MOVING THE EARTH

The Workbook of Excavation

Herbert L. Nichols, Jr.

David A. Day, P.E.

Fifth Edition

McGRAW-HILL

New York   Chicago   San Francisco   Lisbon   London   Madrid
Mexico City   Milan   New Delhi   San Juan   Seoul
Singapore   Sydney   Toronto

The McGraw·Hill Companies

Library of Congress Cataloging-in-Publication Data

Nichols, Herbert L. (Herbert Lownds), date.
    Moving the earth / Herbert L. Nichols, Jr., David A. Day—5th ed.
       p.   cm.
    ISBN 0-07-143058-X
    1. Earthwork. 2. Earthmoving machinery. 3. Excavation. I. Day, David A.,
date. II. Title.
    TA715.N6   2005
    624.1'52—dc22   2004065553

4 5 6 7 8 9 0   DOC/DOC   0 1 0 9

ISBN 0-07-143058-X

*The sponsoring editor for this book was Larry S. Hager, the editing supervisor
was David E. Fogarty, and the production supervisor was Pamela A. Pelton. It
was set in Times Roman by Wayne A. Palmer and Kim J. Sheran of McGraw-
Hill Professional's Hightstown, N.J. composition unit.*

Printed and bound by RR Donnelley.

McGraw-Hill books are available at special quantity discounts to use as premiums and sales promotions, or for use in corporate training programs. For
more information, please write to the Director of Special Sales, McGraw-Hill
Professional, Two Penn Plaza, New York, NY 10121-2298. Or contact your
local bookstore.

CONTENTS

Chapter 3. Rock, Soil, and Mud

Chapter 4. Basements

Chapter 5. Ditching and Dewatering

Chapter 6. Ponds and Earth Dams   6.1

Chapter 7. Landscaping and Agricultural Grading   7.1

Chapter 8. Roadways   8.1

Chapter 9. Blasting and Tunneling   9.1

Part 2   The Machines

Chapter 12. Basic Information   12.3

Chapter 13. Revolving Shovels and Excavators   13.1

Chapter 14. Conveyor Machinery   14.1

Chapter 15. Tractors and Bulldozers   15.1

Chapter 19. Grading and Compacting Machinery 19.1

Chapter 20. Compressors and Drills 20.1

PREFACE
TO FIFTH EDITION

This fifth edition of the classic book *Moving the Earth* is an update to the latest developments in the practical work of handling earthwork. The great value of this excavation workbook has been preserved as far as possible since many of the techniques covered in previous editions have not changed appreciably. Of course, some of those techniques, which were based on human forcefulness, have been augmented by the use of machinery.

Concerns for environmental protection have been extended and included with such techniques as vacuum excavation and horizontal directional drilling. The introduction of remotely controlled equipment as used in microtunneling and some mining operations has been mentioned where appropriate. More coverage of hydraulically operated equipment has been included in this fifth edition.

The outdated forms of equipment, such as cable operated excavators, have been eliminated from this edition. However, reference to older equipment that is still widely used is still in the book. The makes and models of equipment included in the fifth edition are not intended to suggest that they are the best ones to use, but have been included as giving the best example for the points explained.

As mentioned with the fourth edition, it has been a honor and challenge to extend the great value of this book originated by Herbert L. Nichols, Jr., for the benefit of those many contractors and others working to move the earth for the advancement of mankind.

David A. Day, P.E.

ACKNOWLEDGMENTS

Acknowledgment is gratefully made of the assistance supplied by thousands of manufacturers, associations, equipment dealers, contractors, supervisors, operators, engineers, architects, job inspectors, and public officials, who contributed generously of their time and knowledge to make this book possible.

The author sincerely regrets that he has not known the names of most of these helpers and well-wishers, and that the names of so many others are lost among the mountains of paper involved in the preparation of this book, so that it is not possible to thank them personally.

Apologies are offered to those whose contributions have been omitted or wrongly credited.

Particular thanks are now extended to my family and associates who assumed most of the burdens of my work so that I could have time to write.

Many contributors have moved, changed names, gone out of business, or died during the four decades since publication of the first edition of *Moving the Earth*.

INTRODUCTION
TO FIRST EDITION

Moving the Earth has been prepared to supply the first working coverage of the whole excavation industry.

The past twenty-five years have brought a total revolution in the machinery, the methods, and volume of excavation. The rate of development and diversification of machines is steadily increasing, and the techniques of efficient use become more complicated daily. The annual income from their work is now in billions of dollars.

But the literature on the subject has failed to keep up with its advances, and today a very large proportion of the highly specialized knowledge existing about these machines and their proper use has not been put in writing in such a form that it can be used for reference.

The result is that a tragic loss is incurred every year through unwise selection of equipment, damage to machinery through ignorance of its functions and weak points, and waste of time, material, and money in learning by trial and error. While the industry has proved that it is big and dynamic enough to absorb such losses, it would unquestionably be in an even better position today if they had been reduced.

Moving the Earth has been written primarily to fill the needs of those closest to the actual earth moving: the small contractor, the foreman, and the operator. Since their work is basic to all excavation and to the planning and direction of even the largest projects, the know-how for them should also answer most of the needs of the engineer, the architect, the superintendent, and the student.

Some sections of the book have been arranged to meet the needs of the property owner and home builder, and to assist company and public officials whose duties include planning or supervision of earth moving work.

Knowledge of the contractor's viewpoint on problems of site selection, cellar excavation and drainage, backfill and landscaping can result in more satisfactory results at lower cost. The owner or responsible official can learn from these pages what can be done, and how. He can use this information in making his plans, in explaining to his architect and contractor exactly what he wants to do, and in defending his ideas against criticism.

The preparation of *Moving the Earth* has proved a more ambitious project than was apparent at the beginning. Earthmoving is not one industry, but many. It includes hundreds of highly specialized skills, which often have no relation to each other except in their common aim to move dirt or rock. Many of these are important and complex enough to deserve individual reference books.

To drive a tunnel deep underground is a totally different job than to make a fill across a swamp, but the contractor may lump them as a single operation, the waste from one becoming the building material for the other.

The operator of a baby bulldozer grading around a house has nothing in common with the man at the controls of a gigantic stripping shovel, except that they are both classified as operating engineers.

On the other hand, one size and make of bulldozer may be found backfilling a desert pipeline, uprooting jungle stumps in the path of a highway, pushing a scraper on an irrigation land-leveling job, spreading tailings on a mine dump, or placing rock on the face of a dam. One busy hoe shovel may stack logs, pull stumps, pile topsoil, dig a cellar, load trucks, cut ditches, and lay pipe, all on one small job.

It has been difficult to arrange a comprehenisve working guide to an industry with so many varied and separate occupations, which cannot be kept separate because of the way they overlap and interlock. The approach that has been chosen is to divide the book into two distinct sections.

Part One discusses the jobs the excavating contractor is called upon to do, including the preliminary ones of land clearing and rough surveying. The first ten chapters are concerned with the work itself; the basic ways to do it, the problems that arise, and the applications of different types of equipment to the work. Chapter Eleven discusses financing, bookkeeping, estimating, and insurance in regard to the requirements of the excavating contractor.

Part Two is focused on the machines themselves. Every important type of excavating, hauling and grading equipment is discussed and illustrated. Treatment includes mechanical description of the parts and assemblies that go into the machines, the underlying principles of construction, the ways in which special work requirements are met, and suggestions on adjustment and maintenance.

The operation of each machine is explained. In many cases instruction is sufficiently detailed to enable a man to learn from the book alone, although this is less safe and satisfactory than personal instruction. However, because of the complexity of the skills involved, and the secondary importance of the units in general excavation, operation of some machines (oil well drills, for example) has been described for the spectator rather than the operator. Even then, the fundamentals of the work methods are carefully explained.

Special operating techniques, and uses of a machine in various classes of work, will be found both under the main discussion of the unit in Part Two and in appropriate places in Part One. Reference to particular jobs may also be found under several headings. This separation and distribution of subject matter has been necessary to present information in its proper context, to avoid duplication, and to keep an already over-large book from growing out of bounds.

For example, there is a chapter devoted entirely to basements, but further reference to the subject will be found under drainage and landscaping. Crawler tractors and bulldozers have a chapter of their own, but because of almost universal application to earthmoving work, are mentioned in at least nineteen of the other twenty chapters.

In order to provide quick and easy access to all the information on any subject, the subheadings are included in the Table of Contents, and abundant listings and cross references are supplied in an unusually complete index.

Terminology is a difficult problem in this industry, and may well have discouraged attempts of others to write about it. Some words, such as bulldozer and catskinner, have about the same meaning anywhere. Others, such as scraper and ditcher, may be used in reference to several completely different types of equipment. One machine may have many names. An outstanding example is the hoe shovel, which answers to this name, and to hoe, backhoe, pull shovel, drag shovel, ditching shovel, ditcher, and trencher.

This confusion exists partly because of the extremely rapid growth of the business and change in its equipment and methods, and partly because of the complete absence of standard reference material.

All (or almost all) terms used in the text that are peculiar to the construction industry have been defined in the glossary. The definitions chosen are those that seemed most simple, reasonable, or descriptive. Undoubtedly, many interesting and useful terms have been omitted, and in some instances there may be more appropriate meanings than those given. The glossary may therefore arouse more controversy than the text, but it at least represents a first step toward desirable standardization.

Most of the descriptions of equipment are based on specific current or recent models. In general, one representative machine has been selected for careful description, and others in the same class described according to their differences. In no case has an attempt been made to include the whole line of equipment of any manufacturer, nor to judge the respective merits of different makes. Because of changes constantly being made, some information may not apply to current models at the time of reading.

Machines have been chosen for description because of personal experience in operating them, availability of descriptive material and illustrations, or by chance. Such choice does not imply any recommendation of such makes or models over others listed or not listed.

Equipment is usually identified and indexed by manufacturer's name and often by model number as well. This is not only a courtesy to those who so liberally supplied illustrations and information, but also serves to make the discussion clearer and more interesting.

All sections dealing with either the construction or use of particular machines, and with specialized work methods have been submitted to manufacturers and other contractors for criticism and correction. If any errors have survived this checking, or mistakes have been made in final arrangement of copy, the publisher will appreciate being notified.

Moving the Earth is a book about machines and their work. Therefore, comparatively little is said about the men who direct and operate the equipment, or about the hand labor needed to supplement it.

However, I am keenly aware that without the men the machines are nothing. Before there were wheelbarrows, men built roads and dams, quaried and transported rock, and drove tunnels. But without men, the finest machines can do nothing but rust. Without skillful operation, few pieces of equipment can justify their cost. Without maintenance, none can work for long. Without competent direction, all their work is waste. Lack of repeated mention of the men in and behind the machines therefore does not indicate lack of respect and appreciation.

P · A · R · T · 1

THE WORK

CHAPTER 1
LAND CLEARING AND CONTROLS

MACHINES AND WORKERS

Clearing of vegetation is usually necessary and almost always desirable as a preliminary to moving or shaping ground. Any growth makes dirt or rock difficult to handle, and its decay will cause settlement of fills. To satisfy environmental concerns during construction, a building site must be surrounded by silting fencing. This is plastic sheeting several feet high to prevent silt material from being carried off the site by rain or wind onto neighboring property.

Some clearing of growth heavier than grass or weeds is done almost as an end in itself, for agricultural purposes. It makes possible replacing woods and brushlands with pasture, crops, or tree farms.

Clearing is preferably a machine job. It may be done by a wide variety of standard excavators, particularly by bulldozers, front loaders, and backhoes. But if the job is large and/or difficult, it will probably pay to buy or rent one or more of the specialized clearing machines or attachments discussed in Chap. 21.

However, hand labor may be used in addition to or instead of machine work. A small piece of equipment may be able to do the work of a much larger one if occasional oversize trees are cut or stumps are blasted ahead of it.

EROSION CONTROL

In the United States the Environmental Protection Agency (EPA) cites storm water runoff as the most common cause of polluting surface waters and storm water runoff from construction sites has a significant impact on the water quality of streams and rivers. Consequently, the federal government passed regulations called the National Pollutant Discharge Elimination System (NPDES) intended to protect the nation's water supply.

When there is construction activity disturbing the soil on a site, such as excavation or grading during site preparation, there is the potential for runoff that will cause contamination of the nearby waterways. Phase II of the NPDES regulations applies to any site of one acre or larger. Construction projects on sites of this size need an erosion and sediment control plan to apply for a state NPDES permit and also a local permit, if the community has a population of 10,000 or more. Unfortunately, the regulations vary from state to state and also may be different from city to city in the same state. To determine the requirements for a specific project it is advisable to check with the appropriate agency for the project location.

Many municipalities are now mandating the use of readily available erosion control techniques, such as exposing the smallest area of land possible for the shortest period of time or building a retention pond to detain runoff water long enough to allow settling out of suspended sediment.

The erosion and sediment control plan will certainly include silting fencing, mentioned in the first paragraph of this chapter. It may include roughening the soil surface or the use of turf reinforcement mats and straw bales or other forms of check dams to control erosion and hold back runoff flow in ditches. The plan should be prepared by or in the name of the owner-operator of the site. It is generally prepared by the project's consulting engineering or architectural firm. It is not unusual for the contractors, who do earth moving on the site, to be co-permitees and their personnel required to receive instruction about the NPDES regulations before they can do any work on an NPDES-permitted site.

DISPOSAL

Cut or uprooted vegetation must be processed or removed as part of most clearing jobs. Possible ways include burial, allowing time to decay; burning, shredding, or chipping; removal from the area; and various combinations of these methods.

Disposal methods will be discussed below and throughout the chapter. They are considered first because of the extent to which they affect techniques of clearing.

Nitrogen.   Disposal of vegetation by any method other than by burning is likely to be complicated by absorption of nitrogen by decay processes. This element is essential for life of every kind. Although abundant as a free gas in the air, its quantity in fixed or usable form in the soil is limited.

The problem of nitrogen deficiency is usually not a strong objection to disposal of vegetation by any reasonable method such as surface decay, or mixing into soil or plowing it under. Nor does it prohibit the use of wood chips for soil cover. But nitrogen deficiency must be considered in assessing both the immediate and possible long-term effects of the work.

The deficiency can be largely corrected by addition of suitable amounts of nitrogen-rich fertilizer throughout the period of decay or by planting legumes that can obtain nitrogen from the air. The end result is usually substantial enrichment of the soil, as the presence of abundant nitrogen aids the conversion of vegetation to soil humus.

Surface Decay.   Cut or uprooted vegetation is sometimes left on the ground to rot. Soft material such as grass and nonwoody plants may disappear in a few weeks, but trees will make the area unusable for years. Prevention of regrowth is made difficult.

Disadvantages of disposal by surface decay are reduced, and may even be eliminated, if woody material is reduced to small pieces as part of the clearing operation. This may be done by using a shredder or a heavy rotary mower for clearing, or a chipper to grind up pieces after cutting.

This method is suitable for construction only if clearing is done long in advance of removal of topsoil.

Burial.   In agricultural work, burial is often the preferred means of disposal if equipment of the proper type is available, and is big enough to handle the growth density and trunk sizes involved, and if the soil is soft enough to permit it.

Grass, weeds, brush, and sometimes saplings may be buried intact by a brush-breaker plow, or slashed, chopped, and partially or wholly buried by a heavy disk harrow. Rolling choppers can disintegrate and partially bury medium-size trees, including trunks.

Burial of this type is used in agricultural rather than construction clearing, and only when the ground can be left undisturbed (except for planting of a cover crop) until a large part of the material has decayed. This process may take weeks or years, depending on vegetation type and maturity, and weather.

In construction, about the only permissible burial is undisturbed, low-cut stumps under deep fills. When allowed, this exposes the fill and its supported structures (usually road pavement) to eventual settlement as the wood decays. This danger is often outweighed by advantages in prevention of sliding of fill down a slope, and in economy.

Loose stumps are often buried, but the operation is likely to be expensive and unsatisfactory. Attached roots make them enormously bulky; cutting roots back to the buttresses is likely to ruin saw chains by contact with clinging dirt. They are still awkward after cutting back.

Fill including stumps will almost always settle badly. Spaces left in and under irregularities will gradually fill with soil, allowing the surface to sink. Rotting of the wood will cause slow, long-term settlement. Both effects are at a minimum if burial is in permanently wet mud that will flow around them immediately and preserve them against decay.

In any soil, one stump can be packed in more solidly than a number of them. A common practice, when there are just a few big stumps, is to dig out each one with a backhoe, dig the hole deeper, put the stump back in it, and backfill. This shortcut is often not on the plans, and may cause surprise and dismay later, when the surface settles or a small machine tries to dig a trench through the area.

Burning. Where burial is not practical, burning is usually the most efficient method of disposal and does the least long-range damage to the environment. Most of the discussions in this chapter are based on the presumption that fire will be used.

However, since open burning is prohibited or severely restricted in an increasing number of states and localities, it is worthwhile to examine the relative advantages and disadvantages of fire compared with other methods of disposal.

It will be assumed that burning is handled by reasonably experienced crews, with proper regard for safety and confinement of the fire to the vegetation being cleared.

Piles of brush or trees usually contain one-fifth to one-tenth solid matter, the rest being air space. These solids average at least half water, and most of their dry weight is cellulose, lignin, and other burnables. The ash that is left after efficient burning is only a few percent of the dry matter (exact figures are difficult to obtain). A good fire will therefore reduce the vegetation to a small fraction of a percent of the original bulk.

The ash residue is too fine to be good fill material, but its quantity is so insignificant that it can usually be incorporated in other soils, or pushed aside, without difficulty.

It may be reasonably held that efficient burning results in the total removal of cut and uprooted vegetation.

Soil under a hot fire is rendered unfit for supporting growth for 1 to 3 years, but can be restored by plowing or ripping, and fertilizing.

A fire in dry material, with plenty of air, will burn mostly with hot clean flames, which produce carbon dioxide and water vapor, with few pollutants. Green wood and leaves, wet or dirty piles, and most weakly burning fires will give off large amounts of smoke, containing variable quantities of methanol, methane, acetic acid, tars and oils, and carbon monoxide.

Such fires, if upwind from inhabited areas, may create an extreme local nuisance, but its duration is very short. It is doubtful if the pollutants they put into the atmosphere equal those that would have been discharged during the natural lifetime of the plants themselves if they had not been destroyed. The burning concentrates them into a few hours or days.

The pollution problem that faces the world does not arise from country areas or from any activities (including open fires) normally conducted in them. It is a problem of cities, factories, and internal combustion engines.

It is therefore quite unreasonable that burning should come under total or almost total bans, while the real offenders are usually let off with moderate percentage reductions of their offensiveness.

Regulations against burning are costly. Substitute means of disposal, which are discussed in following sections, are more expensive under most circumstances, require vastly increased consumption of fuel, and may create environmental problems of long duration. The extra cost in highway construction alone is probably already in the tens of millions of dollars a year.

Clearing by hand in cold weather may be practical only in the presence of hot fires, for both their emotional uplift and their actual prevention of acute discomfort, including frostbite. Even without need for heat, there is little satisfaction in clearing brushy land if the debris must be left to litter the ground, making it dangerous for people and animals; or heaped into unsightly piles that take years to rot. And loss of unfarmed and unmowed fields to brush is a serious and increasing ecological problem.

The cost of buying or renting shredding equipment, noise, fueling problem, and danger to inexperienced operators put these machines out of reach of most people who wish to do their own clearing.

Chipping. Brush, saplings, and even big trees may be fed into machines that reduce them to chips of small and fairly regular size, by action of a rotating toothed drum (see Fig. 1.1). These chips may be scattered or piled in the work area, or fed through a chute into dump trucks.

A small machine can be towed behind a pickup truck, and often maneuvered on the job by hand. It is hand-fed with bundles of brush, and with saplings up to 3 or 4 inches diameter.

FIGURE 1.1   Brush chipper.

At the opposite end of the scale are monsters which can gulp down entire trees, with trunks up to 20 inches in diameter, without need to even trim the branches.

The chips that are produced may be a definite asset, may be a problem, or may have no importance.

A few modern paper mills are able to digest wood chips that include bark and twigs, and will pay good prices for them. If such a mill is within economical hauling range of a clearing job, chipping pays off both in money and in utilization of the wood. In other areas, chips might be sold for processing into pressed wood, or charcoal and distillation products.

There is also a possibility that chips can be used on the job, either because they are really useful or because it is the best way to get rid of them. They can hold soil on slopes while vegetation becomes established and can add organic material to poor soils. Some applications are discussed in Chap. 7, together with possible problems.

Chips from light to medium thickness of vegetation may be left scattered on the ground, to be incorporated with the topsoil when it is pushed off or cultivated. If the growth is heavy, the chips are likely to make the soil critically short of nitrogen, and difficult to work, and to accumulate in spots as pockets of almost undiluted wood.

Chipping is usually not practical for uprooted stumps, unless they are very small in proportion to machine capacity. Their bulk and shape make it difficult or impossible for the guards to pass them or the drums to grip them, and the dirt and rocks stuck to them damage cutters and make the chips unsalable.

Chips made from stumps in the ground are always contaminated with dirt.

Piling chips, either by keeping a discharge chute in one direction or by dumping them from trucks off the work area, should be permitted only when they are to be reclaimed later. Such piles are likely to remain for many years before decaying enough to permit growth of vegetation.

Chipping machines are expensive, consume large amounts of fuel, are extremely noisy, and may be dangerous to personnel. Their use in mass clearing is justified when there is good use for the chips, in areas of high fire danger, or where smoke cannot be tolerated.

In addition, chippers are valuable in low-volume or selective clearing and trimming, where the cutting would otherwise have to be hauled away.

Removal from Area.   Removal of cleared vegetation from the work area may be anything from a sound and profitable operation to a financial and ecological disaster.

Where lumber or paper mills are within economical hauling distance, it may be possible to sell cut trees profitably. In some cases, the user of the wood may be glad to cut and remove the usable part of the vegetation and pay for the privilege.

The usability and value of trees vary greatly, in both quality to be found on the job and the processing equipment that will handle them. Some users are very narrowly restricted as to species, size, straightness, and soundness. Others will take (usually at a lower price) almost anything that is recognizable as wood. In any big job, it may be worthwhile to invest considerable time in investigating possible outlets.

Firewood is another possibility. Very high prices are often paid for wood cut in 2-foot lengths and split to cross sections averaging 30 square inches or less. Lower but still interesting prices may be paid by cordwood dealers for cut (and perhaps trimmed) trees which they process themselves. But this market is largely limited to the vicinity of cities.

If the vegetation must be removed and nobody wants it, the cheapest disposal is to just push it off the right of way, or out of the construction area, and hope to forget it. Fortunately this practice is usually not allowed. Even if it is, it may have disastrous effects on high-priced surveyors' reference points.

If the contract requirement is off-site disposal at a distance, there may be several possibilities. Both bulk and problems can be greatly reduced, although possibly at considerable cost, by chipping the vegetation and hauling out the chips. Otherwise, trees should be trimmed into lengths suitable for dump trucks or trailers, or flatbeds of either type, with all angles in trunks or branches cut to make them lie flat. Nondumpers require a log-handling crane at the disposal site. Brush may be chipped and loaded, or loaded whole.

Except with medium to large tree trunks, or chips, these loads are likely to be mostly air. Haul cost per pound will be proportionately high.

Unchipped vegetation that is hauled off the job, and is not to be burned, may be dumped in piles over a wide area or stacked in high piles with a log grapple or clamshell. See Fig. 1.2.

The result is almost always an environmental nightmare. Bulk is enormous in relation to the amount of clearing done, appearance is generally a first-class eyesore, and the dumps may be dangerous or impossible to cross for many animals and for people. They may serve as inaccessible infection points for plant insects and diseases.

Depending on the size and variety of vegetation involved, and climatic conditions, these unfavorable conditions may persist for 5 to 20 or more years.

FIGURE 1.2   Clearing with a rake blade.

CONTAMINATED GROUND

Legal Basics. The discovery of contaminated ground is an environmental hazard. Excavating contractors or site developers, who normally move "clean dirt," must be wary about protecting the business from the legalities of environmental hazards. They must do everything feasible to protect themselves from environmental entanglements. An environmental due-diligence survey should be done. It involves three stages: (1) initial assessment (any history or current evidence of contamination), then (2) investigation, and finally (3) remediation.

For instance, assume a basement is being dug in a development and an old, unknown underground storage tank is broken and leaks heating oil into the ground. Who is to be blamed? As the contractor, you may be in trouble if your only insurance is a standard, comprehensive, general liability policy. To cover this special liability, it is necessary to have an "environmental rider" on the basic liability policy.

In addition to adequate insurance coverage, contractors should protect themselves by asking for phase 1 and phase 2 reports at prebid meetings. The phase 1 report involves a thorough investigation of the site's past uses as well as uses of surrounding properties. Phase 2 details the results of soil and water sampling at the site.

Even if the reports indicate the site has a clean bill of health, the construction contract should have a stop-work clause—especially if excavation is involved. This allows the contractor to stop work without penalty if potential environmental hazards are found. The contractor should then contact the project owner, or her or his agent, and report the finding. If the owner refuses to report a situation that poses an immediate hazard to human health or the environment, the contractor may be bound to report the situation to the proper authorities.

Reporting environmental problems can be complicated simply because of the numerous agencies that require notification. If a required one is missed, the fine can reach $25,000. To avoid any oversight, report to all possible agencies—the Environmental Protection Agency (EPA), state agencies, fire departments, local planning commissions, the National Response Commission, and the like.

If a regulatory agency determines that the property's contaminated soil must be cleaned, the owner must carefully select an experienced soil remediation contractor. This is important because the property owner is fully liable for the contractor's actions—and penalties for bad actions start at $10,000 per day.

Treatments. In the early days, most contaminated soil was simply dug up and hauled to a landfill. However, now in the United States the Environment Protection Agency (EPA) encourages alternative treatment methods to actually clean the soil or make its contaminates less harmful. These methods are known as soil remediation techniques. They can be classified into five general categories: biological, physical, immobilization, chemical, and thermal.

A 1993 EPA report "Cleaning Up the Nation's Waste Sites: Markets and Technology Trends" gives insight into the uses of the various technologies. The technique chosen for a given site depends on the contaminants and the site's geology. For example, cleaning up an old gas station may be handled best with vapor extraction if the contamination is mostly gasoline and if the soil is not too dense. In dense soils, biomediation could be a better choice. But if heavier oils are present, thermal desorption may be most effective.

Biological remediation (bioremediation) uses microorganisms, such as bacteria, that eat soil contaminants and turn them to harmless—or at least less toxic—compounds. This method is preferred because it works *in situ,* i.e., is done in place. Nutrients and oxidizers are added to the soil to stimulate the growth of hydrocarbon-eating bacteria. A specialized equipment system, called MecTool, is used for in situ soil remediation and can inject microbial nutrients to depths of 100 feet.

Physical methods for remediation include such processes as drawing a vacuum through wells drilled in the soil to pull out volatiles, or pressurizing the wells with heated air, or in situ steam stripping, a vacuum extraction process with steam injection wells. The vapors captured in all three processes are then treated to remove the contaminants. Another physical technique uses a water-and-detergent solution to wash the contaminants from the soil.

Detection   The use of trenchless technology can be helpful in locating the extent of contamination. Compact directional drilling equipment is used to reach areas of the underground that are inaccessible under a building or other obstruction. Refer to Fig. 20.60.

As stated for the National Ground Water Association in the United States, "horizontally" drilled wells have been installed to perform remediation of subsurface groundwater and soils with pump and treat systems, such as by air spraying, soil vapor extraction, and bioremediation. Horizontal wells also are used to prevent contaminant migration and for characterization, i.e., taking samples for evaluation, under buildings and other areas where surface conditions prevent drilling of vertical wells. Because underground plumes often spread horizontally, fewer horizontal wells are required to treat a site than the numbers of vertical wells that would be needed. The EPA believes horizontal well technology has the potential for significant cost saving on site remediation projects.

The success of horizontal directional drilling (HDD) depends on the drilling machine's electronic guidance system and its correct use. Currently there are two types of systems: (1) the walk-over system and (2) the wireless system. The wireless guidance system uses computers to calculate information necessary to steer the directional bore. This system is generally used for long, deep installations.

The majority of horizontal directional installations use the less expensive walk-over electronics system. There are two basic components: (1) a radio transmitter mounted in the drill head and (2) a handheld receiver operated by a crew member, who walks on the ground surface directly above the drill head as the pilot hole progresses. The transmitter sends signals to the receiver which processes the information and displays it on the receiver. This tells the drill head location and depth, roll angle of the drill head slanted face, and the pitch or horizontal inclination of the drill head. This information is transmitted to the drill machine operator by two-way radio so that he or she can make adjustments in the drilling process.

The use of vacuum excavation or potholing has been useful in locating underground pipelines that could be damaged by HDD running into them. Vacuum excavation, described in Chap. 5, is used to protect against damage to underground utilities when excavating.

Immobilization is a process in which a binding agent, such as cement, fly ash, kiln dust, or asphalt, is mixed with the soil to physically or chemically immobilize contaminants. This method could be called *solidification,* while chemical treatment is more stabilization. Dechlorination is a chemical remediation technique used with soils to react with chlorine in such compounds as polychlorinated biphenyls (PCBs), creating by-products that are less toxic than the original contaminants.

Among remediation methods that rely on heat are the in situ vitrification, incineration, and thermal desorption. Incineration is the most frequently used of any soil treatment method at federal cleanup sites, according to the EPA in 1992. A thermal desorption system puts the contaminated soil through a heated, rotating drum as in an asphalt plant. The temperatures in the drum are sufficient to vaporize the contaminants for collection and treatment.

BRUSH CLEARING

Dozer.   Dozers and loaders are basic machines for clearing, both with regular blades or buckets and with special attachments. They work best when the ground is firm enough for support, and where they are not hampered by holes, gulleys, sharp ridges, and rock. These forms of equipment are described in Chaps. 15 and 16.

Uneven surfaces make it hard to keep the blade in contact with the ground, and lead to burial rather than removal of vegetation in hollows. However, there are few places where a dozer cannot aid hand-clearing crews, by clearing areas where it can work, moving logs and cut brush, cutting roads for supply trucks, or firebreaks.

Dozers have a particular advantage over hand crews where briars and vines are abundant, as these are very tedious to cut but can be readily stripped off by the blade, provided the operator does not take too long a pass and get caught in the tangle.

Unless he or she is completely protected by a cab with windows, an operator should have hand clippers to cut a path out if necessary.

Brush and small trees may be removed by a bulldozer moving with its blade in light contact with the ground. It will uproot or break off a number of the stems and will bend the rest over so that by a return trip in the opposite direction, it can take out more. If the distance is short, it is best to doze the whole patch in one direction, then across or backward.

Individual small trees are first knocked over, then pushed out with another pass in the same direction.

Results will vary with the type of vegetation and the condition of the soil. Hard-baked soils will cause a high percentage of broken stems, while wet or sandy conditions favor uprooting, which is more satisfactory for most purposes. The work can be speeded up by having a laborer cut out or pick up individual bushes that would otherwise require another pass by the dozer.

If the job requires removal of light stumps and roots, they may be overturned in one pass and pushed out in the next. It may be necessary to dig several inches into the soil to get a grip on them, then backblade the soil into the holes.

Brush heaps may be largely freed of dry, loose dirt by rolling them over with the blade and shaking the blade up and down. If this is ineffective, rolling them over backward or pushing them from the side may be tried. A dozer with a blade which can be easily tilted down on either end is very good at this work, as one corner can be used for taking out roots and pushing piles without taking a bladeful of dirt along with it, and the blade can be returned to flat position to skim off surface brush.

Rake Blade.   Rake blades, which are made for the larger bulldozers and loaders, add to clearing efficiency under most conditions. See Fig. 1.2. They allow working below ground level, to take out roots as well as surface material, usually without bringing the soil along with them, if it is dry or sandy.

However, they may be somewhat specialized. A blade with teeth close-set enough to handle brush may bend a tine if it collides with something solid, while one strong enough for impact is apt to have too wide spacing for brush. This type of equipment is described in Chap. 21.

With or without rakes, any mechanical loosening and removal of brush that is to be burned should be done when the soil is dry for best results. If it is wet, it lumps and sticks.

If the loosened material is allowed to dry, much of the dirt can be shaken out while piling, by rolling and shaking.

Burning.   Generally in the United States it is necessary to get a permit from the local authorities to burn for clearing. In general, it is best to burn machine-cleared vegetation at the same time that it is piled. A hot fire, including heavy wood, is prepared, and brush piles are pushed up on it. A new fire is made when the push gets too long.

Best results are obtained if the vegetation is uprooted and allowed to dry at least a few days before burning. This may be done by backing the dozer into the woods from the cleared edge, and uprooting small patches, or individual trees, pushing them clear of the ground, and then leaving them.

The trash dries more rapidly scattered on the ground than in piles. Dirt will tend to dry and break away from stumps, and to sift out of roots and stems. When burning, the brush nearest the fire is put on it first.

Fires fed by a dozer tend to get choked up with dirt. In general, matted light brush is more difficult to clear and to burn than heavy brush or small trees, as it tends to slip under the blade or to bring too much dirt with it.

Fire Box or Trench.   The local authority which issues a permit probably would be more inclined to do so if the burning was to be done in a fire or burner box or trench prepared for the burning. The system for doing this controlled burning is described and illustrated later in this chapter.

Dozer Protection.   When a dozer is clearing dense undergrowth, there is the danger that it will fall into some hole, natural or artificial, whose presence is concealed by the brush. This may be guarded against by scouting the area on foot, and by moving forward in a succession of short pushes overlapping each other on the side, as in Fig. 1.3. This enables the operator to watch from

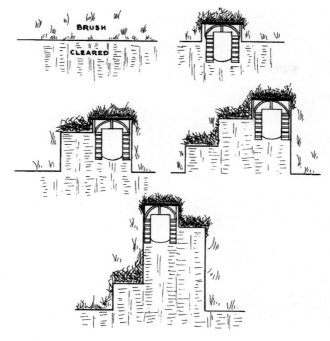

FIGURE 1.3 Clearing thick brush.

one side, without getting branches in her or his face, and to observe the nature of the ground. In addition, it avoids tangling the dozer in branches and vines.

Any dozer used for clearing work should be thoroughly protected with crankcase and radiator guards; the latter include screen with holes not over $\frac{1}{4}$ inch, and accessible for removal of leaves and trash. The engine needs side guards. The operator should carry hand tools to cut herself or himself out of tangles.

Minimum operator protection is a strong overhead structure. See Fig. 1.2. Most new clearing units have complete, extra-strong cabs, often heated and air conditioned.

Accidents have been caused by branches moving throttle and clutch controls.

Other Machines. There are a number of types of equipment that are used for chopping or shredding brush, which may bury it or leave it on the surface to rot or to be removed by other equipment or laborers. Refer to Chap. 21 for the special equipment.

A big rotary mower mounted on the rear of a wheel tractor is highly effective up to its thickness-of-stem limit, which may be $1\frac{1}{2}$ to 3 inches. Vertical rotary shredders may handle double that size. Brush is mostly chopped or shattered into small pieces, but root systems are seldom disturbed. There is no suppression of regrowth.

The sickle bar or hay cutter, in heavy-duty models, will cut brush up to $\frac{3}{4}$-inch, but with considerable wear and breakage. It does not chop the stems and is little used in clearing.

Either a rotary or a sickle bar can be used to suppress regrowth by repeated mowings.

A big moldboard plow, preferably a brush-breaker model, can put brush and saplings underground, cover them neatly with dirt, and leave them to rot. The area is usually harrowed lightly and planted to grass or a cover crop immediately.

A big heavy disc harrow, with discs 24 inches in diameter or larger, chops brush and buries a large part of it. See Fig. 1.4. Big pieces may be loosened, chewed, and pushed around without burying.

FIGURE 1.4   Clearing with a disc harrow. (*Courtesy of Rome Plow Co.*)

Both the plow and the harrow tend to create ridges and troughs in the ground surface, because they move loosened dirt to the side.

A rolling chopper knocks down, tears, and mashes both brush and trees, and cuts near-the-surface roots. A small portion of its cutting is buried.

Carrier.   When the final cleanup of any kind of brush clearing is done by hand, it is helpful to furnish laborers with stick carriers, which may consist of a piece of heavy canvas, 6 or more feet long and 2 to 3 feet wide, with handles on the ends, as in Fig. 1.5. This is laid on the ground, and sticks and branches are piled across it. The handles may then be picked up and several armfuls carried at a time with minimum effort.

TREE REMOVAL

Mechanized Logging.   In some large-scale operations, logging may be almost completely mechanized, with felling, trimming, bucking, and transport (or piling and burning) done by highly specialized machines, some of which are described briefly in Chap. 21.

However, most clearing-for-excavation projects must rely on more standard machines and/or hand labor.

Cutting or Uprooting.   Big equipment can handle small trees in the same manner as brush. But big trees, or any trees too large to be walked down by the equipment on the job, may require special kinds of work.

Pushing a tree over with a dozer or pulling it down with a cable follows the general methods described for stumps later in this chapter. However, a machine can generally uproot a much larger tree than a stump, because of greater leverage from a higher push or pull point and help from the weight of the tree, which tends to tear out its roots as it leans.

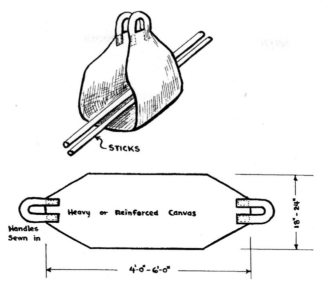

FIGURE 1.5   Trash carrier.

Handling Trunks.   Tree trunks, even in sapling sizes, may be tricky and dangerous to push around. One may ride up over the top of even a big dozer blade during ordinary pushing. It may be put under tension by pushing while an end is blocked, which may result in its whipping with great force against the cab, or into other machines or workers in the area. An operator must be vigilant in avoiding such a situation.

When felled trees are pushed into heaps for burning, the piles contain so much air-space that they may be very fire-resistant. Heavy branches and crooked trunks increase this difficulty.

Branches may be cut off and trunks cut into pieces to make more compact piles. Or the pieces may be placed on an existing fire by a clamshell.

Subcontracts.   If trees are to be removed which are of no value on the job, an attempt should be made to sell them. To contractors desiring to confine themselves to dirt work, the best arrangement is to get the customer, whether sawmill, firewood dealer, or whatever, to buy the trees on the stump and cut and remove them. A danger is that the logger may fail to do the work in the time specified, and so force contractors to do it themselves at the last moment. In making such an arrangement, the disposal of the scrap wood and brush and the height of the stumps should be specified.

A sawmill operator is interested only in large, sound trunks, whereas a pulp or firewood worker can use bulky branches also. The mill will ordinarily pay the best prices but do the least work toward cleanup of the tract, unless it has an arrangement with pulp or firewood users to take its tops and limbs.

No one wants the rotten trees, crooked branches, and brush, but the lumbermen may agree to burn them, if this is a part of local logging practice; or if the contractor accepts a complete cleanup job as partial or full payment for the wood.

Cooperative clearing arrangements may be made in which the logger is assisted by the contractor's tractors or trucks.

Stump Height.   Stump height may be determined by local law or lumbering custom. From a clearing standpoint, high stumps are more easily removed than low ones, and are especially desirable when the machinery is undersized for the job or depends primarily on winches. Low stumps are more difficult to cut, particularly where the trunk flares out widely at the bottom, but do not impede machines as much and can often be filled over and left.

Cutting. If the trees are valuable and lumbermen will not clear them out in time, the contractor may cut and stack them for future sale. This, as a logging proposition, is somewhat out of the field of this book and will be considered very briefly.

Practically all wood cutting is now done with chain saws operated by one worker. Their construction and operation are described in Chap. 21.

There are a number of sizes and models. Thick logs require long blades and increased power. For any log thickness or blade length, increase in engine power means faster cutting, but also more weight to handle.

Most experts recommend use of a bar long enough to cut the average size tree on the job in one cut. Undergrowth should be roughly cleared around the tree before cutting it, to reduce danger to workers and tangles with fallen trees.

A tree should be notched by a V cut up to one-third of trunk thickness, on the side toward which it is expected to fall. The bottom cut of the notch should be made first and should be horizontal, the other sloped down to meet it. See Fig. 1.6.

The tree is then cut through from the other side with a level cut 2 inches or more above the floor of the notch.

If the saw bar is longer than the tree diameter, or a hand crosscut is used, the line of cut is parallel to the back of the notch. With a short bar, procedures shown in Chap. 21 are used to obtain the same effect.

The cut is theoretically finished when the strip of uncut wood (the hingewood) has been reduced to $\frac{1}{10}$ to $\frac{1}{20}$ of trunk diameter. The tree should now fall toward the notched side in a direction at right angles to the length of the hingewood, if its balance has been judged directly. If the hingewood tapers, direction will tend to shift toward the thicker side.

The hinge may be crushed as the tree starts to lean, and may narrow the edge of the cut, binding or even crushing the saw. It is necessary to be alert to pull the bar out quickly. It can be put back in if the fall does not occur.

Direction of Fall. Natural direction of fall of a tree is influenced by a number of factors. Location of its center of gravity, as affected by lean, twist, and limb location, is primary. Direction and velocity of wind may be important.

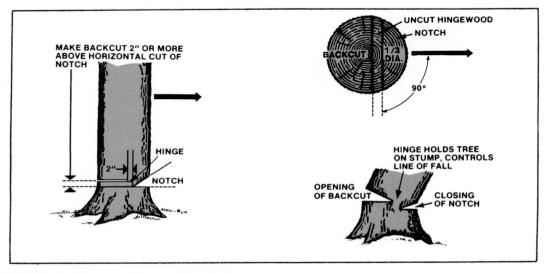

FIGURE 1.6 Felling a tree. (*Courtesy of Homelite.*)

When practical, cutting should be done so as to take advantage of natural factors, to drop the tree in the direction that it tends to fall. This direction can be made positive, or altered moderately, by appropriate notching and cutting.

For greater changes in direction, including tipping the tree oppositely from its normal inclination, wedging or line pull is needed.

Wedging. If there is the slightest question about the direction the tree will fall, a wedge of wood, plastic, or magnesium should be driven lightly into the back of the cut, to prevent wrong-way leaning. This protects the saw against binding, and may influence direction of fall. See Fig. 1.7.

A large tree that has a moderate tendency to fall in a wrong direction can be forced into the right one by wedging. The wedge or wedges may have to be iron or steel, as other materials may not withstand heavy driving. Saw manufacturers oppose any use of hard metal.

The moving saw chain must never touch a hard wedge, as it would suffer immediate and severe damage. The wedge therefore cannot be inserted until the cut is deep enough to provide ample working space for the saw, and the operator must exercise extreme care.

Wedging tends to overturn the tree to the side directly opposite. But the fall will be affected by other factors, such as unbalance to the side, taper in the hingewood, or wind pressure. These factors should be allowed for in placing the wedge.

Wedging may be started with a soft wedge, to keep the cut open until it is deep enough to use the hard one.

One wedge may not be sufficient to tilt the tree. A similar wedge may be driven just above it, Fig. 1.8*A,* or a thicker wedge driven beside it. Since a thin entering wedge is not needed now, a made-on-the-job wood wedge, Fig. 1.8*B,* may be used.

The hingewood may be cut thinner to make wedging easier. But if this strip is made too thin or cut through, or if it is weak because the wood is brittle or decayed, a badly unbalanced tree may fall backward over the wedge.

If tree diameter is too small to safely accommodate a wedge and a chain saw, the chain may be replaced by a hand crosscut, if one is available. Or the tree may be pulled by a line fastened high up, or pushed by a pole or a loader.

Pull or Push. Direction of fall can usually be controlled by a rather light pull on a line fastened high up, the higher the lighter. The first problem is placing the line, which could require either a good climber or a tall ladder, and must be done before starting the cut.

The tree should not fall on the workers or machine doing the pulling. Heights of trees are difficult to estimate, so it is safer, and uses a shorter line, to use a pulley block or blocks and pull from the side. See Fig. 1.9*A.*

The best machine for pushing a tree is a backhoe. It has a high reach in proportion to machine size, and can stand back far enough not to interfere with the cutting.

Pull or push is usually light at first, and is increased as the cut deepens.

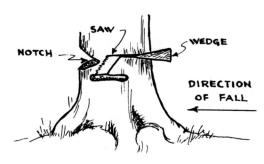

FIGURE 1.7   Use of saw and wedge.

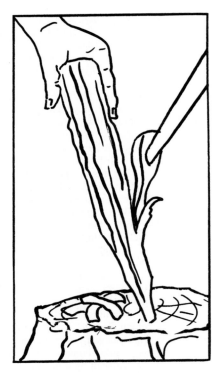

FIGURE 1.8A   Double wedging.

FIGURE 1.8B   Making a wood wedge.

DANGERS

Tree cutting is dangerous work, because of the nature of the tools used and the sometimes unpredictable behavior of trees and their parts.

Direction.   In spite of best judgment and efforts, a tree may fall in an unexpected direction, even backward across wedges. And a fall may occur much sooner than expected, because of concealed weakness in the base, or a puff of wind.

No one should be in a cutting area except the worker(s) at the foot of the tree. They must be alert to move quickly. The critical area should be clear of brush and litter, as tripping over it might have fatal consequences.

A chain saw that is not being used, or any other valuable equipment, should be placed behind a tree or other protection during tree felling.

Overhead Breaks.   Some trees come apart while being cut (Fig. 1.9B). Blows of a hammer on a wedge, rough contact of a pusher bucket, or even the vibration of a saw may loosen dead limbs so that they come crashing down around the base. If the trunk is decayed, it may break as it starts down, with the lower part leaning and falling conventionally, but leaving the upper parts in the air to come straight down.

This will not happen when a tree is live and healthy throughout, but dead branches and decay are not necessarily visible from the ground. It is more common in tall trees than in short ones, and the pieces fall harder.

FIGURE 1.9A   Pulling from the side.

FIGURE 1.9B   Trees may come apart.

Hard hats are not sufficient protection against the weight of pieces that might fall. The top must be watched closely during work, and workers must not be ashamed to run instantly if anything breaks loose. The danger area is usually small and close to the trunk, although this varies with limb spread. But, of course, do not take the direction the tree is expected to go.

Leaners.   A tree may be held from falling by other trees. This may be by comparatively light contact of branches while the cut tree is nearly vertical, or because of fall into a crotch or across heavy limbs.

Such a tree may be brought down by moving the lower end away from the direction of fall. A small or medium tree can be moved by prying the butt up with a pole and swinging the pole. This can be repeated as many times as necessary. See Fig. 1.10. Medium to large trees may be lifted or pulled by a machine, or pulled by a block and fall.

If the trunk cannot be moved, it must be cut in the air. Because of the strain of its position, only shallow cuts or notches can be made in the upper side, with the main cut from below. Pinching of the saw and erratic movements of the trunk can be expected and can be dangerous.

As each piece is cut and knocked or pried out of the bottom, the remainder should slide down to rest on the ground, usually at a steeper angle.

Cutting the supporting tree is too dangerous to be attempted. It is usually under great extra tension and is likely to snap and fall suddenly during work. The leaning tree would then be apt to drop on the workers doing the cutting. See Fig. 1.11.

Trimming.   When the tree is down, the branches should be cut off nearly flush with the trunk, before any other trees are dropped across it to make a tangle. Light branches and any heavier wood which is to be wasted should be piled, burned, chopped up, or taken away by the methods described for brush.

FIGURE 1.10 Moving hung-up tree with pole.

FIGURE 1.11 Never cut the support tree.

Removal. The trunks may be dragged out of the woods by tractors or cut (bucked) into lengths where they fall. Saw logs for small mills may be from 8 to 16 feet long, the size being largely determined by the use for the sawed wood, the capacity of the mill, and the trucks that carry the logs. Piling, which may often be made from thinner trunks than saw logs, is left full length. Cordwood is usually in 4-foot sections, and split to a size that one person can handle. Pulpwood varies in length in different localities and is usually peeled but not split.

Dragging (Skidding). Crawler tractors are the standard equipment for dragging long logs, but with good ground conditions they may be replaced by rubber-tired machines with either four-wheel or two-wheel drive.

Crawlers and four-wheel drives usually carry loaders or dozers. Their efficiency may be increased by mounting winches, and by use of log carriers and various rigs too specialized for description here.

A log is usually pulled by means of a chain or cable fastened around its butt, choker-fashion, and attached to the tractor drawbar or winch. The most important consideration in arranging this is to get the butt off the ground, or riding on the ground very lightly, as digging in will take greatly increased power and will rip up the trail. A short line, particularly to the top of a winch, is helpful unless the log has a greater diameter than the height of the draw point.

The log may also be pulled onto a stoneboat, or other sled, and the line passed through the eye, or two lines used as in Fig. 1.12.

If the tractor is sufficiently powerful, several logs may be pulled at a time by attaching them individually to different lines. If only one line is available, they may be fastened with one choker, which should be fastened well back, as such piles often come apart while being towed.

Two-wheel-drive tractors can drag logs or bundles of logs on dry ground. Loads must be small, but they move briskly. If the tractor has a hydraulic lift drawbar which can be chained to the log to lift its butt off the ground, its efficiency is more than doubled, as the weight on the driving wheels is increased and friction is greatly reduced.

Moving Short Wood.   When trucks can get in the woods, cordwood and pulpwood are usually cut to size and trucked out. As wood is much lighter than dirt, a dump truck can carry several times its body rating, if the pile stays on. Figure 1.13 shows an arrangement of placing planks, poles, or thin split logs vertically along the body sides to permit high piling. These are held in place by the piled logs. If the road is rough, it is wise to pass a chain from the body over each row of logs and to tighten it with a load binder.

If the wood is cut short, and rain or unforeseen mud conditions make trucking impractical, it may be dragged out by tractors. If a stoneboat is available, logs may be piled on it, and the tractor line threaded through the eyehole over the pile and anchored on the back. The eyehole should be beveled so that a chain or cable can slide freely through it, as the tractor pull will then hold the logs to the stoneboat. See Fig. 1.14. If no boat is available, the logs may be piled on the line, which is then looped around it as a choker. Parts of the piles may be chained by snaking the line under them, without repiling.

Storage.   Wood should be stored outside of the work area where it will be accessible both during and after the digging. Poles and logs may be very useful in shoring up banks, making corduroy

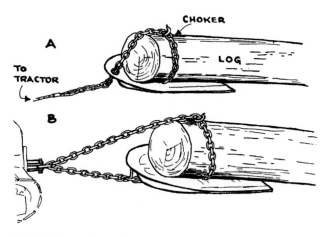

FIGURE 1.12   Skid-pan log hitches.

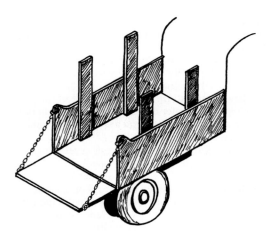

FIGURE 1.13   Sideboards to hold cordwood.

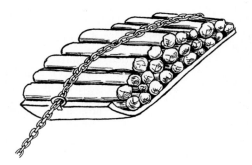

FIGURE 1.14   Cordwood on stoneboat.

roads, getting machinery out of the mud, and other purposes. A buyer might be found for the wood at any time. Stored logs or cordwood should be stacked so as to be off the ground and well ventilated. This makes it easier to remove them later and delays damage from rot and borers. Cordwood is usually stacked in easily measured units.

Personnel.   If possible, experienced loggers should be employed for lumbering. They will be able to do it much more efficiently than equally energetic and resourceful persons not used to the work. The difference may be as much as 10 to 1.

STUMPS

A *stump* is the base of a tree trunk and its attached root system. The trunk part may be anywhere from a few inches to many feet in diameter. It usually flares out near ground level into root buttresses, which connect it to the major roots. Its top may be flush with the ground or several feet high.

Roots form a network near the ground surface. A few species of trees have a taproot, a strong root that extends more or less straight down from the center of the trunk. It makes stump removal much more difficult.

Stumps are a major problem in most clearing that involves trees. They can sometimes be cut low and left under deep fills, but usually must be removed.

Stumps may be broken out by uprooting the whole tree, then disposed of as part of the tree or separately. But it is more usual to cut and remove the tree and then take out the stump.

Stumps may be pushed out by powerful dozers; dug out by somewhat less powerful dozers, rippers, or hoes; pulled out with cables or chains; or blasted. Blasting may be combined with other methods. It is occasionally practical to burn stumps in the ground.

Once out of the ground, stumps present a problem of disposal. As massive pieces of green wood caked with dirt, they are difficult to burn. Nevertheless, this is usually the best way to get rid of them. They are so bulky and irregular in shape that they are hard to bury, and they are an eyesore in piles. Since they rot, they cannot be used in fills under structures.

PUSHING AND DIGGING OUT

Pushing.   Crawler tractors, with dozer blade or loader bucket, or a special narrow stumping blade, are standard for uprooting by pushing. Stumps should be cut high—at least 36 inches—for good leverage.

The blade or bucket is lowered to contact the stump a few inches below the top, and the machine is moved forward in low gear. If the stump yields, forward push is continued until the trunk leans so far that effective contact is lost, or until the roots bulge in front of the tracks.

The tractor is then backed, and the edge forced under the upturned roots. Lifting while moving forward slowly should roll the stump out of the ground, breaking all roots except those on the far side. See Fig. 1.15.

Further pushing to get out the far roots may drop or mire the tractor in the stump hole, or may roll up a too-big ball of roots and soil. It may be better to finish freeing the stump by going to its other side.

A number of stumps may be overturned in one direction, then the tractor turned to finish them up from the other.

The same method is followed to uproot a standing tree, except that a high push point is used for greater leverage, and the capacity of the machine relative to trunk diameter is improved.

Before pushing a tree with any type of machine, the operator should look to make sure that it is alive or at least sound. If a rotten tree is pushed near the base, it may break high up and a top section fall on the dozer. Large dead branches are sometimes dropped with equally disastrous consequences. A dozer to be used extensively for tree pushing should carry overhead guards for the operator.

A tree may bend or split without affecting the roots, in which case the push should be applied lower on the trunk, or from a different direction.

Digging Out.   If it does not yield at all to pushing, it must be dug out. This is done by trenching around it with the dozer to cut the roots. If done systematically, this may follow the pattern in Figure 1.16, but frequently it is unnecessary to go through the whole procedure. Each time the

FIGURE 1.15   Two passes to push out a stump.

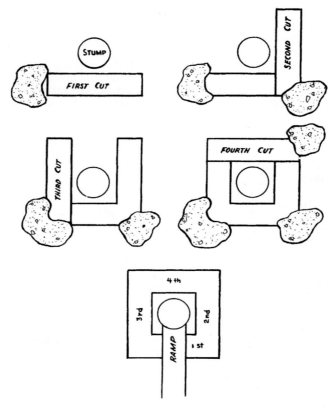

FIGURE 1.16   Digging out a stump.

dozer cuts a big root, it may turn and push the stump to see if it is loosened. The operator will often be able to tell when it has been softened up by the way it shakes as the roots break. Many operators do not bother with cut number 4, but it helps with very heavy stumps, particularly when cut low. The ramp need be built only after an attempt to uproot from a lower level has failed.

Roots should be cut as close to the stump as the power of the dozer permits, but it is a waste of time and power to buck at a heavy root repeatedly when it could be easily broken a foot or two farther out and the stub crumpled back.

Pushing over a stump may leave a hole so large that the dozer cannot cross it to complete the tearing out. In this case the dozer may be stopped at the edge, with brakes locked and the blade holding the stump up. The operator can climb down and block the stump from settling back with stones or a log, then back the machine and push dirt into the hole, or break down its edge so that it can walk into the stump.

If an area is to be excavated after clearing, stumps may be left until digging has undermined them and cut many of their roots when they can be easily removed.

Shovel Dozer.   A shovel dozer or front-end loader can remove stumps in the same manner as a bulldozer, or make use of the hydraulic control bucket in special techniques. A stump of small to medium size may be dug by tilting the bucket floor downward from 30 to 60 degrees and forcing it into the ground close to the stump, as in Fig. 1.17, using both down pressure and forward motion. With the machine pushing forward, the bucket is then flattened and may be driven under

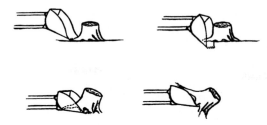

FIGURE 1.17   Stump removal with shovel dozer.

the stump, cutting and tearing the roots, then rolled back, lifting the stump from the ground. If it falls off, the bucket is dropped to contact it and to roll it out of the ground. If it stays in the bucket, it can be carried to a pile or loaded directly on a truck.

The high lift gives the dozer shovel excellent leverage for pushing over trees.

Bucket teeth are desirable for stumping, as they aid penetration, get a better grip on the stump, and are useful for knocking dirt off the ball and for raking out roots.

Methods of handling stumps with a dozer bucket are described in Chap. 16.

Selection of Machinery.   Big machines, and special machines, greatly reduce time, effort, and breakage in clearing work, and should be used whenever a job is large enough to justify their purchases or hire. Stumps that can be knocked right out of the ground may be removed at the rate of one a minute or better; moderately resistant ones may take 2 to 5 minutes; and those which are definitely oversize may take an hour or more. It is easy to see the time that can be saved by applying overwhelming power.

A good clearing team may be made from a heavy tractor with a stumper, assisted by a smaller one with a shovel dozer. If these machines work together closely, the stumper can devote its entire time to breaking out the big ones, while the dozer takes out small stumps, knocks down brush, finishes off loosened stumps, piles and removes them, and smooths the ground.

Revolving Shovels.   The backhoe is probably the best stumper among the shovel attachments. Usually, the operator tries to take the stump by a direct pull first. If it resists, it can be weakened by chopping roots on the far side and by trenching at the sides. Except in the case of very large stumps, digging and removal can be done from one position. It may be necessary to put blocks against the tracks to prevent the shovel from dragging toward the stump when power is applied.

A dragline is used in the same manner but is less efficient at chopping roots and does not have as strong a pull. However, it can often take out a number of stumps without changing position, and is able to backfill the holes and grade the area at the same time.

A power shovel with a dipper stick is used in somewhat the same manner described for a dozer shovel. It has more penetration and can trench to cut roots more readily but is less maneuverable.

The hoe and the dipper are more effective than dozers in rocky ground and among interlocked stumps, as they can apply their power in smaller spaces. It is often necessary to devote time to digging out rocks before the stump can be attacked. If the roots are strongly entrenched in bedrock or oversize boulders, the rock may have to be blasted before the stump can be pulled.

Rippers and Scrapers.   Rippers may be used to cut stump roots and to pull out stumps directly. With two teeth (or one tooth mounted at the side), the ripper can cut roots close to the trunk on all sides. The tractor is then backed with raised teeth to catch the far side of the stump, then forward to pull it over. Another back pass enables the operator to get a tooth under the stump, and lift and roll it out.

Scrapers (nonelevator models) can take out small and low stumps, but this involves complicated maneuvering and danger of getting the stump jammed in the bowl.

Close Quarters.   Stumps are often so located that uprooting them would damage pavement or buildings. Such a stump may be removed in chip form, with little disturbance, by a carbide-toothed wheel driven by a tractor or truck power takeoff. Cutting is usually carried 4 to 6 inches below ground level.

Tree Killing.   Under most conditions dead stumps are easier to remove than live ones, because of the disappearance of the hair roots which bind them to the earth and the weakening of the larger roots. Softwoods in well-drained soils may show perceptible weakening in a few months, while rot-resistant stumps in saturated ground may remain firm for many years. In any case, dead stumps contain lighter wood and hold less dirt than live ones.

It may therefore be advantageous to kill trees well in advance of removal in cases where plans are made early enough. This may be done by cutting, girdling, poisoning, burning, or drowning. The same methods are not equally effective in different localities with different species, so advice should be obtained from local tree experts.

Burning peat soils in the dry season will kill trees and loosen the stumps, but because of smoke and smell nuisance, and danger of spreading, it should be done only under carefully controlled conditions.

If it is possible to dam a stream so as to flood a wooded area for several months during the growing season, some of or all the trees may be killed. Unfortunately this usually is possible only in swamps, and swamp trees are more resistant to drowning than those in dry locations.

PULLING STUMPS

Stumps may be pulled out of the ground by a cable to a power source. This method, although widely used, is less popular than in the past. Bigger machinery and special attachments have made it possible to push or dig out most stumps encountered on construction jobs, without taking the extra time to rig lines.

However, such equipment is not always available, it may cause too much damage, and it cannot work efficiently on soft ground. Pulling will probably continue to be an important method of stump removal.

The pulling line may be a chain, cable, or rope, and the power may be direct pull by a machine or animal, winding in of cable on a winch, either machine- or hand-powered, or a combination of these methods with pulley blocks.

The stump line is generally a choker type which tightens its grip as the pull increases. In smaller sizes, chain is preferred because it is easier to carry, safer to handle, and more resistant to abuse. However, it is much heavier than cable for the same strength, and in large sizes it is too weighty to be practical.

Line pulling is preferred when the ground is too rough or soft to allow machinery to get at stumps directly, and when available force needs to be increased by multiple lines.

Chains.   A more detailed description of chain and fittings will be found in Chap. 21. A standard tow or logging chain is composed of short straight links, carries a round hook on one end and a grabhook on the other. The round hook may be fastened to the chain by a ring, or a ring may be used instead of this hook.

Either the round hook or the ring can be used in chokers. The hook is easier to attach and to detach, but may fall away from the chain when it is slack. The ring may be used by passing the grabhook through it and pulling from the grabhook end; or for stumping, the chain near the ring may be pulled through it to form a loop that is dropped over the stump.

The grabhook fits over any individual chain link, and will not slide along the chain. It is used to adjust the length of chain by increasing or decreasing the amount of double line, by moving it toward or away from the choker end, or by passing the chain behind a tractor drawbar pin, and preventing it from being pulled out again by attaching the grabhook to the slack side, making it too large to be pulled through the space. In this case the surplus chain is slacked, and if it is long, must be hung on some part of the tractor. See Fig. 1.18.

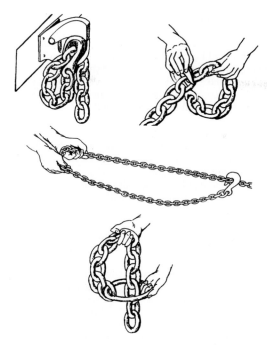

FIGURE 1.18   Grabhook uses, and stump choker.

Grabhooks are used to anchor a chain to a tree that is to be saved. The lack of sliding pressure makes it possible to protect the bark by pads and sticks placed on the side receiving the pull. The grabhook may also be used to make a ring which can be used to make up a choker.

Figure 1.19A shows three ways of fastening a line to a stump. In each case, the stump is shown to be grooved by an axe at the back. This cut is quickly made, and will prevent the chain from squeezing off during the pull and delay its slipping off as the stump leans. Part (A) is the easiest and most usual method, pulling at the center; (B) is a side pull, a little harder to arrange, but it puts less of a kink in the line, so that it can be used with cable as well as chain, and gives the advantage of a twist on the stump; and (C) is the overhead method, which requires an inverted T notch. This gives the greatest leverage but is more likely to slip off than the others.

Care should be exercised not to put loads on a chain that is twisted or kinked, as it will be broken or damaged. It can be readily checked for straightness, as the links which are in one plane should lie in an almost-straight line.

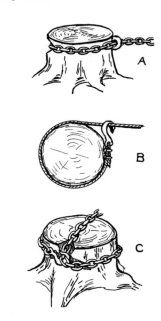

FIGURE 1.19A   Fastening line to stump.

Alloy steel chains weigh only about one-third as much as standard chains in proportion to strength. If a crew is careful enough not to lose chains, and is conscientious enough not to abuse them by kinking or gross overloading, alloy chains will amply repay their much higher cost in reduced labor and fatigue, and by greater efficiency.

As an example, one $\frac{3}{8}$-inch alloy chain, weighing 1.6 pounds per foot, is 30 percent stronger than the same make of $\frac{5}{8}$-inch ordinary chain, weighing 4.1 pounds per foot.

It is recommended that the alloy chain be dipped in bright red paint so that it can be easily recognized, and recovered readily if mislaid.

Cables.   Only the method shown in Fig. 1.19*A*, part (*B*), should be used in pulling a stump with a cable choker, as the sharp bends involved in the others will cause early breakage of the cable.

If a double cable line is used to reduce strain, or to shorten the rope, it should not be bent around sharp angles. A stump is generally round and smooth enough not to cut a cable wrapped around it, and the end hooks or loops can be attached to the drawbar. If the load is angular, it is better to fasten a snatch block to it with a chain or sling choker, and to run the long cable through the block pulley.

If a double cable is so wrapped around the load that it cannot slide around it, great care must be taken to adjust it so that both ends share the strain equally, unless a single line is strong enough to take the entire pull alone.

Root Hook.   A root hook may be used when a stump is too big to pull directly. Enough soil is dug away to expose the lateral roots, the hook is placed to grip one of these, power is applied, and the root is torn out. This process is repeated until the stump is sufficiently weakened to be taken out on one of the root pulls, or by direct pull on the butt.

The root hook may also be laid on top of a stump, with the teeth in a notch on the back. A pull on this gives excellent leverage, but the edge of the stump is liable to tear off. See Fig. 1.19*B*.

Taproots.   The presence of a taproot increases resistance of the stump. If the ground is hard, this root may be broken or pulled apart. If the ground is soft, or the wood very tough or pliable, the pivot point may crush and the root bend so that the pulling power is exerted directly against the length of the root, without benefit of leverage. In such a case, the upper roots of the stump may be torn up sufficiently so that an axe, or a special long chisel, can reach and cut the taproot. The cut should be made while pulling, as tension makes the wood part more easily.

Pulling Clear.   If the force is sufficient to uproot a stump, the roots opposite the pull break first, then those at the side, permitting the stump to be pulled onto its side. If the line does not slip off, the stump may be rolled and dragged out of the ground, but this often takes much more power than overturning the stump, and may be beyond the capacity of the machine that is doing the pulling.

If the stump will not come all the way, the line may be slacked and a log placed or chained against the stump, as in Fig. 1.20. This log will provide a new fulcrum and aid the breaking out. Or the line may be taken off and the tractor moved to pull in the opposite direction, which should

FIGURE 1.19B   Root hook.

FIGURE 1.20 Stump pulling.

free it without difficulty. If a number of stumps are being pulled, all of them may be overturned one way, before pulling the tough ones in the opposite direction.

Half-uprooted stumps are easily knocked out by dozers, and may be left for them to save the trouble of rerigging.

Resistance.   A stump's resistance varies in different directions. If on a slope, downhill pull is most effective. Otherwise it should be pulled toward its strongest roots, as these are easier to bend than to pull apart, and can be dealt with more easily when the rest of the stump is loosened.

The most obvious variable in stump resistance is its height. Greater height means greater leverage and easier pulling. Limiting factors are difficulty of high cutting and of fastening heavy chains at a height, and the trunk breaking under pull.

A buried stump is the hardest of all to pull and usually must be dug out. On filled land, two separate systems of lateral roots may be found, one under the old ground level and the other near the surface, in which case it may be necessary to cut the trunk below the upper roots, in the same way as a taproot.

A stump which yields to pull but will not break loose can often be uprooted by moving it as far as possible, slacking off to allow it to settle back, and pulling again, repeating this process a number of times. This is most effective if done slowly and smoothly, whether with winch or traction. This method is very effective with trees, as the trunk will bend with a whipping motion that exaggerates the force of both the pull and the snapback.

Chopping the roots on the side opposite the pull, while they are under maximum tension, weakens the resistance. A moderate amount of digging will generally expose the main lateral roots.

When a stump has been split by blasting, the pieces are most easily pulled away from the center, rather than across it.

Uprooting Trees.   If trees are so large that their stumps will be difficult to remove, it may be advisable to pull the trees over rather than to cut them down. This gives the opportunity to fasten lines as high as desired and to make use of the weight of the tree. As soon as the tree is pulled toward the tractor, its center of gravity shifts to that side and aids greatly at breaking out the roots. If a large log is chained to its base, on the pull side, the force of the tree's fall will be more effective at breaking roots on that side. The log will also serve to prevent the trunk from digging into the ground where it would be difficult to cut.

If the tree tends to break or split instead of uprooting, additional chokers may be used below the main pull point to distribute the strain and bind the trunk together. This can be done by pulling with two or more machines, or with multiple lines and blocks that will be described later.

If the trunk is smooth, a ladder will be needed to get a high grip. The chain may be held from sliding down by a nail or a notch, when necessary.

Pulling trees is apt to be wasteful of lumber, as the bottom of the trunk may be put under such strain that it will split when cut.

Pulling Small Growth.   Brush and small trees often grow where they cannot be reached by pusher machinery, because of soft or rough ground or nearness to buildings. A landowner may wish to do her or his own clearing without hiring a dozer. Hand cutting may not be satisfactory because of sprouting. In such cases pulling techniques will be applied to small growth.

An automobile has sufficient power for pulling some brush and small, stiff-trunked trees, but the work does not do it any good. Trucks and farm tractors usually put more power on the job and are less likely to be damaged by the exertion.

If the stems are stiff, fastening may be made high for leverage. If they are flexible, height does not matter, and the greater strength of the base may make it the best place.

Chains tend to slide along smooth stems, and they often can be made to grip by wrapping once or twice around before fastening. Light chain with small links holds much better than coarser types. A round hook or ring should be used to make a choker. If stems are close together, it is often possible to pull several at a time by putting a single choker around the group. It will slide up until it can pull them all tight together and then should hold.

Brush tongs get a good grip on small trees and flexible plants, and are easy to attach and to remove, but their weight may outweigh these advantages.

Plants too well rooted to respond to the power available may be weakened by digging out and cutting roots, or pulleys may be used to step up the power.

WINCHES

Power to pull stumps may be supplied by almost any machine or by animals. The crawler tractor is preferred for heavy work, but wheel tractors and trucks may also be used. However, the most powerful, most convenient, and best controlled pull is obtained from winches. These are most efficient when mounted on crawler tractors, but may be on wheel tractors or trucks, or may be portable units operated by a hand crank.

For general clearing work, the most effective tool is a dozer carrying a power winch. The winch consists of a heavy spool drum that is mounted on the back of the tractor and driven by the power takeoff. It is controlled by the tractor main clutch and the power takeoff engagement lever. In addition it may have a transmission, giving rotation of the drum in either direction, and in large machines permitting several speeds of rotation. A jaw clutch or neutral gear is used to disconnect the drum from the driveshaft, to allow it to turn freely when the cable is being removed. A brake is provided to slow or lock the drum when necessary.

The winch may hold 200 or more feet of cable of a size proportionate to its power. Additional cable can be carried on a separate spool and connected to the winch cable by a choker device when needed.

In small sizes, the winch cable generally is fastened at the working end to a short piece of chain equipped with a round hook. Larger cables may be fastened directly, or through a swivel or single link, to a round hook, or a wide-face cable grip hook. The cable is generally underwound on the drum, that is, leads from the work to the lower part of the drum. This gives better stability under heavy load than overwinding.

Stump Pulling.   To winch out a stump, the tractor should be placed facing directly away from it, and both brakes locked on. The winch jaw clutch should be released, its brake set to drag very slightly, and the cable pulled to the stump by hand. If the brake is not used, the drum may continue to spin after being pulled, and unwind and snarl the cable.

If the winch will not freewheel, or the cable is very heavy, the drum is turned backward by the engine to pay it out. It is convenient to have two operators, one to operate the winch and the other to pull the cable. If no helper is available, the operator can stand near the winch while it turns, stripping the cable and coiling it on the ground until there is enough. The operator then stops the winch and drags the cable to the stump. The cable must then be whipped up and down and the twists worked out to avoid kinking when pulled.

The winch cable may be put around the stump directly, may be hooked to a choker chain or cable, or may be run through one or more snatch blocks.

Power is applied to the winch and the cable is reeled in, care being taken to see that it feeds onto the drum properly. The stump may come out or the tractor may be dragged backward. If the latter, the tractor may be anchored by a chain from the blade or front pull hook to a tree. Resistance to pull may also be increased by backing it against a log or bank, or by trying to pull the stump by tractor pull and allowing the tracks to spin until they have built mounds behind them. If the anchoring or blocking is effective, the stump will come out—if nothing breaks, slips, or stalls.

The drum carries a number of layers of cable so that it has a greater spool diameter full than empty. It therefore reels in cable more slowly and powerfully on a bare drum than on a full one. On a bare drum, logging winches will give 50 to 100 percent more pull than the tractor itself; on a full drum, the same pull as the tractor or somewhat less. But with torque converter drive, strongest pull may be with a full drum.

Jammed Cables.   Using a nearly bare drum not only gives the greatest pull but also reduces damage to the cable. If a long cable is wound smoothly onto a drum under moderate tension, and a heavy pull applied when it has built up several layers, the last wrap may squeeze between the wraps below, as in Fig. 1.21(A). This scrapes and wears the cable and jams it so that it will not spool off again. The best way to free it is to turn the drum until the catch is in the position shown in (B), and jerking it, or anchoring the end and driving the tractor away. Or, in the same position on the drum, the cable may be given a couple of wraps around the drawbar, and the winch turned backward as in (C).

If the cable is wound unevenly onto the drum, with the wraps crossing each other at random, it cannot cut down between lower layers readily, but may put severe kinks in sections of cable that cross under it, and this cross wrapping may not entirely prevent it from squeezing in and sticking.

In spite of these difficulties, a long cable is desirable for general work. If reasonable effort is made to spool it in evenly while working, it will usually be rough enough to prevent excessive sticking, without too much bending or crushing.

Two-Part Line.   Where the distance to the stump is less than one-half the cable length, a two-part line may be used by attaching a pulley to the stump and by running the line from the winch around the pulley and back to the drawbar. The useful strength of the cable and the pull between the tractor and the stump are doubled.

The tractor may have to be backed against a heavy log or an outside anchor used in the manner to be described below. The tractor should not be anchored by the front pull hook while using a double line anchored on the drawbar, unless the manufacturer states that it is strong enough to take the strain.

Rocking.   The winch and tractor pulls differ in quality, and it may happen that the pull of the tracks will do jobs that the winch will not. Use of the tractor drive helps in "rocking" stumps or trees out. The line is left slightly slack, and the tractor moved forward in low. As the line tightens, the stump may lean a few inches, then stop. When the tracks start to spin and the clutch is released, the weight of the stump, combined with the spring in the roots and in the line, will pull the tractor back. The clutch is immediately reengaged and held until the tracks spin or the engine lugs down again.

If the stump is within the tractor's power range, repeating this maneuver should gradually break it out. A long cable has more elasticity than a short one, or a chain, and will be more effective at rocking.

This procedure should not be allowed to degenerate into yanking, where the tractor is given a long enough slack run to be brought up with a jerk when it tightens. This will break more tractors and cables than it will pull stumps.

Cable Breakage.   Cable or chain breakage is a serious danger to both operator and helpers. A cable particularly stretches under strain, and if it breaks suddenly, it may whip with great force. The danger to the operator is greatest if the break is fairly near him or her. The cable used should

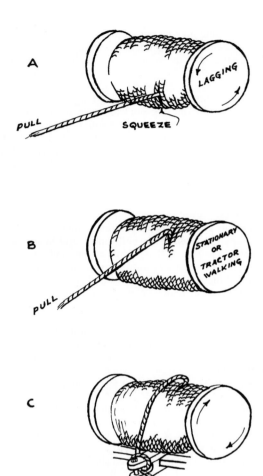

FIGURE 1.21 Freeing winch cable.

be the best quality, in the largest size recommended by the winch manufacturer, if the tractor is to be anchored or used for rocking; and it should be inspected frequently for weak spots.

Another danger inherent in the use of cable is cutting and tearing of the hands and clothing on broken wires. Preformed cable gives minimum trouble of this kind and should be used when possible. Leather-palmed gloves are good protection for the hands.

Either hemp-center or wire-center cable may be used, according to preference or manufacturer's recommendation. Wire center is about 10 percent stronger, size for size, is stiffer, and is not as easily deformed by crushing. It is more difficult to handle, and when kinked or crushed is much harder to straighten. Standard 6×19 constructions are usually recommended.

It is good practice to work a winch at less than its maximum capacity, and to avoid anchoring the tractor unless absolutely necessary. Moderate loads give long life to cables and winch parts, and avoid severe catching on the drum. If the work is heavy, strain can be reduced by the use of pulleys and multiple lines.

Broken cables can be repaired by splicing, but the length of cable used in the splice, and the labor involved, may be too great to justify this method for the short cables ordinarily used in land clearing.

A rough repair may be made by trimming back the broken ends, overlapping them as in Fig. 1.22, and fastening them with two or three cable clamps, for sizes up to $\frac{5}{8}$ or $\frac{1}{2}$ inch, or with three or more for larger sizes. Or two interlocked loops may be made, as in (*B*), fastened with clamps or any type of loop fastening. Cables repaired in this manner are weakened but may last a long time. The patch will not go through pulleys and is inconvenient in other ways.

Winches on Wheel Tractors. If a winch is mounted on a wheel tractor or truck, it is usually necessary to anchor it for heavy pulls. The anchor chain or cable should be attached to the winch frame, or to a heavy member as near to it as possible, to reduce strain on the tractor.

An important consideration in the use of these winches is the fact that the cable will tend to take a straight line between the work and the anchor. In making a high pull, as in Fig. 1.23(*A*), the

FIGURE 1.22 Temporary cable repair.

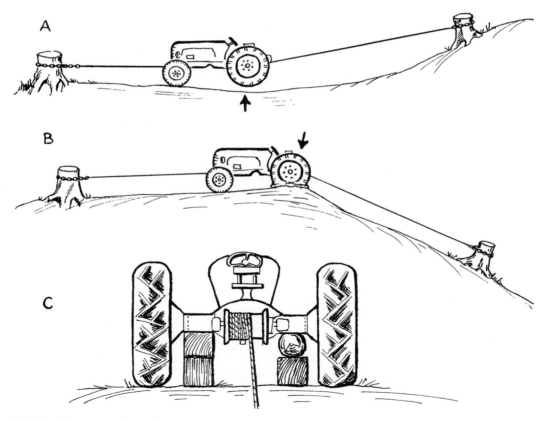

FIGURE 1.23 Vertical effects of winch pull.

tightening cable may lift the tractor and turn it over sideward. In (*B*) the downward pull may blow the tires, unless the axle housing is blocked up, as in (*C*).

If a wheel tractor is not anchored, a rear winch must be underwound, and care must be taken that the machine does not overturn through rising on the front, a danger which is particularly serious if the tractor is driven to move the load.

Truck Winches. A truck winch is usually of the horizontal drum type. It may be mounted in the front bumper, on a flatbed body, or between the body and the cab.

Rotation may be for underwinding, overwinding, or both. Power is from the power takeoff, controlled by the truck engine clutch pedal.

The principal handicap of a truck winch is the difficulty of maneuvering it into position for a straight pull. One or more pulleys may be required to obtain a proper direction of pull and a straight line onto the winch. The truck should have all the wheels blocked, or be anchored by a line from a frame member near the winch.

A gypsy spool or capstan winch (Fig. 1.24) may be mounted vertically on the forward end of a flat body, chiefly for dragging loads onto the truck. A hemp rope is looped around it two or three times, with one end attached to the work and the other end held by the operator. If the operator leaves it slack, the spool will turn inside the rope; if she or he pulls it tight, the working end of the rope will be pulled with great force. The slippage on the spool absorbs shocks that would break the rope and enables it to do very heavy pulling, under exact control. However, the gypsy is not ordinarily used for stumping.

Hand Winches. Hand winches are turned by a hand crank, operating through one or more sets of reduction gears. Under most conditions, it is not possible to make a full turn of the handle because it strikes obstructions, or passes through awkward positions. A large part of the work of winching consists in removing and replacing this handle, and if much work is to be done, a ratchet handle should be purchased, or made up by adapting one from a heavy socket set.

The winch is usually equipped with a friction brake and a pawl that can be engaged to prevent it from turning backward when the handle is released.

Operation of these devices is tedious because of the number of crank turns which must be made to reel in the cable; and exhausting because of the force which must be applied to the handle to develop the rated pull of the winch. It is important that it be thoroughly lubricated.

Hand winches can be used in places inaccessible to power equipment, are comparatively inexpensive, and are surprisingly powerful. Sometimes they can take out tougher stumps than a power winch of the same pull, because the line can be left taut and tightened gradually or from time to time as the stump yields. Their weight, with cable, may be from 75 to 300 pounds, so that carrying one of them any distance is at least a two-person job. It can often be transported in a loader bucket.

Hand winches are sometimes mounted on a truck, in which case they serve largely as a spool to carry cable, most of the pulling being done by the power of the truck. If the job is too heavy for the truck, it may be anchored or blocked and the work done with the winch handle.

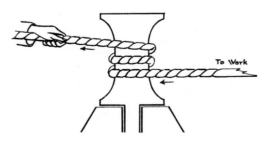

FIGURE 1.24 Capstan winch.

If not mounted on a truck or other carrier, the winch should have a V-shaped towbar, or a subframe by which it can be anchored. Planks should be provided to build up a base in line with the pull, since if the line of pull is high, the winch will be lifted off the ground and will not be steady enough to allow turning the handle.

MULTIPLE LINES

Snatch Blocks.   If pulling stumps takes the full power of the tractor or winch, it may be advisable to use snatch blocks to obtain greater power at slower speed. These devices, also known as blocks and as pulleys, are pulleys set in frames that are provided with one or two round hooks or rings, usually on swivel connections. For most field work, single pulley wheels, with a latch attachment permitting insertion of a cable at the side, are best, as cables usually carry attachments too large for threading; a tedious job even when possible.

These blocks can be obtained in sizes to match any cable or strain. In large sizes they are very heavy, and several workers, or a loader, or light winch or other lines may be used to move them.

Figure 1.25 shows several riggings using pulleys. If the lines are approximately parallel, and the pulley bushings lubricated, each additional line will add about 90 percent to the single line pull. This puts no extra strain on the winch cable, but the chokers holding the blocks must take the combined pull of all the lines fastened to them.

The advantage obtained from the use of a block is decreased when the lines are not parallel, becoming zero when the angle between the lines is 114 degrees. Still wider angles result in loss of power.

The number of lines that can be put on one stump may be determined by the number of blocks on hand, the space available for fastening them, the strength of the available anchorage, or the amount of cable. Light machines may use six or eight blocks on a heavy stump.

Rigging is simplified by the use of series and sling blocks, as in Fig. 1.26A. The tractor line passes around a block pulley to an anchor, doubling the pull at the block frame. This is attached

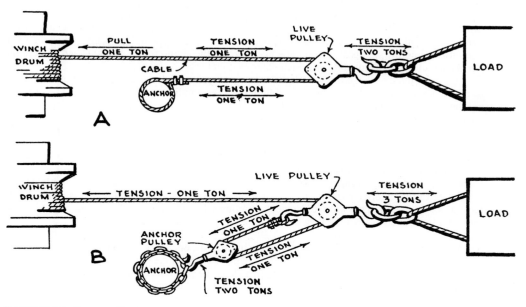

FIGURE 1.25   Stump-pulling layouts.

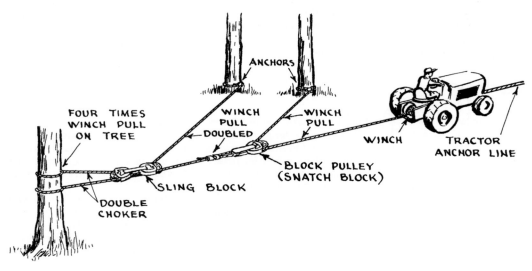

FIGURE 1.26A   Use of sling block.

to a heavier line which passes around another pulley to an anchor. The second pulley is pulled with almost four times the power of the winch.

The sling block makes possible the use of a double choker on the tree being pulled. The choker cable is approximately centered in the sling pulley, and both ends are hooked around the tree. Such a double choker can be of lighter and more flexible cable than a single choker.

Rigging.   Multiple blocks require care in rigging and pulling. Two anchors are better than one, as they spread the cables over a wider space where they are less apt to interfere with each other. Each block is best fastened to a separate choker, but one may be fastened to each end of a chain passed behind the stump if it is strong enough to take a double pull; or one to both ends of a chain given one turn around the stump. It is good practice to notch the stump for each chain used, so that the chains and blocks will not slide into each other as it yields.

Rigging is done with the lines slack. When they are pulled tight, an inspection should be made to make sure that no pulley latches have fallen open, as a pull on an open block will bend it and cut the cable; that no pulleys are jammed with debris, or liable to pull into each other; and that no chain hooks have become disengaged. As the line is wound in, all blocks should be watched to make certain that they do not collide. Lines should not be allowed to drag on each other.

If the winch does not carry sufficient line for the distance or the number of lines involved, extra line can be added by the use of a take-up block, as in Fig. 1.26*B*. A standard practice is to use the winch line from the tractor to near the first snatch block, and the extra line for reeving. The take-up cannot be pulled through a snatch block, and is liable to cause trouble if included in the multiple lines.

The extra line is often carried on a spool supported on a pipe axle and brackets. This should have a drag brake of some kind, to prevent spinning when paying out. The brake might consist only of a log leaned against the face or side, or a worker's gloved hand.

Anchors.   It is often a question of whether the stump or the anchor will yield. Anchor lines should be as low as possible and stump lines high. It may be best to pull the largest stumps first, using several smaller ones for anchorage if necessary. In a clean-clearing job, there is always one last stump for which there is no anchor, and if it is small, it may be pulled out directly; or in any case it will respond to less elaborate artificial anchors than a large one. On the other hand, a large stump will be a dependable anchor, and will prevent the need of frequent rerigging when anchors pull out.

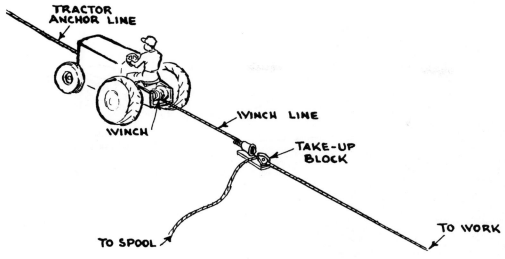

FIGURE 1.26B   Take-up block.

The final stump may be pulled by use of a living tree as an anchor. A choker should not be used under any circumstances on a tree which is to be preserved; padding and blocks should be used with a grabhook loop.

If no anchor is available, one may be made, ground conditions permitting, by digging a T-shaped trench, 2 or more feet in depth, as shown in Fig. 1.27. A log is placed in the crossbar, the cable anchored to it and led up through the sloping trench toward the work. Load and local conditions will determine the depth of cut and size of log. In medium soil, a standard railroad tie 2 feet down should hold a horizontal pull of 5 tons. This is sometimes called a "deadman."

Advantages of Blocks.   Stump pulling with a winch and blocks takes more time and care than direct winch pull, but results are generally more satisfactory. Jerks and jars which are destructive to machinery and cables are largely eliminated. Lighter cable may be used, and a sufficient number of lines will reduce the tension on any one so much that squeezing and crushing on the drum will not occur.

A snatch block may be used to advantage with an anchored winch to avoid shifting its direction for each stump. Figure 1.28 shows a plot in which a number of scattered stumps are to be pulled. A snatch block is anchored to one of them, and the line led through it. The line can then be attached to stumps

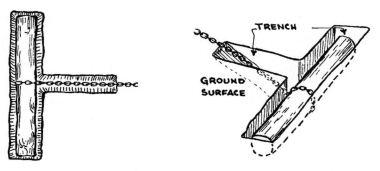

FIGURE 1.27   Artificial anchor.

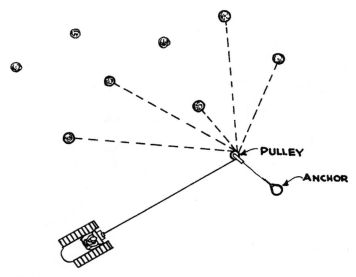

FIGURE 1.28 Use of guide pulley.

straight in line with the winch, or to any on the opposite side from the anchor, as the pulley will lead the line almost straight in, at the price of a small friction loss. The pulley is best placed at a moderate distance from the winch so that the cable can feed evenly onto the drum, instead of tending to pile up in the center. The line from pulley to anchor should be short, to keep shifting of the pulley to a minimum.

STUMP BLASTING

Blasting may be used to break up and remove stumps when machinery heavy enough for the work is not available. Explosives may do the complete job, or just weaken them for removal by equipment.

The standard explosive is 40 percent dynamite, but only because it is most convenient. Slower dynamites and even black powder may be more effective. Charges are generally too small to justify using ammonium nitrate.

Precautions. Stump blasting is dangerous work at best, because of unpredictable conditions underground. Particularly, a rock may be held just over the charge by roots in a position that will cause it to take off like a mortar shell. Split pieces of wood will also fly long distances. Mud or earth packs over charges should be free of stones or pebbles, and personnel should move a long way back from the blast, unless good shelter is available.

All stumps should be accounted for after a blast, as they are sometimes blown up in trees, where they stay until dislodged by wind or another blast, with serious results to persons underneath.

If the blasting is to be done near buildings, logs or saplings chained together should be piled on the ground on the side of the stump toward the building, to stop stones and fragments from flying. Regular blasting mats are safer, but if machinery capable of handling them is on the job, it generally can pull the stumps without the necessity of using explosives in close quarters.

WOOD DISPOSAL

Methods. Some of the methods for disposing of the ground materials in clearing the land have been discussed earlier in this chapter. For many years in the past, brush, tree slash, and stumps to be disposed of while clearing a site have been trucked off to landfill disposal areas. In addition to the loading and hauling costs, the landfill owners might have charged $30, or so, to dump a load on the landfill. Recently, that part of the cost has been climbing steadily and may now be as much as $400 per load. With such a high cost there has been an incentive to develop other means for disposing of the wood growth on the land to be cleared.

Special Equipment.   Brush and limb chippers have been used extensively for reducing small sizes and quantities to wood chips. The material created in this way generally is small in quantity and is left on the site for ground cover. However, this type of equipment cannot handle stumps and other large logs. To handle and dispose of large quantities and sizes of wood, a different type of equipment is needed.

A heavy-duty wood-processing and stump grinder had to be developed. The challenge was met by Morbark with a piece of machinery called a waste recycler. It is a towable wastewood processing machine with a carbide knife cutting wheel in the tub powered by a 650-horsepower diesel engine. The piece of equipment is rigged with a full-circle-rotating, grapple knuckle-boom for self-loading the material to be ground into chips. See Fig. 1.29. Other manufacturers are now making this kind of wood-processing equipment.

One contractor has mounted the waste-processing machine on a crawler track undercarriage so that the machine can be pulled around a large site. This eliminates some of the problems of hauling the timber trash to a central disposal point, saving on labor costs and avoiding a large pile of mulch material.

BURNING STUMPS

In the Ground.   Dead, dry stumps can sometimes be burned without taking them out of the ground, but the process is usually slow and laborious.

FIGURE 1.29   Waste recycling chipper. (*Courtesy of Morbark Industries, Inc.*)

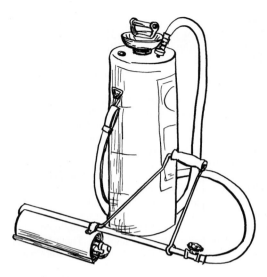

FIGURE 1.30 Flame gun.

A standard tool for stump burning is a large kerosene blowtorch, called a flame gun. See Fig. 1.30. It may be blocked to flame against a stump. One person can operate several, or do other jobs while taking care of one. When the wood starts to burn, the torch may be moved to another stump and brought back if the fire dies down. Green wood will require continuous heat for many hours.

The torch will operate most effectively if directed into a cavity with an opening in the far end so that a draft can move through it. If the flame is aimed into a dead-end hollow, very high temperatures will be attained, but because of lack of oxygen the wood will distill rather than burn and will be destroyed slowly. If the flame is used against the outside of the stump, it should be directed upward to draw a current of fresh air between the flame and the wood.

Dry stumps may also be burned by starting a wood fire alongside them, and keeping it supplied with logs and snags placed to almost touch the stump. The draft and reflection of heat will keep both surfaces burning, but the loose wood must be moved in rather frequently. This method may remove only the top and outside of the stump, leaving a conical core.

Care should be taken to avoid spreading stump fires. Roots may burn underground to start surface fires at a distance. Soils rich in humus, such as swamp peat or forest loam, may burn unless saturated with water, and are very difficult to extinguish.

In a Pile. Because of difficulties in burial, and scenic and environmental damage in piling, the best way to get rid of a pile of stumps is to burn them. However, it is becoming increasingly difficult to get a permit for the work.

The practical side is also difficult. A stump may cling to more than its own weight of dirt and rock, most of which should be knocked off to make it burnable. This may be done by allowing the dirt to dry, then kicking it around with a dozer or other machine, or by picking up with a grapple and clamshell and dropping.

Even if clean, green stumps are difficult to ignite. If they must be piled before burning, the base of the pile should include a substantial quantity of old tires, preferably topped by a layer of logs.

But the surest way is to burn them at the same time they are piled. Build a hot fire by the methods described in the next section, supply it with enough heavy wood to keep it hot for several hours; then roll, push, or lower stumps onto it, in sufficient quantity to make a thick mass that can sustain fire by itself.

A grapple or clamshell on a power crane is the best machine to feed a stump fire, even if the stumps are pushed to it by dozers. It can place them properly, stack them high, and avoid pushing dirt with them.

Stump fires are very hot and may burn for weeks. There is usually very little smoke, and few sparks.

BURNING BRUSH

A great deal of time and effort are wasted in ineffective attempts to burn brush, and for this reason proper procedure will be discussed in some detail.

Even green, wet brush and logs will burn vigorously once properly started, but considerable heat is required to boil off the sap and water, and to ignite the wood. This heat may be obtained originally from a carefully built fire, or by use of flammable chemicals.

Building a Fire. The fire should be on level ground, or on a hump. If built in a hollow or against a rock or stump, inward flow of air will be hindered, and brush added to the top of the fire will be held up away from the heat. All flammable material should be cleared or burned away from around the site, particularly downwind. Fire-fighting tools should be available.

Figure 1.31 illustrates two ways of starting the fire—andirons and tepee. The "andirons" consist of a pair of small logs, or rocks, or ridges of dirt. Twigs and sticks, preferably dry, are laid across the andirons. These should be laid in one direction so that they will lie close together, but should not fit together so well as to prevent air and heat going between them. No leaves or grass should be included.

This pile may be ignited by burning paper, grass, or leaves under it. The material must be dry, and must not be packed tightly, as this reduces the oxygen supply and the heat of the flame.

A self-feeding starter may be made by tearing a section of 10 to 30 pages of newspaper into a strip that will fit easily between the base logs, lying flat. Crumple the top sheet, and light it.

As it burns, the heat will cause the next sheet to curl up and burn. The process repeats for every sheet, keeping a brisk fire going for long enough to ignite dry logs. No kindling is needed, but of course small, dry wood starts faster than thick, green pieces, for which the process may have to be repeated.

When the cross sticks start to burn, more and heavier sticks are added, then partly trimmed branches, and finally, when a good bed of embers and strong flames are present, untrimmed bushes and branches. It is a good plan to put on a few logs or snags at this time to give the fire staying power.

The tepee is similar in principle. The sticks are piled on end around the kindling. As heavier pieces are added, the tepee is crushed, but if it is burning well, this will not matter.

FIGURE 1.31 Starting a fire.

A danger in transition from the hand-tended fire to the roughly piled one is that the untrimmed brush may include so much airspace that the heat cannot cross it effectively. The fire may burn a dome-shaped hole over itself, then die down. In such a case, sticks should be poked into the fire itself to build it up, and the brush over it should be compacted by rearrangement or piling on of heavy sticks. This is tiresome work and may fail. It is better to tend the fire longer before piling on loose material, to be sure it will not have to be worked over afterward.

Artificial Helps.   Old tires provide excellent material for starting a fire. Trimmed brush can be piled on them as soon as they are burning.

A dying fire may be pepped up by use of kerosene, fuel oil, gasoline, or similar fluids. To be effective, these must be applied at the base of the pile. Because of its explosive qualities, gasoline should be applied only as a stream from a blowtorch or similar pressurized device with a fine nozzle, and only when it burns as it is ejected. If it does not burn, it may accumulate in sufficient quantity to cause an explosion.

Putting flammable fluids on the heap itself may produce a fine flame, but it will have little kindling effect, as the evaporating fuel will absorb the heat that radiates downward.

Flame guns produce a hot flame up to 20 inches long. They are effective kindlers when directed into the base of a pile.

A brush burner is a portable unit that combines a heavy fan to direct a strong wind into a pile being burned, with a mist of fuel oil for kindling. Few piles can resist one for long. See Fig. 1.32.

If the fire dies down in spite of nursing, it may be best to build a new fire nearby, with greater care to avoid airspaces and coarse, green wood early in its life.

Transferring Embers.   Once a good fire is burning on the job, its embers may be shoveled out and used for starting other fires. This should be done rather frequently, as a long brush carry adds greatly to labor costs.

FIGURE 1.32   Brush burner.

Four or five shovels of hot embers may be laid on the ground in a pile, and fine brush, or dry twigs and wood, piled on it. Or the embers may be sifted down through piled brush. The embers give a sustained heat and consume little oxygen, so that a strong new fire starts quite quickly.

Feeding.   It usually takes at least two people cutting and dragging brush to keep one fire burning briskly. If it is allowed to burn down, it is good practice to put the unburned ends in the center hot spot, before piling on more brush.

When a dozer is used, ample supplies of fuel can be brought to the fire, and it is usually well packed by the pressure of the blade and the weight of the machine if it climbs up on the pile.

The principal problem of dozer feeding is dirt. This tends to block the fire from spreading into new material, and to smother parts already burning. Every effort must be made to reduce the amount of dirt by rolling and jostling piles, holding the blade high enough not to dig in, and giving the vegetation and mud a chance to dry before bringing it in.

A hot fire will burn through quite a lot of soil, but it will seldom burn clean. After it cools, the remaining stems and stumps can be sifted out by the dozer and used in building the next fire.

Good results in fire feeding are obtained only if most of the new material is placed on top of the flames.

Burning Piles.   If the brush is piled a long time before being burned, dropping a match in it on a hot day may accomplish its complete removal. If it has been piled only a few hours or a few days, a fire may be built on the windward side against it but not under it. This fire may be caused to spread into the heap by keeping it buried under compact brush, so that the fire is fed and the heat reflected into the pile. If the brush has leaves, it is good practice to cover any place where flames show through. A strong fire cannot be smothered with hand-piled brush.

Brush piles may be pushed on top of fires by a dozer, placed by a clamshell, or rolled on by a number of workers using long poles.

If brush is being cut in an area presenting unusual fire hazard, or the cutting is in small, scattered areas, it may be desirable to truck it to a central burning place. A continuous fire may be maintained with incoming loads dozed or hand-piled onto it, or the brush may be piled to dry and burned off occasionally.

Brush up to a few inches in diameter can be reduced to chips by a chopping machine, after which it can either be left on the ground or easily trucked to a dump.

Clamshell.   One excellent combination for heavy clearing and burning is a large dozer, preferably with a rake blade, and a clamshell shovel. The dozer uproots and pushes in brush and trees, and the clamshell picks them up, shakes dirt out, and places them on top of the fire. See Fig. 1.33.

The clamshell can also maintain a fire, moving in unburned ends, and can bury it under dirt at the end of the work day.

A clamshell is also often the best tool for high stacking of vegetation that is to be left to rot, and for burning old piles that need rehandling.

Banking Fires.   If the job is not extensive enough to justify the employment of someone at night to watch the fires, and any flammable material is nearby, they should be buried under a few inches of clean dirt at the end of the work. Humus or rich topsoil should not be used. The soil cover will prevent sparks from blowing, will preserve a hot bed for use in the morning, and, if the cover is not removed, may make a fair grade of charcoal.

FIRE CONTROL

Any contractors burning brush in an area subject to brush or forest fires are subject to heavy responsibility if one of their fires spreads. Also, in the presence of extensive forest fires from any cause, contractors may be required by authorities to use their workers and equipment to control

FIGURE 1.33 A good fire tender.

them. At such a time there might not be experienced fire-fighting personnel available to direct the work. A brief outline of fire-fighting techniques is therefore considered appropriate.

Hand Tools. Where the material burning is largely grass and associated weeds, or thin brush, fire can be beaten out. Household brooms, occasionally dipped in water if possible, are very effective. Shovels or leafy bushes or branches can be used with good effect. Each blow should be directed so that flying sparks are knocked toward the burned area.

The fire may also be starved by scraping away the vegetation just beyond the flames. This may be done with shovels, hoes, rakes, grub axes, or almost any piece of metal. A special type of fire-fighting tool, shaped like a heavy rake and fitted with sickle bar teeth instead of tines, is quite effective. Bushes may be cut with axes, machetes, bush hooks, or pruners.

Extinguishers. Backpack fire extinguishers, which consist of a water tank carried like a knapsack, a flexible hose, a hand pump, and a nozzle, are important pieces of equipment. See Fig. 1.34. If the grass is low or thin, spraying in the path of the fire may stop it. If the fire is strong and moving rapidly, the water may be most effectively used for putting out smoldering spots behind the beaters. Addition of a wetting agent—a small quantity of almost any detergent will do if regular compounds are not available—increases the effectiveness of the water by enabling it to soak through vegetable litter and punky wood.

Pumps. If streams or ponds are available, the contractor's pumps, particularly the light centrifugal type, are very valuable. A welder or machinist can usually make adapters quite quickly that will permit fire hose to be attached to the pump outlet. The high pressures used in regular fire pumps will probably not be developed, but sufficient pressure will be available for wetting down firebreaks or making direct attacks on anything short of a crown fire.

FIGURE 1.34 Backpack extinguisher.

Sprayers.   Tree-spraying outfits make good fire fighters. These usually consist of a tank holding from 200 to 500 gallons; a high-pressure pump driven by a small gasoline engine, or by the power takeoff of a towing tractor, or a carrying truck; a reel of hose; and a nozzle. Those having an engine are generally mounted on a wagon chassis that can be towed by almost any motor vehicle. If the pump is tractor-driven, adaptation to most wheel tractors can be made quickly. The handiest models are those mounted on a motor truck.

Such equipment can generally be rented or borrowed in almost any area. The volume of water delivered through the nozzle is small, but pressure is high and results are usually excellent.

Dozers.   A bulldozer can put out a grass fire by starting behind the fire and straddling its line, as in Fig. 1.35. It may be able to scrape off the grass without cutting much into the ground. If this is not practical, it can skim off the sod until the load is heavy, then swing it into the burned area, or raise the blade and spread the sod over the next few feet of flames, smothering them. An angle dozer can sidecast the sod into the burned area, and a hand beater, or extinguisher, should follow to put out any spots that are missed.

Method of Attack.   Windblown fires should not be attacked directly at the front, as this procedure is both dangerous and ineffective. A new fire running before a wind will assume a shape similar

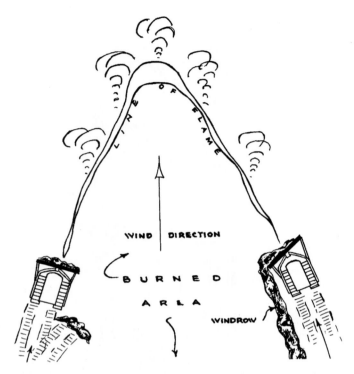

FIGURE 1.35 Fire-fighting with dozers.

to that in Fig. 1.35. A direct attack on the front means fighting flames several feet deep. If these should be put out, fire blowing up the sides could rekindle them in a few seconds. A crippled person or machine ahead of the fire could not escape being burned.

Pinching off the sides is both effective and reasonably safe. The fire is extinguished starting at the back so that the heat and smoke are blown away from the workers. Provided a constant watch is kept behind them for rekindled spots, the fire cannot repossess the extinguished area. When the front is reached, it is attacked from directly behind as well as on the sides.

If the fire is too strong for the force fighting it, the front will continue to advance, but the work on the flanks will limit its width and make easier the task of stopping it with firebreaks or backfires, or after a shift in wind direction. It can sometimes be turned by concentrating on one flank.

Firebreaks. A firebreak is any strip bare enough of flammable vegetation to delay or stop the spread of fire across it. Roads, open water, plowed fields, close-cut lawns, and even footpaths may be used. In addition, breaks may be prepared in anticipation of fire along the crest of hills or mountains, at property lines, or at the edge of the areas being cleared.

Advantage should be taken of any existing breaks when deciding where to place one to stop a fire already burning. A short line is preferable, and valuable property or highly flammable areas should be protected. The break should be far enough from the fire to allow time to finish it and to start backfires; it should be in vegetation least apt to make a spark-producing or a high fire, and on terrain favorable to operation of machinery. A compromise among these features must usually be made.

A bulldozer may be walked along the line of the break, alternately cutting and filling, so as to mix the vegetation with dirt. Hand workers with cutting or digging tools follow, to cut out any spots where fire might cross. If the brush is heavy, the dozer may turn to push heaps of it out of the path. An angle dozer or a heavy grader might be able to make a single clean cut in each direction, turning the sod and brush out from the center.

In grass or light brush, a plow or heavy disc harrow might do a better or faster job than a dozer. The plow makes a rather narrow strip with each trip, and is subject to jamming with brush but does not have to go back over its work. A harrow may require several trips and might not be effective.

The hand tools listed earlier may be used to build a complete break, or to work one over after the machinery has passed.

Backfires.   A strong fire adds to the force of the wind which is moving it, somewhat as a blowtorch builds up its fuel pressure. The combined force may be enough to project a sheet of flame many feet in front of the burning line, and to shower sparks for long distances ahead. For this reason, the fire may cross a break of any practical width and make the area too hot for fire fighters.

The principal use of the firebreak is to provide a line from which backfires can be started. Since the break is made on the downwind side of the fire, a new fire started on that edge burns upwind. The backfire should be made in a continuous strip along the break so that it will not be able to turn and blow back toward it. It will increase in strength as it progresses, but will be steadily farther away from the protected side. When it meets the main fire, there is liable to be a spectacular flareup and heavy production of sparks. If the backfire has been started in time, this should be far enough away from the break that few sparks will cross it, and those can be extinguished by men patrolling the break. If no shift in wind occurs, the sides of the fire can then be put out by the crew working from behind, aided by the firebreak crew.

If the break is made in a forest where the flames might crown (burn in the tops), the trees on each side of the break should be bulldozed or cut so as to fall away from the center.

Since a change in wind direction may occur at any time, care should be taken not to start backfires prematurely, and to keep workers and machines in positions where they can get away if the fire turns toward them. The burnt-over area, ponds or wet swamps, and plowed land or well-grazed pastures are suitable retreats.

Workers on the fire lines must be kept provided with food, water, and tools, and relieved for rest periods. Machinery must have fuel but may be skimped on other maintenance in sufficiently dangerous situations.

Backfiring, and possibly other phases of fire fighting, may be regulated or prohibited by local laws.

Rekindling.   After the spreading of a fire has been checked, it must be patrolled until all danger of its making a fresh start has passed. A grass fire in a clean field may be safe to leave within an hour, while wooded areas containing dead or fallen trees, or rich dry soil, may be dangerous until after several soaking rains.

Dead stumps may burn a long time and are difficult to extinguish unless ample supplies of water are available. Fires burning under and between logs on the ground can often be put out by moving the logs apart, or can be caused to burn out more quickly by piling additional wood on them.

The worst hazards are standing dead or hollow trees, called *snags* by the lumbermen. If close to the line, snags may set fire to the unburned area by falling into it. They frequently produce sparks that may drift long distances. Even thorough soaking may not extinguish them, and it may be necessary to cut them down or maintain an expensive patrol for days or weeks.

Cutting a burning tree is a tricky and dangerous job, as the cutters are in constant danger of being hit by falling pieces, and temperatures at the base may be too high for them or their tools. This job is best left to experienced fire-fighting crews.

Snags may be pushed over by bulldozers, but the tops are apt to fall on the machine. A cab with maximum-strength overhead protection is needed for operator safety.

The best time to check a burned area for hot spots is immediately after a rain, or a heavy dew, as the moisture near the fires will steam.

Underground Fires.   Underground fires, such as occur in rich forest soils and dried-out swamps, constitute a special problem. When fire gets in them, often by smoldering down a dead root, they will burn hot and persistently. Plain water has little effect on such a fire unless applied in such quantities that the area is flooded. Smaller quantities do not penetrate the deeper burning zones, which have sufficient heat to evaporate quantities of water from surrounding peat, and then spread through the dried material.

Special nozzles consisting of pipes long enough to reach the bottom of the fire are helpful. The lower end is plugged, and a fairly large hole is drilled in the plug to wash humus out of the way as the pipe is pushed down, and smaller holes in the side spread a soaking spray. The use of wetting agents will substantially reduce the amount of water required, and may make the difference between success and failure where the water supply is limited.

Such a fire may be confined by trenching down to inorganic or saturated soil. The digging may be quite difficult because of roots, and a backhoe or dragline shovel might have to be used.

Peat fires spread very slowly unless they ignite surface vegetation or litter which set fire to the soil at new points. If equipment is not immediately available to extinguish or ditch the fire, leaves and flammable trash should be removed for 10 or more feet around it, to prevent rapid spreading while arrangements are made to put it out.

Burning Box or Trench   Air Burners' system (see Fig. 1.36) provides a method for burning wood products on site in either a refractory-lined box or an earthen-lined trench. The operating principle of the air curtain within an incineration setup is based on a controlled high-velocity stream of air across the upper portion of the combustion chamber in which the clean wood waste is loaded. For proper operation the air curtain machine has to provide a curtain of air over the fire that has a mass flow and velocity that are in balance with the given mass flow and velocity of the burning wood waste. When done well the ash from the typical wood waste is a very useful soil additive, and as such offers a commodity that can be marketed to plant nurseries and farms as a potting soil additive.

FIGURE 1.36   Incineration box by Air Burners, LLC.

BOULDERS AND BUILDINGS

Boulders.   An area may be so strewn with loose or partially buried boulders that work is difficult, and the removal of these rocks may properly be considered clearing.

If large enough machinery and suitable disposal points are available, the rocks may be turned or dug out and pushed away. If they are too large for easy handling and disposal, they should be broken up.

Breaking may be done by blockhole or mudcap blasting, backhoe-mounted demolition hammers, drilling followed by plug and feathers splitting, or a muscle-operated sledgehammer. See Fig. 1.37.

A plug and feathers set consists of a pair of half-cylinders (feathers) with outer surfaces fitting in a drilled hole, with inner faces shaped for driving a thin steel wedge (plug) between them by air or hand hammer, or by hydraulic pressure.

The very gradual taper of wedge and cylinder halves causes blows or pressure on the plug to be converted into a tremendous sideward pressure, which can split large boulders and break off chunks of bedrock.

Under many circumstances, however, a contractor may prefer to get rid of the rocks by digging and pushing. The dozer is the standard tool for this work. Efficiency can be increased by use of a tilting blade, a dozer shovel bucket, a stumper, or a heavy-duty rake blade.

A dozer can move quite a large rock on firm ground, perhaps several times its own weight. If the stone is too large for direct pushing, it can be pushed first on one side, then on the other, as in Fig. 1.38. If it is rounded, it can be rolled by lifting the blade while pushing. If the blade does not have enough lift to control it over, it can hold it in a partially rolled position, with locked brakes, while the stone is blocked up. The blade may then be lowered and the push and lift repeated.

Partly buried rocks may be pushed or dug out in somewhat the same manner as stumps. The resistance they offer is usually more rigid than that of stumps, and if a rock will shake in the first few

FIGURE 1.37   Boulder split by plug, hydraulic-powered.

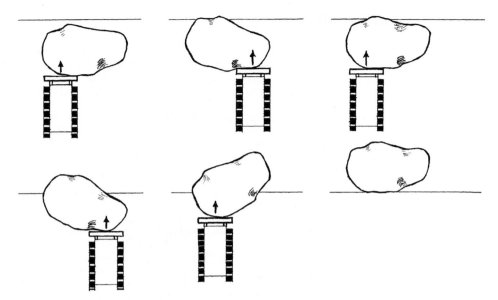

FIGURE 1.38   Pushing oversize boulder.

direct blows of the blade or bucket, it should come out. It is sometimes very difficult to get a grip on smooth sloping surfaces, so that an excessive amount of digging must be done just to get a hold.

When a grip is obtained with a dozer blade, the rock may be raised and pushed. The engine clutch should be slipped, or the converter-equipped engine throttled so as to supply just enough forward pressure to keep the blade in contact with the rock while it lifts it vertically, and, when it is high enough, rolls it out. The rock may slip back into its hole at any time, and it is good to have a helper throw stones or logs under it so the blade can be dropped and a fresh grip obtained. If no helper is available, the operator can lock the brakes to hold pressure against the stone and do the hole filling himself or herself.

If a big stone is rolled out without blocking, it may leave such a large hole that the tractor may be damaged if it falls in it. The danger is more serious than with stumps, as rocks leave sharper-edged and harder holes.

A rock should be pushed from all angles before digging it out, as it may be susceptible to pressure from only one direction. If it is to be dug out, a bowl-shaped crater of considerable size is excavated, working on three sides, if a good grip is available at the top, or all around it if the top is smooth. When it is finally loosened, it may be found that it is so heavy that the dozer cannot get it up out of the hole.

It probably can be pushed out by following the procedure outlined in Fig. 1.39. The dozer builds a ramp out into the crater, shaped so that the machine will be pitched downward when its blade meets the rock. With gravity assisting, it should be able to push the stone a short distance up the opposite slope. The ramp is then built out, and another push made until the stone is out of the hole.

Loose boulders may be pushed out of the work area and scattered, piled, or arranged in walls; or they may be buried, being either used as a fill or wasted in holes. Holes may be dug to bury them.

Where many boulders are pushed into a hole they afford unsafe footing for a bulldozer and may pile up above the desired grade. A moderate amount of dirt, either scraped off the bank or trucked to the spot, will allow the dozer to fill in the holes and stabilize the rocks so that it can walk across them in pushing other boulders to their resting place.

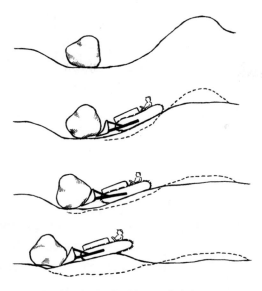

FIGURE 1.39   Getting boulder out of a hole.

If the area is to be finished to a grade, it pays to be liberal in supplying covering soil, for if the layer is thin, the dozer working on it may hook into freshly buried rocks and turn them into high positions. They can seldom be put back in place because of soil and other rocks getting under them, and it may be necessary to knock their tops off with hammers or explosives, or dig them out and rebury them. Digging a boulder out from among others is very difficult, and is likely to turn them up.

Rocks may be trucked away from the job. They may be loaded by lifting in or on a loader or dipper shovel bucket, by clamping in a clamshell or grapple, or between a hydraulic hoe bucket and the stick.

Or a crane may lift them by tongs, chains, or cable slings. Light chains grip well, but are easily damaged.

Ordinary dump truck bodies may be severely battered by oversize rocks. The floor may be protected by an extra sheet of steel, a layer of planks, or just a few inches of dirt.

Stone Walls.   Stone walls built to dispose of boulders removed from farmland are very common in some sections of the country, and may include rocks large enough to present a problem to machinery. The big base stones are often partly or completely buried, interlocked, and bound in place by tree roots. The smaller stones may be valuable for use in masonry, and may be removed by hand before or during the wrecking of the wall.

A dozer of sufficient size can walk right through the wall and scatter it around, but an undersize machine may have to start at a gateway, or find a weak spot to break through and widen the hole by worrying the rocks out one at a time. If the wall cannot be broken from one side, it should be tried from the other.

Foundations.   Old foundations and other masonry structures usually yield readily to heavy machinery. High walls should be pulled down, as they might fall on a machine pushing them.

If a foundation is too strong for available machinery, it may be weakened by blasting along the lines where it meets the floor and other walls, by mudcapping or drilling. Demolishing very heavy or extensive structures, however, is a house-wrecking job out of the field of this book.

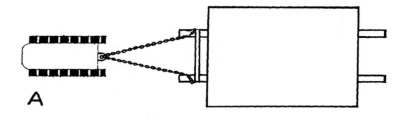

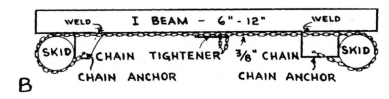

FIGURE 1.40 Bracing tow skids.

Small Buildings. Moving buildings properly is also a highly specialized trade, but an excavating contractor may be called upon to move small structures of minor value out of the work area, or to drag her or his work buildings around on a job.

The easiest method is to jack the building up, or to lift it one end at a time with a shovel or crane, and pull a pair of substantial skid logs under it. These should be beveled at the front so as not to dig in, and notched on the top for the sills of the building. They should be spiked or bolted to the sills. If the building needs additional rigidity, cross logs may be used above the skids and the walls may be braced with diagonal planks.

The skids should be rigidly fastened to each other. If the building is to be pulled by one machine, through a double chain as in Fig. 1.40(A), which is the usual method, the bracing between the front of the skids takes a tremendous inward pressure. Ordinary log or timber braces may not hold, unless very expertly installed, or the pull is light.

It is usually worthwhile to make a steel cross brace such as is shown in (B). The beam itself may be secondhand I beam, or channel or angle. The welded brackets prevent the skids from moving inward, and the chain, pulled tight with a load binder, prevents them from moving out. Notches in the bottom of the skids are necessary to prevent the chain from cutting into the ground and increasing the draft.

The skids should project far enough forward to be easily chained for pulling or lifting. They should be high enough to carry the sills or cross logs over any irregularities in the ground. If this is not practical, rollers, consisting of short logs, may be put under the front of the skids. As the building goes over these, they must be watched so that they will not turn up and injure the structure. When they are left at the back, they may be picked up and carried to the front.

Rollers are also used when the tractor is not powerful enough to pull the skids on the ground.

CHAPTER 2
SURVEYS AND MEASUREMENTS

GENERAL CONSIDERATIONS

Surveying is a profession in itself, and contractors and their employees seldom have time to master it. However, it is possible for a layperson to run levels, to reestablish lines and locations obliterated by construction, and to do rough layout work.

If a job involves as much as a day of work for a surveying crew, it is usually economical to hire professionals. They work more rapidly and efficiently than amateurs, and are less liable to make costly mistakes. Unfortunately, it is frequently not possible to obtain the services of engineers exactly when needed, and there are many jobs which are too small, or too simple, to justify calling them in.

Also, it is sometimes desirable for the owner or contractor to make a rough survey of a project to determine the amount of work to be done, and possible layouts, before bringing in surveyors to provide detailed information. A person can usually obtain a much clearer idea of the problems involved by running his or her own levels than by reading the findings of another.

The methods outlined in this chapter will in some cases be those used by surveyors, but will often be shortcuts and substitutes which can be used by amateurs with reasonably satisfactory results, and which generally are easier to learn, but less accurate, than professional methods.

More detailed information about surveying may be found in textbooks on plane surveying. Work with a survey crew is the soundest training in field methods.

TELESCOPIC LEVELS

The basic surveyors' tool is a telescopic level mounted on a turntable which in turn is usually based on a tripod. The entire unit is often referred to as an instrument. There are a great variety, but most of them may be classified under three headings—level, convertible level, and transit. The difference is partly that in the first the telescope is always used in a horizontal position; in the second the telescope may be lifted out of its frame and reset so as to pivot vertically; and in the transit it is permanently mounted so as to swivel vertically as well as horizontally. However, these general distinctions are not always true in regard to particular models.

Automatic leveling and laser instruments will be discussed later in this chapter.

Builders' Level.   Figure 2.1 shows a type of builders' level which is convenient for general contractors' use. The telescope is held rigidly in a frame that rotates on a vertical spindle, which is perpendicular to the line of sight of the telescope. A spirit level with a graduated glass is mounted on the frame.

The leveling head on which the spindle rotates is fitted with a horizontal circle marked in degrees, in contact with a pointer fastened to the spindle. Many levels do not have this circle, but it is essential for the location work to be described.

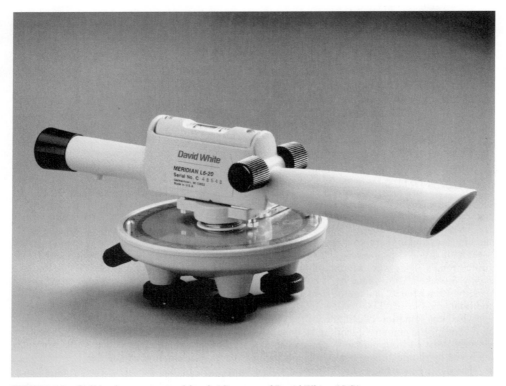

FIGURE 2.1 Builders' or contractors' level. (*Courtesy of David White, LLC.*)

Vernier. The pointer may be expanded into a vernier such as shown in Fig. 2.2. This is a device for reading fractions of a scale, which in the example is calibrated to 30″ (half degree) divisions. The length of 29 divisions on the circle is divided into 30 divisions on the vernier. Each vernier space therefore represents $\frac{29}{30}$ of a space on the circle and is $\frac{1}{30}$ shorter.

In the illustration, if an angle on the main scale is being read from left to right, the zero, or center of the vernier, shows a reading slightly higher than 2°30″. Reading the vernier to the right, it will be found that the tenth division line matches exactly with a line on the circle scale. This indicates that the zero mark was $\frac{10}{30}$, or one-third, of the way from $2\frac{1}{2}$° mark to the 3° mark, as the difference is canceled out in the course of subtracting the $\frac{1}{30}$ difference 10 times.

One-third of 30″ is 10″. The angle is therefore 2°30″ plus 10″, or 2°40″.

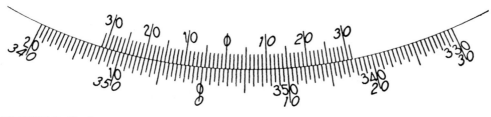

FIGURE 2.2 Vernier.

If the angle were being read from right to left, the main scale would read 357° and a fraction. Reading the vernier to the left, the twentieth division is found to correspond with a line on the circle. The angle is therefore 357° plus $^{20}\!/_{30}$ of 30″, or 357°20″.

The telescope may be locked against swinging by means of a thumbscrew for convenience in reading the scale, or holding it in a certain direction. Another thumbscrew (tangent screw) will then move it slowly for fine adjustments.

Telescope.   The length and power of the telescope and the length of the spirit level determine the range of the instrument and, to a considerable degree, its accuracy. Telescopes range from 10 to 18 inches in length, and from 10X to 35X power in magnification. Spirit levels may be 3 to 10 inches long.

The telescope is focused by means of a knob on the side or top, and possibly by a turning eyepiece also.

The field of view of the telescope is divided into quarters by the crosshairs, shown in Fig. 2.3(*A*), which are held in a frame or diaphragm inside the telescope. Provision is usually made to make these visible or invisible by focusing the eyepiece. The horizontal hair is used for taking levels. If it is correctly placed in the telescope, and the telescope is properly leveled, it indicates the slice of the field of view which is level with the observer's eye. The vertical hair is used to sight a given point or line, and indicates the exact center of the field of view for determining horizontal angles.

Stadia Hairs.   Stadia hairs (*B*) may be fitted into the same frame. These are horizontal and are located above and below the center hair. The distance between the stadia hairs is fixed at a ratio with the telescope, usually 1 to 100, so that if a measuring rod is sighted through the scope, the inches or feet seen between the stadia hairs may be multiplied by 100 to give the distance of the rod from the instrument.

Amateurs are apt to confuse one or the other stadia hair with the crosshair in taking levels, with resultant serious error. If this trouble persists, additional hairs may be installed, as in (*C*) in the form of a letter X, which should make the center hair easy to distinguish.

Base.   The leveling head is mounted on the turntable or base by means of a center pin, on which it can both tip and rotate, and four leveling screws. These screws are threaded into the leveling head and rest on the leveling plate or turntable. They are expanded into knurled wheels for convenience in turning with the fingers, and have expanded feet which do not turn with the screw and which protect the plate.

The turntable base has internal threads by means of which it can be screwed on to the tripod head. It may have a hook on the bottom, at the center, from which a plumb bob (pointed-tip weight) may be hung by a chain and string. The base may be made to slide a limited distance horizontally, relative to the tripod head, for convenience in centering the unit directly over a mark.

Tripod.   A tripod consists of three metal or wood legs, hinged together by a top plate which is threaded for the instrument, as seen in Fig. 2.4. These threads should be protected by a cap whenever the instrument is not mounted. The legs may be one piece, or two pieces sliding on each other and locked by a screw clamp.

The base may be set directly on a flat rock or stump, if it is not possible to set up the tripod, but this is not recommended.

Compass.   Compasses are standard equipment in transits, and can usually be obtained for other types of instruments that have a horizontal scale. They are not necessary for the work to be described in this chapter, although it is often convenient to know the general directions of lines.

FIGURE 2.3   Sighting hairs. (*Courtesy of David White, LLC.*)

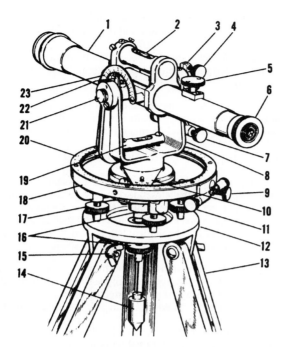

1. Telescope
2. Telescope Bubble Assy
3. Vertical Clamp
4. Vertical Clamp Screw
5. Focusing Screw
6. Eyepiece Cap
7. Vertical Tangent Screw
8. Telescope Support
9. Horizontal Clamp Screw
10. Horizontal Circle
    Vernier Plate
11. Horizontal Tangent Screw
12. Tripod Head and Base Plate
13. Tripod Leg
14. Plumb Bob
15. Tripod Wing Nut
16. Center Screw
17. Leveling Screw
18. Leveling Head
19. Support Level Tube
20. Horizontal Circle Scale
21. Telescope Trunion
22. Vertical Arc Pointer
23. Vertical Arc Scale

FIGURE 2.4   Parts of level transit. (*Courtesy of David White, LLC.*)

Surveys are generally based on the true north, from which the compass north varies rather widely. Part of this variation may be obtained approximately from the map, Fig. 2.5, or exactly from local sources.

If you are in an area of west magnetic declination, the compass needle will point west of the true north by the amount shown on the map.

Another source of error is the magnetic attraction of magnets, iron, and iron ore for the compass needle. It is also affected by the time of day. No confidence should be placed in a compass reading taken near machinery or electrical apparatus. Metal objects in the observer's pockets may cause errors.

Setting Up.   The first step in using the instrument is to set up the tripod. The top should be as level as possible, and the legs pushed into the ground firmly. On a slope, two legs should be downhill. The protecting cap is removed and the instrument screwed on. The telescope frame should be unlocked so that it is free to rotate. The telescope can then be held in one hand and the base screwed on the tripod with the other.

Leveling.   The instrument must now be leveled by means of the four screws. The telescope is turned so that it is over two of them, and those screws are adjusted until the bubble in the level is exactly in the center of the scale. The screws are turned at the same time in opposite directions, so that one pushes the leveling head up while the other makes space for it to come down, as it pivots on its center pin.

The bubble moves in the same direction as the left thumb, as indicated in Fig. 2.6*A*. If the two screws are turned exactly the same amount, the tension on them will remain constant. If the screw toward which the bubble is moving is turned farther, it will jam both screws. If the screw behind the bubble is turned the most, the tension will be reduced and the screws may lose contact with the turntable.

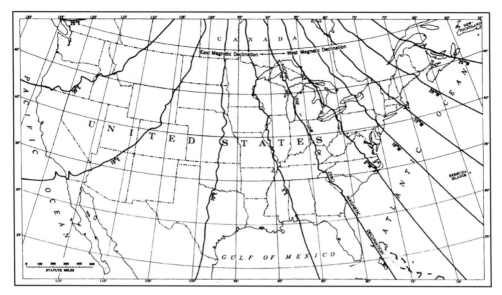

FIGURE 2.5 Magnetic declination map. (*Courtesy of U.S. Coast and Geodetic Survey.*)

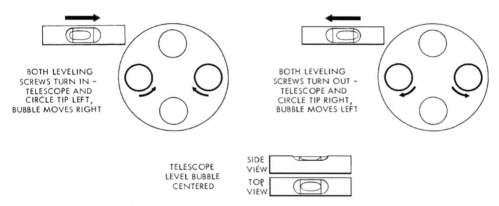

FIGURE 2.6A Leveling screw action. (*Courtesy of David White, LLC.*)

If both screws are turned to the left (clockwise), or one is turned left while the other is stationary, they will be jammed; while if one or both are turned to the right (counterclockwise), they will both lose contact.

The screws should be kept in light contact with the plate during the adjustment and tightened somewhat as it is finished.

When the bubble is approximately centered, the telescope should be swung 90 degrees so as to be over the other pair of screws, which are used to center the bubble in the same manner. This adjustment will disturb the first one, so the telescope must be swung into its original position and leveled, this time more exactly. It should then be checked in the second position, and adjustment in the two positions made alternately until it does not move during the swing through this quarter-circle arc.

The telescope should now be swung through the other three-quarters of the circle. If the adjusting screws are tight and the tripod has not been disturbed, the bubble should not move. If it does move, and the table cannot be leveled so that it will not move when swung, the spirit level is probably out of adjustment.

Releveling.   During use, the spirit level should be checked occasionally and the instrument releveled if necessary. The tripod may settle into the ground, particularly if on some unstable base such as ice, wet clay, or oil road top. Jars from focusing the telescope, or the wind, or other causes, may disturb it. Sometimes it is necessary to put small boards under the legs to avoid settling.

Centering.   Exact placement of the instrument is usually not important in setting grades, but it is essential in most other work. It involves locating the vertical axis on which the telescope swings directly over a marker, often a nail in a 2×2 stake.

The standard procedure is to hang a pointed, balanced weight called a *plumb bob* from the center of the instrument by a light chain and/or string. The tripod is maneuvered until the point of the bob is just over the center of the nail head.

This location may need readjustment after leveling the instrument, and such relocation calls for releveling.

The tripod head in Fig. 2.6*B* allows horizontal shifting a distance of 2 inches in any direction, without disturbing its level. This device saves time and trouble.

The optical plummet, Fig. 2.7, replaces the plumb bob with a low-magnification telescope in the rotating base, which, by means of a prism, looks down the vertical axis. A small bull's-eye and cross line make possible exact alignment with a marker.

Transit Theodolite.   A transit has a vertical swivel and support yoke mounted on the turntable, which permits the telescope to be tilted up and down on a vertical axis in addition to its horizontal

FIGURE 2.6*B*   Shiftable tripod head.

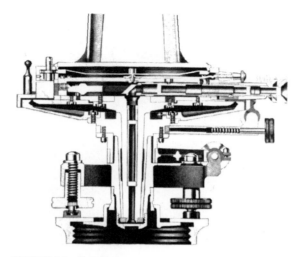

FIGURE 2.7   Optical plummet.

rotation. The angle from the horizontal is shown on a vertical scale with a vernier. When it is used as a level, the reading on that scale should be zero.

A level transit like the one shown in Fig. 2.4 can be tilted in the vertical plane 45 degrees, either up or down. Full transits have an extended support yoke that permits the telescope to turn in a full vertical circle within it. This makes it possible to make a back sight (180° turn) without changing the setting of the horizontal circle.

The latest transit theodolite is computerized for easy, quicker setup and readings. Figure 2.8 shows one of these electronic theodolites for routine layout and measurements on a building site. The compact instrument offers user guidance through the measuring processes and programs. There are only seven keys used for control of the straight forward and fast operation. On the press of a key, both the computed and measured data appear in a clear four-line display. In the measuring programs, the user guidance is supported by a graphical display, making the solution of the problem easier to follow and accept or correct.

SELF-LEVELING LEVEL

The self-leveling or automatic level uses a three-screw leveling base with a circular level vial, for approximate leveling of the instrument. Fine leveling is done by a gravity-controlled device (pendulum or compensator) inside the telescope. See Figs. 2.9*A,* 2.9*B,* and 2.9*C.*

LEVELING ROD

The surveying instrument's best companion piece is the leveling or target rod. This is a measuring stick, marked in feet, tenths of feet, and hundredths of feet; or in feet, inches, and eighths of inches. It may be 8 to 15 feet long, and usually is in two or three pieces which slide on each other, or, occasionally, are hinged or pegged together. The sliding type must be fully closed or fully open to be accurate.

Long rods are very desirable in hilly country.

Spaces may be marked by fine lines similar to those on a ruler, in which case it is called a New York rod. A Philadelphia rod uses the division lines as units of measurement in themselves. See

FIGURE 2.8   Electronic theodolite. (*Courtesy of Carl Zeiss, Inc.*)

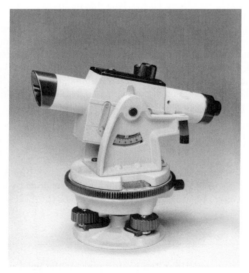

FIGURE 2.9A   Automatic level transit. (*Courtesy of David White, LLC.*)

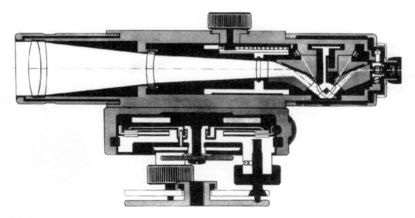

FIGURE 2.9B   Cross section of self-leveling level. (*Courtesy of Carl Zeiss, Inc.*)

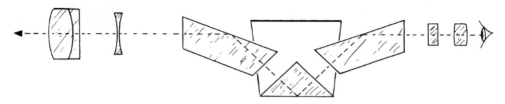

FIGURE 2.9C   Line of sight in self-leveling level.

Fig. 2.10*A*. A rod in the decimal scale has the tenths of feet each divided into 10 equal sections, alternating black and white. If inches are shown, each is divided into eight equal bars of alternating color.

The target, Fig. 2.10*B,* is a metal disc that slides on a track on the sides of the rod. It is painted in quadrants, alternately red and white, with the division lines horizontal and vertical, or in other conspicuous patterns.

The rod operator moves the target up and down on the rod in response to signals from the instrument operator. Readings can be taken in this way when distance or haze prevents reading of figures on the rod.

The target may include a vernier, in which case it can be used for precise work requiring reading of fractions of the smallest divisions of the rod scale.

The rod is used for measuring the distance from the instrument's line of sight down to a point. If the point is almost as high as the instrument, this distance will be short; if it is much lower, the distance will be long.

The elevation of a point is the distance which it is above some standard level. This may be mean sea level—halfway between high and low tide marks—or some local point to which an elevation is assigned arbitrarily.

Elevations are usually positive numbers, measured up from a base point or plane. Rod readings are negative, being measured down from the plane of the instrument.

In taking levels, the positive elevations are obtained from negative rod readings. Care must be taken to avoid confusion. It must be remembered that for any instrument setting, the high readings are low elevations, and vice versa.

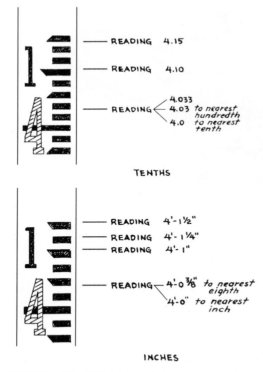

READING 4.15

READING 4.10

READING ⟨ 4.033
4.03 *to nearest hundredth*
4.0 *to nearest tenth*

TENTHS

READING 4'-1½"
READING 4'-1¼"
READING 4'-1"

READING ⟨ 4'-0⅜" *to nearest eighth*
4'-0" *to nearest inch*

INCHES

FIGURE 2.10A Philadelphia rods.

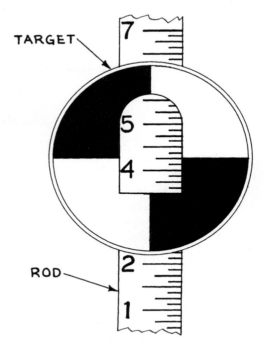

TARGET

ROD

FIGURE 2.10B New York rod and target.

USE OF LEVEL

Use of the instrument as a level depends upon the fact that its crosshair indicates a horizontal plane, level with the observer's eye in all directions. By use of the rod, the amount by which a point is lower than this plane can be measured, and the relative elevation of any number of points within range can be calculated from rod readings. Points above the crosshair cannot be measured, except on vertical walls, without moving the instrument higher and resetting it.

Levels are most accurate over short distances. If the instrument must be used when out of adjustment, readings should be taken as nearly as possible at equal distances.

Converting Readings to Elevations. If the instrument is to be set up only once on a job, and no record is to be made of observations, the rod readings can be used to figure heights. However, since these numbers are negative, the beginner will avoid confusion by calling the lowest point— that with the highest rod reading—zero elevation. The other points will then each have an elevation equal to the difference between its reading and that of the zero point. See Fig. 2.11.

Benchmarks. If the elevations are to be recorded for future use, it is necessary to have some fixed reference point which will not be disturbed and which can be readily identified. A knob on firm bedrock, a nail projecting from a tree trunk, a mark on a building, or a stake hammered flush with the ground may be used. Such a point is called a benchmark, and is abbreviated as BM. A

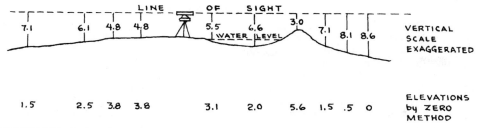

FIGURE 2.11   Rod readings.

reading is taken the first time the instrument is used, and again each time it is set up. These readings will probably all be different, but in each case the elevation of the instrument may be found by adding the rod reading when the rod is sitting on the benchmark to the benchmark elevation.

Since the benchmark is the most permanent point observed, it is good practice to assign an elevation to it, and to calculate all other elevations from that. An assigned number should be large enough that no elevation less than zero will be found on the job, as working with minus figures may cause confusion and error.

If levels have been taken previously in the area, engineers' benchmarks may be found, in which case it is wise to use them. If possible, the elevation assigned to them in the previous survey should be used to facilitate comparison between the two sets of levels.

Even if engineers' benchmarks cannot be used directly in surveying the job, it may be advantageous to run a level to one, and note its elevation in relation to the contractor's own benchmark, so that the two systems can be compared if necessary.

If the job is a type that will involve frequent checks of levels, as on a road where stakes may be knocked out by machinery, it is a good plan to set up benchmarks so that one will be visible from each point where the instrument will be used. This saves time in taking grades on a few stakes and eliminates common errors in moving the instrument, or in taking an elevation from the wrong line stake, or a stake which has been disturbed.

All benchmarks should be figured very carefully and rechecked at least once.

Recording Readings.   Another requirement in recording observations is to identify the spot at which each reading is taken. This is usually done by taking readings at set intervals, such as 10, or 50, or 100 feet. These distances should be marked by stakes, pegs, small rock cairns, or in other ways. The first stake or mark of the series is called the zero stake, and the others are identified by their distances from it. It is customary to give distances in units of hundreds, followed by a plus sign and the other figures of the distance. The zero is written 0 + 0, the 50-foot mark 0 + 50, and the 100-foot mark as 1 + 0. If any points on the line are needed which are not in the series, the distance is measured and entered in the notes with the reading, as 0 + 35.

Important ground features along the centerline, such as crests of rises, bottoms of dips, or beginnings of rock outcrops, should be taken in addition to the stake readings.

Elevations may be taken from the ground, from the top of the stake, or more rarely, from a mark on a stake. If taken from the ground, it should be stamped or cut flat. Such readings are not as accurate as those taken from the top of a stake, and may be very difficult to check back, but they can be used directly in preparing profiles and figuring cut and fill. If the top of the stake is used, it is necessary to measure the stake height.

Tapes.   Measuring is usually done with a steel tape, often called a chain by surveyors. Fifty-foot and 100-foot lengths are standard, and will suffice for most purposes. They should have a nonrusting finish, as it is often difficult to dry and oil them immediately after wet work. Care should be taken not to kink a tape, or to bend it sharply, as such abuse may break it.

If the numbers become illegible, they can be fixed for rough work by measuring and marking the feet, and perhaps some fine divisions, with paint. A broken tape can be repaired by means of a splint and two rivets.

Cloth tapes stretch readily, and are not accurate enough for even rough use. Metallic tapes, composed of cloth with interwoven wires, are variable in quality and resistance to stretching. If used, they should be checked occasionally by a good steel tape.

Steel tapes change length with temperature and stretch under tension, but these changes are so small that they can be ignored in open work.

Tapes must be held level, or very nearly so, on slopes, as engineers' land measurements refer to distances on a horizontal plane. The downhill end of the tape may be placed exactly above the desired point by use of a plumb bob, or by dropping pebbles from the tape end.

Centerlines usually include angles or curves. If the former, measurements must be made to and from the angle point, rather than by a shortcut. Gradual curves may be measured in a series of chords (straight lines beginning and ending in the curve). Sharper curves may require a reduction in the length of the chords, as from 100 to 50, or 25 or even 10 feet. The difference in length between the chord and the arc of the curve may be readily found by laying the tape along the curve from one chord point to another; or measuring a distance along it in very short chords, then measuring the distance between the two points directly. If no significant difference is found, the chords are not too long.

Tapes are best suited to two-person use. However, the loop on the zero end can be anchored in dirt with a screwdriver, and to stakes with a pushpin or thumbtack, and measurements made by one person.

Ground measurements may also be made with the rod, with a short rule, a stick of known length, or for very rough work, by pacing.

Stadia.   If the instrument is equipped with stadia hairs, it may be used to measure distance as well as elevation. If the stadia ratio is the usual 1 to 100, and the rod is marked in feet, tenths and hundredths of feet, each tenth visible between the stadia hairs indicates a distance of 10 feet from the center of the telescope to the rod. Six tenths would mean a distance of 60 feet, a foot would mean 100 feet. This distance may be noted at the same time as the crosshair reading.

If the rod is marked in feet, inches, and eighths of inches, each inch indicates a distance of $8\frac{1}{3}$ feet, each foot 100 feet.

If a distance is to be measured off, the rod is held at increasing distances from the instrument in response to signals, until the proper number of markings shows between the stadia hairs.

The rod should be held perpendicular to the line of sight. The correct angle can be found by pivoting it slowly toward and away from the instrument, until the minimum reading is obtained.

Turning Points.   If elevations are to be taken for any points above the crosshair, the instrument must be picked up and reset at a higher elevation. It must be located so that it can take a reading on at least one point that was taken from the old setting. This point (turning or transfer point) is preferably one of the higher elevations (low readings) taken, and should lie between the two instrument locations. It is best taken from the top of a firm stake, or a knob or a well-marked spot on rock or hard ground, so that the rod set on it will be at exactly the same height at the second reading as at the first. Accuracy in reading at the turning point is very important, as any error made will persist through the rest of the survey. Amateurs are advised to use two turning points with each move, as mistakes in reading or in arithmetic should then show up immediately.

The new instrument elevation (abbreviated H.I. for height of instrument) is found by subtracting the smaller reading from the larger one for each turning point, and, in an uphill move, adding the result to the first elevation of the instrument.

Recording and Figuring.   Figure 2.12 shows some of this work. Part (*A*) shows the slope, the location of benchmarks and stakes, and the two instrument positions used. Part (*B*) is an informal set of notes of rod readings and calculated elevations. Part (*C*) is a profile drawn on cross-section paper from the notes in (*B*). It is made by drawing a baseline, assigning it an elevation lower than

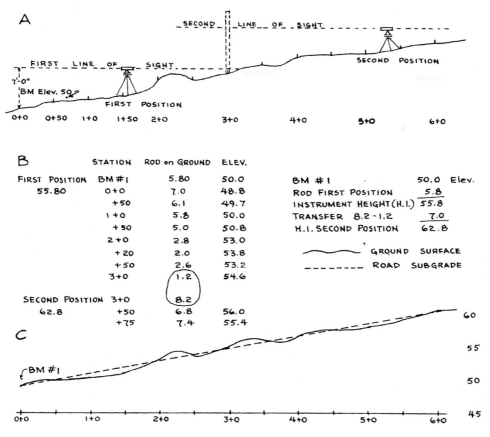

FIGURE 2.12 Recording and figuring.

those of the stakes, and making each square represent a certain distance. In this diagram, each square represents 1 foot vertically and 10 feet horizontally. This vertical exaggeration is necessary to have a large enough scale without making the drawing impossibly long.

The profile is useful in giving a picture of the slope, in determining gradients of roads or ditches, and in figuring the cut and fill necessary to convert the present grade to the new one.

The dotted line is the subgrade for a proposed road. It will be seen that the depth of cut or fill on this line may be approximately determined by measurement with a ruler; the elevation of any point on the road, in relation to the benchmark, can be found in the same manner.

Moving Downhill. If, at the original or at any later location of the instrument, points to be taken are so low that the rod is below the crosshair, the instrument must be moved downhill. A turning point (or points) is chosen with the lowest possible elevation (highest reading), the instrument is moved, and new readings are taken. The low reading is subtracted from the high one, and the result is subtracted from the earlier instrument elevation.

If only one or two points slightly below the crosshair must be taken, the rod may be set on a stake, and the height of the stake added to the reading; or a ruler may be used at either the top or the bottom of the rod to extend it.

Check Runs.   When all the necessary points have been taken, the accuracy of the work may be checked by taking levels back to the starting point. This is usually a faster operation than the outward trip, as it is only necessary to take transfer points and benchmarks. If frequent benchmarks have not been placed, it is advisable to use the same turning points, or to take readings on a few of the grade points, so that if an error is present, it may be localized. It is not necessary or desirable to set up the instrument on the same points for the return trip.

The two elevations found for each point should agree, but a difference, varying with the care with which the work is done, generally exists. Benchmark runs should be held to within a few hundredths of a foot, even in rough work where a difference of several inches on a grade point might be allowable. If any considerable amount of cut or fill is needed, even benchmarks may be left as approximations, until skill or time is available for a more careful run. Any discrepancies found in the check run should be listed in the notes.

If benchmarks are set at the beginning and end of a run, and check properly on the return trip, it will not be necessary to back-check any later run on which these two elevations show correctly. However, if benchmarks have been set by other parties in some previous survey, they should be checked the first time they are used, as they may be wrong or their description misunderstood.

Grade Stakes.   Centerline road stakes are set by instrument and measurements from a prepared baseline. Shoulder, slope, and other side stakes are set from the centerline.

Grading information, that is, the cut or fill necessary at each stake, may be determined in several ways. The preferred method is to take the ground-level elevation at each stake by instrument and rod to the nearest hundredth of a foot or eighth of an inch, figure the difference from the desired elevation, and mark the difference on the stake.

Ground is often so rough that some dirt has to be patted down flat at the foot of the stake to provide a recognizable base for the rod. It may be advisable to put a crayon mark on the stake at ground level, in case anything should change it. This mark, usually a horizontal line, will be needed when grades are marked on it.

Many surveyors prefer to use the top of the stake rather than the ground level. In this way they have a firm base for the rod, and they do not have to be concerned with the possibility that dirt might be kicked away or added. However, this usually requires measuring the stake height to give ground level for yardage calculations.

A horizontal crayon line on the stake may be used instead of the top or ground.

The elevations observed at the stakes are subtracted from those required by the road plans. Plus numbers indicate that fill is needed; minus numbers that the ground must be cut or removed. The symbols written on the stakes are F for fill and C for cut. Each stake must show location.

Another method is to supply the surveying crew with the required grade elevation for each stake. These are subtracted from the H.I. (height of instrument) to show the reading on a rod held with its bottom at correct grade at that location.

If the rod can be held at an elevation so that the reading is seen at the instrument crosshair, the stake is marked with a horizontal line at the base of the rod. This line is marked SG or G. The indication is that ground level must be raised to this line.

If the top of the stake is below the grade, the rod is placed on its top. The calculated reading is subtracted from the actual reading, and the difference written on the stake, following the symbol F. A crayon line is made at the top of the stake, and/or an arrow is drawn from the figures to the top. See Fig. 2.13*A*.

Such a measurement for fill can also be made from a line drawn at any convenient height on the stake. It might be positioned an even distance below the grade, as 2 feet.

If the rod shows less than the calculated reading when it is resting on the ground, a cut is indicated. A line is put on the stake at ground level, or at some other convenient height, and a reading is taken of the elevation of the line. The actual reading is subtracted from the calculated reading, and the difference written on the stake, following the symbol C, as shown in Fig. 2.13*B*.

Stakes placed for the guidance of earth-moving crews should indicate cuts and fills to the grade they are working to produce. This is usually a subgrade, symbol SG, that is lower than the pavement surface or theoretical grade, G. It is very important to make clear which elevation is meant.

Road stakes are discussed in Chap. 8.

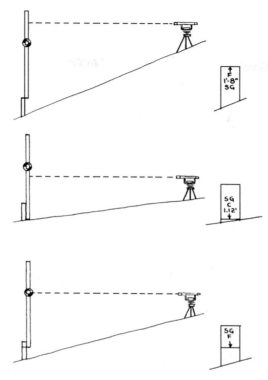

FIGURE 2.13A   Setting grades.

Engineers' grades usually consist of a series of elevations for the finished road. These are plotted on the same sheet of cross-section paper as the profile of the ground surface, and the depth of cut or fill is determined by measuring the distance between the two lines. These figures, if used directly, will not be accurate for most subgrade work, as the thickness of the pavement or gravel and of any special subgrade material must be subtracted to obtain the rough grade elevations.

A misunderstanding as to whether figures on grade stakes are for finish grade or subgrade can be very expensive. Use of subgrade figures for preparing subgrades is usually most satisfactory.

The contractor may obtain from the engineer a list or profile showing subgrade elevation at each station, and information as to the location and elevation of benchmarks. This, combined with sufficient field references to show the centerline, will enable the contractor to replace stakes which have been knocked out, and to find the depth of cut or fill required, by comparing the ground elevation with that required for the road.

LOCATIONS

Turning Angles.   When an instrument is used to turn angles—that is, to measure the horizontal angle between two lines or directions—the axis of revolution of the telescope must be exactly above the intersection of the lines, which may be marked by a nail in a stake driven flush with the ground, a cross chiseled in rock, or markings on concrete or metal plugs.

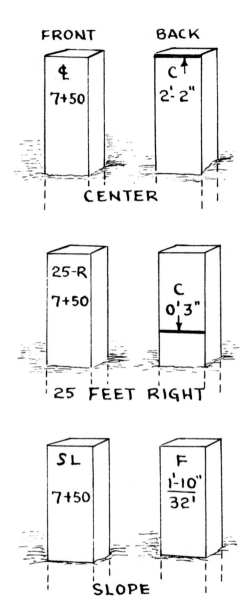

FIGURE 2.13B   Stake markings.

When the instrument is set over a point, a plumb bob should be hung at its center of rotation from a hook or through a hole usually provided. The point of the plumb bob should be just above the mark, and an amateur may have to move the tripod repeatedly before it is placed right.

If the instrument has a shifting base so that it can slide on the tripod, setting up over a point is greatly simplified.

A range pole is a convenient accessory in line and angle work. It is a pole 7 or 8 feet long, equipped with a metal point. It is painted alternately red and white in bands 1 foot wide. It is lighter than a rod and because of its conspicuous pattern is more readily seen at a distance.

This pole is set on one of the lines in question. The instrument is swung so that the vertical crosshair is on the pole. The rotation is locked, and the hair lined exactly on the pole by turning the horizontal tangent screw.

The reading on the horizontal circle and on the vernier is recorded.

The pole is placed on the other line and sighted in the same way. The difference between the two readings is the angle between the lines.

Line and angle work may be done to stake out on the ground locations described on a blueprint or map; or to make a record of ground features or locations on paper so that they may be used in figuring, or replaced or relocated if necessary.

Staking out is best left to surveyors if possible, as accurate work involves trigonometry and skillful use of the instrument, and inaccurate work may result in very expensive mistakes. However, in emergency, or when results need be only approximate, contractors can do it themselves.

Staking from a Map.   An example of staking out from a map is shown in Fig. 2.14. A building, 25×40 feet, is to be erected at the location shown on a plot 100 feet square. Lot corner stakes are at A and B.

The instrument is set up on A and sighted at B. The distances AA″ and A″B″ are measured along the line of sight, and stakes driven at A″ and B″. The instrument is then set up on A″, sighted on B, and turned 90°. The distances A″C and CE are measured and stakes put at C and E.

The transit is now set on C, sighted at A″, and turned 90°, the distance CD measured, and a stake placed at D. Point F is located from the instrument set at E in the same manner. The instrument is now set on F, sighted at E, and turned 90°, measuring the distance to D and B″ for a check on the accuracy of the work. The amount of error allowable will depend on job requirement.

This technique is practical for the amateur only on square or rectangular lots. Another method, which is applicable to any lot for which two widely spaced locations may be found both on the map and in the lot, is illustrated in Fig. 2.15. The house is located on the plot plan, and lines are drawn from the known corners A and B to the near corners of the building, as shown. These lines are measured and converted to feet according to the scale, and the angles they make with line AB and with each other are measured with a protractor. Figures are written on the plan.

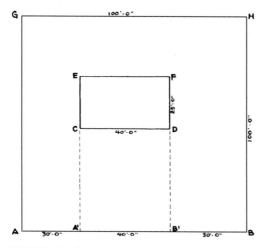

FIGURE 2.14   Locating house site.

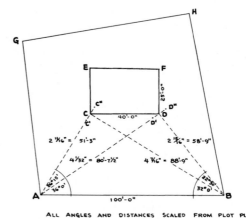

ALL ANGLES AND DISTANCES SCALED FROM PLOT PLAN

FIGURE 2.15   Locating house site in irregular plot.

The instrument is set up on A and a bearing taken on B. A 36° angle is turned, the line AD measured, and the stake D placed. An additional angle of 26°15″ is turned, AC measured, and C marked.

The instrument is now set up at B, sighted on A, turned 32°, and BC measured. The end of this line should be the stake previously driven at C, but if it is not, a second stake C is placed. The instrument is turned an additional 22°30″, and BD measured in the same manner. If the same locations are found for C and D from both A and B, and line CD is the required length, the work thus far is correct. If serious disagreement is found, the work must be rechecked.

The instrument may next be set at C, sighted on D, and turned 90° left. The distance CE is measured and stake E driven. Point F is located by setting the instrument at D in the same manner. This part of the work may be checked by setting up on E, sighting C, turning 90°, and measuring EF.

The accuracy of the location of the house in the lot will depend on exactness of the measurement on the map and the ability to read horizontal angles correctly. Amateurs may be off several feet in such work, and should do it themselves only when such differences are allowable. Under any circumstances, it is necessary to get the building walls of proper length and at the proper angles to each other.

The stakes A and B may be used as benchmarks, and elevations taken at the same time as the bearings and directions.

Recording.   If the location of existing stakes is to be recorded so that they can be replaced if destroyed, the work is the same except that the angles are obtained by sighting the instrument and are copied from the horizontal circle onto a sketch. Distances are measured in the field and noted on the sketch, which is most conveniently made on cross-section paper, roughly to scale. This sketch is used in the same manner as the map in the previous discussion in replacing the stakes. Results are generally much better, as the field figures are more accurate than those obtained with ruler and protractor from the map.

If field observations are to be entered on a map, the baseline or points should be related to features shown on the map, as corner stakes, points measured on a line between diagonal corners, or measured along a boundary. When the baseline is correctly drawn, angles and distances can be marked in with protractor and ruler.

Without Instruments.   Simple location work can also be done without instruments. Figure 2.16 shows the same square building plot. Lines are drawn on the print or tracing prolonging each side of the house to the plot boundaries, from where the distance to the corners is measured. These distances are then measured off on the ground, and stakes are set.

The distances of the house corners from the boundary lines may be scaled from the map and measured on the ground in directions found by sighting between pairs of boundary stakes.

Sighting may be done by placing a thin straight stake, as at L, and another at Q.

A person may stand behind the stake at L in such a position that, when she or he looks with one eye, the stake at Q is centered on L and just above it, as in Fig. 2.16(*B*). Another person, carrying a third stake, measures the distance QE, keeping on line LQ in response to directions from the observer. The measuring is best done by pinning the tape to Q. The stake is set at E so as to be directly in line between stakes L and Q. The distance EC is then measured, and stake C set in the same way. Distance CL is measured for a check.

Stakes F and D may be placed according to sighting from M to P, and measurements similar to the method used for E and C. The four corners of the building are thus located, and in a regular plot such as this no more work would be needed.

However, as a precaution against error, or in an irregular plot, or one with poorly defined boundaries, it is wise to prolong the other sides of the house into lines JO and KN, and to sight and measure the corners again from J and K.

Reference Stakes.   If the corner stakes have been set for a building in a plot without definite boundaries, and the contractor wishes to be able to reset them if necessary, there are several ways in which markers can be set without instruments.

In Fig. 2.17 the house wall lines are shown continued in straight lines out of the digging area. These lines may be established by putting sighting poles on the corner stakes and finding a distant

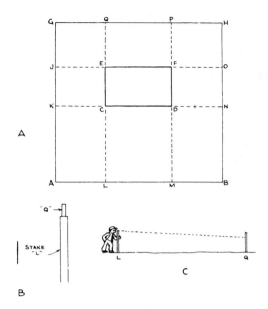

FIGURE 2.16   Staking without instruments.

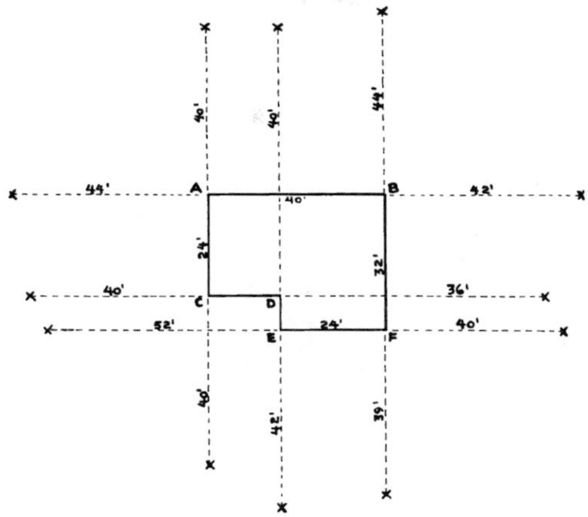

FIGURE 2.17   Cross-reference stakes.

position from which two of these are in line—that is, one partly or completely hides the other. This sighting should be done with one eye and a pole held vertically in line with them. A stake is driven to mark the position of this third pole. The distance from this to the nearest corner stake is measured. This process is repeated for each pair of stakes at the foundation. In the figure, the reference stakes are indicated by Xs and the sight lines by dotted lines. A sketch should be made showing distances.

Any missing stake may be found by sighting from one marker to the other one on the same sight line, and measuring from the nearest marker. Even if the sketch is not available, the point may be found by the intersection of two lines of sight, as described under instrument work.

If each reference stake is set the same distance out from the nearest stake, there is less need of keeping a record.

Locating a Pond.   If an irregular shape, such as a pond, is to be roughly measured and drawn into an existing map, a baseline is first established and two points are measured off. A number of pegs are driven into the shores of the pond at points which will serve to indicate its outline, and are numbered in rotation, as in Fig. 2.18. An instrument with stadia hairs is set up at A, a sight taken on B, or on a more distant marker along the baseline. A sight may also be taken on a corner of the house for a check. Sights are taken on all the stakes in rotation, starting at one, the angle read for each one, and the stadia distance recorded. This information is sufficient to locate the pond by drawing the baseline on the map, and plotting distances and angles. However, to avoid the possibility of gross error, it is safer to set up at B, take a bearing on A, and record the angle and distance of some or all of the points observed from A.

The area of a pond so plotted can be easily obtained by counting squares on cross-section paper, or by the use of a planimeter, which is a small instrument used for measuring areas on paper.

Grids.   If it is necessary to map an area, locating buildings, drainageways, trees, or other features, or to take elevations over a large area in order to prepare grading or drainage plans, a grid

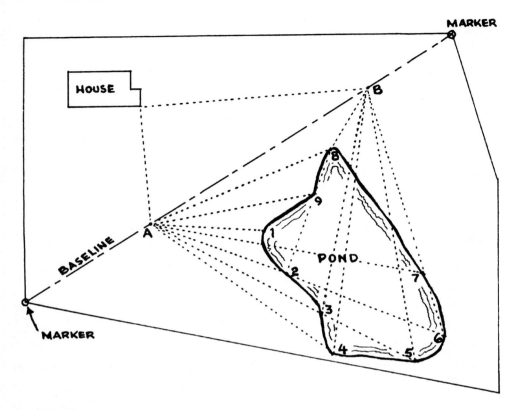

FIGURE 2.18   Locating by stadia.

should be laid out. This consists of pegs or stakes at set intervals. They should be in straight lines, crossing each other at right angles. These lines, intersecting at the pegs, generally divide the area into squares. The interval may be 5 to 20 feet or more.

The grid may be laid out in a number of ways. A baseline should be laid out along an edge of the area. The instrument, preferably a transit, is set up at a corner of the proposed grid and sighted along the baseline. Pegs are set every 10 feet, or at any other desired interval, measured from the instrument, to the end of the grid. Tape measurement is preferable.

The instrument is now turned 90°, and pegs are set at the same interval along the line of sight to the end of the grid. The instrument is set up at the end, a backsight taken, and a 90° turn made. Pegs are set at the same intervals along this third line.

The interior pegs may be placed by the use of a long tape from opposite pegs, or the instrument may be set up over each peg in either the first or third lines, sighted at the corresponding peg in the other line, and pegs set according to its vertical hair and measurement.

Obstacles may make it possible to set all the pegs by any of these systems. Usually, if as many pegs as possible are placed, the rest can be filled in by sighting along lines of pegs, with reasonable accuracy.

The grid should now be copied on cross-section paper with a point representing each peg. Any landscape features may be readily sketched in by estimating or measuring the distance from the nearest peg, and noting the place of the peg in the grid.

Elevations are now taken on each peg, preferably doing them a complete line at a time to avoid confusion. The rod reading may be written just above each point. Readings should also be taken on high and low spots, drain channels, and anything else of interest, and noted in the correct place on the paper.

When the instrument work is finished, the readings are preferably converted to positive numbers that can be penciled below the points, and the rod reading is crossed out.

This grid sheet can be used for reference for any locations or grading estimates which may be required, and in drawing contours, profiles, and cross sections.

Grids without Instruments.   If no instrument that will turn angles is available, a grid may be laid out with a tape, and elevations taken with a hand level. A baseline is decided upon, and a tall stake set at each end. A tape—the longer, the better—is pinned at one end and extended toward the other, and lined up by sighting across it from one stake to another. The intervals are measured, and the tape is moved on and lined up again.

The right angle may be laid out by referring to the ancient engineering knowledge that if the sides of a triangle are in the proportion of 3 to 4 to 5, the angle between sides 3 and 4 is a right angle. The process is illustrated in Fig. 2.19.

First the baseline is laid out, measured off, and pegs are set. The tape is pinned at A, one end of the baseline, and 30 feet is measured off at approximately a right angle. The tape is moved back and forth in an arc that is marked on the ground.

The tape is then pinned to the baseline at C, 40 feet from the end, and an arc of 50-foot radius described, crossing the first arc. A stake is driven at the point where these arcs intersect. Line AE may be located by sighting along stakes A and D, and will be perpendicular to AB.

These figures have been given for the use of a 50-foot tape, but any measurements may be used as long as the 3:4:5 relationship is preserved. Larger triangles will give greater accuracy. If the grid is large in proportion to the triangle, a diagonal should be measured from E to a point on the baseline either ¾ or 1⅓ as far from A as the distance AE, and any necessary correction made if the diagonal is not in the proper proportion.

A very rough grid may be made by sighting along the sides of a building to obtain the right angles, and spotting in the pegs by eye and measurement.

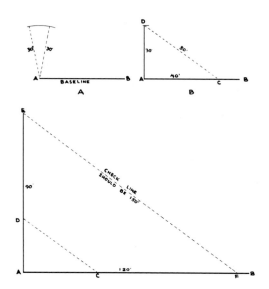

FIGURE 2.19   Right angles.

Obstructions.   Buildings, vegetation, and rough ground interfere seriously with primitive instrument techniques and make it more economical to hire an engineer.

Large permanent obstructions require layout of additional lines and angles to work around them.

Brush clearing for sight lines is laborious and sometimes quite destructive. It is handled by setting up the instrument, pointing it in the desired direction, and directing the cutters so that their work will be kept close to the line of sight.

In heavy undergrowth a mistake in turning an angle may waste hours of cutting work.

Contour Lines.   A grid is frequently used to make a contour or topographic map of an area, as in Fig. 2.20.

A contour line is a line joining points of equal elevation on a surface, or the representation of such a line on a map. In mapping, contours are drawn at set intervals whose size depends on the roughness of the ground and the scale of the map. In nearly flat areas contour interval may be 1 foot, while in steep mountains 100-foot intervals are usual. Every fifth line is drawn more heavily than the others, and its elevation is printed in it.

A contour interval of 2 feet is used in the illustration. The profile shown below the map is plotted along the dotted line. Note that grid points alone would not have shown the depression crossing it.

Since few of the grid points are exactly at a contour elevation, it is necessary to estimate the locations of the lines as they pass between higher and lower points, a process that is called *interpolation*. The 88 contour is placed halfway between point A-3, elevation 87.4, and B-3, elevation 88.6, as it is the same distance above one as it is below the other. The same line passes between B-1 and B-2, elevations 84.1 and 88.1, respectively. In this case the 88 contour is placed very close to B-2.

In irregular land it is very important that the grid observations include the crests of ridges and the bottoms of gullies, as the contours cannot give an accurate picture of the ground if they are omitted. The grid interval must also be close enough to show all important ground features.

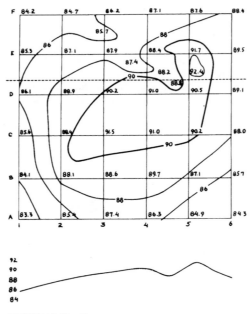

FIGURE 2.20   Contour map.

Placing of contours therefore cannot be an entirely mechanical operation. In the first place, the person making the field notes must understand topography sufficiently to take extra readings where necessary to show up special forms. The valley at the top of this map would not have been shown at all by readings on the grid points only.

The person drawing the contours should have a feeling for landscape forms, to avoid misinterpreting the data from which she or he works.

Topographic Map.   A topographic map is usually a contour map on which both natural and artificial features are indicated. The U.S. Geological Survey, Washington, D.C., has small-scale maps covering most areas of the country that show contours, vegetation, roads, trails, lakes, streams, and much other information. These maps are sold by mail and in bookstores at nominal prices. They are handy for many purposes and essential for some jobs.

Profiles and cross sections may be taken from a contour map by laying a ruler across it, measuring the interval between contour lines, and posting distances and elevations to cross-section paper. The proposed new grades can then be drawn in, and the differences calculated.

The topographic map has a wide variety of uses, including locating highways, haul roads, borrow pits, and dump areas; studies of drainage and stream flow; and estimation of quantities in cuts and fills.

INSTRUMENT ADJUSTMENTS

Surveying instruments are delicate and are easily put out of adjustment by failure of parts, careless handling, or accidents. It is often not possible to have them checked or repaired locally; the return to the factory may mean loss of use for weeks or months.

It is therefore desirable that a person using an instrument be familiar with some adjustments that can be made in the field, without special skill. These include setting of telescope spirit level, and the horizontal and vertical crosshairs.

If these are properly set and the instrument is inaccurate, shop service is probably necessary.

Spirit Level.   The telescope spirit level is usually fastened by a pair of vertical bolts. A single nut holds it in a fixed position at one end, and a pair of nuts, one above and one below, permit moving it up and down on the other end.

This level can be checked each time the instrument is set up. When the turntable is level, the telescope should be able to swing in a full circle without changing the position of the bubble. If no turntable screw adjustment will permit this, the level is presumed to be at fault.

To adjust, the turntable is leveled as accurately as possible and the bubble centered. The telescope is swung a half circle, causing the bubble to shift. The bubble is brought one-quarter of the way back to center by the adjusting nuts, and the rest of the way by using the turntable leveling screws.

The telescope is then swung to its original position, the bubble moved one-quarter of the way to center by adjustment, and centered by the leveling screws. This process is repeated until swinging the telescope does not affect the bubble.

Crosshair.   If the horizontal crosshair is not exactly centered, all readings on the rod will be too high or too low. Readings taken at about equal distances will agree. Greatest errors will be found on long sights.

A reasonably accurate check and adjustment of this hair can be made with the help of a still pond. Two stakes are driven flush with the water surface, about 100 feet apart. The instrument is set in line with them, 10 feet beyond one.

A rod is set on the near stake and a reading taken. This is assumed to be accurate, because the distance is too short for a perceptible error. The target is locked to the rod at this reading, or a note made of it.

The rod is set on the far stake. If the hair is correctly adjusted, the reading should be the same. If it is not, the hair should be raised or lowered until it agrees.

This is done by turning setscrews at the top and bottom of the frame which hold the hair. The screws are unlocked by twisting one or the other a quarter or half turn, after which both are turned in the same direction.

If no pond is available, or a more accurate adjustment is required, two stakes should be driven firmly into the ground, 200 to 400 feet apart, and at almost the same level. The instrument is set up halfway between them, as in Fig. 2.21(A), and leveled carefully.

A leveling rod is held on each of the outer stakes, and an exact reading taken according to the crosshair. These readings will be accurate with reference to each other, as any error in crosshair height is canceled in observations taken at equal distance.

The stake standing on lower ground is assigned an elevation of zero, and has the higher rod reading. The elevation of stake (b) is the difference between the two readings.

The instrument is now set up in line with the two stakes as in (B), about 10 feet beyond stake (b). A reading is taken of (b). Then a reading is taken at (a), which should equal the elevation of stake (b) plus the reading there.

If it does not, adjustment is made in the manner described above.

Vertical Hair.   The vertical cross or direction hair may be checked by driving three stakes exactly in line at 200-foot intervals. The line may be determined by sighting with the vertical hair, and the distances measured by stadia or tape. The stakes should be about on the same level.

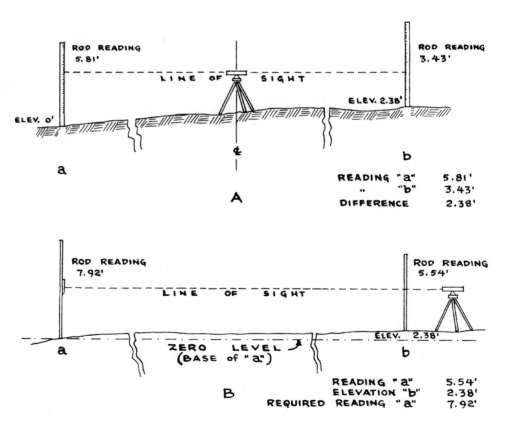

FIGURE 2.21   Checking the crosshair.

The instrument is set up over the center stake, leveled carefully, and turned so that the hair lines up with a rod or pole held vertically over one of the end stakes. The instrument is turned exactly 180°, or, if it is a transit, flipped over vertically. The hair should now line up with a vertical stick on the other stake.

If it does not, the crosshair should be moved one-quarter of the way toward the stick by means of adjustment screws on the sides of the crosshair frame. These work in the same manner as the upper and lower screws.

After the one-quarter adjustment, the telescope is turned until hair and stick coincide. A 180° angle is again measured off and the rod on the first stake sighted. The hair is adjusted to move toward it one-quarter of any distance, then centered on it by moving the scope.

Additional half-circle turns, and adjustments, are made until the hair will coincide with both sticks, 180° apart.

LASER

The word "laser" stands for "light amplification by stimulated emission of radiation" (Fig. 2.22). The type originally used in construction was a tube filled with a mixture of helium and neon gases, stimulated by electric current.

Stimulation causes the gas atoms to emit energy in the form of light. This process produces light with only one frequency, with such intensity that it emerges as a continuous, coherent, narrow beam of red light. Its sides remain almost parallel for considerable distances, instead of diverging and widening as ordinary light.

In recent times the most common lasers used in construction are diodes. A diode is a two-electrode device having an anode and a cathode with marked unidirectional characteristics. This type of laser is an electron-wave tube, which derives its characteristics from the interaction of electrons which are in a beam of initially uniform charged density. The electron wave is focused in a narrow beam of high intensity. It may be in the visible light or infrared (IR) light range. Most commonly used in construction are red or, with certain filters, green which is more intense than the red light.

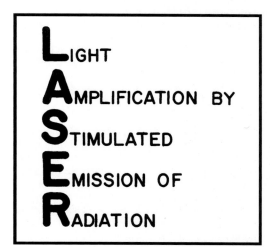

FIGURE 2.22 Meaning of "laser."

The IR lasers are used for heavy construction equipment in the short to medium range of 500 to 1000-foot radius from the laser. There is a laser receiver mounted on the excavator, backhoe, and the like, for elevation indication. With a slope control laser the beam can produce a high-powered IR or barely visible beam for a projection of 1500 or 2500 feet, or more.

The laser tube or cylinder is mounted in an enclosure which may be mounted on a tripod, in a manhole, in a pipe, or in other ways. The design of the instrument and mountings provides threaded adjustments of great delicacy and accuracy, which permit precise regulation of direction (line) and grade.

There are usually two adjustments for vertical alignment. One levels the instrument. The other adjusts the laser tube inside the case for percent of grade.

This percent or gradient is zero for level work. In pipe laying, it is set for whatever plus or minus gradient is required. The setting is indicated by a dial gauge, or by a digital readout.

Power.   The power source is usually a detachable battery pack or a rechargeable 12-volt truck or automobile battery. It is important to use correct polarity, as the laser will not operate if connected to the wrong terminals. Most models will blow a fuse, some suffer serious damage. Others have a built-in electronic safeguard.

The laser beam is generally harmless and is usually invisible when viewed from the side or the rear. It should be visible and harmless when seen in or through a plastic target. OSHA regulations require workers exposed to it to wear goggles designed specifically for protection against the wavelengths and intensity encountered.

Significance.   The laser supplies a visible line-of-sight indicator that can be fixed, or set to move in a pattern, without an operator. It can be used by anyone in its path, by means of simple equipment, to determine grade and/or direction, with a high degree of accuracy. It makes otherwise imaginary reference lines real to workers.

The laser beam is sometimes compared to a taut, no-sag stringline supported at one end only.

A transit or other sight instrument is needed in job layout, and perhaps each time a laser is set up. In addition, there should be a program of frequent checks with a transit to check work being done according to laser guidance.

Maintaining Accuracy.   Once set up accurately, a laser remains accurate unless disturbed. It must be, and usually is, protected against violent disturbance. But a person or object might brush against it or strike it, ground may yield or a support may move, so that it goes off line or grade without the change being noticed.

Devices are available that will shut off a laser if it is disturbed. It can then be turned back on, but must be checked before further use.

With or without such a protective attachment, it is wise to recheck the setting of a laser frequently during precise work, and occasionally in rough work.

PIPE LAYING

For guidance in laying pipe, a laser is set in a manhole or at an angle point in the line, and the beam is directed along the inside of the pipe. Depending on the system being used, the beam may be centered in the pipe, or aligned just above the center of the invert (inside bottom).

The laser location is usually determined by setting a transit directly above the angle point, which (for purposes of this description) is assumed to be approximately centered in a newly built manhole.

The transit may be mounted on a rod based at the bottom of the excavation and supported by an adjustable cross frame, by rods based on a clamp or clamps on the manhole edges, or on a batter board. See Fig. 2.23. It is located exactly over the angle or location point by reference to baseline and hub points.

Measurement is made from the transit down to the grade at which the laser is to be set. A measuring rod support is used directly, being checked with a carpenters' level to ensure that it is vertical. Otherwise, a steel tape and plumb bob are hung from the center of the transit.

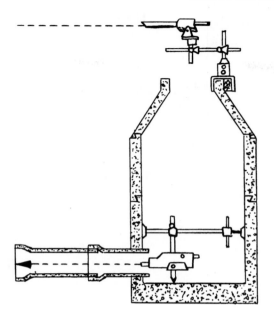

FIGURE 2.23   Transit and laser.

The laser is set inside the manhole, at the level of the pipe center or other measurement level. There are a number of ways of setting up, of which two will be described.

Vertical Rod Mounting.   In this system, both the transit and the laser are mounted on the same vertical column, which is composed of calibrated sections with lockable male-female joints. Cross section is square.

This column may be supported toward the top by an expandable cross brace inside a manhole, or by a batter board. At the bottom, it may be held by a tripod, or by forcing its lower end into soft dirt, or to a support on a steel plate. On fresh concrete, it may rest on a steel pin driven down through the concrete.

By reference to rod calibrations, and/or by measuring with a tape, the laser is clamped to the rod at the desired level, which may be at the pipe center, just above the invert, or at any convenient point between. It faces in the direction in which the pipe is to be laid.

The laser grade indicator is set at the correct slope or gradient for the pipe (see Fig. 2.24), up or down, then the laser is connected to its battery and turned on.

The transit is clamped to the upper part of the column, so that it can look in the direction of pipe laying.

When the ditch has been dug 15 feet or more toward the next manhole, the laser beam should be shining along its centerline, or at least very close to it. A stake or target is placed approximately in the center of the ditch to make the beam visible. The transit is rotated on the column by adjusting a knob, until its vertical crosshair is centered on the beam, as seen in Fig. 2.25.

The transit is now pointed straight down at the laser, and aligned with a reference mark on the barrel by adjusting a different knob than the one mentioned above. The transit is alternately aligned with the beam and with the mark, until it remains exactly centered on both. The vertical line of sight (vertical hair) through the transit now includes the laser beam.

The transit is now aligned with the marker on the next forward manhole by rotating the vertical rod (to which both transit and laser are clamped) relative to its supporting crossbar, with a fine adjustment. Both instruments, being clamped to the rod, remain in alignment with each other.

Therefore, when the marker is bisected by the transit's vertical crosshair, the laser is projected accurately along the line for the pipe.

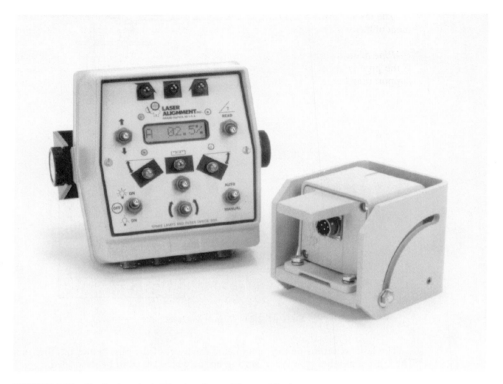

FIGURE 2.24   Gradient readout. (*Courtesy Laser Alignment Inc.*)

FIGURE 2.25   Lining up the beam. (*Courtesy of AGL Lasers.*)

The laser is now rechecked for level and for correct grade setting, and another check of the alignment of the two instruments is made. The level vial should be checked intermittently during the work.

In-Pipe Mounting.   Many lasers are equipped with mountings that permit insertion in a pipe, or if the pipe is very small, just outside one end and looking through. See Figs. 2.26 and 2.27. The support member is a square bar with a cross piece at one end and a threaded adjustment at the other.

FIGURE 2.26   In-pipe mounting.

FIGURE 2.27   In-pipe target and remote control device. (*Courtesy of AGL Lasers.*)

If the pipe has been carefully set to grade and alignment, and the laser is properly mounted, the laser can project the correct centerline or flow line from one pipe through the rest of a straight run.

However, alignment should be checked on a transit-set pole or target 15 to 25 feet ahead.

Refraction.   All light beams, including lasers, may be bent out of a straight line if they pass through air of varying density and/or temperature.

If sun-warmed pipe is set in a cold trench, or cold pipe is put in a warm trench, or if hot sealing compound is used, a mixture of masses of air of different temperatures may occur inside a pipe, causing the laser beam to vary from a straight line.

This situation can be (and must be) corrected by using a blower to direct a steady current of one-temperature air through the pipe. Blowers using the same 12-volt current as the laser are usually supplied with them, and their use is a routine precaution that often avoids costly confusion and mistakes.

Targets.   Targets for use in pipes are usually translucent plastic, or metal with a plastic window. They may be premarked with centering or other information lines, or such indicators may be put on with erasable markers on the job.

One size of target may fit several or all sizes of pipe, or there may be a separate size for each pipe. Some are snap-fitted into the pipe, others have fairly elaborate holding and extension devices, and may include a level vial to check the installation.

Ordinarily, a pipe target is inserted in each pipe at the end away from the laser, and the pipe is shifted until the beam is centered on indicators, usually cross lines, with or without bull's-eye circles. If the laser is set correctly, and the target properly used, accuracy to within $\frac{1}{8}$ or $\frac{1}{16}$ inch is readily obtained. The target is removed as soon as the pipe has been secured in place, and put in the next pipe when it is ready.

Other targets may be range poles or surveyors' rods, with a surface on which the beam can be seen readily, even in brightly lighted areas. There are also electronic sensors that can register the beam even when it cannot be seen.

Changes.   From any one setting, an inside-pipe laser can be used for only one straight section. If the direction or gradient changes, the instrument must be reset at the angle point between the two lines.

If such changes are frequent, it may be better to put the laser on the surface, as described in the next section. It can then be used for gradient only, with direction determined by transit.

A curved pipe must be set from surface instruments.

Ditch Depth.   The laser may be used to regulate ditch depth in between setting pipe sections. Readings may be taken on the bucket of a hoe, the shoe of a ditcher, or a rod held on the ditch floor close to the excavator.

This continual guidance greatly reduces the need for hand grading while laying the pipe.

LASERS ABOVE GROUND

Potential.   The potential of lasers in construction work above ground is very great. A general simplification and improvement in handling a variety of leveling and grading problems may be expected.

Here we will deal with two applications: grade reference within a specific construction area, and guidance of excavating machines.

Tracking Level.   The grade laser tracking level is mounted on a substantial tripod, is leveled, and delivers a rotating beam of laser light. When set for automatic operation, its beam passes over the whole area of a construction site at 6 to 40 revolutions per second (rps).

Each work crew that needs grade checks has a stadia rod, and a target with rechargeable battery that can be clamped to it, similar to that shown in Fig. 2.28.

It is of course very important that the laser beam elevation be accurate. The instrument is set up in somewhat the same manner as a self-leveling engineers' level, by adjustment of three leveling screws. It is turned on, and allowed a 2-minute warmup period before being placed in automatic rotation.

This rotation may shut off automatically if the instrument is disturbed.

A rod and target are set on a benchmark, and a reading is taken. If the elevation is a fractional number, it can be rounded off for greater convenience by lifting or lowering the laser slightly, using a fine vertical adjustment on top of the motor head.

The laser may have a distance range adjustment, which should be correct for job conditions. It may cover an area from 600- to 2000 foot in diameter.

Oscillation. When work areas are typically long and narrow, as in trenching or highway jobs, the laser head may oscillate from side to side, in an arc of 30° or more, instead of rotating.

This oscillation may be quite rapid, and is used more often in automatic control of depth of excavation than as a reference for work crews.

Grade Laser Beacon. The Grade Laser Beacon manufactured by Laser Alignment Inc. is useful in a variety of ways on agricultural and construction jobs. The model shown in Fig. 2.29 is being used for site grading. Seen in that figure are uses with a target on a surveyor's rod and a receiver mounted on a grading machine. The beacon is portable and self-leveling, and indicates when elevation changes have occurred by a disturbance. It is waterproof, can be set for two

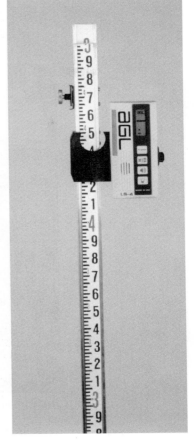

FIGURE 2.28 Target on surveyor's rod. (*Courtesy of AGL Lasers.*)

separate alignment sights, and is workable for an area of 2,000-foot diameter. In addition to site preparation work, it is useful for agricultural land leveling, construction slope control, and ditch excavation.

The laser beam rotation control can be adjusted to 6, 12, 20, or 40 revolutions per second for optimum performance. It would be set at lower rps to save battery life, but the higher speed is needed for use with fast-moving equipment using the laser beam received by hydraulic sensors to adjust for cutting grade. This use with equipment will be explained shortly. The manual adjustments by the equipment operator may be done with a slower laser rps. But the faster the beam rotation, the quicker the grading equipment response, preventing a washboard effect on the graded surface.

For slope control the Grade Laser Beacon is directed up perhaps a 2:1 or 3:1 slope to guide the grading machine or surveyor's rod holder. The slope grade can be set at a range from −4 to + 50 percent. This instrument can be set with two grades set simultaneously on the illuminated grade display. That is useful when two pipelines coming together have different slopes, for example, the feeder lines and the main pipeline. In another case the side slopes for a ditch will probably be steeper than the bottom area running to the water trickle channel.

FIGURE 2.29   Grade Laser Beacon and targets. (*Courtesy of Laser Alignment Inc.*)

Automatic Grade Control.   The basic application of the rotating laser beam is the direct, automatic control of cutting depth (or filling height) of grading machines.

The machine to be controlled must have hydraulic lift for its cutting edge or parts. A vertical mast is erected on a frame resting on the edge, and is kept vertical by a pendulum-controlled valve and hydraulic minicylinders. It has a telescopic adjustment for height.

The mast carries a receiver with a number of cells sensitive to the laser light, which are wired to a solenoid valve that controls the hoist for the cutting edge. This valve opens almost instantaneously when activated by the receiver cells.

Center cells are neutral or on grade indicators, and do not activate the valve. Those just above or just below center will open the valve for 50 milliseconds or $\frac{1}{20}$ second. The top and bottom ones open it for 200 milliseconds or $\frac{1}{5}$ second.

There may be additional cells to activate a control that will swivel the mast to keep the receiver facing the laser.

The mast height is adjusted so that the center of the receiver is in the plane of the laser beam when the cutting edge is exactly on grade, as seen in Fig. 2.30.

The machine is then driven and steered in the normal manner. If the blade drops slightly below grade, the hoist valve will be opened for $\frac{1}{20}$ second to raise it. This action will be repeated 5 times per second until it is on grade.

If the drop is enough to activate the top cells, they will open the valve for $\frac{1}{5}$ second. If the demand for lift is repeated when the beam returns, the valve will stay open continuously until correction shifts the work to the $\frac{1}{20}$-second adjustment, or to neutral.

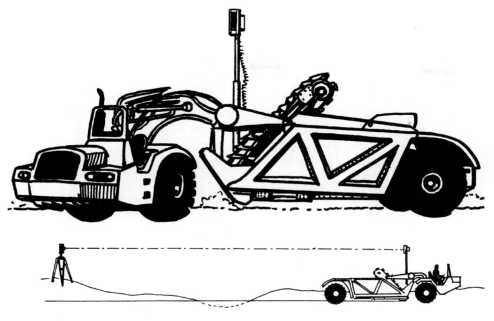

FIGURE 2.30 Laser-controlled scraper.

If the blade moves above grade, similar action will put the valve in the down position. Lights keep the operator informed of the actions of the system.

A refinement for fast-moving machines (scraper or dozer) is a proportional current system, responding to the 10-sweeps-per-second or faster laser. In this, the opening of the valve is proportional to the amount of correction needed.

Repetition of automatic adjustments 5 or 10 times per second results in running a very accurate grade, usually within 0.02 or 0.04 of specification. Accuracy is greatest close in, and declines within these limits to a distance of 1,000 feet, the longest recommended.

With increasing distance, the beam widens so as to lose its pinpoint accuracy, and problems with diffraction become more likely. Errors caused by curvature of the earth's surface are given in Fig. 2.35.

Applications.   This system is well adapted to automatic control of ditchers, landplanes, scrapers, dozers, tube and cable plows, and graders.

The grader requires an additional cross-slope sensor. The laser controls the lift cylinder for the leading edge, while the slope regulator keeps the trailing edge in proper relationship.

Automatic control is largely limited to finishing operations, as it is not designed to sense loads, and may therefore tend to make a machine dig into more material than it can move.

When the automatic control overloads the machine, the operator can override it, and raise the blade. The operator can then return for additional manually controlled passes, finally returning it to automatic for trimming.

Excavating Trench.   To excavate a trench to a given depth and in a specified alignment, Laser Alignment Inc. provides a laser instrument called the Depthmaster, depicted in Fig. 2.31. It can be attached to the dipper stick of a hydraulic excavator and gets direction from a laser beacon, described previously, set up nearby.

The Depthmaster gives readings for vertical alignment and depth of cut. The vertical alignment is noted by the three lights at the bottom of the readout panel. If the middle light shows, the

FIGURE 2.31   Depthmaster—attached to excavator. (*Courtesy of Laser Alignment Inc.*)

excavating dipper is in alignment, whereas the left or right light would indicate that a move to extend or retract is needed. The depth of cut is indicated by the display of light in the strip of lights up and down on the panel. When the dipper is not in the laser beacon's beam because it is too high, or low, no depth light will be lighted. If the light with a horizontal bar on it is lighted, then the bucket in a horizontal digging position is at the correct excavating depth. If one of the upper or lower lights is lighted, then the bucket will have to be lowered or raised to be on grade.

The Depthmaster gives a constant accuracy "on grade" to $\pm \frac{1}{6}$ inch or $\pm \frac{1}{2}$ inch depending on the job site requirements. It has 9 inches of vertical receiving cells and can receive from any laser in all directions, that is, 360° horizontally. The power used for this instrument is 11 to 30 volts direct current, and its operating range is 1,000 feet from the laser. The operator can have a readout panel in the cab of his or her hydraulic excavator in case there is difficulty reading the Depthmaster on the dipper stick.

GLOBAL POSITIONING SYSTEM

The global positioning system (GPS) is a satellite-based system developed for military use but successfully transferred for many civilian uses. GPS makes use of 24 satellites. Commercial uses include land surveying, vehicle navigation and tracking, mining and construction, agricultural systems, and many other applications.

GPS Methods.   Mobile GPS receivers can be mounted on earthmoving or agricultural equipment. The receivers obtain positions by using the signals from the GPS satellites. Two-dimensional positions (latitude and longitude) use the signals from three satellites, and three-dimensional positions (latitude, longitude, and elevation) require signals from four satellites. GPS positions are based on a technique called *trilateration*, which is similar to triangulation but is based on the distances to the satellites, not angles. GPS hardware has very accurate clocks, so timing of these satellite signals is known very accurately. The GPS receiver calculates the distance to satellites, using the signals from them and their location in the sky. Those distances are then used to calculate the receiver's location on earth.

Differential GPS.   A technique called differential GPS (DGPS) is used for guiding earthmoving, mining, and agricultural equipment. DGPS can be used to obtain positions with the accuracy and dynamic positioning needed for vehicle guidance. DGPS requires a base station with a precisely known location. This base station calculates its GPS position from the satellite signals, compares the calculated position with its known position, and generates correction data. The correction data are broadcast by radio in real time to the GPS receivers nearby, for example, on earthmoving equipment. All mobile GPS receivers within range of the radio broadcast can use correction from a single base station. See Fig. 2.32.

Accuracy.   High-quality GPS receivers can receive two types of signals from the satellites: C/A code or carrier phase. With DGPS, the accuracy for C/A code receivers can be improved to $\frac{1}{2}$ meter, and can be up to 10 meters at ranges over 1000 kilometers from the base station. Carrier-phase receivers can have an accuracy to 5 millimeters, and move out to 5 centimeters at ranges of 10 to 15 kilometers from the base station. The important development of real-time kinematic (RTK) allows users to obtain centimeter-accuracy GPS positions in the dynamic mode. With RTK, land surveyors can accurately map property boundaries by simply walking or driving around the property. The location of a piece of earthmoving equipment can be monitored on a real-time basis by the operator.

Mobile Unit Links.   Several types of communication links can be used between the base station and the mobile GPS receiver. The contractor can use UHF and VHF radio modems or cellular phones to establish the link between the base station and the mobile receiver. Near coastlines and navigable waterways, governments broadcast DGPS correction data on frequencies between 283.5 and 325 kilohertz. In addition, in many areas there are commercial services which broadcast DGPS data on an FM subcarrier or by satellite.

FIGURE 2.32   Setting a GPS base station. (*Courtesy of Trimble Navigation Limited.*)

Construction Uses.   In earthmoving, the main uses of GPS for machine guidance are for grade control on dozers, scrapers, and motor graders. Traditionally, the transfer of engineering design from the drawing board or computer design to the field has relied on the accurate placement of stakes by survey teams. With the use of centimeter-accuracy guidance on board the earthmoving machine, the operator can visualize the position of cut-and-fill surfaces in relation to the earthwork design. This allows a complete digital link from design to layout and reduces the time for which expensive machines stand idle or need to move the earth material a second time. Moreover, real-time information can be obtained on the location and productivity of earthmoving equipment, allowing improved management of the construction site.

GRADING INSTRUMENTS ON EQUIPMENT

As mentioned above, instruments with laser controls and GPS referencing are attached to equipment like excavators and graders. An extensive array of these devices is produced by the Topcon manufacturers. Also Leica Geosystems has their make of equipment called GradeStar for 3D grading solutions.

Specific Equipment Applications.   For cutting a specified slope accurately for earthwork, Topcon's Touch Series 5 (TS5) can be used. It involves a Touch Control Panel (TCP) at the operator's station on the excavator, a valve control box on the machine to regulate the hydraulic system of the excavator, and three machine sensors to direct the required angles of the machine parts (boom, dipper stick, and bucket, see Fig. 2.33) to achieve the desired positions as set on the TCP. The automatic controls set the excavator so it does not overcut nor undercut the slope, making for extra work to get the desired result.

The TCP can also be programmed for a fixed depth of basement, footing, or the bottom of a pipe trench. with a display screen to show the bucket position relative to the desired grade, even when the excavator is making "blind cuts," where the operator cannot see the bottom.

Another application is for the TS5 instruments on a grader to automatically grade the subgrade for a pavement. The TS5 consists of a control box, like the TCP, four machine sensors, and an all new High Flow hydraulic valve package. The sensors are on the blade and supports in all their

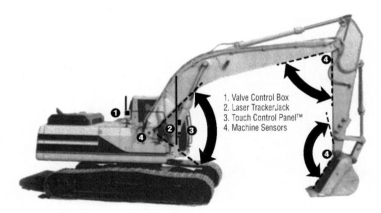

1. Valve Control Box
2. Laser TrackerJack
3. Touch Control Panel™
4. Machine Sensors

FIGURE 2.33   Depth and slope control. By knowing the lengths of each machine member and accurately measuring the angles between them, the control system precisely and automatically positions the bucket to any depth and slope. (*Courtesy of Topcon Positioning Systems.*)

positions for grading. The grader operator uses the control box to enter the desired elevation and slope that TS5 will maintain.

When the machine begins grading, information from each of the four sensors is sent to the control box where comparisons are made to the desired grade entered by the operator. If necessary, the control box sends out correction signals to the hydraulic valves to adjust the blade to the required grade. The TS5 actually performs the measurements and corrections more than 30 times a second so that the grader can move at a relatively high speed.

The Leica Geosystems GPS base station receives time and place information from the GPS satellites in the sky. The base station, which is standard, modem survey equipment, communicates via radio modem to all machines. The precise position is computed on the machine relative to the base station and is then combined with slope and other sensors to determine the exact position and heading of the grader's blade.

By comparing these values to the project data stored in the computer on board the grader, corrections for height and cross slope are transmitted to the SonicMaster blade controller, also on board, freeing the operator to concentrate on steering. Only one or two passes are required to get to final grade. And the result provides accuracy of +/– 0.10 feet (+/– 3 cm) in height for rough grading. A more refined Grade Star TPS solution provides accuracy to +/– 0.02 feet (+/– 5 mm) in height and +/– 0.04 feet (+/– 10mm) for position.

Distance from the roving grader is constrained by the strength of the radio link, from 1,000 feet up to six miles.

DISTANCE METERS

Several battery-powered instruments are available that can read the distance to a target by measuring reflection time of multiple-phase signals.

The short-range Cubitape uses a modulated infrared light source, for distances up to 6,000 feet, with differences in range depending on the type of target and atmospheric conditions. The target is located by means of a built-in telescope.

The distance is automatically displayed in digits for feet (or meters) and their decimal fractions. See Fig. 2.34*A*.

The Electrotape (Fig. 2.34*B*) uses a microwave beam for distances of 100 feet to 30 miles. Two identical instruments are used, with built-in short wave communication between them for the operators. The readings are taken first by one unit, then by the other, as an accuracy check.

These instruments measure direct line-of-sight (slope) distances. If there is a difference in elevation between the points, the horizontal distance must be calculated.

Correction for the curve of the earth's surface is needed if elevations are required. An approximate formula, good up to distances of a few hundred miles, is

$$\text{Correction} = \frac{(\text{distance in miles})^2}{8,000}$$

Figure 2.35 is a table for various distances. The correction is subtracted from the apparent elevation indicated by the instrument. It will be noted that the curvature is very slight within ordinary survey distances, but must be considered in long sights.

MINOR INSTRUMENTS

Hand Level. Rough levels may also be run with hand levels, such as the one shown in Fig. 2.36A. This consists of a sighting tube, in the top of which is a small spirit level parallel with the line of sight. A slanted mirror reflects the spirit level so that it is seen vertically beside the field of view. The object glass is marked with a centerline, and may have two or more stadia lines.

FIGURE 2.34A Cubitape. (*Courtesy of Cubic Industrial Corp.*)

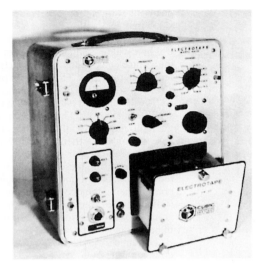

FIGURE 2.34B Electrotape. (*Courtesy of Cubic Industrial Corp.*)

DISTANCE		SUBTRACT FROM READING	
Feet	Miles	Feet	Inches
100	0.019		0.002
500	0.095	0.006	0.070
1000	0.189	0.023	0.279
1500	0.284	0.053	0.633
2500	0.473	0.148	1.774
5000	0.947	0.592	7.096
5280	1.000	0.660	7.920
	2.0	2.64	31.68
	5.0	16.5	
	10.0	66.0	
	100.0	6666.0	

FIGURE 2.35 Corrections for curve of earth's surface.

This level is used by holding it to one eye, and tipping it up or down until the bubble is centered at the centerline on the glass. Any object cut by this line is then on a level with the observer's eye, and nearby elevations may be determined and levels run in the same manner as with an instrument.

Results are much less accurate, but in rough work this may be more than compensated by the ease of use.

The eye height of the observer may be used as a unit of measurement. In taking the height of a hill, as in Fig. 2.36B, the observer holds the level to her or his eye, while the rod operator moves up or down the hill in response to instructions, until the bottom of the rod, a stick, or the operator's

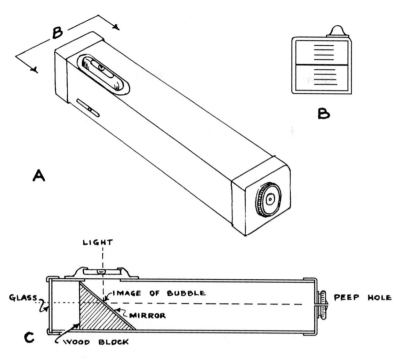

FIGURE 2.36A Hand level.

shoe rests on the ground at eye level. The spot is marked, the observer moves to it, standing with his or her heel on it, and the rod operator moves uphill, repeating the process. When the top is reached, the last observation should be taken on the rod, or a ruler or tape, to show the distance from the hilltop up to eye level. The height of the hill is the height of the observer, multiplied by the number of observations, less the rod reading on the last observation.

In working downhill, Fig. 2.36(*B*), a target fastened at the top of the rod, or a mark on a long stick, is sighted and moved until it is level with the eye. The observer then moves to that spot, while the target is moved downhill until level with the new position. On the last sight, the distance from the ground to eye level is measured.

Each observation covers a drop equal to the height of the target minus the height of the observer, and they will all be equal except the last one, which should be separately figured and added to the others.

An individual's eye height, measured from the heel when she or he is in erect position, will seldom vary more than an inch, which is not too large an error for rough work. Care should be taken that the heel, and not the ball of the foot, is placed on the mark.

Level Clinometer.   The clinometer is a special type of hand level which can be used to measure slopes and vertical angles. The spirit level is hinged so that it can be rotated about 45° in either direction, and a pointer and scale indicate the angle between the spirit level and the line of sight. The bubble will appear at the centerline of the object glass when the hand level is held at the angle indicated by the scale. The scale is usually graduated to indicate both angles and slopes.

The angle of a slope may be measured by setting a target at eye height at one end and sighting it from the other. With the centerline on the target, the spirit level may be adjusted until the bubble is beside the line. The pointer on the scale will then indicate the slope of the hill.

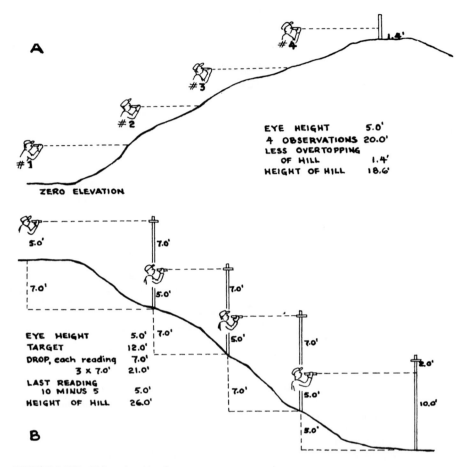

A

EYE HEIGHT   5.0'
4 OBSERVATIONS 20.0'
LESS OVERTOPPING
OF HILL   1.4'
HEIGHT OF HILL   18.6'

ZERO ELEVATION

5.0'   7.0'

7.0'   5.0'   7.0'

EYE HEIGHT   5.0'   7.0'
TARGET   12.0'
DROP, each reading   7.0'   5.0'   7.0'
3 x 7.0'   21.0'
LAST READING
10 MINUS 5   5.0'   2.0'
HEIGHT OF HILL   26.0'   7.0'   5.0'

10.0'

5.0'

B

FIGURE 2.36B   Using a hand level.

It may also be used as an inclinometer by placing it on the slope to be measured, and setting the spirit level until the bubble is centered. It is usually good practice to lay a board on the ground surface and take its slope to eliminate the effect of small irregularities.

String Level.   A string level is convenient to use over short distances. It is a spirit level fitted with prongs by which it can be hung from a string stretched tightly between two marks. Elevations may be taken from the end of the string, or by measuring down from any part of it, as illustrated in Fig. 2.37.

The string used should be strong enough to take sufficient pull to remove all sag. If it is at all slack, it will give false readings, showing slope at the ends of a level stretch and level somewhere near the middle of an inclined string. A special need is depicted in Fig. 2.38.

Most string levels use flexible prongs which are easily bent by light pressure. It is therefore best to check such a level before every use, and occasionally during a job. This is done by leveling a string according to it, slacking the string and reversing the direction of the level on it, and tightening the string to the same marks. If the reading is the same, the level is all right. If it disagrees, bend a prong sufficiently to move the bubble one-quarter of the way to the center, then move the string to center the bubble. Reverse the level on the string, and repeat the procedure as above, until the reading is the same both ways.

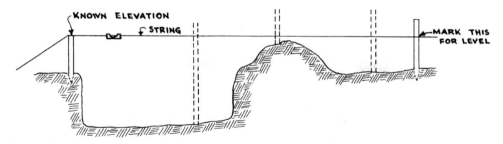

FIGURE 2.37   Using a string level.

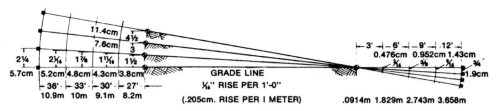

FIGURE 2.38   Figuring string heights on banked curve. (*Courtesy of CMI Corp.*)

Carpenters' Level.   A carpenters' level may be used to level a string, although not as conveniently. A string leveled by a carpenters' level may be used for direct adjustment of a string level.

Altimeter.   Altimeters or barometers which are small enough to be carried, and sensitive enough to react to small changes in elevation, may be used for taking preliminary levels.

   The most convenient altimeters are mountaineers' pocket models, the size of a large watch. Engineers' models, which are rather rare, have a glass for magnifying the scale, for fine reading.

   An altimeter is set for correct or assumed altitude when work with it is started. As it is carried up- or downhill, the hands will point to higher or lower altitudes on the scale, and notes may be made of the reading whenever desired. It is advisable to tap the instrument before each reading.

   Altimeters provide the quickest and easiest means of finding heights and depths in rugged and overgrown country. They are not accurate enough to be used in setting grades, except in very experienced hands.

STRINGLINE

A *stringline* is a string or wire that indicates location, grade, or both location and grade. Its use in construction is widespread and very old. A familiar example is string stretched between batter boards for reference in laying foundations and pipes.

   A stringline may serve as a visual guide for a machine operator to prevent him or her from running off-line between widely spaced stakes. If the machine is equipped with a pointer or a suspended plumb bob on the frame of its grading blade, ditching wheel, or other grade-making tool, it can be adjusted so that following the string accurately with it ensures working to proper alignment and grade.

   The principal and increasing use of stringlines has been in automatic control of grading, excavating, and paving machines. These controls, which are described in Chap. 19, are usually made up of a tracer or wand that stays in contact with the string, and a sensor that reacts to any change in tracer position by actuating hydraulic valves to restore proper relationship with the string.

A machine may need two stringlines, one at each side. However, if it has an automatic cross-slope control, only one line is required.

Accuracy in locating such stringlines is of the utmost importance, as line and grade produced in the automatic operation are absolutely dependent on it.

There are various ways of setting up a stringline. The following description follows a booklet, *Stringline Manual,* published by CMI Corporation.

Apparatus. The string holder is made up of a metal stake (standard length 42 inches), a sliding bracket that can be secured to the stake with a setscrew, and a horizontal rod that slides in the bracket, and can be locked in it with another screw. One end of the rod is notched to hold the string.

The string is usually a special yellow polyethylene selected for strength, stretch characteristics, visibility, and rot resistance. Regular strong white cord may be substituted in an emergency.

The string is carried between jobs on storage reels. When in position, it is tightened by means of take-up reels, one at each end. See Fig. 2.39.

An engineers' steel rule is used to measure the distance from string down to the top of the grade stake (tack line). This distance has been previously determined, and it is the same for each grade stake in a section.

A plumb bob may or may not be used for horizontal alignment.

Location. The stringline is placed outside the strip or lane being graded. In highway work, this usually means in the shoulder, 2 or more feet beyond the edge of the lane. It is preferably 18 to 30 inches above grade, although some machines can use higher and lower settings. It should be at least 5 inches above the ground.

It may be possible to use one setting for up to three processes—subgrade finishing, laying selected subbase, and paving. Clearance from lane edge must be sufficient to accommodate windrowing and removing of waste material, passage of machine tracks if outside the lane, and minimum tracing arm lengths.

Stakes are driven about 1 foot outside of the line proposed for string, alongside marking hubs or grade stakes.

Hubs (Grade Stakes). Hubs are usually 2×2 wood stakes, placed by surveyors along the line of the string, parallel to the roadway centerline. They are driven to grade (bluetops) or to some fixed

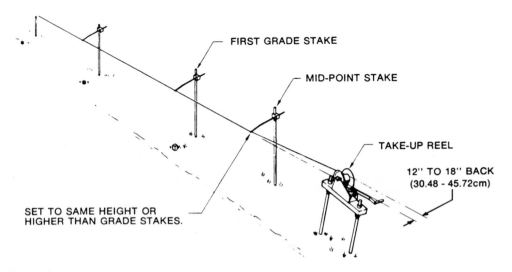

FIRST GRADE STAKE

MID-POINT STAKE

TAKE-UP REEL

12" TO 18" BACK
(30.48 - 45.72cm)

SET TO SAME HEIGHT OR
HIGHER THAN GRADE STAKES.

FIGURE 2.39 Stringline and take-up reel. (*Courtesy of CMI Corp.*)

distance, the same for each stake, above grade. They are usually spaced 50 feet apart for uncomplicated work, and 25 feet for superelevations, vertical curves, and other special situations.

Reference grade is level with the centerline on straightaways. On superelevated (banked) curves, the reference is to the tipped plane of the cross section, so that it will be lower than the near edge on the inside of a curve, and higher than it on the outside (Fig. 2.38).

A tack is driven into the top of each hub for precise location of horizontal alignment.

Setting Up. A string stake is driven alongside each hub, about 1 foot to the outside. It must be vertical, and driven deep enough for good stability. The rod should extend directly toward the stake, and can be slid inward or outward until its slotted end is directly over the tack. The bracket is raised or lowered until the rod is $\frac{1}{4}$ inch above the selected stringline height. Adjustments are locked with setscrews.

A steel engineers' rule is used to measure the height of the rod, and to check its location over the tack.

Take-up reels are then placed, 25 feet before the first stake in the section, and 25 feet past the last stake. These should be offset 12 to 18 inches beyond the end of the line of the string, with the crank to the outside. The reel is first located on the ground, stakes are driven through holes in its base, and it is then raised to a convenient cranking height, and clamped.

A metal midpoint stake is driven in line with the string stakes, halfway between the last (or the first) one and the reel. The rod may be at stringline grade, or higher.

The string itself is reeled off a storage reel (or two reels if there are two lines) held on a bar on a truck. It is anchored by winding about 25 feet onto one of the take-up reels, and is paid out as the truck drives to a few feet beyond the other end. The string is cut and secured to that reel.

The string is pulled taut by hand and by turning the reels. Then it is inserted into each rod notch. It should slip into them easily, but require moderate force to be pulled out. Slot openings can be widened with a screwdriver, or narrowed with pliers. There should be no visible sag between stakes.

Using an engineers' rule, adjust the rod brackets so that the string is at its exact height above the tacks, and with a plumb bob to make sure it is in line. Sight down the completed line, and recheck any break in a smooth flow.

Visual inspection should be repeated just before use, and occasionally as work progresses.

Any breaks in the string can be repaired by relaxing the tension, tying the broken ends together with a square knot, and retightening.

MEASURES OF PRODUCTION

IN-PLACE MEASUREMENTS

Distance. Surface distances may be measured approximately by pacing, stadia, or car speedometer; or more exactly by tape (chaining), the distance meters mentioned earlier, or measuring wheel.

Distances may also be read off job plans or maps. But a possibility of an error in a map, or in interpreting it or its scale, must be guarded against.

A measuring wheel is a convenient and sufficiently accurate means for measuring both short and long distances. Small wheels are hand-rolled along the ground, while large ones may be either hand-rolled or towed slowly by a vehicle. A counter registers the distance covered in feet. Readings down to 2 inches may be obtained. See Fig. 2.40.

Bank. A bank of earth or other material may be measured before and during the time it is dug away. This is usually done to establish a basis for payment to the owner of the land, or from the job to which it is being hauled.

FIGURE 2.40 Measuring wheel. (*Courtesy of Rolatape, Inc.*)

A grid is established before digging starts. Ten- or even 5-foot intervals are used between stakes for close measurement of rough ground, and wider spacing for more even ground and less exact work.

Benchmarks and reference points are established well outside of the digging area. Then the elevation of the ground is taken at each stake. Elevations are also taken of humps or hollows lying between the stakes. A contour map is usually made, and ground elevations are plotted on cross-section paper. Vertical scale may be exaggerated. This original surface, called grade, is a solid line, as in Fig. 2.41. Any desired profile or section that does not follow the stake lines can be made up from the contour map. The new elevations are plotted on the same paper as the originals, and are usually indicated by dotted lines, as illustrated.

Excavated material has been removed from between the solid and the dotted lines on each of the profiles. This area is measured by counting squares, dividing into rectangles and triangles, or mechanical methods. If the horizontal scale is 1 inch to 25 feet, and the vertical scale 1 inch to 5 feet, a square inch indicates 125 square feet or about 13.9 yards. It is best to keep the figuring in feet until the end, for greater accuracy.

The vertical area is determined for each profile, then areas are added and the total is divided by their number to obtain an average. This average is multiplied by the length of the excavation at right angles to the profiles. This gives the amount excavated in cubic feet. Dividing it by 27 reduces it to cubic yards, bank measure.

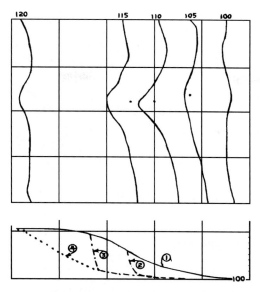

FIGURE 2.41   Measuring a bank.

The same procedure is followed each time the bank is measured. On all calculations after the first it is usual to determine both the amount removed since the previous measurement, and the total amount removed since the job began.

Pile.   An experienced estimator can often judge the volume of a pile quite accurately by inspection, without measurement. However, appearances are often deceptive, and most people are safer if they make at least approximate measurements as a check.

Piles can often be conveniently measured by calculating the volume of regular masses of similar outline, and making plus or minus adjustments for differences.

A pile of clean, dry sand may have a conical shape, or be a ridge with a triangular cross section, ending in half cones. Measurements should be taken to determine base size and height. See Fig. 2.42.

The area of the circular base of a cone is found approximately from the circumference by the formula

$$\text{Area} = \frac{\text{circumference}^2}{12.6} = \frac{(2\pi r)^2}{12.6}$$

and from one-half the diameter by

$$\text{Area} = 3.14 \times \text{radius}^2$$

The volume of a cone is the height times one-third the base area.
The long part of the pile is figured by the formula

$$\text{Volume} = \frac{\text{height} \times \text{width} \times \text{length}}{2}$$

A long pile will have the volume of the center section, plus the volume of one cone, as each of the ends is a half cone.

When the pile shape is flattened, different, or irregular, profiles and cross sections are taken in the same manner described for banks.

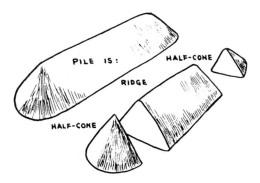

FIGURE 2.42 Pile measurement.

PRODUCTION MEASUREMENTS

In the construction industry the word "production" is assigned the same meaning as "output," which is defined in the dictionary as "the quantity or amount produced, as in a given time." Perhaps a better definition for these words in construction is "useful work accomplished," as in the early stages of a job the chief product is likely to be destruction and upheaval.

Production may be figured in at least three ways. First is on the basis of job requirements. If a time schedule allows 200 working days to move 3,000,000 yards of earth, the contractor's earthmoving machines must move or "produce" 15,000 yards per day. Second, the production of a certain machine is measured or estimated, to determine the number of such machines needed to meet the production required. If a certain scraper can move 1,000 yards a day under this job's conditions, the contractor must keep at least 15 of them working at that rate.

The third way of figuring production is in terms of cost. This is the final and important calculation, as it is the basis on which contracts are let, and on which contractors ride in Cadillacs or go broke. But cost calculations are unlikely to be accurate or useful until job conditions and equipment are known.

Earthmoving quantities may be measured in terms of bulk of material, weight of material, area to be worked, or lineal feet or yards processed.

Cubic Measure. Most earthmoving is computed in cubic yards. A cubic yard is a cube 3 feet long, 3 feet wide, and 3 feet high. It is about 6 inches greater in each dimension than the space occupied by a card table.

"Cubic yard" may be stated and written in full, but is usually shortened to "yard" or abbreviated to cu. yd. or cy. The full designation is used in discussions where the type of yard might be in doubt, as where there is a possibility of confusion with square yards or linear yards. Abbreviations are used in formulas, figuring, and texts where terms are frequently repeated.

Cubic yards may refer to bank, loose, or compacted measure. This will be discussed under Swell and Shrinkage.

Many dimensions in field measurements and contract plans are in feet, so that if they are multiplied together to obtain bulk, results are obtained in cubic feet.

These are usually changed to cubic yards by dividing by 27 (there is 27 cubic feet in 1 cubic yard). Of course it is also possible to divide the original linear measurements by 3 to convert them into yards, and then multiply, but this may lead to working in fractions, decimals, and mixed numbers.

Most countries use metric measures. See pages A.4 and A.19. One meter is 39.37 inches, 3.2808 feet, or 1.0936 yards. A cubic meter is 35.315 cubic feet or 1.308 cubic yards, a bit less than $1\frac{1}{3}$ yards.

Measurements are sometimes given in feet and inches. A cubic inch is too small a unit to be used for dirt (it takes 1,728 of them to make a cubic foot, 46,656 of them to make a cubic yard), but it is used in measurements of engine and compressor displacement and other mechanical descriptions.

It is difficult to multiply feet and inches by feet and inches, so the inches are expressed as either regular or decimal fractions of feet. Regular fractions ($\frac{2}{3}$, $\frac{1}{12}$, etc.) are more accurate, decimals are more convenient.

Weight. In highway jobs, cellar excavation, and site preparation the estimator is chiefly interested in the bulk and digging characteristics of the material to be moved. Its weight has importance chiefly in figuring loading and gradability of haulers, and is only occasionally used as the basis for estimating. See Fig. 3.5.

In mines and quarries weight is likely to be a more important factor than bulk. Products such as coal, crushed stone, cement, ores, and ore concentrates are usually sold by weight, and therefore are measured by weight as they are dug. There are exceptions, however. Some open-pit mines compute their operations on a yardage basis. Others figure ore by the ton and overburden by the yard.

The standard weight measurement is the short ton of 2,000 pounds. There is also a long ton of 2,240 pounds that is used in some mines. It should always be specified as a long ton, and the unmodified word "ton" should mean short ton only.

Other measures are the hundredweight of 100 pounds, and the British stone that weighs 14 pounds. In the metric system a kilogram, often called a kilo, is 2.2046 pounds, say 2.2 for short. A metric ton is 1,000 kilograms, or 2,204.6 pounds, about 10 percent more weight than in a short ton, and a little less than a long ton.

Area. In many operations, bulk and weight do not provide a suitable basis for measuring the amount of work. For example, with any given land-clearing condition, the time and cost of the work vary with the area. Clearing is usually measured by the acre, which is 43,560 square feet or 4,840 square yards. Many estimators do their figuring in acres of 40,000 square feet to simplify arithmetic, and make allowance afterward for the difference of almost 10 percent.

Grading land where cuts and fills are shallow is likely to show costs in relation to area rather than bulk of soil to be moved. But when cuts and fills are deep, bulk becomes more important than area. Exact grade requirements increase grading costs in proportion to area. The estimator must keep track of these factors.

Topsoiling and seeding are figured mostly on the basis of area, although of course the depth of topsoil and the heaviness of seeding are important in determining the rate per square yard.

In the metric system the square yard is replaced by the square meter. This contains 10.76 square feet or 1.196 (say 1.2) square yards. The next larger measure is the "are" that contains 100 square meters, 119.6 square yards, or .0247 acre. Then there is the hectare, that contains 10,000 square meters, 11,960 square yards, or 2.451 (say $2\frac{1}{2}$) acres. The largest measurement is the square kilometer, 1,000,000 square meters, 245.1 acres, or .3861 (say $\frac{2}{5}$) square mile.

Lineal Measurement. Work such as ditching, installing pipe, tile, and fencing is measured by the lineal or running foot, yard, or mile. There are 3 feet in a yard, and 5,280 feet or 1,760 yards in a mile. Metric measures are the meter, 39.37 inches, 3.28 feet, or 1.09 yards; and the kilometer, that is 1,000 meters, 3,281 feet, 1,094 yards, or .6212 (slightly more than $\frac{3}{5}$) mile.

From both a time and cost viewpoint it is of course necessary to know also the width and depth of the ditch, the soil and other special conditions, the size and type of pipe, and the type and quality of fencing.

Drilling is usually measured by the lineal foot, with the size of hole specified.

TYPES OF PRODUCTION

A construction machine may work in an intermittent cycle, in a continuous flow, or in ways that are intermediate between the two types. Figure 2.43 shows the category in which various types of equipment fall.

Cycle	Intermediate	Continuous flow
Revolving shovel, all rigs	Rock drill	Conveyor belt
Shovel dozer	Grader	Belt loader
Bulldozer	Roller	Bucket loader
Scraper	Ripper	Wheel ditcher
Pusher	Plow	Ladder ditcher
Truck	Hopper	Wheel excavator
Pile driver		Paddle loader
Cable excavator		Rock crusher
Concrete mixer		Gravel washer
Pug mill		Screen
		Aggregate spreader
		Compressor
		Hydraulic dredge
		Hydraulic monitor

FIGURE 2.43 Cycle classification.

Intermittent Cycle.   This group includes the most important machines used in primary excavation. They each have a bucket, bowl, or body that is loaded, moved, dumped, and returned to the loading point. One complete set of operations is called a work cycle.

For example, a revolving shovel digs in the bank, swings the bucket over a truck body, dumps it, swings back to the bank, and positions the bucket to dig. A scraper loads in the cut, travels to the fill, dumps, turns, returns to the cut, turns, and gets in position to load. In each case, the set of operations is a cycle.

Production rate depends on the size and efficiency of the earthmoving container, whether bucket, bowl, blade or belt, and the time it takes to go through a complete cycle. The cycle time in turn, depends on the rate at which the container is loaded, moved, dumped, and returned to the loading point.

The distance the load must be moved may be a few feet in the swing of the shovel, or a number of miles in truck haulage. Distance is often the determining factor in the production cycle.

The size of the container is rated by the manufacturer, usually for level and for heaped loads. Container efficiency is the amount of actual load in relation to rated load, and will be discussed later.

The probable production of a machine can be calculated by multiplying its actual capacity by the number of times it can go through its cycle in a given period. Actual production may be found by measurement in the bank, the haulers, or the fill of material removed during a measured time, and/or by measurement of individual loads and timing of cycles.

Continuous Flow.   Continuous flow is found chiefly in equipment using belts, pumps, and/or pipes. It includes machines such as wheel ditchers that dig by means of individual buckets, but which have a number of buckets digging at the same time. The individual bucket cycles overlap, and production is continuous.

Production of a belt machine is found by averaging a number of measurements of the cross section of the load on the belt, and multiplying this figure by the speed of the belt in feet per minute. The cross section is usually measured in square feet, so the result is divided by 27 to obtain loose cubic yards.

Ditchers, bucket conveyors, and other machines using numerous small buckets may be checked by multiplying bucket capacity by the number of buckets per minute, or by measuring the load going through a discharge belt or chute.

Measurements may also be made in the bank, in haulers, or in the fill.

In dredging, the volume of water handled is computed by multiplying a cross section of the stream in the pipe by the rate of flow. This production in cubic feet or cubic yards of water per minute is then multiplied by the percentage of solids in the water—usually 10 to 20 percent—to obtain useful production.

Each belt or pipe might be considered to be a container with a measurable capacity. However, this capacity is related to haul distance rather than rate of production. A conveyor belt carrying a load with a cross section of 2 square feet and traveling 500 feet per minute will deliver 1,000 cubic feet per minute, regardless of whether it is a 10-foot feeder belt or a half-mile-long hauler. The only production difference is the extra time—practically 5 minutes—that it will take to reload the longer belt after unloading it at the end of a shift.

The cost of belt construction and maintenance varies almost directly with the length if width, load, and support requirements remain the same.

Intermediate.   This class of machines requires individual methods of study to determine production.

A grader is a continuous flow machine in sidecasting until it reaches the end of a work area and must turn or reverse. If the area is very short, or it is pushing material bulldozer fashion, it has a cycle. Most grader work is measured by the area processed, either as square yards or as linear feet of roadway of a specified width. Production may be figured on a basis of width processed times speed.

A drill will cut continuously to the end of its stroke, but steels are changed and/or new holes are started frequently. Measurement is by speed in feet per minute or per hour, less nondrilling time.

A hopper often serves to convert intermittent cycle loads to continuous flow, as in a dragline feeding a conveyor belt. Its production rate is usually that of either the excavator or the belt, as it can be adjusted to almost any rate of feed within its capacity.

A special application is in speeding up truck loading. A 2-yard excavator might load sand into a hopper at the rate of 3 yards a minute, and the hopper might be able to fill a 15-yard truck in 1 minute. It serves to convert a rapidly repeated 2-yard shovel cycle into a slower 15-yard truck cycle.

Noncycle Time.   There is almost always a wide gap between production that should be obtained on a basis of cycle speed and capacity, and what is actually produced. Checking the cause of these delays is a drawn-out matter that can usually be best done by studying lost-time reports made out by operators or checkers. Some, however, are frequent enough to be included in brief time studies.

For a shovel there are cleanup and move-up operations, and waits for trucks. Trucks wait to get into loading position, are delayed by traffic, and get stuck. Scrapers wait for pushers, have traffic problems, and sometimes get stuck.

If a breakdown occurs, it is useful to make a note of its nature, and of how long how many machines are put out of action or slowed down.

SWELL AND SHRINKAGE

Swell.   When soil or rock is dug or blasted out of its original position, it breaks up into particles or chunks that lie loosely on each other. This rearrangement creates spaces or voids and adds to its bulk. This increase from bank yards to loose yards is called "swell."

Swell is expressed as a percentage of the bulk in the bank. If 1 bank yard puffs up to $1\frac{1}{4}$ loose yards, the swell is $\frac{1}{4} \times 100$ percent, or 25 percent.

When converting bank yards to loose yards, the measurement is increased by the percentage of swell. In the above example, 10 bank yards could be converted to $10 + 10 \times .25$, or 10×1.25, giving 12.5 loose yards.

Swell Factor.   The swell factor is the percentage of bank yards in loose yards. The contractor uses it to make a conversion on paper of the loose yards being hauled back to the bank yards for which he or she may be getting paid. The swell factor is found by dividing the bulk of a bank yard by the bulk of a loose yard, per the formula

$$\text{Swell factor} = \frac{1}{1 + \text{decimal of swell}}$$

If the swell is 25 percent,

$$\text{Swell factor} = \frac{1}{1 + 0.25} = \frac{1}{1.25} = .8 \text{ or } 80\%$$

The percentage of voids is found by subtracting the swell factor from 1.00.

The tables provided in Fig. 2.44 show average swell, swell factor, and voids for various classes of soil. It also shows the shrinkage factor, which is the reduction in bulk to compacted yards in a fill.

Shrinkage. When soil placed in a fill is thoroughly compacted by rolling, it will shrink, the amount of shrinkage depending on its character, its structure in the bank, the thickness of fill layers, and the weight and type of roller. Blasted rock may retain some swell, while ordinary loam may

TABLE I

	Swell	Voids
Clean sand or gravel	5 to 15%	4.75 to 13%
Top soil	10 to 25%	9 to 20%
Sandy, clayey loam	10 to 35%	9 to 26%
Good common earth	20 to 45%	17.7 to 31%
Clay with sand or gravel	25 to 55%	20 to 35.5%
Clay—friable and light	30 to 60%	23 to 37.5%
Clay—dry, lumpy and tough, with rock	35 to 70%	26 to 41%
Shale and soft rock	40 to 85%	28.5 to 46%
Hard rock—well to poorly blasted	50 to 100%	33.3 to 50%

The loose or aerated part of the load uses space as a percentage of the full heaped pile and is shown above in the second column as "voids."

TABLE II

As a means of simplifying the consideration of these factors, they have been reduced to a representative four as follows:

Sand	10% Voids
Common earth	20% Voids
Clay	30% Voids
Shot rock	40% Voids

TABLE III

Excavation, or place yards may be judged then to represent:

In sand	90% of the heaped maximum capacity
In common earth	80% of the heaped maximum capacity
In clay	70% of the heaped maximum capacity
In rock	60% of the heaped maximum capacity

TABLE IV

Loose yards will represent:

In sand	111% of bank yards
In common earth	125% of bank yards
In clay	143% of bank yards
In rock	167% of bank yards

FIGURE 2.44 Soil swell and shrinkage.

be reduced to 80 or 90 percent of bank volume. Measurement in the fill is described as compacted yards, or yards after shrinkage.

$$\text{Shrinkage factor} = \frac{\text{volume in fill}}{\text{volume in bank}}$$

$$= \frac{\text{volume in fill}}{\text{volume in loose yards} \times \text{swell factor}}$$

$$\text{Shrinkage percent} = 1 - \text{shrinkage factor}$$

Shrinkage may also occur in undisturbed soil on which fills are placed, particularly if 50 tons plus superrollers are used. Such shrinkage may be calculated by the following formula, if more exact information is lacking:

$$\text{Settlement} = \frac{\text{depth of compaction} \times \text{shrinkage percent}}{2}$$

This is just an averaging of maximum compaction at the top and zero compaction just below where the effect is felt. This depth would be about 48 inches in a clay loam soil under a 50-ton roller. Such soil may have a shrinkage factor of .2, so by this formula

$$\text{Settlement, inches} = \frac{48 \times 0.2}{2} = \frac{9.6}{2} = 4.8$$

In wet conditions, or on steep slopes, displacement of subgrade material to the side may reduce its useful bulk much more than its compaction does.

Exceptional Soils.   Soils composed of volcanic ash or pumice, and soils built up by streams on flats in arid regions, may show very peculiar behavior. Particularly, they may show excessive compaction, sometimes to 40 percent of bank volume. In some cases, they even show shrinkage during digging, so that loose yards are heavier than bank yards.

Such soils are found only in limited areas where contractors may be aware of their possibilities. However, they constitute a hazard for the estimator, and any indication of their existence on a job should be carefully checked.

One western state highway department in the United States has adopted a policy of paying for highway borrow by weight rather than by volume to avoid unpleasant surprises for either the budget or the contractor when such materials appear in borrow pits. The necessity of measurement before starting work is avoided also. Portable truck scales that can be set up in a few minutes are used for measurement of pay quantities.

CONTAINERS

"Container" is used here as a broad term to include transporting boxes such as truck bodies, digging and carrying buckets or bowls as in shovels and scrapers, and digging and pushing blades on dozers and graders. But it does not include belts or pipes.

Measurement.   Most buckets and bodies are rated by the manufacturer as to carrying capacity in loose yards. This rating may be water level (the yardage of liquid which could be carried if it did not leak out), line of plate or struck measure, which is water level plus any space between parts of the rim which project above its low point, or heaped.

Shovel dipper buckets and highway truck bodies are normally rated at water level, carrying scrapers and off-the-road trucks at both struck and heaped measure, and clamshells at water level, line of plate and heaped.

Body capacity should be indicated by a printed plate. It is usually in yards, but may be in cubic feet. It ordinarily does not include sideboards, or other removable extensions which increase its volume. The extra load between added boards may be found by the proportion between the heights of the body wall and of the board. For example, if a straight-sided body with 4-foot sides holds 8 yards, a 1-foot sideboard added on each side will increase capacity by one-fourth, or 2 yards.

If there is no plate, size can be determined by measurement of the inside length, width, and height, usually in feet and fractions of feet. These are multiplied together, to obtain cubic feet that are divided by 27 to get cubic yards.

Heaps. The heap on top of a load can be measured only approximately because of its shape. It is usually calculated on the basis of an assumed even slope of the material. The slopes from opposite sides are continued until they meet. In general, the volume of a heap is determined by the slope of the material and the area of the body. A steep slope and a wide body give it maximum yardage.

According to S.A.E. (Society of Automotive Engineers) standards, a heap is figured on a basis of slopes of 3 to 1 (3 horizontal feet to each foot of rise) inward from the top of the body walls. This covers most conditions, but some materials will stack up much higher, particularly if the trip is short and smooth, as in a dragline bucket, which a few will slide or flow to flatter slopes as the container is moved or shaken.

Truck Bodies. Bodies such as are mounted on highway trucks have a rectangular floor and vertical body walls, with provision for increasing the height of sidewalls by adding wood or metal extensions. The rear wall (tailgate) is hinged to dump the load when the body is raised at the front. Most of such bodies are 7 feet wide inside. Some standard heights and capacities are shown in Fig. 2.45.

Bodies of off-the-road trucks are difficult to measure, as they may slope outward from bottom to top and the sloped fixed tailboard ends at a much lower level than the body sides. Ratings generally assume a load struck off by a plank resting on the two sides and moved from front to rear. In the absence of indications to the contrary, manufacturers' ratings for struck load may be accepted.

The more important rating for off-the-road trucks is capacity in tons. This is usually $1\frac{1}{2}$ times the rating in yards. The trucks are designed to carry a certain payload, and the body is designed

Length, inside, in feet	Height, sides in inches	Capacity, cubic yards
9	15-1/2	3
	20-7/8	4
	25-3/4	5
9-1/2	14-3/8	3
	19-1/2	4
	24-1/2	5
10	18-3/4	4
	23	5
10-1/2	20-1/8	4-1/2
	22-1/4	5
11	20-7/8	5
	25	6
12	23	6
	31	8
	38-3/4	10
14	33-3/8	10

FIGURE 2.45 Capacity of standard truck bodies.

to carry that load in material that weighs 3,000 pounds to the loose yard. Lighter materials can be heaped, heavier materials should not fill the body completely. Bodies can be purchased in special sizes for regular handling of material of very different weights.

Capacity Tests.   When a capacity test is to be run, the bank is trimmed and measured and a number of approximately equal loads are taken and counted. The bank is then trimmed off and measured again, and the difference calculated in yards. Then

$$\text{Truck load} = \frac{\text{volume removed from bank}}{\text{number of truck loads taken}}$$

Such tests should be run on the largest practical volume for best accuracy. Minimum amount would be about 200 yards. Counting of excavator bucket loads during the test will supply container and output data for it also.

Scraper Bowl.   Scraper bowls are irregular in shape and in height of rim. Manufacturers' rating on struck load may be accepted if known. If not, inside measurements are taken and capacity is worked out, making allowance for curves and irregularities.

Scrapers are usually rated for heaps with 1-to-1 slopes, 3 times as steep as the S.A.E. standard. Dirt may stand a little more steeply above a scraper because of the way it is crowded up from the bottom, but there is no justification for assuming that it can stand on such a slope, either at the moment of loading or during the shaking and vibration of a haul. Also, such slopes are never carried up to peaks.

A good rule of thumb is to take the scraper's maximum heap at one-half the manufacturer's rating. That is, a scraper rated at 20 yards struck and 28 yards heaped can be assumed to have a heaped capacity of up to 24 loose yards, unless there are definite indications to the contrary. If swell factor were .80, the heaped load would contain 19.2 bank yards. Extensive checking by the U.S. Bureau of Public Roads indicates that scraper loads in bank yards seldom equal their rated struck capacity, so this works out well enough.

Push loading compacts material, so for the same heap a scraper carries fewer bank yards if it is loaded from the top.

Shovel Bucket.   Shovel buckets are rated at struck measure, in yards and fractions of yards. Size can be checked fairly accurately by measuring and multiplying length, width, and height, as in the case of the truck body. Taper, curves, and extension of the lip can be used to increase the result to the nearest $\frac{1}{4}$ yard. Heaps, when digging conditions permit, range between one-eighth and one-fourth of struck capacity.

Hoe buckets are rated and measured in the same manner. Heaps are usually small or absent. Width is an important factor in loading efficiency. Deep narrow buckets may average loads less than one-third of rated capacity.

The open front of the dragline bucket makes it difficult to measure accurately, as the capacity is determined largely by the steepness of slope of the load. A chunky mud that can bulge over the teeth and pile steeply over the top may have almost as much in the heap as there is in the struck load.

Clamshell buckets may be rated water level with the end plates, struck measure along the tops of the sides, or heaped. Another rating that may be important is that of deck area—the length and breadth of the space on a level floor occupied by the bucket when fully open. The same space is required between braces in a trench.

Loader Bucket.   Front-end loader buckets are rated at struck capacity across the sidewalls. If the bucket can be tipped back 20 degrees or more at ground level, and digging conditions and tractor stability permit, it can carry a very large heap, perhaps 50 percent of struck capacity. A bucket that will not tip back seldom can take one-eighth more than struck capacity. However, in hard digging few buckets can pick up even a struck load, and tractor stability may not permit carrying an overload. On the average these buckets are underloaded by loose measurement more often than they are heaped.

Dozer Blade.   The capacity of a bulldozer blade cannot be computed exactly. The blade forms only one of the six sides of a shapeless payload. Since the load must slide or roll along the ground, friction is a very important limiting factor on load size. Materials with low internal friction or light weight, and downhill pushing favor big loads. Pushing through a slot or between windrows, or with the blade almost touching that of another dozer, increases load by reducing side spillage.

A good average load may be said to be 20 percent more than would be measured by the full width and height of the blade, with a forward slope of 1-to-1. The actual load would be higher in the center and skimped in at the corners, averaging out as above. See Fig. 2.46.

The blade on a 25-ton bulldozer might be 12 feet wide and $4\frac{1}{2}$ feet high. The profile of the theoretical load would be a right triangle $4\frac{1}{2}$ by $4\frac{1}{2}$, with an area of 10.25 square feet. Multiplying this by the 12-foot blade width, we have 123 cubic feet. Adding 20 percent for center bulges, this would be 148 cubic feet or 5.5 cubic yards. Different

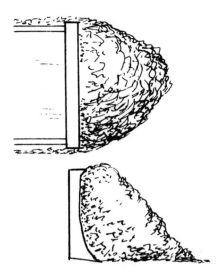

FIGURE 2.46   Bulldozer blade load.

manufacturers rate blades of this size at 5 to $7\frac{1}{2}$ loose yards.

Under very good conditions the load might average 6 inches higher than the blade top, and extend forward on a $1\frac{1}{2}$-to-1 slope. Then the profile would be 5 feet $\times$ $7\frac{1}{2}$ feet or 19 square feet. Multiplying by 12-foot width, we get 230 cubic feet or about $8\frac{1}{2}$ yards loose.

Capacity of a blade is greatly affected by grade. Downhill the load tends to slide along with minimum pressure; uphill its friction rises and it spills more off the sides. Possible load, both from standpoint of power to push it and ability to keep it in front of the blade, increases about 4 percent for each percent of down slope. That is, a dozer might push double its normal load down a 25 percent grade.

Uphill the load falls off at first almost as rapidly, but the decline slows with steeper grades. A half load might be pushed up a 20 percent grade, and a quarter load up 100 percent.

When a job is in progress, dozer blade capacity can be checked by counting the number of full passes required to dig a known yardage of a bank, or by counting loads pushed into a pile and measuring the pile.

The first result will be in bank yards, the second in loose yards.

Container Efficiency.   A machine may go through many production cycles with less than its rated load. The ratio of the container's capacity to its actual load in loose yards is called the efficiency factor of the bucket, bowl, or body.

An excavator may not be able to load its bucket or bowl fully because of hard digging, inadequate power, improper design, dull teeth, traction, heavy material, operator's haste or carelessness, or combinations of these and other unfavorable conditions.

A truck may be run partly empty to enable it to climb steep grades, to reduce strain on rough ground, to cross soft ground, to increase haul speed, or because of poor mechanical or tire conditions.

When material is coarse in proportion to bucket size, chunks may be partly supported by the sides and each other, leaving excessive voids and reducing the actual load below that indicated by the capacity of the container and the bulk of the load. The container efficiency factor (CEF) formula is

$$CEF = \frac{\text{material in container}}{\text{rated capacity of container}}$$

The amount of material in the container may be determined by careful measurement and/or weighing of a number of individual loads, or by measuring either the bank or the fill to determine the amount of material moved in a certain number of cycles.

For example, a ½-yard backhoe might dig a ditch section 2½ feet wide, 6 feet deep and 12 feet long in 25 cycles, removing 180 cubic feet or 6.67 cubic yards bank measure. Dividing 6.67 by 25, we have an average bucket load of .267 yard. Since the rated capacity of the bucket is .5 yard, we have

$$\text{CEF (bank yards)} = \frac{.267}{.5} = .534$$

The efficiency factor in loose yards would be greater by the percentage of swell. In clay loam with 20 percent swell (see below) the loose yards would be 6.67 × 1.20, or 8.0. Dividing by 25 cycles, we would have an average load of .32 loose yard. Then

$$\text{CEF (loose yards)} = \frac{.32}{.5} = .64$$

It is important to always specify whether any container efficiency factor is for loose yards or bank yards. One may be readily converted to the other, the loose yards figure almost always being the larger, as it is increased by the percentage of swell.

OUTPUT

Work Cycles.   The work cycle may be timed as a whole in figuring output for existing conditions. For accuracy it is necessary to take the average of a large number of passes as there may be a considerable variation among them.

If a study is made of a cycle, either to find a way to speed it up, or to use its time intervals as a basis for figuring production under different conditions, it can be broken down into individual operations which are timed separately.

A study of bulldozer and of scraper operation may include some or all of the divisions listed below. Each one should be timed. Digging and traveling distances should be measured. A record should be made of all grades, as these machines are much less efficient going uphill than downhill.

Bulldozer	Scraper
Dig	Load
Shift into second	Shift
Transport	Transport
Dump	Shift
Shift to reverse (raise blade)	Spread
	Shift
Return	Deadhead to turn
Shift to low (lower blade)	Turn
	Return to digging area
	Deadhead to turn
	Turn
	Get to loading position
	Shift into low
	Wait for pusher

Machine Efficiency.   Nonworking time such as delays for moving the machine, minor repair adjustment, rest, getting instructions, or looking at grade stakes is not averaged into cycle time. It is considered separately as the efficiency factor of the machine. It may work out between 70 and 85 percent for short periods with expert operators on a properly run job, with machinery in good condition and weather favorable.

A convenient way of making rule-of-thumb allowance for efficiency of about 83 percent is to consider that 1 hour contains only 50 working minutes. On this basis, a machine with a 30-second cycle cannot be expected to perform it more than 100 times per hour.

If two machines are interdependent, the efficiency factor is usually lower so far as the job is concerned. If two 80 percent efficient machines are down at the same time every time, the efficiency factor remains at 80 percent. If they are down separately each time, the factor drops to 60 percent.

As the period under study becomes longer, efficiency drops steadily, as weather, work sequence stoppages, and major overhauls must be considered. Time losses can be further increased by inefficient management, bad pit layout, poor morale of employees, or other unfavorable conditions.

Output Formulas.   A formula which can be used for figuring production of any machine with a regular cycle is

$$\text{Output, yards per hour} = \frac{Q \times K \times E \times 60 \times f}{C_m}$$

where Q = capacity, either struck or heaped
K = efficiency factor of bucket or body
E = efficiency factor of machine
60 = minutes in 1 hour
f = soil conversion factor
C_m = cycle time, minutes

If the result is to be in bank yards, f has a value obtained from observed swell, or from Fig. 2.44. If the result is to be in loose yards, f equals 1, and can be dropped from the equation. The factor K is not required when full, solid loads can be taken. If efficiency is approximated at $\frac{4}{5}$, then 50 minutes are used instead of 60, for the hour.

Under these conditions, the following simplified formula can be used:

$$\text{Output, loose yards per hour} = \frac{Q \times 50}{C_m}$$

If a 45-minute hour is used, this formula is

$$\text{Output, loose yards per hour} = \frac{Q \times 45}{C_m}$$

When timing machines have a cycle less than a minute, it is more convenient to figure it in seconds. This is done by multiplying the number of minutes by 60, and using C_s—cycle time in seconds—instead of C_m. With these substitutions, the formula will be

$$\text{Output, yards per hour} = \frac{Q \times K \times E \times 3600 \times f}{C_s}$$

Or simplified for a 50-minute hour to

$$\text{Output, loose yards per hour} = \frac{Q \times 3000}{C_s}$$

For a 45-minute hour we would have

$$\text{Output, loose yards per hour} = \frac{Q \times 2700}{C_s}$$

Additional data on output will be found in Chap. 11 and the Appendix, and included in discussions of particular machines.

CHAPTER 3
ROCK, SOIL, AND MUD

SOIL AND ROCK

Soil is loose surface material. Rock is the hard crust of the earth, which underlies and often projects through the soil cover. There is no clear distinction between soil and rock. Geologically, all soils are considered to be rock formations. In ordinary usage, rock is something hard, firm, and stable.

A contractor's definition is that rock is any material which cannot be dug or loosened by available machinery, but this distinction from soil may depend more on the power, size, and digging efficiency of the machinery than on the material itself.

Material to be excavated can also be roughly divided into three classes: rock, hard digging, and easy digging.

Rock is anything that needs blasting or ripping for efficient digging by most machines. Hard digging is compacted, cemented, or rocky dirt, clay, soft shale, and rotten rock, that can be dug directly by heavy machinery, or loosened readily by rippers. Easy to medium digging is any soft or fine, firm or loose deposit.

Soil. Soil is composed of particles of various sizes and chemical compositions. It can be analyzed as to sizes by sifting a dried and weighed sample through a set of testing sieves, such as are shown in Fig. 3.1, and weighing the material retained on each screen.

If further analysis is required for particles passing the smallest (200-mesh) sieve, it is done by hydrometer. This process is based on the fact that the speed of settlement of such particles is proportional to their size.

Figure 3.2*A* indicates the size particles which are included in the common soil classifications. There are several scales, in which the boundaries between different classes may vary. The differences among them are not important to the average contractor. Figure 3.2*B* shows another means of classifying soils.

A plastic soil is one which can be rolled as between the hands, into strings $\frac{1}{8}$ inch in diameter without falling apart. Plasticity is a function of soil character and of moisture content. The minimum amount of water in terms of percent of oven-dry weight of the soil which will make it plastic is defined as the plastic limit of the soil. If no amount of water will allow it to roll into strings, it is called nonplastic, with the symbol NP.

The liquid limit is minimum moisture content, in terms of percent of oven-dry weight, which will cause the soil to flow if jarred slightly.

The plasticity index is the difference between the plastic limit and the liquid limit; that is to say, the range of moisture content in which the soil is plastic.

Soils or other particles may also be classified as to grain shape or hardness, and mineral and organic content. These factors will affect their resistance to weather, stability under load, wear on digging parts, and internal friction.

Most soils are inorganic, and are made up of products of decay and breaking up of rock. Organic soils and organic material in soils are largely humus, which is formed by decay of vegetation and

FIGURE 3.1   Testing sieves. (Courtesy of W. S. Tyler Co.)

has no definite particle size. Organic materials may also consist of lime from shells or from lime-stone originally formed from shells, and animal bones and excrements.

Rock.   Geologically, rocks are classified as to the way in which they were made. Those which solidified out of a molten state are called igneous, and are subdivided into volcanics cooled at the surface, and plutonic hardened deep underground.

Sedimentary rocks are built up of soil or plant or animal remains and have been hardened by pressure, time, and depositing of natural cements.

Metamorphic rocks were originally igneous or sedimentary, but have been altered by extreme heat and pressure.

Figure 3.3 contains tables classifying rocks as to type and hardness. The latter quality is quite variable, even in one formation, and may be made up of different factors, as resistance to penetration, abrasion, or crushing.

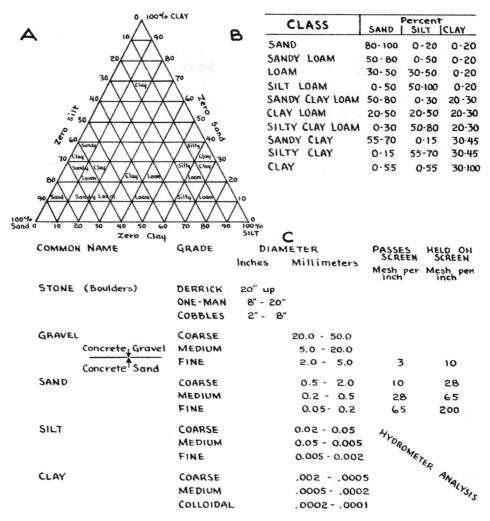

CLASS	Percent		
	SAND	SILT	CLAY
SAND	80-100	0-20	0-20
SANDY LOAM	50-80	0-50	0-20
LOAM	30-50	30-50	0-20
SILT LOAM	0-50	50-100	0-20
SANDY CLAY LOAM	50-80	0-30	20-30
CLAY LOAM	20-50	20-50	20-30
SILTY CLAY LOAM	0-30	50-80	20-30
SANDY CLAY	55-70	0-15	30-45
SILTY CLAY	0-15	55-70	30-45
CLAY	0-55	0-55	30-100

COMMON NAME	GRADE	DIAMETER		PASSES SCREEN	HELD ON SCREEN
		Inches	Millimeters	Mesh per inch	Mesh per inch
STONE (Boulders)	DERRICK	20" up			
	ONE-MAN	8" - 20"			
	COBBLES	2" - 8"			
GRAVEL	COARSE		20.0 - 50.0		
	MEDIUM		5.0 - 20.0		
Concrete Gravel / Concrete Sand	FINE		2.0 - 5.0	3	10
SAND	COARSE		0.5 - 2.0	10	28
	MEDIUM		0.2 - 0.5	28	65
	FINE		0.05 - 0.2	65	200
SILT	COARSE		0.02 - 0.05		
	MEDIUM		0.05 - 0.005		
	FINE		0.005 - 0.002		
CLAY	COARSE		.002 - .0005		
	MEDIUM		.0005 - .0002		
	COLLOIDAL		.0002 - .0001		

FIGURE 3.2A   Soil classification.

Digging Resistance.   The resistance which must be overcome to dig a formation will be made up largely of hardness, coarseness, friction, adhesion, cohesion, and weight.

In digging, hardness is resistance to penetration. It is increased by close packing of soil, or filling of voids with finer particles, or lime or other natural cements. Clay soils are hard when dry, and soft when wet.

Cobbles, boulders, or hard lumps increase the power requirement for penetration. They are most troublesome when they are oversize for the machine, or packed so firmly in place that they cannot slide or rotate away from the cutting edge.

As the digging edge penetrates, friction absorbs an increasing proportion of its force. It is affected by particle size and hardness, by the amount of moisture, and the presence or absence of natural lubricants such as humus or soft clay.

General classification	Granular materials (35% or less passing No. 200)							Silt-clay materials (More than 35% passing No. 200)			
Group classification	A-1		A-3	A-2				A-4	A-5	A-6	A-7 / A-7-5, A-7-6
	A-1-a	A-1-b		A-2-4	A-2-5	A-2-6	A-2-7				
Sieve analysis, percent passing: No. 10 / No. 40 / No. 200	50 max. / 30 max. / 15 max.	— / 50 max. / 25 max.	— / 51 min. / 10 max.	— / — / 35 max.	— / — / 35 max.	— / — / 35 max.	— / — / 35 max.	— / — / 36 min.	— / — / 36 min.	— / — / 36 min.	— / — / 36 min.
Characteristics of fraction passing No. 40: Liquid limit / Plasticity index	— / 6 max.		— / N.P.	40 max. / 10 max.	41 min. / 10 max.	40 max. / 11 min.	41 min. / 11 min.	40 max. / 10 max.	41 min. / 10 max.	40 max. / 11 min.	41 min. / 11 min.*
Usual types of significant constituent materials	Stone fragments, gravel and sand		Fine sand	Silty or clayey gravel and sand				Silty soils		Clayey soils	
General rating as subgrade	Excellent to good							Fair to poor			

*Plasticity index of A-7-5 subgroup is equal to or less than L.L. minus 30. Plasticity index of A-7-6 subgroup is greater than L.L. minus 30.

FIGURE 3.2B Classification of soils and soil-aggregate mixtures. (*Courtesy of Portland Cement Association.*)

ROCK	WEIGHT lbs. per cu. ft.	PERCENT OF WEAR	HARDNESS	TOUGHNESS
Granite	167	4.3	18.3	11
Syenite	171	3.3	18.3	15
Diorite	179	3.0	18.2	17
Gabbro	185	3.0	17.7	14
Peridotite	182	4.0	14.2	11
Rhyolite	159	3.7	18.3	19
Trachyte	170	2.9	18.1	24
Andesite	166	3.9	17.0	18
Basalt	177	3.0	17.1	18
Diabase	186	2.4	18.0	22
Limestone	165	5.0	14.1	9
Dolomite	170	5.5	14.9	9
Sandstone	164	6.2	14.4	10
Chert	159	9.4	18.2	12
Gneiss	172	4.9	17.4	10
Schist	180	4.7	16.6	13
Amphibolite	188	2.8	17.5	19
Quartzite	169	3.2	18.8	18
Eclogite	194	2.4	18.4	22
Marble	173	5.7	13.1	6

CLASS	TYPE	FAMILY
Igneous	Intrusive (plutonic)	Granite, Syenite, Diorite, Gabbro, Peridotite
	Extrusive (volcanic)	Rhyolite, Trachyte, Andesite, Basalt & diabase
Sedimentary	Calcareous	Limestone, Dolomite
	Siliceous	Shale, Sandstone, Chert (flint)
Metamorphic	Foliated	Gneiss, Schist, Amphibolite
	Nonfoliated	Slate, Quartzite, Eclogite, Marble

FIGURE 3.3 Rock types.

Adhesion is the sticking of soil to the digging parts. It may increase the friction load substantially in wet work.

Cohesion is resistance to tearing apart. Firm or hard materials may split readily along bedding or cleavage planes so that they can be dug rather easily from the proper direction. Relatively soft clay banks may be very difficult to dig because of strong and uniform cohesion. A tough formation lacking planes of weakness is described as "tight."

Rippability.   This is a measure of the ease or difficulty with which a rock can be broken by heavy rippers into pieces that can be economically moved by other equipment, usually scrapers.

At least three factors are involved: resistance to breakage of the rock material itself, the extent to which it is weakened by bedding layers (lamination) or by joint cracks or fault movement, and the degree to which the rock has been softened and weakened by weathering.

Many rocks are readily rippable at the surface, but become increasingly resistant with depth because of less exposure to weathering. A generally rippable rock may contain seams or boulders that are difficult or impossible to rip.

Some rocks are nonrippable in their natural condition, but can be ripped after shaking up by light blasting.

In general the igneous rocks such as basalt, granite, diorite, felsite, and lava are not rippable unless they are greatly weakened by weathering. They do not respond well to light blasting unless they have a closely jointed structure.

Massive metamorphic rocks such as gneiss, marble, and quartzite are usually nonrippable, but slate and thinly bedded schist can be ripped.

Some rocks have an absolute resistance to ripping. That is, they cannot be shattered by any amount of pressure that can be exerted by a ripper tooth. More often, rippability is determined by how much force can be applied. Limitations are the strength of teeth and shanks, the power available, and the economics of applying the power.

There are also rocks that will break or tear rather readily, but remain in such large pieces that they cannot be handled economically. Oversize is sometimes broken by walking the ripper-tractor on chunks, by a crane or drop ball, or by blasting.

Ripping is usually not economical unless the product is fine enough for efficient scraper loading, or can be made so by secondary work.

Seismic methods may be used. Rippable rocks seem to carry vibration differently from nonrippable formations, as indicated in Fig. 3.4A.

Seismic testing is the recognized method of testing rippability. It depends on the relationship between the cohesiveness of rock and the speed of vibration through it.

A diagram of the method used is shown in Fig. 3.4B.

Weight.   Figure 3.5 gives the approximate weights per cubic yard of various materials.

Weight may limit the amount that can be dug or carried in a bucket or body, and the speed with which the load can be hoisted or transported. It is a critical factor in selection of dragline and clamshell bucket sizes, and in regard to the length and angle of the boom that carries them.

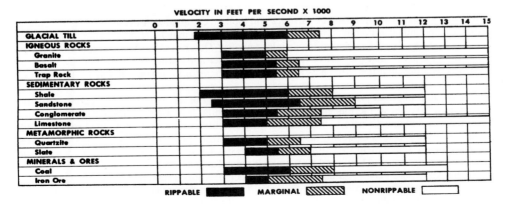

FIGURE 3.4A   Rippability and seismic vibration.

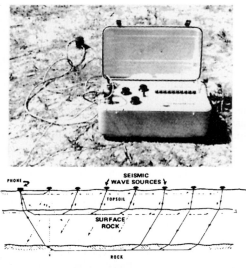

By dividing the distance in feet from the plate to the phone by the time lapse in seconds, the velocity in feet per second is obtained.

$$V \text{ (in FPS)} = \frac{D}{T} \quad \frac{\text{(Distance from phone to plate)}}{\text{(Time lapse in seconds)}}$$

FIGURE 3.4B Seismic recorder and readings.

TRACTION

Soils differ greatly in their ability to support and permit movement of vehicles. An important characteristic is the amount of friction that exists between the ground surface and the drive tires or tracks of a machine on it.

Tractive Efficiency. This is a measure of the proportion of the weight resting on drive wheels or tracks that can be converted into movement of a machine. Soil characteristics and condition are very important in determining it.

It is difficult to estimate and classify the tractive efficiency of ground surfaces, as wide variations may be caused by the shape of particles and their size gradation. In addition, the presence of certain compounds that do not show up in a soil specification may increase traction by improving packing qualities and even cementing particles together. Other substances may reduce traction by lubricating grains so that they move freely on each other and on tire surfaces. For example, salt improves packing qualities, lime acts as cement, and soapstone (serpentine, mostly $H_4Mg_3Si_2O$) acts as a lubricant.

The amount of water present is an important variable. For any soil there is a certain moisture content that gives best traction. In sandy or gravelly soils, less water is likely to allow them to become loose and unsatisfactory, while more water has little effect. In fine-grained soils too little water (once the soil is compacted) may not have any harmful effect except causing dust, but too much water will make them first slippery and then soft. A soaking shower on the dusty loam surface of a heavily traveled haul road may make it temporarily almost as slippery as ice.

Figure 3.6 gives a general guide to traction afforded by various surfaces, expressed as a decimal fraction of the weight on drive wheels or tracks. The variations between low and high values are sufficiently wide to take care of most differences. If the higher values are used, the possibility of having to deal with lower ones should be kept in mind.

This table assigns much lower values to loose and wet soils than most references do.

WEIGHTS OF MATERIALS
(ONE YARD — WEIGHTS IN POUNDS)

MATERIAL	1 YARD 27 CU. FT.
ASHES — PILED DRY	945
BRICK BATS	1485
CEMENT — PORTLAND	2538
CINDERS	1485
CLINKER — PORTLAND CEMENT	2295
CLAY — DRY, IN LUMPS	1701
CLAY — COMPACT, NATURAL BED	2943
COAL — ANTHRACITE	1512
COAL — BITUMINOUS, R. OF M. PILED	1485
COAL — BITUMINOUS SLACK, PILED	1350
CONCRETE — READY TO POUR	3996
DOLOMITE — CRUSHED FINE	2565
DOLOMITE — BROKEN LUMP	2565
EARTH — LOAMY, DRY, LOOSE	2025
EARTH — DRY, PACKED	2565
EARTH — WET (MUD)	2970
GYPSUM — CRUSHED TO 3"	2565
GYPSUM — CALCINED	1620
GRAVEL — DRY, LOOSE	2970
GRAVEL — DRY, PACKED	3051
GRAVEL — WET, PACKED	3240
IRON ORE — 60% IRON	8100
IRON ORE — 50% IRON	6750
IRON ORE — 40% IRON	5400
LIMESTONE — RUN OF CRUSHER	2565
LIMESTONE — FINES OUT	2700
LIMESTONE — 1 1/2" OR 2" GRADE	2295
LIMESTONE — ABOVE 2" GRADE	2160
PHOSPHATE ROCK	2160
SAND — DRY, LOOSE	2565
SAND — WET, PACKED	3240
SHALE — BROKEN	2295
SLAG — BLAST FURNACE, BROKEN	3726
SLAG — OPEN HEARTH, CRUSHED	2835
SLAG — GRANULATED, DRY	1025
SLAG — GRANULATED, WET	1566
SULFUR — BROKEN	1620

FIGURE 3.5   Weights of materials.

Weight Distribution.   The ability of a machine to propel itself on slippery footing is affected by its weight distribution.

For example, if a machine having a weight of 10 tons on the drive wheels can exert a drawbar pull of only 4 tons because of wheel slippage, the tractive efficiency of the ground is 40 percent, or .40. It does not matter whether the whole machine weighs 10 tons or 50 tons, or whether its rim pull is 5 tons or 20, as for this particular calculation we use only weight holding the drive wheels (or tracks) against the ground. However, the total weight of the machine and its load make up the resistance that the drive wheels must move.

Increasing the weight on slipping drive wheels increases drawbar pull in direct proportion, up to the maximum that can be produced by the engine and gears. Increasing weight on nondrive wheels increases resistance. Shift of existing weight to the drive wheels increases potential traction, while shifting weight from drive to nondrive wheels reduces traction. Neither shift affects resistance.

While resistance to movement is in proportion to the weight of the whole machine, ability to move the machine, if power is adequate, depends on the weight on the drive units. All-wheel-drive trucks and most crawlers keep all their weight on the drivers. In other machines weight distribution is important to performance when traction is limited.

Manufacturers usually provide detailed weight specifications for haulers, showing the total weight and its distribution, both loaded and empty.

| | | Tractive Efficiency | | | |
| | | Wheel Factor* | | Track Factor (Grousers) | |
Standard Tables	Type of Surface	Dry	Wet Surface	Dry	Wet Surface
	Smooth blacktop	.8 – 1.0	.6 – .9	–	–
.88 – 1.0	Rough concrete	.9 – 1.0	.8 – 1.0	.3 – .6	.3 – .6
	Hard smooth clay	.6 – 1.0	.1 – .3	.4 – .7	.2 – .4
.40 – .58	Hard clay loam	.5 – .8	.15 – .4	.6 – .9	.4 – .9
	Firm sandy loam	.4 – .8	.25 – .8	.6 – 1.0	.6 – 1.0
	Spongy clay loam	.4 – .6	.15 – .3	.7 – 1.0	.6 – .9
.40 – .44	Rutted clay loam	.3 – .5	.15 – .3	.7 – 1.0	.6 – .9
.20 – .35	Rutted sandy loam	.3 – .4	.2 – .5	.7 – 1.0	.7 – 1.0
.36	Gravel road, firm	.5 – .8	.3 – .9	.7 – .9	.7 – .9
	Gravel, not compacted	.3 – .5	.4 – .6	.5 – .9	.6 – 1.0
	Gravel, loose	.2 – .4	.3 – .5	.4 – .7	.5 – .8
.20 to .35	Sand, loose	.1 – .2	.1 – .4	.3 – .5	.4 – .7
.20	Snow, packed	.1 – .4	.0 – .3	.2 – .6	.2 – .6
.12	Ice, roughened	.1 – .3	.0 – .2	.1 – .4	.0 – .3
	Ice, smooth	.0 – .1	.0 – .0	.0 – .1	.0 – .1

*May be increased by extra wide or extra soft tires.

FIGURE 3.6   Tractive efficiency of surfaces.

If this information is not available, the hauler should be weighed, one axle at a time, both empty and loaded. Different load distributions, to the front or rear of the body, can also be checked on scales.

A highway truck may be taken to scales. Off-the-road trucks and scrapers might have to have scales brought to them, which could prove difficult and expensive.

If there is no specific information on weight distribution and scales are not available, it may be assumed that a rear-drive dump truck carries about 50 percent of its empty weight or 70 percent of its loaded weight on the drive wheels. A front pull scraper, wagon, or rocker usually has 50 to 60 percent of its empty weight or 40 to 50 percent of its loaded weight on the drivers.

On climbs, vehicle weight shifts toward rear drives and away from front drives by about $1\frac{1}{2}$ percent for each percent of grade. This factor increases the already considerable traction advantage of the rear-drive truck in slippery climbs.

Tire Treads.   Tire treads are an important variable in traction. There are a number of tread designs, all of which eventually wear smooth. For most situations high lugs or cleats give best traction, but smoother surfaces are better on dry sand and ice. When tire chains are used, their effectiveness is reduced by high tread blocks.

Since a deep cleated tire gives best traction on soft loam and a smooth tire on loose sand, there are likely to be intermediate surfaces on which they are equally effective, or perhaps ineffective.

Sometimes a well-worn tire will give as good or better traction than the new one ordered to replace it.

FLOTATION

Flotation means the weight-supporting ability of a tire, crawler track, or platform on soft ground. This ability, or lack of it, is the result of a relationship between the weight, the area of contact, and the load-bearing ability of the ground.

The weight divided by the contact area in square inches gives the downward pressure in pounds per square inch (psi). If this pressure is greater than the load-carrying ability of the ground, the machine will sink until it finds enough contact area to support it. Sinking increases rolling resistance. If it is severe, it may prevent the machine from moving under its own power.

Tire Pressure.   When a tire carries enough weight that it tends to sink into the ground, its behavior depends largely on the relationship between its inflation pressure and the load-bearing ability of the ground.

Ground with a bearing strength of 50 pounds per square inch (psi) will allow a loaded tire with 75 pounds of air pressure to sink into it until the same area of tire is in contact with the ground as there would be if the tire had 50 pounds of pressure and flexed to spread out on the surface.

It follows that if a larger tire were used, that could carry the load at 50 pounds of pressure, it would not sink in at all. It does not always work out just this way, but there is no doubt that bigger and softer tires greatly reduce problems with vehicles sinking in soft ground.

The formula for finding the area in square inches over which a tire will contact the ground is

$$\text{Contact area} = \frac{.9 \times \text{tire load}}{\text{inflation pressure}}$$

On a truck with a gross weight of 60,000 pounds, of which 70 percent is on four drive wheels equipped with 14.00 × 24 tires, 20-ply, with 75 pounds of pressure, we would have a load on each tire of $\frac{1}{4}$ of 60,000 × .7, or 10,500 pounds. Then, by our formula,

$$\text{Contact area} = \frac{.9 \times 10,500}{75} = \frac{9,450}{75} = 126$$

If the tires were 16.00 × 24, 16-ply, 45 pounds of pressure, we would have

$$\text{Contact area} = \frac{.9 \times 10,500}{45} = \frac{9,450}{45} = 210$$

Since these loads are within the rated capacity of the tires, they will show only normal flexing, and any additional bearing surface needed will be obtained by sinking into the ground.

To find out whether the ground will support the load, we use the formula

$$\text{Bearing area} = \frac{.9 \times \text{tire load}}{\text{ground bearing capacity}}$$

If the ground has a bearing capacity of 50 pounds per square inch,

$$\text{Bearing area} = \frac{.9 \times 10,500}{50} = \frac{9,450}{50} = 189$$

We find that the necessary ground area to bear this load is 189 square inches. The harder tire has only 126 inches of ground contact, so it will sink until an additional 63 square inches of its surface are in contact. On the other hand, the tire with 45 pounds of pressure spreads its load over 210 square inches, and should not sink at all.

Figure 3.7 shows the bearing capacities that may be expected of various types of soils. The figures should not be trusted very far, as there are many unknowns in this field.

Material	Bearing Power, pounds per square inch
Solid rock	350
Shattered rock	70
Dry clay	55
Medium dry clay	25
Soft clay	12
Cemented gravel	110
Compact sand	100
Clean dry sand	25
Quick sand	5

FIGURE 3.7 Load-bearing capacity.

Other Factors.   For repeated crossings, ground should have two or more times the minimum bearing capacity needed for a static load as above, as impact shock during travel greatly increases effective weight. Once ground begins to yield, truck wheels dropping into holes may deliver blows several hundred percent greater than static weight.

Bearing qualities cannot be measured entirely in square inches. A country road of native soil may easily support a 200-pound person balancing on a 1-inch metal cube, but break down rapidly under 30-ton trucks having only one-third as much ground pressure per square inch.

This may be explained by the formation of a bulb of pressure that is built up underground by the sum of individual surface points or areas of pressure. Figure 3.8 shows a cross section of soil under a rubber tire roller, where each tire produces its own zone of compaction and their combined weights produce another and larger pressure bulb.

On hard ground a high-pressure tire does less flexing and has a smaller area of ground contact, and therefore less rolling resistance than a low-pressure tire. However, if the ground is soft enough for the tire to sink in at all, the low-pressure tire will develop less resistance.

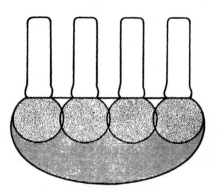

FIGURE 3.8 Pressure bulb.

COMPACTION

The term "compaction" refers to the act of artificially increasing the density of soil. It involves the pressing of soil particles together into closer contact, and expelling air or water from spaces between them.

When the same process occurs in nature, as a result of wetting, drying, freezing, thawing, groundwater movement, and weight of higher soil layers, it is called *settlement*.

The density of soil is measured in terms of its volume-weight, which may be expressed as pounds of wet soil or dry soil per cubic foot, or as porosity in percent of total volume. A high porosity indicates a low density.

The purpose of compaction is to stabilize soil, particularly in built-up fills, embankments, and dams, so that it will show minimum change in volume or shape under influences of weather and time, and under the weight of structures, pavement, and traffic.

Compaction is also useful or necessary to the work of building a fill. Loose soil causes haulers to use excessive power, break drivetrain parts, and get stuck. Rain that would wet only the surface of a compacted fill might sink into a loose one far enough to create several feet of impassable mud that would stop the work.

Background.   The desirability of compacting road fills has been recognized since ancient times. Early methods included driving sheep or cattle back and forth on the fill, and towing weighted wooden rollers with horses.

Reliance on natural settlement and/or simple rule-of-thumb compaction methods is becoming impractical, because of the increasing loads carried by the modern highway. There is constant pressure to revise state and federal laws to permit heavier axle and gross loads, in spite of the fact that the bearing capacity of ordinary soils has already been reached and exceeded in many areas.

Equipment.   Compaction equipment is described in Chap. 19.

Soil may be compacted by pressure, kneading, vibration, impact, or combinations. The steel wheel and sheepsfoot rollers supply pressure, pneumatic tire rollers supply pressure with some kneading, wobble-wheel rollers supply kneading and pressure, and vibratory rollers supply both pressure and vibration.

Steel wheel rollers give good results on all types of soil except clean sands, in layers from 4 to 12 inches deep depending on soil type and roller weight. Clay soil layers should be limited to 4 to 6 inches, to avoid possible compaction of the top of the layer only.

Sheepsfoot rollers compact mostly with the soles of their feet, from the bottom up. As the soil is compacted, the roller rises and walks out of the ground. They do best on fine-grained soils of the plastic groups, and are least efficient in sandy and gravelly types. Excessive weight may have to be avoided, as the feet may shear soil and damage its structure.

Rubber tire rollers are suitable for use in any type of soil, but weight and tire pressure must be proper for the soil type. Results are affected by shape of tires and their air pressure and by total wheel or axle load, not by tire pressure only, as is often supposed.

Vibration is most effective in sand or gravel soils, but may increase the effectiveness of a roller in any soil. It is particularly effective at bringing excess moisture to the surface.

Trench fills and other small areas may be compacted by impact of air, hydraulic, or gasoline hammers, or vibrators. Gravel fills may be compacted by puddling—adding water until the soil is semiliquid—then allowing it to dry and settle. An immersion (concrete) vibrator will speed drying of puddled fill, by bringing its water to the surface.

Moisture Content.   The most critical factor in compaction of a soil is its moisture content, since it can be most thoroughly and conveniently compacted only if it contains just the right amount of water. This quantity is called the optimum moisture content. It must be sufficient to provide a lubricant to allow soil grains to slide on each other as they are pushed together, and not enough to form an incompressible cushion between any of them.

Figure 3.9 shows the relationship between moisture content and compaction for several classes of soil.

A soil that contains too much moisture is likely to become rubbery under a roller, pushing in waves ahead of and behind it, and springing back into its original position when it has passed. This is a very common condition in highway work.

Soil that is too dry may become loose or powdery under pressure, or may be firm but not as dense as it should be. It is standard practice to add water to each such layer by means of sprinkler trucks or trailers. Extra payment may be made for watering.

Stability.   If a road or runway fill is not compacted, it is likely to shrink and settle, injuring or destroying pavements and any other structures on it.

Clay soils provide poor embankment material at best, as they will usually absorb moisture in wet seasons and swell. High, but not excessive, compaction reduces this tendency. However, clay with a high content of the montmorillonite variety is very expansive when exposed to water. It will heave highway pavements and building foundations, even breaking pipes that extend through a wall raised by expansive clay.

Variable compaction may cause the greatest damage. A whole road may settle a few inches with little noticeable harm. But subsidence of a narrow strip of trench fill across a firm embankment will ruin the road surface, as will also subsidence of a poorly compacted embankment on each side of a well-compacted trench fill.

Unpaved road shoulders are more exposed to moisture than the pavement subgrade. If they absorb it, shoulder surfaces may rise noticeably above the pavement, interfering with drainage.

Layers.   Many compaction specifications stress the thickness of layers to be rolled. The best a field crew can hope to do is to come somewhere near the standard. There are some grading bosses who can keep exact control over spreading thickness and areas of a fleet of scrapers, and of the work patterns of rollers. But it is more usual for this work to be somewhat hit-or-miss. The important thing is that the layers should not be thicker than the compaction equipment on the job can handle.

Rock Fills.   Layer specifications reach a point of absurdity in rock fills. Often they are subject to the same maximum thickness regulation of 6 to 12 inches as the soil. In practice, the thickness of rock lifts is often determined by the maximum size rock that the shovel can load.

A rock fill is usually stable if the rock is angular and is well mixed as to size. Such conditions are characteristic of both blasted and ripped rock.

Steel wheel and sheepsfoot rollers are useless on coarse, hard rock and will be damaged. Very heavy (50 tons and more) rubber tire rollers and big hauling equipment provide about as much compaction as is possible or necessary.

Earth Dams.   Discussion so far has been about problems of highway and similar fills. Dams do not ordinarily carry structures, but they must resist penetration by water, changes of shape, and any tendency to slump. There is more about earth dams in Chapter 6.

Fill used in dams is carefully selected for quality, with fine-grained soils used in the core and sand and gravels on the slopes. Methods of compaction are about the same as for highways, but methods and results are very carefully specified and checked.

Specifications.   There are four basic ways in which compaction may be specified:

Method only

Method and result

Suggested method and result

Result only (performance specification)

Specifying the method only is usually the fairest to the contractor, who is told that a certain number of passes with a specified type and weight of roller, on layers of a particular thickness, constitute acceptable compaction. The contractor knows exactly what to do, and the equipment that will be needed.

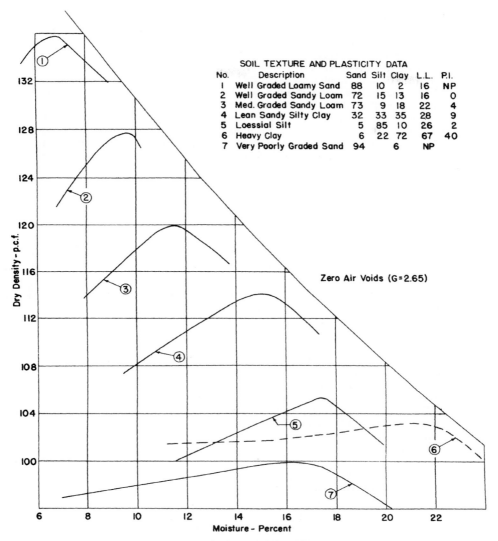

SOIL TEXTURE AND PLASTICITY DATA

No.	Description	Sand	Silt	Clay	L.L.	P.I.
1	Well Graded Loamy Sand	88	10	2	16	NP
2	Well Graded Sandy Loam	72	15	13	16	0
3	Med. Graded Sandy Loam	73	9	18	22	4
4	Lean Sandy Silty Clay	32	33	35	28	9
5	Loessial Silt	5	85	10	26	2
6	Heavy Clay	6	22	72	67	40
7	Very Poorly Graded Sand	94		6		NP

Zero Air Voids (G=2.65)

Moisture-density relationships for seven soils compacted
according to AASHO Method AASHO T99 (in part after "Public Roads").

FIGURE 3.9   Moisture-density relationships.

Unfortunately, most such specifications are archaic, mentioning only steel wheel and/or sheeps-foot rollers, and giving neither the state nor the contractor the opportunity to take advantage of more recent and specialized compactors.

"Method only" relieves the contractor of responsibility for soil moisture content or other conditions beyond her or his control. If brought up to date with a wider range of usable equipment, and selection by the state of proper types for the soils in the contract area, it would provide a good contract basis.

In "method and result" the contractor is told what equipment to use and how, and also the result in density that must be obtained. If the specified equipment will not produce the desired result, there is trouble. The contractor will need a special arrangement to either change the equipment or to get by with less compactions.

The usefulness of "suggested method and result" may depend on the wording and intent of the specification. If the result must be obtained, there is no important difference from specifying result only. However, if the contractor is not held absolutely liable for the result if he or she follows the suggested method, this is a reasonable approach.

A performance specification (result only) allows the contractor to choose the method, but requires a certain density. The state will have established by laboratory tests that this density is possible, but it may develop that the density cannot be obtained in the field because of moisture or other conditions, or can be obtained only by far greater effort than was contemplated by either the state or the contractor. In many jobs it works out very well, but for reasons to be considered below it may be unfair to the contractor unless softened by irregular practices on the part of the state engineers and inspectors. It should include a provision for renegotiation.

Required density is usually specified as a percentage of maximum density produced in a particular test, such as 95 percent modified Proctor.

Laboratory Tests.   Tests are required to determine the optimum moisture content of a soil, the extent to which it can be compacted, and whether it is being sufficiently compacted on the job.

The standard Proctor or AASHO (American Association of State Highway Officials) Standard Test T99 is made as follows:

A sample of soil is moistened and is then compacted in a standard mold 4 inches in diameter with a volume of $\frac{1}{30}$ cu. ft. The soil is placed in three layers of approximately equal thickness, and each layer is subjected to 25 blows of a rammer with a striking face of 2-inch diameter and a weight of $5\frac{1}{2}$ pounds, falling freely a distance of 12 inches. This produces 12,400 foot-pounds of energy per cubic foot of soil.

The sample, which contains a known volume, is then weighed and dried at 105°C for 20 minutes. After that, it is weighed again. The moisture content is computed by the difference between the wet and dry weights. The dry weight is recorded as well as the moisture content. The moisture content is recorded in percentage of weight by the entire dry sample.

By plotting the results of a series of these tests, using the same soil but with different moisture factors, a curve similar to that of Fig. 3.10 will be produced. This curve shows the resulting dry weights obtained in a series of tests on a single sample compacted under a uniform method with varying amounts of moisture.

There are a number of modifications of this basic test, so any reference to an AASHO modified test should specify which one. In general, they employ a heavier hammer or plunger with a longer stroke than in the basic test. The use of modifications is becoming more general because of the higher densities required in construction.

Field Tests.   The standard method of testing actual density of the embankment is to remove a measured sample for laboratory comparison. A sample may be of bulk to leave a hole 8 inches wide and 10 inches deep. Measurement is made by filling the hole with a measured quantity of sand or water. The water is prevented from soaking away by lining the hole with a rubber balloon.

The Proctor needle penetrometer measures the resistance of soil to penetration by a heavy needle, by means of a spring and gauge in the push handle. It is used to compare the density of the embankment with a laboratory sample. It is very useful to check on how things are going, but should be reinforced with regular volume tests. Results in gravel are likely to be misleading.

The most satisfactory and accurate, but also the most expensive, method is nuclear testing. A device using radioactive material directs a beam of gamma or neutron rays into the ground, and counts those that are reflected back into the instrument. This count is compared with a laboratory sample. The higher the count, the greater the density.

This type of instrument is nondestructive, does not interfere with grading operations, does not depend on individual judgment, permits a large number of tests in a short time, and can be used for stony or frozen ground. On the other hand, the instruments are now both expensive and fragile.

A piezometer is a pipe gauge that is placed in the fill as it is built. It will indicate any movement of soil or soil water that would be likely to threaten to damage the embankment.

Too Much Moisture.   The material taken from road cuts often contains too much moisture to meet compaction specifications in the fills.

In dry summer weather, layers of fill may be stirred up with a disc harrow, moved from side to side by graders, or sprayed into the air by a rotary tiller with the hood up. When moisture is sufficiently reduced, the surface is leveled off and the layer is compacted. If it rains during this operation, the result will be mud.

If some fill is wet and some is dry, the two types may be built up in alternating layers, that may or may not be mixed together.

In a wet season, surface drying is impractical, and no dry fill may be found in any of the cuts. Under such conditions it may not be possible to meet density specifications.

This is a very serious problem that is almost completely overlooked in the literature on compaction. It is generally passed off with a paragraph or two on correction by substituting other material or by kiln drying.

Kiln drying of such quantities would involve setting up huge and expensive plants, drastic changes in digging and hauling methods, and lengthening of hauls. It is questionable whether such operations could be included in a highway budget, or even whether improved pavement life would justify the cost.

The most critical point for the contractor is that in many states the terms of the contract require her or him to produce a certain density that can be achieved only at or near optimal moisture content. If the contractor cannot get the soil down to this content, it is unlikely that the density can be obtained. See Fig. 3.10.

Theoretically the contractor should get no pay for the embankment work, and would probably go bankrupt. Actually, this does not seem to occur. Field workers in public works departments are more realistic than the design engineers, and are inclined to pass the work as long as sincere and intelligent efforts have been made to reach the specification. However, this puts both sides in the wrong if their actions become the subject of an official investigation.

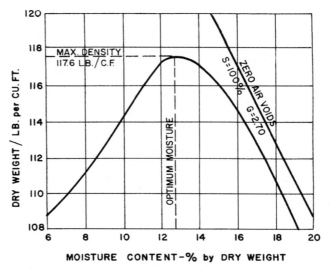

FIGURE 3.10 Moisture-density curve.

MUD

Before proceeding with the details of various types of excavation, it is in order to consider some of the general problems. Mud is one of the most important of these.

NATURE OF MUD

Water Content.   Mud is soil saturated with water to such an extent that it loses its structure and takes on some of the properties of a liquid. Even the driest soils contain some water in very thin films, and moderate additional amounts may give added firmness by acting as a binder. But when the quantity of contained water is sufficient to build up water films around the grains thick enough to serve as a lubricant so that they can move freely on each other, the soil becomes mud.

Particle Size.   The quantity of water necessary to turn mineral soils into mud varies with the size, shape, and arrangement of the particles. Small grains have much less volume in proportion to the thickness of the water film they hold than large ones have, and therefore they form more fluid muds. Sharp angular grains have projections which penetrate the film and interlock, and large grains and pebbles develop high enough contact pressures to cut through the film. If there are enough fine particles in a mixed soil to prevent the coarse ones from touching, the mud will have the qualities of the fines.

A fine-textured soil such as clay will also remain saturated much longer than a coarse one, as the spaces between the grains are so small that water moves through them very slowly.

Humus.   Humus, or peat, which is decayed organic material, absorbs water somewhat as a sponge does, in large quantities, and holds it stubbornly against evaporation and drainage. When saturated, nearly pure humus, as found in some swamps and peat beds, resembles a jelly, fibrous or smooth in texture, and black or brown. It is the most slippery and most treacherous of the muds. It dries very slowly, with shrinkage of 50 percent or more, to a light, fluffy soil. When mixed with inorganic soils, as in topsoil and mucks, it greatly reduces their load-bearing qualities and makes them muddier under wet conditions.

Making Mud.   When undisturbed, inorganic soil is usually quite closely packed, with its grains fitted together closely, and often lightly cemented by mineral deposits. When it is dug up or pushed around, the grains are shaken away from each other into a loose structure. In this condition it can quickly absorb a large quantity of water and become a very soft mud. As it dries, the grains settle together so that less water is absorbed with each subsequent wetting. If it is compacted by rolling, tamping, or vibration before being soaked, it may become even more water-resistant than in its original state.

When a firm, dry soil is covered with water, it gradually absorbs some of it and expands in volume, but it never becomes as soft as if disturbed before wetting.

If a firm, fine-grained soil has a film of rainwater on its surface, and is passed over by a vehicle tire, the water will be forced between the surface particles and the resulting mud will be wiped off and pushed aside, leaving a new surface exposed to the next raindrops and wheel passage. Repetitions of this result in a slippery road, ruts, and mudholes.

Frost.   When soil freezes, the expansion of ice crystals between the particles pushes them out of place. When the soil thaws, it is likely to become a slippery, structureless mud, often resembling toothpaste in consistency. It ordinarily firms up fairly quickly, particularly if vibrated by a heavy rain, but may persist for several months when upward seepage of water prevents settlement. In its extreme condition it will not support loads, and is made more dangerous by its occurrence in places that normally are firm, and under sod which bridges and hides it. Such places can be detected by sounding with a crowbar, and should be avoided or treated with the same precautions as soft swamps.

In northern winters, frost may stabilize a swamp so that it can be worked easily. Ice and frozen earth are liable to be variable in thickness and treacherous because of snow cover and the heat of decay of organic material. But any traveled route will gain in stability as long as freezing continues.

Freezing provides a pavementlike support for machinery, and may stabilize unfrozen mud as well. Ice lenses at the frost line absorb capillary moisture from below. If upward flow is restricted by coarse soil or an impervious formation, the mud will be dried and made firmer.

Mud from thawing is apt to render dirt road surfaces slippery, particularly on sunny slopes. Early mornings such roads are often frozen hard, and work must be done, then, in cloudy weather or at night. Tire chains are useful but may not be adequate.

Geotextile Barrier.   The problem of permanent reinforcement of soft ground for embankments and foundations is now being solved by the manufacturers of sheets of geotextile fabrics. Geotextiles are permeable woven or nonwoven synthetic fabrics—polyolefin, polyester, or fiberglass fibers— as recognized by the Industrial Fabrics Association. They weigh around 5 to 25 ounces per square yard, can drain water and stabilize, separate, and reinforce soils.

The geotextile technique comprises five design considerations, as depicted in Fig. 3.11: (1) the bearing capacity, since the existing soft ground will settle under new load and the fill on top will have to be increased; (2) the global stability in the tensile strength of the fabric; (3) the elastic deformation which defines the central stretching and so settlement of the reinforced soil; (4) the anchorage length to keep the geotextile sheet from pulling out of the soil it is reinforcing; and (5) the lateral spreading of the soil above the sheet of fabric which must be resisted by the friction between the soil and fabric.

Geotextiles have been used to cover a peat bog, to stabilize city streets and roads in new subdivisions, and to cover many bigger areas. The geotextile fabric is made with many sheets seamed together to cover the soft ground of a harbor area with sand fill laid on top, to build up to the desired elevation for heavy industrial equipment to operate. The soft soil consolidates within 6 to 24 months so that the fill can be brought to final design height for permanent use. The runway extensions for Washington's National Airport were made most economically by using geotextile fabrics on the muddy subsoils of the Potomac River.

The Mirafi 140 fabric performs three functions in soil stabilization: separation, filtration, and reinforcement.

In building a road across a swamp, the strip is first cleared of all hard or sharp objects or bumps. Roots, sod, and sometimes soft brush are left undisturbed.

The fabric is unrolled on the surface, with overlapping at any joints. Aggregate, usually bank gravel, is truck-dumped and spread, preferably by light equipment.

There should be a foot or more of aggregate between the fabric and truck tires, to protect it. Any excess can be shaved off after compaction, which preferably is done with a light vibratory roller.

Sand.   Clean sand is as troublesome as mud to two-wheel-drive vehicles. It can be stabilized with pneumatic tire rollers and plenty of water, but the surface will loosen up as soon as it dries. Tires spin and dig down into it rapidly, with a jerking motion that is very damaging to the drive mechanism. All-wheel-drive vehicles ordinarily have less difficulty with it, but it consumes considerable power. The general problem is one of getting traction without digging in, but there is no danger of simply sinking, as in mud. Partially deflating the tires may help; smooth tires will do better than those with tread, as they will not dig down as readily. Mats of brush, wire, grass, or a thin layer of dirt may suffice to give traction.

Tracked vehicles can travel on sand without difficulty, but if equipped with grousers, care must be taken about pulling heavy loads that may cause tracks to spin, as they will then hang up quickly. The silica which makes up most sand is very wearing to the track parts, particularly when particles are angular.

Quicksand.   Quicksand usually is fine sand or silt through which water is moving upward with enough pressure to prevent the grains from settling into firm contact with each other. It

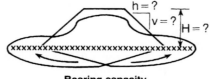

Bearing capacity

Global stability

Elastic deformation

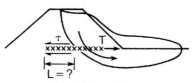

Pullout or anchorage

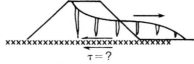

Lateral spreading

xxxxxxxxxx = geotextile

FIGURE 3.11   How geotextile works. (*Courtesy of Civil Engineering, ASCE.*)

provides practically no support for machinery unless its weight is distributed over a large area by platforms.

Work Delays.   Mud is an impediment to work in many ways. Deep mud causes equipment to bog down and to become useless until pulled out, and a film of mud may render firm footing dangerously slippery. Mud sticks to shovel buckets and truck bodies instead of dumping, and builds up in chains and tracks until they jam. It holds objects lying on it by powerful suction so that they become difficult or impossible to lift. When frozen, it can lock together and immobilize the most powerful machines. And it is much heavier than the same amount of dry soil.

EQUIPMENT FOR MUD

Mud trouble can be reduced by using proper equipment. In general, crawlers are preferable to wheels; tracks should be the longest and widest obtainable, tires should be big, soft, and cleated; and units should be the smallest that will do the work. All-wheel-drive is desirable for trucks. Articulated dump trucks with six drive wheels work well in muddy soils.

The ability of a machine to stay on top of soft ground is affected by its ground pressure, which is usually measured in terms of pounds of weight on each square inch of ground contact; shear, which is the load on the edge of the track or tire; and total weight.

Ground pressure is the most important factor in loose soils such as sand or dust. Shear is most important when a soft soil is protected by a harder crust or sod. It is increased when the machine is tipped, and when it pushes a load. Total weight affects deep mud, which may creep or flow from under the machine.

Grousers, cleats, and tire chains cut and churn up their footing but are necessary to get a grip on slippery surfaces.

Some wheel tractors can be fitted with temporary metal and rubber tracks which enable them to work in fairly soft places.

Special vehicles, such as the military weasel and the swamp buggies used for exploration by oil crews, are very useful in supplying fuel and other essentials to machinery working in swamps.

TEMPORARY ROADS

Wheeled Equipment. Wheeled equipment is best kept out of swamps unless they are frozen, dried out artificially or by drought, or have roads built into them, perhaps with geotextile fabric. The use of geotextile sheeting is shown in Fig. 3.12, where a roadway with moderate rutting of the subgrade when dry would become impassable when it rains. The minimum road is a strip in which the soft spots are stuffed with brush or bridged with planks. If poles are used, they should be closely fastened so that they cannot work apart and let wheels down between. Whenever possible, surface poles should be at right angles to the direction of travel.

Pole Tracks. Poles may also be used as tracks to be straddled by dual wheels. They should be straight, free of stubs or sharp projections which might cut tires, and large enough that they cannot pass between the tires toward the hub and small enough that they will not slip sideways out of the groove between the tires. The poles should be overlapped at their ends so that the wheels will not be left without support while passing from one to another.

Front wheels may be placed on skids or runners and chained down. If the mud is not very deep, they may be left to make their own way.

This procedure is for short emergency moves only.

Corduroy. A corduroy road can be built to support heavy machinery on very soft ground. It consists of logs or half logs laid across the traveled way, touching each other. They may be laid directly on the ground, on one or more stringers running lengthwise, or on both stringers and longer cross logs, called sleepers. Several constructions are shown in Fig. 3.13.

A minimum width of 12 feet is recommended for one-way traffic, although it might be possible to get by with 10 feet. Curves should be 2 to 4 feet wider. Cross logs may be extended beyond the road edge for additional stability.

Logs 6 to 10 inches in diameter are generally strong enough for the work, without being excessively heavy. Thinner ones may be used as stringers. Heavier sizes may be split for cross logs, or partly buried for stringers.

The upper surface of cross logs should be smoothed by removal of stubs and bumps, and perhaps by planing down with adze, axe, or ripsaw.

Guard logs are desirable to prevent vehicles from sliding off, to reinforce the road structure, and to retain any surfacing which may be added. Picket stakes may be placed to hold guards in position, and to bind cross logs to the outer stringers. Edges may be bound together by heavy wire,

FIGURE 3.12 Road subgrade covered by geotextile. (*Courtesy of Hoechst Celanese Corporation.*)

cable, spikes, or lag bolts. Such fastenings are important when the road surface is well above the ground, where the mud will not act to hold it in place.

Drainage ditches may be dug on one or both sides to remove standing water. These should be kept from 3 to 10 feet away so that mud will not flow into them from under the logs. In very thin mud their use is inadvisable.

When a drainage channel or small stream must be crossed, the stringers can be increased in number and strength and set on sleepers on each side of the channel so as to serve as a log bridge. If clearance is doubtful, metal pipe may be used to carry the water under the stringers.

Corduroy roads are quick and fairly easy to construct if logs are on the site and usually are very strong. However, their surfaces are extremely rough, and it is advisable to cover them with gravel or other surfacing if they are to have much use. This saves damage to both machinery and road.

A corduroy road made of oak, cypress, or other strong and rot-resistant wood may have a very long life. Some softwoods, such as poplar, will rot out in 2 to 3 years. If mixed species must be used, the inferior ones will give their best service as stringers or sleepers.

Saplings or brush may be wired into tight bundles and used instead of logs for light corduroy.

Plank Roads. Plank road constructions are shown in Fig. 3.14. These are usually more expensive than corduroy, but are easier to lay and provide a smoother surface. Rain or mud may make them very slippery so that sanding will be required.

Nails should be driven diagonally from the sides of the tread planks, as in (*B*), to prevent them from working up and puncturing tires. Even with this precaution, frequent inspections for nails and splinters should be made. If the wood is hard, nails should be soaped or greased before driving. Splitting may be reduced by making thin pilot holes with an electric drill.

Surface Protection. A wet soil with good load-bearing qualities, which might be readily churned into mud, may be protected by a few inches of gravel, spread before traffic uses it. Two feet of broken rock with a skin of gravel may carry heavy traffic for a while over almost any mud. Brush mats or pole corduroy under the rock gives added stability. Between these extremes is every type of condition and cure.

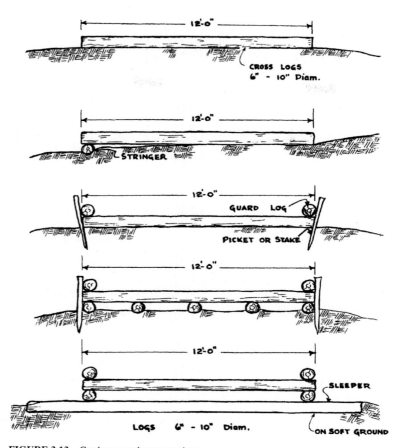

FIGURE 3.13 Corduroy road cross sections.

Fill for mud coverage should be as dry as possible and possess good packing qualities. It should be put on in sufficient thickness that the truck tires will not reach through to mix with the mud. It is strengthened by compacting with a roller, jeep, or empty truck before carrying loaded trucks. A layer of fine-textured, hard-packing soil topped with gravel is often satisfactory, but clean broken rock in coarse sizes up to one-half the fill thickness is longest-lasting.

Flat stones should be placed on edge, whenever possible, so that they will not shift under load. Natural drainage must not be blocked. Iron or steel culvert pipe is most satisfactory on soft bottoms, as it will keep its position and resist separation and breakage.

Swamp Surfaces.   Most swamps are covered with vegetation that gives them some surface stability. Wet swamp sod, in which a person's feet will sink slightly, ordinarily will support a light crawler machine moving steadily across it in a straight line or gradual curves. Crossing should be tried with caution, particularly if the unit has narrow tracks, or high grousers which tend to cut the sod. Sharp turns, if necessary, should be made in the firmest or best-matted sections, or extra support should be provided. Bushes and close-growing saplings which can be walked down by the machine, without cutting, provide excellent natural support.

Although a shovel can be safely walked on quite soft ground, it cannot stand or work on it. If standing, its weight slowly displaces mud beneath it, and as that creeps away, the sod or crust left without support shears and breaks. This process is greatly accelerated by working, as the vibration,

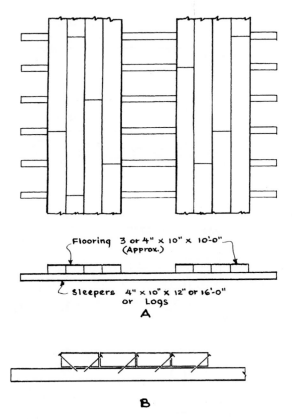

Flooring 3 or 4" x 10" x 10'-0"
(Approx.)

Sleepers 4" x 10" x 12' or 16'-0"
or Logs
A

B

FIGURE 3.14   Plank road.

the variable load, and the twisting reaction to the swing all cause the machine to settle, particularly at the end where the loads are picked up.

PLATFORMS

Platforms, also known as pontoons or mats, are wood supports for machinery working on soft ground. They are most often used under draglines or other revolving shovels.

Their primary function is to spread the weight of the machine over a wide area. They also provide a relatively clean, dry place from which to work, and will act as temporary bridges over ditches and holes.

Construction is not standardized. They are generally made by the contractor who uses them according to local methods.

In general, mats for $\frac{1}{2}$-yard shovels are 12 to 14 feet wide, and for $\frac{3}{4}$-yard, 14 to 18 feet. The length should be about half the overall length of the crawlers, except that they should be shorter if necessary to fit into the truck body that will have to carry them. If they are made for two sizes of shovel, they must be strong enough for the large one, and short enough so as not to be too heavy for the smaller machine.

Oak is quite generally used because of its strength and resistance to decay, and elm because of its toughness. These hardwoods have the disadvantage that they cannot be nailed readily, so it is necessary to drill and bolt. Spruce, although soft, is tough and resists splitting, can be nailed, and is lighter than the hardwoods.

Green lumber can be used. It is tougher and heavier than after curing.

Figure 3.15A shows a sample of very light construction. The loops shown at each end may be of any size cable from $\frac{3}{8}$ inch up, fastened with two clamps. A chain is hooked to both of these and to the shovel bucket when the mats are moved. Three-eighths-inch chain is satisfactory for ordinary service with light shovels.

It is a good plan to make up a special chain about 2 feet longer than the platforms are wide. Round hooks should be used on the ends, preferably the lock-on type. The center should have a ring and a special hook large enough to hook on the bucket, or on the drag chain links next to the bucket.

A mat may be lifted and dragged by one side loop, but this is not as convenient as using two.

The siderails, above the deck, prevent the shovel from sliding off the side. They get in the way occasionally but on the whole are well worth having.

When nails are used, they should be long enough to hammer through the wood. The projecting point should be hammered over flat.

Bolts should have washers on both ends. Heads should be countersunk into the deck.

The laminated platform in Fig. 3.15B is suitable for any but the roughest service and will support light trucks. It is heavier to handle.

The heavy-duty type in Fig. 3.15C will take any abuse a $\frac{1}{2}$-yard shovel can give it, for a while. It is suitable for use on rough or frozen ground, and among stumps and boulders, but it is too heavy to be convenient for ordinary soft ground.

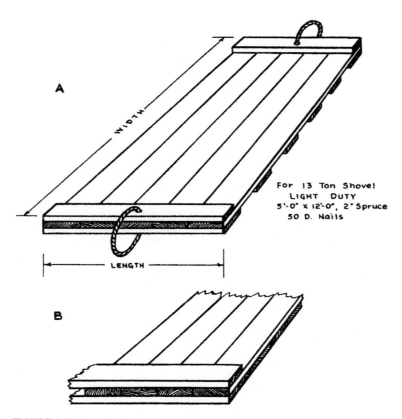

For 13 Ton Shovel
LIGHT DUTY
5'-0" x 12'-0", 2" Spruce
50 D. Nails

FIGURE 3.15A   Butt joint platform.

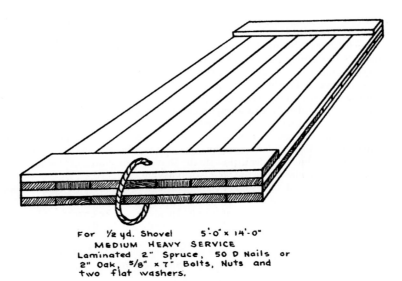

For ½ yd. Shovel   5'-0" x 14'-0"
MEDIUM HEAVY SERVICE
Laminated 2" Spruce, 50 D Nails or
2" Oak, 5/8" x 7" Bolts, Nuts and
two flat washers.

FIGURE 3.15B   Laminated platform.

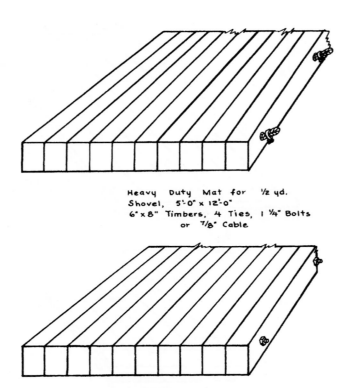

Heavy Duty Mat for ½ yd.
Shovel, 5'-0" x 12'-0"
6" x 8" Timbers, 4 Ties, 1 ¼" Bolts
or 7/8" Cable

FIGURE 3.15C   Timber platform, heavy-duty.

Handling.   Platforms usually are trucked as near to the job as possible, then dragged or carried by a loader or the shovel.

To walk a shovel across an open swamp on platforms, pick up two or more of them and lay them on the mud, as shown in Fig. 3.16(*A*). Move the shovel to the front of these, then pick up the remaining platforms and lay them ahead of the shovel as in (*B*). The minimum number to be used is three—two to stand on and one to move. Four are safer and easier to use, and the maximum is the number the shovel can reach when they are laid in a line behind it. Most draglines use three or four on small jobs, and four to six on large ones.

The shovel then walks to the front of the platforms, (*C*), and picks up those behind it, starting at the rearmost, and lays them in a line in front of itself. In (*D*), the platforms are laid in a curve to avoid a rock. (*E*) to (*G*) shows patterns for laying them in sharp curves. These should be avoided where possible, as the turning of the shovel has a destructive grinding action, and overlapping the platforms causes severe strains, particularly if they are equipped with siderails.

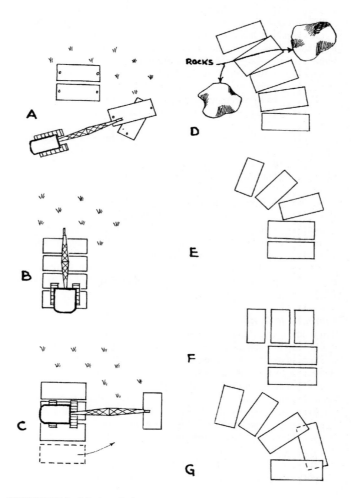

FIGURE 3.16   Moving platforms.

Shovel Rigs.   Techniques of working from platforms vary with the type of shovel attachment being used. The most efficient swamp worker is the dragline. Its long reach enables it to keep well away from the hole it is digging, and to pile spoil far enough away to reduce slumping back into the pit or against the shovel. The sliding action of the bucket during digging and hoisting reduces trouble with suction.

A dragline or a backhoe stands at the back of its line of platforms and digs at the rear or sides. When its work from one position is finished, it walks ahead one or more platform widths, then picks up the platforms behind it and swings and places them in the front; repeating this process as often as a move is required.

Hydraulic full-revolving backhoes are competitive with draglines. They can dig much harder material and to greater depths, and have more precise control in both digging and dumping. They are much better able to get themselves out of trouble, and to work without platforms. But their reach is much shorter in proportion to weight and bucket size. In swamp work, long reach is highly desirable, and weight increases problems, cost, and risks.

The tractor-mounted hoe cannot work on platforms unless another machine places them. Digging reach and dumping reach are poor. But if the tractor also carries a front-end loader, with skillful operation it can work in (and get out of) very soft areas without supports.

A clamshell has almost as much reach as a dragline, and can work without dragging debris, such as roots, stumps, and boulders, up against the platforms and itself. However, it has the severe handicap of pulling the bucket straight up, which in sticky mud requires overcoming suction more resistant than the weight of the bucket and load. Digging is best done off the rear or sides of the platforms, but the ground ahead can be graded before placing the platforms.

The spaces between platforms will vary with the nature of the footing. For very soft conditions, they should be placed in contact with chains fastening them together, or even laid in two layers, as in Fig. 3.17(*A*).

For ordinarily soft or suspicious ground, the widest spacing allowed should be that which will enable the shovel to reach the next platform before its center of gravity reaches the edge of the one it is on, so that the shovel will not tip forward, as in (*B*), so much that the track will push the next platform instead of climbing on it.

Blocking.   Normally, a shovel can be worked on platforms without blocking. There is always a risk, however, that a dragline, or particularly a pull shovel, i.e., backhoe, will unexpectedly hook into something solid and drag itself off the platform before the operator realizes what has happened. If the platforms are slippery with mud or ice, this danger is greatly increased, and the machine is also likely to slide sideways in reaction to swinging.

On a solid platform, wood wedges chained to the platform, as shown in Fig. 3.18(*A*), are effective. For side protection, plank rails, permanently bolted to the side edges, or held there with pins, as shown in (*B*), usually are adequate.

Chaining to Platforms.   For extreme conditions, the shovel may be chained to the platforms. There are many ways to do this, of which one sample is shown in Fig. 3.19. The platforms have a cable loop at each corner, by some of which the two platforms are chained together. To the outer corners of the truck frames are welded brackets holding rings or hooks. Chains are fastened to these and to the corresponding outer corner loops of the platforms, and are drawn up taut with chain tighteners in the rear. The tightness is important, as the momentum of a sliding shovel will break chains or tear out loops which would hold against a direct pull.

All accessories, such as blocks, rails, chains, and tighteners that are to be used in mud, should be kept painted some brilliant color (not green). They are very easily lost and dug under, particularly if some emergency interrupts the routine of their use, and the bright color will greatly increase chances of salvage.

Platforms can be used efficiently only in places where the shovel has room to swing. If trees prevent swinging, they must be cut, or other methods used to support the shovel.

Suction.   Lifting platforms or other objects from mud is greatly impeded by atmospheric pressure (14.7 pounds per square inch), which is felt as suction.

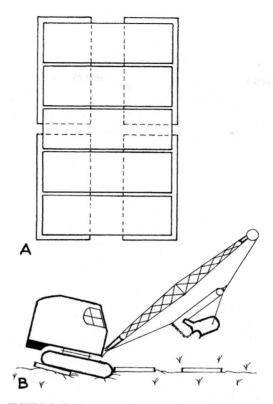

FIGURE 3.17   Overlapping and spacing platforms.

A platform $4\frac{1}{2} \times 12$ feet carries an atmospheric pressure of over 50 tons. When air can flow under it freely, the pressure is upward as well as downward, and is not felt in lifting it. If air is sealed from the bottom by mud, then the lift of the shovel merely serves to reduce the weight of atmosphere and platform on the mud beneath, causing mud at the edges which carries the full atmospheric load to squeeze in under it. A thin mud will flow quickly, a thick tenacious one may stretch but not flow and keep the air locked away from the bottom.

If the suction is too strong to break by a direct lift on the whole platform, the pull should be concentrated on one side or one corner. If the mud seal is broken at one spot, it should shear off from the whole under-surface. The platform might also be lifted and dragged at the same time. Suction presents comparatively little resistance to sliding, and this motion greatly increases the shearing effect on the mud seal. Unfortunately, a dragline is apt to make a pile of mud and debris which would prevent the platform from sliding forward, and the lattice boom is apt to twist or collapse if subjected to heavy side-strain.

A method of releasing is to catch the forward edge with the bucket teeth and hoist while pulling back enough to keep hooked.

Root Mats.   Trees growing closely in swamps usually form a mat of interlocked roots that will support crawler equipment, but will give it a rough trip. Openings may occur where the roots are lacking, which may be crossed by means of green saplings laid across the path of the machine, as indicated in Fig. 3.20. If it is necessary to remove trees to give the shovel space to get through, they should be cut as close to the ground as possible, as the shovel might sink while going over the stump and get hung up on it. Logs laid on each side of the stumps across the line of travel are protection against this.

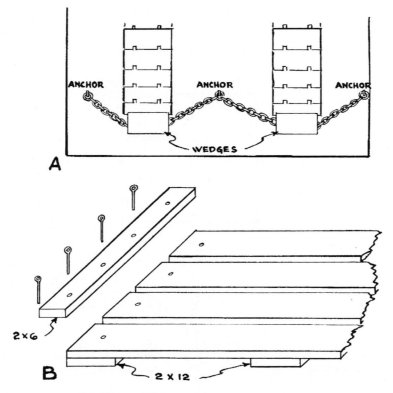

FIGURE 3.18   Blocks and side rails.

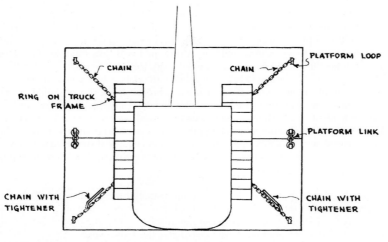

FIGURE 3.19   Chaining to platform.

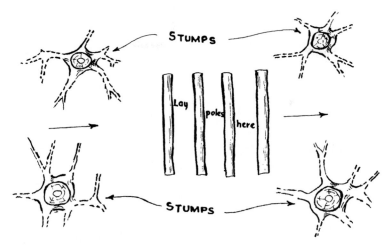

FIGURE 3.20   Support by roots and poles.

In a swamp studded with stumps, or rock, platforms break up very readily. The spaced plank type, which is generally preferred because of light weight and reduced suction, will break if used across a stump or heavy buttress root, and the heavier types will strain and splinter. Some operators prefer to pull the stumps before walking over the spot, thus exchanging the platform-breaking obstacles for a wet and uncertain footing.

It is dangerous to move a machine onto platforms which are placed with one side on firm ground and the other side in loose mud. The mud side may sink enough to cause the shovel to slide to that end and tip over. The soft side may be braced with platforms or logs, or the firm side can be ripped up, before placing the platforms.

Poles.   A shovel may be walked and worked on fairly soft footing by the use of saplings instead of platforms. These should be of firm wood, preferably green, 2 to 10 inches in diameter, and long enough to project 2 or more feet beyond the tracks on each side.

An extra precaution is to place outrigger poles under their ends, parallel to and outside of the machine's path. When the shovel has passed over the poles, they may be retrieved and used again, but the mortality rate usually is high, particularly among softwoods.

When the shovel is working, it will often be found that a pole or two under the front, or where the lifting of the load is done, will suffice to support it. With worse conditions, more cross poles and outrigger poles should be used.

If long poles are not available, short ones may be used, centered under each track, but they are not nearly as satisfactory. There is a danger that they might tip under the shovel and jam into the machinery.

WET DIGGING

In any extensive wet digging, draining or surface pumping is liable to leave a sheet of at least a few inches of water on the bottom. This is a convenience in establishing a flat bottom grade, but unless special precautions are taken, large quantities of water will be dug out along with the soil.

This water may run directly back into the hole, in which case the time and fuel used in lifting it are wasted. If it mixes with soil, either in the bucket or in the pile, it will tend to liquefy so that the spoil will not stand in the high, steep-sided piles which afford maximum dragline production.

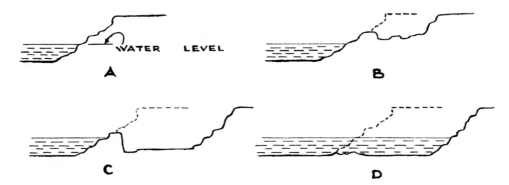

FIGURE 3.21   Compartment digging.

Compartments.   Shallow bottom water may be largely kept out of the bucket by compartment digging, as illustrated in Fig. 3.21. A ridge of clayey soil is left between the digging and the water until bottom grade is reached. The ridge is then dug out and water allowed to flow in. Some mud will be washed onto the digging surface by inflowing water and by the removal of the ridge. Compartments may be large or small.

This method gives waterfree digging for the bulk of the excavation, and can be used, with diminishing efficiency, in increasing depths of water, but not when the whole digging area is under water, as in deepening an undrained pond.

If little surface water is present but groundwater drains into the excavation rapidly enough to be a nuisance, digging can be in two or more compartments. First, the whole area is dug in layers until it gets sloppy, after which the digging is concentrated in one spot until most of the water flows into the hole made. An adjoining area, separated by a ridge, is then dug deeper and the ridge cut.

The water will now flow into the deeper hole, leaving the first one nearly dry and ready to be deepened in its turn. This alternation of digging spots can be continued to the bottom of the cut. If the area is too large to be dug in this manner, the water can be concentrated in gouges dug behind the regular layer cut.

Spoil.   If enough dry or stiff dirt can be dug to build a dike along the excavation edge of the proposed spoil pile, as in Fig. 3.22, it will prevent mud behind it from flowing into the hole. There is

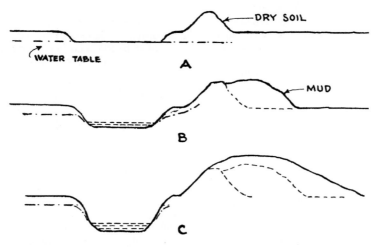

FIGURE 3.22   Dike method of piling mud.

usually ample space behind the pile so that the mud which moves in that direction makes a place for more in the pile, and increases the amount of digging that can be done from one stand. The effectiveness of the dike will of course depend on the quantity and type of dirt available for it.

If the dirt is being loaded in trucks, an effort should be made to get dry material on the bottom to reduce trouble with the load sticking in the body. Most excavating in swamps and mud is done by revolving excavators with dragline attachments.

Bulldozers.   Bulldozers are not suited to wet excavation since they cannot work efficiently on artificial supports. However, they can skim shallow layers of mud off hardpan and dig cautiously in muds compact enough to give them some traction. In skimming work, it is usually best to start at one side of the mud area and make a pass removing the mud cleanly. Each successive pass should be to full depth and cut just enough into the side of the muck to fill the blade, without allowing it to slop off the other side of the blade into the cleared area. The mud should be pushed far back, as the dozer cannot climb up on it to make high piles, and there is danger of its flowing back into the hole.

Wide-gauge dozers are preferred, as they usually have wider track shoes for greater flotation, their weight is spread over a larger area, and they turn more readily on poor footing.

Cleats.   Smooth or semigrouser tracks are unsatisfactory, as they have little traction on wet slippery surfaces. In general, tracks with new and high grousers are better than those which are worn and rounded. However, such grousers on loose soil will dig in very rapidly, and many muds will pack in between them to make a new smooth surface even with their edges so that they become ineffective.

The best solution for straight mud work where rocks or very hard subsoils are not involved is to use flat shoes, with high grousers bolted on every fourth to sixth one. This should put two cleats on each track under the machine at all times, and will give a good grip which will not clog.

A similar effect can be obtained by building up some of the cleats on a set of standard track shoes, or replacing some shoes of a worn set with new ones.

A machine with cleats of variable lengths is likely to be almost incapable of working or traveling on hard ground.

Dragline Road Cut.   A permanent road should not be built over unstable mud unless absolutely necessary. Such muds, particularly if rich in organic matter, will gradually be compressed and displaced by the weight of the road, allowing it to sink unevenly. The mud should therefore be removed down to firm bottom and replaced with clean fill if possible. If it is too deep, or otherwise too difficult to move, measures must be taken to stabilize it, such as with geotextile fabric.

Shallow mud may be removed by a dragline working just ahead of the fill, and piling it to the side. The rate at which the mud pushes back into the hole determines how far ahead the dragline can be operated. A foot or two of mud liquefied by the mixing action of the dragline bucket is usually pushed out by the weight of the fill, but sliding in of the banks must be avoided.

Surface water should be diverted and sufficient pumps used to keep the hole fairly dry. This enables the dragline operator to see to clean out the mud thoroughly without unnecessary digging of firm soil, and will prevent the fill from being turned to mud as it slides down the front of the slope. If the fill is too wide for the dragline to clear, it may be built out in sections.

MUD BLASTING FOR FILL SETTLEMENT

If the mud is too deep for available draglines to handle, or if it slumps into the excavation before fill can be placed, or if there is not adequate equipment to keep the water out, blasting may be resorted to.

Mud is easily blasted out of a limited area and spreads in a thin film over the landscape, leaving no heaps to dispose of. If a nitroglycerin or other sensitive dynamite is used, the concussion from one explosion may detonate other charges of dynamite in the mud nearby, without the necessity of using additional caps. The process, called *propagation,* greatly simplifies mud blasting.

There are three principal techniques: One is to blast a ditch along the right of way, then dump fill in it. Another is to build the road, or a section of it, on top of the swamp and blast the muck from under it. The third, toe shooting, consists of making a big heap of fill and blasting the mud ahead and to the sides from under that.

Ditch Shooting.   The tables in Fig. 3.23 give average loading requirements for various types of open-ditch blasts. Figure 3.24 shows how to load for rather narrow ditches, which are used chiefly for drainage, or to make possible placing of an earth dike to cut off movement of water or mud. It will be noted that the nature of the mud is very important in determining the size ditch which will be obtained from a given charge.

It should also be remembered that water counts as soil in loading calculations.

Blasting techniques for wider ditches, which are used to remove muck to secure a firmer base for road fills, are indicated in Figs. 3.25 and 3.26, which show the cross section of relief methods.

The charges should be 50 or 60 percent straight dynamite, at least 4 inches below the surface, and covered with mud or water. The blasting cap should be in a stick of dynamite at or near the

SPECIFICATIONS FOR SINGLE-LINE COLUMN LOADS

CARTRIDGES PER HOLE	DEPTH TO TOP OF CHARGE	DEPTH OF DITCH	TOP WIDTH OF DITCH	DISTANCE BETWEEN HOLES	POUNDS PER 100'
½	6 - 8"	1½ - 2'	4 - 5'	12"	25
1	6 - 12"	2½ - 3'	6'	15"	40
2	6 - 12"	3 - 3½'	8'	18"	67
3	6 - 12"	4 - 4½'	10'	21"	86
4	6 - 12"	5 - 5½'	13'	24"	100
5	6 - 12"	6 - 6½'	16'	24"	125

CROSS-SECTION LOADING METHOD

CARTRIDGES PER HOLE	DISTANCE BETWEEN HOLES	DISTANCE BETWEEN ROWS	DEPTH OF DITCH
1	15"	30"	2½ - 3'
2	18"	36"	3 - 3½'
3	21"	42"	4 - 4½'
4	24"	48"	5 - 5½'
5	24"	48"	6 - 6½'

SPECIFICATIONS FOR POST-HOLE LOADING

No. of sticks (1¼ x 8") per hole	6	10	20	30	50	100
No. of lbs. per hole	3	5	10	15	25	50
Distance between holes - ft.	3	3½	4	4½	5	6
Depth of ditch "A" - ft.	4	5	6	7	8½	12
Bottom width of ditch "A" - ft.	4	5	6	7	8½	12
Top width of ditch "3A" - ft.	12	15	18	21	25½	36
Depth of load (⅔ of "A") - ft.	2⅔	3⅓	4	4⅔	5⅔	8
Diameter of post-hole - in.	4	4	6	6	8	8
Dynamite per 100 ft. - lbs.	100	142½	250	333	500	833
Material moved per 100 ft. - cu. yds.	118	185	266	363	533	1,067

FIGURE 3.23   Tables for mud blasting.

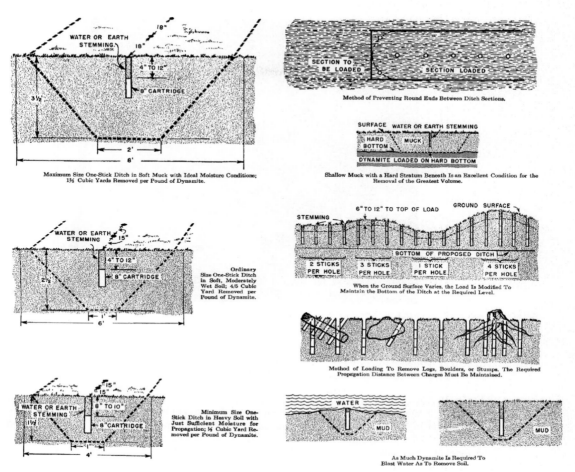

FIGURE 3.24   Loading for ditch blast.

center of the group to be blasted, and with the loads and distances shown in the illustrations, should detonate them all.

However, if the structure of the muck changes between holes, it may shear, and direct the force of the concussion upward or to the side, and charges beyond that point will not explode. In this case, one of the remaining charges should be lifted and primed with a cap, or an additional primed stick should be placed and detonated.

If many reprimings are necessary, it may be better to put a cap in each hole and explode them simultaneously by electricity. As a great many caps are generally used in a single blast of this type, a powerful jolt of electricity is needed.

It is not usually necessary to remove stumps and brush from an area to be ditched by explosives, but trees should be cut and the trunks removed if possible. Extra charges should be placed under stumps, and an effort should be made to get them to throw toward the nearest edge of the ditch. If the stumps are so heavy or so numerous that they prevent accurate ditch work, they may be removed first by draglines, winches, or other machinery, perhaps with some blasting to loosen them up.

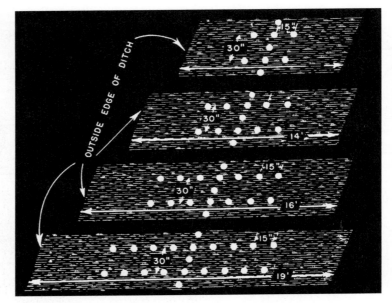

FIGURE 3.25 Cross-section loading.

The dynamite will probably not remove all the mud down to hardpan, but it may liquefy most or all of that which remains, so that the fresh fill will sink through it to firm bottom.

Underfill Shooting. Figure 3.27 shows the second method. Brush, stumps, logs, and other debris are removed from the right of way, and any heavy sod or brush is broken up by machinery or light blasting. Fill is placed in an amount calculated to reach from firm bottom up to the desired road grade. Charges are placed in the mud, in casings driven through the fill. Shooting is by Primacord, or by caps in each hole, rather than by propagation.

The mud explosion is confined by the mass of fill above, and should blow out the sides, driving out most of the mud and liquefying the remainder so that the fill will sink to hardpan.

If the fill is clay or other compact material that might tend to bridge over the blasted cavity instead of settling in it; or if the weight is not considered to be sufficient to resist the explosion; then deep ditches may be blasted on each side of the fill after it has been placed and the charge set off under it afterward. Blasting these relief ditches makes it much easier for the explosion to drive the mud from under the fill, but it also puts a thick layer of muck on top of it, which must be bladed off.

The charges under the fill must be so placed that there will be no danger of their going off by propagation; 30 or 40 percent gelatin is recommended.

Toe Shooting. Toe shooting is illustrated in Fig. 3.28. This is the third method—fill, then blast—done in such short sections that the muck will be driven out to the front as well as to the sides. It has the advantage of allowing the blaster to have frequent checks on the efficiency of the work, so that techniques of loading or filling may be altered readily. A disadvantage is that the muck blown ahead sometimes piles up in a large hill or wave of mud, making continually larger charges necessary to move it. This tendency may be reduced by making the fill with a V-shaped front, so that most of the material will be thrown to the sides. The mud wave may be blasted away in separate operations, but part of it will then have to be cleaned off the fill.

Toe shooting is an excellent method to widen fills, the extra material being piled on the sides and a single line of charges set off in the mud underneath.

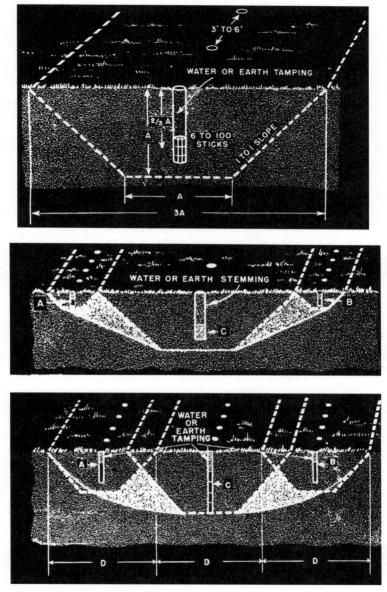

FIGURE 3.26   Posthole and relief loading.

Placing Explosives.   Shallow placement of explosives in mud may be achieved by pushing them down with a stick, or first punching a hole with a crowbar, then using the stick. For deeper work the apparatus shown in Fig. 3.29 is effective. A plain piece of iron pipe of the necessary length is used for a casing, and a round wood pole or a plugged iron pipe making a loose fit serves as a core. The core is set in an iron handle larger than the casing and long enough for a good two-handed grip.

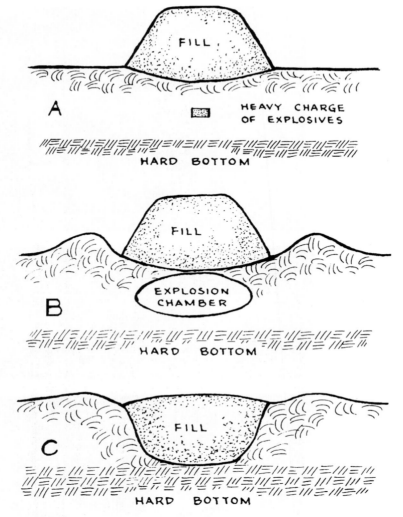

FIGURE 3.27   Bottom shooting.

The core is placed in the casing, and the two are pushed or hammered down into the mud by means of the handle. The core is withdrawn, the casing being held down if necessary by a foot positioned on a bump welded on the side. The charge may then be dropped down in the casing, or pushed down by a long wood tamping stick of the same diameter as the core. This stick is used to hold the charge down while the casing is pulled out. Difficulties experienced will increase with the size and length of the pipe, and it may be necessary to use chain jacks or hoists to pull the casing.

Space permits only this brief summary of mud blasting techniques. Further information may be obtained in other parts of this volume, and from various books and bulletins dealing with explosives, but it is advisable to have an experienced mud blaster on any important job.

Results.   A fault of all mud removal techniques mentioned is that they require placing a considerable depth of fill in a single layer so that it cannot be properly compacted; or the fill is shaken

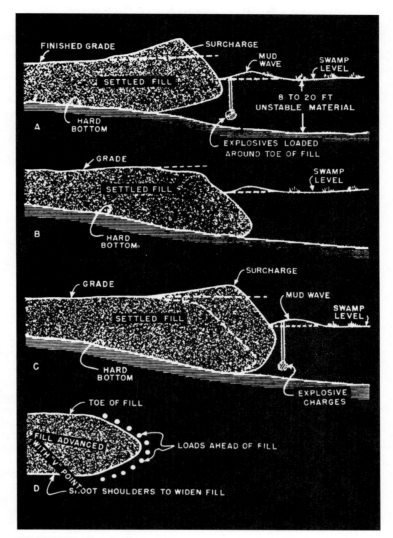

FIGURE 3.28   Toe shooting.

up by blasting and sinking after being placed. This is liable to cause uneven settlement; in addition, there is the danger of pockets or layers of muck getting included in the fill, due to miscalculation, or partial failure of blasts, or rapid slumping after excavation. However, the stability of the fill will always be greater than if it were simply laid on top of the swamp.

STABILIZATION

Dewatering.   One method offers stability equal to that obtained in normal, dry fills, but at high cost. The right of way can be dried up by draining, or well point or sump pumping, as described

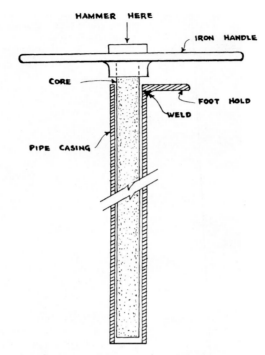

FIGURE 3.29 Cartridge-placing tool.

under Ditching and Drainage, then excavated to firm bottom with machinery. Fill may then be built up to road grade in thin, thoroughly compacted layers and with properly sloped sides. When it is complete, pumping may be discontinued and the muck allowed to settle back against the slopes.

Removal of muck in any way is expensive work, except in the rare cases where it can be sold locally for humus. The expense increases very rapidly with depth, and a point will be reached where it is good practice to stabilize the mud rather than remove it. This is particularly true if the area involved is wide or conditions do not permit side casting or blasting.

Grouting. One method to stabilize ground, as well as to provide seepage control and some rehabilitation of the ground, is by injecting a water-cement mixture which fills the voids and spaces previously occupied by water. The mixture may include some special chemicals that react favorably with the soil to form a hardened mass. Additives may include polymers to stiffen the ground. Many of the grouting techniques were developed in Europe and have slowly been copied in North America.

One technique used in ground made soft by excess groundwater is jet grouting, which originated in Europe but was brought to the United States in the 1980s. Jet grouting has been used not only as a groundwater barrier but also for excavation support, as a bottom-sealing technique to prevent pollutants from entering excavation, to combat scour beneath bridges, to stabilize slopes behind retaining walls, and frequently to underpin existing commercial and industrial foundations.

The jet grouting method involves injecting a water and/or cement grout through jetting pipes under extremely high pressures (generally 4,000 to 6,000 psi), forming a tight mix of grout and native soil. This method can strengthen soft soils or form load-bearing foundations. For example, when the historic structures in the Boston National Historical Park in Charlestown, Massachusetts, were settling, more than 8 inches in some places, due to decades of uncontrolled water infiltration into the loose artificial fill materials, jet grouting was used. A single-fluid jet grouting procedure was used to fill the voids under the buildings and to improve the fill quality. The single-fluid method was used because ground movement during construction is negligible and potential damage to the foundation is minimal.

Sand Drains.   Muck can be stabilized if enough water can be squeezed out of it to convert it from a semiliquid to a solid. The weight of a heavy road fill has a squeezing effect, but the mud and water flow together and very little compaction is obtained for a long time.

Vertical sand drains may be used to dry up the mud so that it will support a load. They consist of columns of sand extending from hard bottom to the top of the mud, and connected at the top to each other, and to an outlet by a sheet of sand or gravel, or other drainage systems. They are discussed in Chap. 5.

Quicklime.   Where the topsoil is so soft that it is impassable by personnel with boots, lime can be spread on the surface and worked in with farm implements such as a disc harrow pulled by a crawler tractor. The quicklime is obtained by cooking limestone at high temperature so it becomes calcium oxide (CaO) which reacts with water, giving off heat and expanding into hydrated lime or calcium hydroxide [$Ca(OH)2$]. Therefore, the quicklime added to the soil sucks up the water, changing the soil from a moldable, soupy mass to a crumbly solid. This makes the ground surface more passable.

Chemicals.   If digging is done in a water-bearing sand or gravel, very soft conditions and extensive caving of banks may be experienced. This difficulty may be avoided by drying the area with well points, or by other methods described in Chap. 5, but it may be more economical to prevent water from moving through the soil. This may be done by means of cement grout, discussed above or by other chemical treatment. The grouting material displaces water, and then hardens or gels into a mass that prevents more water from getting between the soil particles.

One chemical treatment involves pumping a solution of silicate of soda into the ground through drill holes, then pumping in a solution of calcium chloride that causes the silicate to form a dense, hard mass.

American Cyanamid's chemical grout is a thin water solution of two acrylic chemicals and one catalyst, that is mixed with another catalyst solution just before injection into the ground. A chemical reaction causes the mixture to form a permanent, water-impermeable gel, at a time that is controllable between 3 seconds and several hours. These characteristics are particularly useful in sewer rehabilitation work.

Freezing.   Any soil in which pipes may be sunk can be stabilized by freezing, but the presence of salt, or certain other chemicals, may make it difficult.

It is accomplished by sinking a number of metal pipes, rather closely spaced, in the area to be stabilized. Tubing containing a refrigerant, usually ammonia, is placed in these pipes and connected to heavy-duty refrigerating apparatus. The soil water will freeze most rapidly if it is stagnant, but even a steady flow can be checked.

This is an expensive job, mostly confined to a very fine-grained mud, or deep work which does not respond to well points.

FILLS

Compaction.   In fresh fills, the most serious mud difficulties are due to improper compaction. Here it is sufficient to say that excellent compaction may be obtained if a fill is made in layers 6 to 18 inches in depth, and each layer is thoroughly rolled or tamped in all parts. If rollers are not available, trucks may be used, first empty and then loaded. Such a fill usually will be incapable of absorbing enough water to turn to deep mud. The extra expense and nuisance of making a fill in this manner, when specifications do not call for it, may be regarded as an insurance premium against the loss of having to stop a job for days or weeks because of a soft dump.

Clay.   Clay fills are liable to become soft and slippery during rain. The wet surface may be bulldozed off and replaced with dry material, or the dry material may be dumped over the mud in sufficient quantity to carry traffic over it, and may be removed later. If rain is anticipated, the original fill may be left low, to leave room to cover it.

If fill is coming from two or more sources, it is sometimes possible to keep the clay in lower parts of the fill and the better material on the top.

Area Dumping.   If trucked soil is too wet to be stabilized and will not support trucks after being spread, and if the ground (or lower layer of the fill) is passable to trucks, area dumping may be resorted to. A calculation is made of the amount of fill needed in a given area and the amount of each incoming load. Allowance must be made for shrinkage of the soil upon drying and after compaction, and the number of loads required in such an area may be found by dividing the cubic yards required by the corrected yardage of a truckload.

The trucks may then dump their loads in piles which can be left to dry, then be spread and compacted.

Buried Mud.   If it is impossible to truck in the area to be filled, the mud must be covered as it is dumped with a layer of dry fill, gravel, or rock sufficient to support the trucks. Sometimes the mud will support a bulldozer which can roughly grade the soft fill. The dry surfacing is advanced close behind the face of the mud fill, trucks dump the mud at its edge, and the dozer pushes it over the face, digging down sufficiently to allow placement of more surfacing. See Fig. 3.30.

In such mud work, it is advisable to have at least two dozers, so that the one working in mud can be promptly rescued if stuck. Under extreme conditions two dozers may be attached to each other, back to back, by a chain or cable. As one of them pushes mud toward the face, the other backs up toward it, keeping the line slack. When the pusher backs up, the other one pulls it to firm footing. The helper Cat need not have a dozer, and preferably should be larger than the one doing the work. Wide tracks are a big asset in mud.

FIGURE 3.30   Two-layer fill.

STUCK MACHINERY

The next problem to consider is how to rescue machinery that has sunk so far into mud that it cannot move under its own power. The easiest and most attractive method is to stretch a chain or cable to some power source and pull it out. Often there is no such power available; but if there is, it should be used with caution. If the stuck machine has not sunk but has simply lost traction on a slippery surface, power can usually be applied without damage. However if it is sunk in badly, use of moderate power may be useless, and too much power may pull it apart.

TRUCKS

Dump trucks are probably the most frequently stuck type of equipment. If one is loaded, it may be possible and desirable to dump the load, after which the truck may pull out under its own power. But if the truck is tipped sideways, as with one pair of rear wheels bogged down, the other resting on the surface, the raising of the load preparatory to dumping increases the sideward strain, and may overturn the truck (as shown in Fig. 3.31) or tear the body off its base. A winch cable attached to the front top of the body on the high side, and pulling uphill, may permit dumping of a tipped truck, but even with skillful operation, the strain on the body is considerable. Unloading by hand, or partial unloading by power shovel, hoe, or front loader may be necessary.

A shovel is a highly effective rescuer of bogged trucks. A front bucket may be placed under the rear of the frame (or, with some risk of damage, under the rear of the body) and may lift and push at the same time. A front loader can do the same, but might walk into the same mudhole; and its wider bucket may be blocked by tires. A dozer has the same problems.

A hydraulic hoe can get most of the load out of the body, then push and lift with the back of its bucket.

Digging Out.   If the truck cannot be emptied, or is empty and is still stuck, and no powerful equipment is at hand, the next procedure is to dig out the wheels in the direction toward which it is hoped to move. Two-wheel-drive trucks have best traction going forward, and in many situations, such as sinking in a soft shoulder, an attempt to back out will cause the front wheels to get in worse difficulties. Any digging helps, but the best procedure is to go to the bottom of all tires, make a ramp up to the surface with a length of 3 feet or more to every foot of depth, and put a board or boards on this slope, with the lower end against or under the tire. If boards are not available, matted brush, stones, gravel, or anything but mud may be used. If the axles or frame is resting on the ground, an attempt should be made to free them, but this is not often possible. See Fig. 3.32.

Even if this digging does not enable the truck to get itself out, it makes it much easier for a light machine to pull it, and it greatly reduces the danger of damage if pulled by a large machine.

Breakage.   Applying brute force to pulling a deeply bogged truck may result in getting it out minus its rear axle assembly and wheels, a partial victory that brings little satisfaction. In most trucks the rear axle is attached to the frame only by the spring shackles and propeller shaft.

FIGURE 3.31   Dumping on a side slope.

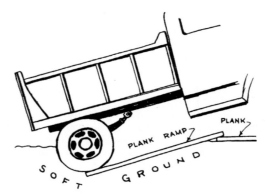

FIGURE 3.32   Digging out of a mudhole.

Lesser and more common damages are tearing off of bumpers and bending of front axles. Many trucks have no satisfactory pull points on the front end. The bumper should be used only at the fastening to the frame. If it is necessary to use the axle, the line should be attached as close to a spring as possible, and care should be used to prevent it from catching any part of the steering when tightened, or when the wheels are turned.

Pull Line.   The most generally effective device for extricating bogged machinery is a winch. It may be mounted on and powered by a tractor or a truck, or may be a portable winch. A power winch may have 200 feet or more of cable which enables it to reach a long way from firm ground, or on shorter pulls, to multiply its power many times by means of pulleys and anchors. The procedure for rigging and operating the winch is the same as that described under stump pulling in Chap. 1. The use of multiple lines is often advisable, even when the machine may be debogged by direct pull, as the slower speed is less liable to cause damage. Hand winches are slow and laborious to wind in, but are powerful and can be used in places inaccessible to larger machines.

Where possible, it is best to pull the truck straight out with the truck wheels driving. If it is necessary to pull at an angle, the truck should be steered toward the pull.

A cable or chain may be stretched from the stuck truck to another truck or a tractor, with or without pulleys, and a traction pull used. Care should be exercised not to use too short a line and get the assisting machine stuck also. See Fig. 3.33.

Push.   A dozer or front-loader tractor can get behind it to lift and shove at the same time, but someone should be stationed to put poles in front of it as it moves so that it will not get stuck in the hole the truck leaves.

FIGURE 3.33   Too much power.

Body Hoist. A dump body usually extends a short distance behind its hinge, so that when the body is raised, the rear edge of it goes down a few inches. Advantage may be taken of this feature by chaining the axle tightly to the frame, perhaps lifting it first by chaining it to the front of the body and hoisting; placing a plank on the ground under the rear of the body, and fitting a stout log between plank and body. If the body can be raised, the rear edge will push down against the log to lift the frame, and through the chain, the axle and wheels. Axle or wheels can then be blocked up, the body lowered, and the log shimmed up to meet the body at its new height, and the process repeated until the wheels are high enough that a support or ramp can be put under them. See Fig. 3.34.

Jacking. If a heavy jack is available, the axle may be chained to the frame, the jack placed on a plank or other support, and frame and wheels raised and blocked in the same manner as when the body lift is used. If it is not possible to chain the axle to the frame, first raise the frame with jack or body hoist, then jack the wheels up with a light jack based on a board and placed under the wheel rim or hub.

Wheel Tractors. A wheel tractor in mud presents similar problems, except there is no load to be removed and it takes more digging to get to the bottom of its big rear tires.

A tractor may have a differential lock, which can be engaged to prevent one wheel from spinning independently of the other. This should be engaged at the first suspicion of bogging down.

Letting some air out of drive tires may be helpful, but if overdone may cause the tire to slip on the rim, tearing out the valve.

If the stuck tractor has a loader, the bucket may be used to push it out of trouble, preferably backward. The procedure is similar to that described for crawler tractors later.

If the tractor carries a hoe, it can lift the rear wheels clear of the ground by forcing the bucket down. It can then push it forward, or swing it to the side.

A tractor equipped with both a front loader and a backhoe can, with skillful operation, work in extremely soft ground.

If a truck or wheel tractor is stuck more because of slippery surface than sinking, tire chains, short chains threaded through wheel openings, or rope or rags tied around the tires will help to give traction.

Wheeled vehicles do best in mud or sand with large tires, or standard tires with reduced pressure. Truck tires break down rapidly if underinflated, but tractor tires usually have thin flexible sidewalls and operate at such low speeds that they can run soft for considerable periods before being damaged. However, they may slip on the rim and tear out their valve stem if too soft.

Churning. A very important item in the sticking of vehicles is to know when to stop struggling. Often a minor tow job is changed into a major project by a driver continuing to spin wheels and buck back and forth until the vehicle has sunk completely. In rescue work, the strongest measures available should be applied early, as every attempt that fails is likely to make the job more difficult, and danger of damage to the sunk vehicle more severe.

FIGURE 3.34 Jacking out with body hoist.

CRAWLER TRACTORS

Crawler machines do not bog down as readily as wheeled ones, but can do an even more thorough job of it.

A crawler tractor may sink in mud too soft to support it, or it may dig itself in while pulling a load, or both. If the tracks are allowed to spin, the grousers act as buckets on a ditcher, digging soil from underneath and piling it behind. On soft ground they can work down rapidly this way, until the frame parts of the tractor are resting on the ground, or on a stump or other object its normal clearance would have taken it over. When the weight of the tractor rests thus on the frame, the tracks churn helplessly in air or loose mud.

Pulling Out. If outside power is available, the machine may be pulled out by a line attached to its drawbar, front pull hook, or other hold. It should be pulled straight forward or backward, if possible, with its own power being used also. The drawbar, or a dozer blade, can take almost any pull, but use of a front pull hook may pull the engine out of some models.

Poling Out. If no outside power is available, and the stuck tractor has no dozer, winch, or other helpful equipment, the first thing to try is digging a shallow ditch in front of (or behind) the tracks, a foot or two wider than the tractor. In this put a green sapling, or strong board as long as the ditch, pressing or wedging it tightly against the tracks, which should then be turned slowly so as to pull the stick or board underneath. When it is well under, press in another stick, and perhaps more. Their effect will be to lift the tractor and restore the weight to the tracks, provide a wide base for support and traction, and cut off or uproot obstacles under the tractor. They are almost certain to get the machine out if it will pull them under.

If the tractor has flat shoes or grousers that will not grip the size of pole available, and no bolt-on cleats can be obtained, or if there is an aversion to digging ditches; planks or heavy angle irons may be drilled and bolted to track shoes on both sides. The effect is then positive, but it is necessary to unbolt and remove them as they come up at the other end. Poles or logs may be fastened to the shoes by loops of cable and used in the same manner. Short sticks should not be used unless absolutely necessary, as they do not afford nearly as much lift or traction as long ones, and they are liable to turn underneath and jam things. See Fig. 3.35.

These stratagems are equally effective at moving the tractor forward or backward, but reverse gear often has less power than low, and backward movement under difficult conditions is a severe strain on the tracks.

Cable Ramps. Another system is to fasten a cable to each track, or the two ends of a single cable to the tracks, perhaps by passing it from the outside through a hole in the shoe, and catching it inside with a loop and clamps. These cables should be stretched parallel or nearly so, straight ahead of the machine, to anchors of some sort. When the tracks are turned to move forward, they will advance on top of the cables, which will prevent them from spinning and provide a tightrope ramp on which they can climb. This technique should be used with caution in reverse, as the strain on the track may break it. See Fig. 3.36.

Chains may be bolted to the track and used instead of cables, but they are much heavier for the same strength and are seldom long enough for the job.

The tractor may also be jacked up on planks, in the same manner as a truck, and the hole filled in or bridged.

Winching Out. If the tractor is equipped with a towing winch, the cable may be fastened to an anchor behind it, and the machine will come out of the mud as the cable winds in. This should be done with caution if the presence of a stump or large stone underneath is suspected, as it might force a track off, or do some other damage. Risk can be reduced by putting blocks behind the tracks so that it will move up as well as back, or by anchoring the cable at a height, as in a large tree.

FIGURE 3.35   Walking out on cross poles.

If the tractor is too badly bogged down to turn, the only available anchor is not directly in line, and the winch is of a type that the cable runs off the spool on an angle pull, then run the cable out until it is slack, then hold it with a crowbar or stick so that it reels on the side of the spool opposite to the direction of the anchor. When the pull begins, it will be off center and will have a tendency to turn the tractor in line with the anchor. If it does not do this, it will wind onto the drum in a spiral making one or more loops before reaching the edge on the anchor side, moving the tractor a short distance. The line can then be slacked and the procedure repeated.

If the angle between tractor direction and anchor is too great, a second anchor must be used and a chain from that adjusted to hold the cable in line. Use of a snatch or pulley block, as shown in Fig. 3.37, will prevent damage to the cable from the chain hook. If alignment is satisfactory but the winch does not have enough power to debog the machine, a pulley may be chained to the anchor and the winch line passed around that and back to the tractor drawbar. This will nearly double the pull.

Dozer Down Pressure.   If the tractor is equipped with a hydraulic bulldozer, but no winch, the blade should be raised; planks, poles or other floats placed, dug or driven under it; and down pressure

FIGURE 3.36 Climbing a cable ramp.

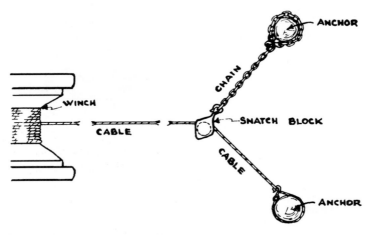

FIGURE 3.37 Changing angle of pull.

applied to the blade. If the hydraulic system is in good condition, this will raise the front of the tracks clear of the mud. Poles may then be forced under the tracks, or the holes filled with rocks, and the tractor can walk out after raising the blade. If the dozer will not raise the tractor, and there is enough oil in the system, try letting the machine cool off, as a worn pump will develop better pressure on thick, cold oil. If it still will not work, or if there is no time to wait, proceed as suggested for a plain tractor.

Cable Dozer.   If a tractor equipped with a cable bulldozer is stuck, and plenty of pulleys and extra cable are available, the blade might be supported by a chain, and the control drum rigged to pull it out. The heavy pull of multiple lines should be through the drawbar or blade.

Front-End Loader.   Front-end loaders equipped with flat or semigrouser shoes get stuck easily in wet holes. The bucket can be used to push it out, preferably backward. Place it on the ground in fully dumped position, apply down pressure, then rotate the floor forward to a flat position, usually turning the tracks slowly backward at the same time. See Fig. 3.38.

Poles may be placed under the bucket and behind the tracks, for extra bearing in very soft ground.

A forward pull may be obtained by placing the bucket flat on the ground, applying enough down pressure to raise the front slightly, and then dumping it slowly. It is not as effective as in reverse.

The bucket can also be used for short-distance self-moving of a machine that cannot travel because of clutch or transmission failure.

The front of the tracks can be raised off the ground by using down pressure with the bucket floor vertical, and supported by poles or blocks. The rear can be raised by shoring up the front and piling the bucket with heavy material.

Hanging Up.   Tractors are frequently hung up on stumps or rocks in the absence of any mud at all, usually through digging in the tracks while moving a heavy load. The first warning is liable to be a failure of the tractor to keep its direction or to steer. The tractor should be stopped immediately and the situation checked over. If the stump (or rock) is under the crankcase guard or other smooth surface, the tractor may escape by backing or turning, or by climbing a rock or a stick placed under one track. If the obstacle has wedged into a hole or raised section, then large rocks or logs under the tracks, or even jacking, may be required to get free.

A crawler tractor may also be hung up by walking over a large flat stone or short log which flips up and catches in the chassis or between the tracks. This situation deserves careful examination before working, as applying too much power in the wrong direction may break a track or do other damage. Movement in the right direction, particularly turning one track only, may free it; or climbing up on blocks or using another power source to pull the rock out the same way it went in may work. Sterner measures are removing the track, or turning the machine on its side in order to break up the stone, or to chop the log.

FIGURE 3.38   Pulling dozer shovel out with bucket.

Overturning.   If a crawler tractor is lying on its side, it may be rolled back on its tracks by pull on a line fastened to the highest substantial structure near the center of the up side. If there is no spot above the center of gravity which will take the pull, the line should be run across the up side and what is usually the top of the tractor until a hold is found. Logs or other blocks under the line should be used to protect the tractor parts against crushing. An improvement on this is to use a pulley in a tree, or in a heavy tripod, so that the line will lift as well as pull.

If the machine is upside down, two pulls may be necessary, one to get it on its side, the other to right it. This is easiest downhill, but blocks must be arranged to prevent it from rolling farther than planned. Power should be applied slowly, to avoid bending or crushing of parts.

If no power is available, hand jacks should be obtained together with planks and blocks. A jack placed on a plank should be used to raise some portion of the tractor, blocks placed to hold it at that height, the jack released, blocks placed under the jack, and the process repeated. One jack will do it, but two are easier. It is good procedure to start jacking the part of the machine which will move the longest distance in resuming upright position, and work the jacks in as space opens up. If the tractor is dozer-equipped, the dozer frame and blade will safely take the strain of jacking. The dozer may sometimes be moved advantageously, using the starter for power.

The type of jack shown in Fig. 3.39 is particularly useful in machine salvage.

FIGURE 3.39   Equipment salvage jack.

SHOVELS

Planks.   Power shovel tracks usually have flat shoes, so that it is difficult to get them to grip poles or planks and drag them under, without excessive trenching, or bolting or cabling them on. Before working a dragline or other shovel in risky places, it is a good plan to get a plank 2 inches × 12 inches, 3 inches × 12 inches or heavier, 2 feet longer than the shovel is wide, and drill holes in it to match the holes in the track shoes so that it may be readily bolted, and carry it for use when necessary as described under tractor rescue. Strain on plank and shovel will be reduced by trenching deeply to the underslope of the idler if possible. If the shovel can climb up on this plank, saplings or other helps can be placed in front of it to assist it to firm footing.

Hoisting Bucket.   If a dragline is down on one end, logs or planks can be forced or dug under the high end, and the low end raised by hooking the bucket or hoist cable to a stump or other anchor, as in Fig. 3.40(*A*) and (*B*). Pulling in the hoist cable, or raising the boom, should tip the shovel forward, lifting the rear sufficiently to allow shoring up with logs.

If the anchor is close, a low boom gives best leverage. If it is distant, a high boom is better. However, a boom angle of over 45° may be dangerous, because if the hoist line breaks or comes off the anchor, the reaction might throw the boom back on the cab.

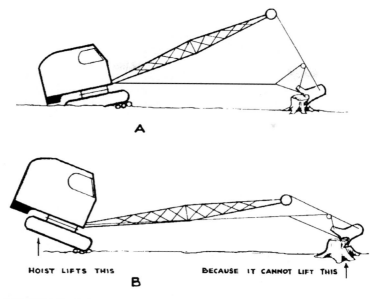

FIGURE 3.40   Raising by an anchored bucket.

The same procedure can be used if one track is bogged and the other one is free.

If the shovel is down all around, trenches can be dug to permit bracing under one end. This is used as a pivot to raise the other end, which is then shored up. The boom is swung to an anchor on the opposite side, and the hoisting and blocking repeated.

Two anchors may be chained together for greater strength, or to secure a better direction of pull.

If no anchors are located in the proper direction, and it is not practical to make one, the bucket can be used as counterweight. The dump cable is shortened to permit holding a load far out. A bucketful of the heaviest dirt available is dug, the boom lowered so that the bucket can be held just above the ground. Rocks or weights are piled on the bucket until the shovel tips. If the rear of the tracks is not high enough, it can be brought up by raising the boom.

In any blocking work a loose track is a nuisance, as it will hang down and impede placing of supports. The slack can be readily taken out by placing a jack on the track frame and raising the upper section.

Drum Line.   A shovel can sometimes pull itself out of trouble by a line attached to one of its own drums, or to a drag bucket. These lines are too fast for a smooth, steady pull, and the drums are so high that there is a tendency to pull the front of the shovel down. Better results will therefore be obtained if the line is passed through a pulley chained to the anchor, and back to the bottom of the dead axle. If several pulleys are available, and all additional lines run to the dead axle, a very powerful, well-directed pull will be obtained.

Jamming Chains.   A special precaution to be observed with shovels is to keep the drive chains free from material of any kind. Stretching or jamming these with mud, gravel, or debris not only is a severe strain on the chain, but also by increasing friction within the chain and with the sprockets absorbs a large part of the drive power needed at the tracks.

Straddling.   It is risky to work a shovel that is straddling a stump or boulder, unless the ground is known to be entirely firm. The rhythm of load, swing, and rebound causes the tracks to work their way down into soft earth, particularly at the load end. If there is a thin layer of sod or hard soil overlying unstable material, the top will give an appearance of solidity, but if it breaks through, will let the machine sink rapidly. Ordinarily, slight sinking can be ignored or checked by poles under the tracks, but if it causes the shovel underparts to rest on a stump, it is a more ticklish situation.

The underparts of a shovel, between the dead axles, are more vulnerable to damage than any part under a tractor. In mud there is often nothing to do but to get out in any way possible and hope for the best; but if possible, any projection reaching up into this vulnerable area should be carefully watched during rescue work.

The power shovel may be freed from an obstruction by anchoring the bucket and hoisting, or by walking it up on logs, or by using a plank bolted to the tracks, or even by using a long crosscut saw to cut off a stump while still under it.

Counterbalancing.　If a shovel sinks so far on one side that it is in danger of overturning, the boom should be swung to the high side and lowered as much as practicable, the bucket extended, and weighted or anchored, then raised to serve as a counterbalance. This should prevent further tipping and may make it possible to shore it up.

Overturning.　If a shovel is overturned or so steeply tipped that it cannot be operated, the swing lock should be set, a line should be attached to the top of the A-frame and to any available anchor on the high side, and drawn taut to prevent further settling. If this line can be attached to a power source of sufficient strength, the shovel may be pulled upright, but this should be done slowly, with careful attention to any tendency toward bending the A-frame or other parts.

If no adequate power is available, a platform must be laid or constructed on the ground, and the shovel raised with jacks and blocks. The best and cheapest blocks are old railroad ties, but any sort of beams or heavy planking may be used. The actual raising of a shovel is so intricate, so dependent on the position and construction of the machine, and so liable to result in severe damage if done improperly that it cannot profitably be described here. The best method is to hire people who specialize in rigging and machine moving, work with them, and remember their methods.

If a shovel starts to tip during a swing, the bucket and load should be dropped, if possible.

Counterweight.　If a shovel is heavily counterweighted to carry a long boom, and the boom is removed to install another attachment, the shovel has a tendency to turn over backward. Removing counterweight is usually too much work to be practical. The shovel may be walked slowly to its other attachment, with a high heavy sawhorse or some other support dragged or pushed along under its tail as shown in Fig. 3.41. Or a line may be rigged from its A-frame to a tractor or heavy truck moving ahead, and kept taut, a device which can be used not only to steady it but also to pull it upright if it does "sit down." Another system is to bring the attachment to the shovel.

FIGURE 3.41　A basic shovel is tail-heavy.

HYDRAULIC BACKHOES

Hydraulic backhoes sink in mud just as readily as other excavators, but they are much better able to get themselves out.

If any solid anchor is within reach, it can be gripped with the bucket teeth, the bucket crowded in to pull the shovel toward the anchor, and down pressure used to raise the near ends of the tracks as they move.

If there is no anchor, or ground firm enough to act as one, poles or other supports may be placed close to the tracks (or wheels). The back of the bucket is placed on them, and down pressure exerted to raise the machine high enough to put supports under it.

The bucket may also be placed against a bank, the ground, or artificial helps, and forced outward to move the shovel back.

If the carrier is a tractor or a truck, the working end of the machine can be moved sideward by forcing the bucket down to raise it, and then swinging. This may also be done with a crawler if the end opposite the bucket is not deeply buried.

FREEZING DOWN

Frozen Mud.   Mud may cause serious difficulty when it freezes, as it sets like concrete and anything in it is likely to stay there. Crawler machines are particularly vulnerable, for if they are muddy, tracks, track wheels, and chassis are liable to be combined into one immovable unit. If they are clean, they are still apt to freeze to the ground.

Precautions to be taken are to clean all mud from the tracks and wheels at the end of the shift, and to park the machine on rock, metal, or wood, with as much of the track length as possible in the air. Mud should be at least scraped off with a trowel, screwdriver, or stick, but it is safer to wash it with a hose also.

Frozen mud is hard, heat-resistant, and tough, particularly when very cold. However, the bond between mud and other materials may be a film of brittle ice, which will resist pull or twist but shatter at a quick blow.

Breaking Loose.   If the tracks are frozen down but the track wheels are free, the machine can often be broken loose by rocking it forward and backward with clutch and gears. Jerky clutch action is much more effective than smooth, but also much harder on the machine. Moving forward tends to loosen the rear pad; backward, the front. The hardest part of the job is getting the first crack in the frost; after that the additional movement makes the breaking away progressively easier. It is sometimes advantageous to apply power to one track at a time, using the steering clutch to disconnect the other.

If there is frozen mud holding the wheels, it is unlikely that the machine can break loose with its own power, and the mud must be chipped or melted away. Chipping is very tedious and often impossible because of lack of space to work. Heat is usually more practical, but if the weather is extremely cold, very large amounts are required.

Melting Out.   A blowtorch—the larger the better—is the usual tool for this work. The flame is moved back and forth over the mud surface, and the thawed material scraped off. The wheels, tracks, and other metal parts will conduct heat to the mud effectively, so that chunks can be undermined and broken off; but care must be taken not to heat the metal enough to take out its temper. If the temperature is very low, a small section should be worked at a time; if it is near or above freezing, the flame may be moved back and forth over the whole area.

Hot water is effective if available in large quantities, as it will wash off the mud particles as they are thawed, exposing new surfaces to the heat. However, it may freeze on other sections of the machine while draining off. A portable heater may be used to blow a current of warm air on it, preferably under a tarpaulin. A flexible tube may be used to conduct hot engine exhaust.

More drastic measures are to erect a tent, or tarpaulin shelter, over the machine and thaw it with a stove; or to drag and push it with locked tracks into a heated building. But often the most practical method is to leave it until a thaw occurs, clean it off, and make good resolutions.

Wheeled Equipment.   Rubber tires on vehicles are not as apt to freeze down or together, as the flexibility of the tires prevents the ice from holding them effectively, and the rotating parts are ordinarily not as close to the fixed parts. If such freezing does occur, the same means may be used to thaw it, except that the rubber must be protected from heat.

The parts most vulnerable to freezing are the brakes, and here water will do as badly as mud. If the vehicle is used in very wet or slushy conditions, then allowed to stand in freezing temperatures, a film of ice will bind the brake linings to the drums. Driving back and forth on a dry stretch of road, even if only a few yards long, with the brakes applied, for a few minutes before parking, will usually dry them out again enough to prevent this trouble.

Frozen brakes may be loosened by rocking the car back and forth with its own power, or can be thawed with hot water or flame.

JUMPING THE TRACK

A weakness of tracked vehicles is the possibility of going off the track. When this happens, it may mean that the track is in a heap alongside the machine, the track wheels are resting on the ground; but more often it means that the track rails are not engaging the wheel flanges properly, being displaced to either side, and contacting the inner surface of the track shoe. Operation in this condition soon leads to the complete separation of machine and track.

Jumping or running off the track occurs most often during sharp turns on uneven ground, and is likely to indicate that tracks are too loose or have a broken link, or track and wheel flanges are worn, or that wheels are out of line. It is usually accompanied by a snapping or grinding noise, and if suspected, the machine should be stopped immediately and inspected, as every inch of movement makes it harder to get back on the track.

If the track is off the truck rollers only, it will usually swing back into place if the track is raised off the ground, by running the bull wheel or idler onto a log or other lift, or by jacking. If the bull wheel is on hard ground, forcing down a hydraulic dozer blade will often raise the track sufficiently.

The principle involved in getting a track back on a bull wheel or idler is similar to that of installing a tight fan belt on a car. It cannot be pulled or pried enough to stretch it over the flange of the pulley; but if it is held in one end of its place in the pulley, and the pulley turned, the wedging action will stretch the belt over the flange, and the part of the belt already in the groove will draw the rest of it in.

Similarly, with the track it is a problem of getting part of the track in line with the flange, and turning the wheel to draw it on. If the track is partially on the wheel, simply turning it in the proper direction does the job. If it is off the flange completely, it may have to be pried into line with a crowbar, jack, or chain, and the track adjustment will usually have to be loosened as well.

If the track is off the upper part of the bull wheel, but still engaged with even one tooth at the bottom, and with the truck rollers, the machine should be moved forward slowly. The sprocket teeth will mesh properly with track appearing from beneath the rear truck roller, and will carry the wrongly meshed section overhead into the slack upper section, where it will be straightened out by the support roller. If the upper part of the bull wheel is correctly engaged, and the lower section off, the track will work into place if the machine is reversed. However, it might be necessary to pry with a crowbar to prevent the track from jumping off the rear roller.

If the track is entirely off the bull wheel flange, but still meshed with the truck rollers, the machine should be moved forward. The bull wheel will roll onto track held correctly by the truck rollers, and, perhaps with the aid of vigorous prying, should mesh with it and pass this correct meshing up around itself. If the track is also off one or more truck rollers, the tractor should be backed so that the bull teeth can mesh with track held correctly by the support rollers. This may require more prying, or a pull in the correct side direction from another power source, or if the disabled machine is a shovel, from a line to its boom.

Should the track be off at the idler, the above methods are still good, with directions of travel reversed. If the track is tight, it may be loosened to facilitate crossing the wheel flanges. If the track is off the support roller, it usually can be replaced by lifting with a crowbar.

If the track is off the bull wheel, truck rollers, and idler but is still on the support rollers, the whole track and wheel assembly should be raised off the ground, and the bull wheel rotated backward. The track can be engaged with this at the top by use of a crowbar, and will be pulled into engagement with the rest of the bull wheel, then the truck rollers and finally the idler. Caution should be used to prevent the track from coming off completely in one place while being meshed in another.

If none of these stratagems work, or if the track is entirely off the wheels, the track adjustment should be loosened and the track "broken" by removing a lock and driving out one of the hinge pins, using the bull wheel as a brace. On most makes, there is only one master pin which can be used. The side of the machine should be raised off the ground, the track placed correctly under the wheels, the machine lowered, and the track wrapped around the bull wheel and idler. The ends should be pulled or forced together at the bull wheel or between it and the top of the idler, by chain tighteners, block and fall, winch, jacks, elbow grease, or any other means available, and the pin inserted and locked. This is a difficult and laborious job, even on a small machine, except for experienced personnel; and on a large one it may require the use of other machinery to handle the track.

It is sometimes easier to replace a track, after opening it, by walking the machine off it onto a plank or beam, aligning the track behind it, and walking it back, than to shift the track around under the machine.

This type of work has been rendered much more difficult by changes in track design and fabrication. A single master pin may be difficult to identify and to get at. When found, it may be so tight that it cannot be driven out until the links are heated with a torch, and the links may be bent by the heavy hammering, during either removal or replacement.

It is common practice among service workers to cut a pair of links with a torch; then take the whole track to a press to be reassembled. A minor incident may be turned into an expensive and time-consuming project in this way.

Split master links make this drastic step unnecessary. They permit opening a track by removing a shoe, then taking out four bolts. Opening the track and closing it are greatly simplified.

CHAPTER 4
BASEMENTS

Excavations for basements may be roughly classified as the dig-and-pile and the dig-and-haulaway types, which will be referred to in this discussion as residential and commercial, respectively, as the larger part of them fall into these categories.

Backfilling around the foundation may be a problem. It must be emphasized here that a wet fill can crumple in a foundation and even move footings, and that this danger is especially severe if the masonry has not had time to cure.

PRELIMINARY WORK

Tree Protection.   In residential excavation, any clearing that may be required is likely to be of the selective type. Large trees, or trees of desirable species, may determine the location of the house, and they must be guarded against damage and burial. It may be advisable to wrap the trunks of such trees in cloth, and protect them by a collar of vertical boards, as in Fig. 4.1. If their bases may be temporarily buried, the original ground line should be marked on the bark with paint, so that the fill may be removed accurately, and burial or overcutting avoided.

Topsoil.   Topsoil is usually present, and if it will be needed for landscaping, it should be saved. This involves taking it off the area to be excavated, and preferably from the areas where spoil is to be piled. This stripping may be a substantial part of the cost of the excavation but is required in most localities.

In places where there is no well-defined topsoil, or the topsoil makes up one-third or more of the spoil and the subsoil will mix well with it, stripping may not be needed.

A method of stripping topsoil which is often most economical in the long run is to remove it completely from all areas to be involved in the digging. Figure 4.2(*A*) shows one method of doing this by placing it in compact piles well away from the two corners of the proposed house. This will usually keep it out of the way of digging, piling, and ditching, and will leave it in a position for straight-line spreading. But it should not be put in corners where a dozer will be unable to get behind it.

Figure 4.2(*B*) shows a more usual method. The digging and piling area is stripped, the topsoil being pushed into two piles, so placed as to be just beyond the spaces for the piles of spoil. If too small an allowance is made for the spoil, the topsoil may have to be moved back farther, or may get buried by the fill and partly lost. In any case it will probably interfere with backfilling.

In (*C*) the topsoil is pushed to the sides and fill piled to the front and rear.

Topsoil stripping is discussed in Chap. 10. It is customarily done by a dozer, which does the cleanest work. Backhoe excavators, and to a lesser extent other shovel rigs, will remove topsoil rapidly, and if the soil is heavy and wet, may be preferred because they do not compact it and cause it to cake. If large areas are involved, scrapers may be used.

If the topsoil is thin and will be required for finishing off, it may be deliberately mixed with some fill while stripping to increase its bulk. If the subsoil has a loose texture, little or no harm will be done to fertility, and regrading will be simplified.

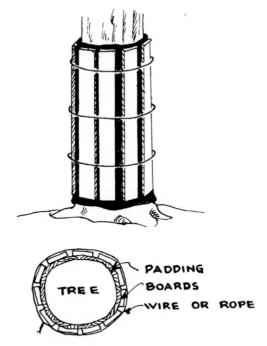

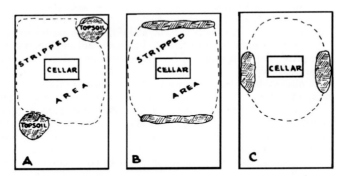

FIGURE 4.1   Protection for tree trunk.

FIGURE 4.2   Topsoil piling.

Artificial Obstacles.   Serious digging difficulties may be caused by obstructions placed by the builder. Batter boards are used for reference in exact placement of foundation walls. They will increase the excavation cost by restricting the movement of machinery, particularly if the house is to be irregular in shape. Piles of building material may also be much in the way.

Corner stakes, with back reference points such as are described in Chap. 2, are adequate location marks for the excavator, and other preparations for building should be postponed until digging is complete.

Depth.   The depth of the excavation depends on the first-floor height relative to original grade, and the depth of substructure below it. Substructure might include floor thickness, rafters, sills, cellar headroom, and thickness of the basement floor and gravel or crusted rock under it. If the full area is to be dug to the bottom of the footings, their depth must be considered also.

On sloping ground, excavation depth will vary at different points.

Factors in determining first-floor level, including a way to probe for rock, are discussed in Chap. 7.

Rock.   Digging in ledge rock or large boulders may increase excavation cost three to five times, or more. Its presence may result in raising the house, or substituting a slab or crawlway for a basement.

If it is to be removed, dirt removal should be completed and the rock surface cleaned before drilling.

Most basement jobs are so located that mats must be used to cover blasts, and particular care taken to safeguard passersby and traffic, and to avoid damage to property.

DIGGING

The three standard machines for digging small basements are the front-end loader, the bulldozer, and the backhoe excavator.

Bulldozer.   A bulldozer can dig a basement very deeply if there is no heavy rock or mud, but it is at its best if the excavation is shallow. This is because it must dig away a considerable amount of the bank to ramp itself in and out, and the whole weight of the machine must come up out of the hole with every pass. It can push much larger loads on a level or a moderate upgrade than on the steep rises from a deep basement.

Digging techniques vary with the operator and the locality. Several methods will be described, but they are presented only as samples and should be followed only where they give satisfactory results.

It is good practice to leave a ramp that will allow trucks to back into the floor of the excavation, for convenience in delivering foundation materials to the point of use.

Example.   The first case to be considered is that of a basement 30 × 40 feet, 4 feet deep, dug in a large, level, treeless lot, from which topsoil has been stripped. The dimensions given are of the outside of the foundation walls, and an additional 18 inches on each side should be allowed for wide footings and working space for the masons, so that the dimensions of the hole should be 33 × 43 feet. Stakes are set 6 inches outside the digging line, as in Fig. 4.3(A) to avoid accidents to them. Any temporary guide pegs the operator needs are set along the edge, or just outside it.

OPEN-FRONT METHOD

Top Layers.   The bulldozer is first worked along the short dimension, inside the stake lines. The blade may be dropped at the front or south line for a fairly deep bite, and when filled, it is lifted to ride the load over the undug ground until the north line is crossed. The dirt may be dropped at the line or pushed a few feet back. The dozer then backs to the south line and takes another bite in a strip adjoining or overlapping the first. It may work the whole width of the front line, as in (A), or only a section of it, before digging the area over which the soil has been pushed. The back edge of the cut is worked north by successive bites until the rear line of the excavation is reached, approximately as shown in the cross section (B).

After completing the removal of the top layer, the dozer may cut out and pile a second layer in the same manner, as in (C). This deeper cut will not extend to either digging line, as the slope down and the slope up are kept inside these lines.

If the soil is so hard that the blade cannot be filled in a short distance, the dozer may be worked over a short digging area only, for several passes, after which the pile of loosened dirt may be

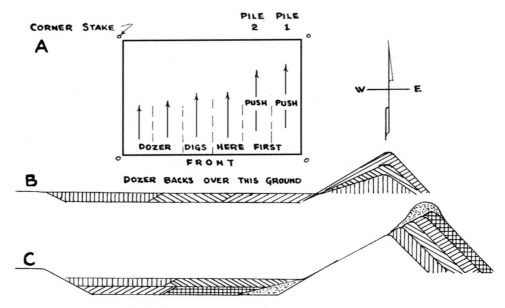

FIGURE 4.3   Basement digging sequence.

pushed to the spoil pile. If a ripper or subsoil plow is available, it might be profitably used to break up layers of dirt in advance of digging.

Ramping Down.   A hydraulic dozer can cut hard soils, but will usually not cut a steep ramp down without special handling. Figure 4.4(*A*) shows a dozer cutting down from a line in soft soil. If digging is good, the blade will penetrate rapidly until maximum depth is reached. It will make a level cut until the machine's center of gravity moves over the cut. It will then fall forward, and the blade will resume cutting until full penetration is reached, when it will level off again. Such a series of steps makes an unsatisfactory ramp.

One way to make a smooth ramp, shown in (*B*), is to start it well back from the edge with a gradual curve that is made steeper at the digging line. Cutting is then regulated so that the full depth of penetration will be reached as the center of gravity crosses the steepened part of the curve. Several passes are made in digging this ramp as indicated. This curve may be made steadily steeper as depth increases, because the tractor itself is at a downward slope.

If the cut cannot be made far enough back into the bank for ramping in, the procedure shown in (*C*) may be followed. Dirt is pushed out of the excavation into a steep pile, the dozer is backed up on this, and is thus pitched down a steep enough angle to cut down sharply.

Piling.   The area to be occupied by the pile should be calculated or guessed at, and the first piles placed at its far side. Successive loads may be dropped toward the excavation, then a load carried over these, pushing the tops off some of the heaps, and dropped at the back. The pile is thus built up in a series of wedges, with their thin ends toward the excavation, as in Fig. 4.5(*A*).

Or the first load may be dropped at the near side of the intended pile, and the next pushed through it with a raised blade so that the approximate upslope is established from the beginning, and successive loads dumped off the back as in the (*B*) series. In either case the pile may show a tendency to build toward the hole, and may have to be dug into severely to cut it back to proper distance.

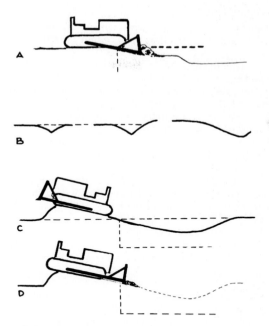

FIGURE 4.4   Cutting down from edge.

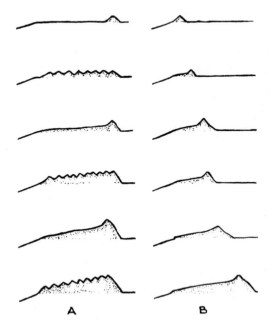

FIGURE 4.5   Piling soil.

The upslope of the pile should be made according to the power and traction of the dozer, the judgment of the operator, and may be between 1 on 5 and 1 on $2\frac{1}{2}$. The easier gradients make it possible to push larger loads, but they must be moved farther and require more ground space.

Lower Layers.   After the digging has reached the half depth, the dozer should be moved out onto the west bank and headed east along the south line of the excavation. An entrance ramp should be started several feet back of the west line, as in Fig. 4.6(A), and a steep ramp cut down toward the 2-foot level. As the blade fills, the machine is swung toward the center of the hole, and the dirt is left on its floor. The dozer returns to the south line, makes another cut, again swinging the spoil out into the center, pushing it somewhat farther. It thus cuts out the full-width ramp by which the dozer entered the pit from the south, and occasionally shifts to pushing the dirt obtained from the ramp up on the north pile. As the east edge is approached, dirt may be pushed up on its bank, cutting a ramp, instead of to the north.

The north slope may then be cut away in the same manner, the spoil first being pushed up the undestroyed section of the ramp, as in (B), and finally onto the west bank. These cuts result in vertical walls along the two long sides of the hole.

The inside ramps may also be removed by oblique passes from near the center (C), steering the dozer so that the blade is parallel with the bank by the time it reaches it. The slope ahead of the dozer can be gouged away in this manner, with the spoil being edged out into the open and pushed up the west bank. With the west section of ramp removed, the dozer can be turned to cut away the east part as in (D).

Finishing.   After removal of the north and south slopes, the floor of the cut may be deepened by pushing to east and west. These ramps up to original grade should have their lower ends at the excavation line, and will therefore be cut back deeply into the bank, as in Fig. 4.7(A). Since a steep ramp means less extra excavation, the slopes should be made as steep as practical. They may be cut all the way through to begin with, for an easy gradient, and steepened as the hole deepens. The

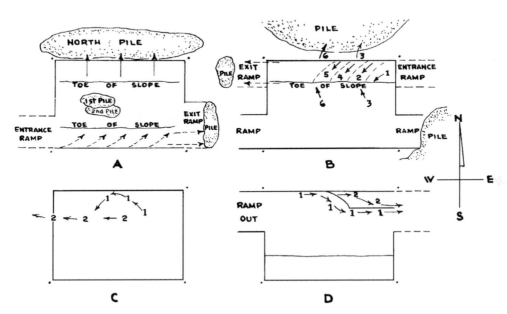

FIGURE 4.6   Cutting lower layers.

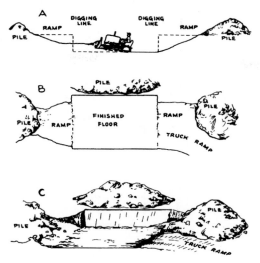

FIGURE 4.7   Finishing excavation.

amount of dirt removed for ramps can be slightly reduced by narrowing them as they go up, as in (*B*). The ramps can be partly filled by the last loads pushed, as in (*C*), except where space is left for supply trucks.

As the bottom of the hole approaches final grade, it should be checked. A 4-foot stick, or a rule, may be used to measure down from the edges, if they are well enough preserved, or from a string stretched diagonally between two corner stakes. The mason contractor generally expects to do some hand leveling and trimming, but will be pleased to find it unnecessary.

It is not practical to dig narrow footwall trenches below the pit level with a dozer.

Results.   The procedure outlined above should produce an excavation such as shown in Fig. 4.7(*C*) with two straight walls correctly spaced. Ramping out of the short sides reduces the amount of extra digging that is one of the drawbacks of dozer work. The whole front is left free for access and for piling building material.

However, the spoil may not be properly placed for backfill and grading. In such a location, a house would usually have the fill spread all around it, with particular attention to building up the front. Here, fill for the front yard would have to be obtained from the sides, which, in turn, might need to be partly replenished from the back. This involves extra pushing, and the presence of a few trees might make distributing it a major problem.

Time.   This basement involves about 220 yards of excavation. It might take an 80-horsepower bulldozer (a good size for the job) from 1 to 3 hours to strip the topsoil, and 2 to 6 hours to do the digging.

Rock (even if not removed), water, irregularities, or obstacles will increase the cost of the work.

OTHER DOZER WORK

Four-Pile Method.   Another pattern for digging this same basement is shown in Fig. 4.8. The soil is pushed in four directions. The east and west ends, for a distance of perhaps 8 feet from the

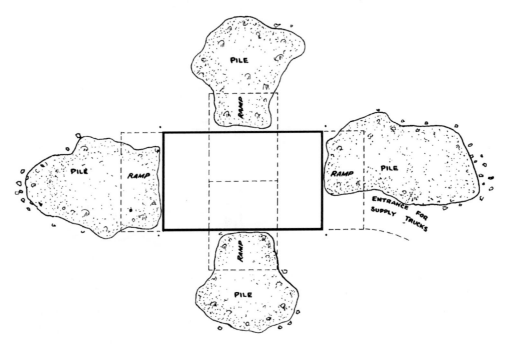

FIGURE 4.8 Four-pile excavation.

line, are pushed up on the east and west piles, respectively, with ramps cut out of the bank and partly refilled as in the previous example. The section between these two cuts is pushed onto piles to the north and south, with ramps cut out of the bank beyond the digging line, each pile being supplied mainly from soil on its side of the center.

By this plan, the whole surface of the excavation can be worked down as a unit, the dozer always pushing dirt to the nearest pile. The four directions of push may be taken in rotation, or varied according to the operator's inclination.

Advantages of this method are efficiency, in that pushing distance is kept to a minimum, and adaptability to grading plans. The four spoil piles are shown to be about equal, but their relative sizes may be changed without varying the method. Access and space for material are not as good as with the open-front method. A larger-amount of dirt is dug out for ramps and must be pushed back later.

Grading. If the topsoil has been pushed well back, it is possible to rough-grade the fill at the same time that it is pushed out of the basement, or immediately afterward. This has the advantage of making all four sides accessible to the builders.

Ordinarily, it is easier to spread several small piles than one big one. For this reason, it is wise to stop excavating occasionally and to spread the piles which have been built up.

The grade should be kept at least 1 foot above its final level, to provide excess soil for back-filling around the foundation.

If the digging is wet, the spoil may be too sloppy to spread immediately.

Irregular Basement. Figure 4.9 shows a basement of irregular shape. This may be dug by the open-front method by cutting back the jog as indicated in (*A*), and by pushing dirt for the small south room into the main basement, to be piled on the east side.

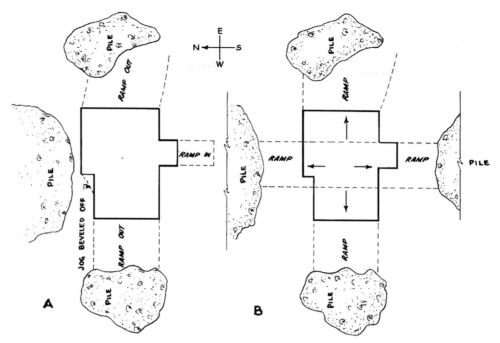

FIGURE 4.9   Irregular basement.

Figure 4.9(*B*) shows the same basement pushed up into four piles. The jogs are handled by simply moving the ramps and piles outward the extra distance.

Limited Access.   Figure 4.10 shows a difficult situation where trees or other buildings permit entrance by the dozer at only three points, and where all the spoil must be pushed out at one of these spots. A detailed description of the work would be lengthy and tedious, and it is hoped that the successive diagrams are sufficiently clear. The dozer movements indicated by the arrows would have to be repeated on each cutting level.

In soft soils it might be possible to cut the corners with the dozer by pushing the dirt to loosen it, then backdragging. However, it may be cheaper to dig them by hand and to throw the dirt out into the dozer path.

This type of excavation takes several times as long as open digging.

Figure 4.11 shows a succession of steps in excavation of a basement in a hillside. This floor base is at grade on the east (right) and requires a cut of 8 feet along the west line. The slope of this hill is about 1 on 3, which a dozer can negotiate going up or down, but cannot safely work sideward, unless it is a wide model.

The easiest way to dig would be to push straight down the hill, but the yardage removed to ramp down would be about two-thirds as much as for the basement itself; and the spoil would be left in such a position that it could not be conveniently used to backfill the ramp.

Dirt may be pushed to the side as well as the bottom by pushing down from the center of the upper line of the excavation to build a step which would be level or tipped oppositely to the hill slope. The dozer may then be operated on this step, pushing to the north and south, along the west line. The step will broaden as it is deepened, and will allow the dozer space to push downhill, first diagonally and then directly, in addition to the back cut parallel to the rear wall. By leaving the two rear corners for hand cutting, the excavation can be completed with small ramp cuts. If the dozer

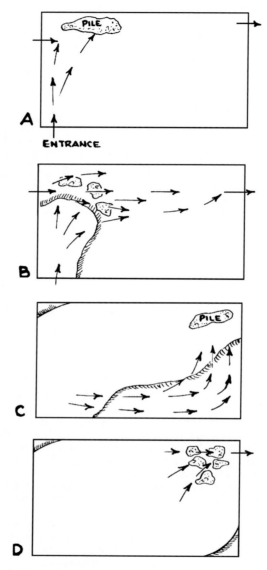

FIGURE 4.10   One-exit digging.

must do the whole job, the back wall cuts can be continued into slot ramps, which should be at least a foot or two wider than the blade to avoid getting the machine jammed against boulders or in slides.

Landslides on Hillsides.   Excavating on a hillside needs to be done with care, because of the potential for a landslide. This could be due to an existing weakness in the soil condition, which might be detected by a gully with an alluvial fan at the end or a scarp indicating a previous landslide in the area. Soil creep could be occurring, but the rate of movement is in fractions of an inch per

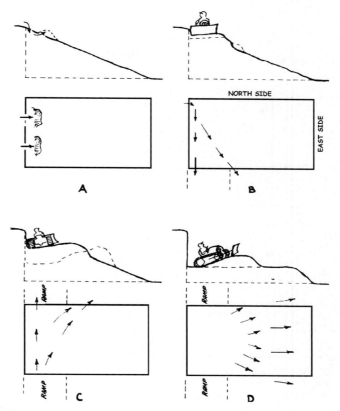

FIGURE 4.11 Sidehill basement.

year, so it is hard to detect. Creep is caused by gravity with seasonal freezing and thawing or alternating shrinking and swelling of clay minerals in the soil.

Building a house with a basement dug into the ground may actually unload a slope and make it safer from a landslide, provided that the soil dug out is hauled away and not left to overload an area where it might be piled. That leaves a decision for the owner or excavator who might need some of the dug material for backfilling. Is it cheaper to take the needed precaution of hauling away material that may be used later for backfill? Or is it worth taking a chance that no slide will occur and do the necessary cleanup if one occurs? The latter choice might be in order for a location in the country, whereas in a built-up urban area where other buildings might be in jeopardy, taking the precaution would be in order.

FRONT-END LOADER

Digging.   A front-end loader, with a standard full-width hydraulic dump bucket, may dig an open type of basement in the same manner as a bulldozer. It should be expected to do a quicker job than a dozer of the same power, because of better penetration in hard soil and ability to push larger loads.

The ability of the front-end loader to cut straight ahead, then back and turn with the load, makes it possible to reduce the amount of excavation outside the digging lines for ramps. If the

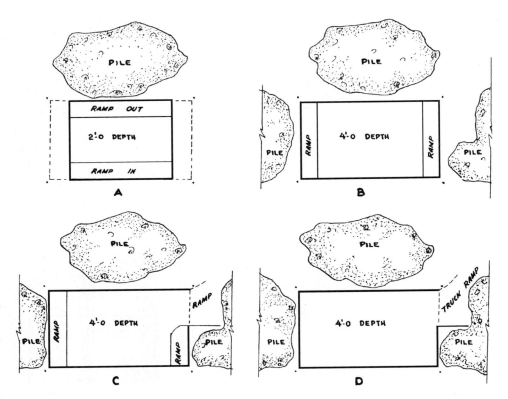

FIGURE 4.12 Three-pile system.

basement is dug by the three-pile or open-front method, the procedure shown in Fig. 4.12 may be followed. The first layer is not cut quite to the ends of the excavation, (*A*), and the ramps built in reaching the bottom are inside the digging lines, (*B*). They are cut back to the steepest practicable grades on the last pushes, and then one ramp, a foot or so wider than the machine, is cut into the bank. All soil left inside the digging line is then removed by being picked up and carried or pushed out the slot ramp.

The machine can carry a larger load up the ramp when moving forward than when reversing. This is because it is nose-heavy when loaded, a condition which is made worse in backing up an incline, both by the shift of the center of gravity toward the front and by the reaction from the driving torque which pushes the front down. In ascending a grade forward, both of these forces tend to raise the front and improve stability. These effects become more pronounced as the incline becomes steeper.

Turning around in a small excavation may be difficult, or impossible, so that it is often better to back out with a small load than to take the time to turn in order to carry a larger one.

Cutting Walls.   So far as possible, the excavation should be finished with vertical walls. Long ones may be cut neatly by digging parallel to the edge, but because of limited space, much finishing must be done by working straight toward the digging line.

Until the line is reached, the bank digging procedures in Chaps. 10 and 16 may be followed.

A vertical face may be cut by allowing bank resistance to move the tractor back as the bucket rises. This may be managed by slipping the clutch with mechanical drive, or slowing the engine that has a torque converter. If the wall is high, the machine must be moved forward again after the bucket lip gets above the push arm hinges.

This back-and-forth motion compensates for the curved path of the rising bucket. The same effect may be achieved by careful regulation of the bucket tilt, as it makes a shallower cut when rolled back than when flat.

Any jogs or irregularities can be cut by digging into the wall from the excavation without making additional ramps. It is easiest to take these out in layers as the floor of the hole is worked down, but they can also be dug after completion of the main work.

Soil carried out of the hole may be spread or distributed to nearby low spots much more easily than by a dozer, and can also be readily loaded into trucks.

Crawler loaders are usually preferred for basement work, but it is done by four-wheel-drive machines also.

Digging under Buildings. A special feature of most small front-end loaders is their ability to dig basements under existing buildings. Particular attention must be paid to bracing the structure over the hole through which entrance is effected, and across any interior pillars which are to be moved. Digging should be done cautiously to avoid damaging the building through collision with rocks, beams, or soil pipes in direct contact with it.

If the building is not large enough that the machine can dig a turning place inside it, it will be necessary to use hand labor to cut back to some of the walls.

Ventilation is very important, and usually requires at least the use of a powerful electric fan to keep the air reasonably free of exhaust fumes.

A machine that is to do much work indoors must be diesel, with a scrubber (fume reducer) on the exhaust.

Hydraulic shovels equipped with telescopic booms can dig under a building from outside, either directly or by removing hand-dug dirt.

BACKHOE EXCAVATOR

The backhoe is a principal competitor for small basement work. There are two types: full-revolving hydraulic or tractor-mounted part-swing hydraulic. Tractor mountings get into difficulties because of restricted swing, and often limited reach. But they can dig a basement efficiently.

Any hoe is capable of shallow digging, but compares most favorably with the dozers when the hole is to be over 6 feet deep, or when unfavorable bottom conditions, such as water, mud, boulders, or ledge, are encountered. It is able to take care of any necessary ditching without change of attachments.

It is recommended that the digging lines be set a few inches outside of the required excavation, although in even-textured soil the backhoe can do a very exact job. In addition to the corner stakes, intermediate guide pegs should be set at short intervals along the digging lines, as the operator cannot sight along these lines without getting down from the shovel, and the finished wall is established with the first cut.

Figure 4.13 shows the 20 by 30 basement with the depth increased to 8 feet.

Lining Up. Accurate lining up of the machine is essential for a clean job. If the cut is to begin along the south line, the shovel is placed as in (A), with the bucket about three-quarters extended and resting a few inches beyond the southwest corner. The boom and the tracks are parallel with the south digging line. Lining it up in this manner is greatly simplified by marking the width of the bucket (including side cutters, if used), centered, on the bottom of both dead axles, with paint, or better yet, with stubs of welding rod. Sighting across the outside pair of these marks from the rear, the outer edge of the bucket should be exactly in line with them, and all three points should be on the digging line.

Digging. A ditch is now dug to bottom grade with its left edge on the digging line, and the spoil is dumped to the south. The far end of the ditch will curve rather sharply inward. When the backhoe

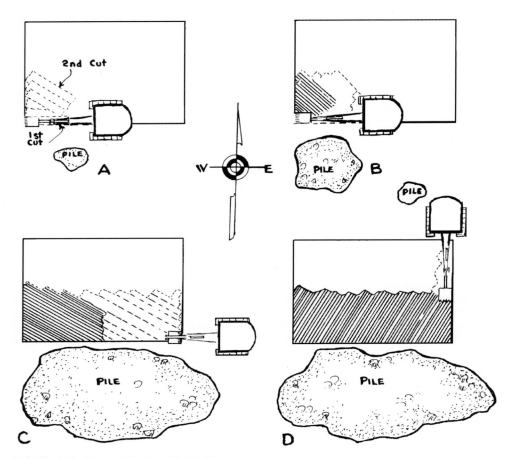

FIGURE 4.13 Basement digging with a backhoe.

has dug as much of the south wall as it can from its position, it reaches to the center of the west line and digs a trench back from there. The triangle included in these ditches is dug in layers to bottom grade, which may be found by a measuring stick, and as near to straight down from the west line as possible.

The shovel is then backed up a few feet to position (B). It can now cut the west end of the ditch to be almost vertical, because of the more extended position of the bucket. The south wall ditch is then extended as near the shovel as possible, and material between it and the center cut down in layers to the bottom. The center-line will be irregular.

The spoil pile will tend to build up too sharply at the edge of the hole unless pushed back. This pushing may be done by regulating the outward swing of the bucket during the dump, so that it strikes the pile at a spot where its momentum will push a considerable quantity of dirt outward, without stopping its own motion. Knocking dirt back should be started early, before the pile gets high. The hoist clutch must be engaged during this operation. The quantity of dirt that can be put in a pile is greatly increased in this way.

Digging is continued in the same manner, with careful attention to a clean, level bottom until the east end is reached. The backhoe can probably cut this to a nearly vertical wall immediately in front of it, but will leave a ragged edge, as (C) farther north. The backhoe is then turned and walked into the unexcavated north section. When its center is a half-bucket width inside the east line, as in (*D*), it stops and shaves the end of the excavation, then trenches to the north edge. It next straddles the north line, and is lined up in the same manner as before, with the bucket resting in the hole in the northeast corner.

The north section is excavated in the same manner as described for the south, and completes the excavation. The west edge may be cleaned up, if necessary, by turning the backhoe to walk parallel to the edge, so that the bucket can dig straight up. The backhoe should not be put in this position, however, unless the soil is firm and is known to have good load-bearing qualities, as a crawler machine is vulnerable to cave-ins or slumping under one track.

This edge may also be trimmed from the north and south banks.

The completed excavation and spoil piles are shown in Fig. 4.14. It will be noticed that the piles are somewhat offset from the hole, so that the south pile can easily be used for fill on the east end, and the north pile on the west end. Both ends are left open for access and storage.

The north cut could have been made in the same direction as the south cut, if the east ditch were shorter, but the fill would then have been concentrated toward the east end.

A more finished hole could be made by starting the digging with a ditch along the west edge, dug from the south. The spoil pile would largely block the access to that side, unless the soil were piled in the basement area for rehandling. If access were not important, this ditch could be widened toward the center, reducing the amount to be piled to south and north. Existence of such a ditch would make it necessary to work the north section toward the east.

Loading Trucks.   A backhoe can load the spoil in trucks instead of dumping it on the ground. Where the piles will be so large that they will have to be dragged back, a truck or trucks may be used to take part of it away, the backhoe continuing to dump on the piles when no truck is in loading position. If grading plans have been prepared requiring use of the fill away from the foundation, it may be cheaper to truck it than to push it later with a dozer. However, enough of a pile should be left by the hole for backfilling between the foundation and the edge of the excavation.

The backhoe can dig footing trenches below the floor level where it is working parallel to the edge, as along the south, east, and north walls in Fig. 4.13. The hydraulic backhoe can dig them anywhere, but the parallel position is easiest.

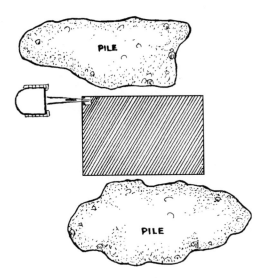

FIGURE 4.14   Finished excavation.

Checking Grade. Cutting the bottom to proper grade is more difficult with a backhoe than with a dozer, as the backhoe operator looks down at the grade rather than along it; has more difficulty climbing down to check it; and cannot move the machine back to grade over mistakes.

It is very helpful to the operator to have someone to check the work, although the operator can manage alone if necessary. The corner stakes, and preferably some other stakes, may be marked at a certain height, as 9 feet above the bottom. In a level field the marks would all be 1 foot above the ground; in a sloping one the highest stake should be marked a little above the ground, and the others with the aid of an instrument or a string level. A stick should be cut 9 feet long.

If the operator is checking the grades alone, he or she may fasten a taut string between two stakes so that it will go over the spot in question and measure the distance from the floor to the string with the stick. Spots which cannot be crossed by the string, or measured directly from the height of the wall, may be checked from a known spot by eye, or with a hand or carpenter's level. See Fig. 4.15(*A*).

If two people are doing the work, a string may be stretched between stakes on one side. Another string fastened to another stake across the excavation may be held in any desired position on the first string, as in (*B*), while the other person holds the stick.

Hand, transit, or laser levels may be used. A long rod is required when the instrument is set outside the hole.

A backhoe cannot cut a perfect floor to a pit, because of the projection of the teeth and the tooth bases. The smoothest grade is obtained when the bottom of the bucket is used for finishing, rather than the teeth.

Irregular Edge. Figure 4.16 shows a basement of the same irregular shape as that in Fig. 4.9. The principal considerations in doing complicated excavations with a backhoe are to avoid digging it into a trap; to avoid surrounding it or blocking it from other work by piles of spoil; and to work either parallel or at right angles to outside edges.

There are several ways in which this basement can be dug. The north side can be dug from the east end, as in (*A*) and (*B*). When the jog is reached it is finished off with a vertical cut, the backhoe backed away, and brought back in position to cut along the inner line. If the start is made at

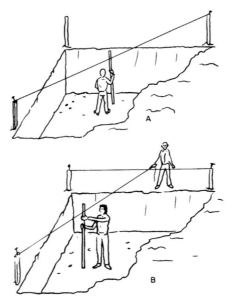

FIGURE 4.15 Checking bottom grade.

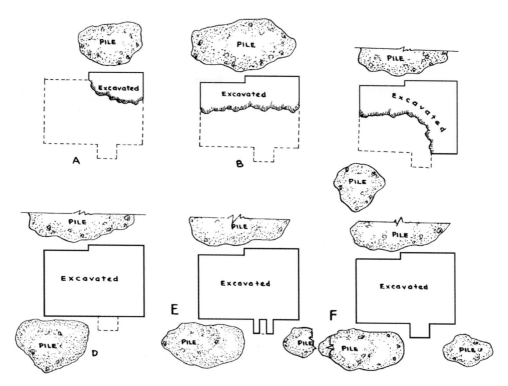

FIGURE 4.16   Cutting jogs with a backhoe.

the west, the cut is brought a little beyond the jog, and the position then shifted to dig along the outer line.

The machine may dig the south side by entering from the west and starting at the southeast corner. Excavation is carried back to the west line, first ditching the edge then digging out the center. Care is taken to begin the spoil pile well west of the south room. The backhoe is then moved off to the south and up to the room. It is first lined up to the west side of this room and makes a cut from the main excavation to the south line of the room. It is then moved so as to cut the east wall of the room, then digs out the rest of the room.

Another way to do this would be in the same manner as the north wall, treating the room as a double jog. However, this would involve extra digging because the bucket needs considerable width in which to cut down.

If the excavation site is a hillside, the work should be managed so that backhoe tracks will head up- or downhill, not across. If the grade is steep, the backhoe should dig from downhill to avoid the danger of being pulled into the hole if the bucket hooks into something solid.

If work must be done from the upper side, the stability of the ground should be checked, and both tracks must be securely blocked against sliding.

OTHER SHOVEL RIGS

Dipper or Front Shovel.   Front shovels (refer to Fig. 13.1) are seldom used in residential basement excavations, but they can do a good job. For satisfactory results, the ground should be firm at bottom grade, and the spoil should build into steep-sided piles.

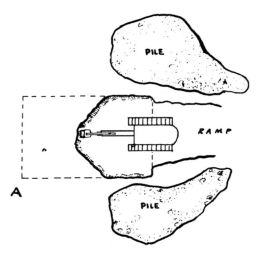

FIGURE 4.17 Large shovel in small basement.

A 30 × 40 basement, 6 feet deep, can be dug as shown in Fig. 4.17. A ramp must be dug outside the excavation line with a slope, usually $3\frac{1}{2}$ or $2\frac{1}{2}$ on 1, which the shovel can climb when the job is finished.

The spoil piles can be pushed back from the edge of the hole to some extent by crowding against them with a closed bucket.

The walls of the hole tend to slope in at the bottom, and to be somewhat jagged because of the different angles at which the bucket cuts them. Both of these features can be reduced or eliminated by careful digging, but extra time will be consumed.

Clamshell. Clamshells are not ordinarily used in this type of excavation because they do not move as many yards per hour as competitive types. However, they turn out as accurate a job as a backhoe, and for small, deep excavations, will do the work more cheaply than a bulldozer. Digging is done from the top so that no ramps are required.

An edge may be cut with the tracks parallel with it, and the tagline chains attached to one jaw, or with the tracks at right angles to the digging line, and the tagline on both jaws. Either of these arrangements will permit cutting straight-sided trenches along the outer lines. The center is best cut in layers, or in sections behind completed edges.

A medium- or heavy-duty bucket should be used. The spoil may be placed in isolated piles, in windrows, or in trucks as desired.

MULTIPLE BASEMENTS

In many residential developments, small houses are built close together in straight rows. Under such circumstances, it may be economical to dig a wide trench straight through the block, and backfill between the houses after the foundations are built.

In Fig. 4.18 the digging is done by a dragline or a large hydraulic backhoe or front shovel which piles the spoil on both sides. Materials for foundations are trucked and piled on the floor of the trench. After the foundations are up, the piles are bulldozed around and between them, and the surplus used to build up the grade throughout the area.

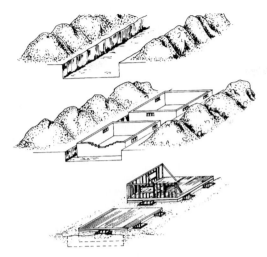

FIGURE 4.18   Multiple-basement cut.

Use of part of the spoil for raising the ground level decreases the depth of digging necessary.

Another method of line digging is to load most of the spoil into trucks, sell or dispose of part of it, and save only as much as is necessary for backfilling spaces between other houses in the same development.

A third method is to haul away all the spoil for use elsewhere, and obtain backfill by dozer cutting on the uphill side of the houses.

It is also possible to dig such a trench with dozers, working across it and piling soil on one or both sides. Or scrapers, working the long way of the cut, can haul to any desired location.

The first method requires at least partial backfilling as soon as the foundations are up, to provide access to building material. The weight of the fill and of the large dozers commonly used is likely to break in uncured foundations, particularly if they are not braced by first-floor beams.

A large dragline might be able to pile all the spoil on one side to allow access to the other.

The other methods give immediate access to the front and back, and allow space for piling materials. Backfilling can be postponed until the building is completed.

Relative cost will depend partly on the value of the spoil removed.

CONTRACTOR'S SETUP

A contractor setup to do basement excavations, one at a time, will probably have a combination of equipment including a backhoe, a small bulldozer, and possibly a dump truck for hauling the spoil material away. The backhoe does the digging, and the dozer moves the material that may be needed for backfill out of the way for the backhoe and then is available for backfilling, when that is to be done. If there is excess material to be hauled away, the fact that the contractor has a dump truck for that purpose means that the owner, or building contractor, does not have to get another subcontractor to do that part of the job. A contractor who does a lot of basement excavations, for either residential or commercial construction, will probably have a fleet of backhoes of various sizes for the variety of excavations to be done, small and large dozers, and various sizes of dump trucks. The fleet may not be large enough to warrant having repair facilities. Instead, the contractor can have extended warranties from the dealers of the equipment, so that they service the equipment as needed. Paying for the dealer's service department is probably going to be more economical than hiring a full-time mechanic and setting up one's own repair facility.

HAULAWAY DIGGING

Trucking Spoil.   In large commercial excavations of the haulaway type, one of the most important considerations is arranging for the disposal of the spoil, except when it is to be used as fill on the same project. It may be possible to sell it profitably, or it might be necessary to pay someone for the privilege of dumping it.

Disposal arrangements may not only determine the price to be charged for the digging, but also the time of starting the work, and the number and type of excavating and hauling units to be used.

The distance to the dump may be much farther for the trucks than for a car because of restrictions on trucking on residential streets. An inspection should be made of the dumping site, and any price quoted for fill should contain provision for additional charges for dumping delays.

Permits.   An excavation contract should specify that the owner, or general contractor, is responsible for obtaining all permits necessary for the work. If the job is stopped because of failure to have such permits in order, the excavator should have the privilege of charging for the tied-up equipment on an hourly basis.

Machinery.   In a medium to large excavation, a crawler front-end loader, a backhoe, or a front shovel unit is usually preferred, unless the bottom is too wet or sandy for trucks. The backhoe needs less space, does not tear up the floor, is better in rock, and can tolerate rough or steep footing.

Increasing size and decreasing depth favor the loader. It can replace the shovel on most jobs.

A hydraulic backhoe of sufficient size does the neatest work. If the bottom is bad, it can load on top, but production is greatest with haulers below it.

Ramps.   In most cases, the power backhoe or front shovel cuts a ramp down, inside the digging lines, which must be of such grade and material that loaded trucks can climb it. The grade may be between 1 on 5 and 1 on 12, depending on the power of the trucks and the loads placed on them. The slope is made as gentle as the length and depth of the hole permit, for larger loads and fewer breakdowns.

If the plot is sloped, the ramp should be cut in from the lowest point on the edge to which trucks have access.

Earth ramps are generally removed immediately upon completion of the excavation they serviced. It is usually necessary to bring in a backhoe or a clamshell for this job, as the yardage is too great for hand labor.

Timber ramps afford less tractive resistance and better footing than earth, and so can be built with a steeper slope. They can be left in place during construction of the foundation for convenience in moving building material. However, timber work is so expensive that these ramps are largely limited to use in excavations that are very deep in proportion to their size.

Pit Floors.   The shovel may dig the floor of the pit exactly to grade, or may dig portions of it below grade, to allow room for disposal of spoil from hand-dug trenches or removal of the ramp.

EXAMPLE

Figures 4.19 and 4.20 show a basement layout for a business building. The excavation is to be 90 by 120 feet for an 88-foot by 116-foot building, 18 feet deep in firm clayey material, in a level plot measuring 100 feet by 200 feet. The structure will be against the sidewalk line, 17 feet back

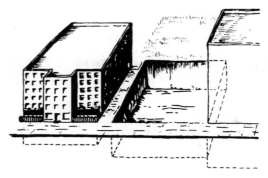

FIGURE 4.19 Haulaway basement job.

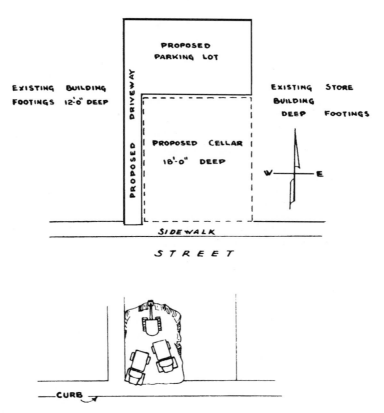

FIGURE 4.20 Starting large basement.

from the curb, will touch an existing store building on the east, a 12-foot driveway backed by another building on the west, and a proposed parking lot. The footings of the east building are below the 18-foot level, those on the west are 12 feet below the surface.

The site has been cleared, but two large stumps have been left. The topsoil is of such poor quality that it will not be separated. The contractor owns a three-quarter yard shovel with dipper, backhoe, and clamshell attachments. This machine is somewhat undersize for the job but may be used unless a premium is placed on speed.

The front shovel is used because of digging and loading efficiency. The ramp is located at the street as there is no access to the other sides except through the driveway, which is too narrow for heavy trucks and will be undermined by the digging. The ramp is next to the driveway, because if the other corner were used, the store building might be damaged by collision or vibration.

Moving In.   Arrangements are made to prohibit parking in front of the work area before machinery is brought in.

The shovel is unloaded from the trailer onto planks laid on the street to protect the pavement, or directly onto the sidewalk which is not protected because the trucking will destroy it anyhow. Digging is commenced in the sidewalk, or at its rear edge, and sloped down at about a 1-on-5 or 20 percent grade, in a cut 30 to 40 feet wide. Trucks are first loaded when standing in the street, parallel to the curb. As the digging progresses, they are backed across the sidewalk and down the ramp, as in Fig. 4.20.

It is necessary to have one person, preferably two, assigned to prevention of tangles between traffic and trucks. These should be police officers or contractor's employees authorized by the police to do this work.

Trucks should be backed into both sides of the ramp, as in Fig. 4.21(*B*), and faced directly away from the shovel while being loaded. At least one truck should always be in loading position.

Trimming.   As the cut progresses, the foreman checks the left edge for accuracy. Because of the angle at which the bucket works, it cannot make flat cuts on the wall, and one or more laborers should trim the face, working either from the top or from the ramp surface beside the boom. Checking may be done by stretching a string along the digging line at the top, and lowering another string to which a plumb bob is attached.

Bank Height.   The front shovel or backhoe would be able to take the full 18-foot depth in one cut, but this would not be good procedure. The shovel cannot cut a straight face higher than the level at which the bucket teeth start to turn away from the bank, and the face above might overhang or break back beyond the digging line. Hand trimming is more difficult on a high face, and caving is more likely and more serious than from a lower one.

The bottom is more likely to be muddy or to contain rock outcrops than higher levels, and it is economical to remove as much soil as possible under good conditions, before tackling the difficulties. In addition, trucks hauling from an upper level have an easier climb to the street.

Under average conditions, this job would be dug in two levels or benches of about 9 feet each.

Support for Bank.   Most deep excavations for buildings in a built-up area need to have support to guard against cave-ins. Certainly this is true when the excavation is alongside a building, as on the east side, even though the foundation for that building goes below the new excavation. There could be a cave-in above the excavation depth, which could affect the basement of the existing building. The building to the west, with foundations above the depth of the excavation, needs to be protected against any settlement that could lead to a lawsuit.

The method of support generally used is either to drive continuous-sheet piling or soldier piles with timber lagging between them. This support system would need to be along the length of each adjacent building and the street. It would not be needed along the north side of this excavation where a cave-in would not cause a serious problem.

There are other methods for supporting an excavated wall of earth. The use of tiebacks, such as with the sheet piling, is a popular method. A variation on tiebacks has been called *soil nailing,* which starts by using about 2 inches of shotcrete, i.e., a mortarlike mix sprayed at high velocity on the earth surface, and wire mesh on the first 3 to 6 feet of excavated depth. Then 4-inch holes are drilled on an incline down into the remaining earth in a set pattern and depth. A lightweight reinforcing bar, perhaps of $\frac{1}{2}$-inch diameter, is placed in the center of each hole, which is grouted up nearly to the top. Another layer of shotcrete is applied, and the excavation is extended for the next lift depth until the full planned depth is reached.

Dozer.   A bulldozer or front-end loader may be used to advantage to dress up the ramp, smooth the pit floor, and be ready to push overloaded or weak trucks up the ramp.

A crawler loader is preferred, as it does better trimming and can help with loading in an emergency. A big one is most useful, but also most in the way.

Trucks.   Trucks with 6- to 8-yard capacity are well matched to this size shovel, but anything from 6 to 14 yards can be used. Small ones can maneuver more easily and work well on certain types of soft ground. Large ones cause less traffic congestion, and under favorable conditions will move soil at lower cost. If the pit is wet or sandy, all-wheel-drive machines are preferable. Trucks must be in good condition to carry capacity loads up the ramp.

First Cut.   In making the ramp, the shovel has worked straight ahead. Upon reaching the floor level of the first cut, it may continue to the back wall, or it may first make a side cut, as in Fig. 4.21(*A*), so that trucks can turn around in the pit. If it works straight through to the back wall, it is then walked to the foot of the ramp again to take another slice toward the back, as in (*B*), as double spotting of trucks is easier when working away from the ramp than toward it.

Once the pit floor is widened enough to allow trucks to turn, the digging may be extended in any direction so long as the shovel may be easily reached by trucks. Sooner or later the bank to the east of the ramp is dug away, and usually part of the ramp itself, leaving it wide enough for one truck only.

Stumps.   When stumps are encountered, the shovel should dig around them before tackling them directly. The depth of this cut is such that they can be undermined enough that their own weight will help to break them out. Roots can be cut easily at a distance of a few feet from the trunk, and the stubs splintered back. Many operators will waste much time, strain their machines, by direct attacks on stumps which would shortly fall out in the ordinary course of digging.

When the stump is loose, it should be knocked around with the bucket to loosen the dirt, then placed in a truck. The tailboard should be folded down or removed, unless the stump is to be lifted off. With skill and luck, the shovel operator may be able to balance the stump on the bucket, then tip it off onto the truck. It may be picked up also by a chain. If it is too heavy for the shovel to lift, the dirt should be dug out of it by hand, and all possible wood removed from it by sawing and chopping.

Barricades.   As digging approaches the sidewalk, barricades must be erected to keep spectators from climbing or falling into the pit, and from using any part of the sidewalk likely to cave in. These barricades may be solid wood fences, or may be perforated so that sidewalk superintendents may watch the work. The contractor may be able to build up local goodwill by encouraging spectators.

Furnishing adequate windows or peep holes reduces the dangerous practice of spectators standing in the truck driveways to watch the work.

Finishing First Cut.   The floor grade of this first drop is approximate, and a foot up or a foot down is not important as long as it is easily passable to trucks. However, the walls must be cut to whatever finish the job calls for. A good operator can cut a straight wall and an almost-square corner with the bucket, but hand finishing is neater and saves machine time.

Bottom Cut.   When the first level is complete, ramp cutting is resumed until the bottom is reached, as in Fig. 4.21(*C*). Trucks will drive down the upper section, turn on the upper level, and back down to the shovel. It is not necessary that the ramp continue in the same direction, but this is the most economical method where the pit is long enough. Any turn must be made very wide for the convenience of the trucks.

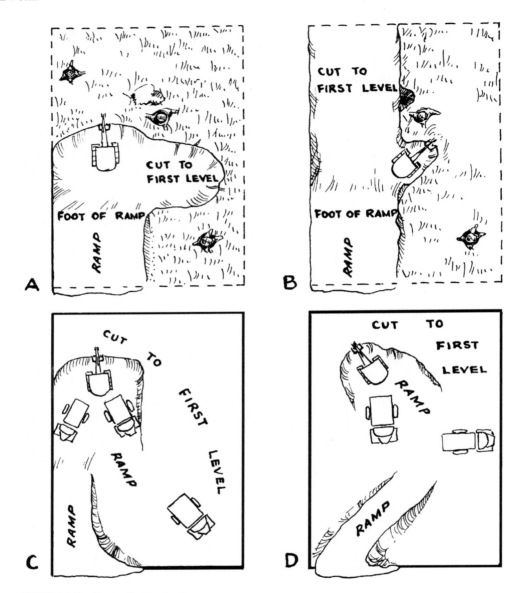

FIGURE 4.21 Cutting first level and ramp.

Cutting of the lower level proceeds in much the same way as for the upper, except that the floor grade must be carefully watched. This is usually checked with a transit or a builder's level. If foot wall trenches below the floor level are required, they may be dug by hand immediately after trimming of the wall is complete, and the spoil moved to the shovel by bulldozer or spread on the pit floor.

A somewhat gentler ramp gradient could be obtained by using a diagonal or a zigzag ramp, as in (D).

SHOVEL TEAMS

Two Power Shovels.   This job is big enough to justify the use of two shovels of three-quarter size or larger. The second shovel might ramp down from the sidewalk along the east side, and cut through to meet the first one at the center. After this, one ramp might be used as an entrance and the other as an exit; or one might be cut away. Or the second shovel might be brought after the first one reaches the upper level, and using the same ramp assist it on that level or ramp down to the bottom.

Traffic.   An external factor which may limit the number and size of shovels in such an excavation is traffic congestion on the street. This may create a bottleneck that would leave a line of empty trucks parked waiting in the street, with the shovels half-idle for lack of trucks to load. In congested areas, traffic may be one of the principal problems of the digging.

Ramp Removal.   When the digging is complete except for removal of the ramp, a backhoe or a clamshell must be employed. This ramp may contain 300 to 500 yards, a sufficient amount to make the use of the faster-digging backhoe better than the clam. If the hoe has an effective downward reach of 16 feet, it will leave a bit of the foundation of the ramp for hand labor; but the clam, in taking the whole ramp out, is apt to require a larger amount of hand labor assistance while working. Another factor may be that the shovel at this job with a crane (clamshell) boom might pick up extra work lowering materials into the excavation. The hoe rig is awkward to handle and to transport when detached, so that it might be more economical to move the shovel to the yard to change over and bring it back than to pick up the hoe attachment, bring it to the pit, and take away the front shovel attachment. The clam rig could be loaded on a truck by a chain hoist or a tractor loader, and moved with little or no blocking.

It might also be good business for the contractor to hire a backhoe or a clamshell rig and move the shovel onto another job.

Whatever machine is used, it will probably stand on the ramp as it tears it up, as the driveway is too narrow for swing space, and for safety at such height.

Teaming Front Shovel and Backhoe.   If two shovels are to be used for the whole excavation, it may be that the larger would be a front shovel and the other a backhoe, although under the ideal conditions considered so far, this is not likely. In such a case, the front shovel ramps down on the building side, unless danger of vibration damage is unusually severe. The strength of the foundation wall might be checked, and permission obtained to brace it from inside if the weight of trucks on the ramp seems to threaten it.

The backhoe is assigned to cutting the north and west walls of the pit because of its ability to make a smooth straight cut without hand trimming. It would preferably start on the north side, cutting from east to west, keeping the line in the manner described earlier, and digging out as much of the center as can be conveniently reached, as in Fig. 4.22. The edge ditch should be made as deep as it can reach, but the rest of the digging only 9 feet. The center digging is discontinued in the last few feet of the north line as the shovel is then backed up against the building beyond the driveway, and turns to get in digging position on the west line. Whether this corner can be cut square will depend on the length and tail swing of the shovel, but in general, it cannot. In any case, it cannot be squared to full depth.

The west line is ditched back to the sidewalk, with some additional material moved from the center as in (*C*). This leaves room to install bank supports. The backhoe can then work a wide cut back from any convenient starting place, taking care that its efforts, combined with those of the front shovel below, do not cut off its exit.

The trucks carrying the spoil from the backhoe may be loaded sideward to the shovel for safety, or from the back for convenience. If the body sides are very high, loading will be inconvenient and spillage excessive. This difficulty may be reduced by loading directly behind the shovel, so that it will have to walk over the spilled material, which will raise it so that loading will be easier. This spillage needs some manipulation to make a smooth ramp, particularly if the soil contains boulders.

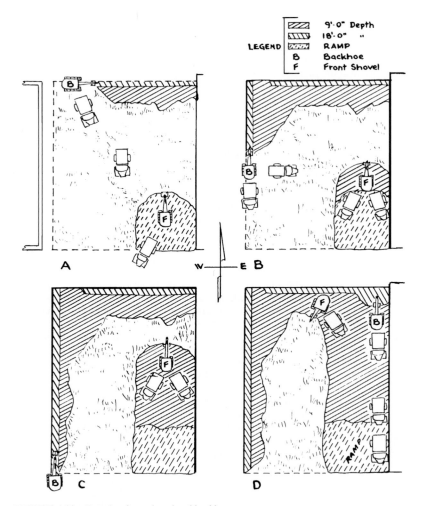

FIGURE 4.22   Teaming front shovel and backhoe.

Another method of loading trucks easily would be to start the cut at the sidewalk, making a ramp down for a truck wide enough to be backed against one part of the face to be loaded, while the hoe digs beside it.

When the combined efforts of the shovels have removed enough of the top cut that there is room for both of them on its floor, the backhoe may be moved down to do the digging to final grade, while the front shovel completes the upper cut, as in (*D*). Both shovels will be working on the same floor, but one will be digging material above, the other below. When the upper layer is finished, the front shovel will leave the job to be completed by the other.

The hoe will move from 10 to 30 percent less dirt each hour than a front shovel of the same size, a loss which may be only partly compensated for by the straight-wall cuts and the ability to take away the ramp without calling in another shovel or rig. However, if certain difficulties develop, the backhoe output will be unaffected, while that of the front shovel will be sharply reduced, and the presence of the hoe is insurance against undue loss of time from such causes.

BACKHOE EXCAVATION

This basement might also be dug entirely by one full-revolving hydraulic backhoe, bucket size $1\frac{1}{2}$ yards or larger. Such a machine may take an 18-foot depth in one layer, and cut foot wall trenches also.

If ground conditions permit, trucks are loaded on the pit floor, for maximum output (Fig. 4.23). The hoe would cut the ramp first, backing away from the street.

This 90-foot width is best taken in three strips, the first from front to back, the others from back to front. Walls are trimmed and trenches cut as parts of the main digging.

The final step is to move around onto the ramp, then dig it, backing up.

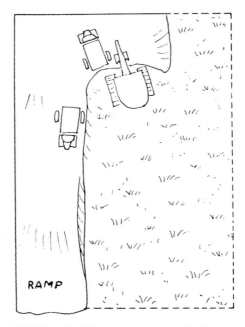

FIGURE 4.23   Full-depth excavation with a hoe.

GROUNDWATER

The most common difficulty is groundwater. It may be in the form of springs or underground streams, or a nearly stagnant water table with capillary water moistening the soil for several feet above it. Wet soils usually turn to mud when loaded or disturbed and impede or bog down trucks.

If the first level should have a firm floor, but water be encountered in the next layer, trucks would not be able to operate on the bottom without expensive aids, so that removal of this bottom layer with a dipper stick would be impractical. The backhoe would not be bothered unless there were sufficient water to hide the bottom, in which case it would have to be pumped out. Special dangers connected with such pumping will be discussed below.

Information about underground conditions may be obtained from test borings or pits on the site; from people who have dug basements or ditches in the neighborhood; and from geologists. Such data may predict with reasonable accuracy the depth at which mud, water, loose sand, or rock might be expected, and digging plans can be made accordingly.

Special conditions might require taking out the ground in three levels, or in one. The pattern should be such that the maximum amount of dirt would be dug by front shovel, on levels which permit trucking. When thin cuts are made, the front shovel can load trucks standing on the upper level, but the extra dumping height slows the digging, and in some materials the bank would not be stable enough to support trucks.

Drainage. Mud can be dried by draining or pumping the water. If the stormwater drain in the street is sufficiently low, arrangements should be made to connect with it before excavating. A ditch is dug from the pipeline in the street to a spot several feet inside the excavation area, and a pipe with sealed joints laid, opening into the storm drain. At the basement end, a vertical pipe of tile or concrete sections with unmortared joints or a perforated pipe is erected, as in Fig. 4.24(*A*) and (*B*). Sand or clean gravel is placed around the vertical pipe as the trench is backfilled, or a wooden barrier is placed to prevent backfill from closing the hole around it.

Each level made during the digging should be sloped to drain to this pipe, which can be opened at any level.

This installation will also serve to remove some groundwater from the site, before excavation.

A general lowering of the water table may be obtained by ditching on the three open sides, as in (*C*), or ditching the center also, as in (*D*). The edge ditches make the digging easier, but the interior trenches complicate it. Heavy wood mats are required wherever shovels or trucks cross them, and these are expensive to build and a nuisance to handle.

If the stormwater drain is not low enough to be useful, similar ditches may be dug and connected with a piped or open sump from which water can be pumped to a catch basin in the street.

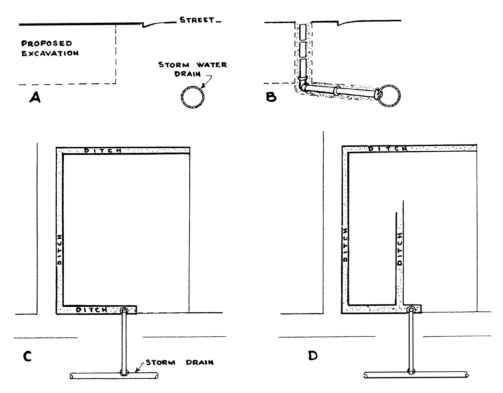

FIGURE 4.24 Drainage for excavation.

An overloaded storm drain may push water into an otherwise dry excavation, unless a check or shutoff valve is provided.

Well Points.   A satisfactory but expensive way of predraining the area is to use well points, which are discussed in the next chapter. Points may be driven outside of the digging line on the north and west, and probably, by special permission, in the sidewalk. Seepage from the east might be blocked by the building. If not, arrangements should be made to put well points in its basement.

Open Pumping.   Digging may be done without predraining and water pumped out of the hole as it appears. If the water is very dirty, and quantities are small or moderate, a diaphragm pump should be used. If the inflow exceeds the capacity of a diaphragm, about 1,500 to 3,000 gallons per hour, several may be used. More often, in holes of this size, centrifugal pumps are employed. Best results will be obtained by locating centrifugal pumps as close to the water level as possible, as their discharge is more efficient than their suction. Holes should be dug so that the inlet will be a foot or more below the water surface. Sucking air in shallow water may be reduced by floating a piece of board over the inlet, where it will block the formation of whirlpools which would conduct air down to the inlet center, or by arranging the hose so that it rises vertically out of the water.

Pumping may be done on a 24-hour per day basis, or only during or just before digging operations. If pumps are to be shut down overnight and on holidays in very wet holes, it may be wise to take them up each time, or to put them on floats for protection against unexpected rises in water level. Other equipment should be moved up to a safe level when work is shut down at the end of the day.

Caving.   Caving of banks and undermining of adjacent structures must be guarded against, particularly in connection with pumping. Caving banks involve hazards to people and equipment, and to adjoining structures, and increase the amount of excavation and backfill necessary.

Some materials, such as dry sand, will not stand in vertical walls, and digging must be figured to include natural slopes from the foundation line outward to the surface, or provision made to drive sheeting, or erect other barriers, to hold it from sliding. Sands or sandy soils containing the right amount of moisture will stand vertically, but they cannot be trusted, as drying will result in surface disintegration and sliding, and heavy rainfall may increase their weight and undermine them by washing grains out at the bottom so that massive caving will follow.

Silts, clays, and loams usually stand well, if not too wet, but if resting on a saturated layer draining into the excavation, may be undermined so as to fall. Vibration of machinery or street traffic may cause clay to creep or flow.

Gravel may stand or may slide, depending on the shape and grading of the coarse particles, presence of cementing material, and the amount of fines. Angular gravel of several sizes, with just enough fines to stick it together, will stand wet or dry unless subject to excessive water flow, or wave action. Very clean gravel, particularly if it includes a large proportion of cobbles and rounded pebbles, may slide in much the manner of dry sand.

Causes of Caving.   Danger of caving continues for days or sometimes weeks after the cut is made. In its natural state the soil is in both static and dynamic balance—static, because of inertia and the manner in which its particles are fitted and stuck together, and dynamic, because the weight overlying soil or structures exerts a sideward as well as downward thrust, which is met by equal counterthrusts from surrounding soil on the sides and below.

When a cut is made, the soil pressure toward it is balanced only by the soil inertia. This may hold it permanently in place, or the pressure may deform the soil and cause breaking apart and rearrangement of its particles, gradually weakening it until it falls. The effect may be likened to the collapse of a building under the weight of snow on its roof, which may occur hours or days after the storm and even after part of the snow is gone.

Groundwater is very effective in both holding and bringing down banks. While in very thin films it serves as a glue or binder. In contact with clay minerals it forms a lubricant, making it easier for particles to change position in response to pressure to such an extent that certain plastic clays will flow slowly. In larger quantities, water will seep or flow through the soil, carrying fine particles with it and cutting minute channels that weaken the structure. The flow of water is much slower through soil than through an open ditch, and it exerts pressure proportional to the restriction of flow.

If the water is allowed to stand in the excavation at its natural level, it will cease to carry particles out of the bank, and will exert a back pressure against the bank that will tend to hold it in place. However, this will not prevent the part of the bank above water from creeping under soil pressure or absorption of capillary water, and wave action set up by wind or dropping of stones or clods will cut into the bank at water level and undermine the top.

In general, where unstable soils or abundant groundwater is expected, open excavation should not be done until preparations have been made to build walls immediately after its completion; and if construction is delayed, it is better not to keep it pumped dry.

Side Effects of Dewatering.   Often the most serious aspect of removing water from an excavation is the effect on adjoining property. Water makes up a substantial part of the bulk of some soils, and its removal, even if it does not carry particles, sometimes causes shrinkage, with settlement of the surface and overlying structures. Damage to structures may also be caused by creeping of plastic soils from beneath them into the pit.

SHORING

Wall Bracing.   Movements of soil into a pit can almost be stopped, and water intake reduced, by the use of timber bulkheads or sheet piling. These are required by law in many cities, and are often good, although expensive, insurance against costly repairs and underpinning.

Installing such bulkheads is a highly technical operation, involving knowledge of soil behavior, engineering calculations, and skilled personnel. There is sufficient space available in this volume for only a brief sketch of one method.

Another method would be to drive interlocking steel sheet piling to several feet below the pit floor and, if needed, tiebacks into the ground the piling is holding back. The tiebacks must be solidly embedded in stable ground, preferably bedrock.

Bracing Stable Soil.   Figure 4.25 illustrates installation of thorough bracing in an excavation where a short section of face will stand for awhile without support. A long section is cut back by the shovel to a slope which is expected to be stable, (A) and (B). Then a short section, perhaps 10 feet, is cut and trimmed to final shape, (C). Sheeting plank, 12 by 3 inches or heavier, is placed vertically against the dirt wall. This plank should be long enough to reach a foot or two below the bottom of the pit, and 1 or 2 feet above the ground surface. Bottom penetration may be obtained by ditching, or by driving the planks down with an air hammer fitted with a special head for the thickness of plank used.

Horizontal timbers, called whalers, are placed along the face of the sheeting, being temporarily supported on cleats nailed to the planks. The whalers should be 6 inches by 6 inches or larger, and should not be more than 5 feet apart vertically.

Beams or plank mats called heels are placed on firm, undisturbed soil in the pit floor, sloping down toward the wall. These are used as abutments to take the thrust of the breast timbers that extend from the heels to the whalers. These should be 10 inches by 10 inches or larger. Each whaler must have two or more breast timbers, spaced 5 or more feet apart. If the heels are firm, the spacing of breast timbers can be increased by using heavier whalers.

While this bracing is being installed, an adjoining section of the wall is trimmed. This is braced in the same manner and the work continued in successive sections.

The sections may be tied together in several ways. The breast timbers may be placed against the whalers where they are butted together, as in Fig. 4.26(A), with or without the plate shown. The joints in different whalers may be staggered, as in (B), or may be overlapped, as in (C).

Nailing is kept to a minimum to avoid damage to the lumber. The bracing is dismantled after the foundation is placed and the material removed for reuse. The sheeting is usually pulled by a crane or pile driver equipped with a special clamp for gripping the tops of the planks.

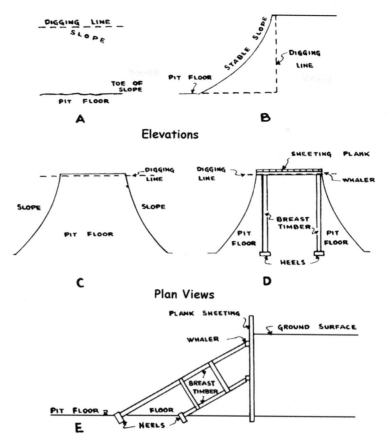

FIGURE 4.25   Bracing.

Unstable Face. If the soil is so unstable that it cannot be trusted to stand even in short sections, the sequence shown in Fig. 4.27 may be followed. A trench is dug with the outer edge at the digging line. This is braced with sheeting, whalers, and sheeting jacks in the manner described in Chap. 5, except that planks are left out of the sheeting on the inner wall at regular intervals.

Additional trenches are now dug into the excavation at right angles to the edge ditch. Heels are placed in them at or below floor level, and breast timbers run from the heels to the whalers on the out wall, through the spaces in the inner sheeting. The dirt between the breast timbers is now dug out, usually with a clamshell with laborers assisting, and the sheeting jacks and inner wall bracing removed.

Steel sheet piling may be driven along the digging line without the ditch. Breast timber ditches are dug in the same manner as described. A ditch is then dug on the inner side of the steel piling, and a whaler and breast timbers are placed. Digging is then carried down to the level of the next whaler, which is placed and braced.

Steel piling does not require as close spacing of the whalers as wood sheeting. A single whaler near the top is often sufficient, and in some cases it is not braced at all.

Cofferdams. When dry excavation is carried a considerable distance below the water table without dewatering the area, the heavy walls constructed to keep out soil and water are called cofferdams.

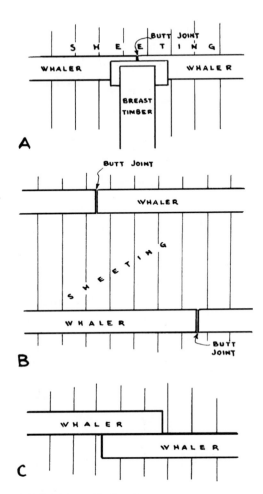

FIGURE 4.26 Detail of bracing.

Cofferdams consisting of single rows of interlocked steel pilings that enclose areas, with interior bracing, have been used for depths up to 60 feet, although ordinary practice limits them to 40. They may be installed by hammering the piling in undisturbed ground until it reaches bedrock, or to sufficient depth below the excavation floor to be considered safe. All the sections should be placed and driven to moderate depth before any of them are driven all the way, to make sure that all joints interlock properly.

Excavation is likely to be done by clamshells. Bracing is placed against the inside of the wall as it becomes exposed.

If the soil is very porous, great difficulty may be experienced getting the water down the first few feet, as the joints between sections leak quite freely until forced together by water pressure. More or larger pumps may be used at this stage of the job than at any later time. It may be necessary to trench outside the wall to place a clay seal partway down, or to partially seal the soil with cement grout.

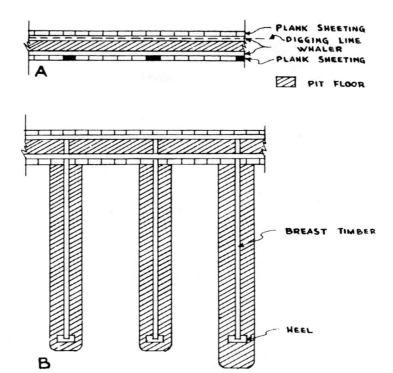

PLANK SHEETING
DIGGING LINE
WHALER
PLANK SHEETING

PIT FLOOR

BREAST TIMBER

HEEL

FIGURE 4.27   Bracing, trench method.

Porous soil under the bottom of the wall may permit excessive quantities of water and sand to boil up in the bottom of the excavation as final grade is approached. If the bottom is in clay, but porous soil is immediately below, the job may proceed easily, and then suffer from a sudden and disastrous blowup of the bottom.

An area protected by a cofferdam may be dug wet, in which case the structure serves to prevent soil from slumping into it.

Caissons.   A caisson is a structure which serves to keep soil and water out of an excavation, and forms part of the permanent structure for which the excavation is made.

A simple type of open-top caisson and stages in its growth are shown by cutaways in Fig. 4.28. A hollow square, ring, or other shape is made of reinforced concrete, with the bottom tapered to an inside edge. If the work starts on dry ground it may be built in a shallow excavation where it is to be used. If the start is underwater, it is made elsewhere with walls high enough to keep out water when it is lowered into place. Transportation is usually by barge.

The caisson is lowered by digging inside to undermine it, and building the top to provide more weight, and to keep it above ground or water as it descends. Most of the digging is done underwater, and it is a very ticklish job to do it accurately enough that the caisson will sink straight. When it comes to the bottom, investigation must be made to determine whether it is on bedrock or boulders. If the rock surface slopes, concrete must be pumped underneath to give it firm bearing on the low side.

The pneumatic caisson has an airtight cap above the bottom, with sufficient air pressure maintained under it to keep water out. Air locks and chambers are provided for entrance and exit of

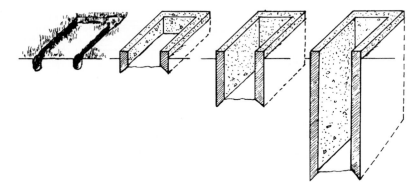

FIGURE 4.28   Sinking a caisson.

personnel and material. Much of the digging is done by hand, and in deep work at high pressures workers may be limited to less than an hour of work at a stretch, with long periods spent in entering and leaving the high-pressure work chamber. Depths up to 100 or 110 feet can be reached.

ROCK

Bedrock.   If bedrock is encountered that is too hard for the shovel to tear apart, it must be blasted. Generally it is best to complete the earth excavation first, to reveal the full extent and as much of the grain and quality of the rock as possible, before going to work on it.

Sometimes, however, drilling and blasting are started as soon as the rock is found, and the shovel doing the earth excavation can be utilized for handling logs or blasting mats. This may save shovel time, as under city conditions it is not often practical to blast rock fast enough to keep a shovel busy, and a shovel whose only duties are handling mats and removing blasted rock is likely to be idle most of the time. On the other hand, earth-hauling trucks will be stopped while the shovel places mats and during blasting.

Backhoes and small dozers are good machines for cleaning the bulk of earth off ledges, but there is almost always need for hand work also.

Procedures for the rock blasting and removal are outlined in Chap. 9. However, it should be emphasized that blasting near streets and buildings is a much more dangerous and specialized job than the same work in a quarry or a country highway cut. Elaborate precautions must be taken to prevent material from flying, and large blasts, or small blasts following each other quickly at regular intervals, must be avoided because of danger of concussion and vibration damage to nearby buildings. Jobs must be inspected in advance by the insurance company in order to set a rate in line with the risks.

Boulders.   Boulders in the soil slow excavation by creating digging resistance and complications, and often by difficulties of disposal after they are dug out.

The hydraulic backhoe, such as a Gradall, usually does the best job in proportion to its size. The comparatively narrow, toothed bucket with wrist action can work around and under large objects, and has great prying force in its closing action.

It is competent at picking them up, but only if they are small enough to be held in or on the bucket, or clamped between it and the stick. Larger ones may be held against the wall of the excavation, or a specially dug slope, and pulled to the top.

The front loader is clumsier at digging out boulders, but does much better at removing the big ones. The wide, deep bucket can pick many up directly, or by crowding against a bank while lifting and curling the bucket. Extra large ones that would fall out can sometimes be chained, or just pushed out of the way to wait for equipment to break them.

Boulders may be loaded into a truck by a hoe or a loader, or more gently by a clamshell, grapple, or crane. For crane work, the rock must be firmly gripped by chains or slings.

Chains should be of the lightest size that will lift the weight, as a thin chain grips rock much more closely than a thick one. Undersize chains break frequently, and spares and repair links and hooks should be kept on hand.

Alloy chains are expensive but are small and light in proportion to strength.

Small cables grip rock well but wear and fray rapidly, so that sharp ends of broken wires make them dangerous to handle.

Slings may be made of several strands of light cable or chain, and combine the grip of small sizes with the strength of large ones.

Boulders may be broken by blasting, but in city areas mud capping is not permissible. Splitting may also be done with sledge hammers, air hammers, or drills and plug-and-feather sets.

HILLSIDE SITES

Downslope. So far we have considered excavation in a level plot. As the basement depth is calculated from street or sidewalk level, a downward pitch to the rear would decrease the amount of excavation, and an upward one would increase it.

If the lot slopes down to the north, as in Fig. 4.29, the natural grade can be cut to the proper slope for a ramp by a bulldozer, and the material removed used to build a flat shelf at the first cutting level on which the front shovel and trucks can start work. If insufficient dirt is cut in making the ramp, the shovel can dig into the hill and sidecast below, to build it up to the desired size.

Excavation is carried back to the side of the ramp and to the south and west digging lines, in any convenient manner, while the bulldozer shapes the bottom level, making a flat space as before.

When the front shovel starts work at the bottom (Fig. 4.30), the excavation and ramp removal are carried out in the manner described earlier.

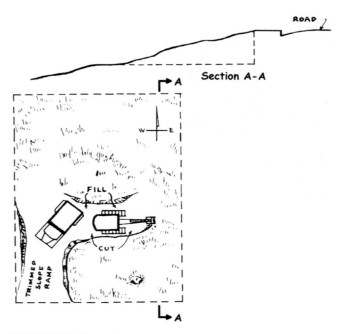

FIGURE 4.29 First cut in a downslope.

If two shovels are used, one can work on each level. The upper one should work across to the east side and finish it first, so that the one on the lower level can work in without cutting it off.

It is unlikely that a backhoe would be used on such a job, except in removing the ramp, unless mud conditions are encountered. Sometimes soft footing can be economically handled by surfacing the truck road with gravel, crushed stone, or dry fill.

Bottom Access.   It may be possible to arrange for the movement of machinery and trucks into the lower end of the lot, as in Fig. 4.31. A bulldozer may then cut a truck road and turn around into one side of the lot. Loaded trucks will now move downhill, and maximum loads can be carried.

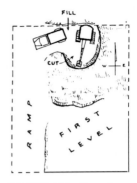

FIGURE 4.30   Second cut.

It may be difficult for the empty trucks to turn on the slope and to back uphill, particularly in sloppy going. If the shovel first digs a wide shelf as in Fig. 4.31, the trucks can turn on it, and another roadway can be graded later for exit so that no uphill backing will be necessary.

Cut and Fill Digging.   Figure 4.32 shows the same sloping lot with a retaining wall built along its back lines. The spoil from the basement is to be used to fill up to this wall for parking area.

A front shovel and trucks may still be effectively used for the digging, but the short haul makes possible the use of other machines.

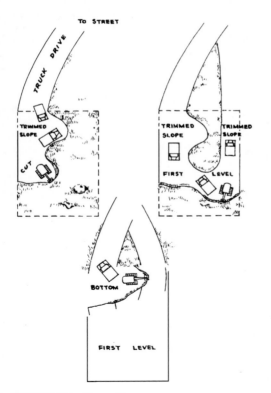

FIGURE 4.31   Slope with rear access.

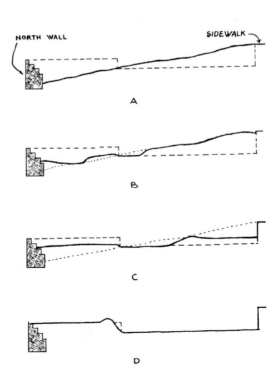

FIGURE 4.32   Filling against retaining wall.

However the soil is moved, it should be spread in thin layers and thoroughly compacted by rolling in open spaces, and tamping where rollers cannot reach. This will prevent serious mud difficulties during the work, possible damage to the retaining wall from pressure or fluid mud after heavy rain, and excessive settling of the finished fill.

The average length of push is about 120 feet, very slightly downhill. This is within the economical range of medium to large bulldozers, or small, self-loading scrapers, but help will be needed from a loader or hand labor to cut out the south corners. The equipment should be small enough to leave by the driveway when the job is done.

A dozer first cuts a shelf, level or sloping opposite to the hill at the top, just below the sidewalk. This is done by digging along the edge line until a bladeful is obtained, then turning downhill, lifting the blade at the same time, so that the fill is built higher than the cut to allow for compaction. Scrapers can be used to cut down this shelf as soon as it is a few feet wider than they are, but a dozer will be needed to keep the walls trimmed back to a vertical. The dozer can also cut much farther into the corners than the scraper can, by the process of gouging and then swinging out.

The front-end loader can square the corners by working against one side, parallel to it, and digging into the bank until the other side of the corner is reached. The spoil is picked up, moved back, and dumped in the path of the scrapers. Best work can be done if the corners are kept cut down within a few feet of the level on which the scrapers are working.

The scrapers may be kept moving in a rotary path, as in Fig. 4.33, digging at the south end, and dumping and spreading along the retaining wall. As the fill rises, it will enlarge to the south. At the same time, a bulldozer can be working down the center section in the soil to be moved the shortest distance. This dozer may also take care of trimming the fill for the scrapers and pushing it into the corners.

A self-powered tamping roller should be kept moving over the fill in both scraper and dozer sections. Hand, gasoline, or air tampers should be used along the wall.

The parking lot fill, and that needed in the rear part of the driveway, cannot be placed on the south side until the foundation of the building is in place. Material needed for this can be piled on the edge, ready to be pushed in place.

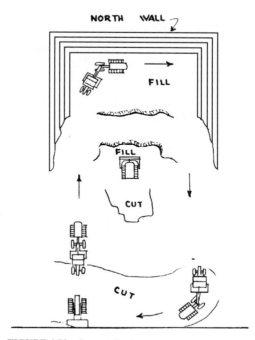

FIGURE 4.33   Scraper digging.

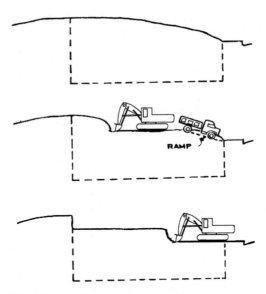

FIGURE 4.34   Basement digging in a hill.

The sidewalk edge of the pit might also be cut by a clamshell standing on the sidewalk. The dirt could either be loaded into trucks or cast out into the pit in reach of the scrapers.

If additional fill is needed in the parking area, it should not be trucked in until the building foundation has set long enough to give support to the driveway.

Hill Removal.   It often happens that the building site slopes up from the street, sometimes very abruptly. The hill must be removed, in layers if it is high enough, before digging down from the street.

Figure 4.34 shows one such situation. The first cut starts above the street level so a ramp is dug up to it. When the top has been removed, digging is started at the street level, and the underground cuts are taken afterward. Two or more shovels can work on the job, usually on different levels.

The upper cuts should be sloped so as to drain toward the street, but not steeply enough to cause gullying and washing of dirt onto the street, as the contractor is responsible for any damages caused by the work.

CHAPTER 5
DITCHING AND DEWATERING

DITCHING

Drainage by ditching is a very ancient type of excavation, and even drainage tunnels were built in prehistoric times. The purpose of this work was generally reclamation of land for agriculture.

Modern advances consist largely in the use of machinery for ditching, some improved types of pipe, and use of pumps to dewater areas that cannot be readily drained by gravity flow.

Most dry ditching is done by backhoes or wheel, ladder, or drag ditchers. In soft swamps, draglines may be best. Clamshells are used for deep and tricky work. Shallow ditches in dry ground may be dug by hoes, ditchers, front shovels, draglines, graders, and dozers of straight, angle, and other varieties.

Investigation.   Before excavation of any kind is begun, the site should be inspected carefully for any conditions requiring precautionary measures. This is especially important when one is working in a developed area with buildings and existing utilities. Survey adjacent properties before excavation and, if possible, before bidding for the work. Record all defects such as cracking and settlement, so that after the excavation any claim of damages can be assessed as having been caused by the work or as a preexisting condition.

The location of existing underground utilities, namely, sewer pipes, electric lines, telephone, fuel, water, and gas, must be determined before excavation. The contractor should contact the utility owners and ask them to establish the location. When necessary, discuss removal, relocation, or service interruption with the utility owner before excavation. If the utility line is paralleling the trench and within the risky triangular wedge by a distance less than the depth plus half of the trench width, extra precautions must be taken to ensure no damage to that utility.

Vacuum Excavation.   In the utility business there is a procedure known as potholing to discover where an existing utility line is located. Many government agencies are adopting regulations that require potholing so that new construction does not damage or knock out existing utility lines. The preferred method for potholing is now by vacuum excavation.

The portable equipment for this is like a large size vacuum cleaner using either high-pressure water or air, up to 1410 cfm, to excavate a hole with the soft ground material being collected in a large tank. Depending on the machine used and soil conditions, a 12-inch square, 5-foot deep pothole can be completed in 20 minutes or less. Most vacuum excavation machines are capable of digging much deeper, but utility potholes usually do not have to go more than six feet deep.

Vacuum excavation was developed more than 50 years ago, but its use has been emphasized by the extensive use of horizontal directional drilling (HDD), which was discussed in Chap. 1 and detailed in Chap. 20. In connection with HDD there are two main functions for vacuum excavation: (1) removal of drilling fluid that escapes from the HDD's pilot hole during drilling, back reaming and product installation and (2) potholing, i.e., uncovering existing buried pipe or cable so HDD crews can visualize their exact location. The vacuum excavation machines are available in many sizes, ranging from compact, self-contained systems, mounted on skids or trailers, to larger truck-mounted models.

DITCHING WITH A BACKHOE

Types.   There are two types of backhoe used in ditching. We have the full-revolving hydraulic hoe, with crawler- or wheel-mounted undercarriages. And there are smaller hydraulic hoes carried on the backs of tractors (or rarely, trucks), whose arc of swing is 200° or less.

Design and operation of these various machines are discussed in Chap. 13.

The hydraulic models provide complete control of most bucket motions, with power for up and down, and pull and push. A "wrist action" bucket can be adjusted in angle during the digging pass for precise control of cutting, for tremendous breakout force on obstructions, for nonspill carrying of bucket loads to the dumping point, and for selection of place and rate of dumping.

Some crawler-mounted hydraulics have separate speed and direction control of the two tracks, a refinement that helps maneuvering in restricted space and reduces cutting up of the ground.

Width.   Ditching with any machine is easiest and neatest if the trench is the same width as the bucket cut. This width is that of the bucket itself, plus any side cutters it may carry. The bucket should be wider at the front than the back, to prevent the sides from binding in the cut and to simplify dumping.

Standard bucket widths, with or without side cutters, are 18 to 42 inches. Small machines may be as narrow as 12 inches, and big ones 5 feet wide or more.

Direction of Work.   A hoe should start at, and dig away from, any obstacle it cannot cross, such as a building. If there are two such obstacles, separate starts are made at each, and an extra-work connection is made between the two trenches.

If the centerline is on a grade, working in the uphill direction makes digging more difficult, by reducing digging force and increasing the tendency of the bucket to pull the machine into the ditch.

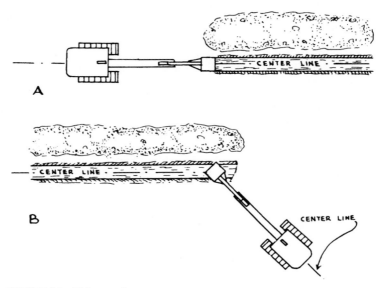

FIGURE 5.1   Lining up a hoe.

While digging downhill is easier, the working end of the trench may fill with water if the ground is wet or the job stands unfinished during a rain. Underwater work is sloppy, inaccurate, and often unstable.

Starting.   The machine is placed so that it is centered on the centerline of the ditch, with the tracks or wheels parallel to it, and the bucket extended to almost its full reach and resting on the starting point, as in Fig. 5.1(A).

Actual digging procedures with the different types of hoe are described in Chap. 13.

Briefly, the soil is taken out in layers down to the required depth. The starting point may be squared off with a vertical face from top to base. The bottom is smoothed off and checked for depth as it is made.

When the desired depth has been obtained along the space the shovel can reach, it is walked away from the ditch from 2 to 12 feet, and a section of that length excavated. Short moves are made in connection with deep ditching, cutting the bottom to an exact grade, or cutting curves; longer moves are feasible for rough, shallow work.

Curves.   Curves are dug as a succession of short, straight ditches, but a skilled operator can bevel the edges to produce a smooth curve. The machine stands with its center a little outside of the centerline, and digging is done in the outer half of the bucket reach. Moves are short.

Angles.   Many kinds of pipe require laying in straight lines and angles rather than curves, and trenches in which they are to be placed are dug accordingly. Angles are made by digging slightly past the angle point, then shifting the shovel to straddle the new centerline, as in (B).

Spoil Piles.   Spoil from the ditch is usually piled on one side, far enough back to allow a footpath or working space between it and the ditch. If a large volume of dirt is being moved, the pile

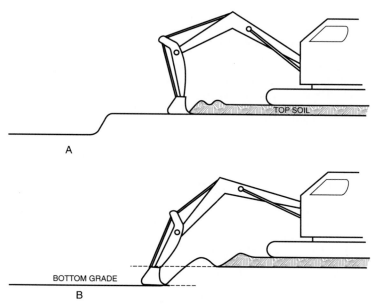

FIGURE 5.2 Separating topsoil.

must be pushed back by the bucket as it is built, and in addition it may be necessary to allow the spoil to come to the edge. Piling on both sides is usually avoided because of backfilling work. It does serve to block off the ditch so that people are less likely to walk into it absentmindedly or in the dark, although it is not adequate barricading.

Topsoil.   If topsoil is to be saved and put back on top of the other fill, it may be piled on the opposite side of the ditch from the deeper digging. If the volume of the spoil is not large, the topsoil may be placed on the same side as the fill but farther back, so that when a dozer backfills, the topsoil will be next to the blade and will reach the ditch after the fill.

Topsoil is salvaged during the digging by scraping it off first, and bringing the bucket as near the shovel as possible on the last bite. The body of the ditch is then dug, with the bucket lifted out short of its closest position. There should then be a separation between the heap of topsoil and that of fill, as in Fig. 5.2. When the shovel backs away, it can dig the pile while stripping the next section.

Sod.   If sod is to be saved, it should be removed ahead of the shovel. It may be dug by hand, or cut in strips by a tractor-drawn or self-powered sod cutter. The strips left by it may be sliced in sections and piled at a safe distance by hand. The sod should be taken out at least 6 inches, and preferably 1 foot, back from the digging sidelines, to avoid damage.

Guides.   Sod removal serves as an excellent and unmistakable indication of the location of the ditch; otherwise a line of pegs low enough to allow the shovel to walk over them may be used. The shovel is lined up over the ditch in the same manner as described for basements in the previous chapter, except that for ditches marked with centerlines, the dead axles should be marked to indicate bucket center, not sides.

If the bottom is to follow ground contour, the bucket stick may be given a paint mark that will be even with the surface when it is straight down at the proper depth. The distance may also be marked on a stick to be used for checking.

A bottom gradient that is independent of surface levels is usually set and checked by instrument from reference points. Anything from a hand level to a laser beam may be used, depending on the conditions and the accuracy needed.

Side Digging.   A hoe should be worked away from the end of the ditch that is blocked. In ditching from a house to the street, it starts at the house and finishes in the open space of the street. However, it often happens that a ditch must be dug between two buildings, or under other circumstances where both ends are blocked.

The simplest method of accomplishing the necessary turnaround is to dig the ditch from one end, then from the other, having them meet at some spot where the shovel can move off to the side. The digging of the second section should be stopped while there is still comfortable room to turn the shovel and get it out, as in Fig. 5.3(A). The shovel is then turned at right angles to the ditch and walked back into the undug space, with its center pin in the centerline of the ditch, as in (C). It then digs as closely to its tracks as possible on both sides and backs away, connecting the ditch sections by digging first on a slant and then at right angles to the trench line.

This method calls for an excavation that is usually two or more ditch-widths with a hydraulic backhoe. (See Fig. 5.4.) Sometimes such a connection can be made where extra width is needed for a manhole, pumphouse, or side ditch.

Tractor Mounting.   The smaller sorts of hydraulic, mounted on the backs of wheel tractors, are usually the most economical and efficient hoes for work around buildings.

They have the advantages of light weight and rubber tires, and thus are less likely to damage lawns and paths than the heavy crawlers, either while working or while getting in and out. Accidental damage to trees and structures caused by operator mistakes is likely to be small.

Another asset is that their small buckets are usually narrower than those of larger machines. For the usual pipe or wire, they take out less soil to put back. However, in this respect they are not nearly as economical as small drag trenchers.

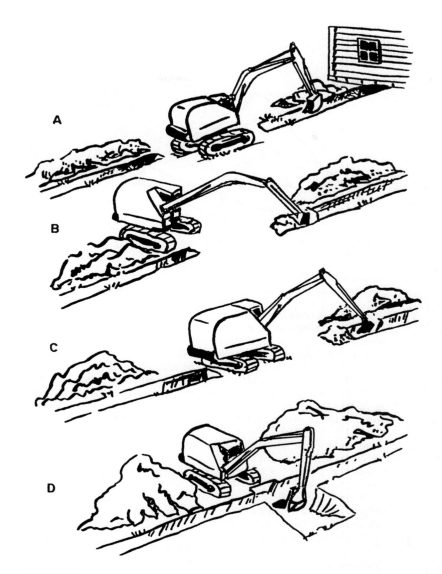

FIGURE 5.3   Connecting trench sections.

Their fast cycle, and the prying power of the wrist-action bucket, often enables them to do work and show production that seems greatly out of proportion to their size.

The tractors usually carry front loaders also, enabling them to backfill their ditches, carry pipe, and do miscellaneous work.

In spite of usually having two-wheel drive, these machines are seldom stuck. The weight of the hoe on the driving wheels supplies excellent traction, and if that fails, downward and outward

FIGURE 5.4 Hydraulics make narrower connections.

pressure on the bucket will lift the rear wheels and push the tractor right out of a mud hole. Or the bucket can pull it out backward, if that is what is needed.

Wide Ditches. When a ditch is to be more than one bucket width, one or both edges will be slightly uneven because the bucket will move inward, toward the center pin of the shovel. Usually one side is made straight by lining the shovel to that side, and by the hacking done on the other side. If neatness is important, the ridges can be smoothed by drawing the bucket in while lightly swinging against the edge.

The full width of the ditch should be taken off in layers if it is to be dug from one position, rather than cutting one side to depth then starting on the other.

If sloping beds of shale are encountered, digging should be arranged, if possible, so that the bucket teeth will cut along the bedding planes. Shale dug in this manner at moderate depths is apt to come up in sheets, so that the ditch will be widened irregularly.

Production. The rate of ditching depends on a number of variables, including depth and width of the ditch, bucket size and efficiency, cycle time of the hoe, digging qualities of the soil, obstacles and hazards both below and aboveground, presence of rock, accuracy of grade required, and need to separate topsoil.

A ditch that is shallow, with soil piled on the edge, offers the fastest digging cycle, but the bucket is not apt to fill well. Deeper digging slows the cycle and means more soil to move, but allows better filling of the bucket. A narrow bucket does not fill as well as a wide one at any depth. A ditch that is wider than the bucket takes much more time.

Boulders and heavy roots slow the digging. Presence of buried pipes or conduits that must not be broken cause serious delays, particularly if their exact location is not known. Buildings or trees that interfere with maneuvering cut down production, as does lack of space to pile spoil.

It takes much longer to clean an irregular rock surface for blasting than to dig a clean trench to grade. Then there is the extra expense of drilling, blasting, and redigging.

The need to keep an accurate grade makes an operator work more slowly, and occasional stops are needed to check gradient or depth, unless a laser-beam setup is used. A smooth bottom finish is produced readily by a wrist-action bucket, but with some difficulty by a rigid one.

Stripping sod and topsoil separately will slow the digging from 5 to 30 percent.

When a trench needs to be braced during or immediately after the digging, production will be determined by the rate at which bracing is set, which almost always is much slower than the digging.

Here's an example of calculating digging rate:

Assume that a $\frac{1}{2}$-yard hoe with a 36-inch-wide bucket, including side cutters, is digging a ditch 3 feet wide and 6 feet deep in common earth, with no special complications.

This ditch has a width of 1 yard and a depth of 2, so its cross section is 2 square yards. There will be 2 cubic yards removed for each lineal yard of digging, or $\frac{2}{3}$ yard per lineal foot.

This machine may have a cycle of 13 seconds, and an efficiency hour of 45 minutes. It should complete 206 cycles per hour.

The soil has a swell factor of .8 (25 percent swell). The bucket averages only four-fifths of a load in loose yards; that is, its efficiency factor is .8. Multiplying the swell factor by the efficiency factor by the $\frac{1}{2}$-yard capacity of the bucket, we have

$$.8 \times .8 \times .5 = .32 \text{ bank yard per cycle}$$

Multiplying 206 cycles per hour by the .32 bucket load, we have a production of 65.92, say 66, yards per hour. Since there is 2 cubic yards to each running yard of ditch, the ditching rate is 33 yards or 99 feet (say 100 feet) per hour.

A 30-inch-wide trench with a 30-inch bucket would come out about the same, as what was gained in handling smaller volume would be lost in poorer bucket efficiency. However, if the ditch were 12 feet deep, either bucket would probably fill well.

OTHER SHOVEL RIGS

Clamshell.   A clamshell ditches best when on the centerline. If the ditch is narrow, the tagline chains are fastened to one jaw, or for a very wide cut, to both jaws. A ditch of intermediate width is made with the chains in the one-jaw position, and the soil is taken out in layers.

Connections between ditch sections are made by attaching the tagline chains to both jaws after completion of the main ditching, and digging the connection from the side. Whole ditches may be done from the side in this manner, but it is harder to keep on the correct line. The side position is desirable in deepening an existing ditch, or in digging beside a wall.

Smooth curves may be dug either by frequent readjustments of the position of the shovel, in the same manner as with a backhoe, or by having someone on the ground twist the bucket into proper position by pushing it by hand or with a stick as it is about to touch the ground.

Dragline.   The dragline is the preferred shovel for ditching in swamps, and for making ditches with sloped banks when the spoil is to be piled alongside. It works along the centerline of the ditch, as in Fig. 5.5(*A*), cutting the bottom and slopes in one operation. If the ditch is too wide for this, two cuts are made from the sides, as in (*B*) and (*C*).

If the fill is to be trucked away, a dragline or a backhoe may be used in this manner. Draglines may have difficulty digging hard earth that the hoe would move easily.

Front Shovel.   The front shovel can dig trenches 4 to 8 feet in depth from the top, or wide trenches from the inside. A neat ditch may be dug from the top by straddling it, if the soil is very firm or if support platforms are used. Trenching may also be done beside and parallel to the shovel's path, but this involves quite a wide cut in proportion to depth and is difficult to trim.

Interior digging conforms in general patterns to that discussed for basements in the previous chapter. Partial swing shovels can dig narrower slots than conventional models, as they do not need space for tail swing, but they cannot load trucks behind them.

Comparisons.   The backhoe is the best machine for ditches of moderate depth and width where boulders or stumps may be encountered. It will break up heavily fractured hard rock and soft or thin bedded shale, and will dig very hard soils if the bucket teeth are long and sharp. It can dig out

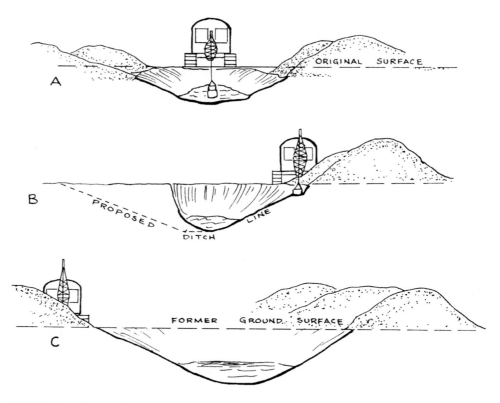

FIGURE 5.5   Wide-dragline ditch.

large boulders by widening the trench as much as necessary, and dragging them up the slope toward itself. The ditch can be easily curved around boulders too large to lift or pull.

The clam is a slower machine but is able to dig to any depth desired, and can work close to obstructions, except those that are overhead.

OTHER DITCHERS

Ditching Machines.   Continuous-type ditching machines offer great advantages in areas where hard bedrock and boulders are rare. They dig by continuous picking and sidecasting, rather than the dig-and-dump cycle of the hoe.

These machines excavate rapidly; make a neat ditch, usually with a curved bottom which is helpful in lining up pipe; can work with less headroom, and do not need space to swing. They can dig certain classes of homogeneous soft rock which a shovel cannot, and seldom tear up banks in shale.

Medium and large machines have a number of small buckets mounted on wheels or double chains, that dump on sidecasting conveyors. Small units have a single chain, fitted with teeth that cut soil and drag it to a surface auger that sidecasts it.

Buckets may be much narrower than in hoes, and drag chains may cut slots only 4 inches wide. Some machines may be fitted with a carbide-toothed wheel that can chew slowly through boulders and bedrock.

Ditchers can be equipped with shoes or reels to lay tile or flexible conduits immediately behind the digging, so that shoring is not necessary.

Design, operation, and applications are discussed in Chap. 14.

Graders and Dozers.   Graders can make shallow ditches with sloped sides rapidly and neatly, by the road-building processes described in Chaps. 8 and 19.

If there is no use for the excavated soil, it is usually spread and blended beside the edges.

The bulldozer can dig a wide, shallow trench from the side, as shown in Fig. 5.6(*A*). The volume of excavation required increases very rapidly with depth because of space needed for ramps.

When the practical limit for side excavation has been reached, the dozer can work in the ditch, pushing dirt into heaps, which it then pushes to the side, as in (*B*) and (*C*).

An angling dozer can excavate by sidecasting in the same manner as a grader, but it may be harder to keep lined up.

ROCK

Stripping.   Ditches frequently contain rock that is too resistant to be dug by the available equipment. Occasionally the line of work may be shifted, but usually it is necessary to blast.

Dirt and rotten rock are removed by conventional methods. Spoil should be piled far enough back to allow space for the drilling equipment, and for the shovel when it returns.

After machinery has removed the soil, the rock surface should be cleaned by hand. If the trench walls are liable to crumble and slide from drilling vibration, they should be shored up, even if depth is shallow.

Open-cut blasting is described in Chap. 9. Trench work differs chiefly in the restricted working space, and in the fact that all shots are tight. Loading must be 50 to 100 percent heavier than on wide faces.

Drilling.   Jackhammers can be used with the operator standing on the rock, or on the surface beside the ditch. Crawler drills are usually on the bank. Special ditching drills may be suspended over the work by cranes.

Several drilling patterns for 3- or 4-foot widths are shown in Fig. 5.7. In each of these, blasting is done back from an edge or face of rock exposed by digging, or by previous blasting.

The distance or length of ditch that can be blasted in a single shot depends on the near presence of buildings, whether it is permissible to overbreak the sides, and whether delay caps are used.

The holes next to the face can throw their burden along the line of the ditch. Any holes behind them, shot at the same time, will tend to expend more of their energy to the side, so that they will need heavier loading. As a result, they tend to break rock outside the digging lines (overbreak) and produce finer pieces.

In (*D*) there are extra edge holes, called relief holes, which are lightly loaded or empty. Their use makes a smoother edge, reducing both overbreak and underbreak.

Delay Caps.   After the blaster has decided on the area to be blasted, the number of holes included in it may be shot at one time; or a much larger number may be fired with delay caps. The first series will be at the face, with the other two groups following in succession.

Millisecond-delay caps give the best results, as they provide thorough breakage with minimum concussion. They are arranged so that the wave of explosion travels back from the face, with such short intervals between rows that each is partially confined by the force of the previous blast. This condition is favorable to good fragmentation and reduced overbreak.

Damages.   Nearby buildings, or more distant installations containing delicate apparatus, may dictate the size and type of shots. The conservative procedure is to fire one row at a time, and muffle and restrain the explosion with dirt and mats. Millisecond delays may permit much larger shots, because the explosion is spread over enough time to reduce its sharpness.

FIGURE 5.6 Bulldozer ditching.

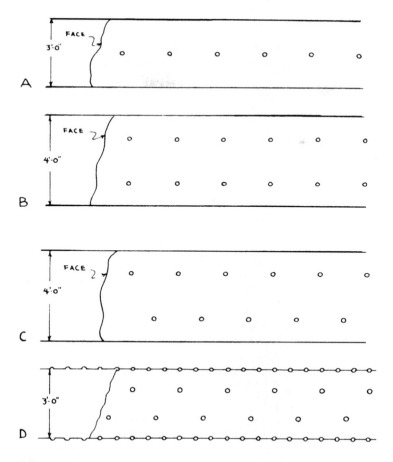

FIGURE 5.7   Rock drilling patterns.

Standard delays permit firing a succession of small blasts with one jolt, but there is danger that they will cause a periodic vibration in some building or object which will damage it more than a heavy single explosion.

Deep shots, or those well buried, produce less damage than shallow ones. If the explosion is so confined that it acts about equally in all directions, concussion is at a minimum. If any part of the force can escape rather readily, its blowout will produce a reaction or back-kick against the solid rock, which will shake the surrounding area severely.

Air waves, which are serious offenders in breaking windows, can be prevented by thorough muffling.

In any blasting near buildings, roads, or people, elaborate safety precautions must be taken. An absolute minimum is to confine the explosion so that no fragments can fly, give ample advance notice of the blast, and block all roads and paths into the danger area. Woven steel mats are the only safe cover for close blasting.

The contractor's liability insurance company usually makes an inspection of the job before any blasting is done. This may include the condition of nearby buildings.

A discussion of blasting damage is given in Chap. 9.

Burial.   When the rock surface is well underground, the ditch may be refilled with dirt after the rock has been drilled and loaded. This confines the explosion, prevents rock from scattering, muffles the noise, and protects the ditch walls from caving. It provides a safe working area for the machine, usually a backhoe or clamshell, that will redig the trench after the blast.

If the dirt fill is shallow, or rocky, it may be necessary to use mats or logs in addition.

Wiring must be thoroughly protected before filling on top of loaded rock. Fine dirt is hand-shoveled over the connecting lines. If the rest of the fill is fine, a thin cushion is enough. If there are heavy or sharp rocks that cannot be kept out, a deep protection is required.

The lead wires may be run along the rock surface beyond the area to be filled, or led up to the surface inside air hose or other tubing.

Two caps may be placed in each hole with duplicate wiring, or Primacord can be used.

Backfill may be pushed in from the side by a dozer, or trucked from an excavator digging another section of the ditch.

Removal.   Blasted rock may be dug from the trench by machine, by hand, or by both methods.

If the ditch is narrow, with hard, irregular walls, projecting stubs may jam a bucket so frequently that hand work may be cheaper. A bucket or container is frequently loaded by hand and hoisted by a machine.

Large rocks are lifted with slings or tongs.

In general, it is good policy to blast sufficient width so as to ensure working space for a clamshell or hoe bucket.

CAVING OF BANKS

Many soils will not stand in vertical walls, so the sides of the ditches must be either sloped back or braced. A few soils will not stand on even a moderate slope, and these will require very heavy bracing.

Groundwater is the most important single factor in collapse of edges of cuts. It acts as a lubricant that enables the soil particles to move on each other readily, and exerts pressure that moves the particles toward the ditch. Sand faces may fall from this cause, or from the drying action of water draining and evaporating, so that the bond between the grains is weakened.

If water is allowed to stand in a ditch, danger of caving from flow of groundwater into it and from other dynamic forces acting in the soil is reduced. However, wave action set up by dropping of stones or chunks of dirt may cut into the walls and undermine them.

It may happen that the upper part of a trench wall will be firm, dry material, but that a shallow layer at the bottom is waterlogged and unstable. The sides will stand for only a short time before becoming undermined by movement of the lower layer.

In general, caving of ditch sides is more apt to happen minutes, hours, or days after the digging than immediately. The usual exceptions are loose, sandy, or semiliquid muds that flow into the excavation, rather than cave or slide into it.

Stabilizing.   Sloping of sides for stability is a technique chiefly used for permanent open ditches, which will be discussed later in this chapter. It is seldom used for trenches for burial of pipes because of the large amount of extra digging, the space required, or the area of pavement, lawn, or other surface disturbed.

Vertical trench walls may be stabilized by bracing, draining, freezing, or chemicals. Bracing is the most common technique, and may be required by law.

Bracing Structures.   Figure 5.8(*A*) shows a light system of bracing or shoring used where danger of caving is slight. Planks are placed vertically in the trench at 4-foot intervals, and pressed against the dirt by means of push-type turnbuckles, called sheeting jacks. Bracing timbers are inserted and the jacks removed. The planks are usually 2 or 3 inches thick, the cross braces 6 × 6 or larger. Wide ditches require heavier cross braces than narrow ones.

A heavier type of shoring is shown in (*B*). The sides of the ditch are lined solidly with vertical planks, called sheeting planks, kept from falling inward by horizontal beams, known as walers, which are braced to each other across the ditch by timbers. These timbers are sprung into place by forcing the walers apart by sheeting jacks.

The weight and spacing of the planks and timbers will be determined by the depth and width of the ditch, and the instability of the soil. It is possible for a usually safe soil to be dangerously soft locally, due to disturbance of underground drainage, leakage of water mains, or other causes, so an ample margin of safety should be allowed.

OSHA.   The Department of Labor in the United States contains the Occupational Safety and Health Administration (OSHA), which oversees and regulates the conditions that could be hazardous for workers, including excavations. Federal law on construction standards (OSHA 2207, issued in 1990) says that employees shall be protected from excavated and other material falling or rolling into an excavation by keeping such material and equipment at least 2 feet from the edge of the excavation.

When a trench is cut, the vertical pressure in the adjacent ground is still resisted, but the horizontal pressure does not have anything to push against. The soil along the trench walls will bulge out slightly. It tries to act as an elastic band to keep the trench walls from bulging. But soil has only so much tensile strength and stretch, and when these limits are exceeded, vertical cracks form along the trench wall. Also when the horizontal pressure decreases, the water pressure in the voids decreases, allowing additional water to be sucked in from adjacent ground. This extra water pushes the soil particles farther apart, decreasing their shear strength. The process leads to trench collapse, which will happen faster in granular soils than in cohesive soils like clayey and silty soils.

Each employee in an excavation shall be protected from cave-ins by an adequate protective system designed to meet certain standards. If the excavation is not in solid rock and is 5 feet deep, or more, there are several options. One option is to slope the sides at an angle no steeper than $1\frac{1}{2}$ horizontal to 1 vertical, i.e., an angle of 34° measured from the horizontal. If the excavation is not deeper than 20 feet, another option is to bench the sides with the bottom lift being 4-foot and above that 5-foot-high lifts, with the slope of the edges to the bottom being $\frac{3}{4}$ horizontal to 1 vertical.

To reduce the amount of excavation, the protective system can function by shoring the vertical sides with structural materials such as mechanical or timber systems, as shown in the Fig. 5.8. Or the shoring can be a metal hydraulic system, as shown in Fig. 5.9. Each of these systems must be carefully designed to fit the earth conditions at the site.

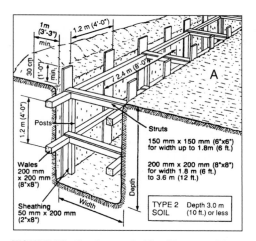 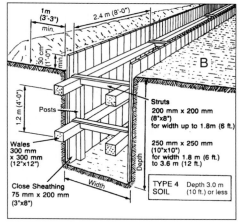

FIGURE 5.8   Bracing trench sides. (*Courtesy of Construction Safety Association of Ontario, Canada.*)

Design.   The OSHA standards contain many design requirements that must be satisfied to meet the regulations for worker safety in the excavation. The standards give the sizes and spacing of timber cross braces (struts), wales, posts, and sheathing for various soil types and trench depths. The timbers are to be oak or the equivalent, with a bending strength not less than 850 pounds per square inch. The soil types are cohesive soils with certain unconfirmed compressive strength (q_u), type A having high q_u as compared to type C which has low q_u. These types are similar to type 2 and type 4 in the above illustrations from the Construction Safety Association of Ontario. Their type 1 is hard ground to dig, in fact, close to rock. There are other standards for aluminum hydraulic shoring that are followed by the manufacturers of that form of shoring. OSHA requires a ladder, stairway, ramp, or other means of access or egress for any excavation deeper than 4 feet. Whatever safe means of access or egress is used, it cannot be located farther than 25 feet away from any employee.

OSHA requires the contractor to sign a Competent Person who is capable of identifying existing and predictable hazards dangerous to employees, and who has authority to take prompt measures to eliminate them. This person does not need to be an engineer, but must have training in, and be knowledgeable about, soil analysis, the use of protective systems, and the requirements of this standard.

Immediate Bracing.   With any bracing the ditch is made sufficiently wider than the bucket that it can work between the struts. The ditch is dug to a depth of about 2 feet, full width, and the top pair of walers placed, and cross braces set with such spaces that the bucket can get down between them. Planks are set vertically touching each other outside the walers.

The shovel, preferably a clamshell, now digs a foot or two below the waler. Workers with hand tools dig the dirt out from under the vertical planks, allowing them to settle, and also remove dirt under the crossbeams which the buckets cannot reach. This dirt is piled in the middle of the ditch and is taken out by the bucket when the laborers are out of range. The shovel then digs more deeply and is followed by hand tool work. At a depth of 2 to 5 feet below the top waler, another pair of walers is set inside the planks and braced across the ditch. Alternate excavation by shovel and hand tools, undermining and dropping of side planks outside the walers, and installation of additional beams are continued until bottom grade is reached. Ordinarily, the walers and crossbeams (struts) are either heavier or more closely spaced with increasing depth as the potential pressure increases.

If the ditch is deeper than the length of available planks, those started at the surface should be of variable length. As each one drops below the top waler, another plank is placed on top of it to follow it down. Mixed lengths make this possible without weakening the structure by having a row of these joints together.

The waler beams are also of different lengths so that both members of a pair do not end together. The joint between any two can therefore be braced against a solid beam on the other side.

Two- or 3-inch sheeting, and 6×6 walers and crossbeams, spaced 5 to 8 feet apart, are strong enough for moderate depths in most soils. If more protection is needed, heavier wood may be used, additional planks can be driven outside the sheeting, or inserted inside by a complicated process of removing and replacing walers. Steel sheet piling is much stronger than wood.

Movable Bracing.   When the work which is to be done in the ditch can be completed in short sections so that the ditch can be backfilled a few yards behind a backhoe, a portable bracing structure can be used. See Fig. 5.9.

It may be made up of steel or wood, and should be equipped with a tow bar or chain at the front bottom, which can be gripped by the bucket teeth. It is lowered into the first section dug, and the pipe laying or other work is done inside it while another section is dug. The shovel drags it along in the ditch whenever sufficient digging or pipe laying has been completed to justify moving it.

Such a device can result in tremendous savings. However, it cannot be used on many jobs because of the necessity of checking the work, or having it inspected. Also, if the sides should close in on it, it might be very difficult to free up for moving.

Backfilling should be done as soon as possible, as allowing the sides to cave may damage the pipe, or shift it out of line.

Flowing Banks.   If the sides are so unstable that they cave or flow immediately upon being cut, the sheeting planks or sheet piling must be driven down by air hammers or pile drivers, and the dirt dug

FIGURE 5.9   Manufactured trench shield. (*Courtesy of Efficiency Production, Inc.*)

from between them afterward. Penetration and control of direction are usually best if the planks are driven only a short distance below the digging. However, mud may flow so readily that the sheeting must be down several feet below excavation level to prevent it from swelling on the bottom. All gradations between this condition and stable banks may be encountered in a short distance.

Washout Failures.   Only wet ground or loose sand exerts very heavy thrusts against the shoring. Water draining down the sides of a braced trench may erode them so the sheeting moves outward, thus loosening the crossbeams or jacks and allowing them to fall, after which the sheeting can be pushed in by any movement of the banks.

Stabilization.   Soft ground may be stabilized by chemical treatment, well-point pumping, drainage, or freezing, by procedures described elsewhere in this book.

STREAM CROSSING

Pipelines must often be built across both large and small streams.

If a wide stream or a pond has a soft bottom and is not used by ships or barges, the pipe may be laid directly on the river bottom. However, it is often necessary to protect it by burying it.

Excavation may be done by a grapple dredge; by a clamshell, dragline, or hoe on a barge made up for the purpose; by a cable excavator on the bank; or from a temporary jetty.

Digging a straight trench from a dredge or barge takes experience. Anchors and winch lines must be so arranged that the barge can both pull itself across the stream, and keep or regain position against pressure of the current. Alignment is checked by surveying instruments that may be on the barge, on the shore, or in both places.

Many streams have sufficient current to fill in a trench as fast as it is dug. At low-water periods it is usually possible to block off a substantial part of the stream channel with a fill or jetty, trench in it or just downstream from it, lay the pipe, and remove the fill. The operation is then repeated in another section. Erosion at the ends of the fill may be severe.

The jetty may be built from one bank to the stream center, and then from the other bank to connect, or a dragline may build itself an island that it moves by digging at the rear and filling the front.

The pipe may be assembled in sections as trench space becomes available; or the whole crossing may be put together at one time, dragged and floated across the water, anchored in approximate position, and sunk into the trench as sections of it are completed. Floating and sinking of the pipe are regulated by the amount of air or water in it, and/or by floats and weights.

Work of this type is subject to disastrous damage if the stream floods. If there is a little warning, equipment can be pulled back from shore-based fills, but a dragline on its own island is likely to be lost. A wire rope connection should be kept to ensure getting back the crew if the water rises too suddenly to permit rescue by boat, and to help locate and salvage the machine if it capsizes. Floating pipe may be pulled back to shore.

Because of flood danger, all possible preparations are made in advance, and the crossing pushed as rapidly as possible once it is started. The urgency depends largely on the past behavior of the stream at that time of year, but few streams are ever entirely secure against flooding.

Small Streams.   It is seldom practical to ditch directly across a brook without taking precautions to keep it out of the trench. This may be done by digging a sump hole upstream, and using a pump or pumps to move the water across the ditch line and back into the stream.

A less expensive method is to confine the stream flow to pipes, and work under them. Sufficient pipe to accommodate the expected flow is laid in the stream across the ditch line. Sectional corrugated pipe should have seams filled with mastic to prevent drip leaks.

Dams are then built across the brook above and below the ditch area, confining the flow to the pipes. There may have to be two sets of dams, a first pair of light temporary ones toward the pipe ends to permit drying out areas where the regular dams can be built of selected soil, carefully tamped around the pipes. Bentonite might be added to the soil to improve water resistance. One dam may be made wide enough to serve as a road for machinery.

The ditch is then dug in the regular manner, under and on each side of the pipes. Water oozing into the ditch can be pumped out, diverted by well pumping, or blocked off by grouting.

When the pipe has been laid in the trench and the path or road across the stream is no longer needed, the dams are dug away and the trench backfilled, and the pipes are lifted out for reuse elsewhere.

If stream flow increases beyond pipe capacity, it will overflow the dams, fill the trench, and stop work. After the flow has subsided, the dams are repaired and the trench is pumped out. Redigging to remove washed-in soil may be necessary.

PERMANENT DITCHES

When a ditch is to be left open permanently, its sides usually must be protected by masonry or rot-resistant sheet piling, or sloped back far enough that they will not slump, cave, or wash into the bottom. If a large volume of water may flow through the trench at any time, the bottom should also be protected against erosion, unless the gradient is so flat, or the water so burdened with silt, that cutting will not occur.

Ditches with low gradients, or which carry dirty water, must be cleaned out periodically by a dragline shovel or other excavator. Masonry, riprap, and in particular vertical stone walls interfere with machine digging and are liable to be damaged. This should be borne in mind in designing any artificial protection.

Sloped Banks.   The most satisfactory bank protection for a country ditch is a stable slope and a good cover of vegetation. This can be reinforced on the outside of bends and other places subject to strong current action, by placing large boulders, walls, riprap, or light piling holding wire or brush mats.

Stable slopes vary in steepness with the character of the soil. Loess may stand indefinitely in vertical cliffs, while certain types of clay may slump if the slope is 1 on 6. Generally, it can be said that slopes of 1 on $2\frac{1}{2}$ or less are advisable if the soil contains much clay or silt; if there is movement of groundwater through it toward the ditch; if there is considerable drainage of surface water running down its face; or if it is in layers that dip toward the ditch.

Vegetation.   If trees or bushes are to be planted or allowed to grow, the slope can be steeper than if it is to be kept in grass. Trees have greater holding power, and maintenance of grass may require the use of mowing machines, whose ability to work on side slopes may determine the grade. However, trees will interfere with access for cleaning.

If the trench is partly or wholly in barren soil, topsoil may have to be spread on the banks to encourage vegetation, although some plants generally can be found that will grow well on the subsoil if encouraged with lime or fertilizer.

Banks of permanently wet ditches may be strengthened by laying willow poles or logs up and down the bank, 2 to 4 feet apart. They should be settled well into the soil for their full length, with the lower end in water or bottom mud, and staked or wired in place.

These poles should grow tops and a continuous root mattress.

Bottom scour may be largely prevented by keeping a gentle gradient. A grass lining will help prevent scour in a permanent ditch, as shown in Fig. 5.10. If part of the trench is too steep, check dams may be built.

This localizes the fall of the water at the erosion-resistant aprons of the dams. Good results can sometimes be obtained with plank dams, or even with heavily anchored brush mats.

FIGURE 5.10   Grass-lined ditch. (*Courtesy of North American Green.*)

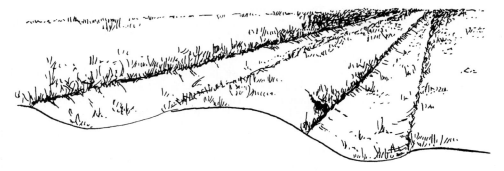

FIGURE 5.11   "W" ditch.

Spoil Arrangements.   The spoil piles from a ditch are apt to interfere with local surface drainage. If the land is flat, spoil may be piled on both sides, but frequent breaks should be made in the windrows so that water will not pond behind them. If the ditch cuts across a slope, these breaks need be made in the upper pile only. Until the ditch slope is protected by vegetation, gullies may form at these spots unless the area is protected by pipe, flumes, or stone.

The "W" ditch seen in Fig. 5.11 is a double ditch, separated by sufficient space to provide disposal area for the spoil. This eliminates any blocking of drainage on either side, and allows maintenance of field grade to the ditch edge.

If the depth of the ditch is determined by the flow capacity required, two can be shallower than one. This construction is only slightly more expensive than the single drainageway, although it usually takes more land out of production.

If depth is determined by a flow gradient, a W ditch will about double the amount of excavation required. If the depth is considerable, the additional spoil is likely to damage an excessive area.

Whenever possible, the spoil piles from permanent ditches should be rounded off so as to permit easy access to the ditch and to make them less prominent in the landscape. This should be done before they are overgrown by trees.

PIPE AND CONDUIT TRENCHES

Most trenches are dug for the purpose of burying pipes or conduits. Conduits, and pipes for gas and water supply, run at more or less fixed depth below the surface. Sewers, storm drains, and other gravity-flow pipes must maintain a minimum gradient from source to outlet or booster pump, and will have a variable depth below an irregular surface,

Fixed Depth.   In cold-winter areas, water pipes are laid below the frost line in the ground. Conduit and wires are laid only deep enough to protect them against accident. In either case, depth may be increased under sharp ridges so as to provide smooth vertical curves.

In fields, the most important menace is the moldboard plow, which penetrates 8 or 10 inches. There is a chance that a subsoil plow, with a penetration of 18 inches to 2 feet, might be used; in addition, land even on gentle slopes gradually washes away, and the surface may be lowered several inches during the life of the conduit. Depths of 2 to 4 feet are usual, and are figured from the surface of the ground regardless of slope.

Gravity Systems.   Close supervision is required to ensure accurate digging for a sewer or other gravity system. A number of methods are used to keep ongrade, following the surveying methods discussed in Chap. 2.

Plotting Profiles.   If there are no plans, or if they merely specify a gradient, the surface profile should be drawn to scale on a sheet or strip of cross-section paper.

In Fig. 5.12 a basement has been dug, and it is desired to lay a pipe on a slope of 1 foot in 50 feet, to take water from a tile drain laid around the outside of the footings. An instrument is set up and the elevation of the basement floor taken. This is arbitrarily assigned a value, say 10, and is used as a benchmark for the rest of the work. Another benchmark on a tree or some surface spot not affected by building work should also be taken for future reference.

A profile is then taken along a downslope, every 25 feet, until points are found well below the floor level.

The figures obtained are plotted on a piece of cross-section paper, which might be 10 squares to the inch. A horizontal scale of 1 inch to 25 feet and a vertical scale of 1 inch to 5 feet are selected. The width of each printed square will then indicate 2½ feet, but its height only 6 inches.

The basement is sketched in, and a base or zero line drawn 20 squares below its floor. The stations (points where elevation is measured) are marked on the vertical lines, one for each inch.

Each of the station elevations may now be marked on the diagram by measuring up from the baseline one square for each 6 inches. These dots are connected by a line which is a picture of the surface slope, with its steepness exaggerated.

The ditch may now be drawn in. A distance of 1½ feet below the basement floor is marked 3 squares down, for the tile. At the last station, 2 + 0 (200 feet), measure 8 squares (4 feet) down from the tile.

A straight line drawn between these points represents a drop of 4 feet in 200, which is the 1 foot in 50 feet that is wanted.

Measurements on this sketch will now give the length of pipe needed, the distance on the ground to the outlet, the elevation of any point on the ditch bottom, and the depth of the ditch anywhere.

A larger scale in which each square would represent a smaller distance will give more accurate readings.

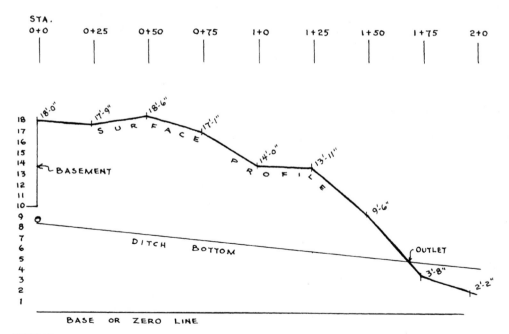

FIGURE 5.12   Determining the gradient.

Finish Levels.   When the digging of a ditch section has been finished, boards may be placed on edge across the ditch at 10- to 25-foot intervals, and staked or otherwise firmly fastened in undisturbed soil, in position such that a tight string may be run over the ditch and adjusted by instrument readings to be parallel with the final grade of the bottom. Extra strips of wood may be nailed on, or notches cut into original boards, if they are too low or too high. Finishing of the bottom, and placement of pipe, is governed by measurements down from this string.

Laser.   Lasers provide means for very exact control over line and grade in trenching, and in pipe placement.
    Lasers are discussed in Chap. 2.

BACKFILLING

General Methods.   Trenches dug for laying of pipes or conduits must be backfilled when the installation is complete. The dirt taken out is pushed or pulled back in. This job can be handled by most earthmoving machines, but the bulldozer is the standard tool for the purpose.
    If the backfill need not be compacted from the bottom up, it may be pushed into the trench in the ways shown in Fig. 5.13. The bulldozer operates at right angles to the trench, taking as large a slice of the pile as it can handle comfortably. Dirt which drifts across the blade is left in windrows that are pushed in a separate series of passes, with poor efficiency, as in (*B*). Or the

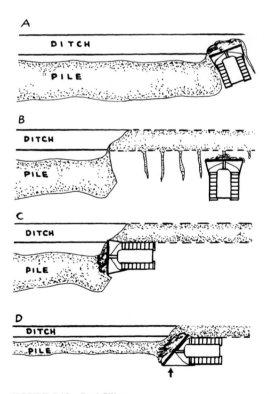

FIGURE 5.13   Backfilling.

remaining soil is pushed parallel with the ditch into the main pile as in (C), or, when the end is reached, distributed along the ditch.

There will usually be too much soil because of the increase in volume of disturbed soil and space occupied by the pipe laid. The excess may be mounded up over the trench and partly compacted by use of a roller, or by driving the dozer or a truck along it. Full natural settlement may take as much as a year, and is liable to leave low and high spots to be graded in.

If the trench is small, it may be refilled by running an angle dozer or a grader through the pile, with the blade set to sidecast it into the ditch as in (D).

Heavy backfill may be done by a front shovel walking through the length of the pile, digging it and dumping in the ditch. A backhoe or a dragline can work from across the ditch, pulling the soil into it. A dragline's efficiency will be greatly increased by fastening a heavy plank or other block across the mouth of the bucket so that it will not fill. Shovels often are used to move the bulk of the backfill, with a bulldozer doing the final cleanup.

When pavement along the sides of the ditch is to be preserved, the best tool is a rubber-tire loader or dozer, but light or medium crawler machines, with semigrousers or flat shoes, may be used.

Special dragline-type backfillers are often used on cross-country trenches.

Compacted Backfill.   If a pavement is to be laid over the refilled trench immediately, the backfill must be carefully compacted from the bottom up. This may be done by dozing or hand-shoveling fill slowly, while workers in the trench compact it with hand or pneumatic tampers. A mechanical tamper may work from the side or straddle the trench. The top layer may be compacted by use of a trench roller with a large wheel that will fit inside the ditch, or by running any heavy machine back and forth along it.

Open-textured soils may be effectively compacted by puddling. Enough water is added to the fill to make it into mud, which, upon drying, will shrink considerably. Fine-grained soils take a long time to dry, and thus are not as readily handled in this way.

Machinery should not be run along wet trench fills as it is almost sure to get stuck in them.

Imported Backfill.   Drainage trenches may be filled with better-draining, more porous material. Spoil removed in the original digging is trucked away or used in grading, and the gravel trucked in for refilling. This may be dumped in piles in and alongside the ditch, and pushed into it by a dozer or grader or hand-shoveled. If considerable work of this type is to be done, a backfilling machine may be profitably used. This carries a hopper that is moved parallel with the ditch by rubber-tire driving wheels. Trucks dump into the hopper, from which a belt carries the soil to the ditch and dumps it. The backfiller can push the truck that is dumping into it, so that the truck driver can concentrate on lifting the body at proper speed.

DEWATERING

DRAINAGE

Both the surface and the subsurface water may be removed by a seasonal drop in groundwater level; by drainage through ditches, pipes, or siphons; by pumping; by walling off or diverting the sources of water; and very often by combinations of two or more of these methods.

The purpose of dewatering may be to promote growth of crops; to dry out swamps or other objectionable wet areas that are not designated wetlands (see Chap. 6); to stabilize slopes, foundations, and road subgrades; or to facilitate excavation for any purpose.

All of these objectives except the last are accomplished chiefly by drainage—that is, causing the unwanted water to flow away through artificial and natural channels or conduits. Pumps may be used to remove water from a sump or low point of a drainage system.

Gradients.   The slope or gradient of a drain will depend on the work it has to do. In tidal marshes and other practically flat swamps, ditches with zero gradient may serve to lower the water level substantially. In general, water will flow through a flat ditch, but is easily choked by sediment, growth of weeds, and dirt falling from banks, as the water flowing through it will have little or no ability to clean it. Too steep a ditch gradient may cause erosion of the bottom, undermining of banks through stream action, and damage from depositing of mud below the discharge point.

The slope must be adjusted first to the necessities of the situation, and second to the relation between the amount of water to be carried and the nature of the soil. A bottom gradient between 1-foot drop to 1,000 feet and 2 feet to 100 feet is desirable under most conditions encountered.

Drainage pipes should not be flat, as costs in cleaning out sediment and debris will be very high. Low gradients can be used when the water is clean, the pipe is short and large enough to allow personnel to work in it conveniently, or there is a sharp fall at the outlet so that water will flow rapidly. Generally, the minimum gradient should be 6 inches to 1,000 feet, and the maximum 2 feet in 100 feet for land title, and 10 feet in 100 for tight joint pipe.

Surface Water.   Surface drainage may consist of disposal of water from rain or melting snow, or lowering the water level in ponds, ditches, or swamps. It may use open channels, conduits, or both. The water is usually led to a natural stream or body of water.

Such drainage may be accomplished by deepening, enlarging, or straightening and protecting existing streambeds; by digging artificial channels; or by installing underground pipes or tunnels.

There is no definite separation between surface and subsurface drainage as they operate on different parts of the same water mass.

Trenchless Excavation.   If it is not practical to ditch to install a drain or diversion pipe, boring or tunneling may be used. Trenchless excavation construction methods include all the methods of installing utility systems below grade without direct installation into an open-cut trench. These methods are differentiated from the large-diameter tunnels by several factors, such as purpose or use and diameter of excavation. Tunneling is discussed in Chap. 9.

Classification.   There are three major categories of trenchless excavation: horizontal earth boring, pipe jacking, and utility tunneling. Horizontal earth boring includes methods in which the bore-hole excavation is done by mechanical means without workers inside the bore hole. See Fig. 5.14. Both pipe jacking and utility tunneling methods require workers inside the bore hole during the excavation and casing installation process. However, they are differentiated by the support structure. The pipe is the structure in one, and the other has liners installed for the structure.

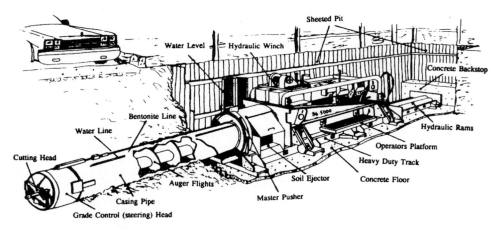

FIGURE 5.14 Trenchless excavation with earth auger borer. (*Courtesy of ASCE Journal of Construction Engineering Management.*)

FIGURE 5.15   Trenchless excavation with earth boring machine. (*Courtesy of Bor-It Mfg. Co. Inc.*)

Earth Boring Methods.   The horizontal auger earth boring method uses the process of simultaneously jacking the casing through the earth while removing the spoil inside the encasement, as shown in Fig. 5.15. There is the slurry rotary drilling method, which uses drill bits and tubing with fluid to remove the spoil earth, instead of augers and cutting heads. The drilling fluid is a bentonite slurry, water, or air which aids in the removal of spoil. This method of horizontal directional drilling is discussed in greater detail in Chap. 20.

There is also the compaction method which compresses and displaces the earth surrounding the casing. This method is restricted to relatively small lines, perhaps 2 to 6 inches in diameter in compressible soil. Another is the water-jetting method, which uses the principle of soil liquefaction to make the bore hole. This is similar to methods for sinking pipe in the ground vertically. There are several other methods for horizontal earth boring, but these should be enough to give an idea of trenchless excavation.

Siphons.   If a drainage line is to be used only occasionally, the expense of ditching or tunneling may be avoided by use of a siphon. This is an airtight pipe or stiff hose across the water barrier, with one end in the water to be drained, the other at a lower level (Fig. 5.16). Maximum rise above water level is about 25 feet. Lower lifts work better.

When the siphon pipe has been filled with water, that which is between the high point and the lower or discharge end moves down the pipe by gravity, while atmospheric pressure, acting through the pond water, pushes the shorter and lighter column of water after it. This water is in turn renewed from the pond so that movement continues until the water level drops sufficiently to outbalance the suction, or to allow air to enter the pipe; air enters through leaks or through the discharge end, or water rises around the outlet to the same level as the intake.

The rate of flow will depend chiefly on the drop between the top of the water being drained and the point where the water loses contact with the outlet end of the siphon. As a pond is drained and its level drops, the flow will become slower.

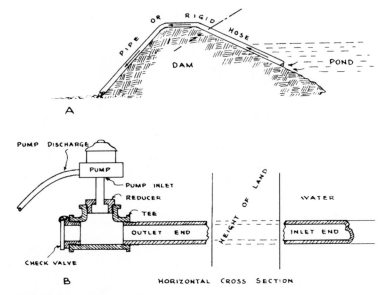

FIGURE 5.16 Siphon and priming pump.

Siphons in which water moves slowly are likely to be stopped by air entering the top of the outlet. This may be prevented by putting the outlet in a box or small pool so that the opening will be under water. A slow current may also allow an air lock to form from the accumulation of air or other gases escaping from the water in the pipe, or leaks from the outside.

Very small siphons may be started by mouth suction, and a medium size by inverting it so that the ends are higher than the middle, filling it with water and holding the ends closed while placing it in position. Or a tee connection in the top may be used to pour water in, keeping the ends plugged until the pipe is full and the tee tightly plugged.

The most satisfactory way to start a large siphon is with a suction pump. A way to connect it is shown in (B). A tee is placed on the outlet end, with the side opening reduced to fit the inlet hose of a small diaphragm or centrifugal pump. Means are provided to prevent air from entering through the lower opening of the tee, by means of a check valve, a screw plug, or a piece of plywood with mud on it. With this stop in place, the pump is started and the air sucked out of the siphon so that water from the pond is drawn through it into the pump. The stop on the main pipe then opens or is removed, and the pump is shut off.

Channels. Channels may consist of natural watercourses; watercourses which have been enlarged, straightened, or paved; or artificial ditches.

A streambed may be dredged to lower its level, to increase its depth or capacity, to keep it from changing its course, or to change its course.

Level may be lowered to drain surface water from a swamp or pond, or to provide better under-drainage for land in its vicinity.

Depth is usually increased to assist navigation, or to provide for more rapid runoff of floodwater. Widening and straightening increase capacity, often at the expense of depth.

Streams normally wander in their courses, cutting away banks in some places and building them in others. When valuable property or structures are threatened by these changes, the channel may be artificially shaped to direct the force of the water away from them. This may involve turning the water back to its original direction, or forcing it to flow in a new one.

Dredging of small streams is generally done from the banks by draglines or clamshells, and of large ones by floating dredges. The material dug may be piled on the banks, or removed by trucks or barges.

When the spoil is used to build banks to control stream direction, it must be protected by paving, masonry, rock, logs, wired brush, sod, or other material. The best emergency protection for a bank that is being washed away is drilled boulders fastened together in groups of three with steel cable.

River dredging may be planned to direct the river current so that it will do most of the excavating in the new channel.

Drainage channels are often paved to protect them from erosion or slumping, to prevent changing of course, and to increase capacity by reducing friction.

Irrigation canal pavements may be used for any of these purposes and to prevent water from leaking out of the canal into surrounding soil.

Check Dams.   When the slope of a channel or gutter is so steep as to make erosion likely, it can be divided into a series of easy gradients, and separated by check dams over which the water falls steeply.

It is important that each dam have a center spillway large enough to prevent water from overtopping the edges and eroding the earth alongside. An apron is also necessary to prevent undermining.

Where elaborate structures are not practical, crude ones made out of brush and logs or loose stones may serve the purpose.

PIPE

Drain and culvert pipe is made in sizes with inside diameters ranging from 3 inches to 15 feet. Materials include concrete, tile (vitrified clay), plastic or PVC, and corrugated iron or steel. Figure 5.17 indicates the names of various parts of a pipe cross section. There are pipe design books to use for selecting drainage pipes.

Ordinary vitrified clay pipe is comparable in strength to plain concrete sewer pipe, and fiber-reinforced plastic to reinforced-concrete culvert pipe. The tile is brittle and fragile to handle.

Concrete.   Concrete pipe may be plain or reinforced; the joints may be butt, bell, slip joint, or gasketed. Size range is from 4 inches in inside diameter and up. Lengths are 2, 3, 4, and 8 feet.

Butt (open) joints are used for land tile.

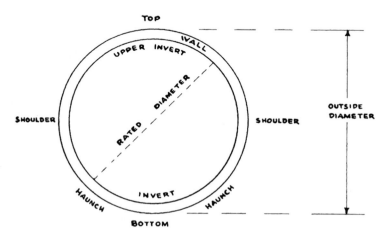

FIGURE 5.17   Pipe details.

Bell joints are resistant to chipping, will hold the pipe against slipping downhill, and, if the joints are open, will reduce flow or seepage of water along the outside of the pipe; but they are difficult to lay.

Slip joints are easier to handle and to lay because of the uniform outside diameter.

Pipes with greater than 12-inch inside diameter are usually reinforced, and this construction is required on most jobs because of its additional strength.

Concrete may be attacked by water carrying certain alkali salts or other chemicals. It is subject to erosion from fast-flowing water carrying abrasive material, and may scale or disintegrate slowly from weathering. Structural difficulties may arise from the comparative weakness of its joints. However, under a wide range of conditions, it is long-lived enough to be considered a permanent installation.

Ductile Iron Pipe.   Some utility companies still use ductile iron pipe because of its inherent strength, toughness, and versatility.

Tile.   Tile may be porous or glazed, and is chiefly made in small and medium diameters, and in 1- to 4-foot lengths. The porous type usually has butt joints and is called land tile. Standard glazed or vitrified tile has bell joints and is called sewer tile.

Tile is lighter than concrete and has excellent bearing strength and resistance to weathering and corrosive chemicals. Its glazing resists erosion. It is fragile, and must be handled with care. In small sizes it is cheap and easy to lay except on unstable ground.

Plastic Pipe.   In small sizes, tile or concrete may be replaced by PVC (polyvinyl chloride) plastic pipe. Standard pieces are 10 feet long, with butt ends that may be linked by sliding collars. Bottoms may be solid to carry water, or perforated to take it in.

This pipe is light, easy to lay, and long-lasting. It is not brittle, and it is seldom broken in handling.

Corrugated Metal.   Corrugated pipe is made in standard, helical, and heavy-duty-constructions. Cross section may be round, elliptical, flattened, or arched.

Standard pipe, illustrated in Fig. 5.18, is made up of galvanized plates of rust-resistant (but not rust-proof) iron or steel, 16 gauge to 8 gauge, which are deformed with parallel corrugations or ripples. These are usually $2\frac{2}{3}$ inches from crest to crest and $\frac{1}{2}$ inch deep. They increase the strength of 16-gauge about 11 times, and of 8-gauge about $3\frac{1}{2}$ times.

The corrugated plates are rolled into cylinders slightly more than 2 feet long, which are lapped and riveted together. Additional cylinders are lapped over the ends and riveted to obtain the desired length.

Inside diameters of standard riveted pipe range from 6 inches to 8 feet. Lengths may be made up in any multiple of 2 feet, but transportation problems usually limit single pieces to 20 or 24 feet. Ends may be beveled or skewed.

Features of arched pipe are lower clearance, greater bottom capacity, and less tendency to settle in soft ground.

Pipe sections are fastened together on the job by band collars. These may be one-piece or two-piece. One-piece bands are usually fastened by compression bolts only. Two-piece may be riveted or bolted to the sections. Because of allowance for overlap, each pipe section is $1\frac{1}{2}$ inches longer than its nominal length. Each joint adds the width of one corrugation.

Under normal conditions, corrugated pipe gives long service, but its life may be shortened by chemicals or electrochemical action, and by erosion of the bottom.

Corrugated pipe is very much lighter than concrete or tile; it is not as readily damaged by carelessness or abuse; it is easily placed, connected, extended, or removed for salvage, and resists movements of fill which would pull short-jointed pipes apart. Its internal flow resistance is higher than that in other types. Its corrugations tend to keep it from moving in the fill, and discourage seepage or overflow from following the outside. It will bridge low or weak spots in its supports.

For subdrainage it can be drilled with $\frac{5}{16}$- or $\frac{3}{8}$-inch holes through the haunches.

Other Types.   Oil or grease drums may have the ends cut out and the cylinders tack-welded together. Such conduit is easy to handle, but compressive strength and resistance to corrosion are poor.

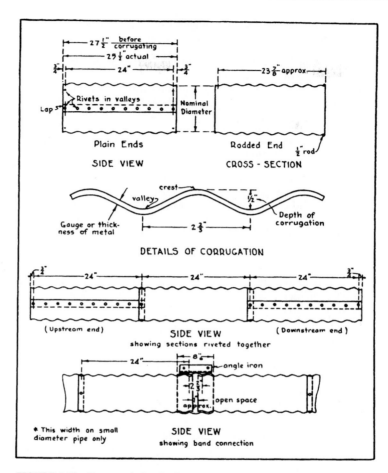

FIGURE 5.18 Corrugated-pipe detail.

It may be strengthened somewhat by stretching it vertically, in the manner to be described for corrugated pipe. Struts may be left in permanently, but are likely to catch debris and to cause clogging. Such installations usually are temporary.

Threaded iron water pipes or well casings are sometimes used for small drains or culverts when they can be obtained second-hand, or no other pipe is available. If the used pipe can be bought cheaply, it is ideal for draining small quantities of water through drives or fills during construction, if the regular drainage system is delayed. It will not pull apart under any natural stress, resists bending, can be cleaned out with a plumber's snake, and is easily salvaged by pulling out from the end, or digging out.

BRIDGES

When a road or other continuous embankment crosses a stream or drainageway, it is usually carried over it on a bridge or a culvert.

These structures may be distinguished from each other on a basis of width of opening. The critical width, or span, at which a bridge becomes a culvert varies from 5 to 20 feet in different localities.

Log. Figure 5.19 shows a type of log bridge suitable for carrying a pioneer or haul road, or a driveway, across a small stream.

Sill logs are set into the bank parallel with the stream, far enough back from it to be secure against being undermined. They may be braced by bolting to stumps or driven piles, or by cables stretched to anchors behind them. This anchoring is very important, as streams often flood sufficiently to float wood bridges.

Logs strong enough to carry expected loads are then placed close together across the stream, resting on the sills and preferably being fastened to them and to each other by lag bolts in drilled holes. Butts and tops should be alternated, so that log taper will not make one side wider and higher than the other. These logs are called stringers.

Next, straight-grained logs of smaller size should be split in quarters or other wedge-shaped fractions and placed, split sides down, between the tops of the big logs. They may be cut into as short sections as necessary to fit snugly and to fill the cracks between the logs. Sections of round poles may be used in the same way.

The split pieces should each be fastened to one log with spikes or lag bolts. Fastening can be on either side, but must not be on both, as that would permit movements of the logs to split them or pull the fastenings loose.

Quarter logs or poles are then fastened to the outside logs, to serve as curbings to prevent vehicles from running off, and/or to retain surfacing. The wood surface may be left exposed for light or occasional use, or covered with gravel or dirt to make it smoother and reduce danger of tire damage if nails work up.

Decks may also be made of planks or split logs nailed at right angles to the stringers.

If the span is long, or the loads are heavy, a stone-filled log or timber crib may be used as a center support. Cribbing may also be used at one or both banks if they are too low or soft to give proper support.

The strength of wood of various species, and in different conditions, varies so widely that individual judgment must be exercised in selecting the logs. If the bridge is to be used over a period of years, resistance to rot may be more important than initial strength.

Greenwood is strong but lacks rigidity, and tends to give too much bounce to a long bridge. It will usually bend and splinter before breaking.

FIGURE 5.19 Log bridge.

If only one side of the stream is accessible to machinery, the logs may be pulled across from that side by the use of a cable through a snatch block anchored on the opposite bank. The block should be placed high, if possible, to prevent the end from digging into the bank.

Concrete.   Concrete bridges consist in general of two abutments supporting a slab. The slab usually includes guardrails, and supporting ribs or stringers which may be flat or arched. The abutments are usually continued into wing walls to direct the stream through the opening and to protect the embankment against sliding or erosion.

Even small structures are quite heavy and require that the abutments rest on solid footings. The flow of the stream should not be restricted, as it might then scour out the material against the abutments and undermine them. Abutments must be strong enough to resist the horizontal thrust of the fill behind them.

Reinforcement should be used throughout the structure, and is particularly important in the slab and its ribs.

The forms for the slab must be supported on a temporary wood or steel bridge of considerable rigidity.

Bridges should be engineered for the site and conditions. Construction should not be attempted without experienced supervision.

CULVERT DESIGN

A culvert may be made of almost any structural material. Reinforced concrete or corrugated metal pipe, and poured reinforced concrete are standard for highways and railroads. Tile and plain concrete may be used for light service. Log and timber construction are usual in pioneer and military roads.

Water passages (barrels) may be round, arched, rectangular, or in special shapes. More than one may be used.

Capacity.   A culvert serves to carry the water from a drainage area or watershed of a certain size. This water includes surface runoff of rain and melted snow and ice, and whatever groundwater comes to the surface within the area.

The size of culvert opening should be determined by the amount of rain which is likely to fall in the watershed within a certain period, and the character and slope of the ground so far as they affect the percentage of water that will run off, and the speed of its flow.

Additional factors to consider are the opening required by normal stream flow before it rains; the extent to which the opening may be restricted by silting; the velocity of water in the culvert; the extent to which water not passed through it can pond against the embankment before overtopping it or damaging property behind it by flooding; and the probable damage from overtopping.

Runoff.   Rate of runoff is determined by intensity of rainfall, the size and shape of the watershed, and the slope, plant cover, and permeability of the soil.

Rainfall is measured in inches, and its intensity in inches per hour, although the period of measurement may be less than an hour. For example, a rainfall of 3 inches might fall at the rate of 6 inches per hour for 30 minutes. In calculating runoff, an adjusted or equivalent rate can be used which makes allowance for variations in rate and duration.

Each watershed has a period of concentration, at the end of which the runoff is assumed to be at a maximum. This is the time required for water to flow from the farthest point in the shed to the culvert. If rainfall is continuous, and ground conditions are unchanging, the runoff at the culvert will increase from the beginning of the rain until it includes water from the whole area, after which it will continue at the same rate.

This period will be longer for long, narrow watersheds than for square or round ones of similar area.

The assumptions involved are not strictly accurate, as runoff increases as the ground becomes saturated, as water penetrating the soil emerges at lower levels, and the rate of flow is more rapid as the volume in channels becomes larger. However, there are so many variables that exact results cannot be obtained, and the average culvert is not important enough to justify an individual study of its drainage area.

The intensity of rainfall will determine the amount of water that will fall on an acre. The ground, slope, and vegetation will regulate how much of that water will flow off, and the speed of its flow. The number of acres in the watershed will determine the total amount of water delivered to the culvert. The period of concentration will determine the length of rainfall necessary to bring the area to the point of full discharge.

There are a number of formulas used in runoff calculations. These may give the volume of water to be expected, or the area in square feet of the culvert or bridge opening required. Information can also be obtained from performance of existing culverts or bridges, and observed heights of floodwater.

The value of results obtained varies with the care with which field studies are made and with a number of factors that are difficult to work out. However, for the contractor who wishes a general guide to culvert size requirements, the simplest method is best.

Figure 5.20 contains two maps showing adjusted rainfall rates in inches per hour for average requirements, and for any installation where overflow or backing up is particularly undesirable. The table supplies the number of square feet of culvert openings required to drain various areas on the basis of 1 inch of rain per hour.

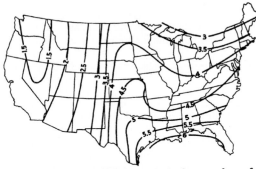

Equivalent rainfall rates in inches per hour for *average* design conditions.

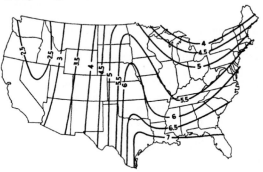

Equivalent rainfall rates in inches per hour for *unusual* design conditions.

Waterway Areas (Sq.Ft.) Required to Drain Different Acreages, M, for Equivalent Rainfall Rate of 1 In. per Hour

M acres	Flat areas not affected by accumulated snow. Length several times width	Rolling farm land. Length of watershed three or four times the width	Rough, hilly watersheds having moderate slopes	Steep, barren watersheds having abrupt slopes
2	0.08	0.14	0.28	0.42
4	0.14	0.24	0.47	0.71
6	0.19	0.32	0.64	0.96
8	0.24	0.40	0.79	1.19
10	0.28	0.47	0.94	1.41
15	0.38	0.63	1.27	1.91
20	0.48	0.79	1.58	2.36
25	0.56	0.93	1.86	2.80
30	0.64	1.07	2.14	3.21
35	0.72	1.20	2.40	3.60
40	0.80	1.33	2.65	3.98
45	0.87	1.45	2.89	4.34
50	0.94	1.57	3.14	4.70
60	1.08	1.80	3.59	5.39
70	1.21	2.02	4.03	6.05
80	1.34	2.23	4.46	6.69
90	1.46	2.43	4.87	7.31
100	1.58	2.63	5.27	7.91
150	2.14	3.57	7.14	10.7
200	2.66	4.43	8.87	13.3
250	3.14	5.24	10.5	15.7
300	3.60	6.00	12.0	18.0
350	4.05	6.74	13.5	20.2
400	4.47	7.45	14.9	22.4
450	4.89	8.14	16.3	24.4
500	5.29	'8.80	17.6	26.4
600	6.06	10.1	20.2	30.3
700	6.81	11.3	22.7	34.0
800	7.52	12.5	25.1	37.6
900	8.22	13.7	27.4	41.1
1000	8.89	14.8	29.6	44.5
1200	10.2	17.0	34.0	51.0
1400	11.5	19.1	38.1	57.2
1600	12.7	21.1	42.2	63.3
1800	13.8	23.0	46.0	69.1
2000	15.0	24.9	49.8	74.8
2500	17.7	29.5	59.0	88.4
3000	20.3	33.8	67.6	101.4
3500	22.8	37.9	75.8	113.8
4000	25.2	41.9	83.9	125.8
4500	27.5	45.8	91.6	137.5
5000	29.7	49.5	99.1	148.7

FIGURE 5.20 Culvert capacity maps and table.

To determine the size of a culvert, the drainage area is measured or estimated. Topographic or airplane maps are particularly useful for this purpose. The number of acres, or the next-higher figure, is selected in the left-hand column. The figure opposite this acreage, in the vertical column whose description best fits the area in question, is taken and multiplied by the rainfall rate shown for the locality by the appropriate map.

This will give the culvert area in square feet. To obtain the diameter of the proper size of round pipe, use the formula

$$\text{Diameter} = 2\sqrt{\frac{\text{area}}{\pi}}$$

(twice the square root of the area divided by 3.14).

The indicated size should be increased if full culvert capacity may not be available, or any local conditions (such as abnormally intense rainfall, or extremely steep and nonabsorbent slopes) indicate the need.

Even generously designed culverts may be inadequate for exceptional storms, as it is seldom economically practical to provide for them.

Sidewalls or Headwalls.   Sidewalls serve to hold embankments from falling into inlet or outlet channels; to direct water into and away from the passage or barrel; to reduce turbulence and prevent undercutting of the embankment; to support the ends of the culvert, and to hold pipe sections against separating inside the fill.

The wall requirement is reduced by lengthening the pipe, as in Fig. 5.21, top. Pipe resting on the original grade, or projecting clear of the bank as in Fig. 5.22, does not usually need a wall at the outlet.

A sidewall is usually of reinforced concrete but may be of stone, wood, or metal. It may be of heavy construction and firmly founded to resist movement in any direction; or it may be light and superficial, so that any settlement will affect it to the same extent as the pipe.

It is most convenient to place wall footings before laying the end pipes.

Metal headwalls (Fig. 5.22), are fastened to corrugated pipe by standard couplings. They can be removed and reused if the culvert is lengthened.

Alignment.   Culvert barrels should be straight under most circumstances.

It is usually desirable to use the original channel for the culvert passage and to have the culvert cross the road at right angles to the centerline.

These two objectives are often in conflict.

The original channel may be undesirable if it is crooked, crosses at a sharp angle or skew, shows rock ridges, is made up of soft mud, or has a strong flow of water. In such cases it may be more economical to dig a trench nearby, lay the pipe, and then divert the stream into it.

Right-angle alignment may be ignored if the natural channel is diagonal, and can be easily prepared for the culvert; or when excessive trenching is required to bring the stream straight across.

Referring to Fig. 5.23(*C*), it will be seen that a slight change in stream alignment under the road can lead to a considerable amount of digging on the side, that the new stream channel will be out of balance, and it may require mats or revetments to protect the outside banks of the curves.

On the other hand, changes may involve comparatively little excavating and produce more satisfactory channels than the original.

If good alignment between stream and culvert cannot be obtained on both sides, the upstream side should be favored. When the capacity of the culvert is heavily taxed by a storm, it is advantageous to get the water into the culvert smoothly.

Gradient.   It is desirable to lay the culvert on the floor of the natural channel, on the original ground surface, or in a smoothly dug ditch. This gives firmer support than fresh fill. Inequalities in channel or ground are smoothed by cutting off ridges and tamping fill into hollows.

The passage should have at least a $\frac{1}{2}$ percent slope, but 2 to 4 percent is preferable. It should not be over 8 or 10 percent, because of probable erosion of the bottom of the lining. The gradient

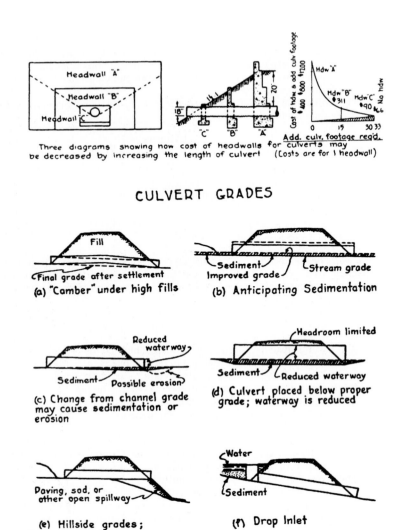

Three diagrams showing how cost of headwalls for culverts may be decreased by increasing the length of culvert (Costs are for 1 headwall)

CULVERT GRADES

(a) "Camber" under high fills

(b) Anticipating Sedimentation

(c) Change from channel grade may cause sedimentation or erosion

(d) Culvert placed below proper grade; waterway is reduced

(e) Hillside grades; erosion prevention

(f) Drop Inlet

FIGURE 5.21 Culvert headwalls and grades.

of the waterway down from the discharge end should be as great as that above the inlet, for a long enough distance that it will not silt up.

If the culvert is on fill or a foundation which is expected to settle, it may be laid in a vertical curve, known as camber (see Fig. 5.21), or so as to include a vertical angle. In each case, a slight gradient is used at the inlet end and a steeper one toward the outlet. Center settlement will tend to straighten out the passage.

Disjointing.   If a fill settles unevenly, or part of it moves laterally, any culvert that is in it will be put under tension. Such forces generally are not sufficient to break pipe, but they are apt to pull apart the joints.

FIGURE 5.22   Metal headwall for pipe. (*Courtesy of Contech Construction Products.*)

Metal pipe is very resistant to such disruptive forces, and can be further strengthened by special joint fastenings. Short-section rigid pipe may be braced at each end by a heavy headwall, founded on underlying stable material, or may be cabled together.

Depth.   Fills over a culvert, to a depth of 4 to 7 feet, protect it by spreading out the weight of vehicles on the surface. Deeper fills have diminishing protective effect and impose the load of their own weight, which at great depths may be sufficient to crush the pipe.

Figures 5.24 and 5.25 contain tables showing the approximate required strength for highway culvert or sewer pipe under various depths of fill. It should be noted that the required strength listed must be multiplied by the diameter of the pipe in feet. This is because the increase in crushing strength with increased size of pipe only partly compensates for its greater surface area, against which pressure is exerted. In short, although large pipe is stronger in itself than small pipe, it is weaker in regard to burial loads.

The table also indicates the importance of pipe bedding in determining strength. Corrugated pipe requires good compaction of fill at the sides, and proper bedding.

It is questionable whether the formed bed shown for the ordinary and first-class installations is often perfect enough to give the calculated support. However, careful tamping of the fill under the curve of the pipe will give similar results.

Restricted Height.   If there is not sufficient space between the channel bed and the embankment surface to install a round pipe of adequate size, two or more smaller round pipes, or low-clearance pipe with a flatter cross section or a pipe arch may be used.

Poured-concrete structures may use a flat rectangle or two or more openings separated by supporting walls.

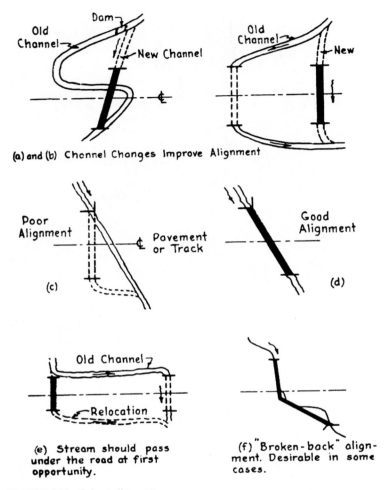

(a) and (b) Channel Changes Improve Alignment

(c)

(d)

(e) Stream should pass under the road at first opportunity.

(f) "Broken-back" alignment. Desirable in some cases.

FIGURE 5.23 Culvert alignment.

Multiple pipes should be parallel, and spaced at a distance of at least half of their outside diameters, to facilitate tamping backfill. They should not be used for streams that carry coarse debris which might choke the openings.

LAYING CORRUGATED PIPE

Handling. Corrugated pipe can be made up in any multiple of 2 feet. Lengths of 8 to 20 feet are usually carried in stock, and longer or shorter ones are obtainable on special order. Shorter pieces can also be made by cutting with a torch.

Small and medium sizes are usually unloaded and placed by hand, and medium to large ones lowered with a rolling hitch or with a crane. The crane may use a hitch around each end or around the middle, with the help of a person to keep it balanced. Soft rope should be used so as not to scratch the galvanizing.

Recommended Gauges for Standard Corrugated Pipe Under Embankments

For Highway Culverts, Municipal Drains, and Railroad Culverts Not Under Track

With Proper Backfilling and Tamping, Using Firm Material

Diam. Inches	Area Sq. Ft.	Fills Up to 15 Ft.	15 to 20 Ft. Fill	20 to 25 Ft. Fill	25 to 30 Ft. Fill	30 to 35 Ft. Fill	35 to 40 Ft. Fill	40 to 45 Ft. Fill	45 to 50 Ft. Fill	50 to 60 Ft. Fill	60 to 70 Ft. Fill	70 to 80 Ft. Fill	80 to 100 Ft. Fill
8	.35	16	16	16	16	16	16	16	16	16	16	16	14
10	.55	16	16	16	16	16	16	16	16	16	16	14	14
12	.79	16	16	16	16	16	16	16	16	16	14	14	14
15	1.23	16	16	16	16	16	16	16	16	14	14	12	12
18	1.77	16	16	16	16	16	16	14	14	14	12	12	12
21	2.41	16	16	16	16	16	14	14	14	12	12	12	10
24	3.14	14	14	14	14	14	14	14	12	12	12	10	10
30	4.91	14	14	14	14	12	12	12	10	10	10	8*	8*
36	7.07	12	12	12	12	10	10	10	8	8	8*	8*	8*
42	9.62	12	12	12	10	10	8	8	8	8	8*	8*	8*
48	12.57	12	12	12	10	10	8	8	8	8	8*	8*	8*
54	15.90	12	12	10	10	10	8	8	8*	8*	8*	8*	
60	19.64	10	10	10	8	8	8						
66	23.76	10	10	8	8								
72	28.27	10	8	8									
78	33.18	8	8	8		Use Multiplate Culverts							
84	38.49	8	8										
90	44.18	8											
96	50.27	8											

The gauges in the first column are minimum requirements for highway embankment conditions and agree with the recommendations of the A.A.S.H.O. specification.

Culverts below the heavy line should be strutted during installation.

Culverts with the mark (*) should be trenched one diameter.

The minimum height of cover should be as follows: For highways with unpaved surfaces, one-half diameter, with 12″ minimum; for highways with flexible and rigid pavements, 12″ minimum.

Gauges heavier than those given here should be used where conditions are unusually severe.

FIGURE 5.24   Recommended gauges for corrugated pipe.

Pipe should not be dragged around on abrasive ground or scratched or banged against anything, as such abuse will damage the galvanizing and shorten the life of the metal.

It is lowered into trenches in the same manner as it is unloaded.

Foundation.   The base should be shaped to fit the lower part of the pipe as closely as possible, by cutting the ground to shape, or building up with well-tamped fill. The work can be checked by placing and removing the pipe, and noting whether it was in full contact.

DRAINAGE & SEWERAGE – LOADING ON PIPES

REQUIRED ULTIMATE STRENGTH, LBS. PER LIN. FT., OF PIPE FOR VARIOUS DEPTHS UNDER H-20 HIGHWAY LOADING.

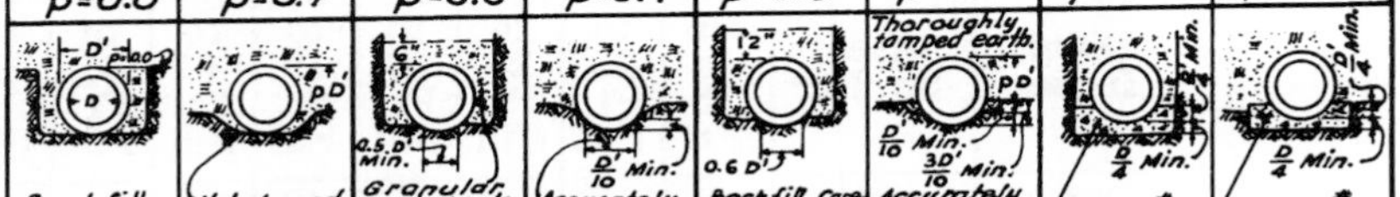

DEPTH OF COVER OVER PIPE (feet)	BEDDING CONDITIONS							
	IMPERMISSIBLE		ORDINARY		FIRST CLASS		CONCRETE CRADLE	
	p=0.0	p=0.7	p=0.0	p=0.7	p=0.0	p=0.7	p=0.0	p=0.7
	Backfill untamped.	Not shaped to fit pipe.	Granular materials shovel placed and tamped.	Accurately shaped to fit pipe.	Backfill care fully tamped in thin layers.	Accurately shaped to fit pipe.	2000# Concrete.	2000# Concrete.
2	2430 D	2490 D	2280 D	2320 D	2250 D	2260 D	2190 D	2190 D
3	2060 D	2190 D	1840 D	1900 D	1770 D	1820 D	1700 D	1690 D
4	1880 D	2040 D	1580 D	1670 D	1500 D	1560 D	1400 D	1400 D
5	1760 D	2180 D	1390 D	1640 D	1290 D	1480 D	1160 D	1240 D
6	1820 D	2220 D	1380 D	1600 D	1260 D	1440 D	1100 D	1160 D
7	1900 D	2380 D	1390 D	1660 D	1240 D	1450 D	1070 D	1140 D
8	2050 D	2550 D	1460 D	1740 D	1290 D	1510 D	1090 D	1160 D
9	2200 D	3030 D	1530 D	1990 D	1340 D	1720 D	1110 D	1290 D
10	2350 D	3240 D	1610 D	**2130 D**	1400 D	1820 D	1150 D	1320 D
12	2650 D	3780 D	1770 D	2430 D	1520 D	2040 D	1210 D	1450 D
14	3020 D	4340 D	2000 D	2830 D	1710 D	2360 D	1350 D	1660 D
16	3390 D	5010 D	2210 D	3170 D	1890 D	2650 D	1500 D	1860 D
18	3780 D	5780 D	2470 D	3650 D	2100 D	3030 D	1640 D	2090 D
20	4170 D	6300 D	2700 D	3970 D	2290 D	3400 D	1790 D	2270 D

D = inside diameter in feet.

EXAMPLE: Assume 24"ø pipe, depth 10 ft., ordinary bedding p=0.7
Table A, above: Ultimate strength per lin. ft. 2130 D = 2130 × 2 = 4260. lbs. per lin. ft.

FIGURE 5.25   Loading on pipes. (*Reprinted with permission from Elwyne and Seelye's Design, John Wiley & Sons, Inc.*)

If the floor is rock, it should be cut from 6 inches to 3 feet below pipe grade, the depth depending on the height of fill to be placed, and backfilled with earth or pea gravel. If it is mud, space should be allowed for enough pea gravel to stabilize the surface. Saplings or wire lath might be laid under the gravel to provide extra stability.

If one end of the culvert is to rest on fill and the other in a cut, the fill under it should be thoroughly tamped to avoid unequal settlement.

Placement.   Each section should be placed with the longitudinal seams at the side shoulders. The cross joints should have the external part of the overlap upstream, so that if the joint is not tight, seepage will tend to move into the pipe instead of out of it.

Joints.   Sections are usually fastened together by a one-piece band. This is placed under the end of the first piece, and the second is laid so that the band will overlap each by the same number of corrugations. The coupler is then drawn tight by turning down the nuts with a wrench.

This joint resists any force tending to pull it apart, being about as strong as the pipe itself. If the pipe will be subject to very severe stress, a two-piece coupler riveted to one or both sections may be used. The pipes are placed so that the collar lines up, and it is fastened with the bolts and nuts.

If the pipe is large enough to work inside (36-inch or more ordinarily, but less if a small worker is available), a one-piece band may be installed in the regular way, and holes then drilled through the band and the pipes, and bolts used to strengthen the joint.

If watertightness is important, an asphalt sealer may be placed inside the band before installing, or a special coupling may be used.

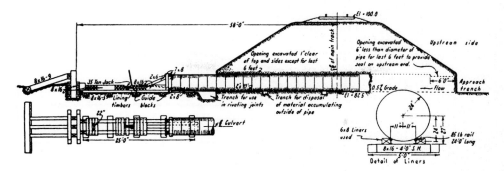

FIGURE 5.26   Jacking pipe through fill.

Jacking.   Corrugated pipe can be pushed through an ordinary fill without trenching.

Figure 5.26 shows a diagram of this process. An approach trench is dug to line the pipe up, and a heavy backstop built similar to Fig. 5.14. A trough is made across the trench at the toe of the fill, for access to the pipe. A wood frame is placed against the pipes to take the push of the jacks.

A person can work inside a large pipe, digging from the face to reduce the pressure needed to push it. Dirt may be loosened ahead of smaller pipes by means of augers or water jets.

Trench drills such as those described in Chap. 20 make jacking unnecessary under many conditions, and can be used for jacking when it is necessary.

LAYING CONCRETE PIPE

Handling.   The standard length of concrete pipe is 4 to 8 feet, except for land tile. The pieces may be unloaded by rolling each pipe so that it will fall from the truck endwise onto soft ground or a couple of old tires. Bell-jointed pipe should be dropped on the straight end.

Pipe up to 2 feet may often be rolled and pried into place by laborers with bars and poles. However, in any size it is more convenient to use some sort of hoist.

Pipe can be handled by any lifting device with power to pick it up, and enough reach and maneuverability to place it easily. A crane is most suitable for placing it in a ditch, and a front-end loader for laying it in the open.

A pipe hook (Fig. 5.27), is very useful. It permits holding the pipe at the free end, and avoids difficulties in balancing it on a chain or sling, and in disturbing the pipe after it has been laid by withdrawing the chain. Figure 5.27(*A*) is alloy bar, 2 inches by 4 inches, for pipe up to 60 inches. In (*B*) 4-inch iron pipe will handle 24-inch concrete safely.

Eight-foot lengths may have a 2-inch hole through the wall at the center. These can be picked up by the means of an eyebolt inserted in the hole and caught inside with a nut and washer.

Foundations.   The base is shaped and tamped in much the same manner as for metal pipe.

If pipe is laid in running water, leveling the bottom may be difficult. Weighted planks or troughs of preservative-treated wood may be placed on the bottom, and carefully leveled and blocked up. Straight-sided pipe can be set on this, and only minor adjustment should be required.

If the load of fill or traffic is to be heavy, or the foundation is partly on fill and partly on rock, or is unstable, a concrete bed may be used. A stiff mixture of concrete, as wide as the trench, or in forms a foot or two wider than the outside diameter of the pipe and one-quarter of the pipe diameter in depth, is placed on a well-compacted earth base. It is roughly hollowed for the pipe, which is settled into it by rocking, or slight raising and lowering.

Bell joint pipe requires cross grooves in the bed to accommodate the oversize end.

Placement.   Pipe is usually laid with the bell ends upstream. If this is the case, placement should start at the downstream end. The first pipe can be lowered level, but the others should have the plain

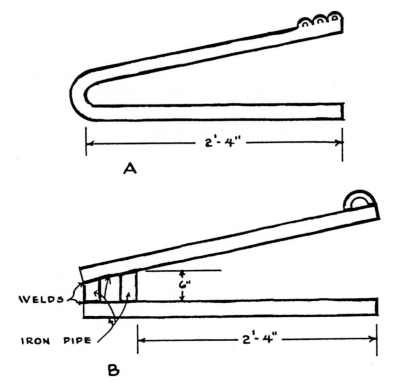

FIGURE 5.27   Pipe hooks.

free end slightly lower so that it can be guided into place without scraping on the bottom. This tip is arranged by inserting the hook only partway into the pipe, or by pushing down on the free end.

If a section is not held in proper position by the bed, it should be chocked securely with stones or blocks until several more sections are laid, or the culvert is completed. This makes it possible to make any necessary readjustments with less work than if fill were tamped in immediately.

It is difficult to get each joint tight without considerable practice. However, it is often possible to lay several loosely, and then push them together from an end. This may be done by a small dozer in the trench, or by a cable threaded through the culvert to a crossbeam.

Joints.   Except in informal or temporary work, joints should be cemented. This is particularly important if water may pond above the outlet so that it will go through under pressure, which might force it out through open joints and cause softening and channeling of the embankment.

Small pipe is cemented by wetting the ends to be joined, and troweling a rich mortar on the upper half of the plain side and the lower half of the bell end. It is desirable to rotate the free pipe slightly, after it is in position, to spread the cement evenly. The outer surface of the upper two-thirds can be troweled off.

If the pipes are large enough for a person to work inside, the whole culvert may be placed dry. Oakum is then hand-tamped into the joint cracks, and cement or bituminous mortar applied from the inside.

Ties.   If foundation conditions are such that the fill may spread and pull the pipe apart at the joints, the culvert may be held together by heavy, deep-based headwalls, or by tie lines.

Tie lines may consist of three rods, cables, or chains, hooked around the end pipes. Turnbuckles or load binders are used to tighten them. They may be internal or external, as shown in Fig. 5.28.

The inside ties will reduce culvert capacity slightly, and may cause jamming of debris and complete stopping. However, they are accessible for inspection and tightening. Outside installations are difficult to service.

A loose cable is sometimes left inside a small-diameter culvert for use if it becomes plugged with silt. Pulling the cable back and forth will make a small hole that can be enlarged by forcing water through.

FIGURE 5.28   Cable ties on concrete pipe.

OTHER FORMS OF CONSTRUCTION

Wood Culverts.   Culverts may be constructed of wood when they are for temporary use, or when time or expense prohibits obtaining more permanent materials.

Construction may be to almost any desired strength. Life expectancy will vary with the type and size of wood, preservative applied, and moisture conditions. In general, the parts that are permanently wet will have a much longer life than those that are exposed to air.

Several designs are shown in Fig. 5.29.

Casual Placement.   There are many situations in which it may be unnecessary or impossible to place culverts with the care required for permanent installations. These would include light-traffic driveways and farm lanes; pioneer or access roads to be used for only a short period; and urgent construction in which it is necessary to get traffic through without delay, even at the cost of possible repair or reconstruction later.

Good standards should be approached as closely as possible, however.

If heavy traffic will ride directly on the pipe, or very closely over it, a strong construction should be used. If silting and trash are not a problem, several small pipes will be better than one large pipe, as they are more resistant to concentrated loads and they can be provided with an adequate depth of cover more readily.

If the foundation is unstable so that a part of the culvert will sink, oversize pipe may be used to provide adequate capacity after settlement and silting. If silting can be prevented, a badly sagging pipe may act as an inverted siphon.

The pipe should be long enough not to require large headwalls, unless they can be easily built with big stones or logs available on the site. Wingwalls, where required, can be made of rocks, of saplings hammered in as piling, or of brush mats.

On wet bottoms, pea gravel or crushed stone should be used under the haunches. Ordinary earth can be used as soon as the gravel has been built up above water level, but it will not consolidate under water.

BACKFILL

Proper placing and compacting of backfill affects both the strength of the pipe and the load it has to carry.

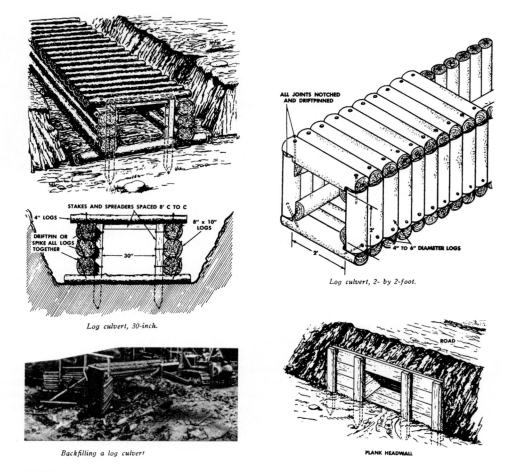

STAKES AND SPREADERS SPACED 8' C TO C

4" LOGS

8" x 10" LOGS

DRIFTPIN OR SPIKE ALL LOGS TOGETHER

30"

Log culvert, 30-inch.

ALL JOINTS NOTCHED AND DRIFTPINNED

4" TO 6" DIAMETER LOGS

2'

2'

Log culvert, 2- by 2-foot.

Backfilling a log culvert

ROAD

PLANK HEADWALL

FIGURE 5.29   Wood culverts. (*Courtesy of U.S. Army Engineers.*)

Load Distribution.   If a round pipe lies on a hard, flat surface and is subjected to load, the entire pressure falls on the line of bottom contact. If the surface is curved, the area of contact is greatly increased and the load per square inch reduced correspondingly. As the haunches curve upward, the amount of vertical support to each square inch of surface decreases until it is zero at the widest point.

Corrugated pipe is flexible and requires horizontal support as well. A normal load tends to flatten the top and spread the sides. If the sides are held in firmly, the arch form of the load-carrying section is retained and strength kept at a maximum,

No part of the foundation or backfill touching a flexible pipe should be rigid, as the whole pipe should be able to change cross-section shape as it deflects under load, and any rigid support will cause excessive strains, particularly at the edges of the contact.

Rigid pipe receives only nominal support from the side fill.

Surface loads on soil masses are ordinarily distributed over increasing areas and reducing pressures on lower levels, as in Fig. 5.30(*A*). If there is a difference in bearing power of the soils within the affected cone, the more rigid soil may carry most or all of the load.

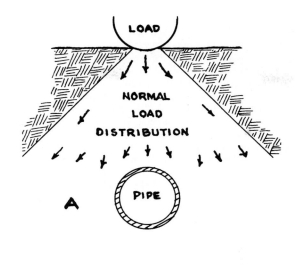

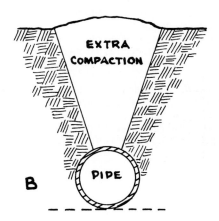

FIGURE 5.30 Load over pipe.

In (B), backfill has been placed loosely over an exposed or projected pipe. Settlement under traffic, or from the effects of weather, will be a fraction or percentage of its depth, so that the thinner fill over the pipe will not sink as far and will project as a surface ridge. This ridge may receive heavier loads, and be more thoroughly compacted, than the soil on each side.

If a trench is loosely filled, settlement will result in a surface trough. Vehicles bumping across this will cause heavy impact loads. If the pipe is near the surface, it may be damaged or destroyed. If deep, the loads will largely be carried by the walls.

A tightly tamped backfill should distribute loads evenly over the whole area and subject the culvert to normal loads for its depth, as in (A).

Whatever load is imposed on the immediate vicinity of the pipe will be shared by it and by the fill on each side. If the fill is tightly tamped, it will bear a larger part of the burden, relieving the top of the pipe. Vertical pressure on the side fill is partly converted into horizontal pressure against the sides of the pipe, providing support for the top load.

Tamping.   Fill must be tamped under the pipe haunches. It should be free of lumps, stones, and trash, and should contain enough moisture to pack, but not enough to make it rubbery. It is placed with a hand shovel in thin layers.

The tamper should have a narrow edge to enable it to get well under the pipe and, if the trench is narrow, may require a curved handle.

Filling and tamping are done evenly on both sides of the trench, to avoid shifting the pipe to the side. It may be necessary to wedge it in place temporarily with rocks or other blocking. Such material may be left and buried if the pipe is rigid, but removed if it is flexible.

Tamping blows should not be so vigorous as to wedge the pipe out of position.

When sufficient fill has been placed that its surface is out from under the pipe, mechanical tampers can be used. Fill should be compacted to the full width of the trench, or, if the pipe is on the surface, for one pipe diameter on each side. Layers should be 4 to 6 inches deep.

When the pipe is nearly or wholly covered, layers of 6 to 12 inches may be placed, and tamping continued, until the trench is filled to grade. A compacting bucket, like the one manufactured by Felco Industries Ltd., mounted on a backhoe, can be useful.

Side Fill.   When an embankment is to be built up on each side of the pipe and above it, much of the compaction can be done by machinery first moving parallel with the pipe, then across it.

A roller working parallel must be kept far enough away so as not to exert a horizontal thrust that will move the pipe. Fill and compaction should be kept even on both sides. Since there is often only one roller available, and it may not be able to get across, the other side may be compressed with a truck or by tamping.

Soil between the rolled strips and the pipe, and between the rolled strips and over the pipe, must be thoroughly tamped.

Successive layers can be rolled closer to the pipe centerline. It is good practice to postpone crossing it until the fill is as deep as the outside diameter of the pipe, to avoid pushing it out of line.

Material is pushed to the pipe by a dozer or grader.

Loosened Backfill.   A compacted embankment may be built up one diameter above the pipe, then ditched over it. This trench is filled in loosely, and layers of compacted fill are built to the top of the embankment in the regular manner.

The soft fill over the pipe causes the load to be transferred to the solid sidewalls.

If any trench containing pipe is backfilled by a dozer, care should be taken not to drop rocks on the pipe.

FORDS AND DIPS

Farm or pioneer roads sometimes cross a shallow stream on its bottom so that vehicles must drive in the water. This type of crossing is called a ford. It may be satisfactory for light or occasional traffic, but it is subject to interruption by high water and ice, and may develop bad bottom conditions that it would be difficult to remedy.

Crusher rock, in mixed sizes from $\frac{1}{2}$ to $2\frac{1}{2}$ inches, makes a good patching and paving material. If bank gravel is used, thorough raking will allow the water to take away excessive fines.

In arid regions, many watercourses are dry most of the time, but will occasionally carry such large volumes of water that adequate bridges would be very expensive. In such cases the road may run across the channel at its natural grade, with no provision being made for passing water under it. The section of road in the channel is usually heavily built, with reinforced-concrete slabs up to 2 feet in thickness sometimes being used. This slab should be sloped on its upstream side, and may have a cutoff or curtain wall extending below its main mass.

Occasionally a culvert may be placed under or beside the dip, to pass small water flows, or a culvert structure may include a spillway for floodwater.

Second-class roads may cross such a streambed on graded local material that must usually be worked over after each flow of water. Roads may also run considerable distances in streambeds, as this may be the only route that is practicable without heavy blasting and grading.

SOIL MOISTURE

Water Table.   Subsurface water exists in three states or zones. The lowest of the series is hydrostatic or free groundwater. Its upper surface is known as the water table, or groundwater level. It follows the contour of the land in a general way, but tends to be farther under the surface in hills and pervious soils than in hollows and fine-grained soils. If it rises to or above the surface, it makes swamps, ponds, or springs.

The actions of this water are controlled by gravity, causing it to seek lower levels by the resistance of the soil to its movement, and by fresh supplies of water reaching it from the surface.

The water table may be static, or fluctuate only slightly, or it may shift up and down widely in response to season or rainfall.

Soil which is saturated with groundwater is usually unstable under load, will turn to mud if disturbed, and does not permit the growth of roots of most plants.

If a hole is dug below the water table, it will fill with water, unless clay seals it off.

Capillary Zone.   The capillary zone lies above the water table. It may be a few inches deep in coarse sand, and 8 feet or more in fine-grained soils. It contains a substantial quantity of water that is held above the gravity surface by capillary attraction and other forces tending to attract and hold it in the finer soil spaces.

The amount of contained water diminishes from the bottom to the top of this zone.

Capillary movement in coarse soils is rapid, in fine ones quite slow. Raising or lowering the water table may raise or lower the capillary zone.

Medium and fine soils in the zone usually contain too much water for stability, and may be subject to frost heaving. In climates where rainfall exceeds evaporation, this zone offers the best conditions for root growth. In arid regions, the water may deposit alkali in the soil and render it unfit for cultivation.

Upper Zone.   The upper or hygroscopic zone contains water which is in very thin films on the particles, or is in chemical or physical combination with them. Some of this water is hygroscopic—absorbed from the atmosphere—and is greatest in amount when humidity is high.

These small quantities of water often give the soil maximum stability, by acting as a cement or binder. Much of the water is too firmly attached to be removed by plant roots, or any method but oven baking.

This zone may also contain varying quantities of rainwater, moving downward by gravity or capillarity, or adhering to soil particles. This is available to plants and may be found in sufficient quantity to make the ground unstable.

SUBSURFACE DRAINAGE

Purpose.   Subsurface drainage lowers the water table. Deep drains, or those in porous soil, will lower the capillary surface also.

Soil must be drained when its water content makes it incapable of supporting roads or other structures on it, or causes frost heaving.

Playgrounds, golf courses, and other recreational areas may require draining to dry up spots that remain wet and soft long after rains.

Farmland drainage may serve to eliminate wet spots that cannot be worked as early as the surrounding land; to speed up the drying and the warming of soil in the spring; to encourage plants to form deep root systems, with resulting increase in vigor and drought resistance; and to leach out harmful substances which may accumulate in the soil.

Methods.   Groundwater level may be artificially lowered by open channels or ditches, or by buried pipe or porous material. Such pipe is generally referred to as tile, even though it might be made of other materials.

In soils that will stand on steep slopes, ditches are the most economical construction down to a depth of a few feet. When wide cuts must be made to produce stable slopes, or when greater depth is required, the open ditch may involve so much excavation as to be more costly than tile.

Ditches, together with any space required for spoil, may occupy rather wide strips of land. If in farms, they cut up the fields and add to the expense of planting and cultivation. They are hazardous when near roads. In any location, they will require occasional culverts or bridges for crossings.

Ditches usually require maintenance. This may include removing silt and cave-ins, repairing erosion damage, and cleaning out vegetation. Neglect may result in general deterioration, with eventual stoppage, or expensive redigging and clearing.

Buried drains do not cut up the fields, or offer hazards along roadsides. However, if one becomes plugged, as a result of poor design, improper installation, or accident, it may be difficult and expensive to locate the difficulty. If the stoppage is due to general silting, it will probably be cheaper to lay a new line than to dig up and clean or repair the old one.

Choice of the type of drainage will depend on local conditions and on individual judgment.

Water slope. The porosity and bedding of the soil largely determine the depth and spacing and to some extent the size of drains required for a given project.

A pool of surface water will assume a slight but measurable slope from its inlet down to an outlet or drain. If the pool is choked with weeds and brush, water may be removed more rapidly than it can flow through the obstructions, so the level at the drain may be several inches below that at other parts of the pond as long as flow continues.

The water table may be considered to be the surface of an underground pond, obstructed in its flow by soil particles. If these particles are coarse and loosely fitted, the spaces will be large enough to allow some freedom of flow, and the slope up from an outlet of the water surface will be gradual. If the soil is fine-grained and compact, the spaces will be so small that flow will be almost stopped and the gradient down to a drain point will be very steep. This slope is called the hydraulic gradient.

Slopes will usually be steeper after rains and in wet seasons than when the surface is dry.

If the drainage is to a single pipe opening in uniform soil, the drained area will assume the shape of an inverted cone, called the cone of depression. See Fig. 5.31.

If the drainage is to a ditch or porous horizontal pipe, the shape will be a trough of roughly triangular cross section. The water surface and movement are shown in Fig. 5.32, and the effects of spacing in Fig. 5.33.

Drainage Layout. Figure 5.34 shows the standard patterns used for subdrainage. Where practical, the intercepting or curtain drain, as in (*E*), is the most economical. Figure 5.35 shows a condition where it should be used.

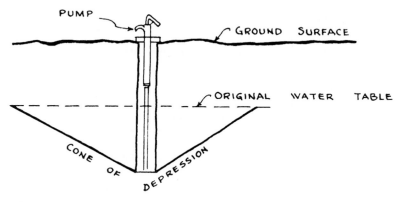

FIGURE 5.31 Cone of depression.

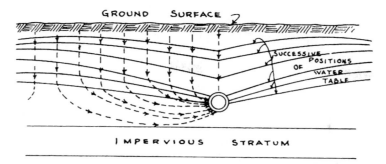

FIGURE 5.32 Groundwater movement.

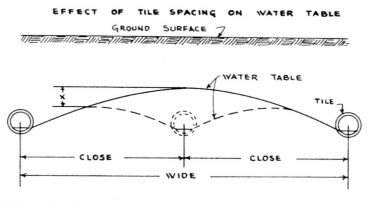

FIGURE 5.33 Effect of tile spacing.

The natural system involves use of the natural drainageways for ditch lines, and involves minimal excavation. Difficulties may be unfavorable surface conditions for ditching; an irregular pattern which duplicates in some spots and is inadequate in others; or excessively crooked lines.

The herringbone, gridiron, and parallel systems are best suited to level or evenly sloping land. The choice will depend largely on which will most readily provide best gradients in the lines.

Depth and Spacing. The table in Fig. 5.36 gives general recommendations for depth and spacing. They cannot be strictly followed in every case because of wide variations in conditions. Tile should at least be below the frost line and out of danger of crushing by machinery.

When a field is first tiled, the widest permissible spaces may be used, and additional laterals added later if they are required.

French Drains. These drains, also called rubble or blind drains, consist of a rock fill in the bottom of a trench, as in Fig. 5.37, with finer material over it, to prevent dirt from working down. The usual practice is to put the large rocks in the bottom.

The more elaborate ones in (*A*) and (*B*) will serve the same purposes as open joint pipe, but the others are not suitable for water carrying sediment, as the lack of concentrated flow will allow the spaces to fill up until the drain is blocked.

Clogging of porous material may be prevented by wrapping it in a geotextile fabric mentioned in Chap. 3. A strip of cloth is unrolled along the top of the completed ditch, then its center is

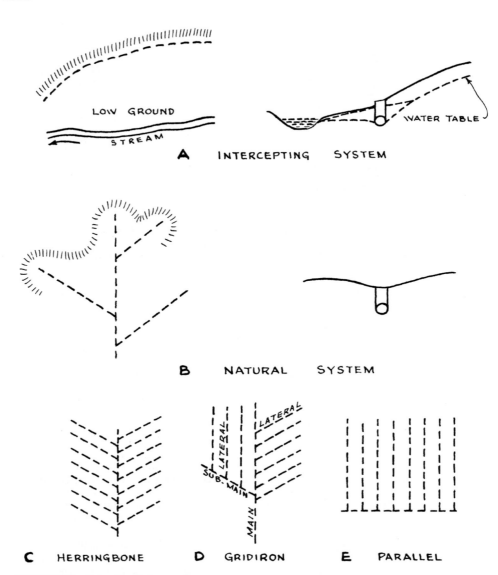

FIGURE 5.34   Subsurface drainage patterns.

pushed to the bottom. The porous material is placed to the desired height, the fabric is folded over the top, and backfilling is completed.

This cloth sifts most dirt particles out of water entering the drain.

French drains are used chiefly where supplies of suitable material are abundant.

Moles.   Certain types of stiff plastic soils may be drained by opening pipelike channels in the soil. This is done by attaching a mole, which is a metal piece shaped like an elongated egg, to the heel of

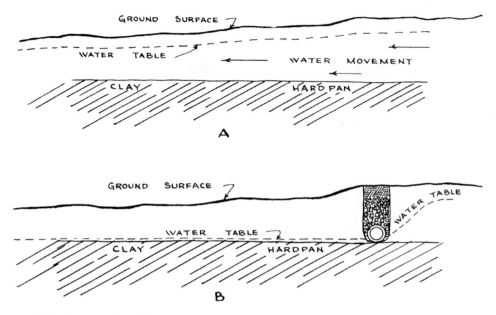

FIGURE 5.35 Intercepting drain.

LOCATION OF LATERALS
AVERAGE AGRICULTURAL PRACTICE

Soil	Depth in inches	Spacing in feet
Irrigated land	60 - 100	125 - 1,000
Peat	42 - 60	75 - 200
Sand	42 - 48	100 - 400
Sandy loam	36 - 48	100 - 250
Loam	36 - 48	40 - 100
Silt loam	36 - 48	35 - 80
Clay loam	30 - 48	30 - 66
Clay	30 - 48	20 - 40

FIGURE 5.36 Depth and spacing of drainage tile.

a subsoil plow, as in Fig. 5.38. The plow is set to penetrate to the desired depth, and the mole is dragged through the ground. It pushes the soil aside and compacts it. Under favorable conditions these tubes will stay open for 5 years or more, and may open permanent channels. In other conditions, they may close immediately or within a few months.

Drainage is usually to a stream, ditch, or hole. The mole is dropped in this and pulled straight into the bank and lifted out at the upper end of the run.

A tile or metal pipe, screened with coarse mesh wire, should be placed in the outlet to protect it against erosion and plugging by entrance of small animals.

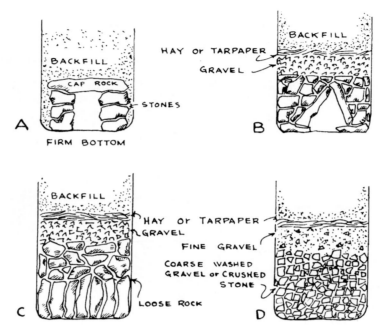

FIGURE 5.37   French drains.

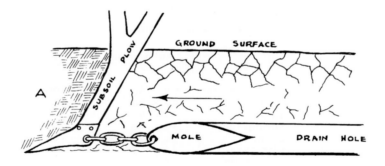

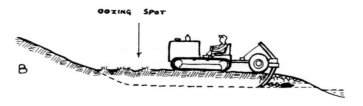

FIGURE 5.38   Mole drainage

TILE DRAINS

The simplest type of drainage by underground tiling is shown in Fig. 5.39 (*A*). A trench is excavated, the bottom smoothed to the desired gradient; butt joint land tile may be laid with ends touching, or with spaces up to $\frac{1}{4}$ inch; and the trench is backfilled.

Water from the affected area drains into the tile through the joints and flows inside the tile to the outlet. Under favorable conditions, such a drain may function for a very long time. However, dirt falling in through the joint and entering with the water may fill it up, or plug low spots left by subsidence of the ditch bottom.

In (*B*) the tile is laid on a bed of gravel. This provides a firmer foundation and a porous space into which dirt can drop through joints from the pipe. This storage space for silt may fill so that the pipe will ultimately fill also; but it may serve to trap all dirt brought in during a period of adjustment, after which little or no dirt will move.

In (*C*) the pipe is surrounded by gravel, which preferably should be a mixture of pieces ranging from coarse sand to $\frac{1}{2}$-inch crushed stone. This serves to filter dirt out of incoming water, keeps loose dirt from reaching the pipe joints, and provides a good bedding.

Tar paper or hay can be used in connection with any of these techniques. It can be laid over the pipe, where it prevents dirt from falling in, particularly when a large opening has been caused by misalignment. The joints may be wrapped individually, or covered by a continuous strip.

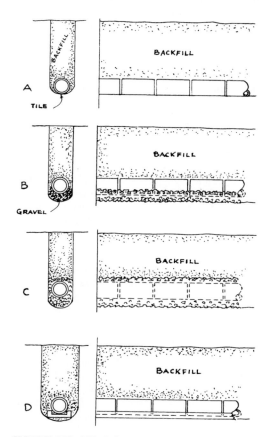

FIGURE 5.39   Tile drains.

It may also be placed over a gravel topping to prevent soil from working down into it. The under surface of the dirt often becomes so well stabilized that it will not cause trouble after hay has rotted out.

When the tile is laid on a curve, the wide spaces at the outside of the joints should be covered with pieces of broken tile.

Connections of branch lines may be made with sewer tile "Y"s or "T"s, or by junction pits which may be made large enough to serve as line cleanouts. A Y provides a smoother flow and larger water capacity than a T of the same size.

Cradling.   If the ground is muck, or otherwise unstable, the tile should be supported by boards, as in (*D*). Cleated boards supporting the haunches are preferable to flat boards because of better support and more permanent alignment.

Corrugated metal pipe with perforations may be used instead of tile and cradles.

Laying Land Tile.   A large part of the land or drain tile used is in farmland. The work is usually on a fairly large scale, on regular grades and with adequate space. Costs must generally be kept to a minimum.

Ditching machines are particularly adapted to a rapid sequence of operations.

Small machines, with buckets as narrow as 6 or 8 inches, may be used for depths up to 4 feet. These involve minimum excavation and ensure lining up of tile. As the maximum depth is approached, it becomes more difficult to place tile accurately, and very difficult to remove stones or earth that may fall from the sides. It is usually not possible to use a tile-laying shoe. It may be inconvenient or impossible to place gravel or tar paper with the tile.

Wider buckets will eliminate these difficulties, but will increase the amount of excavation and backfill.

The tile supply is laid on the field, parallel with the ditch line, just far enough to clear the ditcher, on the side away from the intended spoil pile. Pieces are placed end to end to give the correct number, with a few extra placed at frequent intervals to make up for broken or imperfect tiles.

The tile should be placed on the ditch bottom immediately behind the ditcher to minimize the danger of "losing" the ditch through caving of the sides. The first tile should be plugged with a stone or half brick to protect the line against entrance of dirt or animals. Pieces are usually picked up and placed with an L-shaped rod of light iron. A curved bottom ditch will tend to center them, but they must be checked for alignment anyway.

Tar paper, if used, should be in a narrow continuous strip in a roll, laid over the tile.

If the ditch is wide enough to work in, the tile may be laid in the same manner or by hand. In the latter case, a picker may be used to supply tiles to the ditch worker.

Gravel is sometimes laid under or over the tile by a dump truck with a small opening in the rear gate, similar to that used for supplying automatic sand spreaders. It straddles the ditch. The gravel may pour by gravity, or may be raked or shoveled down the body floor by the person controlling the gate opening.

It is important to smooth off a bottom layer of gravel before placing tile.

Tile Shoes.   If the ground is not firm enough to stand until the pipe and accessories have been laid down, a tile-laying box or shoe towed by the ditcher must be used. A number of varieties are available, many of them of only local distribution.

A tile box should be slightly narrower than the bucket side cutters. It includes the bulldozer or crumber attachment that smooths the ditch bottom behind the buckets, a pair of parallel walls that will slide between the ditch walls, and a chute on which tile may be placed to be fed by gravity or manual control into the ditch bottom.

Figure 5.40 shows a simple type of box that is operated from above. A more elaborate box in which someone can work, and which permits placing of tile, bottom and top gravel, and tar paper, is shown diagrammatically in Fig. 5.41. This is suitable for the greater depths required in irrigated fields.

Two gravel hoppers are mounted on the box, front and rear. A roll of tar paper is mounted on an axle across the inside of the box. The tile layer sits near the bottom, with his or her back to the "bulldozer," a cleaning blade which follows the wheel. The tiles rest on a shelf in front of the layer and are replaced as they are used by someone standing on the box, who picks them up with a rod.

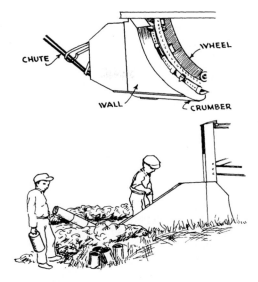

FIGURE 5.40   Tile-laying shoe, fed from top.

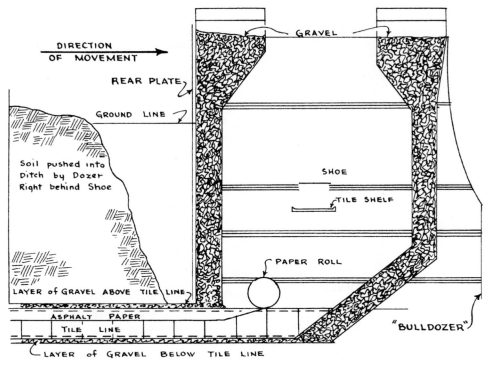

FIGURE 5.41   Tile shoe, worker inside.

As the ditcher moves forward, the blade smooths and shapes the bottom of the ditch. The front hopper spreads a strip of gravel on the bottom. The thickness of this strip is regulated by raising or lowering the hopper spout. The tile is laid on the gravel, with the ends touching. The tar paper rolls off its spool to cover the pipe, and the rear hopper deposits gravel on it.

The bottom of the rear plate may be curved to smooth over the top of the gravel, or may be set to ride several inches above it. A dozer works immediately behind the machine, backfilling. This is necessary, for if the ditch is allowed to stand open an appreciable time, one of the walls might move horizontally and slide the tile out of line.

The gravel may be piled beside the digging line, outside of the tile string, and placed in the hoppers by a small tractor front loader. A hydraulic control clamshell bucket is more efficient than the regular loader bucket, as it picks up the gravel without pushing it around.

A tile box is best adapted to a wheel ditcher, as it does not disturb its digging balance. Operation on a ladder ditcher is possible, but is more difficult as it tends to pull the bottom of the ladder backward and upward.

Backfilling.   In agricultural work, it is customary to backfill with the soil dug from the ditch after placing the pipe and whatever porous material is required. However, if it is to act as an intercepting or curtain drain, imported porous material—such as gravel, sand, or corncobs—may be used to near plow depth. A top layer of native soil should be used to prevent surface water from washing it with its probable burden of silt and trash.

Whatever type of material is used next to the tile, it may be advisable to place it carefully by hand until there is no danger of the pipe sections rolling out of alignment. Compacting it immediately over the pipe (blinding) may reduce silting.

Gradients (Fall or Slope).   Figure 5.42 indicates graphically the most desirable gradients for land tile of various sizes. It will be noted that up to 6-inch diameter a minimum of 1 percent is desirable, over 2 percent is optimal, and that larger sizes require flatter slopes.

Steeper slopes may give sufficient velocity to create eddies that will erode material below the joint, with possible undermining of the tile. This may result in stoppage, in blowing out to the surface, or in washing out of sections of line.

When steeper overall slopes are required by the topography, tiling may be done in a series of benches or levels, connected by inclines of sewer tile with cemented joints.

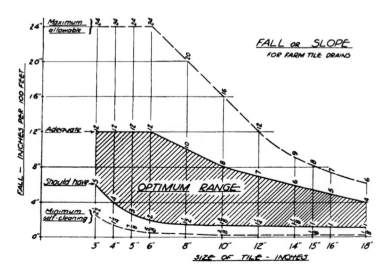

FIGURE 5.42   Tile slope chart.

Flatter slopes increase the danger of silting.

Tile Size.   Figure 5.43 shows the number of acres that will be drained by tile of various sizes, at slopes of up to 1 percent. Some of the sizes listed may not be generally available, and these figures are for average conditions.

It is good practice to use tile of larger than minimum diameter for a line so that effectiveness will not be lost as readily by silting or misalignment.

Surface Inlets.   A tile drainage system may include one or more places where surface water can drain into it. The flow of water should be calculated and the tile increased in size to accommodate it.

Such inlets require careful design. They must be arranged so that dirt cannot be washed in, so that animals cannot enter, and so that hydraulic pressure cannot be exerted on the tile underneath the opening.

For example, if a 6-inch inlet to a 6-inch line is used, water ponding over the inlet due to excessive rainfall may put hydraulic pressure on the underground pipe, causing too fast a flow and probable erosion.

If the inlet is choked down to 3 or 4 inches, or provisions are made for surface overflow, this difficulty should not occur.

Outlets.   The tile line or system may terminate in a drainage ditch; a large drain system; a river, lake, or sea; or a sump. If a sump, an automatic electric pump should be used to keep the water level below the bottom of the tile to ensure free drainage.

A projecting metal pipe may be used, as in Fig. 5.44, or a masonry spillway, as in Fig. 5.45, to avoid bank erosion and undermining. Since outfall lines often lie under surface channels, it may be necessary to make provision for surface flow also, as shown.

The outlet should be protected against entrance of small animals by coarse-mesh wire, a grating, or an automatic gate.

AREA IN ACRES DRAINED BY TILE WITH GIVEN DIAMETER AND SLOPE

PERCOLATION = $\frac{3}{8}''$ IN 24 HOURS — COEFFICIENT OF ROUGHNESS, n = .013

TILE DIAMETER IN INCHES	FALL OR SLOPE — FEET PER 100 FOOT STATION				
	.10	.20	.30	.50	1.00
4	3	4	5	7	10
5	6	9	10	13	19
6	10	14	18	22	31
7	15	22	27	34	48
8	22	32	39	50	71
10	41	58	71	92	130
12	67	96	118	153	213
14	105	149	182	235	333
15	127	179	218	283	400
16	149	211	260	336	473
18	209	295	360	467	660
20	280	393	483	625	880
22	360	510	626	800	1,140
24	453	643	794	1,020	1,440

FIGURE 5.43   Area-tile table.

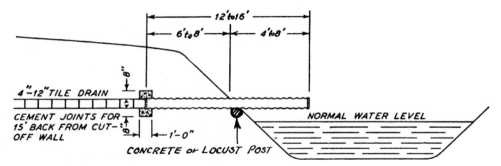

FIGURE 5.44 Projecting drainage outlet.

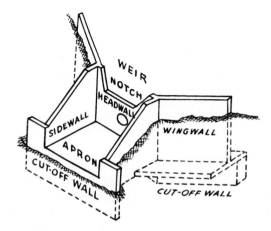

FIGURE 5.45 Outlet for tile, and weir for surface drainage.

A drainage system that ends in salt water between high and low tides, or in a waterway subject to flooding, should be fitted with a check valve or gate that will let the drain water out, but not allow the seawater or floodwater in. To make this effective, the last few feet of tile should be glazed and a concrete headwall used.

Other Pipe. Lightweight, perforated, corrugated-metal pipe may be used for tiling. It has four parallel rows of $\frac{5}{16}$- or $\frac{3}{8}$-inch holes, and is laid so that they are in the haunches. Joints have collars to preserve alignment.

This pipe is too expensive for ordinary agricultural work, but is widely used for road subdrains because of its resistance to crushing and its dependable alignment.

Perforated asphaltic pipe is even lighter and is cheaper, but it is less strong. Both types are made in long pieces and are easy to lay, except where caving soil requires use of a tile-laying box.

Plugging. Tiles may become partially plugged with mud entering with groundwater, particularly when no protecting gravel and paper are used. Flooded streams may also cause plugging, by either backing up the tile or causing the water to stagnate in them.

It is sometimes possible to flush tiles clean by utilizing a surface inlet, or opening the upper end and putting a large volume of clean water through. The flow should be started slowly, as a rush of water might move enough mud to form a solid block, which would necessitate abandonment of the line.

It usually is cheaper to lay new tile than to dig up and clean an old one, unless the stoppage is in a small area and can be located accurately.

SEPTIC DRAINS

Some drains have no open discharge but depend on the porosity of the soil through which they run, or in which they end. A common type is the septic leach field for disposal of domestic waste, a set of standards for which is shown in Figs. 5.46 and 5.47. Septic tank overflow is carried in a sealed pipe to the head of the field, which consists of a number of lines of porous land tile with a maximum slope of 1 inch to 16 feet, with open joints on a base of graded washed gravel that is brought up the sides flush with the top. Building paper or hay is placed over this to prevent dirt fill from sifting into the spaces. Fluids in the pipe leak out at the joints and percolate through the gravel and subsoil down to the groundwater. The gravel serves to widen the area over which the sewage can come in contact with the soil.

Purification is affected by microorganisms in the septic tank and in the soil.

For successful operation, a septic field must distribute the sewage evenly over an area large enough to absorb it completely. If the system is overburdened, or if some part of it carries too large a share of the total load, the fluids may force a channel and flow out on the surface.

Pervious soils absorb the sewage much more rapidly than tight ones, and can handle much larger quantities per foot of tile and square foot of trench.

The size of a system is calculated on the basis of volume of flow and the rate of absorption. For this set of specifications, it is figured there will be 100 gallons per person per day in residences. Day schools have a rate of 15 to 35, and restaurants and business buildings 10.

A percolation test is made on the soil. Holes 1 foot square are dug to the bottom level of the proposed trenches, and filled with water to a depth of 2 feet. The water is allowed to drain until it is 6 inches deep. The time required to fall to 5 inches is recorded.

This process is repeated until the drop from 6 to 5 inches takes the same time on two successive tries. This time is matched to the same or the next-higher figure in column 1, Fig. 5.47. The center column gives the absorption rate, the right-hand one the number of square feet of trench bottom needed.

The tank should have a capacity of not less than 100 gallons per person. Any watertight construction may be permitted. Concrete block, special concrete slabs, poured concrete, and prefabricated tanks are used.

Where a garbage destructor is used, 50 percent should be added to all capacities.

As with most permanent installations, it is good policy to be generous in figuring. The load might increase, or the soil might lose absorption efficiency.

The field should not be heavily shaded, and cannot be crossed by vehicles. It must be at least 25 feet from any pond or stream, and 50 to 100 feet from any drilled well or reservoir. It must not drain toward a surface well or spring used for household water, at any distance.

If the native soil is not suitable, or the water table is high, it may be necessary to install a curtain drain to divert water, or to haul in hundreds of yards of sand to construct a filter bed in which to lay the tile field. This can be expensive enough to be a determining factor in selecting a lot.

Local regulations should be consulted before arranging for sewage disposal, as methods vary in different areas and exact conformance may be required.

Septic systems are used only where public sewers are not accessible, and usually are abandoned if a sewer is laid.

A drain may also terminate in a dry well. The well consists of a hole in the earth, filled with loose rock, rubble, or clean gravel. The pipe discharges into it, and the fluid soaks out of the sides or bottom, or becomes part of the general circulation of groundwater. This system depends on pervious soil in contact with the dry well.

VERTICAL DRAINS

Vertical drains of sand or porous soil can be used to move water up or down. They are often made by drilling or blasting holes in an impervious layer lying over a pervious one. Figure 5.48(A) shows a section of ground in which opening up the hardpan would allow water to drain into underlying gravel, and (B) shows a pond which is able to exist above the water table in a pervious sand,

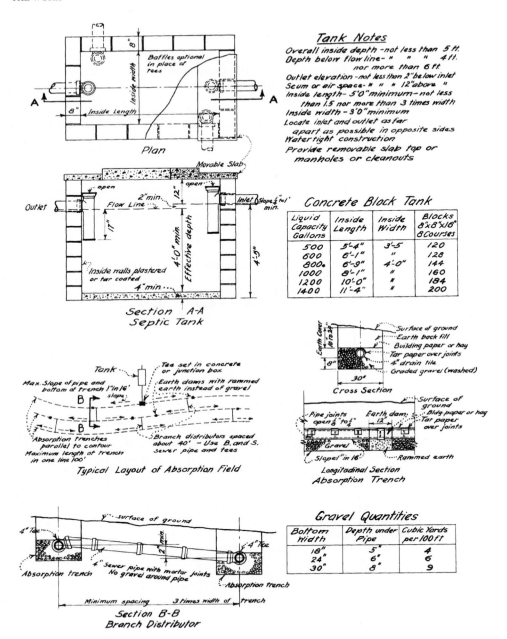

FIGURE 5.46   Details for small sewage disposal systems. [*Reproduced with permission of Westchester County (N.Y.) Department of Health.*]

TIME FOR 1" FALL Minutes	TYPE OF SOIL	ALLOWABLE ABSORPTION RATE Gals. per Sq. Ft. per Day	BOTTOM AREA REQUIRED PER 100 GALS. OF FLOW * Square Feet
5 or less	Sand, gravelly loam	2.5	40 *
8	Light loam	2.0	50
10		1.7	60
12	Loam	1.5	67
15		1.3	76
22	Heavy Loam	1.0	100
30		0.8	125 **
60	Hardpan	0.6	165 **
60	Heavy Clay	None (Use other means of disposal)	

    * 100 sq. ft. minimum per system. To estimate length of trench, multiply by ⅔ for 18 in. trench, by ½ for 24 in. trench, and by ⅖ for 30 in. wide trench.

    ** NOTE: To be used only by special permission.

FIGURE 5.47   Septic field data.

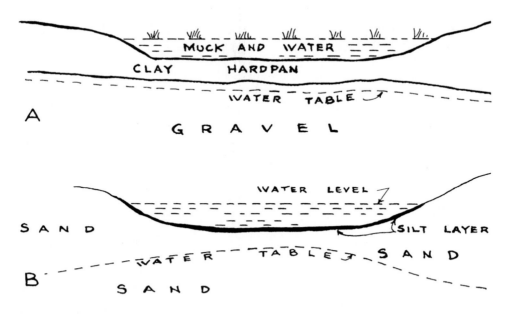

FIGURE 5.48   Perched water tables.

because silt and organic matter deposited from the water have sealed all spaces in a thin layer on its bottom and sides. Breaking this layer by any means will cause the pond to drain, unless sufficient silt is stirred up to seal it again.

Vertical drains, through which water rises because of being displaced by the weight of fill, are used to stabilize deep layers of saturated peat. Such soils may contain 50 percent or more water, and behave as viscous fluids under pressure. Fills placed on them sink, owing in part to water being squeezed out of the mud, and in part to the mud being displaced to each side. This settling of the fill may continue for a great many years, and become so uneven as to make pavements or other structures on the fill become unusable.

The squeezing out of water may be accelerated, and the sideward movement of mud practically eliminated, by making vertical holes in the mud, filling them with clean sand, and connecting their tops with a drainage system, as indicated by diagrams in Fig. 5.49. The fill is then placed, and its

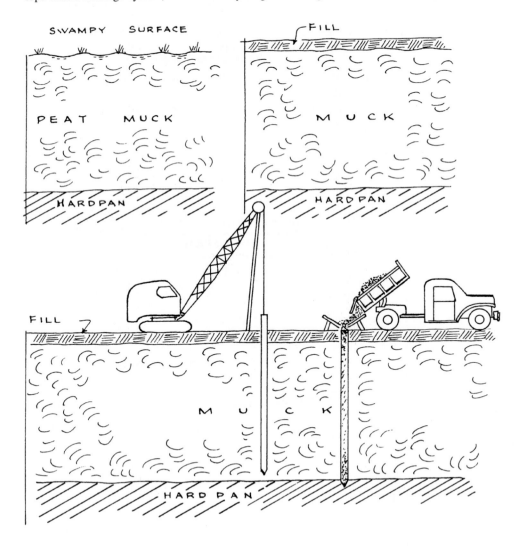

FIGURE 5.49   Vertical sand drain installation.

weight causes the water to enter the sand columns and rise into the drains. If the columns are properly spaced, sufficient water will be removed to stabilize the mud sufficiently to carry the intended load.

Part of the fill may be placed on the swamp, then the holes made by sinking a hollow-walled pipe by jetting with water and compressed air, or by driving a hollow pile with a detachable head. In either case, when the tube has reached the bottom of the soft layer, it is filled with sand and pulled, leaving the sand in the hole. Hole diameters of 16 to 24 inches are commonly used, with spacing varying between 8 and 22 feet.

If the tube is pulled by conventional methods, the sand may stick to it and be raised sufficiently to allow mud to enter crevices in it and interrupt the drainage. This may be avoided by attaching an airtight head to the tube after the sand has been placed, and pumping compressed air between the head and the sand. This raises the tube but exerts an equal downward force against the sand and holds it in place and together.

The tops of the columns may be drained by spreading a blanket of clean gravel or sand, a foot or more in depth, over the whole area; or by connecting the columns with tile or rubble drains. If the fill is all gravel, a drainage layer may not be needed.

Settling may be still further speeded by placing more fill than will be required for final grade. The extra weight will squeeze the water out of the mud more quickly. When settlement is judged to be complete, the extra fill is removed.

EXCAVATIONS

It is desirable to remove surface water and groundwater from areas to be excavated, but the cost may exceed the advantages gained.

Water may be removed naturally by seasonal change, or artificially by diversion, draining, siphoning, or pumping.

Seasonal Lowering.   The seasonal decline in the groundwater level may be quite considerable in areas having dry summers. Some places that are so wet in winter and spring as to be very expensive to work, may become dry to depths of 5 to 30 feet. Permanent swamps may develop crusts sufficient to allow movement of machinery.

Such changes are not uniform, as a wet season may keep water levels abnormally high while an exceptional drought will cause extreme or unseasonable lowering.

When it can be arranged, it is obviously good practice to undertake wet excavation in a dry period, as any reduction in either mud or unwanted water will reduce costs. The economy is greatest in work in marshy areas and in shallow excavations, but may be noticeable even in deep work.

PUMPING

Dewatering of excavations commonly requires the use of pumps.

If the water is small in volume or contains a heavy load of mud or other solids, a diaphragm pump is preferred. A centrifugal pump is needed for larger quantities. A centrifugal pump should be placed as near the water level as possible. More energy is used pumping water over a high bank than over a low one, but the total lift may be largely determined by the job. However, these pumps will push the water more efficiently than they will pull it, and much better output will be obtained by keeping the suction line short.

More details about the pumps can be found in Chap. 21.

Inlet Protection.   The low end of the inlet pipe should be fitted with a screen that will prevent entry of any object large enough to plug the pipe or damage the pump. Water containing leaves or other fibrous matter will clog such a screen readily, and may make necessary the placing of an outer screen of ¼- or ½-inch mesh, or some similar wire. This outer screen is best located far enough from the inlet that water going through it will not have force enough to hold rubbish against it to block it. See Fig. 5.50.

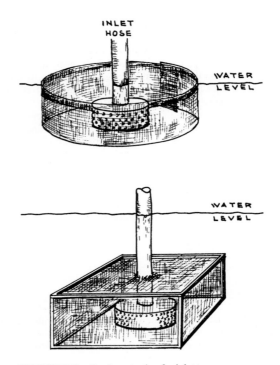

FIGURE 5.50 Trash protection for inlet.

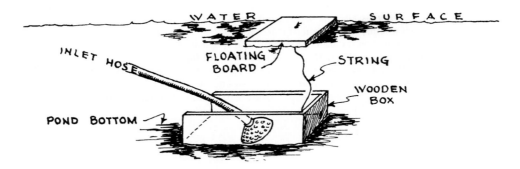

FIGURE 5.51 Mud protection for inlet.

Another inlet problem is that of the pipe and screen sinking into a muddy bottom and becoming blocked, or being buried by soil washed over them. Placing the inlet in a wooden box will prevent sinking and make it easier for the pump to suck up washings. See Fig. 5.51.

In long inlet lines, or large ones, it is good to have a foot valve in the bottom to hold water in the pipe while the pump is not running, unless the pump is tight and the inlet is underwater at all times. This will save time pumping out air each time the pump is started. Unfortunately, foot valves are subject to jamming and sticking, and may need frequent attention to keep them working.

A whirlpool may form over the suction end of the inlet pipe, which will allow air to enter the pipe through several feet of water. This is most apt to happen if the pipe is lying in a nearly horizontal position. It can usually be prevented by arranging the end of the pipe to hang vertically or attaching a shield over the inlet, or by throwing a square or round piece of flat wood in the water, which will tend to center in the whirlpool and block the air passage.

A pump will work best if a foot or two of water is kept over the inlet. In most excavations, the bottoms should be kept as dry as possible. It therefore is advisable to dig a sump pit for the pump hose, and to cut through any ridges that prevent water from flowing into it from the pit.

Dirty Water. If the water flowing into the excavation is dirty, it indicates that soil is being brought in from outside the excavation. Continued pumping may cause caving of banks due to undermining; or even may cause sinking of adjoining buildings or roads. It is wise to keep such pumping to the minimum required for the work and to finish the job as rapidly as possible, even at extra expense. It may be necessary to dry the area by well points, or to block the water off by grout, chemicals, or freezing.

Contractors' liability and property damage insurance ordinarily does not cover damage to structures by undermining, even in the "comprehensive" policies. A special endorsement is necessary, and inspection of the job is usually required.

Well Points. A well-point pump is a centrifugal pump with rather close-fitting parts, and often with an auxiliary air-vacuum pump, and which can work efficiently in spite of a fairly high proportion of air in the intake lines.

A well point is a section of finely perforated pipe that is sunk into the ground by jetting, driving, or drilling. It is attached to ordinary iron pipe, which rises to the surface and is connected by other lines to a pump that usually takes care of a number of points.

When the pump is running, the groundwater in contact with the well point is drawn through the holes of slits into the pipe and pumped away. The holes are so small that only very fine particles of earth will pass through them. The continued suction gradually removes all such particles from the area immediately around the pipe, leaving the coarser ones. This makes a porous screen with an outside area several times larger than that of the point, and improves its gathering efficiency.

Each well point will remove groundwater from a cone of depression around it, the slope of which depends largely on the porosity of the soil.

If well points are placed in a line so that their cones overlap, a continuous band of soil can be dewatered, as in Fig. 5.52.

A ditch could be dug in this band without encountering groundwater, regardless of otherwise saturated conditions throughout the area.

The well points may also be set in a square pattern to dry up the ground for a basement or similar excavation. It sometimes is possible to dewater such an area by using points as a curtain drain where the source and depth of the water are known.

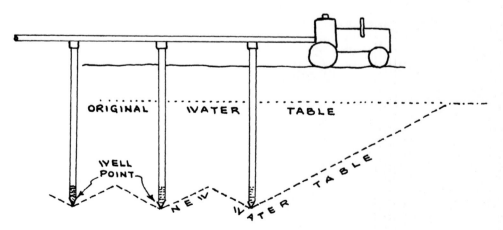

FIGURE 5.52 Well points.

In order to eliminate mud difficulties, the points should be placed deeply enough that the excavation will not reach capillary water standing above the artificially lowered water table.

In a deep excavation, it will be necessary to reset the well points and pumps on successively lower benches as the digging progresses, because of the inefficiency of high suction lifts. This may be done by starting the excavation oversize, so as to leave a shelf at the bottom of each cut for placement of the pumps and lines, as in Fig. 5.53.

Well points are most efficient in porous soils and will ordinarily not give good results in clay soils. In peat, the points are jetted down, and sand is dumped in the hole around them to increase contact area.

Proper use of well points involves considerable work in placing, connecting, and moving points, and pumping usually is on a 24-hour day basis for the duration of the job. In addition, considerable experience is desirable in order to avoid wasted time and possible failure to keep the job dry. It generally is advisable to subcontract this work to specialists.

Deep Hole Pumping. An excavation area may be predrained by sinking a number of shafts, lined with timber or pipe, and pumping from the bottom. The pumps used are usually small, with electric motors. The shafts are more widely spaced than well points and can be used to much greater depths. Drilling and lining are expensive.

Deep well pumps, or the piston or the jet type used for water supply, may be used if equipped with good sand filters.

Sump Pumping. Shallow layers of soil may be dried by digging a deep hole in the area and keeping it pumped out. Effectiveness and promptness of drying may be improved by a system of ditches draining into the sump. These may run along the outside edges of the site and into the interior in any convenient pattern.

This is an excellent and inexpensive way to dewater a swamp before digging a pond, unless the flow of water into the area requires an excessive amount of pumping during the drying process.

Jetting. Jetting with high-pressure water, or less commonly, compressed air, is used in making deep narrow holes for setting piles, installing vertical drains, obtaining soil samples, and various other purposes.

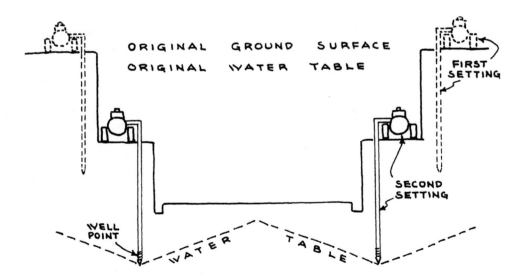

FIGURE 5.53 Deep well-point pumping.

Pressures required range from a few pounds over atmospheric for penetrating loose fine deposits to several hundred for tough clays.

A single pipe with a nozzle or reduction in size at the tip may be used in probing for rock or other obstructions. The tip reduction increases the velocity of the water and makes plugging less likely if it is forced into soil that the water will not cut.

Single pipe holes are irregular in shape, as the exhaust water and spoil rise around the pipe and will erode channels along the path of least resistance.

A better system is to use several water jets around the rim of a pipe so that washings can rise through the pipe to the surface. Water may be supplied through separate pipes, or by welding one pipe inside another, leaving a space between them for passage of water from an upper inlet connection to the bottom jets.

There should be at least three jets, preferably four or more. They must be evenly spaced around the circumference to prevent the pipe from drifting sideward toward the most effective erosion.

The pipe should be handled by a crane or some other type of hoist.

The lower end is sometimes fitted with teeth, and is lifted and dropped to loosen hard materials. The nozzles must be well protected against contact with hard dirt, if this method is used.

BASEMENT DRAINAGE

Excavating contractors are often consulted about the feasibility of having a basement under a house. The problem may be one of the cost of dealing with rock on the site, or a fear of water conditions which would make the basement wet and unusable.

If proper procedures are followed, a basement can be kept dry in any location where water does not spill in the windows or over the top of its wall. The cost ranges from the sometimes nominal expense of installing subdrains, up to more than $1 per square foot for complete waterproofing of floor and walls.

Soils and Locations.   The tight soils such as clay or the various varieties of hardpan, tend to become saturated in wet seasons, even near hilltops. The quantity of water they may carry, which is the basis for deciding on drain size, may be very difficult to determine in a dry season. In general, if the soil contains long streaks or lenses of sand or very fine pebbles, it may be assumed that there is considerable flow through it. If there are spots near the building site which ooze water in the spring, or in which water-loving plants grow, a serious drainage problem is indicated.

Difficulties are sometimes avoided by shifting the building site to a spot with better drainage, or doing only shallow excavation and obtaining depth by filling around the walls. Drains should still be used, as groundwater may rise into the fill.

Subdrainage.   Drainage around the footings is a precaution that should always be taken if there is any lower point to which water can be led. A porous soil such as sand or gravel can seldom hold enough water to wet a basement, but it may be part of a waterlogged lowland or a gradual slope up from one, or have water held in it by layers or lenses of clay.

The standard basement subdrainage consists of a line of land tile laid completely around the outside of the footing, and preferably 1 foot to 18 inches below basement floor level. It should be laid in a fine-crushed stone, protected with tar paper or hay, and backfilled promptly. Such tile has a downward pitch of $\frac{1}{2}$ to 1 percent from a point opposite the outlet.

The outlet may be land tile, but because of the danger of entrance of plant roots, glazed sewer tile with cemented joints is better. It should slope down away from the building at 1 to 5 percent grade to a disposal point. This may be a stormwater drain under the street, a stream, or lower ground.

A stormwater drain complication is that water entering the system at higher levels may back up through the tile and saturate the ground around its walls temporarily.

When there is no stormwater drain, or connection to it is considered unwise, a discharge point on the same property should be sought. It is often easy to get permission to lay pipe through a neighbor's yard, but impossible to get a formal easement to keep it there.

A pipe having an open discharge should always be kept covered with coarse screening to prevent animals from entering it. If the pipe is large, a flap gate can be used, but these are not satisfactory in small sizes.

In many situations an open flow of water from a pipe is objectionable. In such cases the drain may lead to a dry well, or into tile laid out in the same manner as a septic field. An overflow exit may be provided, or the water may be left to work that out for itself.

Standard practice calls for 4-inch tile around the foundation. This is often too small, and its inadequacy is the cause of endless trouble. Even a small building can block a considerable area of horizontal movement of groundwater, which will try to enter it unless it is drained off. After a heavy rain, a previously unnoticed seepage vein or group of small channels may carry more water than a small tile can hope to accommodate, and the foundation wall may cut into a number of them. Six inches is a safer minimum size.

The outlet drain can be the same size if the slope is steeper, or the next larger if it is the same slope.

If the building is on a slope most, although not necessarily all, of the water will be against the upper wall or walls, and may require 8- to 12-inch pipe. If there is a long slope above the building, surface water may constitute a serious problem that is best solved by leading it through gratings and vertical pipes to the footing tile, which may then be as large as 20 inches. The size needed can be figured in the same way shown earlier for culverts, plus a liberal allowance for underground flow.

The floor should be laid on 4 to 8 inches of crushed stone or gravel. This should be connected to the outside subdrain by a tile through or under the footing. If less gravel is used, one or two lines of tile might be laid under the floor. The gravel will serve to catch any vertical seepage of water, and will insulate and strengthen the floor also. The tile serves only for drainage.

Gutter leaders can be tied directly into the footing tile, emptied into nearby dry wells which will ultimately drain into it, or provided with dry wells or outlets at a distance.

If they dump directly into the tile, it must be large enough to carry easily all the water that will enter it from the ground and from the gutters. If the gutter water tries to enter a tile which is already full, it will accumulate in the leader and may build up to a head of 15 or more feet. The resulting pressure inside the tile will force water out of the joints and cause erosion and misalignment that may result in entire failure.

On the other hand, if tile size is ample, the swift current of gutter water will tend to carry away dirt which may work into the tile from the ground.

If dry wells are used, it is best to place them well away from the building.

Where porous fill is available, it should be used for backfill against the foundation to prevent water pockets from forming against the wall. Surface water can be kept out of it by sloping up toward the foundation, and by placing a capping of topsoil.

Porous breather pipes or a hollow-tile outer wall may be carried from the tile to the surface, against the foundation. Air chilled by contact with the ground tends to flow down the drain to the outlet; and when it can be replaced by warm air pulled down from the surface, the resulting circulation warms the soil and the wall, and reduces the problem of condensation inside the building.

A building which has been built without subdrains or with inadequate ones can often be protected by installing a deep curtain drain along the uphill slope. This may be built to cut off underground water only, or to take care of surface water as well.

If a basement is to be built in a hole blasted out of sold rock, it is good practice to provide a sump hole, about 3 feet deep and square, in a corner.

If water difficulties develop, an automatic pump can be quickly installed to remove it from the basement itself, or from the floor base where it might otherwise build up enough pressure to cause heaving and cracking.

The sump can be protected with a manhole cover when not in use. A 36-inch sewer tile, open at the bottom, makes an excellent lining.

This is cheap insurance against the possibility of needing an expensive and damaging ditch-blasting operation, or an elaborate waterproofing job.

Waterproofing.   If there is no downhill point or stormwater pipe to which subsurface water can be drained, and location and soil type indicate the probability of groundwater, the basement walls should be waterproofed. This can be done more thoroughly and economically during construction than afterward.

Condensation.   Before undertaking any considerable expense to stop leaks into a basement, it should be found out definitely whether they exist. Condensation may make substantial amounts of water appear on the walls and floor. If the trouble occurs in hot weather, it is probably condensation; if in the wet season or during heavy storms, it is most likely leakage. If a piece of cardboard is secured against a suspected spot, it will get wet on the wall side if it is leakage, and on the room side if it is condensation.

Condensation may be checked by coating the wall with cement plaster or some other coarse, absorbent material, or by running an electric dehumidifier in the room.

CHAPTER 6
PONDS AND EARTH DAMS

The discussion in this chapter will be limited to ponds varying from a hundred square yards to a few hundred acres in area, which may be built according to the judgment of the landowner or the contractor, rather than according to detailed specifications.

Such ponds may serve as small reservoirs for domestic and industrial use, or to provide water for fire fighting, for animals, and for fishing and other recreation. They may also be useful in swamp reclamation, groundwater replenishment, and flood control.

They may be made by damming streams, digging holes for streams to fill, digging below the water table, or combinations of these techniques. Dry hollows may sometimes be converted into ponds by diverting streams, tapping springs, or lifting underground water by means of windmills or other pumps.

WETLANDS

During these modern times in the United States, before one can think of reforming the earth to build a pond and, perhaps, doing away with a naturally wet area, it is necessary to consider the subject of wetlands. Wetlands are those areas that are inundated or saturated by surface water or groundwater at a frequency and duration sufficient to support a growth of vegetation typically adapted for life in saturated soil conditions. Wetlands are valuable natural resources that provide important benefits to people and the environment. They help improve water quality, reduce flood and storm damages, provide important fish and wildlife habitat, and support hunting and fishing activities.

Clean Water Act.   In 1986 the Congress of the United States included in the Clean Water Act provisions for the protection of wetlands. It laid on the Environmental Protection Agency (EPA), in partnership with state and local governments, the responsibility for restoring and maintaining the chemical, physical, and biological integrity of the nation's waters. Consequently, the EPA is also charged with protecting wetland resources. The major federal regulatory tool for this is Section 404 of the Clean Water Act, which is jointly administered by the U.S. Army Corps of Engineers and EPA.

Types of Wetlands.   There are two types of wetlands, coastal and inland wetlands. The coastal wetlands are linked to estuaries, where seawater mixes with fresh water to make a variety of salinities. The inland wetlands have fresh water commonly in flood plains along rivers and streams, in isolated depressions surrounded by dry land, and along the margins of lakes and ponds. In moving the earth, the wetlands of concern are the inland variety, except where dredging along the coast may have to be done.

Altering Wetlands.   Physically wetlands might be altered by filling, draining, excavating, clearing, flooding, diverting water away, etc. They might be altered chemically by changing nutrient levels or introducing toxics and biologically by grazing or disrupting the natural population. Any of these forms of alteration needs to be planned with approval. Anyone filing with the Corps for a dredge-and-fill permit under Section 404 must first attempt to avoid or drastically reduce the impact on the wetland. Failing that, compensation through mitigation may be required for unavoidable wetland loss, by either restoring an existing wetland or building a new one.

Wetland Evaluation Technique.   The Federal Highway Administration is teaching state departments of transportation (DOTs) and their contractors how to determine the value of a wetland with the wetland evaluation technique. The DOTs require the contractors to take a 4-day course before contractors are allowed to bid on a job involving wetlands, because created wetlands fail when the basics are not being done. If the site for a created wetland is too narrow and the slopes are too steep, the wetland will remain barren. Once built, a newly created wetland must be monitored by a state DOT or developer for 5 years to make sure it is self-sustaining or viable.

SWAMP RECLAMATION

Soil Conditions.   Swamps which are wet all year are logical places to dig ponds. The spoil taken out of the excavation can be used to build up the area around it so that the section worked is changed from a bog into open water and dry land.

Swamps commonly have a top layer of soft peat or muck soil, which may be of any thickness from a few inches to 100 feet or more. This organic material is easy to dig but provides very treacherous support for machinery. Below the muck, any type of soil or rock may be found.

The reader is referred to Chap. 3 for techniques in handling the mud that is one of the usual problems of swamp work. Mud may be reduced or eliminated by working in a dry season; by diverting, draining, or pumping out the water before or during the work; or drying up the area by sump or well-point pumping. These techniques are discussed below and in Chap. 5.

It is very advantageous to get rid of as much water as is reasonably possible. Water prevents the operator from seeing the bottom he or she is cutting, with resultant wasted passes and gouging. It reduces the digging effectiveness of the bucket so that some soils which can easily be dug when dry cannot be penetrated when under 1 or 2 feet of water. Even with skillful operation, water will mix with soil in the bucket, making sloppy spoil piles that reduce the amount of digging at a stand, and which sometimes will flow back into the excavation or cut off the shovel's exit.

Excavators.   For decades the dragline excavator has been the choice of equipment to excavate ponds. It has good reach and can excavate the slopes for the pond very well. More recently, a full-revolving hydraulic backhoe, also known as a hydraulic excavator, has been used to excavate ponds. It does not have the reach of a dragline, but the hydraulic excavator has more power to dig hard material. These forms of power shovels are described in detail in Chap. 13.

For this discussion of pond construction, the use of draglines will predominate so that there is not much concern for the footing of the excavator. Footing is a problem for the hydraulic backhoe, because it has to get closer to the area to be excavated than the dragline does.

Digging Plan.   Figure 6.1 shows a general plan for digging a pond in a swamp and using the spoil to build up the unexcavated parts. A drainage hole is dug at the downstream end, and the water level lowered by a ditch drain or by pumping. A sill may be placed at the entrance to the ditch or the sump hole to hold water back a few inches above the floor of the proposed pond.

If the swamp is fairly dry, and the digging is fast and continuous. the removal of water may be unnecessary as the expanding excavation may keep the water at a low enough level that it will not cause trouble. Surface water may be diverted around the excavation by shallow ditches or dikes as shown, or allowed to flow into the hole.

If no obstructions prevent, the pond is dug from the center toward both sides, with the dragline walking along the longest dimension, which is usually parallel to the direction of water flow, as in Figs. 6.1 and 6.2. The machine keeps back far enough from the centerline that it can reach it with an easy cast. It usually works on platforms or other artificial supports, but if the swamp has

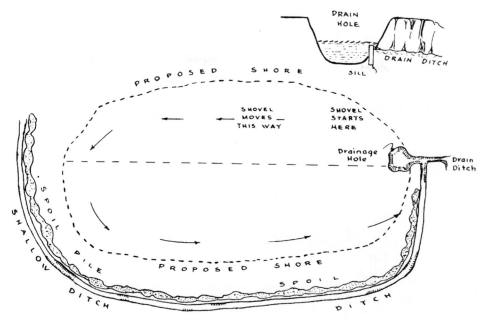

FIGURE 6.1   Digging plan for swamp pond.

been drained enough in advance to be firm, or has gravel soil, or a heavy mat of bushes, these may not be necessary.

The bottom is kept on grade by digging just enough to let the water back over it. If there is not enough water to cover the enlarging bottom, the grade may be checked in the same manner as in a basement excavation.

The length of a pond of this type can be increased indefinitely without change of method. The width, however, is limited by the reach of the dragline and the depth of the hole. The reach determines the width of the strip in which it can dig and pile, and the height of the piles; the depth governs the part of that width which must be reserved for piling spoil.

If a wider pond is required than the machine can dig in one round trip, as illustrated, it must go behind the piles, drag or swing them away from the excavation, and then widen the hole.

Size and Depth.   Calculation of the size and depth of the pond should involve a number of factors. A large shallow pond gives the most for the investment, at first appearance. A deep pond is desirable in that it can be fed by seepage from lower levels, loses a smaller percentage of its water by evaporation, does not lose area by silting as readily, discourages growth of bottom weeds, and is more suitable for fishing and swimming. Against these advantages are increased cost and a possible drowning hazard.

Deep ponds may often be obtained from shallow excavations, or without excavation, by building of dams and dikes; but for the present we will consider results obtained by excavation only.

The pond should be dug to a clean bottom, if possible, and should yield enough spoil to build banks 1 foot or more above the water at the edge, and sloping up away from the pond for drainage. In a limited area that is to be reclaimed, an increase in water surface reduces the area of the banks and the amount of fill needed for them. Fewer yards need be moved for a large shallow pond than a small deep one of the same capacity; although a larger proportion of the yardage may have to be moved more than once.

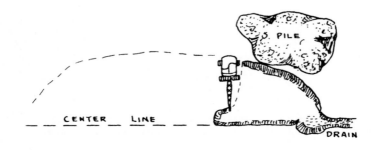

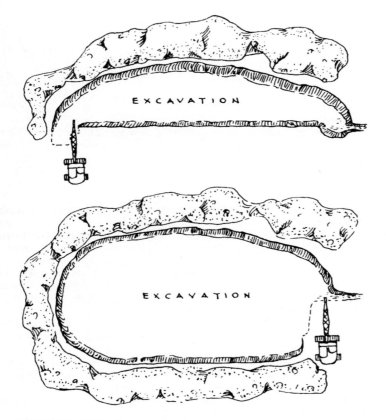

FIGURE 6.2 Making the excavation.

Cut and Pile Relationships. Figures 6.3 and 6.4 show some of the relationships between the cut and the spoil piles. The diagrams show a machine with a 40-foot boom, digging beside and behind itself, and have been simplified by assuming a constant dumping height of 12 feet; an increase in volume of the spoil of 20 percent; nearly vertical slopes in the cuts; a 1-on-1½ slope on the piles, and soft soil, permitting deep digging. Figures on the piles show cross-sectional areas. Muddy conditions cause piles to be lower and wider, reducing the width of cuts.

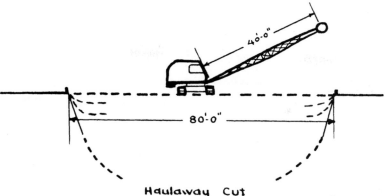

Haulaway Cut
Maximum depth depends on soil
resistance and cable length

CUTTING TO MAXIMUM WIDTH AND PILING

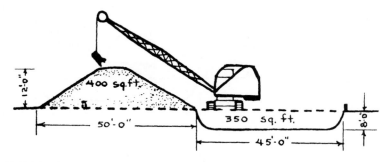

FIGURE 6.3 Dragline cuts.

The dumping height increases with higher boom angles, an advantage partly offset by a shorter dumping reach. A low boom is preferable for deep digging and to obtain a good digging reach without casting the bucket. Extra pile height may be obtained by raising the boom to put a top on it, but keeping it low for the bulk of the digging. A dragline with a live boom can lower it for digging, and raise it for dumping during the swing; but the extra power needed limits this to occasional use.

The swell and the slope of the soil piles indicated in the diagrams might or might not be applicable to a particular job. The regular cross section shown will be found only in sand or other free-flowing, nonsaturated material.

The volume of the piles may be increased by moving the dragline behind them, and pulling

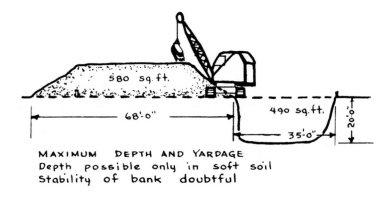

MAXIMUM DEPTH AND YARDAGE
Depth possible only in soft soil
Stability of bank doubtful

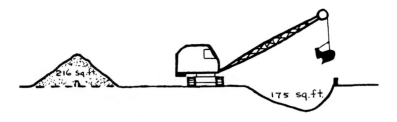

MAXIMUM DISTANCE OF SPOIL MOVEMENT

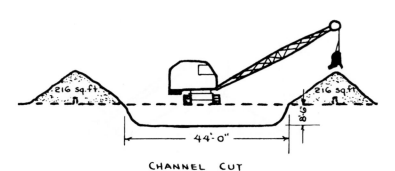

CHANNEL CUT

FIGURE 6.4 More dragline cuts.

them or swinging them back periodically. This extends the pile toward the rear and leaves room for more on top. This is practical if ground conditions permit work without platforms, but otherwise takes too much walking time. A second dragline working behind the windrow and swinging it back makes a very effective combination.

If the excavation is narrow and shallow, and the banks are narrow, a long boom dragline may excavate without building a pile by placing each bucket load in its final position.

Double Cuts. If the width of the pond is to be greater than can be dug in the two cuts described, the dragline may make additional cuts, first removing the soil windrow, then the ground under it.

The windrow is usually moved by standing behind it and digging with a short dump cable, so that the loaded bucket can be picked up without pulling in. It is swung to dump as far back as possible, and a level runway made to permit the dragline to work along the rim of the original cut, as in Fig. 6.5.

If the pile is too large to allow the bucket to reach across it, part of it may be dug away, as in Fig. 6.6, and the balance removed at the same time that fresh ground is dug.

The runway is normally made at the original grade, as vegetation or drying makes it firmer than the soil underneath. However, if the soil is hard to cut, the runway may be lowered to improve digging efficiency. If digging is easy, and the dug soil dry and firm enough to support the shovel, the ramp may be made higher, as in Fig. 6.7, to increase the dumping height. The cross section of a pile increases about in proportion to the square of the height, so the advantage gained is important. The freshly moved dirt is left higher than the undisturbed part of the pile, as it may settle seriously under the dragline. If possible, the machine should be kept on the consolidated part.

If the pile has a wide top, it may simply be leveled by a bulldozer, or by the dragline raking and patting it as it travels along it. However, it should be remembered that the tops of piles are treacherous at best, the machine should be kept back from the edges, and the pile watched carefully for evidence of caving, particularly toward the digging.

As shown in Fig. 6.7, a dragline working on top of a pile is able to not only move the whole pile back but to dig the ground under it in one operation.

For even wider ponds, additional cuts may be made. Each additional slice involves moving all the material which has been dug, with increasingly complicated patterns, and expense mounts rapidly. Usually more than two cuts are made in one direction only when the operation involves cleaning off vegetation and shallow digging. In deep work, trucking the spoil or removing it with conveyors is more economical when several rehandlings with a dragline are necessary.

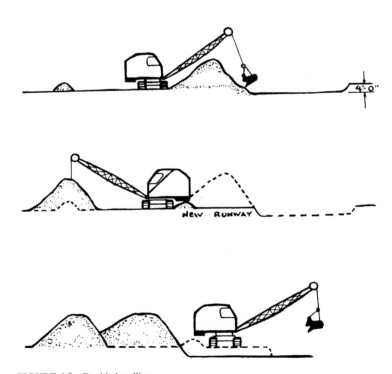

FIGURE 6.5   Double handling.

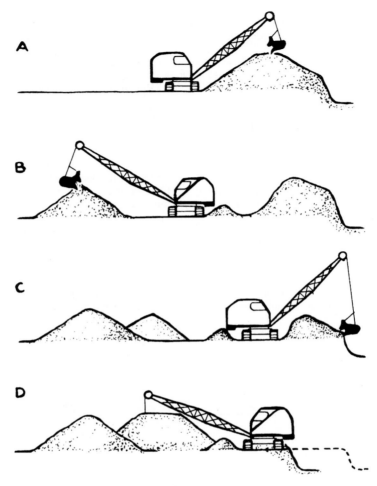

FIGURE 6.6 Working back a heavy pile.

Trucking. A flattened windrow, such as is shown in Fig. 6.7, may be used as a truck road if it is dry and substantial enough, and is connected with dry land at one end. The dragline, or back-hoe, may be started at the far end and worked toward the exit. It may dig to final grade, leave a shelf for further dragline work, or it may load the dry material in the trucks and cast the wet stuff from the bottom to build up a new windrow.

Abnormal delays may be experienced in the trucking. It is usually not practical to build turn-arounds or two-lane roads, so the trucks must back in the full distance from shore, and no other truck can enter until the previous one has left, thus leaving the shovel idle each time. There is also danger of trucks going too close to the edge, or encountering soft spots, so that they get stuck or overturn.

If the windrow can be connected with dry land at both ends, scrapers may be used to remove the dry upper part.

Material may be trucked away economically if the pond is to be large in proportion to the size of the shovel, so that several shovel handlings would be required; if there is insufficient space in the digging area to pile the spoil; if fill is needed elsewhere on the project; or if the spoil can be

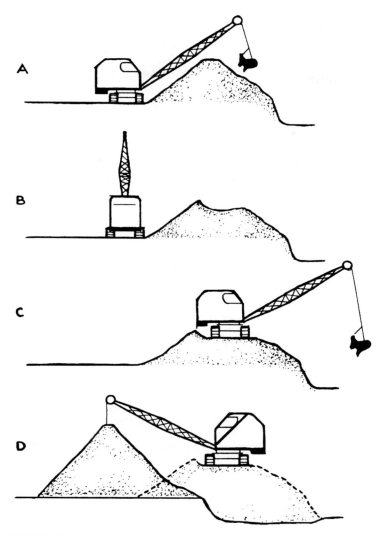

FIGURE 6.7   Working from top of pile.

sold. Under some circumstances, the entire cost of making the pond may be repaid by the sale of the spoil.

Selling Spoil.   Materials that might be sold out of a swamp include the organic earths, such as topsoil, muck, and peat (humus); and inorganic subsoils such as sand, gravel, clay, and loam. Frequently, the values of these materials are destroyed if they are mixed together, as the presence of organic matter makes subsoil undesirable or useless as a structural fill; topsoil or other organic earths are not salable if coarsely mixed with subsoil.

Swamp topsoil and mucks are generally too heavy in texture for use in lawns or gardens, but may improve greatly if left in piles for a year or two, or if mixed with sand or light loam. Nearly pure peats, however, are widely used to enrich soil, and command a fairly good market. These have a very high water content, and may shrink 50 percent or more in the pile.

If the spoil is to be used for the pond banks, or similar unloaded fills, organic and inorganic dirt is generally dug and handled together. If they are to be sold, the topsoil should be stripped and piled to the side, and the subsoil dug afterward in a separate series of operations. As the surface of the subsoil is often below water level, adequate drainage or a dependable pumping system should be provided before putting the shovel on it.

Separation of materials in several layers may be very complex. Examples are digging gravel interrupted by seams of unwanted clay and digging clay seams lying between sandy layers, the decision as to what is desirable being a matter of local demand.

If the dragline is large, or has a long boom in proportion to the area to be dug, these layers may be removed and piled separately, or they may be stripped back separately in the same manner as topsoil. Often, however, a better solution is to truck away the least bulky or sometimes the most valuable of the materials during the digging. Trucking may be possible only during a short period late in the dry season, after a protracted period of artificial drying by drainage or pumping, or only after construction of fills or timber roadways.

Selective digging should be done in the dry, so that the operator can see what he or she is doing. It will be discussed further in connection with borrow pits in Chap. 10.

Predraining.   Whenever the soil is to be trucked out, or when very sloppy conditions are found in a swamp, it is advisable to investigate the possibility of drying up the area before work starts.

Almost any spot can be dewatered, at a price, by diversion of any streams or other surface inflow, and well-point pumping.

Sump pumping may be more economical when surface water can be diverted, and when underground reservoirs are small, or when relatively impervious soils cause groundwater movement to be slow. The soil must be fairly stable.

The sump is a deep hole with a bottom below the proposed digging. One side should be sloped gradually or terraced so that a pump can be set up wherever desired. Outlet hose, or a flume or ditch, must be provided to lead the water out of the area being dried.

For best results, pumping should be started around the beginning of the dry season. If the swamp is underlain by porous gravel or sand, most of the water can be removed from small basins in a few days, and additional pumping will be required only occasionally. In large basins, continuous pumping might be needed during the job. If the soil is tight, less water will be removed at first, and seepage into the hole will be reduced very gradually.

This operation should cause the swamp to dry up, so that difficulties with mud will be reduced or eliminated. However, the effect may be largely lost if heavy rains saturate it again before work is started.

Both the speed and effectiveness of sump pumping can be greatly increased by drainage ditches leading into the sump. A horseshoe-shaped trench enclosing the working area, but leaving an undisturbed space for entrance of machinery, is a convenient layout. Such a trench may involve piling hundreds or thousands of yards of spoil, but should be a good investment if it allows dry digging for the bulk of the project.

Peat will burn except when saturated with water, and a peat deposit drained as suggested might be entirely lost by fire. If this material had no value, this might be the quickest way of removing it. The fire would also loosen any stumps, and if deep, might consume them.

A peat fire might have black or bad-smelling smoke (or be almost smokeless), burn for a long time, and be difficult to prevent from spreading. Such a fire should be started only after consultation with local fire and police officials.

DRAGLINE SIZE

Choosing the right size dragline for pond dredging in swamps involves consideration of a number of factors. Small machines are more easily supported on soft ground, are easier to salvage if they get in trouble, can work in restricted quarters, and usually have a faster digging cycle. However, they must handle material more often to move it the same distance; cannot dig as deep or pile as high or penetrate hard material or take out stumps as effectively as larger machines; and cost more for each yard of digging.

Small draglines may be equipped with long booms to match the reach of large machines, but this reduces their speed and may interfere with stability. Additional counterweight is advisable, and the inertia of this and the reduced leverage on the bucket load cause it to take more time to start and stop each swing, increasing the cycle time by 1 to 5 percent for each foot of boom added. If the swing clutches or engine power is barely adequate to manage a standard boom, loss of time through excessive slipping or lugging down will be much greater than with a high-powered machine of the same rated capacity.

Undersized buckets are often used with extralong booms. These reduce the tendency to tip, and speed up the cycle somewhat, but reduce the payload and the ability to penetrate hard soil.

Digging Cycles.   A three-quarter-yard dragline with a 40-foot boom digging at the end of its reach, and swinging 180° to dump, can move part of the material 70 or more feet at each handling and complete two to three digging cycles a minute. The same machine, in digging up its own track, may move the dirt only 10 to 20 feet at the rate of three to four buckets a minute. The average distance the soil is moved may be found by measuring from the middle of the cut to the middle of the pile.

If the spoil is to be moved back through several handlings by the dragline, the long move is desirable; but if it is being stockpiled, the short, quick cycle is best. Most digging patterns include both long and short swings, to get maximum work out of minimum walking.

Building Up Ends.   If fill is needed at the upper and lower ends of a pond, the dragline may dig and pile in a curve as it rounds the ends, as in Fig. 6.2. Another method is to make straight cuts across the ends, as in Fig. 6.8. Such work at the outlet will block the drainage ditch, or interfere with pumping, so that it should be the last spot to be dug.

Drainage may be maintained by laying pipe in the ditch before covering it. If the spoil is to be spread after the first cut, such pipe may be left permanently, being fitted with a cap or valve so that it may be blocked to fill the pond, or opened to drain it. If two or more handlings of the spoil are necessary, temporary pipe must be placed and then lifted as the digging progresses, for which purpose corrugated steel is most satisfactory because of ease of handling and resistance to rough treatment.

Islands.   Both the quantity and difficulty of digging can often be greatly reduced by building islands. In cases where a shallow cut is made, and the balance of the depth obtained by building a dam, most of the spoil can be disposed of in islands. If cutting is deep, some soil may still be piled on them and their bases do not have to be excavated.

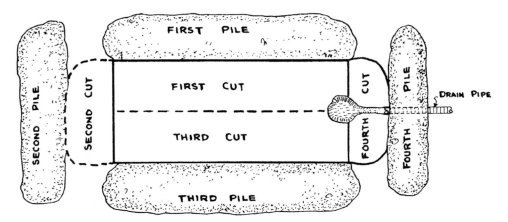

FIGURE 6.8   Straight-end cuts.

Islands may be built at spots most convenient for disposal of spoil, or according to other requirements. A finished height of 2 or 3 feet above water is desirable. In humus, piles 6 feet or more above water level may be needed because of loss through shrinkage, smoothing the top, and slumping under water.

Grading should be done by the dragline immediately after piling, or at least before the pond is filled.

Trees.   Pond sites are usually not as open and regular as the examples discussed above. One of the most common obstacles in the way of systematic work is a tree, or a group of trees. One tree can cause an increase in dragline time of 300 or 400 percent for the digging in its immediate neighborhood, and two trees may make the digging impossible without the use of trucks.

This is one reason why most pond diggers recommend making a clean sweep of all trees within boom reach of the excavation area. Another factor is that if the water level is raised by construction of a dam, or if fill is placed around a tree, even with protection for the trunk, it is liable to die, and large and old trees are particularly sensitive to any such changes around their roots.

Pond construction may cause injury or death to trees at some distance from the work. If a dam is built, the water level may be raised not only at the pond edge but to a decreasing amount for hundreds of feet back. Water level may be raised even when a dam is not built by groundwater backing up behind impermeable fill from the pond, or as a result of raising the grade of the banks. In addition, seepage may drown out trees below a dam. See Fig. 6.9.

Moderate lowering of the water table will only occasionally damage trees. Such lowering may be caused by digging into and draining a water-bearing layer formerly having an outlet on higher ground, or, in the case of a pond without a dam, by digging back into a bank.

Methods of protecting trees during construction, and from the effects of changes in ground and water level, will be discussed in the chapter on landscaping.

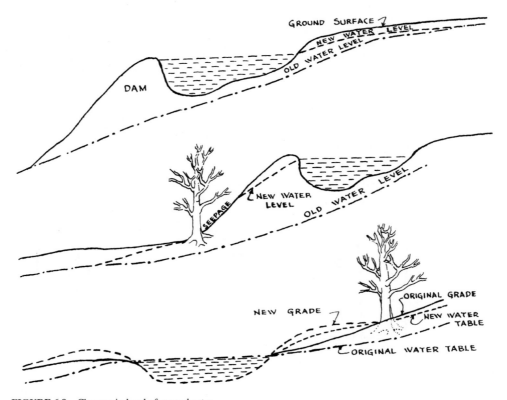

FIGURE 6.9   Changes in level of groundwater.

Removal of small or worthless trees near the pond, and of trees which would shade a beach or float, is desirable from landscaping and recreational standpoints. A strip of grass on the pond edge that can be trimmed will make it easier to keep the shoreline free of bushes and water weeds.

However, none of these factors justify an indiscriminate destruction of trees in the pond area. A tree may be a landscape feature of as much value as the pond, and the cost of digging around rather than through it may be only a small fraction of the entire cost of the project. Also, the cost of cutting and removing a large tree and disposing of the stump may be greater than the cost of operating the shovel for an extra day.

It is a good plan to consult a local tree expert, informing her or him of the scope of the excavation, the changes in water level, and the height of any fill to be placed near the tree, to get an expert opinion as to its chances of survival. This information may be given to the owner with an estimate for the job both with and without removing the tree.

Rock. Ledge rock and boulders cause less difficulty as they interfere only with the digging, not with the swinging. They may often be advantageously used for shoreline or islands, as a rock slope is usually more attractive than a mud or grass bank.

Rock to be blasted must be cleaned off, and should be above water during the work if possible. Standard blasting techniques are used. Refer to Chap. 9.

Bank Preservation. If the pond site includes a dry bank of suitable height, it may be advisable to leave it as one edge of the pond, digging away from it as if it were the centerline in Fig. 6.1 for the first cut, and disposing of all spoil on other edges. This technique can also be used to avoid tangling with trees or landscaped areas. If the pond is wide and without truck access, saving the bank may be too costly.

STREAM CONTROL

Digging Pattern. If a pond is to be excavated in the valley of a flowing stream, special precautions should be taken.

The diagrams in Fig. 6.10 show procedures in excavating a pond similar to that in Figs. 6.1 and 6.2, except that a brook runs through the center of the swamp.

A diversion ditch deep and wide enough to contain the brook is dug, starting downstream at a point lower than the proposed pond bottom, where it meets a ditch from a drainage hole, somewhat deeper than that described earlier. The diversion ditch is continued back to a meeting with the brook on the upstream side of the job. Spoil is piled on the pond side of the ditch to form a dike and to dam the original channel.

The dragline will have to walk across the brook. The banks may be dug away and platforms laid on them or in the water, or the stream may be partly filled with saplings or long logs to permit a crossing.

After diverting the water, one side of the pond is dug. A channel is cut below grade, midway between the centerline and side lines. At the upper end, the shovel cuts a gradual ramp up from the channel bottom, usually after crossing the dry streambed. This ramp should be covered with a brush mat weighted with rocks, or logs or planks should be placed crosswise in the bottom to avoid excessive erosion when the dike is cut and water permitted to flow down it. Sometimes a hole is dug below grade at the foot of the incline to trap most of the dirt washed down, and the protections are omitted. The water should now follow the channel in the pond bottom, and flow over the sill into the drainage ditch, keeping away from the centerline where further digging is to be done.

Another method is to straighten out the stream channel so that it lies on one side of the centerline. The other side is dug and the pile extended well across the stream at the upper end, forcing it to find its way behind the pile and back to the stream below the pond.

Temporary Stream Diversion. It is sometimes possible to put a temporary dam across the stream well above the work area, and to divert it across a low ridge into another valley, or into a

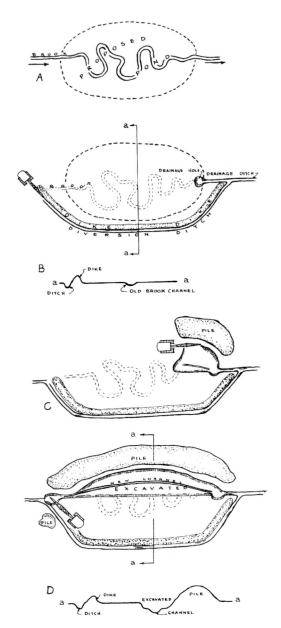

FIGURE 6.10   Digging in streambed—gravity drainage.

trench, flume, or pipe running along a hill slope in the same valley. A large pump may be used to raise the water into such diversions.

Before arranging to pump out a stream, or building a flume or pipeline to carry it, its volume of flow should be measured. This may be roughly done by placing a rectangular wood trough or flume 15 to 20 feet long in the streambed, and packing around the upper end with mud so that all the water will enter it. Chips of wood may be dropped at the upper end, and the time they take to drift through checked with a stopwatch. Mud can be dropped into the water to find if the bottom or sides have a perceptibly slower current than the top.

The flow in cubic feet per second may be found by multiplying the depth of the water in feet or fractions of feet, by the inside width of the flume, and multiplying the product by the distance the wood chip traveled in 1 second.

If the stream has a fairly regular channel, the cross-sectional area of the water in it may be calculated, and the speed of flow measured in the same manner.

Most streams are subject to considerable and sudden changes in volume, and pumps used should have extra capacity, unless it is possible to abandon the job during high water and return to it when the flow is reduced to a volume that the pumps can handle. Diversion channels, pipes, dikes, and dams should also be built to withstand high water.

Digging in Streams. If the stream cannot be diverted, the digging of each strip should start at the upstream end and move downstream. The dirt loosened but not picked up by the bucket will be washed downstream in considerable quantities which might entirely silt up any downstream excavation.

Riparian Rights. Laws relative to stream use and pond construction vary in different states and localities. Where riparian rights law holds, owners of land on a stream below the job must give their permission before the stream can be diverted, even temporarily. They also can collect damages if mud from the excavation work chokes the stream, or is washed onto their property. Excavation permits are often required. It is good to have the law and the neighbors consulted before starting any important pond project.

Permanent Stream Diversion. A pond usually is kept in the best condition and appearance if a strong flow of water goes through it. However, if the pond is to be managed for the production of fish, or if the stream is likely to fill the pond with silt, it may be advisable to make a channel for the stream around the pond, keeping only a controlled flow from it into the pond through a pipe or ditch.

It is difficult to overestimate the power and destructiveness of even a small stream in flood, and it is at flood time that the greatest damage can be done to a pond. It is therefore important to take every precaution to prevent the stream from breaking out of its prepared channel.

The diversion should start, if possible, in a stream section headed in the right direction; and often should be reinforced with heavy rocks or posts driven into the base of the bank on the outside of the curve. If the turn must be in the artificial channel, it should be a gradual one, protected with rocks, posts, or a well-anchored timber bulkhead. A high earth dike should be built between the stream and the pond, planted with sod, bushes, or trees.

Figure 6.11 shows a safe arrangement. A low dam of concrete, masonry, or fitted rocks is placed across the stream to raise its level a foot or so. A pipe leading to the pond is placed below this water level, and a dike of earth or concrete placed over the pipe.

Water in excess of that which will pass through the pipe will flow over the dam into the channel. The pipe may open into a ditch, or continue into the pond. Flow may be reduced by means of a gate valve, or by partially obstructing it with a board, or by placing stones at its mouth.

SHORES

Spreading Piles. After a pond is dug, it is usually surrounded by spoil piles whose size and arrangement depend on the reach of the machine, the digging plan, the shape of the pond, and obstacles to digging or walking. These piles may be left to dry for ultimate sale or removal, but more often are knocked down and graded into banks and slopes. This can be done immediately, but if time is available and considerable yardage is involved, they may be left to dry and shrink. Granular soils and some fine-grained ones should become firm enough that the shovel can work without platforms, saving time and work. Peats and mucks are less likely to become firm, but lose substantially in bulk and weight.

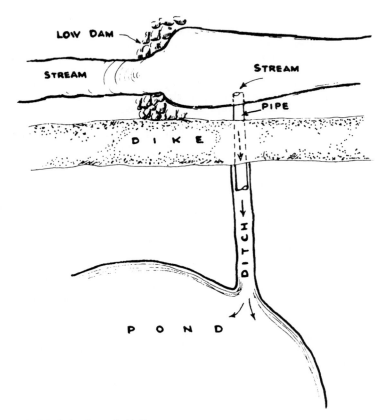

FIGURE 6.11   Controlled inflow.

The dragline or a backhoe are the preferred tools for spreading such piles. A bulldozer can be used if the piles are dry, and is frequently used for finishing after the dragline has knocked them down.

A dragline spreads piles by a combination of dragging down with lifting and swinging. First the machine approaches the pile closely and digs off the top. Each time the bucket is filled, it is pulled closer to the shovel than necessary, sliding several times its capacity ahead of it. It is then lifted, swung, and dumped in a low spot, and the process repeated until the dirt piles against the tracks.

The shovel is then backed a few feet, and the digging and dragging are continued, cutting to somewhere near final grade. The shovel continues to back and dig until the pile is exhausted, when it pulls down the lip in front of it, and walks up on the freshly graded area to work on the portion of the pile that was originally beyond its reach.

In Fig. 6.12 the pile is shown to be on the edge of the pond excavation. The dragline digs this shore to its final slope, widening the pond in the process. It is good procedure to cut banks back to a slope which will be stable under water, as it reduces the accumulation of soft mud at the edge of the bottom from parts of the bank sliding and falling in.

If the spoil is in windrows, the shovel may be walked parallel to the pile, digging and pulling it down until it starts to fall against the tracks, then moving on to wreck another section, continuing until the end is reached. It then comes back, parallel with the windrow but farther back, digging and dragging in the same manner. The ridge pulled against the tracks can be dug and spread behind the shovel. If the windrow is small, one trip may be sufficient.

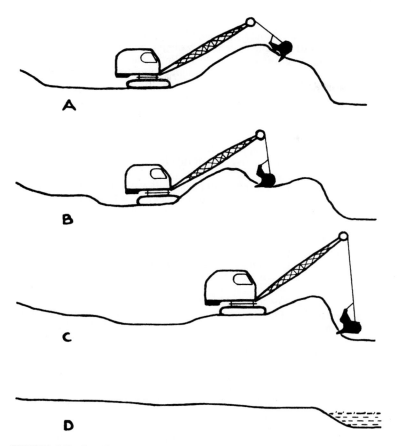

FIGURE 6.12   Leveling piles with dragline.

Grading.   It is sometimes possible to do the final grading while spreading the piles, but more often it is necessary to distribute them roughly to get an idea of the amount of material available, after which the area, or parts of it, is gone over for a light regrading. The finishing is done by raking and patting with the bucket.

There is probably no type of dragline work in which a skillful operator is as valuable as for spreading piles and grading banks. A long boom helps considerably.

Unless the operator is an expert, a better result will be obtained by finishing with a bulldozer or a grader, if the soil will support it. Also, it may be economical to make a quick rough grade with the dragline, and release it for other work while another machine finishes up.

When grading is complete, the area is usually disked and planted. Most peat soils need some time to drain, with the addition of lime, before they can support ordinary field vegetation. Testing outfits may be purchased at garden supply stores, and these may show the cause of trouble if things do not grow well.

Shore Drainage.   The grades of the above-water parts of pond banks are affected by disposal of material, the securing of proper drainage, the nature of the surrounding area, and the personal preference of the person in charge.

A difficult drainage and landscaping problem is presented by a swamp sloping gently up away from the center, changing gradually from wet swamp to dry meadow. Even when tapered to a thin edge, the fill is liable to create a wet spot where it meets the meadow, either because a dike is formed holding surface water on the lowest part of the meadow, or because the whole fill is relatively impervious to water, stopping underground seepage and causing it to overflow at the top edge of the fill.

If possible, the excavation should yield a sufficient volume of spoil to carry it far enough into the meadow to be well above pond level. When sufficiently dried, the lower part of the meadow and the upper part of the fill should be deeply plowed and disked, then blended together with a bulldozer or grader.

If wet spots still appear, they can usually be relieved by mole drainage starting at the pond shore. If this is ineffective, tile or rubble drains may be required.

Shore Erosion.   Freshly built banks will wash and gully badly in heavy rains unless protected. Drainage coming from undisturbed slopes across the fill is particularly destructive. This can often be diverted by shallow ditches made with a plow or by hand. These may be leveled after the banks are anchored by a firm sod.

Disking hay or straw into the surface of the ground increases its resistance to erosion and may supply ample grass and weed seeds. Unless applied with a nitrogen fertilizer, it may delay the growth of vegetation by temporarily absorbing this element from the soil. Such mixing in or scattering hay on the surface is helpful in holding soil that has been graded too late in the year for planting.

Beaches.   If a pond is to be used for swimming, a beach is very desirable. It may also be of use for wading, picnicking, and getting small boats in and out of the water.

A maximum exposure to sunlight is desirable for at least part of a beach. This is best obtained by locating it on the north or east bank, so that midday or afternoon sun, or both, comes over the water. Reflection intensifies its heat, and the slope of the beach is favorable to its reception. In most localities, more swimming is done in the afternoon than in the morning.

If the beach must be located so that its sunlight comes over the land, it may be necessary to cut a number of trees to obtain exposure. If the beach is large, enough trees should be spared to give shade over some part of it, or over a lawn area adjoining it. Sometimes a tall tree that is removed may be profitably replaced with one or more smaller ones to shade a smaller area.

If the pond is being dug, or can be emptied, the beach site can be graded. A gradual underwater slope is desirable for small children and nonswimming adults. Vigorous swimmers are likely to prefer a steep underwater slope, particularly if the water is usually cold. The dry section is usually gently sloped or flat.

A beach must be protected against runoff of water from surrounding land, as this will wash away the sand, spread dirt on it, or do both. A grass-covered ridge immediately behind the sanded area will serve to divert water, and may also function as a very welcome windbreak and heat conserver.

It is desirable for the beach subgrade to be a cut rather than a fill, and be of firm material. If this is the case, 3 inches of sand might suffice for a cover for swimming purposes, but not for building of sand castles. Six inches to a foot is a safer but more expensive depth.

If part of the pond bottom is sand or fine gravel, some of it may be pushed or carried to the proper location during the excavation work.

If the subgrade is soft mud, an attempt may be made to stabilize it with clean bank gravel, pea gravel, or fine crushed rock. A layer of lawn clippings or hay, placed immediately before the sand, may prevent mixing with the mud. This, of course, is not practical underwater.

An attempt should be made to extend the sand blanket to a depth of 4 or more feet below pond level, so that swimmers who are sensitive about walking on mud will be able to take off before they reach it.

Any clean sand that is suitable for concrete or plaster can be used for beach sand. Coarse grades are more attractive than fine, and light colors are better than dark. Where obtainable, white sand from ocean beaches or bars is most satisfactory, but it is apt to be much more expensive than pit or mason's grades. Sometimes the bulk of the beach is made with sand of a cheap quality, and the surface dressed up periodically with a better grade.

A beach will usually require an additional 2 or 3 inches of sand after the first year or two, and occasional freshening up with smaller quantities afterward.

Fire Control. A pond is a valuable asset for fire fighting. Country fire apparatus and many city units have suction pumps so that they can get their water supply from ponds as easily as from hydrants. Even a small pond provides enough water to supply hoses for a considerable time. Many fire crews carry enough hose to utilize water 0.5 mile or more from the blaze. An accessible pond may reduce fire insurance rates substantially.

Suction lines are short, so a rock fill or other firm surface should be provided to allow equipment to get close to the water in any weather. A deep hole should be dug near the shore. A wood or masonry wall to allow the suction hose to enter the water vertically instead of sloping down a bank adds to pumping efficiency. Any shallow bars separating the pumping hole from the bulk of the pond should be ditched.

Such a deep spot also serves well as a location for a diving board and an entrance ladder.

SPECIAL PROBLEMS

Clearing. Land clearing adds materially to the cost of reclaiming swamps, sometimes being more expensive than the earthmoving. Trees, and usually brush, should be removed or burned in advance of digging. Because of soft footing the cutting is usually done by hand, but tree trunks may be dragged out by tractors or winches.

Stumps. The stumps should be cut high in the area to be excavated, and very low where the spoil is to be piled. Height gives leverage which helps in digging them out, and affords a grip for chains for handling them, but makes disposal more difficult.

High stumps in the area to be filled will cause major difficulties during grading by hanging up machines, or tipping or breaking platforms, and by requiring excessive depth of fill.

When a large stump is dug, a slow process of cutting roots, overturning, and dragging out may be required. They are sometimes too heavy to pick up and must be dragged out of the way, cleaned and reduced by hand, or split by blasting.

Lighter stumps, which can be hoisted, usually cannot be held in a bucket and must be gripped with tongs, or other front end attachment, or chained.

Stumps should not be allowed to drag along the ground while being swung, as they will put a serious twisting strain on the boom.

Swamp stumps seldom have taproots, and the lateral roots are very close to the surface, so that they tend to come out as rather thin sheets. These can be most conveniently picked up by a pair of tongs inserted somewhere in the root mat, and chained to the bucket. If the ground is soft, the butt of the stump may be turned down, and driven into the mud in the fill area by patting the roots with the bucket. It is sometimes possible to entirely dispose of large numbers of stumps in this manner.

If this cannot be done, and the dragline or backhoe is working alone, it may be necessary to put the stumps in the spoil pile, where they cause trouble in rehandling. When the spoil is being spread, an effort should be made to rake out the stumps first and put them in the lowest spots, driving them in if possible in order to bury them. If they are too big to bury, they must be piled up to dry for eventual burning, or, more rarely, loaded in trucks and removed.

Digging in stumpy swamps is greatly simplified by the help of a crawler tractor, preferably equipped with a winch, which can stay on firm ground and pull the stumps away as the dragline or backhoe digs them out. These stumps may be winched or bulldozed into low spots for burial under the spoil; scattered around to dry before piling for burning; or piled immediately by passing the winch line through a pulley held high by a tripod, or a stout tree.

The high pulley arrangement decreases the power needed to drag the stumps, but unchaining them on the pile is a messy and somewhat dangerous job. The tripod or tree is usually destroyed when the stumps are burned.

Logging tongs are the preferred tool for gripping muddy stumps for winching. When a chain is used, notching the butt reduces the inclination to slide off.

Boulders. Large boulders also interfere with digging and are very likely to cut the drag cable if lodged in front of the tracks. It is sometimes possible to dig deep holes in which they can be buried, or to line them up along the edge of the pond where they should improve the appearance of the bank. However, they are more difficult to winch out than stumps, because of difficulty in getting a grip on them, and are an even greater nuisance in rehandling spoil. Often it is best to break them with dynamite, or air or hand tools, into pieces small enough to bury or mix with the spoil.

Hard Digging. A small dragline has great difficulty digging hard or rocky soil. It will do its best if the soil is not covered by water; if the bucket teeth are sharp; the boom held at a low angle; and the shovel footing kept as low as possible.

Occasionally a swamp floor may be of cemented gravel or decomposed rock which can be broken up with a tractor-drawn ripper. Such floors, and most hardpans, can be effectively dynamited if charges can be sunk deep enough in drilled or hand-dug holes. It is sometimes sufficient to blast a small area in which the dragline will be able to cut to depth, as it may be able to maintain this depth through the undisturbed material around it. If the whole bottom needs to be blasted, it will probably be cheaper to use other machines. Sometimes a single heavy blast will soften clay throughout the whole area.

Backhoe. A backhoe has quite effective penetration, but is hampered in pond digging by the inability to pile spoil at a distance. This limits the amount it can dig and exposes it to the danger of getting caught in slumping piles. A very good working team is a backhoe digging the hard soil, as in Fig. 6.13, and a dragline taking it away as fast as it is dumped. Best results are obtained if they work together, but because of the exact timing required to avoid accident, it is safer for the hoe to cut as much as it can pile and move on, with the dragline following at a discrete distance behind. At the end of the strip the hoe may turn and work back, building a new pile to be removed by the dragline.

Clamshell. A clamshell with a heavy bucket has good penetration but works quite slowly, and is at a disadvantage in sticky soils because of suction holding the bucket down. If used, it may do the digging and the casting back and spreading; or the rehandling of loosened material may be left to a dragline.

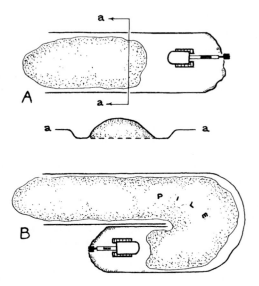

FIGURE 6.13 Loosening hard bottom with a backhoe or front shovel.

Front Shovel.   A front shovel can break up the bottom, provided that drainage is adequate and dependable. The machine can operate on pontoons and ramp itself down to the required depth, and dig as wide a cut as it can reach, or is required, dumping the spoil in a ridge behind it. It may come out of the pond site elsewhere, or turn and emerge near the entrance point, as in Fig. 6.13. Material broken up in this way can be easily dug by a dragline, but may be so soft as to be unsafe even on platforms, until it is well drained.

It rarely happens that a layer under a swamp is sufficiently firm to carry trucks, but in this case regular basement-digging techniques can be used.

Bulldozer digging in softer mud is done by the methods described below for pond cleaning.

Water Level.   The highest water level in a pond depends on the height of the overflow point, whether it is a streambed or an artificial spillway.

The best way to decide upon the new water level is to lay out a grid and take elevations throughout the area. The boundaries of a pond at any level can readily be sketched in, and the amount of cut required for desired depth, and the spoil to be disposed of in the dam and the banks can be roughly calculated.

A less laborious method which is usually satisfactory is to select some spot that would make a good shore and use a transit or hand level to find the corresponding shoreline at other points. This is done by reading a rod set on the selected point and moving the rod up and down any slope in question until the reading is the same, at which place it will be on the same level as the original point.

Readings can be taken on the dam site and on high, low, and normal points in the pond basis, and distances measured with a tape or by stadia.

DAMS

Digging may be done according to patterns outlined previously, with one or more cuts made across the bottom of the pond and piled for the dam. Or the digging may be done parallel with the dam and all the spoil used in its construction.

A dam should fulfill three requirements. It must be high enough in relation to the spillway such that water will never flow over unprotected parts; it should be stable enough not to break, slump, or move under any conditions; and it should not leak.

Usually earth dams for small ponds are given a freeboard, or height above water, of 2 feet. If the spillway is wide, wave action very weak, and the material thoroughly consolidated, 1 foot may be enough; but under reverse conditions, 3 or more feet may be required. For further protection, an earth dam should be covered with rocks, a strong sod, or bushes and trees.

No dam should be built to hold a depth of over 6 feet of water, or any considerable volume of water which would flood an inhabited area if released, without competent engineering advice. In many localities, plans must be filed and permits obtained before building any dam.

Earth Dams.   For stability, an earth dam should rest on a base of firm soil or rock without stratification, dipping away from the pond. It should be well bonded to its base by removing vegetation and plowing or ditching parallel to the axis of the dam.

The dam should be at least 6 feet thick at water level, and slopes should not exceed 1 on 2 on the downstream face, nor 1 on 3 upstream. If the top is to be used as a roadway, it should be at least 10 feet wide.

The soil used should be stable enough to hold itself up, to resist both the push and the softening effect of the water, and to carry any traffic or other loads on the top of the dam. It should also be fine-grained and compact enough to give maximum resistance to movement of water through it.

Stability in the presence of water is best obtained by the use of broken rock or clean gravel, but these materials allow easy passage of water. Clay, and soils rich in clay, is best for sealing off water, but may be inclined to slump and flow when saturated.

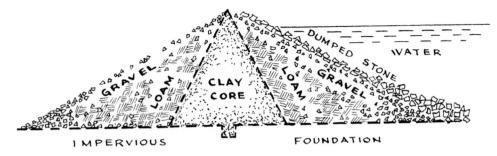

FIGURE 6.14 Hydraulic fill dam.

Large Earth Dams. In large dams, a clay core should be used to stop the water, with loam or gravel faces to support the clay, as in Fig. 6.14. In such dams, the type and amount of each material must be calculated. The dam may be built up in carefully compacted layers from material carried from pits in trucks or scrapers; or the fill may be mixed with water and carried to the dam in pipes laid along its sides. When carried mechanically, the clay core is usually built up a step or two ahead of the faces. The hydraulic method mixes all the soil types together, but as they come out of the pipe, the coarse material is dropped first, at the edge, and successively finer particles as the water flows inward. At the center a pond forms, and fine clay and silt particles are deposited to build the core.

Small Dams. These methods are not well adapted to small dams. The expense of setting up hydraulic equipment can be justified only by large-scale operations. Mechanical transportation and spreading are handicapped by lack of width. Even small trucks, scrapers, and dozers cannot work readily in strips less than 8 feet wide, or in comfort on less than 12 feet. The three sections would accordingly produce a width greatly in excess of that needed. A narrower dam could be built by the use of undersized equipment or hand labor.

A reasonably satisfactory dam may be made of mixed soil dug out of the pond or obtained nearby. This may be piled wet by a dragline or built up in compacted layers in the same manner as a road fill. Dusty soil should be dampened. If much sand or gravel is included, it should be mixed with fine-grained soil, or placed on the downstream face as much as possible.

If the dam is built of dry, uncompacted material, it should be allowed several months and some soaking rains to settle it before it is used to impound water.

If the soil is porous, the dam will leak unless sealed on the upstream side. It is not safe to wait for sediment in the pond to accomplish this, as leakage may liquefy the soil and cause the dam to fail.

The upstream face may be covered with a blanket of clay, heavy soil, or a bentonite mixture.

Bentonite. Bentonite is a volcanic clay which absorbs large quantities of water, changing to a jelly that effectively seals soil against water seepage. It is the expensive variety of clay mentioned in Chap. 3. It is used in many industrial processes and is available in bags in most cities. The pellet size is more desirable for pond work than the powder forms, which tend to float on the water surface for long periods and may be lost over the spillway.

A recommended practice is to mix one part of bentonite with four parts of sandy soil, or six or eight parts of fine-grained loam, and place a 4-inch layer of the mixture over the areas to be water-proofed. When more convenient, the pure material is spread over the ground and raked in. Satisfactory results are often obtained from more economical amounts, applied either in leaner mixtures or in thinner layers.

Either bentonite or the mixture can be shoveled into a pond over leaks, and allowed to settle into them. This is best done when there is no overflow.

Usable Materials.   Small stumps can be used in a dam if the fill is muddy, so that it will form a close bond and fill cavities. It is good practice to cut roots back close to the butt. Boulders may be used in either wet or dry fills if the soil is carefully puddled or tamped around them, and they are not close to each other.

If the fill is rich in organic matter, considerable shrinkage must be allowed for in both height and thickness. Even after years of use, the dam may shrink still further if the pond is dry for an extended period.

Cutoff Trench.   If the soil on the dam site is porous, a trench should be dug down to better material, approximately under the centerline of the dam. This should be filled with clay, well tamped or puddled.

If a deep layer of peat is found at the dam site, it would be best to find another place for the dam. If this is impractical, the peat may be dug or blasted out, or compacted by sand hole vertical drainage. If the budget does not include funds for any of this work, the dam may be built on the peat, and access for machinery provided so that it can be built up later if it sinks. If bulges appear in the peat above or below the dam, they should be left, as they serve to partly counterbalance the weight of the dam.

Settling and Cracking.   In a dam, troubles to be guarded against are settling, cracking, slumping, seepage, erosion, and damage by burrowing animals. Settling is prevented by building on a firm base, using fill low in organic matter, and tamping or rolling it in thin layers if built dry. Cracking may occur in a dam with a high clay content when the pond level is low, and may be avoided by mixing in sandy soils. Such cracking rarely causes dam failure.

Slumping.   Slumping may occur while building a dam with wet fill, and usually necessitates stopping work on the affected section until it has partially dried. Much more serious slumping may occur when water is impounded behind the dam before it is thoroughly consolidated. Wet fills that have not dried, or uncompacted dry fills which have not stood long enough to settle together, are apt to have this trouble. Seepage of pond water into the dam, softening it, and water pressure giving it a push all act together.

Water in a pond exerts pressure against its shores, which tends to balance inward pressure from the water they contain. If the pond is drained, removal of the water support may cause extensive slumping, which may be disastrous if it occurs in a dam.

A dam or causeway separating two ponds is particularly vulnerable if the lower pond is drained. For this reason, it is important to face dams with coarse, self-sustaining material that will resist slumping.

Seepage.   No earth dam is watertight, as there is a slow movement of water even through clay. Water working its way from the pond through the dam is usually called seepage only when it is sufficient in quantity to show on the downstream side, where it may make wet spots on the dam face or marshy patches below it. Aside from the loss of water, such seepage may damage the dam by liquefying it until it slumps; or by making channels of increasing size by washing out particles of earth. Once definite channels have been established, the volume of flow may enable it to tunnel and destroy the dam.

The seepage appearing below the dam may damage it by undermining, but more often merely produces soft wet areas that may detract seriously from the value of the pond area.

Seepage may be largely prevented by cleaning and scarifying the subgrade, careful construction of the dam, using sufficient impervious material, compacting it well, and allowing it to set before raising the water level. If suitable impervious, clay material is not available, a plastic membrane can be used to reduce seepage. A 20-mil-thick PVC membrane placed on the upstream side of the embankment and covered with a protective layer of granular fill will practically eliminate the seepage.

The surest and most expensive cure for seepage in an existing dam is trenching along it, with a backhoe shovel or clamshell, to solid foundation, and building or pouring a concrete core. This may be quite thin if of dense concrete treated with waterproofing on the pond side. Since a leaking dam is liable to have extensive soft spots in its interior, such a ditch may be dug safely only if the pond has been drained for several months, or very heavy bracing is used.

Driving a single line of sheet piling, or tongue-and-groove sheeting down the centerline of the dam, with grouting on the upstream side, is often effective.

The leaks may be stopped by laying a clay blanket on the pond side. The pond should be drained, if possible, to allow inspection. The leakage may be through the upstream side of the dam or in the pond bottom nearby. If the spots cannot be found, clay or fine-grained soil should be laid 6 inches to 2 feet thick on the whole face of the dam, and on the bottom, back about twice the height of the dam, and should be thoroughly tamped. The slope should be gentle, 1 on 4 to 1 on 6, and the clay should be covered with gravel or cobbles where subjected to action of waves. If the leaks are found, dig them out about 2 feet deep, then tamp in clay or heavy earth patches. Bentonite mixture may be used instead of clay.

Impervious patches should never be applied on the downstream side, where the water is leaking out, if the leaks are low on the face. The water will generally work into or around the patch, soften it, and force it out. If the patch holds, the water held in the dam may liquefy parts of it, causing slumping and possibly complete failure.

The best treatment for the downstream slope of a leaking dam is to face it with gravel, with an underdrain below the bottom of the dam opening into the outlet brook, as in Fig. 6.15. The first coating of gravel should be bank run to allow the passage of water, while holding back any soil particles carried with it. Over the bank gravel should be clean coarse gravel or crushed stone, to correct any tendency toward sliding when saturated. If the area is to be planted with grass, stone should be covered with straw, hay, or cut weeds before placing topsoil.

This gravel blanket does not reduce loss of water, but it does stop damage to the dam and eliminates surface wet spots.

Seepage at the foot of the dam may be kept underground by tile and gravel, or stone drains, of the same type used in draining farmland.

If the dam is of pervious material, the methods suggested later in this chapter for stopping seepage into porous soil may be of use.

Overtopping. If the water is allowed to flow over the top of an ordinary earth dam, it may cut a gully to the bottom of it, draining the pond, wrecking the dam, and perhaps causing flood damage below. Freshly built dams are much more subject to damage from overtopping than old established ones that have set and are covered with vegetation.

Overtopping is due to the dam's settling or slumping below a safe height, or an inadequate or too high spillway allowing the pond level to rise too much.

If a dam starts to slump, the water should be drained if possible and the dam allowed to dry, then the dam should be rebuilt with more or better material. If it is not possible to drain the pond, and pumping or siphoning is not practical, the dam should be reinforced by putting first gravel,

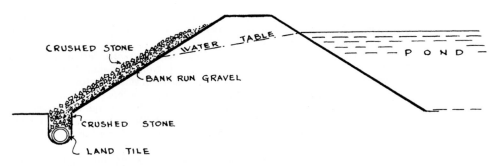

FIGURE 6.15 Seepage apron, small earth dam.

then a heavy fill of coarse rock on the downstream side. An attempt should be made to puddle or blanket the pond side, and the top should be filled to grade. If it settles badly without slumping, the top should be built up, preferably with compacted fill. Sandbags, if obtainable, make an excellent temporary stop.

Sometimes a dam can be saved by partly draining the pond through a trench dug in firm ground nearby. Undisturbed soils can often carry a heavy flow of clean water without severe gullying, particularly if reinforced with roots, boulders, or brush mats.

Repair.   When a gullied dam is fixed, the sides of the break should be smoothed and sloped sufficiently that the fill can be tamped against all parts of them, but it should not be cut into a straight ditch. The bottom should be dried up if possible. Fill should be dumped on the edge and pushed or shoveled down gradually, while workers at the bottom spread it in thin layers, tamping or tramping it thoroughly. If the break is large enough to allow machinery to work in it, it can do most of the spreading and compacting, but the bond with the walls must be done by hand. Dusting bentonite against the sides while filling should prevent seepage along them.

If it is not practical to dry up the bottom, fill should be dumped and kneaded until the water is absorbed into a stiff mud on which a layered fill may be built.

Burrowing Animals.   Earth dams may be damaged by animals burrowing part or all the way through them. Muskrats make holes which run underwater to well under the bank, where they rise above the water. Such tunnels will cause leaks only when they give water access to some line of weakness that did not go through to the pond, or which had been silted shut. Muskrat damage can be largely avoided by using a low dam not containing enough dry ground for home building, or a wide one without porous veins.

Crayfish will at times dig burrows all the way through a dam, creating a water channel large enough to enlarge by erosion, unless a fortunate cave-in should block it. This damage is most apt to occur in soft peat soil, and it may sometimes be cured by injections of cement grout.

Burrowing animals may be discouraged by including $\frac{1}{4}$-inch mesh wire in the underwater part of the upstream slope. This affords fairly good protection for a number of years. It is usually laid on the dam, and 6 inches to 1 foot of fill is spread on it.

Masonry Dams.   Masonry dams may be used instead of earth fills. They are most suited to comparatively narrow sites with firm bedrock near the surface of bottom and sides. Reinforced concrete is the strongest construction, but fieldstone masonry is more attractive and may be less expensive in inaccessible spots.

Earth and decayed rock should be cleaned off the dam site, and the bedrock shaped or gouged in such a way that the dam will not be able to slide on it in any direction. Holes 2 or more feet in depth should be drilled in the rock, and reinforcing steel grouted into them so that it will project into the concrete or other masonry.

If the dam is to be more than a few feet high, it is advisable to have an engineer or a geologist check the ground, as fractured rock can make a leaky and unstable foundation.

The dam should have a bottom thickness of at least 2 to 3 feet for every 3 feet of height.

Masonry Cores.   A masonry core dam consists of a thinner wall, preferably reinforced concrete, with earth piled on both sides. The masonry does not extend much above the waterline, and is ordinarily buried under earth. The core seals off seepage, and the sides support and protect it. It must resist the difference in pressure between the wet and dry earth on its two sides. Thickness is about one-fourth of height.

The core should be founded on a firm, impermeable material, preferably rock. The original surface is ditched for footings. The sides are carried into the banks until they meet rock, or until they are far enough from the water to make seepage unlikely. Rock should be roughened to hold the masonry against shifting.

The core is built and allowed to cure before placing the earth fill. The upstream face should be painted with waterproofing. If its ends are not keyed into rock, they should be fitted with vertical metal baffles sealed to the concrete, and the fill near the baffles should be mixed with bentonite. Failure to take these precautions may lead to serious leakage around the core.

Fills should be placed on both sides of the wall at the same time to avoid unbalanced pressure. If the dam is high, the fill should be carefully compacted. If it is low, this is not necessary unless final grading is to be done immediately.

The masonry core dam is the safest and most satisfactory construction for ponds, but is too expensive for casual use.

Removable Wood Dams. If a small pond is built on a small but fast-flowing stream subject to flood, there is the danger that not only earth or weak masonry dams will be washed out, but that, if the dam holds, the pond may fill completely with mud and debris in a single season, because the slowing and widening of the stream cause it to drop a part of its burden.

A removable wood dam may be used to advantage under such conditions. If the stream is narrow, 10 feet or less, a heavy, well-founded masonry wall is put on each bank, having slots to receive 2- to 3-inch plank, as in Fig. 6.16. A masonry sill, similarly slotted, connects the piers on the stream bottom. Planks cut to the correct size and length are slid down the pier slots, resting on the sill and on each other, until the desired height is achieved. This structure will leak but will impede a brisk stream enough that part of it will flow over the top board, and the desired water height may be maintained. If the stream shrinks, the leaks may be reduced by jamming a tarpaulin in the sill slot, upstream, and pulling it over the dam face and top, and tying weights on the downstream side. Or tongue-and-groove planks may be used to cut leakage, with the top plank fastened down and all the joints packed.

When a flood is expected, or pond use stopped for the season, the planks may be taken out and stored, allowing the stream a clear passage.

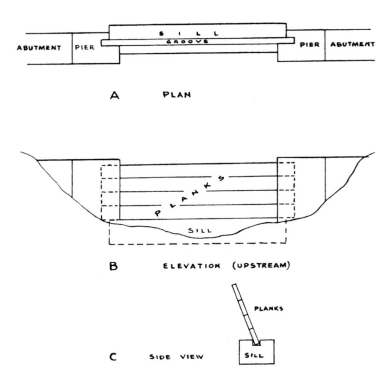

FIGURE 6.16 Removable wood dam.

DRAINS

Gate Valve.   When possible, means should be provided to drain a pond for repair, cleaning, and other purposes. The best, but most expensive, means is to place a metal or concrete pipe under the dam, connected with a gate valve, which may be located in the dam or at either end. Figure 6.17 shows an installation in which the valve is in the downstream face below frost line. To prevent burial and clogging, a vertical 8-inch pipe placed over the valve wheel extends to the surface, where it is plugged or covered. The valve is opened or closed by removing this cover and turning the valve wheel by means of a jaw on the bottom of a rod which can be turned from the top.

If the cover should be left off, and the vertical pipe filled with dirt and trash, it may be jetted out by the use of an engine-driven water pump, delivering water at pressure through a small pipe which is pushed down inside the casing, where it can break up and wash out the debris.

Elbow Drains.   A much less expensive installation, which can be used in climates where freezing is not expected, is shown in Fig. 6.18(A). A metal drainpipe under the dam is fitted with an elbow on the downstream end into which a vertical pipe is threaded. Space is provided so that this pipe can be turned into a horizontal position.

If the open end of the pipe is higher than the water in the pond, no water will move through it. If it is lower, the water will flow through it until the pond level is lowered to the same elevation. The pond level can therefore be adjusted to any height desired by turning the pipe up or down.

In cold climates, the exposed pipe would be subject to breakage because of water freezing in it. This is not likely to occur if the movable pipe is placed in the pond as in (*B*), because of less severe freezing and inward pressure of pond ice. However, the water makes the pipe difficult to

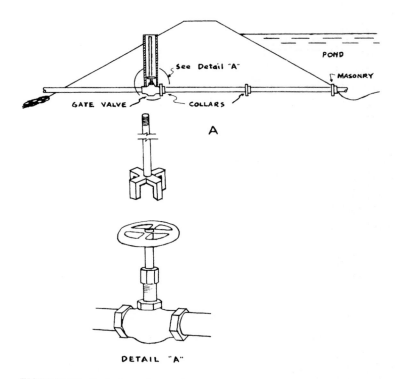

FIGURE 6.17   Drainpipe and gate valve.

get at, so that it must usually be moved by a line stretched to shore, as in (*C*). This will not pull it into a horizontal position and may have difficulty raising it from down position also.

A more satisfactory arrangement for underwater use is shown in Fig. 6.19. The drainpipe is extended by means of a tee, a close nipple, another tee, a short pipe, and a cap. The tees are fitted with pipes long enough to reach the surface of the water. These are set at an angle of about 45° from each other, and the tees welded together. One of the pipes is capped and a ring fastened to it.

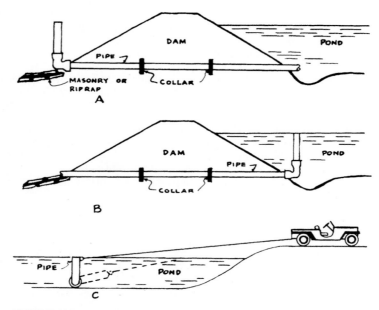

FIGURE 6.18   Elbow drain.

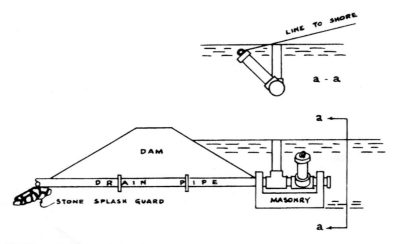

FIGURE 6.19   Elbow drain with pull arm.

This apparatus rests on a small block of concrete, which is cast around the edge of the drainpipe and around tar paper wrapped around the end pipe, but is enough below the tees so that they can turn.

Control is by a rope or cable stretched from the ring, past the vertical drainpipe to the shore. A pull on this line, by hand or machine, should raise the ring pipe and turn the drainpipe down. The drainpipe can be raised by pulling the line from the opposite bank.

With some risk of twisting the end off instead of turning it, the masonry block may be omitted and the outer tee replaced by a street ell welded to the inner tee.

The threads should be treated with waterproof grease or plumber's dope, and wrought iron fittings should be used if possible.

Metal pipe is expensive in large sizes, and 6-inch is about the minimum for a pond drain, except for use in dry seasons only. Considerable expense may be saved by using concrete or tile pipe under the dam, connecting it near the end with metal pipe to the valve or other drain arrangement.

Vertical Tile.   An overflow or trickle drain can also be made entirely with tile. A pipe is laid under the dam, ending on the upstream side in a concrete junction box, as in Fig. 6.20. From this a tile pipe with joints sealed with soft mastic rises to the surface. One of the pipes may have to be clipped short to obtain the proper height. The pond height is limited by overflow into the pipe.

The pond is drained by pulling the top pipe out of its joint and removing the next section when the water has gone down sufficiently, repeating the process until the bottom is reached. Sometimes the whole pipe will pull out of the box and drain the pond all at once. At other times, a pipe may refuse to move and may have to be broken with a hammer or crowbar.

The overflow type of pond drain serves to some extent as a spillway, but a regular or emergency spillway also should be provided for flood conditions, and because of the possibility of the pipe's becoming clogged.

Pipes reaching the surface of the water can be protected against external ice pressure by tying several sticks or boards to the outside.

Drainpipes are a source of weakness to dams and must be carefully installed. It is best to place them before the dam is built, as this eliminates the difficulty of making a proper bond between fill and the wall of a ditch. Pipe joints should be watertight.

One or two collars of metal or masonry should be built out from the pipe, as indicated in the illustrations, and sealed to it by cement or welding. These will discourage seepage from following the outside of the pipe and cutting a channel along it. Clay, or soil mixed with bentonite, should be tamped or puddled around the pipe and the collars.

The first layer of fill should be spread rather evenly along the masonry pipe, as a full load in one spot might push it down enough to open the joints. If the ditch bottom is not firm, the pipe may be set on a reinforced concrete slab the width of the pipe and up to 6 inches thick.

A wood box, 3 feet square or larger, may be built of rot-resistant wood around the upper end of a drainpipe and topped with $\frac{1}{4}$- or $\frac{1}{2}$-inch mesh screen, to keep fish in while lowering the pond level.

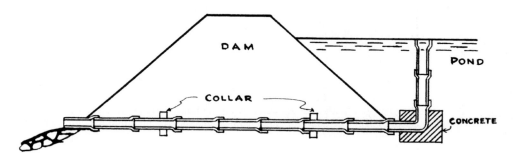

FIGURE 6.20   Spillway drainpipe.

SPILLWAYS

Construction.   Ponds which are made by excavation only, and do not raise the original water level, usually overflow through stabilized streams or channels that do not require any artificial protection against erosion. If an earth or masonry core dam is used, however, an artificial overflow channel, called a spillway, must be prepared.

A spillway may have a surface of any material that will resist the destructive action of the water which might flow across it. A steady flow calls for a structure, usually of stone or concrete, but occasionally wood, metal, or asphalt. A spillway that carries water rarely, as one which is intended to care for floods in excess of the capacity of a masonry or pipe spillway, or to provide for occasional overflow of a normally static pond, may be planted with grass or other well-rooted vegetation.

Spillway size may be calculated on the basis of the area drained, type of land and vegetation, and rainfall records in the same manner as culverts. However, a greater margin for safety should be allowed.

It is good practice to keep the spillway and dam separate if possible, as each is a source of weakness to the other. A recently constructed dam ordinarily lacks the stability necessary to support heavy masonry, and it is difficult to get a leakproof bond between dirt and stone. Any leaking through or around a spillway will be much more destructive to an earth dam than to a long established subgrade. On the other hand, practical and aesthetic considerations frequently require placement of the spillway in the dam.

If the dam including a spillway has a masonry core, the two structures can be combined. However, the core must be widened or buttressed, or the spillway provided with additional foundations as firm as the core wall. If the spillway is supported by a thin core wall and dirt fill, and the fill settles, the spillway will be left supported only at the core, and may break, or may twist and break the core. A preferred method is to extend the core footings far enough to carry piers to support the spillway.

If the overflow is to be carried around the dam, standard practice for masonry structures may be followed. Two more or less parallel walls carrying the water race, which may be a curve or a series of steps, is a standard type of construction. The structure is strongest if built of reinforced concrete well tied together, but stone and mortar make a more attractive appearance. The fill under the water race should be clean sand or gravel with good bottom drainage if the ground freezes in winter.

Settlement.   If the spillway is to be part of a newly made dam, it may be based on the fill material, or may have footings in the native soil underneath the dam. In the first case, any settlement is liable to tilt or break the spillway and to settle away from it, leaving channels for leakage. In the second case, the masonry will stand firm while the dam settles under it and away from it. If the structure includes a core wall long enough to tie into the earth on each side, such settlement may not be serious.

Grouting.   Leakage under a masonry spillway surface, resulting from dirt settling away from it, may be stopped by drilling holes in the masonry and pouring or pumping a cement and water grout into them.

A grout injector may be an air pressure tank or a pump. The tank is provided with an agitator to prevent separation. It is partly filled with grout and tightly closed. Compressed air is piped into the top of the tank, forcing the grout out through a pipe or hose in the bottom. The tank is opened and a fresh batch of grout poured in as often as necessary. Air should not be allowed to enter the outlet hose.

Special pumps may be purchased, or a fluid grease dispenser or a tractor grease gun used. Pumping can be continuous, with extra grout added as necessary.

The holes are drilled or punched to a depth where the leaks are suspected. The grouting tube may be fitted with a rubber collar to fit the holes and held in place by hand, if low pressures are used. For high pressures, a threaded iron pipe is cemented into the hole some days before and the grout pipe coupled to it.

The grout forced underground may penetrate and seal the leaks, may be washed away by water, or may escape to the surface of the ground. If possible, the pond level should be lowered to stop the water flow during grouting. The whole area should be watched for the appearance of grout, particularly at the leakage points.

A very thin grout made with 45 gallons of water to a sack of cement is good for sealing fine porous soil, but will escape readily through small channels. The thickest grout used, $4\frac{1}{2}$ gallons of water to a sack of cement, will escape only through large openings, but does not seal fine passages effectively. Sand mixtures are not recommended for amateur use because of the tendency to separate, but sawdust or fine shavings may be mixed with grout used from pressure containers if the grout is otherwise washed out by water which cannot be stopped.

If grout is applied at a pressure of more than a few pounds, care should be taken that it does not lift or break the spillway, or even split bedrock beneath. A tractor grease gun can develop pressure of thousands of pounds per square inch, and will break up strong masonry with little effort.

All grouting equipment should be thoroughly cleaned immediately after finishing the job, or for any shutdowns of more than a few minutes.

Detailed information on the use of grout for stopping leakage and for other purposes may be obtained by writing to the Portland Cement Association, 5420 Old Orchard Road, Skokie, Illinois 60077.

Wood Spillway.   Trouble from settling under a spillway may be avoided by putting in a temporary structure upon completion of the pond, and removing or destroying it after complete settlement, then building the permanent spillway. Tongue-and-groove plank made into a box is a satisfactory construction. The dam surface on which the wood rests should be coated with bentonite, clay, or other fine-grained soil, and puddled until semifluid. The spillway should be stirred around or vibrated when set, and mud packed in along the sides.

A wood spillway may give satisfactory service for a great many years under favorable conditions.

Horizontal Pipe.   Concrete, tile, or corrugated steel pipe of large size may be used, either, as described, under drains, or laid horizontally through the dam at water level, with the same precautions against seepage.

WATER SUPPLY AND LOSSES

Water Supply.   The ability of a pond to remain nearly full of water through a dry season is to a large extent the measure of its usefulness, except in semiarid sections where it is considered a success if it retains any water at all.

A pond level is kept up by water entering it through rainfall, surface wash, springs and seepage, and streams. It is lowered by evaporation, outflow, leaks, and seepage through sides and floor.

Once a pond is built, little can be done to add water to it except by pumping water from a well, by windmills or engine-driven pumps, or more rarely, diverting water into it. It is therefore important to locate and build it in such a manner as to take full advantage of sources of water.

Ponds dug in swamps may depend primarily on the water table existing before work is started. If possible, fluctuations of this should be watched for a year or two.

A dug pond may cut into active springs or extensive seepage areas which had previously been draining below the site, so that the pond may keep a higher level than the groundwater did. On the other hand, the swamp water might overlie a layer of clay or hardpan, which, when cut, would allow all the water to drain down into unsaturated porous soil, in which case it might be difficult to keep water in the pond.

The best way to estimate the water supply is to measure the drainage area. Figure 6.21 indicates approximate requirements throughout the country.

Seepage into Porous Soil.   Outgoing seepage can be greatly reduced and sometimes stopped altogether by keeping mud in suspension in the pond water for some time. The water in seeping out of the pond takes the suspended particles with it and lodges them in the fine passages through which it travels, thus clogging them up. This process operating naturally over a period of years makes possible the existence of rain-fed ponds and swamps on sand dunes and gravel banks, high above the water table.

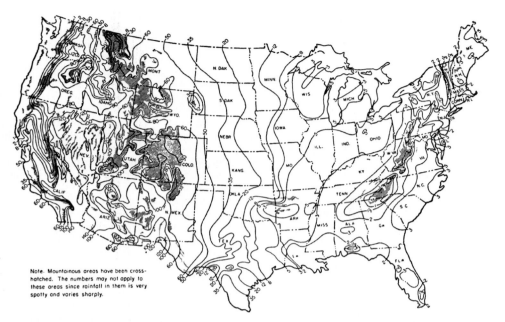

Note. Mountainous areas have been cross-
hatched. The numbers may not apply to
these areas since rainfall in them is very
spotty and varies sharply.

FIGURE 6.21   Drainage area map. A general guide for use in estimating the approximate size of drainage area required for a desired storage capacity in either excavated or impounding reservoirs. The numbers on the chart show the number of acres of drainage area required for 1 acre-foot of water impounded. (*Courtesy U. S. Department of Agriculture.*)

Digging in a pond will keep it muddy, as will driving livestock around in it several times a day. Fine-grained silt, powdered clay, or pellet bentonite may be scattered on the water with hand shovels, preferably when there is no overflow.

If the water is leaking through channels too large to be plugged by sediment, a layer of clay or a soil-bentonite mixture several inches in thickness should be spread over any outcrops of porous veins. If this fails to hold, the pond should be pumped dry and any leakage holes appearing in the clay should be dug out and filled with the blanketing material to a depth of a foot or more.

If the porous vein is comparatively thin and close to the surface, it may be sealed by injections of cement grout in the same manner suggested for spillways.

If leakage is along sod or brush which was not removed before placing fill for the dam, it may be stopped by chopping and mixing. A tamper such as is used in breaking street pavements can drive a narrow tool several feet underground, and repeated blows struck close together will mix the vegetation into the dirt so thoroughly that it will no longer provide water channels. Additional fill can be added to the surface if necessary, and a wide face tamping tool used for compaction.

Seepage along the old ground surface may cease when the vegetation rots, but this cure cannot be depended upon.

Seepage cannot be stopped entirely but will fall to a very small amount in a well-sealed pond, particularly if the water level is not high enough to create a strong pressure toward a nearby low spot.

Movement of groundwater is often nearly horizontal, so that much of the loss from a pond is through the banks rather than the bottom. This is one factor in the excessive shrinking of some small ponds during dry spells.

Evaporation.   Evaporation acts constantly to remove surface water. It varies with heat, humidity, and exposure to sunlight and wind, and may lower the level of a stagnant pond from 5 to 15 feet during a summer. This loss is most pronounced in desert regions.

The rate of evaporation is higher on small ponds than on large, and on shallow ponds as compared with deep ones. A number of factors are involved: The banks heat more readily than the water surface; capillary attraction draws water several feet up on the banks, thus increasing the surface exposed to evaporation; and a large body of water warms more slowly than a small one.

This loss from the water surface may be reduced by shading it with trees, but it is a question whether the trees use as much water as they save. If they are set well back from the edge, they may find a large part of their water supply elsewhere.

Dry Land Ponds.   Losses of water through seepage and evaporation assume their greatest importance in ponds designed to fill with surface runoff in the winter or spring and to hold this water through a dry summer, even though the water table drops many feet below their bottoms.

Such a pond should be so located that the drainage from a large area will flow into it; not only so that it will fill even in years of subnormal rainfall, but also so that it will get the fullest advantage from any freak rains that might fall in the summer. But it should not be placed in the channel of a stream having enough force to fill the pond with sediment during flood time, or to require an unreasonably expensive spillway.

Such a pond may generally be dug in the dry season without any interference from groundwater. Any dry land excavator or set of excavators, down to small tractor-drawn scrapers and scoops may be used. Techniques are similar to those used in borrow pits and basements, except that banks must be sloped, not more steeply than 1 on 1, and it is usual to place a large part of the spoil so as to build up a dam.

For detailed discussion of the locating and building of such dry land ponds, the reader is referred to Agriculture Handbook AH 387, entitled "Ponds for Water Supply and Recreation," issued by the U.S. Department of Agriculture. This can be obtained from the Superintendent of Documents, Washington, D.C. 20402.

POND MAINTENANCE

Silting.   Silting is a problem common to most ponds and reservoirs. Lakes of all sizes are short-lived geologically, because incoming water deposits sediments that fill them, and water flowing out tends to deepen its channel.

The amount of silting will depend largely on the local conditions. Steep slopes, cultivated or bare land, and fast stream flow bring heavy loads of sediment into ponds and cause them to fill rapidly.

Wastage of soil from farmland can be greatly reduced by contour plowing, terracing, and planting steep slopes to permanent grass or trees, with beneficial results to the land, the stream, and the ponds.

If it is not possible to alter watershed conditions, silt traps may be constructed. See Fig. 6.22. These may consist of small ponds built above the main one, or a very deep hole on the upstream end of the pond. Such traps should be so located that a dragline shovel and trucks can reach them for periodic cleaning.

Mud deposits found in ponds and lakes are made up of soil brought in by water or slumping from the banks; dust, leaves, pollen, and other debris falling from the air; and remains of plants and animals living in the pond. A combination of these sources usually produces a soft black mud which dirties and shallows the water. Near inlets and steep banks it may be chiefly silt or sand, and away from shores it is largely organic.

Removal.   A hydraulic dredge removes such a deposit without draining the pond, but its use is often not practical. There must be enough work to justify transporting and launching it, enough water inflow or return flow for its needs, and adequate disposal areas.

Removal by machinery usually requires draining or pumping out of the water to avoid distributing disturbed mud throughout the pond.

After draining, the mud deposit will often be found to be so soft that it will not support machinery safely even on platforms. Given time, it will drain and compact so as to be fairly firm, in which condition

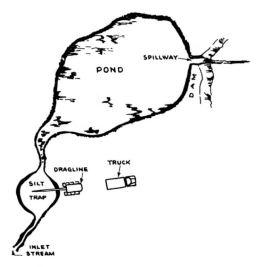

FIGURE 6.22   Silt trap.

it will not only support platforms but will stay in a dragline bucket. This hardening process, which may reduce its bulk as much as 80 percent, can be greatly accelerated by hand ditching into the subsoil for more thorough drainage. The ditching, however, is a sloppy job, and will be very discouraging at first because of mud flowing or slumping into the ditch and blocking it. The first digging should be very shallow and can be gradually deepened as the banks drain.

If the pond is narrow and accessible enough that all parts can be reached by a dragline on the banks, or if the mud overlies firm material that will support a dozer which can push the mud to a dragline, it may not be absolutely necessary to let the mud dry. If it is too thin to be picked up in the bucket, digging the ground under it several times may suffice to get enough of it out. In any event, such undercutting will eventually lower the mud so much that it will no longer be a nuisance.

It is seldom practical to just skim even dry mud off the old bottom. At least several inches of native soil are ordinarily dug with it, and this opportunity is often taken to deepen the pond substantially. In some cleaning methods, it is necessary to take enough subsoil to build firm piles.

The cleaning process differs from the original digging in the usually shallower cuts, the peculiar nature of the mud, the fact that trees and landscaping on the banks often must not be disturbed, and the undesirability of reducing the pond area by piling spoil inside it.

Bottom mud is generally useless for agriculture when freshly dug, but makes excellent topsoil after curing in piles for a year or two. Mixing with sandy subsoil speeds curing and improves its quality. It is often necessary to add lime to correct acidity.

Dragline.   If a dragline can do the necessary cleaning from the banks, the problems are chiefly avoiding or cutting trees, and providing either places to pile the spoil or means of access for trucks to haul it away.

If the width is too great for the boom length, an unassisted dragline must work from the pond bottom, usually on platforms. From there it may pile spoil on the banks to be leveled off later; against the banks, to make a new shore for a smaller pond; load it in trucks on the bank, or build one or more windrows in the pond to be trucked out later.

Trucking windrows must contain enough inorganic soil that they will become firm as they dry, and must be high enough that capillary water will not keep them soft. This height will vary from about 3 feet for a sandy mixture to 7 to 10 for silt or clay. Lower piles, or any piles containing a lot of humus, may require a surfacing of better soil or gravel before they will support trucks. The height of the roadway will be substantially lower than that of the top of the original windrow.

The dragline may roughly level the piles as it builds them, or this work may be left to a bulldozer. It may be advisable to use the lightest dozer that can do the work, as unexpected soft spots may be found, due either to slower drying of sections of fine-grained soil or excessive amounts of humus in spots.

Since cuts are usually shallow near shore, and trees may interfere with maneuverability, it may not be practical to build the piles large enough to make a good land connection, in which case extra fill might be trucked in to bridge the gap.

When the windrows have dried and have been leveled off for a roadway, the dragline or hydraulic backhoe can walk out to the end of one, possibly with the precaution of using platforms or poles, and dig it back from the end, loading trucks backed to it from the shore. It may just dig the piled material, or go down into the pond, either to deepen it or to obtain fill. Sections of roadway may be left to form islands. (See Fig. 6.23.)

Use of this method involves deepening the pond 6 inches to 2 feet or more. The double handling, the trucking, and the volume of material to be removed may make it prohibitively expensive for large areas, although it results in a pond which is better than new.

FIGURE 6.23 Trucking out piled mud.

Dozer.   If the bottom is firm enough to support a dozer, and if the mud is thick enough that a good load will stay in front of the blade, a dozer may provide the fastest and cheapest cleaning job.

The mud can be skimmed off gravel subsoils with little mixing. On softer footings, or wet soils which churn to mud readily, several inches may have to be taken with the mud. In any case, the digging down need not be as deep as with dragline work.

Disposal of the spoil may be a critical problem. It is liable to be too sloppy to pile up high enough for a bank, and to contain too much organic matter to make a satisfactory shore.

It can often be pushed out. The average pond edge is too steep for a dozer to climb with a load, so a ramp or ramps must be cut in it (Fig. 6.24). If the shore is a dam, with low ground beyond, very liquidy muds can be trapped in the ramp entrance and pushed through. Because of light friction, a dozer may push five to ten times its normal yardage on each trip through the slot, but a part of the volume will be water.

The ramp is apt to soften and break down, particularly at the bottom. Also, disposal areas at its head may fill up rapidly. For this reason, a number of ramps are liable to be required, and back-filling these later may be a major project.

The front-end loader, with grousers bolted on every fourth or fifth shoe on each track, is the preferred tool for this work. The widely spaced cleats do not clog with mud. In cutting through heavy deposits, a front-end loader is more adept at side casting than a standard dozer, and can backdrag material out of bad spots. When this is done by filling the bucket, it may be necessary to float it while backing to better ground, as lifting it tends to make the front of the machine sink in. This machine can often unstick itself by using the bucket dump as a pushing or pulling device.

The second choice is a wide-gauge bulldozer. It usually has wide shoes that reduce its tendency to sink, and the width gives extra leverage for turning with loads on slippery footing. If grousers are worn down, a few of them can be built up to provide nonclogging traction.

When the bottom is reliably hard, a large dozer may be used. It is desirable because of greater production in both the volume of mud moved and area left cleaned by a single pass. It can also back into a deeper layer of soft mud without getting hung up than smaller machines with less clearance.

On soft or doubtful bottoms, lighter machines are much less apt to get stuck and are easier to rescue if they do.

Saturated clay, silt, or very fine sand may look and act firm when work starts, but soften under the weight and vibration of machinery. This change will be caused much more quickly by heavy than by light units. However, such soil will often continue to give adequate support to a dozer as long as it keeps moving, even after becoming too soft for comfortable walking. No machinery should be left standing for any length of time, particularly if unattended.

FIGURE 6.24   Pushing mud up slot ramp.

When a dozer is used for swamp digging, means should be provided for prompt rescue in case the bottom proves too soft, or careless operation gets it stuck. If the dozer does not have a winch, a hand or machine winch, or equipment capable of exerting a heavy pull, should be on the bank with sufficient cable or chain to reach any part of the area. Cut green saplings and hand shovels should also be available.

Drain holes in the flywheel and steering clutch housings should be plugged to prevent the entrance of water and mud. Plugs should be taken out periodically to drain any oil that might leak into them.

Fully sealed rollers which are greased on a twice-a-year schedule require no special attention. Other types may require greasing every 2 to 4 hours to prevent mud from working past the seals. Sand in the mud may make it very abrasive so that track wear may be several times as rapid as normal.

Under average conditions, dozer work in a pond bottom offers considerable danger of getting bogged down, and conditions are often found to be so sloppy that little effective work is done. But when it works, it's fine.

Ramps.   The ramp should be at the easiest possible gradient to facilitate pushing large loads and to minimize churning under the tracks.

When the ramp is roughed out, the dozer is backed into the pond mud until a good load is ahead of the blade or bucket. This is then pushed through the ramp to the disposal point, or parked in the ramp to be moved along with an additional load or loads.

The floor of the ramp will usually soften from absorbing water out of the mud being moved over it, and it will be worn down continuously by the push of the tracks and cuts by the blade. These effects are liable to be most severe in the pond at the foot where the dozer turns upward for its climb. A deep hole may be gouged here which will usually fill with a very thin mud. This ordinarily does not bother the machine any more than the same quantity of water, but will eventually reach the fan or other nonsubmersible parts, and the ramp will have to be abandoned or its foot relocated.

Such a hole may be convenient in freezing weather as the tractor may be placed in it overnight, so that the tracks will be underwater and the mud on them will not freeze. This will save a long and messy job of putting it up on blocks and of cleaning and hosing it at the end of the day. It is of course not practical unless the bottom is entirely safe.

The cleared space may be widened by other cuts fanning out from the ramp. This uniform expansion of area is not particularly efficient from the pushing standpoint, as mud tends to spread on each push over ground cleaned by previous passes, and both mud and subsoil become increasingly sloppy from reworking. However, it keeps the dozer close to dry land while bottom conditions are observed. This pattern is shown in Fig. 6.25.

Dozer and Dragline or Backhoe.   Ramp difficulties, or lack of nearby disposal areas, may make dozer cleaning impractical even when the bottom conditions are favorable. In such cases, the dozer may push the mud so that it can be reached by a dragline or backhoe standing on the shore or the dry pond bottom, which can pile it on the bank or load it into trucks, as in Fig. 6.26.

This method can be rather widely applied and is usually more economical than doing the whole job with the dragline.

Pump.   Cleaning by machinery usually mixes some of the mud with so much water that it becomes too thin to be picked up or pushed, but can often be pumped. A diaphragm pump will handle heavier mud than a centrifugal, but the volume moved is much smaller.

If a water source is nearby, clean water can be pumped into a hose line and the mud stirred up, thinned, and driven to the mud pump or gravity outlet by a stream directed from a nozzle. If patience and workforce are sufficient, whole ponds can be cleaned in this manner.

After removal from the pond, the mud may be allowed to flow away from the work area, or to accumulate in natural depressions; it may be held in a settling basin from which it can be dug after it has dried; or it may be placed directly in tight-bodied trucks. The very thin muds which pump most easily are usually the hardest to dispose of. The contractor is liable for mud damage downstream or on adjoining property.

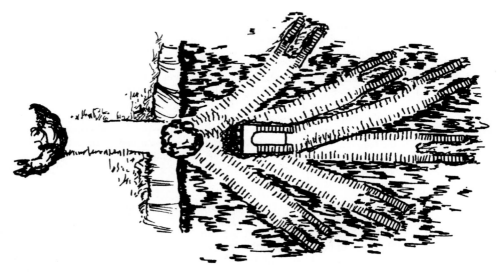

FIGURE 6.25   Gathering mud near ramp.

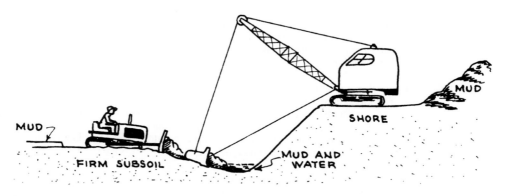

FIGURE 6.26   Dozer-dragline team.

Water Plants.   Vegetation in ponds is of many kinds, few of which are desirable.

The blue-green filamentous (stringy) algae, which grow in surface masses that may have considerable depth, are often a major nuisance. They may be dense enough to make swimming and even boating impossible. These algae grow vigorously (bloom) in the spring, and at irregular intervals throughout the warm months.

In large lakes, these plants can flourish only if the water is overenriched by drainage carrying sewage, detergent, fertilizer, or other nutrients. In small and shallow ponds, however, they often grow very well without such assistance.

Fortunately, such algae are very susceptible to poisoning by tiny amounts of copper sulfate, and die within a few days of contact with it. This chemical may be obtained in the form of coarse blue crystals from large hardware stores, or from dealers in commercial or agricultural chemicals.

The easiest way to apply it is to put in a burlap or other loose-weave bag, then tow it behind a boat, or pull it by hand while wading or swimming.

Two to three pounds to 1 million gallons of water, applied two or three times a year, should keep a pond clean. If it is allowed to become heavily overgrown, applications of two to three times that amount, at 2-week intervals, may be required.

Pond area may be roughly calculated from dimensions found by pacing, or by stadia as in Fig. 2.18. Soundings will indicate the depth. Area in square feet times average depth will give cubic feet of water. One cubic foot equals $7\frac{1}{2}$ gallons.

Copper sulfate cuts down the food supply of fish by reducing the vegetable food that is the basis of all animal life in the pond. Usually a balance can be preserved in which the plants are reduced enough to be unobjectionable but sufficient food is left for a large number of fish.

In the proportions recommended, or even in much heavier doses, this chemical is harmless to swimmers and to fish. However, the abrupt killing of very heavy plant growth may suffocate fish by absorbing the water's oxygen into decaying material.

This danger can be avoided by using light, repeated doses, or by treating only part of the pond at a time.

There are a number of chemicals and compounds used to control algae in swimming pools, which can also be effective in ponds. However, their price is usually higher, and application may be more difficult without special equipment.

Many water weeds that resist copper sulfate may be killed by sodium arsenite. However, it has been found that this very poisonous chemical tends to accumulate in mud bottoms. Its use is prohibited in many areas, and even where permitted, it should be applied by an expert.

Pellets of clay containing 2,4-D or similar materials may be broadcast over the water surface. The pellets break up and release the poison gradually at the plant roots.

Emergent weeds such as cattails and water lilies can be killed by a 1.0 percent spray solution of 2,4-D, or kept from spreading by about one-tenth that strength.

Duckweed (*lemna minor*), a very simple plant consisting of small, flat, floating ovals with roots dangling several inches below, is a problem in many shallow ponds. It often covers the surface completely, giving the appearance of a pale green lawn.

Most of the growth can be killed by chemicals, including sprayed fuel oil spiked with 2,4-D, but it may grow back almost immediately. It may be necessary to remove organic mud deposits and deepen the pond.

State fish and game authorities can usually supply the names of contractors qualified to eradicate or control pond weeds. Permits may be required from state and/or local authorities for chemical treatments.

In small areas the most effective way to get rid of pond weeds is to pull them out by hand and with rakes, as often as they appear. This is a muddy job, but can be enjoyable as a group activity.

HYDRAULIC DREDGES

Hydraulic dredges are very efficient excavators in wetland. However, their use in digging small ponds is limited by the expense of bringing them in and setting them up (see Fig. 6.27), interference of brush and stumps with cutters and suction lines, lack of sufficient water, and problems in spoil disposal.

Pond-cleaning work does not usually involve handling stumps and brush, and most water weeds can be removed by hand or simple equipment. The pond provides some water. The competitive position of the dredge is improved by the great difficulties that land machines have in working pond bottoms too wide to reach from the shore.

This discussion will be limited to pond cleaning. However, the same problems arise in digging a new pond where water is available and vegetation can be handled by other equipment. Dredge construction, work characteristics, and operation are described in Chap. 14.

Water Supply.   A 6-inch hydraulic dredge will pump water out of a pond at a rate between 800 and 1,200 gallons a minute, and will move 1,000 to 2,000 gallons for each yard of soil. An acre

FIGURE 6.27   Loading a small dredge for transportation.

of water contains about 325,000 gallons for each foot of depth (this is a measurement called an acre-foot), or enough water for 4½ to 6½ hours of operation. Because of the slope of banks, each successive foot of water will contain fewer gallons than the one above it. A pond 1 acre in area and 5 feet deep might contain 2½ to 3 acre-feet, say 1 million gallons.

An 8-inch dredge needs about 60 percent more water and moves about 60 percent more soil than a 6-inch model.

If there is no natural inflow into a pond, it is necessary to return most of the water taken by the dredge to be used again. The most economical way to do this is to put the disposal area upstream, so that water will drain back. If this is not practical, a pump is placed so that it can return the water through a hose or pipe. A heavy-duty 4-inch pump can usually handle enough water to keep a 6-inch dredge busy.

Return water is usually dirty so that part of the dredge capacity is wasted rehandling the fines that did not settle out. The amount of circulating soil particles is greatest when the soil is fine-textured, and when settling ponds or areas are small.

The fine organic mud that forms a large part of the deposits on pond bottoms may stay suspended in water for long periods. The experimental uses of cyclones for separating soil particles from dredge water, and of polymers for clotting organic slime, which may reduce this problem, are discussed in Chap. 14.

Dirt in return water increases when the suction strainer of a return pump is too low, or the water is allowed to run on the ground and erode it. It is reduced by having a large settlement pond or area, and/or very sluggish open flow of return water. It tends to increase as the settlement pond fills up with use.

Reusing dredge water has the important advantage of reducing or eliminating the problem of fouling streams or other property below the fill with muddy water.

Any natural flow into a pond reduces the water problem. A pond with no inflow in the summer or dry season may have an ample supply pouring into it in the wet season. If the dredge is operated on a single shift, it could operate on a steady inflow of one-half or one-third of its output.

Disposal Area. Desirable features for a fill disposal area include location close to the pond, drainage back to the pond, good conditions for separation of soil and water, need for fill, and absence of trees and brush.

A 6-inch dredge may be expected to pump spoil from 800 to 1,500 feet, depending on its coarseness and weight, the percentage carried in the water, and the alignment and gradient of the pipe. Maximum distance is obtained with light load, straight pipe, and low lift. A downhill line could be much longer.

Production is reduced by long lines and uphill flow, as these conditions reduce both the volume of liquid and the percentage of solids that can be carried. For maximum output the spoil should be discharged close to the dredge, or downhill from it.

If it is necessary to reuse the water, and two or more dump areas are available, the loss of production from uphill pumping to get gravity flow back to the pond must be balanced against the expense of a return pump at the same or a lower level.

A close location reduces the cost of providing and handling pipe.

Soil settles out most rapidly and completely when water is still. A discharge or settlement pond is usually kept more or less agitated by flow from the pipe. Increasing its area and depth reduces rate of water movement and allows more particles to settle out. The overflow or the sump for pumping should be as far from the flowing water as possible. The size and depth of the pond diminish as it fills during work.

The cost of pond cleaning may be best justified when good use can be made of the material removed. The value of swampland may be greatly increased by building it up to a higher level. Rocky or stumpy fields may be filled over to smooth surfaces.

Fill obtained from cleaning a pond should include both a layer of the original bottom formation and rich black mud built up by the pond water. It is usually of rich fill or topsoil quality, and may be salable after drying.

Mud is held in the area to be filled by putting a dike around it. This may be built by a dozer on firm ground or a dragline where it is soft. A wood dam with movable boards, as in Fig. 6.16, will provide for overflow and height regulation.

Both the owner and the contractor may be liable for any damages caused on neighboring property or downstream by mud or too much water.

However, dredged material has been successfully used throughout the United States for the development of wetlands and aquaculture. It has been used for beach nourishment, shoreline stabilization, and erosion control projects. Also, its use has made for betterment of agriculture, forest, and horticulture. And the U.S. Army Corps of Engineers lists the use of dredged material for open-cast-mine reclamation, solid waste management, construction and industrial projects, and material transfer for fills.

Filling around Trees. Many swampy areas that should be filled are covered with vegetation ranging from brush to big trees. Undesirable brush usually survives partial burial, but trees are very likely to die. The biggest and most valuable specimens have a smaller chance of survival than younger ones.

Trees that will die anyhow should be cut before the fill is brought in. This will allow burial of stumps and unwanted logs. Such logs should be cut in short pieces to avoid later interference with trenching or basement digging.

If a tree is cut after ground is filled, there will be an unsightly stump. This will be extremely difficult to remove because of the depth of the roots under the new surface.

Brush should be cut and burned, or at least knocked down flat, before filling. Afterward it will be difficult to take out and will make grading very difficult.

Wood may decay very slowly under a hydraulic fill.

CHAPTER 7
LANDSCAPING AND AGRICULTURAL GRADING

LANDSCAPING

Landscaping may include the processes of cutting, filling, or grading to change ground contours; retaining or placing adequate topsoil; preserving, moving, or adding vegetation; and planning and installing walls, drives, and game courts.

An important purpose is to produce a pleasing appearance. This may be an end in itself but is usually secondary to the use of the land.

Landscaping is often the final step in jobs which involve earthmoving. It is required in connection with highways, particularly of the parkway or thruway type; to improve the appearance of home or business buildings not surrounded closely by other buildings and paved areas; to beautify parks; and to provide them with suitable recreation areas.

Plans should take into account proper drainage, which may include subdrainage.

Landscaping is often done under the personal direction of the landowner or a representative, but may be finished to grade stakes or left largely to the contractor's judgment.

A large part of the annual landscaping bill is for work around homes and other buildings. Much of this is done during building construction or immediately after its completion, in connection with backfilling around the foundation, disposing of dirt dug for the basement or footings, and restoring surface drainage.

Such landscaping may include construction of terraces, retaining walls, and driveways; moving or planting of trees and shrubs; and making lawns.

The excavating contractor may perform the entire job or only the heavier parts.

CHOOSING THE SITE

Building Elevation. The type of grading close to the building is determined by its elevation relative to the land. The door sills or trim should be at least 4 inches above the finished grade of the topsoil. In general, exposure of more than a foot or two of foundation causes a building to look too high for current styles. The ground should slope down away from the building enough to prevent surface water from standing against the wall.

A building may be set high enough that dirt from the basement excavation can be used entirely in backfilling and grading up to it. If the floor level is determined in reference to the original grade, the bulk of the piles must be "lost" on the grounds, or trucked away.

Grading is also affected by the extent and type of basement excavation. A deep, full basement produces large quantities of fill, while digging for footings and a floor slab may yield little or none. When the building is to have a basement, is to sit low, and is to be built on a plot having a good grade, it will probably be economical to haul away all dirt not required for backfill around the foundation.

Desired depth of the foundation below ground line may be obtained by digging full depth and removing spoil; by putting the basement floor at the original surface and filling; or by an intermediate method. In general, the most economical way is to cut just enough to provide the necessary amount of fill to build the ground up to the building.

Rock and Water.   The presence of rock or water near the surface may make a plot a poor investment, and in any case is important in deciding whether to have a basement, and the depth to place its floor.

Shallow rock can be found with a probe made of 4 or 5 feet of $\frac{5}{16}$-inch stainless steel rod, with a sharp point at one end and a handle at the other. This can be pushed down into any but the hardest soils.

However, it will not tell whether resistance is a cobble or ledge. A long sharp crowbar or prybar can be sunk by repeated dropping and turning. If it is stopped by an obstruction, lack of vibration as it strikes indicates a small stone; vibration only near the hole, a boulder; and a general jarring, a formation of bedrock.

Vegetation will tell a lot about water conditions. Bush willows and bog or bunch grass must have it wet in spring at least. Such water-loving plants on a flat indicate swampy conditions. On a slope they show a spring or seepage, and may warn of ledge rock as well.

If rock or a high water table is found on the site or surface drainage is poor, it is often good practice to reduce the depth of excavation and truck in fill.

No fill should interfere with drainage from adjoining property. If the land must be raised, drains must be placed under or around any dam that is formed.

A septic field on low or impervious ground may have to be placed in a filter bed (pervious fill) which may be quite costly.

Hill or Valley.   A hilltop is almost always well drained, so that the wet basement difficulties discussed in Chap. 5 will not arise. On the other hand, it is much more likely to have rock close to the surface, so that the expense of basement digging may be three to six times greater than for dirt excavation.

Ground drainage can be too good. A person wanting to enjoy lawns and gardens will have difficulty with them in dry weather if they are on a heap of sand or gravel. Topsoil is likely to be poor, thin, and stony.

Building on low ground risks water trouble in the basement, if any, and the possibility of serious flooding from streams or drains. It limits view to the immediate surroundings, provides a higher average temperature but increases danger of frost damage (cold air flows downhill), reduces effect of cooling breezes in the summer and even more cooling gales in the winter, and usually provides rich and moist soil for lawn and garden.

If at all damp, a low site is dangerous to the health of arthritis and asthma victims.

Slopes may offer any combination of features of high and low land. Special factors to consider are that if the land slopes down to the south, it will be warm (or hot), and, down to the north, it will be cold in the northern hemisphere.

View.   A building on high ground may be largely deprived of the enjoyment of a fine view by being set too low or too far back from a slope, or by careless grading or planting.

A common error is to build or fail to remove a high spot which, although lower than the house, blocks the view of nearby down slopes and hollows. See Fig. 7.1.

There is often much conflict between trees and view, which must be decided on a basis of individual preference. In general, ordinary young trees may be quite readily sacrificed while old trees or fine specimens of younger ones should be preserved if possible. Drastic pruning will often serve the same purpose as removal.

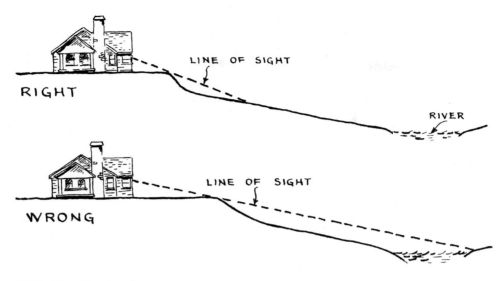

FIGURE 7.1   View downslope.

Shade.   Shading a building and grounds from full sunlight is desirable, but too heavy shade will cause excessive trouble with rot and mildew and create unhealthy conditions, particularly for asthma and arthritis sufferers. Such trouble may be reduced by building in the open, by high trimming of branches of existing trees to permit full air circulation, and by use of discretion in planting.

Noise.   If noise from a highway or railroad is of critical importance in determining building location, it should be remembered that noise travels chiefly upward, partly because of reflection from the pavement or roadbed. Even hundreds of feet up a hillside will not reduce it substantially if the source remains within sight.

   If the river in Fig. 7.1 were a noisy highway, the construction which is wrong from a scenic standpoint would become right when noise only is considered. An earth berm or bank is a more effective sound deflector than a hedge or other planting.

Water Well Drilling.   A substantial portion of both home and industrial building is in areas not reached by water mains. Most farms depend on groundwater for domestic use, and many use it for irrigation also. Factories, theaters, and other large users of water may find that they need a supply in addition to city water. Under such circumstances, the only method of getting a dependable supply of safe water may be to drill for it.

   In sandy or gravelly soils, surface water outcrops, such as ponds and springs, give a rather good indication of the level and abundance of subsurface water. However, a well should go substantially deeper than this level, both for purity and for protection against unusual dry spells.

   Where possible, it is best to get water from rock, or deep down in sandy soil. Danger of contamination is then negligible. Casing is driven down at least far enough to keep surface water and loose soil out of the hole.

   Wells are usually located for convenience, on the first try at least, as prediction of underground water may be highly uncertain. This is particularly so when the soil is too shallow to provide safe supplies and water must be obtained from a rock formation.

   Divining rods of various kinds are used in many sections to locate water. In tests these "dipsticks" have shown a somewhat better record than random drilling, but the difference can usually be accounted for by the good judgment of the experienced person who carries it.

The best place for a well for a residence is just outside the foundation line, so that it can be included in a small extension of the basement or connected by a short pipe, but can still be reached vertically from outside for pulling underground equipment and servicing the underground part of the pump. It is usually drilled and lined (cased) before the basement is dug.

Placing the well away from the building involves constructing a rather costly separate pump house which may pose a landscaping problem and will have to be connected to the building by water and electric lines. It does have the advantage of freeing the building from the noise of the pump and automatic switch, and the possible nuisance of water from leaks.

A well under the building is very convenient, and has become permissible because of improvements in pump design. The flexible plastic pipe and jet pumps, now most commonly used in drilled wells, can be serviced in spite of limited headroom.

Distance between sewage septic fields and wells may be subject to local regulations. Under ordinary circumstances, there is no conflict between having them in the same place if the well is deep, but there is a slight chance that the casing might crack or become disjointed and allow leakage into the water. For this reason, prudence dictates that the well top should be higher than the field, and at least 50 feet away from it.

The truck-mounted spudding or well drill or rotaries and down-hole units have done the drilling. Flow in the well is measured by pumping or bailing.

A flow of 4 gallons per minute is considered adequate for a small residence, but double this is desirable to ensure a generous supply. A small water flow can be partly compensated for by a large storage tank.

SHAPING THE LAND

Backfilling.   In general, it is most satisfactory to backfill around a foundation after the interior horizontal supports for the basement walls are in place but before the upper framing of the building is started. This removes the piles of fill that form an obstacle and a hazard during construction, and provides space for entrance and piling of materials.

Backfill against fresh masonry must be done carefully. A heavy dozer should keep farther away from the wall than the diameter of the largest stone found in the fill, to avoid accidental punching of holes. It should not walk on fresh backfill parallel to the wall, because if it sinks on the side toward the building, it will exert a heavy thrust and be almost impossible to get out without causing damage.

Foundation backfill is seldom tamped when it is placed, but failure to compact it offers the danger of the loose dirt's soaking up enough water during a heavy rain to crush the wall by hydraulic pressure. Good underdrainage around the footings, a proper surface slope away from the building, and compaction of the surface make such a disaster unlikely. Placing floor beams strengthens the foundation.

A foundation of concrete block is subject to damage even after curing. Unless the fill is wet, the weight of the dozer is unlikely to cause damage, but a stone may still be punched through the wall.

A front-end loader is the preferred tool for backfilling and grading around a building. Its ability to back and turn with loads, to cross graded ground with a load without excessive damage, and to place dirt exactly where it is needed enable it to accomplish much more work than a bulldozer of the same size. However, it cannot grade quite as closely to a wall because of the overhang of the back of the bucket in dumped position, and the fact that the bucket is little, if any, wider than the tracks.

Grading.   Grading may be mostly or entirely a problem of disposing of surplus fill to the best advantage. At other times it will consist of arranging for proper drainage, removing objectionable humps or filling gullies, disposing of stone walls or boulders, reshaping to obtain a desirable view or to avoid an undesirable one, or rearranging contours for better appearance. These operations may produce a surplus of soil, or may require bringing in hundreds or even thousands of yards.

Soil in trenches and fills should be thoroughly compacted before the fine grading is done. Unfortunately, it is not common practice to attend to this on small jobs, with the result that an originally pleasing appearance degenerates badly in a year or two. Effects are bad when a level or evenly sloping lawn settles into humps and hollows, and are worse when game courts, stone walls, or paved drives are involved. See Fig. 7.2.

Trench backfill can be compacted by hand; with air, gasoline, or mechanical hammers; or with electric vibrators. If ample time will elapse before grading, ditches can be loosely filled then puddled by flooding with water. Full shrinkage will not occur until they have dried out, a process which takes a few days with porous soils and weeks with heavy ones. While wet, a puddled ditch is a dangerous trap for machinery.

Fills should be compacted by rollers or trucks. If trucks are used, each fill layer (preferably not higher than 10 inches) should be thoroughly rolled, first empty and then loaded. Running a loaded truck on loose fill puts a severe strain on its power train.

A medium-textured fill is more satisfactory for most purposes than either very porous or very clayey soils.

Lawns should not be perfectly flat for any appreciable distance. The maximum slope which it is convenient to mow is about 1 on 6 for long grades, and 1 on 3 for short terraces that are hand-cut. Steeper grades may be left in long grass; planted with vines, shrubs; or fixed as rock gardens.

Old Walls.   In New England and many other sections of the United States, utilizing or disposing of old stone walls is a common problem in landscaping. They often contain huge stones which are so buried and bound that they are a problem to any but the largest machinery. For this reason, and because of the beauty of many of them, it is advisable to leave them in place when possible.

If the wall is to be removed, an attempt should be made to sell it. Weathered field-stone in small sizes is often in demand. Boulders can occasionally be used in deep fills, stream bank riprap, or breakwater construction. Prices obtained for large stone seldom more than repay the expense of handling.

If there is no market for stone, an attempt should be made to bury it. The bulk can be roughly calculated by measuring the length and the average height and width of the wall, including the underground part. If no gully or other natural disposal point is available, a hole or holes should be dug to contain somewhat more than the calculated yardage, allowing for a foot or more of fill over the top.

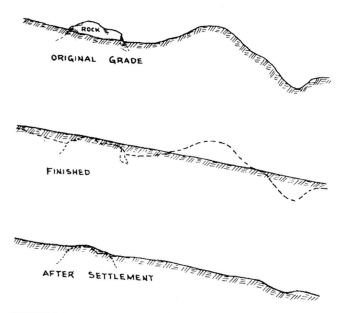

FIGURE 7.2   Irregular settlement after grading.

Excavation is done in the same manner as for a basement. Topsoil should be stripped off the area that is to be dug and regraded. The hole should be deep rather than wide, and might well be dug by a backhoe rather than a dozer, if one is on the job. A backhoe may dig a trench close along the wall, followed immediately by a dozer pushing the stones into it and regrading.

A backhoe is often more efficient than a dozer at breaking up a stubborn wall, as it can work out one stone at a time. However, it cannot transport the stone readily.

The rocks can be trucked away if burial is impractical because of shallow soil, trees, or landscaping. A front-end loader, a backhoe, or a big clamshell can break up the wall and load it. Trucks should preferably have bodies built to carry rock, or be so old and beat-up that damage will not matter.

Loading a wall is slow work. Even small stones may be hard to dig out when in groups, and big ones are hard to get securely in the bucket. Production in yards per hour may be pitifully low.

Retaining Walls. Masonry walls are frequently used to separate different ground levels. They may be required where the slope is too steep for plain earth, or used largely for the sake of appearance. In the first case, the wall may make up only part of the required rise, and an earth slope is continued from its top. Such walls must be strong and well founded if they are to give good service. They are subject to very heavy pressure from the dirt behind them, particularly if it slopes up from the top of the wall, and if it becomes saturated. Freezing will cause a push against the top of the wall and disruptive forces inside it. Tree roots can act to lift and overturn it.

Some cross-sections of retaining walls are shown in Fig. 7.3. The foundation must be adequate, or the wall will fail. If the ground under it is unstable because of its nature, recent placement without proper compaction, or frost heaves, the wall will break up or lean outward. It is therefore

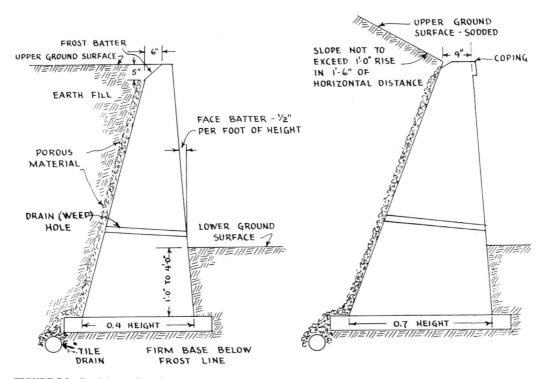

FIGURE 7.3 Retaining-wall sections.

essential to found it below frost level on firm soil or rock. If the quality of the soil is questionable, a wide concrete footing slab may be poured.

The thickness and strength of masonry required for a retaining wall are commonly underestimated. Results of under-strength construction are sometimes satisfactory, but often not. For safety, wall thickness at any point should be between one-third and two-thirds of the height of the wall above that point, and the top should be 6 to 9 inches thick. Minimum thickness is safe when reinforced concrete is used, when height is moderate, and the retained soil is well drained and stable.

Maximum thickness is required when a steep slope rises from the rear of the wall, and when the ground is very unstable. Another consideration is the strength of the masonry. Reinforced concrete is the strongest used. Plain poured concrete is considered stronger than concrete block, brick, or mortared stone. Dry stone walls have little resistance against thrust and should be kept low.

The push from dirt behind the wall can be minimized by keeping it well drained. A layer of gravel or other porous material should be placed along the rear face of the wall. A tile drain should be placed beneath the foundation, and there should also be "weep" holes through the wall itself.

Ground expands when the water in it freezes, and the surface slab formed in this manner can exert a considerable thrust. A slope or batter at the rear corner will deflect this pressure upward so the slab will slide on the wall instead of pushing it.

A vertical wall often has an appearance of overhanging. A backward lean or batter of $\frac{1}{2}$ inch for each 1 foot of height will counteract this. Such batter can be increased to any desired slope with some increase in stability. A face slope in a dry masonry wall may permit outward movement for some years before it becomes vertical or overhanging.

Geosynthetic Walls.   The use of geosynthetics to build high earth retaining walls has been increasing as it is a more economical choice than concrete retaining walls. A comparison is shown in Fig. 7.4. There are two types of geosynthetics. One type is the geotextile-permeable woven or nonwoven synthetic fabric blanket, described in Chap. 3, for stabilizing ground for vehicle passage. The other type is called a *geogrid*, which is like a mesh with openings between strips of high-quality, strong polyethylene material.

Retaining walls built with geosynthetics may be 20 to 40 feet high and can have a nearly vertical face, if that part is built with decorative face blocks or open block to allow viney growth. The geosynthetic reinforcing sheets, which are generally as deep into the fill as the wall is high, are laid horizontally on the earth as the fill is built up. The vertical spacing between sheets is dependent on their required strength according to the loads that will be applied on the ground above and back of the wall. That spacing may be from 1 to 3 feet. Polymers creep at stress levels on the order of

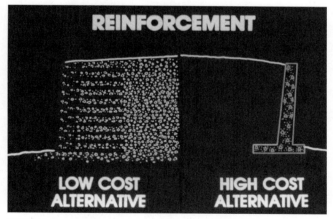

FIGURE 7.4   Comparison of retaining walls. (*Courtesy of Hoechst Celanese Corporation.*)

one-third to one-half their strength. A wall designed for that level of stress will allow creep, whereas a design at one-tenth the strength level should avoid the problem of creep.

Poor drainage is a leading cause of retaining-wall failure. Therefore, the geogrid wall needs various provisions for drainage, as shown in Fig. 7.5. The mortarless, interlocking Keystones drain naturally (1). Surface runoff can be diverted away from the wall (2). Embankment flow can be directed to an outflow pipe (3). And the groundwater that may get into the reinforced zone can be directed to another outflow pipe (4).

Drainage. It is desirable that all areas be provided with sufficient surface slopes, proper sub-drainage, or both, so that water will not stand anywhere and the ground will dry and firm rapidly after saturation. Particular care may be required to subdrain any soil touching basement walls or floor.

When the soil is porous sand or gravel and the water table is low, drainage is usually automatic and mistakes in gradient will show only briefly during rains. Impervious soils, however, demand care in shaping so that they will drain completely, not only when the job is completed but also after settlement of fills.

If pervious fill is placed on a relatively impervious native soil, the surface on the impervious soil should be shaped to drain, to avoid trapping underground water in pockets. If the native soil to be buried is pervious, it need not be graded for drainage, regardless of the type of fill, although shaping to avoid uneven depth of fill is still advisable.

Where areas are large, rainwater flowing on the surface may constitute a serious nuisance even if it does not erode the ground. At a price, such water can be caught in catch basins, then removed through underground pipes. Because of the expense of such an installation it is best to have it

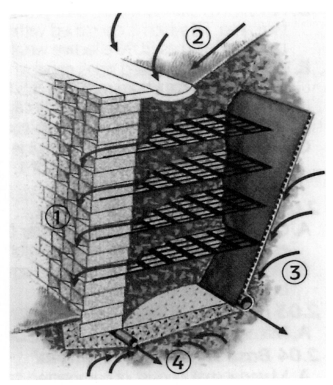

FIGURE 7.5   Drainage for a geogrid wall. (*Courtesy of Keystone Retaining Wall System, Inc. as published by Sweet's Group, McGraw-Hill, Inc.*)

designed by someone familiar with the work. If this is not possible, pipe size should be figured in the same manner as culvert capacity, according to the maps and tables in Chap. 5.

If land tile is used, it will also function as a subdrain. However, care must be taken not to allow more surface water to enter it than it can easily handle, as the hydraulic pressure resulting from water standing in or over the inlets may force channels outside the tile, which will undermine or misalign them, with resultant impairment or destruction. If the important problem is surface water, concrete pipe or sewer tile with mortared joints is preferable.

All inlets should be protected with gratings firmly set in masonry. Lack of these may permit entrance of large objects or masses of material which will plug the drain. Gratings are usually larger in area than their pipe, to allow for partial clogging with leaves. The vertical or steeply sloped pipes up to the catch basin should have tight joints.

If backfill is not tamped in the trench made for the drain, it may settle and leave the grating standing up above the sod. This is unsightly, makes it vulnerable to breakage, and interferes with reception of water.

If a garage is below ground level, a catch basin in the driveway just outside it is necessary. This drain must be adequate, as its failure in a heavy storm will flood the garage and perhaps the basement.

Subdrainage.   Land tile subdrains may be installed under lawns and gardens to correct saturated or oozing conditions, to speed up drying after rain, or to provide better growth conditions for plants.

Subdrains may be tied in with the tiling around the basement and with catch basin systems for surface runoff. They should drain to low areas when possible, as opening into storm water drains exposes them to damage from backed-up flood water.

FINISHING OFF

Topsoil.   Topsoil which has been salvaged in advance of digging may be spread as soon as the fill is graded off, or left piled until the building is finished. Immediate spreading provides a cleaner appearance, which is of particular value to buildings built for sale, but the topsoil is liable to become mixed with various sorts of waste, and to be severely packed by supply trucks.

Two-ton "toy" dozers or small compact loaders are good spreaders, as they are so light that they leave average topsoil in condition to be finished off by hand, where heavier machines compress it so that machine tillage is required. They also can maneuver among trees, retaining walls, and other obstacles with less danger of damage and far less loss of time than larger dozers.

A light wheel tractor with a front bucket or rear grader blade can do light grading.

Freshly spread topsoil or undisturbed field sod which is to be reworked into lawn is often loosened up with a rotary tiller. This machine leaves it soft and easy to work, and if the topsoil is thin will increase its usefulness by mixing in some subsoil.

It is not possible to state a general rule for the amount of topsoil needed around a building. Good topsoil has three important characteristics: it contains humus which absorbs water and doles it out to plants in dry weather, it contains a supply of available fertilizer, and it has a grain size and arrangement that is favorable to plants.

A lawn made with poor or too thin topsoil may be persuaded to grow vigorously by proper fertilizing. However, it will tend to burn out during dry spells unless it is shaded. It will dry out more readily if it is over gravel or sand subsoil than over fine-grained soil. The minimum topsoil depth for a lawn under most conditions is 2 inches, and 4 inches is safer. However, benefits are obtained from greater depths, and it is common practice to use up whatever piles are around. If the original soil was thin, or had been lost, more must be trucked in.

Gardens, flowerbeds, and shrubs like to have about 8 inches, and depths up to 2 feet are recommended for some species.

If peat (humus) is obtainable locally at a low price, it may be spread on subsoil and mixed in by hand or machinery. With the addition of lime and fertilizer it may serve well as topsoil and might be much cheaper.

Converting Field to Lawn.   The original lawn made around a new building may be partly or wholly at the grade of an existing field supporting a growth of mixed grass, wild flowers, and weeds.

One way to make a lawn is to use a rotary tiller or a plow and a harrow to turn in the existing growth, and pulverize the soil for planting of grass seed. It is usually good practice to bury the old vegetation several weeks before planting. Decay of vegetable residues often temporarily deprives the soil of nitrogen to such an extent that the new crop cannot obtain enough for a start. Also, sprouts from roots of undesirable plants can be readily destroyed as they appear on bare ground, and can be reduced or eliminated before planting.

A less expensive method, which is usually satisfactory, is to pull out any brush, then mow the field repeatedly. It may be reduced in one operation from field length to lawn length, then kept short, but better results are apt to be obtained from starting with a high cut as if for hay, and progressively trimming shorter.

The effect of the cutting is to kill or place at a disadvantage plants that prefer to grow tall, and to encourage those which are not damaged by mowing. Most fields contain enough lawn-type grass and clover to take over the whole area within a year when encouraged by repeated cutting.

A similar effect may be obtained by moderate driving and parking of cars and trucks on the field. Field surface may be too rough for a lawn. Large inequalities may have to be cut or filled. Smaller ones can be rolled down, filled with dirt, or both.

A power lawn mower having heavy steel or rubber rolls will tend to flatten ground. Its effect is quite marked when inequalities are small and choppy and the soil is soft, and becomes negligible as the ground bakes in the summer.

A steel wheel roller, weighing from 3 to 5 tons, of the type used in driveway construction and blacktop patching, is a very effective lawn flattener, but must be used when soil is in the right condition. If soil is too soft, it will make ridges, and probably get stuck. Heavy wet soils may pack so hard as to discourage growth.

Hollows may also be filled with topsoil. This may be applied in layers $\frac{1}{4}$ to $\frac{1}{2}$ inch thick so that grass will grow up through them, or the fill made to the level of the surroundings on one lift, then new seed planted where necessary. Sometimes the seed is mixed with the topsoil prior to placing it.

Fills of more than an inch or two on bare ground, and any fill over grass, tend to settle noticeably so that the work will require doing over, although with less material, the following year. Overfilling enough to compensate for this settlement requires expert judgment. Firm tamping is sure to reduce this difficulty and may eliminate it, but might make it difficult for existing grass to push up through the new soil.

Planting Grass.   Soil should be loosened and smoothed before planting. If the area is too large for hand digging and raking, a rotary tiller of appropriate size is the preferred tool. A plow and a disc harrow, or a harrow only, followed by dragging with a plank will often do a good job. Fertilizer, lime, manure, peat moss, or other soil-enriching materials are best mixed in by hand or machinery. If the area is large, a spike-tooth harrow can be used for both smoothing over the ground and covering the seed.

After the machinery is finished, the ground is hand-raked. This looks simple but is not. A certain knack is required to get a smooth surface.

Soil acidity can be tested with litmus paper and color scales that are obtainable at garden supply and drug stores. Most lawn grasses like pH 6. If the soil is pH 5, about 75 pounds of ground limestone should be added to each 1000 square feet. A pH of 4 will call for 100 to 200 pounds for the same area.

Slaked or hydrated lime has a higher calcium content, 100 pounds of it being equivalent to about 135 pounds of ground limestone. Quicklime is still more concentrated, but is inconvenient and dangerous to handle.

It is good practice to add some fertilizer, but the amount varies widely with circumstances and individual judgment. If vegetation has been mixed into the soil within the last two or three weeks, either on the spot or at the source from which it is being hauled, some nitrogen fertilizer must be added, to make up for that borrowed by the decay processes. In general, an addition of a moderate amount of general fertilizer is cheap insurance for the work and expense invested in preparing and planting the lawn.

Any of the lawn seed mixtures are good, and the differences can be explained by the store selling them. A mixture should almost always be used instead of a single kind, as each variety has its own special and often obscure likes and dislikes in soil conditions. If several kinds are planted, there is a good chance one of them will like the circumstances, and its vigorous growth will compensate for any sulking by other varieties.

Seed scattered on the ground surface in very early spring may be worked underground by freezing and thawing so that no effort is required to cover. Seed lying on the surface will sprout and take root successfully during a long wet spell, but hot sun may kill it, or birds eat it. Raking after seeding reduces these dangers.

The grass will come up faster and better if the ground is rolled lightly after seeding. This may be done with a muscle-powered lawn roller, a hand tamper, a power lawn mower of the roller type, or the smallest size of steel wheel gasoline roller.

Erosion.   A sloping lawn is vulnerable to severe damage from flowing water from the time the topsoil is spread until the grass has made a good root growth. The probability of such damage should be figured into cost estimates by both the owner and the contractor.

Danger of erosion can be reduced by mixing straw or lawn clippings into the surface. This necessitates a heavy addition of nitrogen fertilizer, and makes it harder to cover the seed.

If a seeded surface is sprayed with asphalt emulsion, it will be held against ordinary erosion, without preventing growth of the seedlings.

It is sometimes possible to divert drainage into other areas. One section may be fixed up and seeded first, and water routed through the part which is only rough-graded. After grass is firmly established, drainage may be shifted to go over it, so the rough area can be smoothed and planted.

Sod.   Drainageways and steep slopes can be protected with sod. This is cut out of existing lawns or mowed fields by means of hand tools or an engine-driven sod cutter, laid on freshly loosened and smoothed topsoil, and tamped into firm contact. It is sometimes fastened in place by driving pegs or thin stakes, or by pegging chickenwire firmly over the whole area.

Sod may be cut in strips 12 or 15 inches wide and 6 to 10 feet long or in squares or rectangles of any convenient size. A depth of $1\frac{1}{2}$ inches usually suffices to get practically all the roots.

It is essential that newly placed sod be thoroughly watered and tamped to establish its contact with the ground. It should be watered as necessary until it shows that it can take care of itself.

TREE PROTECTION AND REMOVAL

Trees are liable to destruction or damage from various causes during construction work. Trunks or branches may be broken or scraped by accidental contact with machines; roots may be dug away by ditching or lowering of grade, lessening the tree's ability to obtain food and water and rendering it more vulnerable to uprooting by wind; its trunk may die because of dirt piled around it; or its roots may be drowned or suffocated by placing of fill.

In general, the larger and more valuable trees are less subject to fatal damage from collisions, although the scars they do get heal more slowly; but they are much more likely to die from root cutting or suffocating than younger and more adaptable specimens.

Bark Damage.   Trees can be partly or wholly protected from collisions by wrapping with burlap or other cloth, and tying thin wood strips around the trunks and any particularly exposed branches. If used for anchors in pulling out machinery, trunks must have very heavy padding and thick wood pieces between the bark and the chain; and the chain loop should be fastened with a grab hook, bolt, or knot that will not slide. The choker effect obtained from round hooks or rings can readily crush bark and wood all around a tree so that it will be fatally injured.

If a tree is girdled by removing even a narrow ring of bark around the whole trunk, it will probably die. However, if the sapwood is not injured and the damage is kept shaded so that the wood will not dry out, young and vigorous trees may repair the cut by growing several inches of callus and new bark across the injury from top and bottom.

If the gap is wide, a skillful worker may be able to graft strips of bark across the injury. If circulation is established through these, they will serve to keep the tree alive and they will widen out so that the damage may heal over entirely.

Scars on trunks or branches should be promptly covered temporarily or left bare for self healing.

Ditching.   Ditching on one side of a tree ordinarily does not injure it severely. However, it is best to keep the cut as far from the trunk as possible, thus reducing the number of roots lost, minimizing the danger of tearing the trunk, and making the digging easier.

If a hoe or dozer digs within two trunk diameters of the tree, the roots should be uncovered, and then cut by hand to avoid danger of splitting the trunk while tearing them up.

A close cut weakens the tree's resistance against a wind that tends to tip it away from the ditch. If uprooting in that direction would cause it to fall on a building or across a highway, a tree expert should be consulted about the advisability of providing cable support, Fig. 7.6, or removing some of the upper branches.

Burial.   A tree's reaction to having its trunk buried varies with its species, health, and the nature of the dirt. Burial is fatal to the majority if the fill is deep enough or of such a nature that it will smother the bark and support organisms which will destroy it.

The fill also changes the air and water content of the topsoil and subsoil around the roots. Such changes may damage or kill the roots directly, or indirectly by changing the nature of the soil population.

The best defense a tree can muster is to put out new roots near the surface. The willow does this automatically, but the majority of temperate-zone trees do it with difficulty or not at all.

Trunk damage can be avoided by building a stone wall around the tree on the original ground, at a sufficient distance to allow free air circulation. See Fig. 7.7. The space inside is called the well. Sometimes the fill is made and part of it dug away by hand to make space for the wall. Or the wall may be built first, and the dirt placed around it.

The first method is expensive and offers some danger of damaging the tree with the digging tools. The second is subject to the danger of knocking over the wall while placing fill, or accidentally spilling dirt over it that will fill up the well.

In general, the most satisfactory technique is to build the wall first, and fill the well with easily removed material such as stones, wood scrap, or crumpled newspaper. Such items will prevent any appreciable amount of dirt from entering the hole, and are easily taken out when grading is complete.

Sometimes pebbles or crushed-stone collars are used to avoid unsightly or dangerous holes. These will usually allow sufficient air circulation when new, but are likely to plug up with dirt.

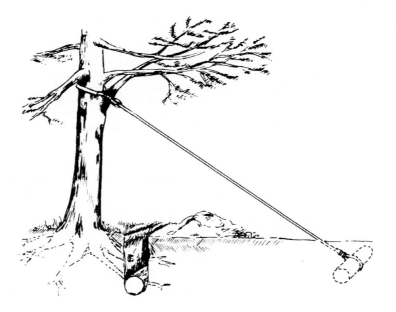

FIGURE 7.6   Temporary tree brace.

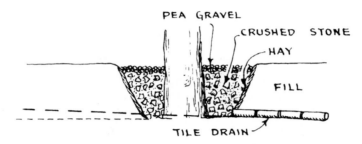

FIGURE 7.7   Tree trunk protection in fill.

The fill should be pervious enough that water will not stand in it and in the holes. Tile may be laid to drain the tree wells, but a saturated fill is liable to kill the roots anyhow.

Root protection is more complex, and the results are less certain. Land tile is laid on the old surface of the ground or slightly below it, with lines 3 to 6 feet apart. A 4- to 6-inch blanket of crushed stone is laid over the area and covered with hay. Pipe openings at each end of the fill or into wells should allow enough air circulation to preserve favorable conditions long enough to enable the tree to adjust to the changing conditions. Wire mesh must be placed across openings to keep animals out.

If it is not economically feasible to take these precautions, the fill should be made of clean bank gravel or coarse sand. Trees may survive heavy additions of such open textured material.

Removal.   Landscaping work may involve removal of trees. If they are to be destroyed, the job resembles the land clearing described in Chap. 1, except that interference from buildings, wires, valuable trees, and other obstacles is much more common.

If the ground is to be filled, trees may be cut as nearly flush with the ground as possible. This may also be done if the grade is not to be changed and the presence of the stump is considered less objectionable than the cost of removing it. If the grade is to be cut, stumps must be uprooted.

DRIVEWAYS

Most home landscaping involves the planning of a driveway. It may be a straight connection to a street a few feet away, or a long roadway involving considerable problems.

Short, straight drives can be as narrow as 8 feet for use by passenger cars only, but 12 feet is more comfortable. A long drive, one in a slot between walls, or any drive to be used by trucks, should be at least 12 feet wide.

Curves should be 1 to 3 feet wider than straightaways, the sharper turns requiring the greater width.

The entrance from the street should be 30 feet wide at the curb, in order to permit turning into it from the near side of the street. An effort should be made to avoid entering the road through a deep cut, between large trees, or in or very near to a curve, as any of these features add to the danger of accident.

Sidehills.   A long driveway of the farm or estate type may have to cross a hill slope. It is notched into it in the same way as a pioneer road. However, since it is a permanent improvement which should have a pleasant appearance, special procedures are followed.

Such a drive serves not only as an automobile route but also as a drainageway. Unless diversion ditches are made above it, whatever water flows down the slope will land in the driveway cut, and flow down it or across it. Unless an ample channel is provided, the drive may often resemble a streambed.

A driveway crossing a hillside below a long slope should be 14 to 16 feet in width, including drive, shoulder, and gutter, but not the slopes. The gutter should be at least 3 feet wide and deep enough to carry all the water. It can be relieved at frequent intervals by diagonal cross drains to the lower slope. The drive cross-section may be crowned or sloped oppositely to the hill, but should never slope with it. See Figs. 7.8 and 8.1.

The cheapest gutter is sod, and it may hold on quite steep slopes if well established. Temporary diversion ditches can be made with a plow to keep much of the water off until it is well established. Stone, concrete, and blacktop are water-resistant, but are subject to frost heaving unless on a stone or gravel base.

The slopes of the cut and fill should be topsoiled and seeded. They are often too steep for mowing.

Garage Level.   Driveways offer minimum trouble in use and maintenance if they are nearly level. Unnecessary expense and inconvenience may be caused by placing the garage so that grades are created or exaggerated.

If the driveway is long, the garage should be at about the grade of the ground around its entrance after landscaping. If the drive is short, the garage should be at about street level.

It is unusual for a garage to be higher than the grade around it, but it is very common practice to place it under the building at basement level. In this case the driveway often must enter through a deep cut bordered by steep slopes or retaining walls, and usually descends more or less steeply as well.

A descending drive must level off several feet outside of the doors, and must not do it so abruptly as to cause a car's bumpers to scrape when entering or leaving. Extra width is needed between walls. One or more grating drains should be provided outside the doors, with plenty of capacity. Drainage from a long drive or from the lawn should be diverted to other drains or channels. Failure to observe both of these precautions may result in a flooded garage or basement. It is also necessary to keep the area clear of leaves and trash that might block the drain.

If the drive is short, its slope will be determined by the difference in level between the garage and the street. Driveways as steep as 30 percent grade are in use, but they are both difficult and dangerous.

Any steep drive requires care in designing the vertical curves at each end so that the center of the car will not scrape on a convex curve, nor its bumpers hit on a concave one. There should be a parking place which is moderately level if possible.

In climates where freezing weather occurs, steep grades may become dangerous or impassable. Also, drifting snow may entirely fill a cut to a low garage, from which it will probably have to be dug or blown rather than plowed.

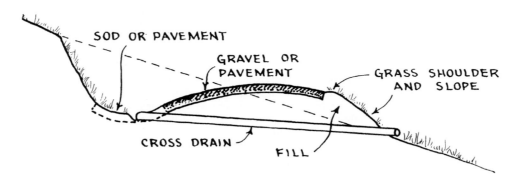

FIGURE 7.8   Hill slope driveway section.

Snow Melting.   Heating pipes can be installed under driveways, walks, and outside stairways to prevent snow and ice from resting on them.

The preferred method is to lay wrought iron pipe or copper tubing in the pavement slab or immediately below it, and circulate an antifreeze solution heated by a heat exchanger in the building's steam or hot water boiler. It can be turned on and off by hand, or by automatic controls operated by the weight of the snow, or by its interference with a light beam reflected off a polished surface to an electric eye.

The system must have ample capacity, or will occasionally do more harm than good. If it does not quite keep up with the snowfall, at the end of the storm it may leave the area covered by a layer of slush, which might then be frozen by an extreme drop in temperature and kept frozen until the weather has warmed slightly.

Electrically heated wires, available in hardware stores, may be placed on ice for emergency melting. Covering with cloth or paper increases effectiveness.

Turnarounds.   A driveway which does not include a turning place requires that a car be backed out of it or into it. This is entirely impractical on long or curving drives, and is a nuisance and a danger in any case. Wherever property size permits, a turnaround should be provided.

The best way to lay one out is to have the people who are to use it make some trial turns, add a few feet to the space they require to allow for carelessness or a bigger car, and build the drive accordingly. If there is no opportunity to practice, any of the layouts shown in Fig. 7.9 should prove satisfactory.

Allowance should always be made for car overhang. This may be 2 feet front, up to 4 feet rear, and 8 inches at the sides. It is desirable to have a curbing that will keep the wheels about a foot away from vertical walls to protect the car from scraping at the side.

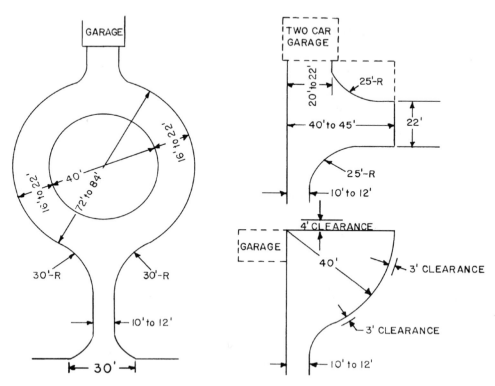

FIGURE 7.9 Turnarounds.

Extra space may be provided in a turnaround for parking, or to supply peace of mind to uncertain drivers.

Surfacing. Four inches of good bank gravel, crushed rock, shell, or similar materials should be stable enough for a building drive on well-drained soil. Under average conditions 6 inches is safer, and when the ground is soft and wet, 8 inches to 1 foot or more may be required.

A stone fill underneath can be used to reduce the gravel requirement. Any flat stones near the surface should be set on edge so that they will not rock and disturb the top dressing.

If the driveway is long, it may pay to try to get by with a minimum depth and add more material to any soft spots as they develop. However, it is often necessary to dig away the softened gravel, as it mixed with mud underneath. If the driveway is short or the budget liberal, it is good practice to put down a safe depth in the first place.

These materials may be used for both the bulk and the surface of the drive, may be given a surface treatment, or may serve only as a base course.

If used alone, varying amounts of difficulty may be found with loose stones, gullies, dust, tracking small particles into buildings and cars, muddy surfaces, ruts or mudholes, or scattering on the grass, depending on the kind and quality of material used and the circumstances.

Calcium chloride, either scattered on the surface or mixed in, will prevent the drive from becoming dusty, and will help to hold it against washing and scattering. It should not be used in the immediate vicinity of the building, where it might do damage if tracked in.

A thin layer of loose pebbles or fine crushed rock makes an attractive surface for light and slow-moving traffic, but scatters under fast traffic. Snowplows are likely to move a large part of it to the lawn.

AGRICULTURAL GRADING

NEED FOR GRADING

It is often necessary or desirable to regrade land in order to use it for farming. In arid regions, land is leveled to permit even distribution of irrigation water. In semiarid climates, sloping land may be terraced to hold rainfall behind dikes so that it will soak into the ground instead of flowing off.

Where the rainfall is adequate or excessive, terracing may be necessary to reduce washing of soil from cultivated slopes. Under any conditions of climate or soil, leveling may be desirable to allow use of large or high-speed machinery. Alone or in conjunction with underdrainage it may increase yields by eliminating burning out of crops on ridges and drowning in hollows.

Agricultural grading differs from other types of earthmoving in the large areas to be treated in proportion to the money available, the flexibility in engineering requirements to suit conditions and cost factors, and the problems relating to the handling of topsoil.

Cuts and fills are typically shallow, vertical movement of soil is slight, and horizontal movement is relatively great.

TERRACING

Terracing land is the grading process of interrupting slopes with ridges, channels, or benches, or combinations of them, in order to slow or stop the flow of rainwater and to prevent harmful soil erosion.

Terracing may serve to hold water on the slope so that it will soak into the ground; allow water to flow off it while keeping the loss of soil to a minimum, or to reduce slopes so as to make them more readily workable or irrigable.

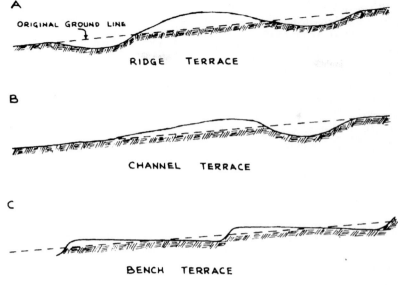

FIGURE 7.10   Terrace types.

Terrace Types.   Three principal types of terrace are used. Each is constructed along level or contour lines. The ridge terrace, Fig. 7.10(A), is a ridge built of soil obtained from both sides. The channel terrace, (B), is a ridge constructed of dirt from the upper side only, and the channel formed by this excavation is an essential part of the structure. The bench terrace, (C), has a stair structure with steep risers separating relatively flat cultivated areas.

Ridge and channel terraces are usually built with sufficiently gentle slopes to allow farm machinery to work along or across them. Best results are obtained if farming operations are done parallel with their centerlines.

Ridge Terraces.   The ridge or absorptive-type terrace is used primarily to conserve water in regions of deficient rainfall. Each ridge serves as a dam for a pond, which is deepest in the excavated area immediately above it. Water may also be impounded in the trough formed below this ridge by borrow of material.

A larger area and quantity of water can be held on slight gradients than on steep ones, by any one size of ridge. Not only is more water retained per yard of dirt used in the ridge, but also its distribution over the land is more uniform.

Too great a depth of water may drown out crops immediately above the ridge.

It is ordinarily not economical to construct terraces for water conservation alone on slopes over 3 percent, and structures for reducing soil erosion are more often of the channel or intermediate types.

Overflow channels may be provided to carry off rain in excess of that for which the system was designed. These should be protected like channel terrace spillways.

Channel Terraces.   Channel or drainage-type terraces are essentially shallow diversion ditches which catch water flowing down a hill and lead it off to drainageways that have been protected against erosion.

The channel depends on the ridge of excavated material for much of its capacity. Its grade is flat, or nearly so, so that only extremely fine soil particles can be carried by the water it discharges.

Bench Terraces. The principal application of bench terraces in the United States is in connection with irrigation. If the original slope of the land is greater than that of the graded fields, each field will constitute a terrace, separated from fields above and below by comparatively steep slopes.

Benches may also be made in steep cultivated land by leaving narrow contour strips in grass or other permanent vegetation. Soil washing from the wider strips between will be caught by the grass and will tend to build up the low side of the cultivated piece, while its top is lowered by erosion. This process, often accelerated by plowing so as to throw dirt downhill, will ultimately result in gentle slopes separated by steep banks.

This work is ordinarily done by farmers without assistance from contractors.

Surveying. A terrace system must be carefully surveyed and planned before construction starts. The interval between terraces may be taken from the chart in Fig. 7.11, or better, determined after conference with soil conservation specialists.

Stakes are placed from top to bottom of the field at the selected intervals. From each of these a level line is run the full width of the field or area to be processed. These lines, known as contours or contour lines, will bend toward the high side of the slope in hollows, and away from it on ridges.

Each level line may be found by setting a level transit or laser at a point, then measuring its height above the ground. A marker or laser receptor is set at that height on a rod.

At various distances, the rod is moved up and down the slope until the marker is at instrument height when the rod is on the ground. A marking stake is placed.

A series of such readings indicates a level line. Rechecking is advisable.

Each line indicates the location of a terrace. However, sharp angles and extremely irregular lines are not desirable and can often be reduced or eliminated by minor adjustments, as in Fig. 7.12. Farming is simplified when adjoining terraces can be made parallel.

If an angle in a gully is eliminated by moving the line downhill, the terrace ridge will have to be built higher above ground level to preserve its grade, and water will be ponded behind it.

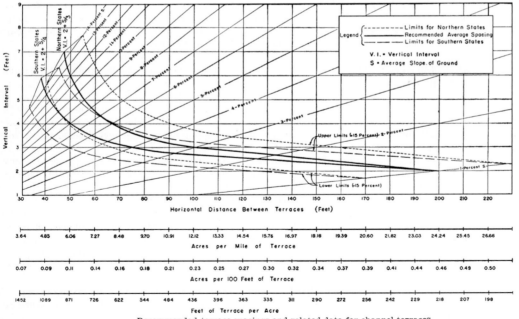

Recommended terrace spacings and related data for channel terraces.

FIGURE 7.11 Channel terrace data.

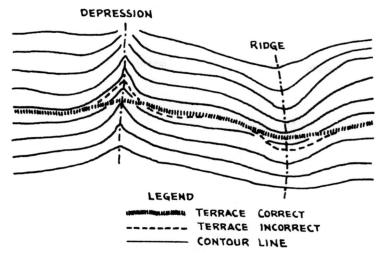

FIGURE 7.12   Terrace line correction.

If the line is moved uphill to cut off a bend on a ridge, the channel will have to be dug more deeply to allow flow of water through it. However, such a ridge may be used as a divide or drainage head, in which case little water will be present.

Stakes are ordinarily used only for location guides, but may be marked with grades where the terrace is to be higher or the channel deeper than standard.

It is desirable that the top of the terrace system be also the top of the drainage area. The top terrace should serve the same width of ground as the terraces below it. If a larger area must be served because of flow from higher fields, the channel capacity must be increased proportionately, or some other type of intercepting drain used.

Grades.   Ridge terraces usually have a level grade. Channel terraces may be level for the section most distant from the outlet, and slope increasingly to about a ½ percent grade at the outlet. Drainage is normally from ridges toward hollows. Short terraces require less maintenance than long ones.

Outlets.   The discharge from a terrace should be into a waterway that is capable of carrying the water directly down the slope, without eroding. Shallow depressions carrying a permanent sod are often satisfactory. These should have enough drop from side to center to ensure gathering of all water discharged from the channels, but should not concentrate enough flow at the center to cause erosive velocities.

The strength of the sod, as affected by fertilizing and grazing, and the condition of the soil will determine the maximum safe gradient for a meadow outlet. This is ordinarily 6 percent or less.

Sod should extend several feet above flow lines on the sides of the waterway. Steeper or narrower channels may be protected by ungrazed and uncut growth of grass, weeds, bushes, or trees.

For extreme conditions, a channel protected with check dams (Fig. 7.13), pavement, or other artificial structures may be required. Steep banks require sod or artificial flumes for the terrace discharges.

Permission of owners of land below the farm must be obtained if the terrace system alters the path or concentration of water on their properties.

Outlets *must be* completed and protected before terraces are built.

Construction.   Terrace construction is primarily a matter of sidecasting. The work is commonly done with graders of either the powered or towed types. Bulldozers working at right angles to the terrace are also effective, particularly in the channel type.

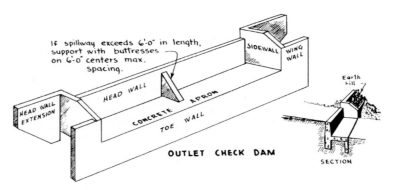

FIGURE 7.13   Check dam.

Belt loaders, of both standard and special terracing models, give excellent results.

If the channel or cut depth is greater than that of the topsoil, a barren strip will be left which will yield poor crops. This damage may be reduced by overcutting the channel so as to leave it below grade, and blading some topsoil from above to cover it.

Maintenance.   Terraces will serve their purpose best if plowing and cultivating are done along contour lines—that is, parallel with the terrace lines.

A terrace can be enlarged by plowing so that dead furrows are in the channels and lands on the ridges, as in Fig. 7.14.

Any accidental blocking of a channel or damage to a ridge should be repaired immediately, as it might cause the terrace to fail in a heavy rain, with possible destruction of the terraces below it.

GULLIES

Characteristics.   So far as this discussion is concerned, a gully is a drainage channel that has become so deepened or enlarged that its banks are unstable and tend to extend destructively into surrounding land (Fig. 7.15).

Control of gullies is largely an agricultural problem, but may also be required to protect highways or structures.

Gullies are a sign of the beginning of a new cycle of erosion which tends to dissect smooth slopes or high levels of ground into tablelands separated by steep-walled channels or canyons. Unless controlled, it will eventually narrow such tables into peaks. Geologically, they are small examples of the type of stream erosion which carves rising land into mountains.

The new erosion cycle may be started by land rising so that steepening channels add to the velocity of flowing water; by lowering the outlet of a stream with the same result, or by reducing the resistance to erosion of the land.

Gullies are caused most frequently by the destruction of the vegetation which protects the land surface, although they may also be started by lowering of outlets, due to highway or stream channel work, or land slips.

The majority of gullies contain intermittent streams which flow only during or immediately after rains, or in wet seasons. Permanent streams are less often affected as their beds are unsuitable for agriculture. These are most apt to be raised and choked by silt deposits resulting from bad farming.

Growth.   When a slope is covered with vegetation, whether sod, bushes, or forest, rainwater tends to move downward as a flowing sheet, and is only gradually gathered in definite drainageways. Its eroding action on the ground is slight, as it is held from contact with the dirt and its velocity is lowered by stems and roots that form a protective mat.

FIGURE 7.14   Plowing to preserve terrace.

FIGURE 7.15   Active gully. (*Courtesy U.S. Department of Agriculture.*)

If the vegetation is removed by plowing, disking, close grazing, or fire, the water comes in direct contact with the soil and tends to remove the surface particles. This effect is usually rather uniformly distributed at first, and is called sheet erosion. It can be reduced to slight proportions by proper farming, including contour plowing and cultivating, and terracing or returning to sod when necessary.

Erosion is most active where the amount or velocity of water is greatest, or the soil is least resistant. Such places tend to wash out more than the surrounding area, and then, being lower as a result, will catch the runoff from a larger area, increasing the quantity and velocity of water, and its eroding effect. The deepening of the channel therefore tends to build up forces which will make it deepen more rapidly.

In its early stages such a gully may be destroyed by plowing or harrowing, so that it is choked by clods and some of its water is diverted elsewhere. However, unless close-growing vegetation is planted, or weeds are allowed to grow, new storms will reform the channel or create new ones nearby, and they may eventually become too deep to be choked by plowing or even to be crossed by a plow.

Once a gully is formed, it enlarges by three separate processes. One of these is channel erosion— the scouring action of the water deepening the bottom. This is accompanied by the falling in of the sides as they are undermined.

The upper ends (heads) of gullies advance into the land by waterfall erosion, both along the main drainage line and branches which are acquired. Subsoil is often less resistant to erosion than topsoil. Water pouring into the gully will cut it into steep banks, undermining the topsoil and causing it to fall. The impact of the waterfall on the bottom gouges holes which accelerate channel erosion.

Waterfall erosion usually produces a gully with a U cross-section. It becomes less active as it approaches the head of drainage and the quantity of water is reduced.

If the subsoil is equal or superior to the topsoil in resistance, waterfalls will not develop, but the gully can progress by extensions of channel erosion.

The third major factor in extending gullies is sloughing off of soil due to alternate freezing and thawing, or following saturation by heavy rains. This process is most active on southern and eastern slopes, and will eat through a field with little regard to slopes or drainage lines.

Continued progress of either waterfall or sloughing erosion depends on sufficiently active channel erosion to remove the loosened dirt.

Once well established, a gully will continue to enlarge even if the surrounding land is planted in erosion-resistant vegetation, as the head and side slopes will undermine the surface.

A gully may advance downstream by channel erosion. More often, it deposits debris in a delta fan at or near its mouth, so that the land is built up. This process is destructive also as it buries topsoil under subsoil.

Damage.   The damage from gullies includes actual destruction of farmland, cutting up fields so that they cannot be worked economically, lowering the water table so that crops dry up, undermining buildings, roads, and bridges, burial of lower lands under barren subsoil, and choking of streams with silt.

An individual gully can do damage amounting to many thousands of dollars, and the national loss from them is in the hundreds of millions. Their control is therefore of great importance.

CONTROL

Control measures after gullying has started may include diverting water to other drainageways, planting, breaking down walls, building check dams, and proper use of the affected land.

Diversion.   Water entering the gully can sometimes be diverted by plowing an arc around its head, as in Fig. 7.16(*A*). The slice should be turned toward the gully to make a dam to back up the furrow.

More often it is necessary to build dams or to dig ditches. Dams are safer, as a new ditch or even a plow furrow may start a new gully unless watched and controlled.

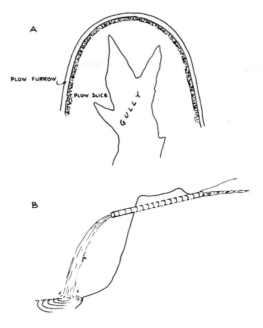

FIGURE 7.16 Diversion of water from gully head.

Where diversion is not practical, waterfall erosion may be checked by conducting the water into an overhanging pipe or flume, as in (*B*).

If water can be permanently diverted from a gully, the walls will eventually break down into stable slopes, which can be covered and held by vegetation.

Planting. Gully growth may be checked by planting. Small gullies may be held by pasture plants or vines; larger ones require shrubs and trees.

The plants should be of types that can grow well in poor soil, have extensive root systems, and will not be injured by partial burial. If the ground is damp most of the year, willows are probably the most efficient, particularly as they will grow from poles or logs secured horizontally in the floor across the direction of flow, thus providing mechanical control while roots and stems are sprouting.

Black locust grows vigorously in poor soil, has a widespread root system, and yields a crop of fencepost material. Many of the pines do well in barren soil, but their roots are not as strong. Vines such as kudzu and honeysuckle are used successfully, though kudzu may take over all vegetation. Whatever plants are chosen, enough soil should be loosened to give them a good grip. Whenever possible, fertilizer, manure, or topsoil pockets should be provided to give them a good start. Animals should be fenced out of the area.

It is usually not practical to plant very steep walls. When vegetation is growing well on the bottom, soil caving from the walls will be held so that the slopes will gradually become less abrupt, and will allow growth of self-seeded plants.

It is often necessary to divert water or to reduce its velocity in the channel before planting can be successful.

Breaking Down Walls. The healing of a gully scar can be greatly accelerated by breaking down the walls into slopes gentle enough to support vegetation. When practical, it is desirable to have the slopes such that farm machinery can cross them in any direction.

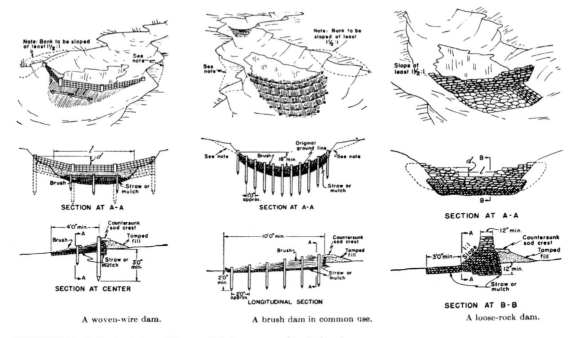

FIGURE 7.17 Gully check dams. (*Courtesy U.S. Department of Agriculture.*)

Very small gullies can be broken down with plows and other farm implements. Somewhat larger ones can be reduced by graders or angle dozers working parallel with the edge. Big ones require dozers pushing the bank more or less straight into the gully, or a dragline pulling it from below.

Gullies may be of such large size—depths of 50 feet or more—that it is not economically practical to grade them in, even with the largest machinery.

The new slopes produced are planted in somewhat the same manner as road banks.

Check Dams.   In order to obtain permanent control, it is usually necessary to slow, stop, or reverse the process of channel erosion. An actively cutting channel will steepen and undermine its banks and the new grading or planting they support.

A channel interrupted by dams or other obstacles will tend to silt up to a higher gradient, reducing the height of its walls and encouraging plant growth. If flow is only occasional, close plantings of bushes in the channel and on its banks may be sufficient. These are often planted in lines across the gully, and protected against washing out by wire stretched across deeply driven posts.

Sod, combinations of cut brush and wire, or stone or logs can be used in building check dams. Some constructions are shown in Fig. 7.17.

In general, any structure which water cannot get under or around, which is tight enough to prevent dirt from going through it, and which is strong enough not to be washed out, will serve as a check dam.

IRRIGATION SYSTEMS

The design of an irrigation system is an engineering problem out of the field of this book. It is in order, however, to mention briefly some of the factors.

Wells.   Water may be obtained from wells, occasionally by natural flow but more often by pumping. Such water, if used in the immediate vicinity, involves minimal piping and distribution difficulties. The most serious problem is the likelihood of using too much water, so that wells must be deepened or new ones drilled to keep in touch with the falling water table. In many localities such overpumping will eventually reach salty or alkaline water unfit for agriculture.

River water may be pumped up to the land to be irrigated, but more often flows onto the land from a higher point in the river. A ditch or canal is cut from the stream, and run horizontally or with a very slight gradient along the side of the valley. The land between the canal and the river is irrigated by gravity flow from the canal into ditches, and from them onto the fields.

Most rivers have a wide variation in level, so that such a canal might be subject to being left dry at times and flooded at others. A more constant level can be obtained by damming the stream below the canal entrance to keep the water high enough to enter it, and by providing gates and a protective dike to prevent flooding. A dam and control gates above the inlet will permit regulation of the river level and inflow.

In dry climates, streams may flow only during the winter and spring or in occasional floods. Water from such a source may be useful for crops which will mature in the wet season or as a supplement to water obtained from wells, but the cost of handling it may not be justified.

Reservoirs.   Such streams may be utilized by building one or more impounding reservoirs upstream. These will permit storing of the heavy flow of some months to be released in drier periods. Large reservoirs of this type may be ideal sites for hydroelectric plants, as the irrigation water can be used to turn turbines on its way through the dam.

Most of the large irrigation systems of the United States depend on storage reservoirs.

Lakes may be used as sources of water. It may be pumped up onto adjoining land or conducted through siphons or canals to lower land. Occasionally, a lake may be tapped by means of a tunnel and used for stream regulation in the same way as an artificial reservoir.

Sediment.   Direct river flow into an irrigation canal carries varying quantities of sediment with the water. The flat gradients and slow motion of the water in the system will cause extensive silting, which will ultimately reduce water capacity, may cause continuous trouble in operation of gates and other devices, and will clog pipes. The design should take this quality of the water into account. Very dirty water requires oversized waterways and steepened gradients, particularly in pipes; cleanout traps or flushing devices for pipes; and access to all ditches for cleaning by machinery or by hand.

Upstream reservoirs catch most, or sometimes all, of the nonsoluble sediment. Any residue may be still further reduced by settling basins.

Dirty water has certain advantages. In most localities it is good for the land, carrying topsoil and minerals which help to replace natural losses. It also tends to seal leaks in canals and ditches so that wastage by seepage is reduced.

Its disadvantage is the greatly increased cleaning and maintenance work required, which is so important a problem that most irrigation engineers will secure clean water whenever possible.

Canals.   The main canal may be an unlined ditch or a ditch lined with special impervious soils or a concrete or other artificial lining. See. Fig. 7.18. If it is carried well below natural grade at any point, it is customary to use a lined tunnel or concrete pipe at such a section.

The unlined ditch is the most economical to construct, and may be used where soils are impervious, where the water carries enough sediment to seal leaks, or where the supply of water is so large that seepage losses are less important than the expense of preventing them.

A ditch may be lined, either in sections or as a whole, by placing a layer of clay or impervious silt on the bottom and sides. This may be placed in a dry ditch, or fine clay may be added to the irrigation water until it has coated the canal.

A ditch must have bank slopes which will not cave or slump into the water. Vegetation may be used to hold dirt banks and will permit steeper slopes, but its consumption of water is often excessive. Banks should be protected from erosion by surface water.

Concrete lining is expensive construction, and if properly done is most satisfactory. Such leaks as occur will generally be localized and comparatively easy to find and repair.

FIGURE 7.18 Concrete lining in irrigation canal. (*Courtesy U.S. Department of Agriculture.*)

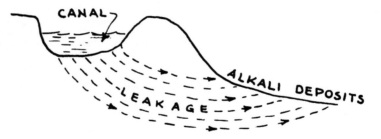

FIGURE 7.19 Damage to land from canal seepage.

Damage from leaking canals is not confined to the water loss and the breaking down of the canal structure. Farmland near the leaks may be rendered unusable by excessive water and resulting alkali deposits, unless subdrainage is provided. (See Fig. 7.19.)

Pipes. Either ditches or pipes may be used to distribute the water from the canal. Ditches are more economical to construct, but cut up the land, are expensive to maintain, form breeding spots for insect pests, are a hazard for children, and may allow substantial water losses by seepage and evaporation.

Pipes may be concrete, glazed tile, plastic, steel, or some composite material. The first two sometimes separate at the joints when filled with cold water, either after lying empty or being used to carry warm water. Iron pipes may corrode rapidly due to alkalies in either the soil or the water.

Various methods have been developed of joining concrete and of treating iron pipes, which reduce the difficulties mentioned. On the whole, pipes require less maintenance and waste much less water than ditches. However, they must be laid on steeper grades to prevent silting. If they become blocked, they are very expensive to clean.

Large farms or ranches may control an entire irrigation system, but water is usually distributed to a number of users by a water district or other government agency, or by a water company. One water gate or weir is usually provided for each land unit, and distribution within that unit is cared for by the farmer.

This may be handled by running a pipe line along one edge of the field, with standpipes and valves at 10- to 20-foot intervals. If the field is long, several distribution lines may be used, connected by a main.

Ditches may be used instead of pipes, but they require more maintenance and may interfere with tillage.

The *direction of irrigation* is the direction the water flows on the surface of the field from the distribution pipe or ditch. It is generally at right angles to the pipe.

LAND LEVELING

Slope Patterns.   Land leveling may be divided into six classes, according to the result obtained:

1. Spot grading
2. General downward slope away from water supply—for sprinklers
3. Uniform grade in direction of irrigation
4. Uniform grade in direction of irrigation and at right angles to it
5. Uniform grade in direction of irrigation and exactly level at right angles to it
6. Exact level

Spot grading consists in removing humps or filling hollows, without establishing a uniform grade in any direction. It is sometimes done in advance of better leveling for irrigation, and is of general use to make possible faster tillage and more even production.

If water distribution is to be by means of sprinklers, perfectly uniform slopes are not required. For water distribution, it is only necessary that the land have a general slope down from the source of water. In climates where deep freezing of soil occurs, the slope should be uniform enough to make possible drainage of sprinkler pipe laid at a fairly regular depth.

When the water reaches the individual plants by flowing on the surface of the ground, it is necessary to have an almost uniform slope in the direction of irrigation. The steepness of slope may be determined by the character of the soil, the crop to be planted, the original grade, and the rate of water use.

Economies may be affected on many plots by leveling only in the direction of irrigation, and following the original profile at right angles to it. This type of job is used chiefly in orchards which can readily be cultivated into ridges that will regulate water drifting across the field.

Choice between the fourth and fifth methods will depend largely upon economies in working over the natural grade. In very large fields, the two-way slope will facilitate movement of water through the cross-distribution pipes. The cross grade should be so slight that even light ridging will prevent sideward drift of the water.

Entirely level plots are usually limited to rice fields, and alfalfa and other crops which can tolerate flooding.

Flow and Absorption.   The rate of water flow and absorption should balance, so that water will reach the lower end of the slope in sufficient quantity for a crop without flooding or running off. In practice this balance may be difficult to achieve, and provision is made for draining off excess water when necessary.

Increase of gradient will accelerate the flow and decrease the rate of penetration. Light sandy or gravelly soils absorb water rapidly and require steeper grades than clay, which may be almost waterproof unless freshly loosened.

The maximum gradient would be below that which will cause the soil to wash and gully during irrigation or heavy rains. The minimum is flat.

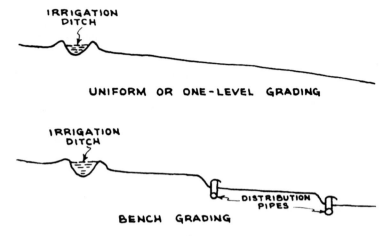

FIGURE 7.20   One-level and bench grading.

If the maximum practical gradient is not sufficient to move the water the length of the field, additional distribution lines can be installed. See Fig. 7.20. Very porous soils require pipes and sprinklers.

Figuring Gradients.   Earthmoving should be kept to a minimum to save money and to conserve topsoil. If conditions justify the use of any one of several slopes, the one will be chosen which conforms most closely to the natural topography.

If the steepest possible gradient is so much flatter than the original grade that excessive earthmoving will be required, because of deep cuts at the top and high fills at the bottom, the field may be divided into two or more levels or benches. These will have the desired slope and will be separated by steep banks. A separate waterline is required for each bench.

The high corner or end of the field must be below the level of water in the ditch or its head in the standpipe.

The new gradient must usually be placed at a level, or levels, where cut and fill will balance, as cost may be greatly increased by bringing in borrow or dumping surplus soil. After the surrounding areas are largely under cultivation, borrow or disposal may not be possible.

Topsoil is not a problem in many arid valleys where soil is fertile to a considerable depth. However, when topsoil is thin and rests on layers of soil which are infertile or hard to work, or when the surface soil differs sharply in character from that underlying it, the cuts should be kept shallow.

Stakes.   Stakes set in a square grid at 100-foot intervals, and on high and low spots, are used in measuring the original surface and for marking the new grade. One or more benchmarks are set outside the grading area, as discussed in Chap. 2.

The new grades are marked on the stakes in any convenient manner. On fills, time may be saved by tying strips of cloth on grade lines to enable operators to see them without getting off their machines. Where soils are loose, two stakes may be used to advantage—one hammered down to ground level, the other left to project 2 or more feet. Cut and fill are figured from the top of the lower stake and marked on the upper one.

Clearing.   Much of the growth in arid regions is of a light and brittle character which breaks up during grading and mixes with the soil, and has value as a binder and a source of plant food that is more important than the slight difficulties it causes during finishing. This group includes the various kinds of sage, tumbleweed, and many of the smaller cacti.

However, larger shrubs and trees such as mesquite, greasewood, and acacia, the presence of which often indicates good soil, are tough and deeply rooted, and their removal requires heavy machinery. In thick stands of these plants, clearing is more expensive than grading.

Such growth can usually be piled and burned immediately after removal. The leaves and sapwood are resinous, and the dry soil sifts out of the piles so that they burn readily. Heavy trunks are more difficult to ignite, and if only a few are present, it may be easier to haul them to a dump.

Clearing methods and machinery are discussed in Chaps. 1 and 21.

Savings may be effected by cutting the trees flush with the ground wherever the fill will be deep enough to permit tilling over them. However this is not recommended, as it will prohibit the future use of pan breakers or other deep tillage tools and will add greatly to the expense of installing underdrains if they should become necessary. The same objection applies to burying logs in the fills.

Wind Damage. Clearing and grading should not be started until irrigation water is available. The native vegetation, even when very sparse, has some power to break the wind and hold the soil. The weathered ground surface usually has a crust which resists wind scour. These natural protections are destroyed by the work, and unless water can be put on and a holding crop started immediately, the best part of the soil may be blown away or piled into dunes that may be more costly to level than the original surface.

Wind damage during the work may often be avoided by choosing a season in which windstorms are infrequent. If this is not possible, the final leveling and planing should follow immediately behind the rough grade, as a perfectly smooth surface is much more resistant to scour and dune formation than one having ridges or tracks of machinery on it.

If such a planed surface becomes roughened by wind, it should be replaned before the next storm, and kept flattened until a crop can be grown.

Machinery. Dozers are used to clear, to take the tops off ridges and dunes, to bevel steep slopes, and to fill in pits; for cut-and-fill work on short pushes; and for pusher work with scrapers.

A drag leveler can be used to smooth out rough spots wherever it is possible to walk the tractor over them. It can transport soil long distances, although its efficiency diminishes rapidly over 200 feet. Compared with the dozer, it has the advantage of making a wider cut, with little tendency toward scalloping, and has a greater transporting capacity and speed. Compared with scrapers, it has greater stability against overturning, smooths a wider area with each pass, cuts down and fills more quickly, and can be dumped promptly if the tractor gets stuck. It has a smaller transporting capacity and generally will not make as smooth a grade.

On any large area, the bulk of the dirt moving is most efficiently done with scrapers. Because of the width of most of these grading jobs, and the small slope of the land, these can be used in almost any pattern preferred by the supervisor or operators.

Grading. One technique is to produce a rough finish grade in the high corner of the field and to expand this grade as continuously as possible. Where necessary, spot grading operations are done beyond this area in order to secure fill or to dispose of surplus.

Economies may be effected by loading the scrapers toward adjoining depressions so that the soil pushed along in their efforts to load will fill them. If the soil is very loose, this may be more important than loading in the direction of the dump, but it is often possible to do both.

Fills are usually made in thin layers in order to get maximum compaction from hauling equipment, as rollers are seldom used. Tamping rollers will produce a more permanent grade where fills of more than a foot or two are required, but close competition for the work may not permit the necessary increase in price.

Rough-graded sections may be settled by flooding before doing finish work.

It is often possible to keep the short-range equipment, dozers and drag scrapers, working all the way through the job if the area contains a number of adjacent humps and hollows. The drag level can also do the light finishing more economically than the scrapers.

Finishing Off. As soon as any considerable section is rough-finished, grades are rechecked and additional stakes may be placed. These can be on 50-foot centers both ways, or 50 feet one way and 100 the other, or at the intersection of diagonals across the original squares. Any changes of grade necessary at this stage can usually be made by the drag level.

If a small error has been made in balancing cut and fill, the gradient in the lower part of a field or bench may be increased or decreased slightly to change the quantity needed to balance. Large errors may require lifting or lowering the whole field, or one of its benches, or making arrangements for disposal or borrow outside the plot.

When the entire field, or a large section of it, has been brought as near the finish grade as is practical, all stakes are removed and the job is finished with a land plane. See Chap. 17. This will flatten ridges and hollows around stakes, plane off spill windrows, piles, and track marks, and even local inaccuracies at hitting the grade.

Planing also serves as a maintenance operation and is sometimes repeated after each harvest.

Distribution pipes are usually laid immediately after completion of leveling. Ditches should not be dug until the pipe is on the job, as drifting soil can fill it very rapidly. It is usual to lay and cement the pipe, to partially cover it by hand, and to allow it to cure. This fill is removed by hand where standpipes are installed. The ditch is then backfilled and graded over by machinery.

IRRIGATION DRAINAGE

Alkali.   All soils contain some salts, in both soluble and insoluble forms, and these are necessary for plant growth. In rainy climates the soluble salts tend to be leached out of the soil and carried away in underground water about as rapidly as they are made soluble by plant action and weathering, or added as fertilizer.

In dry climates where there is not sufficient rain to leach them effectively, they accumulate in the soil, accounting for the great richness and productivity of the land. However, in flat low areas and some other places, the salts, then known as alkali, may be concentrated so heavily that they kill plants instead of aiding them. Alkali may also appear as a surface crust where groundwater comes to the surface and evaporates.

Underground Pools.   Where soaking rains are rare, natural underground drainage tends to be poorly developed or nonexistent. If such an area is irrigated, water absorbed in excess of that required by the crops will accumulate in a stagnant underground pond whose top may rise close to the surface.

This water will dissolve minerals on its way down and while lying underground, and usually becomes so alkaline that it injures or kills plants which absorb it. If it does not become loaded enough to do this, it still may injure plants by drowning their lower roots. Also, when the water table is near enough to the surface that capillary attraction will lift it to the surface—a short distance for granular soils, a long one for fine-grained soil—its evaporation will form an alkali crust.

When such a stagnant or semistagnant pool forms, the land above it usually becomes unfit for crops; and even if irrigation is stopped and the water slowly drains away, the alkali deposits in the soil may render it unusable. Artificial leaching would reestablish the underground water.

Drainage.   The area can usually be put back in production by the installation of an adequate system of drains. These will serve to lower the water table below the trouble line, or to give the water enough flow toward the drains that it will not stay in the soil long enough to become alkaline.

Such drains preferably are deep, 6 to 7½ feet being usual, and spaced from 75 to 800 feet. Close spacing is for impervious soils, wide for granular ones. However, for subdrainage purposes, the porosity of the soil cannot be judged from casual inspection, or even by analysis of samples. Heavy impervious clays often respond readily to tiling because they are filled with fissures, either open or sand-filled, which conduct the water. Many really tight soils will not require drainage because of their refusal to absorb the irrigation water.

Some irrigated lands are composed of alternating layers of heavy and porous soil, which are in the form of lenses tapering to nothing on each end, so that natural drainage must move through both types of soil. Ditching cuts and drains the porous lenses.

The tile lines which do most of the work of draining are called laterals, and the larger pipe into which they empty is called the baseline. Four- or 5-inch laterals and 6- or 8-inch baselines are usual. Sizes are ordinarily much smaller in proportion to acreage than in nonirrigated fields, but layouts are similar.

If the problem is water leaking from an adjoining canal, an intercepting drain should be placed parallel to the canal, at a distance of 50 to 70 feet. Water may leak under it if it is too close or too shallow.

It is best practice to lay all drainage tile on gravel, and under tar paper and gravel, as described in Chap. 5, as the effective life of the system will be many times that of plain tile. Since tiling is generally done in saturated land, a tile box should be used to avoid danger of cave-ins.

Moles.   A mole or mole-ball, Fig. 7.21, is a ripper accessory that will open a drainage tunnel in a plastic soil. The type illustrated is a torpedo-shaped piece of iron, attached to the heel of the standard. As it is pulled through the ground, the mole presses the soil outward with great force, leaving an open tunnel with a firm lining. Seepage from the surface is aided by soil breakage.

FIGURE 7.21   Mole ball. (*Courtesy of John Deere.*)

Gradient is determined by surface slope and the route taken across it. It should be between 6 inches and 2 feet to 1,000 feet, and in no case should be steep enough to permit erosion in the tunnel. A piece of tile should be placed in each outlet to protect it from erosion or stoppage.

This device is effective only in soils which are damp and plastic enough to be molded, and not soft or loose enough to flow or cave into the passage. An uneven surface or stones in the soil make it difficult or impossible to run it at an even gradient.

Mole drains may not work at all, may stop up in a couple of weeks, or may give satisfactory service for years. Occasionally their good effects are accomplished as much or more by the incidental breaking of an impervious hardpan than by the drains themselves. The work is much more economical than tiling.

Leaching.   After a field has been subdrained, alkali can be leached out of it by repeated soaking with irrigation water. This dissolves the chemicals and removes them through the tile lines.

CHAPTER 8
ROADWAYS

Roads are of many kinds, from cart tracks to superhighways. The importance of the simpler types is often overlooked. They are essential in their own places, and they show principles that are basic to the more elaborate highways.

ROAD TYPES

Pioneer Roads.   Pioneer roads are access roads built along the route of a highway, pipeline, or other heavy construction project to allow the movement of equipment to and between different sections of the job. If such a road is required, it should be the first work undertaken; and any delays in cutting it through will slow the starting of the job and may keep workers and equipment idle.

It is best to locate it sufficiently to the side that it will not be blocked or cut off by the main work, and if it must cross the construction strip, it should do so where it is close to subgrade.

The importance of the pioneer road decreases as sections of the main road become passable for trucks, but it often retains at least emergency or detour value until the job is finished.

If it is to be used only for moving in equipment, it may be narrow, crooked, and steep for the sake of economy or haste. Specifications written, and the route surveyed or walked through for it, serve as guides rather than instructions, and the job supervisors are usually given wide latitude in altering them for the sake of speed or economy.

Pioneer roads are most often needed in mountainous and timber country where severe obstacles hinder cross-country travel. Where fill is available, trees are cut flush and the stumps buried; otherwise they are uprooted and the holes graded in. Topsoil is handled as fill.

Rock is avoided as much as possible in the layout of the road, and when found is often buried instead of blasted. If an excessive amount of rock must be moved, it may be economical to place the pioneer road in the route of the highway, as the cost of the separate blasting may outweigh the advantage of the independent road.

Grades follow the land contour as closely as possible. The maximum grade will depend on the use. Shovels, tractors, and lightly loaded trucks should be able to negotiate grades up to 30 percent, but serious delays can be caused by stalling of weak units, or as a result of skidding. Ten to 15 percent grades are more practical.

Curves should be wide enough to enable the longest units to get around them somehow, and the machines in steady use should be able to make them without backing. Attention should be paid to the lane width needed, so that inside rear wheels will not run off the road. Width requirement increases with length of wheelbase and sharpness of turn.

The road width is determined by its intended traffic, construction problems, and haste. It is desirable that it be two lanes wide, but often this is not practical. On steep slopes, two one-way roads may be constructed, one above the other.

Two-way traffic on one lane will require turnouts at 100- to 500-foot intervals. It is best to make these of two-lane roads the length of two vehicles, but deadend turnoffs may be easier to build and will serve the same purpose. A vehicle may turn into one, and back out on the road again when opposing traffic has passed.

Small streams are best bridged with corrugated metal culvert pipe and fill. Occasionally, bottoms may be hard enough to permit easy fording.

Fords are the most economical means of crossing larger streams. A soft bottom can sometimes be made safe by a rock fill. Its downstream edge should contain heavy boulders. Its surface can be crushed rock or clean gravel. It may be placed over a culvert pipe that will handle normal stream flow but not high water.

If a ford is not practical, multiple-culvert pipe, log or timber bridges or trestles, or prefabricated steel bridges may be used.

Roads built for use in a dry season may be so constructed that they will be washed out when the rains come, if the contractor believes they will have served their purpose by then.

The bulldozer, or angle dozer, is usually the primary tool for cutting a pioneer road. Methods are described in a later section.

In sidehill cuts, the road surface should slope down to the bank or inner side, and may have a berm (ridge) along the outer edge. This shape allows for fill settlement, reduces washing of the fill slope, and decreases danger of sliding off the road.

Drainage from the road surface and the hill slope is carried along the inside bank to culverts, or to outward-dipping sections of road reinforced with rock or blacktop. Overhangs or sluices must be provided to carry the water across the fill.

One of the constructions used by the U.S. Forest Service is shown in Fig. 8.1.

Access and Farm Roads. A pioneer road is an access road for each otherwise isolated piece of the job it services. However, the term access road usually means a road by which a whole job is connected to a highway system, and is generally used in connection with pits and dams.

The quality of construction is variable. If the project is small, or to be quickly finished, and no substantial amount of raw material is to be trucked in or products taken out, rough pioneer construction may suffice. More often, it must be built as a haul road. Occasionally, a first-class highway will be required.

Farm roads are usually graded native soil, two lanes wide, with gravel, dirt, or other low-cost surfacing.

Haul and Logging Roads. There is no sharp distinction between these two types. Both must carry heavily loaded trucks at a good speed, and are ordinarily located according to a favorable terrain, rather than property lines. The logging road is likely to be longer, to climb to much greater elevations, and, under modern lumbering practice, to be permanent. The haul road will carry a much greater traffic for a limited time, and then often will be abandoned.

As compared with the pioneer and access types, these roads differ in that grades are limited. Ten percent is the usual maximum for the logging road, and in haul roads grade is sometimes kept as low as 3 percent of climb in the direction of load movement. Culverts and bridge capacities are designed according to the period of use, and the comparative expense of large openings, or repairing washouts over smaller ones.

The long climbs needed on log roads in mountainous country are best ascended at even grades, which can only be attained by careful survey of possible routes. Where the direct distance along a valley wall is too short to provide the ascent at the required grade, the road may be run back into spur valleys instead of crossing them on trestles, or may ascend the slope in a series of switchbacks, or hairpin turns. The turns require a wide space, which, for economy, should be placed where the grade is flatter than ordinary, or where excavation will require minimum blasting.

These factors limit the route rather closely to that originally surveyed, although occasionally, if the contractor runs into unexpected difficulties, he or she can have the road shifted to avoid them.

The haul road seldom has long ascents and descents, but switchbacks and side wanders must often be used to get them out of a deep pit or over a massive ridge.

Logging roads are surfaced with local material where possible, from cuts or borrow pits along the road. Any fairly hard and porous material, such as gravel, disintegrated granite, or broken

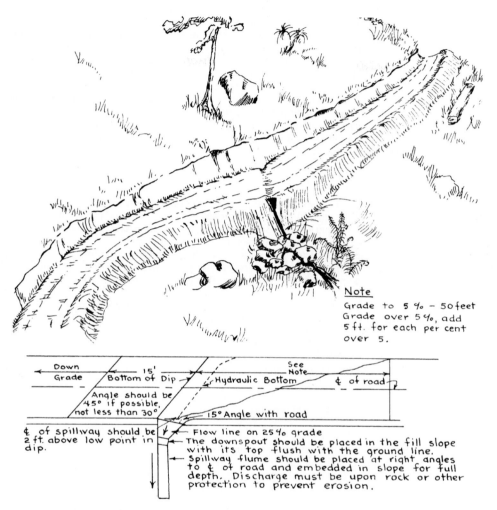

FIGURE 8.1   Hillside roadway. (*Courtesy of U.S. Forest Service.*)

shale can be used, as traffic is ordinarily not long sustained. Haul roads may be oiled to control dust and speed traffic.

Trouble with snow or ice is minimized by locating on the north or east slopes of valleys, in the northern hemisphere.

Development Roads.   Roads built for real estate subdivision vary in quality from the crudest pioneer type to city streets. Differences depend on the type of development, local regulations, value of land, capital available for improvement, terrain, and other factors.

Rural subdivisions are seldom regulated, but those in and near cities may have to have roads built to high standards. However, the developer may be allowed latitude in locating roads, or shifting them to avoid obstacles, to run cuts through banks of desirable fill or gravel, to change lot lines, or to obtain a more attractive appearance.

Subdivision roads may be financed partly by sale of topsoil, gravel, fill, and other surplus material. Construction costs may be reduced and swampland "reclaimed," except where area must be kept as a wetland, by using such areas as dumps for quarry waste or other clean, solid fill.

Fills made over swamps are subject to severe settlement. For a good-quality road, mud should be removed for use in landscaping lots, so that road fill may rest on a firm bottom. Or geotextile blankets can be used as described in Chap. 3 and depicted in Fig. 3.12.

City Streets.   City streets are built to exact specifications, often under circumstances which do not allow maximum output from either machines or workers.

All operations are likely to be impeded by traffic, which will probably require working the job in sections limited to a few blocks, and frequently to half the street width. Provisions must often be made to pass traffic on intersecting streets through the work. In addition to direct interference with work schedules, congestion will probably delay trucks and machines entering and leaving the job.

Removal of old pavement is usually the first construction step. Asphalt paving, on a gravel or crushed-rock base, can be dug by most front loaders, or backhoes. Occasionally it is hard enough to require preliminary breaking with a ripper or scarifier, or direct loading with a large excavator.

A backhoe can dig close to manholes, but care should be taken not to hook into them, or into a widened masonry base, as these are easily broken or crushed. Pavement chunks sliding up on the manhole cover may be thrown into the bucket by hand.

Concrete pavements and bases are tough, particularly if reinforced. They may be bonded to the manholes or their bases so as to require breaking away by air hammers, ahead of backhoe digging. They break out in big slabs which are difficult to pick up in the bucket, and to dump out of a small or medium truck.

Soil beneath the pavement is removed with it to required depth. It may be native soil, or rock, dirt, or even garbage fill. It may be honeycombed with pipes and conduits that may belong to the city, or to various utility companies.

If the grade is to be lowered, some of the pipes may have to be dug in more deeply. In any case, extensive repairs, enlargements, or relocations of piping are liable to be done between the removal of the old pavement and the laying of the new. This will involve a lot of ditching and probably considerable delay.

The subgrade is graded and compacted according to specifications. Because of interference with manholes, and the need for working in short sections, a large amount of handwork will probably be required.

Highways.   Highways make up the bulk of the excavating contractors' road work. Modern standards of width, grade, and alignment require heavy cuts and fills in rough or rolling land, and grading and compaction of subgrades involve heavy work on any terrain.

Contracts may be let on a basis of a fixed price for a job; a fixed price plus specified extras, such as allowance for overhaul, rock blasting, slides, or other difficulties whose extent cannot be conveniently estimated in advance; or on a price-per-yard basis. Less frequently, they are constructed on a cost-plus or equipment rental arrangement.

Highway earthwork jobs may involve widening and straightening of roads, building a new road in the approximate location of the present one, building a new road which will run along or cross the old one only occasionally, or building a totally new road crossing undeveloped country. There are of course no definite lines of distinction among these types.

A requirement of most highway construction is to provide for continuance of traffic along any roads running along or crossing the job. This may be a controlling factor in job sequence.

Airports.   An airport runway is essentially a very wide, short, straight road. It is usually located on the flattest land available, but deep fills are often required.

Banks of cuts must be graded back to very gentle slopes to avoid choppy air currents. Borrow is frequently obtained from the glide areas at the ends of the runway. It is standard practice to cut away any ridges which might be hit by a plane climbing slowly off either end of the runway.

The runway may have a level centerline, crowned up from the sides slightly for drainage, or have a flat cross section and a longitudinal slope. In either case drainage slopes are very slight, and the surface must be exactly on grade to avoid puddles.

Taxiways and plane parking areas are roads surfaced to an ample width to carry the wheels of a plane running on the ground. Additional areas on each side must be cleared and lowered to allow clearance for the wings.

Airport subgrades and pavement may have to exceed standards for heavy truck highways if maximum size planes are to be carried.

Road Markers. A construction or mining road should be plainly marked as such, to prevent accidental entrance by motorists. Cars and big machinery do not mix well, and also drivers who are lost or confused may get in the way of a blast or run off a cliff.

Warning signs should be placed on highways at least 400 feet on each side of a haul road crossing. If either road is a busy one, the intersection should be protected by a flagman or a traffic light.

One-way haul roads should be marked plainly and frequently with direction signs. There should be a sign wherever any vehicle could enter, and additional signs along the roadways to warn drivers going the wrong way. If the road is paved, arrows should be painted on the pavement, pointing in the direction of travel. A conventional dashed stripe line down the center could have arrowheads painted on some of the dashed lines.

Signs at entrances to one-way sections are not enough, as they may be destroyed by accident or vandalism, or obscured.

Failure to provide sufficient notice and warning of traffic direction is the cause of many head-on crashes on divided state and federal highways. There the blame is put on the wrong-way driver, but in a private construction road the contractor is likely to be held responsible.

ROADWAY LAYOUT

Roadways are planned and staked out with consideration for horizontal alignment, vertical alignment, and cross section.

Horizontal Alignment. Horizontal alignment is the route as it would appear on a map, with detail enough to enable field engineers and contractors to lay out and build the road exactly as it was planned. It is figured in terms of the location of the centerline.

Curves. Curves are laid out as arcs of circles. Each point on such an arc is equally distant from the center of the circle that would be formed by continuing the arc on the same curve. See Fig. 8.2.

A curve is described or defined by its degree of curve or by the length of its radius. Its degree of curve is the number of degrees in the angle at the center that is made by drawing lines from the center to points on the curve that are 100 feet apart. A sharp curve will have a higher degree of curve than a gradual one.

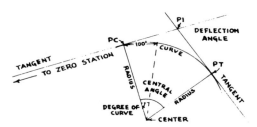

FIGURE 8.2 Curve and tangents.

Curves may also be defined according to the length of the radius, that is, the distance from any point on the curve to the center. The radius of a 1° curve is 5,730 feet. A 2° curve will have half that radius, or 2,865 feet. A short radius means a sharp curve.

Either measurement may be converted into the other by using one of the formulas:

$$\text{Radius} = \frac{5,730}{\text{degree of curve}}$$

$$\text{Degree of curve} = \frac{5,730}{\text{radius}}$$

If someone were in a hurry, or trying to work the problem out in his or her head, the person could change the 5,730 to 6,000 and be less than 5 percent off.

A highway curve may be compound. A compound curve includes two or more arcs having different degrees of curve, and may include some short straight lines also.

Tangents.   In highway work and in many other surveys, straight lines are called tangents. This is because they are tangent to the connecting curves. This one word makes clear that a straight line forms a smooth continuation of an adjoining curve.

Tangents vary in length from a few inches up to many miles.

Whenever two tangents are joined by a curve, lines are drawn on the plan continuing them until they cross each other outside the curve. The point where they meet is called the point of intersection, or PI for short. The tangent distance is measured between the point where the tangent meets the curve (called PC at the beginning of the curve, or PT at its end) to the PI. The side or deflection (def.) angle formed by the intersection of the extended tangents equals the central angle of the simple curve between the tangents.

Baseline.   In making a road survey, the engineers first lay out a baseline that follows the general route of the road, but that may be partly or wholly outside of the right of way. This line is often made before the exact location of the highway has been decided.

The baseline is very carefully surveyed and marked. Some points on it, called hubs, are more important to the engineers than any of the road line stakes. They may be conspicuously marked, or concealed with leaves or rocks.

No contractor or employee should ever destroy, move, or otherwise interfere with any stake or marker on the job, regardless of whether it seems to fit in with the markers he or she recognizes.

Centerline.   The centerline is the basic location reference for the highway itself. It is the center of the pavement in a single road, or the center of the median division of a dual highway whose two roadways are a fixed distance from each other.

The engineers set the centerline according to angles and distances from baseline points. Measurements are made along it with a steel tape (an operation often called chaining), and stakes are set at 100-foot intervals.

All distances are measured along the centerline, and structures and stakes are located in reference to it. It is also the basis for grade calculations for single roadways.

There are a number of construction lines that run parallel or almost parallel to the centerline. These include pavement, shoulder, gutter, and slope edges. They are usually measured off from the centerline, at right angles to it on straight stretches, and along radial lines on curves.

Profile.   The profile of a road is the vertical alignment of the centerline or of a theoretical grade line. It is a representation of its rise and fall, without indication of whether its route is straight or curved.

Two profiles are prepared, one of the existing ground surface, the other of the proposed pavement surface. Both of these are drawn on one sheet or roll of cross-section paper marked off in squares of $\frac{1}{10}$ inch, with inch squares indicated by heavier lines.

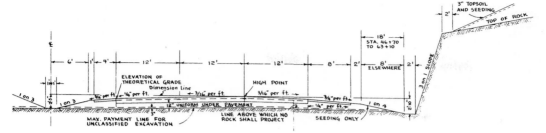

FIGURE 8.3   Typical cross section.

The usual scale is 100 feet to 1 inch horizontally and 10 feet to 1 inch vertically. The exaggeration of the vertical scale is necessary because the ups and downs are usually quite small compared to the horizontal distances, and would be hard to measure accurately on a small scale.

The road profile is made of a series of straight lines or grades connected by curves. These vertical curves are usually arcs of parabolas, not circles. Plus grades go up as they go away from zero station, and minus grades go down.

The ground profile is prepared from topographic maps, often made by stereoscopic photography from the air. Profiles may be prepared for several possible routes, and highway profiles sketched along them. A ground survey is made along the selected route to serve as the basis for final plans.

A rough estimate of the volume of cuts and fills can be made from the profile, but accurate determination usually requires cross sections showing side slope of the ground, slopes proposed for highway cuts and fills, and other details.

Cross Sections.   There are two types of highway cross section. The plans usually include a set of "typical road sections" that show the details of pavement width and thickness, shoulder and gutter width, crown or side slope, and other construction information. These typical sections serve as guides in staking out and building the road. See Fig. 8.3. The cross section for the roadway can be achieved without numerous stakes by using the TS5 instruments, described in Chap. 2, to enable the operator on the grader to get the required shape.

An ordinary cross section is a profile taken at right angles to the centerline. It is at least long enough to include the full width that will be graded. It is usually taken with a transit, but for rough calculations a hand level may be satisfactory.

The number of cross sections taken depends chiefly on the irregularity of the ground. In hilly country they are taken at each 100-foot station, and at additional points where the ground surface changes. On perfectly flat land only one or two might be taken on a whole project.

The cross section of the ground is drawn on cross-section paper with the vertical and horizontal scales the same, or the vertical scale exaggerated. Then the proper typical road section is selected and its subgrade line is drawn in, on the same scale and in proper location. Wherever present grade is above proposed subgrade, material must be "cut" or dug and removed. Where present grade is lower than subgrade, material must be added or filled in.

Such cross sections provide data to figure the cut and fill for the road.

STAKES

Stakes are used to guide the contractor and employees in following the engineer's plans. They also assist inspectors in checking up on the contractor's performance. (See Fig. 8.4.)

The first working stakes on the job may be the centerline, showing depth of cut and height of fill, and slope stakes that show the outer limits of the area to be cleared, grubbed, and graded, and usually the cut and fill information also. Some of these stakes would not be necessary if the TS5, mentioned above, is to be used.

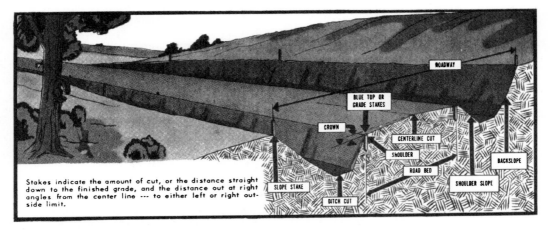

Stakes indicate the amount of cut, or the distance straight down to the finished grade, and the distance out at right angles from the center line --- to either left or right outside limit.

FIGURE 8.4 Stakes for cut.

When heavy cuts and fills are required, most of the work may be done with guidance of only slope stakes, both the originals and others that are set up or down the slopes as work progresses. (See Fig. 8.5.)

When the working levels approach the subgrade, additional stakes are needed. Centerline stakes will be restored, and lines of shoulders and gutters may be staked.

Finishing may be done with blue tops, which are stakes driven until the tops are at the grade desired, usually subgrade, and/or string on shoulder stakes.

Centerline. The centerline is usually staked at 100-foot intervals in preliminary work, and sometimes as closely as every 25 feet in narrow, winding roads or in finishing operations.

These stakes are called stations. The first one, the zero station, is at the beginning of the road or other project. The distance in feet from zero is marked as a double figure.

Stakes at 100-foot intervals are called full stations, others are called plus stations. Station numbers are made up of the distance from zero, with the hundreds divided from the last two figures by a

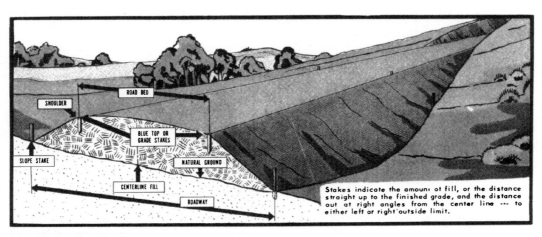

Stakes indicate the amount of fill, or the distance straight up to the finished grade, and the distance out at right angles from the center line --- to either left or right outside limit.

FIGURE 8.5 Stakes for fill.

plus sign. For example, a full station 500 feet from zero is 5 + 00, and the part or plus station at 545 feet is 5 + 45. A distance of 3,456.2 feet from zero is station 34 + 56.2.

If changes in plan should cause the project to be extended to the other side of the zero point, minus stations would be used. A stake 180 feet beyond zero would be station −1 + 80.

Location measurements refer to distances on a horizontal plane, unless specified otherwise. As a result, 100-foot stations will appear to be more than 100 feet apart if measurement is made along steeply sloping ground. On a 1-on-3 slope ($33\frac{1}{3}$ percent grade) the surface distance would be about 105.4 feet. But if the stakes were extended upward or downward, any horizontal line between them would be 100 feet long.

Centerline stakes are in the middle of a single roadway, and usually in the median of a double one. They are marked L, C, or /c. They show depth of cut or height of fill needed, and usually carry information about location of culverts, structures, and other features.

Slope.   Slope stakes are set where the outer slopes of the cuts or fills meet the original grade, usually at 100-foot intervals along the roadway, and also at other points where ground slope changes or special structures affect the slope. They are always at points of no cut and no fill.

Slope stakes are usually set with a transit or dumpy level and a 100-foot steel tape. They provide the first markers the work crews need, as they show the outer limits of the area to be cleared.

Each stake should show the cut or fill necessary to make the ground level with the centerline at that point, and the distance to the centerline. It should also show the steepness of the slope, but it often does not. If this is the same for the whole job, the grading supervisor can carry the information in her or his head.

Slope and other side stakes are usually marked with the station number, the distance from the centerline, and the direction of the centerline. Direction is indicated by the letter R for right or L for left. Such directions for plus stations are read looking from the zero stake. In some localities 25-R means 25 feet right of the centerline, in others that the centerline is 25 feet to the right of the stake.

A slope that is not known can be figured by subtracting one-half the width of the road (including a gutter, if there is one) from the distance from slope stake to center, and dividing the remainder by the cut or fill measurement.

For example, if the distance from slope stake to center were 28 feet and the half width of the road were 24 feet, the width of the slope would be 4 feet. If its height were 2 feet, the slope would be 1 on 2.

As a cut deepens or a fill is built, it is usually necessary to check the slope with new stakes. These may be set from the original with a string or carpenters' level, a rule, and a plumb bob. (See Fig. 8.6.)

Reference (Offset).   Stakes on areas to be cut will be dug away, and those inside fill lines will be buried. In shallow cuts, stakes can be left temporarily in islands; and in shallow fills, long stakes may be used which will project from the top unless they are knocked over. Slope stakes are liable to be undercut or buried. Any stakes are apt to be moved by accidents, particularly if the ground is stony or frozen.

It is therefore desirable to set reference stakes well outside the work lines to simplify resetting of the work stakes.

Such stakes may be set on one side or on both sides. They are marked with the station and the distance from the centerline, and may be identified by lettering such as OFF or REF.

If the road strip is narrow or of moderate length, reference stakes on both sides will permit replacement of the working stakes simply by measuring between the offsets.

Where trees or heavy rocks are near the road, nails may be driven into trees, or marks chiseled on rocks, on opposite sides of the road, and a tape stretched between them. The reading at the centerline and ends and the station are noted. With these notes it should be possible to find the center again quickly and accurately.

When a few centerline points can be found from side references, it is often possible to sight in the rest of the missing stakes by eye with reasonable accuracy.

Grades may be marked on offset stakes, or a separate system of benchmarks may be used.

Surveyors often set a line of offset stakes instead of a centerline, leaving the center and other stakes to the contractor's crew.

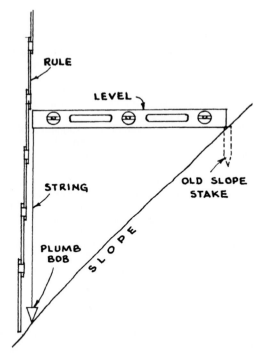

FIGURE 8.6   Checking a slope.

Grade.   Grade stakes show the distance that the ground surface is above or below a desired elevation or grade. Vertical distances to grade are marked on the stakes in feet, inches, and eighths of inches, or in feet, tenths, and hundredths of feet. Figures are preceded by the letter C, for cut, if the ground is high and must be cut or dug away, or by F if it is low and must be filled.

Cuts and fills may be figured from the base of the stake (ground level), from its top, or a line drawn on it. Any basis except ground level is confusing to operators and may cause serious mistakes. However, ground level should be marked in case soil falls away or is added without disturbing the stake.

If the fill is less than the height of the stake, the grade may be marked directly on it with crayon. It is an excellent practice to tie a rag around the mark to make it readily visible to the operator.

Shallow cuts may be marked temporarily with rags a specified distance, such as 1 or 2 feet, above grade, so that operators will not have to dismount to read the figures.

A great number of rags can be made of one old sheet by tearing it in narrow strips. If none is available, unsterilized 1-inch bandage can be bought quite cheaply for the purpose. These cloths are easily dyed.

Original centerline stakes are usually marked to show finish grade, that is, the surface of the pavement, since it is the line that forms the basis for engineers' calculations. The letter G indicates that reference is to finish grade.

Subgrade is the surface of the native soil after cutting, filling, grading, and compaction. See Fig. 8.7. It is lower than finish grade by the thickness of the pavement and any pavement-base and/or subbase that may be required. The combined thickness of these layers may be almost nothing, if the surfacing is to be oil or cement stabilization of native soil; or 3 or more feet for very heavy construction.

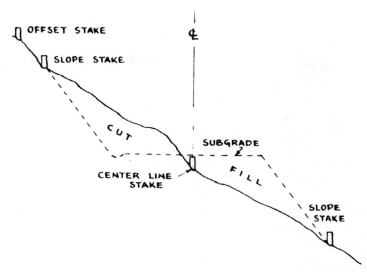

FIGURE 8.7 Slope stakes.

Since the cut and fill operations are directed toward reaching subgrade, much confusion will be avoided by making all such stake markings refer to subgrade, using the symbol SG.

A misunderstanding as to whether figures on grade stakes are for grade or subgrade can be very expensive.

Centerline stakes are quickly lost in most heavy grading, and the work is checked mostly by slope stakes. However, as sections are brought near the correct subgrade, the center stakes are replaced, and additional lines of markers are used to show the crown or cross slope of the road, and shoulders and gutters.

Stakes at the edge of any grading area should be set back about 6 inches to 1 foot, so that they can serve as a guide without interfering with the work.

It is frequently necessary to remove spilled dirt or level around a stake by hand, so that the operator can read it and see whether the grade is high or low in reference to it.

Blue Tops.   The final or fine grading operation is often guided by blue tops. These are usually 2 × 2 grade stakes driven down until tops are at subgrade. The tops are often colored with blue crayon to make them more visible. Any that are driven below the surface are marked by a light stake alongside.

An expert grader operator can work over blue tops without disturbing them. However, it is necessary for a person on the ground to remove spill piles that hide them, and to expose them if they become buried. Even with this precaution, varying numbers of these stakes are caught by the blade, or rolled over by tires, so that they have to be reset.

To set blue tops, a telescopic level of any type is set up, and its height (HI) figured from a benchmark. The correct rod reading for grade is figured for each stake location from the center or the theoretical grade profile, with allowance for crown or banking where necessary. The rod worker starts the stake and holds the rod on top of it, and is told by the instrument operator how high it is. The rod worker drives the stake the approximate distance, another reading is taken, and the process is repeated until the top is within a few hundredths of feet of grade.

It is common practice to set blue tops 50 feet apart along the length of the roadway, and at 12- to 15-foot spacing across it. The 50-foot spacing sometimes produces a low, wavelike effect, as a grader operator may get the grade perfectly at the stakes but have a tendency to run consistently high or low between them. This may be prevented by reducing spacing to 25 feet.

A grader equipped with automatic blade control can grade widths up to 40 feet from a single line of blue tops, resulting in substantial savings in grading and staking time.

Care.   Operators should be very careful when working around stakes as they are valuable, both as guides to correct work and in relation to replacement cost. In general, an occasional stake can be replaced readily, sometimes without instruments, but a group of them may involve considerable work for surveyors.

Errors.   A new set of stakes may not agree in grade or location with the missing ones. This difficulty could arise from an error in the original settings or in replacing them. A satisfactory road can often be built according to an error, but seldom when right and wrong markings are mixed together.

Stakes are accepted as correct until discrepancies are noted. If any stake appears to be out of line, or badly off grade, it may have been moved or disturbed; it may be a baseline or other marker; or a mistake may have been made in placing or marking it.

When possible, the surveying crew should be recalled to check it. If this is not practical, the supervisor may be called upon to use his or her judgment as to whether it should be remeasured. It should not be disturbed, however, unless absolutely necessary, as the suspected stake may be right and others wrong.

GRAVEL ROADS

Surfaces of bank gravel and other low-cost materials are so frequently required for haul, access, and other work roads that a brief discussion of them is in order.

Bank Gravel.   Bank gravel is a natural mixture of pebbles and sand. For road-building purposes, it should contain some fines that will act as a binder. Most deposits contain cobblestones and boulders.

Specifications for road gravel vary greatly. The following spread includes most of them:

Sieve	% passing each sieve
2 inch	80–100
1 inch	60–100
$\frac{1}{4}$ inch	40–85
10 mesh	15–70
200 mesh (fines)	5–25

In general, gravels with over 10 percent fines are not suitable for roads that will be subjected to freezing. Less than 5 percent may lead to loosening up in hot, dry weather. However, an increase in the percentage of coarse particles will lessen the softness caused by too much binder. Variations in particle shape and material will also affect results considerably. Increase of depth may make up for weakness.

There is no consistent difference between the parts of gravel banks which are above and below the water table. Water levels usually are different at the times when the material is deposited. However, there is very often a difference in color due to above-water oxidation of certain pigments.

Engineers frequently write ideal specifications for gravel that is not obtainable, and contracts are let to use practical grades on a price or availability basis.

Screened Gravel.   Specifications may call for screening gravel to be used in the top course or in the full road depth. Maximum size stones may be limited to 1-, 2-, or 4-inch diameters.

Screening is desirable to obtain a smooth, easily worked surface, but it often involves wasting of an excessive amount of stone which could be worked into the road. The resulting loss of strength may affect the road stability, particularly in crossing soft or wet ground.

In general, most oversize stone can be eliminated during the spreading and grading processes at less expense than pit screening, except in patching work.

Crusher Gravel. Bank gravel which is short of pebbles and long on stones may be run through a crusher to reduce the oversize to pebbles. The result may be superior to run of bank of similar size distribution because of the angular shape of the crushed pieces.

Blasted rock which is run through a crusher, without separation of the product, will often produce a material similar in size, distribution, and performance to the best of bank gravels.

Crusher gravel is usually more expensive than run of bank because of the extra processing.

Similar Materials. Any hard material which is broken into particles of the gravel size range may be used in its place. The breakage may be from blasting, rooting or digging, burning, or the effect of heat and cold. Such materials include shale, soft limestone, fine blasted rock of any kind, scoria, red dog, slag, disintegrated granite, cinders, and shell.

Exposures of shale rock are frequently soft enough to be dug by a small shovel without blasting. The broken shale has the appearance of excellent road material, but breaks down readily into mud. Some very expensive road failures have been caused by allowing traffic to use a shale subgrade, then putting a concrete pavement over it after its usefulness had been destroyed.

Soft limestone is the "coral" of the island military bases. It is often dug from the solid by loosening with heavy rippers, or hydraulic dozer blades fitted with teeth. It is easy to drill, but blasting may require as heavy loading as hard rock.

Such limestone is used as it comes from the pit. It should be rolled promptly after placing, as rain can make a soggy mess of it when loose. After compaction, wetting sets it into a hard surface that requires less maintenance than gravel.

Rock from tunnels (muck) is well suited for road fills, as the tight, heavily loaded shots cause fine fragmentation.

Scoria comes from clay beds that have been cooked by the underground burning of adjacent coal seams. It resembles broken brick. Red dog is a similar material that is produced by the burning of piles of waste bituminous coal with a high clay content. Both of these substances may break down into mud under traffic unless protected by some other surfacing.

Disintegrated granite is the standard low-cost road material in many parts of the southwest. It is a rough, coarse sand with excellent compaction and drainage characteristics.

Slag is a by-product of hot refining of metals, which may be poured molten onto dumps where it hardens into rock, or may be cooled and broken up by a water spray.

Cinders are of two kinds—refuse from steam power plants burning lump coal, and aerated rock blown from volcanoes.

Those from power plants are light, easily worked, and free-draining. However, they pound into mud quickly under traffic, and are useful for light-duty footings and emergency surfaces only. They are becoming rare because of power plants changing over to powdered coal and other fuels.

The use of fly ash mixed with lime makes an economical slurry to stabilize a wet base. The fly ash is a by-product of coal-fired power plants. The slurry might have a mix of 4 parts of fly ash with 1 part of lime. It is injected into a depth of up to 4 feet of base. The fly ash drives out the moisture by filling the voids in the subgrade, and the lime locks the fine-grained soils into a firm, stabilized mass.

Weight of volcanic cinders and ash ranges from 60 to 120 pounds per cubic foot, compacted. The difference is chiefly in entrained air, the lighter qualities being so full of bubbles that they have little strength.

Most volcanic cinders from 90 pounds up make good road material, but care may be needed in selecting them in the pit. They are too resilient for use under rigid pavements, but have good frost resistance because of their air content.

Shells are dredged in enormous quantities from bars along the Gulf of Mexico, and serve locally for the principal low-cost road base and surface.

Preparing Subgrade. The subgrade should be finished as accurately as possible. Ridges or hummocks of subsoil which extend up into the gravel weaken it. If the subgrade is clay or silt, it is good practice to place a blanket of clean, coarse sand to interrupt capillary flow and add to road stability.

The subgrade should be compacted if it is practical to do so. However, temporary gravel roads are often put across wet spots that are not workable. Rock fill or extra depth of gravel is used to make up for lack of subgrade preparation.

The use of soil cement is a means for making a subgrade of granular material with some fines into a stabilized surface. The soil and cement are blended in a portable mixer and conveyed over the surface like any paving material to make the improved subgrade. More about the uses and methods for soil cement will be found in the coverage of Soil and Cement Mixtures at the end of this chapter.

Cross Sections.   Three cross sections in common use are shown in Fig. 8.8. The feather-edge construction in (A) calls for a flat subgrade. Its advantage is ease of construction. Disadvantages include poor drainage of water out of the center gravel, deficient strength at the edges, and the necessity of blading fill from gutters or shoulders into the road during maintenance.

The trench section (*B*) provides center drainage and strength to the outer edge of the gravel. However, frequent bleeder drains through the shoulders may be needed to prevent water from ponding in the edges, soft shoulders may be a hazard in wet weather, and maintenance work will put dirt over the gravel.

The full-width surfacing in (*C*) is the best construction, and is to be recommended wherever the price of gravel is not a controlling factor. It saves the trouble and expense of edging, provides hard shoulders and good drainage throughout the surface, and minimizes maintenance difficulties.

Placing Gravel.   On good subgrades, gravel may be very thin, but it is the best practice to use 6 to 8 inches compacted depth, and to spread it in two layers. On soft ground, the depth may be 12 inches or more. The greater part of deep gravel is usually in the bottom layer.

The best gravel available should be in the top layer. It should not contain many stones larger than 1 inch, or at the most, 2 inches in diameter. It should be coarse enough to resist the action of tire suction, water, and wind, and should have enough binder to hold it in dry weather but not enough to make it sloppy when wet or thawing.

In the bottom, stones up to two-thirds of the layer thickness can be tolerated. Clean sand without stone may serve, if the top layer is thick and well bound enough to hold it together.

Gravel is ordinarily trucked in and spread by a dozer or grader. Occasionally, hauling and spreading can be done by scrapers.

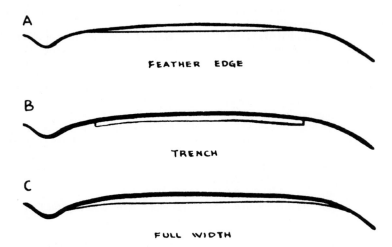

FIGURE 8.8   Cross sections of gravel roads.

Oversized stone may be bladed to the side, or picked out by hand. Small loose stone may also be taken off, or it may be left to be pushed back into the gravel by the roller.

Oversize stones that remain in the gravel after spreading and smoothing may be pulled out by a rake grader blade or a spike-tooth harrow.

Hand-picked stones may be thrown directly into trucks, or placed alongside the road for later removal. Second handling is easier if they are piled rather than scattered along the edge.

Compaction.   Each layer should be thoroughly compacted by pneumatic tired or steel wheel rollers, or by traffic. A heavy steel wheel roller will work back into the gravel all small stone pulled out by spreading work, and it gives a well-finished appearance. See Fig. 8.9.

The edges should be rolled first, and strips should be overlapped. This preserves the crown of the road, which should be at least 4 inches for a 20-foot road.

Proper compaction is impossible if the gravel is entirely dry, and difficult if it is too wet. However, dry gravel can be watered, and wet gravel will usually drain quite quickly. Full compaction on a gravel surface is not as important as in subgrades for pavements.

Dust Control.   Drying out, with resultant dust nuisance and aggravation of washboarding, may be prevented by use of calcium chloride. This is spread by hand shovels or machinery on the surface, and absorbs air moisture that soaks into and dampens the surface.

Recommended application is $1\frac{1}{2}$ pounds per square yard at the beginning of the dust season, and $\frac{1}{2}$ pound a month or two later. However, satisfactory results may be obtained with much lighter applications where summers are not entirely dry.

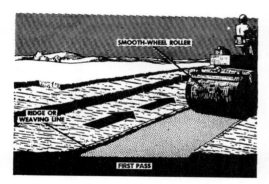
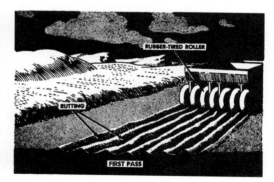
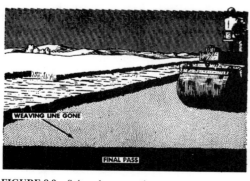
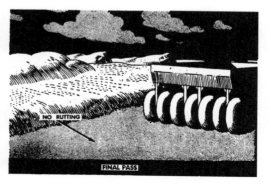

FIGURE 8.9   Subgrade compaction.

Lignum sulfonate, a by-product of paper manufacture, is used for the same purpose. It is sold in 50 percent solution form in drums, diluted to 10 percent, and spread by watering trucks.

Gravel road maintenance is discussed in Chap. 19. For haul use the surface should be kept free of loose stone and coarse gravel pieces, as these may cut tire mileage by as much as 50 percent.

Quantities of gravel needed for various areas and depths are given in the Appendix.

SIDEHILL CUTS

In hilly or mountainous country, roads are largely notched into slopes so that the land rises from one side of the road and dips away on the other. Such a road may be constructed by digging on the high side and using the spoil to build up the low side, as in Fig. 8.10(A); by cutting only, as in (B); or, less commonly, by building a shelf of fill as in (C).

Difficulties of design, excavation, draining, and stabilizing increase rapidly as hill slopes become steeper.

Stripping.   Removal of topsoil, stumps, and logs may or may not be required. This matter will be decided by the job specifications, or by the judgment of the engineer or contractor.

In general, stripping of topsoil becomes both more difficult and less important as the slope increases, as deep cuts in steep hills increase the proportion of subsoil in the dirt moved.

When stripping is required, the topsoil can most economically be pushed straight downhill by dozers to form the toe of the fill, as in Fig. 8.11(A), or a windrow below it, as in (B). Such a windrow may be moved by scrapers to a stockpile, or left to be pushed or pulled back up the slope to cover the completed fill.

If the hill is too steep to back up, the dozer may be equipped with a towing winch, and the line anchored above the work so that it can pull itself up the slope. It may also be helped by winch or direct pull from a tractor above it, or by a line around a pulley anchored above it to a tractor on its own level or on a lower one.

Loose stumps can be used in pioneer road fills but are unsuitable for highways. Their use in intermediate classes of roads will depend on job conditions, the estimated useful life of the road, and the rate of decay of the stumps.

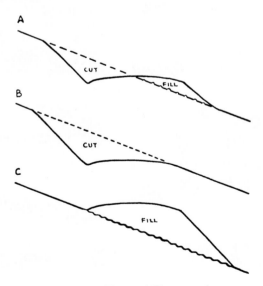

FIGURE 8.10   Sidehill cut-and-fill cross sections.

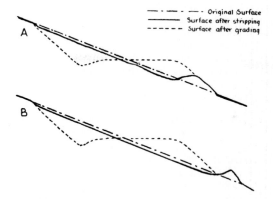

- - - - Original Surface
——— Surface after stripping
- - - - Surface after grading

FIGURE 8.11   Topsoil stripping.

Logs placed at the toe of the fill are useful in catching rolling boulders and checking slides. These, or more elaborate precautions, are often required to protect roads or structures below. They may be held by their own weight, by resting against stumps, or by cables and anchors. They may be used as temporary expedients in most work.

Stumps left intact in steep fill areas may serve to prevent the completed fill from sliding downhill as a mass. Specifications often permit leaving them if they will be covered 2 or 3 feet deep.

If a sidehill is cleared and stripped, the areas to be filled should be plowed or roughened across the slope to reduce the danger of slides.

Dozer Digging.   If the side slope is gentle, the road shelf may be cut by pushing downhill. Steeper slopes may be started in the same manner and finished by working along the road line, as in Fig. 8.12.

In general, when the upper bank becomes so steep that the dozer cannot back up it without assistance, it is more economical to work from the side. However, if the line of cut is interrupted by rock ribs, which are not to be blasted until the softer parts of the road are made, a dozer with a helper cable may be used to cut benches in each section, at least long enough to permit it to start a sidecasting cut.

Pushing from above, where practical, is faster than sidecasting.

Sidecasting.   The standard method of notching a steep sidehill is to sidecast with a dozer. A wide-track, close-coupled dozer with a blade that can be tilted to cut low on the uphill side is most efficient. An angling blade, set with the uphill side low and angled to cast down the hill, is useful,

FIGURE 8.12   Starting sidehill cut with dozer.

particularly in light soils and shallow cuts. The advanced position of the blade may make it difficult to turn with heavy loads.

Work is started near the upper slope stakes, at a spot naturally or artificially level enough to permit the dozer to work parallel with the road centerline, at the upper edge of the cut. A blade full of earth is dug along the upper cut line, then the blade is lifted and the machine turned downhill at the same time. After dumping, the dozer is backed until parallel to and touching the upper line. Another scoop is dug and swung downhill.

One or several layers may have to be dug in one spot to obtain enough fill to build out the shelf wide enough to carry the dozer. The steeper the slope, the more passes are needed.

The blade is raised sufficiently during the dump to keep the fill higher than the cut so that the notch will slope oppositely to the hill. This keeps the dozer tilted for efficient cutting, allows for compaction of the fill when it is walked on, and also provides the proper cross section for a pioneer road.

When the shelf is wide enough to hold the dozer, further procedures are varied to suit the slope, the soil, the machine, and the operator's preference. The cut can be lengthened to the end of the slope, then cut in successive layers to grade and width, or it may be developed to full size in a single cut.

Layer cutting involves more rehandling of the dirt, as the loads dropped from the first cut are moved again as it is deepened. However, it is easier for a dozer to make shallow cuts, and the angle blade sidecasts most effectively and puts minimum strain on the tractor when the cut is light.

Deep single cuts make it difficult to trim the bank and may have to be avoided for that reason.

Rock Slopes.   If the slope is composed of material that the dozer cannot dig, or that it can dig only with difficulty great enough to reduce production and increase repair costs substantially, the material should be softened ahead of the dozer.

Hard clay and soft rock on moderate slopes may be loosened with a tractor-mounted ripper. But if the rock is hard or the slope is steep, drilling and blasting will probably be necessary for the pioneer cut.

Engineering geologists say slope failures are increasingly common, particularly where heavy blasting takes place during construction, where natural fractures undermine rock face stability, or where the slope has long been exposed to weathering.

Very steep or broken slopes call for hand drills. Holes may be drilled along the top line of the cut or horizontally at the first-floor level. It usually pays to drill closely and load heavily, as ribs or poorly fractured rock will delay the dozer operation out of proportion to any saving in costs.

When the rock is patchily covered with loose overburden, it may be necessary to dig holes before drilling. At other times only the exposed rock is drilled, and secondary work done on the parts that are uncovered during dozer work.

If there is a considerable amount of pioneering in rock slopes, best results may be obtained by use of light self-propelled drills on crawler mountings. They can reach and work in very difficult places, and can tow their own compressors except under extreme conditions.

After blasting, dozer sidecasting proceeds in the same manner as in naturally loose soil.

Once the pioneer bench has been established, the character of the rock will determine whether rippers or drills and explosives should be used to loosen it.

Lower levels may utilize dozer sidecasting, backhoe sidecasting, scraper hauling, or shovel and truck hauling, depending on the job plan. If the material is used elsewhere, it is of course desirable to take it away immediately, rather than sidecast it first and then redig and haul it.

Belt Loaders.   Once a cut of sufficient width has been made between two areas that are wide and level enough for turning, a belt loader (see Chap. 14) is sometimes used for widening and deepening the cut by sidecasting if the soil is suitable. This machine may work in only one direction. It may be followed on each trip by a dozer grading off the spoil.

Excavator.   The front shovel or backhoe can be used instead of a dozer for notching a slope. It can usually do the rough work in one trip, as in Fig. 8.13, but if the bank must be trimmed or the cut is very deep, it may be done in layers.

When the width of the cut will allow it, it is good practice to keep the shovel on its floor rather than with one track on the fill. For narrow roads and deep cuts, a small backhoe with a short rear overhang is desirable. The cut should be kept sloped into the bank to keep the weight off the edge.

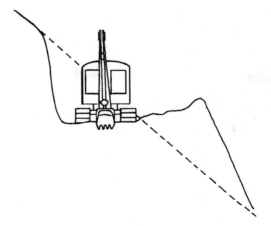

FIGURE 8.13   Sidehill cut with excavator.

The fill is kept higher than the cut, particularly if used for footing. Poles or platforms can be used for extra support under both tracks, or under the outside track only.

When the ground is soft or wet, the slope is very steep, soil layers slope with the hillside, or smooth bedrock is just under the cut, the smallest backhoe which can handle the digging should be used. The weight of a large machine, together with the vibration of its work, may cause a slide.

Shovel spoil can most conveniently be sidecast, but also can be loaded into trucks backed up to it. If the road is long and narrow, trucking out all the spoil will be very slow work.

Rock exposures along the road line should be blasted, as a shovel cannot be readily moved up and down steep slopes to bypass them. Use of a shovel is indicated when soil is too soft or rocky for effective dozer work, when cuts are deep, and when spoil is to be used at a distance.

The work is ordinarily left rough to be finished off by a dozer or grader.

Side Cuts.   When the notch is to be largely or entirely a cut and the spoil is to be used nearby on the job, dozer sidecasting is used only until the shelf is of ample width to hold the machinery. The material is then pushed or carried along the shelf to the fill area.

Big dozers can be used for pushes up to 200 feet on the level, and farther downhill with fair efficiency. When the cut is too narrow to allow machines to pass each other, their production can be stepped up, at some additional cost, by using two or more dozers in relays. One, working from the back of the cut, will push a load partway to the fill and spread it a bit in dumping it. The dozer below it will back over the heap and push it to the end of its beat.

Scrapers.   The possibility of using scrapers should be considered. Their use on short runs is discussed later.

Tractor scrapers are impossible to back, so in a narrow road they require an additional road to bring them back from the fill. This may have to go back to the beginning of the hillside, or enter it at some intermediate point. In either case, the scraper's travel distances are apt to be much greater than those of the bulldozer.

If some spoil is being sidecast and some hauled away, a dozer can work on widening and serve as a pusher.

In the first stages of enlarging a notch, it may be difficult to keep the road sloping into the hill because of scrapers sinking and gouging into the loose fill. This pitch may be preserved or restored by running a grader or an angling dozer close to the wall, and casting out. As the cut widens and enters solid ground for its full width, it will become possible to keep it trimmed on the bottom by proper manipulation of the scrapers.

When a steep hill contains boulders, stumps, or ledge, sidecasting is preferred to hauling, and dozers will probably be both safer and more economical than scrapers if short or medium hauls are required.

Pioneering is done by dozers. If the cuts are in material they can handle, scrapers take over the job. In very rugged terrain, crawler tractors with standard or undersized scrapers are preferred, as they can work on the steepest grades and need minimum turn space. After grades are reduced to 20 percent, self-powered scrapers can take over, if the haul is long enough to justify their use.

For rubber-tire jobs, turnarounds should be kept nearly level, and machines should be driven directly up and directly down the grades. Overhung scrapers are in their most vulnerable position in regard to overturning when turning downhill on a side slope.

Compaction.   When a wide road is notched into a hillside by cut-and-fill methods, it may be difficult or impossible to compact the fill if it is sidecast.

If compaction is required, two pioneer notches may be made (Fig. 8.14), at the top and the bottom of the cut. Scrapers are then used to cut the top down and build the bottom up. Compaction of the fill can be handled by rollers following the scrapers, until sufficient width is obtained to permit them to pass the scrapers on the fill, after which they can operate in both directions.

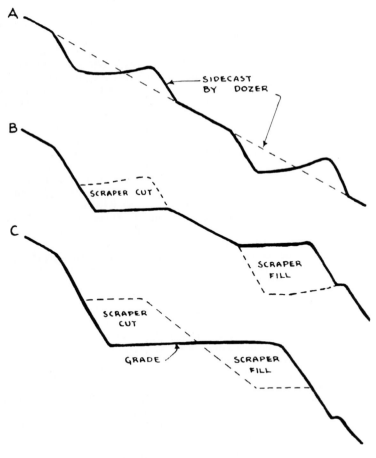

FIGURE 8.14   Parallel cut and fill.

BANK SLOPES

Angle.   The angle at which bank slopes will stand in cuts and on fills is an important factor in the cost, and sometimes in the feasibility, of sidehill construction. It is also a limiting factor in the depth of through cuts.

There are two approaches to determining how steep a cut slope may be left. One is the behavior of the same or similar material in cuts and on natural slopes, the other is soil analysis and calculations. They are frequently used in conjunction, a tentative slope being determined by field observation and then checked by engineering research.

Natural slopes are seldom steeper than those that can be used in the same material in a highway cut. The exceptions usually involve groundwater problems or the binding effect of vegetation.

However, natural slopes may not be nearly as steep as the soil qualities permit. In general, a hill whose foot is being vigorously eroded by a fast-flowing stream will approach maximum steepness, and one rising above meadows will tend to have a flatter slope.

Old cuts give a much more accurate indication. However, before dependence can be placed on them, it would have to be ascertained that the material is actually the same, that it is subject to the same weather conditions (freezing and thawing loosen faces more actively on the shaded side than on the sunny side walls of canyons), to the same dip of strata, as in Fig. 8.15, and that groundwater conditions are similar.

It is desirable to cut back to entirely safe slopes, but this may not be possible. In notching along the side of a mountain, the cut wall must be substantially steeper than the natural slope to avoid excavating tremendous yardages. Also, a fill slope must be steeper than that of the sidehill on which it rests if it is to support a road.

Some soils, such as loess in the midwest and lightly cemented gravels in the southwest, will stand for long periods with almost vertical faces. In general, slopes can be steeper in arid climates than in wet ones.

There are varieties of clay that will stand steeply when first cut, but under influence of surface freezing and thawing, groundwater pressure, and vibration will slump to a 10 percent (1-on-10) grade unless stabilized by topsoiling and planting.

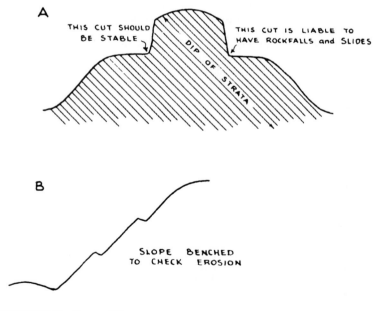

FIGURE 8.15   Slope stability.

In general, cut slopes range from vertical or even overhanging in rock and 5 on 1 in the most stable soils down to about 1 on 6. It is the engineer's responsibility to decide where in this wide range the requirements of stability, safety, and the roadway budget can best be compromised.

Slides.   The most serious problem associated with deep roadway cuts is that of landslides. These may occur during the work, or at any time after completion. Dangers include loss of life or injury among those building or using the road; destruction of excavators, trucks, cars, and other equipment; and loss of use of the roadway for long periods.

The likelihood of slides increases with height and steepness of banks, but slides are caused by internal conditions.

In rock, the cause is usually a dip of seams or joint structure that provides an inclined slide for the cut layers, as in the right side of Fig. 8.15(*A*). This structure, when well lubricated by groundwater, may produce anything from a series of minor rockfalls to a 100,000-yard roadblock.

Much the same effect is produced in unconsolidated soils that have sloping layers of pervious and impervious material, or where a slightly pervious soil mass rests on a sloping base of clay or rock.

The existence of such a condition may be revealed by preliminary borings, or be shown by a line of springs as the top of the impervious layer is excavated. In this case a slide is most likely to occur after heavy rains, when the loose soil is heavy with water, and water movement on the base formation provides both a lubricant and pressure.

Slumping and sliding may also occur in seemingly uniform soil masses, because of water seepage or saturation zones.

The forces involved in deep-seated, large-volume slides are usually too great to be controlled by braces or anchors. However, most can be prevented by cutting to a flatter slope in danger areas, and/or diverting or draining the water that starts them moving.

The likelihood of slides may be a determining factor in selecting a route. However, there are many areas in the world where mountain roadways must be constructed for long distances along steep soil slopes, and where money is not available for engineering investigations or drainage works.

Under such conditions a pioneer road may be cut by a dozer sidecasting down slopes and pushing fills across canyons. The road is widened and improved by cutting down its floor.

When slides occur, the dozers simply cut new road shelves across them, repeating as often as necessary. When fills wash out they are replaced, cutting new slots on the slopes above the road cut where necessary.

In the course of a few years it is usually found that the larger part of the route is reasonably stable, and that other sections stabilize with repeated working. The true problem spots that remain can then receive detailed engineering investigation and corrective measures, at a fraction of the cost of similar work for the entire route.

Checking.   A technique is available for testing the stability of slopes with the Seismitron, described briefly at the end of Chap. 9. The instrument's probe or receiver may be placed on the slope surface, but results are more accurate if it is placed in a drilled hole in the bank. It picks up tiny sounds of ground movement called microseisms, and amplifies them so that they can be heard in earphones.

An increasing or high frequency, say over 25 or 30 microseisms per minute, indicates danger of a slide. A low or decreasing rate is an assurance of stability. It has been found that slopes that failed when wet had given warning of possible failure while they were dry.

Clay, mud, or fine sand may not produce warning noises that can be recognized. Most other formations do.

Stabilization.   When the bank is high, it may be necessary to drill long holes into the toe to prevent water from causing the face to slump. Since this trouble usually occurs in soil that is firm enough to leave in high banks but not hard enough to resist percolating water, augers may be the preferred drilling tool. Perforated metal pipe is inserted in the holes while drilling or immediately afterward.

Long slopes may be benched, as in Fig. 8.15(*B*), to break the flow of surface water. Each bench has a reverse slope so that it acts as a diversion ditch, with water flowing along the back. A gentle grade spills the water toward one or both ends of the cut.

Benches may also serve to catch falling rock. Their effectiveness for this is increased by a berm of dirt along the outer edge. This construction is particularly useful in banks of cemented gravel, 1 on 1 or steeper, from which surface cobbles are released.

Slopes can be stabilized by growth of vegetation. Most types will provide surface protection, and types with deep or interlocking roots may hold against some internal pressure as well. To help the revegetation after fire had destroyed grass cover on a steep slope, straw and coconut-fiber blankets, as seen in Fig. 8.16, were used.

Artificial protections include supporting walls, drainage systems to intercept or remove groundwater, and fences to catch rolling pieces.

Walls may be of masonry, interlocked concrete, or metal bins. Strength of the last two constructions depends on their being filled with coarse, pervious fill. Any of these must rest on a solid footing that can resist both weight and thrust.

Logs can be used for temporary retaining walls and to catch boulders rolling during work.

Drainage. Freshly worked embankments should be protected against surface water flowing from adjoining ground. In cuts, a diversion ditch may be dug a few feet back from the upper edge. Unless its gradient is gentle, its bottom may need protection to prevent it from developing into a gully that would damage land below it and eventually break out through the bank.

Such protection may include establishment of a strong sod, construction of a series of check dams, paving with resistant materials, diversion of some of the natural drainage at higher points, or use of discharge flumes down the slope.

If the slope is threatened by softening or washing by groundwater, subdrainage may be required also. Land tile may be laid under the surface channel if its floor is impervious enough not

FIGURE 8.16   Revegetation on a steep slope. (*Courtesy of North American Green.*)

to allow excessive surface water to enter the tile. Underdrainage may be required in the gutter at the foot of the slope, and in or behind wet spots in the slope to catch the seepage.

In areas of rapid runoff, a roadway may be protected along its entire uphill side by a system of diversion ditches that channel all drainage into culverts or access dips. For economic reasons this type of work is limited to diggable soils on slopes that are accessible to machinery.

Fills usually have less drainage across them, but because they are not as well bonded together, they are more subject to surface erosion than cuts. Water may flow onto them from the road and from slopes above the road. They can be protected by berms along the outer edge of the road shoulders, which will prevent water from going down the side of the fill, except at points protected by pipes, flumes, or pavement.

Fills which are built on sidehills have a tendency to slide along the old surface, unless it is well roughened. Leaving of stumps and boulders, roughening by plowing, and placing of subdrains to stop seepage of water along the joint are common methods of reducing this danger.

Any soil, whether original bank or fill, which rests on smooth, steep rock slopes is liable to slide. The most important step in preventing slippage is to divert groundwater moving down the surface of the rock.

Grading.   Steep side slopes should be finish-graded as they are made, as it may be difficult and dangerous to work them afterward. But if it becomes necessary, a wide-track dozer may work a long slope in strips, from the top down or diagonally.

Horizontal trimming by use of graders or dozers on steep side slopes may be made safer by cabling to another machine moving parallel to it on the top of the bank. Two cables are used, attached to the front and rear of the lower machine.

It is not safe to operate unsupported heavy equipment along slopes that contain rocks, soft spots, or frozen ground.

Topsoiling.   The best protection for a dirt slope is a good cover of vegetation. Grass, weeds, bushes, and trees are all effective controllers of erosion. See Fig. 8.16. The type selected will depend on the locality, soil, and season.

On most jobs, it is necessary to place a layer of topsoil over the fill or exposed earth in order to get a good growth. Occasionally plants will grow well enough on raw earth, or with the aid of some lime or fertilizer.

Deep topsoil is favorable to growth, but it may discourage plants from rooting into the subsoil, and absorb too much water so that it will slide off during rains. For this reason, and for economy, topsoiling of steep slopes is usually limited to a depth of 2 to 4 inches.

The fill surface should be roughened so as to bond with the topsoil. A sheepsfoot or tamping roller is one of the best tools for accomplishing this. If the slope cannot be worked, the roller may be operated by a dragline at the top. The drag cable is used to pull the roller up and to let it down, and the walking of the shovel moves it along the slope.

Topsoil may be pushed up a slope from stockpiles at the bottom, pushed down it from piles trucked to the top, or distributed over the surface by a clamshell working from either top or bottom, and the resulting piles shoveled or raked out by hand.

The Gradall is an excellent tool for final shaping of the subsoil and spreading topsoil on any area it can reach.

Freshly spread topsoil gullies readily and needs protection on slopes. A thin coat of sprayed asphalt emulsion will carry off rain, and still allow grass to grow through it.

A layer of hay or straw may be mixed into topsoil by a tamping roller to hold it. The hay should be well cured, as rapid decay would make its useful life too short. It is apt to absorb so much nitrogen from the soil as to interfere with growth of seedlings. Use of barn straw that contains some manure, or adding nitrogen fertilizer, cures this difficulty.

Some hay and straw contain enough grass and weed seeds to establish a good cover. Other types are deficient and require that the ground be seeded. Seed can be mixed with water and sprayed onto slopes.

On small areas, topsoil may be held by adding straw, and holding it with chicken wire firmly pegged down. Horizontal wood slats are sometimes used. Placing and tamping cut sod in drainageways, in horizontal strips on slopes, or on the whole surface are very effective, but the cost is high.

Rock Faces.   Rock cuts can be left with very steep or vertical faces, and occasionally are allowed to overhang. Such faces usually cause a hazard of rockfalls to the roadway, but the expense of cutting rock back to completely safe slopes can seldom be justified.

Some rock formations tend to break up into gravel or small stones at the face because of temperature changes, and will at times subject the road to an almost continuous bombardment. Such faces should be cut back sufficiently to permit a wall or fence to be put beside the road, with space behind to catch falling stones.

More massive cliffs may present the danger of occasional falls of larger rocks or of whole sections. These may be checked in the danger season by a worker with a bar, supported by a rope held at the top. Loose pieces can be pried out.

Long expansion bolts, similar to those used to secure tunnel roofs, can be placed to fasten a whole slope into a solid and safe unit. They are particularly efficient in shale beds parallel to the slope.

Vegetation tends to break up rock faces, so artificial planting should not be attempted.

THROUGH CUTS

A through cut has a high wall on each side, as shown in Fig. 8.17, so that little or no material can be excavated by sidecasting.

If it is on a sidehill, one edge will be higher than the other. The part of the cut which is above the low wall is actually a sidehill cut, and may be handled as one or as a through cut.

Through cuts are seldom used in building pioneer roads, except where borrow is needed to cross a ravine. When roads are narrow, and the sharpness of curves is not an important consideration, sidehill work is faster and less expensive.

FIGURE 8.17   Cross section of scraper cut. (*Courtesy of U.S. Army Engineers.*)

SCRAPERS

Scrapers, which are described in Chap. 17, are the standard excavator for alternating cuts and fills, where the soil is soft or fine enough for them to work, or can be made so by rippers and explosives, and where the haul is too long for dozers.

Preparation.   The first requirement is to smooth over the cut and the fill areas so that scrapers can work them. This is usually a dozer job. The ground is cleared of vegetation and boulders, holes and gullies are broken in and ramped over, sharp ridges beveled off, side slopes notched, and turning places graded off.

It is not absolutely necessary to prepare the whole area in order to have the scrapers move in. Their work can start on the high part of the cut and the low part of the fill, while the dozers are clearing and smoothing the balance of the area.

If the cut has a high side, it is cut to a passable driveway by straight pushing or sidecasting. The bottom of this cut is sloped oppositely to the hill.

If the hill is high in the center of the cut, the hump is graded off sufficiently to afford good footing for scrapers.

It is sometimes economical to make small fills in areas which are to be lowered, and small cuts under future fills in order to smooth out working areas quickly.

When a dozer is not available, a scraper can smooth moderately rough ground by driving through it with the knife held low enough to cut off the bumps and high spots. If the tailgate is held near dumped position, it will act as a dozer blade.

Scraper work on side slopes is simplified by first cutting a shelf with a dozer. If no dozer is available, the scraper can be taken uphill to the start of the cut, the blade dropped, and the scraper turned to dig along the upper cut line. The turn will cause the edge to cut most deeply on the uphill side, and if done repeatedly, will level the digging area, or slope it oppositely from the hill.

Cutting Ridgetops.   If the slope up from the fill is too steep for the scraper to climb, it may be broken down into a ramp by dozers, or the cut made with an excavator.

If the slope away from the fill is too steep for scrapers, the top can be lowered by the combined work of scrapers and dozers, as shown in Fig. 8.18. Full-trailer scrapers will dig across the cut as they turn, as in (*A*). Semitrailers can be backed up to the edge, as in (*C*), and if a snatch tractor is available, can be backed over it. Digging is then done straight toward the fill.

The undug lip left by the first method is pushed over the edge by a dozer, as in (*B*). This filling, and the cutting into the slope, will extend the floor and allow scrapers to work farther back.

Eventually the bank will be lowered sufficiently to make it practical to break it down with dozers (*D*), so that scrapers can go through to dump on the far side, or turn to continue hauling in the original direction.

Shaping.   The outer edges of the slope should be determined before starting work so that steep banks may be cut to final grade from the first. They are taken down in a series of steps. If the slope is 1 on 3, and the scrapers are taking 6-inch slices, each new level cut should be 18 inches farther from the bank than the previous one, as shown in Fig. 8.19.

The slope should be checked frequently by engineers for correctness, and trimmed off by a grader working on the floor of the cut, as it may become very difficult to reshape when the floor has been cut too far down.

The floor of the cut should slope down toward the edges. This slope may be originally established by an angle dozer or a grader, after which it will tend to perpetuate itself, as the weight of the machine will be greater on the down side so that it will tend to cut low there. If the slope becomes too great, the upper part may be readily planed off.

Machinery may not be available to shape the original surface, or the crown may be lost because of oppositely sloping strata or by careless operation. A scraper can cut a crown by taking advantage of the fact that the oscillating tractor part does not affect the side-to-side tilt of the knife, which is determined almost entirely by the rear axle. A gouge taken heading up a slope can be

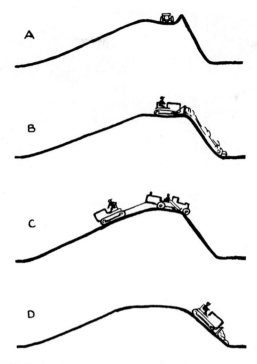

FIGURE 8.18   Breaking down steep slope.

FIGURE 8.19   Cutting slopes with scraper. (*Courtesy of U.S. Army Engineers.*)

used to tilt the rear axle so that the knife can cut on the uphill side when turned along the slope. See Fig. 8.20.

Whenever possible, the cut should be arranged for digging downhill and toward the fill. The first factor is usually more important. The grade of the cut is most important when it is or can be made steep, and when power and traction are small in proportion to the size of load desired.

If no bulldozer is available, scrapers can pioneer side-hill cuts. At the point where the cut should be started, turn sharply toward the row of stakes above you.

When tractor has completed a full turn, drop the cutting edge to start loading. As the scraper turns, the hillside corner of the blade will cut into the high side of the hill.

In this manner a bench will be started. Repeat this procedure whenever loading next to the bank. Result will be cuts that always slope toward the high side. This prevents scraper from drifting.

FIGURE 8.20 Scraper pioneering a sidehill cut.

To facilitate rapid movement and easy loading, it is important to keep the pit from getting too rough or ridged.

Hard Digging. Scrapers can penetrate fairly hard soils, as the cutting edge is sharp and held at an effective angle. The machine cannot be overbalanced by suction, as the knife is carried between the axles, but a plastic soil may pull the edge a few inches deeper than intended by flexing the tires or causing them to sink.

There are hardpans and rocky soils which the knife will not cut and many others which can be dug only by the expenditure of so much power in penetrating that little force is left for the loading. In such cases, the use of rippers ahead of the scrapers is advisable. Single or widely spaced teeth give best results, as coarsely broken ground is usually easier to load than that which is reduced to very fine pieces or pulverized.

In hard digging, a straddle loading sequence is often helpful. Parallel cuts are made, leaving a ridge between which is narrower than the bowl, as shown in Fig. 8.21. The ridge is then taken out on a third pass, and it will be found that the digging resistance is more nearly in proportion to the shallow cuts at each side than to the deep one under the ridge.

This method should be used with caution near the edges of cuts as it may destroy the crown.

If bedrock needing blasting is found in the cut, a dozer could strip the overlying soil and push it out to be picked up by the scrapers. A scraper is more vulnerable to damage from contact with rock, and its loading will be slowed by any effort made by the operator to avoid such damage.

FIGURE 8.21 Scraper straddle loading. (*Courtesy of U.S. Army Engineers.*)

Trimming Banks. Successive cuts are set back from the edge to provide proper slope. Scrapers will not cut vertical walls but will leave very steep faces.

Slopes between 1 on 1 and 1 on 4 are usual in soil cuts. If the scraper takes a 6-inch slice, a 1-on-2 slope would require each pass to be 1 foot inside the last. If 1 on 4, it would be spaced 2 feet.

The steps are best trimmed to a smooth slope by a grader working on the floor. The excess material is cut and slides to the bottom to be removed by the scrapers.

If trimming is done with a scraper, one rear wheel should be on the bottom, the other on the slope. If it is steep, the tailgate should be carried well forward so that loosened dirt will slide downhill rather than enter the bowl.

The cut should not be deepened so far between trimmings that the grader cannot reach all the steps. This is particularly important when the slope is so steep that it cannot be worked by machinery later.

As the cut deepens, new slope stakes are placed. They are often set from the originals with a string level, rule, and plumb bob. If they are driven in flush and marked with light sticks, a good grader operator can trim the bank without knocking them out.

Finishing Subgrade.   The bulk of a deep cut can be made without staking except for the slopes. As it approaches bottom, however, grade stakes should be set, and digging done with sufficient care to avoid overcutting and the resulting need for patch fills.

Good scraper operators can hit a grade within a fraction of an inch if the soil is smooth, but it is often more economical to have them run a rough grade and go on to other work while a grader finishes up. The grader will probably be required to cut and shape gutters, in any event.

The road, or the shoulders, is sometimes overcut to allow space for spoil from ditches. If this is not done, ditch cuttings may be windrowed on the road for later removal.

Any areas that are cut and refilled in this manner must be thoroughly compacted. A few patches may be rolled by trucks or scrapers, but regular rolling equipment should be brought in for any extensive areas. Failure to compact may lead to local settlement and pavement failure.

Selected Base.   A layer of porous soil with high bearing strength is usually placed between the native soil and the pavement, in both cuts and fills. If it is obtained from nearby pits, it will probably be economical to bring it in and spread it with scrapers.

The pit wall or floor may be shaped for direct scraper digging. If this is not possible, hauling may be done with either scrapers or trucks loaded by excavators or other equipment. The scrapers offer the advantage of doing their own spreading.

Self-loading scrapers are good in this work, where quantities are relatively small and pit shape is irregular, as they do not have to be part of a balanced spread which may be difficult to organize.

PUSHER

The pusher is a separate tractor that pushes the scraper while it is loading. It is almost always required for efficient full loading of single-engine self-powered scrapers, and is usually desirable with two-engine scrapers and with crawler-drawn scrapers that are oversized for the tractor or that are in hard digging. See Fig. 8.22.

Types.   Pushers are usually crawlers with tractor weights of 20 tons or more.

There are also four-wheel-drive pushers with similar or greater total weight. These have the advantage of much greater speed. They can often get a scraper through the cut much more rapidly and with less scraper strain and wheel spin than a crawler of similar weight can, and they can make fast moves from one scraper to the next, and between jobs. However, they do not have nearly as much push at stall speeds; they may lose their speed advantage by requiring shallower and longer cuts. Tires lose traction badly on wet, slippery surfaces.

Crawler tractors with torque converters have maximum pushing power and traction, and have some speed flexibility.

Several thousand dollars may be saved by buying a pusher with a fixed push plate instead of a center-reinforced dozer blade. However, the fixed plate will not permit the tractor to do cleanup or backripping work in the pit, and it is likely to find little use outside of it. It may cause difficulty by not meeting the scraper at an efficient angle, or by losing contact on rough ground.

A good push plate needs little maintenance, but even the most heavily reinforced dozer blades are likely to cave in when used for scraper pushing.

FIGURE 8.22   Scraper and pusher.

There are also small push plates that mount on the C-frame of a dozer or angle dozer. This installation is more expensive than the fixed plate, but permits accurate lining up with scrapers, can be used for pushing big boulders, and can be easily replaced with a dozer blade.

Either type of plate is much less likely to damage scraper rear tires than a dozer blade is.

Loading Effect.   Soil and slope conditions being equal, loading time and the size of load in a particular scraper are determined by the power applied to it, regardless of whether that power comes from one, two, or three tractors.

The rule of thumb is that 1 pound of push puts 1 pound of dirt per minute in the scraper. This "push" includes the scraper's own effort.

Tandem Pushers.   It follows that size of scraper loads and speed of loading can both be increased by increasing pusher power. Additional power may be obtained by using bigger tractors, using two or more tractors together, or in both ways.

The standard arrangement for pushing with two or more tractors is to line them up behind the scraper. The front dozer pushes the scraper, it is pushed by the machine behind, and sometimes that is pushed by still another. However, two pushers are much more usual than three.

These tractors must have push brackets at the rear, fastened to the side frames so as to put the thrust directly against the dozer arms.

Tandem pushing involves extra delays in getting the machines in contact with each other and loading, but it usually more than makes up for this in extra speed and depth of slice in the cut.

Snatch Tractors.   A few contractors use a pull tractor instead of or in addition to pusher tractor(s) for loading scrapers. Coupling may be by means of a short tow cable, or preferably by a coupling that locks automatically on contact and can be opened by the puller operator from her or his seat.

The tow cable requires a full stop and the services of a ground worker; the automatic coupler, a high degree of operating skill.

Snatch tractors show good results with experienced crews, but most contractors are better off if they stick to pushing.

Scraper-Pushers.   Scrapers may be equipped with pusher plates or blades so that they can push each other to help in loading. This arrangement is usually not as efficient as using a separate pusher. A scraper cannot deliver nearly the push of a tractor of the same price. There is a chain loading effect that requires a long cut or better-than-average supervision.

There are a few jobs on which scraper-pushers may work out well. They may save or postpone a big investment in a crawler tractor, or fill in time while a pusher is out of service.

Graders.   A heavy grader equipped with a dozer blade may make a good pusher, and may be used for this purpose in an emergency. Under most conditions, however, it is much more valuable as a first-rate finishing and maintenance machine than as a second-rate pusher.

Loading on Grades.   The effect of gravity is to pull a load going downhill to the extent of about 20 pounds per ton of weight per percent of grade. An uphill load will be held back to the same extent.

This will cause a pusher weighing 25 tons to gain or lose 1,000 pounds in net drawbar pull on a 2 percent grade, while a 50-ton loader scraper will be affected doubly.

The difference between a 2 percent upgrade and 2 percent downgrade will be thousands of pounds of pull, maybe 20 percent of the pusher's power. Using the rule of thumb that 1 pound of push produces 1 pound of dirt per minute in the scraper, direction on this grade would mean a difference of 3 tons or 2 yards of earth in the load, or a proportionate amount of time in obtaining the load.

It is therefore profitable to keep the loading down grade even on a gentle slope, whenever job conditions permit.

Pushing.   The pusher is driven up behind the scraper in low or sometimes in second gear, and contact with its bumper is made as smoothly as possible. The scraper unit, in low gear, applies as much power as it can without spinning the wheels or drawing away from the pusher.

The pusher moves as fast as it can without making the scraper jackknife or either unit twist sideward. Twisting or off-center pushing is likely to cause a dozer blade edge to rip into a tire.

Most pushing is done in low gear because it is important to keep going right through the pass. Second gear may be used to advantage when the pusher is overpowered for the size of the scraper or the load that it is to carry. There is an extra hazard of damage from making contact too roughly. Cushion dozer blades may be used to reduce the shock.

Patterns.   The simplest pusher loading pattern is called backtrack or shuttle loading. See Fig. 8.23(A). The scraper drops its bowl at the beginning of the cut, and the pusher makes contact. The scraper pulls and the pusher pushes until the desired load is obtained. The scraper bowl is raised, informing the pusher operator that he or she is not needed, if the operator has not already learned this from watching the load. The scraper tractor then shifts into higher gear and departs.

The pusher returns to the beginning of the cut. It may make this move in reverse, or by turning and using forward speeds. The length of the cut and the relative speeds the machine can make in the highest usable reverse and forward speeds determine which should be used.

If the next scraper gets in loading position before return of the pusher, it will start to load itself, as it can pick up part of a load readily, and will thereby decrease the distance the pusher must come back in order to get behind it.

Another pattern, shown in (B), is suitable only for long cuts. After loading the first scraper, the pusher waits for the next empty one to come up alongside it, pushes that until it is full; pushes the next from its stopping point; and so on until the end of the cut is reached. The pusher is then turned and run back to the beginning of the cut to start another series.

A third system, which is useful where there are more scrapers than the pusher can readily handle, or where the dirt can be moved in two directions, is outlined in (C). Each scraper is loaded moving in the opposite direction from the previous one, so that the pusher need only turn around to be in position for the next push, instead of having to move back to a starting place.

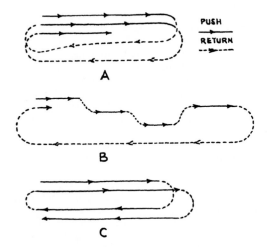

PUSH

RETURN

FIGURE 8.23   Pusher patterns.

A pusher should have as many scrapers as it can conveniently handle, but it is difficult to maintain a proper proportion because of changes in the length of haul. Two scrapers might keep a pusher busy on a very short haul, whereas a dozen might not work it steadily on a long run.

Where there are more units than a pusher can service, so that one or more are waiting, and it is not possible to shift any to longer runs, it may be wise to have the more powerful tractors, or those with the shortest runs, load without assistance, so that all can be kept moving, even if with a smaller average yardage.

Time and Distance.   A scraper loading pass may take from 20 seconds to 2½ minutes. It is best to keep it down to 1 minute or less.

Speed of both machines in the cut is determined by that of the pusher. Low-gear speed of most gearshift crawler tractors is about 1½ miles per hour, say from 2.2 to 2.5 feet per second. In heavy pushing, speed may be reduced below 1 foot per second by track slippage. Torque converter tractors in low gear may go up to 4 miles per hour, or 6 feet per second, when loads are light, but in heavy pushing may be slower than clutch-type units.

Length of the pass should be kept between 90 and 125 feet, as long runs are wasteful of both pusher and scraper time. The pusher usually has to go back the same distance it worked forward, and the scraper should shift up and start its haul.

Rubber-tire pushers may have low-gear speeds of 2½ miles per hour or more, and usually have the speed flexibility of the torque converter as well. The higher pushing speed synchronizes better with the low-gear speed of scrapers. Very fast travel gears are available for the return trip.

The higher speed of the four-wheel-drive pushers makes it practical to use longer runs in the cut, perhaps up to 200 feet. But the extra length in the pass largely cancels the speed advantage.

Quickest loads, in both time and distance, are obtained with easy digging, downhill loading, and small loads in proportion to total digging power.

When job study figures are not available, an estimator may assume that an average push is 1 minute long. This push represents about one-half the pusher cycle, as it will spend about as much time getting to and contacting each scraper as it does pushing it.

This gives a rule-of-thumb pusher cycle time of 2 minutes. With good operation and supervision it should be much shorter, but under field conditions it is just as likely to be longer.

Scrapers Serviced.   The number of scrapers that can be serviced by a pusher depends on the relationship between the length of the pusher and scraper cycles. The formula is

$$\text{Scrapers per pusher} = \frac{\text{scraper cycle}}{\text{pusher cycle}}$$

If a scraper cycle were 5 minutes (300 seconds) and the pusher cycle 2 minutes (120 seconds), the pusher could take care of $2\frac{1}{2}$ scrapers. This could be managed on a big job by using two pushers and five scrapers, but on a small one the contractor would run either three or two.

Three scrapers would mean waiting time for scrapers; two of them would not keep the pusher busy.

If pushing could be speeded up by more efficient patterns, better operating skills, or a faster machine, so that the pusher cycle were reduced to 100 seconds, one pusher could take care of the three scrapers.

The pusher cycle might be shortened to 90 seconds by keeping the scraper in the cut only 45 seconds and sending it out with a lighter load. Scraper cycle would be shortened by 15 seconds in the cut and by about 10 seconds in the haul and dump, by faster acceleration and faster spreading. On this basis one pusher could take care of three scrapers.

Figure 8.24 shows how these various arrangements would work out on a basis of a 45-minute hour, an hourly cost of $50 for a scraper and $45 for a pusher in the United States, an 18-yard load in 60 seconds, and a 16-yard load in 45 seconds.

Other Work for Pushers.   In ordinary backtrack work, the pusher uses one-half its travel distance and about one-third of its time moving from loaded scrapers to empty ones, usually in reverse. In addition it will spend a variable but often considerable amount of its time waiting for scrapers. The thrifty contractor will wish to make profitable use of this nonproductive time.

Backup time of dozer pushers may be used either to smooth the surface of the cut by back-dragging, or to loosen the ground with backripper teeth hinged to the blade.

This ground smoothing requires no extra equipment on a dozer pusher, and uses only a little extra time to vary return paths. It is quite effective in loose or sandy soils, but results may be poor on hard or stony ground.

Grading that must be done in forward gears reduces the time available for pushing scrapers, and the effort to combine duties is apt to lead to inefficiency in both assignments.

Ripping.   If the ground is too hard for good scraper loading, it may pay to rip it. This work can often be done by the pusher.

Backrippers are economical to buy, and permit doing most of the ripping while the pusher is backing from one scraper to the next. This usually does not work out so that the whole area can be loosened, but if the scrapers are able to load the material anyhow, whatever work the teeth do

Number of scrapers	3	2	3	3	5	5
Number of pushers	1	1	1	1	2	2
Push time, seconds	60	60	60	45	60	45
Scraper cycle time, seconds	300	300	300	275	300	275
Scraper waiting time for pusher	60	–	–	–	–	–
Pusher cycle time, seconds	120	120	100	90	120	90
Size of load, cubic yards	18	18	18	16	18	16
Loads per 45 minute hour	22.5	18	27	29.5	45	49
Yards per 45 minute hour	405	324	486	472	800	784
Scraper cost @ $50 per hour	$150	$100	$150	$150	$250	$250
Pusher cost @ $45 per hour	45	45	45	45	90	90
Equipment cost per hour, total	195	145	195	195	340	340
Cost per cubic yard, cents	48.1	44.8	40.1	41.3	42.5	43.4

FIGURE 8.24   Costs in scraper and pusher cycles.

is so much to the good. On the other hand, the operator may take ripping so seriously that she or he will let scrapers wait while finishing a strip. This can result in considerable loss of production.

Soils too hard for backrippers, and many rock formations, can be ripped for scraper loading by a rear-mounted ripper with hydraulic down pressure.

Methods of use and rate of production of these tools are discussed in Chap. 21. In most soils a big tractor should be able to keep three teeth in the ground, and break soil much faster than an ordinary fleet of scrapers could remove it. In frost or rock one tooth would be used, and output might be so reduced that one scraper could handle it.

A pusher can do heavy ripping only if it has enough time between scrapers to prepare enough soil for them. If it does not have this much time, the ripping will be incomplete or the scrapers will have to wait, or both. Each case must be judged on its own merits, but under such conditions it is usually more efficient to use one tractor for ripping and another for pushing.

If this arrangement leaves the ripper with idle time, it can be used as a tandem pusher when it is available, with the regular pusher carrying the whole load during ripping periods.

Economics of Ripping.   Many soils become so loose when they are broken up that they are harder to load, or provide smaller loads, than when in solid state. Unless loading results are clearly and definitely favorable, ripping should not be done.

Ripping rock for scrapers pays only when a fairly fine and uniform breakage is obtained. Scrapers load badly in coarse rock, and unless specially reinforced, may sustain severe damage from repeated contacts with big pieces and unbroken ledge.

WORK PATTERNS

Scraper work patterns should be arranged to allow for as many of the following as possible:

1. Digging downhill
2. Digging in the direction where load will be dumped
3. Utilization of pushed soil
4. Efficient turns with minimum deadheading
5. Cuts that start at high points, and fills at low ones

Direction of Digging.   A favorable grade increases the speed and the effectiveness of loading and reduces wear. The advantage becomes more marked as the downgrades get steeper.

Figure 8.25 shows three ways to make a deep scraper cut. Figure 8.25(A) is inefficient because the downgrade is used in transporting where little power is required and does not assist the digging. Figure 8.25(B) takes full advantage of the downgrade but may create an inconveniently sharp angle at the beginning of the cut.

In (C) the digging is started on the upgrade, just before the crest. The power requirement for the first few yards is small as resistance increases with load. The machine is rounded into the downgrade for the bulk of the load. This keeps the crest cut down without sacrificing much of the advantage of the slope.

Digging in the direction of the work is desirable. A loaded scraper moves more slowly, wears more, uses more fuel, and may be less stable on turns than an empty one. If the load is picked up heading toward the fill, it is able to take the shortest path to the dump, and to make the turns and the longer run between them empty.

However, there is often sufficient reason for digging away from the dump. Digging downhill is more important than direction. Occasionally a pusher can be best utilized if scrapers are loading in both directions, which, in a single cut and fill, would require about half the unit's load before turning to go to the fill.

Pushed Dirt.   The scraper knife usually pushes some dirt ahead of it, the amount increasing with the size of the load. Loose material such as sand may be moved in considerable quantities. This is left in low piles when the bowl is lifted.

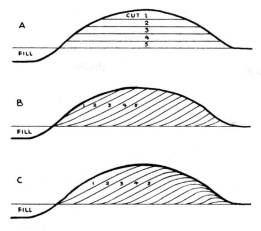

FIGURE 8.25   Taking off a hill.

This dirt can be utilized to build up the fill where it meets the cut, by allowing the bowl to drag slightly until the fill is reached. However, dragging may cause a loss of speed which outweighs the importance of the dirt moved.

Care should be taken not to cut below grade at the junction with the fill unless it is necessary to make a ramp.

Turns.   The time consumed in making a U turn with a scraper may vary from 5 to 60 seconds or more, depending on space available, ground conditions, type of machine, traffic, and operator. Fifteen seconds is a fair average.

Time consumed deadheading from the working area to the turn and back may be considered part of either the turn or the haul, but it is better practice to consider it a separate part of the cycle.

On short hauls, turns and deadhead time have an important effect on production. As hauls become longer, their significance decreases.

There are four major patterns of scraper operation, which are shown in Fig. 8.26. In the first, (*A*) and (*B*), there are two turns to each dig-dump cycle; in the second, (*C*) and (*D*), one; and in (*E*), one-half. Figure 8.26(*F*), with no U turns, is only practical when a very wide area, such as a field or runway, is being graded.

When both cut and fill are wide enough for easy turns, the (*C*) and (*D*) layouts may be most efficient, particularly when work areas are long and tractor speeds low. The advantage is that the scraper can turn to start a new cycle immediately after digging or dumping. The length of haul can therefore be figured between the centers of mass of the cut and the fill, as the longer and shorter runs will average out.

The diagram and arithmetic in Fig. 8.27 indicate the advantage of operating one cut with one fill under low-speed conditions, and combining them when travel speeds are higher.

If turns cannot be made immediately at the end of the work, the time required to travel the average distance from the ends of spreading runs, and from the beginning of cuts to their turns, must be added to the cycles, as in Fig. 8.28.

Through-travel highway patterns, seen in Fig. 8.26(*E*), have their greatest use where the graded area is too narrow for turns, and cuts and fills are rather short and closely spaced. Their efficiency depends on the extra time required for through travel, compared with that used for turns and deadheading.

Turn patterns are of course subject to the overall plan for the job, which may include some very complex factors of distribution, work sequence, and separating of different types of soil.

These examples are somewhat oversimplified for demonstration purposes.

Turnarounds.   The location of the turnaround in a narrow one-way cut is affected by the difficulty of making it. For efficiency, it should be slightly across the hillcrest from the fill, so that the

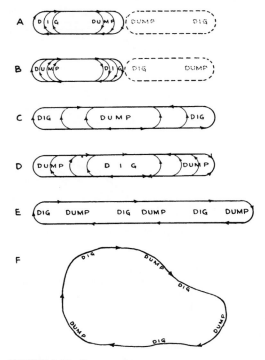

FIGURE 8.26 Scraper patterns.

scraper can be straightened out to load just before it crosses the crest. However, the digging will work the crest back and destroy the turnaround quite soon. It may therefore be wise to locate it well back from the crest.

Whenever possible, a turning place should be wide enough for the machines using it to get around without backing. Space requirements vary greatly in different sizes and types of scrapers. Time may be saved, and accidents reduced, by providing more space than the minimum requirement, particularly for sharp-turning models.

A big scraper will need 30 to 50 feet to make a U turn (180 degrees) on good ground, up to 80 feet on bad footing. If it is not possible to grade enough space for a full turn where it is needed, the machine must back one or more times, or go on until space is available.

The bottom of a cut for a four-lane undivided highway will usually allow somewhat more than tight turn space for a scraper; two-lane cuts may not permit a nonstop turn. Wider roads and the upper parts of any deep cut will allow ample space.

Deadheading. If the cut and/or the fill are too narrow for turns, or there are traffic difficulties, the scrapers will go past the work areas to a turnaround. The extra travel is called deadheading when traveling empty and is usually for only short distances.

At the cut the trip to the turnaround will be made in one direction at full travel speed, except for any necessary slowing for traffic, while the return will be at the speed of the turn or only slightly better, because of lack of space for acceleration. Beyond the fill the turn will be approached at dumping speed; the return from it will be a part of the empty return and will be at full acceleration if traffic allows.

Deadheading distances are often increased unnecessarily by careless dumping or loading that leaves the fill or cut too rough for a turn, or by placing too many grade stakes to permit turning in the work area.

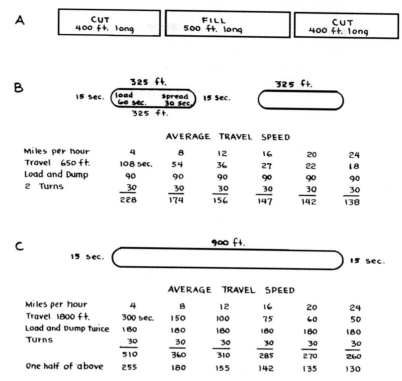

FIGURE 8.27 Turns may save time.

Intermediate Hauls.   Some hauls are too short for normal scraper use and too long for dozers.

A crawler dozer's output falls rapidly as length of haul increases. It becomes uneconomical somewhere between 100 and 200 feet on level ground, although it may be used for much longer distances. Because of complications of turning and distances required to load and to spread, scraper use generally starts at 200 to 400 feet. These figures are for large machines. Small bulldozers lose effectiveness at lesser distances, and small scrapers are operated on much shorter runs.

Because of their higher speed, rubber-tire dozers may be used for substantially longer pushes than crawlers. They are less efficient in a short run as, except in the easiest digging, the loading part of their cycle is longer.

Scrapers can often be used more efficiently on very short hauls than is generally believed. They can waste considerable time deadheading to turns, and carry undersize loads because of short digging runs, and still move much more dirt than a dozer.

When the fill is too short for proper spreading, part of the load can be carried around the turn and dropped on the way back to the cut.

Two-wheel drag scrapers, such as are used for land leveling, can be used shuttle fashion and in easy digging will move about double the load of a dozer blade. Large models are very wide so as to be inconvenient to transport and unable to work with other machines in ordinary roadways.

Haul Speed.   Scrapers have advertised top speeds from 20 to 30 and even 40 miles per hour. However, the average construction job does not permit much use of these high speeds, and even when they are used, the average speed of straightaway hauling is very much lower.

Top speed is limited by the space needed to attain it. A loaded scraper or other hauler in its highest gear has little surplus power for acceleration, after taking care of the rolling resistance of

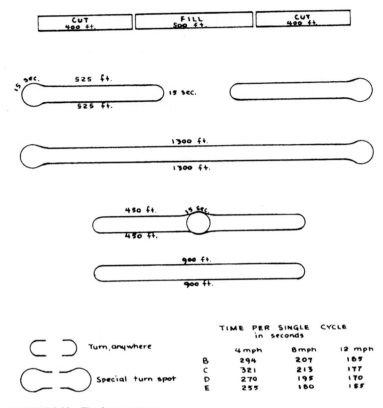

	TIME PER SINGLE CYCLE in seconds		
	4 mph	8 mph	12 mph
B	294	207	165
C	321	213	177
D	270	195	170
E	255	180	155

FIGURE 8.28 Fixed turn patterns.

big tires and dirt roads. It might take such a machine a half mile of level road to reach top speed in its top gear, and this distance is longer than the average scraper haul.

A favorable grade will encourage rapid acceleration, but it is likely to make high speed dangerous. And for every favorable grade in a round trip there must be an unfavorable one to balance it.

Scrapers are rough-riding vehicles. While the big tires absorb bumps, no way has been found to fit them with shock absorbers, and bouncing will prohibit full speed unless exceptional attention is paid to maintaining the haul road.

Scrapers with overhung, two-wheel tractors are likely to develop a rhythmic galloping effect even on smooth roads, unless equipped with a cushion hitch.

Hauler speed is discussed under Travel Resistance in Chap. 12. When there is no time to work out level ground performance in detail, figure 40 percent of top speed in the highest gear to be used for hauls up to 1,500 feet, 50 percent to 2,500 feet, and 60 percent from there on.

SHOVEL CUTS

Through cuts are made by excavators or front shovels and trucks when the original surface cannot be readily leveled for scraper operation, when the ground is rocky or wet, when the fill is too soft or too narrow for surface dumping, and when the haul is too long for scrapers.

In general, shovels do best in banks that are about 15 to 25 feet high. Soft, sliding banks of sand or gravel provide best digging when they are very high, but there may be danger from slides.

If the cut is considerably deeper than the favorable bank height, it may be taken in two or more layers or benches. On through cuts with a fairly level cross section, as in Fig. 8.29(A), the top is removed first. On sidehills there is an option of taking the top first, as in (B), or cutting the toe, then the top, then the floor of the upper cut, as in (C).

It is first necessary to build a haul road between the start of the shovel cut and the fill. Most of this may be already provided by highways or construction grading. It should be wide enough for two trucks, although for short or small jobs, one lane with turnouts may be adequate. Slope up to the shovel should not be over 15 percent, although in special cases, grades up to 35 percent have been used.

If no natural turnaround exists at the start of the excavation, and the grade is easy, trucks may be backed in at first. As excavation progresses, the pit floor will provide turning space.

The roads in (B) and (C) are usually cut by dozers. They allow a rotary movement of one-way traffic past the shovel on its first cut on each level. They eliminate the necessity of turning at the shovel in cramped quarters. However, they may be too expensive or inconvenient to build.

When there is no through road, the shovel starts each level by taking as wide a cut as it can reach, as in Fig. 8.30 (top), to allow space for two-way traffic, turning, and spotting two trucks at a time. Subsequent cuts are made about half as wide to facilitate loading and truck movement.

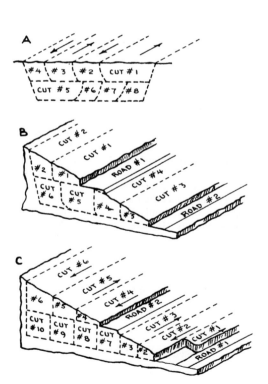

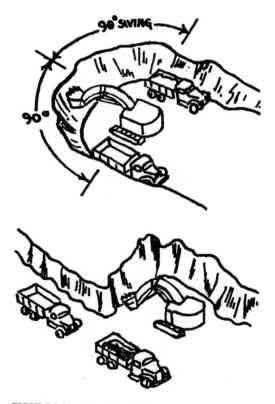

FIGURE 8.29 Sequence of shovel cuts.

FIGURE 8.30 Through and side cuts.

If the cut floor is soft, it is the best practice to use a dragline working from the top. However, if none is available, or the digging is too hard for it, the front shovel may work from supporting platforms and have a gravel, stone, or other road built behind it for the trucks. If it is working on the bottom level, the road can be left to facilitate surfacing work.

A busy shovel should have at least occasional help from a dozer, which can level the pit floor, clean up spilled dirt, get boulders out of the way, and assist stalled trucks. Even if the shovel can handle the operation without assistance, it will produce more if it needs only to dig.

A shovel cut should be started on the low side of the grade, and worked uphill so that it will drain. The floor should be shaped carefully to avoid excessive working over.

LOADER CUTS

Cuts made with front-end loaders, of either the crawler or wheel-mounted type, are similar to shovel cuts in many basic features. They are moderately deep, usually between the height of the push arm hinges and the maximum lift of the bucket edge. They should be sloped to drain, to avoid difficulties with rain or groundwater.

The loader is not as efficient as the shovel if the ground is muddy, soft, or loose; or if the bank is extremely hard. While it can work in narrow places, production may be greatly reduced by lack of generous space in which to turn.

The loader can keep its working area smooth and free of rocks. However, if there are enough trucks to keep it busy, it is often better to send in another machine for cleanup.

If the distance is 1,200 feet or less, large rubber-tire loaders may carry their loads to the fill more economically than trucks can haul them. See Chap. 16.

Belt and wheel loader cuts will be discussed in Chap. 14.

FILLS

Fills are made to bring a road or area up to a desired grade, to elevate it above water or drifting snow, to bury stumps or rocks, or to add strength to ground too unstable to support road surface or traffic.

Fill may be obtained by the removal of high spots or banks along the same road or project, by digging gutters or ditches alongside or near the fill, or by hauling from necessary excavation on other jobs, from commercial pits, or borrow pits opened just to obtain the fill.

Nearby cuts on the same project are usually the cheapest source, as the digging costs and part of the hauling can be charged against the excavation. Also, excavation in adjacent hills will lower the grade, and thereby decrease the volume of fill needed to carry the road across hollows.

Roads in hilly country are often engineered to balance the cuts and fills, so that all the material cut out of high spots is just enough to build up all the low spots. However, where the road crosses ridges of hard rock close to the surface, good borrow is available nearer fills, or snow removal problems are severe, it may be advisable to keep cuts to a minimum and haul in dirt.

Where very heavy hill cuts must be made without a corresponding need for fill, the surplus may be wasted in dumps off the road. This may be preferred to raising road levels to absorb the fill, because of the economy of a waste dump compared with a compacted highway fill. Also, high fills may require purchase of extra land width to avoid the need for steep and dangerous side slopes.

Cuts and fills on a road are sometimes so far apart that combining them would cost more than wasting spoil from the cut nearby and getting the fill from borrow pits.

Types of Fill. Any type of mineral earth or rock can be used as road fill, but clay and silt are generally undesirable. They soften when wet, frequently with changes in volume, and may act as a wick to bring groundwater to the surface. Humus is avoided, particularly in its pure state, because of lack of bearing strength and excessive water absorption. Topsoil, a mixture of mineral soil and humus, may or may not be permissible, depending on its qualities and its location in the fill.

Sand and loose, clean gravel have excellent bearing power but afford poor traction, are hard to compact, and must be held in by other materials.

The most desirable fills are mixtures of two or more simple types. Varying proportions of clay, silt, sand, gravel, and stones are found in loams, boulder clay, and glacial till. Sand and gravel are most desirable when mixed with enough clay or silt to bind them together. Various soil mixtures are described in Chap. 3.

Granular soils with a high percentage of sand or gravel are desirable when work must be done in rainy places or seasons. They absorb and drain off large quantities of water, and do not get slippery easily.

Moisture Content.   The water content of soils largely determines their behavior in a fill. Each soil has a best (optimal) water content which favors compaction. Less water will allow the grains too free motion in relation to each other, and more will permit soil to bend or creep away from pressure.

A soil which contains too much moisture will develop a rubbery quality. It will move away from the roller, and when its weight has passed, spring back into nearly its original position.

A loose soil may hold too much moisture for best compaction and still appear fairly dry. When the grains are squeezed together, water films between them are displaced and tend to work up toward the surface, rendering it wet. This condition may be cumulative through a number of layers of fill.

Some compaction is accomplished by rolling a rubbery soil, and the operation warms the ground and brings moisture to the top so that drying is speeded up.

The problem caused by soil that is too wet for specified compaction is discussed in Chap. 3.

If the soil is too dry, it is watered by sprinkler-equipped tank trucks or trailers while being spread and rolled.

Swell and Compaction.   Undisturbed soil has generally been in the same position for long periods. The particles are well settled against each other, leaving little space. Natural cements may bind them together.

When such a soil is dug or disturbed, it breaks up into chunks or grains which are thrown against each other in a disorderly arrangement, leaving airspaces or voids between. This increases the bulk of the soil, and increases its ability to absorb and conduct water. Such a loosened soil will turn to a very soft mud if soaked.

The process of soaking and then drying will settle the grains together somewhat, reducing the voids. Repeated wetting and drying will cause it to shrink to about its original bulk. Freezing and thawing will accelerate this settlement, as will also the weight of traffic or additional fill.

Compaction by Hauling Units.   Considerable packing down of fill can be done by hauling and grading equipment. Ground pressure under loaded scraper tires may be 30 to 40 pounds per square inch, and the kneading effect of these tires and/or the vibration of crawler tracks are quite effective.

However, compaction tends to decrease with distance from the cut, as all the fill material must pass over the near portion, and only a small fraction over the far end. In addition it is difficult, sometimes impossible, to get the operators to vary their routes enough to give systematic rolling to the full width. Routes may have to be shifted by stationing one or more operators along the way to tell or signal the operators where to go, or by the use of movable obstacles.

It is usually inadvisable to have a heavily loaded unit break a new path in soft fill, as the power requirement and strain on the machine are excessive. Trail breaking should be done on the empty return trip, and loaded units then turned into those tracks.

These difficulties can be avoided by compacting the fill as it is placed. Rollers of various types are used on thin layers of fill to squeeze the grains into even closer contact than they had in the bank. They are aided by the weight of grading and hauling units. Loam soils may be reduced to 90 percent of their bank volume by thorough compaction.

A properly compacted fill should not shrink on exposure to time and weather, so that it is theoretically possible to put a permanent surface on it immediately. In addition, it has the highest bearing power possible to its particular soil type, so that wheels and tracks will not sink into it much, and speed and capacity of hauling equipment on it are increased.

A compacted fill will not absorb rainwater readily, so that the fill should remain hard enough to work even after heavy rains. Whether the surface will become greasy depends on the clay content and the possible presence of a layer of dry uncompacted dust before the rain.

Rollers.   Rollers may be smooth steel wheel, tamping sheeps-foot, or grid, or rubber-tire. Some have rubber on two drive wheels, and a steel roll, rough or smooth. Rollers may depend entirely on weight and shape for effectiveness, or may have vibrators also. See Chap. 19.

The smooth steel-wheel models are just known as rollers. They are usually self-powered, and may have either two or three rolls. Weights range from $1\frac{1}{2}$ to over 20 tons. They are primarily finishing machines, used more often on surfaces than on subgrades. Tops of thick layers may be better compacted than bottoms.

These machines have little traction, particularly with tandem construction, and are not suited to rough ground. Grid or segmented drive rolls overcome this difficulty.

Towed tamping or sheepsfoot rollers are steel drums 4 or 5 feet long and 40 to 60 inches in diameter, fitted with projecting lugs (feet or legs) 7 or more inches long. There may be three lugs to every 2 square feet of drum surface.

Feet may have expanded soles (which may kick up soft dirt), or taper from a wide base to a flat end. Either way, they penetrate soft fill until weight is carried on the sole, where compaction begins. As the ground is compressed on successive passes, the feet do not sink as far and start to "walk out" of the ground.

Drums may be filled with sand or water. Sole pressures range from 250 to 750 pounds per square inch. The drums are mounted in box frames fitted with drawbars. Up to three may be mounted side by side, and two pairs may follow each other. Working speed is 2 to 3 miles per hour. Power requirements are high, particularly in soft fill.

These units are being replaced by self-powered units with one or two tamping drums, and a pair of rubber-tire drive wheels. This is in line with a general move away from towed equipment.

Pneumatic tired rollers are ballast boxes supported by wheels with smooth-tread tires. The wheels may roll straight, vibrate, or move up and down or wobble as they revolve. They compact by a combination of weight and kneading action of the soft tire walls. Weights vary up to more than 80 tons. They can compact single fill layers as deep as 24 inches.

Fill Bases.   It is desirable that a fill be firmly bonded to the surface on which it rests to prevent formation of saturated zones, water channels, and possible sliding downslopes. This is usually accomplished by removing vegetation and topsoil, and plowing ridges across any slopes.

Methods of removing humus and other muds from the location to be filled, and of stabilizing such muds when removal is impractical, are discussed in Chap. 3.

When the area to be filled is wet, rough, or otherwise impassable to machinery, the first layer is built by dump trucks and dozers to a height at least sufficient to carry the hauling units over the soft spots or obstacles.

If the surface is uneven but passable, low spots may be built up first with compacted layers, or high spots removed, before the main fill is placed.

Rock Separation.   Handling and compaction of fill material are rendered difficult by the presence of loose stone. Rocks of even small sizes interfere with grading. If their diameter is greater than that of the fill layer, they will project from the top. If two or more rocks are in contact, they are liable to prevent even distribution and compaction of fill under their adjoining edges.

For this reason, the size and number of rocks present in thin or layered fills are often limited. This may be done by using selected borrow, or by putting bouldery material through a grizzly.

The arrangement shown in Fig. 8.31 represents a minimum of equipment for screening. A truck on a high level dumps on a sloping grizzly, dirt falls between the bars into a truck parked below, and boulders roll to the side.

Oversize material may be allowed to roll directly into trucks, be loaded from beside the grizzly, dozed away from it to a stockpile, or, if the grizzly can be located on the edge of an abandoned pit, allowed to accumulate.

FIGURE 8.31   Screening oversized rock from fill.

One or two workers are needed to free oversize stones stuck between the bars and to coordinate the trucks. If the stones are a substantial part of the bulk of the soil, smaller trucks may be used under the grizzly than on top of it.

If sticking of stones or sliding off of chunks of earth is much of a problem, a vibrating grizzly, or a standard grizzly with a vibrator bolted to it, may be desirable. The flat slope illustrated is suitable only for loose soil and large openings.

Rock Fill.   Various results are obtained from all-rock fills. If the largest pieces are smaller than the depth of the fill, and sizes are mixed, including a good proportion of fines, a solid fill with a good surface may be obtained by pushing piles off an edge with a bulldozer. Large pieces tend to move ahead and over the bank, while smaller ones drift under the blade to form a topping.

If there are not enough small pieces to provide a working surface, finer material should be brought to fill surface holes and even off the top.

Rocks too large to fit in the fill can be rolled ahead of it until a hole is found or is made to bury them, or they are reduced by splitting or blasting.

Rock fills are generally almost incompressible, exceptions being when rock is soft or fissured, and very heavy weights are used. However, they are apt to be subject to only minor and local settlement, where fines are shaken or washed into spaces between rocks below them.

Rock is desirable fill material for the bottom layer in crossing water or mud, as it is not softened by contact with water and spreads surface loads over large areas of the base. In such locations it may settle due to displacement or compression of the ground under it.

The volume of fill is greater than the unbroken rock in the bank. The difference will vary with the quality of rock, type of fragmentation, and amount and kind of compaction. And 50 percent is a rule-of-thumb average that can be used except where there are indications to the contrary.

If the rock must be used in the fill, it is best placed at the bottom. Unfortunately, rock is ordinarily the last material to be taken from a cut as soil is stripped prior to blasting.

SCRAPER FILLS

The standard method of building a fill with scrapers is to start with an area sufficiently leveled to allow the scrapers to travel on it, and to build it up in thin layers, starting at the outside edges or at low spots.

Spreading depth may vary from 2 inches to the maximum lift of the bowl—8 inches to 1 foot. Thin layers favor compaction (see Fig. 8.32), particularly if the scrapers are depended on for the rolling and facilitate smooth building up of the grade. Chunky, sticky, or rocky fill will not spread thinly, or can be made to do so only by very slow travel during the dump.

FIGURE 8.32   Scraper spreading fill.

Thick spreading is liable to flow out of the bowl more smoothly, can be done at higher speed, and reduces the dump time. However, it tends to make a rough fill which will require slower travel speeds, or smoothing work with a dozer or grader.

Edges.   If a fill is high, the edges may be troublesome and dangerous unless carefully made. The problems are to keep it at the correct toe alignment, proper slope, at full density or compaction, and not rolling any machinery off it. These problems are affected by the nature of the fill and by its height and slope.

Loose fills of sand, clean gravel, or too-dry dirt tend to cave under the weight of machinery close to the edges. Finer-grained fills may have excellent bearing power if well compacted and not too wet. However, while being compacted, they tend to squeeze outward, and an allowance for this creeping must be made when placing the first fill, so that it will not move out past the toe stakes.

The behavior of the fill on edges may be anticipated by making soil analysis or by consulting with contractors or machinery operators who have worked with the same formation.

Except for allowance made for creeping under load, or spillage from above, which seldom should be more than a foot or two, the fill is started at the toe line and built up of layers, usually not over 6 or 8 inches, loose. Each layer should be rolled with a tamping roller that is allowed to project slightly beyond the edge. For this purpose, two or more rollers should be fastened in a single yoke so that their width will be substantially greater than that of the tow tractor, which should not have to walk on the edge. This is particularly important with high banks and wheel tractors.

If watering is required for proper compaction, application may be somewhat heavier at the edge to allow for side evaporation. However, it should not be sufficient to make it soft or muddy.

The fill should slope up at the edges in order to incline the center of gravity of the machinery toward the center and minimize the danger of caving. If the fill is narrow, it will have a trough shape, and if wide, it will be flat with raised sides.

This slope is most easily started by a grader or an angle dozer working over the first layer or two left by the scrapers. Once made, it will tend to preserve itself as the tilt will tend to make the inside wheels of the scraper sink more deeply. If it becomes too steep, it is readily reduced by filling toward the center.

If the job is shut down during any period when rain is expected, it may be wise to build the fill up to a crown in order to allow it to drain. This involves resloping the edges on resumption of work, and, if the work is done under exact compaction specifications, may cause confusion in the treatment of the tapered layers required.

Another treatment is to preserve the trough but so grade it that all water will flow to selected low spots. Here ditches are dug through the raised edges, and troughs of metal or wood placed to lead the water down the slope. This is readily done in hilly country, where most of the road is on definite gradients, but not in level country.

Such drain ditches may be made wide and gentle, so that they can be dug, backfilled, and compacted by machinery, or may be hand-dug, refilled, and tamped.

If the trough shape is left without precautions, a center gully may be scoured by a heavy rain, a pond formed in low spots, and damage done to edges by overtopping and concentrated runoff.

Scraper distance from the edge is determined by depth of spread and slope. If a slope is 1 on 2, spreads are 6 inches deep, and compaction is one-third, each pass will be 8 inches inside the previous one.

If the edge is not firm enough to support scrapers at the proper distance to dump loads, they should be spread farther back and dirt cast out to the edge by a grader or angling dozer.

Additional slope stakes should be set as a high fill is built up to maintain the correct width.

DRAINAGE

Drainage is an important factor in the construction of most roadways. Groundwater must be kept far enough below the surface so as not to damage it, or weaken the subgrade directly or by supplying capillary water. Water falling on the surface of the road must be conducted off it, and runoff or streams crossing the road must be provided for.

Raising the Grade.   In swamps and lowlands, the only practical method of getting the road well above groundwater is to build a high fill. If the base course can be made entirely of rock, it will break any contact between the water and the balance of the fill. Clean gravel or coarse sand may serve the same purpose.

Proper quality of fill can reduce the required height substantially. However, it is often more economical to make a higher fill of inferior material obtained from roadside ditches, as in Fig. 8.33(A), and it is often possible to lower the water level by the same operation. Draglines or excavators are generally used, but dipper dredges may be preferred when much of the land is under water, or it is intersected by numerous channels.

Either machine may work along the ditch lines, piling spoil toward the center, as in (B). The dragline will work away from the cut, as shown, but the dredge will float in it.

If a dragline has a sufficiently long boom, it can travel on the road centerline, and dig both ditches and pile the spoil in one pass, as in (D).

Road fill may also be obtained by ditching in dry flatlands where the road is to be raised above floods or snowdrifts. In such circumstances elevating graders may be used as shown in Fig. 8.34, or dozers or scrapers.

Tiling.   In sloping land, it is usually more economical to lower the water table by drainage. The standard method, Fig. 8.35, is to put shallow ditches to carry surface water through cuts (A), and, if necessary, to place porous tile or other drains (B) 2 to 3 feet deeper in loam soils. Silt or clay deposits may require a drain depth of as much as 7 feet, but in such a case, better results may be obtained by a normal drain depth and by the use of a layer of pervious material under the road.

The design of subsurface drains must be carefully adapted to the requirements of the particular job. The ground may drain naturally so that no work is necessary. There may be a saturated condition that could be relieved by providing a drain through an impervious barrier (C), or by cutting

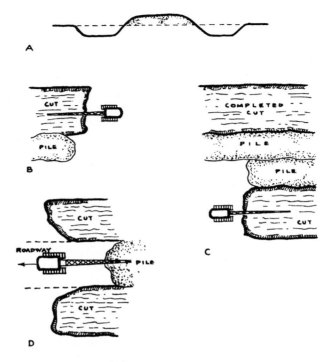

FIGURE 8.33 Road fill from gutters.

FIGURE 8.34 Borrowing road fill from sides.

off the source of water (*D*). There may be springs or seepage rising under the road, which would require center or lateral drains, (*E*) and (*F*). Such drains could also be needed to take off water soaking through a porous road surface.

When the ground is generally dry and firm but has local springs or seepage, the wet areas should be dug well below the intended drain and backfilled with stones and clean gravel, topped

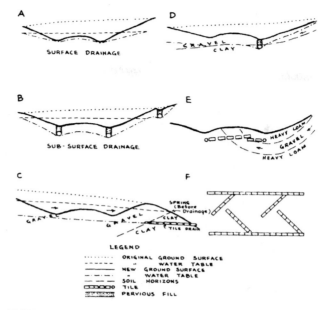

FIGURE 8.35 Roadway subdrainage.

with sand. The drain itself may be any type of pipe, laid at the lowest convenient level, and opening into side drains, a catch basin, or a gutter.

The rock fill directs the water toward the pipe and reduces or eliminates softening of adjacent areas.

JOB STUDY

Road construction may involve clearing vegetation; stripping and storing of topsoil; excavating soil and rock to cut natural levels to road grades; hauling the spoil to road fills or waste dumps; building culverts, bridges, and drainage systems; raising low areas to road grade by fill obtained from roadway cuts or borrow pits and finishing, topsoiling, and seeding of slopes; and cleaning up the work area.

Usually, this work must be accomplished within a time limit. It is desirable to get the maximum number of machines and workers on the job as soon as possible after the start, but it is more important to keep them efficiently employed once they are there.

Sequences. When time permits, it is often desirable to perform complete operations in sequence. If an entire work area is cleared, it will usually be easier to arrange dirt moving sequences than if the excavators have to be limited to a few small sections. Culvert construction should be completed before fills are raised high enough to go over them, unless they are to be installed by ditching the completed subgrade.

Liberal areas of rock should be cleaned before drilling starts. Pioneer bulldozer work should be well advanced before scrapers operate.

If the schedule is close, delay in one operation will delay others that have to wait for it, which may be more costly in machine and worker time. These secondary delays are much more serious when the maximum amount of equipment is crammed into a job than when a few units are doing it over a longer period.

Basic Factors. Basic factors to be considered in figuring grading for a road may include

1. Clearing costs
2. Topsoil stripping, storage, reclamation, spreading, and planting
3. Amount and type of soil excavation in cuts or borrow pits
4. Amount and type of rock excavation
5. Availability of suitable borrow and cost of purchase
6. Haul road construction and maintenance, and length of hauls
7. Quality of fill required, and processing required of material from cuts and pits
8. Fill compaction, shrinkage, and disposal of surplus
9. Slope finishing and protection
10. Groundwater conditions and drainage requirements
11. Structures such as bridges, culverts, and retaining walls
12. Possession or availability of proper machinery, with necessary parts and supplies; extra costs of using second-choice or beat-up equipment
13. Availability of construction supplies such as pipe, forms, etc.
14. Labor supply
15. Weather—rain, snow, ice, dust, frozen ground, frozen equipment, mud
16. Time of completion of related structures such as bridges, being built under separate contract

In roadway work, the amount, kind, and location of cut, borrow, and fill, and the length of haul, may be specified. Haul may be described as *normal* or free, up to a certain distance, which may be 300 to 1,000 feet, and longer hauls called *overhaul.* Excavation may be described as *unclassified,* or divided into rock yards and dirt yards.

In less formal jobs, these factors may be indicated only approximately, or may be figured by the contractor from grade or route plans.

Casual Estimating. Where cuts and fills are shallow, and side slopes lacking or moderate, grading can often be estimated fairly accurately by inspection of centerline stakes. The exact yardage is sometimes not of primary importance, as stripping topsoil and working over a piece of ground represent an amount of machine time that may be only moderately increased by the cuts and fills.

Several errors must be watched for, however. Cuts and fills on the stakes may be figured from the top of the stake, from ground level, or from a line on the stake. The grade indicated may be subgrade, in which case it is taken at face value, or finish grade when the depth of base courses and of surfacing must be added to the cuts and subtracted from the fills. The width to be figured on is not only the road and shoulders, but also gutters and slopes. The depth of topsoil to be stripped is subtracted from the cuts, added to the fills, and considered separately as an important cost factor.

When cuts or fills are deep, side slopes exist, topsoil need not be stripped, or when the job is large, yardages should be carefully calculated. If this is not done on the plans, the contractor can do it.

YARDAGE CALCULATION

Center Profile. The minimum staking for a road is the centerline. When this is done, a profile is taken, showing the elevation of the ground at each stake. These elevations are plotted on cross-section paper, usually with the vertical scale 10 times the horizontal, and the points connected by a line. A profile for the road is then sketched in according to the standards of grade and vertical curve required, or from some previously formed plan. This line should represent the subgrade before the addition of any imported material.

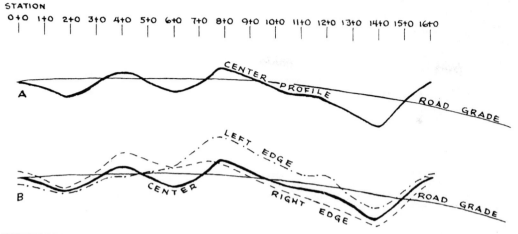

FIGURE 8.36   Center and side profiles.

Distances measured from the road line to the ground line will indicate the depths of cut and fill required to establish the road grade. If topsoil is to be stripped, its depth should be added to the fills and subtracted from the cuts.

If the ground does not slope across the line of the road, this type of profile, shown in Fig. 8.36(A), should give a reasonably accurate picture of the relative volume of cuts and fills, and the distances they are to be moved. However, to obtain yardages, cross sections usually must be calculated, as described below.

Side Profiles.   If the road is laid out on side hills, side stakes and slope stakes may be set. The side stakes may be at the edge of the pavement, at the outer edge of the shoulder, or the far side of the gutter, if any. In general, the shoulder or the gutter locations are preferable. Slope stakes are placed where the intended cut in a bank reaches its top, or at the outer, base edge of a proposed fill. These are not placed until cross sections are calculated.

If the side stake elevations are plotted in the same manner as the centerline, two additional profiles can be drawn, as in (B). These will give additional information about the bulk of material to be moved, but since they often do not include cuts for gutters, and cannot show the volume which must be dug or filled for side slopes outside the road lines, they are not an adequate basis for careful calculation.

Cross Sections.   A cross section is a profile taken at right angles to the line of the roadway. It is at least long enough to include the full width that will be graded. Such profiles are sometimes taken with hand or string levels. They may be taken at each 100-foot station, plus points where the ground surface changes, or, in smooth terrain, less frequently.

This cross profile is also drawn on cross-section paper, preferably on the same vertical scale as the center profile. Horizontal scale may be the same as vertical, or at any convenient proportion to it. The cross section of the road subgrade is drawn in.

A number of such cross sections are shown in Fig. 8.37, together with the cut and fill for each.

Wherever the ground line is above the road line, there will be a cut; and where the road line is higher than the ground line, there will be a fill. If topsoil is to be stripped and saved, it may be advantageous to lower the ground line by the depth of the topsoil to save confusion.

Figure 8.38 shows a sample cross section with the arithmetic involved in computing its area. The road and gutter surfaces have been simplified in the calculation, as this reduces the work without introducing too large an error for rough figuring. The problem is also simplified by a horizontal ground surface.

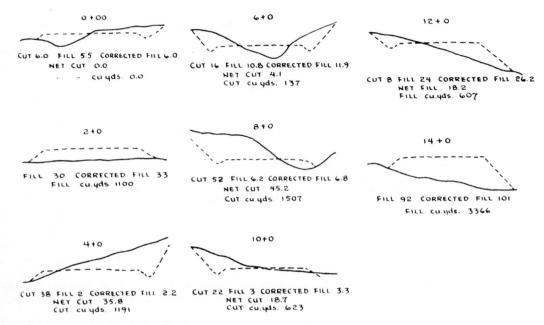

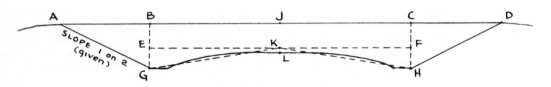

FIGURE 8.37 Roadway cross sections.

Ground Surface AD Level

Cut at ℄ JL   4'-0" given

" " gutter BG   6'-0" "

GK & KH are simplified subgrade lines

JK, BE & CF – each   3.3' scaled

BC = EK +KF =   35'-0" given

AB = CD =   12'-0" scaled

EG = FH =   2.7' scaled

Road & Gutter Cut = ▭BEFC + △EGK + △KFH

Area ▭BEFC = BE × BC = 3.3 × 35 =   115.50'

Area △EGK = △KFH = $\frac{EG \times EK}{2}$ = $\frac{2.7 \times 17.5}{2}$ =

$\frac{47.25}{2}$

Area △EGK + △KFH = 2 × $\frac{47.25}{2}$ =   47.25

9 | 162.75

Area road cut cross section (sq.yds.) = 18.08

Area side cut = △ABG + △CHD

" △ABG = $\frac{AB \times BG}{2}$ = $\frac{12 \times 6}{2}$ = $\frac{72}{2}$ = 36

" △CHD = $\frac{CD \times CH}{2}$ = $\frac{12 \times 6}{2}$ = $\frac{72}{2}$ = 36

9 | 72

Area slope sections (sq.yds.)   8.00

Total cut area (sq.yds.) 26.08

FIGURE 8.38 Figuring a cross section.

The cut is divided into slope triangles; and a road section, which in turn is divided by line EF into a rectangle and two triangles. The data given by the engineer are labeled *given,* and those measured off the diagram as *scaled.* The areas of the triangles and the rectangle are readily computed, their measurements in feet being used for convenience, and the result is converted to square yards by dividing by 9.

The areas of a succession of cross sections are obtained in this manner and averaged by adding together and dividing by the number of sections added. The result is multiplied by the length in yards of the area in which the sections were taken, giving the number of cubic yards of excavation required. Figures from in-place measurements are in bank yards.

Where the ground slopes irregularly, the ground surface is simplified by drawing straight lines, and the cut and fill areas are divided into triangles.

The road and gutter cuts could be figured by averaging the width and average depth at each cross section, then multiplying the product by the length of the sectioned area. The slope sections cannot be averaged, as their areas vary with the square of the depth of cut, and use of average depth would indicate a much smaller yardage than is actually required.

The most convenient way to measure the areas of cut and fill is by counting squares and fractions of squares. If a lot of work is to be done, areas can be measured by means of a planimeter.

Fill Shrinkage.   When fills are rolled to the compaction required in modern roadways, the material is often compressed into a smaller space than it occupied in the bank. This shrinkage should be allowed for in figuring cross sections. Loam soils often shrink 10 percent, clean sand 5 percent or less, and blasted rock, not mixed with other dirt, will show a minus shrinkage, or swell.

Compaction by hauling equipment without rolling is variable and will seldom cause shrinkage.

The examples in Fig. 8.37 use a shrinkage factor of 10 percent, but the figure selected should depend on job conditions.

Net Cut or Fill.   On side hills, one station is likely to include both cut and fill. The smaller amount is subtracted from the larger, giving net cut or net corrected fill.

Converting to Cubic Yards.   The net square yards of the cross section is converted to cubic yards by multiplying by the length of the road it represents. If sections are taken at 100-foot intervals, each will represent a piece 100 feet long, that is, halfway to the next section, on each side. If a special section is taken 40 feet from a 100-foot station, it will cover 20 feet on one side and 30 feet on the other—a total of 50 feet. The adjoining sections will be reduced proportionately.

When the 100-foot interval is used, it represents $33\frac{1}{3}$ yards. It is easier to multiply the section in square yards by 100, then divide by 3, than to multiply by $33\frac{1}{3}$.

The net cut and net fill figures, when converted to cubic yards, are used in making a mass profile. The gross cut figures are converted to cubic yards in the same manner to determine the total excavation, exclusive of topsoil.

Topsoil volume is figured by multiplying the length of the road, the average width to be stripped as indicated by the cross sections, and the average depth.

Cubic yards of net cut are added and compared with the total of net fill yards, to determine whether extra fill will have to be obtained from pits or whether fill will have to be wasted outside the road area.

Mass Profile.   A mass profile is prepared by drawing on cross-section paper a straight line to indicate the road grade, dividing it into stations, and posting cubic yards of net cut above it and net corrected fill below it, on any convenient scale. It is sometimes helpful to draw in blocks representing the fill at each station, as in Fig. 8.39(*A*).

A curved line, the mass profile, is drawn connecting the station points. The amount of net cut or net fill at any point along the road can now be scaled off, as well as the haul distance between cuts and fills.

The haul distance is measured between the centers of mass, or centers of gravity, of the cut and fill. The longer and shorter hauls should average out.

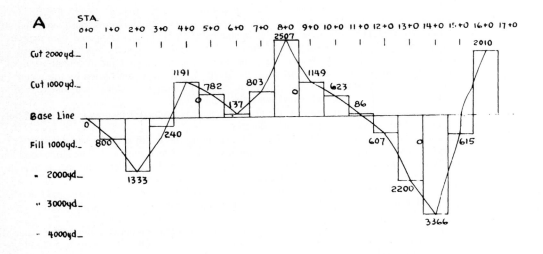

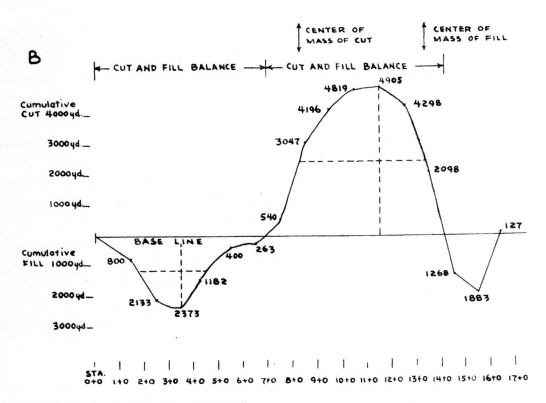

FIGURE 8.39   Mass profile (*A*) and mass diagram (*B*).

Mass Diagram.   Many engineers prefer to use the mass diagram shown in (B). A straight baseline or zero line is drawn on cross-section paper and marked off for road stations, and plus and minus yardages in the same manner as for the mass profile.

Points are plotted for cumulative or total yardages, starting at zero station. Points are placed one-half an interval farther to the right than the station they represent, as the full yardage figured for each station is not accumulated until the end of that station block.

At station 1 + 50, the minus yards of fill for station 1 + 0 is entered. At station 2 + 50, the total of the fill for stations 1 + 0 and 2 + 0 is posted; and at 3 + 50, the total fills for stations 1, 2, and 3.

When a cut is reached, at 4 + 0, the cut yardage is subtracted from the accumulated fill so that the line turns up. This line, called the mass curve, crosses the baseline when the accumulated cut equals the accumulated fill, and continued cuts raise it above that line until a fill is reached and pulls it down.

In short, wherever accumulated fill, starting at zero station, exceeds accumulated cut, the mass curve will be below the base. If there is an excess of cut, it will be above.

The mass curve line does not show total yardages of either cut or fill.

The points of loops which are farthest from the baseline indicate changes from cut to fill, or fill to cut. They also represent the total net yardage to be moved from cut to fill along the road line, but disregard sidecast material.

Any horizontal line drawn on the diagram is called a balance line. The yardages between any two places at which it intersects the curve have a balance of cut and fill. The baseline will often serve as a balance line, as in the illustration.

The centers of mass of a cut or fill can be found by drawing a vertical line from the outermost point of a loop to the balance line. A horizontal line is drawn through the center point of this vertical. Its points of intersection with the sides of the loop are approximately at the center of gravity of the cut and the fill for that balance line.

A single balance line may be used for the whole road. Any part of the mass curve which extends beyond the last balance point to the first or last station of the road will represent yardage to be borrowed, if it is below the base; or to be wasted, if it is above.

A loop above the balance line indicates fill movement to the right in the diagram, and below it to the left.

Any number of balance lines can be used as long as they end in points on the mass curve and do not overlap. The vertical distance between two balance lines represents borrow or waste in the part of the curve connecting them.

The mass diagram is a very flexible and useful aid in studying yardage distribution. However, it is so confusing to persons not accustomed to this type of computation that the average contractor working out such a problem may do better to use a mass profile.

SOIL AND CEMENT MIXTURES

Types.   There are three general types of soil and cement mixtures: compacted soil-cement, cement-modified soil, and plastic soil-cement. The first of these is by far the most important, and will be discussed in some detail in a following section.

In all of these, portland cement is mixed with native (in-place) soil or with borrow from nearby sources, to produce a low-cost, consolidated material. For reasons of economy, cement content is usually held to the minimum to produce a specific result.

Cement-Modified Soil.   This is the most economical and casual of these cement mixtures. The cement used may be only 1 or 2 percent of the soil volume.

The purpose is usually to improve a soil that is unfit for a pavement base or subgrade. The modified product may be caked or slightly hardened, but the principal object is to reduce plasticity, water-holding and volume-change capacities, and increase its load-bearing strength.

Soft spots encountered in underlying soil in building secondary roads are sometimes stabilized by mixing in small quantities of cement. Modification is applicable to a wide range of soils, including expansive clays. Mixing is usually done on the road, in the same manner as soil-cement, but less attention is paid to compaction.

Plastic Soil-Cement.   This is a mixture in which sufficient water is included to make it soft, like plastering mortar. It does not require compaction, and therefore can be used in places inaccessible to road-building equipment.

Plastic soil-cement is used to line or pave ditches, slopes, canal banks, and other places that are subject to erosion.

It may also be made by mixing high early cement into the natural material in mudholes, for a temporary emergency patch. Hand shovels and rakes are the usual tools.

COMPACTED SOIL-CEMENT

Compacted soil-cement is usually referred to just as soil-cement, as the other two kinds of soil-cement mixtures have comparatively little use.

This soil-cement is a mix of pulverized soil and carefully calculated amounts of portland cement and water, compacted to a high density. The resulting material is a rigid slab, with moderate compressive strength, which is resistant to the disintegrating effects of wetting and drying and freezing and thawing.

Soil.   Practically all substandard soils can be improved for use as structural material by mixing with portland cement. However, many of them require excessive quantities of cement or are difficult to work, and therefore are seldom processed in this manner.

Stabilization is most efficient with sandy or gravelly soils with 10 to 35 percent silt and clay, with not over 45 percent of the pieces larger than $\frac{1}{4}$ inch.

Sandy soils with few fines, or none, require somewhat more cement, and may create traction problems for rubber-tire processing equipment.

Silty and clayey soils make satisfactory soil-cement, but clay may have a high cement requirement and may be unusable if it cannot be pulverized. Both season and weather are important when working with these soils.

Figure 8.40 gives average cement requirements, by both volume and weight, for various soil types and miscellaneous materials.

Stones over an inch or two in diameter are considered highly undesirable. Old blacktop can be included in the mix if it can be broken into fine enough pieces.

Organic material in the soil usually has a very unfavorable effect on soil-cement, so the use of topsoil, or soil contaminated with topsoil, should be avoided whenever possible. However, the effect depends partly on other soil qualities, and satisfactory results have been obtained with organic content as high as 3 percent, without unreasonable increase in cement.

Mixture Design.   Presumably because of its appeal as a low-cost road material, soil-cement is not mixed to obtain maximum strength or durability, but to use the minimum amount of cement that will enable it to pass two standard laboratory tests.

The wet-dry test involves samples containing varying proportions of cement that have cured in high humidity for 7 days. They are weighed, submerged in tap water at room temperature for 5 hours, then placed in an oven at 160°F (71°C) for 42 hours. They are then given two firm strokes on all sides with a wire brush to remove material loosened by the wetting and drying. Reweighing and subtracting the new weight from the old indicate the amount of disintegration that occurred during the cycle.

This process is repeated 12 times. The passing grade ranges from 14 percent loss for sandy or gravelly soils down to 7 percent for clayey soil.

Normal Range of Cement Requirements for B- and C-Horizon Soils†

AASHO SOIL GROUP	Percent by vol.	Percent by wt.
A-1-a	5-7	3-5
A-1-b	7-9	5-8
A-2-4 A-2-5 A-2-6 A-2-7	7-10	5-9
A-3	8-12	7-11
A-4	8-12	7-12
A-5	8-12	8-13
A-6	10-14	9-15
A-7	10-14	10-16

†A-horizon soils (topsoils) may contain organic or other material detrimental to cement reaction and may require higher cement factors. For dark grey to grey A-horizon soils, increase the above cement contents 4 percentage points; for black A-horizon soils, 6 percentage points.

Average Cement Requirements of Miscellaneous Materials

MATERIAL	Percent by vol.	Percent by wt.
Caliche	8	7
Chat	8	7
Chert	9	8
Cinders	8	8
Limestone screenings	7	5
Marl	11	11
Red dog	9	8
Scoria containing plus No. 4 material	12	11
Scoria (minus No. 4 material only)	8	7
Shale or disintegrated shale	11	10
Shell soils	8	7
Slag (air-cooled)	9	7
Slag (water-cooled)	10	12

FIGURE 8.40 Cement requirements for soils. (*Courtesy of Portland Cement Association.*)

In the freeze-thaw tests, 7-day specimens are placed on moist blotters, refrigerated at $-10°F$ ($-23°C$) for 24 hours. They are then thawed in a moist atmosphere at $70°F$ ($21°C$) for 23 hours. They are brushed, and half-loose scales are pried off with an ice pick. After 12 cycles, the specimens are oven-dried and weighed. Permissible, or perhaps desirable, loss is the same as in the wet-dry test.

Compressive strength, in spite of its great importance in the pavement, is not necessarily tested directly, but is assumed from these two. The 7-day strength ranges from 300 to 800 pounds per square inch. After long curing, it may be 1,200 to 3,000 psi.

Water. Water requirement is figured out very carefully for optimum compaction, without regard to the quantity needed for hydration of the cement. Air-dry, pulverized soil is weighed in a laboratory, then blended with the proper amount of cement. Water is mixed in, in small measured quantities, and compaction tested after each addition. A moisture-density curve is plotted from these results.

Moisture content of soil is averaged from a number of spot-check tests, and subtracted from the optimal moisture. That is, if the soil has 10 percent water by weight, and contains 4, then 6 percent must be added. An arbitrary assumption of combined absorption by the cement, and loss from evaporation, is made, perhaps 2 percent. The result is that this soil-cement mixture needs 8 percent of its dry weight in additional water.

The weight of material going through the mixer per minute is calculated and corrected to dry weight. Assuming that this worked out to 9,000 pounds, the water requirement would be $.08 \times 9,000$, or 720 pounds. At 8.33 pounds per gallon, this would be 86.4 gallons per minute.

This water is supplied to on-the-road mixers by tank truck. The mixer should have a small tank to keep it working during truck changes.

When weather is dry, additional water may be added almost constantly to the surface during processing, as a light or fog spray from a distributor truck. The amount of such addition should just compensate for evaporation, and depends upon the judgment of the foreman or inspector.

Pavement Design.   Soil-cement is used in two principal ways. In one, it is the structural part of a pavement, but is protected against surface abrasion by a bituminous coating or pavement. In the other, it is a base (called subbase) for a concrete pavement.

Methods of construction are almost identical. The pavement use will be considered first.

Soil-cement usually has a compacted depth of 6 to 8 inches. On firm subgrades and under light traffic conditions, this may be reduced to 5 or even 4 inches, but then its advantage over granular bases becomes doubtful.

Twelve or more inches may be placed under special circumstances, but 8 inches is the most that should be put down in one layer.

To develop its full strength, the correct proportions of cement, water, and soil must be present, and compaction must be 95 to 100 percent or (more) of maximum density.

Preparation.   If the original road surface is to provide the soil for the new pavement, it must be carefully shaped to final grade. Very little displacement of material occurs during mixing, and any inaccuracies in original grade will cause extra work at the end and probably result in too-thin and too-thick areas in the pavement.

If the roadbed contains oversized stone, or is so hard that the mixing machines do not reach design depth, it should be scarified or ripped until loose enough to work, and until all stones have been pulled to the surface and removed. Grading, or regrading, should follow this operation.

Any soft spots that develop during this work should be repaired, as proper compaction cannot be obtained on a yielding base. It may be necessary to dig out and replace wet material. It may be possible to stabilize limited areas by mixing in high early cement. Permanent drainage should be installed where indicated.

Clayey soils that are in dry, hard lumps usually are easier to pulverize if water is added a day or so in advance. If the lumps are wet, they should be partly dried by scarifying or disking repeatedly to expose them to air and sun.

If the mix soil is to be hauled in, the road must be shaped and made firm, but scarifying is not required unless to remove projecting stone.

Cement.   On a small job, cement may be placed on the road in bags in a regular pattern. (See Fig. 8.41.) A marked tape or string can be used as a guide in distributing them properly. The bags are opened by hand and spread in windrows across the work strip. These windrows are then spread lengthwise by a small tractor pulling a spike-tooth harrow or some other drag. (See Fig. 8.42.) Several trips back and forth should distribute the cement in a fairly even layer.

But in most of this work, the cement is brought in trucks and distributed by a mechanical spreader. (See Fig. 8.43.) Ordinary dump trucks, fitted with a body cover to prevent blowing, can be used. There are bulk cement trucks that discharge by means of an auger, and agricultural lime trucks which might be able to both haul and spread.

FIGURE 8.41   Distributing bags of cement. (*Courtesy of Portland Cement Association.*)

FIGURE 8.42   Preliminary mixing of cement. (*Courtesy of Portland Cement Association.*)

The spreader should be towed rather than pushed to avoid picking up of cement by the truck tires. It should be kept at least half full, to avoid changes in rate of flow.

The rate of cement delivery is rather critical. There must be enough to keep the mixer working, but cement is vulnerable to loss by wind and spoilage by rain, so to spread it far in advance is risky.

Mixing.   Two types of in-place mixers are used, single-shaft and multiple-shaft. A single-shaft machine is shown in Fig. 8.44. As the machine moves forward, rapidly revolving tines strike downward at the cement and earth, pulverizing and mixing with repeated blows as the mixture is carried around inside a covering hood. Water may be added by a spray bar in the hood, or it may have been applied to the soil ahead of the mixer.

FIGURE 8.43 Mechanical cement spreader. (*Courtesy of Portland Cement Association.*)

FIGURE 8.44 Single-shaft mixer. (*Courtesy of Portland Cement Association.*)

This machine might accomplish the whole mixing operation in one pass, but two or more are usually made to be sure.

Multiple-axle machines have either two or three axles, each equipped with tines. The first set pulverizes and mixes; water is then added by a spray bar supplied by a tank truck; the second and possibly the third set complete the job. These machines are often called single-pass.

If the rear plate of a mixer hood is lifted, the material is ejected a considerable distance to the rear. On a sunny day this, as a preliminary operation, will produce quick drying of too-wet soil.

If the soil is trucked in, it is dumped, then bladed by a grader or shaped by a towed proportioner into a symmetric windrow for the width being worked, with cement usually added on the top. Another type of mixer picks up the windrow and passes it through a revolving drum. The first few paddles blend the dry material, then water is sprayed in and mixing completed by the remaining paddles.

The material is discharged to the rear, and spread by an attached strike-off blade, by a following grader, or by both.

The width that can be processed in one pass is limited by the width of the in-place mixer, or the capacity of the windrow machine. This is usually around 10 feet, or one-half the width of a two-lane road.

If traffic must be maintained, a quite long section of one lane is completed, traffic turned onto it, and the other lane worked. This method creates several problems, the most serious being the center joint between the two strips.

When traffic can be detoured, the two strips are worked alternately, leapfrog fashion. First 300 to 500 feet is done on one side, then 600 to 1,000 on the other, so that the second lane can be placed before the first one has set.

Premix.   If mixing is done at a central plant, the blended and moistened material is brought by trucks and dumped into aggregate spreaders. (See Fig. 8.45.) These should not be allowed to run empty, but should be stopped during truck changes. If the subgrade is dry, it must be moistened before the mix is placed on it.

Compaction.   A variety of rollers may be used for compaction of soil-cement. The sheepsfoot or tamping roller, either towed or self-powered, is favored for all except the most granular soils. Weight of ballast is adjusted so that there will be initial penetration to near the bottom, followed by walking out to the surface in a few passes. Unit pressure may be as low as 75 pounds per square inch for friable soils, and as high as 300 for heavy clay.

If the feet will not penetrate and pressure cannot be increased, the material may be loosened by scarifying or dry running through a mixer.

Pneumatic-tired rollers are favored for coarse, loose mixes. Models that permit varying tire pressure during work are the most efficient. Three-wheel steel rollers in the 12-ton class may also be used, for the whole job or just for final smooth-up rolling.

Grid rollers, segmented rollers, and various types of vibratory rollers all have their backers for soil-cement compaction. The most important factor is that compaction be started immediately after spreading, and continued steadily until completion. Hydration of cement starts as soon as it is wet, and best strength will be obtained with least work if all processing is finished quickly.

Finishing.   For various reasons, the compacted surface will not be entirely smooth, and will usually need some reshaping with a motor grader. This involves trimming high spots, ridges, and other irregularities, and either using the material to fill depressions or wasting it at the road edges. (See Fig. 8.46.)

Differences from trimming a gravel or dirt road include effects of partial setting or drying of the surface and the possible presence of compaction planes. These are smooth or dense surfaces, usu-

FIGURE 8.45   Placing premixed soil cement. (*Courtesy of Portland Cement Association.*)

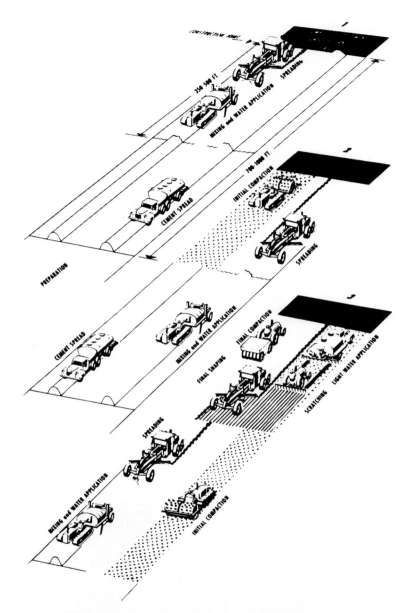

FIGURE 8.46 Windrow processing sequence. (*Courtesy of Portland Cement Association.*)

ally tire prints, close to final grade. They are found chiefly in heavy soils. Material bladed onto them may not form a good bond, so that loosening and spalling will occur in the finished pavement.

Compaction planes are eliminated or made harmless by scratching the surface with a spike-tooth harrow, weeder, or steel broom pulled by a small tractor. Final grading may then be done in a conventional manner, except that the surface must be kept damp during the work.

If compaction planes are at or very close to final grade, or if the surface is partly set, it may be economical to waste any cuttings at the side, rather than to try to use them in the road.

Final grading is followed immediately by moistening, and compaction with a pneumatic-tired roller. (See Fig. 8.47.)

If another layer of soil-cement is to be built on top of this one, exact surface grade is not critical, and compaction planes need not be roughened. (See Fig. 8.48.)

Joints.   Joints occur at the end of each day's construction, and between parallel work strips.

A vertical joint at the end of a day is made by cutting straight down into the mixed and compacted material, either immediately after stopping work or before starting the next morning. The toe of a grader blade, a special disk attachment on a blade, or hand grub axes or shovels may be used. (See Fig. 8.49.)

When work is started again, the cut surface is brushed clean of loose material, dampened, and new mix placed against it. Compaction may be by a roller or grader operated across the road, or by a hand vibrator or tamper. Excess material should be placed at the fresh side, to be pressed into the joint by regular longitudinal rolling. Remaining excess, if any, can be bladed off when finishing.

A longitudinal joint between two work strips is made by overlapping the mixer a few inches, if the material is still fresh and soft, or only partially hardened. Test holes may be made with a pick to locate the firm edge of the original strip, and stakes or string placed as a guide for the operator.

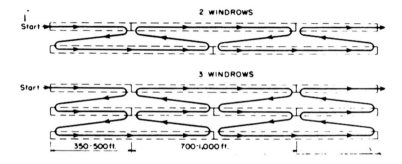

FIGURE 8.47   Another processing sequence. (*Courtesy of Portland Cement Association.*)

FIGURE 8.48   Multiple-shaft mixer and compactor. (*Courtesy of Portland Cement Association.*)

FIGURE 8.49   Cutting a joint. (*Courtesy of Portland Cement Association.*)

Fully hardened soil-cement is shaved back to solid material with a grader blade or disc, with hand-tool assistance as necessary, cleaned of loose material, and moistened, after which the new mix is placed against it with some excess.

Curing.   The freshly finished soil-cement surface must be protected from drying. The standard method is to cover it with a thin bituminous coating, as soon as possible after completion.

The surface should be damp, and free of any loose material. Cutbacks RC-250, MC-250, and emulsion RS-2 are suitable curing materials.

Sanding is required if traffic is to use the surface before the asphalt has set. The soil-cement can then safely carry loads that are not greater than the compaction equipment used, but low speeds are advisable for a few days.

Asphalt is the most convenient curing material, but anything that will hold moisture will do. Plastic or waterproof paper sheets, wet cloth, straw, or sand, or membrane spray can be used.

Surface.   A soil-cement surface is ordinarily too friable and absorbent to stand up under exposure to traffic. It is therefore customary to apply a regular bituminous surface on top of the curing coat. This may be placed immediately, or delayed a month or more to allow shrinkage cracks to develop.

For light traffic, a single-seal coat may be sufficient. A double-seal coat with a total thickness of ¾ inch is suitable for general use, except in snow areas. There, 1½ inches of plant mix blacktop is recommended, for assurance against peeling by snow plows.

A single-seal coat may be applied immediately, and a second coat or blacktop put on several months later, preferably in cool weather, to bridge over shrinkage cracks.

These cracks are inevitable in soil-cement, as the hydration process causes shrinkage. However, breaks are so irregular that the edges support each other, so structural weakening is slight. The principal damage is to the appearance of the surface, which can be minimized by delaying the top coat until after they have formed.

SOIL-CEMENT BASE

Soil-cement base for concrete pavement is constructed in the manner just described, except that curing is likely to be done with some nonbituminous material.

Getting correct surface grade is critical, and may be somewhat difficult. If a gravel or other base is found to be off-grade just before concrete is laid, correction is made rather easily by removing material or bringing it in. With hardened soil-cement this is not practical, so great care must be taken to get it right.

CHAPTER 9
BLASTING AND TUNNELING

For the purposes of this chapter, rock is defined as material which requires loosening by explosives or rippers in order to be dug economically by machinery.

PURPOSES OF ROCK EXCAVATION

Surface excavation of rock is done chiefly for the following purposes:

1. *Stripping*—the removal and wasting of any type of rock or dirt in order to uncover valuable layers.

2. *Cutting*—removal primarily to lower the surface. In road and airport construction the spoil is generally used for fill elsewhere on the project. In ditching, it is often used for backfill after installation of pipes.

3. *Quarrying or mining*—excavation of rock which has value in itself, either before or after processing. A rough distinction can be made between these two, in that quarries are ordinarily concerned with the physical characteristics of the stone, and mines with its chemical composition. However, the terms will be used interchangeably in this discussion.

One excavation can involve all three classifications, as in a heavy road cut where some material is wasted, other is used for road fill, and the best rock is crushed and used for aggregate.

Blasting may be divided into a primary operation in which rock is loosened from its original position in bulk, and secondary work which consists of reducing oversize fragments and breaking back ridges and spurs. The latter is done in the same manner as other light blasting, such as breaking boulders and chipping out ledges.

Rock work may also be classified as to the type and fineness of breakage required. Quarrying of building or dimension stone involves loosening large solid pieces from the parent rock, while blasting for fill or crushed rock requires pieces small enough to fit in the shovel bucket, the fill layer, or the crusher.

Stripping. In most stripping work the spoil has no value, so that the cheapest way of handling it is the best way. It is often possible to dump it in excavated areas from which pay material has been removed.

It is common practice to shoot and dig overburdens over 100 feet deep in a single layer, and the use of the largest shovels and draglines is required for such work.

Drilling may be done horizontally from a face, as in Fig. 9.1(A), vertically from the top, as in (B), or in combination, as in (C).

Horizontal drilling has its best use when the mineral deposit is immediately under soft rock. Auger-type drills with extensions 6 to 10 feet long, and diameters of about 5 inches, are used. These have a tendency to drift downward, and since distances of 30 to 75 feet are commonly drilled, it is necessary to start them several feet above the deposit, or to start them at a slight upward slant. Spacing may be 10 to 30 feet.

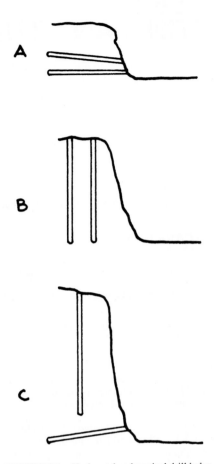

FIGURE 9.1   Horizontal and vertical drill holes.

For harder bottom rock, crawler-mounted (track) air drills may be used. The hole is usually started several feet above the bottom. Lower holes are slanted downward. Higher ones, if drilled, may be horizontal or slant upward.

If thick, higher layers of the overburden are hard rock such as sandstone or limestone, holes are drilled vertically, as in (*B*) or (*C*), using rotary or downhole drills with bits from 4 to 12 inches in diameter.

Burdens (distances from face) from 25 to 35 feet and spacings in rows from 25 to 35 feet are common in high faces. Track drills are used generally to depths of 50 feet, and occasionally to 100. (See Fig. 9.2.)

Scrapers or dozers are sometimes used to remove most of the loose soil before drilling. This saves a considerable drilling footage, makes castings unnecessary, and, on low and medium faces, simplifies the use of track drills.

If the pit area is to be regraded when work is completed, the scrapers can be used to fill the hollows between the piles left by the shovels, or to place topsoil over regraded areas.

Because of the tremendous size and power of the excavators used in large pits, the blasting need only shake up, crack, and loosen the overburden without producing fragmentation comparable to that required in a cut or quarry. Wide-spaced holes and light charges can therefore be used successfully.

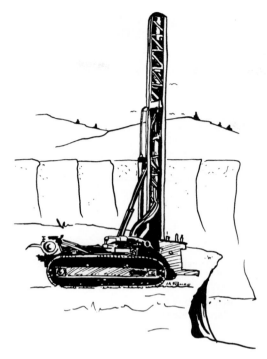

FIGURE 9.2 Drilling near a high face.

Highway Cuts.   Rock cuts for highways may be of the through type as in Fig. 9.3(*A*), and the sidehill (*B*). Material from a sidehill cut may be thrown down to make a fill, as in (*C*).

The area to be cut should first be cleared and stripped of loose soil, and preferably of rotten rock. This may be done with dozers, scrapers, or shovels, depending on the conditions and the equipment available.

If the rock is soft, its upper surface may be loosened with rippers and removed, along with any dirt pockets it may contain. If it is hard and irregular, extensive cleaning by hand and with small equipment may be necessary. It is desirable to remove all loose dirt, particularly if the rock is to be used as crusher aggregate for road topping.

When water and disposal areas are available, hydraulic cleaning with contractors' pumps and fire hose or with special equipment may be used.

If cleaning is not practical over the whole area, the spots to be drilled can be cleaned individually. The top layer may then be drilled, shot, and removed for fill, and any clean rock required can be obtained from lower levels.

If the cut is 20 feet or less in depth, it may be taken in a single layer, but depths of 12 to 15 feet are generally considered most satisfactory for track drills, and digging by 1- to $2\frac{1}{2}$-yard shovels or medium-large front loaders.

In a through cut, the full width is used as a face to provide maximum space for machinery. On a sidehill, the same technique or one or more bench faces parallel with the centerline may be used.

Degree of fragmentation required is determined largely by the depth of fill layers where the spoil is used.

Mining and Quarrying.   Pit operations are largely conducted to obtain certain classes of rock or earth. The general aspects of this work will be discussed in the next chapter.

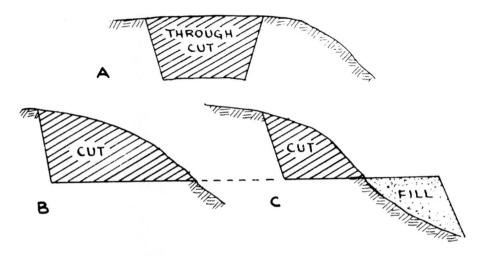

FIGURE 9.3   Through and sidehill cuts.

Rock excavation may follow highway techniques in exploiting comparatively narrow or irregular veins; or large-scale stripping work may be necessary to make pay rock accessible to surface digging units.

Pits are often distinguished by the use of high and wide faces; or holes sunk below surrounding grades, with access by ramps or inclined or vertical hoists.

It is advantageous to have the face wide enough that several operations can be carried on in different sections with minimum interference.

The fineness of fragmentation which must be obtained by blasting is generally determined by the size of the hoppers or grizzlies on the crushers or processing machines.

BOREHOLE PATTERNS

The simplest type of drilling pattern is a straight line of vertical holes parallel to a vertical face. The distance from each hole to the face is called its burden, and the distance between holes their spacing.

The holes are drilled somewhat more deeply than the face so that any ridges left between them will not project above the new grade.

Blasts tend to overbreak at the top and not shatter completely at the base. As a result, faces tend to slope back. The projection of the bottom beyond the vertical line from the top is called the toe.

The extra burden at the toe may be handled by bottom drilling, or heavier loading (more powerful explosive or tighter packing) in the bottoms of vertical holes.

Holes may be drilled at an angle so that they are parallel to the slope of the face. This angle may be from 5 to 30 degrees.

Angle or slope drilling keeps the burden at the toe from being greater than at the top, so that no especially heavy charge is required at the bottom. Improved bottom breakage reduces the need for drilling below the floor, so that the total length of hole is usually not greater than in vertical drilling.

Cleaning.   Small and medium air drills are not suited for working through soil overburden. It tends to choke up their air passages and to fall in on top of the bit. If this falling occurs after penetration into rock, it may make pulling the steel difficult or even impossible. Soil drills readily only when frozen.

The whole area is usually machine-stripped, and the actual spots to be drilled are cleaned by hand. At other times the soil is cleaned off only in slots, and drilling is done in lines along them.

Face Height.   A rock mass may be taken out in a single layer, or in a series of benches. Highway contractors usually prefer benches, as do most open-pit miners, but quarry operators may take single slices 100 or even 200 feet high. This is partly because quarry rock is often sound enough to stand at great heights without much danger of collapse.

High faces are usually developed by pushing low or moderate ones back into a hillside, as in Fig. 9.4(*A*). Where low faces are preferred, such a cut may be made in a series of benches, as in (*B*) and (*C*). This is the safest method in most formations. The final slope should not be so steep that it cannot stand by itself.

Face height affects the method of drilling and the size and placement of holes. Accessibility of the top may be a determining factor in taking off the top layer.

In general, hand drills can be used efficiently for holes up to 10 feet deep, and they can be carried anywhere a person can go. Their comparatively light air hoses can be strung over long distances, although at the cost of reduced pressure.

Track-mounted drills work well from 1 to 50 feet deep, and may go to 100 or more. They can climb and work on slopes over 30 degrees. However, risks and delays are involved in working them on very steep or rough ground, and it may be more economical to take a first cut with hand drills.

Depths over 30 feet call for blast hole drills, of the rotary, downhole, or churn types. These heavy and expensive machines should be kept on safe, fairly even ground. They can dig through any depth of overburden readily, and can start their work from pioneer roads notched into soil slopes by dozers. See Fig. 9.5.

Hole Size.   Hand drills will produce hole diameters of 1 to 2 inches, wagon and track drills 1 1/2 to 5 inches, and blast hole drills 4 to 12 inches. Pneumatic drills of all sizes make tapering holes with steel bits, as the hole gets smaller toward the bottom as they wear. Carbide bits and rotary and churn drills produce holes with little or no taper.

Borehole Loads.   Figure 9.6(A) shows the cubic foot capacity of holes of various diameters for each linear foot. For example, to find out how much bulk of explosive will be needed to fill a hole 3 inches in diameter and 9 feet deep, find the value 3 in the first column. Next to it is the capacity of 1 linear foot, .05 ($\frac{1}{20}$) cubic foot. Multiplying this last figure by the 9-foot depth (below stemming) of the hole, we find that the capacity is 0.45 or $\frac{9}{20}$ cubic foot.

The rest of the columns indicate the weight of explosive of various densities that it will take to fill the hole completely. The actual load will usually be somewhat less, except with free-running

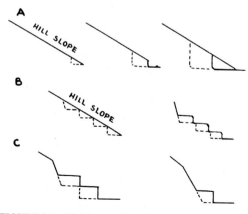

FIGURE 9.4   High face and bench cuts.

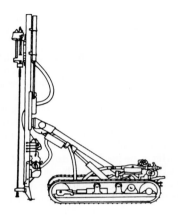

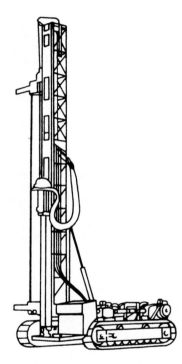

FIGURE 9.5   Track and blast hole drills.

materials, because of waste space around cartridges. The amount of difference will depend on the efficiency of tamping.

To find the amount of dynamite with a rated density of 164 sticks to the 50-pound case that can be tamped into this hole, follow the horizontal line in the table from the 3-inch line to the right to the fourth column, under the values 47 and 164. We find that a 3-inch borehole will hold 2.3 pounds of this powder per lineal foot. Multiplying this by the 9-foot length, we have a possible load of 20.7 pounds.

A

BOREHOLE CAPACITY

Specific gravity of explosive		.64	.75	.80	.96	1.12	1.28	1.44
Pounds per cubic foot		40	47	50	60	70	80	90
Number 1-1/4 x 8 cartridges in 50 lb. case		192	164	154	128	110	96	85
Hole Dia., Inches	**No. Cu. Ft. Per Linear Ft.**				Capacity, pounds per linear foot			
1-1/2	.012	.48	.6	.6	.7	.8	1.0	1.1
2	.022	.88	1.0	1.1	1.3	1.5	1.8	2.0
2-1/4	.027	1.08	1.3	1.4	1.6	1.9	2.2	2.4
2-1/2	.034	1.4	1.6	1.9	2.0	2.4	2.7	3.1
3	.05	2.0	2.3	2.5	3.0	3.5	4.0	4.5
3-1/2	.06	2.4	3.0	3.0	3.6	4.2	4.8	5.4
4	.09	3.6	4.1	4.5	5.4	6.3	7.2	8.1
4-1/2	.11	4.4	5.2	5.5	6.6	7.7	8.8	9.9
5	.14	5.6	6.4	7.0	8.4	9.8	11.2	12.6
5-1/2	.16	6.4	7.7	8.0	9.6	11.2	12.8	14.4
6	.20	8.0	9.2	10.0	12.0	14.0	16.0	18.0
7	.27	10.8	12.5	13.5	18.2	18.9	21.6	24.3
8	.35	14.0	16.4	18.5	21.0	24.5	28.0	31.5
9	.44	17.6	20.7	22.0	26.4	30.8	36.2	39.6
10	.54	21.6	25.6	27.0	32.4	37.8	43.2	48.6
12	.78	31.2	36.9	39.0	46.8	54.6	62.4	70.2

B

CUBIC YARDS DISPLACED PER FOOT OF BOREHOLE

SPACING OF BOREHOLES

Average Burden on Boreholes	6	7	8	9	10	11	12	13	14	15	16	17	18	19	20
6	1.33	1.55	1.77	2.0	2.22	2.44	2.65								
7	1.55	1.81	2.0	2.33	2.7	2.85	3.11								
8	1.77	2.0	2.37	2.65	2.96	3.26	3.55								
9	2.0	2.33	2.65	3.0	3.33	3.66	4.0								
10	2.22	2.7	2.96	3.33	3.7	4.1	4.44	4.81	5.18	5.55	5.92				
11			3.26	3.66	4.1	4.48	4.88	5.3	5.7	6.11	6.52				
12				4.0	4.44	4.88	5.33	5.77	6.22	6.66	7.11				
13					4.81	5.3	5.77	6.26	6.74	7.22	7.70				
14					5.18	5.7	6.22	6.74	7.26	7.77	8.30				
15					5.55	6.11	6.66	7.22	7.77	8.33	8.88	9.44	10.0	10.55	11.11
16							7.11	7.70	8.30	8.88	9.48	10.07	10.66	11.3	11.85
17							7.55	8.18	8.81	9.41	10.07	10.70	11.33	11.96	12.59
18							8.0	8.66	9.33	10.0	10.66	11.33	12.0	12.66	13.33
19								9.15	9.85	10.55	11.3	11.96	12.66	13.37	14.07
20								9.63	10.37	11.11	11.85	12.59	13.33	14.07	14.81
21										11.66	12.44	13.22	14.37	14.77	15.55
22										12.22	13.03	13.85	14.66	15.48	16.30
23										12.78	13.63	14.48	15.33	16.18	17.03
24										13.33	14.22	15.11	16.0	16.88	17.77
25										13.88	14.81	15.74	16.66	17.60	18.51
26										14.44	15.74	16.37	17.33	18.30	19.26
30										16.66	17.77	18.88	20.0	21.1	22.22

NOTE: — To reduce to tons: For limestone, shale, granite, etc., multiply by 2¼; for trap rock, multiply by 2½; for sandstone, multiply by 2¾.

FIGURE 9.6   Borehole capacity and burden.

Untamped Cartridges. If cartridges are not slit and tamped, capacity is figured on the basis of the length and weight of the cartridges, and the number that can be placed side by side. Three of the standard 1¼ × 8 inch cartridges would fit in the hole if tied together. It would take 13 of these bundles to fill the hole within 4 inches of the stemming line. There would be 39 cartridges, each weighing ¹⁄₆₄ of 50 pounds, or about .3 pound per stick. The total load by this method would be 11.7 pounds, just a little more than half of the possible load.

Burden. The explosive in each hole is supposed to break out a section of the rock between the line of holes and the face. Only experience with the particular rock and explosive will indicate exactly the amount and type required.

In a general way, however, it may be said that 1 pound of 40 percent dynamite should break up and move 2 yards of soft rock, or 1 yard of medium hard rock, on an open face. In soft, layered, or rubbery rocks, 20 percent dynamite might move more per pound; while in very hard rocks, even higher-strength dynamites might have smaller production. In tight holes, at edges, and in corners heavier loading is required.

Figure 9.6(*B*) is a table showing the yardage of rock to be moved per foot of hole for various burdens and spacings. As an example of its use, consider that the powder that packed 2.3 pounds to the foot of 3-inch borehole would move 1½ yards of rock per pound, or 3.45 cubic yards to the foot of borehole.

According to the table, 3.33 yards of rock would be moved if spacing were 9 feet and burden 10, or the other way around. An 8-foot spacing and 12-foot burden would yield 3.55 yards.

Pull. If a hole fails to move its burden, it is said that it did not "pull." This usually occurs at the bottom of the hole, and most often in edge or corner holes where the rock is held back on two or three sides. Such failures may be due to too heavy a burden or too wide a spacing, to improper stemming of a shallow hole, use of the wrong explosive, explosive not reaching the bottom of the hole, or a partial misfire. It is generally necessary to remove the blasted rock, check for any unexploded charge, then drill and shoot the bottom again.

Measurement. In order to drill and load holes accurately, it is necessary to know the height of the face and the amount of the toe. With low faces, or in casual operations, or in working upper lifts to temporary grades, depths may be estimated, although this is always risky.

Faces between 10 and 70 feet may be measured by the device and method shown in Fig. 9.7. A 45° right triangle is carefully made of two-by-twos or two-by-fours, with the sides of the right angle equal and from 2 to 4 feet long.

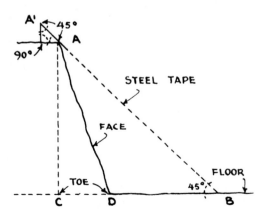

FIGURE 9.7 Measuring face height.

This is placed on the top edge of the face, as shown, and the bottom carefully leveled. A sight is taken along line A"AB and the spot B marked on the quarry floor by an assistant. The distance AB is then measured with a steel tape. Multiplying its length by .71 (the sine of 45°) gives the height of the face and the distance BC. The distance BD is then measured, and when subtracted from BC, gives CD, the projection of the toe.

This measurement may be repeated at various points along the face.

If the face is high, its top irregular, or considerable accuracy is necessary, it may be preferable to make a transit survey of the site, establish benchmarks and location points on all levels, and calculate from direct measurements from these points.

Each drill hole may be marked according to the cut to bottom grade, or by the drilling depth desired. For convenience in loading, the projection of the toe may also be noted on the marker.

Bottom Grade.   The bottom grade should have a slope for drainage that may be away from the face or toward the sides, but not toward the face. If natural drainage is not possible, adequate pumps should be provided.

If blasting and excavating are accidentally carried below grade, hollows can be readily filled with fine shot rock. If the floor is too high, and the rock is soft, it may be possible to take it down with rippers and dozers. If it is too hard for machinery, a tedious job of shallow drilling and blasting, or of noisy and inefficient mudcappings, is required.

Spacing.   In general, large drill holes are increasingly prevalent in faces over 30 feet high, with proportionate increase in spacing and burden.

With most rock types, a point will be reached where enlarging and spreading the holes will result in poor fragmentation in the center of the blocks.

Faces 50 feet or higher are usually shot with a single row of holes at a time. Low faces use additional rows, fired simultaneously or in short-interval succession. In any height, holes in the same row are now fired in sequence.

Each blast should supply enough rock to keep the shovel busy for at least half a day; therefore, the lower faces must be shot back more deeply than the high ones, particularly if the shovel is large and its working area narrow.

The entire group of holes may be drilled on a rectangular pattern, as in Fig. 9.8, or they may be staggered to improve fragmentation.

Best results from multiple rows are obtained if there is a free cleavage plane at the new grade. If the bottom is very hard to pull, heavier loading may be required than when firing single rows; or burdens may have to be reduced progressively toward the back, resulting in higher costs.

Tight Holes.   When blasted rock must be sheared away on two or more planes, the shots are called "tight." In Fig. 9.8 the holes marked *a* are open, those marked *b* are tighter, having to shear off the back or side as well as the bottom, and the *c* holes must shear along back, side, and bottom. In general, the tighter the hole, the greater the likelihood that it will fail to pull. It is usually cheaper to take special precautions with tight holes in the original blast, than to do secondary drilling and blasting.

The tightest blasting found in open work is the start of a cut down from the surface: The first rock blasted can move upward only. If the whole set of drill holes are shot together, each of them will be very hard to pull. However, if the center holes are made oversize, loaded more heavily, and fired first in a delay sequence arrangement that progresses toward the outer limits of the area, the adjoining holes can throw into the space left, then the next holes throw into their space.

Buffers.   Blasted rock may be entirely dug away before the next blast, or varying quantities may be left against the face. Complete cleanup is required if the toe is to be accurately measured or drilled horizontally, and is considered good practice with high faces.

If the working space is narrow, and the face low, it may be desirable to leave some shot rock as a buffer or "blanket." This confines the force of the explosion, and may prevent blocking of the work area and aid fragmentation. However, the buffer should be small enough that some scattering occurs, as this makes it easier to find and blast oversize rock before the shovel gets in the heap. Heavier loading is required.

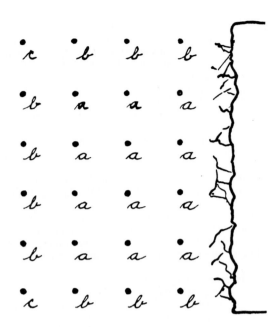

FIGURE 9.8   Open and tight holes.

Snake Holing.   Sometimes a face can be most economically blasted from the bottom only. This may be when a cliff (face) is being cut into a steep hill with poor access to the crest, the rock has a vertical cleavage so that it will break away without dangerous overhang when the bottom is blasted out, and is brittle enough to shatter when it falls. Such an operation may be very economical of powder, as the breaking out of the crest and the fragmentation upon hitting the floor are accomplished partly by gravity.

Coyote Holes.   Coyote holes, illustrated in Fig. 9.9, are used for heavy undermining blasts. They may be used alone to topple a cliff, or to break out heavy toe burdens in conjunction with well drill holes loaded from the top. Faces should be at least 60 to 80 feet high to justify their use.

Coyote holes are used when material is difficult to drill, a large yardage is required at one time, and the rock fractures readily.

They consist of small tunnels, 3 or 4 feet in diameter, driven horizontally into the face at floor level, and one or two cross tunnels parallel to the face. Explosives are stacked in the cross tunnels, and the entrance is securely blocked with stemming. Firing is best done by Primacord.

CONTROLLED BLASTING

Overbreak.   Rock usually tends to overbreak at the top of the bank, and special drilling or loading may be required to avoid leaving a hard bottom rib at each blast junction.

It may be necessary to set back the first row of holes for the next blast for more than the normal burden, as their burden may be partly shattered, so that the shot is not confined as well as in the other holes. Drills may not be able to work close to the edge, or may not be able to penetrate shattered material.

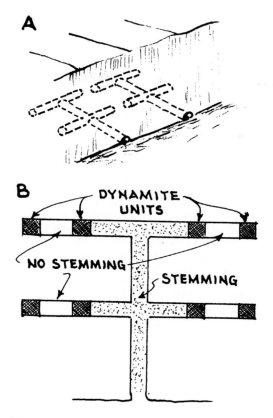

FIGURE 9.9   Coyote holes.

In such conditions the front row may not pull the bottom unless it is drilled considerably deeper than the others, some horizontal bottom snake holes are drilled, or a denser or faster explosive is used.

However, shovels are frequently able to scrape away a few feet of unblasted soft rock, without excessive wear, in which case ribs may be ignored in the blasting pattern and dug out when found.

Edges.   In large-area blasting, control of overbreak and underbreak is mostly a matter of getting the most efficient use from equipment and explosives.

In road cuts, and other work where remaining rock is a permanent part of the finished work, accurate finishing may be required by the contract, either directly as a specification, or indirectly in the form of nonpayment or penalties for excess removal.

In most rock formations, good to precise faces may be cut by line drilling, cushion blasting, preshearing, and various combination techniques. However, selection and refinement of method for a particular formation may be difficult.

For success, the rock must be reasonably sound and cohesive, and capable of standing at the slope to which it is trimmed. In starting a new project, it is highly advisable to obtain advice from your explosive company's representative, and from local sources.

It is essential that the drills and the drillers be able to keep the holes in accurate alignment, both to obtain the desired edge and because the holes often show in the completed work.

Line Drilling.   In line drilling, a single row of holes, usually 3-inch or less, is drilled along the edge, either from top to bottom, or in benches. Spacing is very close, from 2 to 4 hole diameters.

These holes are left empty. They create a line of weakness along which the rock should be sheared by the preliminary blast.

The back row of primary holes may be closer to the line than the normal between-row interval, and may be more closely spaced and lightly loaded than other rows.

This method is best suited to formations with a minimum of bedding planes, joints, and other weaknesses which may affect breakage more than the line of holes. Weak-bedded rock may respond satisfactorily if its planes are nearly at right angles with the proposed slope, or exactly parallel to it.

Cushion Blasting.   Drilling for cushion blasting is similar to that for line drilling, except that larger holes and wider spacings may be used. (See Fig. 9.10.)

The holes are loaded with light charges, with undersize and/or spaced cartridges, often strung out on detonating cord. All space not occupied by explosive is usually filled with stemming, although some blasters prefer to stem only the top. See Fig. 9.11.

Closer drilling or heavier loading, or both, are needed in cutting angles in the face. (See Fig. 9.12.) Extra unloaded relief holes may be put between the loaded ones.

Standard procedure is to blast and usually to excavate the main cut back to within a few feet of the final line, either before or after drilling the cushion holes, and fire this final row separately. Detonating cord or instantaneous caps are normally used. If shock and noise are problems, close-interval DuPont MS delay caps may be substituted.

Preshearing.   Preshearing resembles cushion blasting in drilling and loading, but firing is in advance of the main blast. Since the explosive force has nowhere to go (except up), it may be expected to produce finer and more even fragmentation between the holes, and a totally effective shear plane for the primary blast. The two blasts may be fired together, using delay caps.

EXPLOSIVES

General Properties.   Explosives are chemical compounds that can decompose quickly and violently. The original solid or liquid chemicals are largely changed into gases, including steam, that

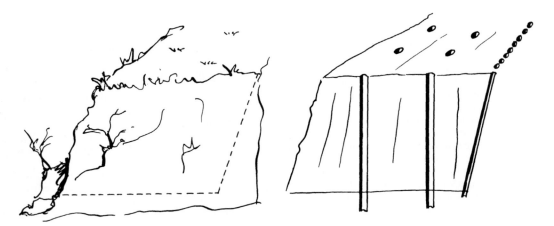

FIGURE 9.10   Preparing for a cushioned blast.

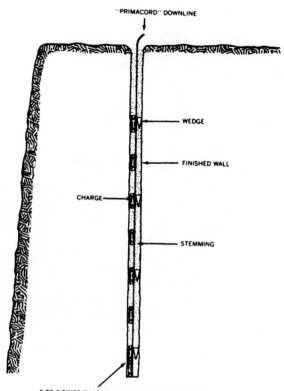

"PRIMACORD" DOWNLINE

WEDGE

FINISHED WALL

CHARGE

STEMMING

2 TO 3 TIMES CHARGE/FT IN BOTTOM TO INSURE SHEAR AT FLOOR

FIGURE 9.11   Charge placement for cushion blasting.

have a much greater volume. Heat is generated by the change, and serves to expand the gases greatly.

Explosion by rapid burning is called deflagration, and by almost instantaneous decomposition is called detonation. High explosives detonate.

Handling explosives requires experience. A widely used reference is *Blaster's Handbook,* published by International Society of Explosives Engineers.

Properties to be considered in selecting an explosive include sensitivity, density, strength, velocity, water resistance, fumes, price, and availability.

Sensitivity is a measure of the ease with which a substance can be caused to explode and its capacity to maintain an explosion through the length of a borehole. It is also a measure of safety—the higher the sensitivity, the greater the risks of handling.

Nitroglycerin is so sensitive that it must be mixed with other substances before it can be used in commercial explosives. Compounds such as fulminate of mercury and lead azide that are used in detonators are so sensitive that they will explode at a light hammer blow or when exposed to moderate heat. At the other extreme, ammonium nitrate is so insensitive that few precautions and no permits are required for shipping and storing it.

Density is the volume of an explosive in proportion to its weight. It is measured in pounds per cubic foot, or in the number of $1\frac{1}{4} \times 8$ inch cartridges in a 50-pound case for wrapped dynamites. Such a cartridge contains about 9.78 cubic inches. A count of 100 sticks to the case is roughly equivalent to a density of 77 pounds per cubic foot. A cubic foot of water weighs 62.4 pounds.

FIGURE 9.12   Angle drilling.

Strength is the energy content of an explosive in relation to its weight. In general, maximum explosive power can be obtained from a given borehole by using a high-density, high-strength explosive in it.

Velocity is a measure, expressed in feet per second, of the speed at which the burning or the detonation wave travels through an explosive. It varies from 1,000 to 3,000 feet per second for black powders to 23,000 feet per second for blasting gelatin.

Low-velocity explosion has a heaving and separating effect, while high velocity crushes and shatters.

Water resistance is an important factor in wet rock, and varies with not only the character of the explosive but the manner in which it is packed and wrapped. Manufacturers are able to put water resistance in the explosive rather than in the wrapper.

The gases resulting from explosions vary in toxic and irritating qualities. This is very important in underground work, particularly if ventilation is poor. Explosives are rated by the manufacturers according to fumes as excellent, good, fair, and poor.

Explosives vary widely in the length of time they can be kept under various conditions before deterioration makes them dangerous or useless. Dynamite was formerly damaged by freezing, but this difficulty has been entirely overcome. Spoiling may be a serious factor if use is subject to delay, particularly in hot, wet weather.

Different explosives vary widely in price. The most economical one for a certain use may be one with a high cost. It is important to select an explosive on the basis of all the related factors, rather than purchase cost alone.

In many areas, very few types of explosive may be available, and because of the complications of shipping, delivery of special orders may be delayed weeks or months. Under such conditions, use of the standard dynamite may be advisable even if it is not exactly suited to the job.

If special explosives are purchased from a contractor or a quarry, it may be necessary to handle the transaction through a dealer in explosives to comply with state laws.

Permissible dynamites are those approved by the U.S. Bureau of Mines for use in gassy and dusty coal mines. Their most important feature is minimum flame in the explosion.

Black Powder. The explosives which explode by burning are called low explosives. Black powder is the only commercially important member of this class, and is the oldest explosive known.

Black powder is ordinarily composed of sodium nitrate, sulfur, and charcoal, finely ground and combined in grains of various sizes. The grains are then coated with graphite or other glazing to make the powder free-running. A more expensive powder for special purposes uses potassium nitrate instead of the sodium compound.

Fine-grain powders burn and explode faster than coarse-grain ones. Somewhat more powder can be packed in a borehole by mixing two or more grain sizes.

Black powder can also be obtained in pellets, which are short cylinders of compressed powder with a center hole. They are wrapped into 8-inch cartridges resembling dynamite in appearance. They are more convenient to use in small boreholes than the loose powder, and are somewhat safer to handle.

Black powder may be ignited or exploded by flame, heat, sparks, or concussion, and requires more careful handling than most dynamites. A special hazard is that powder spilled on the ground or on the magazine floor may ignite if stepped on or scuffed.

The blasting action will depend on the degree of confinement, the bulk, the grain size, and the closeness of packing. Unconfined powder will flash-burn, without explosion; and poorly confined powder will waste much of its energy along the path of least resistance.

Black powder produces considerable smoke and quite toxic fumes, the quantities of which vary considerably in different blasting procedures.

Black powder can be used to advantage when large, firm pieces of rock are desired, or when the material being blasted is soft and resilient enough to absorb the shattering blow of high explosives.

It cannot be used underground where ventilation is poor, or where the air may contain inflammable gas which may be ignited by the flame from the powder, or in wet holes. It has been replaced by other explosives in most applications.

Dynamite. Dynamite is the best known and one of the most widely used commercial high explosives. The name includes several different chemical groups, wrapped and marketed in about the same manner.

The "straight" dynamites consist primarily of a mixture of nitroglycerin, sodium nitrate, and combustible absorbents such as wood pulp, wrapped in strong paper to make a cylindrical cartridge. Although a wide variety of sizes are available, the most popular are 8 inches long and $1\frac{1}{8}$ or $1\frac{1}{4}$ inches in diameter.

The percentage of nitroglycerin by weight contained in the mixture is used to identify it, according to strength. From 15 to 60 percent may be used.

Strength does not increase in proportion to the percentage of nitroglycerin because the other ingredients also contribute gas and heat. For example, a 60 percent dynamite is about $1\frac{1}{2}$ times as strong as a 20 percent.

Higher percentages are faster and more sensitive. Speed is desirable in hard rock and where the explosive is not confined, as in mud-capping boulders. Sensitivity is necessary when blasting mud ditches by the propagation method.

Straight dynamites have fair water resistance. Their fumes are poor, however, and they are never recommended for underground work.

Any type of dynamite of the general-purpose 40 percent strength will explode if subjected to sharp concussion, such as explosion of a blasting cap; from impact of a rifle bullet; from excessive heat, whether produced by fire, friction, or impact; and from sparks.

When dynamite is burned—usually to destroy surplus or deteriorated stock—it is spread in a thin layer on straw or other combustible material, which is ignited. All personnel should keep a safe distance from the fire. Dynamite will usually burn without incident, but there is always a chance that it may explode.

Spoiled dynamite may soak into its containers, and render them explosive. The cases and wrappings should therefore be burned with the same precautions as would be taken with dynamite.

The "ammonia" dynamites use ammonium nitrate as the principal explosive, in combination with some nitroglycerin. They do not catch fire as easily as straight dynamites, and are less sensitive to shock and friction. Water resistance is generally inferior, but fumes are less objectionable.

These are rated on a percentage strength basis, but the figures do not indicate anything of their chemical composition but simply that performance is comparable to that of a straight dynamite of the same rating.

A third type of explosive used in commercial blasting is gelatin dynamite. This is based upon a jelly made by dissolving nitrocotton in nitroglycerin. Various other ingredients are added.

The gelatin dynamites are dense, plastic, cohesive, and practically waterproof. Fumes are excellent in all but the highest strengths, which vary up to 90 percent.

Shot with a standard cap, and when not confined, ordinary gelatin dynamites will explode at a velocity of about 5,000 feet per second. If confined, or shot with a straight dynamite primer, velocities of 13,000 to 22,000, depending on the strength, will be obtained. Certain types may also be obtained which will always detonate at the higher velocity.

Gelatin dynamites are relatively insensitive to shock, and often will not explode by propagation from adjacent holes. Their plasticity makes it easy to load them solidly in boreholes, and to pack tightly in cracks for mudcapping. The velocity of the higher grades and their high density recommend them for hard, tight blasting; and the waterproof qualities for any underwater work not requiring propagation.

Ditching, stumping, and agricultural dynamites are usually one of the standard strengths best suited for the purpose, with a special designation.

Blasts are initiated or set off by timing devices, by remote electrical controls, or by a combination of these methods.

AMMONIUM NITRATE

Ammonium nitrate (NH_4NO_3), called AN for short, is a nitrogen fertilizer that has largely replaced dynamite in medium and large borehole blasting.

At ordinary temperatures AN is generally a stable compound. If heated to 300 to 400°F, it will decompose without exploding into water and nitrous oxide, a brownish red gas with a pungent smell.

If subjected to great heat under confinement, or direct detonation of high explosives, AN decomposes explosively into water, nitrogen, and oxygen. The formula is

$$2NH_4NO_3 = 2N_2 + 4H_2O + O_2$$

Ammonium nitrate is so insensitive that it is not rated as an explosive. It is called a "blasting agent," and it can be shipped and stored free of special regulations and permits. However, it does burn, and since it supplies its own oxygen, it cannot be smothered. The only way to put it out is to use plenty of water.

If burning AN is confined in rooms, or if it is piled in such bulk (over 100 tons to a pile) that heat and pressure can build up inside it, it may explode as destructively as dynamite.

AN should never be transported or stored with high explosives.

Additives. The sensitivity and explosive power of AN are greatly increased by mixing or blending it with organic materials that absorb oxygen when they burn. Several disastrous explosions have occurred in AN fertilizer that was treated with a small amount of wax to prevent sticking.

For use as an explosive, additives such as lampblack, powdered coal, sawdust, and fuel oil have been tried. Fuel oil is very successful, of which more will be said below. Any of these provide material to combine with the surplus oxygen to produce additional heat and gas volume.

Prills. Most AN is prepared in prill form. Prills, sometimes miscalled "pellets," are globular, porous particles obtained by spraying a 95 percent solution of AN into a rising current of warm, dry air. They are usually coated with about 3 percent by weight of kieselguhr (diatomaceous earth, about 80 percent finely divided silica, called guhr for short) to prevent sticking. There are also uncoated prills, and prills coated with minute amounts of other materials. Prills vary in density, size, and size mixtures according to brand and specification.

Other processes produce flakes, grains, or pellets that usually are denser. These have not been as successful in direct blasting as the prills up until now.

History. AN has been used as an ingredient in dynamite and other explosives since 1867. In 1935 a canned AN mixed with a small quantity of fuel was introduced by du Pont under the name Nitramon. Because of its low sensitivity it could not be detonated by caps, and special primers of amatol, a TNT and AN compound, were provided.

Nitramon and similar products are safe and clean to handle, economical, excellent in wet holes and coyote holes, but are now declining in use because of competition from fertilizer grade AN.

Nitrex, a canned AN, was used in a 3,300,000-pound blast to remove Ripple Rock above Vancouver.

Fuel Oil. In 1955 it was discovered that No. 2 fuel oil is an almost ideal material to mix with AN to make a practical explosive. The mixture is called AN-FO for convenience.

Maximum explosive power is obtained when the mixture contains about 94 percent AN and 6 percent fuel oil. This is about 1 gallon of oil to a 100-pound bag of fertilizer. This is also the proportion of oil that the AN absorbs most readily.

Mixing may be done at the factory, at a fixed plant near the job, by mobile equipment or by hand at the borehole, or in the borehole. Relative cost depends largely on the size of the job.

Factory mix is the most thorough and the most expensive, and is the least trouble on the job. The mixed AN-FO, called nitrocarbonitrate by the ICC, is more dangerous to handle than the unmixed AN. Various mild precautions must be taken, and vehicles carrying it must be marked "DANGEROUS."

AN-FO should not be handled or transported in the original AN bags unless the change in contents is plainly marked.

AN-FO may be made dangerously sensitive, or even caused to explode spontaneously, after long standing in stemmed boreholes, by contamination with unidentified, naturally occurring chemicals.

Local plants should mix as well as the factory, and mobile equipment almost as well. Hand mixing is much less effective, and mixing in the borehole is rather hit-or-miss. A thorough blending is important for highest blasting efficiency, but it may cost more than it is worth.

Poor mix, or poor detonation for other reasons, is likely to be indicated by "Kodachrome clouds" of yellow, orange, or brown smoke from the explosion. The color is caused by nitrous oxide mixing with steam and other gases.

Borehole mixing is done by pouring in a bag or two of AN, adding the correct quantity of fuel oil, then putting in more AN and more oil alternately until the full charge is placed.

Hand mixing outside the hole may be done by pouring 3 quarts to a gallon of fuel oil into each 80-pound bag of the fertilizer, or 2 to 2½ quarts into each 50-pound bag, moving the bag around to help the liquid to distribute itself evenly, and allowing it to stand for 5 to 20 minutes before pouring it into the hole.

Somewhat better results may be obtained by putting the AN in a mixing trough such as is used to hand-mix concrete or plaster, spraying the fuel oil over it, and mixing with a hoe or shovel.

Unopened bags may have fuel oil injected into them under high pressure by the use of an engine-driven pump and a sharp-pointed nozzle that can be pushed through the wrappings. This method, pioneered by Monsanto Chemical Company, is called the Needle Fuel Injector System.

Mechanical mixing may be done with engine-driven cement mixers. Oil should be added as a spray rather than a solid stream.

A great deal of specialized equipment has been designed, both in the factory and in the field, for mixing quantities of prills with fuel oil, and discharging them rapidly into boreholes. Augers may move prills from a truck body into a stream of compressed air, in which they are mixed with oil and blown into the borehole through flexible hose.

Loading prills by compressed air may build up charges of static electricity in the borehole and its vicinity. Such charges create danger of premature discharge of electric caps, and even possible hazards with fuse caps. Minimum precautions are grounding of the pneumatic loader and use of semiconducting hose. Caps should be of types least sensitive to stray currents.

This problem, and a number of important precautions, are discussed in a U.S. Bureau of Mines Report, IC 8179, titled "Safety Recommendations for Sensitized Ammonium Nitrate Blasting Agents."

Priming.   Most AN-FO cannot be detonated dependably by blasting caps or regular Primacord. Even when detonated, it may not maintain full speed of explosion, or even any explosion, for the length of the hole.

It is therefore necessary to use primers or boosters that can be exploded by caps or Primacord, and that will produce sufficient explosion to detonate the AN-FO at high velocity; and to use enough of them in the hole that the explosion wave will not have enough space between them to weaken.

Such primers may be made of one or more sticks of gelatin dynamite of 60 percent or higher strengths, or of special "cast" explosives. The cast boosters are somewhat more expensive than dynamite, but are safer to handle and are more powerful.

Figure 9.13 shows a cross section of a cast booster. The sensitive explosive that is detonated by Primacord or a cap is completely surrounded and protected by another high explosive that cannot be detonated by a cap or any probable accident, but can be detonated by the core, and has sufficient strength to set off any AN-FO mixture.

Borehole diameter is the most important single factor affecting propagation of the explosion, as this dies out much more quickly in small holes than in large ones. Although AN-FO has been successfully used in holes as small as $2\frac{1}{2}$ inches, variable results, and cost of extra boosters, limit its use in holes smaller than 4 inches.

Specification.   AN-FO is a passive explosive material. Its performance is governed by such factors as confinement, diameter, and fuel oil content. Under defined conditions of density, confinement, diameter, and particle size distribution. AN-FO will have only one velocity at which it can be detonated. The maximum velocity at which it can detonate is 14,600 feet per second in a borehole 10 inches or larger in diameter. In a 4-inch borehole, the steady-state velocity of AN-FO is 12,500 feet per second, and in a 6-inch borehole it is 13,900 feet per second. If the primer has a lower velocity, it may not be able to detonate the AN-FO. Ideally, the primer will have a higher velocity than the AN-FO, and there will be a downward change in velocity known as overdrive. Overdrive velocities have been recorded only in boreholes up to 6-inch diameter.

The other key factor in detonation of AN-FO is the diameter of the primer charge relative to the AN-FO explosive. As the diameter of the primer is reduced below that of the explosive, the effect on the initial velocity of the AN-FO is considerable. An extreme low velocity is developed in the AN-FO when a 1-inch-diameter primer is used. This suggests using a slurry-type primer that will fill the borehole as the AN-FO does. If that is not practical to overcome the effect of diameter mismatch, it is necessary that primers have a detonation pressure of at least 80 kilobars (more than 1,000,000 pounds per square inch). Primers satisfying that requirement include Aquaram,

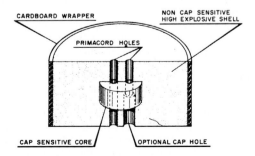

FIGURE 9.13   A cast booster.

Aquanal, Power Primer, Pentolite, gelatin dynamite, and perhaps TNT or water gel, but not regular dynamite.

To summarize: The efficient initiation of AN-FO requires a primer that has sufficient diameter to nearly fill the borehole and has a minimum length of at least one borehole diameter; the detonation pressure of the primer should be at least 80 kilobars.

Specifics.   The du Pont company originally developed a family of high-explosive primers known as Detaprime, which stress safety, high performance, economy, and ease of handling. These primers have a velocity of 24,000 feet per second and a density of 1.50. They are insensitive to mechanical impact, have no metal-to-metal contact between the primer and cap, and do not detonate when exposed to open flame. There is no sensitivity to temperatures from −40 to 225°F. Their water resistance is excellent, including resistance to moisture throughout shelf life. And storage life is 3 to 5 years at ambient temperature. The Detaprime primers are about ½ inch in diameter and 2 to 3 inches long, with a hollow core for detonating cord, fuse, or cap insertion.

Poor confinement may also cause the detonation to slow or stop. A soft rock or mud seam between layers of hard rock may not confine the AN-FO sufficiently for it to carry the explosion across.

In general there should be a primer at the bottom and at the top of the hole, and one at least every 20 feet. However, 50-grain Primacord and a single primer at the bottom may prove adequate.

Loading.   Dry holes are loaded by placing the primers with the detonating cord or wire, then pouring the mixture or the two ingredients separately until the proper amount has been placed. Stemming is then added, as described later under LOADING.

There is a tendency to load holes higher with AN-FO than with dynamite because of its lower cost, and to use less stemming. Overdoing this is likely to be a waste of the explosive, and to add to the hazard of high flying rocks.

Unprotected AN-FO cannot be expected to perform dependably in wet holes. It can be protected by putting it in plastic bags. It may be bought this way, or the bags made up from sheet plastic with the help of a sealing machine.

Bagged AN-FO tends to float on any water in the hole, and must be forced down by the weight of explosive above it. Even with firm tamping the bags will reduce the amount that can be placed because they will not conform perfectly to the walls. Bags may tear and allow water to ruin their contents, perhaps cutting off part of the blast.

Density.   The density of AN ranges from 47 pounds per cubic foot for the prilled variety to 64 pounds for fine-grained types. This compares with dynamite densities of 37 to 90 pounds.

In loading calculations the low density of prilled AN-FO compared to heavier dynamites is offset to a variable degree by complete filling of all hole space by the free-running material. That is, if a dynamite with a 60-pound density filled only 80 percent of the bore space in spite of slitting and tamping the cartridges, then its effective density in the hole would be .8×60 or 48 pounds per cubic foot, about the same as the prills.

The finer and denser types of ammonium nitrate are not well suited for use as blasting agents with present techniques. They are more sensitive and less powerful than the prills, and are more difficult to mix with fuel oil. Some increase in density of prills may be obtained by mixing two or more sizes, but the heavier charge may still be offset by slower detonation.

SLURRY (WATER GEL)

Composition.   Slurries, also known as water gels or dense blasting agents (DBAs), are usually mixtures of a "sensitizer," an oxidizer, water, and thickener.

The sensitizer may be any of a number of reducing (oxygen-hungry) chemicals. This is usually the explosive TNT (trinitrotoluene), but may be (or include) finely divided aluminum and/or other substances that may or may not be explosives themselves. They may be in sand-size granules, very fine powder, or other forms.

The oxidizer is an oxygen-rich chemical such as ammonium nitrate and/or sodium nitrate.

Characteristics.   Consistency is regulated by the amount of thickener or jelling agent (often guar gum) that is used. It varies from that of pancake syrup to soft (flows if jiggled) jelly at room temperature. Stiffening occurs at low temperatures, but most formulas are resistant to damage by freezing.

Water resistance varies from good to excellent, unless water flow is sufficient to wash it away. Loading water gel in its sealed plastic packages is recommended for severe water conditions.

Packaged slurries are generally jelled in cylindrical shapes slightly smaller in diameter than the boreholes in which they are to be used. They are shipped in polyethylene bags, protected by burlap or by cardboard containers.

The bags are usually soft enough to allow slump to fill the borehole almost completely, and can be obtained in even softer, expandable types. However, it is normal to slit the top 6 inches of each bag, and drop it into the hole. Some water gels have unique gellant systems which are designed to permit them to be poured into the hole.

It does not mix readily with water, and its weight (specific gravity is 1.1 to 1.5) causes it to flow to the bottom, displacing any water and filling all borehole space. But if it is very thick, it may bridge over and retain water pockets.

For large-scale use, slurries may be shipped and even mixed in tank trucks and pumped into holes. If the hose can be extended to the bottom of the hole, danger of bridging is eliminated. Most slurries are insensitive. They cannot be detonated by blasting caps or Primacord, and require special booster-primers.

Recently developed small- and intermediate-diameter grades of water gel are cap-sensitive, so that they require shipment and storage as a class A explosive. This type is finding wide acceptance as a replacement for dynamite in bottom loads.

Applications.   Slurries are less expensive than dynamite, but cost more than ammonium nitrate–fuel oil mixtures (AN-FO). They are now used chiefly in open-pit mines where rock is hard and/or holes are wet.

The high density of slurry (1.37 to 1.68) compared with AN-FO permits use of smaller-diameter boreholes, or wider spacing, to obtain the same explosive power and fragmentation. The higher price of the slurry may be more than offset by reduced drilling cost.

In wet holes AN-FO is not practical, so the choice is between dynamites and slurries. Here the slurries have a smaller advantage in density, but have a price advantage as well.

Some slurries have been employed successfully in boreholes down to $1\frac{1}{2}$-inch diameter, using a special small-diameter booster.

INITIATORS

Safety Fuse.   The original timing device is a fuse made up of a black powder core, surrounded by layers of protective wrappings. Two speed ranges are available, with burning speeds of 90 seconds and 120 seconds to the yard. These speeds must be considered approximate, as they are affected by altitude, weather, storage conditions, and possible damage to the fuse.

Fuse is water-resistant except at the cut ends, which are immediately spoiled by contact with moisture. It should not be used unless it can be shot the same day it is loaded.

It is manufactured in lengths of 50 feet or more, and wound in coils or on spools to be cut to the desired length on the job. As short an interval as possible should be allowed between cutting and using.

Squibs.   Electric squibs are devices for igniting charges of black powder, and may be used instead of fuses. They consist of copper or aluminum tubes with powder, an electric firing element, and wires sealed into them. They are embedded in the powder charges, and when sufficient current passes through them, they catch fire and ignite the charge.

Blasting Caps.   Figure 9.14 shows construction of blasting caps for both fuse and electric firing. Fuses are inserted in the hollow shell of the cap, and fastened in by crimping the metal with special

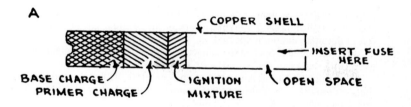

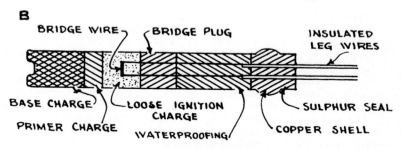

FIGURE 9.14   Detonating caps.

tools or machines. The fuse should be cut off square, preferably by a razor blade or other very sharp edge, which will not pinch the wrappings together.

The electric firing device consists of a very thin wire lying in a highly combustible mixture. Passage of electricity causes the wire to become white hot and ignite the mixture, which explodes the primer and high explosive.

Delay Blasting.   This is used to improve rock fragmentation and rock movement. Rock fragmentation is dependent on the stress wave set up in the rock by the explosion. That may be between 10,000 and 20,000 feet per second—faster in harder rock. The rock fractures radiating out from the borehole travel from 0.15 to 0.4 times the stress wave in the rock, so they travel from 1,500 to 8,000 feet per second. In other words, the radial cracks travel from 1½ to 8 feet per millisecond.

As a guide to when rock movement starts, the common expression of 1 millisecond per foot of burden is reasonably correct. The rate of movement is used as a guide so that rock broken by the first row of holes will move out of the way before the second holes fire, or material from the second holes has no place to go. The delay firing may be obtained by use of delay electric blasting caps. Time interval averages about 1 second in "standard" caps. A more effective delay system will be obtained by using MS connectors with 5- to 25-millisecond delays. The very short intervals give the finest breakage but give reduced efficiency and increase throw, making mucking of the round difficult due to widely scattered rock.

Wires may be obtained in almost any desired length, and should be long enough to reach the wires from adjacent charges or to connect with the lead wire to the electrical source. They may be copper or iron, and are protected by a plastic insulation. The ends of the wires are fastened together into a bridge or shunt, to prevent premature firing through contact with stray electric currents.

Instantaneous and delay caps may be used on the same round. If they are in the same series, it is good to increase firing current by one-third. Caps made by different manufacturers should never be fired together. They are almost certain to have different current requirements, so that one brand of cap will fire and break the circuit before the bridge wires in the other brand are heated enough to fire.

Caps contain explosives which are more sensitive than dynamite, and they must be handled with great care. Heat, friction, and shock must be avoided. In the original package, electric caps are usually cushioned in their own folded wires, and are often protected by a cardboard or paper wrapper in addition.

But the greatest danger of accidental firing of electric caps comes from electricity. Even with their wires shorted by a soldered shunt in the original package, a very powerful nearby current might detonate them unexpectedly.

After the shunt is opened to connect to other wires, there is danger from any stray electric current, even from radios.

The content of a cap is small, but one can blow off fingers and toes, and flying particles of the copper case may cause injury to personnel within a radius of 30 feet. The most serious danger in an accidental explosion of caps, however, is that of setting off primers or nearby heavy explosives.

Caps should be buried before exploding for test purposes.

Primacord.   Primacord is a detonating (exploding) fuse, made up of a core of an insensitive high explosive, pentaerythritol-tetranitrate, that is called PETN for short, surrounded by a protective wrapping. Primacord is detonated by means of a blasting cap. The explosion travels along it at a rate of about 21,000 feet per second, and detonates any cap-sensitive explosives with which it is in contact. See Fig. 9.15.

It is produced in a number of types that are classified according to explosive content in grains per foot, and/or the type of protective covering around the explosive core.

Standard types usually have 50 to 60 grains per foot. Lighter grades, including "E-Cord" with 25 grains, are used chiefly for secondary and very shallow blasting. Heavier ones, with 100 to 400 grains, are used for continuous column initiation of ammonium nitrate–fuel oil mixtures, or for cutting into short lengths for use as primers.

Wrappings are rated on the basis of strength and water resistance. All are water-resistant, but plastic-coated ones are essentially waterproof except at cut ends. The plastic is also resistant to oil, an important point when using an AN-FO mixture.

E-Cord and Primacord with 45 or more grains will fire when wet with water or oil if it is initiated from a dry spot. Wet Primacord can be initiated only with a very powerful primer, such as 80 percent gelatin dynamite or a special booster. It will not maintain detonation through a knot connection.

Cut ends of Primacord will pick up some moisture from capillary attraction, usually only a few inches in. If the cord is lower than the water level, it may in time become soaked all the way through.

Textile-reinforced Primacords are usually used in ordinary downholes, wire reinforcement in rough and jagged ones.

The individual lines in the holes, called branch or down lines, are usually connected to the blasting cap by a trunk line of Primacord. Fastening is done by simple knot connections. It is

FIGURE 9.15   Primacord is supplied in boxes and rolls.

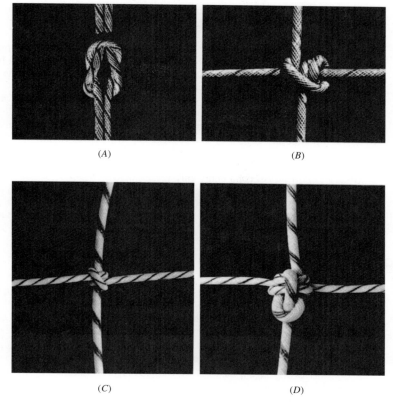

(A)   (B)

(C)   (D)

FIGURE 9.16   Primacord connections. (*A*) Trunkline spliced with square knot. (*B*) Double wrap, half-hitch knot connection between a reinforced trunkline and downline. (*C*) Trunkline clove hitched over a downline. (*D*) Special clove hitched trunkline to prevent slippage from subsidence of the charge.

important that the lines be at right angles to each other. (See Fig. 9.16.) When very stiff cord is used, it may be necessary or convenient to use plastic connectors instead of knots.

Electric caps may be used to combine quiet surface wiring with Primacord downlines. Connection is made by taping a cap (or preferably two caps) tightly to the Primacord. This may be a short piece, called a tail, which is tied to the down line with a square knot just before firing. The caps are wired into a conventional blasting circuit.

There are also nonelectric delay caps, Primadets, that are made up attached to a light detonating cord, Primaline. They are strong enough to explode AN-FO and most slurries directly, without a primer.

The cap is placed near the bottom of the hole, with or without a primer, before pouring in the main charge of explosive. The Primaline is knotted to the Primacord trunk line. It does not have sufficient strength to detonate AN-FO or slurry, so timing is regulated by the delay feature in the cap. It is not used with dynamite or other explosives that are sensitive enough to be set off by the cord.

On all large blasts it is customary to arrange Primacord so that each hole may be reached by the explosion from two directions. Large blast holes, or holes of any size with deck loading, may have two cords strung on opposite sides of the hole to ensure firing.

Primacord has become the standard method of setting off large blasts, because of its exceptional safety. As an explosive it is quite inert, and is less likely to be detonated by accident than the main charge of explosives. Particularly, it cannot be exploded by stray electric currents, a serious hazard for electrical hookups. An entire blast can be prepared with detonating cord, and a fuse or electric cap attached at the last minute.

Low-Energy Detonating Cord.   The noise made by the explosion of Primacord trunk lines aboveground may be objectionable. This noise can be considerably reduced by covering it to depths from a few inches up to a foot with sand, dirt, drill cuttings, or other stone-free material, or almost eliminated by using low-energy cord.

LEDC, a low-energy cord that is relatively noisefree, was developed jointly by Ensign Bickford Company and E. I. du Pont de Nemours Co. It contains 2 grains of PETN per foot, protected in a tiny lead tube covered by wrapping of cotton and plastic. The noise made by exploding 150 feet of it is equal to that caused by one blasting cap or 2 inches of reinforced Primacord.

Special connectors are available to fasten LEDC to Primacord and to itself. When the explosion travels from LEDC to LEDC or to Primacord, a booster charge is needed. This may include a 0-, 10-, 15-, or 25-millisecond delay unit.

The preferred method of firing is to fasten an electric blasting cap to one of the down lines of Primacord, and to fasten the LEDC to this down line with a nonexplosive connector called a trunk line adapter. The LEDC is connected to other down lines by booster delay caps.

PRIMING

Primers.   A primer may be a stick of dynamite that contains a blasting cap; or is any other heavy explosive which has been fitted with a device for setting it off.

Since these primers combine the power of the dynamite with most of the sensitivity of the cap, they must be handled with greater care than any other units of explosives.

They are ordinarily prepared at the borehole immediately before being placed, but may be made in some central place and delivered to the loaders as required.

The essentials of a good primer are that the cap must be powerful enough to produce detonation, there must be intimate contact between cap and explosive, they must be fastened together so that they will not separate while being placed, the cap should be shielded from shock or friction, and the wires or fuse should not be kinked or strained.

Black Powder.   Black powder may be primed by placing a fuse in the hole and pouring the powder around it. This method may be improved by tying a knot in the end of the fuse to anchor it, and making several slits into the core above the knot where they will be in contact with the powder. A paper cartridge may be prepared to hold powder closely around the knot and slits.

If the powder is to be exploded by an electric squib, a similar cartridge is made up to enclose the squib with some powder.

Blasting caps may also be used to explode black powder.

Fuse Caps.   Preparation of dynamite and fuse cap primers includes two jobs—attaching the fuse to the cap and the cap to the powder. The fuse ends *must* be dry.

The fuse should be cut squarely with a clean, sharp blade, preferably a razor blade in a suitable mounting, and pushed into the cap until it is seated against the explosive compound. The copper shell is then crimped firmly onto the fuse with a hand or a bench crimper.

If the fuse is cut on a bevel, it may fail to make proper contact, Fig. 9.17, top, or the end of the casing may curl over and prevent contact. A good contact is shown at the bottom.

The interior of the cap should be clean. Any foreign matter in it should be tapped out or removed with a straw or toothpick. Blowing into it may dampen it and cause it to fail. If the cap is suspected of being damp from any cause, it should not be used.

Figure 9.18 shows two ways to place and fasten a fuse cap in dynamite. In each case a hole or slit is made in the cartridge and the cap inserted. The primer can be held together by lacing the fuse through another hole, or by tying it with string.

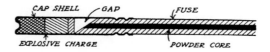

FIGURE 9.17   Insertion of fuse in cap.

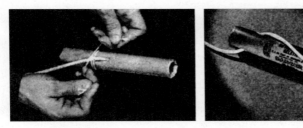

FIGURE 9.18   Priming dynamite with fuse cap.

In shallow holes, and in blockholing or mudcapping, it is practical to simply insert the cap in the cartridge end, without fastening, as the primer need not go out of reach. Friction will hold it in place against a moderate pull, but not against yanking.

Electric Cap Primer.   Figure 9.19 illustrates the most common method of priming a small-diameter stick of dynamite. The cap is pushed into the navel-like end of the wrapper, and the wires are caught in a half hitch around the center. If the dynamite is hard, a hole may be made in the end with a wood peg to make it easier to insert the cap securely so that it will not slip out during loading.

Large-diameter sticks are best primed by this sequence: the cap is inserted in the top, and a loop of wire is pushed through a slanting hole and is caught around the middle.

In each case the cap is entirely inside the dynamite, cannot work into a position where it might scrape the side of the borehole, and its direction of explosion, away from the wires, is directed into the dynamite.

Detonating Cord.   To make up a primer, Primacord is fastened to a large stick of dynamite by being threaded through and secured with tape and to a small one by being tied tightly. The resulting primer is usually the first explosive to be put in the hole. This puts the detonating cord in contact with the entire charge. See Fig. 9.20.

Placing Primers.   If one primer is to be used in a borehole, it is best to place it at the top, or one cartridge down from the top. This keeps to a minimum the danger of damage to fuse or wires while

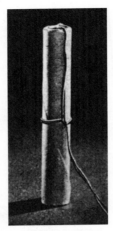

FIGURE 9.19   Priming small dynamite stick.

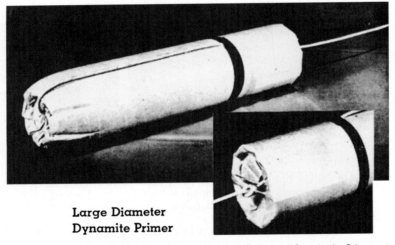

**Large Diameter
Dynamite Primer**

Punch a hole diagonally through the top end of the cartridge, tie the Primacord on to the lowering tape allowing 8 to 12 inches plus the length of the cartridge for Primacord length below the knot to complete the assembly. Thread this length through the diagonal hole, then down the side of the cartridge and insert the end of the Primacord into a hole punched in the bottom of the cartridge. For cartridge reinforcement, wrap electrician's tape around the cartridge as shown.

FIGURE 9.20   Primacord and a large dynamite stick. (*Courtesy of Ensign-Bickford Co.*)

placing and tamping the charge. Use of one stick above the primed one cushions the primer against jars from overzealous tamping, and from contact with abrasive stemming material.

If the hole is long, or the charge heavy, it may be considered a good precaution to use two or more primers. The second one is liable to be in the bottom, and any additional ones are spaced throughout the column.

With all types of delay firing the primer must be in or near the bottom. This largely avoids the danger of throwing the primer out in the muck pile, where it would be likely to detonate during digging, with a possibility of disastrous results.

Spaced charges are more likely to require additional priming than solid ones. Deck charges need a primer in each level.

Correct placing of primers, and even correct direction of the cap in the primer, is of importance under some circumstances and makes little difference in others. Because of the speed and destructiveness of blasts, exact analysis of the mechanism and results is difficult.

Speed of Explosion.   There are four classes of speed concerned in the firing of explosives. There is the slow burning of a fuse, explosive burning of confined black powder, extremely rapid but somewhat variable detonation of high explosives, and almost instantaneous passage of electricity.

Black powder is little used, but still serves to illustrate the effect of point of ignition on explosive performance in a blast hole. Its slow-explosive burning speed may be 1,000 to 3,000 feet per second, so that it would take $\frac{1}{5}$ to $\frac{1}{15}$ second for a 200-foot borehole to fire. If ignited at the top, the upper rock might be moved a considerable distance before the bottom fired. In one way, this would act to lighten the bottom burden and help it to pull, but it might also serve to "uncork" the borehole and allow the bottom of the charge to blow upward rather than horizontally. On the other hand, the force of the upper part of the explosion, reacting against a heavy burden, might press down on the unexploded charge with great force and seal it.

If the column were set off from the bottom by electricity, the toe would be well kicked out before the explosion reached the top. If several caps were placed at intervals in the column and fired together, the whole burden would be moved out at approximately the same time.

The same action is found in high explosives, although the rapid detonation makes it less effective. (See Fig. 9.21.) A 200-foot column of a 40 percent dynamite with a velocity of 10,000 feet per second, fired from the top, would explode at the bottom $\frac{1}{50}$ second afterward. If Primacord were used to the full depth, the detonation would take only $\frac{1}{100}$ second. If electric caps were used at top and bottom, the time at the bottom would be about $\frac{1}{5,000,000}$ second after the top, but the lag to the center would be $\frac{1}{100}$ second.

At first glance, these small fractions of a second might not seem significant. In many cases they are not, but sometimes they do have an important effect on both the performance and the concussion of a blast.

If long-borehole blasts do not give the desired effect in fragmentation, throw, or any other way, it may be advisable to change the location, number, or type of primers to try for better results.

Cameras can be obtained that will take pictures of the various stages of the explosion for study. These may be high-speed streak cameras, which can take 550,000 pictures per second, or instantaneous pictures, as seen in Fig. 9.22.

Precautions.   Precautions to be observed in regard to a primer in placing and leaving it in the hole include: placing it in such a manner that it is not subject to shock or jar, and is not penetrated by rock splinters or other sharp objects; that it is not to be wet for a longer time than the powder, the cap, the fuse, or the wiring can stand; that the fuse and wires lead to the top without kinks, and are held so as not to be damaged by placing and tamping of additional charges and stemming.

Water Resistance.   This is the resistance of an explosive to penetration by moisture, and/or its ability to explode when wet or damp. It is important when holes are wet, or when there is a long time between loading and firing.

Gelatin dynamite and slurry are good to excellent. Other dynamites are good with intact wrappings, fair to poor if cartridges are torn or punctured.

Ammonium nitrate has very poor resistance, but can be kept dry in plastic bags.

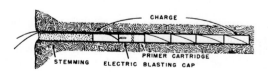

FIGURE 9.21   Loading with instantaneous cap primer.

FIGURE 9.22 A 700-ton blast recorded by camera.

Electric caps are waterproof. Fuse and Primacord are practically waterproof except at cut ends, where fuse has no resistance. Cut Primacord absorbs water slowly. Damp sections are difficult to detonate.

HANDLING EXPLOSIVES

Transporting. Large quantities of explosives should be transported in special vehicles marked in accordance with state or interstate laws. Smaller quantities may be carried in an ordinary car or truck, with any required warning signs made so that they can be removed when not in use.

Caps and explosives should be carried in different trips or vehicles unless quantities are small, in which case they may be carried in one vehicle if kept well separated and if permitted by law.

ICC regulations are accepted by most states for intrastate transportation, but some have more restrictive laws.

Storage. Different classes of explosives should be kept in separate magazines. These should be far enough apart that an explosion in one would not affect the other; and should be surrounded so far as possible by earth barricades or higher ground so that the force of an explosion would be deflected upward.

Magazine areas should be as far as practicable from roads, railroads, or structures, and should be posted with warning signs and fenced if possible.

Magazines should be constructed of cohesive fire-resistant material, such as sheet iron, or soft material which will tear or crush rather than separate into flying fragments. Ventilation and protection from grass fires and from excessive heat should be provided. Doors should be heavy and provided with strong locks.

Portable magazines to hold a few cases of powder or boxes of caps are most easily made from large metal tool or packing boxes fastened with padlocks. When properly marked these are legal in most states, although many laws and regulations recommend more complicated units. When a portable magazine is to be left on a job, it should be chained and locked to a tree or other anchor.

Handling.   Dynamite may cause severe headaches. This is especially apt to occur if it is unwrapped and handled with bare hands. Different brands and strengths differ in headache-producing qualities, and individual reaction is highly variable.

Persons handling explosives should not smoke and preferably should not carry matches. A complete list of safety precautions recommended by the manufacturers will be found in each box of dynamite. A complete list of the "don'ts" issued by the Institute of Makers of Explosives is included in the Appendix.

Boxes.   Dynamite is usually packed in 50- or 60-pound boxes, although 25 pounds may be available. The wooden box, which was inseparable from blasting operations for many years, is no longer used.

The standard box is now made of fiberboard, with a full lift-up cover whose overlap provides double sides for the container. Sealing is by means of tape.

The box is lined by a polyethylene bag, which is an effective moisture and chemical barrier, and can be readily opened and reclosed in the field.

A small possibility of damage to both the dynamite and the liner, with resultant contamination of the fiberboard with explosive, is the reason that these boxes should not be burned, except with the same precautions as with dynamite itself.

LOADING

Holes may be loaded in a number of ways. These may be classified as solid, string, spaced, deck, and spring.

Solid.   In solid loading as much explosive is crammed into the hole as possible. Free-running explosive is poured into the hole, or blown in pneumatically from bulk carriers or from portable hoppers. The air tends to build up static electricity, so the unit must be grounded, and other precautions taken. (See Fig. 9.23.)

Water gels are usually loaded in their plastic containers, which may be intact or slit, but may be poured in. On large-scale operations it is economical to use a truck pumper, which may mix the ingredients also.

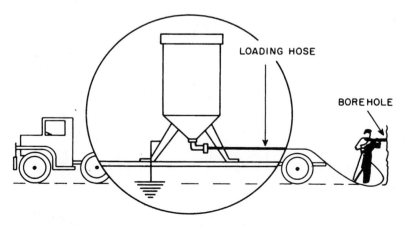

FIGURE 9.23 Grounding a loading hopper. *Note:* Mounted Du Pont Airloaders should be grounded to the ground with cable or wire. DO NOT use chain.

Cartridges smaller than the hole are slit up the sides in two to four places, so they will spread when tamped. Slitting is better than unwrapping, because of reduced danger from spilled powder or of headache from skin contact, and the wrapper ends prevent powder from sticking to the tamper.

Some cartridged explosives have special perforated wrappers that do not need slitting, as they tear and unwind under heavy end pressure.

Tamping. Tamping is a process of compacting explosives in boreholes by comparatively light blows, and/or pressure, of a stick or weight. This tool must not have exposed metal of any kind.

For best compaction, tamp each cartridge or layer separately with a firm, pressing stroke. Sharp blows are less effective, and should be avoided.

A tamping stick should be of round wood, with slightly smaller size than the smallest part of the hole, with a straight cut across the working end. If the hole is too deep for the use of a single tamping stick, several sticks should be drilled lengthwise and strung together with a cord. When the cord is slack, the stick will fold and can easily be fed into the hole. If any stick is held, and the cord tightened above it, the joints below the pull will be made rigid.

If the lower end of the stick wears to a taper, it should be cut back. The taper may punch holes in the tops of cartridges that would not be filled by pressure on the next one placed, it may grind some of the powder against the sides, and it may stick in cartridges and pick them up.

Large blast holes made by rotaries are usually tamped with a block on the end of a rope. The block should be of hardwood to resist abrasion, be slightly smaller than the bore of the hole, and have a flat end.

If weight is needed for heavy tamping, or working in wet holes that would float wood, the block may be drilled and weighted with lead or other heavy metal plugs, which should be covered with wood.

This type of block is not adapted to ramming down cartridges which have stuck in the hole above the bottom, as it may cause excessive side friction. A special block with a chisel-point stake that will break up the stuck cartridges is better.

These blocks are shown in Fig. 9.24.

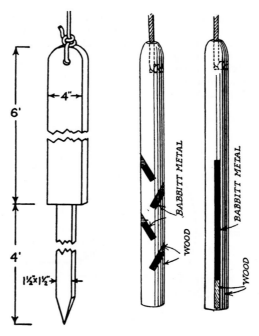

FIGURE 9.24 Tamping and cutting blocks.

Deep Holes.   It is a good plan to check a deep large borehole before loading it, by inspecting it with a flashlight or sunlight reflected from a mirror, or sounding it with a tamping block, to make sure it is not obstructed. The block can be used to knock obstructing pieces or scale to the bottom.

Cartridges may be dropped into shallow holes or deep smooth ones. (See Fig. 9.25.) If the hole is deep and rough, and there is a possibility that they may stick partway down, they should be lowered. All explosives manufacturers will provide cartridges equipped with means to attach a lower rope. A band of special Scotch tape and a readily disengaged hook are used.

The impact of the cartridges on the bottom and the weight of the column above frequently compress charges well enough so that tamping is not necessary.

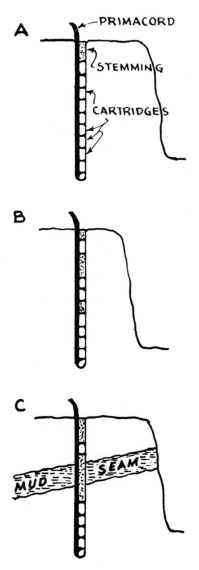

FIGURE 9.25   Loading deep holes.

If the hole is ragged or partially caved so that it is not practical to load it with cartridges, a free-running explosive may be poured down it. If it blocks the hole and starts to build over an obstruction, it should be poked down with a long jointed pole or a dislodging block.

If such a hole is wet, it may be necessary to use a water-resistant slurry or water gel, either placed or poured.

A tamping block used for unwrapped dynamite should be kept clean by resting it on a box or some sacks when it is not in use.

String.   If the borehole is wet enough that slit or unwrapped dynamite would be spoiled, or if solid loading would make too heavy a charge, cartridges somewhat smaller than the borehole, but not small enough to fit side by side, may be dropped in one after the other without tamping, or after having tamped the bottom cartridge or two.

This is the easiest way to load, and is satisfactory for small or occasional blasts. However, it is inefficient. More rock must be drilled than is necessary to hold the charge that is used. Part of the strength of the explosive is wasted in the air cushion around it.

Spaced.   Spacers may be used to string cartridges out along a hole that is not to be fully loaded. These may be square, round, or hollow pieces of wood, tile, lean concrete, or rolled cardboard. They are usually made up ahead of time, in lengths of 8 to 10 inches. There should be sufficient airspace around them to allow for detonating cord or wires, without squeezing or rubbing.

Spacers may be alternated with cartridges or pairs of cartridges in the parts of the borehole that are not to be fully loaded. The primer cartridge should have at least one additional stick in contact with it.

Decking.   In large boreholes, charges which are to be strung out are usually separated by solid plugs of sand or other stemming material, and each section of the charge primed separately, unless fired with Primacord or other detonating fuse.

Stemming.   Stemming is inert material such as dirt, sand, or finely crushed rock that is used to fill parts of a borehole that do not contain explosive.

Its primary use is in filling vertical holes from the top of the powder to the surface. Its use improves breakage by confining the force of the explosion, and adds to safety by preventing accidental igniting of the charge before it is fired.

Stemming is also used to space out charges, as in Fig. 9.25.

The minimum depth of top stemming should be about 1 foot in a $1\frac{1}{2}$-inch hole, and 12 feet in a 12-inch hole. Deeper stemming is used where the top will be shattered simply by having its base blown out from under it. About two-thirds of a very high face may be broken in this way, with explosive in only the bottom third.

Carrying the load too high in a hole is at best a waste of powder, and at worst can mean excessive and damaging rock throw and noise. On the other hand, too little explosive and too much stemming will give poor top breakage.

Drill cuttings are often the best source of stemming, particularly from rotary drills or air drills equipped with dust collectors. It is important that stemming contain no sharp pieces or sizable stones that might cut wires or fly in bullet fashion. Moist material usually is more effective than loose dry sand.

Mines and quarries that have rock-crushing plants and that use large-diameter holes often use fine crushed rock (screenings) for stemming. This may be hauled in Dumpcrete trucks that can dump directly in the hole through chutes. Ordinary dump trucks may leave a pile of screenings near each hole or group of holes, for distribution by hand shovel or wheelbarrow.

LIGHTING FUSES

The length of fuse determines the time which will elapse between lighting it and the explosion. Regular sequences of firing can be obtained by varying the lengths of fuse in different holes, and lighting them at the same time.

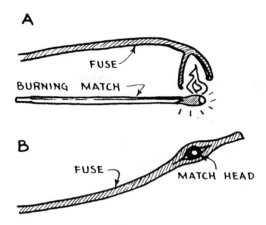

FIGURE 9.26 Lighting a fuse with a match.

Fuse does not light readily with a match because of the small area of powder exposed, and the likelihood of wax from the coverings being spread over the end while it is being cut.

If only a few fuses are to be lit, good results can be obtained by splitting each fuse with a knife or razor blade, as in Fig. 9.26, bending the fuse so the opening is down. It may then be lit with a match. Care should be taken to keep the fingers out of a line with the end of the fuse, as it will spit out a jet of flame.

The split fuse can be ignited more readily by having the opening horizontal or upward, placing a broken-off match head between the halves, and squeezing them together, as in (*B*). The match head gives a much hotter flame than the stick.

It is possible to buy a number of devices which simplify the lighting of fuses. The match lighter is a short paper tube which fits over the end of the unslit fuse and is coated on one end with a compound similar to that of a safety match head. This is readily ignited with a match or the edge of a match box, and subjects the fuse end to intense heat.

This lighter throws a jet of flame which resembles that caused by ignition of the fuse. To tell whether the fuse is actually burning, it is necessary to observe a moment later whether a thin stream of smoke is issuing from it. If it is not smoking, the lighter should be removed and another applied.

The pull-wire lighter is of similar material, but clamps on the fuse and is ignited by pulling a wire.

The lead spitter is a coil of thin lead tubing containing black powder. A piece is cut off, lit with a match, and the resulting hot flame used to then light the fuses. The lead melts back as the powder burns.

The hot wire is similar to a fireworks sparkler. It burns slowly with a very hot flame, and is the safest and most dependable of the devices listed.

A burning fuse will light black powder by contact. When used to explode dynamite, it must be connected to an explosive cap.

ELECTRICAL FIRING

Electrical firing requires a complete circuit from the power source through all the caps and back to the power source. (See Fig. 9.27.) One cap, or several hundred, may be used.

A cap may detonate the charge through a primer in the borehole, or initiate a detonating cord hooked up with a number of holes.

Source of Current.   One standard source of electric energy for blasting is a blasting machine. This may be a generator which delivers high voltage when a handle is pushed or twisted vigorously.

These devices are rated according to the number of caps in series which they can fire at one time. Rating is usually conservative. Their efficiency should be tested occasionally, particularly

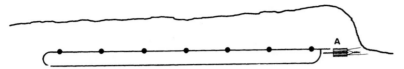

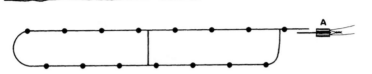

(A) Single row of holes connected by detonating cord and detonated by electric blasting caps or caps and fuse attached to detonating cord at A. Note detonating cord loop is tied to trunkline with right angle connection.

(B) Double row of holes connected by detonating cord. Note detonating cord from back line is tied to trunkline with right angle connection. Detonated by electric blasting caps or caps and fuse attached at point A.

FIGURE 9.27   *Detonating cord layouts.*

if they are not in steady use, as they may deteriorate rapidly in damp storage. Testing is done by means of a special rheostat which sets up a resistance in the line; and from one to four blasting caps. If the machine will overcome the rheostat resistance of its rated capacity, and fire the caps in addition, it is in good condition.

Newer-type blasting machines operate on the capacitor or condenser principle. Current is supplied by flashlight-type dry cells, or by a 12-volt rechargeable nickel-cadmium unit. This modest current is built up to high voltage in condensers, then discharged into the blasting circuit.

The high voltage may be built up when required, by pressing a Charge button or switch, then used by depressing a Fire button, after a buildup time of 5 to 30 seconds.

Some models automatically build up voltage between uses, and can fire instantly.

Storage or dry cell batteries may be used in emergencies for shooting a small number of caps, but they are not considered to be adequate or safe for regular use.

High Lines.   Where electric lines (called high lines) are available, 220 to 440 volts may be used for firing. Special switches are made for connecting into such lines. They automatically shunt or short out the firing lines until the moment the switch is pulled.

Series.   There are three basic types of circuit—series, parallel, and parallel series.

When the caps are arranged in series, Fig. 9.28, the current must have enough force, or voltage, to overcome in succession the resistance of the lead wire, the caps and their wires, and the return lead wire, in addition to the variable resistance offered by connections between wires.

The voltage required can be calculated by Ohm's law. This basic law states that the current, in amperes, in an electric circuit will be equal to the potential, or voltage, of the power supply divided by the resistance, in ohms, of the circuit. That is, if sufficient current is supplied at 110 volts to a circuit with 10-ohm resistance, the flow of current will be 11 amperes. If the voltage is 6, the flow will be only .6 ampere.

A single cap requires a current of about .5 ampere. A series of caps takes 1.5 amperes, with sufficient voltage to overcome all resistances in the circuit.

The tables in Fig. 9.29 indicate the resistance of caps and wires commonly used. The supply of current should be well over the calculated need, however. Minute differences in the bridge wires in caps may vary their resistance, so that a weak current might burn some of them through and break the circuit before all are exploded. Be specially liberal if the series includes both regular and delay caps.

Series circuits are easy to lay out, to hook up, and to test.

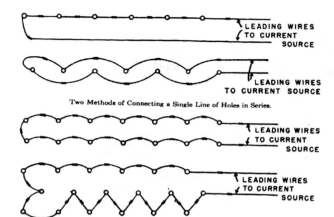

FIGURE 9.28   Series wiring.

Resistance of Electric Blasting Caps in Ohms per Cap

Length of wire, feet	Copper wire		Iron wire	
	Instantaneous caps	Delay caps	Instantaneous caps	Delay caps
4	1.26	1.16	2.10	2.00
6	1.34	1.24	2.59	2.49
8	1.42	1.32	3.09	2.99
10	1.50	1.40	3.59	3.49
12	1.58	1.48	4.09	3.99
14	1.67	1.57	4.58	4.48
16	1.75	1.65	5.08	4.98
20	1.91	1.81	6.08	5.98
24	2.07	1.97		
30	2.31	2.21		
50	2.42	2.32		
60	2.69	2.59		
80	2.71	2.61		
100	3.11	3.01		
150	4.11	4.01		
200	5.12	5.02		

Resistance of Copper Wire

AWG Gauge No.	Ohms per 1,000 feet
6	0.395
8	0.628
10	0.999
12	1.588
14	2.525
16	4.02
18	6.39
20	10.15
22	16.14

FIGURE 9.29   Resistance of caps and wires. (*From the 16th Edition of Blasters' Handbook.*)

PARALLEL

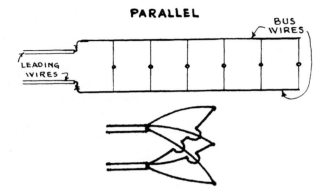

FIGURE 9.30   Parallel hookup.

Caps made by different manufacturers must not be used in a series, because of variation in current requirement for detonation.

Parallel.   In a parallel circuit, Fig. 9.30, the current does not go through the caps one after another but goes through all of them at the same time. A poor connection on a cap wire affects only that cap. The voltage requirement is lower than when the same number of caps is shot in series, but more amperage is needed.

Two caps in series have twice the resistance of one cap. Two in parallel have only one-half the resistance of one, as less potential is required to force the current through a large conductor than a small one. But where 1.5 amperes was sufficient current to shoot a whole series of caps, .5 ampere is required for each cap placed in parallel.

Parallel wiring is therefore preferred where a source with low voltage and high amperage, such as a storage battery, is to be used. It is not suitable for blasting machines, and the results with dry cell batteries are doubtful. High lines are equally efficient with either arrangement.

The most common simple parallel hookup is the second one shown, and it is not recognized as such by many who use it. It is the most convenient way to fire a small irregular group of blasts.

In figuring a parallel circuit, the resistance on one cap is divided by the total number of caps, and in a large blast, may be so small a figure that it can be ignored. The resistance of the two bus wires, between the leads and the last cap, is approximately one-half the resistance of the same length of the same wire used for a lead. Lead wire resistance is the same as in a series circuit.

The lowered resistance of the bus wires is due to the fact that some of the current is diverted at each cap. Full current is present at the beginning, and zero current at the end, so that it averages out to about one-half current for the full length of these wires.

Parallel Series.   This layout, Fig. 9.31, makes it possible to shoot large numbers of caps without requiring excessive voltage or amperage.

There is some disagreement about the balancing of the size of the different series. Technically, each series should have the same number of caps. Many blasters, however, claim better results when the series differ from each other by a set amount. This is said to be particularly advantageous when firing an excessive number of caps with a blasting machine.

When the series are equal, and juice is put in the line, all caps are equally heated and should detonate simultaneously. If any series gets current but less than its share, due to a poor connection or other defects, it may not fire. However, as the other series fire, current will cease to go through them, and all of it will go through the remaining one, unless the wires are broken, until it fires also.

If the series have different numbers of caps, at first the current will flow most strongly through the series with the fewest caps and least resistance, and having fired that, will concentrate on the

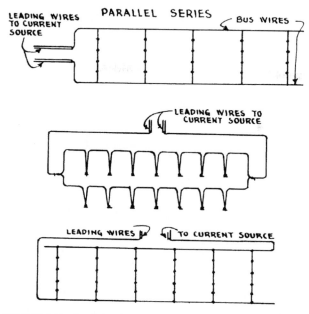

FIGURE 9.31   Parallel series.

next longer string, and so on through the whole group. If the current is weak, there may be a brief but definite time interval between the series, so that short strings should be at the face. If it is strong, explosions may be simultaneous.

As the current from a blasting machine flows very briefly, it is possible that in a graded pattern only the shorter series would fire.

In most blasting patterns, the equal series hookup will be very much simpler than the graded series, which is generally considered obsolete.

Making Connections.   It is most convenient to have the cap wires of such length that they meet each other with a moderate amount of slack between the holes. If they are short, an extra piece of wire must be spliced in at each connection; if they just reach, insulation or primers may be strained while splicing; and if much too long, they will need cutting, or will make a tangle of loose wire that may lead to mistakes in connecting, and accidents.

While connections are being made, the power or far end of the lead wires should be fastened together to ground out any induced current. The electrical source should be locked or at least removed from the immediate vicinity of the wires.

In a series circuit, the current runs from one lead wire through all the caps and their wires, to the other lead wire. The insulation on each lead is stripped back with a knife or plucked off with pliers for an inch or two, and the wire is bent to a tight loop.

The cap wires are pulled apart at their soldered connection, and each wire is connected to one from an adjoining cap. When all the caps are connected in this manner, the two end wires are connected to the lead. Figure 9.32 (*A*) to (*C*) shows connections commonly used between caps, and Fig. 9.32(*D*) shows two cap-wire-to-lead-wire hookups.

It is usually possible to arrange the circuit so that it is convenient to hook both end caps to the lead. However, in a one-row straight-line layout, an extra wire must be used to connect the last cap to the lead.

If cap wires are long, such connecting wires are usually made up of scrap wire left from preceding blasts. It is good practice to extend the leading wires some distance with a lighter and less

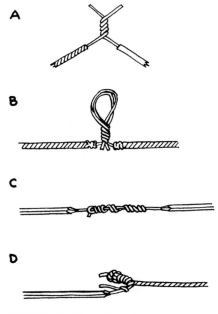

FIGURE 9.32 Connecting wires.

expensive wire, to reduce damage to the ends. If not enough surplus wire is available, connecting wire may be brought on the job on a spool and cut as needed.

The use of scrap wire involves a risk of misfires due to breaks inside the insulation, which is not justified by the small value of the wire saved.

When the last cap has been connected, the whole series should be rechecked, to make sure that no hole has been omitted, that no loose ends of wire are lying around, and that connections are tight. It is good practice to squeeze each connection with pliers at this time.

Bare connections may be propped up on sticks or rocks where necessary to keep them out of water, or from contacting wet ground. If any connections are unavoidably wet, they may be tightly taped or smeared with water-resistant grease.

If the shot is only of a few caps in a limited area, and the electrical source is of ample power, precautions against bare wet joints are necessary only if the water has a high mineral content.

Electrical Hazards. Electric caps are supplied with a connection between the ends of the wires which is called a shunt. This prevents the accidental buildup of opposite electric charges in the two wires, which might pass enough current through the cap to explode it.

Such charges may be caused by the near presence of electrical machinery or transmission lines, a radio transmitter, stray currents in the ground, thunderstorms, or static electricity from dust storms or escaping steam. Such currents in lead wires may often be detected by inserting a No. 47 radio pilot lamp in the circuit instead of the cap. If it glows, conditions are unsafe.

The best precaution to take in blasting near an electrical hazard is to use fuse caps and Primacord. However, certain precautions can be taken which will reduce the danger of using electric caps.

Lead or other wires should not run parallel to electric lines.

The shunts should be left on cap wires until they are connected into the blasting circuit, and the circuit should be shorted until ready to fire.

The two-cap layout in Fig. 9.33 illustrates the method to be used. The lead wire is shorted at each end. A wire is connected to one lead, and to one leg of a cap. A connection is made from the

FIGURE 9.33 Protecting caps against stray currents.

other leg or the shunt, to one leg of the next cap which has its other leg connected to the return lead. Making cuts where indicated will include the caps in a firing circuit which is closed until the leads are separated at the battery end.

Even with these precautions, however, blasting should be discontinued if there is a thunderstorm within 5 miles, or other severe hazards exist.

Crackling in a portable battery-type radio (not FM), left at high volume, provides warning of approach of a thunderstorm.

Testing.   A circuit tester should be used for checking before attempting to blast. This device consists of a galvanometer and a silver chloride dry cell, which produces a current too weak to fire a standard blasting cap. The lead or cap wires are fastened to its terminals, and the action of the indicator needle shows the condition of the circuit.

If the circuit is good, the indicator needle will move an amount inversely proportional to the resistance offered by the caps and wire used. If the needle does not move, there is an open break. If it moves only slightly, a loose connection, or a break with wires just touching, is indicated. If the needle moves farther than it should, a short or a ground is present.

Each hole may be tested before hooking into the circuit, and a single test may be made from the power end of the lead wires when wiring is complete.

Any trouble in the system can be spotted by making a series of tests with a long connecting wire. In Fig. 9.34, the connecting wire N is fastened to one lead and to a tester post. The other

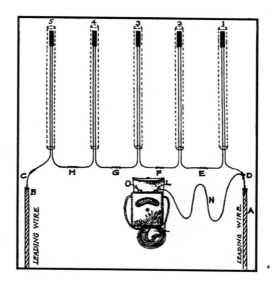

FIGURE 9.34 Testing for break in blasting circuit.

post, or a wire securely connected to it, is touched in succession to connections E, F, G, H, and C. The bad reading will show up whenever the difficulty is inside the circuit being tested. As an example, if normal readings are shown at E and F, but an abnormal one is found at G, the trouble is in number 3 cap or its wiring.

After the blasting circuit tests properly, an additional test is made at the power end of the leading wires.

Warning.   Warning must be given of intention to blast. The type of warning may be determined by either law or custom. For large blasts, particularly in pits employing numerous workers and machines, blasting should be done at specified times, as at 12 and at quitting time, and should be preceded by definite and well-understood signals, such as horns, sirens, whistles, or yelling, long enough in advance to notify all persons and give them time to prepare.

A usual blast signal is a call of "Fire" or "Fire in the hole," repeated two or three times at intervals of 10 to 30 seconds. Signals should be arranged by which workers detailed to watch entrances to the area can stop the blast if necessary.

It is the responsibility of the blasting crew to make sure that all personnel are out of the way, machines are protected, and no visitors or trespassers are where they can get hurt.

The area, and particularly any roads or paths leading into it, should be marked with warning signs. If any public roads are within the danger area, traffic should be stopped at a safe distance.

Firing.   During the signaling, the wires are connected to the blasting machine or switch, or if a battery is used, one wire is connected to a post. To fire, the blasting machine handle is slammed to the bottom, or the switch is closed.

There is a wide divergence of opinion as to what constitutes a safe distance from which to fire a blast. Some experienced people will take shelter behind a nearby rock or tree, whereas others consider 500 feet a bare minimum. No one should stand in front of a rock face, at any distance.

Proper barricading may be as safe as distance and more convenient. Full protection requires some sort of roof or overhang. A very safe spot is in the tucked-in bucket of a big front shovel which is turned away from the blast. The shovel itself may be protected by wood lagging on the rear of the cab.

The return to the blast should be slow for several reasons. The fumes, which dissipate in a few minutes in the open, are toxic and may cause severe headaches and nausea. If more than one hole has been shot, one of them may fire late. Rocks are occasionally thrown so high that they take a long time returning. Rocks or debris may be lodged precariously in trees.

LIGHT BLASTING

Light blasting, Fig. 9.35, includes loosening up of shallow or small outcrops of rock and breaking boulders. It may constitute the entire job, be done in connection with dirt excavation, or follow heavy blasting which has failed to cut to grade or slope lines or has left chunks too large to load.

Chip Blasting.   Shallow rock outcrops are most conveniently broken up by drilling and blasting. Unless the rock breaks readily along planes more or less parallel with the surface desired, it will be necessary either to drill much deeper than grade or to space the holes closely. It is often good practice to blast each row before drilling the next.

Loading may be light, or very heavy, but in general it is necessary to use more powder per yard of solid rock than in heavier work.

Laminated or jointed formations may be shaken apart by light charges. Fragments may be thrown long distances, and mats used to confine them are more subject to damage than with deeper blasts.

The amount and direction of throw can often be controlled to a large extent by drilling and loading procedures. A vertical hole causes scattering in all directions. A sloped hole tends to leave the lower slope in place and to throw the upper one away from it. Throw is reduced by increasing the number of

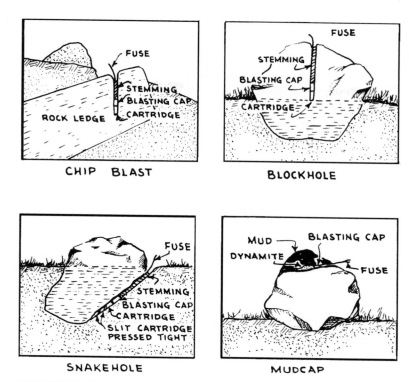

FIGURE 9.35   Spot blasts.

holes, reducing the charges, or drilling deeper than required by the breakage line. These last two place the powder deeper, where more of its power is applied to breaking and less to scattering.

When breaking must be done exactly to a line, holes are drilled closely along the line and a variable number left without charges, as discussed on page 9.12.

Blockholing.   Boulders and oversize pieces of blasted rock may be broken by drilling a hole slightly more than halfway through, and exploding a small charge of dynamite in the hole.

Fragments may be thrown for long distances, so that protection should be provided for the blaster and other personnel. High-velocity explosive, or large charges, will produce finest fragmentation.

Chip blasting may be called blockholing also.

Mudcapping.   Ledges may be chipped and boulders broken by mudcapping instead of drilling. Heavy charges of dynamite, preferably of the highest-velocity type that is obtainable, are laid on the surface of the rock, primed, and covered with a few inches of mud. The explosion acts as a giant hammer blow and should split or crush the stone.

Knowledge of the grain and jointing of rock is important in successful mudcapping. The charge is placed in the same place which, in hand breaking, would be hit with a hammer or opened with a wedge. In general, hammerlike crushing is most effective on loose boulders, and splitting on ledges.

It is a common error to suppose that the force of black powder is chiefly exerted upward, and that of dynamite downward. In each case the explosion acts equally in all directions, but when it acts slowly, it can find and follow paths of least resistance, where the quicker-acting dynamites

DIAMETER OF BOULDER IN FEET	APPROXIMATE NUMBER OF CARTRIDGES, 1¼ × 8" — IN AVERAGE HARD STONE — REQUIRED FOR:		
	Mudcapping	Snakeholing	Blockholing
1 ½	2	1	¼
2	3	1	¼
3	4	1 ½	½
4	7	4	¾
5	12	6	1

FIGURE 9.36 Charges for boulder blasting.

deliver such a rapid blow that they will crush objects under them, even when not confined. However, a study of the table in Fig. 9.36, showing quantities of dynamite used for blockholing and mudcapping, will show the waste involved in open explosions.

The mud pack over the charge is usually 2 to 6 inches thick. It serves to confine the explosion slightly, increasing the force exerted on the rock and reducing noise and airborne concussion. Mud is preferred to any other substance. It is much more effective at confining the explosion than dry or damp dirt or sand, as it packs and sticks together better. It should be free of stone or pebbles that would create a hazard by flying long distances.

Charges can be fired on bare rock, but they are less efficient and even noisier.

Mudcapping is wasteful of powder, excessively noisy, and less certain in effect than drill hole blasting. However, it causes less rock scatter than other methods of shallow blasting, and does not require the presence of a compressor.

Snakeholing.   Boulders are most readily broken if they are lying on the surface of the ground. If partly buried, the earth or other rock around them provides a support and a vibration-absorbing cushion that may prevent or reduce breakage.

Embedded boulders may resist machinery that can handle them readily once they have been loosened up.

Snakeholing consists of making a hole beside or under a boulder, and firing a charge sufficient to roll it out of the ground, and preferably to break it also. Any further breakage required can then be accomplished by mudcapping.

Snakeholing is more laborious than mudcapping, but is more economical of powder and is much less noisy.

MECHANICAL BREAKAGE

Drop Ball.   Balls of hard steel, familiarly known as drop balls or skullcrackers, may be carried by a crane, and used for breaking loose rock. Weights range from 1,000 to 8,000 pounds new. They wear down in use, and are replaced whenever they become too light for the work they are expected to do.

The ball is carried on the hoist line of a revolving crane. It is positioned high over the rock to be broken, the brake is released, and the ball falls almost freely to strike the rock. The brake must be put on as it hits, to avoid spinning out of cable. Long booms held high increase striking power, but may reduce accuracy, particularly on windy days.

Skullcrackers eliminate the danger and nuisance of secondary blasting, but they cause problems of their own. Rock chips fly from 50 to 200 feet, and they make the area around the operation one of continuous danger. The operator's station on the crane must be protected with wire mesh or bulletproof glass, and signals must be arranged to stop the work when the area must be entered by others.

Some rock breaks very well on impact, and others do not. If breakage is good, the rock can be readily reached by the crane, and other operations are not unreasonably delayed by it, this is a good method.

Some quarries or pits keep a drop ball stored on a crane at all times, and may use it continuously or only occasionally. Gravel pits may push boulders aside until there are too many around, and then put a ball on one of the draglines or clamshells for a day or two. Rock breaking may be done on days when the pit is shut down, to avoid interference.

Plug and Feathers. Rock may be broken by first drilling, then inserting a device in the hole that can be caused to expand until it breaks the rock. The Egyptians used dry wood plugs, which they supplied with water until they swelled and broke the rock. Now we use wedges or jacks to obtain the same result.

The plug and feathers (Fig. 9.37) is a three-piece wedge set with a very gradual taper. The outer pieces, called feathers, are placed in the hole, and the plug is forced between them by blows from a sledgehammer or air hammer, or by hydraulic pressure up to 7,000 pounds per square inch, using a tool known as a hydraulic splitter, as in Fig. 9.38.

On large or extremely hard boulders or concrete blocks, which might be impossible to break with one tool alone, two or more tools can be hooked up in parallel, doubling the force and greatly increasing production. Granite boulders over 20 feet in diameter have been broken in half with this technique.

Pressure may be developed by an attached hand jack, or by an engine-driven hydraulic pump. Expansion force of 80 to over 400 tons may be developed by the wedging. Two or more sets may be used in adjacent holes. It is probable that any rock strong enough not to crumble, and that has a free face to move toward, can be split in this way. Solid rock faces are taken in shallow cuts.

In general, these devices are used where explosives are particularly dangerous or annoying, are prohibited, or the job is very small.

Backhoe. A hydraulic or air hammer, mounted on a hydraulic backhoe in place of the bucket, is an effective boulder breaker and is convenient to use.

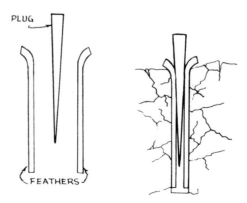

FIGURE 9.37 Plug and feathers.

On large or extremely hard boulders or concrete blocks, which might be impossible to break with one tool alone, two or more tools can be hooked up in parallel doubling the force and greatly increasing production. Granite boulders over 20′ in diameter have been broken in half with this technique.

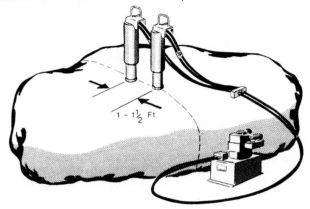

FIGURE 9.38   Hydraulic rock splitters. (*Courtesy of Elco International, Inc.*)

ROCK MECHANICS

Rock is broken or fractured by an explosion in three ways: compression, shear, and tension. Compression is obtained by the direct, hammerlike blow of the explosion against an unyielding rock mass. An explosive that is so deeply buried that it cannot break out to the surface breaks by compression only. This is the least effective way to use it.

Shear is movement of pieces or blocks of rock along lines of weakness. Tension is produced by reflection of the explosion back from an unconfined surface or face of the rock.

In a hard rock, maximum effectiveness of explosives is in tension. Tensile strength is only about one-tenth of shear strength, and shear strength is only one-tenth of compressive strength. This means that a blast that can break out to a free face at an efficient distance may produce 100 times the rock breakage of one that is completely confined.

The importance of tension has been demonstrated by setting off a tightly confined explosive far enough below a horizontal surface that the explosion could not break out, but near enough that surface rock was shattered in a cone-shaped crater, separated from the small explosion chamber by solid rock.

It is very fortunate that rock breakage is not produced mostly by the outward movement of gases from an explosion, as in this case rock throw would be many times greater than it is. In a well-engineered and successful blast, most of the rock moves less by explosive force while being broken than it does afterward by slumping under the pull of gravity.

The improved fragmentation obtained by using millisecond-delay caps is partly due to creation of a series of free faces from which waves of succeeding explosions reflect to produce tension in the rock.

DAMAGE

One of the contractor's problems in connection with blasting is the possibility of real or imaginary damage being done to structures in the vicinity.

An explosive, if properly used, will expend most of its energy in shattering the rock immediately around it. The remaining energy will set up waves or vibrations in the ground, and sound and concussion in the air.

Noise.   The noise of an explosion may cause most or all of the neighborhood difficulties. Mudcaps, shallow blasts, overloaded holes, fractured rock, and other conditions that allow the explosion to break out into open air before expending its energies, are the cause of complaints all out of proportion to the amount of explosive used.

In the first place, the noise attracts attention to the fact that blasting is going on. It causes the householder to concentrate on trying to feel the jar or shake of the blast, to look for cracks in plaster, and to speculate about other damages that might be done. In many cases, the sound of the blasting will annoy sensitive people so that they will invent or exaggerate physical effects. The contractor or quarry operator's first rule is therefore to blast as quietly as possible in any area where there is a possibility of complaint.

This means a first rule of *no mudcapping.* This technique is not only wasteful of explosive, but a sure way to lose the goodwill of the neighborhood and of the insurance adjuster.

Boulders and oversize blast fragments should be drilled before blasting. The noise is tremendously reduced, and it will usually be found that the saving on explosive and the better fragmentation obtained will more than outweigh the cost of the drilling. When there are only a few pieces, the nuisance of clearing the pit for blasting may be avoided by plug-and-feather splitting. In brittle rock, a crane with a skull-cracker steel ball may be the most economical solution.

To minimize noise, do blasting when weather conditions are more favorable, that is, clear to partly cloudy skies and relatively warm daytime temperatures. Unfavorable weather conditions are those when the air is relatively still. These occur when the days are foggy, hazy, or smoky and the temperature is constant from the ground up to a high altitude. About the only general rule is that the blaster should consider avoidance of noise one of the important objectives.

Sound travels rather slowly. Its distribution is affected by winds, as shown in Fig. 9.39; by reflection from hills, clouds, or atmospheric layers; and by temperature and humidity.

Characteristics of Damage.   In most cases, there is an observable difference between building damage caused solely by vibration, and that resulting only from structural defects and/or settlement that change the shape of the building. Most of the distinction is in the pattern of wall cracks.

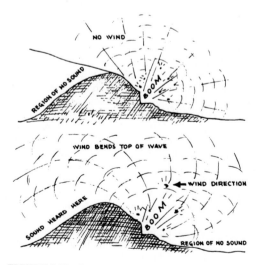

FIGURE 9.39   Sound travel from blast.

Cracks caused by vibration are usually fairly uniform in width with little or no displacement of the sides relative to each other. Vibration cracks tend to fan out diagonally from corners of windows and door frames.

Either type of stress frequently causes walls to crack along their junctions with a chimney or column. Settlement or tilting of the chimney only, or the building only, usually results in an opened or spread crack on one side, and a closed or even pressure-squeezed crack on the other. Vibration more often produces equal-width cracks on both sides.

A distinction can sometimes be made on an age basis. New cracks show clean surfaces of material; older ones become progressively grimier, the rate depending on conditions. Dust, spot deposits, and light-bleaching of color pigments are indicators. Recent cracks should be clean and unfaded.

Concussion. Airborne concussion is responsible for a large share of the damage in bombing, and in accidental detonation of explosives, but is rarely a factor of importance in blast damage. It consists of one or more waves of highly compressed air moving outward from the explosion.

Sufficient explosive to cause concussion more than a few feet away should not be used in mud-capping. Even very heavy blasts in solid rock cause little or no concussion if they are laid out and loaded properly.

Any damage caused by concussion is usually obvious. Glass breakage in closed windows at right angles to the path of the waves is the most common result. In the absence of extensive glass breakage, it is very doubtful that any other parts of a structure could be damaged.

If a building is to be endangered by blasting, windows should be opened or removed, particularly those on the facing and far sides of the house. Store windows may be braced as a routine precaution. Careful check should be made of the condition of plaster and masonry, so that claims need not be paid on preexisting defects.

Rock Throw. Unexpected damage may be done by rock or other material thrown through the air by blasts. In general, shallow blasts, overloaded holes, shots in rock with irregular resistance, and block-holed boulders give the most trouble in proportion to the amount of powder used.

Thrown objects may cause injury or death, and their control is therefore of first importance. Property damage may or may not be severe, but at least claims filed on this ground are usually sincere.

Danger of damage from rock throw may be reduced by increasing the number of holes so that smaller charges may be used, by sloping holes to throw rock away from danger points; by reducing the quantity or strength of powder, and handling any resulting oversize fragments by blockholing under mats or by the use of larger machinery.

Covered Blasts. Throw can be closely controlled by working downward, using small blasts and covering them with mats or chained logs. If the cover is large and heavy in proportion to the strength of the explosion, it will prevent any scattering of fragments. If the charge is heavy enough to lift the cover, it will move somewhat less than the average distance of throw to be expected from an uncovered blast, and fragments with higher-than-usual velocity will be held in.

It is important that the cover extend several feet beyond the area being shot, particularly if the charge is heavy enough to lift the mat, as fragments might escape under its edges.

When a power shovel is used to remove the shot rock, it is advantageous to use a woven steel mat as it is easily handled with chains, and provides a quicker and more secure cover than logs. The mat is lowered over the holes, or dragged in such a manner that it will not damage the wiring and cause misfires.

Logs are used when no mat is available, or when there is no machinery on the job which can handle one. They should be long enough to overlap the blast at both sides, and light enough to allow the crew to carry them by hand. Two chains should be laid on the ground first, the logs piled, and the chains fastened over them, preferably by wired square knots.

Chaining is important, as unfastened logs may be thrown farther than rocks.

Neither mats nor logs should be laid directly over mudcaps as they are liable to be thrown long distances and be severely damaged as well.

Blasting mats should be used wherever there is the slightly possibility of fragments reaching people or property. Even a scattering of sand or fine pebbles on their property will make people nervous and resentful, and is an indication that loading should be reduced or the technique changed to a safer one.

Reasons for Complaints.   Studies indicate that much greater vibrations are produced in house structures by slamming doors, running, and often street traffic, than by even severe blasting. It would appear that at ordinary distances the ground and air vibrations set up by heavy blasts are so weak as to be incapable of affecting any structure. Yet complaints and claims for damage pour in on every blasting job. Why?

There are a number of reasons. One is that the ability of rock, soil, and water to transmit vibration varies much more than is indicated by the relatively superficial testing that has been done.

It would appear from an old table of recordings that 600 pounds of explosive would not produce sufficient ground waves to damage a house on average overburden 100 feet away (!!). But there are records of high five-figure awards paid for damages done to a village 2 miles away from an underwater blast of this size, indicating a difference of over 10,000 percent between theory and fact.

Contractors frequently blast much more heavily than is indicated by their records and statements, particularly when a job gets behind schedule. Mistakes in loading can occur. Variation in the strength and quality of explosives can be a factor.

Most of the checking of blast damage to date has been done by representatives of mining and insurance interests, who are more interested in disproving it than in making an impartial study. Some of the instruments used for measurement leave much to be desired.

There are also psychological reasons for exaggeration of blast damage. Bomb damage received much publicity in modern wars and recent terrorist attacks and has made people overly conscious of the dangers of explosives. There is also fear and resentment of the unusual, which makes blast vibration appear more significant than that from a truck.

Prestressing.   One of the principal defenses advanced by defendants in blasting damage suits is criticism of the condition of the structure before the blast. If it is in a condition of stress due to unequal settlement, warping or shrinking of timbers, or overloading, it will change in shape and its plaster will crack.

If a blast vibration is within "safe" limits, an overstressed condition may cause cracking from the blast. The theory is that if the blast had not been set off, the same cracks might have developed shortly from natural causes.

In general, the poorer the quality of construction, the greater the probability that stresses will develop, plaster crack, and misalignment occur. But this is not always so.

The prestress argument is unpleasantly reminiscent of the whitewash given the Donora smog by a group of doctors. They said in effect that it was nothing to fuss about, as only people with a previous history of respiratory disease had died.

It would be unjust to allow reckless blasters to evade payment of damages on these grounds, or to make property owners go without recompense because their building standards fall short of those set up by the U.S. Bureau of Mines. However, blasters should not be compelled to subsidize substandard construction. It is likely that most cases where prestressing is actually proved should be subject to compromise settlements.

Water Supply.   Blasting sometimes causes springs and even deep wells to go dry. The vibration causes underground movements that may close water passages or open new ones. However, explosives probably are responsible for only a fraction of the difficulties for which they are blamed.

Underground water circulation is under constant change. Old seepage veins become plugged with mineral deposits, new ones are opened by solution and erosion. Changes in rainfall pattern, in conversion of forest land to farms, or back again, may alter the quantity and location of underground water over a wide area. Overpumping will lower the water table.

A new well may tap into an underground reservoir of limited size, which once pumped out will not refill. Such a well may show a very high yield on its first test but decline markedly after long use, when it comes to depend on circulating water only.

Keeping Out of Trouble.   Under all ordinary circumstances, blasting should be kept light enough not to damage buildings. The job should be figured on a basis of conservative blasting, and the work done the same way.

Short-period delays are a real friend to the person who wants heavy blasts, but is surrounded by structures. Up to 70 percent of the charge of an instantaneous blast can be used with *each* of 10 or 12 short delay periods, without increasing the vibration. Or to look at it another way, the same loading can be used as for an instantaneous shot, and 10 periods used to cut the damage potential by four-fifths.

Even with ultraconservative procedures, inspections should be made of nearby buildings before blasting. If property is valuable, vibration-testing devices will be supplied by the insurance company or by a blasting consultant to measure the disturbance caused. Such instruments should be used to check the next blast in any building or area from which complaints are received.

As detailed before, noise should be kept to a minimum. If there are few people in the area, it should be possible to notify them before blasts, so that they will not feel it necessary to be tense all day waiting for an explosion. Another method, applicable to heavily populated areas also, is to set a definite time or times each day for shooting, and stick to it.

If a claim is made and is justified, it should of course be paid. But if it is clearly unjustified, it probably should not be paid even if apparently too small to be worth arguing about. One paid claim is likely to bring in a dozen or a hundred others, and the contractor might be forced to replaster and decorate a whole town before he or she knew it. Payment of any claim makes any other much harder to defend in front of a jury.

Of course, contractors should protect themselves with insurance, and usually do so. But in the long run the premiums they pay are based on what the company pays out for damages, so their interests are identical.

DIGGING UNDERGROUND

TUNNEL WORK

Tunnels are underground passageways of any size, and may be natural (as in limestone caverns) or made by animals or humans. Those discussed in this section are made by people. They serve a variety of purposes, including mining, water supply and drainage, laying sewer and other pipes, railroad and vehicular shortcuts or water crossings, and air raid protection.

Rock tunnels are driven through solid material that usually requires blasting and may support itself permanently, or at least long enough to allow setting up of bracing after digging out a short section. Soft ground tunnels involve digging or pushing aside soil, and the roof (called the crown) and the walls may require support before removing the soil. Mixed-face tunnels go through both types of ground, either together or in different sections.

People have driven tunnels since prehistoric times. They usually worked in rock, because the difficulty of digging it was more than compensated for by its ability to hold itself up. Cutting was done with hand tools, or by heating the face with wood fires, then throwing cold water or cold water and vinegar on it to cause sections to crack off.

The vinegar technique, with little or no ventilation, must have been really rough on the workers. A rough approximation of the atmosphere might be obtained by building a good blaze in a fireplace, shutting off the chimney damper, then putting out the fire with vinegar.

Layout and Problems.   The methods used to drive a tunnel vary tremendously with the nature and water content of the material to be penetrated, depth and size required, surface conditions along the route, time allowed, and background of the workers doing the job. There is space in this section to indicate only a few of the problems most often encountered.

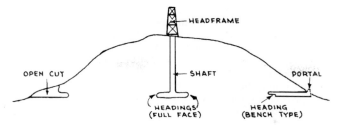

FIGURE 9.40   Tunnel layout.

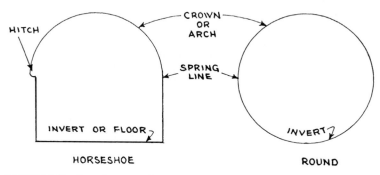

FIGURE 9.41   Tunnel cross sections.

The diagrams in Figs. 9.40 and 9.41 show the layout of a simple tunnel job, and the names for some of its parts. If it is driven more or less horizontally into a hillside, the opening is called the portal. The working face, where the digging is done, is the heading. Vertical access tunnels descending from the surface to the main tunnel level are known as shafts.

In the tunnel itself, the floor is the invert, and the roof is the crown. The spring line is the meeting of the vertical sidewall with the curve of the roof arch. A supporting shelf cut at this line is the hitch. A small pioneer or accessory tunnel is called a drift. Standard cross sections are rectangular, round, and horseshoe-shaped.

There are a great many special problems connected with even a simple tunnel project. To the open-cut worker, one of the most impressive is lack of space. Many tunnels have been driven with cross sections as small as 4 × 4 feet—not even big enough to stand in.

MICROTUNNELING

For a relatively small pipe installation, such as a 36-inch sewer pipe, and adverse conditions, such as high groundwater, the microtunneling technique may be used. This technique eliminates the need for open trench excavation and possibly tight sheathing with the need for a dewatering system. It is sometimes called trenchless technology.

The technique makes use of a microtunneling machine that uses a pipe jacking apparatus that guides the pipe through the ground as it advances. The apparatus is remotely operated with laser guidance to keep it in line. The microtunneling machine has a sealed head in the cutting face to keep the soil pressure balanced between the slurry injection system and the removal rate of the soil dug by the cutting head. The slurry injected through hoses from the back end of the machine and through holes in the front end of the pipe is generally a bentonitic mixture of earth material that lubricates the movement of the pipe and seals the outer sides of the pipe when it becomes stationary in its final position.

There are several problems that occur for the microtunneling machine. Because of its sealed head arrangement it can not deal with obstacles in the way of the advancing pipe the way a horizontal boring machine can. But the boring machine needs to be operated with manpower inside and other complications. Also there is a limit to the toughness of the ground that the pipe is driven through using microtunneling. Generally, if the ground is no harder than 5,000 psi resistance, then the microtunneling machine can operate.

A microtunneling machine performs better as the moisture content of the soil increases and the soil type changes from a stiff clay to a more granular material. In good conditions this method may be able to advance from 20 to 80 feet per day.

LARGE TUNNELS

Twenty- to 30-foot-diameter tunnels are big, yet they provide a floor width that would be considered skimpy for a haul road on top.

Equipment to be managed at and near a tunnel heading may include a drill jumbo (a movable frame almost as big as the tunnel, carrying a battery of drills), a machine for loading muck (the below-ground name for spoil) and railcars or rubber-tire trucks to remove it; the same or other cars or trucks to bring drill steel, bits, explosives, and other supplies to the heading; a locomotive to push and pull cars; and a switching or passing device to permit hauling units to get past each other, although often there is room only for a single width of track or roadway.

There will be high-pressure air pipes to supply the drills, and often large low-pressure ducts for ventilation. Overhead wires or ground cables carry electricity for light, power, and blasting juice. Water under pressure may be supplied for wet drilling. A system of drainage, pumping, or both may have to handle tremendous volumes of water.

In addition to the regular equipment there may be need for a diamond drill to make test and grouting holes, grouting equipment to seal off leaks and solidify wet ground, and/or a movable buffer to confine rock throw from blasts.

If the tunnel is to be timbered for support, or lined for support or for permanent use, the crews and materials for this work may follow the digging closely, and in any event will have to work in and over the single entranceway.

If driving is from a shaft, its bottom is another crowded point. Haulage equipment may be lifted to the top to dump, or it may dump into containers at the bottom. Supplies must be unloaded from the elevator cages and reloaded for hauling to the face. Crews leaving work, and their supervisors and inspectors, wait here for transportation to the surface. Pumps, compressors, and even drill and repair shops may be located in skimpy quarters excavated near the shaft.

Sequences are very exacting. The tunnel cycle (the succession of drilling, shooting, and mucking) must keep the largest possible number of workers and machines usefully employed, and the time interval from any operation to its repetition should not vary. Whenever possible, two or more operations should be performed simultaneously, as drilling the top of a face while digging the bottom, and installing lining a few feet back at the same time.

When two headings driven from one shaft are close together, one may be drilled while the other is shot and mucked. With increasing distance, the advantages of this arrangement are reduced.

Most tunnel crews are of the universal type, and perform all operations in the cycle. This saves the contractor from paying a crew waiting time because of a delay in a prior operation.

Speed.   Under favorable conditions, tunneling may progress very rapidly. The Owens River Gorge Power Tunnels in the Los Angeles water system were driven as fast as 104 feet in a day, and 2442 feet in the best 31-day period. Other tunnels excavated by using tunnel-boring machines, called TBMs, working in fairly ideal ground conditions have made even better progress.

Gold mines at Kimberley, South Africa, hold the depth record at 9000 feet. These tunnels must be air-conditioned, as otherwise the heat would make it impossible to work in them.

The 12-mile Simplon tunnel in the Alps is 7,000 feet beneath the surface at one point. Temperatures up to 131°F were encountered in drilling it.

A record for its time was established in twin power tunnels at Niagara Falls in Canada. These are each 51 feet in diameter and 5½ miles long. Together they required over 5,000,000 yards of excavation.

There is such constant improvement in tunnel-driving techniques, and increase in confidence to undertake bigger projects, that some or all of these records may have been surpassed by the time this book is in print.

Plant.   The plant at a tunnel may include the tower, hoist, and hopper; compressors, low-pressure ventilation system, water pumps, electric transformers or generators, change rooms with showers and lockers, provision for emergency treatment of injuries, a blacksmith, forging, and bit dressing shop; welding and repair equipment; and telephone or radio communication systems.

Compressors are usually at the surface. They are usually of the two-stage type, and have an after-cooler as well as an intercooler, to avoid transporting any heat of compression into the heading, which is often too hot already.

Alternating current is used. When possible, it is purchased from a utility. It is usually stepped down to 220 or 110 volts at the entrance, but on some jobs is taken in at several thousand volts, in armored parkway three-wire cable. Dry transformers (oil-filled ones are a fire hazard underground) are set about 1,000 feet back from the faces, and advanced in long jumps as progress warrants. This system avoids the power loss and voltage drop associated with long-distance transmission of low-voltage current.

There may be three electric circuits in the tunnel, a 220- or 440-volt for power, a 110-volt for light, and a high-voltage line for firing explosives. Some operators standardize on 220 for both lighting and power. Sometimes 220-watt bulbs are a nuisance to get in the United States, but they have the advantage of being useless in an ordinary lighting circuit, so they are seldom pilfered.

No drill dust can be tolerated. It may be suppressed by detergent or foam, or drowned in wet drills with water supplied through pipes from outside the tunnel.

Surveying.   Tunnel sections meet each other far from their portals or shafts, sometimes after curves, with uncanny precision. Differences usually vary from a small fraction of an inch up to several inches. These are too small to be noticed on the walls, but are measured at the surveyed centerline (axis).

An underground direction is obtained by establishing a baseline at the surface, running close to the line of the tunnel. This is very carefully done, and it is marked at frequent intervals by permanent monuments, with exact points pricked into metallic bolts embedded in concrete.

Two plumb bobs weighing 20 to 30 pounds each are suspended close to the bottom of the shaft by piano wire from the surface. They are as far apart as shaft width permits. Vibration and tendency to swing may be dampened by hanging them in pails of water. Very careful observations are taken of the wires at the surface, relative to the proposed tunnel centerline. Direction is identical with that of the same wires at the bottom.

Careful observations are taken of the bottom part of wires, using a very accurate instrument and special sighting devices. Readings are taken over and over, and the results averaged. The tunnel line is then established in the correct direction, by reference to surface readings.

This work must be done at a time when workers and equipment are not working, as ventilating currents and vibration can disturb the wires.

The line is extended through the tunnel by laser beam and/or transit, and marked on spads (markers) driven into holes drilled in the roof.

Exploration.   Tunnels are seldom driven blind. Preliminary drilling is done along the route to determine the type of rock, the amount of water to be expected, and the danger of mud slides. Test holes are drilled from the surface, usually with diamond drills that can bring up cores for inspection.

Diamond drilling may also be done from the heading, where dangerous conditions are expected. This precaution has often revealed the presence of such quantities of water or unstable soil ahead, that disaster might have resulted had it been broken into by a full-face blast.

Figure 9.42 shows extensive core drilling that was done for a sewer tunnel under the East River, New York City, in order to find a way to avoid a dangerous seam of decayed rock.

Dangers.   Underground work naturally is very dangerous, and it is greatly to the credit of tunnel workers and labor departments that there are so few accidents.

The most evident danger is that of collapse. Most soils and many rock formations will slump rather quickly into any hole cut under them. In any given material, this tendency increases markedly

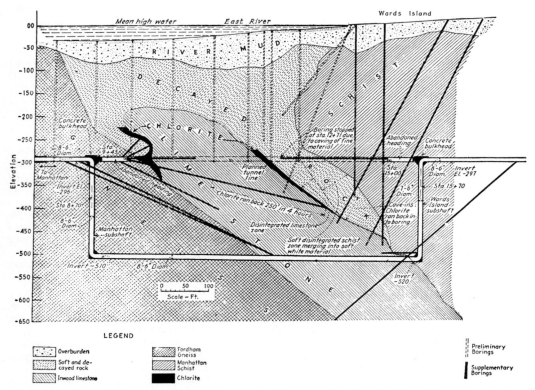

Extensive core drilling on Wards Island sewer tunnel under the East River, New York, showed the way to get under a bad fault that stopped driving on original upper level tunnel.   Preliminary borings failed to reveal true conditions.

FIGURE 9.42   Underground exploration and tunnel detour. (*By permission from "Practical Tunnel Driving" by Richardson & Mayo, McGraw-Hill.*)

with depth. Below 500 feet even apparently firm rock may creep, and break off slabs with explosive violence. There is always danger of loose pieces falling.

Caving and breaking off are combatted with compressed air, timbers, steel and concrete linings, and holding rock bolts.

If the soil will not stand at all without support, bracing must be installed ahead of the digging; or the heading protected by a movable shield.

Water, with or without accompanying soil, may break into a tunnel in such volume as to flood it completely within minutes. Escape of workers may be difficult, machinery is apt to be abandoned, and an expensive and tedious job of sealing off the water and pumping out the tunnel is often required before work can be resumed.

Fire must be carefully guarded against, particularly on jobs using compressed air and/or timbers.

Air conditions are difficult to keep healthy. Drills produce rock dust, and most air-powered machines have foul, oil-charged exhausts. Explosives produce fumes. Some clay and rock formations give off unpleasant or poisonous vapors. The increasing use of internal combustion engines underground makes tremendous demands on ventilation systems that try to clear out dangerous or irritating exhaust gases.

Silicosis, a lung disease caused by breathing dust from rock drills over long periods, was formerly one of the greatest health dangers of tunnel work. It now can be almost entirely avoided by wet drilling and good ventilation.

The Delaware Aqueduct used 10,000 feet of air per minute at each heading. A standard estimate is 1,500 feet per drill.

Shallow rocks are cool, but in deeper work an increase of 1°F for every 50 feet of depth can be expected.

Noise is deafening, particularly during drilling, as the sound echoes back and forth in the confined space. Diesel engines add to the uproar.

MSHA.   The Department of Labor in the United States includes the Mine Safety and Health Administration (MSHA), which oversees and regulates the conditions that could be hazardous for workers in gassy metal and nonmetal mines. Although they were not originally intended to apply to tunnels for highways, railroads, water pipelines, and utility lines, the MSHA rules should be and are frequently applied in these tunnels.

MSHA defines a *blowout* as a sudden, violent, and unplanned release of gas or liquid due to reservoir pressure in petroleum mines and an *outburst* as a sudden, violent release of occluded gases and solids under high pressures from a geologic formation. Either a blowout or an outburst could happen in the types of tunnels discussed in this chapter. For the sake of reference, the *mine atmosphere* is tested at any point at least 12 inches away from the face, rib, back, and floor of the tunnel.

Requirements.   The MSHA standards contain operational requirements that must be satisfied to meet the regulations for worker safety in the excavation. The standards stress tunnel ventilation to control airborne contaminants, including natural methane gas and equipment exhaust. There are minimum airflow quantities for ventilation. If air passing a particular input point has a mine atmosphere with more than 0.25 percent methane, it cannot be used for ventilation. Testing for methane must be done at least once every work shift. The volume of air ventilating each face at a working place shall be at least 20 feet per minute multiplied by the open cross-sectional area in square feet of the entry. According to that minimum, if it is a 10-foot-diameter tunnel, there needs to be at least 1,570 cubic feet per minute (20 × 3.14 × 25) of ventilation air flowing to the face. However, the required minimum at the face is 2,000 cubic feet per minute.

The tunneling equipment is to be "permissible," that is, it does not emit gaseous exhaust or electric sparks that could ignite a gaseous mixture. Therefore the best equipment to use is electrically or compressed-air powered. Within 100 feet of the tunnel face or bench, the equipment used must be permissible, except that nonpermissible front-end loaders and haulage trucks equipped with methane monitors may be used at the face or bench after blasting. It is recommended that workers be removed from a blasting area for at least 30 minutes after a blast. MSHA requires that methane-monitoring devices be installed on continuous mining machines, and that should apply to tunnel-boring machines also.

SHAFTS

Shafts—vertical passages between the tunnel and the ground surface over it—are required for the majority of tunnel jobs. They are sometimes the only access. Even when there are portals, shafts shorten the time required to do the work, as each makes it possible to work on two extra headings. In addition, underground hauling is a headache, and runs should be kept as short as practical. Some shafts are part of the permanent tunnel project.

The advantages of shafts must be balanced against the considerable expense of sinking and equipping them.

Shaft location may be chosen to keep depth to a minimum, as in the troughs of valleys over the tunnel; to take advantage of an easily worked or stable formation; or on the basis of surface conditions such as good access, nearby dumping areas, cheap land, or distance of shielding from populated areas.

Size.   Shaft size is highly variable, depending largely on the volume of material it must handle and the size of objects that must be lifted and lowered. A minimum size, about 11 × 13 feet inside the lining, accommodates a single hoist and supply elevator and a ladderway. The ladderway includes a ladder, electric and high-pressure air lines, water pump discharge pipe, and ventilator ducts, all of which must be protected from swinging loads or falling chunks.

The headframe is a tower of prefabricated steel, as in Fig. 9.43, or may be built with timbers. This carries the hoist sheaves, dumping mechanism, and the discharge chute or hopper. The hoist engine and winch are ordinarily in a separate structure nearby.

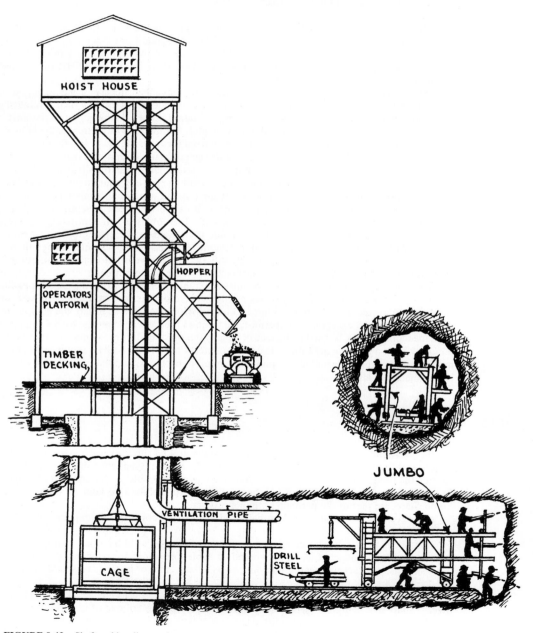

FIGURE 9.43 Shaft and heading equipment.

Soil Excavation.   Digging is started with a clamshell, which can dig soft soil unaided, and remove hard soil and rock after they have been loosened. One or two signalmen direct the operator's movements, as he or she cannot see the bottom, and any wrong move with the heavy bucket might be disastrous to the workers. The clamshell is ordinarily not used below a 25-foot depth.

The next stage may be to replace the digging bucket with a light bucket or container that is lowered to the floor, and loaded by hand or by equipment suited to the cramped work space. The container is raised out of the shaft by the hoist line, swung to the side, and dumped by a trip device or by hand. This may be used to a depth of 100 feet, or a direct transition may be made from the digging bucket to use of the headframe hoist.

A special small clamshell may operate from a platform close to the bottom, loading the containers that are lifted past it to the top by the main hoist.

Blasting.   In shaft rock blasting, all the holes are tight—that is, there is no open face to permit sideward throw of the rock—so that close drilling and heavy loading are the rule. It is necessary that the rock be cut back cleanly to the digging lines and important that overbreak be kept to a minimum, because of the high expense of removing muck and the frequent requirement of filling all spaces outside the lining.

Figure 9.44 shows typical drilling patterns for shaft and tunnel work. A set of two or more converging angle holes (wedge holes) is drilled, and other sets of straight or slightly angled holes next to them, until the rim has been reached. The wedge holes are heavily loaded, so that they crush and kick out the rock between them, making an opening into which the rock around can move sideward when the next ring of holes is fired. These in turn make space for the next set. Firing is best done by short-period delays.

In (*B*) the floor is lowered on only one side in each shot. This permits drilling to be resumed on one side while muck is being loaded from the other.

Figure 9.45 shows a burn hole shot, with the center holes parallel instead of angled.

The blast is fired from the top, after all workers and equipment are out of the shaft, except that in very deep work some equipment might be merely raised far enough to be out of immediate danger.

After the explosion the bottom will be full of fumes, which would take a long time to dissipate naturally. These may be blown out by lowering the tool air lines with the ends open, or extending

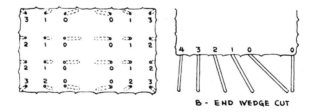

B - END WEDGE CUT

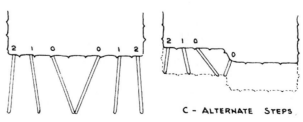

A - CENTER WEDGE CUT

C - ALTERNATE STEPS

FIGURE 9.44   Wedge drilling in shaft.

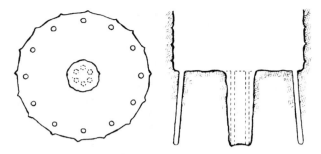

FIGURE 9.45   Burn cut.

low-pressure ventilating ducts to the bottom. (These have to be dismantled or pulled back a considerable distance to avoid damage from the blast.) A suction line (foul air duct) is more effective at cleaning the air than a blow or pressure line, as the fumes tend to settle.

Some shafts are large enough to provide space to load the muck by machinery, but in many of them it is tossed, rolled, or hand-shoveled into buckets or skips, that are removed by the hoist when filled. The best fragmentation for this type of loading is usually one-person stone, that is, pieces that one worker can handle conveniently.

Drilling can be resumed as soon as part of the bottom is cleared. Six- or 7-foot steels giving a 5-foot penetration are often used, but longer or shorter ones may be better in particular circumstances. Hand and wagon drills are standard, although special jumbos have given good results.

Working Up.   When a large shaft is required as part of the finished job but is not needed for the early tunnel work, it may be more economically cut from below. With this method the blasted rock falls to the tunnel floor, and is removed through the portal.

The glory hole method is to sink a small pilot shaft to the tunnel, then to dig the large shaft from above, blasting or pushing the muck into the small shaft so that it will fall to the tunnel. See Fig. 9.46. Alternatively, a raise boring machine can enlarge a circular shaft to full diameter by operating through the pilot bore. In that case the pilot bore must be centered in the shaft's diameter. See Chap. 20.

Shaft Lining.   In soft ground the shaft is protected from caving by setting sheeting planks or sheet piling, held by whalers, in much the same way as described earlier for ditches and basements. The whalers are interlocked at the corners to hold each other in position, and additional divider beams may be run across between the ladderway and the hoistway. See Fig. 9.47.

If the soft soil is too deep to be held by sheeting driven from the top, successive layers can be driven from inside the shaft, inside the upper ones. If the sheeting is driven with an outward batter, shaft size will be preserved. If driven straight, the inside diameter of the shaft will decrease, so that the top would have to be oversize to allow it to be full size at the bottom.

Shaft lining or timbering is required for the hoist and to a lesser extent for the utilities, even when the soil or rock is self-supporting. When not needed for wall support, timbering may follow 20 or more feet behind the digging, to avoid interference with the work, and damage to the lower section in blasts.

The lining must be supported vertically to prevent it from slipping down. Each new set is fastened by hanging bolts to that above, and every 100 feet or so horizontal notches or shelves are cut into the walls to provide fresh support.

Timbered shafts are usually rectangular, while metal or concrete linings call for circular cross section. Steel ribs are made up, curved to the proper arc, and divided into two or three pieces that are lowered endways and then supported by hanging bolts until fastened into a full circle. The actual lining or lagging may be sheet piling or similar material, or the ribs may be built as liner plates, with curved flanges which butt against those above and below to make a continuous sheet.

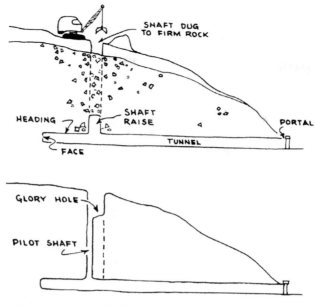

FIGURE 9.46   Raise and glory hole shaft excavation.

A continuous lining or lagging is used in soil that might squeeze between ribs or timbers, or in rock that scales or breaks off so that falling pieces would endanger workers. Unlagged walls may often be kept intact by spraying with shotcrete or a bituminous mixture.

Drainage.   Most shafts are wet. If there is only a little water, it can be bailed into the bucket and hoisted with the muck. More often it is removed by a pump with a discharge line reaching to the surface or, if the height is great, to one or more pumps that help push the water out of the shaft. All pumps used in deep shaft work should be able to develop very high discharge pressures, so that a good lift can be obtained between boosters.

If water conditions are severe, the area may be predrained by sinking 4- to 12-inch holes with rotary drills, and pumping from them. Depth is too great for ordinary well point work from the surface, but in flowing ground well points may be sunk from the shaft bottom or sides, and the water rehandled by the regular pumps.

A deep, wet shaft should have gutters and sumps at intervals, to catch water running down the sides. Pumping to the top from intermediate points may be more efficient than allowing it to get down to the bottom and raising it from there.

Shaft Sinking Machines.   A large and increasing percentage of new shafts are cut in a single or dual operation by gigantic drills, described in Chap. 20.

HEADINGS

A heading is a digging face and its work area. Conventional tunnel driving is discussed here. Tunneling machines (borers or moles) are described in Chap. 20.

When the shaft has reached the level of the proposed tunnel floor, two headings are started, one in each direction along the line of the tunnel. In addition, the foot of the shaft may be greatly expanded for storage and maneuver space, and one or more rooms may be built to house compressors, pumps, and other plant equipment.

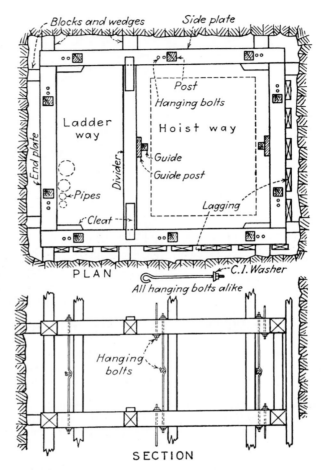

FIGURE 9.47 Timbering for small shafts. (By permission from "*Practical Tunnel Driving*" by Richardson & Mayo, McGraw-Hill.)

At first only a single set of tunnel-driving equipment may be used, as there will not be space enough for two, and greatest efficiency will be obtained by drilling at one face while mucking at the other.

Drilling patterns may be similar to those described for shafts—wedge or burn holes, and successive rings breaking into the crushed-out area. The whole face is usually drilled and blasted in one operation (full-face attack), but a small tunnel (drift) may be drilled full face, blasted, and cleaned out, then enlarged by radial drilling; or the top may be kept ahead of the bottom (bench-and-heading method). See Figs. 9.48 and 9.49.

Pilot Tunnel. Shafts may be partly or wholly replaced by a small pilot tunnel, driven parallel and close to the main tunnel. Crosscuts are driven from this to the main tunnel, wherever new headings are to be started. The main tunnel is opened up with a center drift, and enlargement started after it has been cut through enough that both tunnels can be used for traffic.

The extra tunnel may be used for ventilation, both during the work and afterward. It permits a great many operations to be performed at the same time, and may save considerable expense in sinking shafts. This method has been used chiefly for long railroad tunnels through mountains where depth was too great for shafts.

FIGURE 9.48   Tunneling with big equipment. (*Courtesy of Atlas Copco.*)

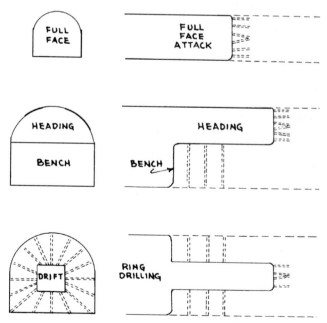

FIGURE 9.49   Tunnel headings.

Drilling.   The standard tool for small tunnel drilling has been the drifter, a medium-weight hand drill with a hand or automatic feed, mounted on a vertical column or a horizontal bar of such length that it can be secured between the floor and roof, or between the sides, by screw-jack ends. Because of the weight of the columns, they become impractical for full-face work in tunnels of greater cross section than 10 × 10 feet.

The drifter permits the drill crew to resume work on the top of the face as soon as blast fumes have cleared away, with the drill operators standing on the pile of muck until it is dug away. They can drill the bottom after it has been cleared.

Larger tunnels may be done by the heading-and-bench method shown in Fig. 9.49. This permits the use of drifters on short columns for the advance, and approximately vertical jackhammer or wagon drilling for the bench. Sometimes the heading is extended far ahead of the bench, and has its own hauling equipment that dumps over the bench face into other cars, or into a pile to be dug away.

For larger tunnels to be drilled and blasted, the standard method is to use a drill carriage (jumbo) on which power feed drills can be mounted so as to reach all parts of the face at correct angle and to correct depth. Each drill usually does several holes. It can be positioned by hand, or by mechanical, air, or hydraulic controls. Such jumbos may be so constructed as to straddle hauling equipment, so that it need not interfere with removal of muck. They may also carry a cherry picker crane to pick up empty cars to switch loaded ones through. They are backed away from the face before each blast.

On very large tunnels jumbos may be used on both levels of heading-and-bench work.

Usual drilling depth is 10 to 12 feet, but in any case is seldom deeper than two-thirds the smallest dimension of the tunnel.

Figure 9.50 shows the typical full-face drilling patterns.

Bits.   Recently tunnel drilling has been partly standardized to use steels threaded to carry detachable bits. These may be multiuse types that can be sharpened by grinding, or sharpened and reshaped by hot milling; one-use or throwaway bits that are discarded when dull; and carbide insert bits. The carbide insert bit has caused a spectacular advance in speed and ease of hardrock tunneling. Carbide outwears steel at an average of about 100 to 1, and gives much more rapid hard rock penetration. The time of handling, transporting, and processing bits is reduced from a major to a minor problem.

Loading.   Water-resistant explosives with good fume characteristics are desirable in underground work. These qualities are found in gelatin dynamites.

When all holes in a face have been drilled, each is blown out with a high-pressure air jet to remove loose cuttings and water. Cartridges are slit (unless the explosive has been damaged by water and the hole is wet) and tamped firmly with a wooden pole. It is common practice to place the primer after the first cartridge, with the cap pointed toward the collar of the hole.

Stemming may be taken from the drill cuttings. It is most convenient to use if wrapped in paper bags of the same size as the cartridges. If this material is very high in silica, its use as stemming might increase the silica in the air enough that prewrapped blanks supplied by powder manufacturers might be preferred. There are also wood and rubber plugs that are very satisfactory.

It is good practice to place a wad of paper between the explosive and the stemming, so that the powder can be easily and safely located in case of a misfire.

There is danger of premature explosion from stray currents. A common precaution is to take down or "kill" all electric wiring within 500 feet of the face before starting to load. Safety flashlights, of hand or cap models, or headlights from a battery locomotive can be used. It is sometimes a question whether the poor lighting obtained offers as much of a hazard as the electricity would.

Even the complete absence of electricity on the job would not guarantee a tunnel face against currents, as underground water is often highly mineralized and will conduct a charge for long distances. Metallic ores may be excellent conductors.

The precautions described earlier for blasting in the presence of electrical hazards should be followed.

Firing.   Any wiring hookup can be used—series, parallel, or parallel series, depending on the preference of the blaster. If 440-volt electricity is available, it is preferred for firing, although 220 or even 110 will do. Regular blasting machines are also used, but they should not be kept in the tunnel when not in use, because of possible damage from dampness.

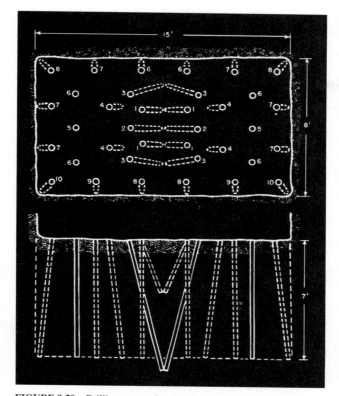

FIGURE 9.50 Drilling pattern, large tunnel.

All equipment is moved 500 to 1,000 feet back from the face, as rocks caroming off the walls can travel long distances. Compressor pipe can be left fairly close to the blast, but ventilation conduit must be stripped way back.

Checking. It is important that a thorough check be made after the blast for misfires. Tunnel work brings a large number of workers into close contact with the heading, and any accidental explosion during mucking or drilling would be disastrous. The best check is inspection by experienced people.

If an unexploded hole is found, and the wires are intact, they can be hooked up and fired. If the wires are missing, the stemming can be washed out by a water jet, and a new primer inserted and fired. Or a parallel hole, about 2 feet away, can be drilled, loaded, and fired. The muck must be inspected for unexploded cartridges.

MUCKING

Loading. In small tunnels, blasted rock may be dug by hand, although the excellent mechanical loaders adapted to work in tight quarters that are now available, and the rising price of labor, are steadily reducing the practice. Output for the loading gang is generally figured at about $\frac{1}{2}$ to $\frac{2}{3}$ yard per hour per person, although one person may load up to 2 yards under favorable conditions. The difference lies in the work of loosening, and of handling cars, and in other delays.

The swell or "growth" of rock in passing from the solid to the blasted state averages about 50 percent. In tunnels, mucking is usually calculated in terms of loose yards, in mines in number of tons loaded.

Slick sheets should be used in connection with hand loading. These are thin steel plate, $\frac{1}{4}$ or $\frac{5}{16}$ inch, in pieces about 4×6 feet, with holes punched for convenience in picking up for moving. They are laid to cover the tunnel floor for 10 to 25 feet back from the face before each shot. Large rocks are picked up and thrown into the cars individually, while the finer material is dug by shovels that slide easily along the metal surface.

Mechanical loaders include full-revolving shovels with short booms and proportionately larger buckets, that move on either crawlers or rails.

Special tunnel-mucking machines are available in large variety. Most are rail-mounted, although crawlers are gaining in popularity. The bucket can be swung from side to side to reach the full floor area, and is filled by pushing into the pile.

It is then lifted, in some models over the machine to discharge into a car or conveyor belt behind; in others it loads a built-in conveyor that discharges to the rear. In either case, the car may be coupled to the mucker so that it is always in loading position.

BORING

A tunnel may be cut to full size in a single operation by boring with a tunneling machine, sometimes called a mole. These machines are described in Chap. 20. See Fig. 9.51.

A tunnel borer grinds, chips, or digs its way through formations, by rotary or oscillating motion of cutter teeth, and deposits the muck onto a conveyor belt for discharge into haulers at the rear.

FIGURE 9.51   Bored tunnel face.

In addition, it may and often must provide for placing steel or concrete linings around or behind itself. Such lining can be used to take the thrust of its crowding force on the face.

These machines can usually be disassembled sufficiently to be brought down a large shaft, and assembled in a conventionally dug tunnel section at its foot. However, whenever possible, they start work at a portal (outside entrance).

HAULING

Almost any type of hauling unit may be used in a tunnel, from a wheelbarrow to an off-the-road ten-wheeler. It is a matter of tunnel size, speed of driving, ventilation, and preferences of the management.

The traditional system is small muck cars pulled along narrow-gauge tracks by electric locomotives. The locomotives can take power from either batteries or high lines, and start at 4 tons. There is an increasing use of diesel locomotives with exhaust conditioners in well-ventilated tunnels. See Fig. 9.52.

Cars are usually side-dump types, although many special constructions are found. The width is governed by the tunnel and the gauge of the track, and should be small enough to allow passing in the tunnel. Car width is generally about twice the track width.

The capacity of the car may be limited by switching arrangements. If they are pushed by hand, capacity is limited to 1 or 2 yards, as heavier cars will need to be pried along the tracks, rather than shoulder-pushed. The car must be low enough to go under the discharge of the mucking machine being used. If hand-loaded, it must not be over 4 feet high.

The loaded muck cars are hauled to the shaft and run into hoisting cages, in which they are lifted to the top, where they are dumped by side tipping. There are also special cars that can be lifted directly, without entering a cage. Or they may be dumped at the bottom into a hoisting skip.

FIGURE 9.52 Diesel-powered mine locomotive.

The perpetual problem in tunnel haulage, which becomes more acute as size decreases, is bypassing the empty cars (or trucks) going to the face around the full ones coming away from it. Empty cars may be switched to the side; or if they are small, lifted or pushed off the track by hand, where there is space for only one track. Larger ones may be handled by a cherry picker. In either case the spotting arrangement shown in Fig. 9.53 may be used.

The locomotive pulls a string of empties into the heading and stops to let the cherry picker take up the rearmost car and set it aside. The locomotive then backs far enough that the car can be replaced on the track in front of it; then pushes that car up the loader. While it is being loaded, it backs so that another car can be picked off.

When the car is loaded, the locomotive couples to it and backs past the cherry picker, which places the empty in front of it to be pushed to the face. While it is loaded, the rear empty is again set aside, to be pushed in on the next cycle. When all the cars are filled in this manner, the locomotive pulls them to the shaft.

In a tunnel of sufficient height, a movable framework called a Grasshopper, Fig. 9.54, can be used. This allows the empty cars to be moved over the loaded ones, and can be pulled up to the face by the loader.

A conveyor belt may be set up so that a full train of cars can be backed under it, and loaded one by one from the front to back.

Conveyor belts can also be set up to haul from the face to the shaft. No switching arrangements are required, but this unit cannot be used readily to bring supplies from the shaft to the face; considerable work is involved in dismantling or protecting it for a blast, and there is constant work adding sections to keep it in touch with the digging.

Diesel-powered trucks are used for large tunnels. They carry much bigger loads than mine cars, and if sufficient width is available to make passing possible, they get past each other with fewer complications than rail-mounted carriers. The shuttle types, such as the Dumptor, which are

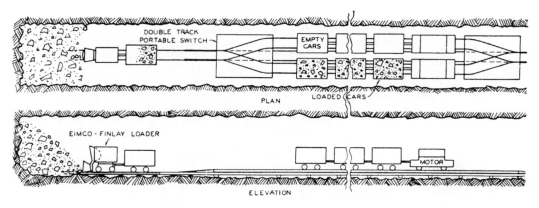

FIGURE 9.53   Portable switch.

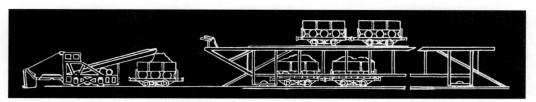

FIGURE 9.54   Grasshopper overhead switch.

equally comfortable going backward or forward, are often better adapted to the work than those which have to be turned in the tunnel.

The use of internal-combustion engines fouls the air, so that very good ventilation is required.

Exhaust Gas.   The exhaust from a gasoline engine contains carbon monoxide, an odorless but poisonous gas that soon makes any closed-in place deadly to life. Amounts of carbon monoxide that are not sufficient to cause unconsciousness or death may temporarily damage judgment and reasoning power, causing an increased danger of accidents.

Diesel exhaust contains little carbon monoxide, but it is rich in various chemicals that smell badly, are irritating to eyes and throat, and that fog up the air so that visibility is dangerously reduced. This last difficulty is increased by the usually bad lighting in a tunnel.

The danger from gasoline engine exhaust has largely prevented use of this type of power underground. Diesels are used in spite of the irritation and danger they cause. Their presence is partly compensated for by increasing the ventilation, but conditions do become very bad. They are often made worse by an astonishing lack of care in adjustment of the engines. Diesel trucks sometimes emerge from tunnels belching black smoke, presumably caused by defective or souped-up injectors.

Various types of scrubbers using water and chemicals to dissolve and neutralize gases, and secondary catalytic oxidizers that serve also as mufflers, are used to make internal-combustion engines acceptable underground. These are described in Chap. 12, under Exhaust Conditioners.

Good ventilation and lots of it are a basic requirement, even when such devices are efficient. The most they can do is reduce the exhaust to carbon dioxide and water. Carbon dioxide is not poisonous or irritating, but in sufficient concentration it has a suffocating effect that can cause impairment of judgment, unconsciousness, and death.

WATER

Groundwater is a problem in most tunnels, and may be the principal one in some. Many mining tunnels, some of them miles in length, are made solely to lower the water table. There may be seepage all along the line, adding up to a considerable volume to be drained or more often pumped away. Gushing springs may be exposed by any blast, or may open up from seepage points well behind the face. Underground lakes or rivers may be encountered that are capable of flooding the work in spite of continuous pumping. Veins of soft, water-soaked soil may be found in hard rock, that may break into and fill the tunnel.

The first necessity is adequate pump capacity. The tendency is to underestimate requirements, largely because pumps and lines are expensive, partly because even careful exploration from the top seldom reveals the full quantity and pressure of water that may be encountered.

If a tunnel runs uphill from a portal, drainage may be by natural flow through a ditch cut along the side. If an upgrade from a shaft, it can be drained to a pump inlet at the shaft foot. This arrangement is easy and inexpensive but seldom satisfactory, because of repeated blocking of the ditch by rockfalls from walls or from hauling equipment, resulting in water running over the floor, making it sloppy and often undermining the track or spoiling the road surface. The ditch also takes up more space than a pipe, and there has not yet been a tunnel with floor space to spare.

The conventional arrangement is to pump all water. A small centrifugal pump, usually air-driven, is kept near the face, and takes from a sump and discharges into a pipe running back toward the portal or shaft. Another sump is provided every 500 to 1,500 feet back to collect local water for another centrifugal, usually electric-powered pump. Each pump may discharge into the sump behind it, which is kept down by another pump, usually of a larger size. Another arrangement is to have all pumps discharge through check valves into a common discharge line. A powerful electric pump of the piston or centrifugal jetting type is installed at the shaft bottom, and as many boosters as are required for the lift installed at intervals in niches in the shaft.

Pipelines vary from $1\frac{1}{2}$ to 10 inches in diameter.

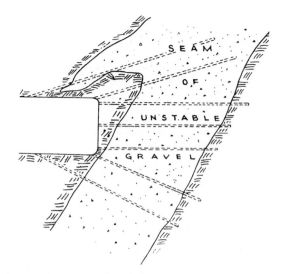

FIGURE 9.55 Exploration and grouting holes.

The pump or pumps at the base of the shaft are sometimes placed in a sealed room, with power and control directly from the shaft top. In other cases the pumps are in the open, but are of the submersible type. These arrangements permit use of the units along with emergency pumps if the tunnel should be flooded.

Grouting. Water inflow can often be checked by grouting. This may be done by drilling deep into the rock in the direction of the supposed source of the water, sealing in pipes with cement, and then pumping cement and water grout through them, either straight for seepage or mixed with sawdust or shavings for gushing flow. This may be done in advance of the tunnel driving in very wet areas, by fanning the grout holes out from the face and edges of the heading, as in Fig. 9.55.

Grouting is also done through completed linings, either to check water or to fill in spaces between it and the wall. Grout pipes may be cemented into a concrete lining when it is poured.

Successful grouting of a wet seam sometimes merely diverts the water so that it enters the tunnel at another point that was previously dry. This also may be grouted, but a point may be reached where the contractor either installs a complete concrete lining or gives up the effort to seal off and relies on pumps.

The aboveground uses of grouting were discussed in Chap. 6.

ROCK SUPPORT

Ground pressure in rock tunnels is difficult or impossible to estimate. In firm formations there will be little or no pressure until depths of greater than 500 feet are reached.

However, there are soft, joined, or laminated formations that will scale off or fall from a flat or moderately curved roof, until a Gothic or pointed arch develops, after which it will be self-supporting. If bracing is done only to support the roof, it is a question whether it will be more economical to cut up to a stable roof line, and avoid placing of supports. See Fig. 9.56.

In any roof problem, width is a very important factor, as wide spans will drop pieces or fall in much more readily than narrow ones.

Many rock tunnels are perfectly safe without any bracing. Others get by without accidents. But very often is is necessary to place supports directly after the digging, or within a few days. Also,

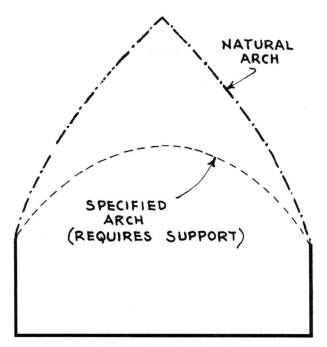

FIGURE 9.56 Cutting crown for self-support.

the majority of tunnels outside of mines are more or less permanent in nature, and except in very firm rock, will require lining to prevent deterioration and to reduce or eliminate maintenance.

Support or lining may be wood timber, steel ribs, plates or bolts, or concrete. Concrete is frequently placed inside one of the other types of support.

Timber is the oldest material used, and is found in ancient tunnels. Concrete was used to some extent by the Romans, and has become the standard for permanent installations. Steel liners and roof bolts are quite modern developments, and are rapidly replacing timbering.

Timbering. Figures 9.57 to 9.59 show some designs for timbering. The square-set framing is confined to small tunnels, and various forms of arch construction can be used in quite large ones. The arch may be supported on posts supported in the floor, or rest on a springline shelf (hitch) cut in the sidewalls. Support may vary between these methods with changes in ground, or in shape of the edges.

Posts should be fastened to the wall plate by dowels, lag bolts, or scabs (nailed-on pieces) so that they cannot fall if relieved of weight.

The weight of timbering varies with expected ground pressure. Sometimes it is merely a light roof to catch light rockfalls, at other times a high-strength lining designed to resist squeeze from all directions, including the bottom.

Where timbering ends at a portal, or at an enlarged shaft base, it must be securely braced by diagonal beams, as in Fig. 9.60, so that any compression developing in the tunnel will not squeeze it out.

Packing. In rock or soil that tends to push in, it is important not to leave any space between the lagging and the wall or roof, as any inward movement will increase the instability of the ground, and may cause it to exert tremendously more pressure than if it had been held in its original position.

An exception is found in swelling or squeezing ground that is allowed a limited space for movement.

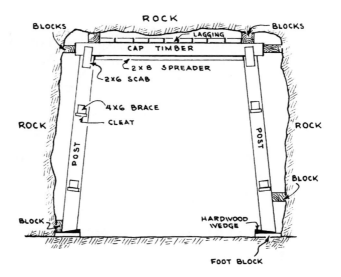

FIGURE 9.57 Timber set, small tunnel.

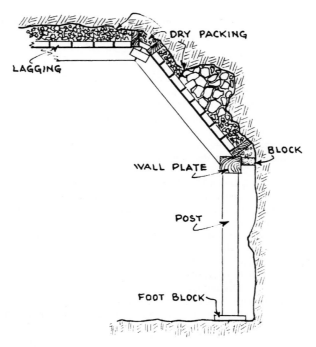

FIGURE 9.58 Timber arch on posts.

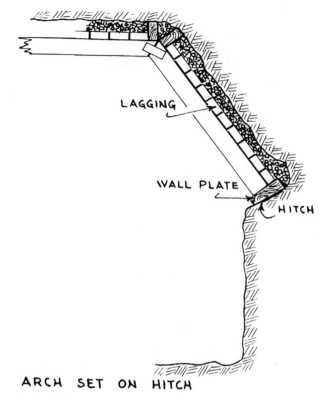

FIGURE 9.59 Timber arch on hitch.

Initial movement is prevented by packing the space between the lagging and the rock. The most economical system is to use a dry packing of fine muck, which is shoveled behind the planks as they are placed. At the crown it must be thrown in from the end and securely rammed—a tedious, disagreeable job that is seldom well done.

Dry packing may also be done with pea or birds-eye gravel, shot into place with pneumatic guns either through holes in the lagging or from the end. Its use is more common in soil than in rock.

Lean concrete, with a cement–sand–small stone mix of 1:3:6 or 1:4:8, can be shot into the arch with a pneumatic placing tool. This must be very dry so that it will not leak through cracks between the planks. This is done after the set has been erected and securely blocked, as the fresh concrete may impose very heavy stress.

Lagging is usually set closely (skintight) in the crown. On the walls it may be widely spaced or lacking, as even if the rock squeezes in, the spans between timbers are too short to allow bulging. Under very heavy conditions the timbers may be set skintight, so that lagging is not needed.

If the tunnel is to be concreted, the lagging may be placed inside the timbers, to provide a smoother outer form that saves concrete yardage. The disadvantages are that much more packing is needed and that the fastening is under tension rather than compression, so that heavy pressure may make the lagging pop off the timbers. The effect may be cumulative, as yielding of one fastening increases the strain on the next, so that a considerable length may give way at one time.

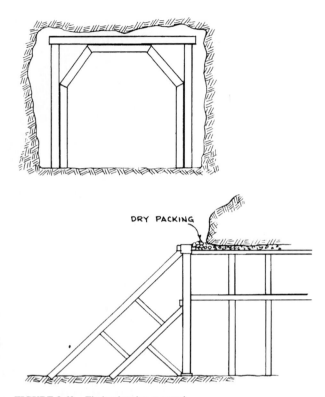

DRY PACKING

FIGURE 9.60 Timber bracing at portal.

Steel Ribs. Steel supports are standard in tunnel work. They are easier to handle, and allow substantial saving in excavation. This is because for a given strength, they are only half as thick; and the projections of ribs into a concrete lining are counted as reinforcing. In timber construction, the outside line of the concrete is figured as the inside line of the timbers, and the concrete used to fill out to the lagging is largely figured as waste. On small tunnels the saving by use of steel in excavation may be 30 percent and in concrete, 50.

However, steel liners are more vulnerable to blasting damage, and do not give warning of impending collapse under load by groaning, as timbers do.

The steel ribs are made in two pieces, occasionally more. They are brought in endways and set up individually. The lagging may be wood planks or steel liner plates. If the former, the ribs must be well strutted to each other to keep them in line.

As in the case of wood, steel lining may be only a roof or crown support based on shelves at the spring line in the sidewalls, or a complete tunnel enclosure.

Roof Bolts. It has been found in mining and tunneling operations that unsafe rock will often support itself safely over wide spans if it is reinforced with steel bolts. (See Fig. 9.61.)

In laminated (thin-bedded) formations, the effect is similar to that obtained in plywood and other layered-wood constructions. Several weak and thin layers may be very strong when bonded together. In jointed and fissured rock, the bolts, if used properly and in sufficient numbers, restore to the rock the massive strength it had before it separated into blocks and pieces.

Expansion bolts are used, rather similar to those that fasten wood framing to masonry. The type shown in Fig. 9.62 is made in $\frac{5}{8}$-, $\frac{3}{4}$-, and $\frac{7}{8}$-inch diameter. The $\frac{3}{4}$ has a minimum breaking

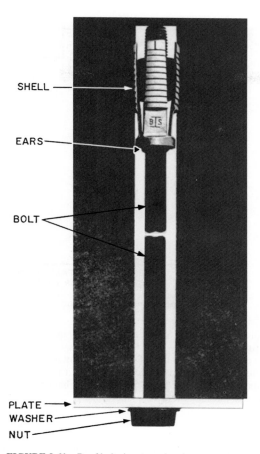

SHELL

EARS

BOLT

PLATE
WASHER
NUT

FIGURE 9.61   Roof bolts in a tunnel arch.

load of 22,500 pounds in regular strength, and 32,000 in high-strength steel. Lengths are 2 to 8 feet, in 6-inch steps.

The bolt threads into the plug of an expansion shell that fits into a $1\frac{3}{8}$-inch-diameter hole. Ears on the bolt prevent it from sliding too far into the shell, so that tightening pulls the plug down into the shell, expanding it against the sides of the hole. (See Fig. 9.63.)

Roof ties may be used to support the roof between bolts. Wire mesh can be used in addition to the ties, or instead of them, where the problem is separation of small pieces.

Flat or dished (reinforced) plates of $\frac{3}{8}$- or $\frac{1}{4}$-inch thickness and 6-inch diameter are used where ties are not needed. They are usually made with a $1\frac{3}{8}$-inch hole. A hard steel washer prevents the bolt head from pulling through.

The drilled hole must be at least as deep as the bolt is long, and may be deeper. The bolt is usually provided with a hard steel washer and assembled to the shell at the factory. The shell is pushed through a plate and as far as it will go into the hole, and the bolt head is tightened with an impact air wrench, usually to a torque of 150 to 200 pounds-feet.

The bolt head is held outside by the plate and washer, so the threaded plug is pulled outward as it is tightened, squeezing the shell against and into the walls of the hole. The grip of the shell in hard shale, sandstone, or limestone is usually greater than the breaking load of the bolt. In soft shale, grip is usually less than bolt strength. In rotten rock, grip may not be adequate.

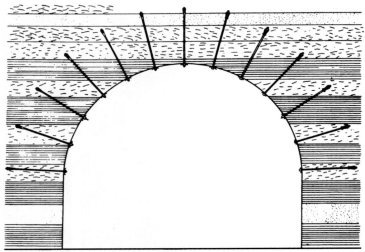

**TYPICAL INSTALLATION OF
ROOF BOLTS FOR TUNNEL WORK**

FIGURE 9.62 Square-head mine roof bolt.

Holding power can often be increased by injecting cement grout into the space around the bolt, after tightening.

A very strong grip is obtained by anchoring with resin. A capsule containing resin, fine aggregate, and a tube of hardener is placed in the hole, then broken and mixed by turning a bolt into it. After the mixture has set for 30 to 60 minutes, a bearing plate and hard washer are tightened onto the bolt with a nut and a torque wrench.

The washers offer the advantage of producing greater bolt tension with the same effort, or the same tension with less effort. The more uniform tension provides greater security.

Bolting requires from one-fifth to one-tenth the steel required for ribs and lagging, and under many conditions is equally strong. In addition, it saves the need of excavating space in which to set the steel structure, and reduces the amount of concrete required for permanent lining.

Elimination of all ribs and timbering makes a tunnel easier to work in, as there are fewer obstructions, and it provides for a smoother flow of ventilating air.

Another important advantage is that the economy of the work causes it to be done on roofs that might be judged to be self-supporting if bracing were time-consuming and expensive. The bolts can also be installed right up to the face immediately after blasting, so that protection is available to the heading crew. As a result, their use in the rather wide range of conditions where they are applicable results in a marked decrease in roof-fall accidents.

Heavy wire mesh may be used to prevent falling of small fragments in between the bolts. In some instances gunnite is used to minimize air-slacking and spalling.

Rock anchor bolts, which are similar to the slotted mine roof bolts, are used along highway and railroad cuts to prevent rock falls and slides.

Concrete Lining. Installation of concrete lining is construction rather than excavation work (however necessary it may be to the excavation) and will be only briefly considered.

There are two general procedures—soft ground technique, in which it is placed immediately behind the digging and is necessary to the driving of the tunnel, and hard ground. Under the second heading comes work in rock that is self-supporting, and requires lining for permanence, scaling protection, or waterproofing; and unstable soil or rock that is adequately held in place by timber or steel.

1. Drill hole to required depth. Install bolts perpendicular to bearing surface. Pattern of bolt holes is predetermined according to conditions of particular strata.

1. Drill 1¼-in. diameter hole (for 1-in. slotted bolt) to a depth 3 in. less than the length of the bolt.

2. Insert assembly —square-head bolt, hardened steel washer, plate washer, and expansion shell and steel plug—into hole.

2. Insert steel wedge into slotted end of bolt. Place bolt through opening in both the hardened steel washer and the plate washer and drive unit into hole.

3. Tighten bolt head with power wrench. As bolt turns, plug is drawn down on threads. This action expands the serrated section of the shell.

3. Driving dolly must be used to protect threads. Wedge expands slotted end of bolt providing anchorage.

4. Shell is now anchored securely. Steel roof plate furnishes additional support. Steel ties, connecting a series of bolts, may be used in addition to steel plates.

4. Tighten nut with impact wrench. With nut drawn up snug, the steel anchor plate bears against the rock surface.

HEADED BOLT   **SLOTTED BOLT**

FIGURE 9.63   Installation of headed and slotted bolts.

The soft ground technique is to follow the heading closely, with some resulting interference between operations. Perhaps the most serious is maintaining a track for muck cars through the pouring operation, and across the freshly laid invert (floor). Steel beam bridges may be used to carry the track in this section.

The invert may be laid about $1\frac{1}{2}$ inches low, protected with planking, and brought up to grade with a top dressing of thin concrete as a finishing operation after the tunnel is complete.

Traveling forms of various types are used. For fast schedules, it is essential to have telescoping forms that can be folded up and moved through other forms supporting more recently poured sections. On other jobs, forms are used that can be collapsed just enough to break away from the concrete surface so as to be moved ahead to the next section. In either case the forms are carried on carriages that may move on steel wheels and tracks or on rubber tires, depending largely on the muck haulage method used.

Breakthroughs.   Sometimes, in spite of precautions, there will be a sudden rush of water or mud into the tunnel. This most often occurs at the face immediately after a blast. Sometimes the source is a limited underground pocket which will give no trouble after it has once drained off. At other times a stream or large body of water will keep up a continuous flow. If the water is muddy, or the flow is partly or wholly mud, an unstable soil formation has been reached which may give increasing rather than diminishing trouble.

In any case the first step is to seal off the face with a bulkhead (wall) as quickly as possible. Timber, sandbags, or sandbags with timber may be used. Occasionally timber may be backed with concrete.

The bulkhead must not be used as a dam while being built. Pipes should be built into it large enough to take the water flow until the structure is complete. Otherwise water pressure will tend to destroy the bulkhead as it is being erected, and conditions will be very dangerous to personnel. With water discharged through pipes, the structure can be properly and strongly made and keyed into the tunnel rim. The water can then be controlled by valves on the pipes.

The bulkhead should also be fitted with pipes for grouting and concrete placement. After the water has been shut off, grout can be injected into the space between the bulkhead and the break, and will sometimes work back along the water seam and stop or reduce the flow. Grouting may also be done through exploration holes drilled through the bulkhead and into the rock beyond. Over 90,000 bags of cement have been used to control one water pocket.

Further tunneling through such a spot is first in the form of drifts (small tunnels) each of which serves as a base for further grouting, until the ground is consolidated enough to drive the big tunnel.

Tunneling machines, as described in Chap. 20, have an important effect on methods and costs of solid rock tunneling.

SOFT GROUND TUNNELING

Soft ground is divided roughly into the following subclasses, description of which is abbreviated from *Practical Tunnel Driving* by Richardson and Mayo, McGraw-Hill.

Running ground: Must be instantly supported. May be dry sand or gravel, quicksand, silts, and muds.

Soft ground: Roof must be instantly supported, but walls will stand vertically for a few minutes.

Firm ground: Roof will stay up unsupported for a few minutes, and the sidewalls and face for an hour.

Self-supporting ground: Will stand unsupported while the entire tunnel is driven a few feet ahead of the timbering.

The standard methods of driving through soft ground are forepoling with wood or steel, or working in a shield. The plenum method is keeping out soil and water with air pressure, with either forepoling or shield.

Forepoling.   The use of plank forepoles was formerly the standard method of driving a tunnel through soft ground. While this technique has been largely replaced by steel liner and poling plates, it is still widely used on jobs too small to justify obtaining steel.

In forepoling, the tunnel is protected by timbering, and by breast boards set against the face. Planks are driven through slots cut in the breast board and supported cantilever fashion to make a temporary roof, under which dirt can be dug and permanent supports installed.

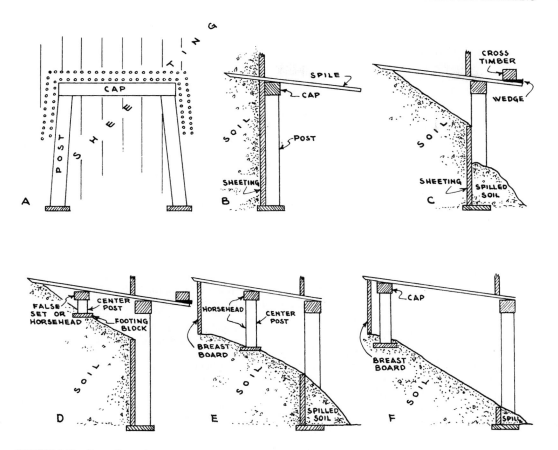

FIGURE 9.64   Forepoling soft ground.

Figure 9.64 indicates the terminology of the parts and something of the method.

Starting from a shaft lined with plank sheeting, a bent (cap or roof timber and two post supports) is set and securely braced close to the sheeting. Close-set holes are drilled through the sheeting in a double line just above the cap and a single line about 18 inches below it. Double vertical lines are drilled just outside the posts (A).

A set of light forepoles or spiles are made of 2 × 6 planks 5 to 6 feet long, sharpened to a chisel point on one end. A piece of sheeting is knocked out between the lines of drill holes above the cap, and a forepole is rested on the cap and driven through the hole, at an upward slant of about 2 inches per foot, for about half its length (B). Another bit of sheeting is cut or knocked out, and another forepole driven in the same manner, parallel to and touching the first. This process is repeated until the full width of the cap has been covered.

Spiles are then driven into the sides, flaring out about 2 inches per foot, to a penetration 6 or 8 inches deeper than the roof pieces. These may be driven horizontally, or at an upward slant to keep contact with the roof.

A timber is now placed across the shaft, immediately above the free ends of the roof forepoles, which are then forced downward slightly by driving wedges under the timber. The poles are now supported on the cap and held down tightly by the timber and wedges at the rear, so that the front is supported cantilever fashion.

The sheeting is then broken out from the spiles down to the lower line of holes, and the ground allowed to run into the tunnel until it assumes its natural angle (*C*). The resulting slope will normally not extend back to the points of the forepoles, but will end at some intermediate position.

Next a horsehead, or false set, is placed under the poles about 2 feet beyond the sheeting. This consists of a cross piece under the spiles, and a center post set on a small supporting block in the dirt (*D*). The spiles are then driven to their full penetration, substituting the support of the horsehead for that of the cap rear timber.

Earth is then raked in until the points of the spiles are almost uncovered. A board the width of the cut and about 18 inches high is set vertically immediately under them. This serves as a breast board to keep more dirt from flowing in, and supports the spiles (*E*).

A cap timber is then set to line and grade, and is temporarily supported by a single center post. A "bridge" of 2 × 6 planks is fastened to the top of the cap but separated from it by 4-inch blocks (*F*).

The remainder of the side spiles are now driven. Some of these are tapered and are used wide end forward, so as to reverse the upward slope of the roof spiles and the upper few wall spiles.

The forward cap is now supported by a pair of beams resting on short temporary posts and wedged down from a cross timber. The remainder of the sheeting below the first cap is now broken out from the top down, and the dirt pulled into the tunnel and hauled away. Additional breast boards are set under each other as space becomes available, and held in place by cleats nailed to the side spiles.

When the floor is cut to grade, side posts (legs) are set on below-grade blocks and wedged up until they take the weight of the cap. Wedges are driven between the posts and the side spiles to tighten them. Sometimes a trench jack must be used to force the side spiles out while setting the legs.

The next set of roof spiles is entered through the bridge slot on top of the second cap, and is driven at the same upward angle. Space for side spiles outside the legs is obtained by knocking out the wedges as the spiles are placed for driving.

All spiles should be driven skintight (touching throughout their length) except at the corners, where 1-inch boards (lacing) are tacked on. When necessary, cracks are stuffed with excelsior, salt hay, or other packing to prevent inward leakage of soil.

Each timbering set must be braced securely to that behind it, as any shifting will severely weaken the structure. Spiles are usually driven with a sledgehammer or air hammer. Sometimes they are jacked in—a very tedious job—to avoid jarring the soil.

If the tunnel floor tends to get muddy, it should be floored, for convenience of workers and to avoid possible shifting or settling of the foot blocks. Sometimes floor spiles are driven if the bottom tends to boil up, but compressed air is a better way to combat this and other difficulties with excessively soft ground.

This method is relatively easy to follow in many soils, but it takes an experienced crew to get through boulders, flowing mud, and other difficult conditions.

Forepoles may also be used with steel ribs instead of timber sets.

The standard soft ground tunneling hand tool is a short-handled, round-pointed shovel, aided when necessary by a grub hoe (mattock), pickaxe, or crowbar, and often by paving breakers. Special grub hoes have one hammer face for use in driving wedges. In soft clay a curved two-handled draw knife can be used to advantage. It is pulled by two workers or a power winch, and slices the clay off in strips.

Liner Plates.   Corrugated steel liner plates, curved to match the tunnel rim and supplied with drilled bolt holes in flanges or overlaps for fastening to each other, are increasingly used for soft ground tunneling. They are made in various sizes, with 16 × 36 inches in common use. A plate of this size made of ⅛-inch metal weighs about 27 pounds, and if ¼-inch stock, 53 pounds. Short plates are available for fitting into the tunnel circumference.

Stiffening ribs are used when the tunnel is over 10 feet in diameter, and for heavy loads in any size opening. They are generally not used when the same strength can be supplied by a heavier-gauge plate.

Liner plates are usually a temporary support to hold the tunnel until a concrete lining has been installed, usually a matter of hours or a few days after the digging. They are sometimes "robbed" for reuse immediately before the concrete is placed. The safety of this practice depends on the character of the soil, which is a matter for engineers to pass on in each case.

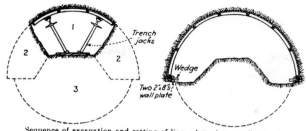

Sequence of excavation and setting of liner plates in a small tunnel.

FIGURE 9.65   Setting liner plates. (*By permission from "Practical Tunnel Driving" by Richardson & Mayo, McGraw-Hill.*)

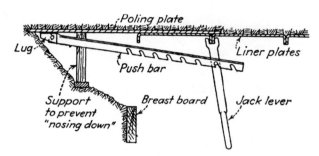

FIGURE 9.66   Steel poling plate and jack. (*By permission from "Practical Tunnel Driving" by Richardson & Mayo, McGraw-Hill.*)

Liner plates are placed from the top down. A small section is dug ahead, the center plate placed and braced with a post or jack; and then the sides dug away to place the adjoining plates. These are supported radially on cleated center blocks, as in Fig. 9.65. When the spring line or base of the roof arch is reached, two 2 × 8 planks, called footing boards, are placed on each side. Wedges are driven between the two boards until they lift the arch of liner plates enough to take the weight off the jacks. The lower plates are then nailed to the boards to prevent slipping off them.

If the ground is too soft for this method, interlocked poling plates (Fig. 9.66), can be placed outside and forward of the completed liner, and jacked forward from inside.

Shields.   Shields have become the standard equipment for driving major tunnels in soft ground. A schematic view of one is shown in Figs. 9.67 and 9.68. It resembles a tin can with an open back and controlled openings in the front. The front may be open, with grooves to allow setting a breast board or plates if necessary, or closed by a bulkhead with controlled ports. The back or tail is large enough to permit placing the tunnel lining inside it.

The shield is forced forward into the dirt by jacks based on tunnel lining. Doors in the front are opened to allow soil to flow in, or to be shoveled. In very soft ground where bulging of the surface will cause no damage (as under rivers or swamps), no dirt need to be taken into the tunnel, as it will be pushed aside by the pressure of the shield.

A primary lining, which is most often of bolted cast-iron segments but sometimes of cast steel (for unusual stress), fabricated plates, concrete blocks, or timber, is constructed in the tail, which is long enough to protect a complete segment. This lining must be strong enough to not only resist full soil pressure, but also to take the thrust of the jacks that move the shield forward.

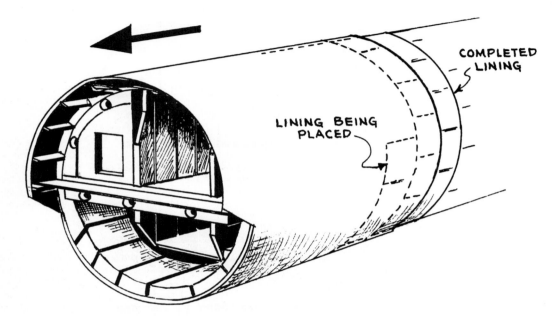

FIGURE 9.67 Tunnel-driving shield.

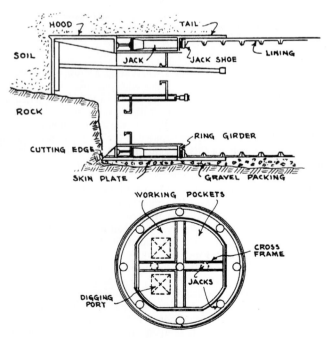

FIGURE 9.68 Shield in mixed face heading.

The outside diameter of the shield tail must of course be larger than that of the lining built inside it. A few plastic soils can be manipulated so as to close in smoothly on the lining as the tail moves away from it, but under most conditions the space must be filled. Failure to do so will leave the lining without proper side support, so that the arch will tend to sag.

Grout was originally the standard filler for this space. Grout plug holes were built into the liner pieces. When the tail cleared them, grout was forced into the bottom hole, with the next above used as an air vent. When grout appeared at the upper hole, the grout hose was transferred there, the bottom plugged, and injection continued. The full circumference was worked in this manner from the bottom up.

The amount of cement used makes this operation costly, and in addition the grout has a tendency to move forward along the outside of the lining and flow under the tail into the shield. Also, grout may work up to the surface, cause heaving of pavements, or break into sewers or conduits.

Gravel filling is used to avoid these difficulties. Bird's-eye gravel (uniform size, passing $\frac{1}{4}$-inch screen, 33 percent voids) or similar sizes of slag or screenings can be blown by an air gun into the grout holes, also starting at the bottom. This will not leak into the shield or travel far from the tunnel, but it may not fill spaces uniformly. It is therefore usually followed by regular grout. The quantity required is greatly reduced, and its tendency to travel is checked by the presence of the gravel.

Compressed Air.   The compressed-air or plenum method of tunnel driving makes it possible to work with relative safety in soft mud and under bodies of water. The principle is that the inward and downward pressure of water and of soils can be counteracted by increasing the outward pressure exerted by air in the tunnel.

The rule of thumb is that each $\frac{1}{2}$ pound of air pressure over atmospheric will support a 1-foot height (head) of water. Actually, the pressure required is often far less, because of the stability of the soil, restriction of water passages, and other factors.

The extra pressure is built up by low-pressure compressors (converters) at the surface, and piped through a retaining bulkhead into the tunnel. Workers and materials are passed through this bulkhead through one or more locks. Air in the tunnel may leak out through the soil as fast as it is supplied, or may be exhausted from the heading through a blowline.

A lock is a passageway between two airtight doors. The outer door of the tunnel is opened to admit entering personnel or materials. It is then closed, air pressure is raised to match that in the tunnel, and the inner door opened to complete the passage.

When exiting, valves are opened to bring pressure in the lock up to that in the tunnel. The inner door is opened, the traffic moved into the lock, and the door is closed.

Air is allowed to escape from the lock until it is at atmospheric pressure. The outer door can then be opened.

This device permits maintenance of pressure in the tunnel, and limits traffic air loss to the relatively small amount in the lock at each use.

It is best practice to have at least two locks, one for people and one for materials. The human lock should be large enough for the whole crew, and must have valves by which pressure can be closely controlled so that it will drop gradually for minutes or hours while crew leaving the tunnel are in it. This process of gradual reduction, called decompression, is necessary to prevent nitrogen dissolved in the blood from being suddenly liberated to cause a painful and sometimes fatal ailment called the bends.

There may also be a small emergency lock, high in the bulkhead so as to be the last place flooded. This is left open to the high pressure, so as to be ready for immediate use. One or more cross partitions may be placed in the crown to hold air pockets in case the tunnel should be flooded.

The materials lock should be long enough to accommodate hauling units of the size used. It may be small, so that one car at a time is pushed by hand in one end and then pulled out the other; or it may be as much as 80 feet long, to accommodate a train and a locomotive. Lock construction is expensive, but a liberal size speeds work greatly.

Fire danger under high air pressure is severe. The extra supply of oxygen in close contact may cause even wet wood to burn vigorously. Smoking and other fire hazards must be avoided, and there should be a liberal supply of fire extinguishers, and fire hose connected to high-pressure water.

Clay reacts most satisfactorily to compressed air, as it is so nearly impervious that it is well supported by the air and seals it in. Primary bracing may not be required before placing the permanent lining.

On first exposure to compressed air, silt acts as clay, but it then tends to dry out and crumble off at the top, and to turn to mud and flow at the bottom. The higher the tunnel, the greater the differences between top and bottom behavior.

This is because the air pressure is the same on all parts of the tunnel rim, but the head of water that tends to force water into the tunnel, or resists its being forced out of the lining soil, is much less at the top than at the bottom. A partial cure for the difficulty is to excavate the upper or arch section first under low pressure, install liner plates or other support; then increase pressure and dig the bottom. Once a full lining is installed, the unbalanced condition becomes unimportant.

In sand the air penetrates several feet at the top, and leaves the bottom wet enough that boards have to be stuffed with excelsior to stop sand runs. The best cure for this condition is to drive well points ahead of and below the face, and keep the lower sand dry until lining is placed.

Air will escape in any formation except tight clay, and will reach the surface by following porous veins, old wells, or even sewers. It is best conserved by getting the lining in immediately after the digging. Airtightness of the lining is not automatic, however. Grouting outside it (which is necessary for firmness also) and painting the inside of concrete with cement and water greatly reduce leaks.

Liner plates may be made airtight by spreading wet clay along the joints. Building paper can be used on wood lagging.

About 20 cubic feet of atmospheric air per minute is required for each square foot of face area, with an additional allowance for losses through the locks.

Blowouts.   Sometimes the compressed air in the tunnel blows out the surface. This is particularly likely to occur in shallow tunneling in soft underwater mud. Any outward leak must be immediately plugged with any material on hand, valuable or otherwise. From the outside a blowout can be prevented or stopped by dumping enormous quantities of clay from barges.

The blowout can be disastrous in itself, hurling workers and equipment up into the water. The immediate drop in pressure allows water and mud to enter the tunnel, threatening those in it with drowning or suffocation. A job "lost" in this manner is expensive and tedious to resume, and sometimes driving can more easily be done on a different route.

In addition to this below-ground work, a variable amount of surface construction is required. The following is a list of major items of this nature:

- Access roads
  Construction
  Maintenance
- Power supply
  Installation of lines
  Construction of generating plant if necessary
- Surface buildings
  Change and washroom facilities
  Blacksmith shop
  Machine shop
  Compressor building
  Powder magazine
  Cap magazine
  Miscellaneous buildings
- Construction camp (if needed)

- Portal excavation
- Water supply
- Sewer system

MINING

A mine is an excavation made in order to obtain (recover) material that has valuable chemical or physical characteristics. If it is an open-cut project, it may be called a pit or a quarry. Strictly speaking, a quarry is usually concerned chiefly with a material desired for its physical characteristics, as trap rock for road aggregate. However, if a quarry goes underground, it is called a mine.

The problem of mining is to get the highest possible percentage of the pay material out, at minimum expense. In some cases the best system is to confine excavation almost entirely to pay dirt, even if it requires a maze of small and irregular tunnels. In others it is more efficient to blast and remove 100 feet of overburden so as to expose only a few feet thickness of ore.

The first step in deciding upon an approach is to find how much ore or other pay material there is, exactly where it is located, the extent to which it is interrupted by other materials, and its physical and chemical characteristics. This information may be required to determine not only whether the deposit is worth mining and the method, but also the type and size of any processing plant required.

Mineral Deposits.   Exploration is very complex, because of the number of factors that influence distribution of minerals. Sedimentary rocks develop in more or less horizontal layers, but may then be folded, twisted, or even turned upside down. Faults are breaks that extend across the layers, and are made by movement of whole blocks of the earth's crust. They may be a single clean-sliding plane, or a width of hundreds of yards in which the rock is smashed up. Movement along a fault may be a fraction of an inch, so that the same formations are found on each side, or several miles, so that one section of a deposit may be at a great distance, or lost entirely. Movement may have taken place in the ancient past, or might occur from time to time during mining.

Most metallic minerals are associated with invasion of formations by molten rock from below. If fluid rock reaches the surface, it becomes the lava, ash, and other usually valueless materials associated with volcanoes. If it stays far below, it hardens gradually into granite or other coarse-grain rock, and while cooling may give off great quantities of minerals, in fluid or gaseous form, that penetrate the surrounding rock for miles. The weak or porous streaks, through which they move and in which they are deposited, are the miners' pay veins. Parts of the main mass may become mineralized, often resulting in an extensive but low-grade ore body with rather uniform composition.

One area may undergo several successive periods of mineralization, the later ones reworking, removing, or enriching some of the earlier deposits. When the area cools sufficiently, groundwater becomes active at dissolving, transporting, and redepositing material.

The result of these factors is that underground structure is often extremely complex, and while exploration can give a general picture of what to look for, only very extensive (and expensive) diamond drilling, or the actual removal of the ore, will give the complete story. Access to formations is often obtainable only by the hardest and most costly type of digging—hard rock tunneling.

Exploration.   There are a great many methods of exploring an area for valuable minerals. Until rather recently most deposits were found by surface inspection and sampling of the ground. People on foot or horseback found pay outcrops, pieces of them below or downstream, or formations associated with them. Major finds are still made in this manner, but complex techniques have become more important.

A search may start with studies of geologic maps indicating more or less completely the rocks to be found in a region. Inspection and photographs from planes may reveal promising areas, which can then be scouted on foot. Test holes can be made with almost any tool from a pickaxe to a diamond drill. Radiation-sensitive instruments such as the Geiger counter are used to locate radioactive deposits. Local changes in gravity may indicate metallic ores.

A very interesting method is seismic prospecting. A deep hole is made, usually with a diamond drill, and a heavy charge of explosives fired in it. Sensitive instruments at selected spots in the area record the time and pattern of the resulting earth waves, which indicate the nature of the ground through which they pass. The information may be used directly in locating oil and some other deposits, or in working out underground structure to indicate the location of veins whose outcrops are confusing.

Following the Ore. When mining has started, with or without benefit of thorough exploration, the digging is kept in the ore whenever it does not make too complicated a pattern. In general, large, well-financed operations are more inclined to place their haulageways and shafts with an eye toward long-term efficiency, where the small operator keeps in pay rock as much as possible, sometimes with most unfortunate effects on later operations.

Figure 9.69 shows some of the tunnels that might be included in a mine. The main route in from the portal is the haulageway. A drift is an approximately horizontal tunnel of small size, a stope is excavation of a room. A raise is a shaft worked from the bottom. Drainage tunnels, sometimes miles in length, are driven to save the cost and danger of heavy pumping inside wet highlands. The glory hole is a shaft enlarged from the top, with the muck descending by gravity to a floor from which it can be loaded by machinery. This loading area is a draw point.

Any of these except the glory hole may be in either ore or nonpay rock. They are contracted to minimum dimensions (which for a haulageway may still be quite large) when in country rock, and are expanded and supplemented by side drifts when in ore. If a vein is too narrow or too low for working space, excavation may include enough other rock to give head or side room.

The "arch" of the rock (the span of roof that can be allowed between walls or pillars, with or without timbering) and the height of the veins are important factors, as they determine the yardage (called tonnage underground) that can be taken out at one stand. Larger volume and working space permits use of bigger and more efficient machinery.

Development. Development is that part of mining that prepares an ore body for removal. It is likely to include the sinking of shafts, the driving of tunnels, and the installing of chutes and transportation and drainage systems.

If development is done in the ore or other pay mineral, expenses may be partly or wholly paid by the value of the product. In some procedures, as in room and pillar mining where the pillars are left, the development may be the entire mining operation.

Shafts. Shafts are similar to those used for other tunneling. They provide for entrance of personnel and supplies, removal of ore and waste, drainage, and ventilation. They vary in size upward from a drill hole with a casing 8 inches or more in diameter that serves as a chute for concrete or fill,

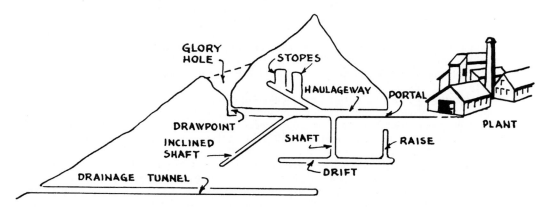

FIGURE 9.69 Mine shafts and tunnels.

to openings large enough for four divisions—two for a pair of skips for bringing up ore and waste, one for an elevator for personnel, and one for pipes and ladders. Supplies may be lowered in human elevators, in ore skips, or by a separate system. There may be an additional system of ventilator shafts, that may be either dug or rotary-drilled.

Haulageways and Drifts.   Tunnels that are made primarily for hauling ore or other materials from digging points to shafts or portals are called haulageways. Their size varies with that of the haulage units to be used in them. Old time human-and-wheelbarrow methods could get by with a width of 4 feet and a height of 6½ feet. Two-way haulage with off-the-road trucks requires a width of about 30 feet and a height of 20. (See Fig. 9.70.)

The majority of underground mines use track haulage, although rubber-tire trucks and shuttle cars are also used. A few still use hand-push cars on 18-inch-gauge tracks. Gauge is the spacing, center to center, of the rails. Locomotive haulage may call for track gauges from 24 to 42 inches. Cars may project 18 inches beyond the track on each side. An additional space of at least 18 inches on one side is required so as not to crush workers against the walls. There is usually a gutter or pipe for drainage water, piped high-pressure air for power and low-pressure air for ventilation, and electric wires or cable.

Haulageways that are to be used for a long time are given strong and permanent linings. Concrete without reinforcing is used because of its compressive strength, comparative simplicity of placement, and the fact that when broken, it is readily replaced or removed. However, steel lining, timber sets, roof bolting, or sometimes just shaping and scaling of a natural rock roof are all used under suitable conditions.

Drifts are tunnels made during exploration and development. They are usually smaller, shorter, or are expected to carry less material than a haulageway, but there is no clear distinction.

Remote Control of Loader/Haulers.   In recent years a System for Integrated and Automated Mining (SIAM) has been developed by Noranda for their Brunswick mine. It consists of a multimedia communications backbone, a video-assisted teleoperation system, a system for automating load-haul-dump

FIGURE 9.70   Diesel mine hauler.

(LHD) bucket loading, and an automated guidance system for LHB machines. In other words, the remote control is operated from a control room at a safe distance from the mining face. There the operator with joysticks maneuvers an LHD into the muck at the face of the mine to scoop up a bucket full of muck, then moves the LHD to carry the muck to a dumping point. The bucket is filled by an automated system based on hydraulic pressures, cylinder extensions, and load on the front axle. With such automation the system ensures that the bucket will be essentially full every time.

This remote control system has proven to be successful at consistently filling the bucket of each LHD at the face and displaying the load to the operator in the control room, improving productivity, and reducing costs of the mining operation. The reduction of costs is made possible by having a single operator handle several machines at a time. The system uses visual navigation and each vehicle is equipped with a series of sensors, an on-board computer, and two cameras, one at each end. The cameras transmit images through an optical tape that runs along the roof of the drift.

Extraction.   Extraction is the operation of removing ore and other minerals that have already been made accessible by development work.

ROOM-AND-PILLAR EXCAVATION

Minerals in extensive beds of fairly uniform thickness may be mined by a method known as "room and pillar." Coal, limestone, and salt are frequently dug in this way.

A number of parallel corridors (tunnels or drifts) are cut into the formation, and are connected by series of parallel corridors crossing at right angles. There must be at least two of the main corridors to provide for ventilation, and there may be as many as five in a set—two haulage, two ventilation, and one spare. The corridors are the rooms, and the blocks or walls between them are the pillars.

Figure 9.71 shows a layout during the development stage, as the rooms are being cut. The rooms and pillars are the same width. This could be as little as 6 feet with a weak roof, or as great as 50

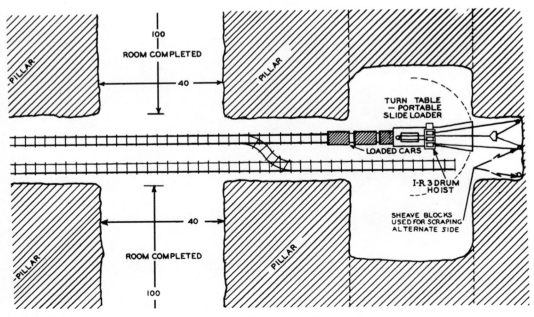

FIGURE 9.71   Room-and-pillar excavation. (*Courtesy of Ingersoll-Rand Company.*)

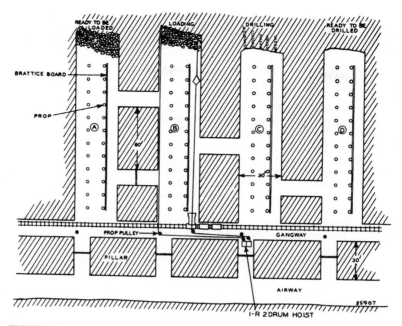

FIGURE 9.72   Four-heading cycle. (*Courtesy of Ingersoll-Rand Company.*)

feet under very favorable conditions. With such an equal-width layout, the development work would remove 75 percent of the mineral formation, if a clean cut were made to the floor and the roof.

Another method, Fig. 9.72, is to leave the pillars as continuous walls except for occasional ventilation connections between the corridors. This is often done when the corridors will be back-filled to support the roof, and the pillars then dug out.

Leaving the Pillars.   The pillars may be left permanently to support the roof. This may be done because the supply of the mineral is so large that it is more efficient to waste the pillars than to do the extra work of extracting them; when roof conditions make extraction particularly dangerous; or to prevent damage by subsidence of the ground above. Worked-out mines of this type are being used for storage of records, and for underground factories.

Pillars left in this manner may be the full size left by the development, or they may be cut away (robbed) to the point where they are barely adequate to support the roof.

Limitations.   Room-and-pillar work of this type is usually limited to beds whose thickness is not much greater than the safe span of the roof between pillars. Tall pillars must be thicker than short ones to support the same weight, and in thick beds with narrow corridors they will contain too high a proportion of the mineral.

A mineral body must also be fairly regular for efficient room-and-pillar work. Interruption of the pattern by barren or soft ground, or sudden changes in pitch, increases costs and decreases recovery.

STOPING

The term *stoping* is used so differently in different mining areas that it has little specific meaning. In general, and in this discussion, it applies to any underground digging of valuable minerals that does not create a passageway. It covers most forms of ore extraction other than room-and-pillar and block caving.

A stope is the cavity in which stoping is being done or has been done. Refer to Chap. 20.

Gravity.   Almost everything is against economical work underground: restricted working space, tight blasting, need of roof and sometimes wall supports, cost of hoisting to the surface, darkness, limit on size of machinery, and drainage, pumping, and ventilation requirements.

There is one helpful force that the miner puts to use very effectively: gravity. The miner employs it to help break rock, or sometimes to do the whole breaking job, and to convey the broken material through chutes to loading points.

Chutes.   Underground operations in extraction or stoping are very complex, and almost any cut-away view and brief description are likely to be incomprehensible to the surface worker. For this reason, the subject will be opened with an illustrated and oversimplified description of procedures.

If ore lying above a development tunnel is to be mined, it might theoretically be loosened by a bar or by blasting to fall directly into a mine car parked below, as in Fig. 9.73(A).

However, some of it would fall beside and between the cars, and it would be difficult to control the amount loaded. For this reason it would be better if the loosening were done somewhat higher, as in (B), and the broken ore fed down through a chute with a control gate.

But a vertical chute dumping chunks of rock out of the ceiling is dangerous, and the gate will be very difficult to work, as a great weight of rock can rest directly on it. Therefore the chute is made a sloping one coming in at the side, as in (C). And since a tunnel has two sides, two chutes can be cut into it at one point, as in (D). Several chutes or pairs of chutes may be spaced along the tunnel so that each can load a separate car of a parked train at one time. In this manner, a 12-car train could be loaded by two chute locations with six stops or by three chutes with four stops.

Gate-controlled, car-loading chutes may be as small as 2 feet square, are unlikely to be over 5 feet wide. Small sizes may be operated by hand or compressed air, larger ones by air only.

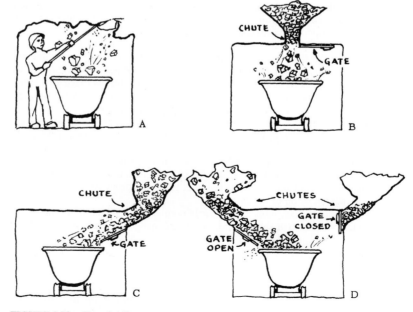

FIGURE 9.73   Chute loading.

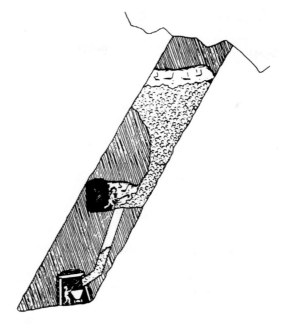

FIGURE 9.74 Chutes.

This type of chute may be very subject to jamming by oversize pieces that may be difficult and expensive to remove.

This problem is met by having a sorting and screening point between the stope and the chute, as in Fig. 9.74. A tunnel is cut above the haulage drift, into which broken ore can flow by gravity so that it will spill into the chute, or can be raked or shoveled into it. The top of the chute is protected by grizzly bars. Any piece too large to go through them is broken up, or pulled out of the way by a miner.

Chutes that will handle a considerable tonnage of ore may be lined with timber, metal, or concrete. Wear on concrete is reduced by using a stepped underslope instead of a smooth slide.

Draw Point.   Jamming may also be reduced or eliminated by using a much larger chute ending in a draw point instead of a gate. The one shown in Fig. 9.75 opens into a side tunnel or room where it spills on the floor, and is loaded into cars or onto a belt by an overhead shovel or some other mucking machine. The ore is held by the floor and by its natural slope, and feeds down automatically as the toe of the pile is dug away.

Another type of draw point or free-flowing chute empties directly onto the tunnel floor, putting the ore in the working path of a drag scraper (slusher).

This machine, the two versions of which are shown in Fig. 9.76, is a bucket, usually bottomless, that is pulled to the digging area by a cable attached to its rear, and then pulled forward through its digging, transporting, and dumping cycle by the other end of the same line, attached to its front.

The line operates between a winch at or near the dumping point, and a pulley block anchored behind the digging area, several feet above the tunnel floor. (See Fig. 9.77.)

One scraper may service several draw points.

Square-Set Stoping.   This method does not depend on broken ore for a working floor. The ore is cut in a succession of identical blocks, that may be between $5 \times 5 \times 7$ and $6 \times 6 \times 8$ feet, the

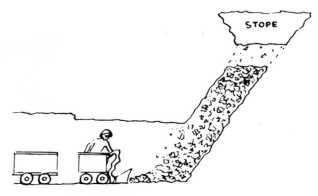

FIGURE 9.75   Draw point.

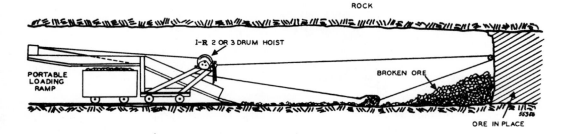

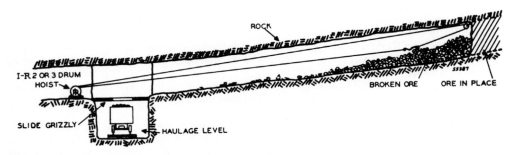

FIGURE 9.76   Drag scraper installations.

largest dimension being vertical. As each block is removed, a squared timber frame is erected in the space and roofed with planks. The stability of the material will determine whether the blocks must be dug and shored one at a time, or if a number can be opened and then shored at one time.

This work may be started at the bottom of the ore body or of the section being worked, and the bottom level (or slice) is followed in sequence two or three sets behind by higher slices, so that the cross section of the front resembles the underside of stairs. The roof of each frame serves as the floor of the one above it. The line of advance may assume any one of a number of patterns, the straight wall from side to side of the ore or section, or a center advance followed by stepped backsides.

The strength of the timbering in a large stope depends largely on keeping the sets in exact alignment in all three directions. If any part starts to move, diagonals and bracing plates are installed, sections are rebuilt, or the worked-out sections may be filled with waste rock.

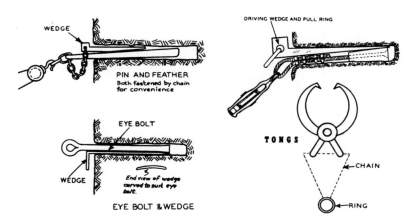

FIGURE 9.77   Drag scraper anchors.

If the ore is mixed with country rock, a large amount of filling may be done by separating ore from waste while mucking, and moving the waste into the bottom of the stope. The convenience of doing this is a great advantage of square sets over shrinkage stoping and other caving methods that require taking everything, good or bad, to mechanical separating equipment that is usually in the mill aboveground.

Backfilling.   Square sets and other types of stope may be backfilled as a regular part of the extraction operation. This may be done to strengthen any bracing structures and support the walls and overhangs during extraction operations, to provide elevated surfaces on which miners can work, and/or to provide against long-term subsidence that might affect other parts of the mine and its equipment, or surface buildings (frequently including the mill).

The backfill may consist in part of waste separated in the stope or at the crusher, of tailings from the mill, or of granulated slag. When operated in connection with open-pit or open-stope mines, or when the proportion of waste is very high, this material may be all that is needed. Otherwise sand, gravel, or other suitable material is dug in the vicinity and hauled to the mine.

The waste fill is usually handled by a system of shafts and passages separate from that used for ore removal. Sand, boulder-free gravel, and dry tailings may be dropped through chutes made from cased drill holes. Wet tailings may be pumped down similar bores, or through pipe rigged in the main shaftway. Cement may be mixed with backfill gravel to make a very lean concrete, in order to avoid settlement and slumping into adjoining pillar extraction work.

BLOCK CAVING

If a large enough area of rock or ore is undermined, it will cave in. Many formations will break up in collapsing so that most of the pieces are small enough to go down ore chutes. This behavior is utilized in block caving.

The minimum width for sure caving is around 100 to 150 feet. Under various conditions block dimensions may vary from 100 × 100 to 200 × 300 feet. Height of blocks varies from 200 feet up to perhaps 400, and a substantial weight of overburden may (or may not) be helpful. The block is sometimes cut off from adjoining blocks, or from separating walls by shrinkage stoping or a network of raises or drifts.

Development work prior to caving is fairly complicated. Three levels of tunnels are needed: the undercutting level where the block is undermined, the grizzly level below it for sorting and feeding into chutes, and the bottom or haulage level. See Fig. 9.78.

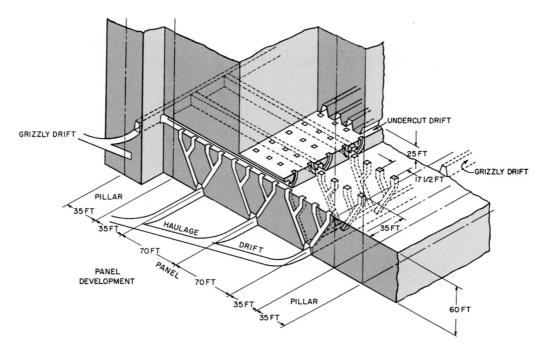

FIGURE 9.78   Block caving.

The levels are connected by raises. The upper set, from the grizzly to the undercut passages, are called finger raises. Their upper ends may be belled out into funnels. These terminate beside and above grizzly bar entrances to the control raises (chutes), so that oversize pieces may be broken or set aside.

When the development work has been completed for a block or a large enough part of one to start caving, the pillars separating the undercut drifts, and usually their ceilings as well, are drilled and blasted. The broken ore spreads and settles into the drifts, leaving the roof entirely unsupported, so that it breaks under its own weight and crumbles into the raises. This process may start immediately, or after a delay of several days.

The number of control raises can be greatly reduced by using drag scrapers (slushers) to pull ore away from the bottoms of the finger raises to a few large raises.

A number of problems arise in block caving. First, there is the blocking of finger raises by oversize pieces, that must be broken by "bombs" of dynamite that are pushed up the raises on poles to the obstruction, and exploded. Then there is the arching of the ore over a raise or group of raises so that it will not feed, usually a tricky and sometimes a dangerous proposition to correct.

Another difficulty is that uneven breakage and settlement may put great pressure on certain sections, crushing even heavily reinforced control and haulage tunnels. If the blocks are separated by pillars (walls), and the barren capping rock is stronger than the ore, settling in the blocks may result in these walls supporting large areas of overburden, with resultant heavy pressure on passages under them. Another possibility is that a large mass of rock in one of the blocks may be more resistant to breakage than the ore, so that it rests more and more heavily on the ground beneath it as the ore crumbles and is drawn away. One block-caving operation lost a haulageway for 18 months as a result of such a mass crushing it repeatedly.

In spite of these and other complications, block caving is the most economical method of mining large, fairly uniform ore bodies that will break into small pieces under their own weight, and that are too far underground for open-pit recovery.

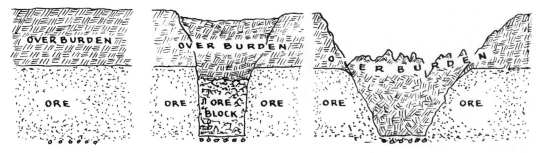

FIGURE 9.79   Surface subsidence.

Surface Subsidence.   On the planning side, any mining method that allows the roof to cave as the ore is taken must allow for the effects on the surface. A thin, even bed such as a 6-foot coal seam may let the surface down evenly, with comparatively little damage to structures. But any substantial and irregular removal of underground material will result in subsidence pits. These may be hundreds of feet deep and thousands of feet wide, and are likely to destroy any road or structure within their reach.

The subsidence of a caving block may be nearly vertical at first, but the cleavage in the over-burden is likely to fan out widely. The ultimate effect, in a wide variety of rock formations, is crumbling and sinking inside a slope line of 40° to 45° outwardly from the bottom of the block or caving operation, with additional slip faults that might extend to a 38° slope. See Fig. 9.79.

All too often ore treatment mills, shaft head structures, and towns are involved in such subsidence, so that very expensive relocation is required.

FORECASTING ROCKFALLS

One of the principal safety problems of underground work is the danger of rockfalls and rockbursts. Knowledge of where such incidents are likely to occur usually makes it possible to install additional supports, or if that is not possible, to remove personnel and equipment from the danger area.

The cracking and popping of rock, often heard by personnel working underground, has long been recognized as a warning of moving, unstable ground. It has now been discovered that sounds of the same type, so slight as to be heard only by very sensitive instruments, are made by ground that is under even moderate stress. Increasing frequency and intensity of the sounds, called microseisms, indicates an increasingly unstable condition that is likely to result in collapse.

Microseisms occur within the audio frequency range and, when amplified, sound like "clicks" akin to the creaking of wooden stairways, floors, and diving boards. Almost any hard material, such as salt, rock, shales, sandstone, quartz, brick, glass, wood, steel, concrete, sand, and sugar, will produce microseisms. Microseisms by their clicking nature can be easily distinguished from sounds caused by foot shuffling, drill rigs, trucks, mucking cars, talking, and other means. Microseismic apparatus can never be used when vibrations are so great as to overwhelm the equipment.

The Seismitron (Fig. 9.80) is an instrument based on the microseismic principle and has been in use since 1951. Its primary purpose is for the prediction of rockfalls in tunnels and mines as well as for the determination of stability of earth dams and slopes at the sides of mountains and highways. The apparatus is made by the Walter Nold Company of Natick, Massachusetts, and is certified (*permissible*) for use in coal mines by the U.S. Bureau of Mines and by the Department of Mines and Technical Surveys for Canada.

Receiving sensors (geophones) are cylinders $1\frac{1}{4}$ inches in diameter and 9 inches long. They contain synthetically grown crystals which, when stressed, generate low-level electric currents. A

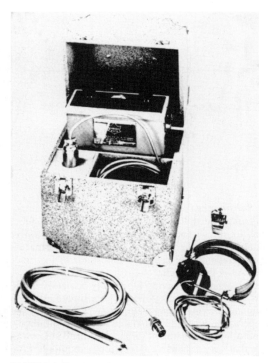

FIGURE 9.80   Seismitron.

portable battery-powered amplifier increases the level to a condition that enables the microseisms to be heard by means of simple earphones.

The number of microseisms per minute which occur when the rock is at the failure point, or about to burst, is determined by monitoring the heading (of a tunnel, for instance) immediately after a blast. The microseismic rate of decay under such a state is the same as the rate of increase prior to a rockfall. This is the failure point.

Six-foot-deep holes are drilled and spaced every 100 feet apart in the suspected area. Each of these stations is monitored each week for microseismic activity. This is accomplished by placing the geophone in the hole and sealing the opening with waste. The microseisms heard are manually counted for a 15-minute period and the rate per minute versus dated time plotted on graph paper at a later time.

A history is thus built up for each station. A low, little-changing rate of microseismic activity is indicative of stable conditions. A continually increasing rate indicates suspect conditions, especially when 120 to 180 microseisms occur per minute. A doubling of the rate within any 24-hour period indicates imminent collapse.

An excellent prediction of collapse, sometimes as much as 45 days ahead of time under ideal conditions, can be made by extrapolation of the plots on the graph to the rock failure point. The curvature becomes hyperbolic as failure approaches.

If the active rock, which is sometimes hidden by gunnite, should become stabilized either naturally or by roof bolting or timbering, the decaying microseismic rate would immediately indicate the tendency toward safe conditions.

The instrument in this manner becomes a very useful tool for judging the necessity for, or the effectiveness of, expensive roof bolting. The Seismitron is used in earth drilling to determine depth of bedrock. It is also used to determine leaking running water in earth dams, evidence of life in blocked tunnels and earthquake rubble, stability of overhanging cliffs, large ice masses, concrete structures during quiet periods of construction, underground foundations while under-pinning is being replaced, communication through rock or pipe, robbing pillars of mineral-bearing ore, and so forth.

With the Seismitron, timbers about to break sound just as one would expect wood to sound under these conditions. Rock about to fall sounds like rock about to fall. Earth about to slide sounds just like earth about to slide. In other words, just like the subject of physics; natural and normal.

CHAPTER 10
PIT OPERATION

This chapter contains an outline of some of the principles of pit layout and operation. The term *pit* is intended to cover any open excavation made to obtain material of value, whether it be coal, mineral ore, quarry rock, gravel, or fill.

Because of the complexity of the subject, treatment will have to be brief, and many operations omitted entirely.

Techniques of drainage, road building, and blasting have been discussed in previous chapters.

Most pit operations are started with the removal of soil or rock lying over the deposit to be mined. The problems involved in stripping will therefore be considered first.

STRIPPING OVERBURDEN

Overburden may include topsoil, subsoil, sand, gravel, clay, shale, limestone, sandstone, and other sedimentary deposits.

The depth of overburden that may be removed depends on its character and accessibility, the value of the underlying formation, the comparative cost of underground mining, and the extent to which the spoil can be sold or utilized.

Need for Stripping. Stripping overburden may be a very large part of the cost of mining, and a number of factors should be considered before undertaking it.

First, is it necessary to strip it? It may be possible to mix it with the product, or to separate it at lower cost during processing.

If the pit has gravel or stone screening, separating, or washing equipment of adequate capacity, soil may be dug or blasted down with the pay dirt, and separated as part of the regular processing. In this case, thorough clearing is necessary as sod and brush clog screens and crushers.

Utilization of Spoil. If the land must be stripped, the next question concerns possible profit or use to be obtained from the material. Near large cities, topsoil can often be sold at a higher price than the regular pit product. This may also be true of peat deposits.

Any substantial layer of clay, or fine earth, should be sampled and analyzed. Clays which are superficially similar in appearance are used in widely different products, such as fire, paving, and common brick; tile, pottery, portland cement, flux, mud for rotary drills, and specialized functions in chemical processing.

Limestone is often found in overburden, and it is extensively quarried for crusher rock, building stone, and cement manufacture.

If it is not possible to get good prices for any part of the spoil, it may still be possible to get enough for it as fill to repay some of the stripping cost. The stripper is in a good competitive position because he or she has to pay for digging, and often for dumping as well, and is better off selling for a fraction of the excavation cost than not selling at all.

The limiting factor is the additional cost in making the spoil salable or delivering it. If selective digging that will slow operation, or finer fragmentation, is required, or if the stripper must truck material which otherwise could be cast, the salvage may cost more than it is worth.

When the spoil is being removed in trucks or scrapers, and no market exists for it, and it is not practical to set up any plant to process it, it might be used for real estate improvement or for goodwill.

Swampland along railroads or highways can often be bought very cheaply, and can be converted into valuable industrial sites by filling and grading, if the laws about preserving wetlands are not being violated. Filled land near towns may be sold for residential purposes.

Pits are frequently unpopular with the neighborhood because of blasting, heavy trucking, or dust. Local authorities may impose restrictions which would make operation more expensive or impossible. Under any circumstances, cultivation of local goodwill is sound business practice.

Fill which is being hauled and wasted can be utilized to reclaim land for parks or parking areas, building road or airport fills, and blocking gullies. Topsoil can be used to enrich farms, gardens, or lawns.

However, there is danger in such work of getting into much greater expense for extended hauling and grading than anticipated, so a careful study should be made in advance. The people or community benefited will often be willing to pay at least these extra costs.

As a general proposition, however, the pit operator must figure that the spoil from stripping will not be salable or of any use, and when it must be hauled away, it may be a problem to find a disposal area. This is particularly likely in large operations.

Regrading.   A number of states now require grading of spoil piles and breaking down of walls of worked-out pits. This work should be included in cost calculations.

Replacing of topsoil and/or replanting may be required by law, or by private contract between the pit operator and the land owner.

Pit floors, slopes, and spoil piles are usually quite barren, and are often very rocky as well. They may be entirely lacking in nitrogen, and their potash and phosphorus supplies are usually in insoluble forms that become available to plants only after long weathering.

Reclamation involves grading to smooth contours; planting with grass, hardy trees, or other vegetation; and protecting the graded land against erosion until the plants have taken hold.

A reclamation plan must usually be filed and bonded before work starts. If the pay seam is thin, restoration of original contours will probably be required. If it is thick, or underlies rough land, new topography may be permitted.

Removal of coal often makes groundwater highly acid. Treatment of this condition within a project, and prevention of contamination of streams flowing from it, may be a major reclamation problem.

SIDECASTING

When the spoil cannot be utilized, the cheapest way to move it is by sidecasting—that is, to pile it alongside the cut within dumping range of the excavators. This disposal is possible when the pay formation has only a narrow exposure, or can be worked in narrow strips, so that the overburden can be economically moved across the pit, from the unopened to the worked-out area.

Sidecasting is generally not practical in working thick layers of quarry rock or mineral ore, because of need for wide working space below the face. Its best-known application is in the strip mining of bituminous coal.

One-Spot Strip Mining.   The diagrams in Fig. 10.1 show the basic process of stripping, mining, and backgrading a narrow lens. It consists of three operations aside from the drilling and blasting frequently necessary. Top material is cast out of the way, pay material is dug and trucked away, and the top pushed or cast back in.

Progressive stripping of a wide deposit, as in Fig. 10.2, resembles the action of a gigantic moldboard plow, taking slices off the high wall and laying them against the spoil heap. Mining is done between trips.

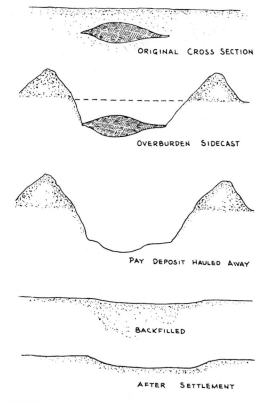

ORIGINAL CROSS SECTION

OVERBURDEN SIDECAST

PAY DEPOSIT HAULED AWAY

BACKFILLED

AFTER SETTLEMENT

FIGURE 10.1   Steps in strip-mining a lens.

If the spoil piles are smoothed over where they fall, without substantial removing, the original grade will be raised in a ridge parallel with the beginning of the work, and lowered at the last cut, these changes corresponding to the land and the dead furrow of the ploughed field.

Bulldozers.   Where the deposit is shallow and the pit narrow, bulldozers make good stripping tools, particularly when the land slopes across the pit. Figure 10.3 shows a sequence of operations. This can be repeated in successive strips across the deposit if conditions remain the same.

Such shallow overburdens seldom require blasting, but use of a rooter may speed the digging.

It will be noted that this machine backfills the face of the pay layer so that work must be done from the surface of the seam.

Bulldozers are also essential auxiliary tools in the heavier stripping work.

Scraper.   Scrapers, or pans, are used in sidecasting, haul-away, and combined stripping. Their best application is to soils that can be dug easily with the power that is available, or that can be broken into fine fragments by ripping or blasting; and that are not deep enough to justify the use of shovels with enough reach to give the pit the width it requires.

Figure 10.4 illustrates a method of scraper use. These are considered sidecasting since the spoil pile is placed immediately behind the pit, but otherwise the work is identical with haul-away with the same machines. A particular problem that becomes more important as depth increases, is maintenance of haul roads across the pit and up the spoil bank.

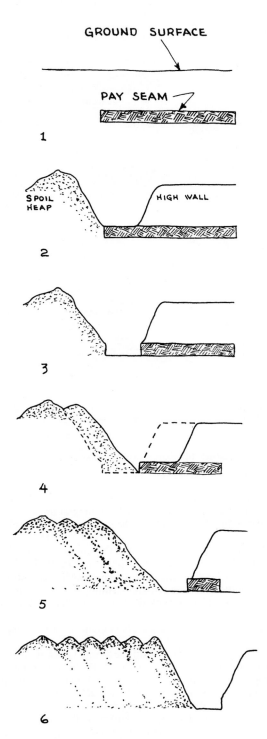

FIGURE 10.2 Strip-mining a horizontal bed.

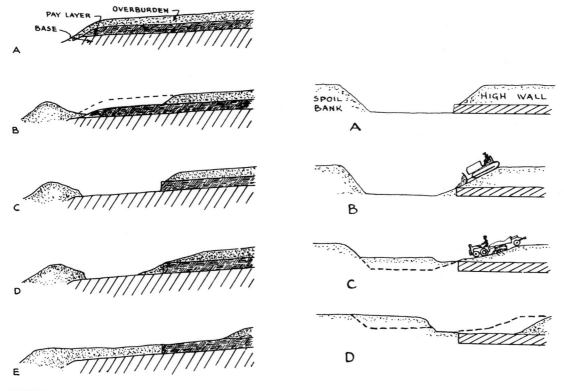

FIGURE 10.3 Bulldozer stripping.

FIGURE 10.4   Stripping shallow overburden with scrapers.

In Fig. 10.4 the high wall is smoothed off and the outcrop of the pay formation covered by bulldozers working down the slope. The scrapers then dig the high wall, working downhill, and build their fill on the full width of the empty pit.

Lengthwise digging is well adapted to combined operations. If only part of the overburden were to be removed by the scrapers, and the balance by shovels, a narrower fill would be made. The scraper work would thus serve to reduce the depth to be handled and the distance it would have to be thrown.

Scraper cuts are frequently made to dispose of loose or soft overburden on rock which requires blasting. This reduces the drilling footage, eliminates the need of casing large blast holes, and permits use of wagon drills on the exposed rock.

Front Shovel.   For sidecasting work, front shovels are equipped with extralong booms and sticks to increase reach. These are compensated by extra counterweight and power or smaller buckets.

Crawler-mounted stripping shovels are made with buckets from $2\frac{1}{2}$- up to 140-yard capacities. In strip work they are generally used for sidecasting, but except in the largest sizes they may also be used for loading trucks or trains that stand in the pit or on comparatively low walls.

Small- and medium-sized stripping shovels are generally more or less standard units that can be readily transported from job to job. Large ones are likely to be custom-made, shipped to the job in pieces, and erected at considerable expense. Their high cost is only justified when enough work is available in one area to repay it.

For casting, reach must be increased with the depth of overburden, as the slope of the pile progressively narrows the work area in proportion to top width of the cut. Increases in power and bucket capacity are required for digging harder and coarser material, and for greater output.

A shovel may make one, two, or more trips to clear a working area (Fig. 10.5). It may be worked from the coal or pay seam, or from the floor left after its removal. Maximum swing required is 180°.

The shovel is followed by one or more cleanup bulldozers. The toe of the high wall is scraped back, and the pay seam is cleaned of material left or dropped by the shovel, or which has slid or rolled from the pile. This debris is pushed into the spoil pile.

Coal may be injured by the grousers of heavy bulldozers. Rubber-tire dozers, or crawlers with the grousers trimmed back or covered by street shoes, can be used to avoid damage.

Dragline. Stripping draglines are being manufactured to carry buckets up to 220-yard capacity, and booms over 300 feet long. Diesel and diesel electric power are standard up to $3\frac{1}{2}$ yards, and are sometimes used in machines as large as 8 yards.

Crawler mountings are standard in small machines, and walking bases in large ones. The walker rests directly on the ground and moves by eccentric movement of shoes on each side. It is safer for use on high or loose banks because its ground pressure is low (5 to 12 pounds per square inch, as against 50 or 60 for comparable crawler machines), and it can change its travel direction without exerting any side thrust. The mechanism is described in Chap. 13.

Walkers are not adapted to use on pit floors because they can only walk away from the boom so that they may be trapped if unable to swing freely. In addition, they are usually wider than crawlers of the same weight.

The dragline strips by moving along the high wall, parallel with the pit, digging behind it and casting onto the spoil pile, as in Fig. 10.6. Maximum swing is about 90°. Standard practice is to dig within the radius of the boom point.

Large draglines can dig hard and coarse formations but are somewhat less efficient in them than power shovels.

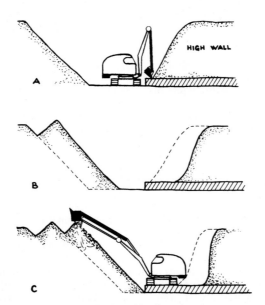

FIGURE 10.5 Stripping with a front shovel.

FIGURE 10.6   Stripping with a dragline.

    Draglines can load part or all of the spoil into trucks or trains, on the wall or in the pit. They can dig selectively from the top down, handling different formations in succession from one stand. This ability makes possible a rough division of the bank into select material to be hauled away, and waste to be sidecast.

    Cleanup bulldozers follow immediately after the dragline and push their loads within reach of its bucket.

Stacker.   Stackers are mobile elevating belts. The belt is carried in a boom which can usually be raised, lowered, and swung while operating, and may be adjustable in length as well.

    The hopper is usually crawler-mounted and self-propelled. It is protected by a grizzly, so that pieces too large for the belt will be rejected onto the pit floor.

    Figure 10.7 illustrates the use of one of these machines in conjunction with a standard-boom shovel. The shovel digs the bank and dumps in the hopper, from which the belt carries it to the spoil bank.

    Dug soil can usually be moved more economically by a conveyor belt than by swinging a shovel heavy enough to carry a boom of the same range. However, the stacker is not as flexible and will not handle as coarse material as a stripping shovel or dragline.

    The stacker is also used in the pit for loading trucks up on the bank.

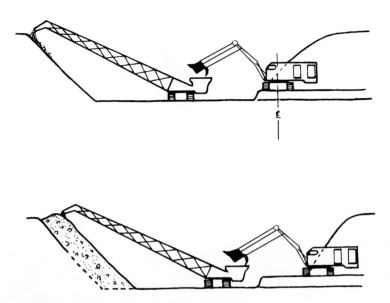

FIGURE 10.7 Front shovel and stacker handling overburden.

Wheel Excavator. The wheel excavator is the newest big machine to appear in the coal fields. It is a self-propelled, crawler-mounted unit that carries a cutting wheel on the end of a long boom that can be raised, lowered, and swung; it also has a stackerlike conveyor belt for discharging the spoil, that may or may not be separately controlled. (Refer to Chap. 14.)

One of these machines was used by Peabody Coal Company in its River King mine in southern Illinois. It was teamed with a 70-yard front shovel to handle overburden from 65 to 135 feet deep.

The wheel excavator took the top third of the overburden, which was mostly soil. Where the cut was 90 feet, the wheel took off 30 feet. The bottom 40 feet contained hard rock that was drilled and blasted; this work was done from the shelf left by the wheel. The shovel dug this and piled it across the pit in the space from which coal was removed.

On the next pass, the wheel spoil was deposited above and behind the shovel piles, as in Fig. 10.8. This made an excellent division of work, as the wheel could not handle coarse rock, and its soft spoil was held from sliding back into the pit by the shovel's rock pile. On the high wall side, the wheel shelf provided a better surface for the drill than the original ground, and drill holes were much shorter. The shelf also caught bank slides that would otherwise go down on the coal.

The coal-loading shovel followed the wheel, making a 45-foot cut. Overall pit width at the bottom, from high wall to the toe of the spoil bank, was 130 feet to provide space for these big machines to pass each other.

Cable Excavator. Overburden can sometimes be economically stripped with slackline cableways. A long pit will require a mobile or portable tower and means for easily moving the tail tower.

A good practice is to locate the heavy head tower on undisturbed ground, and the tail tower on the old spoil, as in Fig. 10.9. The bucket and its mechanism are rigged to dump at the low end, and digging is done toward the head tower to avoid dragging spoil over the face.

Slacklines are best adapted to wide cuts. They may be assisted by rooters or by blasting in hard deposits.

If access to the face is not required, a power drag scraper can be used to pull the overburden into the old pit.

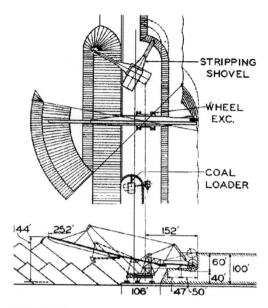

FIGURE 10.8   Wheel and shovel team.

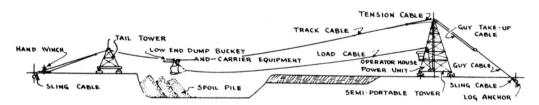

FIGURE 10.9   Overburden removed with drag scraper.

Double Casting.   If a space is required which is wider than that which can be stripped by available shovel and dragline equipment, and the spoil is too coarse for scrapers, a shovel may be used to swing the spoil out onto the pit floor, and a dragline to pile it on the bank. The illustration in Fig. 10.10 is of a quarry requiring a wide working area below the face.

For best results, the two machines must work together as, if the backhoe is allowed to build a substantial windrow, the dragline will need a longer reach to get to its far edge.

A dragline may also be operated on the previous spoil bank, to take off the top of the pile being built by a shovel and thereby to increase its disposal capacity.

The extra expense of double casting limits its use to special conditions.

Deep Stripping.   Strip operations often run into areas where depth of overburden is too great for efficient sidecasting with available equipment. This may be handled by two trips of a shovel, by a shovel and wheel, or by a shovel and dragline.

Two trips with excavators may be made for drilling and blasting convenience rather than to make up for a lack of reach. Taking a high bank in two levels may result in worthwhile savings.

The strata in the upper and lower lifts may differ in quality so that different drilling and blasting techniques are used. In some cases one will require blasting whereas the other will not.

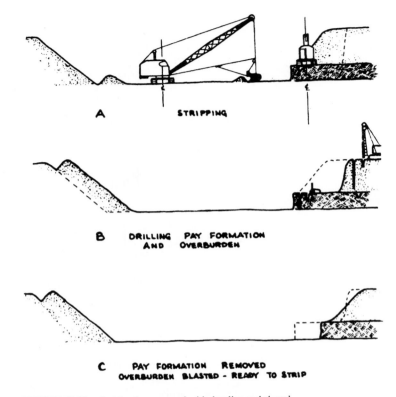

FIGURE 10.10   Overburden removed with dragline and shovel.

Very large draglines may be able to dig solid shale, but not sandstone. In such formations only the sandstone layers would need to be drilled and blasted.

Grading Spoil.   The spoil banks from light stripping operations can be leveled by bulldozers, scrapers, or graders. However, the huge machines used in heavy stripping leave such large, coarse piles that the biggest dozers cannot reduce them economically.

The problem is largely a new one, as prior to legislation of the subject, most of the stripped areas were left as permanent wastelands.

If a front shovel is moved short distances frequently, the high peaks of its spoil ridge are greatly reduced, as indicated between Fig. 10.11(A) and (B). The more even ridge top is the more readily broken down by a bulldozer working along its crest.

If the boom and stick are longer than required by the height and slope of the bank, the ridge and trough surface can be eliminated by building out level with the old bank, (D) and (E), instead of heaping toward the pit, as in (C). Accurate grading cannot be expected in such overhead dumping, but the bulk of the irregularities can be eliminated.

A dragline with extra boom length can build a flat-top pile by dumping against the bank near the top at full reach, and building out the edge by shorter or longer swings as required. Approximately the same result can be obtained with a slightly shorter boom by throwing some spoil.

Another dragline procedure, requiring extra clearance at the pile foot, is to drag the top of the completed pile onto the slope to the pit, as in (F) and (G).

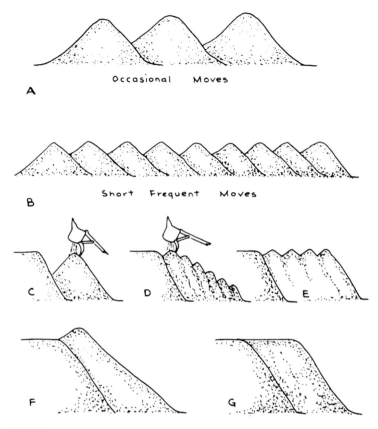

Occasional   Moves

A

Short   Frequent   Moves

B

C   D   E

F   G

FIGURE 10.11   Piling spoil.

If the machine can be reeved so as to have a live boom, grading off spoil heaps may be made much easier as the reach can be varied by raising or lowering the boom.

Another system is to cut rough haul roads through the troughs, and build them up with spoil hauled from fresh stripping. Scrapers usually do this more economically than trucks as they can spread as well as dump. The troughs may be entirely filled, or just brought up far enough to enable dozers or other machines to take down the tops.

HAUL-AWAY STRIPPING

When sidecasting does not provide a wide enough work area for removal of a pay formation, or there is no place for sidecast spoil to be piled, the overburden is loaded and hauled to a dump.

Depth of overburden may be a few inches to more than 600 feet, and the area may be only a few acres or several square miles. In general, the method is adapted to thicker pay deposits and deeper overburdens than sidecast stripping.

Mines in which stripping is chiefly done by sidecasting are called strip mines, while those whose overburden is hauled away are called open pits. Most sand and gravel are obtained from open pits. Open-pit overburden may be any material, from soft, lake-bottom mud to the hardest varieties of rock.

Stripping Area.   The size of the area to be stripped for full recovery of a deposit is determined by the area of the pay material, its bottom depth, the thickness of the layer of waste, the slope of the natural ground surface, and the steepness of the safe slope of cuts.

As a practical matter, this ideal area may be limited by property lines, the presence of valuable buildings, access or drainage requirements, the money available for stripping, or combinations of these factors.

Property lines may not be absolute barriers. Permission may often be obtained, at a price, to extend excavation across them. In the Mesabi range where open-pit mines adjoin each other, different companies often exchange both overburden and ore, balancing such accounts over a period of years.

If the mineral is valuable enough, whole towns and mills worth millions of dollars may be moved or demolished to permit pit expansion.

Cut Slopes.   It is seldom possible to limit stripping to the area immediately above the pay deposit, unless the overburden is very shallow or very firm. Cutting must be kept at a slope that will stand during the period of the work.

Even the firmest rock formations should not be left vertical for over 200 feet. The majority of rocks are not safe until slopes are reduced to 38° to 45°. The only guide that is reasonably accurate is behavior of similar rock in other pits, as laboratory tests for shear and other characteristics are unlikely to include all the characteristics of large masses.

Unconsolidated soil has the same general slope requirements, except that a combination of fine-textured soil and groundwater may reduce possible slopes to 10 percent or less. Water is the most important single factor in the stability of soil slopes.

While natural cliffs may be vertical for thousands of feet, they almost always have rock heaps at their feet that show falls that would have been disastrous in a busy pit.

In benched cuts there are three slopes: that of the individual faces, the average working slope, and the final or residual slope that is left when the work is completed. The average and final slopes are taken from crest to crest. See Fig. 10.12. The final slope is usually steeper than the working slope.

Rate of Removal.   It is usually necessary to remove some overburden before the pay material can be dug. When money is short, extraction work may start before there is really room for it, with resulting inefficiency. A well-financed project may remove much or all of the waste, perhaps at an expense of millions of dollars, before digging any ore.

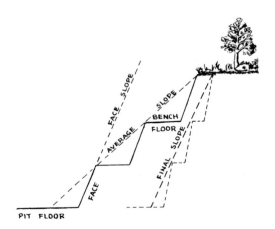

FIGURE 10.12   Benched slope.

It is usual for both stripping and extracting operations to be carried on through most of the life of a pit, either at the same time or alternately. In the Mesabi iron mines the same equipment that digs ore in the summer loads overburden in the winter. Gravel pits do their stripping in slack seasons, whenever they occur.

Some mines blast and load their own ore but let contracts for stripping off waste, so that the two jobs are active side by side through the whole working year.

The relative rates of stripping and mining are an important factor in mine efficiency. When stripping is well ahead, the mine has plenty of working space, with wide bench floors. If mining is faster than stripping, benches narrow and space becomes crowded, reducing efficiency.

The size of the pay formation may change during mining. Additional reserves may be located, or the cutoff point between usable material and waste may be changed by variations in price, processing methods, or market. Any extension of mining will probably make it necessary to extend the stripping. However, abandonment of part of the project cannot unstrip the affected area, or recoup any part of its substantial cost.

A lag in stripping is often due to lack of funds, or to reluctance to invest them any sooner than is absolutely necessary.

Stripping Costs.   Stripping costs are affected by the type and amount of material to be removed, the distance it must be hauled, grades on the haul route, and efficiency of the operation. These costs would include all loading and hauling costs as well as supervision, but not drilling and blasting.

Scrapers can move suitable material at 50 to 60 percent of the cost of shovels and trucks, if conditions favor them. But in coarse and rough material their production goes down and costs go up rapidly.

Hydraulicking and dredging are limited to soft overburden, abundant water supply, and dumps suitable for ponding hydraulic fill. However, if conditions are right, they can remove overburden as cheaply as any method.

Rail haulage is likely to be cheaper than truck haulage if large volumes are to be moved more than 2 miles.

Small pits may be able to sell or use their overburden, medium sizes may have little difficulty finding places to dump it, and large operations may find considerable difficulty in finding adequate disposal areas. For this reason, large-volume stripping may show a higher cost per yard than small-volume, because of the longer hauls needed.

Working at the Edge.   A pit in a deep sand or gravel deposit usually expands rather slowly, and stripping work is done at intervals, often by the pit machinery in slack periods. Stripping may be postponed too long, until the face is pushed back against the overburden, as in Fig. 10.13(A).

Stripping may then be done by either pushing into the pit and loading from the bottom, or throwing back with a dragline, backhoe, or clamshell. Consideration should be given to the question of caving of high banks.

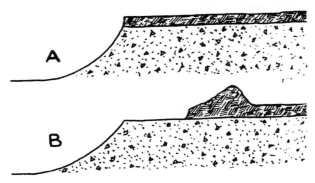

FIGURE 10.13   Stripping edge of expanding pit.

Once the edge is cleared, as in (B), the burden can be moved back farther by recasting with the dragline, pushing with a dozer, or carrying in trucks or scrapers.

It is vital to pit efficiency to keep overburden stripped far enough ahead to be out of the way in rush periods. Failure to do this often results in spoiling or losing valuable material, and in inability to fill important orders.

Dump Location.   A dump for waste from stripping should be as near the pit as possible to reduce haul costs, but it should not be within an area that might be dug away because of any conceivable extension of the stripping. These two considerations may be opposed, and deciding between them may be difficult. It is a matter of regret that initial economies have often resulted in disproportionate later expense in redigging and moving a dump pile.

For example, loading overburden might cost 90¢ per yard, hauling it 1 mile about 25¢, and additional miles 15¢ each. A single move of 2 miles would cost $1.30, while moving it 1 mile, redigging, and moving another mile would cost $2.55 per yard. Double handling is always expensive.

If it can be managed without substantial extra expense, different types of spoil should be placed in separate dumps in such a manner that they will be accessible for redigging. Changed conditions may make previously worthless material valuable. Examples are the reclaiming of mine tailings and slag heaps.

Haul Grades.   Grade of haul roads is important to economical hauling. A level run from cut to dump is desirable for speed and economy. An adverse grade (upgrade in the direction of haul) will cut both the speed and the load-carrying capacity. The extent of this loss depends on many variables. A rule of thumb is that production will be reduced about 5 percent for each percent of adverse grade.

The adverse grade will increase fuel consumption, tire wear, and maintenance costs. Wear on the truck engine and drive train is increased disproportionately on grades over 6 percent.

A downgrade in the direction of haul (favorable grade) is helpful up to about 2 percent, but steeper grades may reduce production about as much as an adverse grade. Downhill speed must be limited for safety reasons, and even empty trucks are slowed by upgrades.

Favorable grades over 2 percent and adverse grades over 5 percent call for special retarding devices in torque-converter-equipped trucks.

Grades may change considerably during a stripping operation. The floor of the cut moves downward, but its edges move outward and often upward. Dumps may stay at the same level, but if space is restricted, they usually build upward.

Haul Routes.   Two-way roads for heavy hauling should be from 4 to $4\frac{1}{2}$ times as wide as the vehicles using them. That is, highway trucks should have 32 to 36 feet between gutters or banks, and 11-foot, off-the-road haulers from 44 to 50 feet. Hauling can be done on much narrower roads when necessary, but liberal width pays whenever large volumes must be moved. Even wider roads are made for some mines.

A haul route that crosses a public road is subject to serious traffic delay. For example, an automatic traffic light that is set against pit traffic, but trips within 10 seconds when a truck reaches it, will delay the hauler as much as an extra 1,000 feet at 20 miles per hour. A full stop sign will cause the same or greater delays, depending on the density and speed of highway traffic.

A signal worker at the intersection reduces delays to a minimum if the signaler is allowed to favor the pit traffic.

Hillside Dump.   The easiest way to dispose of stripping waste in trucks is to dump it off a bank, that is high enough so that it grows outward quite slowly. Height may be anything from 10 feet to several hundred.

Such a dump may be started by flattening off a hilltop enough to give trucks space to turn, or by cutting a pioneer road along a slope and dumping from it.

Capacity can be figured in two ways. Annual or daily capacity depends chiefly on the length of the dumping face, and to a smaller extent on its height. Total capacity is the volume that can be dumped without spilling beyond the boundaries. It depends on the area that can be used, the height of the fill, and the slope of the dumped material.

Blasted rock dumped off a bank usually has an angle of repose of about 1 on 1, or 45°, and is likely to stay at its original slope indefinitely unless the native soil beneath it moves outward.

Dumped soil tends to assume a somewhat flatter angle, which will depend on the size and shape of its particles. Wet soil flows, and it is important that no drainage from other areas flow into the dump. Rain falling on the turning area on the top may gully the slope and spread mud over a large area below it. This can be prevented by sloping the surface up toward the edge, keeping a berm or ridge of dirt at the edge, and providing another escape for the water.

Both rock and soil slopes offer a hazard of rounded and oversize pieces rolling far beyond the toe of the fill. In empty country this may not matter, and brush and trees check such objects naturally. But dirt or log barriers may have to be built to prevent rolling onto paths, roads, buildings, or other property.

The slope of dumped fill is likely to be between 27° and 35°, with coarser material having the steeper slopes. For rough calculations, assume that it will be 1 on 2. If the waste is already being dumped or piled, the slope can be measured for more accurate figuring of areas and quantities.

Dump Operation.   Trucks are backed square to the edge and dumped. Methods of keeping them from backing over the edge vary widely. Sometimes the driver is just supposed to stop in the right spot and at the right distance, with or without a spotter. Or a dump log may be placed to both indicate the dump spot and protect the truck against backing too far. The chief problem here is that a log heavy enough to stop 30 to 60 tons of loaded truck is difficult to move.

The simplest and best protection for trucks is an 18-inch to 2-foot ridge of dirt left at the edge by the grading dozer. If the edge is very soft, a ridge may be built at a distance from the edge.

Off-the-road trucks with standard spill chutes instead of tailgates will dump their loads clear over a ridge at the edge, so that piling up does not occur at the top until it builds up from the bottom. Tailgate bodies may or may not spill part of their loads on the top.

It is good practice to maintain an upward slope from the truck entrance to the dumping edge. One-half percent is sufficient on porous fills, and $1\frac{1}{2}$ percent is sufficient for any material that is kept well graded.

Fills tend to settle under weight of traffic, and with time and weather. This sinking is most rapid toward the edges that are being built out. It is necessary to correct the resulting downslope with wedge fills as downslopes are dangerous to trucks and may cause gullying.

A wedge fill is made by dumping on the surface at the back or thin edge of the settlement, and grading with a dozer to restore or increase the original upslope, as in Fig. 10.14. The dozer first pushes toward the edge to establish the slope, then parallel to it so as to smooth and compact it and to leave an even windrow along the edge.

TOPSOIL

Topsoil is frequently the only material sold from a temporary pit. At other times, it may be a highly profitable sideline, or a costly stripping and wasting problem.

In this discussion, topsoil is defined as any layer or layers of soil containing sufficient humus (organic matter) and plant food to support a good growth of grass or other desirable vegetation. It is ordinarily on the surface but is occasionally buried by flood deposits or slides.

In the eastern United States, topsoils are predominantly brown, with a humus content between 3 and 20 percent by weight. Depth varies from zero on ridges to many feet in river bottom lands, but is usually between 4 and 10 inches. Division from lower soil layers is usually definite. These soils will be the basis of most of the following discussion.

In arid and semiarid sections of the west, topsoil tends to be deficient in humus and rich in minerals. It is often difficult to distinguish from subsoil. It may occur in deep layers or deposits and in general does not obtain as high a price as in the east.

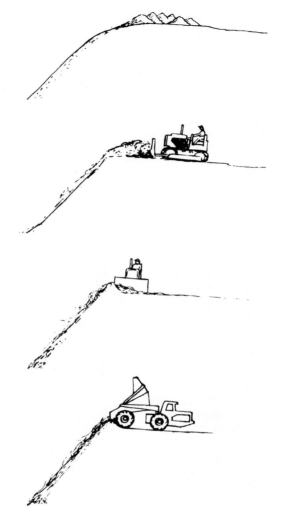

FIGURE 10.14 Building up a dump edge.

The prairie topsoils range from 8 inches to several feet in depth, may be brown or black, are rich in both humus and minerals, and generally have an excellent texture.

Swamp topsoils, in any section, tend to be gray to black and may contain up to 85 percent organic matter. Depths vary from a few inches to hundreds of feet. The richer deposits are not topsoil as defined above, and will be discussed separately as peat.

In general, the salability of topsoil is determined more by appearance and texture than by the ability to grow crops. The average topsoil buyer will seldom have soil tested, and tests are often not as reliable as good judgment.

Topsoils with high percentages of clay or silt will be heavy, slow-draining, and inclined to pack into hard lumps if disturbed when wet. Increase of humus content will soften the lumps.

Sandy or gravelly topsoil is loose in texture, drains readily, tends to dry out, and can be worked when wet without caking. Most soils are of an intermediate structure, with variable draining and lumping.

Testing.   One test of topsoil is the observation of the type and condition of vegetation it supports before stripping. The vigor of weed growth on piled topsoil is an excellent index to its quality.

Laboratory or field tests can be made for humus content, grain size, acidity, and available plant food. Humus is measured by the ignition test to be described for peat, and grain size on screens used for testing sand and gravel.

The test for acid-alkaline balance is commonly made by pressing litmus paper against the damp soil, and comparing its new color with a chart. If the soil is dry, distilled water should be used to dampen it, as tap or pond water may give a false reading.

Acidity is expressed in terms of pH (percentage of free hydrogen ions). A reading of 7.0 is neutral, lower readings increasingly acid, and higher ones alkaline. Most plants will grow under quite a wide range of conditions. A slightly acid condition is desirable for most of them.

Excessive acidity is readily corrected by the addition of lime, which can be spread on the field before plowing or disking. Soils are made more acid by mixing with humus, oak leaves, or aluminum sulfate.

Kits obtained at garden supply stores can be used to measure available or soluble nitrogen, potash (potassium oxide), and phosphorus or phosphorus oxide. It should be remembered, however, that these chemicals are often taken up by plants, or leached out by rain, as fast as they become soluble. The real measure of prolonged fertility is the insoluble reserves that are gradually made available by soil organisms, plants, and weathering. Except for humus, which is rich in nitrogen, and often in the other plant-supporting chemicals, such reserves are difficult to measure.

Preparation for Stripping.   The cost of properly preparing a field from which topsoil is to be sold is usually a small part of the total expense, and should increase the value of the soil so that it will either command a higher price or be more readily salable at a standard price.

Aside from clearing, field preparation may not be necessary if the soil is to be left piled long enough to rot the vegetation—usually 4 to 6 weeks of warm weather for sod—or if digging is to be done by a chain bucket loader. Plowed land is more easily and cheaply piled by a bulldozer than solid fields.

The field should be plowed to the full depth of the topsoil, if possible, or at least deeply enough to turn up most of the roots of the grass or crop. However, turning up of subsoil should be kept to a minimum, particularly if it is of a conspicuously different color. Therefore, if the topsoil depth is variable, or plow depth hard to control, it may be necessary to plow very lightly.

After plowing, the field should be thoroughly disked so that the vegetation is chopped up and well mixed with the soil. It is then ready for stripping.

If soil is to be removed in a wet season, it may be necessary to leave some strips of sod intact to support trucks.

Noxious weeds can be reduced or eliminated by planting and turning under one or more vigorous, close-growing cover crops. Buckwheat is particularly effective at smothering out.

Nitrogen Deficiency.   When vegetation is turned under and mixed with soil, there is an immediate and rapid increase in the number of the microorganisms which cause it to decay. For their growth they need nitrogen. If the crop is a legume, such as clover or vetch, they will be able to obtain it as they break down the plant material; otherwise they obtain it, or as much as they can get, from the soil.

This is likely to result in temporary total exhaustion of available nitrogen. When the vegetation has decayed so that it no longer provides sufficient food, most of the organisms die and their nitrogen is largely returned to the soil.

During the interval, which is 2 or 3 weeks for fresh green material in warm weather, and longer when the material is dry or coarse, when soil moisture is deficient, or when the weather is cold, any crop which is planted will be starved for nitrogen and make little or no growth. This effect is most severe when conditions favor rapid decay.

Packing Soil.   The value of topsoil is reduced or lost if it is packed into lumps. Except for a few very light and friable soil types, varying degrees of damage will be done if it is worked wet, or trucked over except when thoroughly dry.

Wet working breaks down the soil particles into a structureless mud that dries into lumps and sheets. The probability of damage increases with the amount of contained water, and its severity decreases with increased proportions of sand or humus.

Packing under trucks or other heavy machinery may produce the same result by bringing water out of small spaces between particles so that it will make a mud. Trucking on rather dry soils will produce compression cakes, which are usually softer than the mud lumps.

A rough test of condition may be made by rubbing a sample of soil between thumb and finger. If it smears, it is too wet, and if the particles remain separate, it is probably ready to work. The dirt turned by the plow or dozer blade should be watched for smearing, which indicates that soil is too wet.

Topsoil should not be trucked over, but it is often impossible to avoid doing it. The stripped areas may be too soft, or they may not offer enough space for maneuvering. If the latter, trucks may be routed to drive empty across the topsoil and run on the subsoil with loads.

It is usually better to completely ruin a narrow strip of soil by using it as a haul road than to damage a large area by allowing trucks to wander around on it.

Soil lumps are completely broken down by freezing and thawing, and usually disintegrate slowly in wet seasons. If still on the field, the roots of a cover crop, or sometimes only a thorough rolling or disking, will reduce them. If absolutely necessary, a hammer mill shredder of the type used for humus can be used to pulverize them.

Piling.   The standard tool for piling topsoil is the front-end loader. The old standby, the bulldozer, does the same work but rather less efficiently, as it cannot make high piles without walking on them. But it may separate topsoil more cleanly.

When piling is done in advance of loading, the standard practice is to heap the soil in windrows (long piles). These may be run up and down the slope so as not to interfere with drainage; or across it to keep in uniform soil types, or for convenience in trucking. It may be necessary to make occasional breaks in windrows to prevent ponding of water above them. Piling should be started at the entrance to avoid trucking over unstripped areas.

Windrow size will vary with loading requirements and soil depth, and the size of the bulldozer. Small, closely spaced ridges are most easily piled, but loading machines work best in large, high piles. Large dozers and deep soil favor building big piles. Building of the piles is described in Chap. 4.

Careful separation of the topsoil from subsoil is a requirement of most stripping. Inclusion of even considerable amounts of loose fill in topsoil does not usually damage its usefulness, but it is a type of adulteration that is unpopular with the buyer. The damage to appearance and value is especially severe if the subsoil is a conspicuously different color, or texture, or is in the form of lumps or sheets. It is generally better to leave a thin sheet of topsoil on the field than to mix topsoil and fill.

However, if the stripping is done in order to get clean dirt or gravel for roads or other purposes, it may be better to concentrate on cleaning the surface, even if some fill mixes with the topsoil.

Topsoil and subsoil are separated chiefly on the basis of color. If the difference is prominent, the distinction is easy to make, but the results of a mistake are painfully obvious.

Light conditions may obscure the color difference. When the sun is very low, in the morning or evening, or high in a clear sky, subsoil and topsoil may look the same. Cloudy days and intermediate sun elevations give the clearest distinction.

Any shovel rig can be used for piling topsoil. The backhoe and dragline are best at it, and the toothed clamshell slowest. Shovels work rather slowly in shallow soil. The hoe is adept at salvaging soil along a wall or fence.

Shovels loosen and aerate the soil, and cause minimum damage in handling it when wet. Fill dug with the topsoil can be concealed by mixing in. Teeth prevent absolutely accurate work, and it is good practice to clean up afterward with a dozer.

Scrapers are used for piling topsoil chiefly when it has to be completely removed from a large work area. They generally make wide, low piles that are more readily rehandled by scrapers than by other excavators. They pack the soil heavily even when it is in good working condition.

If topsoil is stripped from one part of the job at the same time that it is spread on another, the scraper can combine the two operations very efficiently.

Scrapers drawn by fast wheel tractors may be used to dig topsoil and deliver it by road to local customers, who will usually appreciate having it spread.

Loading from the Pile.   Topsoil is comparatively light and normally has a low digging resistance. However, it tends to push ahead of narrow buckets, instead of entering them, and this factor, coupled with the small size of the usual pile, reduces production to below that of hard digging in a bank.

Difficulty in filling the bucket may be reduced by thorough chopping of sod and weeds before piling, building large piles, and building piles on undug areas so that the bottom of the bucket will work in firm soil.

The front-end loader is the best equipment for loading because of flexibility, ability to load from either pile or field, to clean up as it works, to pile when not loading, and high production in relation to purchase price. Trucks are backed into the end of the pile, in variable positions to keep them at an angle of about 45° to the digging, as shown in Fig. 10.15. These wide buckets get heaping loads until the end of the pile is reached. The remnants can be pushed to the next pile.

The excavator is also used. It may have trouble filling the bucket, but its principal difficulty is in cleaning up. If the pile is narrow, as in Fig. 10.16(A), it can walk down the center and scrape in the sides by swing dragging. If the pile is heavy, as in (B), it can dig the bulk from one side, easily cleaning as it goes, and come back through the small remnant. In either case, it is best to have a bulldozer work with it, at least part-time, cleaning up. A light rubber-tire dozer is generally adequate.

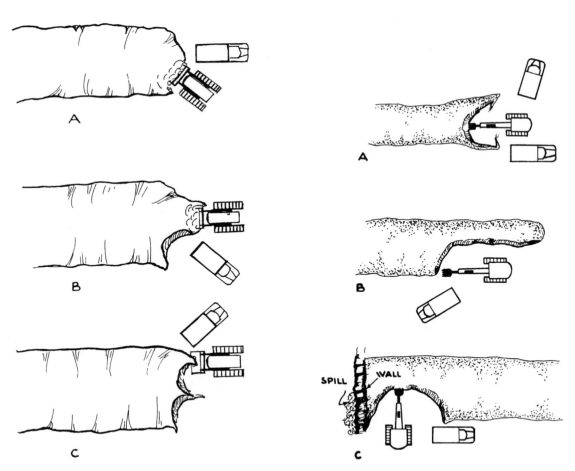

FIGURE 10.15   Loading topsoil with front-end loader.   **FIGURE 10.16**   Loading topsoil with shovel.

When digging a windrow ending in a wall or property line, as in (C), the shovel should be started at that end and worked in, to minimize working soil over the end and losing it.

Draglines and backhoe shovels can be worked from one side of the pile, as in Fig. 10.17(A), or preferably from the smoothed-off top, as in (B). The top of the pile is particularly suitable for backhoes, as in (C), as they load more rapidly and easily if higher than the trucks. These rigs clean up the edges without extra work but can strip closer if aided by a dozer.

Any shovel cutting to a grade will occasionally go below it so that fill will be dug. The mistake is immediately shown by the color of the bottom. The bucket should be dumped off to the side to be wasted, or on the pile, to break up and mix the slice of fill. Subsoil may be deliberately dug and mixed in this manner if the topsoil is rich and the price is low or the buyer indifferent.

Loading from the Field.   Topsoil is often loaded directly from the field, with no systematic preliminary piling. The efficiency of the operation depends on the depth of the soil, the machinery used, and the output required.

Best results will be attained in deep soils, using the shovel excavator or backhoe, as increase in depth reduces the proportion of time spent in trimming the bottom and in moving into the digging.

Unless the soil is very hard or rocky, the front-end loader gives good results, particularly as vegetation does not have to be disked ahead of it. Front loaders dig easily, but may roll up big masses of sod. Almost any loading machine can be used unless the soil is shallow or rocky.

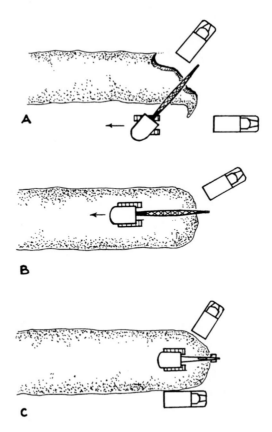

FIGURE 10.17   Loading topsoil with dragline or backhoe.

High production cannot be expected of any machine, except possibly the front-end loader, in shallow stripping. However, if the trucks are scheduled so that the front-end loader has free time between them, it can stockpile enough to enable it to fill the truck quickly.

Sometimes a bulldozer is used to boost output by feeding soil to the loader. This may be done on the whole job or just on thin spots. Dozer cleanup after digging is usually desirable.

Screening. Topsoil may be coarsely screened on a portable grizzly, which is placed on the truck body prior to loading, or may be up on legs so that the truck can back under it. The latter type requires large piles for convenient use.

These are used either to market topsoil which is otherwise unsalable because of presence of rocks, roots, and trash; or to obtain a premium price. It is usually necessary to have a worker clean stuck material off the grizzly. Best results are obtained with dry or sandy soils. Cohesive soils require a coarse mesh and must be put on a small quantity at a time, or too large a proportion may be rejected.

Square openings from $\frac{1}{2}$ inch up to 4 inches are used, or similar bar spacings. A much finer product, with less waste in rejections, can be obtained by using a revolving screen on the discharge of a front-end loader.

Storage. Topsoil is often left piled for long periods. Texture is generally improved by standing over the winter, and quality does not seem to be damaged much, even by years of storage. In hot, dry seasons, it may bake out to colorless dust, which is unpleasant to load and hard to sell, but a wet season will restore its original texture.

However, piled topsoil will support a luxuriant growth of grass and weeds, which quickly forms a sod that is very undesirable. The irregular shape of the piles makes cultivation with any ordinary tools almost impossible. Flame guns are tedious to use and ineffective, and chemical weed killers are selective and uncertain in action.

Such growth can be kept down, at a price, by cultivation with a bulldozer, as in Fig. 10.18. The machine is run along the crown of the windrow, cutting off the top and allowing the dirt to flow down the sides. This uproots or buries a large part of the weeds. Later the bulldozer pushes the sides back up to the top, as in (*C*), destroying the rest of them. Most economical results are obtained by allowing a week or more to pass between the two operations. The job is done over as often as necessary.

This serves not only to keep the soil clean, but also to enrich it by the decay of the weeds.

Restoring Vegetation. When topsoil stripping is not followed by other work, areas are left denuded of vegetation and the ability to grow it. They tend to cause a dust nuisance and to erode badly, becoming a mass of unsightly gullies, and often silt up streams and block roads with the waste.

As a result of such nuisances, many communities now forbid the stripping of topsoil, or impose restrictions on the work. The least of these is to require a guarantee to restore the vegetation on the stripped areas.

The process is similar to that required to restore worn-out and eroded farmland. The ground is loosened with rippers and plows, loose rocks are picked up or pushed off, a liberal application of manure or fertilizer and perhaps of lime made, and a crop planted. Fertilizing and seeding should be heaviest in drainageways. This crop should be one that will grow readily on poor land. The local Farm Bureau will be able to advise one or more suitable for the conditions and season.

In many localities, buckwheat or soybeans make a good summer crop, and a rye and vetch mixture is suitable for fall or early spring. When the plants are in flower or beginning to seed, they should be plowed or disked under, and additional fertilizer supplied to spots that did not grow properly. After an interval of 2 or more weeks, depending on local custom, plant another crop. A self-seeding legume such as sweet clover is good. This seed should be inoculated with a culture of the proper nitrogen-fixing bacteria. Some patch fertilization and replanting may be necessary later, and any tendency to gully can be checked with topsoil patches, heavy fertilizing and planting, or brush mats.

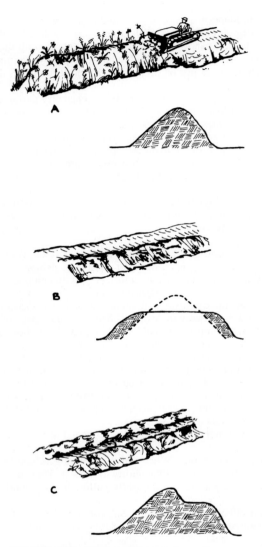

FIGURE 10.18   Weed control on piled topsoil.

With luck and good farming, the second crop may hold the land so that no further plantings are necessary. Decay of plants and nitrogen absorbed from the air will enrich the soil, so that the native vegetation of the area will soon be able to reestablish itself.

Reclamation probably gives best results on glacial till soils, which will develop a new topsoil cover in a surprisingly short time.

Floors of deep pits will respond to the same methods but much more slowly. Deep subsoils rarely have proper tilth, or plant food in quickly available form for crops, so more "green manuring," or turning under of vegetation, is required.

Animal manure will give better results than chemical fertilizers, particularly on floors of deep excavations, largely because it contains many organisms essential to a healthy soil. However, in many localities it is difficult and expensive to obtain.

It is advantageous to plant as soon after stripping as possible. If an interval is to elapse between piling and loading, it is good practice to plant a cover crop between the piles to prevent the development of a dust nuisance and loss of topsoil remnants.

PEAT (HUMUS)

Characteristics. This is a light, soft, absorbent, organic substance formed by partial decay of vegetation, which accumulates in more or less pure form in swamps or shallow ponds. It is usually brown or black, and may be a structureless jelly, or fibrous, or lumpy with recognizable remnants of wood or leaves.

It may be mixed with varying quantities of soil which increase its weight and will make it lighter in color, particularly when dry. Water content is 50 percent or more by volume. When pure, from 70 to 85 percent of its dry weight is organic and the balance mineral ash.

In some countries it is used extensively as fuel, when partly or wholly dried. If allowed to dry in stockpiles or dewatered deposits, it will catch fire rather easily and burn slowly and persistently. It can be put out by complete soaking, or controlled by cutoff trenches. This problem is discussed in Chap. 1.

Peat is usually very acidic. It may contain no immediately available plant food, but reserve supplies in insoluble form are good, particularly of nitrogen. It may be sterile, or able to support only limited growth, when first dug, but on exposure to weather it will gradually become fertile. It will make good topsoil when mixed with soil or sand, and lime and manure.

It is used to enrich lawns and gardens and occasionally farms. Golf courses use large quantities. It will soften cohesive soils so that they will not bake hard; and make granular soils absorbent so that they will not dry out readily. Use of large quantities at one application, particularly without thorough mixing, may unbalance the soil so as to make it temporarily unsuitable for some crops. Lime should be used liberally with it unless acid-loving plants are to be grown. Manure, or to a lesser extent fertilizers, may prevent unbalance.

Humus may be tested for water content, organic matter, and acidity. A quantity is weighed, baked dry at low heat, and weighed again. The difference is the water which it held.

The dry humus is then raised to a red heat, stirred occasionally, and kept at this temperature until it has been reduced to ash. The ash is weighed. The difference from dry weight is the organic content.

Burning humus may produce noxious fumes, so good ventilation should be provided.

Tests for acidity are the same as used for topsoil. Extremely acidic conditions are to be expected.

Digging. Peat is normally a water-level or underwater deposit. It is often readily drained or pumped dry, and is soft and light to dig. However, it is difficult to recover, particularly in large deposits, as it is sometimes too unsubstantial to make safe footing even for crawler draglines on platforms, or to support ordinary haul roads.

Methods of digging and getting it out of the pit are discussed in Chaps. 3 and 6. In general, the cable excavator mines it at lowest overall cost, and with fewest complications, but operations are often too small to justify the necessary investment.

Curing. A particular problem is the high water content. Half to three-quarters of the water will drain out of a pile in a month or two of dry weather, reducing its bulk 50 percent or more. The balance will stay in it unless baked out, and is a normal part of bulk or screened humus.

Before draining, the humus is too wet for handling as it becomes sloppier each time it is moved, and it will not respond to processing. Provision must therefore be made for curing between the pit and the plant or customer, unless the plant is equipped with drying apparatus.

A drag scraper or slackline may dump into a hopper, from which one or more movable conveyors carry it to piles, as in Fig. 10.19. The conveyors may have to use cleated belts, or buckets, to prevent the peat from flowing back along them. Conveyors, loaded by dragline or clamshell, can reclaim the piles to the plant; or the material may be loaded into trucks for sale as "raw" humus as soon as it has drained.

Draglines may pile it up at the pit edge and return to load it.

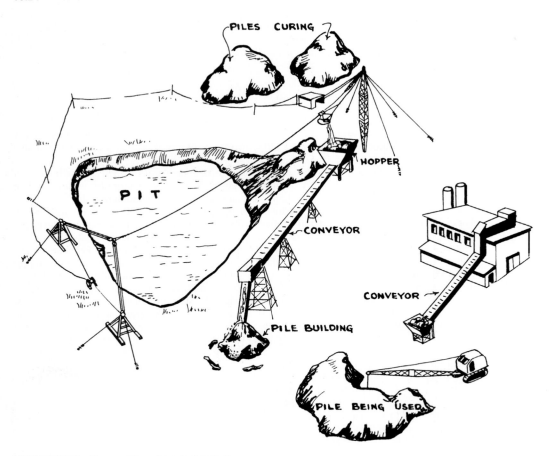

FIGURE 10.19   Humus pit layout (scale distorted).

Machines handling dry or partly dried humus can be equipped with oversize buckets, or oversize drum lagging, or can be run in a higher gear than it can when handling heavier soils.

Processing.   Humus is processed to break up lumps, to mix together different colors or grades, to mix in soil accidentally or deliberately added, and to increase bulk by fluffing and aerating.

When lumps are a serious problem, or the product must meet rigid specifications, a shredder, often a special type of high-speed hammermill, is used. It is lighter and less costly than rock-crushing models, but operates in the same way. Toothed drums may also be used.

In small operations, the shredder may be supplied with hand shovels, and the product hand-shoveled into trucks or left in a windrow to be gathered by a loader.

A portable conveyor belt may take the humus from the discharge opening and load it in trucks, or on a pile. Larger portable mills may have a high hopper opening supplied by a hand-loaded conveyor belt. The shredder can also be fed mechanically by any kind of loader, without hand work.

Portable crushing and gravel plants are sometimes successfully adapted to processing humus.

Fixed plants should be on firm ground outside the pit, and may be supplied by cable excavator, conveyors, or trucks. They are capable of much higher production, and may turn out a finer or more uniform product. They can be equipped with drying and packaging machinery so that their product can be sold in stores.

In any calculation about humus it should be remembered that 1 yard in the wet deposit or pile means $\frac{1}{2}$ yard or less in the cured pile.

GRAVEL, SAND, AND CLAY

Bank Gravel.   Bank gravel is a useful and highly varied material. It consists chiefly of sand, pebbles, and cobbles, but may also contain clay, silt, and boulders, mixed in or in accompanying layers or pockets. The gravel proper is the pebbles and cobbles in sizes from $\frac{1}{4}$ to 2 inches.

The specifications which gravel must meet to do certain jobs, and the proportions found in deposits, vary widely.

Bank gravels consist mainly of deposits laid down by fast-running streams, often of glacial origin, but they are also formed by waves on the seashore. The quality depends on the original stone, the proportion of sizes, and the angularity of the particles. Wave-formed gravels are predominantly rounded, glacial ones subangular, and product of other streams variable.

Talus gravels, formed at the foot of cliffs by falling and sliding, may be coarse and angular, but are often weak stone.

If gravels are not sufficiently angular for their job, and contain oversize stones, they may be run through a crusher which will produce angular fragments.

Fines in bank gravel act as a cement or binder, holding it together when dry. Gravel without binder becomes too loose for road use in hot, dry weather.

Fines in excess of 8 or 10 percent may cause a gravel to become sloppy after repeated freezing and thawing when wet. Fines over 15 percent may cause it to soften under prolonged soaking. Softening is made more likely by a high proportion of fine sand in the mixture, and less likely if thorough compaction precedes the freezing or soaking.

Any gravel will become sloppy if soaked when freshly dug, but if of good quality, should drain and firm quite quickly.

Gravels derived from continental glaciers are largely of hard rock. River and mountain glacier gravels are derived from upstream formations, and occasionally include too much shale or other soft rock for some purposes.

There are a number of tests for gravel, for field and laboratory use. A sample, with stones over $\frac{1}{4}$ inch removed, can be shaken up with water in a glass jar, then allowed to stand. The pebbles will form a layer in the bottom, with coarse and then fine sand on top. Silt and clay will settle out more slowly, and may take an additional day to compact. The relative amounts of the different-size particles can then be determined by inspection.

In the laboratory, gravel is dried, weighed, and put through a vibrating screen with many different meshes. The particles caught on each tray are weighed. Any lumps have to be broken up. This operation gives a classification of the specimen for size gradation.

Gravel can be tested for abrasion resistance by rolling in a cylinder with steel balls or other hard weights. Resistance to breaking up by freezing can be tested with cold, or with chemicals which duplicate its effect.

Clean bank gravel of the proper sand-gravel proportions is frequently mixed directly with cement for concrete.

Sand.   Most bank gravel deposits are more than half sand. In addition, sand deposits occur in many areas where no gravel is found.

Ocean beaches are typically sand, and river deposits usually contain high proportions of it. If the river flows slowly, the sand may be mixed with silt and clay, which usually must be separated before use.

Most sand is largely particles of silicon dioxide, best known in the form of quartz. It is very hard and withstands the abrasion of water working, which reduces other minerals occurring with it to fines. Calcium carbonate, mica, feldspar, gypsum, and many other minerals may also occur as sand.

Many sandbanks are clean enough for use without processing, but in most cases it is safer to screen and wash before using in concrete.

Occurrence.   Sand and gravel deposits occur in all parts of the world, and with special frequency on or near past or present shores, glaciers, and mountains. They may be thin, irregular deposits, or in heavy masses. In general, gravel is more variable than sand in size and type of particles, and thickness and shape of beds.

Running water needs higher velocity to carry large pieces than small, and in general, gravel is deposited nearer the source than sand, or at times of heavier stream flow. However, a stream which is building up a deposit alternates, bringing in materials and cutting parts of it away. Channels wander over the whole area. Oversize material beyond the capacity of the water to carry may be rolled long distances along the bottom. Clay and silt may be deposited in temporary pools and cut-off and stagnant channels.

The result of these factors is that gravel, sand, and clay deposits are often extremely variable and uncertain. When this is the case, mining them requires constant good judgment in deciding which horizons should be combined and which separated; and what can be used and what must be wasted.

Processing.   Sand and gravel may be processed to clean out dirt; to separate into different sizes; to combine different sizes and materials; to remove or crush oversize stones; and for combinations of these purposes.

In variable formations, the primary processing is selection at the bank as discussed under selective digging.

The processing plant proper may consist of a washer, a screen, a crusher, or multiples or combinations of these units, together with feed hopper, and transfer and discharge conveyors. These plants, available in both mobile and portable types, are described in Chap. 21.

By the use of units of proper size, any desired reduction, combination, or separation can be secured. It should be remembered, however, that no plant can produce a coarse product from fine particles. Deficiencies in gravel content must be made up by mixing in stone of proper size, or oversize up to the crusher capacity, in addition to the run-of-pit material.

Clay.   Clay, like sand and gravel, may be found in massive deposits or in irregular layers and lenses. It is often interbedded or mixed with other materials in very complex ways.

Underwater clay may be soft enough to be dug with a small dragline, or quite hard. Dry clay grades from hard shovel digging to shales requiring heavy blasting.

Pit operators usually find it economical to loosen up dry clay with at least light blasting, to facilitate digging. Electric or gasoline-driven augers are extensively used for drilling, and slow to standard velocity explosives for blasting.

When valuable clay is in narrow and confused beds, it is often blasted, then separated by hand into piles which are loaded by machine.

LOADING OUT OF THE BANK

Most primary pit excavation is in formations deep enough to be loaded directly from the bank. The material may be in its natural state or loosened by blasting.

Bank Height.   In free-flowing material, such as loose dry sand, the only limit to bank height is that imposed by safety. This will be discussed below.

If a formation will stand in vertical or overhanging walls, and is dug from the bottom, the face should not be higher than the machine can reach, as it may be necessary to dislodge overhanging pieces with the bucket to avoid danger from falls. Half this height is usually more convenient and may allow greater production if the top of the bank does not keep falling as the lower face is cut.

For example, excavators of $2\frac{1}{2}$-yard capacity may do their fastest loading in banks in which the dipper teeth do not have to be lifted above 12 to 15 feet. However, they can trim banks up to about 25 feet.

When working in rock, it is the height of the blasted rock heap that counts, not the height of the face. The amount of settlement depends largely on the proportion between the height of the face and the depth to which it is blasted.

A 200-foot face blasted back 20 feet may yield a muck pile only 15 or 20 feet high. A 20-foot face blasted back 200 feet might produce a 30-foot-high rock pile.

The height of a rock face may be limited by the length of a drill feed. Changing steels or adding drill rod takes time. This time can be saved by fixing the face height somewhat lower than the length of the feed. Rotary drills with 50-foot masts are used extensively on 40-foot banks.

Low faces require frequent moves. A height of less than one-half of the shipper shaft height may make it difficult to fill the bucket.

Benching. Whenever a noncaving formation is too deep for convenient digging, it is removed in layers. In shovel work these are called benches. They may be anywhere from 6 to 200 feet high. In highway work 12- to 20-foot faces are common, being suited both to the mobile light drills favored by contractors, and to the ¾ to 3½-yard power shovels they use.

In very massive rock cuts in highways, and in large-scale quarry and mine operations, heights of 30 to 60 feet are usual, with drilling by rotary or down-the-hole drills, and loading by 2½- to 8-yard shovels or the biggest front loaders.

The best place to start benching is at the top. This simple fact is often obscured by other considerations, so that hillside work may be started at the bottom or the middle, and the pattern straightened out later.

The width of a bench, from the edge to the toe of the unblasted rock, should be at least 50 feet. Greater widths are better.

Types of Machinery. Loading machinery used for pit excavation can be roughly divided into tractor loaders, which depend on traction on the pit floor for digging power; revolving shovels with dipper, clamshell, or skimmer front ends, which stand on the floor while working; revolving shovels, with dragline, backhoe, or clamshell rigs, which load from the top of the bank; and scrapers and bulldozers which work down the bank slope.

Selection of machinery will depend on the location and digging characteristics of the formations, the volume of output required, the type and importance of other work that must be done by the same machines, the type of haulage or conveyor units, and the costs involved.

Production Factors. Big machines are suited to hard and coarse formation and to high production requirements.

Practically all excavators are available in different sizes. Production usually does not increase in direct proportion to power and weight, as the more massive construction of heavier units may require lower speeds, and space may be lacking for convenient operation.

Manufacturers' data on output should not be accepted without careful study. Some firms deliberately underestimate production to avoid arguments, while others exaggerate it to make sales. Others base it on time-motion calculations, with little reference to field conditions.

Production ratings based on loose yards, or on a 60-minute hour, will be higher than for bank yards, or a 50-minute hour.

Also, there is room for honest difference of opinion about whether a formation is hard or soft, and conditions average, poor or ideal.

A rough index to output can be obtained by timing a machine at work in various materials. A stopwatch should be used and the results written down. The cycle time is the elapsed time between a certain movement, such as entering the bank, and the repetition of that movement. Average number of cycles per minute, from a number of observations, multiplied by the average bucket load in yards, will give the production rate in yards per minute in simple work such as sidecasting. Extra passes made to trim the bottom, or to break out or avoid boulders, may be averaged in or considered separately.

If the machine is loading, the loss of time in spotting trucks and trimming up their loads should be observed.

Data for calculating production are included in Chap. 2.

Big machines can almost always dig hard material better than small ones of the same type, but this factor is even harder to calculate. A rough index to penetration in material of even texture can be obtained by dividing the force which can be applied to the bucket by the width of the edge, or the combined width of the teeth. The extra resistance to the thicker teeth of the heavy bucket may be negligible in brittle formations and important in resilient ones.

In poorly blasted rock, or boulder-filled banks, the gain in penetration is much greater, as nearly the full power is often applied in succession to points of greatest resistance.

A wide bucket may be at a disadvantage because of inability to get between obstacles to attack them separately, or benefit from its capacity for large chunks.

In any digging, sharp cutting edges are essential to best work. In hard formations, teeth of proper spacing will give better results than straightedges.

Mobility is an important factor for machines which may dig for short periods from a number of different bank sections, or are used for loading from storage piles as well. Ability to do several types of work is liable to be useful, particularly in small pits.

It is good practice, although not always essential, to match the size of loading and hauling units. If large shovels are used with small trucks, time is wasted centering the bucket and material will be spilled off the sides. Truck tailgates may be jammed by oversize pieces and trucks damaged by impact. If the trucks are too large for the shovel, they must spend too much time being loaded; the shovel may be unable to fill them from one stand, and high body walls may hamper it. Generally, the loading equipment should govern the production of the operation, but to be cost-effective the number of haul units should not lead to noticeable idle time for the haulers.

Revolving shovels are usually teamed with trucks which will carry between five and ten bucket loads. Capacity is not as important for tractor loaders, but body walls should be low enough to permit easy placement.

Tractor Loader.   Tractor front-end loaders include crawler types, which can do quite hard digging and heavy bulldozing; four-wheel-drive, rubber-mounted units suited for medium-hard banks, and two-wheel-drive loaders for soft or loose material and firm ground operation.

Crawler-mounted loaders are easy enough to move around pits of moderate size, but the wheel mountings are superior in speed and cause less wear to themselves and to the roads while traveling.

Four-wheel-drive loaders may be obtained with standard buckets up to 24 yards in capacity. They can often replace an excavator-and-truck combination in filling hoppers, and in short to medium hauls.

Front-end loaders are described in Chap. 16. Almost all of them now have an open-front bucket that rotates forward to dump and backward into a cupped position for slicing upward in a bank and for carrying a load.

Multipurpose or four-in-one buckets are easier to dump into high trucks, and are invaluable in handling bulky objects. However, their greater weight is a disadvantage in ordinary digging and loading.

A good truck loading pattern for a front-end loader is shown in Fig. 10.20. While these machines are flexible and can dig under very awkward conditions, best production is obtained if both angle of turn and walking distance are kept to a minimum. Best height for noncaving banks is somewhere between the height of the push arm hinges, and the maximum upward reach of the bucket edge.

Loaders retain a high degree of efficiency in very low banks, but are seriously exposed to slides or cave-ins in very high banks.

In spare time or in emergencies, either crawler or wheel-mounted front-end loaders can do almost any bulldozer work except scraper pushing. They are used for tidying the pit, smoothing haul roads, shifting heavy machinery, carrying heavy or bulky objects, and rescuing or starting stuck trucks and other equipment.

Excavator.   Hydraulic excavators or front shovels are excellent machines for bank excavation. Although fastest loading is in soft material that will heap on the bucket, they can maintain good output

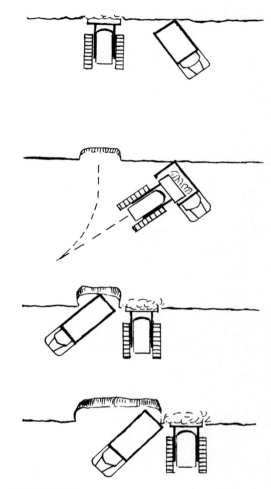

FIGURE 10.20   Loading from bank with front-end loader.

in very hard or rough material. They are more costly in proportion to capacity than the tractor front-end loaders, but have lower repair requirements as the tracks do not move during the digging cycle.

They will dig from any graded floor that will support trucks. Best production in noncaving material is usually obtained when the bank is about as high as the shipper shaft (dipper stick hinge).

A short arc of swing is important in getting maximum production. The bucket can usually be moved from break-out position at the bank, to correct height and distance for dumping in a truck, during 30° to 45° of swing. Any longer swing required by truck position slows the digging cycle.

When a shovel is walked straight into a wide bank, the initial swing required is about 60°. As the machine works in, the swing becomes longer, finally approaching 180°. See Fig. 10.21(A) to (C).

Except in the case of a through cut being made in just this width, it is obviously inefficient to penetrate so deeply into the bank on one path.

If the shovel is walked parallel to the bank, as in (D), trucks can be spotted ahead, at a slight angle to the bank, with a minimum swing of about 40° and a maximum of 140°. If trucks are also placed behind, the maximum swing can be reduced somewhat. As long as the shovel is kept slightly outside

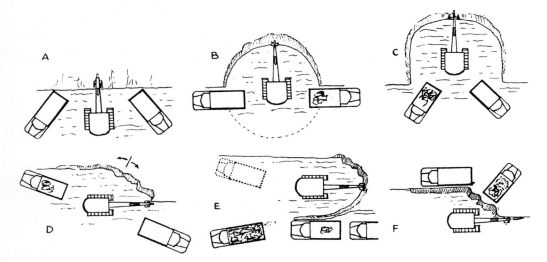

FIGURE 10.21 Loading from bank with excavator.

the toe line of the bank, as shown, it can do its own cleanup. However, production is increased if the shovel operator can dig roughly, depending on another machine to smooth out the floor.

If the shovel is kept deeper in the bank, as in (E), a ridge will be left near the toe line, reach to load across it will be longer, and a dozer will be needed. However, production from each stand will be greater, which in a low bank may be an important factor.

When the pit floor is narrow, sandy, or wet, so that trucks must keep to beaten paths, a drive-through pattern, as in (E), can be used. The shovel again works along the toe of the bank, and the trucks run parallel to it, at a convenient loading distance. Only one truck can be spotted at a time, but it can be moved into position much more rapidly than when backed in.

If the pit floor is too soft for trucks, and only a front shovel is available for loading, trucks may be put on top of a low bank. This works well only with a big shovel and low trucks.

In banks offering a danger of slides, the shovel should be worked straight in, so as to be able to back out directly if partly buried. Cuts should be kept shallow by frequent moves to different parts of the face.

Clamshell. The clamshell is so versatile that it is difficult to set up patterns for it. It can stand at the foot of a fairly high bank and dig from the top, or stand on the top and dig from the foot, or can work at any intermediate level. It digs straight down, gathering in its load, without pushing or pulling the surplus. These features make it very valuable in selective digging.

The clamshell is adapted to various types of digging by changing buckets or bucket plates. Heavy-duty buckets of great weight and reduced capacity will dig very hard dirt and even soft rock. Rehandling buckets are larger, light, and often lack teeth. Medium-duty or general-purpose models are intermediate in weight and have teeth.

The clamshell has a smaller output than other shovel rigs and is more often used in stockpile, rehandling, and hopper feeding than in primary digging.

Dragline. The dragline is the best machine for loading from the top of the bank if it can dig the material. Small draglines usually are quite helpless in tight or rocky soil, but very large ones will dig even tight, unblasted shale.

Difficulty of penetration increases with depth. For deep work, the boom should be long and digging done well out. This minimizes the upward pull of the drag cable, which decreases the effective weight of the bucket.

A dragline can dig harder material from a face than it can cut vertically. If a wide ditch is started by other machinery or by blasting, it can be continued back into the bank by a dragline. If hard, it will tend to narrow down and become shallow.

The most efficient arrangement for hard digging is with the machine digging from the top of the bank and the trucks in the pit.

A dragline's reach enables it to stand well back from treacherous banks so that it can usually make deep cuts safely.

The dragline loads best if it digs inside the boom point, at a medium depth, with a swing which takes no longer than the raising of the bucket, and the haul units are on the pit floor beside the bucket.

"Throwing" the bucket, that is, casting it so that it digs beyond the boom point, adds substantially to the area a dragline can reach from one stand or one line of work. But it is a practice that is best reserved for special situations, such as small cleanup jobs where access is difficult.

Throwing the bucket may add from 10 to 50 percent to the cycle time, since it usually requires that the bucket be pulled in, swung out, then dragged in with a full load until it can be picked up. If digging is done close in, the bucket is simply dropped and raised as soon as it is filled.

The extra reach is rarely more important than the time consumed. Also, careless casting causes or increases damage to bucket and cables.

When the loading or piling area is on the top of the bank, digging becomes slower as the reach becomes deeper, because of the time required to reel in the additional hoist cable required. At usual hoist line speeds, an extra second is needed for each $2\frac{1}{2}$ to 3 feet of depth. However, if the swing is long and unobstructed, the time of raising the bucket may not affect the length of digging cycle. If trucks are in the pit, the bucket may be raised only a few feet, regardless of depth.

If the dragline is not overloaded, it should have power enough to perform simultaneously the three functions of raising and braking the bucket, and swinging without lugging down the engine. If the bucket is lifted to dumping position before the swing is completed, it is the length of the swing which determines the loading speed. If the swing is delayed in order to raise the bucket, it is the hoist, and therefore digging depth, which regulates it.

Carrying Scraper.   Scrapers are not ordinarily considered to be bank-digging tools, but they may give lowest cost on combined digging and hauling.

Self-loading models, whether elevator, crawler-drawn, or two-engine, are most suitable for pit use, as they can work alone.

The bank is first shaped to a slope that may be between 10 and 25 percent if the machines are to climb it, and steeper if they reach the top by a haul road. It is desirable that the top of the bank be flat or have only a gentle grade to reduce the danger of tipping while turning. The scrapers are loaded by driving straight down the slope, which should be long enough to give them plenty of space to load. Rippers and pushers should not be needed.

The scraper then hauls its load away to dump into a hopper, build a storage pile, or deliver it to a job. On its return, it is driven up the bank, or a haul road, turned on the top and again loaded coming down. The cycle is illustrated in Fig. 10.22. Semitrailer machines may be backed up the slope if turn space is lacking.

Once the bank is properly sloped, a single scraper may perform all the functions of digging, hauling, and storing without help from other machines or personnel. Such scrapers can also be used for digging in the pit floor, building haul roads, and grading.

Bulldozer.   Bulldozers can load trucks from banks high enough to permit the machine to push into or over the body. For occasional loads, this may be done directly from the bank, as in Fig. 10.23(A) to (D). Considerable material is usually lost in building the bank out to the truck, and repeated loading extends the bank out into the pit, requiring a longer push with each load. The truck may get stuck in the spill.

A retaining wall and platform, as in (E) and (F), will eliminate this difficulty. Many other constructions are used. The platform should have steel strips or rails to keep the blade from digging into the timbers. These should be spaced so that the tracks will not have to walk on them. If they are raised above the wood, they will cause a dirt cushion to be built up which will protect the wood from the grousers.

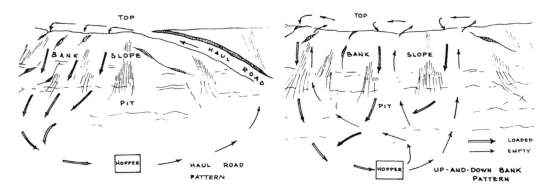

FIGURE 10.22 Loading from bank with scrapers.

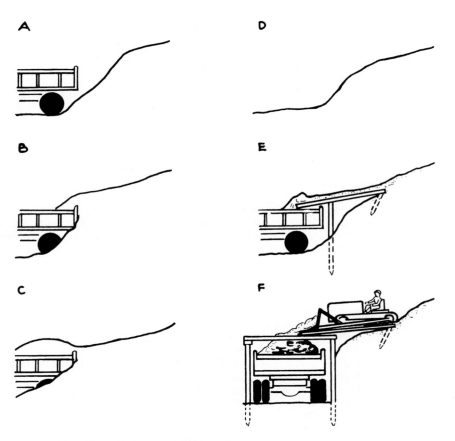

FIGURE 10.23 Loading from bank with a bulldozer.

Dozer push loading is most effective in high banks with a slope steep enough to allow pushing of large loads and to still allow the tractor to back up easily. Much steeper banks can be used in clay and hardpan than in loose sand or gravel.

Spoil can be pushed in the same manner into a hopper and conveyor, which may be a light homemade arrangement such as that in Fig. 10.24(A), or a factory-built, high-capacity belt loader such as that in (B). In either case, when the material within efficient range is exhausted, the hopper should be moved.

It is common practice to postpone moving for much too long. Dozer loading on level or slight grades is inefficient and should be avoided. No part of the push should ever be uphill.

Bulldozers are also used to push bank material within reach of excavators which are stopped by rock outcrops in the toe, and to keep high banks sloped to prevent undermining and caving.

In "glory hole" excavation, which is usually in rock, a tunnel is driven in from the toe and a connecting shaft run to the surface. Rock blasted from the sides of the shaft feeds by gravity to a conveyor, drag scraper, or railcars, which haul it out of the tunnel. A bulldozer is not required until the pit has widened its slopes so that rock will no longer slide.

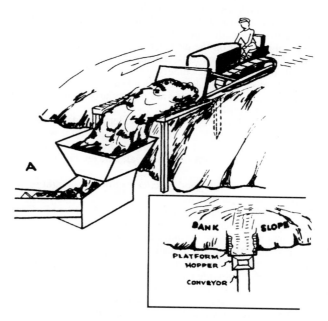

FIGURE 10.24   Loading a hopper with a bulldozer.

Bank Slides.   Most materials will rest temporarily at a steeper slope than their natural angle of repose. Some sand and gravel may stand in vertical or overhanging banks when freshly cut, but eventually fall or slide to slopes between 1 on 1 and 1 on 2.

The danger of undercutting high, noncaving banks is obvious. It is less apparent when a bank caves and slides steadily when dug at the bottom, preserving a fairly uniform slope. Such a formation may gradually become too steep to be stable, without giving any indication of its condition. A change in moisture content, a blast, thunder, or the dropping of a rock may start a slump, which reduces the slope of the face by lowering the crest and advancing the toe into the pit.

The danger from such slides increases rapidly with bank height and steepness. In many cases, workers have been killed and machinery buried in them.

Changes in moisture content affect both internal friction and weight, and either drying out or becoming soaked may create or intensify unbalanced conditions.

Aside from this danger, a high, sliding bank offers the best possible bottom-loading conditions. Because of the constant supply of fresh, loose material, digging is easy, and the excavator has to move forward only at long intervals.

Damp clay will usually stand vertically when cut, but will slump or fall eventually. Vibration from passing machinery or nearby drilling is liable to break down its structure so that it will flow. If the movement starts at the top, a dangerous collapse may be caused.

A face of clay or silt exposed to alternate freezing and thawing or internal water pressure may liquefy and flow out on the pit floor, eventually assuming a gradient as low as 10 percent. This action is usually too slow to be dangerous, but should be allowed for in figuring clearances for haul roads or in parking machinery.

High, steep rock walls should be checked for fissures running parallel to the face, which would allow sections to fall off. These are particularly dangerous if filled with dirt that might absorb water and exert a push.

No high face of any kind should be undercut widely without adequate bracing.

The safest way to dig high, steep banks in general is with a drag scraper. Sometimes a dragline with a very long light boom is used to pull the crest down to the excavators.

On lower slopes, and on firm material, a dozer can be used.

Benching.   It is usually good practice to limit the height of shovel cuts by taking the materials in a series of layers or benches.

Two methods of benching a hill slope are described in Chap. 8. Pits are liable to take much larger areas and require many more benches.

Benching may be done from the top down, or from top and bottom, as in Fig. 10.25. A boundary cut is frequently carried down below the pit floor when the higher parts are exhausted.

A large number of benches may be worked at the same time, or in rotation. Each bench should be large enough to provide ample space for shovel and trucks. If it is accessible at each end, one-way traffic can be maintained and the need for turnaround space avoided. However, narrow roads are often blocked by stalled vehicles, slides, rockfalls, or overbreaks.

If the bench is accessible from only one end, the shovel should work from that end so that the width of the new cut will be available to traffic.

When layers are taken from the top down, starting at a hillcrest, or a back line which will stand steeply, the working area may be made about as wide as desired. The excavating done on the top widens the area available for the next cut.

When cuts are worked up from the bottom, width is largely determined by slope gradient and face height; if the slope is 45°, a 1-foot height is required for each foot of bench width.

Top benching is preferred for steep slopes whenever immediate access can be had to the top.

LEVEL DIGGING

Material is frequently obtained by sinking the pit floor, or a part of it, in thin layers without developing a bank except at the boundaries of the excavation. The material may be piled before loading, in

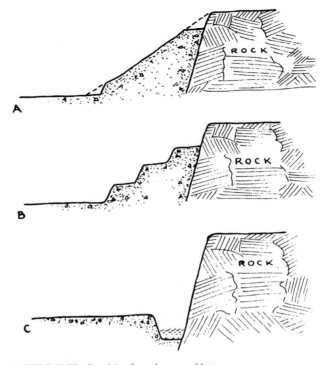

FIGURE 10.25   Benching from the top and bottom.

the same manner as topsoil, but it is usually more convenient to dig directly, with carrying scrapers or cable excavators.

Rooters may be used to loosen the ground for scrapers and for cable excavators except in wet digging.

The wheeled scraper is more flexible in digging, can vary the dumping spot readily, and by change of the number and size of the machines can excavate at almost any rate desired. The machine may also be used in other pit work or outside jobs when the cut is idle.

Under favorable conditions, the cable excavator can dig at lower cost per yard. The fact that it is difficult to move is sometimes in its favor, as it will be there when needed.

SURFACE WATER

Rain.   Rain will usually stop excavating and hauling operations. In addition, it may soften stock-piles and turn pit floors and haul roads into swamps or ponds so that work may not be resumed for days or weeks.

Some pits are in such porous soil that any volume of water will soak away quickly, and neither mud nor standing water will delay work more than a few hours. Others are in such dry climates that it is better to run a small risk of water delay than to spend the necessary thought, time, and money on arranging for drainage. The majority, however, are so situated that at least routine precautions should be taken to keep them usable.

A first principle is to shape pit floors so that they will drain. In cutting into a hill, the floor should slope slightly upward toward the face so that water will flow away from this line of greatest activity.

If this rear drainage is not practical, the slope may be made to the side or to channels or drains in the floor.

If the pit is sunken, drainage can sometimes be arranged into a deeper portion which will take it off the working floor and allow it to soak away gradually. This will often be a pond dug under the water table to supply the plant with water. See Fig. 10.25(C).

If pumping is necessary, it should be done from a sump that will hold a large volume of water, and in which the pump can be protected against being covered if heavy rain falls while the pit is shut down. Such a sump may serve as a storage reservoir for plant supply.

Runoff. A pit may be troubled by water running off surrounding areas, either during rains or in the form of permanent streams, which should be diverted around the working areas or channeled through them in such a way that it will cause minimum interference.

The best practice is to dig diversion ditches to lead the flow in other directions. However, if the pit is expanding, these ditches will require relocating and may cost more than they are worth. This is particularly likely if the ground is steep or rocky so that ditching is difficult.

Also, the water may be needed in the pit for plant or dredge supply.

If the water flows only occasionally, it can be led through the pit in wide shallow channels which can be crossed by machinery and trucks at any point. If it flows often or continuously, it should be in a ditch and taken through haul roads in pipes or on rock-paved fords.

GROUNDWATER

Layout, machinery, and methods in a pit may be affected by the water table, or level of groundwater.

This unseen water surface may be practically level, evenly sloped, or irregular. The water generally appears to be stagnant, but it is almost always in slow motion and will slope down from the source to the outlet. The angle of this slope is the hydraulic gradient, which is determined by the resistance of the material to the passage of water, the pressure and volume of the supply, and the relative heights of inlet and outlet.

Porous materials, such as gravel and sand, have low gradients, and tight ones, such as silt or clay, steep ones.

The water table tends to follow the slope of the land, but at reduced grades so that water will be farther from the surface in hilltops than in valleys.

Above the true water table is the so-called capillary table, which is kept wet by water rising in the spaces between soil particles. The finer spaces cause greater rise in fine-grained soils than in porous ones. Capillary water gives comparatively little trouble in clean sand and gravel, but causes serious softening of other soils.

Capillary water may come up higher when ground is compressed, as under a haul road.

Underground water ranges in quality from fine spring water to solutions of salt or chemicals unfit for any use.

A shallow pit may be kept well above the water table, which is then of little interest except as a source of water for processing. A pit carried sideward into a hill may cut into the table and have drainage problems of varying severity. A downward cut will get into water eventually, except in arid climates, or very tight ground.

Surface water falling into a depressed pit as rain, or flowing into it from adjoining areas, must also be taken into account in coping with the groundwater.

Permanent plants and all-year haul roads should be above the highest water levels.

When it is necessary to use materials lying at or under the water table, they may be obtained by wet digging, digging in dry seasons only, draining, pumping, and combinations of these methods.

Digging underwater. Any machine which can dig from the top of a bank can dig underwater to some extent. However, there are a number of special difficulties, including inability to see the work, weaker penetration because of decreased bucket weight and interference of water currents, loss of material carried out of the bucket by water, and sloppy condition of the spoil. Wet banks are also more liable to cave under a shovel.

Loose underwater material, such as sand, is efficiently dug with a clamshell, as it is securely held in the bowl while being lifted. Draglines, or long-boom hydraulic hoes, have highest production. For large areas, hydraulic dredges are preferred.

If a lake or river borders a pit, or a large enough pond is dug in it, a hydraulic dredge can be trucked in, assembled, and floated. The largest dredges may be able to cut 100 feet underwater, but 40 feet is the maximum depth usually recommended. The spoil is usually pumped through a pipeline directly into a processing plant, but storage piles can be built. It is also possible to put the plant on the same hull as the dredge and discharge processed material into barges or conveyors.

The dredge can also enlarge the pond by undermining banks so that they cave within reach of its suction. High banks that do not slide should be lowered by land machinery to avoid danger of damaging the dredge when they come down.

Unless the pond is very large, or has a large inflow of water, the dredge may depend on a prompt return of water pumped out with the spoil. If the product does not contain many fines, water may be returned directly to the pond through a pipe or sluice. Fines can be filtered by allowing the water to drain back through gravel or sand, or by holding it awhile in a settling basin. In either case a larger water supply may be required than for direct return.

When the dredge has cut the working area to the maximum depth allowed by its ladder, it may be possible to reach further supplies by lowering the pond level. This may be done by partial drainage, by diverting the wastewater from the plant into other drainageways, or by combining diversion with pumping.

If the material is soft enough not to require cutting, suction pipes can be extended below the ladder to the desired depth, but recovery will probably not be complete.

Care should be taken not to locate sunken and wet pits where they will interfere with the orderly development of higher layers. It should be remembered that drainage through a bank may cause it to settle to a very gradual slope, which might run it much farther back into the pit floor than was intended. This type of spreading is particularly apt to occur when the pond is used as a water source, and wastewater returned by being allowed to soak into the ground.

Drainage.   The costs of wet digging are generally higher than those of doing the same work dry. If the ground surface slopes down far enough in the vicinity of the pit, it may be more economical to undertake even a large drainage project than to dig wet or to pump.

Draining is particularly feasible when spoil from an open cut ditch is of the same type as that which is being mined in the pit.

Where possible, provision in the original work should be made to drain to the full depth of the outlet or of the deposit. This may involve a very wide top cut, stable side slopes, and trucking out of spoil. If the spoil cannot be used, trenching, installing of drain pipe, and backfilling may be more economical.

Sometimes it is practical to run a tunnel, or one or more drilled holes, from the low spot into the edge of the pit area where a connecting surface shaft can be sunk. Each level can then be connected in turn to drain into the shaft. Precautions must be taken against the entry of dirt or trash that might block the tunnels.

Sand and gravel will generally drain into either a shaft or trench with little further attention, but other deposits may require trenching of various types. The problems are similar to those involved in digging and draining a large basement, and are discussed in Chaps. 4 and 5.

When a sidehill cut gets into water, a curtain drain may be required at the toe of the bank to avoid damage to the floor from seepage or flowing water.

A high water table may be supported by an impervious layer separating it from well-drained formations, as in Fig. 10.26. Such a perched water table can be drained by drilling or digging through the impervious clay layer below.

Surface ponds on sand or gravel deposits will often drain if the silt layer on the bottom is opened. A few sticks of dynamite exploded on the pond floor, or a shallow dragline cut, may do the work.

Pumping.   When drainage is not practical, water may be removed by pumping. This may be the preferred method if the water can be used in the processing plant, then discharged outside the pit.

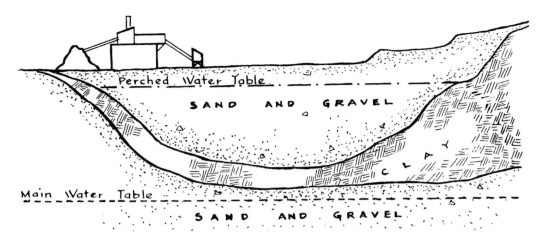

FIGURE 10.26   Perched water table.

Pit pumping follows the practices described in Chaps. 5, 6, and 21. It usually consists of removing open water standing against or over the deposit being dug. A small sump is generally made by digging part of the hole more deeply and placing the suction hose in that.

Success in pumping depends on the relation between the volume of water in the hole and the rate of inflow, and the capacity of the pump or pumps. Costs per gallon are usually smaller, and work can be started or resumed more promptly if the pumps are oversize and can handle many times the volume of inflow, so that a large part of their capacity can be used to lower the open water.

Extensive gravel layers may contain billions of gallons of water over areas of many square miles, which will drain into the sump. If no rain falls, the rate of flow gradually declines as the continued drainage flattens the hydraulic gradient, but this situation requires handling so much water that expenses are usually too high.

If the gravel is of limited depth, much of the inflow might be sealed by cement grout forced down into the gravel at pit boundaries, or where flow channels are suspected. Pumping should be stopped before and during the grouting.

In gravel formations of limited size, and in tight materials such as clay and peat, the original rate of inflow usually declines rapidly, and the underground reservoir may be exhausted so that inflow will stop until it rains.

Each time a wet pit is enlarged by working, the pumping job of reopening it becomes greater because of the increased pond volume.

Pumping should be done in dry seasons when the water is lowest and interruption from rain least likely. In general, it is better practice to pump out, dig a large volume as quickly as possible, and allow to refill than to maintain pumping and a slow digging rate over a long period. This is particularly true in the porous, quick-filling formations.

Both surface and underground water are sometimes removed by pumping out of deep wells drilled in the pit floor or near its boundaries. This method is particularly suited to plant supply.

PIT PLANNING

Pits may be opened casually by digging in a roadside bank, or large sums may be spent on investigations, plans, road building, and site preparation, before work is started. Most of them start with small equipment and output and increase in scale if they prosper.

If a large-scale operation is planned involving plant and other equipment bought specially, the building of haul roads, clearing and stripping, or if the deposit is of such value that every yard must be removed, then the formation should be carefully investigated for extent, quality, accessibility, and water conditions. An option to buy or to develop should be secured before this investigation.

Zoning.   In many areas, zoning regulations absolutely prohibit opening a pit as such. Frequently, however, if it can be shown that the land will be improved by the work, other types of permits can be obtained which will allow limited or complete operations.

Favorable conditions include deep road cuts through hills and the removal of underwater deposits to make a lake. Other projects include taking away ridges that block view or drainage, or leveling of land for residential or industrial development.

It is probable that restriction of mining has been carried too far in many areas. Where acute shortages of sand and gravel exist, there are sometimes millions of yards of these materials made permanently inaccessible by reserving the land for houses.

On the other hand, there is no question but that a pit in a residential area is a dust, noise, and traffic nuisance, and is often an eyesore as well. A pit operator who takes care that no machinery is operated without mufflers, or on Sunday; that calcium chloride or oil is used freely to keep down dust; that the floors and banks are kept trimmed and reasonably neat; and that finished areas are promptly topsoiled and planted, will encounter minimum resistance to expansion or to opening other pits in the vicinity.

Wherever possible, operations should be screened from the public view by leaving or planting trees and shrubs, or by natural or artificial ridges covered with vegetation.

Permits.   In areas which are not zoned against digging, it still may be necessary to get both state and local permits to operate, and to put up bonds to guarantee that the area will be smoothed and planted afterward.

Heavy fines may be imposed for any failure to abide by such regulations, and the operation may be closed down, perhaps permanently.

Investigation.   The only certain way of finding exactly what is underground is to dig it out. However, inspection and mapping of surface indications, study of other pits or holes in the area, and talks with geologists and local old-timers can provide at least an idea of where to look.

Next may come some test digging and/or boring. A tractor-mounted hoe can dig inspection pits 10 or more feet in depth. If the soil does not cave, quite deep holes can be sunk by a clamshell of sufficient bucket weight. But do not go down in them, unless protected by very heavy bracing. Get your information from what came out, and possibly by inspecting the walls with a light and a mirror.

Rotaries and downhole drills grind or pound soil and rock to chips, sand, and dust. Spudders (churns) and some rotaries complicate the problem further by mixing the spoil with water. The resulting mud may contain material eroded from the walls of the hole.

Interpretation of such samples requires knowledge and skill. Augers give a fairly accurate picture of formations which they can penetrate, except that the material is loosened and mixed.

Core drills give excellent and reliable samples of rock, but may not pick up soft or loose material satisfactorily.

Any hole or shaft should eventually indicate the water level. In porous soils this takes a few minutes, in tight ones as much as a month. If a record of the seasonal rise and fall of the water is to be kept, the hole should be lined with perforated casing, unless it is in rock.

Market Analysis.   The next step is to figure the extent and durability of the market for the products to be mined, and the likelihood of competition from similar projects or from small workings with less overhead.

If the total consumption of the products within shipping range is small, investment in a big plant would be inadvisable unless new outlets could be opened by lower prices, superior quality, or building of consumer plants. If the demand is large but is already adequately supplied, the question will be whether the new pit can supply better material or service, or cut prices, or create additional demand. The efficiency of existing pits and the extent of their reserves should be studied.

If the potential market is large, and the supply limited, the question of future competition will depend on the availability of similar material to others, and whether the intended plant will retain its relative efficiency long enough to repay its cost.

The amount of processing required to fit sand or gravel to specifications of highway departments and other wholesale users may be an important factor in costs.

Capital.   A contractor already operating in other lines may not need to make any extra investment in a pit until business is brisk enough to justify it. In general, however, the minimum capital required consists of down payments on machinery and land or digging rights, and money to carry payroll, operating expenses, and installments until there is income to take care of them.

If the pit must build up its market gradually, or if the demand is seasonal so that stockpiles are accumulated throughout the year, to be sold during a short period, substantial capital will be needed to carry through the slack periods.

The necessity of selling on credit, so that short to long periods elapse between loading the material and getting the money, will sometimes tie up more capital than all the other investments combined. This problem is discussed in the next chapter.

The risks of pit operation, as of any business venture, are greatly increased by a lack of surplus funds or borrowing capacity reserved for emergencies.

Selling without Loading.   A pit may be opened or operated without capital by the owner of the land or digging rights, by selling material "in the bank" to customers who will dig it themselves. Such arrangements are usually based on a price per yard, measured either in the bank or in the trucks. Occasionally, a certain portion of the deposit is marked off and sold, or the buyer may be allowed to dig all he or she needs for a certain job, for a lump sum.

A customer taking a substantial amount is usually expected to do her or his own stripping of overburden, keeping this section of the pit orderly and doing any required pushing back of topsoil after completion.

Sales of bank yards involve measuring the ground surface before and after their removal. This is usually done by surveyors, who work out a grid or a series of profiles. The original measurement may be fairly expensive. Later ones are much cheaper if the bench and location marks have not been disturbed. This method is best adapted to large yardages.

It is often necessary to limit a buyer to a small area so that he or she will not wander around picking out pockets of especially good material, making unsightly holes, and leaving substandard remainders.

Working Space.   A pit which sells directly from bank to customer may require only enough working space to back in a truck. However, it is usually desirable to have a flat area in which to park idle machinery, to pile topsoil or other good material not immediately salable, and to place boulders and stumps until they can be disposed of.

If a crushing, screening, or other processing plant is used, space requirements are greatly increased. It is unusual to be able to sell products in the same proportions in which they are produced. There is usually a surplus of one or more grades that must be stored for future sale.

Since it is often easier to ship directly from the plants than to reclaim from storage, piles may grow when demand for their particular item is weak, and remain untouched when demand is good. This may result in steadily increasing storage requirements as the pit is enlarged.

If a pit is started in a small way and preserves a more or less level floor while being dug into a hillside, storage area may increase automatically with requirements. But if a big operation is to be started full scale, level land outside the pit area must be obtained or built up with waste overburden.

The plant is almost always located at or near the original ground level, and hauling products down to any of the cut benches for storage, and back out to sell, would be uneconomical. This type of pit will require storage space outside the digging area, which is best provided at the beginning of the work.

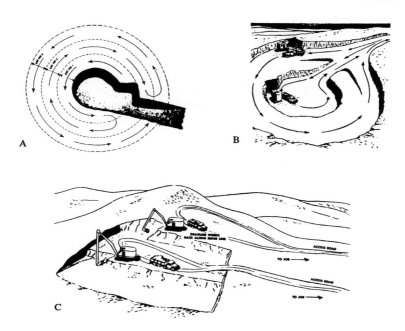

FIGURE 10.27 Pit patterns.

Excavating Patterns. Hill pits may be opened by a straight cut-in or by benching. After reduction to the level of the surrounding land, they are dug as sunken pits.

Subsurface workings, called sunken or dig-down pits, may be opened with front shovels and ramps, or by dragline or backhoe work from the top, in much the same manner as a haul-away basement excavation. The circular pattern shown in Fig. 10.27(*A*) and (*B*) is also widely used, for both subsurface and slopes with gentle gradients.

Backhoes can take gentle slopes in a series of benches, as in (*C*). It is necessary to level a strip for walking, as accurate loading is difficult on a slant.

PROCESSING PLANTS

This heading includes screens, crushers, and washers with their feeding and discharge mechanisms. These units will be described in Chap. 21, and are discussed here only in relation to pit layout and other operations.

Portable Plants. The simplest screening equipment is that described earlier in connection with topsoil. These pickup or skid grizzlies can be used wherever a shovel can work, but require ramps if they are to be used with tractor loaders. Their product is not well graded, as narrow oversize pieces pass through readily.

Mobile plants having screening, usually crushing, and occasionally washing, equipment mounted on one or more wheeled trailers, require from a few minutes to several hours to move up to a bank and start work. Short moves in the pit require less downtime than highway transportation as conveyors and other projecting parts need not be removed or folded in.

One of these units is usually able to eliminate primary hauling or to reduce it to a single truck shuttle, or a short conveyor. For direct loading from a low bank, particularly by a short-range excavator, it may be desirable to keep a tractor constantly on hand so as to move the unit up.

The use of portable units allows almost as much flexibility as in direct-from-the-bank selling. However, highway requirements keep their maximum size and weight below those required for many jobs; and the necessity of packing everything neatly in minimum space causes them to be harder to service than fixed plants of the same capacity.

They can often be used profitably for handling the more distant banks, or for filling special orders, in a pit that has a fixed plant. They may also be used for outside jobs or subsidiary pits. Hauling is a major cost factor in gravel and crushed rock, and the ability to open and process banks near the job may be valuable.

A mobile screener and crusher will reduce the risk involved in opening a new pit. Although not very readily disposed of, they are more or less standardized, and if found to be of the wrong size and type, can be sold or turned in, with a far smaller loss than a fixed plant.

Also, the use of one or more of these units may permit the pit to be developed until adequate space is dug for a fixed plant. If the pit is in a hillside, with steadily increasing bank height, getting the permanent plant well back in it will result in cheaper primary hauling over the whole job.

Fixed Plants. A fixed plant should be the right type for the material to be processed; it must be large enough for the job and within the capital budget. The first consideration is generally the most important, for if business is good, the plant can be expanded although at relatively higher cost; and exceeding a budget may be less damaging than getting the wrong equipment.

Plant manufacturers are ready to supply good engineering advice on every aspect of plant layout. It is a sound plan to get at least general recommendations from two or more companies, and to compare their findings with local practice. Even with these precautions, no person without a good working knowledge of the business should make a heavy investment in machinery from catalogs.

Plant Location. A permanent or semipermanent plant should be as close to the digging as it can be without being in danger from blasting and slides. If the pit is wide, or includes many sections, the plant should be near the center, or the side which produces the greatest yardage.

Dig-down pits may supply the plant by means of ramps and trucks, trucks with cable assists, vertical bucket conveyors, clamshells, or elevators. If supply is vertical, the plant should be located on the pit edge, or as near to it as firm footing can be found. This method is ordinarily used only in rock pits, but not all rock will support a factory on the edge of a cliff.

If the spoil is trucked up, it is best to locate the plant well back from the pit to allow for ramps, storage, and room for expansion.

Wherever a fixed plant is placed, the cost of hauling to it will increase as the digging progresses, whether laterally, downward, or both.

Taking down, moving, and setting up a permanent plant are usually tedious and expensive operations, particularly if it is of a large size. Even if it is of prefabricated, knock-down construction, rust, wear, and patching may make it hard to handle and foundations are generally left behind. Many millwrights consider it best to salvage only the operating units, and to order or build new frames to carry them.

It is usually sound policy to charge the entire cost of such a plant against the material that can be handled at its original location. If the pit area is definitely limited by property boundaries, zoning restrictions, or change of ground, and the depth of the deposit is known, the yardage can be calculated. It is best to make a liberal allowance for occurrence of unexpected masses of unusable material.

HAULING

Pit hauling includes the movement of material from the bank to the plant or to storage, and between the plant and storage in both directions. It also involves delivery from these three locations to the job, although a variable amount of the product may be hauled from the pit in the customers' trucks.

The principal hauling units for pit use are trucks, including semitrailers and full trailers, and conveyor belts. Railroad freight trains, of either narrow or full gauge, are used in big pits, the latter particularly in taking raw material from banks to a distant market. Digging units such as scrapers and cable excavators may also do a substantial amount of hauling.

Conveyor belts and cable excavators and, to a smaller extent, scrapers are largely confined to work inside pits. Trucks are equally adapted to inside hauling and outside delivery. On very long hauls, heavy materials are more economically moved by standard gauge rail.

Conveyor belts may be considered either hauling units or part of the plant itself. They move and elevate material with minimum effort, but are usually difficult to set up and locate. They may be used instead of haul roads and trucks for delivery of a heavy volume of material to a single point many miles away.

Trucks are excellent flexible, general-purpose units. They are available in a wide range of standard sizes and can be adapted to different-size loaders or production schedules by varying the number on the run.

Scrapers, to operate as such, need ground they can dig and hoppers which they can drive across, or storage areas giving them room to maneuver. Banks which they cannot dig can be loaded into them. However, scrapers are more costly and are usually slower than dump trucks of the same size, so it is not good practice to use them steadily under shovels.

A scraper can dump beside a sunken hopper which is kept filled by a dozer.

Truck hauls may be kept short by adding conveyor belts to the plant. The new belt will dump on the receiving end of the previous belt. Such installations may be quite long and are justified whenever considerable yardage will be handled.

Hoppers which are built so that the truck can drive straight across, instead of backing to dump, are more expensive to construct but will allow a faster truck cycle. Such hoppers can also be used for scrapers.

SELECTIVE DIGGING

Selective digging may be done to separate, at the face, two or more materials of value and to remove them; to remove one or more formations, leaving unwanted material; or to dig two or more materials so as to combine them.

Any or all of the spoil from these operations may be hauled away or sidecast.

Layers. If the different formations are in vertical sheets, as in Fig. 10.28(A), any machine which is accurate enough to work the narrowest vein can be used. If they lie horizontally, as in (B), any excavator can move them if they can be cut as separate banks. If they are horizontal, and two or more must be removed at once, the excavator should be able to work from the top down. If divisions run in several directions, and separation must be exact, a Gradall excavator, with assistance from hand labor, could be used.

When horizontal layers are separated by a dragline, as in (C), it should have a boom at least twice as long as the bank is high. The boom angle should be low and the dump cable short to make possible picking up the bucket at a distance.

A clamshell can do the same work with a shorter boom as no allowance need be made for space to drag the load.

Inclined strata fall into any of the above classes. In general, it is bad practice to remove enough of any layer to leave the one above it overhanging.

Selective digging is quite commonly required in stripping overburden, and in gravel and clay pits. The operator may have the responsibility of choosing the section of bank most suitable to plant or customer requirements, and supplying deficiencies by mixing different sections or layers.

Mixing. A good way to mix at the bank is to build a stockpile by dumping several materials on one spot. A conical pile will be built, with each bucket load separating and sliding down the sides. A succession of very thin layers will be made which, upon redigging to load, should mix together quite smoothly.

Such a pile will tend to concentrate round or coarse pieces at the bottom, but these will be remixed in handling by a machine working from the bottom.

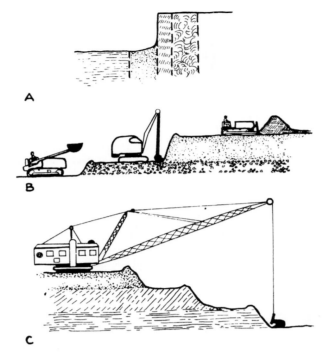

FIGURE 10.28   Selective digging.

Mixing is also done directly at the bank by digging from one formation, dumping on another, and then loading the two together.

If the layers are horizontal, digging in slices from the bottom up will mix them. Scrapers may bring to the top of the bank material to be mixed into it by loading.

Pockets.   When an irregular deposit, with sloped or vertical edges and numerous interruptions, is dug from the bottom, a cluttered pit is left. All parts of it are accessible to digging equipment and trucks, but there may be little room to maneuver and none for storage. If the digging is done from the top, the area may become a badland, with little or no access and probably deficient drainage.

Such areas should be dug out, or smoothed down, at the first opportunity, particularly if any further work is to be done behind or under them. Many pits have strangled themselves out of business while they had ample reserves, because leaving obstacles to orderly digging has forced haphazard development with increasing excavation and hauling costs.

Cutting a pit floor by pursuing a good vein across it is a bad practice that is sometimes difficult to avoid. In general, it should not be done unless its value is sufficient to cover the cost of backfilling. If floor and walls of the cut are smoothed off, it may be filled with material to be stored which can be recovered when the whole floor is lowered. If no further digging is to be done in the floor, the slot can be filled with waste of a type which will not become too soft to carry pit traffic.

Cuts down from a floor, whether pockets or the start of new levels, should be near the outer edges of the pit.

Boulders.   A common problem in pits which are dug without blasting is the occurrence of boulders too large for the loading or processing machinery. These are found in glacial and stream deposits, in disintegrated rock, and near steep slopes.

In pits selling only directly from the bank, there is no convenient way of utilizing boulders or of disposing of them. Blasting will reduce them to a size which can be loaded, but the market for coarse rock is so limited that they may have to be sold as second-grade fill, or wasted. The pit operator will generally prefer to allow them to accumulate along the bank, or will have holes dug to bury them. Occasionally, an abandoned pit is close and deep enough to permit disposal by pushing them over the edge.

If allowed to remain where they fell out of the bank, or pushed into occasional piles, they will present obstacles to orderly development similar to those left by pocket digging. In general, the nuisance value increases with the size of the pieces, relative to the power of the dozer which must handle them.

It is occasionally possible to sell boulders for use in jetties or breakwaters, at a price high enough to justify hiring a machine big enough to load them.

STOCKPILING

Stockpiling is most efficiently done on hard, flat, clear areas. Dumping may be done on the flat, off piles, or from side banks. The location should be convenient to the face, the plant, and the market, the relative importance varying with the use of the material.

Trucks.   If available space is very large compared with the bulk to be stored, trucks may dump piles against each other, as closely as possible, without further grading or heaping. Large trucks make high piles and place maximum yardage in an area.

This method takes a lot of space, forms a bank too low for efficient loading by many machines, and causes maximum danger of mixing the stored material with the floor.

If packing by trucks will not cause damage, such a piled area may be smoothed off by a dozer, and one or more additional layers added. Factors limiting the maximum height are the slope in from the edges and the more gradual grade for the truck ramp, which steadily cut in on the area available at the bottom.

Figure 10.29 illustrates the building of a stockpile by backing trucks up on the dump and building it up in layers. Ramp grade should not be so steep as to strain the trucks or prevent them from dumping cleanly.

At any time the building of the top can be discontinued, and loads dumped off the end. The trucks are then usually driven up forward and turned on the pile.

These two methods can be alternated so that both height and area can be increased to the limits of the space available.

If the material is somewhat too soft or loose to support trucks, the road may be strengthened by the use of wire mats, or small quantities of screenings, soil, or other binders, if their use will not spoil the value of the stockpile. However, it is usually easier and safer to use other piling methods with such material.

Trucks can build a stockpile by dumping off a high bank. This involves a minimum of dozer work. The special problems are loss and contamination because of mixing with the bank.

Reclaiming may be done by a loader at the foot of the pile, or a dragline or backhoe at the top.

Trucks may be kept off a pile, either for safety or to avoid packing, by dumping on a level and piling by machine.

Heaping up may be done by dozer or by loader, on either tracks or wheels. The front loader is more efficient, as it can combine lift with push for higher, steeper piles with shorter moves and less power consumption. In discussions of jobs where either machine can do the work, "dozer" may refer to either of them.

If the stockpile can be laid out as a windrow, and the dumping kept close to it, the operation is almost as efficient as immediate heaping, allows the use of the dozer on other jobs, and makes hauling and piling work largely independent of each other.

The dozer heaps up stockpiles rapidly, is entirely flexible in placing them, or varying their size or shape, and can be used for a variety of other work. However, it must move its entire weight up

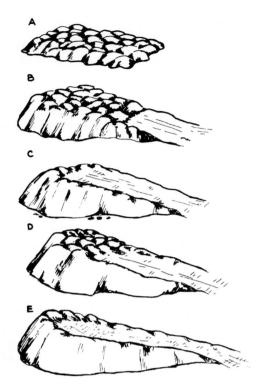

FIGURE 10.29   Truck-and-dozer stockpile.

the pile with each load, has constantly working tracks which may be subject to severe wear in sand or other abrasive piles, and may pack or crush soft materials so as to reduce their value seriously.

Choice of tracks or wheels depends on availability of equipment and the type of material. Wheels provide more compaction and cause less breakage into fines. They wear less in sand or gravel, but become ineffective under slippery conditions.

Loaders of either type can be used to reclaim the pile, loading it into trucks or carrying it to hoppers or to the area of use. If the carry becomes longer than 50 feet, the tires have an increasing advantage in speed and economy.

Clamshell.   The clamshell is commonly used for building stockpiles from shallow dumps, barges, and railcars. A light rehandling bucket, with toothless lips, is used for loose material on a floor.

It can be used at the same time for loading trucks from either the dump or the stockpile. It often alternates hopper loading from the stockpile with the heaping work.

It is flexible in regard to placement and shape of piles, although not as much so as the dozer. It does not crush or pack.

Drag Scraper.   If the storage location is fairly permanent, it may be more economical to heap up the truck dump with a drag scraper than with the trucks and bulldozer. The mast should be high enough to allow for piling the maximum amount to be stored, and the tail tower should be readily moved so that a wide area can be utilized.

The drag scraper may also be fixed to reclaim the material with a reversed bucket, and dump it into a hopper from which it is conveyed into the plant, dump bins, or directly into trucks.

If several classes of material are to be handled, both head and tail towers may be mobile.

A scraper installation does not have the flexibility of dozers and is not available for other work. However, it is very efficient in that only the weight of the bucket needs to be pulled up with the load, it does not pack or crush the pile, and only the bucket is subject to abrasive wear.

Carrying Scraper. The rear-dump scraper builds a pile in about the same manner as a truck. It can operate on a steeper ramp and in much tighter places, particularly if towed by a crawler.

The bottom-dump scraper is most efficient at building a long pile, in the same manner as a highway fill. The sides are built up, starting at the outer edges, as steeply as the material will permit. The entrance ramp at one end and the exit at the other are started at gentle slopes and steepened as necessary. The pile may be started full length, or made short, and extended when additional capacity, or an easier ramp grade, is needed.

A dozer or a motor grader should be available for at least occasional trimming of the surface. If the material tends to get soggy when water-soaked, or is unstable at the edges, compaction with sheepsfoot or rubber-tire rollers might be advisable. Ordinarily, however, the scrapers themselves provide sufficient compaction for stockpiling purposes.

Scrapers may also make shallow piles for rehandling by drag scrapers.

Conveyors. Stackers and other conveyors of the boom type will build high piles with material dumped into a hopper. If these machines are wheel- or track-mounted, they can be towed away from the pile so that it will build into a windrow. Width of pile can be increased or separate piles made by pivoting the boom or the whole machine.

These are most easily fed by a drag scraper or revolving shovel, but a skid-mounted ramp that will allow trucks to dump into the hopper can be made out of heavy timber.

Long conveyors, of either elevating or horizontal types, may be made so that a dumping device can be inserted at any spot desired. This makes possible the building of a windrow the full length of the conveyor, or making a series of separate piles of different grades. Lengthening the dump conveyor will increase the area which can be used.

SEGREGATION

Different-size pieces in a material being piled or distributed have a tendency to separate from each other, so that a disproportionately large amount of coarse pieces will be found in one part of the pile, and finer ones in another. This process is called segregation or separation.

Whenever size gradation is important to the use of the material, the system of handling must be checked to make sure that it either will not cause segregation or will include adequate remixing.

Even if aggregate is to be rescreened before use, as in a mix plant, the screens must be supplied with a proper assortment of sizes.

Dry. In dry materials, a principal cause of separation is sliding and rolling down slopes. When a pile is built from the top, material falls or settles onto the pile, picking up greater or less momentum from the downward movement.

Small particles develop little energy, and tend to come to rest almost immediately. Larger ones have enough momentum to slide or roll down the surface, the biggest tending to reach the ground, while intermediate sizes stop on the slope.

The extent of the separation increases as the range in sizes becomes greater, and when the material has low internal friction. Fine or sticky material tends to pile up to unstable slopes and then fall or slide in masses, minimizing separation. For example, damp sand does not usually separate, dry sand or fine crushed rock separates a little, while coarse-and-fine rock and rounded river gravel may segregate almost completely.

When sliding in chutes, or subjected to random movements or strong vibration, large pieces tend to be wedged upward by small ones working under them, with a layering effect opposite to a top-built pile.

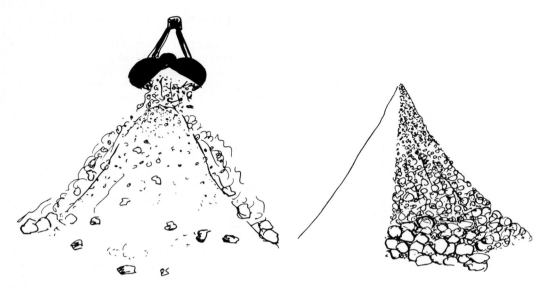

FIGURE 10.30 Segregation in a dry gravel pile.

The segregation caused by building a pile of assorted-size pieces from the top, with a clamshell or a fixed-discharge conveyor, is shown in Fig. 10.30. The pile is always surrounded at ground level by a ring of the coarsest pieces. As the pile expands, its bottom layer is made up of these. Above it is a zone of somewhat smaller particles, with a fairly even gradation to mostly fines at the top.

Such separation is almost never complete, as big pieces will get trapped in the top and small ones get to the bottom, but it is often sufficient to make the pile unusable until recombined.

The clamshell operator can largely avoid layering by estimating pile area in advance, and building it in layers. Each successive layer must be enough smaller than the one below it to prevent sliding over the edges.

Remixing. A dragline working from the top of a pile, or a long-boom front shovel at the bottom, can make long shallow cuts from bottom to top, to get some of each layer in each bucket load. A clamshell should dig at various layers in turn, trying to provide a good average mix in the hopper or hauler.

Layers are more or less mixed when loading is done from the bottom by a loader. Undermining causes the finer upper part to fall and slide, mixing with the bottom.

A hopper gate or other fixed-in-place reclaiming unit under a pile must take what it gets, so it is essential to keep material properly unsorted above it.

Thin Liquids. Some processes, such as hydraulic dredging, move mixtures of water with heavier solids. These behave more or less as true fluids as long as they are in sufficiently rapid motion, but separate into liquid and solids when slowed or stopped.

If speed of flow is marginal, several zones may develop in a pipe or stream. Coarse pieces slide or roll along the bottom, somewhat finer pieces are partly pushed and partly carried, and still finer pieces remain in suspension and function as part of the liquid (Fig. 10.31). The margins of the zones are moved up and down by turbulence.

With decreasing velocity, coarse pieces no longer move and medium sizes slide or roll sluggishly, until only fines are moving with the water. Buildup of the coarser parts is likely to plug the pipe.

FIGURE 10.31   Layers in dredge pipe.

On the other hand, if velocity is increased sufficiently, all pieces will be swept along as part of the fluid, although there is usually a higher percentage both of solids and of their coarse portion toward the bottom.

If such a high-velocity fluid is released from confinement, as at the discharge end of a pipe, it will lose speed and force abruptly. Coarse pieces will be dropped immediately, finer ones carried somewhat farther, and only very fine particles kept in suspension. This behavior is used to advantage in building hydraulic fill dams. See Chap. 14.

In a washing plant, it means that an open-pipe discharge to a stockpile will build a coarse center and top, and finer edges. As it grows, it may tend to have a fine layer at the bottom, and coarser ones toward the top.

However, if the material builds in steep slopes, this effect may be balanced or reversed by the gravity separation described for dry material.

REFUSE DISPOSAL

A pit near a city may yield a much larger profit from being refilled than from the original digging. Disposal of raw garbage, incinerator ash, scrap, and industrial waste is a severe and increasing problem in many areas; and a worked-out pit or pit section may provide an ideal dumping area.

The pit owner may charge the city or the garbage contractor on the basis of each load dumped, by the cubic yard of dumped and settled or compacted waste, on a monthly or yearly service basis, or at a flat price for use of a certain area. Leveling, compacting, covering, and/or burning may be done by either the refuse hauler or the pit owner.

Arrangements for refuse disposal in or near settled areas should include detailed coverage of the methods of operating the dump, as it can easily become a nuisance, and in any case it would be subject to regulation, interference, and possible closing by state or local authorities.

There are a great many ways to dispose of refuse, ranging from open dumping of mixed garbage and scrap in fields, to filling of prepared holes with thorough compaction and immediate coverage with clean fill. If the dump is kept free of litter, and all garbage is promptly compacted and covered by dirt, it is a sanitary fill.

Segregation.   A dump may be expected to handle a wide variety of material, from old automobiles and heavy steel scrap to semiliquids. Both of these extremes must be avoided on the main area of a commercial dump, as they make it difficult or impossible to handle rubbish or garbage efficiently when they are mixed with it.

Burning.   Some or all of the material on a dump is usually burned. This reduces its bulk substantially, and often makes it more manageable by destroying long or awkwardly shaped pieces. Such burning may be done in a series of separate fires on the top, by keeping the forward part of the dump itself continuously burning, or by both methods.

Surface fires of wood, dry paper, tires, and other flammable scrap seldom cause disagreeable smoke or odors. However, fires in mixed materials including garbage may be offensive to workers and drivers on the dump, and to nearby communities. Such fires may be started by a surface fire working down, by hot ashes or spontaneous combustion in the trash, or by deliberate setting. They may go out by themselves, or persist in spite of determined efforts to extinguish them.

Spreading. Refuse is dumped from trucks and spread by a bulldozer or front-end loader, usually a crawler-mounted model. Rubber tires are too apt to be damaged by sharp pieces of metal. Tracks may get badly snarled with wire, cable, or bedsprings, but can be freed by wire or bolt cutters, hacksaws or other tools with little or no damage. A dozer operator on a dump should develop a special knack to avoid track tangling.

The dump may be built out in a single high face, in a series of thin layers, or in compartments. A high face is the least work, makes the worst mess, gets the least compaction, and is likely to cause maximum difficulty with rats and insects.

A dozer crushes and compacts the surface of the trash as it moves over it. A bull clam or a rolled-back loader bucket may be used to flatten it ahead of the machine. Compacting effect usually goes down 1 to 3 feet.

Compaction is desirable to increase the amount of refuse that can be put in the available space, to make a stable fill that will be able to support graded surfaces and buildings, and to reduce the number of rats.

Rats. Garbage dumps usually provide abundant food and shelter for large numbers of rats. They are a major nuisance. They spread disease. They kill or drive away many species of desirable birds and animals in adjoining wild areas. When a dump is closed, thousands of starving rats are likely to desert it and pillage the neighborhood.

Rats may be controlled to a limited extent by poison, inoculation with disease, and encouragement to use them as targets for .22 rifles. But the most effective way to suppress them is to compact the garbage so thoroughly that they cannot tunnel through it, and to cover it with dirt before they can feed on it. The same work eliminates breeding places for flies and mosquitoes.

Litter. Papers and other light trash articles blow around a dump and its neighborhood, producing an untidy appearance that may be one of the chief factors in local hostility to the operation.

Litter is controlled by enclosing the area with a high fence, putting wire or weights over papers as soon as they are dumped, and prompt burning or burial of all trash that might blow. It is often necessary to spend considerable time retrieving pieces by hand that have escaped in spite of one or more of these control methods.

Smell. Objectionable odors around a dump may be caused by decay, burning, or chemicals, or various combinations of these sources.

Decay odors can be almost entirely prevented by prompt burial in sanitary fills. Separate burning of clean trash, and avoiding fires in the dump itself will eliminate burning odors. Dumping of chemicals may have to be controlled or prohibited.

PIT LININGS

A pit that will contain any sort of contaminated material needs to have a lining to prevent that contamination from migrating to the surrounding earth and groundwater. This lining could be an impervious, compacted clay layer, a flexible membrane, or other geosynthetics. The latter are used in the United States to satisfy the Resource Conservation and Recovery Act regulations for liner systems.

Liner Quality. To ensure that no contaminants leak out of the pit, there needs to be well-executed construction quality assurance. That may call for a double geocomposite liner system. The system may have a primary liner and a secondary, high-density, polyethylene liner along with two compacted clay layers. The contractor would begin liner construction by preparing the earth base covered with a nonwoven geotextile. This provides confinement to sandy, base soil, allowing better compaction of the clay layer above it. Then the second liner of polyethylene membrane can be placed, and to ensure tightness, the seams are sewed together instead of merely overlapping the joints. The final compacted clay layer is laid to protect the membranes from cuts or tears that would destroy the containment quality of the pit lining. This liner system may be considered the ultimate lining, and one less complicated might be accepted as sufficient for site conditions.

SANITARY FILL

A sanitary fill is usually a garbage dump in which the rubbish is thoroughly compacted in thin layers and is promptly covered with clean fill, and that is therefore free of persistent bad odors, large rat populations, and severe litter nuisance.

Dug-Cell Method.   The cell method usually depends on obtaining clean fill from the dump area itself. Procedure is probably not exactly the same in any two dumps.

Figure 10.32 illustrates the parallel or double-trench method. First, a wide flat bottom trench is dug, usually by the dozer that spreads the garbage. The soil is piled to the side for use in final grading. Garbage is dumped at one end of the trench, where a ramp is provided to permit trucks to back in. Garbage is spread in one to six layers (two shown), depending on trench depth and thickness of layers. They are compacted by the back-and-forth and turning motions of the dozer.

Another trench is dug alongside the first one, in the waiting time between trucks. Fill obtained from it is pushed or carried across to cover the surface of the garbage, either the top layer only, as shown, or each layer as it is made. Thickness of dirt layers in between garbage lifts may be only 4 to 6 inches, but the final cover should be 18 to 36 inches.

A single trench may be used in progressive fashion, where fill for the day's garbage is obtained from digging another section for the next day. Or the trench may be made by sidecasting, and backfilled from the side to cover rubbish.

Trench cell methods are efficient only in areas that have a good depth of firm dry soil, preferably with a sand or loam texture. Wet conditions would require using a dragline or backhoe for excavation and a pump to keep the trench dry. Mud difficulties would be expected until the bottom

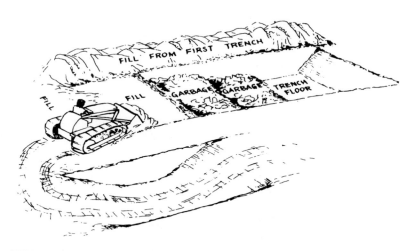

FIGURE 10.32   Sanitary fill, dug-cell type.

layer of garbage had been placed. Pump failure or heavy rains might stop work, and float rubbish out of the area. These conditions involve extra expense and nuisance.

Hard, coarse, or boulder-filled ground, such as makes up the floor of many old pits, will make cell digging impractical.

Ground level can be raised only a few feet by this method, because of the inefficiency of dozers in deep trenching.

Dug cells are therefore poorly suited to land improvement, as they can be made only in land that is fairly good to begin with. A community would use its rubbish to better effect by filling swamps or burying rough, rocky, or stump-filled areas.

Covered Fills. A sanitary dump for improvement of poor land may be made up of one or more layers of garbage covered by imported fill. The bottom layer should be deep enough to keep trucks and spreading equipment safely above water and obstructions. The dozer should make enough spreading passes to compact it thoroughly. The layer should then be sealed off with a 4- to 6-inch soil cover, and another garbage layer started. (See Fig. 10.33.)

Maximum density is obtained if layers are limited to a compacted depth of about 2 feet. High density permits putting more rubbish in the same space, and prevents future settlement. Some types of trash compact satisfactorily in lifts as high as 5 feet, but this is unusual.

A garbage fill can be called sanitary only if clean cover is put on within a few hours of spreading the garbage. Fill trucks or scrapers may haul at the same time as garbage trucks, or stockpiles may be made from time to time within reach of the dozer.

At the end of the day the cover should be extended over the face of the layer, so that no trash is left exposed. Except in the first layer on wet ground, the dozer should walk down the slope of the face while spreading fill, to compact and seal it.

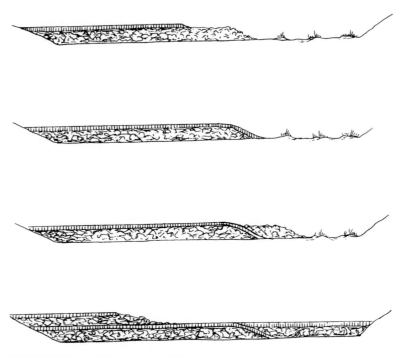

FIGURE 10.33 Building up sanitary fill.

When a first layer must be built in standing water, the operator should try to work large, heavy pieces of trash out into it, to minimize the floating away of light pieces.

As with all fills, the surface should be given a slight grade preferably toward the face, to prevent ponding of water on it. The project should be arranged so as not to interfere with natural drainage either during filling or after completion.

Artificial Hills.   If a community runs out of low places to be filled with garbage, it may decide to build it up into a hill. This idea usually does not create popular enthusiasm. In flat country, the dump may become the most conspicuous feature of the landscape, for example, the more than 50-foot-high sanitary landfill north of Glenview, Illinois.

But as long as it is built carefully by sanitary fill methods, its appearance need never be distressing. And upon completion, it can become a very pleasant park or a golf course, as is planned for the Glenview landfill.

Recreational Park.   A landfill site in Cambridge, Massachusetts, was studied carefully to evaluate site conditions, development possibilities, and potential public health risks. The studies investigated settlement, combustible gas migration and generation, air and groundwater quality, radioactivity, storm water drainage, and revegetation.

After all was evaluated and methods were declared satisfactory, the landfill was made, creating a park including softball and soccer fields, children's play areas, and others, along with 2.5 miles of trails for walking, jogging, and biking. And there are about 20 acres of slopes planted with wildflowers and more than 800 trees. A 2-acre wetland meadow for storm water control and restrooms and parking facilities complete the landfill park.

CHAPTER 11
COSTS AND MANAGEMENT

BOOKKEEPING

Adequate bookkeeping is a basic necessity both for intelligent estimating and profitable operation. Most earthmoving contractors start in business with some knowledge of how to get work done, but with little or no understanding of how to keep track of what they are doing.

Fortunately, it is not necessary for the contractor to keep the books personally. Large organizations have their own bookkeeping departments with full-time employees. A very small operator can hire an accountant or bookkeeper for part-time work, even for one evening per week or month, for a fraction of the money that will be saved. Even if the contractor can do the figuring, he or she will be wise to have a trained person check the books regularly.

A usual procedure for the small contractor is to hire a bookkeeper to make up a system, and to train him or her or an employee in using it. Daily entries and rough work are done by the contractor, and the bookkeeper makes periodic inspections of the records, posts items to the proper accounts, balances the books, and calls attention to mistakes and omissions. The frequency of the bookkeeper's visits will depend on the volume of work, and upon the care and competence with which the contractor keeps the books.

Books should be kept on a double-entry system, in which a record is made of two sides of each situation or transaction. For instance, a sale might result in the receipt of cash; therefore both the receipt and the sale are entered, and the books are "in balance."

The checking account is usually the basis for the books and records. Entries are made on the stubs and/or in a separate book, to correspond with both bank deposits and checks written. The figures are reconciled with the bank statement monthly. In this manner each month is put into balance.

Balance Sheet. The balance sheet shows what the business owns and what it owes. An individual owner of a business that is not incorporated may include the nonbusiness property and debts. It is better practice to keep them separated as much as possible.

A contractor's balance sheet might include the items in Fig. 11.1.

Net worth is the amount left after subtracting total liabilities from total assets. It is listed as a liability in order to balance the two columns, and because it may be said that the business owes this amount to its owner or owners.

Day Book. Every contractor should keep a daily record in a book of what he or she does. It should show jobs worked, labor time, machine time, services provided, and materials used. Definite figures in feet, yards, tons, hours, and/or dollars are best. Such a record is easier to use than a collection of sales and job tickets, that are likely to get mixed up or lost. However, these tickets should be kept also, at least until payment for the work is received.

The day book may also serve as a diary for nonbookkeeping matters, such as important contacts with customers; promises of work and material to customers or from subcontractors or suppliers; important difficulties with weather, footing, breakdowns, or employees. It should record money spent, at least if it is in cash.

<table>
<tr><th colspan="2">ASSETS</th><th colspan="2">LIABILITIES</th></tr>
<tr><td>Cash on hand and in bank</td><td>$ 3,640</td><td>Accounts payable</td><td>$ 4,430</td></tr>
<tr><td>Accounts receivable</td><td>13,830</td><td>Notes payable on equipment</td><td>16,880</td></tr>
<tr><td>Notes receivable</td><td>–</td><td>Notes payable, other</td><td>12,000</td></tr>
<tr><td>Work in progress</td><td>16,400</td><td>Interest, taxes, wages due</td><td>1,360</td></tr>
<tr><td>Depreciation reserve</td><td>11,200</td><td>Advance payments not earned</td><td>1,100</td></tr>
<tr><td>Construction supplies</td><td>1,420</td><td>Mortgage on real estate</td><td>112,000</td></tr>
<tr><td>Machinery</td><td>159,860</td><td></td><td></td></tr>
<tr><td>Land and buildings</td><td>118,700</td><td>TOTAL LIABILITIES</td><td>147,770</td></tr>
<tr><td>Processing plants</td><td>–</td><td>NET WORTH</td><td>177,280</td></tr>
<tr><td>TOTAL ASSETS</td><td>$325,050</td><td></td><td>$325,050</td></tr>
</table>

FIGURE 11.1   Balance sheet.

Such a daily record provides data for settling disputes about work done, payroll, and other matters, keeping track of work in progress and materials used, for obtaining adjustments in insurance rates, and backing up income tax returns.

Other Records.   Contractors, like everyone else in business, must fill out forms for income tax for themselves and for withholding and social security for the employees. The contractor will probably have to keep track of sales and use taxes, fuel taxes, compensation insurance, and perhaps truck mileage.

Other records will depend on the volume and variety of business, and how much he or she believes in paperwork. Records can get too numerous and too detailed, but in the construction field they are usually too few and too carelessly kept.

DEFINITION OF COSTS

It is customary to divide contractors' costs into overhead and operating expenses. Overhead, often miscalled "fixed cost," may be divided into overhead and job overhead.

Overhead.   Overhead is made up of costs which do not vary immediately or directly with volume or type of work. It may include the following items:

Drawing accounts, or living expenses, of owner or partners

Management and supervision—salaries of executives, engineers, superintendents, and foremen

Office rent, payroll, and supplies

Interest paid on loans, or charged against capital investment

Insurance for fire, theft, and liability if paid on the ownership of equipment and premises

Ownership taxes on land, equipment, and other capital assets

Depreciation

Job Overhead.   This heading may include any of the overhead items which are increased to take care of a particular job. When a contractor takes on a big project, the office and supervision force may be enlarged several hundred percent for its duration. This increase, arising from the one job, can justifiably be charged against it.

If job conditions require providing guaranteed pay, meals, rooms, or services to field employees, such expense may be labeled overhead, operating, or job overhead.

Job overhead may also include a proportion of home office overhead.

Operating Costs.   This heading includes

The field payroll of employees hired by the hour or day, or for the job
Payroll taxes
Liability and compensation insurance based on payroll, work, volume, or job conditions
Machinery fuel, lubrication, maintenance, and repair
Machinery rental, delivery, and changing rigs
Expendable supplies

Borderline Costs.   It is often difficult to classify particular expenses, to decide just which account should carry them. As long as the contractor is consistent, he or she can list them very much as desired. However, following accepted practices makes it easier to keep bookkeepers, and to explain matters to banks or bonding companies when it is necessary to do so.

Personal expenses.   The contractor who runs his or her own business should keep books sufficiently to distinguish between business and personal expenses. However, she or he should bear in mind that these come out of the same pocket, and that living costs are part of business overhead to the extent that it is up to the business to provide money to cover them.

It is common practice for owners to draw a fixed amount, and to consider this to be the only personal charge on the business. However, if personal expenses are in excess of the drawings, and the difference results in running up bills, these will ultimately have to be paid by the business, and might better be considered a charge against it from the first.

If personal expenses are not closely accounted for, a one-person business which is profitable in itself may go steadily downhill, without the proprietor's ever understanding why.

RECEIVABLES

Importance.   An important consideration for a contractor or a pit operator is the amount of capital required to carry customers' accounts. In most localities it is difficult or impossible to work on a cash basis. Even when the primary business is selling a commodity in great demand, as gravel in a gravel-scarce area, and operations are started successfully on a cash-for-each-load formula, good customers have a way of working away from it through a series of steps, such as pay after several loads, at the end of the day, at the end of the week, and at the end of the month, to a regular charge account, perhaps tying up thousands of dollars for long periods. Losses on jobs, or difficulty in collecting accounts, may change a well-heeled customer into a slow-paying one.

Credit granted to one makes it triply difficult to refuse it to others.

It takes more backbone, or perhaps uncooperativeness, than is possessed by the average contractor to resist this technique of opening and increasing accounts. Also, it is often true that an enterprise cannot maintain a profitable volume except on credit, particularly if competition is severe.

The contractor who does small and medium-size jobs for a number of different customers has no choice but to extend credit. Insistence on cash in advance or even on payments during work usually means the loss of too many jobs.

Receivables not only tie up a large amount of working capital, but include a probability of bad-debt losses. These can be minimized by good judgment in extending credit and skillful collection methods, but they cannot be entirely avoided.

A bank is usually willing to lend money on receivables. If the account has a good credit rating or local reputation, it may advance the full amount, less a discount which serves for an interest payment, on the understanding that any money received from that customer goes directly to the bank. Or a certain portion of the total amount of receivables may be lent on a regular interest-bearing note.

The cost of such discounts or interest, and an allowance for uncollectible accounts, should be figured into the prices charged for material and services.

Offering discounts directly to the customer for cash or prompt payment is helpful in bringing quick money from good accounts, but is not very effective with those who are really hard up, and who constitute the major problem.

If a business is run partly with owned machinery, and partly with units rented from others, an over-large or doubtful account with a contractor may be tactfully collected by hiring the customer's machinery until he has worked it off. Sometimes an arrangement is made to pay the customer partly in cash to enable him or her to keep up with the payroll, and to apply the balance to the bill.

Liens.   A contractor or subcontractor can usually file a lien for an unpaid bill against the property on which the work was done. Such a lien stays in force for a number of years, and must be paid when the property is sold, mortgaged, or remortgaged.

In most states it is necessary to file a lien quite soon after completion of the work. The contractor or supplier must be aware of the local time limitation, and not allow an unpaid account to run until it is too late.

An old account can sometimes be brought up to date for lien purposes by making one more shipment of material, or performing one more service for which a charge can be made, as it is the date of the last item that determines the last date for filing.

Filing a lien does not prevent a contractor from taking other collection action, such as a lawsuit or attachment of other assets.

Bonds.   Government and government agencies can usually be depended on to pay their bills, although they are sometimes slow and they may dispute amounts. But a subcontractor may have to be careful that the general contractor does not collect without paying out.

In almost all public jobs, and in many large private ones, the general contractor must put up a bond. This usually means that a responsible insurance company guarantees that the contractor will complete the work and pay all suppliers and subcontractors. If the contractor fails to pay any bill incurred on the job, the creditor can collect from the bonding company.

However, claims under bonds must be filed very promptly, often within 90 days of the date of the work, or protection is lost.

Most losses due to late filing of liens and claims under bonds are due to originally friendly relations between the parties, so that collection of the account is not handled in a businesslike way.

Work in Progress.   A contractor may tie up substantial amounts of cash and credit in jobs before he or she is able to even ask for payment. On small jobs he or she may have to wait until the work is finished, on big ones there is usually an arrangement by which the contractor is paid in installments as the work progresses.

Installments may become due on completion and approval of parts or stages of the work, as a building contractor may receive a first payment when the foundation is completed, another when framing is done, and so on. The owner, in turn, may receive installments on the mortgage loan at such times.

In highway and other large heavy construction projects, a number of different stages may be worked at the same time. Rough grading and even clearing may be in progress on one part of the job while fine grading is being completed on another.

For this reason payments are made at regular intervals on the basis of measurements and estimates of the amount of work completed. Five or 10 percent of the amount due is usually held back until the end of the job. Payment may be made 1 to 20 days after the end of the work period, which is usually a month, but may be at shorter or longer intervals.

A schedule involving frequent and prompt payments reduces the contractor's need for working capital. However, the contractor must have money on hand to keep going if a payment is delayed by disagreements or other difficulties.

The extent of such delays is largely dependent on the policies of the owner, and the owner's reputation may enable the contractor to make proper allowance for them in advance.

Cumulative Cost and Income.   The graph in Fig. 11.2 shows a simplified example of the drain on a contractor's resources during an installment-payment contract. The job is assumed to use no

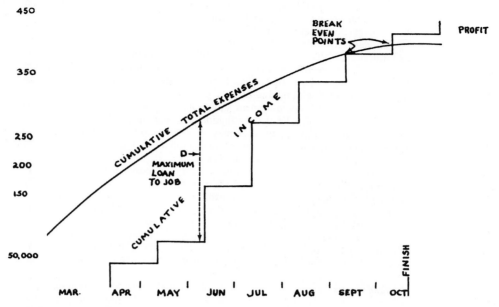

FIGURE 11.2   Cash requirements of job.

more than the contractor's regular equipment, so that none need be purchased specially. Costs are actual expenditures, plus calculated machinery depreciation.

The cost curve shows the approximate amount spent at any time during the job. The stepped line indicates the total received in payments. The vertical distance between these lines first shows the amount "loaned" to the job, and later the profit.

The line "Maximum Loan to Job" is the greatest distance, and indicates the minimum amount of cash and credit that will be required to carry the job under normal conditions. If the May payment were smaller or the June payment were delayed, the line would be longer, indicating a need for more money.

MACHINERY PURCHASE

Purchase of a machine, whether new or used, involves consideration of the type and amount of work in hand and expected, price and availability of suitable models, as well as operator skills, work habits, and personal preference. With the establishment of large national equipment rental companies offering lower rental rates, the alternative of renting construction equipment versus purchasing it should be considered.

Size.   The arguments about machine size can appropriately be restated here. A big excavator is more costly to buy and to move, and requires more working space. It will dig more dirt in a given time, will handle harder and coarser formations, and will show a lower cost per yard if it has space to work and is teamed with other equipment of proper size. It is harder to service and repair because of volume of fuel and lubricants used, and weight of parts. It gets stuck more easily and seriously in soft spots, but seldom hangs up on rough ground.

When space is restricted, ground is soft, or other conditions are unfavorable to the large unit, a small machine may not only work at a lower cost per yard, but may handle a larger volume as well.

Under conditions of equipment shortage, the large unit often has a proportionately higher resale value than the small one. There is a steady trend toward the use of bigger equipment, resulting in reduced labor expense per unit of production.

New or Used?   Some successful contractors buy nothing but new equipment, while others buy only used pieces. Before the Tax Reform Act of 1986 in the United States there was a tax incentive to buy new capital equipment. Now, with the ever-increasing cost for new equipment, more contractors resort to buying used equipment, or rent (short-term) or lease (long-term) new equipment. In general, but not always, a new machine will have less mechanical trouble, and will receive better service from the dealer. It is more costly in purchase price, and in percentage of loss when sold. It has advertising or prestige value. It may be difficult or impossible to secure in the make, size, and model wanted within a reasonable time.

A purchaser of used equipment should have a good knowledge of mechanical condition and current values, and must be alert for liquidations, auctions, and other forced sales where good values can be obtained. Considerable time may be required to find a particular make and model at a good price, and haste may make it necessary to pay too much. On the average, repairs will be more costly and service less satisfactory than on new units.

The expert buyer of used machinery is often able to sell the purchases at a profit, sometimes obtaining considerable work from them first. The average buyer, however, will seldom accomplish this, and is liable to be stuck with worthless machines now and then.

Primary production machines such as excavators and wheel loaders are replaced about every three years with a new generation of equipment that significantly outperforms its predecessor. Firms competing in a market that endeavors to constantly improve quality at low cost can not afford to hang on to key machines through two or three rebuild cycles. If they do not upgrade production units regularly, their competitors using new machines will be able to outproduce them.

Rubber versus Tracks.   Rubber mountings usually provide more mobility and less traction and flotation than tracks. They offer the advantage of working over pavements and hard obstruction without damage, and can move over public roads without the use of trailers. With some exceptions, they are not as maneuverable in close quarters, and they are more readily slowed or put out of action by soft or slippery footing.

Tracks are often better in the cut or at the bank, but tires are superior on the move or the haul.

Big rubber tires are given credit for adding to operator comfort. This would be true at crawler speeds, but fast-moving wheeled equipment on uneven ground can be very rough on operators.

The contractor planning to use such machines must figure on grading equipment to keep haul routes relatively smooth.

If rubber-tire equipment is selected because of its ability to travel on roads without a trailer, the cost of licensing and insurance should be investigated. Some states license heavy equipment for a set fee of a few dollars. Others charge the same rates per pound or per horsepower as for trucks, which, on heavy equipment, may be a large sum.

Technicalities involved in obtaining permits to move overwide machines may be so tedious as to interfere with their use. This is a question not only of the law but of the attitude of local authorities toward its enforcement.

Cost.   A contractor should figure the cost of an intended equipment purchase in two ways—total outlay of cash and credit involved in buying the machine and putting it to work, and the relationship between its cost of ownership and operation and the money it can earn.

The expenditure, particularly the cash down payment, is the most important figure to the contractor with limited capital, but may be merely a factor in considering long-term costs for the large or well-financed operator.

Care should be taken to include in the estimated cost all expenses involved. These may include list price, taxes, delivery to the freight station and then to the job or yard; extra front ends or other units to adapt the machine to different types of work; accessories such as cabs, lights, spare tires, parts, and special tools; repairs or alterations necessary immediately; and allied equipment required to get full use of the machine.

Some of these items are self-explanatory. Immediate repairs are required only on used machines and include such items as replacement of worn tires or tracks, mechanical repairs, engine work, or complete overhaul.

	New	Used
Track loader, complete with standard equipment	$170,000	$75,000
Diesel engine	20,000	–
Spare parts	0	2,000
Sales and use tax, 4.5%	8,550	3,465
Rail freight	1,600	–
Trailer haulage	150	300
Install center pin, overhaul engine	–	12,500
PURCHASE PRICE	200,300	93,265
Turn-in allowance on old loader	30,000	–
Sale price of old track loader	–	20,000
NET PURCHASE COST	170,300	73,265
Finance charges, 18% of $160,000	28,800	–
Finance charges, 10% of $ 65,000	–	6,500
NET COST PLUS FINANCING	$199,100	$79,765

FIGURE 11.3 Example of purchase cost, 2-yard track loader.

Alterations may be changes made to adapt to overloads or special work, or may be necessary to correct mistakes or omissions of the manufacturer. They can include fishplating and other types of reinforcement, building up wearing surfaces with hard steel, and adding safety guards.

Allied equipment might be a trailer to carry the machine, ramps for loading it, and different sizes of excavator or hauler to match its size.

It is also advisable to add up the interest or finance charges that will be incurred in making the purchase. As these are not actually part of the price, they should be added after the original cost figures are determined.

For example, if a contractor decides to replace an old loader with either a new or a used 2-yard machine, he or she might study the comparative costs by setting up figures such as those in Fig. 11.3.

DEPRECIATION

When contractors buy a piece of equipment at a fair price, they do not "spend" the amount they pay. They invest it. They exchange their money for something of equal value.

But the value of machinery starts to decrease as soon as it is delivered, because of use, wear, weathering, and passage of time. This decline in value represents the true spending of the invested money, and it is entered in the books and deducted from taxable income as an expense—depreciation.

It would be expensive and unsatisfactory to have a machine appraised every year to determine how much it had depreciated. It is also necessary to estimate in advance the rate of depreciation, as it is an important factor in establishing the price to be charged for the machine's work.

Depreciation, also known as accelerated cost recovery, is therefore calculated in advance according to various formulas. Each of them provides a basis for balance sheets, profit and loss statements, and income tax. Annual depreciation is converted to an hourly figure for estimates and cost records.

More simply, hourly depreciation is the cost of a machine divided by the number of hours that it is expected to work.

Useful Life.   Depreciation schedules must be based on the number of years the equipment is expected to be in service. Its useful years depend on the type of equipment, the class of work it will do, how much of that work it does, the care it receives, industry standards, and income tax decisions. At best, the time selected represents only an informed guess.

Bulletin F was the Internal Revenue Service's guideline to equipment life for many years. Extracts are shown in Fig. 11.4. It has been retired as a guide, but may still be used as a reference and basis for discussion.

Backfillers, power, light	3	Hoists, air, electric, or steam	8
Medium or tractor	5	Gas	6
Heavy	6	Hand power	8
Breakers, pavement, pneumatic	3	Jacks, hydraulic	8
Buckets, cableway, clamshell,		Pile drivers, barge	8
Orange peel, scraper, and dragline	6	Steam, on skids	10
Concrete, elevator, and bail	5	Track	12
Bulldozers, tractor	4	Pile hammers, steam or air, heavy	10
Gradebuilders	8	Light	4
Compressors, gasoline, portable	6	Medium	5
Motor-truck unit	5	Pipe lines and fittings for floating	
Conveyors, belt, elevating, portable	3	dredges	10
Stationary	6	Pit and quarry plants	6
Bucket	6	Pumping units, gas or electric	
Cranes, crawler, gas, 2–1/2, 5 tons	5	Centrifugal or diaphragm	6
10, 15 tons	7	Highway contractor's	4
20 tons and over	12	Rollers, road, gas	10
Dragline	10	Scarifiers, attachments	4
Universal, gas, on 10 ton truck	6	Scrapers, blade, carryall	6
Crushers, rock, portable	8	Rotary	4
Stationary	10	Wheel	5
Drag lines, gasoline,		Shovels, electric or gasoline,	
1/2, 3/4 cubic yard	6	crawler or wheel, 1/2, 3/4 yard	5
1, 1–1/4, and 1–1/2 cubic yards	9	1 to 1–1/2 cubic yards	6
2 cubic yards and over	12	2 cubic yards and over	8
Dredges, clamshell	16	Tanks, gasoline, storage	6
Dipper	8	Water or air, storage, steel	10
Hydraulic	20	Tarpaulins and tents	3
Drill points, well	5	Tongs, chain	4
Drills, airdrifter	3	Tower excavators	12
Jackhammer	3	Towers, steel	6
Tunnel carriage	5	Tractors, gas or steam, 3–ton	4
Well	10	5–ton	6
Engines, gas	10	10–ton	8
Excavators, cableway complete	4	20–ton	10
Trench, gasoline, 7 to 12 foot depth	6	Trailers, dump, steel	10
18 foot depth	8	Washers, gravel	3
Wheel or ladder type	5	Wagons, dump, steel	6
Graders, blade, road, 7, 8 foot blade	4	Welding outfits, acetylene or	
9, 10 foot blade	5	electric	10
Over 10 foot blade	8	Winches, electric and	
Rooters, wheel	5	pneumatic	10

FIGURE 11.4   Depreciation periods from Bulletin F.

Class· Property	Recovery Period, years
General purpose trucks	5
Agricultural machinery and equipment	7
Mining equipment	7
Contract construction other than marine	7
Marine equipment	10

FIGURE 11.5 Guideline for depreciation periods.

There is now a class life setup, with 5, 7, or 10 years as life periods for the contractor, shown in Fig. 11.5.

Schedules calling for more increased early depreciation than is allowed under the declining-balance method are likely to be disapproved.

Used equipment may be given the same depreciation period as if it were new, or any shorter period that appears to be reasonable.

Fast Writeoff.   It is considered to be good practice to depreciate equipment at the fastest possible rate. This is called fast writeoff. It permits charging the largest proportion of costs against the machine when it is new and best able to carry the burden, and when it is doing the specific work for which it was bought.

Fast writeoff also keeps the book value of equipment down near its real value.

The most important advantage of fast writeoff is related to income tax. The faster the depreciation, the greater the deduction that can be made *now,* and the less to be left for an uncertain future. However, this expected advantage might turn out badly, as a contractor may waste the heavy depreciation on unprofitable years, and not have the deductions in later profitable periods.

Capital Gain.   If a machine is sold for more than its depreciated value, the profit is a capital gain that is generally taxed at less than the rate of ordinary income, while depreciation is deductible at the full rate. A substantial tax saving may therefore result from a fast writeoff that overdepreciates equipment. The U.S. Internal Revenue Code now allows an equipment exchange provision that saves money in the turnover of a piece of equipment for similar replacement.

Salvage Value.   This is the value of a machine after it is fully depreciated. It may be the actual sale price, or the value it might be assumed to have to the contractor when it is theoretically overage and worn out.

Salvage value varies greatly with the type of equipment, its condition, its scarcity, and the local prosperity of the construction industry. Sometimes it is only a few dollars per ton for scrap, at other times, but more rarely, as much as 60 percent of new cost. It is usually estimated at somewhere between 5 and 20 percent of purchase price.

The declining-balance depreciation method leaves a small salvage value automatically. With other methods, any estimated salvage is subtracted from purchase price before figuring depreciation.

Amounts allowed for salvage can be adjusted to simplify arithmetic. For example, if a machine with a 5-year life cost $16,346.93 and might be expected to bring $1,000 to $1,500 salvage, the salvage value could be taken as $1,346.93, leaving an even $15,000 to be depreciated.

Internal Revenue sometimes insists on deducting salvage value before figuring depreciation, and sometimes does not. It is better to depreciate the full purchase price when possible. This simplifies bookkeeping and makes an allowance for a probable increase in replacement cost, a matter that will be discussed below under Price Increases.

DEPRECIATION SCHEDULES

There are two official tax methods of computing depreciation for periods of 3 years or longer. These are known as the straight-line and the declining-balance methods. For short periods of less than 5 years only straight-line is used.

Information in regard to taxes may become obsolete while it is being printed. It should always be checked before use.

Straight Line.   Straight-line is the simplest method, gives a uniform basis for figuring machine costs, and avoids complications in reserve for depreciation. The only thing it lacks is fast writeoff.

The cost of the machine, less any salvage value, is divided by the number of years it is expected to be useful. The resulting figure is the annual depreciation. It is the same amount each year.

Declining Balance.   This method is based on the total cost of the machine. The maximum depreciation rate is twice that allowed by the straight-line method, but it is applied only to the value at the beginning of the year, which is the original cost less all depreciation that has been deducted.

For example, a $20,000 piece of equipment with a 5-year useful life would depreciate 20 percent or $4,000 each year under the straight-line method. With declining-balance, depreciation the first year would be 40 percent of $20,000, or $8,000, the second year 40 percent of $12,000, or $4,800, the third year 40 percent of $7,200, or $2,880. At the end of the fifth year a salvage value of $1,555.20 would remain.

If the machine's life expectancy were 8 years, the depreciation each year would be 25 percent of the value at the beginning of the year.

Sum-of-the-Years Digits.   This method, formerly recognized by the Internal Revenue Service, is based on cost less estimated salvage value. The number of years of useful life is taken as the first figure in a descending series, which for a 5-year period would be 5, 4, 3, 2, 1, and for 8 years 8, 7, 6, 5, 4, 3, 2, 1. The series is added together, giving 15 for the 5-year period or 36 for 8 years.

A fraction is made by placing the number of years of life from the beginning of the year over the total obtained by adding all the numbers in the series together. This is multiplied by the cost to give the depreciation for the year.

On a $20,000 machine with a 5-year life, first-year depreciation will be $5/15$ (or $1/3$) times $20,000, or $6,666.67. In the second year, machine life from the beginning of the year is 4 years, so the fraction is $4/15$, or $5,333.33.

The whole series of deductions would be $5/15$, $4/15$, $3/15$, $2/15$, and $1/15$, totaling $15/15$, or the entire cost.

On an 8-year basis the first-year depreciation would be $8/36$ (or $2/9$) of the cost, or $4,444.44. The next year would be $7/36$, and so forth.

Choice of Method.   Declining-balance and sum-of-the-years-digits formulas are designed to put more of the depreciation at the beginning of the life. They provide the fast writeoff that is liked by industry, and they conform most accurately to the actual loss of value of equipment in normal markets.

However, they cause problems in converting to an hourly basis for use in figuring job costs. Using a different rate each year would be difficult. If there were several machines of the same model but different years, the attempt to charge different prices for them would be confusing to the bookkeeper and aggravating to the customers, and would make accurate pricing of a job almost impossible.

The contractor should use straight-line depreciation in figuring hourly costs, regardless of the method used for income tax and annual reports.

Figure 11.6 shows graph lines for these three types of depreciation, and Fig. 11.7 gives annual depreciation figures per $1,000 of cost. This table can be applied for calculations on any price of

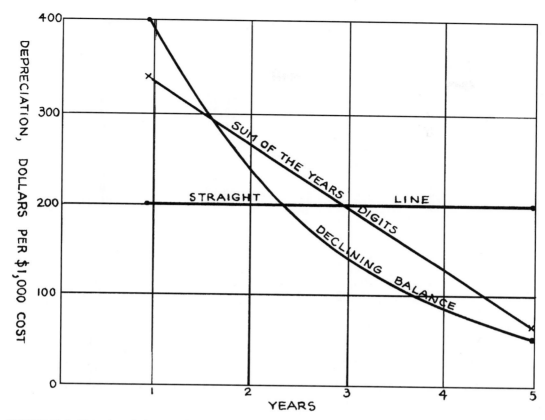

FIGURE 11.6 Three depreciation methods.

machine, by multiplying by its cost divided by 1,000. That is, for a machine costing $15,500, the table figures are multiplied by 15½.

Hours of Use. A contractor may elect to depreciate machinery on an hourly-use basis, without regard to calendar time. The contractor may buy a bulldozer for $25,000 and expect to use it 5,000 hours. He or she will charge $5.00 per hour against its jobs, and at the end of the year depreciate it by $5.00 times the number of hours it worked. If it were busy 600 hours, depreciation would be $3,000; if working time were 1,400, the year's depreciation would be $7,000.

Units of Work. A machine's production may be used as a basis for depreciation, if it can be measured accurately. A mine may buy a 5-yard shovel for $900,000 in the expectation that it will work 20,000 hours and load 8,000,000 tons of rock and ore before it is scrapped. If business is good, it may work 6,000 hours per year; if it is very poor, the machine might be entirely idle.

Under such conditions annual depreciation would not be appropriate. Instead, the $900,000 value of the shovel might be divided by the 8,000,000 tons it is expected to handle, giving a depreciation figure of $0.1125 (11¼¢) a ton. Each year it is depreciated on the basis of the number of tons loaded.

Tires. It is common practice to deduct the value of tires from the purchase price of equipment, charging them as operating expense and depreciating the balance as a capital investment.

Depreciation Schedule

Term of Years	Year	Straight Line	Declining Balance	Sum-of-the-Years-Digits
2	1	$500.00	–	–
	2	500.00	–	–
3	1	333.33	$666.67	$500.00
	2	333.34	222.22	333.33
	3	333.33	74.00	166.67
4	1	250.00	500.00	400.00
	2	250.00	250.00	300.00
	3	250.00	125.00	200.00
	4	250.00	62.50	100.00
5	1	200.00	400.00	333.33
	2	200.00	240.00	266.67
	3	200.00	144.00	200.00
	4	200.00	86.40	133.33
	5	200.00	51.84	66.67
6	1	166.67	333.33	285.71
	2	166.67	222.22	238.09
	3	166.67	148.15	190.45
	4	166.67	98.77	142.83
	5	166.67	65.84	95.21
	6	166.67	43.90	47.59
8	1	125.00	250.00	222.22
	2	125.00	187.50	194.44
	3	125.00	140.62	166.66
	4	125.00	105.47	138.88
	5	125.00	79.10	111.10
	6	125.00	59.33	83.32
	7	125.00	44.50	55.54
	8	125.00	33.37	27.66
10	1	100.00	200.00	181.80
	2	100.00	160.00	163.62
	3	100.00	128.00	145.44
	4	100.00	102.40	127.56
	5	100.00	81.92	109.08
	6	100.00	65.54	90.90
	7	100.00	52.43	72.72
	8	100.00	41.94	54.54
	9	100.00	33.55	36.36
	10	100.00	26.84	18.18

FIGURE 11.7 Annual depreciation on $1,000 cost.

The advantages and disadvantages of this approach will be considered later. It is subject to disapproval by the Internal Revenue Service unless the contractor can prove from records that such tires usually last 1 year or less.

Repairs as Capital. Repairs are considered an operating expense as long as they do not add greatly to the value of a machine. But major overhauls, particularly if done late in the depreciation years, may be considered to be a capital expense that must be depreciated over several years.

For example, if a contractor spends $4,000 rebuilding a $10,000 machine in the last year of its depreciation schedule, he or she may have to list the expense as a capital investment, and set up a new schedule for it.

Fully Depreciated Equipment. If a machine is kept beyond the end of its depreciation period, no further depreciation is charged against it. However, the hourly price for it should remain the same. The part of its earnings that formerly paid for depreciation becomes profit, one of the "hidden profits" that help to keep contractors in business.

However, this extra profit may easily turn into a loss because of high repair costs and too much downtime. It is not good business to run old machines unless they are in good condition.

Short-Term Use. Many contractors buy machines for particular jobs, and sell them as soon as they finish. Others have a policy of turning in equipment after a certain amount of use, to reduce maintenance and job delays and to have the prestige value of up-to-date machines. Cost estimates are then based on the difference between purchase price and estimated sales price.

For example, a contractor may buy a fleet of $240,000 scrapers for use in two seasons of about 1,200 hours each, after which they will sell for one-third of their cost. Each of them will depreciate $160,000, and that cost per hour will be $33.33. This would compare with no-salvage depreciation of $24.00 for 10,000 hours or $48.00 per hour for 5,000 hours.

When periods of use are to be very short, it may be cheaper and/or less risky to rent equipment. This will be discussed later in the chapter.

DEPRECIATION RESERVE

The contractor who intends to stay in business should set up a depreciation reserve, in which funds can accumulate to replace equipment as it wears out or becomes obsolete. This reserve may be a separate bank account, a fund maintained inside the regular account, or perhaps only a page in the ledger.

As depreciation is charged against a machine and deducted from income, it should be paid into the reserve. If emergencies prevent saving the actual cash, the amount should at least be entered as a liability so that it will not be forgotten.

Need for Reserve. Machines wear out and must be replaced. Money is needed for the replacement, and it should be provided by the machines as they work. Otherwise the capital invested in them is consumed and destroyed.

If whatever money it made has been eaten up, the contractor may not even have the down payment on new equipment. Without adequate books the contractor will find it hard to understand why he or she should finish a number of busy and apparently successful years without money to replace machinery.

Inventory and Reserve. When a machine is purchased, its value is listed as an asset under Equipment Inventory or some such heading. The depreciation is deducted from this each year, and added to the Depreciation Reserve. On a $16,000 machine with a 5-year life, the straight-line method would work out as in Fig. 11.8.

Inadequate Depreciation. Most manufacturers recommend that construction equipment, aside from big loaders and special units, be depreciated on the basis of 10,000 hours of use in 5 years. But as we will see later, most contractors are doing all right if they work 1,000 hours per year.

	Machinery Inventory	Depreciation	Depreciation Reserve
Purchase date	$16,000	$ –	$ –
End of first year	12,800	3,200	3,200
End of second year	9,600	3,200	6,400
End of third year	6,400	3,200	9,600
End of fourth year	3,200	3,200	12,800
End of fifth year	–	3,200	16,000
	$48,000	$16,000	$48,000

FIGURE 11.8   Inventory depreciating into reserve, 5-year basis.

On the recommended basis a $100,000 bulldozer would depreciate $20,000 per year and $10.00 per hour, and its price on jobs would be set accordingly. But at the end of a year, it might have worked only 1,100 hours because of weather, job delays, and repairs.

The jobs would owe the depreciation reserve $20,000 for the year, but the machine would have earned only $11,000 for this purpose. There would be a deficit in the reserve of $9,000, which would have to come out of profits, or if there were none, out of other funds.

If this machine use had been more realistically figured at 1,000 hours per year for 5 years, depreciation would be $20,000 per year and $20.00 per hour. In 1,100 hours of work in 1 year, the machine would have been able to provide the full $20,000 for the reserve, plus a $2,000 surplus for profit. The extra $10.00 an hour might make jobs harder to get, but if that is the actual depreciation, the contractor must charge accordingly or lose money.

If experience shows that the machine will last 10 years at the 1,000 hours per year rate, its schedule could be set up for 10,000 hours in 10 years, and $10.00 per hour. On this basis the 1,100 hours of work would pay the $10,000 depreciation charge, with $1,000 left over. But if the machine had to be scrapped at the end of 5 years, the reserve fund would be short $50,000.

It may seem to the reader that this is just a matter of juggling figures. But the figures are very real, and understanding them and arranging them properly may mean the difference between prosperity and bankruptcy.

A contractor who can really use a machine for 10,000 hours is justified in basing estimates on long use. But if he or she is getting only 3,000, 5,000, or 6,000 hours out of the equipment now, basing costs on longer use is foolish and dangerous.

Price Increases.   Contractors share with all other users of modern machinery the problem of price increases. The price of any equipment is likely to increase during its life, so that the replacement cost is more expensive than the original cost. This arises both from a general rise in prices, and from improvements in equipment that add to its cost.

A properly kept depreciation account for a single machine will seldom contain enough money to buy a new one if the original unit is scrapped at the end of its calculated life. Other funds or loans have to be added to buy a replacement.

This difficulty may be partly or wholly overcome by figuring that the machine has no salvage value. Since it almost always has some salvage value, even if only for junk steel at $10.00 per ton, and is often worth a substantial amount if in good condition, its value plus the depreciation reserve may provide fully for a replacement of the same type and size.

Another hedge against price inflation is to increase the depreciation charge against the machine whenever its replacement price is increased, so that it is the same as for a new model.

For example, if a $20,000 machine were depreciated on a 5,000-hour basis, the hourly depreciation would be $4.00. If after 2 years the price of similar machines were increased by the manufacturer to $23,000, the depreciation charge against the old machine would be increased to $4.60. This is for internal bookkeeping only, and cannot be used on income tax returns.

This has the advantage of partially providing for replacement at an increased price, and of keeping prices uniform with new units that might be added.

Advance in prices to cover rise in replacement costs is particularly important for firms that obtain a substantial part of their income from renting out machinery.

Improvements may be made in equipment so that a new model is not strictly comparable to one 2 or 3 years old. However, the only point of importance in regard to upgrading the old model in price per hour is whether the changes produce an important increase in production. They often do not.

INVESTMENT

There are at least three different ways to consider an investment in equipment. They are: initial, total, and average annual investment.

Initial Investment.   This is the net cost—the total of cash paid and debts incurred to buy a machine. It causes a shift in the balance sheet, adding to machinery inventory by reducing cash and/or increasing liabilities. This could be called the market value method for covering the equipment cost.

At the beginning of each year the value is estimated and covered by the estimated hours the piece of equipment will be used that year. This method results in a higher charge in the earlier years but lower in the years nearer the end of the machine's life

Total Investment.   It is customary, although not particularly reasonable, to charge equipment with the interest on any cash invested in it. Also, property taxes and loss insurance are paid on the basis of machine value. This method could be called the amortization method which uses a discounted cash flow annuity calculation. It is the method banks and finance houses use. It will likely result in a charge slightly higher than with the average annual investment method.

Since a certain part of the initial investment is amortized—that is, paid off in depreciation charges—each year, the investment on which such charges are figured is reduced in a series of steps. For the first year it will be the purchase price, the second year purchase price less one year's depreciation, and so on.

Such charges are most easily worked out for the whole life of a machine by using the total investment. This is found by adding the machinery inventory value for every year of the unit's life. In Fig. 11.8 the total of the first column, $48,000, is the total investment for that machine.

Total investment (TI) may also be found by adding 1 to the depreciation period in years (DP,yrs), and multiplying by one-half the original cost. Stated as a formula, this is

$$TI = (1 + DP,yrs) \times cost/2$$

For a $16,000 machine used for 5 years,

$$TI = 6 \times \frac{\$16,000}{2}$$

$$= 6 \times \$8,000 = \$48,000$$

If interest were to be charged against the unit at 6 percent, the total interest for its life would be 6 percent of $48,000, or $2,880.

Average Annual Investment (AAI).   This is a more realistic figure that averages the purchase cost over the years of machine life. It can be found by dividing the total investment by the number of years.

Average annual investment may also be obtained by multiplying the cost by the years of depreciation (DP,yrs) plus 1, and dividing by twice the years of depreciation. Stated as a formula,

$$AAI = \frac{cost \times (DP,yrs\ plus\ 1)}{2 \times DP,yrs}$$

Figure 11.9 gives the average annual investment for $1,000 purchase cost for the most used depreciation periods up to 25 years. To use this, multiply the figure appearing after the number of years of life by the number of thousands of dollars in machine cost.

Useful Life, Years	Average Annual Investment	Useful Life, Years	Average Annual Investment
1	$1,000.00	8	$ 562.50
2	750.00	9	555.55
3	666.67	10	550.00
4	625.00	12	541.67
5	600.00	15	533.33
6	583.33	20	525.00
7	571.42	25	520.00

FIGURE 11.9   Average annual investment for $1,000 cost.

INTEREST

Rates.   Interest rates vary greatly with different types of loans, with the risk and bookkeeping involved for the lender, and with the general level of interest rates at the time the loan is made. They can be very confusing.

If a person borrows $100 and pays it back at the end of a year, plus $6.00 interest, the interest rate is 6 percent. If he or she keeps the money for 2 years and pays only $6.00 interest at the end of that time, the rate is 3 percent.

If the $100 is borrowed on a discount basis with the interest paid at the beginning, the borrower will receive $94 when the loan is made, and pay back $100 at the end of the year. Here the real rate is 6.38 percent (100/94 − 1).

On an installment loan with an advertised rate of 6 percent on the unpaid balance, the borrower will receive $100 and pay $106 in 12 equal monthly payments. The real interest rate is about 11 percent, as the average indebtedness for the year is only $54.16.

True interest rates can be found by dividing the amount borrowed into the interest paid, and multiplying by a fraction made of 1 over the time in years, or 12 over the time in months.

If $5.00 in interest is paid on $100 borrowed for 2 years, we have

$$\text{Rate} = \frac{5 \text{ (interest)}}{100 \text{ (loan)}} \times \frac{1}{2 \text{ years}}$$

$$= \tfrac{1}{40} = 2\tfrac{1}{2}\%$$

If the term of the loan were 4 months, then

$$\text{Rate} = \frac{5 \text{ (interest)}}{100 \text{ (loan)}} \times \frac{12}{4 \text{ months}}$$

$$= \tfrac{3}{20} = 15\%$$

Interest on Equipment.   As mentioned before, interest should be charged against a machine even if it is bought for cash. If the purchase is financed at a rate of more than 6 percent, the higher rate is charged, while if interest is less than 6 percent or if none is paid, the charge is kept at 6 percent. Interest is figured on a yearly basis on the average annual investment.

This custom does not conform to good accounting practice, and is in conflict with methods of treating money tied up in other ways.

Equipment Debt.   There is good reason for charging interest on equipment purchase loans against the equipment, although a good case could also be made for charging it to general overhead.

Carrying it as an equipment expense has the advantage of simplicity and of automatically identifying the source of the charge.

General Debt.   A second case is to charge bought-for-cash equipment with the same interest rate being paid on open loans for general purposes. It should be considered that money borrowed for general business use is a general overhead item, and responsibility for it is shared by field, shop, and office equipment, materials on hand, cash in the checking account, accounts receivable, and work in process.

The value of owned equipment is usually much greater than the amount of the general debt. If a machinery inventory of $100,000 were charged with 6 percent interest because of a debt of $20,000, it would be paying $4,800 more than the cost of the interest.

If this equipment were charged with the actual interest, the $1,200 paid would be applied to the whole $100,000, so that interest would be at the rate of 1.2 percent. If the debt were carried by all the money-consuming items mentioned above, which might easily total $150,000, the effective interest rate would be only .8 percent.

No Debt.   A third situation is to assign an interest charge of 6 percent to equipment, even though its owner pays no interest on any kind of debt. The idea is that the contractor could invest money elsewhere if he or she did not buy equipment. But the only investments that a contractor can make and still keep the money available in the business are savings accounts and short-term bonds, that might pay 5 percent or less.

It is of course both possible and likely that funds invested outside the business would fail to return 6 percent interest, and might be partly or wholly lost.

Equipment is not a bond or mortgage that justifies itself by paying interest. It pays its way by production. To saddle it with interest charges is to ask it to produce a profit before it goes to work. In this highly competitive business, such an arbitrary increase in its cost basis may make the difference between getting a job and losing it.

However, since assuming of interest charges is now a widespread practice in the industry, it will be taken into account in some of the cost calculations that follow.

Installment Interest.   The interest rate on an installment contract is found by dividing the total debt into the total interest.

Total debt (TD) is a figure that is similar to total investment. It is found by adding 1 to the number of monthly installments (1), multiplying by the loan, that is, the amount borrowed before interest, and dividing the result by 24. The formula is

$$TD = (1 + 1) \times \frac{\text{loan}}{24}$$

The amount of interest is found by adding all the installments together, or multiplying their number by their amount, and subtracting the amount of the loan.

For example, $12,000 of a $16,000 purchase is financed in 36 monthly notes, 35 for $393.33 each and a final one of $393.45, totaling $14,160. Subtracting $12,000, we find that the interest totals $2,160. By the formula above, the total debt is $18,500. Dividing $18,500 into $2,160, we have .117, or an interest rate of 11.7 percent.

Some contractors want to know the interest rate of finance charges on the whole purchase price. If the above machine had 5-year depreciation, its total investment would be $48,000. This would be divided into the $2,160 interest, showing a rate of 4.5 percent on the machine.

If the machine is to be charged with 6 percent interest on the nonfinanced part of the investment, the total debt is subtracted from the total investment, and the remainder multiplied by .06. In this example, we would have $48,000 minus $18,500, or $29,500, multiplied by .06 to give an interest charge of $1,770. Added to the finance charge, this would give a total interest cost of $3,930 and an average rate of a little less than 8.2 percent.

It is worthwhile for a contractor to understand interest rates. The contractor will then have a clear picture of the extra expense involved in financing equipment, and may be able to save substantial amounts by being able to detect mistakes or fraud in papers.

FINANCING

The cheapest way to finance the purchase of a piece of equipment is to borrow from a bank on a straight time note at the regular rate of interest. Such a loan may be obtained by pledging collateral such as stocks, bonds, or accounts receivable. A substantial contractor may be able to obtain such a loan without putting up security.

Installment Plans.   Most equipment financing is done on a straight-line installment basis, with a down payment of 20 to 40 percent (usually 25 percent) and the balance plus interest paid in equal monthly installments over a period of 1 to 5 years, with 18-month to 3-year terms the most common.

The finance or interest charge may be 10 percent per year on the original amount of the loan. That is, if $1,000 is borrowed to be repaid in 12 monthly installments, the interest is $100.00. If installments extend over 3 years, this charge is $300.00. It works out to an actual rate of around 19 percent, the higher cost being found on the longer terms.

Installment payments are secured by a chattel mortgage on the equipment, that is recorded in the town or city records. The borrower must be sure to have this canceled by filing a release from the finance company or bank when she or he has completed payments.

When the value of the equipment and/or the contractor's ability to make the payments is questionable, the lender may ask for additional security, such as an endorser on the notes, or a mortgage on additional pieces of equipment that have no debt against them.

If loan installments are not paid on time, an extra charge may be made for each one that is delayed. The lender also has the privilege of demanding immediate payment of the whole sum if even one payment is unreasonably delayed, and may seize and sell the equipment to collect. Machinery sold in such proceedings is not likely to bring its full value, and the contractor may still owe a balance even after the equipment is lost.

Schedules may be made up to allow omitting payments in off seasons, usually three or four winter months. Such a provision will either make the other payments larger, or stretch them over more years. Most contractors manage to make regular winter payments with surplus from working months, collection of accounts, or short-term borrowing from banks.

OTHER OWNERSHIP COSTS

Property Tax.   The contractor must pay a variety of taxes, including real estate, personal property, excise, and payroll levies. Here we are concerned wit the personal property taxes payable on the assessed value of equipment.

This is entirely a local matter. In some states the local governments are permitted or required to tax machinery and other movable property in the same manner as real estate. This tax may range from 2 to 5 percent of the assessed value of the equipment, depending on the type of equipment, its costs, age, and condition. In other states or localities there are no property taxes whatever on construction equipment.

It is customary for estimating advice to suggest using the nationwide average tax of 1.5 to 2 percent of value in figuring ownership costs. However, in this case, average costs have little bearing on particular costs. Contractors must find out what taxes, if any, they will pay before they can use them in figuring.

The tax is usually low in country districts and high in cities, but it varies with local financial policies. A high rate with a low assessment may mean a lower tax than a lower rate and full-value assessment.

Assessments may or may not follow the depreciation schedule of the contractor. But it is a general practice to assess a machine for at least 20 percent of its cost as long as it looks as if it might run.

Registration.   Highway vehicles must have registration plates. The cost is moderate for cars, pickups, and jeeps, but may be very heavy for big trucks.

In most states this tax is based on weight and/or capacity. In some there is an additional mileage charge. There is no close relationship to purchase price, so that it cannot be handled on a percentage basis.

Registration is an overhead expense, mileage an operating item. Both are added to other costs in setting a price on a truck's services, but this must be done on an individual basis.

Liability insurance.   Highway vehicles are not covered by a contractor's general liability and property damage insurance. They have special coverage at much higher rates.

This is another ownership expense that is not related to purchase cost. Its amount is affected by vehicle weight, type of use, accident record of the owner, and miles driven.

Loss Insurance.   The cost of insurance against fire, collision, upset, and theft is an ownership cost that is charged against each piece of equipment in proportion to its value. There are equipment theft prevention systems available that should reduce the insurance cost if one is installed.

The charge for insurance of this type is known in insurance circles as a judgment rate, as it is set for each locality or contractor according to the insurance companies' judgment of the risks involved.

The rate for fire, collision, and upset in a combined extended-coverage policy is usually about 1 percent of the actual value of the equipment, for the small contractor with a few machines used in miscellaneous work. Very large earthmoving or construction projects, such as the St. Lawrence Seaway sections, may be given a rate as low as $\frac{1}{2}$ percent. This is in spite of the fact that some of the machines work under very dangerous conditions, as the extreme risk positions are outbalanced by many behind-the-lines units working under safe and stodgy circumstances.

The highest rates for this coverage may be $1\frac{1}{2}$ to 2 percent. These are charged where the job conditions are more dangerous than average, or where the contractor is considered to be careless or reckless in management.

Theft insurance may be written into these policies as an extra coverage. With a $50.00 deductible clause it may be free in country districts where stealing is rare, and up to $\frac{1}{2}$ percent of value in cities. One-quarter percent is a usual charge. Companies may refuse to issue theft coverage at any price in certain cities or areas.

Premiums are usually charged on the basis of the contractor's valuation of his or her equipment, as long as the contractor follows any reasonable and consistent system of depreciation. Each year, or at more frequent intervals, the contractor sends the insurance company a list of equipment, showing date of purchase, original cost, and present value. The premium is charged as a percentage of the total.

If a unit must be replaced because of insured loss, payment is made on the basis of the actual value of similar equipment in the locality at the time of loss. However, the company has the right (which it may not use) to refuse to pay more than the value of the machine stated in the policy schedule.

Therefore, if the value stated in the policy schedule, which should be the same as in the equipment inventory, is more than the actual value, the premium on the excess might be wasted money. If schedule value is less than real value, the equipment is not fully protected against complete loss. However, complete loss is rare except in very small units, and most payments under these policies are for repairs.

Storage.   It is unusual for there to be any storage cost directly chargeable to a piece of equipment. Most contractors have at least one home lot, often near their repair shop. This has room for a number of pieces of equipment. The rest are kept out on jobs, where they must be to earn their keep. They usually can be left on or near a job until they are moved directly to the next one.

Ownership and maintenance of a storage yard are strictly a general overhead expense, as this facility is not expanded and reduced with purchase or sale of machines.

However, a contractor who wants to charge it against individual machines can do so by finding the annual cost per square foot of the yard, and charging each machine according to the number of square feet it occupies when it is there.

For example, a piece of industrial land in outskirts of a city might cost $75,000 per acre, including a graded and stabilized surface. It might be assessed at full value, with a tax rate of 4 percent. As this is not an income-producing investment, 6 percent interest might be charged against it. Cost per square foot might be worked out as in Fig. 11.10, to $0.413.

A large scraper might occupy a space 50 feet long and 12 feet wide, or 600 square feet. Allowing 400 more feet for maneuver space, its requirement would be 1,000 square feet. Annual cost would then be 1,000 × .413, or $413. This machine might cost about $50,000, and have an average annual investment of $90,000, so storage would be nearly 0.5 percent of value. A shovel of similar value would need less than half as much space.

It is unusual to store large pieces of equipment indoors. If it is considered necessary to do so because of vandalism, extreme cold, or other conditions, the cost may be as high as 5 percent of investment.

Summary.   Figure 11.11 shows the normal range in ownership costs or carrying charges, on a per year per $1,000 of average annual investment basis.

There is a wide range, from .5 to 23.5 percent. Most estimating advice recommends using 10 to 13 percent. Ten percent is an easy figure to remember and to use, but 8 percent is likely to be more accurate if interest charges are limited to those actually paid.

The contractor or estimator should not rely on any general average of costs, but should find out what they really are for her or his own situation.

Purchase price - $75,000

Interest on investment, 6%	$4,500
Taxes, 4%	3,000
Maintenance	1,500
Night watchman	9,000
	$18,000
Annual cost	

Area, one acre, or 43,560 square feet

$$\text{Annual cost per square foot} = \frac{\$18,000}{43,560} = \$0.413$$

FIGURE 11.10   Cost of storage yard.

	Minimum	Ordinary	Maximum
Interest on investment	$ 0.00	$ 45.00	$ 115.00
Loss insurance	5.00	15.00	25.00
Property taxes	0.00	20.00	45.00
Storage	0.00	0.00	50.00
Total	$ 5.00	$ 80.00	$ 235.00
Per cent of average annual investment	.5	8.0	23.5
Per cent of cost, with 5— year depreciation	.3	4.8	14.1

FIGURE 11.11   Ownership costs per $1,000 average annual investment, without depreciation.

EQUIPMENT WORK HOURS

Annual depreciation and other ownership costs are converted to an hourly figure as a basis for charging out equipment time. At first glance this appears to be easy. It is only necessary to divide the annual costs by the hours worked per year, or the total work hours by the total costs.

For example, a machine whose fixed costs during a year are $3,600 and that worked 1,200 hours in that year will show fixed cost per hour of $3.00. Or if its total costs for life are $18,000 and its total hours of work are 6,000, the figure is still $3.00.

But it is difficult to settle on the number of hours that equipment can be expected to work, as that is affected by a number of variable factors.

This discussion will be limited to the problems of those contractors who work a single shift and are subject to delays in weather, getting jobs, and keeping equipment running—which includes most of them. It will also deal chiefly with the contractor's first-line equipment that has work most of the time.

Maximum Use. Estimating advice from manufacturers usually recommends a basis of 2,000 hours per year, and a 5-year life. But most construction work is done in 8-hour days and 5-day weeks, with shutdowns for a minimum of six holidays. The maximum number of hours that can be worked in a year is 2,040 on this program. It is usual to lose an extra 5 days in special holidays or shutdowns, reducing the year to 2,000 work hours.

Bad Weather. Weather often makes outdoor heavy construction work impractical or impossible. One New England state highway department estimates that weather and ground conditions permit the following number of days per quarter:

January–March	35
April–June	55
July–September	60
October–December	55
	205

These figures are on the optimistic side for the area, and they are based on working Saturdays when necessary to make up for rained-out weekdays.

A 5-year survey by the U.S. Bureau of Public Roads indicates that the nationwide average of shutdowns on highway jobs that are due to weather amount to about $\frac{1}{5}$ of working time.

The southeast and south central states do not have to stop work for snow and ice, but they do have rain that may have equally bad results. Only in certain areas in the southwest can the 2,000-hour figure be even closely approached on a permitted-by-weather basis.

Maximum working hours are affected by job conditions. Work in rock, gravel, or sand, or on surfaced haul roads, can continue under conditions that would make a job in loam or clay impossible. Pressure of a deadline can make it worthwhile to work under very unfavorable conditions, just in the hope of making some progress.

The type of equipment also affects lost time during weather. A dragline piling wet soil may not be affected by rain unless it is flooded out. Crawler equipment keeps going after rubber-tire types give up. Vehicles may carry part loads where full ones would make them bog down.

If there is a lack of any specific information to the contrary, the estimator should allow for a loss of 20 percent of annual working time because of unfavorable weather.

No Work. Equipment can work only when there is a job. This means not only work in general, but for the specific machine under consideration.

Some contractors find little difficulty in keeping busy all working season or all year; others must get through frequent or prolonged periods of insufficient work or no work. The differences depend on construction activity in the area, the specialties of the contractor and the demand for such specialties, the aggressiveness and reputation of the contractor, and a factor of luck in bidding and in selling services.

Even when a contractor has a job, it may not be for all the equipment. The contractor may even have to leave his or her own machinery idle and work with hired equipment at a job that is outside the regular field.

As a general average, a capable contractor may hope to keep the first-line equipment busy on jobs about 80 percent of the time that weather permits working.

Downtime. Even when weather is good and work is available, a machine may not be able to work because of need for repairs to itself or to another unit whose operation is necessary to it, or as a result of shortage of materials, strikes, or other causes. This nonworking time on the job is called downtime.

Studies conducted by the U.S. Bureau of Public Roads, now the Federal Highway Administration, show that equipment downtime on the job is likely to be between 20 and 65 percent of working time, with age and condition of equipment and competence of management being the most important factors in the variation.

Most of this downtime is considered to be working time (if the machine were rented, rental would be charged), but the owner must take its loss into consideration in figuring the work gotten out of the machine.

Such downtime is in addition to the small delays that are taken into account by using a 45- or 50-minute work hour.

Work Hours Summary. A rule of thumb for the hours that heavy equipment will work is to assume a one-shift, 2,000-hour year; take off 20 percent for bad weather, leaving 1,600 hours; take off another 20 percent for lack of work, leaving 1,280 hours, and another 20 percent (an absolute minimum) for lost time on the job, leaving a net working time of 1,024, or say 1,000 hours. This is the Rule of the Three Twenties.

Like all rules of thumb, this can be way off. But before it is discarded, the estimator should study his or her own conditions carefully to see if they are really better, or quite possibly worse.

This rule does not apply to mines and pits, that may work three shifts on a 7-day week, and have up to 8,600 scheduled machine work hours in a year. They do not ordinarily lose as high a proportion of this time.

A number of machine cost computations in this book use a 1,200-hour year as a basis. This is due partly to the fact that many contractors consider the lost time on the job to be working time, and partly to the longer-than-5-year life enjoyed by many machines. That is, the hourly costs come out nearly the same whether the machine is used 1,200 hours per year for 5 years or 1,000 hours for each of 6 years.

Equipment Life. The useful life of construction equipment varies depending on how it is used and maintained, also how long the contractor wants to keep using it as opposed to replacing it with a new piece. A study conducted for Construction Equipment magazine in the 1990s found that the average life of major equipment kept in a contractor's fleet was about 7,000 hours. Figure 11.12 shows the range of useful hours to replacement for key types of equipment according to the study.

A contractor might use a fleet information system (FIS) computer program to help decide on the equipment life for a piece or set of equipment he or she owns. The program calculates what-if costs based on current information. The computer software projects ownership and operating costs of a machine being analyzed for replacement. Estimated downtime is calculated based on the reliability and life averages for similar equipment. Cost and life data is drawn from the database maintained with the firm's fleet management records.

The FIS system then balances costs with the expected revenue the machine will earn, based on its past averages of usage and revenue earned. The system also estimates residual value expected at the time of replacement, including any repairs that might have to be made.

The result is an expected cost per hour for a rebuilt machine, which can be compared to the costs for buying and operating a new machine. If the decision is to retire the old machine, the equipment life of that machine has been determined.

Type of equipment	Lighter weight	Heavier weight
Hydraulic excavators	To 20 tons 8,000–14,000 hrs	More than 20 tons 10,000–18,000 hrs
Crawler loaders	To 100 hp 7,500–13,500 hrs	More than 100 hp 12,000–18,000 hrs
Wheel loaders	To 5 yards 10,000–20,000 hrs	More than 5 yards 14,000–25,000 hrs
Backhoe-loaders	To 75 hp 8,000–10,000 hrs	More than 75 hp 10,000–15,000 hrs
Skid-steer loaders	6,000–8,000 hrs	8,000–12,000 hrs
Crawler dozers	To 100 hp 10,000–15,000 hrs	More than 100 hp 15,000–20,000 hrs
Scrapers	Conventional 10,000–22,000 hrs	Elevating 10,000–18,000 hrs
Graders	Rigid-frame 10,000–25,000 hrs	Articulated 10,000–20,000 hrs

FIGURE 11.12 Useful life targets for key machines.

Equipment Life Based on Repair Cost. It has been shown that the life of a piece of equipment depends on cost of repairs to the machine. The cost of those repairs per hour is at a minimum after 7,000 to 10,000 hours of use depending on the type of equipment and its use. Then they jump because of requirements for a new set of tires, hydraulic pump, rebuild of the transmission, or maybe a new engine. That time of minimum cost of repairs in the life of a piece of equipment might be called its sweet spot.

Auxiliary and Emergency Equipment. Most contractors own a certain amount of equipment for which they find little use. For example, a general contractor who seldom does rock work may keep a compressor and drill to have them immediately available if required.

   If a contractor owns a compressor or pump that is used only 50 hours per year, the depreciation per working hour is 20 times as much as that of another contractor who uses hers 1,000 hours. The depreciation on such a unit must be charged to general overhead, as the machine cannot hope to earn it.

Equipment Investment Analysis. The total investment analysis for a piece of construction equipment is an involved process. Major factors to take into account are the selling price, resale value, financing costs, accelerated cost recovery (depreciation), insurance, and a variety of taxes. These must account for the estimated market value, the finance period, the finance charge or add-on interest rate, and the residual book value. To make a satisfactory analysis is more involved than can be shown properly in this book. It can be done with the help of a professional financial person or using a guide like the one produced by Caterpillar, Inc.

OPERATING EXPENSE

The expense of operating a piece of equipment is likely to include the following:

   Fuel: both fuel and handling
   Lubrication: cost of oil, grease, lube equipment, and labor
   Maintenance and repair: parts, supplies, shop equipment, and labor
   Labor: operator, oiler, helper, ground men, supervision

Fuel.   Fuel cost varies widely with the power, type, and condition of engines; the type and condition of equipment; type of work; and the grade of fuel.

Fuel consumption in relation to horsepower and load is discussed in the next chapter.

Fuel costs vary with the prices of crude oil, distance from the source, quantities delivered, seasonal demand, and taxes imposed.

The delivery quantity may be very important. The contractor with a tank of 275- or 550-gallon capacity may have to pay up to 5¢ per gallon more than a big competitor who can take 2,000 or 3,000 gallons at a time. However, this difference can be reduced or eliminated if the small order can be filled on the same trip as others in the locality.

There are state taxes of 9¢ to 36¢ per gallon that apply to fuel used in highway vehicles, and the federal gasoline tax in the United States is nearly 20¢ per gallon. Generally, any vehicle that is registered for highway use must be charged with the state tax, even if operation is off the roads.

Taxes may be paid by the distributor at the highest rate and passed on to the contractor, who then must report the amounts used at lower tax rates to obtain a refund. Or the fuel may be delivered tax-free, and the user required to make monthly or quarterly statements of use, with payment of tax due. Payment by the distributor is usually most convenient.

These taxes are substantial enough that it pays the contractor to keep careful account of her or his use of fuel. A tally sheet must be kept at the pump or in the distribution truck, showing quantities, type of equipment, and class of use. The bookkeeper needs the information on these sheets to make up reports, for either tax payment or refunds.

Lubrication.   There is considerable variation in lubricant prices and applications, with resulting confusion to the purchaser. In general, the best quality and most suitable lubricant is the most economical regardless of its price per gallon or per pound, as the cost of labor in using it, and the expense of repairing wear and damage resulting from poor lubrication are vastly greater than the price differences.

Oil.   Equipment manufacturers recommend that engine oil be changed at regular intervals, that may vary from 75 to 200 hours in different makes or models. The time between changes may be shortened under dusty or extreme temperature conditions, or lengthened where work is light, air is dust-free, and/or a special type of filtering or reclaiming apparatus is used.

Crankcase capacities vary widely with size and design of engines. They may hold a quart of oil for every $3\frac{1}{2}$ horsepower, or only a quart for 13 horsepower.

While oil consumption may be negligible in new engines, it may be as high as $\frac{1}{20}$ of fuel consumption in engines that have badly worn piston rings and/or external leaks. However, no properly run job would tolerate oil loss of more than $\frac{1}{50}$ of fuel use, as pumping oil into cylinders is accompanied by losses of fuel and power, and leaks are likely to allow dirt to get in.

Oil in transmissions, rear ends, and final drives is usually changed twice per year, the most important change being in the fall. Loss between changes is usually negligible, but may become severe because of failure of seals and gaskets, or cracks in housings. Any type of leak may allow dirt to enter, so prompt repair is important.

In general, an allowance of 3 times the reservoir capacity per year will take care of two changes and losses by leakage or accident.

Grease.   Equipment varies tremendously in its requirement for grease. For example, a 20-ton crawler tractor may use from 1 to 5 pounds of grease in old-fashioned track rollers every 8 hours or less. A similar machine having positive seals may need lubricant in the rollers only at 1,000-hour intervals or when the rollers are rebuilt.

Here records are the only indication of what to expect. Even if they only indicate the pounds of grease bought and the total equipment work hours in a year, they will at least provide an average requirement for the fleet.

Small equipment is carried in the tools account and is difficult to separate, while big units are depreciated in the same manner as other equipment. Lacking information to the contrary, a 6,000-hour life may be assumed for them.

Lube Labor.   The pay of the people who operate a grease truck or a stationary rack is definitely charged to lubrication. But an oiler on a shovel, in addition to taking care of oiling and greasing, is likely to assist the operator with other maintenance, repair, moves, and in many other ways. It is usual to carry their pay in the same account as that of the operators.

A great deal of lubrication is done by the operators themselves. They may be paid for $\frac{1}{2}$ hour overtime a day to take care of this and fueling, or may do it during the shift in pauses in the work.

A grease truck crew can take care of about three machines per worker hour. This figure is an average of daily lubrications that may take 5 minutes or less, periodic thorough jobs where all points are reached and all reservoirs checked, and complete lubes including oil change.

Rule of Thumb.   In view of all the variables and borderline costs, the estimator is justified in accepting and using the rule of thumb that costs of lubrication equal one-third of the cost of diesel fuel or one-quarter of the cost of gasoline. There will usually be some error as a result, but it is likely to be less than that resulting from a superficial attempt to work out the actual figures.

The important thing about keeping track of these costs is to decide on a system and stick to it. A contractor who uses a different method each time he or she thinks of one will not be able to make comparisons between different jobs and different years.

Always keep in mind that the biggest lube expenses are the failures—the breakdowns that are caused by improper or neglected lubrication.

MAINTENANCE AND REPAIR

There is no definite line of division between maintenance and repair. It is usual to say that maintenance includes items such as cleaning, inspection, adjustment, routine replacements, and hard face or other build-up welding, while repair consists of fixing or replacing worn or broken parts.

Lubrication is often treated as a maintenance expense, and it is probably the most basic and important of all the maintenance operations.

Many contractors and most equipment rental firms divide repairs into two classes—major repairs, overhauls, and painting; and small repairs and maintenance. The first class may be called shop work, as it should be done in the repair shop even if it actually has to be done in the field, and the second class is called field repairs and maintenance.

In rental arrangements, the shop repairs are usually done by the owners, the others by the lessee, although the contracts do not say so specifically.

A repair, whether in the shop or the field class, serves simply to fix or replace a defective or broken part, together with any associated parts that have caused the breakdown or have been affected by it. An overhaul involves thorough inspection and all necessary rebuilding of an entire unit.

For example, a transmission with a broken gear may be repaired by simply replacing the gear; but if it were overhauled, it would be completely disassembled, all parts would be cleaned and checked, and any defective ones replaced.

Whatever classification is used, there is nothing that is more important to the contractor's success than careful maintenance and prompt repair, as it will save equipment and money.

Estimating Repairs.   A contractor must have a fairly accurate idea of the future cost of maintaining and repairing a machine, before a price can be put on its use. If good records have been kept, the contractor can check his or her own experience, and use it as a basis on which to allow for future expenses.

If there are no records, or if new equipment and/or new jobs are so different that old records do not apply, estimating must be done on the basis of reports of other people's records or ideas. These must be modified to suit particular conditions.

Most manufacturers and estimating books recommend setting total nontire repair cost during the life of the machine at 60 to 100 percent of depreciation. However, most of these same sources set machine life at 10,000 hours of use in 5 years. But we have seen that the contractor usually does not get over 5,000 or 6,000 hours of machine time in 5 years. This leaves the question of whether these authorities really expect life to be 10,000 hours or 5 years.

There are so many variables in this field that experience records can be found to support almost any estimate. Records can be obtained at any time in the field by handheld instruments reading data in bar code form from the equipment. The bars may be set to give: hours of operation, hours since last oil change, and other such data. Costs are affected by the quality of the machine, accessibility of its parts, standards of lubrication and maintenance, skill of mechanics, work conditions, hours and years of use, and quality of supervision and operation. There is also an important factor of luck.

The contractor who has just one important machine may be made or broken by different combinations of these factors. But possession of a number of machines will usually cause good and bad features of individual machines to average out, and a succession of jobs is likely to smooth out the ups and downs of work conditions.

Equipment Monitoring.   Remote control monitoring systems are now available in heavy equipment. The systems monitor machine conditions such as operating hours, temperature of coolant and various oils, oil pressures, and shaft speeds. They use a global positioning system (GPS) to fix the current position of the machine, then they communicate detailed location and the performance data over a wireless link to the company's computer, pager, or cellular phone. The cost of the hardware for this monitoring system in a machine is only in the thousand dollar range and monthly fees may be as low as $20 per machine.

The monitoring, that can be done on a continuing basis, is able to pick up danger signals from a piece of equipment that suggest it be stopped for maintenance work and repair, if necessary. The information is received in a timely manner before a major breakdown would occur and allows for preventive maintenance to be done. The system can be set up to tie in with the on-board computer so that the operator can be aware of the need for maintenance.

Commercially available monitoring packages usually include the GPS tie for location and the hours of use. But then the user can choose to monitor: coolant temperature, hydraulic temperature, engine or hydraulic oil pressure, or other functions of working parts. For instance, on harddriving machines, like bulldozers and scrapers, the maintenance manager may want to monitor the transmission oil temperature.

In the beginning of equipment monitoring systems the hardware bounced data off satellites, but satellite air time can be expensive. So vendors have worked out the system to transmit data using commercial radio frequencies and cellular phone systems. That way the charge may be as low as $15 per month per machine. A major problem for the monitoring system is the massive volume of data, so it needs to be stored in short term memory and delivered to a maintenance decision maker for early action as needed.

Repair Factors.   Figure 11.13 gives a table of repair factors that may be useful in determining probable repair costs over the life of a machine, in adjusting experience records to new conditions, or in explaining expenses that have already been incurred.

In using this table, the estimator selects the description under each heading that most nearly represents the conditions expected, and takes the figure that follows it. These figures are multiplied by each other to produce a combined repair factor, that is then multiplied by 1/10,000 of the purchase price of the piece of equipment.

Unless special conditions have an unusual effect on tire life, these factors may be used for the whole machine, including tires. When tires have exceptionally short or long life, these factors should apply only to the nontire part of the equipment, and the factors in Fig. 12.137 used to determine tire life.

For example, a contractor may buy a crawler-mounted front loader for $190,000. It is a top-quality machine, maintenance is expected to be good, work conditions are heavy, temperature is normal; experience, work pressure, and operation are average; and the machine is expected to be used a total of 6,000 hours in 5 years.

1. TYPE OF EQUIPMENT		2. TOTAL HOURS OF USE		3. YEARS OF USEFUL LIFE	
Crane, revolving	.5	1,000	.5	1	.6
Compressor, air	.8	2,000	.5	2	.7
Truck, highway dump	.8	3,000	.6	3	.8
Dragline and clamshell	.9	4,000	.7	4	.9
Shovels, dipper and hoe	1.0	5,000	.9	5	1.0
Truck, off the road	1.0	6,000	1.0	6	1.0
End loader, 4–wheel drive	1.0	8,000	1.3	7	1.1
Scraper, all types	1.1	10,000	1.6	8	1.2
Dozer, crawler	1.2	12,000	1.9	9	1.3
End loader, crawler	1.4	15,000	2.3	10	1.4
End loader, 2–wheel drive	1.6	20,000	3.0	15	2.0
Crawler and mounted ripper	2.5				

4. TEMPERATURE, FARENHEIT			5. WORK CONDITIONS		6. MAINTENANCE	
Very hot	Over 100°	1.3	Mostly standby	.4	Excellent	.6
Hot	85 to 90°	1.1	Light	.8	Good	.8
Normal	32 to 84°	1.0	Average	1.0	Average	1.0
Cold	0 to 31°	1.2	Heavy	1.4	Poor	1.5
Very cold	Under 0°	2.0	Rough	2.0	None	3.0

7. TYPE OF SERVICE		8. OPERATORS		9. EXPERIENCE (LUCK?)	
Mine or large pit	.5	Exceptional	.8	Excellent	.6
Small pit	.8	Good	.9	Good	.8
Contractor	1.0	Average	1.0	Average	1.0
Rental to others	1.4	Rough	1.2	Poor	1.5
		Cowboy	2.0		

10. EQUIPMENT QUALITY		11. WORK PRESSURE	
Top	.8	Leisurely	.9
Average	1.0	Average	1.0
Poor	1.5	Desperate haste	1.5

FIGURE 11.13 Repair cost factors.

For the given example see the following table:

1. Type of equipment	1.4	7. Type of service	1.0
2. Total hours of use	1.0	8. Operators	1.0
3. Years of life	1.0	9. Experience	1.0
4. Temperature	1.0	10. Equipment quality	.8
5. Work conditions	1.4	11. Work pressure	1.0
6. Maintenance	.8		

Dropping the 1.0 factors because they do not affect the multiplication, we have

$$1.4 \times 1.4 \times .8 \times .8 = 1.2544, \text{ say } 1.25$$

We multiply this by $19.00, which is 1/10,000 of the purchase price of $190,000. Then

$$\text{Hourly repair cost} = 1.25 \times 19 = \$23.75$$

These factors should be used only by persons experienced in heavy equipment use, as judgment is required in selecting the correct factors, and in deciding whether the results obtained are reasonable.

If equipment is not used in its proper jobs, in relation to its size and its design, repair costs will be affected. For example, a ½-yard shovel used in coarse blasted rock would be in the rough-conditions classification, even if the bank were average digging for a 2½-yard machine.

A highway-type dumper used in off-the-road work will suffer severely. Repairs are likely to be those of rough conditions, even if the job were average or light for an off-the-road hauler.

End-of-Period Cost.   Repair cost increases as equipment ages, and the increase is faster than is indicated by whole-life averages. It is desirable to estimate its actual rate at the end of possible life periods, to determine whether it is likely to be so high as to make it uneconomical to keep using the machine.

Average repair cost can be converted into the end-of-period rate by multiplying by the proper one of the following factors:

Hours of use	Factor	Hours of use	Factor
2,000	1.0	8,000	1.8
3,000	1.3	9,000	1.8
4,000	1.5	10,000	1.8
5,000	1.6	12,000	1.9
6,000	1.7	15,000	1.9
7,000	1.7	20,000	1.9

Repair Cost.   Figure 11.14 is offered for those who prefer taking a quick approximation from a graph to working out an answer with a set of factors. It shows hourly repair costs for a piece of equipment in average and heavy work conditions on a $1,000 cost basis. Both average and end-of-period figures are included.

This graph takes care of groups 2 and 5 in the factor table. Its figures can be adjusted for any other of the groups simply by multiplying by a factor in that group.

Percentage of Depreciation.   Repair costs are often calculated simply as a percentage of depreciation, varying from 60 to 100 percent. Depreciation periods also vary, from 5,000 to 10,000 hours or more, so here the estimator can select from a wide range of possibilities.

Percentage of Cost.   The Associated General Contractors of America publishes a table of ownership expense of construction equipment, compiled from reports from members. This lists a combined item of "Overhauling, Major Repairs, Painting" that is 12, 15, or 20 percent of the purchase price for most items of earthmoving machinery.

Heavy repairs are considered as an ownership expense for the purpose of computing charges for renting equipment to others where the owner pays the major repairs and the user the smaller ones. They make up from 50 to 80 percent of total repair and nonlubrication maintenance expenses.

Using these figures, the hourly cost for heavy repairs alone for each $1,000 investment would be

Percent	Annual use, hours	Heavy repairs per hour
12	800	$.15
	1,000	.12
	1,200	.10
	2,000	.06
15	800	.19
	1,000	.15
	1,200	.125
	2,000	.075
20	800	.25
	1,000	.20
	1,200	.167
	2,000	.10

As we will see later, repair expense on rented equipment tends to be higher than normal.

Usefulness. Tables, factors, and percentages are all based on averages from large numbers of machines, and do not necessarily hold good for any one machine. Use bar code data for a machine where possible.

The increase in costs with longer use is also an average. Any machine, and even most fleets of equipment, will go through good periods of little expense, and bad periods of frequent breakdowns.

In the same manner, July should be a good month for earthmoving and January a bad one, but occasionally the reverse condition occurs. But the contractor still must figure on working the next July and losing time during the next January.

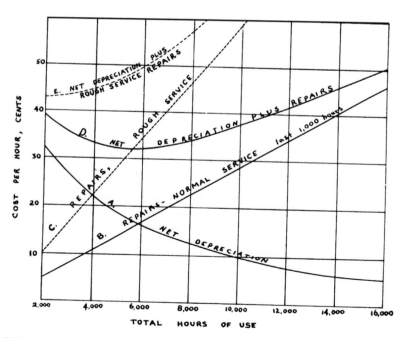

FIGURE 11.14 Depreciation and repair.

Rising Repair Costs.   As a machine gets older, its repair costs increase. Fuel and lubrication bills also increase unless held down by first-class maintenance.

Depreciation cost per hour decreases steadily with longer use. This serves partly to offset rising repair costs. Figure 11.14 shows a curve, (A), for net depreciation, and lines (B) and (C) for normal and extraheavy repairs. Curves (D) and (E) give combined net depreciation and repair.

The low parts of curves D and E show the most economical life period in hours. For average medium repair costs this is 5,000 to 6,000 hours, with little difference shown from 4,000 to 8,000 hours. For extraheavy repairs a 2,000-hour machine life appears to be less costly, with moderate increase to 4,000 hours and then a steep rise in expense.

Such very heavy repair costs with resulting short economical life can be combatted by using bigger and stronger equipment, reducing work pressure and loads, stepped-up maintenance with frequent overhauls including bracing and substituting heavier components when they are available, and by various combinations of these methods.

Light Duty.   The heavy repairs usually incurred by using a machine late in its useful life can sometimes be avoided by assigning the unit to light or standby duty. A bulldozer may be assigned to cleaning up loose dirt around a shovel or shovels, a tired wheel tractor may pull a water wagon, and an old shovel be kept in soft digging. Such assignments are more easily made in open-pit mining than in general earthmoving, and they help to explain the large number of over-age machines that continue to render satisfactory service in pits.

Downtime.   An item that does not appear on the bill at all is often the most expensive part of equipment repair. That is the lost time on the job while mechanics and parts are being located and the repair is being made. This cost is usually at its highest rate during the first few minutes or hours of the breakdown, while operating costs are still at a maximum and before the work program has been changed.

If the shutdown is a short one and only one machine is affected, for example, if a hose bursts in a self-loading scraper working alone, it causes loss of only the production of that one unit. The same difficulty in a loader might stop the loader, a string of trucks, a dozer, and compacting equipment also.

Conditions involved in downtime are so variable that they cannot be put into graphs or tables. Losses are kept down by alert supervision, new equipment, and expert maintenance.

TIRES

Tires may represent an important part of the cost of new equipment, and they have several characteristics that make them difficult to fit into the same cost calculation as the rest of the machine.

Noncapital Treatment.   Some contractors follow the practice of deducting the cost of tires from the price of a new machine before setting up its depreciation account. If this deduction is made, it should be on the basis of the actual cost of replacing such tires, but list price is sometimes used.

The chief reason for subtracting tire cost is to obtain the fastest possible writeoff. The excuse used is that tires are not physically part of the machine and wear at a faster rate, so that the ordinary long depreciation period does not apply.

A hauler may go through two or four sets of tires before it wears out itself. If the original tires are capitalized, the owner may be still deducting depreciation on them years after they have been scrapped. He or she will have paid the cost of replacing them before having been able to get a tax reduction on all of their original cost.

Another reason for keeping tire accounts separate from the machines that carry them is that they wear at different rates and are affected by different conditions. Mechanical parts may be little affected by differences between sandstone and shale, but they may cause a 3-to-1 difference in tire wear. Extreme cold may increase equipment repair costs and prolong tire life.

None of these factors entirely justify separation of tires from the rest of the machine in bookkeeping. The same arguments could be advanced for separate accounts for crawler tracks and rollers, blade and bowl edges, batteries, and even shovel buckets. And the double bookkeeping does exaggerate the cost of tires.

If earthmover tires have an average life of a year or less, the Internal Revenue Service agents will approve their deduction from capital investment. If average life is over a year, they have the right to disapprove. In general, they are against any type of separation of a unit into various pieces for separate depreciation treatment.

Operation Cost.   The major operating expense of a tire is its replacement. The actual cost divided by the number of hours it operates gives this cost on an hourly basis. Since many tires reach an early and sudden end through accident or abuse, the life of a number of tires must be averaged to obtain a fair figure.

Tire maintenance and repair are assumed to cost about 15 percent of replacement.

LABOR

In spite of mechanization, labor accounts for a big part of every heavy construction dollar. No estimator can afford to overlook any of the labor costs.

Current pay rates on a national basis are shown in Fig. 11.15.

Operator pay is often left out of equipment costs in estimating advice, because it varies so widely from place to place. Estimators must be very careful not to leave it out of their figures.

Almost every piece of construction equipment has an operator. Many shovels have an oiler too. Laborers are needed to handle supplies, spot trucks, direct traffic, pick up rocks, trim banks, scrape sticky soil out of bodies and buckets, and do many other tasks.

Supervision at the superintendent level is considered an overhead expense. Foremen may be charged to overhead also, but it is more usual to enter their pay as an operating expense. It may be charged against the job in a lump, or divided by the number of workers supervised, and added in as part of operator payroll.

An operator is usually paid for a good many more hours than her or his machine works. He or she may get a full day's pay just for reporting, whether the machine runs or not. The operator is certainly kept on the payroll during short delays for adjustment and repair, and when standing by during various job delays. The operator may be paid for extra nonoperating time in which he or she greases and services the equipment.

If an operator is paid on an annual salary basis, the wage should be divided by the number of hours the equipment works or is expected to work during the year to obtain an hourly rate.

Hourly wage rates: base rate + fringe benefits, U.S. dollars

| | 20-city average | | | | | |
Classification	Price	Percent change from 9/02	Boston	Birmingham	Chicago	Cincinnati
Laborers						
Heavy, highway	27.72	+3.8	34.50	18.06	34.89	25.45
Operating engineers						
Cranes, shovels	36.87	+4.0	43.44	20.86	46.63	33.17
Heavy equipment	36.52	+3.6	43.44	21.51	46.08	31.75
Small equipment	33.19	+4.4	39.66	21.36	44.03	29.15
Teamsters						
Truck drivers	30.53	+6.0	33.95	17.65	34.78	na

FIGURE 11.15   Representative pay rates, September 2003.

Mine workers are usually paid on a portal-to-portal basis, that is, from the time they check in at the gate or the main building until they get back to the time cards. They receive full pay for time spent between the entrance and the place of work. This may make a substantial reduction in the time actually worked during a shift. To look at it another way, it means a higher per hour cost for the time they are working.

Construction workers usually check in close to the work, and are usually expected to have equipment running and ready to go at the start of a shift.

To find the full cost of labor, it is necessary to add in payroll taxes, both for Social Security and unemployment compensation insurance, and fringe benefits such as paid holidays and sick time, reserve for pensions special travel or subsistence allowances, and pay for nonworking time on temporary job shutdowns. These extras may increase base pay from 10 to 30 percent or more.

Shift. A shift is the continuous (except for breaks for meals) time worked by one crew in one day. It is usually 8 hours, but it may be 7, 10, or even 12. The longer shifts usually include an overtime pay rate.

Multiple Shifts. Work may sometimes be speeded by working two or three shifts. Three shifts are commonly 8 hours each, one crew taking over from another without any shutdown. The day shift is from 8:00 a.m. to 4:00 p.m., the "swing" until midnight, and the "graveyard" until the day gang takes over. Pay time is 8 hours, but a "lunch" period, and time lost in the changing of the shifts, reduces work time to less than 7½.

Two shifts may be of either 8 or 10 hours each. The job is usually shut down after each shift, except for lubrication and repair crews.

Night work is less efficient than day because of the need for artificial light and the lessened accuracy and usually lower mental and physical vigor of the workers.

Multiple shifts may work at cross-purposes, or at least with insufficient understanding of what has been done. This difficulty is somewhat less when the new crew arrives before the other leaves. If there is no contact, the supervisors should meet in the idle period to discuss the work and coordinate their efforts.

There should be a system for rotating workers among the shifts, but it should be administered intelligently. Night shifts are generally unpopular, but some individuals prefer them and they should be left in them. Swapping of shifts among equally qualified workers should always be allowed. Rotation should be at rather long intervals to enable the workers to adjust to changes in sleep and work hours. Two weeks is the shortest period which should be considered.

Overtime. Where the industry operates on 5 working days of 8 hours each per week, additional hours worked on any of the 5 days and any time worked Saturdays are called overtime, and paid at 1½ or 2 times the regular hourly rate. Sunday and holiday work may be time-and-a-half, double, or even triple time.

A contractor who must finish work before a contract deadline, or before bad weather, or who wishes to take advantage of a busy season, may ask workers to work overtime; hire additional personnel to work two or three shifts; or may buy or rent additional equipment to work one extra shift.

In general, it is profitable to work large machines overtime, as the extra wages are more than offset by the drop in hourly ownership costs caused by spreading them over a greater number of hours. Small machines may show either increased or reduced profits.

It should be remembered that payroll insurance premiums increase in direct proportion to the amount of pay, although some payroll taxes do not apply over a certain amount.

The small contractor whose machine operators are frequently able to work for extended periods without help or supervision, is more likely to work overtime than the large-scale operator.

When work is being done on a fixed price contract, or at a fixed hourly rate, overtime costs must be carefully watched. If the work is on a basis of cost plus a fixed fee, overtime will merely require a larger investment to obtain the same profit. If payment is cost plus a percentage of cost, overtime, as well as any other extra expenses, will increase the contractor's profit.

Cost-plus contracts may require that a contractor obtain written permission before incurring overtime or special expenses.

WORKING TIME

Time can be used as a measurement in direct clock and calendar divisions, or on a basis of working times that are fractions or combinations of them.

Efficiency Hour.   A second and a minute always have the same time meaning in construction as they do on a clock. But an hour may contain the regular 60 minutes, or may be an "efficiency hour" of 50 minutes or less. This special hour allows for lost time in a way that can be easily included in calculations.

For example, a machine may be able to move 2 yards of earth per minute in steady digging. This would represent an output of 120 yards in a full hour. But no machine can be counted on for absolutely steady work because of delays from such causes as need for adjustments and minor repairs, changing positions, lack of supporting equipment, cigarette time, digging obstacles, and so forth.

The actual production of the machine, averaged over many hours of work, may be only 90 yards per hour. This is 75 percent of its maximum potential output, and is called 75 percent efficiency. It means that the machine is working to capacity only 3 out of every 4 minutes.

It is customary to express the reduced ability to produce by reducing the number of minutes in the hour, rather than by deducting a percentage from production. In this example the work hour would be $\frac{3}{4}$ of 60, or 45 minutes. Multiplying 45 minutes by the maximum production of 2 yards per minute, we get an hourly production of 90 yards, which is where we started.

In the excavation industry it is usual to assume an average efficiency of 83 percent, so that many calculations are based on a 50-minute hour. This efficiency is not unknown, but the average is very much lower. Most examples in this book will use a 45-minute hour.

Day and Week.   A workday includes all the hours of work during a regular 24-hour period. This is usually 8 hours, but may be $8\frac{1}{2}$, 10, or some other time. If there is one shift, the shift and the day are the same time measurement.

In construction, a workweek is usually 5 days, with overtime pay for working additional days. Weeks in which holidays occur are shorter. In mines, 6- and 7-day weeks are common.

If business is poor, work may be stretched out by shortening the work week to as little as 1 or 2 days. This practice is rare in construction, but is common in mines.

Job.   The time that elapses between starting a job and finishing it is known as job time. The completion date may be stated in the contract, either with or without penalties for not meeting it, or may be set by the contractor's own schedules. It may be measured in hours, days, weeks, or years. The number of workdays must be carefully distinguished from the number of calendar days.

Job time gives an excellent check on progress. Daily or weekly plottings of accomplishment against percentage of time used will indicate whether the job is on schedule.

Progress charts.   A form that indicates the percentage of work intended and accomplished in each time period is shown in Fig. 11.16. Time is shown on the horizontal scale, percentage on the vertical. Dashed-line curves are drawn in for the schedule, and a solid line is plotted in week by week according to progress made in the field, through the week of May 5 in this example.

The dashed curves usually have somewhat the shape of a letter S, as work starts slowly, speeds up as workers and equipment become adjusted to it, and slows again as the workforce is reduced for finishing operations toward the end.

Taking the heavy grading work for an example, we can tell from the graph that it started a week late, but in one week was a little ahead of the two-week schedule. But it then fell behind, as there was no work the next week because of rain and mud, and progress was poor the following week because of mud. After that progress was good, and another week at the present rate should reach or pass the scheduled output.

The next set of lines indicates that finish subgrade work started early and is running well ahead of schedule.

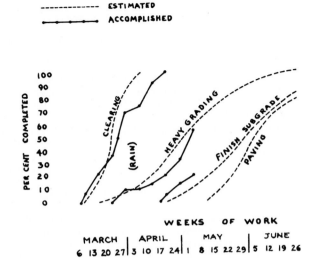

FIGURE 11.16 Work schedule graph.

A highway job involves many other items, and if each is to be followed, it will be necessary to use several graphs to avoid confusion from crossing lines. Colored pencils are always helpful in making graphs easy to understand.

A contractor seldom expects the work to correspond closely to her or his ideal curve, but it is important to know how far off it is, and why.

EQUIPMENT SELECTION

For many earthmoving operations there are alternative equipment selections that can be considered. The object is to select the piece of equipment, or the combination of equipment, that will produce the lowest overall cost per yard moved, assuming that the earthmoving operation is the primary one that governs the total work to be done. This is a tall order, but one that must receive careful attention. Side-by-side analysis in the field probably is the most conclusive way of answering most of the questions for a specific job. However, computer simulation is a valuable tool to cover all the "what if" variations that should be considered.

The variations that might be considered for an earthmoving operation include crawler or rubber-tire equipment, weight-to-horsepower ratios, tandem or single-engine power, large-capacity or smaller, more maneuverable equipment, self-loading scrapers or scrapers with pushers, scrapers or top-loaded haulers, and other alternatives. These differences are more evident when the variations of job requirement and conditions are included in the selection process. The tabulation given in Fig. 11.17 suggests possible equipment selections for differing job requirements and job conditions.

EQUIPMENT RENTAL

There are many different types of equipment rental arrangements. The one that we will discuss here is the renting of a machine owned by a contractor, distributor, or rental agency to a contractor who will use it as his or her own during the period for which payment is made.

Job requirement	Job condition	Possible equipment
Haul length		
<1000 feet	a. Relatively level	1. Rubber-tired dozer 2. Front-end wheeled loader 3. Single-axle drive elevating scraper
	b. Adverse grades	1. Crawler dozer 2. Tandem elevating scraper
1000–3000 feet	a. Relatively level	1. Single-axle scraper with wheel pusher 2. Loader with single-axle hauler
	b. Many grades	1. Tandem drive scraper
>3000 feet	a. Relatively level	1. Excavator with bottom-dump wagons 2. Belt loader with large haulers
	b. Grades and turns	1. Excavator with articulated haulers
Haul route condition		
	a. Compacted earth	1. Rubber-tired equipment 2. Single-axle drive equipment
	b. Soft and muddy	1. Crawler-mounted equipment 2. Excavator with track haulers 3. Dragline with dual drive haulers

FIGURE 11.17 Possible earthmoving equipment selections. More about these alternatives is discussed in Part 2 of the book.

Rental of equipment with fuel, maintenance, operator, and supervision will be considered later under Contracts.

A contractor may decide to rent part or all of the machinery needed on a job because of short period of use, availability, lack of confidence in future work, lack of capital, and/or other reasons.

Cost.   Equipment rental is often casual in nature. A contractor who has a machine that is idle or nearly so will rent it to another contractor who has need of it. Price is likely to be strongly affected by the amount of demand for the unit, its condition, and the financial positions of the parties. Rental rates under these conditions may vary widely.

Rates may be based on the following references or some other formula. They are usually for the base machine, with separate rates for some buckets and equipment. Delivery and operator, fuel, lubrication, or service, if available, are extras.

Rates are usually based on one shift of 8 hours per day, 40 hours per week, and 176 hours per month of a 30-consecutive-day period. Contractors may find that it pays to work a 10-hour day or use other overtime arrangements to get full time on expensive rented equipment.

Overtime on the machine is paid at the same hourly rate as regular work time. Time may be taken from the contractor's records, inspectors' reports, hour meters, engineer revolution counters, or combinations of these methods.

Unless other arrangements are made, the rental period starts when the machine leaves the owner's yard, and does not end until it is back in the yard, or is taken to or by another contractor by arrangement with the owner.

Even if the machine does not work the full number of hours, or any hours at all, during the rental period, full charge will be made except under special conditions. Most firms renting equipment will make allowances for time lost through long breakdowns, excessive bad weather, or strikes or material shortages; but the conditions under which such allowances will be made should be clearly understood in advance.

AED.   The Associated Equipment Distributors (AED) used to publish every other year a Rental Compilation. Now it is published by K-III Directory Corporation as the AED Green Book. A warning in the compilation reads

> The rental rates and terms set forth in this compilation are for informational purposes only and not to suggest or to influence the rates or conditions of rental of any item of equipment, as this is a matter which must be determined by the lessee and the lessor of the equipment....
>
> For any distributor, or any other person, to enter into any agreement, understanding, combination or concerted action with one or more distributors, or with one or more other persons, to adhere to the rental rates shown in this Compilation, or to refrain from charging less than such rental rates, would be a violation of the Federal and State Anti-Trust Laws.

K-III Directory Corporation also publishes *Rental Rate Blue Book for Construction Equipment.* This is also a compilation. It is more detailed in that it provides rates for specific make-and-model items of equipment, and has information on regional variations.

The two references differ widely on many items. However, either (or both) may be very useful as a quick guide to relative costs and values.

Repairs.   A definite understanding should be clear about repairs in connection with every rental agreement. Policies of owners vary with local custom and the type and condition of equipment.

The owner should take care of overhauls, major repairs, cleaning, and painting. The owner should do this conscientiously enough that the equipment is able to work through its rental period without major breakdowns. The renter is supposed to take care of small field repairs, all damage from abuse or accident, replacement of cables, cutting edges, and other fast-wearing parts, and to return the machine in as good condition as he or she got it, except for normal wear.

Field repairs should be the responsibility of the owner if the breakdown is in parts known to be defective at the start of the rental period.

Some rental arrangements distinguish between nontractor equipment, for which the owner assumes responsibility for wear and tear, and tractor and rubber-tired scrapers and haulers, on which the owner expects the contractor to pay all expenses, including those resulting from ordinary wear and tear in normal use.

Misunderstandings often arise as to the responsibility for major repairs needed during the rental period, and the amount of wear that can be said to be normal. Adjustment of these differences can mean a cost variation of several dollars per hour, so a clear understanding in advance is important.

The owner usually reserves the right to pull equipment off a job where it is being abused.

Ownership versus Rental or Leasing.   The contractor who keeps machinery from job to job and takes good care of it operates at lower cost than if the equipment were rented. Rental prices include an allowance for greater-than-average major repairs because few people are as careful of rented equipment as they are of their own, and the owner's profit is of course added in.

However, for short jobs with no sure usefulness for the machinery after completion, renting is cheaper than purchasing for the job.

Contractors whose work is scattered over the country usually rent machinery at each job, instead of owning and moving it. This saves heavy transportation expense, reduces hostility to a "foreign" contractor, and makes it easier to hire and control local operators.

The following will serve as an example of figuring comparative costs on one job.

Assume that a job requires a front loader to work 600 hours during a 4-month period, and that the new price of such a machine is $100,000 including incidental expenses of purchase.

If the contractor buys the loader, and keeps it employed for 1,200 hours per year for 5 years, then junks it, the average hourly cost for depreciation will be

$$\text{Depreciation, } 100,000/6,000 = 16.67$$

If the contractor buys the loader and sells it for $75,000 on completion of the job, the costs will be

$$\text{Depreciation } 25,000/600 = \$41.67$$

If the contractor rents the machine at the rate of $3,000 per month, he or she will pay $12,000 plus a one-way delivery charge of $300.00. Then

$$\text{Rent, } 12{,}300/600 = \$20.50$$

Other costs for each alternative, such as ownership or insurance, repairs, fuel, and lubrication, should be added to each, but they will be relatively small.

The advantages in renting equipment versus owning it include: no capital expenditure and modern efficient and well-maintained equipment are nearly always available. Leasing equipment is a form of renting long term with the option to buy the equipment. True leases are not included as debt on a contractor's balance sheet which helps in assessing his or her financial stability.

A study in the United States in 2001 showed that the percentage of contractors who were buying equipment outright versus financing or leasing was decreasing so that the alternatives are practically equal. The percentage of contractors leasing versus renting short term was about the same. However, the percentage of contractors buying new heavy earthmoving equipment was slightly higher than the percentage of those renting but twice as high as the percentage of those buying used equipment and much higher than the percentage of those leasing this equipment.

The probable reason for more renting of equipment, particularly when the economy was poor and rental rates have been lower, was that local rental companies have been taken over by larger, national companies, who can buy equipment at better discount prices. When renting a piece of equipment the contractor must be careful to understand the time periods for the rental rates, e.g., some rental companies may base the monthly rate on 28 days instead of 30 and the extra days of use would be prorated.

It would be possible for the contractor to partner with the rental company, which would help if there is downtime because of equipment failure to be replaced, if training assistance is needed for the operator, and to improve on productivity of the equipment. With the availability of a wide selection of equipment the contractor can always have the right equipment for his or her job. To find just the right piece of equipment to rent there are Internet Web sites that can lead to the desirable machine anywhere in the country.

ESTIMATING

A contractor is usually called upon to estimate the time, material, and expense involved in a piece of work. This estimate may involve careful calculation of all factors, may be made up from records of similar work, or the memory of them; or, in bidding on a small proposition, be only an informed guess.

An estimate may be used as a basis for making a fixed price bid, or simply to give the customer an idea of cost while performing the work on an hourly or cost-plus basis.

The first requirement for most estimating is practical experience with the work involved. In large organizations, this experience may be only in handling cost, production, and time figures. In small firms, the figuring is often done by the same person who does or directs the work. That person should be familiar not only with excavation in general, but with the specific type or types of work to be done.

Checklist. Every estimator needs a checklist of the items involved or possibly involved in the job being figured. For simple work or rough estimates he or she may keep it in his or her head, but it is better practice to have it in writing and to refer to it frequently.

The principal use of the checklist is to remind the estimator of items he or she might forget. An experienced person might feel that he or she no longer has need for such artificial helps, but anybody can forget something.

Records of state highway departments indicate that careless mistakes are common even in multimillion-dollar estimates produced by experienced people. Errors in arithmetic are the most common failing, and leaving out operations is the next. A contractor may estimate concrete at

$55.00 per cubic yard and put it in a bid at $5.50. Or the contractor may figure out to four decimal places what it costs to drill, blast, and shovel load a rock ledge, and entirely forget the haul cost.

An estimator, whether a contractor or hired by one, should work up his or her own checklist for each type of work, refer to it, and add to it whenever necessary. It can be one of the estimator's most valuable assets.

Round Numbers.   An estimate is an informed guess. No matter how solidly it is founded in experience and knowledge, it deals with future work in which unexpected conditions can upset the most careful calculations. It is also often the basis of a competitive bid that must be lower than that of any other qualified contractor in order to get the job.

Since the figures themselves may prove to be inaccurate, and because they may be changed in the bid to meet a price, it is usually pointless to work them out to several decimal places. Excessive detail adds greatly to the time and labor of making up a bid, and the estimator may become so lost in complicated figures that he or she will overlook errors in arithmetic, or whole items that ought to be included.

A sense of proportion must be preserved. A per-yard cost of moving dirt might well be carried out into several decimals if there is 1 million yards to move. But as final figures are approached, pennies should be dropped, and dollars rounded off to the nearest 10, 100, or 1,000, depending on the size of the job. The rounding off should be indicated at the point where it is done to avoid confusion, by writing in a word such as *say* or *approximately* or an abbreviation such as *approx.*

For example, an engineer's calculation may indicate that there is 16,828 yards of soil in a bank, and excavation cost is figured at 51¢ per yard. Cost of digging the whole bank would be

$$16,828 \times .51 = 8,582.28, \text{ say, } \$8,600$$

If the engineer had simplified the figure to approximately 16,800, the calculation would be

$$16,800 \times .51 = 8,568, \text{ say, } \$8,600$$

Estimating Excavation.   The gross factors in estimating excavations are the quantity of material to be dug, its digging qualities, the distance it must be moved, haul conditions, and the manner of its use or disposal; all in relation to the equipment to be used.

Quantity, which is usually measured in bank yards, should include anything that must be dug, quarried, or moved in the course of the work. Material which is stored and reclaimed must be added in twice.

Digging qualities will include not only the hardness and coarseness of the bank, but water or sand conditions on the pit floor, danger of slides, etc. It will largely determine the type of excavators to be used, and whether blasting will be necessary or not.

The distance to be moved will dictate whether it is more economical to push or to carry it, and the types of hauling unit to be used. In general, haulage is figured from the center of mass of the cut to the center of mass of the fill, but the lengths of the longest and shortest hauls must also be considered.

Haul calculations should include attention to the type of ground to be crossed, its probable carrying capacity and tractive resistance, grades to be climbed, and the cost of making and keeping it passable.

Spoil can be dumped over a high bank more economically than it can be spread and compacted in a fill. Operations will be slowed unless there is space for equipment to maneuver and dump in, and unless the fill will support and give adequate traction to the hauling units.

Fill requirements can be greatly increased by a soft base that will compress or shift under its weight.

Digging Factors.   The digging qualities of a soil are of great importance in estimating. If blasting is required, expenses are increased 5 times or more, with the extra costs per yard increasing if the quantities are small, or if precautions must be taken against damaging property.

Hard soil that can barely be dug without blasting will also prove expensive, in the terms of slower production and increased breakage of equipment. It may require the purchase or rental of special or larger machines.

Wet digging requires working from above with shovel backhoes or draglines, results in partial loads, may call for expensive drainage or pumping, and will cause mud difficulties at the dump. Operation on wet or muddy pit floors may require the use of tracked hauling units instead of rubber-tired, with a resultant drop in speed; or substitution of all-wheel drive for conventional trucks.

Fills.   Trucked fill placed in thin layers requires more or larger dozers for spreading than when in high lifts. Even if no rollers are used, compaction and rain resistance will be improved because of better vertical distribution of the weight of the hauling units. If rollers are used, the thin layers will have more total surface to be treated, but compaction may be secured with lighter machines, or with fewer passes on each level.

Wet clay may require sandwiching with layers of sand or gravel to make a stable fill.

Specifications for compaction may be impractical, except for a highway, airfield, or earth dam fill, and compliance may be very costly in time and effort.

Sequences.   Excavation or grading projects often involve a sequence of two or more operations. Sufficient delay in one of them will slow or stop work on those which follow it. Increase in the number of operations makes the final ones more subject to delay. If each step in a series is followed closely by the next, through physical necessity, or haste, the possibility of continuing some work after a breakdown is reduced.

As an example, in laying subsurface drains, a ditch is dug, tile is laid in it, and the ditch is refilled. If the tile is laid and the ditch backfilled immediately behind the ditcher, it cannot even stop for fuel without making the tiling crew and the dozer idle. Any delay in the supplying or the placing of tile will shut down the dozer and, if the ditch is likely to cave, the ditcher as well.

If tile is supplied by truck as required, or a little ahead of use, truck breakdown will stop work quickly. On the other hand, if several hours' supply is laid out along the ditch line, work can continue while the truck is repaired or replaced.

In shovel loading, the sequence is digging, hauling, and spreading. If the shovel stops, the job stops. If a truck stops, shovel and dozer work is usually slowed. If the dozer quits, work may shut down after a few loads, or continue for some time, depending on dumping conditions.

Slowing or stopping of a job increases the contractor's cost, especially when there is no other work to which machinery can be shifted for the time involved. Fixed expenses continue, and part or all of the payroll. The effect on contracts involving penalties for failure to finish on schedule may be even more serious.

Bottlenecks are another hazard of sequences. Any machine, or any operation, which is slower than those preceding and following it will set the pace, or the lag, for the whole job, until the condition is corrected. This situation may arise through improper selection of a machine, delivery of the wrong size or type, mechanical or digging difficulties, labor shortage, lack of skill, or mistakes in figuring.

In making an estimate, sequences should be studied carefully and allowance made for the probable delays.

Rush Jobs.   Rush jobs usually involve very close sequences to such an extent that machines and workers are so on top of each other that a great deal of time is wasted, even if no breakdowns or serious tie-ups occur. An extra charge should be made to cover this inefficiency.

Another type of rush which is frequently experienced is that a customer, often an owner or building contractor, will demand that machinery be sent over immediately to backfill and grade around a building, dig ditches, or perform other work required to obtain a payment on a building mortgage, or to make the house look attractive to possible buyers on a weekend.

If such a call is answered promptly without investigation, it will often be found that the site is not in workable condition. Perhaps the whole area is cluttered with piles of sand, gravel, bricks, and lumber; or the foundation has not been painted with waterproofing, or the scaffolding removed; or neither the boss nor the plans can be found.

Owner Delays. An extra amount may be allowed on an estimate for excessive job delays caused by inadequate or contradictory plans, expectation of changes during the work, and owner meddling with work methods.

Such an extra charge may be based on inspection of plans, on the owner's reputation, or both.

Some owners are poor credit risks, and work may have to be slowed or stopped during the job because of lack of money.

Public highway contracts may have a provision that excavation must be stopped immediately in any area where prehistorical or historical ruins or objects are encountered, until the objects are checked and possibly removed by experts. Such stoppages can interfere seriously with orderly work on a project.

Other contractors may have jobs in the construction area, installing or relocating utilities, that may cause confusion and delay.

Production. Most estimators are familiar with the output of the machines to be used on jobs that they figure. If they are not, production can be determined from field studies, taken from manufacturers' charts, or worked up on paper from discussions of various classes of equipment in Chaps. 13 through 21, and from other sources.

There is a learning effect at the start of a construction operation that causes cycle times and costs to be higher than anticipated. This must be taken into account when estimating. It has been reported in *Journal of Construction and Management* of ASCE that the accuracy of predicting future performance gets about as good as it is going to get at about 25 to 30 percent of activity completion. After this point the difference between the predicted total remaining cost and the actual remaining cost is within plus or minus 15 to 20 percent.

Allowance must always be made for special conditions that will affect machine performance. These are usually on the bad side—water, mud, cramped working areas, high altitude, steep grades, and so forth. But there are also favorable possibilities, such as light, easily dug soil, rock with good fragmentation, or expert operators.

Cost of Production. The cost of owning and operating the job equipment must be known, so that its production can be converted into cost figures. If a shovel can load an average of 100 yards per hour after allowance for average delays, and all costs including operator are $80.00 per hour, the loading cost is 80¢ per yard.

The time the machine will be on the job is found by dividing its production into the volume of work. This same shovel would take 1,000 hours to move 100,000 yards of dirt. This is a year's work, $80,000 worth. Total yards divided into total cost gives unit cost again.

It is important to figure all side expenses such as supervision, spotting, pit maintenance, and incidental labor into each part of a job.

Total Quality Management. The latest pitch is for total quality management (TQM) to improve company management running the business. Applying TQM to a contractor's equipment maintenance department simply requires finding what the project or operations people want from the equipment and encouraging the workers, such as mechanics and lube crews, to help figure out how to deliver it most efficiently.

Overhead. When each part of a project has been figured, the costs are added together. Overhead expense must then be added. It is made up of the part of general overhead that will be devoted to the job, and any additional overhead costs that are incurred for it. This may be figured out separately for each bid, or an arbitrary percentage of the cost total may be used. Ten percent is usual.

Profit. Profit is what the contractors get out of their work and risk if they have estimated properly, get the job, and do it successfully. It is usually figured as a percentage of total estimated cost.

The contractors must decide on this amount for themselves. If they put it too high and don't get the job, there won't be any profit on this one. If they put it too low and get the job, they may wish they hadn't. Five to 10 percent is often used. A combined figure of 15 percent for overhead and profit is standard in many areas.

Jobs are sometimes bid on a no-profit basis to keep money turning over so that bills and installments can be paid, or to keep an organization together in hope of profitable jobs in the future. But it should be remembered that in this business, the person who breaks even is usually losing money after hidden and delayed costs are counted up.

CONTRACTS

Small jobs may be done on the basis of verbal agreements, that may be quite specific and definite (or very vague). Big jobs should always have a written agreement, that is usually in the form of a contract.

The contract describes the work that is to be done and the price that is to be paid for it. This may be done in two paragraphs up to hundreds of pages. There are usually drawings or plans, that may be one sheet or several hundred. Standard forms should be used when possible.

If good faith exists on both sides, it is usually easy to arrange a simple contract between contractors, or between a contractor and someone who is familiar with the work involved.

In making arrangements with persons having little or no knowledge of excavating procedures, the greatest care should be taken to explain both what will be done and what will not be done.

Payment.   Payment basis may be a lump sum or fixed price for the whole job, unit prices that vary with quantities, cost plus, or combinations of these methods.

Any type of contract may call for either a single payment, or installments based on the contractor's investment, work, and/or accomplishment. Monthly payments based on a percentage of work completed are usual in large jobs.

Lump Sum.   In a straight lump-sum or fixed-price contract, the owner agrees to pay an agreed price for a certain piece of work. This is a good arrangement when all the factors that will affect the job are known, but it must be based on a thorough understanding of the nature and finish of the work by both the owner and the contractor.

In a fixed-price contract the contractors are on their own as long as they keep to the job specifications and time schedule. They can reduce measurement, classification, timekeeping, and bookkeeping to what they need themselves. While the prudent owner will still have an inspector on the job, she or he has a minimum of measurement and timekeeping to do.

However, unless advance engineering work and site study are very complete, such contracts may result in disagreeable surprises for either party. The owner may have paid blasting price for a volume of rock that is readily broken out by a shovel; or for removal of valuable material that could have been dug at a profit. The contractor may be digging rock where he or she looked for loam, or running pumps 24 hours a day where the contractor thought he or she would be high and dry.

A fixed-price job that is turning out disastrously for the contractor can sometimes be renegotiated, but unless the provisions for possible change are written into the contract, the contractor is largely dependent on the goodwill and generosity of the owner for such relief.

However, the contractor can demand extra payment if the unfavorable conditions were known to and concealed by the owner, or if the owner withheld information that would have enabled the contractor to anticipate the difficulties.

Unit Prices.   When quantities have not been determined exactly, or when they may be subject to considerable change during the job, parts of a contract or a whole contract may be let at unit rates.

For example, an owner might ask for bids on removing a hill of approximately 30,000 yards of dirt. The job is let to a contractor who bids 60¢ per yard. The hill is measured before work is started, at intervals during the work, and after the job is complete. It is found that 37,000 yards has been moved. Payment to the contractor is .60 × 37,000, or $22,200.

If the contract were let on a lump-sum basis of the estimated yardage of 30,000 times .60, payment would be $18,000. But the contractor would be likely to claim that the 30,000 figure was not honest, and disagreements and even lawsuits might ensue. On the unit basis the owner pays for just the volume that is moved, whether it is more or less than his or her estimate.

Unit prices reduce the requirement for careful prejob investigations that can be very expensive if underground conditions are involved. On the other hand, measurement of quantities is difficult and sometimes inaccurate when cuts are shallow and the ground is irregular.

Quantities can also be measured by truckload or by measurement of fill.

Unit prices for earth and rock are usually based on cubic yards, trenches on linear feet or occasionally linear yards, and clearing on acres.

A typical unit price bid schedule has a number of work items for which there are quantities estimated by the owner or the owner's engineer. The contractor's bid is the sum of the price for each item multiplied by the estimated quantity. The bid is said to be unbalanced if the prices for some items are higher than they should be and others are lower. The contractor may do this because he or she thinks the estimates are wrong or wants to get more money in the early time of the contract. Some owners may think this is an unethical practice.

Classified Excavation.   On a big job that involves various digging conditions, excavation may be divided into a number of different classifications. These may be separated according to the type of work, as road cut, borrow, shallow trench, culvert, and deep trench. Or the classification may be according to difficulty of digging, as soil or rock, or dry or wet.

The most important classifications in regard to total money involved, and problems in estimating, bidding, and working, are earth and rock. The practical distinction is that soil can be dug directly by shovels of normal size for the job, while rock must be blasted or ripped before it is dug. The pay difference may be made on this basis, or according to geologic definitions.

It is fairly standard practice for the contractor to remove all or most of the soil over rock, then send for the owner to inspect and measure the rock for payment. Sometimes the two parties will agree on the amount of rock before excavation, depending on inspection of outcrops and depth of soil in test holes for the amounts.

Boulders are measured after they are freed from the bank, and before they are broken or loaded out.

In tunnels, and in some trenches and road cuts, payment for rock may be varied according to its position. Full price may be paid inside the bore, side, or slope lines, a lesser price for moderate overbreak, and nothing for excessive overcutting.

Numerous problems arise in connection with identifying and measuring rock. Many engineers and public works departments prefer to avoid them by letting excavation work on an unclassified basis. The contractor is given access to whatever boring and test hole data are available, allowed to look over the ground, and makes an estimate or perhaps guess about how much rock will be found. This method diminishes risk for the owner and increases it for the contractor.

Excavation prices usually include hauling to the fill and compacting. In highway work in some areas the haul distance is limited to a few hundred yards, beyond which an item called overhaul or paid haul calls for additional payment.

Cost-Plus.   If conditions are such that the contractor cannot readily tell the amount, kind, or conditions of excavation; if the amount of work to be done has not been determined; if it is not practical to clearly define the extent of the work and the condition in which it is to be left; or if the job is to be done a little at a time, as equipment or funds are available, the cost-plus or hourly basis will probably be the most satisfactory.

On a cost-plus arrangement the contractor will have all the costs in doing the work repaid, and will receive either a fixed fee or a percentage of the costs in addition. This type of contract is most often let in government or other work where haste prevents thorough investigation of the site, or plans are subject to change during operations.

The fixed-fee basis is appropriate where the total amount of work can be estimated with fair accuracy, and the percentage where changes and extras can make up a substantial part of the job. The latter system is subject to grave abuses, as mistakes which add to the cost will increase the profit, so that inefficiency is rewarded.

A serious cost-plus difficulty is that it is apt to lead the customer to interfere with the contractor's policies and management on the job. This effort to lower costs is liable to be of the pennywise, pound-foolish variety, and increases expenses more often than it reduces them.

It is important for the contractor to include all indirect as well as direct costs in this type of bid.

A variation of cost-plus is a bid listing hourly or daily rates for all machines, services, and personnel to be employed on the job. The contractor figures the profit on each unit into the price charged for it.

Hourly Work.   Working by the hour is almost the standard practice on small jobs in which the expenses of investigations by the owner and estimating by the contractor are not justified by the money involved. It is also common in subcontracts and other arrangements between contractors.

A working hour may be considered to be the time that the machine is present on the job, the time it is present and ready to work, or only the time it is actively working, depending on the arrangements made.

In effect, the contractor who owns the equipment is renting it to the customer, but the contractor usually retains the right to supervise, and pays all expenses, including the operator, fuel, lubricants, and repairs. Occasionally, the customer may furnish fuel or other items if he or she can do so more conveniently than the contractor can.

If equipment is rented to a job without an operator, it is a rental rather than a working agreement.

Pay for time during which the machine is stuck in mud is usually on the lessee, as it is a mishap caused by job conditions. However, if the fault lies with a disobedient or careless operator, or if the owner has warranted that the machine will not get stuck on that job, payment may be withheld.

The machine is not paid for time lost because of mechanical failure or absence of the operator. However, stops for adjustments, minor repairs, fueling, lubrication, or cigarettes, which average less than 10 minutes per hour, may be considered working time if agreement is made to that effect.

Timing.   Working or pay time may be taken from readings of electric hour meters, which register the time the engine is running; from mechanical counters which register engine revolutions in terms of hours of wide-open operation; from special checking by a foreman or timekeeper, from the lessee's job time sheets; or from the contractor payroll records.

On operator work, it is good practice to check time daily and have the customer sign a ticket for it.

Timing by hour meter leads to the equipment owners' operators keeping the engine running, whether it is needed or not. Many jobs involve substantial amounts of waiting time, during which noise and wear would be reduced by stopping the engine, but this action by the operator would penalize her or his employer and possibly herself or himself.

Hour meters should be checked frequently as they may become disconnected.

Engine revolution counters are more accurate, and seldom get out of order, but when used as a pay basis, offer the added disadvantage of placing a premium on running the engine at full throttle at all times. This may make it difficult to do precise or fine work and will cause excessive wear, waste, and noise.

Jokers.   Many contracts are tricky and can be used to make the contractor do the work for part pay, or to take the responsibility for conditions beyond her or his control.

It is customary to make payments on account on jobs which take over a few weeks to complete, so that the contractor will not have to scratch for payroll and immediate expense money. In many types of work, it is usual for the owner to withhold a percentage, which may range from 5 to 50 percent of the value of work performed, as security for completion of the job and fulfillment of any guarantees.

Such contracts may leave the owner the option of not completing the job, thus withholding final payment indefinitely. For example, a contractor might bid on a development job of installing sewers, backfilling the ditches, laying gravel roads, and then blacktopping. This is one job, and 10 percent withheld from payments is not due until the blacktop is completed.

But the developer can sell houses when the gravel is in. He or she may lack the money to put on blacktop, or just figure it is a good idea not to do it, and hold onto the percentage. This may be possible under the contract, unless it specifies that each operation calls for a separate final settlement, or that all work may be performed in proper sequence, without specific authorization.

On contracts which call for penalties for failure to complete by a certain date, the contractor should be protected against delays caused by the owner. These may include failure to complete prior operations or to remove surplus material left from them; or not supplying plans, grade stakes, work permits, or access to property on time.

The contractor should also protect against shortage of materials, by means of delay-caused-by-circumstances-beyond-control clauses, the option of substituting available for unavailable items, or both.

However, the greatest losses to contractors occur because of their forgetting to leave a loophole for possible underground conditions. The big three are rock, mud, and flowing water, and one or all of them can crop up in the most unexpected places.

Highway contracts may call for compaction of fills to a specified density that can only be reached if the soil contains just the right quantity of water. It may be impossible to reach this density if both the cuts and the weather are wet.

CRITICAL-PATH SCHEDULING

Critical-path scheduling permits visualizing projects, study, and working out sequences, time, and costs more readily than is possible with bar graphs.

Most projects include one or more jobs or job sequences that must be completed before another phase of the work can be begun. For example, one sequence may be first clearing, then trenching a culvert site, another procuring and bringing in specially designed pipe. Completion of these two sequences is necessary before pipe can be laid.

If one job or one sequence takes longer than the others leading to the same result, its time determines the time for achieving that result, whether it is starting the next phase or finishing the project.

Because of its important effect on work scheduling, the operation or sequence taking the longest time is called the critical path.

Vocabulary.   Critical-path scheduling has been set up in a somewhat formal manner as to vocabulary and format, to enable its users to understand each other, and to permit solving its more intricate problems by means of computers.

For purposes of this work, the following words are limited to the meanings listed for them:

Chain: a sequence of jobs following each other

Crash: speedup or rush work

Duration: the time required by a job

Event: the start or finish of a job or jobs, as at *a, b,* etc. in Fig 11.18

Float: time available for a job, minus job duration

Job: one small activity or single class of work, as clear site

Arrow Diagram.   Critical-path schedules are worked out in arrow diagrams, the simplest form of which is shown in Fig. 11.18. Each arrow represents a job or activity. It may be labeled by description as in this illustration, by code letters, or by an event numbering system. The first arrow is usually for lead time, getting ready to start work. An arrow may be added for cleanup.

Arrows are made in any convenient length, and may be straight or curved. They indicate only the sequence of the jobs in the pattern of a project, and have no scale.

Three questions should be asked and answered about each arrow:

1. What immediately precedes this job?
2. What immediately follows this job?
3. What can be concurrent with this job?

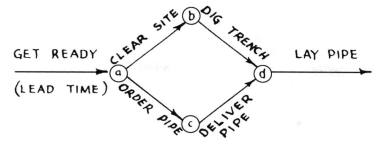

FIGURE 11.18   Simple arrow diagram.

The arrow diagram must be worked out logically and thoroughly in regard to sequence and interdependence of jobs. Omission of any item gives a false picture and may lead to mistakes in scheduling. On the other hand, the simple act of placing each activity in a frame of reference with other jobs helps in building up an intimate knowledge of the project.

Overlapping Jobs.   It is usual for construction projects to have overlapping sequences. Brush clearing comes long before laying pavement, yet the two operations may go on at the same time in different parts of a highway section.

As an example, let us take laying a pipeline. On a simplified diagram this may be broken down into three activities that must be done in succession: trenching, laying pipe, and backfilling. A work section of a pipeline may be many miles long, but it may be possible to start each job as soon as a few hundred feet of the previous job are completed.

Each of these jobs may be considered to be done in three sections: the initial, continuing, and finish. Initial work must be completed before the next job can start, while the other two run at the same time in different areas. Figure 11.19 shows the arrangement of arrows to indicate this situation. Note that all are the same length, although initial work may take a day or less, while continuing work may go on for months.

The pipeline may require a pumping station. Building it would be a very different type of work that might be subcontracted, or diagrammed separately. However, the line is not usable without it, so it should be represented in this master diagram as the lower arrow, D.

Events.   The start and finish of every job is called an event. Therefore every arrow begins at an event and ends at one. These events are numbered, starting with 1 or 0 at the beginning of the project and continuing through an unbroken sequence of numbers to the finish. However, because of concurrent or parallel job chains, the numbers are not necessarily in sequence in any one chain.

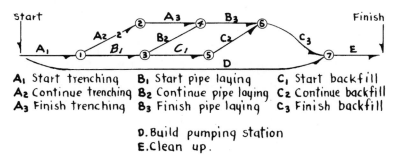

A_1 Start trenching B_1 Start pipe laying C_1 Start backfill
A_2 Continue trenching B_2 Continue pipe laying C_2 Continue backfill
A_3 Finish trenching B_3 Finish pipe laying C_3 Finish backfill

D. Build pumping station
E. Clean up.

FIGURE 11.19   Overlapping jobs.

The only absolute rule in assigning these numbers is that the number at the head of an arrow must always be larger than that at its tail.

Event numbers are used to identify the arrows between them. In Fig. 11.20, A is 0-1, B is 1-2, C is 1-4, and so forth. Since diagrams often include enough arrows to use the alphabet many times over, identification by event numbers is more practical than letter codes. It is also necessary when problems are to be handled by a computer.

Sometimes two or more jobs will begin and end at the same event. A borrow pit may require clearing, testing, and measuring before digging starts. In Fig. 11.21 the three arrows B, C, and D

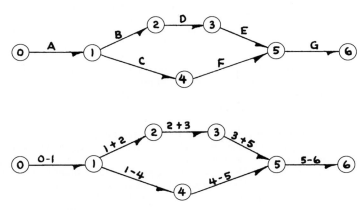

FIGURE 11.20   Event numbers.

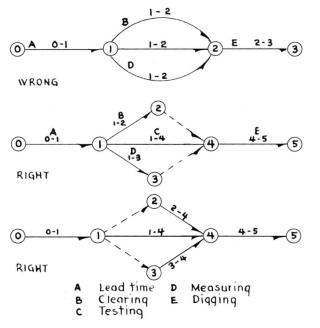

A	Lead time	D	Measuring
B	Clearing	E	Digging
C	Testing		

FIGURE 11.21   Dummy arrows for identification.

in the top diagram would each be designated 1-2. This duplication is avoided by introducing a dummy, as shown in the two lower illustrations. The dummy may be either before or after the arrow. The junction between the arrow and the dummy is given an event number.

The rest of the illustrations in this section will use letter codes instead of event numbers, to avoid confusion with other figures. A real working diagram is drawn on a scale large enough to put in all necessary figures without crowding them.

Duration. When an arrow diagram has been completed and checked, the time that the job is expected to take is written under each arrow. This may be in hours, days, months, or any appropriate unit, but the same measurement must be used all the way through a diagram. If days are used, they must be working rather than calendar days, to avoid confusing calculations with holidays and weekends.

The duration assigned is first a normal or average time, taken from experience, job studies, or an estimator's figures. In an ordinary construction project it would involve one-shift operation and use of equipment on hand or readily obtained, without either rush pressure or deliberate waste of time.

Dummy arrows are dashed or dotted lines and always have a zero duration, as they are only symbols to show connection between jobs. The top diagram in Fig. 11.22 shows duration times.

Event Times. An event time is the sum of the durations of the jobs that precede the event, and represents the time that will elapse between the start of the project and that event. If two or more

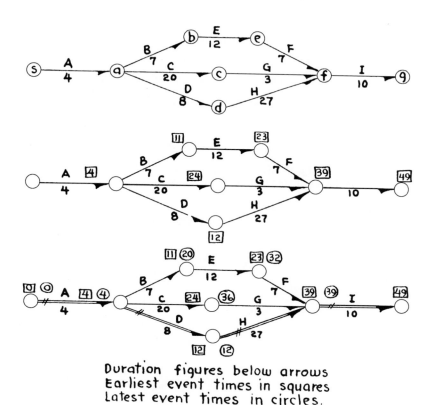

Duration figures below arrows
Earliest event times in squares
Latest event times in circles.

FIGURE 11.22 Earliest and latest event times.

chains or sequences of jobs are needed to make the event possible, its time is determined by the slowest path.

The earliest event time, abbreviated EET or sometimes e.t., is defined as the earliest finish of the event by the slowest path. It is indicated on arrow diagrams by a number inside a square, and is worked out for each event in a diagram, as in the middle drawing.

Working out the EET is simply a matter of addition until a junction of two arrowheads is reached. Here both preceding chains are figured, and the larger number is used. In the illustration, the three chains leading up to the beginning of I add up to 30, 27, and 39, so 39 is used.

The earliest event time calculation shows how long the job will take under ordinary conditions.

The latest event time, abbreviated LET or l.t., is the latest time at which an event can be finished without delaying completion of the project. It is found by working backward from the earliest event time for completion, along the slowest path. It is written in a circle, alongside the square containing the EET, as shown in the bottom diagram of Fig. 11.22.

This is simple subtraction, working backward from the finish, except at a junction of arrow tails, where the smaller number is used. In the illustration for the finish event of job A, the three chains show LETs of 13, 16, and 4, so 4 is selected and written down.

The latest event times provide a quick method for determining what chain of jobs sets the time for the whole project, and is therefore its critical path.

Critical Path.   The critical path is a sequence made up of one or more jobs or job series, whose duration is the determining factor in the length of the whole project. In order to be critical, a job or series must conform to all of the following requirements:

1. EET and LET must be equal at the start.
2. EET and LET must be equal at the finish.
3. The time available for the job must be equal to its duration.

The time available for each job is found by subtracting the starting EET from the finish LET.

In Fig. 11.22 the critical path is ADHI. Arrow F is not a critical job because the starting figures, 23 and 32, are different. Arrow G begins with an EET of 24 and a LET of 36. One noncritical job prevents a series from being critical.

As critical jobs are identified, they are marked with a double hash stripe. When the critical path is worked out, it may be emphasized by making the arrows heavier or with double lines, or by color.

There may be two or more critical paths, in which case each of them is marked.

The critical path ADHI determines the overall time of 34 days. Any efforts to save time and shorten the job should be first concentrated on it.

Float.   The spare time in the quicker jobs, series, or paths is called float. In Fig. 11.23 the critical path is ADEF, with a total duration of 34. The alternate path, ABCF, would allow completion in 24 days if jobs D and E were not needed or could be speeded up. This also means that B could be started 10 days after the completion of A, without delaying project completion. This 10 days is float.

From a scheduling standpoint, float time may be regarded as waste time. It indicates a possibility of idle time for the workers and equipment doing the jobs, and it presents a problem in utilizing them to speed up critical work.

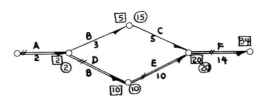

FIGURE 11.23   Critical path.

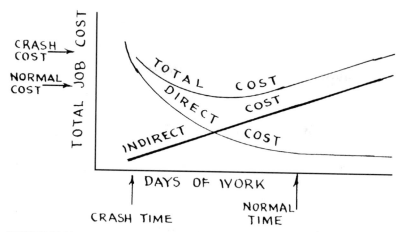

FIGURE 11.24   Speedup and cost relationship.

For example, our illustration might represent grading for a highway, with a small cut, B clearing and C digging, and a larger one, D clearing and E digging. The original durations might have been assigned on the basis of an equal force in each cut. By taking personnel and equipment from the small job and assigning them to the critical path, the two cuts might be done in 13 days each, permitting blue-top work, F, 5 days earlier.

This is one way of using float time. It often happens that the same crews cannot work on two jobs. Then an effort is made to shorten the critical-path durations in other ways, to squeeze some or all of the float out of the faster series.

Crashing.   Rushing a job through by an intensified effort that involves a substantial increase in costs is called crashing. Some jobs can be shortened by a big percentage at moderate cost, others respond poorly to unlimited extra expenditure.

A usual relationship between job duration and job cost is shown by the lower curve in Fig. 11.24. Extending time beyond normal duration saves little money, and pouring in money after a saturation point saves very little time. The greatest gains in time for extra dollars spent occur in the first few days saved, and the smallest gains per dollar are found near the minimum time end of the curve.

One curve is for direct costs only. Overhead or indirect costs are likely to be about the same per day whether the job is crashed or allowed to sleep. The straight line shows these, and the upper curve shows a total cost against project duration. The low point in this upper curve occurs at the most efficient time for the job.

The shape and pitch of the direct cost curve and the steepness of the indirect cost line vary with each contractor and job. In general, the highest proportion of overhead cost to direct cost is found in the very large companies and in the very small ones. A big organization must carry many salaried people, a one-person outfit must meet her or his daily living costs out of a small work volume. In either situation, the best return will often be obtained from crashing jobs, rather than letting them just plod along.

CAUSES OF FAILURE

Every year many excavating and general contractors fail, or sustain losses that force them to operate on a reduced scale, or give up. Most of the failures arise from one or more of the following causes:

Unforeseen price rises

Abnormal labor cost

Abnormal equipment breakage

Death or disability of owner or key people

Fire not adequately insured

Liability or property damage not adequately insured

Poor accident record

Failure of subcontractors

Adverse weather

Unforeseen subsurface difficulties

Faulty credit judgment

Sudden restriction or withdrawal of credit

Unavailability of materials

Taking on too much work for financial resources

Taking on too much work for adequate supervision

Speculation

Diversion of funds to nonbusiness use

Embezzlement by employees

Some of these subjects have been discussed previously; others are of a general business nature and are too complex for discussion here. Two subjects of particular importance to the excavator, however, are accidents and insurance.

ACCIDENTS

An accident may be defined as an unforeseen sudden happening, or as an unintentional and damaging interruption in an orderly process.

The important accidents are those in which persons are injured. However, this is often a matter of chance rather than the character of the happening, and an accident in which no one is hurt should be taken seriously, and steps taken to prevent its recurrence.

Employees should be protected by workers' compensation insurance. This coverage is usually required by state law, but in any case is a *must* for any employer interested in the welfare of his employees, and in his own. Nonemployees and property of others should be protected by liability and property damage insurance, lack of which can wipe out a prosperous business overnight. A contractor can protect his or her own equipment and property with fire and damage insurance.

However, the possession of full insurance does not justify the slightest negligence in regard to accident prevention. For one thing, the best insurance will only pay the more obvious costs. In small accidents that are most common, indirect uninsurable costs may run 5 times as high as the payments under compensation.

Some of these expenses and losses are

1. Increase in insurance rates
2. Payment to injured employee of wages for period too short for compensation
3. Loss of time of other employees who stop work at the time of the accident and because of it
4. Time spent by foremen and supervisors in assisting injured person, investigating the cause, selecting and briefing or training another person for the job, and preparing accident reports and attending hearings

5. Slowdown of job, with possible failure to finish by deadline

6. Paying full wages to employees who return to work before being capable of performing full duties

7. Loss of chance for profit on an operator and the machine

8. Lowering of morale of other workers on the job

9. Possible interference with work methods by public officials

10. Unfavorable newspaper and other publicity

Prevention.   The first rule in accident prevention is to use common sense—in laying out a job, assigning machines and personnel to their duties, providing adequate supervision without fussiness, and setting up sensible and reasonable safety rules.

Too many safety rules may be worse than none. Every one of us has a limit to the amount of good advice we can absorb, and that limit is often painfully low. It is better to take a few important points at a time, and hammer them home, than to prepare long lists that will neither be read nor remembered.

Enforcement of safety rules should not be so strict as to cause workers to fail to report for first aid for minor accidents, or to lie about the way in which they occurred.

Excellent posters and leaflets can be obtained from insurance companies and safety councils, and when used in moderation bring very good results. Only those that have some bearing on the work should be selected.

Workers' suggestions, on both safety and other matters, should be encouraged and acted upon.

A worker's skill should not be taken for granted. In an emergency an unfamiliar machine might cause an experienced operator to make the wrong move. Judgment should be used in giving out ticklish assignments. Training and refresher programs should be given periodically and whenever needed, and reference material on proper operation and procedures should be available.

Good housekeeping is important. Piles of junk, material, litter, boards with projecting nails, carelessly piled bags of material or heavy parts, and accumulations of grease and dirt cause accidents directly, and also indirectly by encouraging sloppy work attitudes.

Crowding causes accidents. On a rush job a boss tends to jam as many machines and personnel into the work area as it will take without bulging. That may mean collisions, and collisions lose time. One person can dig a ditch faster alone than with a helper who hits him or her on the head with a pick.

Piling Materials.   High piles are dangerous piles, with the exception of loose material lying at its angle of repose.

In excavating, even a shallow ditch can injure someone seriously by caving, and deep ones are killers. High vertical faces around a cellar excavation might stay up, but it is safer not to trust them. Shore them up, and make sure the shoring is strong enough. Do not just guess; have it designed and inspected by an experienced and careful person.

Barricades.   It is not only the workers who must be kept out of accidents, but also the public. There are sidewalk superintendents who like to watch the work, and are apt to be foolish enough to fall into it if they have the chance. If there is an attractive danger spot, like a basement excavation in a city, they must be fenced out. Such a fence should be strong, and at least 7 feet high.

The fence or barricade must be secure itself, so that it will not fall into the excavation, or be left partly in space by a slide. It should have windows or peep holes in it. These build goodwill for the contractor, and make spectators less likely to move into the very dangerous truck drives that penetrate the fencing.

Barricades, signs, and flares can hardly be overdone on roadways. Any excavation that extends into a road, and particularly into a high-speed highway, is just asking for trouble. And it is not enough to mark it so well that only 1 in 1,000 would fail to notice it—10,000 cars might pass while it is open. And the police, the lawyers, and the newspapers will not be interested in the 9,999 who did not crack up in it. Just in the one who did.

Insect Stings. Clearing and excavating bring workers into painful contact with hornets, yellow jackets, and other stinging insects so often that it is one of the special risks of the business. While in most cases no serious injury results, such stings can be more dangerous than is commonly realized, and they cause a number of deaths every year. They respond excellently to proper and early treatment.

There are three dangers:

Allergy to the injected poison, which will cause exaggerated reactions, and if very severe may result in shock or death from a single sting.

Stings close to the eyes or other vulnerable parts, which may disable a normally sensitive person.

Multiple stings from a swarm of insects, which may produce serious poisoning.

Most trouble comes from unexpected contacts. Preliminary scouting of an area on foot may reveal the location of nests, particularly of hornets on branches.

When possible, such nests should be destroyed in advance of the work. This can be done at night with little danger, as the insects are then sluggish and nearly blind. Also, as they are all nested, a 100 percent kill may be effected.

Ground nests are eliminated by pouring $\frac{1}{4}$ or $\frac{1}{2}$ cup of insecticide down the hole, then tamping dirt in the top.

Paper hornet nests should be wrapped in wire screening, mosquito netting, or cloth; cut off the branch; and burned on a hot fire or kept under water for at least 48 hours.

The worker doing this job can be protected by heavy clothing, gauntlet leather gloves, a hat or helmet, and a head-protecting mosquito net. The last item is the most important, as face stings are painful and dangerous.

If it is necessary to work among ground nests that have not been treated, they should be completely destroyed by pushing out or deep burial on the first approach. The insects then are disorganized and less likely to attack, particularly if the machine is kept in motion.

Minimum protection for operators in a danger area is a head net.

Laborers known to be particularly sensitive to stings should be kept on safer work until they can be desensitized to the poison by a series of shots.

Treatment. Treatment consists of stopping the swelling, slowing absorption of poison into the system, and stimulation to help to overcome its effects.

Three minims (a minim is $\frac{1}{15}$ cubic centimeter) of epinephrine, divided among two or more shallow injections at the edge of the swelling, will constrict the blood vessels, stop enlargement of the swelling, and wall off the poison. This treatment should be a routine precaution for any sting near the eyes.

A dose of the same size injected in the upper arm rallies the system for defense. If no "lift" is felt, the arm injection can be repeated in 10 minutes, or sooner if the patient is unconscious.

These injections are made much more effective by addition of equal amounts of Chlor-Trimeton (strong solution) or some other injectible antihistamine to epinephrine before injection.

Ordinarily, injections can be made only by a doctor or a nurse. Sometimes it is possible to obtain bee sting kits including automatic injectors, for lay use in emergencies before medical help can be obtained.

The most vital factor in treatment is quick action. Every minute of delay increases the extent of the injury, and the danger of shock. Even single stings in sensitive people, and multiple stings in anyone, should receive prompt attention.

In the absence of other remedies, absorption of the toxin may be slowed by an ice pack on the stings and/or a tourniquet above them. Danger of shock may be reduced by strong black coffee, taken by mouth if the patient is conscious, rectally if not.

Surface applications of mud or ointments may relieve pain, but have little or no effect on swelling or systemic reactions. Use of such remedies should not be discouraged, however, as they satisfy the person's desire to "do something."

INSURANCE

Every contractor needs insurance. The only questions are, What kinds and how much? There are two types of insurance. One protects property owned by the insured, who is paid if it is damaged or lost. The other protects against claims for damage to other people because of the insured's negligence. They are both important, but the second much more so than the first.

Much of the insurance protection a contractor needs is required by the majority of business people, but there are special angles.

To the layperson, insurance policies are complicated and confusing. There are many kinds of coverage, some of them overlapping; and many circumstances that affect each type. It is important to go to a good broker or agent who can explain in detail the purpose of each policy and what it covers, and even more important, what it does not cover.

Self-protection.   To protect her or his own property, a contractor should have fire insurance on the buildings and their contents, and separate all risk "floater" insurance on the equipment. Cars and trucks may be covered under the floater, or under separate motor vehicle policies for fire, theft, collision, and other damages.

The building insurance is made more complete by extended coverage, added at moderate additional cost, that protects against damage from wind, storm, hail, aircraft, vehicles, smoke, and certain other causes. Vandalism, earthquake, and some other coverages may need special endorsements on the policy. It should be remembered that these, and flood damage, are not included in extended coverage.

A good tools and equipment floater policy will protect a contractor against most damages to the machines—fire, theft, overturning, tornado, upset, and collapse of bridges. But riot, vandalism, malicious mischief (increasingly important), and "loss while waterborne" are probably included only on payment of an extra premium.

Such a policy may list all pieces of equipment covered, or list the large units and lump the smaller ones. Another method is to declare a gross value for all the machinery, and pay a premium on that. If equipment is listed individually, there is usually automatic coverage of new machines for a short period after purchase.

Compensation.   Workers' compensation insurance, required of employers by law in practically all of the United States, and by common sense and self-interest in all of them, pays medical expenses, part wages (as disability benefits), and damages to employees injured on the job. Usually there is a period of time, such as a week, in which workers' compensation pays no wages unless the disability extends over a longer period. There may also be gradations from partial to full compensation for time lost, as the no-work period lengthens.

Premiums are based on the type of work and the amount of the payroll. Rates and requirements differ in various states, and a contractor working across state lines must take care to be covered on both sides.

The cost of workers' compensation insurance has risen sharply in the closing decades of the 20th century. In the United States between 1985 and 1993, premiums for the construction trades have increased an average of more than 10 percent. This is in spite of the fact that injury rates decreased. The premium increases were due to soaring medical costs and widespread abuse of benefits.

Liability and Property Damage.   Liability insurance pays for injuries to people caused by acts of negligence for which the insured is liable. Property damage pays for similar injury to property.

A contractor is neither a good business person nor a good citizen if he or she is not well insured for injuries and damage to others. The equipment and the nature of the work both make it likely that claims will be brought against the contractor. He or she cannot afford to be put in bankruptcy by an operator's carelessness, nor should the contractor risk causing damages for which she or he could not settle.

All too many contractors, and other business people also, think they are completely insured until an accident shows a hole in their coverage. This section will point out a few of the pitfalls, but the best precaution is to be friendly with a good insurance agent and talk to him or her freely about jobs and work methods.

Most liability policies have a minimum coverage of $25,000 for injury to one person, and $50,000 for injury to two or more in the same accident. The policy covers each of a series of accidents in the same amounts, until it expires or is canceled.

In addition to the face amount of insurance, the company pays for investigation and for legal and trial costs, bonding fees, and release of attachments, which may add up to substantial costs.

Exact coverages of policies vary from company to company and state to state, so the following discussion is only a general guide to what might be included.

First there is motor vehicle insurance, on personal and business cars, pickups, trucks, trailers, and equipment that travels under its own power or is towed on public roads. This includes wheel tractors, graders, and self-powered scrapers.

Rates on trucks increase with their gross weight. Rates on wheel tractors and other heavy, slow-moving equipment are prohibitively high. Arrangements can sometimes be made for coverage on job-to-job moves under the general contractors' liability. Careful investigation should be made of this point.

Towing a trailer of any kind may invalidate car or truck insurance, unless provided for in the policy, or the trailer is separately insured. If such towing of an uninsured trailer is rarely done, the company insuring the vehicle should be willing to issue a special endorsement or binder to cover the combination for a specific trip or time period, at little or no cost.

If there are a number of motor vehicles, economies may be affected by insuring them together in a fleet policy, and by keeping some of them on low mileage and therefore low-rate local errands.

Contractors' Liability.   There are a number of classifications of liability risks for the contractor that can be insured separately. It is good business to lump as many as possible in a comprehensive policy, to avoid extra payments on overlapping coverage, and to avoid confusion.

A comprehensive policy may cover

Ownership, use, and operation of buildings and premises

Construction machinery, as above

Completed work (products) having defects causing injury or damage

All contractual work of kinds specified in the policy

Operations of subcontractors, except in maintenance of insured's property

It probably will not cover

Dogs, animals, boats, aircraft, or vehicles

Blasting

Damage to subsurface pipes, conduits, and wires

Collapse of structures caused by excavation or underpinning work

Tunneling and bridge construction

Obligations assumed for others

Damage to rented or controlled equipment

The first exclusion in the above list is made because these risks should be covered by other types of policy. The next four are high-rate risks, and losses incurred under them can be more justly paid under special endorsements or other policies, by those who do such work, than by the larger number of contractors who do not.

These risks can be covered for specific jobs, usually only after inspection by the company so that it can see what it is letting itself in for, and set the premium accordingly. It is in the contractor's interest to have such inspections made to obtain the necessary coverage; not only for her or his own protection, but because it is only the most experienced of supervisors who will not benefit from talking over a job with a good inspector.

Employers often feel that inspectors are a threat and a nuisance, but they perform invaluable services both as safety engineers and job consultants. Contractors who will listen to their discussions of methods used on other jobs will often find that they will save more than the cost of the premiums charged and the safety procedures required.

"Obligations assumed for others" is a tricky one that has caused many painful surprises. It is a too-common practice for an owner to write up a work contract specifying that the contractor assumes all liability for everything that happens on the premises while he or she is working on them. This may extend the contractor's risks far beyond the premium paid for his or her own activities. It is much better for the owner to take out an owner's risk policy for work in progress, and ideal if the owner can place it in the same company that insures the contractor.

If this is not possible, the contractor can show the contract to his or her own company, and pay an extra premium for an endorsement to cover any obligations assumed under it.

If such precautions are not taken, the results of the owner passing responsibilities to the contractor may be disastrous to them both, as neither of them is insured for the owner's risks and both are responsible for them.

"Damage to rented or controlled equipment" is another joker on which many a contractor has tumbled, although the amounts involved are usually modest. Liability policies are designed to protect against claims from others. If a contractor hires a machine, it is his or hers for the period of use, and may be the subject of a claim against him or her.

Coverage to protect such equipment can be obtained by endorsement of the liability policy. The extra premium is usually based on the rental cost.

Rates.   Insurance is priced so that each class of risk will bring in enough money in premiums to pay sales, administrative, and legal costs, and the claims that have to be paid, and to leave a surplus for reserves, and dividends to stockholders or policy holders.

An increase in losses automatically results in an increase in rates, although this effect may be delayed. The increase may be applied generally to all those having the particular type of insurance, or specifically to those whose accidents have piled up the claims.

Most insurance is written on one or more basic rates covering a general class of risk, with upward or downward revision depending on local conditions and experience with a particular risk or a particular customer.

Fire insurance premiums are affected by how likely the property is to take fire, how readily and completely it will burn, and the availability of firefighting equipment and water.

Contractors' liability and property damage rates are extremely variable. They are based first on experience with a particular type of work, so that blasting will have a higher rate than landscaping. Again, coverage for blasting in the country may be at nominal cost, whereas in a city it might be as high as 50 percent of the payroll.

A small contractor may be just carried at the average of the industry. A larger operator will be assigned an experience rating, based on the number of accidents and how expensive they have been. This rating may make the contractor's insurance more or less expensive than that of the competitors, and may thus affect his or her position in competitive bidding.

The premium for compensation insurance basically consists of a percentage of the payroll expressed in terms of dollars per $100 of wages. At the start of the policy term, the company and the insured define the risks that are to be covered, estimate the payroll for 6 months, and set the premium on the basis of the estimate. Then every 6 months the company makes an inspection of the insured's books, or perhaps only of the payroll tax returns, and an additional amount is charged or a credit issued for any difference from the estimated charge.

If a contractor has a number of different activities, and does not keep separate payroll records for them, the rate of the most expensive coverage for all of them is used. It is therefore in his interest to keep the different classifications at least roughly divided.

Liability insurance may be assessed according to the payroll, or by the value of the work done during the period. Here also a separation should be made between jobs carrying different rates.

The contractor must pay liability premiums on all work done by subcontractors and by hired machinery unless he or she obtains and shows to the company certificates of insurance coverage from the subcontractors.

BONDS

The excavating contractor shares with other forms of business the danger of serious loss through dishonesty of an employee, or employees. For a contractor, the loss is as likely to be in property taken or sold "over the fence" as it is in money.

Fidelity bonds of various types are available for protection against losses of this nature.

Construction contract bonds are required of contractors performing work for federal, state, and local governments. There is an increasing use of them in contracts with private owners.

A bond is a three-party agreement, made by the contractor and the bonding or surety company to protect the owner. It usually covers all obligations that the contractor assumes on the job, including completing the work to specification, and paying subcontractors and employees so that no liens or actions can be brought by them against the owner.

Three types of bonds may be involved. The first, the bid bond, accompanies a bid or proposal on a job, and guarantees that the bidder who is given the job will enter into a formal contract to complete it and will supply bonds to complete the contract.

The bond supplied for the work itself is made up of two bonds, which are separate, but seldom if ever written separately. One is a performance bond, covering fulfillment of the contract, the other a labor and material payment bond, guaranteeing payment to personnel, suppliers, and subcontractors.

These last two are drawn separately so that no question of priority can arise when claims are presented by both the owner and those who have supplied services and materials. In the early days of bonding, the government had to be paid or satisfied first, and the others got what was left. This meant at least long delays, and in cases where the bond was too small, losses for the small claimants.

In order to obtain a bond, a contractor must convince the company that he or she is competent to do the job, and financially able to carry it. The contractor pays the premium, usually not over 1 percent of the contract price, figures it as part of the cost, and passes it on to the owner in the bid or estimate.

Performance and payment bonds traditionally were in the full amount of the contract. However, surety companies that provide the bonds have become concerned over the size of large contracts, running into the hundreds of millions of dollars, which may provide a bond for less than 100 percent of the contract price. Often a general contractor will require that subcontractors provide performance and payment bonds.

Substantial all-around benefits are sometimes obtained from writing of construction bonds. The owner can let the contract to the lowest bidder without having to inquire into the question of whether he can complete it, as the bonding company guarantees performance. The contractor may save money by driving hard bargains with subcontractors who cut their figures a little closer because they know they will be paid.

If a contractor fails to complete the job or to pay the subcontractors, the bonding company takes over, lets a new contract to finish, and pays up the bills. Quite often, the new contract will be let to the contractor who defaulted, as her or his equipment is on the job.

The contractor is legally obligated to repay to the surety company everything that it has spent to finish the work. The company makes a more cooperative and intelligent creditor than a combination of an enraged owner and starving subcontractors, and in most cases the contractor is able to work out the difficulties and avoid a failure that might have been inevitable without the protection of the bond.

Unfortunately, there is another side to the picture. Many contractors who are thoroughly competent and reliable and have adequate resources for a job cannot get a bond to cover it. Potential low bidders may thus be weeded out, and work concentrated in the hands of a favored clique.

Inability to get a bond may result from a poor background, lack of resources, too many jobs already in progress, or other reasonable causes.

THE MACHINES

CHAPTER 12
BASIC INFORMATION

In studying the construction of machinery, it is necessary, or at least convenient, to have an understanding of its basic principles. This chapter therefore includes a brief review of a few of the laws of physics involved in machinery construction and operation.

There are a number of parts or assemblies such as gears, brakes, and clutches that are common to machines of widely different types. These are described together for convenience of reference, and to avoid complicating the descriptions in the following chapters.

PRINCIPLES OF LEVERAGE

Speed and Force. An engine usually turns its crankshaft too fast and with too little force (torque) for the machine it powers. This speed can be reduced and the force increased in the same proportion by the use of gears, levers, pulleys, and other devices. Explanation of how this is done can best be started by referring to some basic laws of physics.

Conservation of Energy. The first law is that of conservation of energy—that energy cannot be created or destroyed, only changed from one form to another. In machinery, speed and force are applied by the engine. Some of this energy will be transformed by friction into heat and wasted, but most of the rest may be converted from speed to power, or power to speed, and back again.

Levers. The oldest and simplest converter is the lever, which may be divided into three types. Figure 12.1(*A*) shows a first-class lever, consisting of a crowbar with its business end under a rock, a stone 2 feet away serving as a pivot or fulcrum, and a person pushing down on the handle end, 6 feet from the fulcrum. The man is three times as far from the fulcrum as the rock is, so that if he pushes the bar down 3 inches, the end under the rock will rise only 1 inch. This loss of distance (distance is a function of speed) is balanced by an increase in force—the 100 pounds of pressure he puts on the bar becomes a lift of 300 pounds under the rock.

Figure 12.1(*B*) shows the same example except that the fulcrum has been moved to a spot 1 foot from the rock, increasing the fulcrum-to-handle distance to 7 feet. The handle end of the lever is now 7 times as long as the working end; it will have to be pushed down 7 inches to raise the rock 1 inch, but the up-push against the rock will be 700 pounds.

If the rock is too heavy to be lifted, and the fulcrum stone is pushed down into the ground by the force applied, the bar becomes a second-class lever, the end under the rock becoming a fulcrum and the lower stone becoming the load.

The same distance and force ratios apply. It is apparent that the difference between first and second class may be just which of the two points is weaker.

Increase of force in a lever or other device is called the mechanical advantage, abbreviated M.A.

A third-class lever has the force applied between the fulcrum and the work so that it can be used only to increase speed or distance at the expense of force. Figure 12.2 shows a man lifting a board 20 feet long so that one end will rest on a wall-top 10 feet above him. He places one end under a weight, making it the fulcrum, and picks it up 5 feet from that end. As the other end is

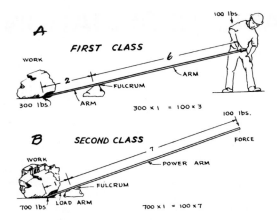

FIGURE 12.1   First- and second-class levers.

20 feet from the fulcrum, it will move 4 times as fast and as far, so that he need raise the board only $2\frac{1}{2}$ feet to get its end to the wall top. The 100-pound lift he is exerting is reduced to 25 pounds at the upper end, but since in this case the board is both the lever and the load, the lift need not be calculated at the end, but at the center of gravity 10 feet from the fulcrum. His effective lift is thus reduced only to 50 pounds.

All parts of a lever move in curves rather than straight lines. The curves are arcs of circles centered at the fulcrum.

A lever must be rigid for the distance-force ratios to apply accurately. These are always figured in straight lines from the point where force is applied to the fulcrum, and from the fulcrum to the workpoint. The lever itself may be straight, angled, curved, or offset.

If a lever has a sharp angle at the fulcrum, as in Fig. 12.3, it is called a bell crank. The space between the arms may be braced for greater rigidity, producing the triangular bell crank.

Wheel and Axle.   A wheel and axle, Fig. 12.4, is a specialized type of lever. For example, consider a wheel 3 feet in diameter rigidly fastened to a 6-inch axle, revolving at 10 revolutions per minute (rpm). Any point on it will describe a circle every 6 seconds, and the farther the point is from the center, the larger the circle will be, and the faster the point must move to get around it in 6 seconds. A point on the axle surface moves 19 inches with each turn, and on the outer edge 113 inches. As with all mechanical devices with a fixed amount of power, greater speed means

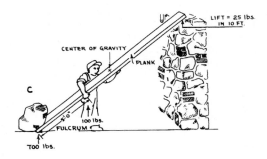

FIGURE 12.2   Third-class lever.

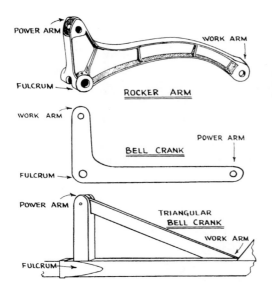

FIGURE 12.3 Rocker arm and bell crank.

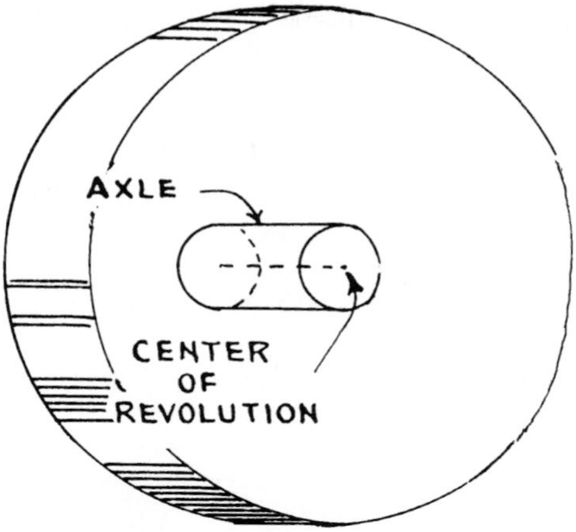

FIGURE 12.4 Wheel and axle.

less force. This wheel may be considered to be a number of third-class levers, with the center of rotation the fulcrum, the axle surface the power point, and the wheel perimeter the work.

If power is applied to the axle and work is done at the outer edge of the wheel, the result is an increase of speed and decrease of force in a 6:1 ratio. If the power is applied to the wheel rim, the wheel acts as a number of second-class levers, with the fulcrum again at the center of rotation but the work at the axle surface; power will be multiplied by 6 and speed (distance) cut by the same amount.

GEARS

Types of Gears.   If the axle is locked to two wheels of different size, it will deliver different proportions of power and speed to their circumferences because of the different leverage ratios. If the wheel edges are notched into teeth, which can be meshed with other toothed wheels, they are called gears and have almost unlimited capacity for changing the power-speed relationships between the input (driving) and output (driven) shaft.

Figures 12.5 and 12.6 illustrate some of the more common types of gears, which may be roughly described as follows:

A spur gear has a rim so notched that the teeth are nearly the same size as the space between them, and the tooth edges are parallel to the axle. This is a simple, rugged type of gear that tends to develop noise and backlash more readily than the next two to be described.

A helical gear is wider, with teeth set diagonally. The teeth are usually comparatively small in cross section, and may be curved. It is usually quieter and smoother in operation than the spur but exerts considerable side thrust.

Gear teeth are often not flat on their contact surfaces, but have a definite bulge or crown that reduces friction.

A herringbone gear is a double helix, with teeth on the two sides sloping oppositely to each other. It has the advantages of the single helix and has no side thrust.

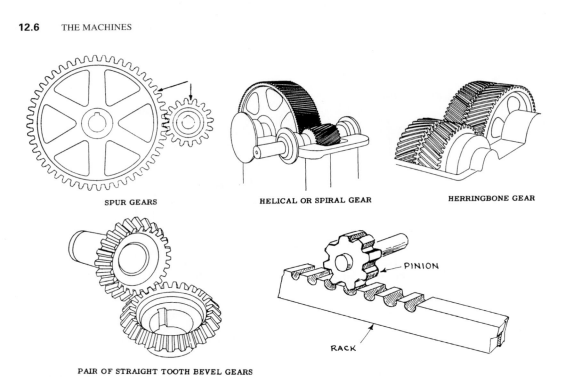

SPUR GEARS HELICAL OR SPIRAL GEAR HERRINGBONE GEAR

PAIR OF STRAIGHT TOOTH BEVEL GEARS

FIGURE 12.5 Gears.

FIGURE 12.6 Worm gears.

A rack or straight gear is a bar with teeth cut in one side. It is used where the forces are too heavy to depend on friction between a wheel and a track, or where an exact relationship between turning and moving must be preserved. A very large circular gear may be called a rack, also.

A worm is a cylinder with spiral teeth cut in it, and has a resemblance to a coarsely threaded bolt. It turns a worm gear or wheel gear, which is a type of spur gear with teeth cut in a curve to mesh with the worm. This accomplishes a very great speed reduction. It is usually made irreversible so

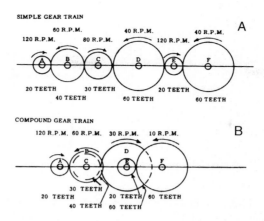

FIGURE 12.7   Gear trains.

that the gear cannot turn the worm, or semi-irreversible so a light automatic brake can prevent it from doing so.

Meshed Gears.   Figure 12.7 shows simple meshed gears. Any two that mesh with each other must rotate in opposite directions. If three are meshed in series, the two outer ones rotate the same way. The center counterrotating gear is a reversing idler.

The speed ratio of two gears is proportional to the number of their teeth. In a simple set, the speed ratio of any two gears is proportional to the number of their teeth, regardless of whether they may be connected through larger or smaller gears.

Figure 12.7(*B*) illustrates a compound gear set in which two gears of different sizes are locked on the same axle. Such sets permit great changes in ratio in a comparatively small space.

When it is important that gears be quiet, they should be "matched," that is, ground or run together by the manufacturer so that the teeth match each other perfectly. This reduces noise, friction, and wear.

If a gear set of any type is taken apart, the gears should be marked so that they can be meshed together exactly as before, so that the advantages of their wearing in will not be lost.

Bevel Gear Set.   Bevel gears can carry rotation around a right angle and change the speed-power ratio at the same time. An important additional use is to reverse direction of rotation of a shaft.

Figure 12.8 shows a set of three bevel gears. The two upper gears are the same size and revolve freely on a rotating horizontal shaft. The lower gear is larger and rotates more slowly on an axis at right angles to the horizontal shaft.

In the figure, the left-hand gear is shown revolving in the same direction as the shaft, so that its lower teeth are moving toward the observer. The lower gear is thus caused to rotate from left to right, and turns the right-hand gear so that it turns oppositely from the shaft, and the lower teeth move away from the observer. Two right-angle turns have been made in the gear set, and the two upper gears, meshed with each other through the lower gear, turn in opposite directions at the same speed on the same shaft.

If the horizontal shaft were fitted with two clutches, by means of which the two upper gears could be separately locked to the shaft, engagement of the left-hand clutch would rotate the gears as described above. Disengagement of this clutch, and engagement of the other, would cause the gears to rotate in the opposite direction. The lower gear, and any machinery driven by it, can therefore be rotated in either direction by the engagement of the proper clutch, or left idle by disengagement of both of them.

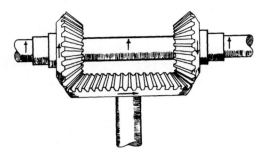

FIGURE 12.8 Reversing set of bevel gears.

Differentials. A more complex set of bevel gears is the differential used in cars, trucks, and other machines, illustrated in Fig. 12.9. It consists of a ring gear driven by a pinion or a reducing gear set, and a pair of horizontal shafts (axles), driven by the ring gear through a set of six bevel gears. Two of these bevels are splined to the inner ends of the axles; the other four, called spiders, are mounted in opposed pairs in a case attached to the ring gear. The spiders rotate with the ring gear and turn the axle gears, the whole system revolving on the axis of the axles. If the load on the two axles is the same, the spiders do not revolve individually, but only with the case around the axles.

If both axles should be locked, none of these gears could turn. If one is locked, each opposed pair of spider gears, in addition to revolving around the main axis, will revolve in opposite directions around their own axis (which turns end over end with the ring gear), rolling on the locked axle gear. The other axle continues to be driven by the rotation of the case, and is also turned by the turning of the spider gears, so that its speed is doubled. If one axle is slowed, the other will be speeded up proportionately.

A differential is essential for delivering power to a pair of wheels which may have to revolve at different speeds, as in a car or truck going around a curve, where the outside wheels go farther than the inside ones. It is also used to steer machines and change gear ratios by proper application of brakes. A disadvantage is that it will deliver most of the power to the axle or unit which is disconnected or

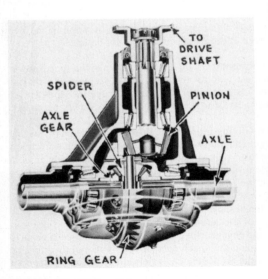

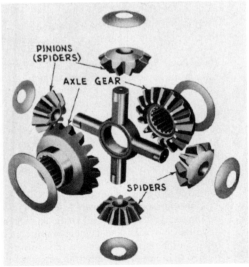

FIGURE 12.9 Differential gears.

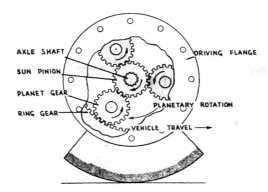

FIGURE 12.10   Planetary final drive.

running free, as to a car wheel on a piece of ice, which will spin while the other wheel on dry pavement is not turned at all.

There are many no-spin differentials in which this difficulty is overcome by the use of automatically locking spiders, or of worms and worm gears instead of bevel spiders, in which differential action in rounding corners is permitted, but which automatically lock against spinning.

Planetary Gears.   A planetary or sun gear set, Fig. 12.10, includes a central sun gear or pinion splined or keyed to a shaft or axle, two or more planet gears, and a stationary ring gear with internal teeth. The planet gears turn on shafts fastened to a carrier or driving flange. For reduction in speed and increase in torque, power is supplied to the axle and work is performed by members attached to or driven by the flange.

The planets are caused to rotate by the turning of the sun gear. This rotation forces them to walk around the toothed track formed by the ring gear, pulling the carrier or spider around with them in the same direction the axle rotates.

The torque multiplication is found by dividing the number of teeth in the ring by the number of teeth in the sun gear and adding 1. For example, a ring of 30 teeth and a sun pinion of 10 teeth would produce a torque multiplication of 4.

The final drive shown in Fig. 12.11 has a sun pinion driven by the live axle, a ring gear splined to the end of the axle housing, and a flange which is bolted to the wheel hub.

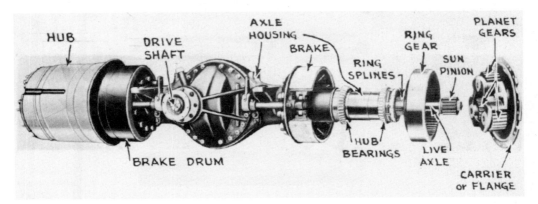

FIGURE 12.11   Axle housing and final drive.

A planetary gear set with a fixed ring such as this provides a compact and sturdy means of multiplying shaft torque. If the ring and the spider can be either held or allowed to rotate, brake and clutch action can be obtained.

Power shift and automatic transmissions usually include planetary gear sets, controlled by disc clutches and/or brakes.

SHAFTS

Keys and Splines. Gears may be connected to the source of power or to the work by means of shafts or axles, or they may revolve freely on shafts that simply hold them in proper position.

A rigid connection can be made to the shaft by means of keys or splines. If keyed, a square-sided slot called a keyway is cut in both shaft and gear hub as in Fig. 12.12. The key, a square strip or specially shaped piece of very hard steel, is inserted in the shaft slot and the gear forced over it. The key should be a tight fit in both slots so that the gear and the shaft cannot twist independently of each other.

A splined shaft, Fig. 12.13, has a number of grooves similar to keyways cut at evenly spaced intervals around it. Matching slots are cut in the gear hub, and it is slid or pressed on to the splines in the shaft. If it is a sliding fit, its position on the shaft may be controlled by a clutch-type yoke or collar. If a tight or press fit, it is held in place by friction, set screws, snap rings, thrust washers, or other means. A splined connection is much stronger than a keyed one.

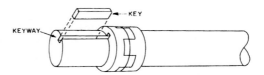

FIGURE 12.12   Keyed shaft.

FIGURE 12.13   Splined shafts.

Universal Joints.   Shafts carrying power from one unit to another seldom can stay on a single axis of revolution determined by the bearings in both units. Even if the original installation is exactly in line, wear of bearings, bending of frames, or setting of foundations is apt to cause misalignment. Often the units are deliberately installed out of line for convenience, or their position in regard to each other varies.

Even a very slight misalignment will cause a shaft to impose excessive strain and wear on its bearings, as exaggerated in Fig. 12.14(A), and may also cause it to whip or break. It is therefore standard practice to place one or more (usually two) flexible couplings or universal joints in shafts connecting separate units, to permit the shafts to adjust to differences in line.

Misalignment may be angular, as in (A) and (B), or parallel as in (C). If parallel, it is usual to handle it as two angles at the ends of an intermediate hub or shaft, as in (D). A simple type of flexible coupling which carries light loads through angles is shown in Fig. 12.15. This also absorbs vibration and shock.

If the differences are slight, roller chain or silent chain couplings such as are shown in Fig. 12.16 can be used.

A Spicer universal joint is illustrated in Fig. 12.17. It consists of a journal cross, two yokes, and four needle bearings. The driving yoke holds two opposite arms of the cross, and the driven yoke the other pair.

Each pair of arms rotates on the same axis as the yoke which holds it, and the two axes intersect at the center of the cross. The twist caused by revolving in two planes is absorbed by oscillation of the arms inside the yoke bearings.

Change in length of shaft is permitted by a sliding spline (slip joint) in one of the yokes.

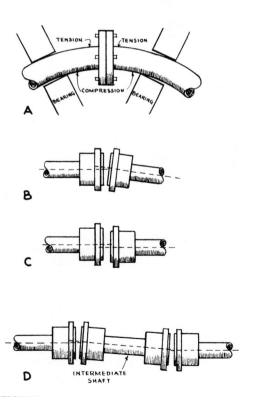

FIGURE 12.14   Shaft misalignment.

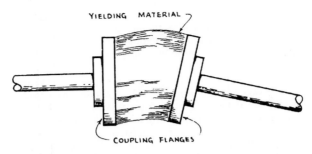

FIGURE 12.15 Simple flexible coupling.

FIGURE 12.16 Roller chain coupling.

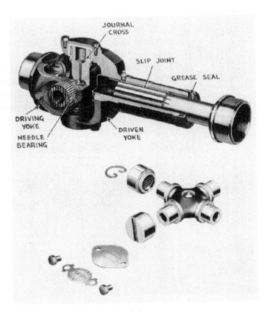

FIGURE 12.17 Universal joint.

Journal cross universals are usually made for normal angles up to 8° or 10°, and momentary angles up to 15° or 20°. Higher angles require special construction.

Universal joints are generally used in pairs. This allows them to accommodate sideward as well as angular displacement, and it makes it possible to cancel out any pulsation caused in the driven shaft by joint action.

BUSHINGS AND BEARINGS

Where a gear or wheel turns upon a shaft, or the shaft turns within its supports, provision must be made to reduce friction and wear. Some premium U-joint and slip member assemblies are sealed and need no lubrication, but most driveshafts still need the right grease at the right time to keep working reliably.

Bushing (Sleeve Bearing).   There is no definite distinction between sleeve or solid bearings and bushings. They are sleeves or hollow cylinders installed between a shaft and its supports, or between a gear, pulley, or wheel and the shaft on which it turns.

A bushing is usually made of some metal alloy that is softer than steel, can retain an oil film that will prevent direct metal-to-metal friction, and is easily worked. Bronze, brass, babbit, and oilite are some examples.

When such bushings are kept free of dirt and are properly lubricated, they reduce friction and absorb most of the wear caused by the rotation of the shaft or wheel. A bushing may be replaced several times before the more expensive shaft has to be repaired, and they protect the supports (often called pillow blocks) for the life of the machine.

There are also hard steel bushings whose only function is to protect the support from damage by the shaft. They are usually found where shafts do not make complete revolutions (hinge pins of a front loader, for example), or where they turn very slowly. They may or may not be lubricated, and usually do not have seals against dirt.

Bushings in pillow blocks are usually split, that is, cut lengthwise into two pieces. This permits easy installation, as they do not have to be slid along the shaft to get in place.

A bushing is always fastened to the outer member, whether it is stationary or rotating. It is easier to lock them against turning on the outside, and the inside turns a little more slowly than the outside, because of its smaller diameter.

Main and rod bearings in an engine have split bushings called bearing inserts. See Fig. 12.18.

If a bushing is in sight, or can be readily seen after guards are removed, wear can be observed. If the space between it and the shaft changes with differences in load or direction of rotation, or if grease squeezes out and is drawn back in as it works, there is looseness that may or may not be serious.

The friction area in a bushing is large enough to convert an important amount of power into damaging heat if lubrication is not adequate. Heat is quickly followed by scoring (deep scratching) of both the bushing and the shaft. A worn bushing will not hold the shaft in alignment, and more trouble may follow.

Most bushings have a drilled passage to a grease fitting or oil line, and grooves that distribute lubricant through the full width. There is a slow motion of the lubricant toward the edges of the bushing, where it is usually lost. Supply must be fast enough or frequent enough to maintain a film of oil over the whole friction surface.

Antifriction Bearings.   Antifriction bearings are used instead of bushings at most rotary friction points where exact alignment and low friction are required, and the shape of the shaft permits their use. The two principal types are ball bearings and roller bearings.

Ball bearings consist of two rings of hard steel, one of which fits closely on the shaft and has its outer surface machined into a smooth track; the other fits closely into the casing or hub and has a machined inner surface. Between these rings, known as the inner and outer rings or races, is a circle of hard steel balls held in place by a light cage. See Fig. 12.19.

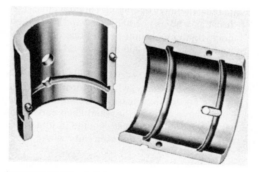

FIGURE 12.18   Split bushings (bearing inserts).

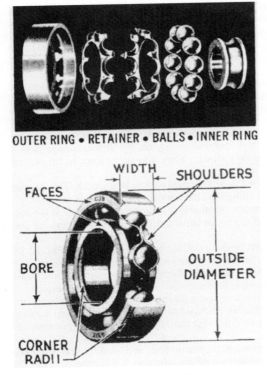

OUTER RING • RETAINER • BALLS • INNER RING

FIGURE 12.19 Ball bearing construction.

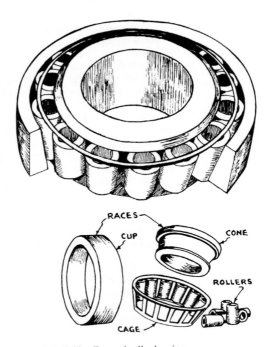

FIGURE 12.20 Tapered roller bearing.

The inner race turns with the shaft, the outer race stays stationary with the casing, and the balls keep these races at an exact distance from each other and revolve so that all friction is changed from the sliding to the rolling type. Pushing a heavy piece of furniture, first without casters then with them, will give a somewhat exaggerated idea of the difference. The space around the balls is filled with grease or oil which is either sealed in or renewed periodically.

Roller Bearing. Here the balls are replaced by steel cylinders which roll between two races in the same manner. An important variation is the tapered roller bearing illustrated in Fig. 12.20. The races are tapered, the inner one being called a cone, the outer a cup. This shape enables them to resist side thrust as well as normal loads, and since they are used in pairs with the small ends of the cone facing, tightening a nut behind one of them will force the cones nearer the cups, thus compensating for any wear in the roller or races.

Value. Bushings and bearings serve several functions. They reduce friction, provide for lubrication, and are replaceable individually, when worn, at less expense than replacing a whole shaft or casting.

For most uses antifriction bearings are superior. They have less friction, can be more effectively lubricated, have a longer life, and keep parts in more exact alignment. However, they are more expensive, are subject to rust damage in idle machines, require more space than bushings, and since they cannot be split, are used only where they can be slid onto a shaft.

Cleaning. Used antifriction bearings must be cleaned before their condition can be judged. They are usually washed out in a can or pan of kerosene, fuel oil, or gasoline. They should not rest on the bottom of the pail, as they are likely to pick up dirt there. Hang a wire basket inside the pail, or block up a piece of coarse screening above the bottom.

Open bearings can be scrubbed with a natural bristle brush (synthetic fibers may dissolve in the kerosene), rubbed with the fingers or a cloth, or blown out (but not spun) with compressed air. A number of cleanings and rinsings may be required. It is a good plan to dip them in oil when they are finished, to prevent corrosion if they have to stand around.

They should be packed with grease if they are to be stored. It should be worked well in by turning the races.

Inspection.   Tarnish or stain on any surface of a bearing, or corrosion on nonworking surfaces, will probably not interfere with the usefulness of a bearing. Slight roughness or tendency to stick may only mean that it needs to be recleaned. A good test for real roughness is to press the two races toward each other while turning them or to lay the bearing flat on a clean surface and rotate it with heavy pressure from the palm of the hand.

The following defects should be looked for, and if found they are cause for scrapping the affected part if it is separate, or the whole bearing if it is not.

Broken or cracked rings

Dented seals or shields

Cracked or broken separators

Broken or cracked balls or rollers

Flaked areas on working surfaces

Brownish-blue or blue-black color indicating overheating

Races indented by pressure of balls or rollers

Looseness may be a characteristic of that model bearing, or may indicate extensive wear. Compare with a new bearing, if possible, after packing the old one with thin grease.

TRANSMISSION

A transmission, gear set, or gearbox is a set of gears and shafts which provide a change in the speed-power ratio. It may be a single-speed or gear reduction type in which a small gear driven by the engine meshes with a larger gear that turns the working parts; or it may be a selective or sliding gear type with several ratios.

Two-Speed.   Figure 12.21 shows the simplest selective type, an auxiliary transmission having two forward speeds. The engine-driven jackshaft carries a main drive gear which meshes with a larger gear splined to the front of the countershaft. The countershaft also carries a smaller fixed gear which may mesh with a sliding gear on the main shaft. The main shaft is a continuation of the jackshaft and may revolve separately from it in low position (shown), or may be connected to it in high by a jaw clutch, a device described later under Clutches.

A fixed jaw is on the jackshaft, and a sliding jaw on splines on the main shaft. The sliding jaw is expanded into a gear which may be meshed with the countershaft gear. This combination gear and clutch jaw is controlled by a fork and shifting collar which can slide it forward to connect the clutch and disconnect the gear, to the center so neither will be engaged, and to the rear to engage the gear only.

When the jaw clutch is engaged, power is transmitted straight through the transmission as if on a single shaft and the countershaft turns without doing any work. This is called direct drive or high gear. When clutch and gear are both disconnected, the transmission is in neutral and no power goes through it. When the gears are meshed, the drive goes from jackshaft to countershaft and from countershaft to main shaft, involving two pairs of gears which can be made in ratios to give any desired gain in power through reduction in speed.

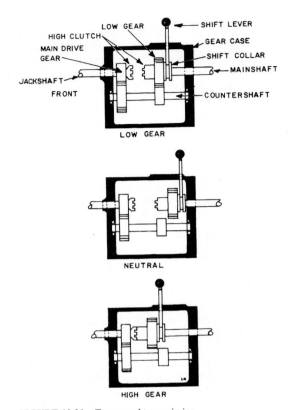

FIGURE 12.21 Two-speed transmission.

Figure 12.22 shows a three-speed forward and one reverse speed transmission. The countershaft gear is enlarged to a cluster gear consisting of three gears made in one piece, the forward one being the largest and the rear the smallest. The center section is meshed with a gear on the idler shaft.

A combination sliding jaw and gear, similar to that described in the two-speed transmission, slides forward along the main shaft to mesh with the jackshaft jaw for high gear, and back to mesh with the large part of the cluster gear for second. Another sliding gear on the main shaft is moved forward to mesh with the small member of the cluster gear for first or low gear, and backward to mesh with a gear on the idler shaft for reverse.

The jackshaft and mainshaft revolve in the same direction whenever they are connected directly or through two countershaft gears, regardless of their ratios. Putting an extra (idler) gear in the series reverses the rotation of the mainshaft. See Fig. 12.7 for this effect.

Compound Gearing. The larger trucks and tractors often have two transmissions in series, the engine power going first through one, then through the other. The smaller one is called an auxiliary. In tractors it may have one high and one low gear, or a forward and a reverse. In trucks it is likely to have from two to four forward speeds.

Trucks may have two-speed axles. This means that there is a high-low shift in the differential, that has the same effect as an auxiliary two-speed transmission.

The total number of gear speeds may be found by multiplying those in the two transmissions. For example, a jeep has a main transmission with three forward and one reverse, and a high-low auxiliary. This gives it six forward and two reverse speeds.

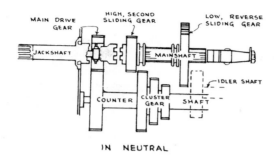

FIGURE 12.22   Three-speed and reverse transmission.

   The two transmissions may be in separate cases with separate levers, or both in one case, with one or two shift sticks. One stick usually has an interlock that permits starting only when it is in neutral.

Shift.   Shifting gears is made possible by a friction clutch between the engine and the transmission. When under load, meshed gears stick together because of pressure against the teeth. When in neutral, a pair of gears cannot be engaged unless the two sets of teeth to be meshed are moving at about the same speed. To shift with the machinery stationary and the engine going, the engine clutch is disengaged, the jackshaft allowed time to slow or to stop its spinning, and the desired shift made. If the teeth will not mesh, the clutch is engaged slightly to turn the jackshaft to another position.
   To shift while the machinery is moving, the engine clutch is disengaged to take the load off the gears, and the sliding gear is shifted into neutral. If the shift is from low to high, the jackshaft will be turning too rapidly for quiet engagement. Since it is no longer being turned by the engine, it will soon slow to proper speed and can than be engaged. If the engine throttle is closed during shifting, this waiting period can be reduced by engaging and disengaging the clutch (double-clutching) while in neutral, as the engine loses speed more rapidly than the free-spinning shaft.
   If the shift is from high to low, the jackshaft will be moving too slowly. After shifting into neutral, the clutch should be reengaged, the engine speeded up, the clutch disengaged, and the shift completed.
   A clutch brake may be installed on the jackshaft which will stop it from spinning when the clutch is fully out. This makes shifting easier from low to high, but might make it more difficult from high to low.

Constant Mesh.   In a constant-mesh type of transmission, the main shift gear would be permanently engaged with the countershaft gear, but would spin freely on the main shaft unless keyed to it by a sliding jaw. This shifts more quietly because of special tooth design, and the slow speed of the gear hub as compared with its teeth.
   A synchromesh transmission is a constant mesh type in which leather collars on the two jaws touch before the teeth do, providing just enough friction to slow or speed the jackshaft so that the jaws will synchronize and mesh quietly.

Power Shift.   Transmissions that can be shifted while transmitting full engine power to the wheels are called power shift or shift-on-the-go units.
   Power shift transmissions generally use sets of multiple disc clutches to control constant mesh gearing of either planetary or countershaft design.
   Power shift units are usually teamed with torque converters, which effectively absorb drive train shock loads caused by changes in gear ratios. The transmissions may be designed specifically for cycling applications (loader or dozer) or for hauling (truck or scraper).
   Two speeds are provided automatically by the twin-turbine torque converter, in which each turbine is connected to its separate output gear set, Fig. 12.23.

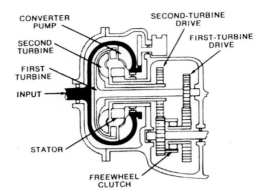

FIGURE 12.23 Twin-turbine torque converter.

These gear sets are connected to the range gearing, which consists of a reverse and a low planetary gear set, plus a high gear clutch.

Range shifting is done by means of a control valve that directs hydraulic pressure to multiple disc clutches, which hold or release parts of the gear train to accomplish speed changes.

When vehicle motion is started, oil flow within the converter causes the first turbine to turn, driving a low-speed combining gear, to deliver high torque to the range gears.

As vehicle speed increases, the second turbine assumes the drive and, through its high-speed combining gear, delivers lower torque but higher-speed drive to the range gears. The first turbine and its combining gear freewheel when the second turbine is operating at higher speeds.

The result is automatic two-speed performance from the torque converter, which combines with two speeds in the forward range gearing to provide four forward speeds.

Reverse, having just one range, provides only the automatic two-speed performance.

Automatic Shift. Transmissions that change gear ratio without action by the operator, while transmitting full engine power, are termed automatic. They generally use sets of multiple disc clutches to control constant mesh gearing. See Fig. 12.24.

Automatic shifting is effected by hydraulic pressure or electric signals that are proportional to vehicle speed, but are modulated or biased by a signal of throttle position.

These transmissions are teamed with torque converters, whose shock-absorbing and torque-multiplying properties are important for their success.

Performance and reliability of heavy duty automatic transmissions depend on the quality of the fluid in the sump (as written in *Construction Equipment* magazine). Also too little fluid starves splash-lubricated parts, while too much fluid tends to aerate, causing erratic shifting and overheating.

Transmission Failure. *Construction Equipment* research suggests that most transmissions need repairs in 5,000 to 10,000 hours of use. Failure becomes more likely when there is a drop in shift performance, when the equipment operator says that the machine hesitates when up-shifting, or when the transmission slips under heavy loads and drops into neutral at low idle. The hesitation or slippage might be caused by worn plates and discs, linkage out of adjustment, wrong oil used, low fluid level, or an incorrect pressure setting.

The use of oil monitoring (discussed later in this chapter) can be helpful in analyzing the problem. For instance, elevated tin contamination in the oil points to a failing babbitt bearing in the oil pump of the tramsmission.

Heat is a major obstacle to longer transmission life. This is particularly true with crawler dozers and other heavily loaded tractors because of the high torque passing through the clutch. This can sometimes be detected when there is discoloration of the paint on the case.

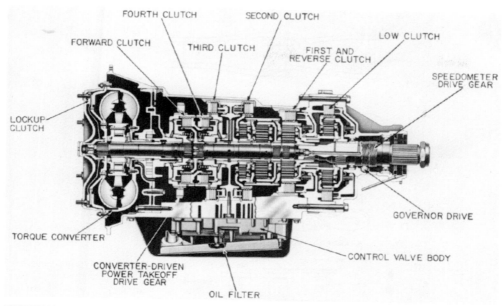

FIGURE 12.24   Automatic shift transmission.

CHAIN DRIVES

Roller Chains.   Most gears transmit their power through direct meshing of teeth, but sprocket gears are used to drive through roller chains.

Roller chains have been used in revolving shovels in the track drive, crowd mechanism, and deck machinery; in ditching machines in track and digging drives; in grader final drives; and in many other places.

Both offset and "standard" types are used. Figure 12.25(A) shows offset construction.

Standard construction is shown in (B) and (C). There are two types of link, each with straight, parallel side plates. The wide ones have removable pins fastening the plates together at each end; the narrow links are fastened together with sleeves (bushings) inside rollers. These links alternate in the chain.

Sprockets connected by one side of a chain or belt revolve in the same direction. If connected on reverse sides, they revolve oppositely. See Fig. 12.26.

A chain or belt system needs a movable sprocket or pulley to permit adjusting tension.

Crawler tracks, described in Chaps. 13 and 15, are special types of roller chain. Their load-carrying shoes may be bolted to chain links, or may be links themselves.

Silent Chain.   "Silent" chain, Fig. 12.27, is made from a series of flat metal links which are tooth-shaped at each end. The ends are pierced for long cross pins. A center guide groove in the sprocket teeth, and retainer plates in the center links, may be used to hold the chain securely on the sprocket.

This chain is quieter and has less vibration than roller chain and can be run at higher speed. Its links are small in proportion to its strength so that it can be used on smaller-diameter sprockets. It tends to cushion shocks and even out irregularities.

It is more expensive.

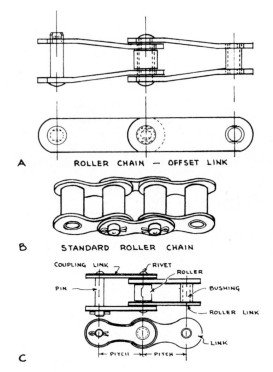

FIGURE 12.25   Roller chain.

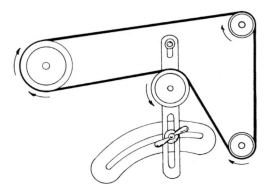

FIGURE 12.26   Adjustment pulley.

Alignment.   It is important for either type of chain that sprockets be properly lined up with each other. If they are not, the chain will tend to climb the sides of the sprockets and jump off them.

There are two types of misalignment—shafts not parallel and sprockets that are offset or not in line. If shafts are not parallel, it is the angle between them that counts, regardless of the distance between them. This condition is common in construction machinery. It is indicated by a one-sided pattern of bright spots on the chain or the sprocket. Figure 12.28 illustrates sprockets

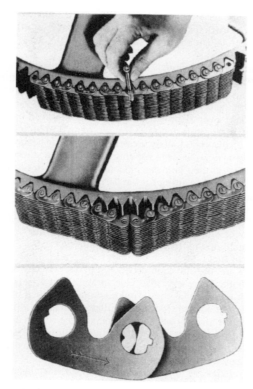

FIGURE 12.27   Silent chain.

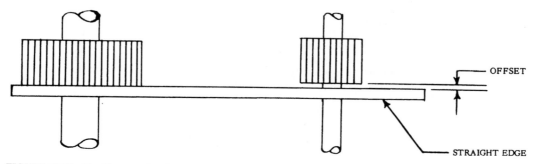

FIGURE 12.28   Checking sprockets for offset.

that are offset out of line. This condition may result from poor installation, from the sprocket's shifting on the shaft, or from end play in the shaft. It is checked by putting a straightedge or stretching a string along one side of the two sprockets. They should line up exactly.

Wear.   Both roller and silent chains stretch as they wear. The case is often small enough so the chain will slap against it when it becomes loose. This slapping is a danger signal. It means that it is time to adjust the chain, or if there is no adjustment or the adjustment is fully extended, that the chain must be rebuilt or replaced. Otherwise the chain will probably start to jump, that is, to allow one of the sprockets to turn inside it. This will cause damage to the chain and the sprocket, and will prevent dependable delivery of power.

Repair and Replacement.   A roller chain can usually be shortened by taking out a link in offset types. In standard chain a pair of links must be removed. If this shortens it too much, a half link, which is just an offset link used in a standard chain, may be substituted for the pair.

Each link gets a little longer as the chain stretches, so that they no longer mesh accurately with the sprocket. Even if proper tension is restored to a badly worn chain by removal of links, the sprocket will be subjected to excessive wear, and jumping may occur.

Roller chains occasionally break links. This may be the result of a defective link, a shock overload, natural wear, or wear from rubbing against some stationary object. A few spare links should be kept on hand, as their help can put a machine back in service in a few minutes that otherwise might be delayed a week waiting for the parts. Links can be obtained from the equipment distributor, or from a sales agency for a chain manufacturer.

Lubrication.   A roller chain operating at low speed where it is likely to get mixed up with dirt should not be lubricated. The grease and oil will simply pick up sand and clay and make them into a grinding compound. Such a chain should be run dry, and occasionally taken off and washed in kerosene or fuel oil.

Engine and transmission drives run in covered cases and are lubricated by partial submersion in an oil bath. There may also be a spray of pumped oil. Silent chains are always run in covered cases with lubricant, as they would be quickly ruined by exposure to dirt and dust.

The oil that does the work on a roller chain is not what is visible on the surface, but that which is between the sleeves and the rollers, and the rollers and the pins. Since it has to soak in, a thin oil is usually best, but good penetrative qualities are sometimes obtained with thicker oils and light greases. Be careful about this, however, as it is no use keeping the outside of the chain looking well lubricated if its insides are dry.

Roller chains are not suited to high-speed work because there is no way of forcing oil continuously between the pin-bushing and bushing-roller surfaces. If the chain is run too fast, or the oil is too thick to penetrate or too thin to lubricate, sliding friction between pin and bushing may create enough heat to actually weld the joints of the chain, a difficulty that is called galling.

Silent chains are somewhat tricky to lubricate when they run at high speeds. Fortunately the contractor does not have to worry about the complicated business of designing a lubrication system— he or she just has to know that the one that is there is working. But the contractor must know what it is. For example, on a combination of oil bath and pumped spray, lubrication is not adequate if the pump fails and the chain is being lubricated by the bath only.

Silent chain lubricant is good-quality, nondetergent, petroleum-base oil. It may have inhibitors against foam and rust, and to improve film strength.

Tracks.   The crawler track is a specialized type of roller chain. There are two principal types.

Linked shoe tracks, described in Chap. 13, are made by fastening shoes or pads into an endless belt by means of hinge pins. They are used in the majority of machines that do not depend on thrust against the ground for most of their working force.

Crawler tractors and some other machines have separate shoes bolted onto roller chains. This construction is described in Chap. 15.

BELT DRIVES

Flexible belts made of fabric and rubber, or more rarely of leather, are widely used to convey power between parallel shafts. Such drives are easy to design and are usually cheap to make. They absorb shocks, are cheaper, and are easier to install and service than chains, but will not last as long, or carry as heavy loads, and cannot be used where exact timing is required.

A belt will stretch under a heavy pull, and resume almost its original size when released. This characteristic makes a belt drive a shock-absorbing one that gives some protection to the machine. An overloaded belt will usually slip, acting as a safety clutch to prevent damage. However, this slippage makes it impractical to use belts for most heavy power applications.

Pulleys.   A belt runs around two or more pulleys. Those in contact with the inside of the belt will all turn in one direction. Any pulley in contact with the outside of the belt will revolve in the opposite direction, in the same manner as with roller chain.

One pulley in any belt system should be adjustable, Fig. 12.26, as belts wear and stretch, and must be tightened by moving the pulleys apart. If the pulley can be moved rapidly and locked in tight and loose positions, it can be used as a clutch. If it is loose, the drive pulley will turn inside the belt without moving it, but when tightened, friction will cause the two to turn together.

If two pulleys in a belt system are different sizes, the larger one will turn more slowly. The driven shaft can be made to turn faster or more slowly than the driving or engine shaft by proper selection of pulley sizes.

A belt drive can be made to provide several speed ratios. Several pulley sizes are mounted side by side on the same shaft, and the belt is shifted from one set to another as required, like changing gears on a bicycle. Or a V pulley may have a movable side which when slid inward will cause the belt to ride higher in a larger circle.

A belt changes shape as it goes around a pulley, as it is thicker on the slack side than on the loaded side. See Fig. 12.29.

Flat Belts.   Flat belts are of two types: the endless, which is made in a closed circle; and cut, which is fastened into the circular form by a pin passed through wire loops crimped into its ends.

A center bulge in the flat pulley helps to keep the belt on. Centrifugal force tends to throw the belt outward so that it tends to move to its line of greatest tension across the pulley, and stay there. A bulge puts more stress on the center of the belt than on the sides. Unless the pulleys are in good alignment with each other—their axes of revolution parallel to each other, and perpendicular to the direction of travel of the belt—the belt will climb the tighter side, and will probably slip off under heavy load. Flat belts require more tension than V-belts, to avoid slippage.

Flat belts are cheaper than V belts and are usually longer-lived because their shape reduces internal friction, but they need much more attention. They are preferred when the work is done by a different machine than that which supplies the power, such as for a pulley on a tractor driving a cement mixer, a buzz saw, or a pump.

Conveyors.   Conveyor belts are wide, flat belts that transport dirt or other material. They will be discussed in Chap. 14.

V Belts.   The V belt is the standard means of driving permanently mounted, light-load accessories, such as fans, generators, and water pumps.

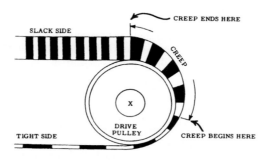

FIGURE 12.29   Changes in belt shape (exaggerated).

V pulleys have such high steep flanges that the belt cannot climb up them readily. The driving friction is entirely on the sides of the belt, and is very effective because of the wedge action of the belt taper.

V belts are kept small in cross section to minimize internal friction and heating. When too much power must be transmitted for the strength of one belt, two or more are used, working in parallel grooves in the same pulleys. Such belts must be the same size and dimensions, or the tighter one will carry too much of the load, with resultant strain, slippage, and rapid wear. Since belts stretch in service, this means that if one belt breaks, the whole set must be replaced.

Lugged belts have a series of slots in the inner surface. This gives them greater flexibility and better grip on the pulleys and reduces fatigue so that they last longer. They are particularly good for small-pulley heavy-duty work. They are recommended whenever a chronic slippage problem is encountered.

A belt drive may slip because of a slippery glaze on belt and pulley surfaces. This can sometimes be removed by wiping with gasoline, or dusting with fullers earth.

Fit.   V belts are made in several side slopes. If one flares out more sharply than its pulley groove, contact will be made only at the upper belt corners. This gives it a poor grip, causing damaging distortion of the belt structure, excessive wear on the corners, and damage to the pulley groove.

If the belt does not flare out as much as the pulley groove, it will ride hard on the bottom corners, with about the same results.

If the groove is too wide or too shallow the belt will ride on the bottom, so that it will lose the wedging effect of the sides. Such a belt will require excessive and damaging tension to reduce or prevent slippage.

If the belt is so wide at the top that it rides high in the groove, it will tend to turn over, and the cover will wear quickly along its top contact.

A rough or irregular surface in the pulley groove will give the belt a poor grip and is likely to cause rapid wear. Smoothing can often be done with a fine file, grinding stone, or emery paper.

Projections on the pulley surface cut into a belt very rapidly, because it creeps and changes shape as it goes around the pulley, causing it to slide along the surface.

Installation.   To install a V belt, move the adjusting pulley inward as far as it will go. Put the belt in the groove around the least accessible pulley or pulleys, or the hardest one to turn. Then try to slide the belt onto the last pulley.

If it will not slip on, pull or push it on as far as possible, then turn the pulley to wind on the installed part. If for any reason the pulley cannot be turned, the belt may be pried onto it with a screwdriver, taking care not to damage the fabric.

When the belt is in the grooves of all pulleys, the adjustment pulley is moved outward until the belt is under light tension, and is then locked in place.

Correct tension for light service allows pushing the center of a section of belt inward about 1 inch for each 1 foot of unsupported span between the two pulleys. For heavy service such as a traction drive in a small machine, the yield should be only $\frac{1}{2}$ inch to 1 foot of span.

An engine with a crankshaft-driven, front-mounted power takeoff may require a major disassembly job to install an ordinary fan belt, or it may have a spring-loaded jaw that can be pushed forward or back to open the shaft just far enough to slip a fan belt through its opening.

It is sometimes possible to obtain belts with a coupling that can be opened for installation, and closed after putting around the shaft.

INCLINED PLANES

Ramps.   The inclined plane is applied generally in reducing force needed to lift heavy weights. Figure 12.30 shows a 150-pound roller that is to be placed in a truck body 3 feet above the ground. Two workers could lift it onto the truck, but one can push it more easily up the ramp shown. A long ramp has less of an upgrade than a short one for the same height and will require less force

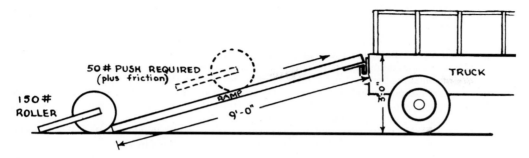

FIGURE 12.30   Mechanical advantage of a ramp.

to move an object up it, but the push must be exerted for a longer time. Friction may be very important in determining whether an inclined plane should be used, as a sliding load on a rough ramp might cause more friction resistance than its own weight.

The mechanical advantage (MA) of an inclined plane (ignoring friction) is found by dividing its length by its height. In the illustration it is three, which means that the force requirement of rolling up it is one-third of that needed to lift it vertically.

Threads.   The jack shown in Fig. 12.31 is an application of inclined-plane mechanics. The 2-inch center shaft of the jack carries a spiral thread which fits within a similar thread in the jack body. Turning the shaft by means of a bar through a hole in its top causes the shaft to move upward on the circular ramp provided by the threads. A full turn on the shaft raises it the vertical distance between two threads.

The $\frac{1}{4}$-inch pitch gives a 4-to-1 ratio. The bar acts as a second-class lever, with the work at the thread. If the bar is 20 inches long, its power end travels in a circle with a 40-inch diameter, a distance of 40 × 3.14 or 125.6. Since 4 turns are needed to raise the load 1 inch, the level end

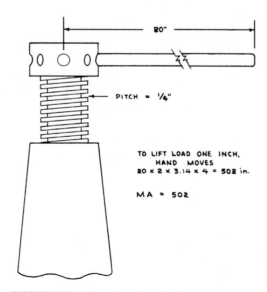

FIGURE 12.31   Screw-type house jack.

travels 4 × 125.6 or 502.4 (say 502) inches, giving the jack a leverage or mechanical advantage (MA) of 502.

Bolts and Nuts.   Bolts and nuts are the best known applications of this method of turning rotary motion into a straight push or pull. Two threads are in general use.

Coarse thread is the most used. It is cut more deeply into the bolt, each thread is wider and heavier, and it is more steeply pitched so that it tightens with only two-thirds the revolutions needed for the same length of fine thread. As a result, it does not get as much tightening from the same wrench force as with fine, but it tightens faster.

Coarse threads, when compared with fine, are easier to start without cross threading, particularly in hard-to-reach areas, and are less likely to strip or to be ruined by rust or careless handling. Fine threads will develop greater bolt tension for the same wrench pull, are more suitable for adjustments, take less away from bolt strength, are somewhat less likely to work loose, and their heads and nuts are narrower so that they require less space.

A fine-thread bolt is about 20 percent stronger than a coarse thread of the same size and material. However, the thread itself is much weaker.

In general, a coarse thread will have enough strength to allow breakage of the bolt by turning the nut, but a fine thread will strip before the bolt breaks.

Three grades of steel are in common use in bolts in heavy equipment. They are rated by minimum tensile strength (resistance to permanent stretching) in pounds per square inch, abbreviated psi.

Grade 1 is a low-carbon steel rated at 55,000 pounds per square inch. It is used in machine bolts that have square heads, square nuts, coarse threads, and a black finish. Their use is limited because of low strength and the inconvenience of square nuts in places that are hard to get at.

Grade 2 is a somewhat stronger low-carbon steel rated at 64,000 pounds per square inch. It is used in bolts and nuts of natural steel color (bright finish), with six-sided heads and nuts, in both fine and coarse threads.

Grade 5 is a medium-carbon steel strengthened by tempering to a rating of 105,000 pounds per square inch. Finish is black, and threads are usually coarse. It is identified by three radial lines on the bolt head, and may have a manufacturer's symbol too. Nuts are higher than standard.

Bolts must be tight to hold. Figure 12.32 shows how looseness may increase the leverage against them. Looseness also encourages nuts to loosen further.

Other Threads.   Pipe thread is a different coarse thread, used only on pipes and fittings. Thin-walled tubing usually has a fine thread. Pipe and tubing thread will not mesh with each other, and the tubing will be damaged if an attempt is made to force them.

Wedges.   The wedge is an allied device that is capable of building up enormous pressure at an angle to the direction of the force, and which can convert hammer blows into steady pressures. A wedge hammered into a log to split it is a familiar example. Friction losses are very large in most wedging work.

Cutting Edges and Teeth.   The cutting edges of most excavating machines are wedge-shaped, at least when new. This shape is strong and has high penetrative and disruptive ability.

A wedge tooth entering material which resists equally on both faces will tend to move in a direction midway between the slopes of its faces. If resistance is unequal, the wedge will tend to slide on the face contacting the greater resistance. Figure 12.33 shows the actions of a wedge tooth on various slopes while being moved horizontally. Held at the pitch (*A*) it will penetrate steep slopes; at (*B*) it will cut into any kind of slope but will not dig down; whereas held as in (*C*) it will penetrate upslope or level. It will be noted that it is the angle between the underside of the tooth and the ground that determines its penetration. Figure 12.33(*D*) shows the same angle as (*C*), but the point and part of the underslope of the tooth are worn away, destroying its "suck" and therefore its penetration.

Sufficient down pressure will make any of these pitches or conditions dig down to some extent, but this is an expensive and inefficient substitute for properly positioned and sharpened teeth. However, in hard materials, weight or down pressure greatly assists the digging by forcing the point of the wedge into the soil and keeping it there.

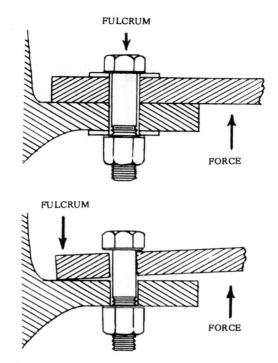

FIGURE 12.32   Leverage against loose bolt.

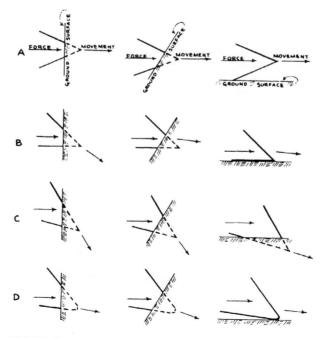

FIGURE 12.33   Penetration of a wedge.

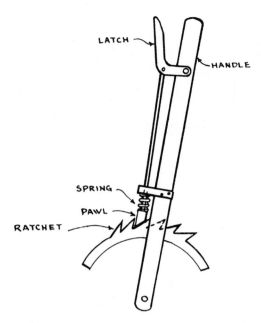

FIGURE 12.34   Pawl and ratchet.

Separated teeth cut into hard earth and rock much better than a continuous cutting edge because they concentrate the force in small areas, allow the material to be displaced to each side as well as above and below, and usually so shatter or weaken material between them that the lip at the tooth bases can easily penetrate it.

A moldboard is a continuation of the upper surface of the tooth or cutting edge, and its shape is important to the efficiency of many machines. In a farmer's plow it turns the slice of earth on its side or over and presses it into place. In a grader or dozer it rolls the earth upward until it falls forward, lessening friction in the load being pushed, and facilitating drifting of the material side-ward if the blade is angled.

Ratchet and Pawl.   The ratchet and pawl, one type of which is illustrated in Fig. 12.34, permits motion of a lever one way, and resists reverse motion unless a catch is released. The ratchet is a plate with hard steel teeth in its curved upper surface. The teeth are sloped on one side, straight on the other. The pawl is a single tooth held in a slide on the lever and pressed against the ratchet by a spring. It has the same shape as a ratchet tooth but is turned backward. In the illustration, if the lever is moved to the right, the pawl will move freely across the ratchet teeth, being wedged up and over by the sloped sides. But if moved to the left, its straight edge catches on the straight edge of the first tooth and remains until pulled up by a latch.

Overrunning Clutch.   An overrunning clutch, or freewheeling unit, may be a circular ratchet, or the types illustrated in Fig. 12.35. In the first example, cylinders of graduated sizes lie in tapered hollows between inner and outer hubs. Rotation of the outer hub clockwise (or the inner one coun-terclockwise) will move the cylinders into the small end of the slot, where they will jam the two parts together so that they will rotate as one. Reversing the rotation of either shaft will move the cylinders to the large end of the slot, where they can revolve freely without preventing the inner and outer hubs from rotating independently.

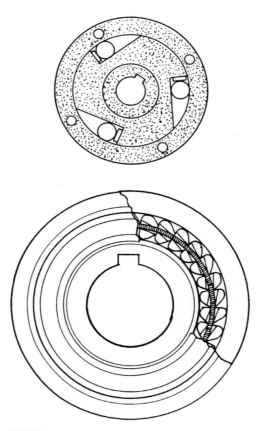

FIGURE 12.35   Freewheeling units.

In the other type shown, rotation in one direction will cause the cams to lock the two hubs, and in the other will release them.

CONTROLS

Levers.   Clutches and brakes may be operated by foot pedals or hand levers. The pedal is pushed down to release the clutch or apply the brake, and is returned to position by springs that may be in the unit, in the linkage, or in both.

Hand levers may be pushed or pulled. A single lever may control two clutches (for example, the swing lever on a shovel) and move both ways from a center neutral or released position. On single controls, the linkage can sometimes be changed to reverse the direction of operation.

A lever is ordinarily returned to neutral by springs and/or compression of lining and parts in the unit. It may be held engaged by a hand-operated pawl and ratchet, by a ball or wheel pushed into a socket by a spring (ball detent), or more commonly by locking over center.

Over-center locking may be obtained in different ways in various linkages. A simple type is shown in Fig. 12.36, in which a sliding collar causes two arms attached to it to move outward. Maximum pressure against the arms, and through them against the friction lining of the attached

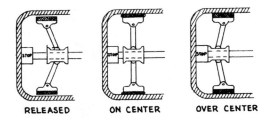

RELEASED   ON CENTER   OVER CENTER

FIGURE 12.36   Locking over center.

shoes, is obtained when the arms are in line with each other. Moving the collar farther, against the stop, relaxes the pressure slightly. The collar is now held between the stop and the pressure required to move it back across center, and it will stay put until force is applied to move it.

Rod Linkage.   Connections with mechanical controls may be of lever, clevis and rod construction, or stiff wire in a flexible casing.

An example of a pedal and rod linkage is shown in Fig. 12.44.

This type of system is often difficult to design, but once built is sturdy and durable. Difficulties include friction in numerous pivot points, and the fact that slight wear in each of many spots adds up to lost motion that may make control difficult. Careless reassembly after an overhaul, particularly failure to get arms nearly perpendicular to the direction of pull, may make the system ineffective. A loose or bent support will allow the cross shaft to twist and bind.

Hydraulic.   There are at least three types of hydraulic control. One has valve-directed flowing oil between a pump and working parts. This may be called a power or dynamic system.

A second type uses controlled slippage, as in a torque converter, described in detail later.

The third, discussed here, involves application of pressure to one part of an enclosed body of fluid, which transmits the pressure to other units in the system, moving them against working resistance and springs.

The most important and best known application of this static type of control is in automotive hydraulic brakes. For convenience, it will be called a hydraulic brake system here, although it has many other uses in machines.

A hydraulic brake system includes a master cylinder with a built-in reservoir, that is sometimes called a compensator; lines, and wheel or acting cylinders.

Pressure on a pedal or pull on a lever moves a piston in the master cylinder that forces fluid through the lines to the wheel cylinders, where the pressure forces other pistons to move and apply brakes or do other work. When pressure is released, springs move the wheel pistons back to their original positions, forcing the fluid back into the master cylinder.

Hydraulic brake systems are nearly frictionless, and usually require less force to operate than mechanical systems do. Equal pressure is supplied to each wheel cylinder, but differences in action may be caused by design differences in wheel cylinder sizes or brake linings, or accidentally by different, dirty, oily, or wet linings.

Maintenance consists chiefly of keeping the system full of fluid and free of air. Low fluid shows up by free play at the top of the pedal, and a corresponding lowering of the point at which the brake becomes effective.

Air may enter because of low fluid, leaks, or disconnecting lines during repairs. It makes the pedal feel spongy, and the brakes become unresponsive and may not work at all.

Air is removed by bleeding. This is done in the field by filling the reservoir, putting gentle pressure on the pedal, and opening bleed screws in the wheel cylinders, one at a time. Fluid is allowed to flow out of the bleed opening until it is free of bubbles. The screw is retightened, and another opened in its turn. Several fillings of the reservoir may be necessary.

The wasted fluid may be caught in a container by means of a special bleeder tube, soaked up with rags, or allowed to drip, depending on conditions.

Leakage of fluid onto linings destroys their smooth-friction quality.

Brake systems use rubber in their sliding and sealing parts. It may swell and rot upon contact with any petroleum product, which includes most "hydraulic fluid." They can use only genuine brake fluid, whose main ingredient is glycerin.

Hydraulic brakes may have a vacuum booster. It must be bled in addition to the wheel cylinders, and may have as many as three bleed connections. Leaking lines must be replaced. Leaking cylinders can often be rebuilt.

Power Controls.   Clutches, brakes, and other machinery controls can be operated by hand or foot pressure, or partly or wholly by power. In general, small and light machines will respond easily to manual controls, while heavy ones often need some sort of booster action.

The power may be mechanical, vacuum, air, hydraulic, electric, or electronic.

Mechanical boosters include brakes which are not strong enough to stop the drum against which they work, and which will be shifted by contact with the drum so as to pull or push on a linkage that will apply a larger brake.

Vacuum employs the intake suction of a gasoline engine, or a separate vacuum pump, to pull a piston in a cylinder. The movement of this piston may operate a brake through a mechanical linkage, or combine with pressure from a hydraulic master cylinder, to apply hydraulic brakes.

Air controls use air supplied by an engine-driven compressor to operate brakes or clutches by moving short mechanical linkages, expanding flexible tubes so that they press against the bands, shoes or discs, or pushing pistons in hydraulic brake systems. The last is called "air over hydraulic."

The vacuum-hydraulic and air-mechanical systems used in trucks are described in Chap. 18.

The degree of response from an air brake depends on the amount of pressure admitted to the lines, and a well-balanced valve is required for smooth action.

Hydraulic power systems are described later in this chapter, and in discussions of bulldozers, graders, and heavy trucks.

Electric controls operate by supplying current through a rheostat or other regulator to electric motors or magnets attached to mechanical linkages. Increase of current increases the force of application.

An electronic control box, found in the operator's cab, monitors and compares input and feedback signals against a programmed memory. With that electronic information it produces an output signal that initiates a shift. The control box may receive signals from four sources: (1) A switch below the transmission selector lever tells the control box what gear has been selected; (2) a transmission switch tells the control box what gear the transmission is actually in; (3) a magnetic speed sensor on the transmission relays ground speed by measuring output shaft rpm; and (4) on a scraper the transmission hold pedal, when depressed, sends a signal to the control box to prevent the transmission from shifting. In the automatic mode, a shift is initiated when input and feedback signals meet the programmed conditions. A signal is then sent to the upshift or the downshift solenoid. The solenoids momentarily open valves, permitting hydraulics to position a rotary selector valve which physically shifts the transmission.

Onboard electronic systems deliver information to the electronic control unit (ECU). When the computer senses a system is operating outside an acceptable range, such as the coolant temperature higher than a preset limit or engine speed faster than a recommended maximum, the ECU flashes a fault code to the operator.

The computers used in today's machines give service personnel with a laptop computer plugged into the ECU an insight into how operations people use their equipment. Laptop diagnostic menus give the user the opportunity to display output from the various sensors working in the equipment. One nice feature of the new electronic system is a menu option that allows the technician to test the continuity of individual electrical circuits. This feature isolates each circuit the user chooses and confirms there are no shorts in the wiring harness between the controller and the sensor.

Power controls often eliminate a complicated mass of mechanical linkage which is hard to service. They permit using brakes or clutches of adequate size and rugged design without much consideration of the effort required to apply them.

ELECTRIC DRIVE

Accessories. A familiar example of electric drive is a generator on an automobile engine supplying electricity to a heater motor or a windshield wiper. Another example is the electric starter. In this case the generator's output must be stored in a battery, and released to the starter when needed.

In construction equipment, electric drive may be extended to motors for steering and other applications.

Either alternating current (ac) or direct current (dc) may be used. AC motors usually run at a set speed, and are likely to be damaged by low voltage or lugging down under load. DC motors will run at a wide range of speeds, depending on voltage and load, and are not easily damaged by lugging down.

Figure 12.37 shows a weatherproofed ac electric motor with built-in gear reduction designed for winches. It is equipped with a spring-loaded brake that automatically locks the unit whenever current is cut off.

This motor is said to be able to accelerate from a standing start to its full speed of 180 revolutions per minute, in $\frac{1}{5}$ second, and to reverse for full power in the opposite direction in $\frac{1}{4}$ second.

Such motors are installed at the point where power is applied, and substitute electric wiring for drive shafts.

Diesel-Electric. In this type of drive, most or all of the engine power is used to turn a generator that supplies electricity to motors that do the heavy work of the machine. This type of drive is found in medium to large shovels and in dredges. It is standard in most railroad locomotives and in big scrapers, off-the-road trucks, and rollers.

Full Electric. Power for electric motors may be obtained from a ground cable or trolley wires connected to a stationary source of current. This may be a "high line" or regular industrial supply system, or a diesel generator unit set up for the job.

Full electric drive is simple and quiet, and avoids fueling problems. It is necessary where the power requirement of machines exceeds that of diesel engines they can carry, and is desirable under many other conditions.

A machine may be made so that it can operate on either its own diesel-electric system or outside power.

Electric Wheel. Diesel-electric may be applied to haulers by the electric wheel, a self-contained unit that includes a dc motor of 40 to 400 horsepower, reduction gearing, a wide-base rim, and a tubeless tire. Figure 12.38 shows a cutaway view of one.

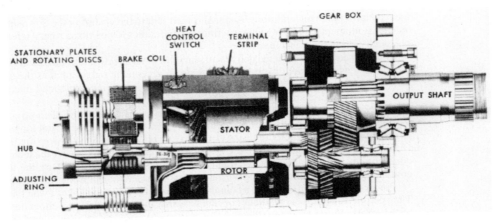

FIGURE 12.37 Electric drive for winch.

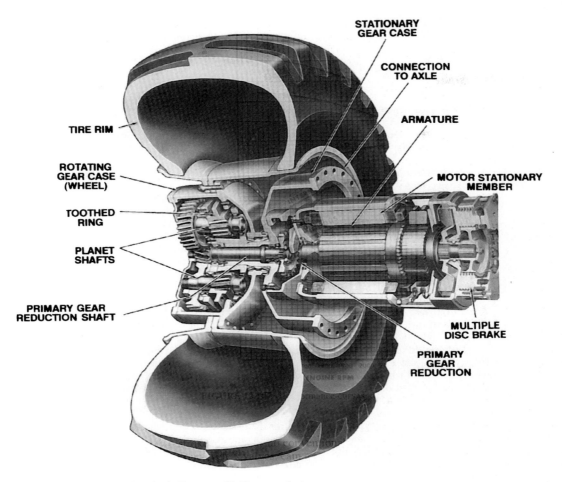

FIGURE 12.38   Electric drive wheel. (Courtesy of LeTourneau, Inc.)

The stationary member or gearbox bolts to the outer end of a dead axle. The pole ring of the motor is bolted to it. The rotating armature is geared down to turn the primary reduction shaft. This drives a compound set of planetary gears in a stationary carrier. The large driven planet gears turn short shafts carrying smaller inner gears that are meshed with an internal toothed ring that is bolted to the rotating gear case that makes up the wheel of the machine.

The multiple-disc brake is mounted on an inner extension of the armature shaft. It is applied by springs, and released by an electromagnet. It is applied automatically whenever there is no current in the brake circuit, either because the controller is in the off position or because of a general power failure.

The motor also provides regenerative electric braking that is sufficient to slow the machine on downgrades and stop it in normal operation. The disc brakes are used for sudden stops and for parking.

Power is usually provided by a diesel-driven generator mounted on the machine. It may also be supplied by overhead trolley wires. A single machine may be designed to use current from either source.

Current is controlled by hand switches and rheostats. Since the motors are reversible, the same performance is obtained backward and forward, and the operator's station may swivel to face either way.

The electric wheel is finding its principal construction applications in large all-wheel drive scrapers, rollers, and trucks. Further references will be found in the sections about these machines.

WHEEL STEERING

Wheeled equipment uses many different steering systems. They include manual, booster, hydraulic, and others.

The manual type that was standard on all cars years ago transmits power from wheel through a worm gear to the steering linkage. Its operation is too familiar to justify description here.

Power Booster.   This mechanism is similar to manual steering, but it is geared higher so as to involve less spinning of the wheel, and includes in the linkage a hydraulic piston and valve.

Operation is similar to power steering in a car. The maximum effort exerted to turn the front wheels under the worst conditions is the same as that required while rolling on a hard roadway. In each case it need be just sufficient to compress a spring.

If the front wheels are jammed so that they will not turn, a relief valve will open in the case. This may make a noise and/or cause the wheel to pulsate. A properly designed booster will not exert enough pressure to damage the linkage, but an overstrength one might, so it is best not to try to force it.

Steering may also be stiff when the oil is cold. The booster does not work at all if the engine is not running.

Brake-Assisted.   If a brake is applied so as to slow one side of a vehicle and not the other, the vehicle will turn toward the braked side. One-side braking is the only steering means in most crawlers, and in a few skid-steer wheel tractors.

Most rear-drive wheel tractors have a combination, with sharp automotive-type angling of the front wheels, and individual braking on the rear wheels. Most steering is done with the front wheels only. But when pulling heavy loads, and/or making very sharp turns, the front wheels may skid ahead instead of turning the tractor. This tendency can be corrected by applying the rear brake on the inside of the turn.

Full Hydraulic.   This type of steering turns the front and/or rear wheels directly or through simple linkages by means of one or more hydraulic rams. Control is by a valve operated by a lever.

In this arrangement the position of the steered wheels depends directly on the position of a piston in a ram, as the height of a hydraulic dozer blade depends on its ram pistons. There is no direct relationship between the position of the control lever and that of the wheels.

When the wheels have reached the desired angle, the steering lever is moved back to center or neutral, holding them at the angle. When the turn is completed, the lever must be moved to the right and held until the wheels are straight.

Full hydraulic steering of wheels is limited to machines such as graders and rollers whose wheels are easily visible to the operator. But it can be adapted to general use by a follower valve that enables it to follow the movements of a steering wheel in the same manner as a booster system.

Articulated.   Full hydraulic may also be used to turn a whole section of a machine on a massive, vertical hinge. One or two double-acting cylinders are employed. This construction, called articulated steering, has been standard in self-powered overhung (four-wheel) scrapers since they first appeared. It has also become the preferred construction in large four-wheel-drive tractors and some trucks.

In the scraper, Fig. 12.39, the hinge, (kingpin) is near the front of the machine, and it is always the front or tractor part that swings when steering, with a result close to (although not exactly like) turning the front wheels.

In the tractor loader, Fig. 12.40, the hinge is midway between the axles, so both parts share equally in the swinging. This action produces the effect of the four-wheel coordinated steering described below. Front and rear wheels run in each others' tracks, backward and forward, and the whole weight of the tractor lines up efficiently behind any pushed load.

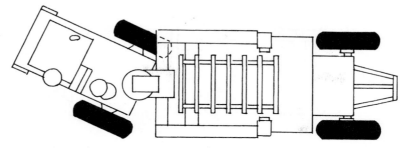

FIGURE 12.39   A scraper's articulation is forward.

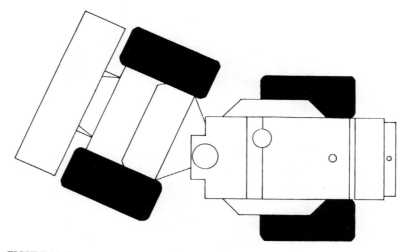

FIGURE 12.40   A tractor's articulation is centered.

Electric.   Some scrapers have used electric motors, operating through reduction gearing, to control turning around the kingpin.

There were two fingertip switches, one for right and one for left. Operation followed the full hydraulic pattern, requiring return by power to center position.

Front and Rear Steering.   Wheeled vehicles may be designed to steer by angling the front wheels, the rear wheels, and/or both the front and rear wheels. See Fig. 12.41.

Front-wheel steer is the standard method. It is the safest and most satisfactory for most conditions, both because operators are used to it from driving cars and because the effect is the simplest. The vehicle follows the angling of the wheels. The rear wheels do not go outside the path of the front ones, but trail inside it.

Rear-wheel steer swings the rear wheels outside of the front-wheel tracks. As this happens behind the operator, it may give some unpleasant surprises in close quarters. The principal advantage is greater effectiveness in handling off-center loads at either the front or rear, and in preventing "drift" down a sideslope. It was widely used in front loaders in preference to front-steer, as it kept machine weight squarely behind the bucket on turns, and kept the front wheels tracking inside the rear wheels while backing from a bank, close to a truck. But it has been replaced by articulation.

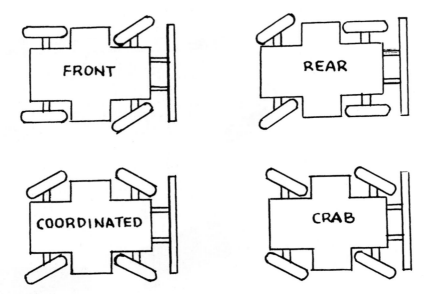

FIGURE 12.41   Front-and-rear steering.

A vehicle with both front and rear steer can use it in two separate ways. In coordinated steering the front wheels are turned one way and the rear wheels to the same angle in the opposite direction. The wheels will then track; that is, the training wheels will always move in the same tracks as the leading wheels, whether the machine is moving forward or backward. This feature lessens rolling resistance on soft ground, as one pair prepares a pathway for the other.

Coordinated steering also provides maximum control of direction under load, enabling the machine to keep on a straight course in spite of side forces, and to force loads to change direction. It also permits accurate cutting along curved lines, and short turns in proportion to the maximum angle of the wheels.

In all-wheel crab steering, both sets of wheels are turned in the same direction. If both sets are turned at the same angle, the machine will move in a straight line at an angle to its centerline. If it carries a straight dozer blade, this will meet dirt at an angle and sidecast in the manner of an angle dozer. Each wheel makes it own track, giving maximum flotation.

Special results can be obtained with either coordinated or crab steering by using different angles of turn on independently controlled front and rear wheels.

Skid Steer.   Skid steering is effected by separately controlled power trains in the two sides of a four-wheel-drive tractor.

In some models, the mechanism is similar to that in a crawler tractor. One side is driven while the other is slowed or stopped, causing a turn toward the slow side.

However, most skid steer machines have separate forward and reverse drives to the sides. One side can be driven forward, the other backward, for a spin or spot turn, usually within the machine's length. In addition, any lesser degree of turn can be obtained by varying the relative speeds of the two sides.

BRAKES

A brake is a device for slowing, stopping, or holding a machine. A friction brake performs all three functions, but a tooth or jaw brake is intended only to hold.

Friction brakes turn vehicle or machinery momentum into heat. They usually consist of metal bands or shoes with composition lining on one side, with linkage by which the lining can be forced against a smooth, narrow, cylindrical drum. Bands are ribbons of slightly flexible steel; shoes are rigid pieces shaped to a drum. Brakes that are outside the drum and squeeze inward are called external contracting; those that are inside pushing out are internal expanding.

An internal expanding two-shoe brake, Fig. 12.42, has two rigid shoes hinged at one end of the backplate. There is a spring which holds the shoes together in the released position, and a cam between the two shoe ends which is turned by the brake linkage and forces the shoes apart and against the drum to apply the brake. Adjustment is made in the linkage or by moving the hinges.

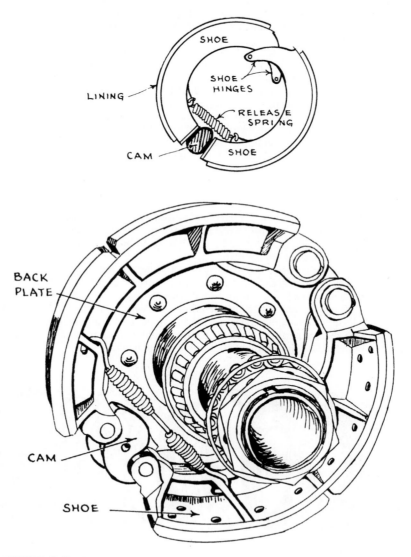

FIGURE 12.42 Two-shoe brakes.

In hydraulic brakes, the cam is replaced by a wheel cylinder, which is a cylindrical chamber in which a pair of pistons move out equally and oppositely when pushed by pressure from the master cylinder.

Disc.   Disc brakes are similar to the disc clutches described in the next section, except that half the discs do not rotate.

The disc brake construction shown in Fig. 12.43 has a set of alternately placed lined and unlined steel discs. The lined discs are splined to the brake hub and turn with it. The unlined discs are splined on the outside to a stationary housing.

The discs are held apart by springs while released, and are applied by air pressure that forces them together. Braking is obtained from the friction between the rotating and nonrotating surfaces, and is proportional to the force with which they are squeezed together.

There are also single-disc and shoe-type disc brakes. Application may be mechanical, hydraulic, or air.

Adjustments.   There are at least four types of adjustments that may be made on brakes. Examples of these are shown in Fig. 12.44. First, pedal height may be adjusted in the brake lever for convenience of the operator. This particular adjustment has no effect on brake performance unless a low position could make the pedal rod strike the floorboard.

Then there is the clevis adjustment at the left end of the operating rod. Shortening the rod will tighten the brake, lengthening it will loosen the brake. In some machines this adjustment might be for pedal position only.

The most-used adjustment on this type of band brake is the brake adjuster bolt and nut. Tightening the nut shortens the working part of the rod, pulling the band closer to the drum, and reducing free movement of the linkage necessary to make contact. In this example the adjustment affects only one end of the band directly, but it brings the whole circumference closer by pulling the band around.

Chatter.   Some brakes (and clutches too) will chatter under certain conditions, often when only partially engaged. This is a real nuisance, often interfering with slow, delicate work. It is hard on the operator's nerves, and even harder on the equipment. It causes crystallization and early failure of shafts, cases, and brake parts.

Chatter is basically caused by defective design, but usually does not occur until something else is wrong to bring out the weakness. This may be gummy lining, the wrong type of lining, a bent band or warped shoe, out-of-round drum, wrong hookup, or looseness in the linkage.

It may be very difficult to determine just what is causing a chatter, and how to stop it. The best adviser is a manufacturer's service representative. Sometimes he or she can provide a package of replacement parts designed to cure the trouble, or can recommend a harder lining, a softer lining, or perhaps a different length of lining on the band.

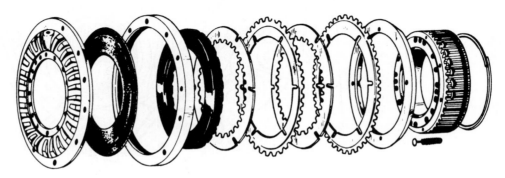

FIGURE 12.43   Disc brake, exploded view.

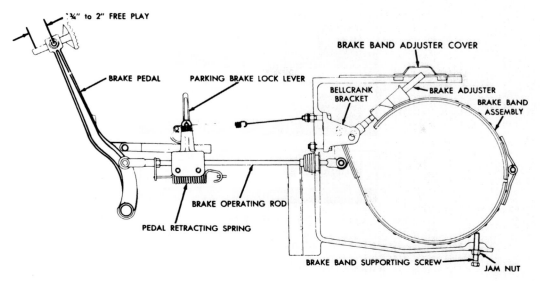

FIGURE 12.44 Steering brake linkage.

Linings. Most brake linings are of the molded type or woven lining. Molded-type linings are made of metal shavings and fiberglass or ceramic material mixtures which are carefully manufactured to meet the use of the particular lining, and are hardened in molds by heat and pressure. Such linings are brittle, and it is difficult to adapt them to drums that are more than 10 percent larger or smaller than those for which they are shaped.

Woven lining is a fabric of composite material and metal threads. It is quite flexible and can be cut to the proper length as needed. It will adapt to different-diameter drums and will not break while being handled. But it is much more easily damaged by contact with oil or brake fluid, as these material soak into it.

Variability in friction coefficient among different linings offers a way to cure some deficiencies in brakes. A grabbing brake might be smoothed down by using a lower friction lining, while an inadequate or heavy-pressure brake might be improved by a higher friction lining. See Fig. 12.45.

Most linings are fastened to shoes or bands by means of brass or copper rivets. Holes the size of the rivet shanks are drilled through the lining, in position to correspond exactly with the holes through the band or shoe. A larger hole, the size of the rivet head, is drilled about two-thirds the depth of the lining from the face or friction surface, the rivet placed in it, and expanded at the back of the band.

Lining must be replaced as soon as it is worn down to the rivet heads. Rivets are harder than lining and wear down more slowly, cutting grooves in the drum. New lining working against a grooved drum will not make full contact with it, the brake will not work properly, and the roughness will cut the new lining.

Bonded or rivetless lining is glued under heat and pressure to brake shoes. When such a lining wears out, the shoe is replaced with another that has been lined at the factory or in a shop. Bonding permits more economical use of lining, as it can be worn down nearer the shoes than a riveted lining before it must be discarded. Efficient braking takes place only when lining material meets drum walls correctly.

Condition. Most linings are made to work dry, and they will hold only when they are clean and dry. If oil, grease, brake fluid, or other liquids get on them, their friction characteristics will change. Usually they stop working, but occasionally friction will be increased so that they will grab and chatter with unpleasant effects on the operator's nerves and the machine's performance.

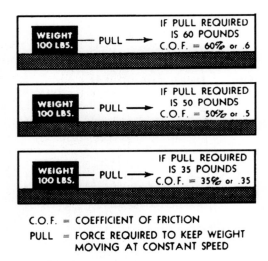

C.O.F. = COEFFICIENT OF FRICTION
PULL   = FORCE REQUIRED TO KEEP WEIGHT
         MOVING AT CONSTANT SPEED

FIGURE 12.45   Coefficient of friction.

If you act promptly you can often save the lining by washing it in naphtha, benzene, or white gasoline, either on the machine or after taking it off. You can use carbon tetrachloride if there is plenty of ventilation. Do not use leaded gasoline, kerosene, or fuel oil, as they leave harmful deposits.

Grease and other substances may leave a hard glazed or gummy deposit on brake lining that is difficult or impossible to remove with solvents. Its effects may be that the operator will have to push or pull harder to get the original braking result, it may prevent the brake from working sufficiently with any pressure, or it may grab and slip alternately, causing the brake to chatter.

If neither solvent and wire brushing nor fullers earth will make the brake work properly, it should be relined.

Wet Linings.   Some linings are made to work in oil, which cleans and cools them. It is aggravating to see how well these hold when oil-soaked, whereas a dry lining is made useless by a few drops. These "wet" linings wear out very quickly if used dry.

Maintenance.   In maintenance work for brakes, it is unwise to do repair work of the wearing parts on just one wheel because the brake action of all wheels need to be balanced to act together. It is also suggested that the reuse of wear parts should be avoided.

CLUTCH

A clutch is a device by which two shafts turning on the same axis—that is, in line with each other—can be connected and disconnected. Clutchlike action can also be obtained by the use of a moveable pulley in a belt system, by engaging and disengaging gears, and in other ways.

Jaw Clutch.   Figure 12.46 shows jaw clutch constructions. Each unit consists of two toothed rings, called jaws, with the teeth facing each other. One jaw is keyed onto the drive shaft, the other is moveably splined to the driven shaft. A circular groove is cut in the back of the driven jaw. In its hollow a brass ring or shifting collar rides, held from rotating by a yoke that is connected to the clutch control linkage. It pulls the jaw back to disengage the clutch, or pushes it forward to engage its teeth with those of the lower jaw. When engaged, the two shafts will rotate as one, being bound

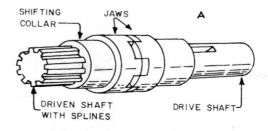

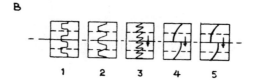

FIGURE 12.46   Jaw clutches.

together through the clutch teeth and the key and splines. When disengaged, one may rotate without affecting the other.

This type of clutch has the virtue of positive action as it cannot slip, although it might disengage itself or break. It is cheap to build, occupies little space, particularly as one jaw might consist merely of tooth sockets cut in the hub of a gear serving other purposes. It will operate dry or wet.

However, it is rough and inconvenient in action, hard to engage at any time, and impossible to disengage under load. If engaged at speed, it may give damaging shocks to shafts and gears. If both jaws are stationary, tooth may strike against tooth, so that one jaw must be turned until tooth is against hole to enable them to mesh.

In (B), (1) is the same as (A) and (2) can be engaged at low speeds. It will show a tendency to slip out of engagement when overloaded, particularly as it becomes worn. It is the only design shown which can be disengaged under load.

The other three are one-way clutches which will be disengaged automatically by reversing rotation. They can be engaged at moderate differences in speed, although (3) may break tooth points; (5) is the strongest construction.

For best results, a friction clutch should be placed between a jaw clutch and the power source.

Single-Plate Clutch.   The dry disc or single-plate clutch is the friction type most used in cars, trucks, and many other machines. If an engine clutch, it is built into the flywheel; if used elsewhere, it has a similar housing.

Figure 12.47 shows a typical specimen. The flywheel turns with the engine crankshaft, and at its center is the pilot bearing in which the front of the jackshaft rests. The purpose of the clutch is to connect and disconnect these two shafts which are in line with each other (have the same centerline of rotation).

The pressure plate has a machined front surface and is mounted on bolts threaded into a cover plate bolted to the outer part of the flywheel. It can slide backward and forward on these bolts. Each bolt carries a coil spring which pushes the pressure plate forward.

The clutch-driven disc, or plate, is between a machined rear surface of the flywheel and the pressure plate, but does not extend out from the hub as far as the bolts and springs. It has friction lining on both its front and rear, and is movably splined to the jackshaft.

The springs push the pressure plate against the disc, pushing it forward against the flywheel plate. The disc is squeezed between two plates which are turning with the engine, and the friction between the machined surfaces and its lining is sufficient to turn the disc, the jackshaft, and any load driven by it, up to the power of the engine.

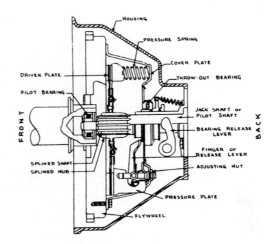

FIGURE 12.47   Single-plate clutch.

The clutch is disengaged by pulling the pressure plate back against the springs by means of three levers, called fingers. These levers are pivoted on rear extensions of the flywheel and are attached to brackets on the back of the pressure plate.

This clutch is standard for use in engine flywheels and finds many other applications. It is usually smooth in operation and long-wearing. However, except in special constructions, it should not be slipped for extended periods because of excessive wear and possible heat warping or scoring of the plates.

Double-Plate.   A double-plate has two discs splined to the jackshaft. They are separated by an unlined plate driven by the flywheel. Extra friction surface is obtained without increasing diameter.

Heavy-Duty.   There are two types of special construction used for disc clutches in extraheavy service, as in tractors that carry front-end dozers or rippers.

Caterpillar uses a double-plate clutch that operates in oil that is kept in circulation by a pump. Linings are a special type that grip when oil-soaked, and the circulation carries away the heat generated by slippage.

Dry clutches may carry ceramic discs bonded to the plate instead of regular lining. The ceramic is made of sintered metal and clay or similar substances. It is highly resistant to both heat and wear, and will outlast regular lining several times under extreme conditions. In any construction, generous design size will tend to provide long life.

Multiple-Disc.   The multiple-disc clutch consists of an engine-driven splined hub, a driving drum or spider with a splined inner surface, and a number of driven discs splined to the shaft and free of the drum, alternating with driving discs splined to the drum and rotating freely on the shaft. Also splined to the hub on each side of the discs are the fixed backplate and the movable pressure plate, which have polished surfaces toward the discs. The clutch is engaged by moving the collar against the pressure plate, causing it to press the discs together so that friction between them transmits power from shaft to drum. To release, the collar is slid away from the plate, and springs force the discs apart.

The discs may be all metal, or alternately metal surface and friction-lined. Wear is taken up by means of a threaded collar.

The multiple-disc clutch is as smooth in action as the single-plate, is somewhat longer but much smaller in diameter for the same holding power, and may be run dry or in oil. It is somewhat more complicated and expensive.

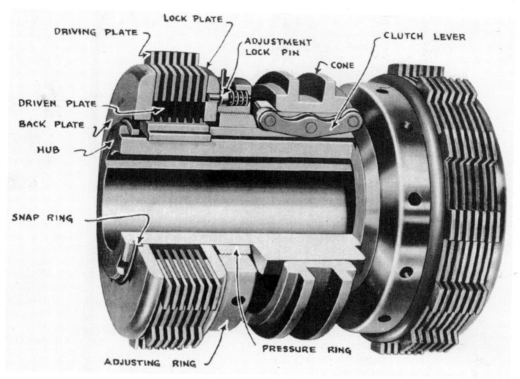

FIGURE 12.48   Multiple-disc clutch, reversing control type. (Courtesy of Twin Disc, Inc.)

The unit shown in Fig. 12.48 includes two multiple-disc clutches, and is used both to disconnect power and to reverse its direction. The control lever has forward, neutral (release), and reverse positions.

Shoe Clutch.   Heavy-duty equipment often uses internal expanding band or shoe clutches. These are similar to band or shoe brakes, except that the bands and linkage revolve, and therefore have to be operated through a throwout collar. This type of clutch is most used under severe conditions where it must operate partially engaged at times, and is subject to rapid changes in load. See Fig. 12.49.

Linkage.   Clutches may be operated by a foot pedal, as in cars and trucks. In this case foot pressure releases the clutch, and it is reengaged by the pressure plate springs and usually by a spring on the clutch pedal also, after foot pressure is removed. Clutches may also be operated by hand levers, as in shovels and crawler tractors, in which case the pressure plate springs are comparatively weak. The lever is pulled back so that is locks over center and holds the clutch in engagement until it is pushed forward, where it is held in the released position by a light spring.

Adjustment.   A newly adjusted clutch not equipped with a booster will require a moderate to heavy pull to lock over the hand lever. As the clutch is used day after day, the pull needed will gradually diminish, perhaps to such an extent that it will not lock.

This easing results from wear of the facing, and too-easy operation usually indicates a clutch that is either slipping or will slip under severe strain.

FIGURE 12.49   Internal expanding band clutch.

Adjustment is usually made on the basis of free play and "feel" in the clutch lever or pedal. When it is moved from fully engaged position, there should be a free travel space, $\frac{1}{4}$ to 1 inch on pedals, 1 to 3 inches on levers, before the resistance of the pressure plate springs is felt.

Failure to leave this free play will cause excessive wear to the clutch collar and fingers and may cause slippage as well. It has the same effect as riding the foot on a clutch pedal.

A clutch hand lever should lock over center in fully engaged position. Tightening the clutch increases the amount of pull to required to lock it, and overtightening may prevent it from locking.

Instruction books usually specify the number of pounds of pull that indicate a correctly adjusted clutch, and this may vary from 25 to 60 pounds, although 40 is quite usual. An experienced operator can judge the adjustment on her or his own machine by feel.

The important thing about adjusting a clutch is that it must not drag when released or slip when engaged.

Dragging.   Drag is tested by disengaging the clutch, then moving the gear lever slowly from neutral toward any gear position. If it grinds continuously when the teeth are held in light contact, there is a drag.

A drag that occurs only when the engine is cold is not serious. This condition may be chronic in oil bath clutches. At operating temperatures, however, a drag is an extreme nuisance to the operator, and indicates that the clutch has too little clearance, so that loosening or repair is needed.

Slipping.   A slippage test is made by holding the machine either with its brakes or against an obstacle, and then engaging the clutch slowly in high gear. If the engine does not stall, and there is no fluid drive or safety slip clutch, the clutch is too loose.

If lever pull or pedal position indicates correct adjustment, slipping shows that servicing is needed.

Safety Clutches.   There are cushion clutches which will slip a certain distance under a shock load before reestablishing a solid connection, and safety clutches which will slip rather than transmit enough strain to break parts. These units save damage to machinery and cables due to sudden increase of load or the hitting of obstacles.

Centrifugal.   A centrifugal or automatic clutch is usually a small friction clutch whose linkage is operated by a pair of weights flexibly mounted on the drive shaft. See Fig. 12.50.

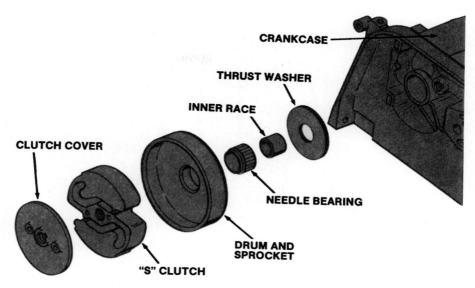

FIGURE 12.50 Centrifugal clutch. (Courtesy of Homelite.)

Rapid rotation of this shaft will cause the weights to move outward by centrifugal force, moving the linkage to engage the clutch. Slowing the shaft will allow them to move back to center, disengaging it.

A well-known application is in the chain saw. The clutch is fully engaged at full-throttle cutting speeds, and disengages when the engine is cut back to low speed. This avoids the hazard of motion in the cutting chain when it is not in use.

FLUID DRIVES

Fluid drives or couplings usually transmit engine power through fluid, without mechanical linkage, but may drive through gears in such a way that hydraulically controlled slippage occurs and similar cushioning is achieved. See Fig. 12.51.

The fluid drives discussed in this section are the hydraulic coupling and the torque converter. In both of these, the driving and driven members are inside the same case, and have the same axis of rotation.

Pump-and-motor and hydrostatic drives, which are quite different, will be covered in a later section in this chapter.

Most fluid drives are designed for use with engines having speeds of 1800 rpm or more. If used with lower-speed engines, a larger size must be used in proportion to power. At engine cranking speeds they transmit no torque, but in cold weather they may constitute a drag which will increase starting difficulties unless a conventional engine clutch is used also. At engine idling speeds they transmit enough power to turn the output shaft against light loads so that the power train tends to "creep."

An engine protected by a fluid coupling cannot be stalled by a load, but it may stall while idling because of defects in ignition and carburation.

Because their effects are achieved through slippage, fluid couplings may increase fuel consumption by an amount which varies with the type and design of the unit, and the class of service. Some models may be provided with lockup clutches, which will permit changing to a mechanical connection under conditions where cushioning is not required and slippage is undesirable.

FIGURE 12.51 Fluid coupling, exploded view. (*Courtesy of Twin Disc Inc.*)

The slippage generates heat. In torque converters this is usually dissipated through an outside radiator and cooling system. In most other installations the heat is radiated from the outside of the case. Cooling problems become more acute when the coupling is small in proportion to the power of the engine, when heavy continuous loads are carried, and when air temperatures are high. Capacity of a unit can often be increased by an auxiliary cooling system.

Means may be provided to drain the fluid into a reservoir (dump the clutch) so that no power will be transmitted through it.

The smooth application of fluid power improves starting traction of wheels against dirt or pavement, and of pulleys on belts. Engine lugging ability is increased because engine speed can be maintained in spite of slowing of the power train. Torque converters have this effect, and they also multiply the force exerted by the engine.

Properly engineered fluid couplings reduce maintenance and repair costs, increase production, and relax the operators, advantages that more than offset increased fuel consumption in many types of service.

Hydraulic Coupling. The hydraulic or fluid clutch or coupling consists of an oil chamber which contains a set of pump vanes driven by the engine, and a turbine set connected to the driven machinery. These vane sets, somewhat resembling half grapefruits with the fruit removed, are set close together with the flat sides facing, and they turn on the same axis.

The Twin Disc coupling, Fig. 12.52, uses two impellers and two turbines, thus providing increased capacity in proportion to diameter.

Rotation of the pump vanes by the engine-driven input shaft causes the oil to spin, and the oil rotates the turbine in the same direction as the pump. The high rotation speed and close clearances permit transmission of very heavy loads.

The force exerted against the turbine is slight at engine idle, and increases with speed until the two members turn almost as a unit. Load on the turbine or output shaft will cause increased slippage so that it will turn more slowly while the engine continues to work at full speed and power.

A fluid clutch is run about 85 percent full of oil. The proper level is indicated by the location of the filler plug. If it is on the rim of the coupling, the coupling is turned until a plate reading "Top for Filling" is at the top. If the filler is on the side, it is turned into its highest possible position, and fluid poured in until it runs out of a check hole.

The airspace in the coupling provides for expansion of fluid as it heats in service. Provision for further expansion may be made in the shaft seal by a spring and bellows arrangement on one side

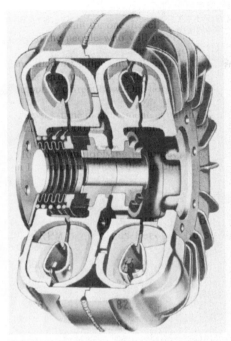

FIGURE 12.52 Fluid coupling, cutaway. (*Courtesy of Twin Disc Inc.*)

of a pair of matched seal rings in a positive-type seal. High pressure will compress the bellows spring and allow air or fluid to escape between the rings.

The fluid coupling is little used in construction machinery, having been almost entirely replaced by torque converters.

TORQUE CONVERTER

A torque converter is a fluid coupling so constructed that the flow caused in the oil by slipping under load is converted into additional force acting on the turbine vanes.

Single-Stage.   One form of torque converter, a hydromechanical transmission adapted from European railroad locomotives, essentially replaces clutches in the power transmission. See Fig. 12.53. The converter pump is bolted to the flywheel, and together they form an oiltight housing within which the turbine and stator (reactor) operate. The turbine is splined to the output shaft at the front. The rear of this shaft is surrounded by a stationary casing called the ground sleeve.

When the pump is turned by the engine, it picks up oil near the center and throws it outward and backward against the turbine vanes at a predetermined angle. The turbine has sharply curved blades which catch the speeding oil from the pump and extract maximum force from it by changing the direction of flow. This causes the turbine to rotate and the oil to move from the outer circumference to the center of the turbine, which it leaves traveling in the opposite direction from the rotation of the pump and turbine. Outlet passages are smaller than the inlet so that when the turbine is moving at low speed under load, the oil emerges with greater velocity than it enters. It is still capable of exerting force, but will tend to exert it against the pump unless its direction is changed.

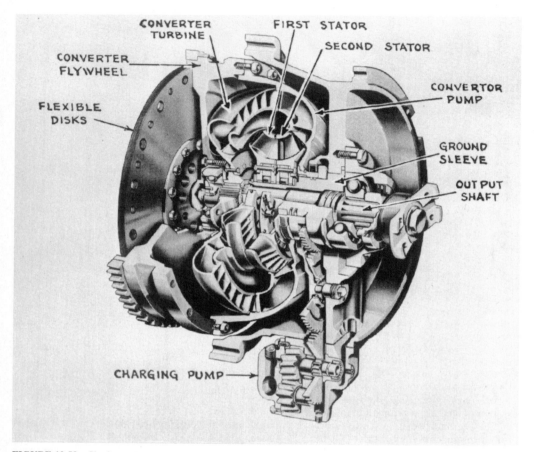

FIGURE 12.53 Single-stage torque converter. (*Courtesy of Detroit Diesel.*)

The stator is mounted on the ground sleeve and includes a freewheeling unit which allows it to spin freely in the direction of pump rotation, but locks when backward pressure is applied. When the stator is struck by oil speeding back from the turbine, it locks against the ground sleeve and provides curved passages that change the direction of oil flow so that it enters the pump hub moving in the same direction the pump is rotating. Its velocity is added to the velocity of the oil developed in the pump so that the total force of the oil leaving the pump is correspondingly greater. This regenerative action is the key to the torque multiplication developed in the converter, and increases automatically with slippage caused by increase in load on the output shaft.

When slippage decreases so that oil leaves the turbine vanes slowly, the stator turns in the direction of pump rotation, and the converter acts as a fluid coupling.

A total of five converter series are available, with torque ratios ranging from 2.5 to 1 up to 4.0 to 1. Capacities cover engines from 60 to 600 horsepower.

Twin-Turbine. A twin-turbine converter has a single pump with two turbines, one inside the other. Each turbine drives a different gear train in a transmission. If load on the drive shaft is high, as in starting to move a load, the first turbine turns, driving a low-gear system. If the load is reduced, higher-velocity oil reaches the second turbine and turns it, along with its higher-speed gear connections.

This results in automatic two-speed action, which is usually supplemented by an additional choice of ranges in the transmission.

Three-Stage.   The Twin Disc, Figs. 12.54 to 12.57, is a three-stage converter with fixed reactors. The fluid leaving the pump or impeller impinges on the first-stage vanes in the rim of the turbine, is diverted by a set of reactor vanes and passages in the case to exert push on the second rim set of turbine vanes.

From these it passes through a second reactor and spends its remaining force on an inner set of turbine vanes. Most of it then goes to the pump, but a small part is diverted through the cooling system.

The impeller of the converter is a centrifugal pump and, therefore, must obey the following basic laws:

1. Input torque varies with the square of the input speed.
2. Input horsepower varies with the cube of the input speed.
3. Input horsepower varies with the fifth power of the diameter of the impeller.

To handle a complete range of engines from 40 to 580 horsepower, Twin Disc manufactures three sizes of three-stage converters: the 10,000 series, the 11,500 standard, and heavy-duty 11,500 series. The 11,500 series impeller wheel is approximately 15 percent larger than that of the 10,000 series, and since $1.15^5 = 2$, the horsepower capacity range is twice that of the 10,000 series.

In each Twin Disc series there are torque converters available with different specific torque ratings, each covering a portion of the broad capacity range of the series. This enables the user to match the

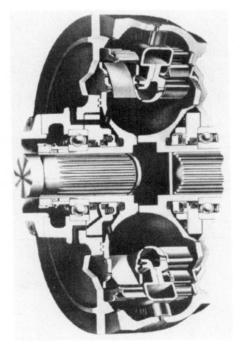

FIGURE 12.54   Three-stage torque converter, cut-away. (*Courtesy of Twin Disc Inc.*)

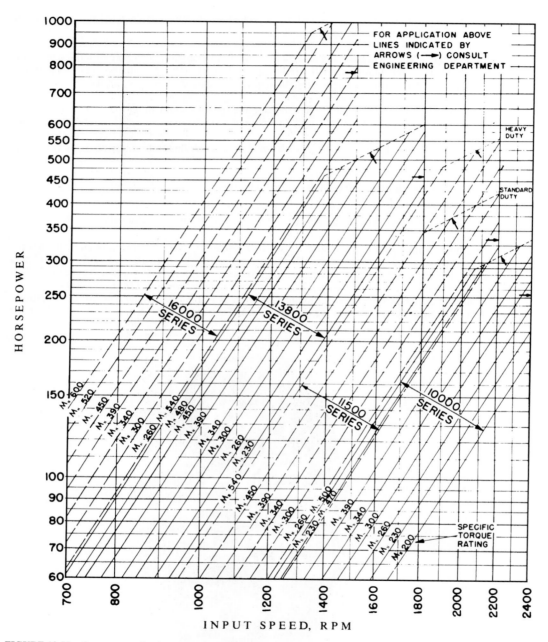

FIGURE 12.55 Converter application chart. (*Courtesy of Twin Disc Inc.*)

FIGURE 12.56   Three-stage torque converter, exploded view. (*Courtesy of Twin Disc Inc.*)

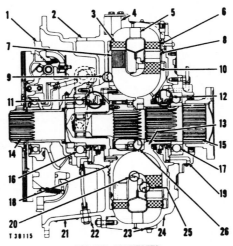

TORQUE CONVERTER
(CROSS-SECTION)

1–Clutch assembly. 2–Inlet from fluid cooler. 3–Turbine blade (first stage). 4–Outlet to fluid cooler. 5–First stage reactor blade (converter housing). 6–Turbine blade (second stage). 7–Impeller blade. 8–Second stage reactor blade (converter housing). 9–Labyrinth seal (impeller to clutch housing). 10–Turbine blade (third stage). 11–Seal assembly (impeller). 12–Seal assembly (turbine). 13–Turbine assembly. 14–Input shaft. 15–Output shaft. 16–Converter front bearing. 17–Converter rear bearing. 18–Impeller assembly. 19–Rear seal drain hole. 20–Labyrinth seal (impeller to turbine). 21–Clutch housing. 22–Front seal drain. 23–Drain plug. 24–Converter housing. 25–Labyrinth seal (turbine to reactor ring). 26–Freewheel assembly.

FIGURE 12.57   Cross section of a three-stage converter.

converter exactly to the engine, thus fully utilizing all available horsepower. Maximum torque ratios range from 2.0 to 1 to 6.0 to 1. These torque converters are used in bulldozers, trucks, and other heavy equipment. Figure 12.55 provides information on their speed and power capacities.

The choice of a converter depends on engine characteristics, the type of machine in which it is to be installed, and even expected machine operating conditions.

Cooling and Circulation. Slippage inside a fluid drive unit creates heat. The torque converter usually makes so much heat that it requires an outside cooling system, just as an engine does. The type of system varies with the make, model, and type of machine. Figure 12.58 is a three-stage converter with a stationary housing and a cooling radiator. Oil pressure in the converter is higher at the rim than near the hub, so circulation is obtained by tapping the line to the radiator on the rim and the return line near the hub.

Cooling circulation in a single-stage converter with a rotating housing depends on the charging pump's keeping pressure in the converter. Some of this fluid enters the pipe to the radiator or heat exchanger, and cooled oil goes into a sump to be picked up again by the pump.

Proper temperature is very important to a converter. Good operating range is between 180 and 220°F, with absolute minimum 160 and maximum for steady operation 250°F. Operators should be reminded that they must watch their gauges, and should stop and report if these indicate trouble.

The basic circulation in the converter is from impeller to turbine to stator to impeller, over and over. In addition there is the circulation for cooling, mentioned above, and the bleed-and-charging circuit.

In the three-stage unit, the charging pump draws fluid from a reserve tank, that may be the machine's diesel fuel tank, and pushes it through a filter and into the converter. A pressure relief valve built into the pump regulates the amount of pressure it can build up in the converter. This pressure varies from unit to unit, but it is very important that it be correct for each one. Low-pressure systems, used chiefly in shovels, may require only 20 or 30 pounds, whereas a tractor converter may need from 45 up to 75 pounds.

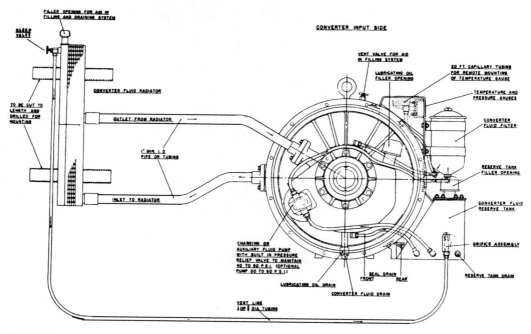

FIGURE 12.58 Cooling circulation, three-stage converter. (*Courtesy of Twin Disc Inc.*)

The high point of the fluid system, located at the top of the radiator in our illustration, has a bleeder opening and tube into the reserve tank, which excess oil enters through an orifice (small, precision opening) that will not allow passage of as much fluid as the charging pump can deliver. This restricted opening enables the charging pump to keep the pressure in the converter and cooling system as high as the setting of its relief valve.

Air.   Torque converters can function properly only if they are filled solidly with fluid. Air mixed with the oil will cause poor performance, overheating, and possibly serious damage. It may enter the system if the fluid in the reserve tank is low enough to permit the charging pump to suck air (a moderately low level might permit sucking air if the machine is working on very steep slopes), or there is a leak in the suction line.

Suction leaks may be too small to show up by outward leakage of oil.

Of course, air enters the system when oil or filter elements are changed and when the lines are opened for any reason.

Moderate quantities of air can be taken off by the bleeder system, as it will rise to the top and then be forced through the tube and orifice by pressure of liquid behind it. Partly opening the bleeder screw on top of the converter housing while the unit is running will also permit some air to escape.

Check for the presence of air by stopping and starting the engine, while watching the pressure gauge. The hand should move immediately from operating to zero pressure when the engine stops and right back up when it starts. If it hesitates in either direction, there is air in the system.

Overheating.   Overheating is a major problem in converter operation. Its frequency and extent are affected by design, the type of work, the operator, the air temperature, and the condition of the unit. The causes for overheating are given in Fig. 12.59.

Use of too low a gear and high engine speed may also result in some overheating. This type of converter will also heat badly if left connected to an idling engine. Under this condition the input shaft turns at several hundred revolutions per minute, while there is likely to be enough drag on the output shaft to keep it stopped or hardly turning. The lost power is turned into heat, and the cooling system works sluggishly at low speeds because there is little pressure difference between the inner and outer parts of the converter.

A converter should not be left idling. If it has an input clutch, release it. Otherwise, stop the engine.

Selection of Ratios.   The torque converter automatically proportions the amount of torque delivered to the output shaft to the load through infinite gradations. However, it does not necessarily do this according to the desires of the machine operator, or even for maximum efficiency.

The accuracy with which a converter selects ratios depends on its internal design and the conditions of its use. Its efficiency usually drops when the speed of the output shaft is less than one-half that of the input shaft. For these reasons it is often desirable or necessary to use a transmission in conjunction with it. This may have standard construction, or be automatic or semiautomatic in action. A converter with a narrower range of action and snappier response can then be used, and a much wider range of ratios supplied.

Cleanliness.   Torque converters are precision-made and depend on very close tolerances. Some parts are lapped smooth with irregularities allowed only up to 12 millionths of an inch. Oil in these units circulates at high velocity, and any foreign material it carries will wear down the edges and pit the hollows in vanes, changing the effective shape. This is in addition to bearing and seal damage always caused by dirty oil.

A converter for a medium-size off-the-road truck may cost thousands of dollars, and single parts for it cost hundreds of dollars. It is certainly worth the trouble of seeing that every drop of oil that goes into it is clean, and that its filters are inspected and cleaned and replaced regularly.

The recommended interval for changing converter oil is every 1,000 hours or every 3 months. Since construction equipment has an average use of around 1,000 hours per year, the actual interval in your shop is something to be discussed with the equipment dealer. If the unit uses fuel oil from the engine supply tank, changing may be automatic.

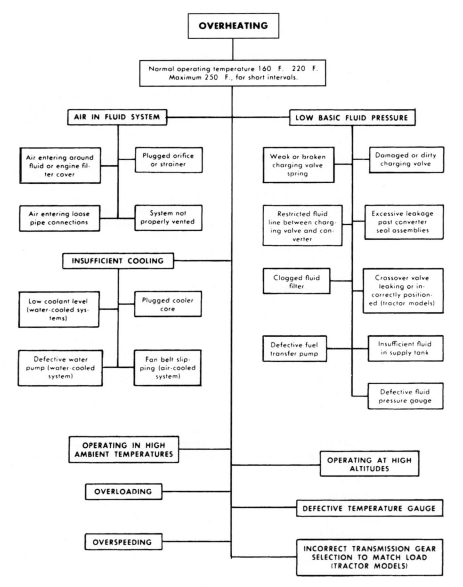

FIGURE 12.59 Causes of converter overheating.

The three-stage converter has four drain points. They are the bottom of the converter housing, the reserve tank, the filter, and the radiator. Air vents in the top must be opened to ensure rapid and complete draining. The converter should be warm or hot so that the fluid will flow freely.

The filter element is replaced, and its case cleaned and rinsed, whenever the oil is changed.

To refill, pour fluid in the radiator until it comes out the converter housing vent. Close the vent, add fluid to the filler opening, and replace the radiator cap. Fill the reserve tank 1 inch below the full mark.

Then start the engine and run it at half throttle. Check fittings for leaks, and check the gauge for proper operating pressure. Keep checking the reserve tank dipstick, and add fluid if the level goes down.

Many torque converters are lubricated by the converter oil, as this is pumped through its bearings as part of the circulation system. The three-stage type, however, is separately lubricated by dip and splash from a lubricating oil reservoir, or by grease from outside fittings.

This separation of fluid and lubricant requires the use of seals to keep each substance in its place. These are in double sets, with a drain between them that will take any leaking fluid into the reserve tank. This makes the unit safe against damage from diluted lubricant that might occur if only a single internal seal separated them.

Converter Clutches.   For efficient use in construction equipment it is usually desirable to supplement a torque converter with one or more clutch or clutchlike devices.

An input clutch is placed between the engine and the converter. It can be released to protect the converter against being overheated by an idling engine, and to prevent "creep" of the output shaft and the machine. It is used chiefly in equipment that does not have a gearshift, or that does not require gear shifting on the move.

An output clutch is a heavier and more expensive unit, as it has to carry the multiplied torque. It is needed for mechanical shifting with the machine moving, as it relieves the jackshaft of the inertia of the converter. It will stop creep, but may not prevent overheating when idling.

A lockup clutch locks the input and output shafts together, so that the converter is put out of operation. It prevents slippage and the resulting loss of efficiency during operations that do not require either torque multiplication or fluid cushioning. For example, an off-the-road hauler might need the converter in climbing out of a pit, but would perform better without it on a flat haul at the top. This device may also be called a direct drive, and may operate automatically.

The so-called free wheel produces the opposite effect from freewheeling in a car. It is an over-running clutch that locks the shafts together whenever the output tries to move faster than the input. This permits use of the braking effect of engine compression in going down grades. Converter drag also helps in slowing the machine. This device, or a lockup, is necessary in any downhill hauling operations.

Another advantage of the lockup clutch and the free wheel is that either will make it possible to turn and start the engine by pushing or towing the machine. This is often very important.

Hydraulic Retarder.   Hydraulic retarders are used to reduce speed of trucks and other haulers on downgrades, and to slow them down preparatory to stopping with the friction service brakes. See Fig. 12.60.

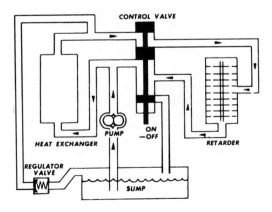

FIGURE 12.60 Hydraulic retarder.

A retarder consists of a paddle or vane-type rotor turned by the drive shaft or the converter output shaft, a fixed casing or stator fitted with vanes, and an oil circulation system.

When the casing is empty, there is practically no friction. When oil is admitted, it sets up a drag or resistance that slows the shaft, turning its energy into heat. The degree of retarding or braking depends on the amount of oil in the unit.

Oil may be supplied and cooled by a torque converter system, or by being forced out of a hydraulic cylinder and through a heat exchanger. The surplus heat is useful in keeping the converter or engine warm on long downgrades.

A hydraulic retarder greatly increases the safety of hauler operation in hilly country, and usually permits higher average speed as well.

Output Shaft Governor.   Torque converters sometimes produce excessive output shaft and vehicle speeds under light loads, as compared to the speeds under full-rated loads.

It is usually desirable to maintain, within limits, a constant output shaft speed in spite of changes in the load placed on the converter. To obtain this control, it is necessary that the governing device be driven by the converter output shaft, so that it will respond to variations in output shaft speed rather than engine speed.

A dozer, for example, will need full power while cutting hard material, and little power while spreading it. If equipped with a torque converter and a conventional engine governor, it would move slowly in the cut at full throttle, then speed up as the load diminished until it would move too rapidly for accurate grading.

If engine speed were controlled from the output shaft only, the speed of the dozer would be the same through cut and fill, except when working beyond capacity. However, engine speed might be increased too much under heavy load conditions, unless an engine governor were used also.

The output shaft governor installation therefore consists of two separate governors, one driven by the engine and one driven by the converter output shaft, either of which may override to reduce fuel to the engine. Speed may be under the control of either, depending upon the load and speed settings.

At output speeds below the set value, the shaft governor will permit the engine to run at full speed, but the engine governor will hold it down to whatever maximum it is set for. When shaft speed reaches the set value, the shaft governor will close the throttle sufficiently to prevent speeding, even if engine rpm is less than that allowed by the engine governor.

AIR MOTORS

Rotating motors powered by compressed air are used in wagon drill feeds, in hoists, and in many other machines which operate near a supply of compressed air in conditions where exhaust fumes would be a problem, and where a light motor developing high torque is required.

A stalled air motor will resume turning if the load is reduced within its power range, without needing to be restarted. It will not be damaged or rendered inefficient by lugging down. It is lubricated by oil mixed with the incoming air.

Both piston and vane types are used. A four-cylinder piston (radial) motor is shown in the cutaway, Fig. 12.61. All cylinders operate off one throw of a counterbalanced crankshaft. A rotary valve releases compressed air into each cylinder as its piston passes dead center on the upstroke. The air expands, driving the piston down, and is exhausted on the next upstroke.

Air motors are supplied in both reversible and nonreversible models. Speed is controlled by an air throttle. Gearing reduces speed and increases torque.

The vane motor, Fig. 12.62, has a rotating cylindrical hub set off-center inside a casing cylinder. Flat vanes are set in longitudinal slots in the hub. Air pressure in the hub and centrifugal force move them outward into light contact with the cylinder wall.

The curved space lying between the rotor and the more distant section of the casing has an inlet at one end and an outlet at the other. Compressed air entering through the inlet pushes against the nearest vane, moving it and rotating the hub until the air can escape through the outlet. The five vanes provide for a smooth continuous rotation.

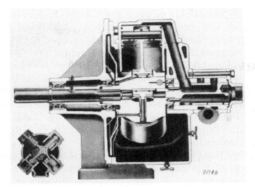

FIGURE 12.61   Radial air motor. (*Courtesy of Ingersoll-Rand Company.*)

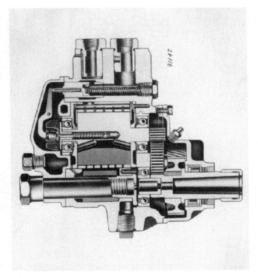

FIGURE 12.62   Vane air motor. (*Courtesy of Ingersoll-Rand Company.*)

CABLE (WIRE ROPE)

Cable, or wire rope, is one of the most important materials or parts used in excavation machinery. There are many types for different uses, but most are made up of carbon-steel wires wound into strands, and strands wound with each other to make cable. The strands are wound around a center or core, which may be an additional strand, a miniature cable, a rope made of sisal or manila, or plastic. The wire core is stronger and more resistant to crushing, but is less flexible and resilient than the hemp or plastic.

A cable is designated by its size, by the grade of steel wire used in it, as to whether it is preformed, by its lay, the number of strands not including the core, and the number of wires in each strand.

A widely used construction is the 6 × 19, that is, six strands of 19 wires each. The wires may be all of one size, or of two or more sizes. Additional wires, to a total of 25 per strand, may be added without changing the 6 × 19 designation. See Fig. 12.63. Each construction has a name, often the name of the designer of the particular type. Variations in flexibility and in resistance to crushing and to abrasion are obtained. Small wires are desirable when the cable is subjected to sharp bending; large outer wires when it may be rubbed and chafed.

Lay. The lay of a wire rope is the direction of twist of the wires in the strands and the strands in the cable. Four standard lays are illustrated in Fig. 12.64. In practice, the difference is that the Lang lay has better fatigue resistance because of the flatter exposure of the wire, but it has a tendency to untwist unless both ends are held.

Right lay is the usual construction and is recommended for overwinding on a drum when the anchor is on the left, and for underwinding when the anchor is on the right. Left lay is preferable for overwinding from the right, or underwinding from the left. This is because the cable, when relieved from strain, tends to twist slightly as if to unwind its strands, and if used as advised above, this twisting will cause the wraps on the drum to hug each other, instead of loosening and spreading apart.

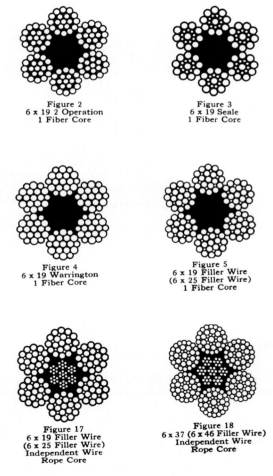

Figure 2
6 x 19 2 Operation
1 Fiber Core

Figure 3
6 x 19 Seale
1 Fiber Core

Figure 4
6 x 19 Warrington
1 Fiber Core

Figure 5
6 x 19 Filler Wire
(6 x 25 Filler Wire)
1 Fiber Core

Figure 17
6 x 19 Filler Wire
(6 x 25 Filler Wire)
Independent Wire
Rope Core

Figure 18
6 x 37 (6 x 46 Filler Wire)
Independent Wire
Rope Core

FIGURE 12.63 Wire rope cross sections.

REGULAR LAY ROPE
A regular lay rope is one in which the direction of lay of the strands
in the rope is opposite to the direction of lay of the wires in the strands.

LANG LAY ROPE
A lang lay rope is one in which the direction of lay of the strands
in the rope is the same as the direction of lay of the wires in the strands.

RIGHT LAY ROPE
A right lay rope is one in which the path of the strands in the rope is
from left to right in a direction away from the observer. A right lay rope
may either be regular lay or lang lay.

LEFT LAY ROPE
A left lay rope is one in which the path of the strands in the rope is
from right to left in a direction away from the observer. A left lay rope
may be either regular lay or lang lay.

FIGURE 12.64   Wire rope lays.

To determine whether a drum is overwinding or underwinding, stand behind it, looking along the outgoing cable. If it takes off from the top or near side of the drum, it is over; if the bottom or far side, under. See Fig. 12.65.

Grades.   Several grades of steel are used in wire rope. The two most suited for excavation machinery are plow steel and improved plow steel. The improved variety is about 15 percent stronger than the plow steel, which in turn is considerably stronger than the mild plow and traction steels used in elevator cable, stationary guy ropes, and highway guards. Weight and strength data are given in Fig. 12.66. Cable size is found as shown in Fig. 12.67.

There is also a higher-grade, extraimproved plow steel, which is 15 percent stronger than improved plow.

Wire.   Cable wire is stiff and springy. In ordinary nonpreformed construction, if a wire is cut or broken, the ends will straighten and project from the rope surface at an angle. These ends cause extra wear to sheaves, drums, and other wraps of cable, and will cut unprotected hands.

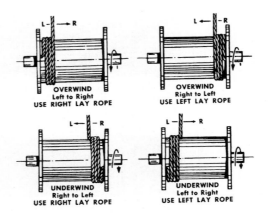

FIGURE 12.65   Winding on drum.

If such a cable is cut or broken, the wires and strands will untwist for several feet or yards on each side, unless bound (seized) or clamped.

Preforming.   Preformed cable is made of wires which are shaped so that they lie naturally in their positions in the strand and the rope. They show little tendency to stand out from the surface or to unravel when cut.

Preformed cable is safer to handle than the straight wire type and is more resistant to fatigue caused by working over small sheaves, or around sharp angles. It is recommended for use in most excavating machines, and is replacing the older type.

Seizing.   Wire rope is manufactured in long pieces that are cut into shipping or working lengths by the manufacturer, dealer, or user.

Cable formed from straight wire should be firmly bound (seized) with soft iron wire in two to four places on each side, before cutting. This process is illustrated in Fig. 12.68. A tighter wrap can be obtained by holding the loose end of the wire under some tension and twisting it onto the rope with bar, as in Fig. 12.69.

Large cables require more and tighter seizing than small ones.

Seizing wire should be thin enough so that the cable end will go through a socket or clamp easily. If difficulty is experienced, the cable and binding can be flattened with a heavy hammer.

Preformed cable is usually bound only when it is to be placed in a poured socket, or is to be stored or roughly handled before use.

Cutting.   The best tool for cutting is an oxyacetylene torch, as it is quick and easy and will weld the ends, prevent raveling, and reduce the strain on the seizing. The next choice is a regular cable cutter, which is a concave chisel in a vertical guide. The cable is placed in a groove under the edge and the chisel struck several times with a heavy hammer. An ordinary cold chisel may also be used.

Standard bolt cutters will be damaged by the hard wires. A hacksaw is rather tedious to use. It will work best if fitted with blades designed for use on hard steel.

Bending.   Wire rope is very flexible, but repeated bending causes the metal to lose its resiliency and break, usually first in individual wires, producing a weak spot that finally breaks under strain. Sharp bends, as around small drums and sheaves, are more damaging than gradual bends around large ones.

Wire rope manufacturers recommend that drum or sheave diameter be 45 times the diameter of the cable to minimize bending stresses. Unfortunately, it is impractical to build excavating machines up to these standards, and ratios vary from 30 down to 10 or below. This results in a generally short life for cable, and in some cases it is more economical to use a small rope which

CABLE SIZE	APPROX. BREAKING STRENGTH	FACTOR OF SAFETY & ALLOWABLE LOAD							WEIGHT LBS/FT
		7	6	5	4.5	4	3.5	3	
1/2"	* 21,200	3,300	3,600	4,300	4,700	5,350	6,050	7,150	.46
	+ 26,700	3,700	4,450	5,350	5,900	6,700	7,600	8,900	
5/8"	* 33,200	4,800	5,550	6,700	7,300	8,350	9,400	11,150	.72
	+ 41,200	5,900	6,900	8,300	9,100	10,400	11,700	13,900	
3/4"	* 47,300	6,700	7,750	9,400	10,500	11,750	13,500	15,700	1.04
	+ 58,900	8,450	9,850	11,800	13,000	14,800	16,800	19,700	
7/8"	* 64,300	9,150	10,650	12,800	14,200	16,000	18,300	21,200	1.42
	+ 79,600	11,300	13,150	15,800	17,600	19,700	22,700	25,650	
1"	* 83,300	11,900	13,950	16,700	18,500	20,900	23,800	27,900	1.85
	+ 103,300	14,700	17,150	20,300	22,900	25,400	29,500	34,200	
1-1/8"	* 105,100	15,000	17,500	21,000	23,300	26,100	30,000	35,000	2.34
	+ 130,000	18,600	21,300	26,000	28,800	32,300	37,100	43,200	
1-1/4"	* 129,100	18,400	21,250	25,900	28,600	32,200	36,800	43,000	2.80
	+ 168,900	24,100	28,100	33,900	37,500	42,200	48,200	56,200	
1-3/8"	* 155,200	22,100	25,900	31,000	34,400	38,900	44,300	51,600	3.50
	+ 192,000	27,200	32,000	38,200	42,600	48,000	54,800	64,000	
1-1/2"	* 184,000	26,200	30,700	36,900	40,800	46,100	52,500	61,200	4.16
	+ 228,000	32,500	38,000	45,500	50,600	57,000	65,100	76,000	

*6 x 19 I.W.R.C. improved plow steel hoisting cable (upper line) 1/2" through 1" diameter

+6 x 19 I.W.R.C. extra improved plow steel hoisting cable (lower line) 1/2" through 1" diameter

*6 x 37 I.W.R.C. improved plow steel hoisting cable (upper line) 1-1/8" through 1-1/2" diameter

+6 x 37 I.W.R.C. extra improved plow steel hoisting cable (lower line) 1-1/8" through 1-1/2" diameter

The above chart is primarily intended to determine safe loads for hoist load lines and guy lines. If used for slings, the values must be reduced by the loss for the type of end fitting.

FIGURE 12.66   Strength and weight of wire rope.

is not strong enough for shock loads, rather than the recommended size which is so damaged by bending over small sheaves that it fails quickly.

Crushing.   Wire rope is damaged when wound unevenly on a drum and placed under strain. The wraps of cable crush and kink each other. In many machines the working end of the rope does not stay in line with the drum so that a device of pulleys or rollers, called a fairlead, must be placed between the drum and the work to line the cable up for smooth winding on the drum.

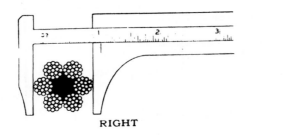
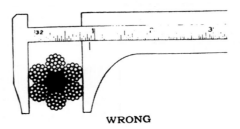

RIGHT   WRONG

Wire Rope is usually manufactured slightly larger than the nominal diameter. The diameter of a new rope may exceed the nominal diameter by the amounts as shown in the United States Federal Specification for Wire Rope.

FIGURE 12.67   Measuring wire rope.

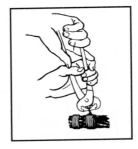

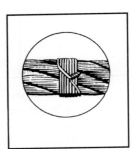

FIGURE 12.68   Seizing wire rope.

Fleet angle, Fig. 12.70, is the maximum angle made by a rope with a line perpendicular with the drum. It should not be more than $1\frac{1}{2}°$ if the drum is smooth, or $2°$ if it is grooved. It can be reduced by increasing the distance between the drum and the guide sheave.

Kinking.   Cable is severely damaged by kinking which occurs when it is pulled tight while it has a loop or twist in it. To avoid this damage it should always be unwound, not lifted, from a reel or coil. A reel should be set so that it can revolve as the cable is pulled off it. A coil should be untied, the outside end laid on the ground, and the coil rolled away, leaving the cable extended behind it.

FIGURE 12.69   Tight-wrapping seizing wire.

Lubrication.   Wire rope needs a lubricant to minimize friction between the wires and avoid fatigue of the rope. The manufacturer lubricates it thoroughly, and for short cables in continuous use this is sufficient. However, if it is to lie idle for any length of time, it should be oiled or greased to avoid rust damage. Also, if it wears shiny in use, it should be greased for the sake of the sheaves as much as that of the cable.

Safety Factor.   The safety factor in rope is the ratio between the pull required to break a new cable and the load it carries in service. Safety factors of 3, 5, or better in relation to normal pull are recommended. A shovel with a 5-ton pull on a drag cable should use one with a tested strength of 15 to 25 tons. This safety factor allows a margin for shock loads, which may be several times normal load for a moment or two, and permits the rope to continue working after some of its wires have weakened and broken.

Capacity.   Drum capacity may be calculated according to Fig. 12.71.

CABLE SYSTEMS

Fittings and Anchors.   In general, cable ends require some sort of clamp or fitting to attach them to the power source or work. A variety of these is shown in Figs. 12.72 and 12.73.

The connection between the wire rope and the fitting may be secured by clamps, wedges, or filters.

The standard clip, or cable clamp, consists of a U bolt, a saddle, and two nuts. The cable is doubled over on itself, and the two thicknesses squeezed between the U and the saddle by tightening nuts. The grooved inner surface of the saddle has a better grip than the U, and is therefore used on the live or working end of the rope.

The wedge socket jams the cable in too-small grooves in the outer surface of the wedge and in the inner surface of the socket. The cable is wrapped around the wedge, which is tapped into the socket. Pull on the cable pulls the wedge farther, tightening the connection.

This device is best suited to excavator and other cables whose ends are more or less fixed, and which have to be removed at intervals. It is too bulky for many applications.

A poured (filled) fitting consists of a conical socket attached to an eye, loop, hook, or other device. It is installed by putting the end of the rope through the small end of the socket, fraying it

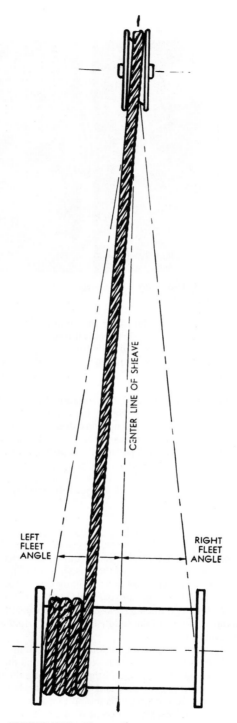

FIGURE 12.70 Fleet angle.

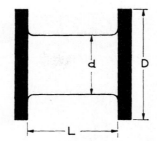

TO FIND THE WIRE ROPE CAPACITY OF A DRUM

Length of wire rope in feet that a drum will hold = $(D^2 - d^2)$ LX

D = Diameter of flange in inches
d = Diameter of drum in inches
L = Inside width of drum in inches
X = Rope Factor—See Table below

Rope Factors

$\frac{1}{8}$ inch	4.250	1 inch	.0655
$\frac{1}{4}$ inch	1.018	$1\frac{1}{8}$ inch	.0516
$\frac{3}{8}$ inch	.466	$1\frac{1}{4}$ inch	.0418
$\frac{7}{16}$ inch	.341	$1\frac{3}{8}$ inch	.0347
$\frac{1}{2}$ inch	.262	$1\frac{1}{2}$ inch	.0292
$\frac{9}{16}$ inch	.2064	$1\frac{5}{8}$ inch	.0248
$\frac{5}{8}$ inch...	.168	$1\frac{3}{4}$ inch	.0214
$\frac{11}{16}$ inch	.138	$1\frac{7}{8}$ inch	.0186
$\frac{3}{4}$ inch	.116	2 inch	.0164
$\frac{13}{16}$ inch	.099	$2\frac{1}{4}$ inch	.0129
$\frac{7}{8}$ inch...	.085	$2\frac{1}{2}$ inch	.0105

FIGURE 12.71   Rope capacity of winch drum.

out like a brush, removing the hemp center, if any, as well as any lubricant or other foreign matter. Molten zinc is then poured in the socket. It hardens, holding the individual wires in their expanded position so that they cannot be pulled through the small end.

Babbit and lead are too soft for use in these fittings unless the load is to be extremely light.

Drums.   Power is usually imparted to a cable by anchoring it to a spool-type drum, rotating the drum so that the cable is wound in around it. This pulls the cable, and the cable pulls anything to which it is attached.

The surface onto which the cable winds is called the lagging. There are three main types: smooth, cut with grooves to guide the cable into position, and tapered or conical with the small diameter at the anchor side. The tendency of the cable to slide down the taper keeps the wraps tight against each other. This tapered lagging should not be used if there is sufficient rope to wind in a second layer.

If a cable is attached to a reversible drum—that is, one which can be rotated by power in both directions—it is usually not anchored to it, but is wrapped around it two or more times. Whichever way the drum turns it will wind in one section of cable and pay the other out in the same amount. The friction between the wraps of cable and the lagging provides the necessary traction.

Sheaves.   The direction of cable pull may be changed by running it over a pulley or sheave (pronounced *shiv*). This is a steel wheel revolving on or with an axle, and having a groove in its rim to hold the particular size of cable to be used. Too large a groove will not support the cable properly and puts extra flexing strain on it; too small a groove will pinch and wear it.

Sheave grooves for $\frac{3}{8}$ to $\frac{3}{4}$ ropes should be $\frac{1}{32}$ to $\frac{3}{64}$ inch wider than the cable.

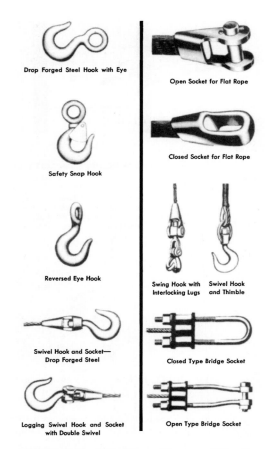

FIGURE 12.72   Cable fittings. (*Courtesy of Wire Rope Institute.*)

Multiple Lines.   A pulley or set of pulleys may be used to change the power-speed ratio of a cable system. Figure 1.25(*A*) shows a drum with a line attached to it passed around a pulley in a snatch block and brought back to an anchor near the drum. The drum is pulling in the rope with a force of 1 ton. This force, less friction losses in the block, is exerted on the anchor. There is tension of 1 ton on the line between drum and pulley, and on the line between pulley and anchor also. Since both pulls are exerted on the pulley, the pull against it is 2 tons. If it should yield to this force and move inward, it would move only 1 foot for each 2 feet of cable wound on the drum. The 2 feet is pulled out of the drum-to-pulley section, but as the block moves, 1 foot from the other section crosses it toward the drum, so the drum-to-pulley section is shortened by only 1 foot. It will be noted that the ratio checks—the force is doubled and the distance (speed) halved. Actually, there will be a loss of 5 percent or more in the pulley, and a slightly less than 2-to-1 advantage if the drum line and the anchor line are wider apart than the width of the sheave. The mechanical advantage of this system declines as the angle between the two cables increases, and it becomes zero at 114°.

Figure 1.25(*B*) shows the same drum and pulley with an additional block attached to the anchor. The rope is now run around the first pulley, around the anchor sheave, and anchored to a hook on the first sheave. If the drum winds in 3 feet of cable with a force of 1 ton, the first pulley will be subjected to a force of 3 tons, 1 ton for each line, and in response to this force it may move

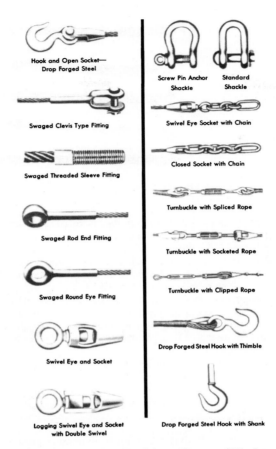

FIGURE 12.73   More cable fittings. (*Courtesy of Wire Rope Institute.*)

1 foot. The pull on the anchor pulley is 2 tons, and whichever of the three units is least able to take the pull to which it is subjected will move inward.

The drum is said to be attached to a one-part line, the anchor to a two-part line, and the first pulley to a three-part line. A four-part line could be rigged by putting an additional pulley with the first one, and reeving the cable around that and back to the anchor, by which means a 4-ton pull (less friction) could be obtained.

The number of parts of line may be increased up to the space and pulleys available. The force exerted on any part of the system will be the force at the drum multiplied by the number of lines; and the distance any member will move is the length of cable wound in, divided by the number of lines acting on the member.

CHAIN

Chain is used in work where the line is subject to scraping, kinking, twisting, and other abuse. It is heavier and bulkier than cable of similar strength. Except in small sizes it is not used on drums or sheaves.

FIGURE 12.74   Logging chain.

It is not damaged by sharp bending, is resistant to abrasion, and can be easily attached, detached, lengthened, shortened, and repaired.

Chain is made up of a series of interlocked links. Each of these may be a single piece of rod bent to shape and welded, or two half-links butt-welded together. Chain size designation indicates the approximate diameter of the rod stock used. For example, $\frac{3}{8}$-inch chain is composed of either $\frac{3}{8}$-inch or $\frac{13}{32}$-inch round rod. See Fig. 12.74.

The strength of chain is affected by the size and shape of the links. Short links are stronger but do not move on each other as readily as larger ones, are more inclined to kink, and are harder to repair.

The most important strength factor, however, is the quality of the metal used. A good alloy steel chain is more than 3 times as strong as one of the same size and weight but of only standard, hardware store quality.

Chain and associated tackle are discussed in Chap. 1.

HYDRAULICS

Hydraulic systems of the dynamic or flowing type are of great importance in construction machinery. See Fig. 12.75. Refer to the variable flow hydraulic system for a backhoe excavator shown in Fig 12.76. Such systems depend on the fact that liquids cannot be compressed except under extreme laboratory conditions, so that pressure exerted on any part of a confined fluid will be exerted by the fluid everywhere on the confining surface.

Lines carrying fluid from the pump to working parts may be very short, very long, or anything in between. Power loss in the lines is slight if they are large enough for the flow. There is no need for alignment among the separate parts of the system. This is an important advantage over mechanical drive.

Fluid.   Hydraulic fluid for flowing systems is a petroleum-base oil, similar to but not identical with lubricating oil for engines. It must be distinguished from the glycerin-base hydraulic fluid used in brakes and other static systems. The two liquids are absolutely incompatible, and cannot be substituted for, or mixed with, each other.

Viscosity is measured in the same way as motor oil, with 5W very thin and 40 fairly heavy or thick. Too-thin oil tends to leak back through pumps and out through seals. If too thick, it consumes extra power and makes operation sluggish.

Viscosity index (VI) is a measure of a fluid's change in thickness as it heats during work and cools when idle. If the changes are slight, the VI is high and the oil is good. If the index is low, it may be possible to raise it by adding a substance called a viscosity index improver.

Anti-wear (AW) hydraulic fluid is the top-recommended fluid by pump manufacturers. AW fluid will demulsify or separate water from oil for daily removal. And high VI fluids (over 100) will handle extreme temperatures and pressures in hydraulic systems.

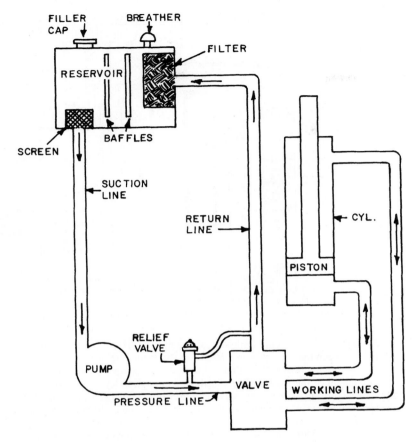

FIGURE 12.75 Hydraulic system.

A good fluid should lubricate the system. It should be able to penetrate fine spaces, resist being wiped off completely by pressure, keep friction to a minimum, and act as a cooling medium. It may contain an extreme-pressure additive to ensure these characteristics.

The oil should resist breakdown by oxidation, and by reaction with parts of the system. It must be carefully refined to obtain this characteristic, and have one or more special chemicals added. Such special oil will also keep damage to metal from rusting or corrosion (eating away) to a minimum.

To retain its stability and neutrality the hydraulic fluid must be kept filtered to remove water, and particles of every kind. It is important to change filters as recommended by the manufacturer.

Good hydraulic fluid may dissolve small quantities of air without harm, but it must not carry it in the form of foam or bubbles. These will compress in working parts, causing mushy, unreliable performance. Good oil resists foaming, but here again an additive may be needed.

The fluid should remain separate from any water that enters the system, so that it can be filtered out. Combining in the form of an emulsion allows the water to stay, and causes oxidation, deterioration of fluid characteristics, and deficient lubrication.

The cylinders of backhoes typically wear out fastest. Leaking oil means a seal has failed. That also means dirt is entering the hydraulic system by clinging to the oil film on the cylinder rod.

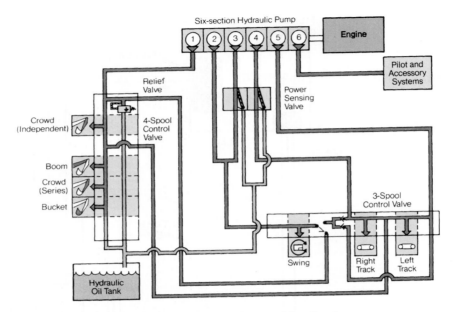

FIGURE 12.76 Variable-flow hydraulic system. (*Courtesy of Case Corp.*)

Pressure. The potential pressure at a working part is the same as that at the pump, as long as the line between them is open and adequate. But if the rate of flow is reduced below demand by a line that is too small or a fluid that is too thick, pressure will fall at the work end, unless (or until) movement of the working parts is slowed enough to let the fluid catch up with them.

Pressure inside the system is measured in pounds per square inch (psi). Changes in force and speed, equivalent to mechanical leverage advantages, are made by varying the relative areas of pistons and other movable parts.

High pressure increases efficiency and promptness of response, and reduces bulk and weight. It intensifies problems of leakage, breakage, and wear.

Efficiency. Hydraulic systems, if properly designed and in good condition, may be highly efficient but are subject to many variables.

There are always some losses in friction, and in most systems there is some slippage and internal leakage. Movement of fluid and these losses create heat, which absorbs power. The heat is often sufficient to require removal by a cooling system.

System efficiency may be as low as 70 percent, but is often between 85 and 95 percent. Contamination is most frequently blamed for lost productivity and downtime for repairs.

Leverage, Hydraulic Jack. The hydraulic jack cross section in Fig. 12.77 shows the way in which difference in area provides leverage. Its rigid casing includes a large cylinder containing a piston which raises the car or other load, a small cylinder containing a piston, pump, and a reservoir.

When the pump handle is raised, it draws the small piston to the top of its cylinder, permitting oil from the reservoir to flow under it. The piston is pushed down, first closing the hole that admitted the oil, then pushing the oil through the check valve into the bottom of the large cylinder. This check valve consists of a ball spring held against a carefully machined seat. Pressure from the pump forces the ball back against the spring, opening the passage and allowing oil to pass through. The spring and pressure from the large cylinder force the ball back against its seat when the handle is raised so that no oil can pass back through it. Each stroke of the pump therefore adds liquid and builds up pressure in the large cylinder, thus raising the piston and whatever load may be on it.

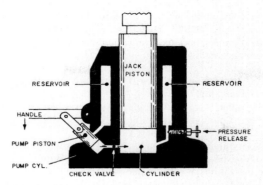

FIGURE 12.77   Hydraulic jack.

The operator of the jack may apply 50 pounds of pressure to the handle. If the leverage of the handle is 6 to 1, then 300 pounds of force is exerted on the pump piston, which has an area of $\frac{1}{4}$ square inch. Pump pressure is therefore 1,200 pounds per square inch, which pressure passes practically undiminished through the check valve, passage, and cylinder to the bottom of the large piston which has an area of 5 square inches. The fluid presses against the whole area, so the total pressure on the piston is 6,000 pounds, or 3 tons. If the person applies a force of 100 pounds to the handle, the lift will be increased to 6 tons.

The pump cylinder's effective length is 1 inch, so each stroke will deliver $\frac{1}{4}$ cubic inch of oil. In the lift this is spread out over an area of 5 square inches and will raise the piston $\frac{1}{20}$ inch. Inside this particular jack the mechanical advantage (MA), or change in force-distance, is 20 to 1, the force being multiplied by 20 and the distance moved divided by the same amount. Including the leverage of the handle, the jack has a mechanical advantage of 120, with a friction loss of less than 1 percent. This ratio could be increased up to any force desired simply by reducing the size of the small piston or increasing that of the large one.

However, the useful power of any jack is limited by the strength of the casing and other parts.

The jack is lowered by opening a return passage by means of a thumbscrew. The weight of the piston and its load then pushes the fluid in the cylinder back into the reservoir.

PUMPS

The pump in the jack is a one-cylinder reciprocating type, hand-operated. In most cases, however, pumps in excavators are driven by rotary shafts. There are three principal types of rotary pump, known as gear, vane, and piston. Each is made in several designs.

Pumps may also be classified as fixed displacement and variable displacement. The displacement is the amount of liquid it moves in 1 revolution or cycle. If it is fixed, the output should always be the same at any one speed of rotation. But output can of course be changed by varying the speed of revolution.

The variable type can be adjusted internally to change or (usually) to stop its output, without changing its speed. The hydrostatic piston pump usually has this feature, and is also reversible.

A pump may have a built-in pressure-limiting device, or such a relief valve may be a separate unit in the system.

The pump may be permanently attached so that it works whenever the engine is running, as such, a tractor front-end pump; whenever the engine clutch is engaged, as in a rear-mounted tractor pump; or to be connected specially each time it is to be used, as in a truck dump hoist. In any case, the pressure developed is utilized through one or more valves.

Even when pumps are new, they are likely to allow a slight movement of oil from the pressure side back toward the inlet. Such losses occur along the sidewalls and between meshed teeth, around the ends of vanes, and, to a limited extent, past pistons. The rated capacity of pumps takes such losses into account.

As the pump wears, the losses will increase, although not necessarily at a steady rate. The result will be a loss in volume that will be greatest at high pressures, and a drop in the maximum pressure it can develop.

A pump that is small for its job lacks reserve capacity and will show results of wear much sooner than a bigger pump with the same work. Also, what seems to be a rather sudden pump failure may be just the final stage of a long, slow decline.

A worn pump will give much better performance with thick oil than with thin oil. One of its symptoms is brisk performance only until the fluid is well warmed up.

The indications of a pump that is defective because of either wear or breakage are slower than normal hoist or other movement and less than normal power; both conditions being more noticeable when the hydraulic oil is hot than when it is cold.

Gear Pump. The gear pump, Fig. 12.78, consists of two accurately meshed spur gears between side plates, turning in a chamber shaped so that the teeth are in close contact with a housing, for about half the circumference of each gear.

Oil enters the inlet side by gravity and/or atmospheric pressure, is caught in the hollows between the teeth in both gears, and is carried to the other side of the housing. Meshing of the teeth in the center prevents it from returning between the gears, so it is forced into the outlet passage. A continuous flow is generated.

Very close clearances are required, to prevent back leakage. The side plates are sometimes movable, and are then kept in contact with the gears by conducting pressure oil behind them.

This is the simplest and most economical of the rotary pumps, and is perhaps the most widely used. But it is not adapted to extreme high pressure.

A rotary may be the only pump in systems up to about 2000 pounds of pressure, or a charging pump in a higher-pressure system.

Vane. Figure 12.79 shows a vane pump of the unbalanced type. This is not widely used, but is easier to explain than the balanced model in the next illustration.

A cylindrical, engine-driven rotor turns off-center inside a circular rotor ring or case. This rotor has a number of radial slots that hold flat vanes, in a sliding fit. When it turns, the vanes are pushed outward by centrifugal force until they reach the ring, against which they then fit closely.

The vanes divide the crescent-shaped area between the rotor and the case into compartments which expand and shrink alternatively as the rotor turns.

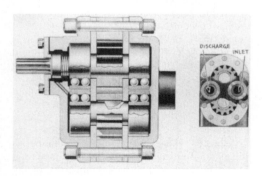

FIGURE 12.78 Gear-type hydraulic pump.

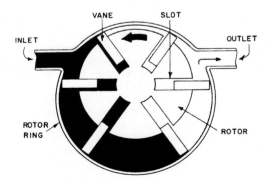

FIGURE 12.79 Unbalanced vane hydraulic pump.

Fluid feeds into the inlet at the point where the chambers start to expand, creating a suction, and is discharged into the outlet as they contract. Beyond there, the chamber shrinks to the dimension of a lubricating film, so that pressure air cannot follow the vanes back to the inlet.

This pump does most of its work on one side, resulting in off-center loads that tend to cause rapid wear. For this reason, most vane pumps are the balanced type, Fig. 12.80, with two port sets on opposite sides. The ring is oval instead of circular, with a centered rotor.

The same pumping action is accomplished in half a revolution instead of a full revolution.

The balanced vane pump is somewhat more expensive and complicated than the gear pump and may develop somewhat higher pressure.

Piston Pump (Hydrostatic).   Hydrostatic is a popular name for these pumps, the almost-identical motors that are driven by them, and the drive systems in which they are used.

The term is inappropriate. Strictly speaking, *hydrostatic* means hydraulic pressure without motion, as in the push of water or wet soil against bracing in an excavation. However, industry has accepted the term for these systems, and it will be used here.

The axial piston or hydrostatic pump, Figs. 12.81 and 12.82, consists of a number (five or more) of pistons in a rotating cylinder block. The pistons are parallel to the axis of revolution, and have their heads in sliding contact with a nonrotating, tilted plate, known mostly as a swashplate but occasionally as a cam.

The casing has two ports, which carry low-pressure fluid from a supply or charging pump, which goes through a check valve, an open tube, and an outlet to the working parts.

In one pumping position, the swashplate is tipped away from the upper port, which becomes an inlet. Charging pressure pushes pistons deep into their cylinders, until their ends press against the plate.

As the block turns, the inlet port closes, and the plate's slope forces the pistons back into their cylinders, putting a high wedging pressure against the fluid in them. At about half a turn from the inlet, this fluid is forced through the outlet. It goes through a motor circuit, then returns to the pump inlet.

The pistons are closely fitted in the cylinders. No fluid except a film for lubrication should get past them. For any one angle of the swashplate, the displacement (the amount of fluid pumped

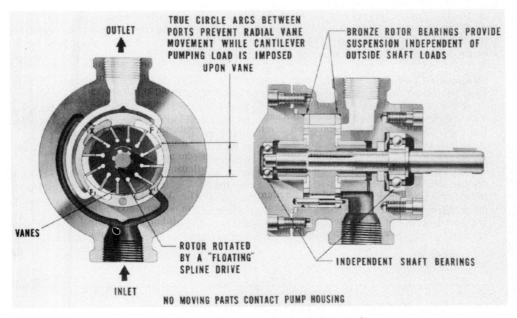

FIGURE 12.80   Balanced vane hydraulic pump. (*Courtesy of Vickers Incorporated.*)

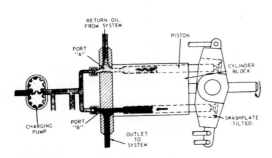

FIGURE 12.81   Hydrostatic pump in operation.

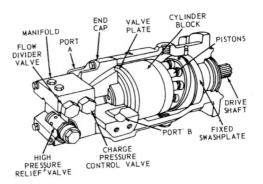

FIGURE 12.82   Fixed-displacement piston motor.

with each revolution of the shaft) and the potential pressure are the same, whether the engine is idling or turning at maximum speed.

In a fixed-displacement hydrostatic pump, the swashplate angle cannot be changed.

Gear and vane pumps also have fixed displacement, but it is usually not so exact, because of leakage from the pressure to the inlet side. Such leakage is increased by thin oil and by pump wear, and is proportionately greater at low speeds than at high ones

Variable Displacement.   The displacement and pressure of the hydrostatic pump depend on the angle of the swashplate. If its slant is increased, the stroke of the pistons is lengthened, increasing the volume of fluid handled. The steeper angle reduces the strength of the wedging effect, so that less pressure can be developed.

Flattening the angle of the plate has the reverse effect, reducing volume and increasing pressure.

If the plate is flat, at right angles to the axis of rotation, the pistons will not move in the cylinders, no fluid will move, no pressure will be produced, and no power will be consumed.

When the plate angle is reversed, the pump will take in oil at the port which was the outlet, and discharge it at the former inlet. This reverses flow in the system, so that a motor driven by this pump can be caused to rotate in either direction, by proper setting of the swashplate.

Swashplate Control.   The angle of the swashplate in variable-displacement units may be regulated by manual control, an automatic device, or both.

A manual control may be a lever to provide a neutral center, then gradually increasing speed with lever movement in either direction away from neutral. In a travel drive, moving the lever forward would cause forward travel; to the rear would reverse the machine.

In this arrangement, the control lever provides the effect of both a gearshift and a clutch, supplying infinitely graduated changes in speed-power ratio and a disconnect feature.

One automatic control in an excavator depends on engine speed. If engine rpm drops below its throttle setting because of hard digging that slows the bucket, a flow valve in a control circuit reduces the swashplate angle, thus decreasing the rate of flow to the bucket cylinder, increasing the pressure in it, and relieving load on the engine so that it can resume speed.

This effect is somewhat similar to that obtained by shifting a transmission into a lower gear, but adjustments can be made smoothly under power to any of an infinite number of ratios.

VALVES

Check Valve.   The check valve is a simple and important device that allows passage of fluid in one direction only. A small ball check is used in the jack in Fig. 12.77.

The principle is simple. A fluid passage is made with a machined collar. A solid plunger, which may be a ball or a plug, is shaped to a leaktight fit against the collar, and is held against it by a spring. Pressure from behind, that is, in the same direction as spring pressure, holds it more firmly on its seat, keeping the passage blocked. Pressure from the other side pushes the plunger back by compressing the spring, so that fluid can flow past it.

Relief Valve.   Most hydraulic systems, and most engine lubricating systems also, depend heavily on a relief valve for proper operation. The pump is built to deliver a much higher pressure than the system can use so that it will have reserve capacity to compensate for damage or wear. The relief valve is set for the pressure that the system should use, and spills any excess into a bypass back to the reservoir.

The pressure-relief valve in Fig. 12.83 is similar to the check valve in the hydraulic jack, except that it is on a larger scale, the spring is quite heavy, and the spring tension, and therefore the opening pressure, is adjustable.

If the valve opens at too low a pressure, because of either a weakened spring or an improper adjustment, the hydraulic part of the machine will operate sluggishly and with less than normal power. If the valve is set for too high a pressure, action will be brisk and snappy, and usually jerky, power will be greater than normal, and bursting of hoses is likely to be frequent.

Control.   Control valves for high-pressure systems usually have a movable core, with slots or holes that can be brought into and out of register with slots or holes in a case or body. The matching surfaces of the core and body are machined to a close sliding fit that prevents leakage between ports.

The quarter-turn petcock, Fig. 12.84, is a simple, familiar type of rotary valve. It consists of a tapered bore with a hole drilled through it, held tightly in a similarly tapered case by a spring. The case has a hole through it, one end of which is attached to a tube or a tank. The core may be turned by means of a handle so that its hole will coincide with the casing hole and fluid can flow through it; or so that its hole is at right angles to that in the casing, and its solid part blocks the passage and prevents fluid from passing through.

Figure 12.85 shows operating positions of a schematic four-way spool (sliding) control valve, of a type used in bulldozers. The case and the core are each drilled or cut away for a number of passages, which are opened into each other, or closed, as the spool slides in the case.

The passages in the casing are connected with four outside pipes—one bringing pressure from the pump, two leading to the working part of the system, and one emptying into the reservoir. The two working positions of the valve admit pressure from the pump and direct it to one or the other of the working lines, and connect the other working line to the reservoir. The "hold" position closes both working lines and opens the pump line to the reservoir. The "float" position opens the pressure and the working lines to the reservoir. In the hold and float positions, the pump develops only sufficient pressure to overcome line friction on the way to the reservoir. Its pressure in the working position is determined first by the degree of resistance, second by the setting of the relief valve, and third by the ability of the pump.

This four-way valve might also be used in a dozer hoist, or a loader with a separate dump valve. Most excavator circuits have three-way valves with no float position.

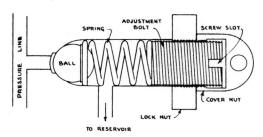

FIGURE 12.83   Pressure-relief valve.

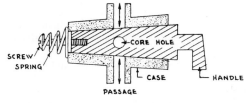

FIGURE 12.84   Simple rotary valve.

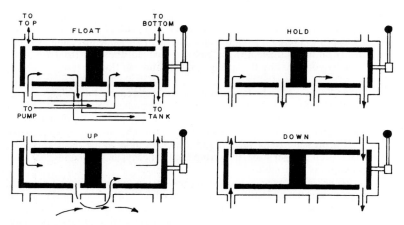

FIGURE 12.85   Schematic control valve.

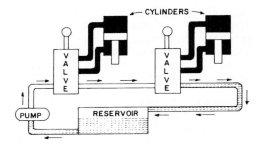

FIGURE 12.86   Series hydraulic circuit.

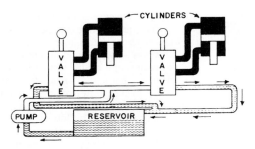

FIGURE 12.87   Parallel hydraulic circuits.

Sequences.   When two or more working circuits are powered by one pump, they must share its flow, which tends to always go to the area of least resistance. The designer must figure out where the power is needed in what proportions, and include appropriate devices to distribute it.

There are two basic ways to route the power flow: series and parallel. They are shown in Figs. 12.86 and 12.87 for systems in which two circuits might operate together or separately.

In a simple series, fluid goes through the two valves in succession, and then returns to the tank. If the first circuit is used, it takes all the flow, but its exhaust goes through the second valve. If that is put in working position, resistance to its piston will cause back pressure on the first circuit. Pressure will become balanced in the two circuits, and the one with lower resistance will get most or all of the fluid, "robbing" the other.

In a parallel system, the two valves have separate pressure and exhaust systems, but are designed so that putting one valve in working position will block the open center in the other also, or so that in neutral, one valve has open center, the other closed center.

In either case, one active valve will receive full flow, two active valves will compete with each other for the intake, and the one with lower resistance will get the larger part, or all.

There are flow-divider valves. One might be placed at the division of the pump line into parallel circuits. It could be fixed to favor one or the other in timing, in volume of fluid, or in both; or to ensure equal promptness and fluid to each.

The routing of fluid may also be affected by mechanical arrangements. With equal loads, a small cylinder would need higher pressure than a large one, and so lose out in a parallel arrangement.

Or with cylinders of equal size, leverage might be made more favorable for one than the other, so it could operate on lower pressure.

Providing a separate pump for each circuit is costly, but under a wide range of conditions it is justified by superior performance.

LINES AND FITTINGS

Lines.   Hydraulic pressure systems may use plumbing pipe (but not galvanized, it flakes) for rigid lines, but tubing is preferable. It is usually stronger, more resistant to vibration, and less likely to leak at connections. It can also be bent at the factory to fit the machine.

The wall strength of any conduit, whether rigid or flexible, must be increased with increase in diameter, due to enlargement of area exposed to internal pressure.

A rigid line must be firmly clamped close to any connection to a flexible line, so that its movements will not have sufficient leverage to strain and eventually break the pipe at the clamp.

Most flexible lines in construction equipment are high-pressure-type hose woven of fabric, rubber, and metal. Their bursting strength is far greater than the pressure they are required to hold, but they become weakened by repeated bending or rubbing, so that they eventually blow out.

Useful life varies with the quality and pressure resistance of the hose, its flexibility, and pressure in it, the tightness of its curves, the sharpness and frequency of bending, and the way it is installed. A wrong-length hose can change a good arrangement into a bad one.

A hose shortens as pressure inside it increases so that a correct curve during installation may mean a tight bend under pressure, and most hose damage occurs under pressure.

Hose should be marked with a dashed line or some other pattern by which you can tell whether you have it straight. Be sure to check this as you finish an installation, as any twist will damage a hose that is under pressure.

Fittings may have either coarse pipe thread or fine tubing thread. Never try to force mismatched threads together, as they will strip.

Bolted flanges may be used instead of threaded fittings in many places in very high-pressure systems.

A swivel fitting, Fig. 12.88 top, has a threaded collar which turns freely on the fitting when it is loose, but pulls into a rigid nonleaking connection when tightened. A pipe union is another example of this type of fitting.

In a hydraulic system, each hose and each tube or pipe that both begins and ends in a solid connection must include a swivel. This permits installation and tightening at the fixed end or ends, then using the swivel for the final assembly and tightening without twisting the line.

The swivel is the first joint to take apart, and the last one to put together.

Two wrenches must be used in working on swivels, to avoid twisting the lines. Open-end or adjustable wrenches are best, but if a Stillson is used carefully, it should not damage the fittings.

RESERVOIR

The reservoir of a hydraulic system serves basically as a supply tank, feeding fluid into the pump as needed, and receiving low-pressure oil exhausted or bypassed by working parts. It must hold sufficient oil to take care of pump demand in spite of tipping of the machine, surges in the system, normal leakage, and spills from broken lines.

Sometimes a capacity of hundreds of gallons is required, and since excavators often have little spare space, location of the tank may be difficult. It may be beside a tractor seat, in loader columns, in an enlarged transmission case, or almost anywhere.

The reservoir normally includes a filler cap and drain plug, an outlet to the pump and an inlet from discharge lines, an air breather (unless in a sealed system), a baffle or baffles to reduce sloshing

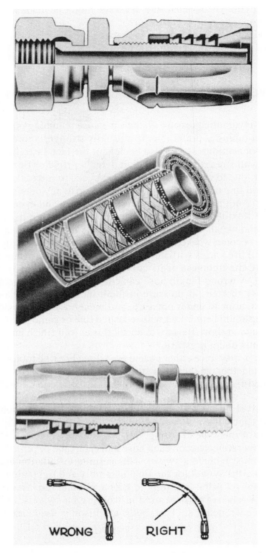

FIGURE 12.88   Pressure hose and fittings. (*Courtesy of Aeroquip Corporation.*)

and to prevent inlet oil from going directly out of the outlet, an oil level gauge, an outlet screen, and perhaps one or more filters.

Another function is to cool oil. Radiation from the sides of the tank may be enough for a system with moderate pressure and activity, but a separate cooler is needed in many high-pressure systems.

Open or Closed.   A hydraulic system may be more or less open to the air at the reservoir, or it may be sealed.

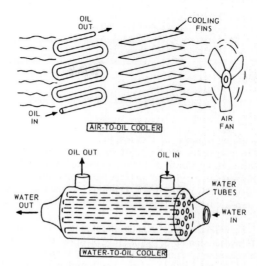

FIGURE 12.89 Oil coolers.

In an open system with a properly filled tank, the resulting surges that raise and lower level simply push air out or draw it back in. In a closed system, they build up pressure or pull it down to the vacuum. A large air capacity may be needed to cushion these effects.

An over-full open tank may squirt oil out the vent, sometimes giving the operator a disagreeable hot shower. If underfilled, the system may suck air, causing chattering, poor response, and other troubles.

The sealed construction is best suited to the modern, ultrahigh-pressure systems, because of their increased sensitivity to dirt damage, and possible danger in too-forceful sprays of oil.

Coolers (Heat Exchangers).   Hydraulic oil and seals and other parts tend to deteriorate if system temperature rises over 180 to 200°F.

If radiation of heat from the reservoir and exposed parts is not sufficient protection, a cooling system must be installed. See Fig. 12.89.

A widely used system is to place an oil-cooling radiator in front of the engine water-cooling radiator, with alterations in fan size and other engine cooling details.

Oil is circulated through this cooler at a pressure of about 100 pounds per square inch, which might be obtained by a constriction in the return line to the reservoir, or from a special pump.

CYLINDERS AND MOTORS

Pumps convert mechanical energy into fluid energy. Cylinders and motors turn the fluid energy into work. Cylinders push and pull in straight lines or arcs; motors rotate.

Cylinder, Single-Acting.   The single-acting or one-way cylinder may be a ram type or a piston-and-rod design, Fig. 12.90.

A steel plunger or piston slides in a cylinder, both machined to be a close but not a wiping fit. The piston has one or more rings to seal the contact with the cylinder, so that only a thin film for lubrication can get past.

The cylinder has a single port at its attachment end, through which pressure fluid enters to push against the face of the piston to move it, the rod if any, and any attached load, away from the cylinder.

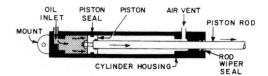

FIGURE 12.90 Hydraulic cylinder, one-way.

When the oil pressure is released, the weight of the load, or sometimes a spring, moves the piston back to its original position, forcing oil out of the cylinder back to a reservoir. There is also a hold position, in which oil is trapped in the cylinder and will keep the piston and load in position.

The single-acting cylinder has comparatively limited use in excavating machinery, as two-way power is usually advantageous. Exceptions include the lighter dump truck hoists, three-point tractor hitches, part-swing devices where two rams push against each other, and other specialities.

Cylinder, Double-Acting. The double-acting or two-way cylinder is the workhorse of excavators, for functions that can be performed by pushing and pulling. See Fig. 12.91.

The mechanism is similar to the one-way type with piston and rod, except that an oil port and line are substituted for the air vent, and a high-pressure seal replaces the rod wiper.

Cylinder seals wear over time of use, leaking pressurized fluid to the drain or into the environment. If the leaking seals are not replaced, eventually the cylinder will not lift a load, than the cylinder needs to be rebuilt.

The most costly reseal job on construction equipment will probably cost less than $1000. If the resealing is not done, to refurbish the cylinder will cost at least 6 times as much.

A control valve directs pressure oil into either end of the cylinder while it permits flow out the other end. The piston moves away from the pressure.

The valve may also block both ports, locking the piston in place, or open a passage between the two ends, allowing the piston to move freely under load, or float.

Many cylinders have a snubbing device, to prevent the piston from moving at full speed and pressure against a cylinder end, with resultant shock to the system. The snubber may be a tapered or double port which is almost closed by the piston as it nears the end of its stroke. The reduced opening restricts outflow of fluid, creating back pressure to oppose and slow the piston.

Most double-acting cylinders are unbalanced, meaning that they have a stronger push (expansion) from the piston's open side than pull (contraction) from pressure on the rod side, because of difference in area exposed to fluid pressure. See Fig. 12.92.

A cylinder with a 5-inch piston attached to a $2\frac{1}{2}$-inch rod, operating on 2,000 pounds per square inch pressure, would have a push of about 39,000 pounds and a pull of 29,000. A 3-inch rod would reduce pull to about 25,000 pounds.

When volume of flow through lines is a limiting factor, the pull is faster than the push for light loads.

Mounting. A hydraulic piston rod is strong only for straight push and pull, and has little resistance to bending stresses from the side. The cylinder and the outer end of the rod are usually mounted hinge-fashion, and movements in the plane of these hinges impose no bending stresses.

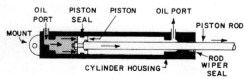

FIGURE 12.91 Hydraulic cylinder, two-way.

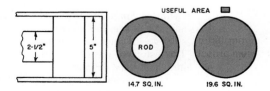

FIGURE 12.92 Effective piston areas.

If there is any possibility of side strains bending or twisting the working parts out of this plane, one or both ends of the unit must have a universal or spherical mounting, to permit sideward pivoting to align with the changed direction of the stress.

Hydraulic Motors.   The hydraulic motor is a hydraulic pump in reverse. The pump is driven by a rotating shaft and produces fluid power. The motor uses that power to rotate a shaft.

Their use is a paying proposition because when two shafts are in different places, fluid lines may be an efficient and convenient way to carry power from one to the other. Such transfer may include a change of direction, a reversal of rotation, and/or shift in speed-power ratio.

Motors of either gear or piston type may be almost identical in appearance and description to pumps of the same type. However, they are seldom interchangeable unless specifically designed for dual use. They differ in details of fluid flow and in points of stress, and may use different types of bearings and seals.

A vane motor differs from a vane pump in having springs or clips to hold the vanes out against the housing when at rest, as it cannot depend on centrifugal force (as the motor does) until the fluid starts to turn it.

Figure 12.93 is a cross section of a rotor or internal gear motor. Not actually gears, the moving parts are called the rotor and the rotor ring. The rotor is driven inside the rotor ring, which is held by the rigid case.

The rotor is mounted eccentric to the ring. The ring has one more lobe than the rotor, so that only one lobe is in full engagement at any time. This causes the rotor's lobes to slide over the outer lobes, making a seal.

In operation, high-pressure fluid enters the motor, striking the rotor and rotor ring lobes, forcing both to rotate. As they rotate, a seal is formed, then broken, as each inner lobe engages a cavity in the outer ring. The fluid is discharged at low pressure at the outlet port in the case.

Hydraulic motors may be used for almost any kind of rotary drive, and can be designed to slow or stop their output shafts when input stops. However, because of the possibility of slight internal leakage under sustained static pressure, they cannot be depended upon to keep a load locked in place.

If locking is necessary, it is usual to put a friction brake or clamp in the mechanical part of the system. This is held engaged by springs when the motor is idle, and released by hydraulic pressure when it is working.

The efficient speed for a motor is often much higher than the desired speed for the driven unit. Motor shafts may then be part of a final drive type of gearbox, as in Fig. 12.94. The gear set can then include the holding brake mentioned above.

Hydrostatic Circuit.   A hydrostatic motor is very similar to the in-line axial piston pump described earlier. High-pressure oil from the pump enters cylinders on the high side of an angled swashplate. The pistons slide down the slope, causing its cylinder block to revolve, turning an attached output shaft.

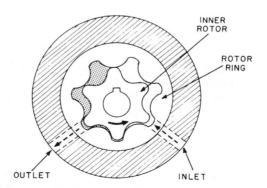

FIGURE 12.93   Internal gear motor.

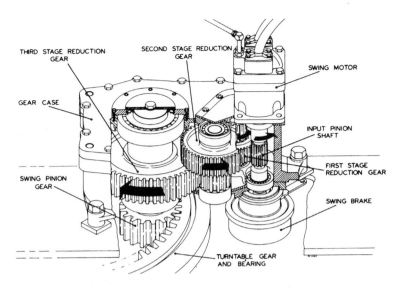

FIGURE 12.94   Hydraulic motor with reduction gearing.

The exhaust line from the motor is the inlet line of the pump.

It is usual to have an adjustable swashplate on the pump and a fixed one in the motor. However, when two motors with different requirements are driven from one pump, they may have the adjustable members.

Pumps in hydrostatic transmissions are usually the axial piston type with variable displacement provided by a movable swashplate or "rocker cam" that tilts back and forth to vary the stroke of the pump's pistons.

To get long life from hydraulic pumps and motors operations, they must be operated with care. Avoid loading the system until it warms. Cold hydraulic fluid flows slowly and can starve the pumps, allowing them to pump air which can result in scoring and pitting. Conversely, high temperatures in a hydrostatic transmission are deadly. Excessive heat thins the fluid so that its lubricating properties are reduced and wearing of precision surfaces is accelerated. An operator can create excessive heat in a hydrostatic transmission by stalling a crawler loader by pushing too hard into a pile when the tracks do not slip.

The fluid oil used in a hydrostatic transmission is crucial. It is best to follow the machine manufacturer's recommendations. A great way to monitor the fluid's performance is by oil sampling to test for wear metals, viscosity, and water. Sources of contamination can include bad seals in hydraulic cylinders, pitted metal in cylinder rods, or damaged fittings and control valves.

A hydrostatic circuit is frequently a closed loop, involving one pump (with charging pump), one motor, and a pair of connecting lines. The pump and motor may be built into a single block with drilled passages for fluid, or be at considerable distances, with connection by piping and/or flexible lines. One reservoir and filter may serve one circuit, or several,

In-block construction may serve simply as a transmission with variable speed and direction, or provide change in direction (angle drive) also. Separated units also allow replacement of more or less complicated layouts of shafts, gears, and/or chains with more adaptable hydraulic lines.

SEALS AND FILTERS

The proper functioning of a hydraulic system is completely dependent on having effective seals that will keep fluid and pressure in and dirt out. Design, manufacture, and installation of such seals become increasingly critical as pressures increase.

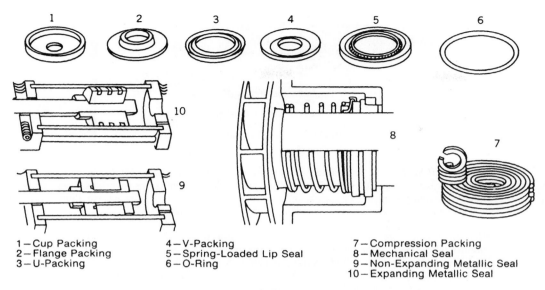

FIGURE 12.95   Types of hydraulic seals. (*Courtesy of Deere & Company.*)

1—Cup Packing
2—Flange Packing
3—U-Packing

4—V-Packing
5—Spring-Loaded Lip Seal
6—O-Ring

7—Compression Packing
8—Mechanical Seal
9—Non-Expanding Metallic Seal
10—Expanding Metallic Seal

There are two distinct types. Static seals, which include gaskets, seal fixed parts together. Dynamic seals are for parts that move against each other. The motion may be rotation of shafts, or sliding of pistons and rods. See Fig. 12.95.

Compression Packing.   Compression packing consists of string, or shaped inserts, that are wound or placed around a shaft, and tightened against it by a packing nut or gland. The sealing material may be cotton string, graphite compounds, plastic, rubber, or flexible metals.

These seals will hold only a few hundred pounds pressure, and are found mostly on old equipment.

U- and V-Packing.   U- and V-packing, often called chevron packing, is made of various materials compressed and molded into rings that are concave in one direction and convex in the other. The rings may be intact, for slide-on installation, or cut for installation from the side. Several of them are placed around a shaft with their open or concave sides toward the pressure, so that it will tend to expand them, forcing the inner lip against the shaft and the outer lip against the case.

They are rather lightly compressed by being held in a gland or case, with a flange or plate tightened by bolts. The bolts are often secured against loosening by a wire passed through drilled holes in the heads.

These packings are satisfactory for low to medium high pressure.

O-Ring.   The O-ring is probably the most popular seal for sliding shafts such as piston rods, and is important for static use as a gasket.

It is composed of a slender ring of synthetic, oil-immune rubber, which is usually quite soft and flexible. It is used only between smooth, carefully shaped metal surfaces, as it does not have the bulk to take care of irregularities, and will wear rapidly or tear immediately on sliding parts that are rough, jointed, or angled. It is also worn rapidly by rotating shafts.

The O-ring is compressed moderately, about 10 percent, in a specially shaped groove between the two parts. For gasket use it may be reinforced by a backup ring of stronger material. See Fig. 12.96.

An O-ring must be the right size, both in ring diameter and in gauge, to do its work. It must also be in perfect condition. Tiny nicks or scrapes, loss of shape, or hardening from heat exposure is likely to make it useless.

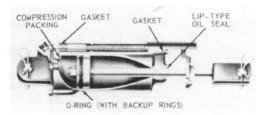

FIGURE 12.96   Seals in hydraulic cylinder.

Mechanical Seal.   The mechanical or positive seal is usually a pair of hard washers, lapped to make a perfect rotating fit with each other, pressed together by diaphragms or springs. These are used chiefly for rotary seals at low speeds and low to moderate pressure. They are excellent to protect lubricated units, such as track wheels, that work in dirt and mud.

Metallic Seals.   Metallic seals for pistons and piston rods are quite similar to piston rings in engines. However, engine piston rings are of the expanding type: cut at one point, and shaped to push outward against the cylinder walls for a smooth, sliding fit.

Metallic piston seals may be of the same type, or may be nonexpanding. The latter needs extreme precision in manufacture, to avoid serious leakage.

Piston rod seals may be fitted into the cylinder housing, and contract around the rod.

In general, metallic seals are subject to more leakage than other types. This may not matter on pistons, but for rods may call for an additional wiper seal.

Gaskets.   Gaskets are flat sheets, formed to fit between flat pieces of metal, and cut away wherever necessary to allow for passages, bolt holes, or moving parts. Shapes for hydraulic components tend to be fairly simple. Gasketed parts are usually bolted together.

Gaskets are made from a number of quite different materials, which may be metallic, nonmetallic, or both. They must be yielding enough to mold into all irregularities of the mating surfaces, but strong enough to resist penetration or blowing out sideward.

Bolts for gasketed connection should be tightened firmly, up to the torque recommended by the manufacturer; and then retightened after some hours of use.

Filters.   Every hydraulic system needs one or more filters to remove fine particles. This is in addition to the comparatively coarse screen that is customarily put at the entrance to the pump inlet line.

Bypass filters are arranged so that only a part of the circulating fluid goes through one on each cycle. The greatly preferred full-flow type filters all the oil every time, at least until it clogs. See Fig. 12.97.

There are surface filters made of very fine wire mesh, porous paper in various shapes, or even of stacked discs of metal or paper. The surface may be made very large in a small container by elaborate folding.

Depth filters are packed with absorbent material such as cotton waste. Since oil can take many different paths through them, they tend to pick up particles at all levels, instead of just on the surface.

Adsorbent filters are chemically treated to attract and remove certain contaminants, which may be not only particles, but also liquids and sludge from deteriorated oil.

A standard for many filters is removal of particles down to 10 micrometers in diameter. This is about 1/4,000,000th inch. A few systems trap much finer bits, some down to 2 micrometers.

Even a new filter impedes oil flow a little, so that pressure is slightly higher on the upstream side. As it works, holes clog and the pressure difference increases until the filter is a serious obstacle to flow. A bypass should then open, filtration will cease, and a warning signal of some kind should appear or be heard.

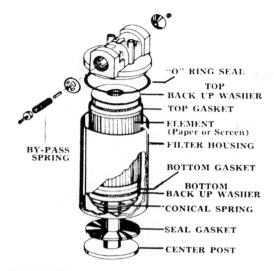

FIGURE 12.97 Hydraulic filter.

MAINTENANCE

Hydraulic systems are rugged and may operate for long periods with little attention. However, it is urgent that they do receive the servicing they require, as failures can be very expensive.

The first and most urgent requirement is clean oil. Filters must be the proper type and size, with cleaning or changing as often as necessary. Oil should be changed according to manufacturer's directions. However, it pays to have maintenance intervals based on fuel used rather than hours worked.

Leaks should be fixed promptly. They make the machine dirty and hard to service, and any hole that lets oil out might let dirt in, in spite of the great pressure difference.

All hoses should be inspected periodically, and sharp bends and rubbing areas corrected. Where rubbing is chronic, protection should be arranged.

Some contractors with large fleets of equipment have found that increasing the number of service trucks—providing lubricants, oil, etc.—in the field improves equipment uptime considerably and increases their component life.

LUBRICANTS

A fundamental necessity for machine operation is proper lubrication. The lubricant provides a slippery film between surfaces rubbing, turning, or scraping on each other. This film greatly reduces friction and the wasted power, wear, and heating that friction causes. Lubricant may also serve as a cooling medium and as a barrier or cleaner to keep abrasive material from getting or remaining between moving parts.

Oil and Grease.   Lubricants are called oils or greases. Oils are fluid and vary from the extreme thinness of penetrating oil to the slow-flowing transmission oils, which are more often called greases. The term grease may be said to include these thick oils, but more specifically means the semisolid and solid mixtures of oil with special soaps or fillers which give the combination the qualities of body (flow resistance), adhesiveness, pressure endurance, water resistance, and melting point on the basis of which greases are selected.

Maintenance mechanics make a mistake trying to service all drives with a single oil because they find that shortcut will lead to shorter life for some components. They should follow the manufacturer's instructions. One way for detecting wear of parts is by inspecting the magnetic plug at the oil port. If there is a thin film of iron fillings on the plug that may be normal. But if there are flakes, cuttings, chunks of metal, or shiny bits of bearings that show up on the magnet, that shows something has failed.

Dip Lubrication.   Transmissions and other gear boxes usually are partly filled with oil or fluid grease. Some of the gears are partly immersed in this lubricant and carry it on their teeth to the higher gears with which they are meshed. Other gears, bearings, and splines are lubricated by splash, by gravity flow of oil carried to higher points, or by both.

The dip method is best suited to heavy lubricants, which cling to parts enough that adequate quantities will be picked up and transferred to higher levels. Rotation should be slow and construction simple enough that local hot spots will not result from uneven distribution of the lubricant.

When engines are lubricated in this manner, the crankshaft usually has projections which dip into the oil and splash it around so that it reaches all surfaces requiring it.

An engine or a gearbox may be lubricated partly by dip and splash, and partly by pumped oil.

Pump Systems.   A pump may pick up oil from a reservoir, usually the crankcase or oil pan under the engine, and force it through the crankshaft and camshaft in drilled passages which have openings in each bearing. The amount of oil which escapes at each point is regulated partly by the size of the outlet, but chiefly by the closeness of bearing fit. Connecting rods may also be drilled to carry oil to wrist pins. Parts not reached directly by pumped oil, such as cylinder walls, are lubricated by an oil mist generated by leakage out of the bearings, and by dipping and splashing as well. All oil returns to the reservoir to be picked up again by the pump.

A weakness of many of these systems is that the pump moves only sufficient volume for normal requirements. If bearings wear or the oil becomes too thin, through error in selection or because of dilution or too much heat, an excessive amount of oil will escape at the bearings. This will lower the oil pressure and make it likely that the last bearings in the series will receive too little lubricant, with resultant damage.

Dirty Oil.   It is difficult or impossible to keep foreign materials out of oil. Dirt can enter an engine through outside contamination of oil in cans or funnels, by the oil dipstick that often is so located that it is very difficult to avoid touching it to dirty parts when checking oil level, through an inadequately protected or improperly serviced air intake and then past the piston rings, or through an improperly protected crankcase breather.

Carbon may work down from the combustion chamber and metal particles may appear from anywhere. A machine whose engine pan gets in the dirt, such as a tractor, may take in some of it through holes in the oil pan or past a defective seal on the rear main bearing.

Pump systems may be protected by filters. There is usually a screen at the pump intake. More important are the line filters which contain replaceable elements of fiber, cloth, or paper, or permanent ones of closely spaced metal discs or porous stone.

All serious sources of dirt in engine oil put it in the crankcase first. A particle may be put through the system several times before it happens to get in the filter. If it escapes through a bearing, it may recirculate dozens of times, each time taking a little metal out of the bearing, the shaft which rides in it, or both. Small particles, fine-sand-size and smaller, are likely to stay active much longer than coarse ones. If highly abrasive, like sharp silica particles, a fraction of a teaspoonful may cut a big engine to pieces before it is filtered out. More damage may be done in a few hours or even minutes than in years of normal operation.

The answer to this is the full-flow filter, in the line from the pump to the working parts, which is large enough to easily filter all the oil going into the engine passages. The pressure gauge, if tapped into the line after the filter, will give warning of any clogging sufficient to reduce oil flow.

The most serious contamination comes from metal filings. These are produced very slowly if the unit is in good condition, and more rapidly as bearings wear and shafts and gears get out of line. A large quantity can be permanently taken out of circulation by using magnetic drain and check-level plugs, which will hold them until removed for cleaning.

The only cure for dirt in a dip or splash system is to change the oil. It is a good plan to follow draining by putting in a thinner oil, running long enough to give it a chance to wash all parts, then draining that.

Oil Sampling.   Caterpillar Inc. has developed a scheduled oil sampling (S.O.S.) program to test for contaminants in the oil. The sampling provides three specific types of diagnostic tests:

1. *Wear analysis* monitors machine wear by detecting, identifying, and assessing the amount and type of metal wear elements found in used oil.
2. *Chemical and physical tests* detect the physical presence of unwanted fluids (water, fuel, antifreeze) in the oil.
3. *Oil condition analysis* identifies loss of lubricating properties by quantifying combustion by-products (soot, sulfur, oxidation and nitration products).

Oil contamination and degradation can create conditions leading to excessive wear and eventual machine failure. Regular S.O.S. identifies and measures these various contaminants and sources of oil breakdown. Each indication leads to the problem area and causes and then identifies the potential wear and what needs to be done to alleviate that.

There are several ways to collect oil samples for testing. The least desirable way is collecting drain oil because that is most susceptible to inconsistency. The best way is to install a pressure valve sampling port, just ahead of the filter, in a pressurized oil circulation system, such as for the engine, hydraulic system, or transmission. Another method makes use of a tube inside the tube for adding oil into the component. This method uses a vacuum extraction device.

To save on oil sampling costs it is best to do the testing at the time of regular oil changes and not start the sampling until the oil change just before the predicted time of major part failure, which may be at 5,000 to 6,000 hours of use.

Diesel Lube Oil.   Diesel engines tend to produce sludge and varnish-depositing compounds as by-products of combustion. Special heavy-duty oils have been developed that contain detergents that keep these substances in suspension, rather than making harmful deposits. It is absolutely necessary that these be used instead of ordinary motor oil.

Such oils can also be used to advantage in gasoline engines. On the first filling they may pick up so much accumulated sludge and varnish as to need to be changed very quickly, but this cleanout is very beneficial to the engine.

Injection Systems.   The solidified lubricants are usually pumped into bearings and other parts requiring them at intervals calculated so that there will always be sufficient grease sticking to the rubbing surfaces. The pumping may be done by hand guns, air guns, or mechanical injectors. The bearing, or a tube that opens into it, has a grease fitting that matches the nozzle of the gun and permits grease at pressures as high as 10,000 pounds per square inch to pass from the gun into the bearing.

It is a mistake to overgrease, since 60 percent of all seal failures or failures associated with bearings are the result of overlubrication. Overgreasing will damage and rupture the seals allowing contaminants in.

Some antifriction bearings are greased and sealed at the factory and require no further lubrication at any time. More often bearings are greased periodically, the intervals depending on the speed, the load, and the danger of dirt's getting into them. Bushings may require very frequent lubrication, often every 4 to 5 hours of operation.

Dirt.   If sand or other abrasive particles are allowed to mix with grease, they form a grinding compound that will cut away the hardest metal rapidly. The rate of wear increases with speed and with load.

Hinges which must work in the dirt, such as unsealed track pins and bushings, last longer if run dry than lubricated.

CABS

Machines may be fitted with cabs to protect the machinery, the operator, or both. Protection for the engine only is called a hood.

Revolving shovels, highway-type trucks, and some other machines carry cabs as standard equipment.

Cabs or canopies may be fitted on tractors, rollers, graders, and most bare machines that are large enough to carry them. They are used to protect the operator against rain, sun, cold, or wind.

Rain protection requires a roof, a glass windshield with a wiper, and walls, windows, or curtains at the sides and rear. Means should be provided to roll, swing, or fold all the vertical panels out of the way or to remove them when they are not needed.

An increasing number of cabs are entirely enclosed, with glass windshield, and glass windows, some of which can be raised and lowered. This construction makes it possible to provide effective heating in cold weather, and air conditioning where heat is excessive.

A major problem with air conditioning (AC) systems is their tendency to leak refrigerant. Electronic leak detectors have become standard devices. To be accurate the AC system must be pressurized to at least 50 psi.

The entire cab should be readily removable in case it becomes uncomfortable or interferes with operation or repairs.

A cab will increase operator comfort under many conditions, and often permit work in weather too severe for an open machine. It is essential for snow plowing and will pay handsome dividends in work when weather is very hot, cold, or rainy.

A cab will interfere more or less with the operator's view of the surroundings so that it may reduce accuracy of work or cause an accident hazard. It may make it difficult to enter and leave the seat.

ROPS (rollover protection standards) now make it obligatory to equip a wide variety of machines with cabs that feature members (usually tubing) strong enough to protect the operator if the machine overturns.

POWER PLANTS

The majority of excavating and hauling machines use gasoline or diesel engines. These are called internal-combustion engines because fuel is burned inside the same unit that turns the shaft. They were given the name to distinguish them from steam engines, which burn fuel to make steam and then pipe the steam to an engine that converts it to usable power.

The latest entry into this field is the gas turbine, which so far is used only in very large off-the-road trucks. It has the advantages of simplicity, light weight, and high power. But it is efficient only at full speed.

Gasoline and diesel engines have many things in common. They burn a mixture of air and fuel, turning the heat of their explosive combination into pressure against a piston on a crank that turns a shaft. They must use clean air and clean fuel, must keep a film of oil on all moving parts, should be kept at an even temperature by a cooling system, and usually have a throttle to regulate speed. Industrial engines such as are used in excavators have a governor that automatically opens and closes the throttle to maintain proper speed.

The exhaust from both engine types is extremely noisy, and requires muffling to avoid operator ear damage, and to keep job noise down to endurable levels.

Air Filter.   Dust must be filtered out of air taken in by an engine to prevent it from wearing moving parts by scratching and grinding them and from building up gummy deposits by combining with the lubricating oil. Excavating machines work on dusty jobs so often that it is particularly important that they have good filters that are properly cared for.

A Donaldson dry cleaner is shown in Fig. 12.98. The precleaner is replaced by simple wire mesh, which serves to prevent entrance of coarse particles. A straight intake pipe brings air to a chamber in the top of the cleaner section.

The air enters the top of the outer chambers of the 32 to 42 nylon cyclone tubes. Vanes cause it to spin. At the bottom the air makes a sharp U-turn to go up through the center tube into the filter

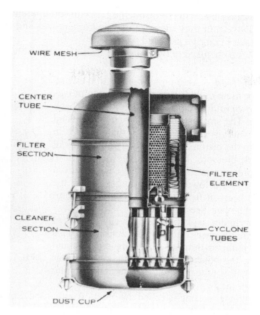

FIGURE 12.98   Dry-type air filter.

chamber. Dirt, thrown out of the air by centrifugal force and the sharp turn, drops out of the bottom of the tubes into the dust cup. This section takes out about 95 percent of the dirt.

The air moves from the outside of the filter chamber inward through a resin-impregnated, pleated-cellulose element that removes the remainder of foreign particles from the air. Filter efficiency ranges from 99.8 to 100 percent.

The dust cup is removed, emptied, and replaced daily, or as necessary. It is very important that no oil be put in it. The cyclone tubes need no routine attention, as they are self-cleaning.

The filter may be serviced by using a jet of clean, dry air at a pressure of 40 pounds or less. The air is directed first against the inside of the element, then the outside, alternating until it is clean. Clean water may be used instead of air, at the same pressure. The air or water is directed along the complete length of each pleat.

Oil or sooty soil deposits may be removed by washing with warm water and a nondsudsing household detergent.

Excessive engine exhaust smoke and/or loss of power may indicate that the filter needs servicing. If the trouble continues after cleaning, and is not due to other causes, the filter element must be replaced. Its life may be as short as 125 hours or as long as 3,000 hours. Life is affected by the quantity and nature of the dust and other pollutants, and by care or lack of it in emptying and cleaning.

Engines with more than one intake pipe will have a complete air cleaner for each.

Altitude.   High altitudes reduce the power of all internal-combustion engines. This is because the air becomes less dense as height above sea level increases. The engine therefore draws in fewer molecules with each stroke, develops lower compression, and has less oxygen to combine with fuel. Two-cycle engines suffer a smaller decline up to about 10,000 feet, because their blower feed packs more air in.

Installation of a supercharger on an engine to be used at high altitudes may restore or increase its sea-level power. However, the same engine with the same supercharger would show increased power if taken down to sea level.

Many diesels now have turbochargers (turbo-superchargers) as standard equipment. The units are expensive, but they increase engine horsepower and altitude adaptability substantially, without increase in engine size and weight.

Engines to be used entirely at high altitudes should be specially constructed and adjusted.

Block and Head.   An internal-combustion engine has one or more cylinders in which a piston moves up and down or, in the rare horizontal engine, back and forth. The small space between the piston and cylinder wall is sealed by rings that are set in grooves in the piston and push out against the wall with gentle pressure. The upper rings are shaped to prevent pressure in the combustion chamber from blowing by; the lower one distributes oil on the wall, and keeps it from working up into the combustion chamber. See Fig. 12.99.

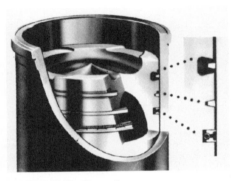

FIGURE 12.99   Piston rings.

The cylinder is kept cool by circulation of water through spaces (jackets) surrounding it. This water is pumped through a cooling radiator and returned to the jackets. Air-cooled engines have fins on the outside of the cylinders that radiate heat into a stream of air.

The cylinder casting is called the block. Another casting, the head, is bolted to the top of it. The two pieces have flat machined surfaces facing, with a gasket or seal rings to ensure an air- and watertight fit.

Lubrication is usually by pressure from a pump to all bearings, and by a splash and mist to cylinder walls.

Each piston is attached by a wrist pin (hinge pin) to a connecting rod. This is attached to the crankshaft by a lubricated bearing (a babbit bushing, called a rod bearing). Up-and-down motion of the piston, powered by fuel-air explosions above it, rotates the crankshaft.

This shaft is usually connected to the working parts of the machine by a friction clutch or torque converter, and a transmission.

If the stroke—the distance the piston moves up and down—is short, the piston speed at any given engine speed will be less than if the stroke were long. The compression ratio need not be affected, as a smaller space above the piston will compensate for its smaller displacement.

Four-cycle engines have intake and exhaust valves opened and closed by the camshaft, which is driven by the crankshaft through timing gears or chain.

Water Cooling.   Water is forced by a centrifugal pump throughout the block and head, where it takes heat from walls of cylinders and combustion chambers. It then flows through small tubes in a radiator, where it delivers the heat to air that is pulled or pushed past the outside of the tubes by a fan. See Fig. 12.100.

The cooled water flows into the pump, and returns to the engine.

A thermostat, usually located in the upper radiator connection, closes to prevent or reduce circulation whenever the engine is too cold for efficient operation.

The fan is mounted on and driven by the engine. In vehicles and haulers, it pulls air from front to rear. In tractors, it may pull, push, or be reversible.

The airstream of a pull fan is aided by the wind created by vehicle motion forward. The pusher is helpful in keeping dust away from the operator, and is very suitable for loaders. Reversing may help to free the radiator from accumulations of dust and fine trash.

Cooling water should contain additives to reduce corrosion of metal parts. It requires addition of ethylene glycol antifreeze wherever there is a possibility of freezing. Alcohol is not suitable because of its low boiling point.

Air Cooling.   Engines may also be cooled directly by flow of air around the outside of the cylinders, which are formed with fins to increase their heat-radiating surfaces. See Fig. 12.101.

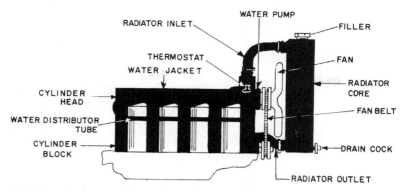

FIGURE 12.100 Water cooling system.

FIGURE 12.101 Air-cooled diesel engine. (*Courtesy of Deutz Corporation.*)

The air is forced through passages by a blower-type fan, with multiple blades turning in a case. The passages are designed to distribute the airstream equably, so that all parts of each cylinder are kept at a similar temperature. See Fig. 12.102.

In air cooling, there is a much greater temperature difference between the metal parts and the cooling medium, than in water cooling. The engine may be run much hotter, with temperature limited chiefly by the necessity of maintaining an oil film on the cylinder walls.

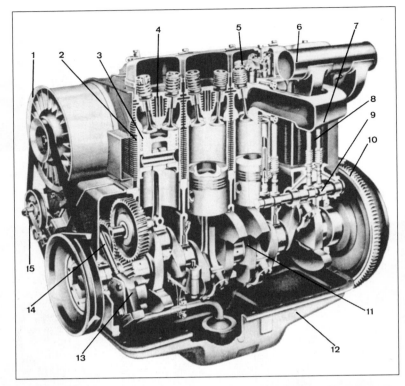

FIGURE 12.102 Cutaway view of air-cooled diesel. (*Courtesy of Deutz Corporation.*)

1) cooling air blower
2) cylinder liner with cooling fins
3) piston with combustion chamber (direct injection)
4) light-metal cylinder head with inlet and exhaust valves
5) rocker arm
6) air intake manifold
7) exhaust manifold
8) pushrod with cover tube
9) camshaft
10) starter ring gear on flywheel
11) crankshaft with counterweights
12) oil sump
13) lube oil pump
14) timing gears
15) V-belt idler

Air cooling uses a smaller volume of air than liquid cooling, eliminates the need of filling and maintaining a cooling system, has fewer parts and connections, and eliminates danger of (and precautions against) freezing and boiling.

GOVERNOR

Most heavy equipment is required to operate at a fairly steady speed, regardless of load changes that would tend to slow the engine or speed it up. Engine speed is kept at the proper level by a governor. There are three kinds in common use.

Mechanical.   In a mechanical governor there are weights that revolve with an engine-driven shaft. They are held in toward it by a spring or springs, and moved away from it by centrifugal force which increases with speed. When the shaft turns rapidly, the weights move out; when it slows, they move inward. An arm controlled by the position of the weights regulates engine speed by controlling fuel injection in diesels or the air valve in gasoline engines. For simplicity, we will discuss only the application to diesels. See Fig. 12.103.

When the engine is required to move a heavy load, it slows down, causing the weights to drop inward and increase the fuel supply, as an automobile driver pushes the accelerator to keep speed up a hill.

If the load is reduced and the engine starts to race, the higher speed moves the weights outward, reducing the fuel supply and keeping the engine slowed to proper speed.

A mechanical governor cannot keep the engine running at exactly the same speed with changes in load. An increase in load causes the steady speed of the engine to reduce slightly; and loss of load will increase the steady speed slightly. This is because a return to the original steady speed while carrying a load would reduce the fuel so that the load could not be carried, and speed would fall off again.

Hydraulic.   Hydraulic governors are operated by revolving weights and springs in the same manner, but the linkage controls a hydraulic valve in a pressure system, which in turn controls the fuel supply. This makes it possible to obtain very prompt response, and to reduce or eliminate speed droop.

Velocity.   Velocity-type governors for gasoline engines may be built into a carburetor or may be a self-contained unit installed between the carburetor and intake manifold, Fig. 12.104. Operation depends on the velocity of air-fuel mixture impinging on a throttle plate normally held open by a spring. When the accelerating engine reaches the desired maximum speed, the air velocity will overcome the spring, causing the throttle plate to close, thus maintaining the engine at that speed. When the engine speed decreases, the spring will pull the throttle plate open.

This type of governor is not as quick to respond as a mechanical type, allowing the speed to drop more under full load. Velocity governors are less expensive to purchase and service, and

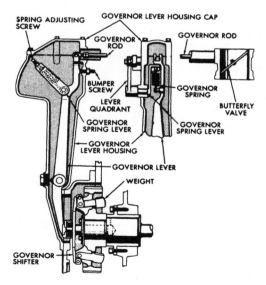

FIGURE 12.103   Governor details. (*Courtesy of Waukesha Motor Co.*)

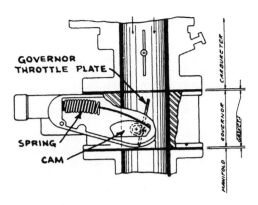

FIGURE 12.104   Velocity-type governor.

operate very satisfactorily where immediate speed response is not absolutely essential or where the engine is geared down so far that this characteristic does not show in performance.

These governors are the accepted installation for over-the-road vehicles and operate satisfactorily on revolving shovels. They are not the proper installation for tractors or road graders.

GASOLINE POWER

Figure 12.105 is a cutaway view of an industrial four-cycle (actually four-stroke cycle) gasoline engine. The cycle starts with the intake stroke. The piston is pulled from the top to the bottom position, pulling an air-fuel mixture through the open intake valve.

When the piston reaches bottom, the intake valve closes, and the piston moves up, squeezing the air mixture into the combustion chamber. This movement is called the compression stroke. The amount that the air is compressed is called the compression ratio. If the ratio is 7 to 1, it means that the charge in the cylinder is compressed to one-seventh of its former volume.

As the piston nears the top of the compression stroke, a spark jumps across the electrodes of a spark plug and sets fire to the air-gasoline mixture, so that it explodes and drives the piston down on the power stroke. The timing of the spark is regulated by a rotating contact in the distributor, that is also driven by the timing chain or gears.

As the piston reaches the bottom and starts up, the exhaust valve opens, so that the gases resulting from the explosion can be pushed out into the exhaust passages. This is the exhaust stroke, and it finishes the cycle. The next move of the piston pulls in a fresh charge of air and gasoline, which is where we started.

Carburetor.   Air and gasoline are mixed in the carburetor, shown schematically in Fig. 12.106. This contains a pool of gasoline supplied by a low-pressure fuel pump, and kept at a constant level by a float that shuts it off when too high and lets it in when low. As air is pulled through the carburetor by the intake strokes of the piston, it picks up a spray of gasoline through metering jets. If the jets pass a lot of gasoline, the mixture is called rich; if only a little, it is lean.

There are two valves to control the airstream through the carburetor. One, the throttle, is between it and the engine. When it is closed, it allows just enough air-fuel mixture through to allow the engine to idle. When it is open, the engine pulls in all it can hold. In any position except wide open, the throttle causes a partial vacuum in the manifold, as the piston tries to pull more air out of it than the throttle lets in.

The other valve, called the choke, is between the carburetor and the open air. When that is closed, the vacuum pulls on the gasoline supply so that raw gasoline is pulled into the engine, and

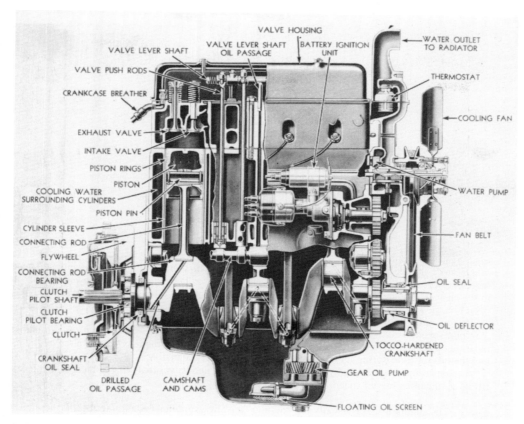

FIGURE 12.105   Cutaway view of industrial gasoline engine.

such a high vacuum is caused that much of it changes to vapor, even if it is cold. This rich mixture of gasoline helps to start a cold engine and keep it running, but is too much in a warm engine.

Whenever the throttle is open to allow more air-fuel mixture to get into the cylinder, the explosion becomes more powerful and tends to make the piston move faster and turn the crankshaft with more power.

The gasoline engine requires a fuel that turns easily into an explosive vapor, and an electric spark to set off each power explosion.

In some engines, the gasoline is injected into the cylinder in about the same manner as in a diesel, so that the carburetor is eliminated.

Compression.   The compression ratio is important to the efficiency and performance of the engine. When the molecules of air and gasoline are squeezed tightly together, they will deliver a higher proportion of their explosive energy to the piston, and waste less in heating the chamber, than when compression is low. However, many practical problems are created by high compression. Industrial engines usually have a ratio between 7.0 and 8.0 to 1. Some automobile engines have still higher compression ratios.

One limit to compression is that air is heated when it is compressed, and a very high ratio will make the mixture hot enough to explode without waiting for a spark. This heat is what makes a diesel go, as even its relatively inert fuel burns automatically when compression reaches 15.0 to 1. See Fig. 12.107.

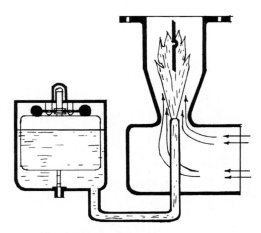

FIGURE 12.106   Simplified carburetor jet.

Of more immediate importance in gasoline combustion is uneven burning. As engine speed increases, it is able to take in less of a charge on each stroke, as the incoming air is slowed by friction in the cleaner, carburetor, and manifold passages; and the interval during which the intake valve is open becomes shorter.

Up to a rather high limit, the faster an engine goes, the more horsepower it produces; but after a certain speed, usually somewhere near one-half to two-thirds of top speed, its torque or actual twisting power decreases, as shown in Fig. 12.108, because of the smaller charges in the cylinders and the briefer time during which the explosion can affect the piston.

A "souped-up" engine usually has bigger air passages, and valves that open farther, so that it can fill its cylinders at higher speed.

Fuel.   The standard fuel is usually the "regular" gasoline supplied for automobiles. It includes a small quantity of tetraethyl of lead, a poisonous compound that reduces tendency to knock, and/or other chemicals to improve performance; together with a dye to give warning of the poison, or to identify make or quality.

Gasoline is classified by octane rating. Octane is an excellent fuel in antiknock characteristics, and is given a 100 percent rating. Heptane is a poor antiknock fuel and is given a 0 percent rating. The octane number of a gasoline is the octane percentage in a mixture of octane and heptane which it matches in antiknock value.

Commercial range now is approximately 85 to 90 for regular unleaded gasoline with additives and 91 to 95 for premium quality. Leaded or doped gasoline cannot be used in engines equipped with the catalytic type of combination muffler and exhaust scrubber.

Gasoline engines can also be run on compressed butane gas, and on various special fuels available in oil production centers. Special tanks, carburetors, and other equipment are needed.

Two-Cycle.   The two-stroke-cycle gas engine fires each time the piston comes to the top. Figure 12.109 contains a diagram showing how it is able to do this.

At the "bottom" of its stroke the piston uncovers a port in each side of the cylinder, one opening into the crankcase, the other into an exhaust passage. The piston has a ridge that prevents direct communication between these ports. Incoming air goes through a conventional carburetor and into the crankcase under the piston.

As the piston moves down in the power stroke, the fuel mixture in the crankcase is compressed. When the piston reaches the "bottom" of the stroke, it opens both ports. The compressed charge enters through one opening while the burned gases go out the other. The piston goes up on its compression stroke, a spark plug fires the charge near the top, and the cycle is repeated.

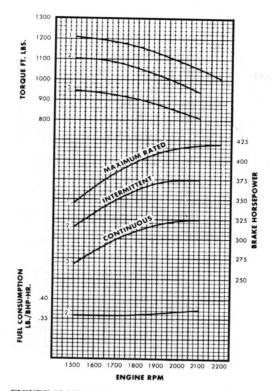

FIGURE 12.107   Performance curves, diesel engine.

Since the two-stroke engine has no conventional valve train, it eliminates both the complexity and weight of the four-cycle's valves, cam, rocker arms, springs, seals, etc. Therefore, it does not have an oil sump. Instead, the two-stroke engine relies on a small amount of oil mixed into the fuel for lubrication. The ratio of gasoline to oil varies by engine and manufacturer, but typically ranges from 16 parts gasoline to 1 part oil (16:1) to 50:1.

The two-cycle gasoline engine is used chiefly in small, high-speed units, such as chain saws, and is of little importance in excavating equipment.

DIESEL ENGINES

In a diesel engine, plain air is drawn in on the intake stroke. It is compressed so tightly that it becomes very hot, and a spray of fuel oil injected in it near the top of the stroke is ignited just by contact with the heated air.

The diesel therefore needs no carburetor or ignition system. But it does need a method of metering the fuel correctly and injecting it at just the right time. Since the fuel is a nonvolatile liquid, precautions must be taken to make sure that it mixes with the air thoroughly enough to burn cleanly and completely in the very brief time available.

A diesel must be more strongly and finely built than a gasoline engine, because of the strains imposed by the higher compression and temperatures. For this reason diesels cost more than gas engines in small models, but this difference diminishes with increase in size. See Fig. 12.110.

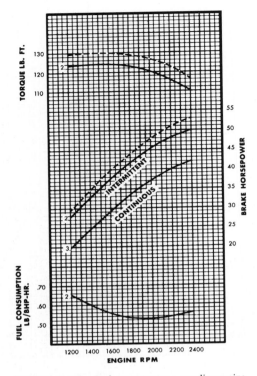

FIGURE 12.108   Performance curves, gasoline engine.

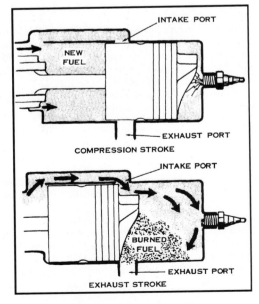

FIGURE 12.109   Two-stroke cycle, gasoline.

Fuel Injection System Bosch in-line injection pumps are standard equipment on the Series 40. Gear driven, the system provides high-pressure delivery up to 17,000 P.S.I, enabling the Series 40 engine to meet emission standards without sacrificing performance or economy. Series 40 engines have a lengthened fuel filter interval to 24,000 miles (900 hours).

Serpentine Belt Drive The serpentine belt is the main drive for the alternator, water pump, and cooling fan. The single, 8 groove, poly-vee design with automatic tensioning results in a virtually maintenance free belt drive.

Piston Design The newly designed pistons include a high technology 3-ring pack. The top fire ring is a keystone design, and is faced with a plasma spray-coating of molybdenum, nickel, and chromium. This new design provides exceptional oil control and maximizes piston life.

Wet Cylinder Liners The Series 40 wet-type replaceable liners are spin-cast, induction hardened, and plateau honed. This wet liner design provides maximum heat transfer, extended piston and ring life, and facilitates rebuild to original factory specifications.

Gear Train Series 40 engines provide the ability to gear drive an engine air compressor and power steering pump from a single location. This location may alternately be used for other accessories to 55 horsepower.

Crankshaft The Series 40 crankshaft is made of micro alloy forged steel which is body hardened. Journals are induction hardened for maximum strength. This design also allows for multiple regrindings which reduces rebuild costs.

Cylinder Block The new cylinder block consists of high strength alloy cast iron with an integral deep skirt design for maximum strength and lower noise levels. This same block is used in both Series 40 displacements—7.6L and 8.7L—providing common installation requirements over a wide horsepower range.

Series 40 Features and Benefits

The Series 40 has been designed in 2 displacements, while maintaining a high degree of component commonality. Components that are shared among the two displacements include: cylinder block, cylinder head, front cover and gear train, flywheel housing, and intake and exhaust manifolds.

The premium design features that this family of engines shares allow Series 40 engines to provide:

- Reliability
- Durability
- Low Emissions
- High Performance
- Low Maintenance Cost
- Optimum Fuel Economy
- Gear Driven Accessories
- Light Weight

Cylinder Head The one piece cylinder head incorporates helical inlet ports designed to provide enhanced swirl for efficient combustion. Optimum cylinder head to block clamp load and maximum gasket sealing are ensured by 26 equally spaced bolts.

Combustion System The Series 40 engine design incorporates the latest technology in direct fuel injection, turbocharging as well as air-to-air charge cooling. This technology enables the Series 40 to continue to meet increasing on- and off-highway emission standards. Current automotive Series 40 engine models are certified to meet EPA and CARB emission standards without the use of exhaust aftertreatment including a catalytic convertor.

Lube Oil System An engine mounted oil cooler maintains oil temperature within the optimum range, while two oil jets per cylinder cool and lubricate the piston to prolong engine life. The oil system includes a thermostatically controlled temperature relief valve for fast warm up. The high pressure relief valve provides cold start engine protection. The new Series 40 has lengthened oil and oil filter change intervals to 12,000 miles (450 hours).

FIGURE 12.110 Cutaway view of diesel engine. (*Courtesy Detroit Diesel Corporation.*)

Special oil containing detergent compounds must be used to prevent deposits of sludge and varnish from forming. This diesel lube oil can also be used to advantage in gasoline crankcases, as both a lubricant and a cleaner.

Diesel is more economical than gasoline for several reasons. The higher compression makes it much more efficient. Compare Figs. 12.107 and 12.108. The fuel has a higher heat value. A greater proportion of the available power in the fuel is set to work turning the crankshaft. This means less fuel to buy and to handle. Fire danger is greatly reduced.

Diesel engines used in excavators have a size range from approximately 60 to 1,600 horsepower. They are available in four-cycle and two-cycle construction.

Four-Cycle.   In the four-cycle (four-stroke cycle) diesel, Fig. 12.111, a downward stroke of the piston with the intake valve open fills the cylinder with air. The upward stroke against the closed valve compresses it, injection and burning of the fuel occur at the top, the burning drives the piston down on the power stroke, and the next upward stroke pushes the products of combustion out past the open exhaust valve.

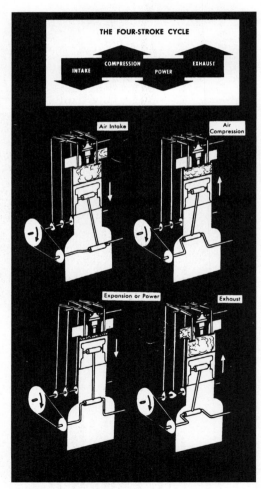

FIGURE 12.111   Four-stroke cycle, diesel.

While the piston-and-valve action is the same as in the four-stroke gas engine, the power principle is quite different. The air drawn into the cylinder is just air, and its flow is not regulated by a butterfly valve as in a gas engine.

Compression is very high, the minimum ratio being about 16 to 1 and the highest at present being over 20 to 1. Diesel (nonvolatile) fuel is injected into the combustion chamber when the piston is at or near the top of the compression stroke, and is ignited by the heat of compression. Engine speed is regulated by controlling the amount of fuel injected on each stroke. Pressures and temperatures in the cylinder are much higher than in gas engines.

Two-Cycle.   The two-cycle (two-stroke cycle) diesel, Fig. 12.112, has one power stroke for each two strokes of the piston; that is, every down stroke produces power. A blower in the intake passage, Fig. 12.113, pushes air at low pressure into a chamber in the block that opens into the cylinder through a ring of holes or ports that are covered by the piston except at the bottom of its stroke.

The power stroke starts by injection of fuel into hot compressed air when the piston is at the top of its stroke. This burns and expands, forcing the piston down. As the piston nears the bottom, the exhaust valve opens and the burned gas starts to escape. Further movement of the piston uncovers

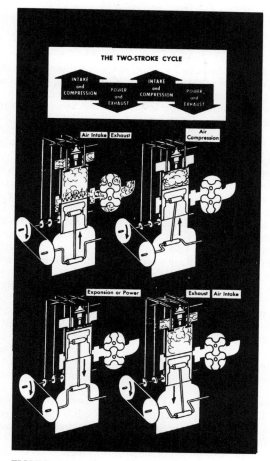

FIGURE 12.112   Two-stroke cycle, diesel.

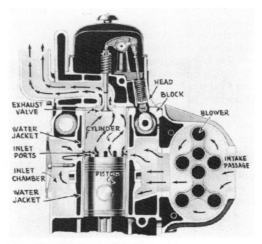

FIGURE 12.113   Cutaway view, two-stroke cycle diesel.
(*Courtesy of Detroit Diesel Corp.*)

the intake ports, so that an inrush of clean air forces the remainder of the burned gases out. The piston then rises to compress its new filling of air, and fuel is injected at the right moment to start the next power stroke.

Since each down stroke, instead of every other down stroke, is a power stroke, as few as two cylinders can be used without excessive roughness.

Fuel.   Ignition quality of diesel fuels is rated by cetane numbers, which may be between 35 and 60. Higher numbers indicate easier starting.

Most diesels in construction equipment use No. 2 diesel oil. This is a little heavier in body and less flammable than kerosene, evaporates very slowly, and has a slight lubricating value.

Some engines may permit substitution of No. 2 furnace oil, used widely for domestic heating. Diesel oil cetane number is between 40 and 50, as against 37 to 40 for heating oil.

In any grade of fuel, the most important requirement is cleanliness. Many fuels contain sulfur and other corrosive chemicals in sufficient quantity to damage pumps and injectors. Any of them will have more or less foreign matter that absolutely must be strained out, as the close fits in a diesel fuel system will not tolerate any solids. It is customary to have both primary and secondary filters, and often a final one at each injector.

Second-grade fuel usually increases downtime and maintenance expense.

Fuel Supply.   In order to get fuel into each cylinder in the right quantity and at the right time, four separate functions must be performed. The fuel must be measured, directed to the proper cylinder, timed to reach it at the right time, and put under sufficient pressure to enter it in a vaporized spray. As might be expected, a considerable variety of methods are used in different makes.

Distributor Pump.   A number of engines combine all four functions in a high-pressure pump.

Pump and Distributor.   The pump and distributor mechanism may be separate, as shown in Fig. 12.114. The single plunger pump puts the low-pressure fuel from the transfer pump under high pressure, and also meters the amount by a spiral relief groove and a governor-controlled mechanism for turning the plunger.

The high-pressure jets of fuel from this pump enter the distributor passage. Four poppet valves, lifted by cams, direct the fuel through lines to the injectors in accordance with engine firing order. One of these will be open for each push of the high-pressure pump, and the measured fuel will

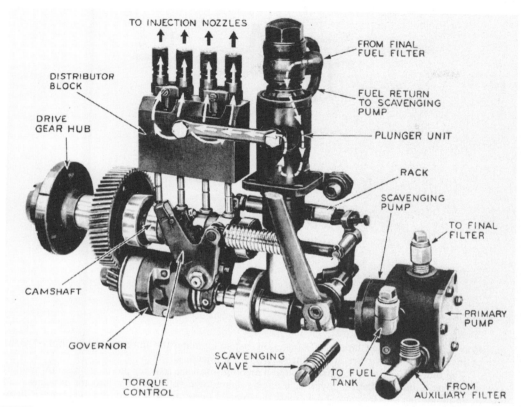

FIGURE 12.114   Injection pump and distributor.

pass it into high-pressure tubing that will conduct it to an injector. There it will unseat a valve and spray into the precombustion chamber.

Precombustion Chamber.   A precombustion chamber is a small chamber connected by an open passage to the main combustion chamber. Fuel is injected into it and is ignited by compression heat. There is not enough air to burn it fully. The burning mixture expands and rushes into the main chamber, where it mixes with more air to burn completely.

This arrangement makes possible the use of relatively low pressures in the fuel injection system, promotes efficient performance on standard fuels under a wide range of load and speed conditions, permits low idling speed, and reduces exhaust pollutants.

The prompt and thorough mixing of fuel and air is very important, and there are a number of designs using different shapes of chambers and piston tops.

Unit Injector.   In the unit injector system, the whole job—pressure, timing, distribution, and measurement—is handled by the injectors.

The General Motors diesel, Fig. 12.115, has a low-pressure fuel pump, located between the primary and secondary filters, that delivers fuel directly to the injectors in the head. These are actuated by the main camshaft acting through rocker arms. The throttle and control rack regulates the amount of fuel dispensed by each by turning the injector plunger, the camshaft determines the distribution and the timing, and the pressure is produced by a plunger in the injector. This pressure can be much higher than can be carried by outside tubing, and makes it possible to inject the oil as a very fine spray directly into the main combustion chamber.

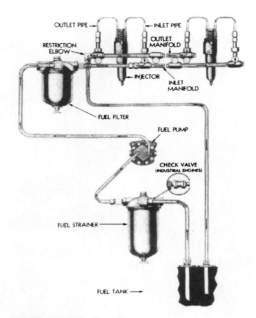

FIGURE 12.115   Unit injector system.

Pressure-Time.   The Cummins PT injection system regulates the amount of fuel supplied to the cylinder by varying the pressure in the supply to the injectors.

The fuel flow is from the tank through a medium-pressure transfer pump, a pressure-regulating valve, a throttle shaft, a governor plunger, a shutdown valve, tubing, injector passages, and drain tubing back to the tank. Fuel enters the injector cups on the upstroke of the plunger, in an amount determined by its pressure.

Figure 12.116 shows the pump and regulating mechanism in a cutaway view. The throttle shaft regulates the flow of fuel between the pressure regulator and the governor. When idling, it cuts off the fuel flow through the main passage and forces it to pass through an idling hole to the governor idling port. The governor controls the fuel pressure at idling speed and when the engine reaches maximum speed, the governor shuts off the operating fuel supply.

The shutdown valve is used for stopping the engine, and should remain closed until the engine is to be restarted. This shutdown may be either a manual or electric solenoid type.

The injector, Fig. 12.117, has no throttle connection and meters the fuel according to its pressure, which is responsive to the position of the throttle.

Starting.   Diesel engines are started by cranking, in much the manner of gasoline engines. There is usually an electric starting motor of the type used for automobile engines, but of much heavier construction. The battery or battery set may be 12 to 48 volts.

A heavy diesel may be started by a compressed air motor, supplied from the machine's air brake system. This does not have to rest until it runs out of air. It is important that there be no leaks that might empty the system overnight.

An air starter needs a larger reservoir (air receiver), and a connection by which air can be supplied in emergencies by another machine or system.

Smoke.   The diesel is normally a clean-burning engine, as the cylinder is charged with more than enough air to burn the maximum amount of fuel injected. It has a big advantage over gas engines in that the exhaust is almost free from carbon monoxide. However, it does exhaust some bad-smelling, irritating, and moderately toxic gases, that prohibit its use in poorly ventilated places unless an efficient exhaust scrubber is attached.

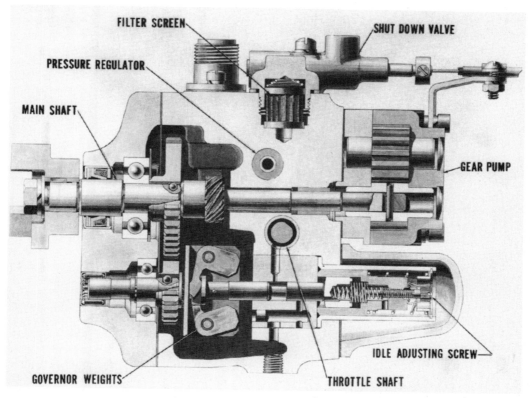

FILTER SCREEN

PRESSURE REGULATOR

SHUT DOWN VALVE

MAIN SHAFT

GEAR PUMP

IDLE ADJUSTING SCREW

GOVERNOR WEIGHTS

THROTTLE SHAFT

FIGURE 12.116   Pressure-time pump, cutaway view. (*Courtesy of Cummins Engine Company, Inc.*)

In view of its generally clean-burning characteristics, it is unfortunate that so many diesel trucks trail clouds of black smoke behind them, to the annoyance of everyone on or near the highway. This nuisance is caused by injecting more fuel than the engine is designed to use, so that the mixture is too rich.

Excavating machines and off-the-road haulers almost never show black exhaust smoke, while too many highway trucks put on a good imitation of a coal-burning locomotive firing up. Smoking is an indication of fuel being wasted, oil being contaminated with sludge, and exhaust valves and mufflers being damaged by contact with still-burning gases. For this reason, any alert foreman or operator will send for a service representative upon seeing a dirty exhaust.

Superchargers.   A supercharger is a blower or pump that forces air into the intake of an engine at higher pressure than the atmosphere. As a result, more oxygen is packed into a cylinder, so that more fuel can be burned and more power is produced by the same size engine, or equal power by a smaller engine.

A supercharger will overcome friction of air passages so that cylinders can fill completely at higher speeds than when atmospheric pressure alone pushes the air. It can also compensate for the reduced density of air at high altitudes.

The turbosupercharger, Fig. 12.118, includes a set of turbine vanes driven by the high-velocity exhaust gases, that spin compressor or blower vanes that force air into the engine.

The exhaust gases are channeled through nozzles in which their velocity is built up to exert maximum force on the turbine vanes, and the impulses from individual cylinders are blended together into a smooth stream. This process creates some back-pressure, but less than buildup of intake pressure.

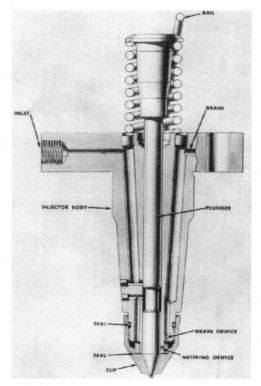

FIGURE 12.117   Injector, cutaway view. (*Courtesy of Cummins Engine Company, Inc.*)

The turbosupercharger adds to engine efficiency by converting waste energy in the exhaust gas into intake pressure that increases compression, cleans out burned gas completely and may even convert the intake stroke into an air pressure power stroke.

The turbocharger may rotate at 80,000 rpm or more. One side is kept hot by exhaust gas right out of the engine; the other is cooled by incoming air. Proper care includes idling the engine 5 to 15 minutes between a hard pull and shutting down, to equalize temperatures. Neglecting this precaution is likely to ruin the unit.

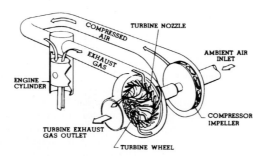

FIGURE 12.118   Turbocharger.

EXHAUST CONDITIONERS

Gasoline exhaust contains substantial quantities of carbon monoxide, an odorless, poison gas, together with small quantities of other chemicals. Diesel exhaust has little carbon monoxide, but a wealth of irritating gases of varying degrees of toxicity. These include hydrocarbons, aldehydes, and oxides of nitrogen.

Because diesel-contaminated air usually becomes intensely irritating before it is a serious health danger, diesels are safer than gasoline engines wherever ample natural ventilation is lacking. No matter how efficient protective devices may be, there is always a chance that they will fail unexpectedly.

The amount of irritating and poisonous chemicals exhausted can be kept at a minimum by a proper air-fuel mixture in the cylinder. In a diesel, 15 to 22 parts of air to 1 part fuel is the best range. If there is too much air, as when the engine is idling or running at low throttle, there may be excessive production of irritants. If the mixture is too rich, there are many products of incomplete combustion, and particles of solid carbon as well.

The noxious exhaust gases are soluble in water, and combine with it to produce acids. In the presence of heat and a catalytic agent, they will combine with oxygen to form carbon dioxide and water. The carbon dioxide has a suffocating rather than a poisoning effect, and is comparatively easy to control by ventilation.

When internal-combustion engines are used in buildings or underground, it is essential that both exhaust conditioners and good ventilation be provided.

Exhaust Scrubber.   One method of cleaning up exhaust gas is to pass it through alkaline water. The poisonous and irritating chemicals dissolve in the water, and the resulting acids are neutralized by contact with crushed limestone or specially selected salts and minerals. See Fig. 12.119*A*.

A venturi tube is built on the outlet to mix the exhaust with 5 times its volume of air, so that any remaining deleterious substances will be thinned out before being breathed.

The scrubber is contained in a case that is mounted in any convenient place on the machine. Crawler tractors usually carry it on the rear.

The conditioner must be carefully chosen to fit the particular engine and its load requirements and thereby provide sufficient capacity and minimum back-pressure.

Cooled ERG Engines.   Cooled exhaust gas recirculation (ERG) systems have been used since 2002 by major engine manufacturers to reduce emissions of smog producing nitrogen oxides

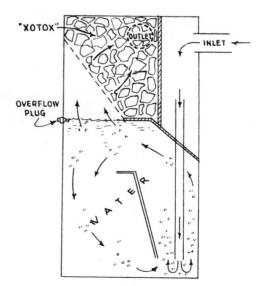

FIGURE 12.119*A*   Exhaust scrubber, schematic diagram.

(NOx) from diesel trucks. These systems divert 5 to 30 percent of an engine's exhaust stream through an air-to-water cooler, then back into the combustion chambers, where the cooled gases reduce peak temperature and thus retard NOx formation. See Figure 12.119*B*.

Cooled ERG systems include a number of sensors and actuators that are controlled and monitored by the engine's electronics. These systems are virtually maintenance free.

Catalytic Exhaust.   The OCM Catalytic Exhaust, Fig. 12.120, is a replacement muffler containing sticks of ceramic porcelain coated with a platinum compound. When heated to 500°F, this catalytic film will cause the carbon monoxide and hydrocarbons to combine with the fresh air supply drawn through the venturi, producing water and carbon dioxide. The heat of the exhaust, assisted by that generated by this secondary combustion, keeps the catalyst tubes hot and the reaction going.

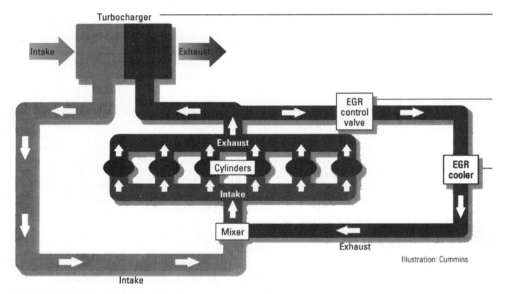

FIGURE 12.119*B* Schematic of ERG system. (*Illustration: Cummins.*)

FIGURE 12.120 Catalytic exhaust muffler. (*Courtesy of Oxy-Catalyst, Incorporated.*)

A pyrometer in the exhaust tube indicates whether the unit is working at correct temperature, and may also give information about the fuel mixture in the engine.

It takes about 5 minutes of full-load operation to create sufficient heat to start the reaction. With gasoline engines it will continue even at idling speed, but diesels must be kept at about 60 percent load to maintain it.

These conditions are used successfully with white gasoline, diesel fuel, and LP (propane) gas.

FUEL CONSUMPTION

Manufacturers often show fuel consumption per engine brake horsepower as an extra line on torque and horsepower graphs, as in Figs. 12.107 and 12.108. The line is usually a flat curve, with highest consumption rate at the lowest revolutions per minute, and the lowest rate near the optimum rpm.

Unfortunately, consumption is usually given in pounds. This may be satisfactory to engineers, but it is a nuisance to contractors, who must measure and pay for fuel in gallons. To convert to gallons, divide pounds of gasoline by 6.0, and pounds of diesel fuel by 7.0.

Load.   Fuel curves on engine graphs are for full-load operation. Few excavators or haulers work continuously at full load, and their engines could not maintain performance if they did. The power shovel may coast on its swing and has little load while lowering its bucket; the dozer backs up without load, and the truck and scraper return empty and may descend grades. Average load may run from less than $\frac{1}{4}$ to over $\frac{3}{4}$ of maximum. An engine running without load uses about 30 percent as much fuel as it would under full load at the same speed. Any reduction in load results in a saving of fuel, so that in most operations fuel consumption is well below rated levels.

Work in hard digging or hauling on soft ground or at high speed will usually consume more fuel than under more favorable conditions. But there are exceptions. A shovel scraping at a hard bank may use less fuel per hour (although more per yard) than when in easy digging that permits deep penetration and heavy bucket loads.

Efficiency.   Many engines do not operate anywhere near their top efficiency. Dirty air cleaners, wear and faulty adjustments in the engine and its fuel system, compression leakage past worn or stuck valves or rings, and the drag of constantly engaged accessories such as generators and hydraulic pumps all add to fuel consumption. A new or newly rebuilt engine that has been properly broken in will use less fuel than its full-load rating on practically any job, but as it approaches the end of its useful life, or is allowed to deteriorate, it may use more at part load than it is supposed to at full load.

Engines that are governed or altered to produce less than their rated horsepower will use fuel in proportion to the power that they are permitted to develop, rather than to their rating.

High altitude and high temperature increase fuel consumption unless special adjustments are made.

The table in Fig. 12.121 indicates fuel consumption that may normally be expected under three classes of conditions. "Light" is for equipment in excellent condition with light loads, "heavy" covers old engines and equipment in difficult working conditions, while "average" represents the ordinary run of equipment on jobs.

		Type of Service	
Fuel	Light	Average	Heavy
Diesel	.02 to .03	.04	.07
Gasoline	.06 to .07	.08	.10

FIGURE 12.121   Fuel consumption, gallons per horsepower hour.

For fast, rough figuring, use 80 percent of rated full-load fuel consumption as a basis. But be sure to change pounds of fuel to gallons, to avoid major confusion.

The contractor who has records of fuel consumption should use her or his own figures, as they are likely to be much more accurate than average figures such as those above.

POWER

Power may be measured in terms of force—push, pull, twist, or lift, as a tractor may have a drawbar pull of 30,000 pounds. It may be measured in terms of work, as if the tractor pulled a 30,000-pound load 6 feet, it would have done work amounting to 6 feet × 30,000 pounds, or 180,000 foot-pounds.

Small loads and small distances may be measured in inch-pounds. That is, if a 20-pound weight were lifted 7 inches, the work would amount to 140 inch-pounds. Dividing by 12 will convert inch-pounds to foot-pounds.

The general formula is

$$\text{Work} = \text{force} \times \text{distance}$$

The third measurement is in terms of work and time. If the tractor pulled 30,000 pounds 100 feet in 1 minute, it would have a work rate or production of 30,000 pounds × 100 feet, or 3,000,000 foot-pounds per minute.

Torque.   Torque is defined as the turning power of a shaft. It is the measure of twisting power.

Engine torque is rated in terms of pound-feet, which is the amount of force exerted at a distance of 1 foot from the center of the crankshaft.

A dynamometer may be used to measure engine torque. A simple type called the Prony brake is shown in Fig. 12.122. A brake acts on a drum that revolves with the flywheel. The brake carries a force arm that has a scale at its outer end. Engine speed is measured in revolutions per minute (rpm) by a tachometer.

When the brake is applied, its drag against the revolving drum turns it and the arm, and the force is registered by the scale.

Whenever the brake holds engine speed to a lower level than its throttle setting calls for, the scale indicates the amount of force the engine is exerting at that speed against the resistance or load of the brake.

If the force arm were 1 foot long, the scale reading would indicate the torque. If it were 2 feet long, its reading would have to be multiplied by 2 to obtain the torque.

With a 2-foot arm, the throttle wide open, and the brake holding the engine to a speed of 1,000 rpm, the scale might register 401 pounds. Multiplying by 2, we would have a torque figure of 802 pound-feet (pounds for short) at that speed.

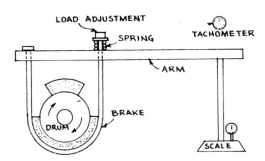

FIGURE 12.122   Prony brake or dynamometer.

The brake can be used to hold the engine at a number of test speeds, perhaps at 200-rpm intervals throughout its speed range. The result for one diesel engine is shown in Fig. 12.123.

If the test is to be of value to the people who will use the results in estimating, the engine should be fully equipped with water pump, fan, generator, governor, and muffler. It will carry these accessories when it is installed in equipment, and each of them takes some power.

Most engine testing is now done on dynamometers, that use an electric generator (dynamo) to apply the load, and measure power output by current flow. One horsepower equals .746 (say $\frac{3}{4}$) kilowatt.

Horsepower.   Horsepower is a work-time measurement, equalling 33,000 foot-pounds per minute. The tractor that worked at the rate of 3,000,000 foot-pounds per minute was using about 91 horsepower.

The work of an engine is found by multiplying its torque (force) by the distance it moves or can move a 1-foot-long arm in 1 revolution. This is the circumference of a circle with 1 foot radius, or 6.2832 feet.

Engine RPM	Measured Torque, Pounds-Feet	Brake Horsepower
800	725	114.2
1000	802	152.7
1200	825	188.5
1400	815	217.7
1600	790	240.7
1800	756	259.0
2000	710	270.0

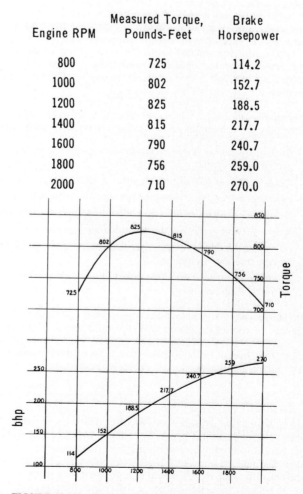

FIGURE 12.123   Engine speed, torque, and horsepower.

Horsepower can be obtained by multiplying this result by the rpm (time) and dividing by 33,000. The basic formula is

$$HP = \frac{torque \times rpm \times 6.2832}{33,000}$$

However, since the two figures 6.2832 and 33,000 are always present, the formula can be simplified by dividing one into the other. The formula will then be

$$HP = \frac{torque \times rpm}{5252}$$

Applying this formula to the torque readings, we obtain the horsepower ratings in the third column of Fig. 12.123. A comparison of the two sets of figures shows that horsepower continues to rise with increasing speed long after torque is dropping.

Torque and horsepower figures such as these are usually posted to graph paper and used to plot curves, such as those in Figs. 12.107 and 12.108.

When an engine is slowed down by load (lugs down), this increases the force it is putting into the work until speed drops below the peak of the torque curve. While working in the range between full speed and maximum torque it is said to be lugged down.

The ability of an engine to lug down is very important in certain types of work, particularly for excavators. However, operating an engine at a lugged-down speed for more than a few seconds is inefficient, and puts heavy strain on the engine. The load should be reduced or the gear changed to allow it to regain full speed.

Horsepower Ratings.   There are a number of different kinds of horsepower ratings. Knowing what they are sometimes makes it easier to understand specifications.

Brake horsepower is engine horsepower measured as described above, or in any way that would give similar results.

Net horsepower is brake horsepower with the drag of accessories deducted.

Flywheel horsepower is engine horsepower with or without accessories.

Drawbar horsepower is engine horsepower less friction losses in the power train.

Maximum horsepower is the most that an engine can develop for 5 consecutive minutes.

Intermittent horsepower is power that can be developed for changing loads such as are provided by construction equipment.

Continuous horsepower is the rating for long life performance under steady loads, such as deep-well irrigation pumping.

Displacement.   A manufacturer's specification sheet should state the cubic inches of displacement. This is the sum of the piston areas times the length of the stroke, both in inches.

An efficient internal-combustion engine should develop at least $\frac{5}{8}$ (.625) pound-feet of torque for each cubic inch of piston displacement. This relationship varies, but can be used as a general check. To find the approximate torque, multiply the displacement by 5 and divide the result by 8. That is,

$$Torque = \frac{5 \times displacement}{8}$$

Altitude and Temperature.   Air becomes thinner as altitude increases or its temperature is raised. The thinner air contains less oxygen to combine with fuel, and provides less substance for compression.

As a result, engine power diminishes with increases in altitude and/or temperature. Standard tests are made at a sea-level barometer pressure of 29.92 and a temperature of 60°F. If test conditions differ, the report should specify what they were.

In general, increase in altitude will reduce the power of a four-cycle engine about 3 percent for each 1,000 feet about sea level. A two-cycle diesel will keep full power to 1,000 feet, and then lose about .9 percent for each 1,000 feet above that. The better showing is due to the low-pressure blower used in the intake.

A barometer drop of 1 inch, as for example from 30.0 to 29.0 as a storm approaches, will reduce engine power about as much as an increase in altitude of 900 feet.

Increase in temperature reduces engine power at a rate of about .9 percent for each 10°F. Cooling has the reverse effect.

The Appendix gives tables of power change with altitude and temperature.

Such power losses have a considerable effect on performance and production at high altitudes, and must be considered in estimates. For example, a loaded truck might be able to climb a 10 percent grade in San Francisco in third speed, but could manage only 8.5 percent in that gear in Denver. If it went up to work on a pass in the Rockies at 12,000-feet altitude, the grade ability would be reduced to 6.4 percent. Lower gears would have to be used for all climbs.

Fuel consumption increases as power is lost with altitude, as the thinner air does not provide an efficient mixture. Special carburetor, injector, or pump adjustments may be made to save some of the fuel, but they cannot restore the power.

Supercharging.   Superchargers, usually of the turbo or exhaust-driven type, are widely used to provide more than normal power in proportion to displacement, and/or to retain normal power in engines used at high altitudes.

Most supercharged engines will also lose power if altitude is increased, or gain it if brought nearer sea level. But such losses are unlikely to equal the additional power obtained from supercharging, and the charger can usually be adjusted to restore the altitude losses.

HAULER POWER TRAIN

In most construction machinery, engine power is delivered to working parts through gears, chains, belts, converters, and/or pulleys and cables that multiply torque and reduce speed.

Gearing.   Except for considerable losses in friction and slippage that will be discussed later, such stepping up of power or torque is always exactly proportional to the loss in speed. If an engine produces 300 foot-pounds of torque at 1,800 revolutions per minute, a 3-to-1 gear set will convert this to 900 pounds of torque at 600 rpm. If gearing is 25 to 1, output will be 7,500 pounds torque at 72 rpm.

A standard truck transmission may drive through two sets of gears (main shaft to countershaft, and back to main shaft) in all speeds except one, which is direct through the mainshaft. An auxiliary transmission may have two additional gear sets. A rear end has one set of gears if it is single reduction, two if it is double reduction. It may have two speeds.

An off-the-road truck or scraper may have these gears, and another reduction set in the final drives in the hubs.

The total torque multiplication, or total gear reduction (TGR), is found by multiplying all gear reductions being used in the power train by each other. If reductions in low gear were 6 to 1 in the transmission, 4 to 1 in the rear end, and 6 to 1 in the final drive, then total reduction would be $6 \times 4 \times 6 = 144$. It is usual to simplify figuring by combining the rear end and final drive ratios into an axle ratio, in this case 24 to 1. The total reduction would then be found by multiplying this by the transmission ratio, $6 \times 24 = 144$. In direct drive in the transmission the total reduction would be 24.

Step-up gearing to increase speed at the expense of power may be used in the highest transmission gear or gears. Such gears are called overdrives. If the truck above used an overdrive with a ratio of .7 to 1, the total reduction would be $.7 \times 24 = 16.8$.

Converter.   Torque multiplication in a torque converter is infinitely variable from 1 to 1 up to its capacity, that may be anywhere from 2.1 to 1 to 6 to 1. A hauler similar to the above, equipped with a converter with a maximum ratio of 2.7 to 1 and a low transmission gear of 4.0 to 1, would have a total reduction of 2.7 × 4 × 24 = 259.2. This would be at stall speed, and the reduction would automatically decrease as it gained speed, up to a minimum reduction at no load in low gear of 1 × 4 × 24 = 96. In direct gear maximum reduction would be 2.7 × 24 = 64.8, and reduction at light load 1 × 24 = 24.

Mechanical Efficiency (ME).   The mechanical efficiency of a mechanism is the percentage of the power put into it that it delivers to its work. If a hauler engine has 300 pounds of torque (T) at 1,800 rpm and has a total gear reduction (TGR) of 100 to 1, it should deliver 30,000 pound-feet of torque to the rear hubs. If the hub torque (HT) is found by testing to be 24,000 pounds, the friction loss is 6,000 pounds. To find the percentage of engine power, we use the formula

$$ME = \frac{HT}{T \times TGR}$$

and we have

$$ME = \frac{24,000}{300 \times 100}$$

$$= .8 \text{ or } 80 \text{ percent}$$

The loss of torque in a mechanical power train varies with the number and type of gear sets, the design, the perfection of finish, the condition of the unit, and the amount of torque carried. It may range from 2 to 5 percent for each gear set.

Losses are greatest in the lowest gear, in which the greatest number of reductions is used, and in which the greatest strain is placed on the drive line. Friction is increased by high pressures on gear teeth and bearings, and by shafts changing shape under twist. These effects are greatly increased as equipment wears.

A rule-of-thumb figure for the overall efficiency of an off-the-road hauler is 80 percent for new equipment. An old and badly maintained machine may go down to 70 or even 60 percent efficiency. Highway trucks have higher efficiency than off-the-road models. Tandem drive consumes extra power.

Torque converter machines usually have lower mechanical efficiency because of slippage. An average figure may be taken as 75 percent.

Mechanical efficiency should not be confused with work or job efficiency. It is simply a measurement of power lost between the engine and the rear axle.

Loaded Tire Radius (LR).   The loaded tire radius is the distance from the center of the wheel hub to the ground when the tire is carrying its full rated load. It is the length of the power arm through which the torque at the hub acts to drive the machine. Lengthening this radius by putting on bigger tires increases vehicle speed and decreases power.

Figure 12.124 gives the static loaded radius for some tire sizes. This radius is increased somewhat by centrifugal force when the truck is in motion, and is greater if load is less than that for which the tire is rated.

Loaded radius is reduced by low tire pressure and by tread wear. A big tire may have a change of 3½ inches from new tread down to fabric. It can be measured on the vehicle by parking it on hard, level footing, placing a carpenter's level at the center of the hub, leveling it, and measuring from its other end to the ground.

The loaded radius is always less than the unloaded radius, which is one-half the overall diameter. The difference is about ½-inch in a 6.00-16 and more than 5 inches in a 30.00-33.

The variables in loaded tire radius alone are enough to make it important to field-check the calculated performance of haulers.

Rim Pull (Tractive Effort).   Rim pull is the torque that a machine is capable of exerting at the contact of its drive tires with the ground, that is, at the distance from the axle represented by the loaded radius of the drive tires.

The term is not a very clear one, as the rim is actually the outer part of the metal wheel, and except in this particular use it does not mean the tread surface of the tire.

Rock Grip Excavator H.D. Tire
Static Loaded Radius

Size	Plies	Radius	Size	Plies	Radius
7.50–20	12	17.6	13.00–24	18	23.7
8.25–20	12	18.1	14.00–24	20	25.2
9.00–20	12	19.1	16.00–24	20,24	27.0
10.00–20	14	19.7	16.00–25	24	27.0
11–22.5	14	19.8	18.00–24	20	29.4
10.00–24	14	21.7	18.00–25	16,20	29.4
11.00–20	14	20.5	18.00–32	24	33.4
12–22.5	14	20.5	18.00–33	32	33.4
11.00–22	14	21.5	21.00–24	20,24	31.2
12–24.5	14	21.5	21.00–25	16,24	31.2
11.00–24	14	22.5	24.00–25	24	33.6
12.00–20	16	21.0	24.00–29	24,36	35.7
12.00–24	16	23.0	27.00–33	30,36	40.1

FIGURE 12.124   Loaded radius of some tire sizes.

Tractive effort is a term that may be used instead of rim pull. It comes closer to expressing the meaning, as this is the effort that the machine can exert to move itself. However, the term *rim pull* is used in manufacturers' specifications. It will therefore be used here.

Rim pull is stated in foot-pounds unless some other measurement is specified. It is different in each gearshift position, being greatest in the lowest and slowest gears.

Figuring Rim Pull.   Rim pull (RmP) is figured as the hub torque divided by the loaded tire radius (LR) in feet. The hub torque is the engine torque (EnT) times the total gear reduction (TGR) times the mechanical efficiency (ME). Combining these, we have

$$RmP = \frac{EnT \times TGR \times ME}{LR, \text{ feet}}$$

Since tire radius is always given in inches, it may be less confusing to change the formula to

$$RmP = \frac{EnT \times TGR \times ME \times 12}{LR, \text{ inches}}$$

Estimating rim pull for on-road, i.e., hard surface, conditions using the loaded radius is reasonably accurate, but off-road, where the tire sinks into soil, LR will be too large and needed torque will be greater than on the road. See Figs. 12.125 and 12.126.

It often happens that the estimator does not know either the gear reduction or the loaded radius. But even one-page flyers will usually give the engine horsepower and the speed in each gear. The estimator can use them to get an approximate value for rim pull by using the arbitrary formula

$$RmP = \frac{300 \times \text{engine horsepower}}{\text{maximum speed, mph}}$$

Figure 12.127 gives power and speed data for a simplified 300-horsepower hauler that will be used as the basis of sample calculations.

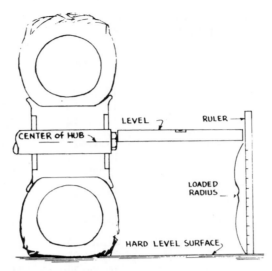

FIGURE 12.125   Measuring loaded radius.

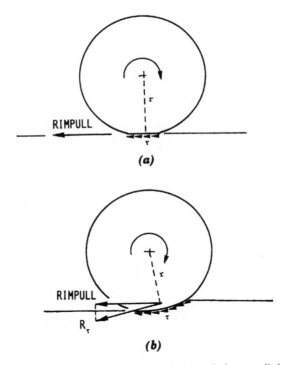

FIGURE 12.126   Determination of rim pull from applied torque, (*a*) On hard surface (pavement); (*b*) on deformable surface (soil) (T = torque; r = rolling radius; τ = shear stress; and R_τ = resultant of shear stresses), where RmP = T/r. (*Courtesy of ASCE Journal of Construction Engineering and Management.*)

Gear	Total Gear Reduction	Speed, mph	Pounds Rim pull, full governed speed*	Rim Pull, pounds per ton** Vehicle Weight	
				Loaded	Empty
1st	180	2.5	36,000	720	1,440
2nd	90	5.0	18,000	360	720
3rd	60	7.5	12,000	240	480
4th	36	12.5	7,200	145	290
5th	22.5	20.0	4,500	90	180
6th	15	30.0	3,000	60	120

*Maximum rim pull with the engine lugged down may be 10 to 15% greater.

**These two columns not ordinarily supplied in specifications, but are easily obtained by dividing rim pull by vehicle weights.

FIGURE 12.127   Power in gears, simplified hauler.

TRAVEL RESISTANCE

Rim pull must overcome travel resistance before it can move a machine, and in order to keep it moving.

Travel resistance is rolling resistance, plus or minus any hindrance or help from grades. It is figured as a percentage of machine gross weight, without reference to engine power.

Rolling Resistance.   Rolling resistance is the power consumed by the ground surface and the tire as the vehicle moves. It is expressed in pounds of pull per ton of vehicle weight, or as a percentage obtained by dividing the pounds of pull per ton by 2,000.

On a paved level road, rolling resistance usually is 40 to 60 pounds per ton or 2 to 3 percent of vehicle weight. It includes power absorbed in flexing of tires and by friction in wheel bearings and between the tire tread and the road.

Rolling resistance may be as high as 600 pounds per ton. Figure 12.128 shows the amount that may be expected on various surface. Such figures are approximations, but serve as a convenient basis for working out power requirements and speed of haulers under various conditions.

Rolling resistance increases above the 40-pound minimum chiefly because of the tendency of tires to either sink into a yielding surface or depress the general area on which they travel, so that it is as if they were constantly climbing a grade to carry the truck along on a level. There is also a factor of roughness of surface.

On rough ground, large-diameter tires will meet less rolling resistance than small ones. This is because the big tire bridges over hollows and absorbs humps, and the higher center of revolution tends to roll it over humps instead of trying to push it through them.

Type of Haul Road Surface	Pounds per Ton of Gross Vehicle	Per cent of Gross Vehicle Weight
Concrete and asphalt	30 lbs.	1.5%
Smooth, hard, dry dirt and gravel. Well maintained. Free of loose material	40 lbs.	2%
Dry dirt and gravel. Not firmly packed. Some loose material	60 lbs.	3%
Soft unplowed dirt, poorly maintained	80 lbs.	4%
Wet muddy surface on firm base	80 lbs.	4%
Snow—Packed	50 lbs.	2.5%
4″ Loose	90 lbs.	4.5%
Soft, plowed dirt or unpacked dirt fills	160 lbs.	8%
Loose sand or gravel	200 lbs.	10%
Deeply rutted, or soft spongy base	320 lbs.	16%

FIGURE 12.128   Rolling resistance.

Hard tires show least friction and rolling resistance on hard surfaces, and the most sinking and highest resistance on soft surfaces.

Grade Resistance.   The extra power required to overcome gravity in moving a machine up a grade is called grade resistance. It amounts to about 20 pounds per ton, or 1 percent of loaded weight, for each percent of grade. That is, a 5 percent grade would offer a resistance of 100 pounds for every ton of vehicle weight.

The estimator adds grade resistance to rolling resistance for uphill work, and subtracts it for downhill hauls. If assistance from a favorable grade is greater than the rolling resistance, the machine should be able to coast down. It is then time to consider the power of its brakes or retarders, rather than that of the engine and the gear reductions.

The grade at which a machine starts to roll freely provides a rough field check on the accuracy of estimates of rolling resistance.

Figure 12.129 gives the travel resistance for combinations of rolling resistance from 40 to 200 pounds, with various grades up to 25 percent.

Gradeability.   Travel resistance is compared with pull in various gear speeds to determine which one will be used.

Vehicle weight is taken from its specifications, or by weighing in the field. The figure is rounded off to the nearest 1,000 pounds or the nearest ton for convenience in arithmetic. This weight, multiplied by the proper figure selected from Fig. 12.129, will give the pounds resistance that must be overcome for that grade and ground condition. It is compared with the gear speed and pull chart, as in Fig 12.127, and a gear with power greater than the resistance is selected. Haul time is figured on the basis of speed in that gear.

For example, our sample hauler with a gross weight of 50 tons must climb a 7 percent grade on a haul road with a rolling resistance of 4 percent (80 pounds to the ton). From Fig. 12.129, we take 220 for the combined or travel resistance. Then

$$\text{Required rim pull} = 50 \times 220 = 11,000$$

According to Fig. 12.127, this truck has a rim pull of 12,000 pounds in third gear, and only 7,200 pounds in fourth. Third gear is selected. The 1,000-pound margin gives security against minor errors in measuring grade or selecting the soil factor.

Rolling Resistance, Pounds per ton	Grade per cent																
	-10	-6	-4	-2	-1	0	+1	+2	+3	+4	+5	+6	+8	+10	+15	+20	+25
0	-200	-120	-80	-40	-20	0	20	40	60	80	100	120	160	200	300	400	500
30	-170	- 90	-50	-10	10	30	50	70	90	110	130	150	190	230	330	430	530
40	-160	-80	-40	–	20	40	60	80	100	120	140	160	200	240	340	440	540
60	-140	-60	-20	20	40	60	80	100	120	140	160	180	220	260	360	460	560
80	-120	-40	–	40	60	80	100	120	140	160	180	200	240	280	380	480	580
100	-100	-20	20	60	80	100	120	140	160	180	200	220	260	300	400	500	600
120	-80	–	40	80	100	120	140	160	180	200	220	240	280	320	420	520	620
150	-50	30	70	110	130	150	170	190	210	230	250	270	310	350	450	550	650
200	–	80	120	160	180	200	220	240	260	280	300	320	360	400	500	600	700

FIGURE 12.129   Travel resistance.

Maximum speed in third is 7.5 miles per hour. Whether it will attain this speed depends on its start up the grade and its length, as it does not have enough surplus power for rapid acceleration.

Traction.   Tire treads must be able to grip the ground surface to use the power that is applied to them. Their ability to do so is called tractive efficiency, which is discussed in Chap. 3.

A radial tire gives better traction than a bias tire because its sidewall is able to flex without causing distortion in the tread face.

A machine cannot use more power than its traction permits. If travel resistance is greater than traction, wheels will spin without moving the machine.

Acceleration.   In any haul or haul section where the equipment has to start from a stop, or enters the section at a speed lower than it can attain, the average speed is affected by the time required to accelerate.

Acceleration is rapid when the rim pull is high in proportion to the travel (rolling plus grade) resistance. Since the rim pull is less in the higher gears, more acceleration time is needed in them, particularly when a gear is used that is just able to pull the load.

The rule for computing acceleration is: For each pound of rim pull per gross ton of machine and load that is in excess of that required to move the machine, an acceleration of .01 mile per hour in each second can be expected.

That is, if a machine can exert a rim pull of 300 pounds for each ton of its weight, and travel resistance is 80 pounds, the excess or accelerating rim pull (AccRmP) is 220 pounds.

The acceleration formula is

$$\text{Accel, mph per second} = \text{AccRmP} \times .01$$

This does not work well for the lowest gears where there is a very great amount of excess rim pull. Acceleration in them is very rapid, but not as fast as is required by the formula. This difference arises partly from sluggishness within the engine, and partly from other causes.

Results will be more accurate if arbitrary reductions are made in high excess rim pulls per ton, as follows:

Over 300 pounds, subtract 75%

From 200 to 299 pounds, subtract 50%

From 100 to 199 pounds, subtract 25%

In a hauler with 640 pounds per ton of excess rim pull in low gear, this would work out:

Accelerating rim pull		640
Subtract 75% of 340	225	
50% of 100	50	
25% of 100	25	330
Corrected accelerating pull		310
Acceleration, mph per second		3.1

Starting from a dead stop with a load may take another 2 to 4 seconds. Shifts with sliding gears average almost 4 seconds, synchromesh shifts 1 to 2 seconds, and power shifts no time. Start and shift time are added to acceleration time.

If this machine has a top speed of 2.5 mph in low, it will take 2.5/3.1, or .8 second, plus perhaps 2 seconds starting time, to reach full speed in this gear. In a high gear with only 10 pounds excess rim pull, accelerating time would be 75 seconds.

Production tables for scrapers often figure in acceleration and deceleration as part of "fixed" time.

Accelerating time may also be figured by multiplying the maximum speed by the appropriate factor from Fig. 12.130. But much more accurate results may usually be obtained by working out figures for the hauler and the job conditions.

A hauler may require 1,800 feet or more to reach full speed in a gear in which it has very little excess rim pull.

Length of Section, Feet	Starting From Stop	Headway when entering
0–250	.25 to .4	.5
250–500	.4 to .5	.6
500–1,000	.5 to .6	.7
1,000–2,000	.6 to .7	.8
2,000–5,000	.7 to .8	.8

FIGURE 12.130   Factors, maximum to average speed.

TIRES

Importance.   Tires make up a substantial fraction of the purchase price of rubber-mounted machinery. They usually wear out or break up more rapidly than the rest of the machine, so that they require replacement one or more times during its life. In addition, they may need repairs.

Repair and replacement costs are greatly affected by care and conditions of use. Proper tire selection, and good maintenance and work practices, often add substantially to profits.

Good tire management includes

1. Buying tires that have the proper size, strength, tread, and speed rating for the job
2. Using tires only for purposes, loads, and speeds for which they are designed
3. Maintaining tires at proper pressure and tires and machines in good condition
4. Keeping excavation areas, haul roads, and fills smooth and free of spillage

Construction.   All tires used on construction equipment are the pneumatic type, in which air or fluid is held under pressure inside a flexible, ring-shaped container.

There are many different sizes and types of tire, but they all have the same principles of construction and parts as shown in Figs. 12.131 and 12.132.

The typical tire is made up of the following:

1. *Beads.* The tire beads are bundles of strong steel wire that prevent any change of shape that would interfere with fit on the rim.
2. *Casing, cord body, or carcass.* The fabric body of the tire gives it strength to hold the internal pressure that supports the load.
3. *Plies.* The plies are the individual fabric layers in the cord body. They are usually woven of cotton, rayon, or nylon cords, or steel wires, surrounded by rubber and looped around the beads.
4. *Tread.* The part of the tire that contacts the road. It must provide traction, long wear, cushioning qualities, and cut resistance.
5. *Sidewalls.* The sides of the tire, between the tread and the bead. Technically, may refer only to the protective rubber outside coating on the sides.
6. *Tread plies or breaker.* Part plies that are only under the tread, and that serve to resist and distribute road shocks, and to resist penetration by sharp objects.
7. *Chafer plies.* Part plies that protect the tire from rim damage near the bead.

In addition, a tubeless tire has an airtight inner layer on the casing that seals in the pressure. The rim must also be airtight. Other tires have a soft inner tube to contain the air. This is protected from possible damage from the rim by a cloth flap.

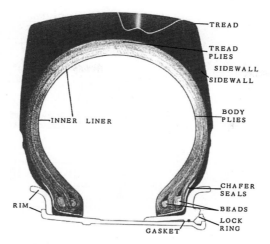

FIGURE 12.131   Cross section of tire.

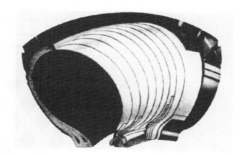

FIGURE 12.132   Cutaway view of tire.

Steel wire has increasing application in heavy-duty tires. Heat conduction is excellent, so that the tire does not overheat readily, and has great strength to resist impacts and cuts.

Weight.   Tire weights range from about 40 pounds for the smallest truck tire, up to about 1,700 for a 27.00-33 off-the-road unit. For tube-type tires, add about 10 percent for tube and flap.

Measurements.   Figure 12.133 gives accepted names for the various tire dimensions. Tires are identified as to size by the approximate width of a section of inflated tire that is not under load (section S in the diagram) and by the diameter of the rim where the tire bead rests on it. A tire with a width of 10 inches on a 20-inch rim would be called a 10.00-20 tire. See Fig. 12.134.

Most tires are nearly round in cross section, so that this tire would have an outside diameter slightly larger than the rim size plus twice the width, or $20 + 2(2 \times 10) = 40$ inches. This size relationship does not exist in wide-base tires.

Tire Selection.   The Caterpillar *Performance Handbook* notes that proper tire selection, application, and maintenance are one of the most important factors in earthmoving economics. Off-the-road tires must operate under a wide variety of conditions ranging from dry "potato dirt" through wet, severe shot rock. Speed conditions vary widely, and the gradient may vary from 75 percent favorable

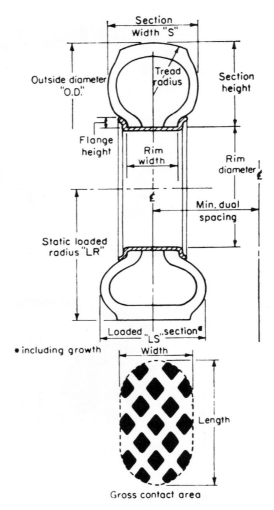

FIGURE 12.133 Tire dimensions.

to 30 percent adverse. Climate conditions, operator skills, and maintenance practices all may have a profound effect on tire life and unit costs.

No one tire construction can meet all requirements on any one machine and, in many instances, not even on one job. The optimum tire selection for a specific machine on a given job should be a joint decision between the user and tire supplier.

Tire Types.   The *Handbook* says off-the-road tires are classified by application in one of the following three categories:

1. *Transport tire*—for earthmoving machines in transporting material

2. *Work tire*—applied to tractive-type earthmoving machines such as wheel tractors and loaders

3. *Load and carry*—for wheel loaders doing transporting as well as digging

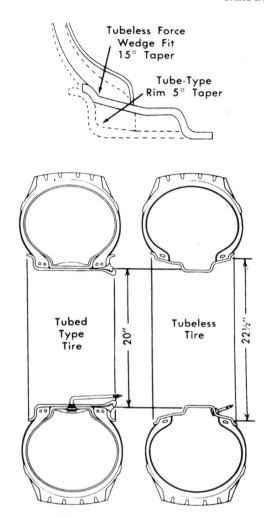

FIGURE 12.134   Rim shapes for tubed and tubeless tires.

With these variations and those mentioned in the first paragraph under Tire Selection, the selection process is difficult. Add to those variables the ones involving nomenclature.

There are also solid tires used for the most severe, abrasive ground conditions, especially for skid steer equipment. They are called segmented tires which are puncture proof and resistant to cutting, heat, and chemicals. However, these tires are heavy; it is hard for the machine to get moving or stop; the tires provide less traction than pneumatic tires, and the solid tires give the operator a rough ride.

Tire Size Nomenclature.   Tire size is derived from the approximate cross-section width and rim diameter, with various systems available:

1. A wide-base tire is designated, for example, as a 29.5-35 with the approximate cross-section width (inches) being the first number and rim diameter (inches) the second number.
2. A standard base tire is designated, for example, as a 24.00-35 with the approximate width (inches) being the first number and the rim diameter the second.

3. A low-profile tire is designated, for example, as a 40/65-39 with the approximate cross-sectional width (inches) being the first (40) number and the rim diameter (inches) the third (39). The second number (65 actually 0.65) is the aspect ratio, i.e., the section height divided by the section width.

The wide-base tire has an aspect ratio of approximately 0.83, and the standard base tire approximately 0.95. A wide-base tire gives better floatation in mud because it distributes the load, i.e., causes less pounds per square inch on the ground, keeping the vehicle on top of the sloppy ground instead of being mired down in it.

Wide-base tires look much bigger and stronger in their measurements than they really are. You subtract from 4.5 to 6.5 from their width to find the equivalent size and load capacity in conventional standard tires.

The greater tread width and lower pressure of wide-base tires give them much greater ground-contact area, and therefore better traction and flotation on soft ground than standard types.

Ply Rating.   The strength of tire bodies formerly varied directly with the number of plies or fabric layers, so that the load a tire should carry might be figured according to the number of plies. Modern practice is to use cords of different strength and type, such as cotton, rayon, nylon, or steel, and to use different weights and spacings of cord in the fabric.

As a result, the number of plies no longer has much meaning by itself. However, since plies were a convenient way of rating tire strengths, the term has been retained as "ply rating." This term indicates the strength of the casing in relation to plies of standard strength. The amount of load recommended by the Tire and Rim Association for each ply rating is included in the specification table in Fig. 12.135. When there are two or more ply ratings in one size, the higher number can take higher pressure and carry heavier loads.

OFF-THE-ROAD HAULAGE SERVICE—30 MPH MAXIMUM SPEED

WIDE BASE DIAGONAL (BIAS) PLY TIRES

DISTANCE — Up to 2.5 miles (one way)

TIRE SIZE DESIGNATION	TIRE LOAD LIMITS (LBS.) AT VARIOUS COLD INFLATION PRESSURES (PSI)						
	25	30	35	40	45	50	55
15.5-25	5720(8)	6370	6970(12)				
17.5-25	6900(10)	7680	8410(12)	9090(14)	9740(16)	10360	10950(20)
20.5-25	8660	9630(12)	10540	11400(16)	12210	12990(20)	
23.5-25	11170(12)	12430	13600(16)	14710(20)	15760	16760(24)	
26.5-25	14240	15840(16)	17340(20)	18750	20080(24)	21360	22590(28)
26.5-29	15220	16930(18)	18530	20040(22)	21470	22830(26)	
29.5-25	17880(16)	19890	21770(22)	23530	25210(28)		
29.5-29	19040(16)	21180	23180(22)	25060	26850(28)	28560	30200(34)
29.5-35	20720	23050	25230(22)	27280	29230(28)	31080	32870(34)
33.25-29	22750	25310	27700(26)	29950	32080(32)		
33.25-35	24660(20)	27430	30020(26)	32460	34780(32)	36990	39110(38)
33.5-33	25270	28120	30770(26)	33270	35650(32)	37910	40090(38)
33.5-39	27220	30290	33150(26)	35840	38390(32)	40840	43180(38)
37.25-35	30050	33440	36590	39570(30)	42390(36)	45080	47670(42)
37.5-33	30830	34300(24)	37530	40580(30)	43480(36)	46240	48890(42)
37.5-39	33080	36810	40280(28)	43550	46660(36)	49630	52470(44)
37.5-51	37400	41610	45540	49240	52750(36)	56110	59320(44)

Figures in parentheses denote ply rating for which bold face loads and inflations are maximum.

FIGURE 12.135   Off-the-road wide-base tires. (*Courtesy of Rubber Manufacturers Association.*)

At the same air pressure, the recommended load is usually the same for a tire size, regardless of the ply rating. The difference is that the heavier construction can take more air pressure, and the greater pressure will carry a heavier load.

However, the heavier tire is more resistant to carcass damage under lighter pressure and loads. Where damage from bruising is an important factor, the heavier tire is likely to be the better investment even if its extra load capacity is not needed.

Load.   Load is probably the most critical factor in tire costs. Load capacity increases with the size of the tire and with its ply rating. Both size and plies are expensive.

In a single size of tire, the price range between the lightest and the heaviest construction may be from 10 to 35 percent. The range between the largest and smallest tire that can be readily adapted to a single vehicle may be over 100 percent.

The most economical time to select a tire size is before the hauler is shipped from the factory. This is particularly the case if optional tires are enough larger to require special rims, wheels, and/or a change in axle ratio.

Tire Sizing.   Haulers are usually factory-equipped with tires that are barely adequate for the vehicles' recommended loads. Offering larger tires as standard equipment would increase the advertised price of the unit, putting it at a competitive disadvantage with others offering minimum equipment. If the hauler is to be used only for rated or smaller loads, such tires may be satisfactory, but they leave no margin of safety.

Occasionally heavy trucks are offered with tires that are definitely too small. It is worthwhile to look up the truck's loaded weight and weight distribution in its specifications to see whether it will carry the load.

For example, an off-the-road truck with a loaded or gross vehicle weight of 90,000 pounds, with 64,000 pounds on the rear axle, is equipped with four 18.00-25, 24-ply drive tires. They are rated at 16,000 pounds at 30 miles per hour, 17,920 pounds at 10 miles per hour. Maximum vehicle speed is 31 mph.

At full rated speed, the drive tires have a combined capacity of 64,000 pounds, exactly equal to the rated load. There is no margin of capacity for accidental or deliberate overload, shifting of load to the rear while climbing steep grades, or weight shift on curves. To keep any margin of safety, the owner would have to underload or keep speed down to 20 mph or below.

For average work involving moderate to high speeds, full loads, and some road roughness, it would probably be wise to specify 28-ply tires in the same size, with a load capacity of 17,500 each, or a margin of 1,500 pounds per tire.

If still greater capacity were desired, different wheels, rims, and rear-end gearing could be used, and 24-ply 21.00-25 tires used.

A contractor might make either of these extra investments to reduce tire damage or to permit overloading the truck without tire damage.

Tread.   In general, high, widely separated diagonal cleats are used for drive tires on soft earth; wide, closely spaced bars or lugs for rock, button or diamond treads for free-rolling tires, or drivers on sand or dry humus; straight ribs for grader and farm tractor front wheels, notched ribs for highway travel, and smooth treads for compactors and loose sand.

Diagonal thin cleats tend to break up on rock, and the wide spaces between them do not have enough tread to protect the body against bruising. Rock treads tend to fill up on soft earth or mud. Any design using cross or diagonal cleats will wear irregularly on pavements, and will probably be very noisy. Designs that lack cleats or lugs may not have sufficient traction in soft ground.

Many tread designs are intended to serve reasonably well under two or more different conditions, rather than very well under one.

Rims.   Small truck tires are mounted on drop-center rims that may be part of the wheel, as in a car, or demountable by removing lug nuts.

Tires larger than 12.00 are mounted on three- to five-piece rims held together with a lock ring. This ring can be very dangerous if a tire is inflated before it is properly secured, as it can blow out with explosive force.

Tapered rims are best for low-pressure tires, as the tire bead is wedged against the sloping floor of the rim, reducing the tendency of the wheel to revolve inside the tire.

TIRE WEAR

A tire parts with a little of its surface on every revolution, partly because it changes shape with a scuffing effect as it contacts the ground. This changing of shape is called deflection. It is increased if the pressure is decreased (less air to hold the tire stiff) or if the load is increased (the extra weight pushes down harder on the tire).

The amount of wear that will result from any given amount of deflection depends largely upon how abrasive or sharp the road or ground surface is. Surfaces such as new brush-finish concrete, sharp gravel, and crushed or blasted rock take tread off rapidly, while old concrete, blacktop, fine sand, and clay are less wearing.

Wet rubber cuts more easily than dry rubber, so damage from cuts and scratches is likely to be increased by rain. But sometimes water acts as a protective lubricant and reduces wear.

Spinning the wheels will take tread off, sometimes as fast as a buffing machine would. When the tread shows long scratches running part or all the way around the tire, the wheel has been spun. Using brakes increases tread wear, and the harder they are applied, the faster the tread comes off. If the wheel locks and slides, a flat spot may be made on the tread, that will tend to get worse with ordinary use. A high spot on a brake will concentrate wear on one spot also.

Alignment.   If an axle sags, tires will wear on the inner edge. If a wheel leans outward, the wear will be on the outer edge of the tread. Excessive toe-in or toe-out will produce feathered edge wear on one side.

Alignment is usually more troublesome in highway vehicles than in off-the-road equipment.

Cupping.   Tires often wear irregularly in a pattern of alternating hollows and ridges, an effect that is called cupping. This may be caused by wheel alignment, but in earthmover tires it is usually caused by driving on paved roads.

Widely separated diagonal cleats are particularly subject to this difficulty, but any coarse tread not including at least two around-the-tire ribs is likely to cup.

Cupping occurs even at very low speeds, but it becomes more severe as speed increases. The tread may become so rough that the tire must be buffed down, retreaded, or scrapped.

Inflation.   If a tire has too much pressure for the weight it is carrying, it will round out so that only the center of the tread will be on the road. If it has too little pressure, the center will fold up and in and it will ride on the edges. And whatever it rides on, it wears. If there is center wear, the tire has traveled a long way with too much air; if it is worn on both edges, it has probably been too soft or overloaded too long. See Fig. 12.136.

Air pressure in a tire should be sufficient to carry the load without either of these effects. The load carried determines the correct pressure for each type of service. The table in Fig. 12.135 shows that with each increase in pressure the allowable load is increased, until the maximum pressure for the tire is reached. If the load exceeds that amount, a tire with more plies or a bigger size must be used. This table should be read both ways, as it also means that with decrease in load, tire pressure should be decreased.

Dump truck and scraper tires are inflated according to the weight they carry when the vehicle is loaded.

When pressure is higher than that permitted by tire design, it tends to put too much stress on the cords. This tension makes the tire easier to break and puncture. For an example of this effect, push a pencil point lightly against a toy balloon that is only partly inflated, and then against the same balloon when it is blown up tight.

Tire pressure must be checked when the tire is cold, before it has been run. This is because the heat generated in it by use increases the air pressure. If it has the right air pressure when it is hot, it will be too soft when it cools off. A tire is built to stand the extra pressure that is caused by heat as long as it is not carrying more than its proper load.

Section of overinflated tire showing how center of tread is subjected to excessive wear.

Underinflated tires flex at every turn of wheel, resulting in high internal heat, fabric breaks.

Correctly inflated tires permit all of tread to contact road, are not soft enough to flex excessively.

FIGURE 12.136   Effects of inflation on tire shape. (*Courtesy of Rubber Manufacturers Association.*)

Heat.   An underinflated or overloaded tire flexes too much, causing internal shear and eventual separation in the fabric. This mechanical effect is damaging enough, but the heat that it creates is far more destructive. Bending of any material creates heat. For example, bend a wire coat hanger rapidly back and forth a few times, then cautiously touch the spot where it bends. It may be hot enough to burn your finger.

Rubber is a poor conductor, so the heat created by flexing builds up in the tire. If it passes 250°F, the rubber will lose strength and some fibers will be weakened. At 280°F, the temperature at which many tires are cured and retreaded, plies may separate, and the tire may blow out.

Overloads.   There are two types of overloading. One is carrying a load that is within the rated tire capacity, but that is too much for the pressure in the tire. This is really underinflation, and has the same effects.

The other type of overload is placing more weight on the tire than it can carry at its maximum recommended pressure. Such overloading usually results from heaping loads and heavy materials. It produces the results of underinflation in a much more severe form.

Speed.   Heat is generated in each part of the tire as it flexes in passing under the wheel, and heat is radiated constantly from the tire surfaces and through the wheel. Speeding up will increase the generation of heat but will not affect the rate at which it is radiated away.

A tire with a 12-foot circumference will turn about 37 times per minute at 5 miles per hour, and 220 times at 30 miles per hour. Six times as much heat will be produced at the higher speed.

This tire might run moderately underinflated or with an overload at 5 miles per hour with little damage, as heat could be radiated away as fast as it was produced. But at 30 miles per hour, a high speed for earthmover tires, extreme overheating and a blowout would be likely.

Also, injuries from impact are more likely and more severe as speed increases. The force of the blow increases greatly with speed.

Retreading.   Retreading is used here as a general term covering both retreading and recapping. Strictly speaking, a tire is retreaded by buffing the worn tread down close to the fabric, and vulcanizing on a new tread with part sidewalls that extend halfway to the bead. Recapping, a slightly cheaper and sometimes less satisfactory operation, replaces the tread only. However, most construction workers use the terms interchangeably.

Retreading costs about half as much as a new tire, and is subject to the same fleet discounts. There are few places that can retread the very large tires, so they may involve extra cost for freight, and considerable lost time.

Some contractors and many mines find that they can retread most of their tires again and again, perhaps as many as four times, and that this is the most efficient and economical way to buy rubber. Others report that they have had no success whatever with retreading, and have adopted a policy of using only new tires. The in-between majority do some retreading or recapping and a lot of replacing.

TIRE LIFE

Differences of Opinion.   There are sharp differences in the published literature about tire life. Manufacturers of earthmoving equipment assume it to be 5,000 hours. Most estimating books accept these figures. On the other hand, tire manufacturers state that such tire performance is most unusual, and that 1,000 to 3,000 hours is a more reasonable estimate.

It is likely that the tire industry is much closer to an accurate estimate of tire performance than the equipment people are. However, there is enough evidence on both sides to justify their claims.

Wear Factors.   Several tire companies have published tables of factors by which probable tire life can be calculated. Figure 12.137 is based in part on such tables, and in part on additional information obtained from contractors and mines.

Maximum normal life of off-the-road tires under favorable conditions is assumed to be 6,000 hours or 60,000 miles, with the hourly basis being more convenient for us to use.

Each of the groups of operating conditions has a series of decimal fractions indicating the part of full tire life that can be expected under each condition. For example, in group F, trailing tires that neither propel nor steer the vehicle, and therefore have a minimum of wear, are assigned a factor of 1.0, indicating that in this position full life may be expected.

In the same group, front wheels that have tread rubber scuffed off as they resist sideward push as the machine is steered are given a factor of .9, meaning that life should be nine-tenths that of a trailing tire under the same conditions. Scraper driving wheels, that operate under severe conditions including spinning, are given a factor of .6.

Neutral conditions are given a rating of 1.0, and disappear in the arithmetic. Factors larger than 1 show better than ordinary conditions and increased life.

To find the effect of wheel position on tire service, we multiply the proper factor by the maximum tire life of 6,000 hours. We find that if all other conditions are entirely favorable, a tire is likely to run 6,000 hours on a trailing wheel, 5,400 on a front wheel, and 3,600 on a scraper drive wheel.

It is of course unlikely that all other conditions are excellent. A factor is selected from each group, representing as closely as possible the conditions on the job, These factors are multiplied by each other, and their product is multiplied by 6,000 to give the average life in hours of tires on the job being studied.

These figures assume that a tire's life is finished when the tread is worn off. If the casing is good, it can be retreaded; but it then starts a new life on a different cost basis. The saving is taken into account as a reduction in replacement cost.

Miscellaneous Conditions and Combinations might include extreme heat as a factor by itself, or in combination with speed and/or overloads, some rock on road in combination with speed and overloads, cowboy behavior of operators, frequent ripping of scraper tires by dozer blades, or favorable factors such as low-friction road surface of borax or serpentine.

If tread wear is extremely rapid, the tire life may depend entirely on its rate. On the other hand, fast driving on coarse rock with overloads might land every tire on the scrap pile with thick tread still on it. A ruinous factor like blasted volcanic glass determines tire life by itself, without multiplying by other factors.

Therefore, while this table of wear factors is a helpful guide, it cannot stand by itself. It should be used by people who are familiar with tires and with the conditions under which they operate.

Predicting Tire Life.   An aid to help choose the correct tire for earthmoving and mining operations is the ton-mile-per-hour (TMPH) concept. Basically, TMPH is an indication of heat buildup, which can cause separation and tire failure. It is a rating of a tire measuring its work capability at the tire's maximum safe operating temperature. Tire manufacturing requires heat in the vulcanizing process that converts crude rubber and additives to a homogeneous compound. The heat required

FACTORS IN EARTHMOVER TIRE LIFE
To be multiplied by 6,000 hours or 60,000 miles

Group A — MAINTENANCE
INCLUDES INFLATION

Excellent	1.1
Average	1.0
Poor	.7
Very bad	.4

Group B — MAXIMUM SPEEDS

10 miles per hour	1.2
20 miles per hour	1.0
30 miles per hour	.8
40 miles per hour	.5

Group C — CURVES

None	1.1
Moderate	1.0
Severe, single wheels	.8
Severe, dual wheels	.7
Severe, tandem wheels	.6

Group D — SURFACE

Snow, packed, no road exposed	3.0
Earth	
Hard packed earth	1.0
Soft earth or sand, maintained	1.0
Gravel road, well maintained	.9
Soft earth, some rock	.8
Mud, ordinary	.8
Gravel road, poorly maintained	.7
Mud, abrasive or with rock	.5
Blasted rock	
Soft coal	.9
Soft shale or limestone	.7
Granite, gneiss, trap, basalt,	
hard shale or limestone	.6
Slate or schist	.4
Lava, hard surface	.3
Obsidian, volcanic glass, flint	.1
Blacktop	
Clean, wet	1.4
Cold weather	1.2
Hot weather, 75 to 100° F.	.8
Very hot, over 100° F.	.5

Group E — LOADS

Recommended by Tire and Rim Assn.	1.0
50% underload	1.2
20% underload	1.1
10% overload	1.0
20% overload	.8
40% overload	.5

Group F — WHEEL POSITION

Trailing	1.0
Front (non-driving)	.9
Driving	
Rear dump	.8
Rear dump tandem	.7
Bottom dump	.7
Scraper, self propelled	.6

Group G — GRADES, DRIVE TIRES ONLY

Level	1.0
Firm surface	
6% maximum	.9
10% maximum	.8
15% maximum	.7
25% maximum	.4
Loose or slippery surface	
6% maximum	.6
10% maximum	.6
15% maximum	.4

Group H — MISCELLANEOUS CONDITIONS AND COMBINATIONS

Favorable or counteracting	1.5
None	1.0
Unfavorable	.8
Very unfavorable	.6

FIGURE 12.137   Factor table, earthmover tire life.

is above 270°F (132°C). A tire generates heat as it rolls and flexes. Heat generated faster than it can be radiated into the atmosphere gradually builds up within the tire.

Caterpillar Inc. was instrumental in developing the TMPH method of rating tires. The TMPH calculation, simply stated, is the average tire load multiplied by the average vehicle speed over the work shift. For example, if a truck axle is carrying an average total load, between empty and loaded travel, of 28 tons, and there are two wheels with tires, the average tire load is 14 tons. And if the truck works 8 hours, making 16 trips each of 10 miles round trip, the average shift speed is 20 mph. Therefore, the TMPH is 280 ton-miles per hour. This indicates that the tires for this truck's axle need to have a TMPH rating of 280 or higher to avoid heat problems.

Tests have shown that the TMPH formula does not apply when tires are loaded 20 percent above capacity, or on hauls longer than 20 miles one way. Under either of those conditions, the work capability factor (WCF) should be used for rating the tires. The WCF formula is the average tire load in tons, multiplied by the maximum average speed.

Dozer and loader tires are thicker and stronger than other off-road designs. Heat can therefore build up faster for the tires on this equipment, if haul distances exceed 50 feet and speeds are above 5 mph. The WCF provides a way to select tires to handle the work. For example, if a wheeled front-end loader with full load weighs 84,000 pounds and travels at an average of 5.5 mph, then the WCF for each tire is 57.75 (84,000 divided by 4 × 2,000 then times 5.5). Therefore, the tires for this equipment should have a WCF rating of 57.75 or above.

TIRE BALLAST

Liquid Ballast (Hydroflation).   Drive tires may be 75 to 100 percent filled with fluid (hydroflated) to obtain additional weight, in order to increase traction and drawbar pull used. See Fig. 12.138.

The most effective fluid is a solution of 5 pounds of commercial-grade calcium chloride ($CaCl_2$) in each gallon of water. (See Fig. 12.139.) This solution weighs 10.6 pounds to the gallon. Freezing point is 52°F below zero. Added weight for 75 percent fill ranges from 395 pounds in a $15\frac{1}{2}$-38 farm tractor tire to 5,060 in a 33.5 × 33.

Plain water may be used where freezing is not a problem, but it weighs only 8.3 pounds to the gallon. Some contractors use a 3-pound solution with a weight of 10.1 pounds to the gallon and a freezing point of −3°F.

Fluid may be put in a tire by gravity flow from an elevated tank, but it is much more efficient to use a special pump. A number of models are available, for volumes from 240 to 600 gallons per hour.

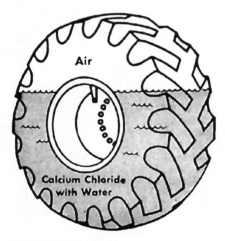

FIGURE 12.138   Hydroflated tire.

Tire Size	Gallons Water 75%	*3½ Lbs. Calcium Chloride Per Gallon Water			†5 Lbs. Calcium Chloride Per Gallon Water		
		Gallons Water	Lbs. CaCl₂	Total Weight	Gallons Water	Lbs. CaCl₂	Total Weight
WIDE BASE AND GRADER SIZES — 75% FULL OR VALVE LEVEL							
15.5-25	46	40	139	470	37	187	500
17.5-25	60	51	180	609	48	243	647
20.5-25	90	77	269	910	72	362	967
23.5-25	118	101	354	1198	95	478	1274
26.5-25	159	136	477	1614	129	643	1716
26.5-29	174	149	521	1764	141	703	1875
29.5-25	207	177	618	2090	167	833	2223
29.5-29	224	192	673	2275	181	907	2419
29.5-35	251	215	753	2547	203	1015	2708
30-33	167	143	500	1690	134	670	1785
33.25-29	296	253	885	2985	238	1190	3170
33.25-35	319	274	958	3242	258	1292	3447
33.5-33	328	281	983	3326	265	1325	3536
33.5-39	363	311	1089	3684	294	1470	3917
37.25-35	355	304	1065	3610	287	1435	3820
37.5-33	423	362	1268	4290	342	1710	4562
37.5-39	466	399	1397	4729	377	1885	5020
37.5-51	552	473	1656	5603	447	2235	5958
43.5-43	291	249	873	2915	235	1175	3126
10.00-24TG	25	21	74	254	20	101	270
12.00-24TG	34	29	103	348	28	139	370
13.00-24TG	41	35	122	412	33	164	438
14.00-20TG	45	39	135	458	36	182	487
14.00-24TG	51	43	152	513	41	205	546
16.00-24TG	72	62	216	730	58	291	775
18.00-26TG	113	97	339	1147	91	457	1220

*3½ Lb. solution provides anti-freeze protection to −15°F, slush free.   †5 Lb. solution provides anti-freeze protection to −53°F, slush free.

FIGURE 12.139   Hydroflation table for wide-base tires. (*Courtesy of Rubber Manufacturers Association.*)

If a tire is to be hydroflated, the load is taken off it by jacking or by removing it from the machine. It is kept in a vertical position, with the air-water type valve at the top. An adapter is used to permit fluid to be pumped in and air vented at the same time.

Some manufacturers recommend 90 to 100 percent filling, but this makes the tire too rigid for riding comfort. Pressure should be checked after a few hours' operation. Checking is done with the valve in its highest position. Corrosion-resistant gauges are needed to resist attack by the calcium chloride salt.

Liquid tire ballast is particularly useful in tractor, tractor-loader, and grader drive wheels, as these machines usually have more power than traction. It is sometimes used in both drive and free-rolling tires, both for traction and to lower the center of gravity in rubber-mounted cranes, loaders, forklifts, trenchers, and drills. This type of ballast serves to improve traction and stability, reduces tire wear resulting from slippage, reduces the bounce of big tires, and is said to increase puncture resistance. These advantages far outweigh the moderate cost, and the messiness that may accompany tire repair.

TIRE CHAINS

Traction (Skid) Chains.   Small and medium-size earthmover tires, and highway tires of any size, may be equipped with automotive-type tire chains, provided there is enough clearance to accommodate them. Each chain usually consists of a pair of circular chains smaller than tread diameter, resting against the sidewalls and connected across the tread by a number of short cross chains. See Fig. 12.140.

These chains are essential equipment for most two-wheel-drive vehicles or machines that work in snow or on ice. They may be valuable for occasional mud conditions, but tend to increase their severity by cutting and churning the ground. They increase the danger of getting stuck in sand.

All-wheel-drive machines function well in snow and mud without chains, but may need them on ice.

FIGURE 12.140 Tire protection chains on loader.

This type of chain has a very long life if used on dirt, ice, snow, or loose rock; or on pavement that is largely covered with snow or ice. But it wears badly on bare pavement, with the rate of wear of both chain and pavement increasing rapidly with speed. It causes no significant tire damage.

Installation. In well-equipped shops, chains may be installed by lifting the wheel clear of the ground, placing the opened chain on top, and linking it at or near the bottom. If no adequate hoist or jack is available, the chain is laid on the ground and the machine moved until the wheel is resting on the cross chains, a foot or two from one end. The other end is lifted and dragged over the top of the tire, and brought down to be connected.

Chains may loosen substantially in the first few turns of the wheel. They should be checked, and tightened if necessary.

Protection Chains. Drive tires on loaders and scrapers working in abrasive blasted rock may have a very short life, due chiefly to grinding and gouging the tread as wheels spin. There may also be many premature casing failures because of damage from sharp rocks and momentary overloads.

Tread wear and casing damage may both be greatly reduced by covering the tire with chain mesh. Use of these chains permits the operator to ignore the possibility of tire damage, and to apply the full power of the machine to its work. On the other hand, there are ground conditions under which the chains may reduce traction seriously.

Steel Treads. Another approach to loader tire protection is found in Loegering's Trail Blazers, designed for skid steering. It is claimed they provide a smoother ride and reduce rutting and compaction. See Fig. 12.141.

OPERATOR RESPONSIBILITY

The operator of excavation equipment must be a responsible person, for at least two reasons. This machinery is heavy and powerful. It may be capable of enormous destruction to property, and injury and death to people (including the operator) if it is operated improperly or carelessly, or if it goes out of control because of mechanical failure.

FIGURE 12.141   Tire crawlers for skid steering. (*Courtesy of Loegering Mfg. Inc.*)

This equipment is both very expensive, and extremely vulnerable to damage arising from improper maintenance. In order to do its work, it must be lubricated according to schedules that vary widely from machine to machine. There are many parts that deteriorate rapidly if used when worn or out of adjustment, and are likely to cause rapidly increasing damage to other parts if their repair is postponed. Mechanical neglect greatly increases danger of accidents.

The operator should check his or her machine before, during, and after work, and correct or report improper conditions. Such checking always includes level of lubricant, hydraulic oil, and coolant, and indications of unusual wear.

PRESTART CHECK

On arrival on the job, the operator should walk around the machine, looking at it carefully. The parts deserving particular attention vary with the type of machine, the work it is doing, and often the service history of this particular unit.

Tires.   If it has tires, proper pressure is of the first importance. It should take only a couple of minutes to check the whole set with a gauge, but an experienced operator may be able to check them adequately by eye.

Tires should be inspected for cuts, tears, and worn tread. A pattern of scratches and grooves on drive tires indicates wasteful spinning. If the tread is smooth, traction will be poor except in sand, and the tire will soon be worn past the possibility of retreading.

FIGURE 12.142   This track should be reported.

Tracks.   Track tension can be checked only if the machine is reasonably clean, and was moving forward when stopped for parking. Then there should be a slight sag on each side of the carrier (upper) roller. Too tight a track will wear rapidly, while if too loose it will also wear rapidly and may come off. See Fig. 12.142.

Bright patches of scraped metal usually indicate that something is loose or out of line, although work in loose gravel may produce a similar effect. Prompt inspection by a mechanic is called for.

Low cleats and bent shoes on a tractor mean poor traction, and reduced safety on side slopes, but this is a long-term situation.

A flat spot or spots on the carrier roller show that it does not turn freely. If it is jammed by surface debris, free it. Otherwise it is damaged inside, and must be repaired or replaced promptly.

Working Tool.   The bucket, blade, edge, tooth, or other part that is forced through dirt or rock during excavation is subject to both wear and breakage.

Wear is usually gradual, but may be quite rapid in abrasive material such as sharp gravel or sandstone. It usually makes the cutting edge thicker and duller, so more power is needed to do less work. It may change a straightedge into scallops. The operator should be alert to report excessive wear.

Breakage of cutting parts may occur without warning, but it is often preceded by cracking. Detection of cracks usually makes it possible to patch-weld and reinforce the unit, to postpone or avoid expensive repair. A break usually tears and bends the metal so as to make rebuilding difficult.

Loader and hoe buckets are particularly vulnerable to breakage following cracking.

Linkage.   Most excavating tools are hinged to their power units. Linkage may be very simple, as in a direct-lift dozer blade or ripper, or quite complex, as in a front loader.

In any case, each hinge is a point where lubrication is probably required, and where a pin might loosen and come out. Loosening of a pin is rare, but can be very damaging, so it pays to look at each one to see that it is properly in place.

Looseness due to wear of the pin and bushing is most readily detected during operation, but it can usually be studied in the parked machine by prying with a crowbar.

Steering linkage acting on the wheels is subject to damage from collision with rocks or other objects. Steering action in an articulated machine might be completely blocked by a careless mechanic leaving a safety brace attached.

Leaks.   Surfaces of the machine, and the ground under it, should be inspected for evidence of leakage of lubricating or hydraulic oil, or coolant.

The seriousness of leakage is variable. Any opening that will let oil out may let dirt in. In spite of the efficiency of modern filters, this is a danger. It may show that a seal is beginning to fail, or worse, that a case has cracked. Or it may be normal outflow from a pressure-greased fitting. Each leak must be judged individually.

In the hydraulic system, leaks are usually at joints. If the joint is tight, the leak probably indicates deterioration of the hose inside the joint clamp, and may (or may not) be the warning of a coming blowout at that point, or elsewhere in that hose.

If the hose itself starts to exude oil, it should be replaced immediately.

Radiator.   A number of fluid levels should be checked at the beginning of each shift. The radiator should be filled to within a few inches of the filler cap. Check it by unscrewing the cap (always with a cold or at least not-hot engine) and looking. The cooling system is sealed, so any loss of fluid is cause for alarm.

Leakage may occur externally through hoses or their connections, a worn water pump, or a punctured radiator. Such leaks should be visible.

More serious loss is through the cylinder head gasket into the cylinders. Then the coolant might work its way down into the crankcase oil, or be blown out the exhaust.

Worse yet, leakage of hot gas from the cylinder may heat the coolant to boiling.

The fluid is usually a mixture of water and ethylene glycol antifreeze, possibly with other additives. In summer it can probably be brought up to level with plain water, but in freezing weather, losses must be replaced with antifreeze mixture.

The air passages through the radiator may become plugged by trash, such as leaves, straw, and seeds. This is particularly likely during mowing or clearing operations. Interference with cooling increases rapidly with the area affected.

Reservoirs.   Almost any piece of equipment has one or more fluid reservoirs whose level is checked by a dipstick (or more rarely, by a sight gauge), usually while the engine is shut off.

There is always a dipstick for the engine. The operator must be sure to use it, and to add oil if necessary, before starting. In addition, there may be dipsticks for a torque converter, the transmission, and other drive units. It is necessary to know where each of these is, and to remember to check it. Availability of the proper fluid may be a problem.

Standard transmissions, differentials, and final drives often are checked by removing a screw filler plug. Lubricant should be up to its level, or slightly above. A daily check is not customary for these, unless there are signs of leakage.

If there is an air pressure tank, open the drain cock (or drain cocks) in the bottom to let out any fluid, and retighten.

Check the fuel level, unless there is a gauge on the dash. If drain cocks are provided for removal of water, use them.

Air Cleaner.   The air precleaner usually has an easily removable container (dust cup) for coarse trash. This may have a sight gauge on the dash, with a red strip to warn that it is full. This may not register unless it is tapped.

If the indicator shows that trash is present, open and dump the container. If there is no indicator, open it to check.

STARTING THE ENGINE

Interlocks.   Most new machines, and many old ones, have one or more safety devices, called interlocks, that prevent the starter from operating if controls are not in proper position.

To avoid delay and embarrassment (not to mention possible accident), it is wise to follow all instructions in the book and/or on the dash, before and during starting. The following list may not be complete.

If there is a seat belt, fasten it.

If the brakes can be locked down, press and lock them.

Put all gearshift levers in neutral. With power shift, there is usually an additional shift lock lever that must be engaged.

Put all equipment control levers in neutral, Hold, or Float.

Connect the battery to the electrical system. The disconnect, if there is one, may be a lever or handle in the battery case, or on or under the dash, or a key-operated switch on the dash.

The starter switch itself may be a pushbutton or pedal in old equipment. Now it is usually a turn switch, with or without a key. It may have three positions—off, run, and start.

Diesel.   Most excavation equipment is diesel-powered. The diesel is stopped by simply shutting off the fuel. It is started by opening the fuel passage and cranking with the starter. In cool or cold weather, starting aids are needed.

The fuel shutoff may be a separate lever or a dashboard knob that is pushed in to run, and pulled out to stop. Or it may be a position on the hand throttle or foot accelerator. When either of these speed controls is moved from a speed position to low idle, an additional pull past a stop puts it in shutoff position, where it will be held by a detent.

To start in warm weather above 60°F (16°C), put the separate lever in the on position, or move the throttle out of the detent into idle position, and turn the switch to Start. Release the switch when the engine starts.

A diesel in good condition should start within 5 or 10 seconds in warm weather. If it does not start within 15 seconds, throttle position and fuel supply should be checked.

A starter should not be used continuously for more than 30 seconds, after which it needs about 2 minutes to cool off.

Cold Weather.   Many diesels will start on a warm weather basis down to about 40°F (5°C), but this varies with make, model, and individual engines, and cannot be depended upon.

There are three principal methods for helping a cold diesel to start—ether, preheater, and glow plug.

Ether is sprayed sparingly from an aerosol can into the intake of the air cleaner while the engine is being cranked. Do not use it either before cranking or after starting. This volatile, highly flammable fluid ignites readily in the cylinders. Some machines have built-in ether dispensers, which may involve supplying it in can or capsules. The aerosol is generally more convenient and dependable.

Ether is effective all the way from moderate temperatures down to extreme cold.

Ether is very poisonous, flammable, and explosive. It must not be used in an unventilated room, or near heat, flame, or sparks. It is likely to explode in the manifold if used for starting at the same time as a preheater.

A preheater is often supplied for cold-weather diesel starting. This may resemble a miniature furnace-type oil burner inside the intake passage. There is a spark plug, a nozzle, and a dashboard hand pump. The spark is switched on, fuel is forced out the nozzle and is ignited, and the engine is turned with the starter. The burner heats the air going into the cylinders. It consumes some oxygen, but usually not enough to interfere with combustion in the cylinders. The principal drawback is that mechanical difficulties, particularly shorting out the plug with fuel oil and failure of the pump to operate, may prevent it from working when most needed.

A glow plug is a small electric heater in a precombustion chamber or intake passage. At moderate temperatures it is switched on for 1 minute before using the starter, when below freezing for 2 minutes. It may be used alone, or in addition to ether injection. See Fig. 12.143.

Gasoline.   A gasoline engine is stopped by turning off the ignition, usually with a key. If it is warm or hot, it is started by turning on the ignition and using the starter, which is often on the same switch.

An exception to this statement is that a very hot engine may boil the gasoline in the line to the carburetor, creating a won't-start condition called vapor lock. Sometimes you can cool the line by wrapping it in a damp cloth, but usually you just wait.

If the engine is cool or cold, it will need to be choked. A choke is a valve that cuts off almost all the air flowing into the carburetor. It is usually operated by a dashboard knob that you pull for choking (closing the choke). Cranking the engine then creates a vacuum in the carburetor, pulling extra gasoline through the jets. Partial choking creates a very rich fuel-air mixture; full choke (choke closed) draws liquid gasoline also.

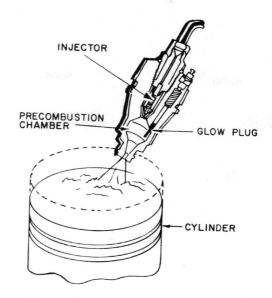

FIGURE 12.143   Glow plug.

Engines are quite individual in their response to choking, but in general, the colder it is, the more they need.

When the engine starts, the choke should be partly opened immediately, but partial choking may be necessary for several minutes. Not enough choke causes stalling and/or misfiring or hesitating during acceleration. Too much choke causes rolling, or pulsation in speed, and may stall the engine if not corrected. Overchoking also may dilute lubricating oil and foul spark plugs.

If the engine does not start with the starter turning briskly and the choke full out, it may be overchoked. Many units can go directly from a not-enough to a too-much choke state without catching, and sometimes without even kicking. Test for this by opening (pushing) the choke and continuing to crank. A wide-open throttle is helpful in getting rid of extra gas, but starting is easier when it is partly closed.

If the engine has an automatic choke, simply open the throttle halfway, use the starter, and hope for the best. You can usually check the action of the automatic by taking off the air cleaner and looking into the carburetor intake. Full choking can be obtained by entirely blocking the opening with your hand.

Extra gasoline may be supplied by pumping the accelerator.

Poor starting usually means that an ignition tuneup is needed.

Alternative Fuel.   There are alternative fuels being tested and used in engines for vehicles and other construction machinery. One is natural gas, which requires a remodeling of the fuel supply mechanism and the engine for the equipment. Another is called water-based fuel. Both of these alternatives have the advantage over gasoline and diesel fuel of reducing the undesirable emissions into the atmosphere.

The water-based fuel, which was initially known as A-21, is a combination of 30% water, 65% naphtha, and 5% additive. The surfactant additive is needed because naphtha, like all oils, is immiscible with water, and it forces the two to combine if they are sufficiently mixed together. The combustion of A-21 leads to a significant reduction in the emission of nitrous oxides (NO_x) in the exhaust. An advantage of the water-based fuel is that a major remodeling of the fuel supply mechanism and engine is not needed.

Oil Drag.   There are four factors that make cold-weather starting difficult—the extra heat required to raise fuel-air mixtures to the ignition point, the slower vaporization of fuel (particularly important with gasoline), the drag of thick, cold oil on all parts, and the lowered efficiency of cold batteries.

The oil drag can be very serious. In general, thinner oils should be used in cold than in warm weather, both to reduce drag and to supply better lubrication. If conditions are severe, or particular machines are hard-starting, it is a good plan to put a small quantity of gasoline, 1 or 2 cups, in the crankcase oil filler tube just before shutting down. The engine should be turned over afterward just enough to mix the oil and gas in all its parts.

This will thin the oil so that drag will be greatly reduced the next morning. As soon as the engine warms up, the gasoline will evaporate rapidly, returning the oil to its proper viscosity. The gas vapor will escape through the crankcase breather pipe, and if it has a filter element in it, danger of fire from this source is negligible.

Dry Fuel.   A basic all-year precaution for good starting is to keep fuel oil and gasoline dry. Moisture gets in them from condensation in the tank above the fuel level, when the machine cools at night, and it may be present in the fuel supply.

Condensation can be prevented by filling the tank at the end of the work shift.

Water in the fuel can be made harmless by adding a small quantity of methanol (wood alcohol) daily. This prevents freezing, and enables water to mix with fuel so it seeps through filters harmlessly.

Booster Cables.   If the battery is too weak for starting, you may be able to use another battery, or set of batteries, of the same rated voltage, either on the ground or in another machine. See Fig. 12.144.

Heavy insulated copper cables with spring clamps on each end are used. One is usually marked in red, the other in green. Hook up negative to negative, and positive to positive, putting the clamps on the battery posts or cable clamps. It may be necessary to twist or wiggle them to get a good connection.

If the auxiliary battery has a good charge, the starter should work normally. Remove the cables when the engine has started.

Push-start.   Many engines start best if turned by rolling, pushing, or pulling the machine rather than by using a starter. It is a very good plan to leave equipment in such position that it can be

FIGURE 12.144   Bring a booster battery.

readily towed if necessary, as this may save hours of monkeying around with adjustments that might better be left for a slack time. Coasting downhill is a good means of starting, and a crawler machine may be backed up a steep pile, and started the next day on a run of a few feet.

Unfortunately, this can seldom be done with newer machines, as torque converters have too much slippage unless equipped with lockup clutches, and power shifts are likely to be damaged.

Warming Up.   After starting, a cold engine should be run at one-third to one-half speed for 3 to 5 minutes, with no load or light load. The primary purpose is to get the lubricating oil warmed and thinned, and distributed to all parts needing its protection. If there is a turbocharger (see discussion under Stopping), it is particularly important that it turn slowly until warmed and lubricated.

The dash gauges should be watched. The temperature indicator should gradually climb into proper operating range. Oil pressure should be correct or a little high within a few seconds. The ammeter should show a high charging rate for a few minutes.

Equipment having air brakes should not be moved, nor should the brakes be unlocked, until pressure reaches normal range, usually 100 pounds.

Running the engine will turn the hydraulic pump or pumps, and circulate fluid through the valve and reservoir. After a couple of minutes, all valves should be manipulated to move hydraulic-powered attachments through their full range of action, without load. This serves to warm the fluid in their lines, cylinders, and motors, and is an important check on their proper operation.

The steering should be checked twice through its full range, for the same reason.

None of these warmup precautions are needed in restarting after stopping for servicing, lunch, or job delays, provided that operating parts are still hot to the touch. If they are merely warm, a minute or two of running may be advisable.

Toward the end of the warming-up period, make any dipstick checks which are supposed to be done with the engine idling. These may include engine, torque converter, and transmission levels.

Many machines have points that require once-a-shift greasing, and these are likely to be the operator's responsibility. These can be attended to at this time also. Usually, you pump in enough fresh grease with a hand gun or bucket pump to make some old grease come out. If there is resistance, check to make sure grease is supposed to come out, before forcing it.

STOPPING

Idling.   It is a general rule that an engine that has been working under load should run half-speed with no load for 5 minutes, then idle ½ minute before being shut off. This permits gradual readjustment of temperature between hot spots and cooler sections. It is particularly important for a turbocharger.

In short stops, as for lunch, some manufacturers recommend allowing the engine to idle. Others never want it idled for more than a minute or two. This is because some engines are adapted to idling, others are not. Consult the instruction book or the foreman.

The transmission should be put in neutral whenever the engine is run without the machine working.

Turbocharger.   A turbocharger is a double blower on a single shaft. One side is rotated by exhaust gas, the other crams fresh air through the intake passages into the cylinders. With more air, the cylinder can burn more fuel, and power is increased without adding to engine size.

The blowers may turn at 100,000 or more revolutions per minute, while the diesel is revved up to 2,400 or less. The exhaust side is very hot, the other side is cool. It is lubricated by pressure from the engine oil pump.

When the engine stops, the flow of oil to the turbocharger bearings stops immediately. If it continues spinning at high speed for only a few seconds, the bearings are likely to burn out. The hot side may tend to cook the cool side.

When the engine runs at part speed without load, the turbocharger turns at moderate speed without picking up much heat. Idling slows it down still more, and prepares it for a no-damage stop.

FIGURE 12.145   These tracks will not freeze down.

Just one quick engine shutoff from full speed at full load may put the turbocharger out of action. And it is very expensive to repair or replace.

The charger may also be damaged by a quick speedup of a cold engine, before lubricating oil reaches it. It may spin in the wind, so the intake pipe must be plugged or covered if the machine is transported, to prevent this unlubricated rotation.

Parking.   The machine should be parked on reasonably level ground. Mud spots, or areas subject to flooding, should be avoided.

In freezing weather, tracks must be cleaned, and kept from extensive ground contact by being driven onto planks, logs, or stones. (See Fig. 12.145.)

Most parts supported by hydraulics or by cables, such as buckets, blades, and bowls, should be lowered to the ground, and controls left in the Float or Hold position. Shovel and crane booms, however, are usually left up.

The transmission (or transmissions) should be placed in neutral, and locked if power shift.

Shut off engine by turning the key, pulling the shutoff knob, and/or moving the throttle below its idle position.

Disconnect electrical system, if equipment permits.

Turn seat upside down, or cover it, to keep it dry, if it is exposed to weather.

If there is a cab, put up windows, close doors, and lock it. If the machine may stand in hot sun, leave windows open a crack at the top.

SAFETY

To operate safely, a person should be alert, observant, reasonably cautious, and willing to invest a little extra time and trouble in doing things right.

One elementary precaution is to use handholds, steps, ladders, or whatever helps are provided for getting in and out of the seat, and reaching service and inspection areas. An agile person can skip many of these, but should not. Surfaces are often slippery with water, ice, or oil; and long steps and reaches make sliding more likely. See Fig. 12.146.

Warning devices should be kept in working order. These include intermittent blowing of a horn whenever reverse gear is engaged, flashing "lollipop" lights when working along roads, and bright headlights and rear lights for night operation.

Restricted vision is a problem that increases with machine size. Before resuming work after parking, an operator must walk around the machine to make sure no one is eating or dozing in its shade. When moving, keep aware of blind spots.

FIGURE 12.146   Use ladders and handholds.

Rollover protection structures are important for safety, and cabs may add greatly to comfort. But both of them restrict vision even under good circumstances, and call for increased alertness. A cab with misted windows is very dangerous, and they must be wiped off as many times as needed, before and during operation. All glass must be kept clean.

Attachments should be on the ground when you leave the machine. An opened connection or an accidentally tripped valve may otherwise lead to disaster. However, it is customary to leave crane booms up when they are cable-supported.

After shutting off the engine, move all operating levers back and forth, to make sure you have not left anything up that ought to be down.

These are only samples of various safety practices that must be kept in mind by the careful operator.

CHAPTER 13
REVOLVING SHOVELS AND EXCAVATORS

BACKGROUND

Revolving shovels were the first important power excavators. Part-swing, steam-powered dipper models mounted on railroad cars or barges were in use over 150 years ago.

Other models—clamshell, dragline, and backhoe—were gradually developed. In the first 20 years of the 1900s, steam power was largely replaced by internal-combustion engines and electric motors.

Also, full-swing replaced part-swing, and crawler self-propelled mountings became standard. More recently, rubber-tire carriers have become common.

Until around 1950 practically all these machines manipulated their buckets by means of cables (wire ropes), winding onto mechanically or electrically driven drums. Swing and travel depended on gear arrangements. Controls were mechanical, air, hand or foot pressure hydraulic, or combinations of these systems.

By 1975, production of cable and mechanical drives in backhoes had been largely replaced by hydraulics in all but large mining excavators. Hydraulic-operated cranes and booms are standard among smaller machines, and the hydraulic clamshell has gained acceptance. Also, the hydraulic excavator has replaced the dipper shovel in most kinds of work.

As a result, the importance of the cable-operated excavators has declined greatly, particularly among small machines. However, they are dominant as draglines, clamshells, and big cranes and dipper models.

BASIC SHOVEL

Structure. A shovel has three structural divisions. Anatomically, the top or revolving unit is the head and torso, the mounting or travel unit is the legs, and the various attachments are the arms and hands. A revolving unit and a travel unit together make up a "basic shovel."

There are five attachments (also called rigs or fronts) which have primary importance. These are known best under the names front shovel (or dipper stick), backhoe, dragline, clamshell, and crane. They are shown in their age-old forms in Fig. 13.1.

A basic shovel has three sets of machinery. One of them is made up of deck-mounted drums which are fitted with spools for cable or hydraulic pumps and hoses to rotate and stop the deck by clutches and brakes. These also control the in-and-out and up-and-down movement of the bucket for digging and dumping.

The second set rotates (swings) the deck, upper machinery, and front end around a center pin. The upper unit is supported by rollers or balls on a circular track or turntable, and is rotated by a vertical pinion gear that meshes with bull gear teeth in the turntable. It "walks" around the bull gear, swinging the shovel as it does so. The gear is controlled by a reversing clutch (a unit consisting of two friction clutches and a set of bevel gears), which can turn it in either direction.

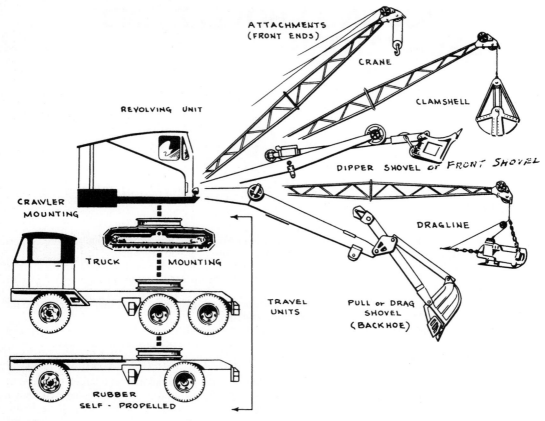

FIGURE 13.1   Shovel rigs and mountings.

This mechanism is an important factor in the efficiency and adaptability of the shovel. It enables it to face in any direction for digging and dumping, and to move loads quickly anywhere within its reach. Comparatively few wearing parts are involved, none of them work in the dirt, and friction losses are slight.

The third power train provides means to walk or propel the shovel. A vertical shaft extends downward through the deck and the hollow center pin to drive a horizontal axle, which has a clutch and a brake on each side. These may be of either jaw (tooth) or friction construction. A pair of sprockets on the ends of the outer axles drive the tracks through roller chains.

The propel mechanism is controlled by a reversing friction clutch set, usually the same one that controls the swing. A few machines have an independent travel shaft and clutches. Truck mountings depend entirely on the truck for travel.

When one set of friction clutches controls two or more functions, jaw clutches are shifted to connect the power train that is to be used, and to disconnect the other.

Small shovels use gasoline, diesel, or electric engines, with mechanical drive to all moving parts. Large shovels may use diesels, with mechanical or electric drive; or several electric motors supplied from "high lines" through a cable.

The engine is fitted with a disc clutch, so that it can be cut off from the machinery. It may also have a torque converter, that serves to cushion it against shocks, to reduce lugging down, and to give additional power in hard digging.

There may be a two-speed engine transmission, chiefly to allow faster travel.

Size.   There have been light shovels with $\frac{1}{4}$-yard buckets, and $\frac{3}{4}$-yard is a very useful and popular size weighing around 20 tons. The largest models, used in strip mining, have dipper buckets up to 140 or more yards, or draglines over 200 yards, and weigh thousands of tons.

REVOLVING UNIT

The revolving unit, or superstructure, is built on and around a heavy steel deck or bed plate. This carries the machinery frame, which may be welded to it.

The deck is supported by the swing rollers which rest on the turntable of the travel unit. It carries the engine, transmission, and operating machinery.

A diesel (or, less commonly, gasoline) engine or an electric motor is mounted across the back. Drive from its clutch, converter, or transmission shaft, with mechanical connections or hydraulic pumps and tubing, run the mechanism. In either case, it has a reduction ratio that makes the deck machinery turn more slowly than the engine.

In Fig. 13.2 the deck machinery for a shovel outfitted for a clamshell or dragline operation is shown with various alternatives. The power unit is either an internal-combustion (IC) engine or an electric motor. The drive chain from the motor is a silent chain, which turns the swing shaft (reversing shaft), which also serves as a transmission shaft, driving the intermediate and drum shafts through a small gear on its opposite end.

In Fig. 13.3 the excavator is controlled by a hydraulic system. There are two joystick-type hand levers and two foot pedals. These control raising or lowering the boom and other movements required in excavating operations.

Propel Shaft.   A layout with hydraulic power is shown in Fig. 13.4. The swing shaft is used only for swinging.

At the lower end of the vertical swing shaft is a pinion (small spur) gear which meshes with the large internal toothed bull gear in the travel unit (see Fig. 13.14) which gear is centered on the vertical propel shaft. The propel shaft terminates in a bevel gear that meshes with a bevel (ring) gear on the live axle, which drives the tracks through chains.

Figure 13.5 shows a deck layout with separate propel and swing.

If a live boom is not required, independent swing allows any rig but a front shovel to walk without any shifting of jaw clutches, thus saving most of the time in move-ups. With any rig, it will allow simultaneous control of walk and swing, which saves time and effort in restricted spaces.

Some large draglines are equipped with a walking shaft that is independent of all other functions. Such machines cannot be readily converted to shovel use.

Large Mine Excavators.   The revolving frame design for a large (20- to 80-cubic-yard bucket sizes) excavator with electric motor power is more involved than the smaller shovels described previously. The deck frame has multiple deep-section vertical members welded to heavy top and bottom cover plates. This forms a rigid mounting deck for the hoist and swing machinery. The machinery deck arrangement may be as shown in Fig. 13.6.

A distinct difference from the smaller shovels is the use of two motors for the hoist and two for the swing motion. These provisions are to give quick reactions to load change capability. The electrical inertia, or resistance to change of motion, of the motor armatures is proportional to the square of their diameters. Therefore, smaller-diameter armatures will make changes considerably faster than one large one. That means two motors with smaller armatures will have a quicker cycle time than if there were just one larger motor for each of these functions.

A-Frame and Gantry.   The deck A-frame consists of a triangular frame of angles or tubing on the deck on each side of the machinery, and a cross bar slightly above the cab roof which carries sheaves for the boom support line. See Fig. 13.7.

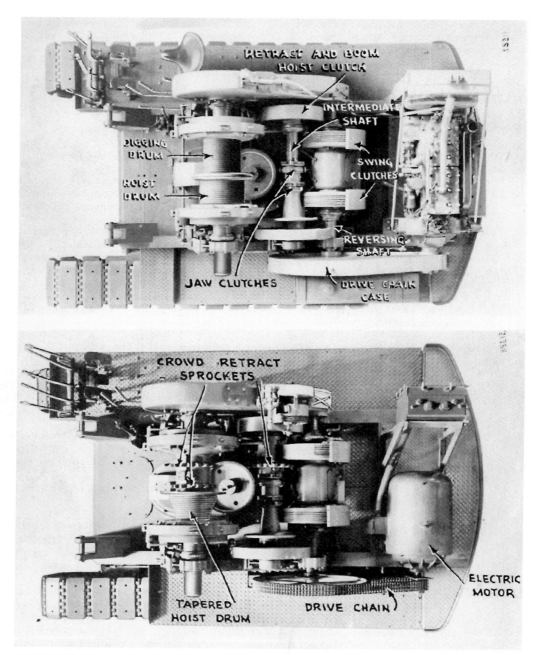

FIGURE 13.2   Deck machinery for cable operation.

FIGURE 13.3   Hydraulic controls in an excavator. (*Courtesy of The Gradall Company.*)

Larger shovels require higher supports for the boom cables. When the A-frame projects high above the roof, as in Fig. 13.8, it is called a gantry. Gantries, and some A-frames, are of folding or collapsible construction so that they can be lowered by hand or power for shorter booms, or to go under obstructions.

Boom Foot.   The boom foot is a pair of hinge brackets on the reinforced front edge of the deck. Booms are attached by horizontal pins running through these brackets so that they can pivot up and down, but are held from side movement.

Cab.   The cab is a shell of sheet metal and glass, designed to protect the machinery and the operator from the weather, but also to give the operator as much visibility as possible to the work area.

Controls.   The operator's seat may be on either side of the deck, forward of the center. Levers and pedals or pushbuttons controlling all functions of the machine are near it. A representative set of controls is shown in Fig. 13.9 for a mechanically driven cable machine.

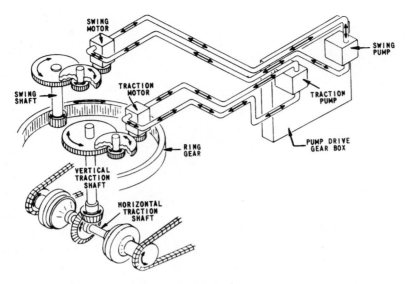

FIGURE 13.4   Hydraulic swing and traction circuits.

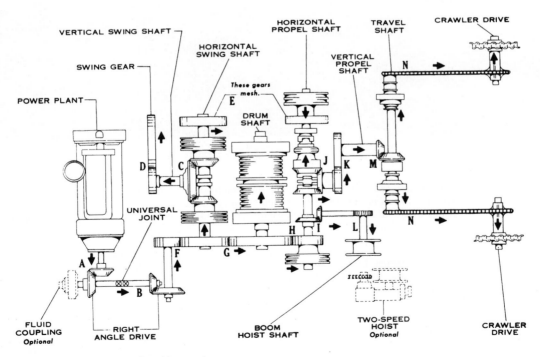

FIGURE 13.5   Deck layout for cable operation.

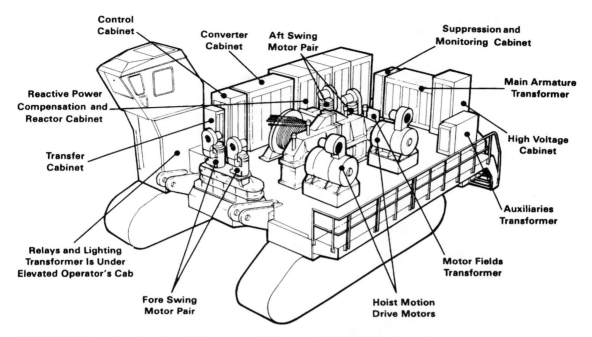

FIGURE 13.6   Large shovel machinery deck arrangement. (*Courtesy of P&H Mining Equipment.*)

FIGURE 13.7   Circle vision cab and A-frame.

   In the cable-controlled machine, the hoist lever engages the hoist clutch when pushed forward. The lever is held back by a light spring, but when pushed forward beyond the point necessary to engage the clutch, it locks over center, and requires a pull to disengage it. Use of this level rotates the hoist or front drum so as to wind in the cable attached to it.
   The digging, drag, or crowd lever, when pushed forward, engages the clutch of the same name, winding in cable that forces the bucket into material being dug. If used with a dipper stick, it can be pulled backward from center also, engaging the retract clutch. These movements are made against spring tension, and both clutches lock over center.

FIGURE 13.8 Gantry boom cable support. (*Courtesy of P&H Mining Equipment.*)

The boom hoist lever, when pulled back, engages the boom clutch, causing the boom cable to wind in and raise the boom. In center position, it releases this clutch and allows an automatic brake to lock the boom drum. When pushed forward, it releases the brake and allows the weight on the cable to turn the boom drum as fast as the drum shaft turns. There is also a ratchet which can be set to prevent the boom cable from paying out at all.

The swing lever, when pushed forward, engages the front swing clutch and causes the shovel to swing left or, in walking position, causes it to move forward. When pulled back, this lever engages the other swing clutch, with reversed results.

The swing lock, when pulled back, engages a toothed lock with the bull gear and prevents the shovel from swinging. The swing-propel shift lever, when forward, engages the jaw clutch for swinging. When back, it connects the walking jaw. In the middle, both are disengaged.

The hoist brake locks or slows the rotation of the hoist drum; the drag brake does the same for the drag drum. These brakes are held in released position by springs, but can be locked down by engaging with a ratchet. Pushing on the top of the pedal locks it, on the bottom releases it.

The throttle regulates the speed of the engine by changing the governor setting. The steering lever is used only when walking. Forward turns to the right, backward to the left, center straight ahead, in forward motion.

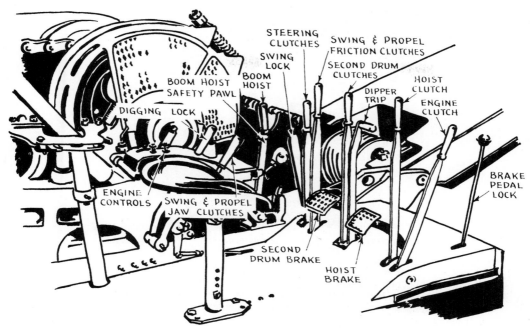

FIGURE 13.9 Operator's mechanical control station.

The engine clutch is used to disconnect the machinery from the engine, for safety when adjusting or lubricating with the engine running, and for starting when cold.

The digging lock will lock the shovel tracks when pushed forward, will release them when pulled back, and in the center will allow the shovel to move forward only.

Controls are different in each make of shovel. Many use vacuum, air, or hydraulic controls, with shorter operating levers. Steering brakes are generally used to hold for digging also, and may be applied by hand or automatically. The direction of throw of operating levers can often be reversed.

The two-lever "joy stick" controls in a hydraulic-controlled shovel are shown in Fig. 13.10.

In a large, electrically controlled mining excavator, the operator may have a programmable message display (PMD) module, as shown in Fig. 13.11. This is true with the P&H Electrotorque system, which automatically converts incoming high-voltage ac power to the controlled dc power that operates the motor armatures. Once the module is programmed, the operator would only have the buttons on the PMD panel to push to control the operations of the shovel.

Swing Rollers. The whole weight of the revolving unit and attachment rests on rollers. These are tapered so that they can follow the circular turntable without any part dragging. Bushings or antifriction bearings are used.

In Figure 13.12 the swing rollers carry the load only, and hook rollers stabilize the machine. Adjustment is made in the hook roller brackets.

Figure 13.13 shows a single row of balls with each race forged in one piece. This is assembled without the balls, that are then fed with their separators through a hole in the inner race. The hole is then plugged. Dirt is kept out of this oversize bearing, and lubricant retained in it, by upper and lower neoprene seals.

Counterweight. The weight of the attachment and its load are balanced by putting the center of rotation in the front center of the deck, so that the engine and most of the deck machinery are behind it.

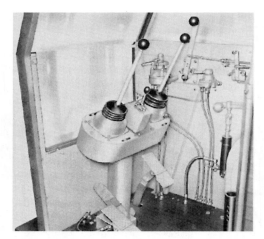

FIGURE 13.10   Hydraulic control station.

In addition, counterweight may be placed in the rear of the revolving unit in pockets behind the engine, or in a band around the rear of the deck, outside of the cab. The exposed counterweight makes an effective bumper and guard for the rear of the machine, but adds to its length, or tail swing.

Counterweight can be added or taken away to suit different applications.

Lubrication.   Most of the deck machinery may be enclosed in oiltight cases, and lubricated by gears dipping oil out of one or more reservoirs and carrying it to distribution points, from which it flows over the other gears and bearings and back to the reservoirs. The oil level should be checked often through inspection plugs.

The turntable gear is greased with asphalt-base lubricant. This may be thinned and pumped through a deck-mounted dispenser, or thinned or heated and poured through a hole in the deck, while revolving the upper frame. The gear should be checked at least once a day, and greased whenever bright spots appear on the teeth.

Pressure gun fittings throughout the shovel are greased in accordance with a schedule supplied with the machine. In general, bushings require more frequent attention than antifriction bearings, some of which may be sealed so as not to require lubrication. The crowd chains for the dipper stick should be kept painted with oil or light grease, but the tracks and drive chains should be left dry.

CRAWLER TRAVEL UNIT

Many shovels may be purchased with three different types of travel unit—crawler (track or cat mounting), truck mounting with two engines, and rubbertire self-propelled, as shown in Fig. 13.1. The crawler is the most widely used and will be described first.

General Construction.   The crawler chassis, shown in Fig. 13.14, is made up of a turntable welded to a frame consisting of two heavy I beams, called dead axles, that connect two heavy truck frames which rest on the track wheels and are surrounded by the track. The track usually consists of flat shoes hinged and pinned together at their ends, as shown in Fig. 13.17.

The travel unit is considered to have the bull wheels in the rear and the idlers in the front, regardless of whether this corresponds to the front and rear position of the revolving unit.

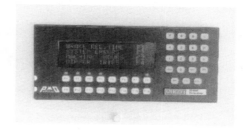

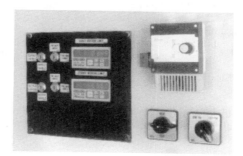

FIGURE 13.11   Electric controls for large mining excavator. (*Courtesy of P&H Mining Equipment.*)

Live Axle.   The three-piece live axle, Fig. 13.15, runs across the upper frame, parallel to the frame axles. Near its center is a bevel gear, which is driven by the vertical propel shaft. On each side of this are steering clutches, with the inner jaws fixed to the center section of the axle.

Steering.   If both axle jaw (steering) clutches are engaged, rotation of the vertical propel shaft will turn both tracks equally. If the right clutch is disengaged, the left track only will drive. The effect on the machine's direction of travel will be determined by the steepness or roughness of the footing, and friction in the right track, Fig. 13.16. On firm level ground the machine will move straight forward, or a little to the right; on steep upgrades or on rough ground, it may turn quite

FIGURE 13.12   Separate swing and hook rollers. (*Courtesy of Harnischfeger Corporation.*)

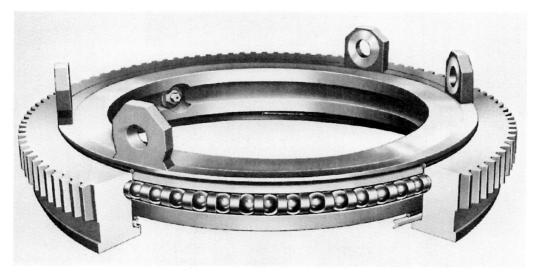

FIGURE 13.13   Ball race turntable.

sharply to the right. If the right axle is locked by the brake, the right track will stop, and the left one will walk in a circle around it, turning the machine very abruptly. Turning in soft material is difficult, as the locked track digs in. On slippery surfaces one or both tracks may skid, preventing accurate steering.

Turning to the left involves engaging the right-hand clutch and disengaging the left one, and perhaps locking the left side if a sharp turn is desired. A gradual turn may be made by use of the clutch alone, by partial application of a friction brake, or by a succession of short, sharp braked turns with straight runs between.

If the machine is moving down a grade steep enough that its engine is holding it back rather than pushing it forward, the steering action of the clutches will be reversed; that is, disengaging the right clutch will cause the machine to turn to the left, as the disconnected track can roll faster than the one connected to and held back by the engine. When the brake is used, steering is about normal regardless of grade.

The machine should be stopped before engaging or disengaging the jaw clutches or brakes. While they are transmitting the engine power to the tracks, they are almost impossible to disengage; and when apart, they might be damaged by being forced into engagement while the jaws are

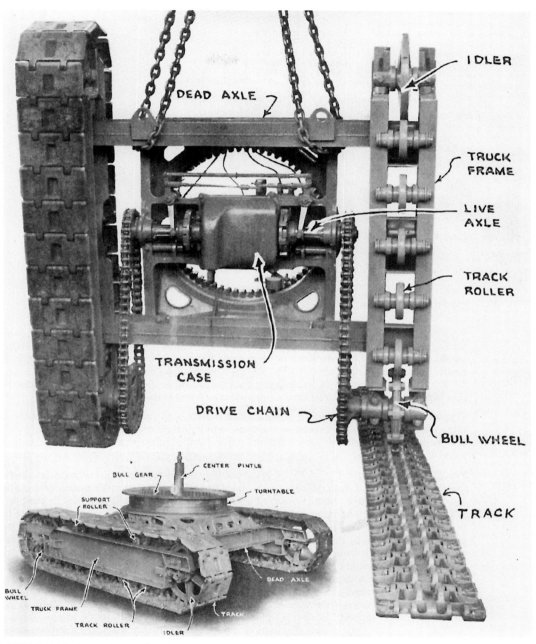

FIGURE 13.14 Lower frame and tracks.

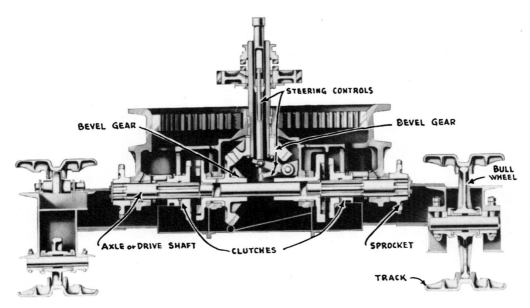

FIGURE 13.15   Propel power train in base.

turning at different speeds. The machine is stopped by partially engaging the friction clutch that would move it in the opposite direction or by using friction brakes. In an emergency a ratchet lock or jaw brakes can be used for stopping, but this is very jarring to the machinery.

The steering method is the same going backward (toward the bull wheels), except that for a turn in the same compass direction the opposite clutch and brake are used.

Tracks.   Figure 13.17 shows part of a track that is used, with variations, on most excavators and other machines that do not depend on the tractive pull of their crawlers for their ability to work. It is made up of a number of identical shoes cut and drilled at their ends so that they can be fastened together by pins.

Track Behavior.   Refer to Fig. 13.18. If the machine moves forward, the bull wheel will turn so that the teeth will mesh with the pads under and ahead of it, picking them up one by one and passing them around its rear and then forward over its top. As the bull wheel moves forward on the track, it pushes the truck frame and the idler ahead. The idler, pressing against the track in front of it, pulls on both top and bottom sections, but since the bottom is pinned down beneath the shovel, and is pulled tight by the bull wheel, the idler pulls in the track from above about as fast as the bull wheel supplies it from behind, and turns it back to the ground, where it stays, supporting the truck wheels and weight of the shovel until picked up by the bull wheel again.

If the machine is walking backward, the mechanism may behave in two ways. If the footing is good, and the grade nearly level, the bull wheel will pull in some slack from the top of the track and pass it underneath, moving backward and pulling the truck frame and idler after it. The track it has turned underneath supports the shovel and is then carried up by friction with the idler.

If the footing or grade is such that there is insufficient traction in the pads directly underneath the bull wheel to move the shovel, they will be skidded forward against the rear truck roller, the bull wheel will pull the upper track tight, and it will pull the idler backward. The idler moves truck frame, shovel, and bull wheel backward.

There are several reasons why a shovel or other tracked vehicle should be walked forward when possible. In forward motion, the track is laid smoothly along the ground, and its slack hangs

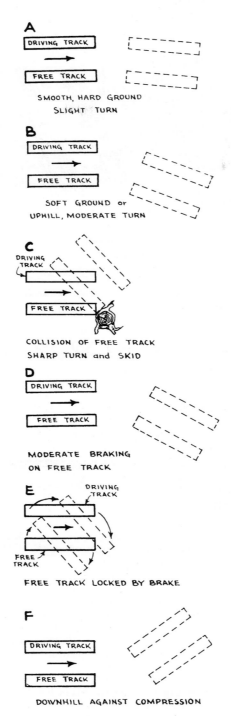

FIGURE 13.16 Steering with clutches and friction brakes.

FIGURE 13.17 Excavator track.

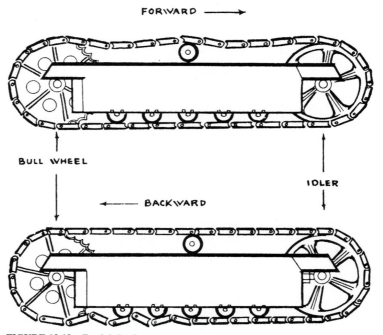

FIGURE 13.18 Track behavior.

harmlessly in gradual curves in the top section. The heavy strain is on the short piece from the bottom of the bull wheel to the first truck wheel supporting a good share of the shovel's weight. In reverse movement, the track may be kinked in one or more damaging angles which the truck rollers must climb over or push down, and the traction stress is on the whole distance from the top of the bull wheel to the front slope of the idler, causing excessive wear and danger of breakage.

Multiple Crawlers. Very large stripping shovels may weigh several thousand tons. They usually travel and work on the surface of coal seams that have limited resistance to crushing and shearing.

They may be supported on four crawler units, one at each corner of the lower frame. Each unit includes a pair of crawler frames and tracks, and an ac drive motor.

Universal mountings, that allow oscillation of individual tracks and swiveling of the whole unit for steering, connect the crawler units to massive hydraulic jacks that support the lower frame, and automatically keep it level on uneven ground.

Travel and steering controls are located on the lower frame.

RUBBER MOUNTINGS

Two-Engine Truck Mounting. The revolving unit may be carried on a turntable fastened to a truck chassis. The truck engine is then used for traveling, and the shovel engine for digging. Ordinarily, the shovel controls for walking, steering, and traction braking are disconnected or missing. Occasionally, however, arrangements may be made to connect the shovel engine to the wheels through the truck transfer case, and to steer and brake the truck from the shovel cab. This mechanism is used for short moves at low speeds in the working area.

Any standard truck chassis of sufficient rated capacity can be used to carry a shovel, although considerable extra bracing is required, and better service should be obtained from a chassis specially engineered for the shovel. Tandem drive gives best support, and all-wheel drive is advisable in work where mud or sand may be encountered.

The truck should have the right-hand side of the cab cut away, to make it possible to carry the boom forward at a low angle. The frame and the rear axle should be a single rigid unit while digging. Springs may be eliminated, locked during work, or replaced by stabilizer bars in the tandem construction.

The truck-mounted shovel can ordinarily swing in a full circle, but with most attachments can work through only 270° because of interference of the cab and the truck front.

Figures 13.19 and 13.20 indicate various power trains used in the trucks.

Outriggers. Outriggers are used to increase stability. These may be beams which can be slid or folded out of the bumpers, usually only at the rear, and which are supported by blocks or jacks. They provide a much larger and more rigid base than the tires. When they are used, lifting capacity is greater than that of a crawler of the same size, particularly when working off the back so that the truck engine acts as a counterweight.

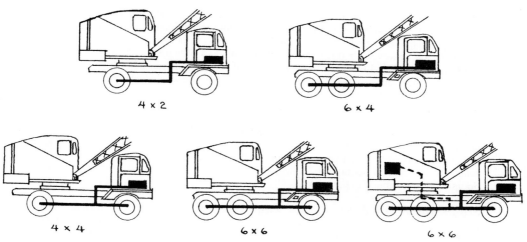

FIGURE 13.19 Power trains in truck mountings.

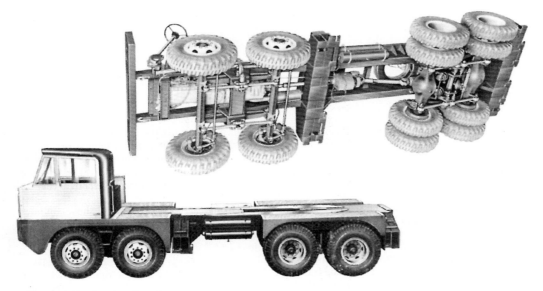

FIGURE 13.20   8 × 4 truck carrier.

Platforms may be put under the blocks to give support when working on soft ground, and the jacks may be used to lift the wheels out of mud or sand pockets.

There are also power outriggers, as in Figs. 13.21A and 13.21B. A curved arm with a hinged ground shoe is moved outward and downward, or inward and upward, by a hydraulic two-way telescoping ram. The set of four can be used to steady, level, and even to lift the machine.

These outriggers are individually controlled from the truck cab. Automatic locking wedges relieve the hydraulic rams of strain during work, and prevent settlement that might be allowed by fluid leakage.

Uses of Truck Mounting.   The advantage of truck mounting is its capacity for rapid and inexpensive movement from one job to another. With many models, the shovel can be placed and locked in traveling position in less than a minute, and then moved along roads at 20 to 35 miles per hour.

This is a pleasant contrast to the slow and laborious job of loading and securing a crawler type on a trailer, and unloading it at its destination.

On the job, however, the truck-shovel suffers from lack of maneuverability. Instead of turning in nearly its own length, like the crawler, it must have considerable space in which to turn or sidestep.

Its most important weakness for excavation work is the ease with which it can get stuck. Even with all-wheel drive, its ability to get in and out of soft spots is inferior to that of a crawler, and with rear drive only, constant care must be exercised to keep it out of trouble in soft ground and during rains.

Two-Piece Boom.   For an excavator on wheels designed for travel on roadways, a two-piece boom is advantageous. See Fig. 13.22. It tucks up much tighter than the one-piece boom for travel between job sites. The two-piece boom uses a hydraulically adjustable base section that is pinned to the second boom section, which is also hydraulically controlled. In turn the second boom section is pinned to the stick.

Self-Propelled.   The rubber-mounted, self-propelled shovel is a third type. It is, in effect, a compromise between the crawler and truck mountings. It can utilize any front-end attachment.

The revolving upper is similar to those found on crawler mountings, except for different controls for steering, brakes, and stabilizers, which in this machine are power hydraulic.

FIGURE 13.21A   Cutaway of hydraulic outrigger.

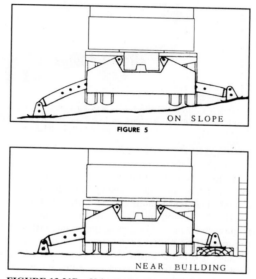

FIGURE 13.21B   Using hydraulic outriggers.

It is manufactured as a short-chassis, two- or four-wheel drive carrier, with dual wheels front and rear. There are no springs. The front axle is connected to the frame by an oscillating mechanism that can be leveled and locked by power hydraulic stabilizers. The rear axle is bolted solidly to the frame.

Maneuverability on the job is subject to the same limitations as the truck, except that the short wheel base, and in some models four-wheel steering, allows it in tighter places. It is permitted to travel on highways, but for long trips it is better carried on a trailer.

HYDRAULIC FULL-REVOLVING EXCAVATORS

Like its cable-operated counterpart, the full-revolving hydraulic excavator may be said to be made up of three structures: the revolving deck unit, the travel base, and the attachment. See Figs. 13.23 and 13.24.

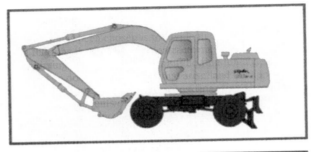

FIGURE 13.22   Comparing booms for mobile excavators. This illustration shows how much tighter the two-piece, hydraulically adjustable boom (bottom) tucks in for driving, compared to the standard one-piece boom (top). *Illustration courtesy of Hyundai.*

In hydraulic machines these distinctions remain clear only while considering the frames, arms, and other mechanical parts. The hydraulic system is operative in both deck and attachment, and often in the base as well.

This discussion will first cover the mechanical arrangements, including the placement of the principal hydraulic units; and will then deal with the generation and distribution of the hydraulic power.

The revolving unit consists essentially of a heavy rectangular steel deck, formed and reinforced to carry engine, pumps, attachment, controls, and cab; and to rest and revolve on a turntable. Some parts are heavy steel plate and specially shaped ribs and bases, others may be open girders.

The center of rotation is usually forward of the center. The area of greatest strain, and therefore of heaviest construction, includes this spot, the mounting hinges for the boom just forward of it, and the hinges for the hoist cylinder(s) on the front edge of the deck.

The forward location of the rotation center places the major part of the deck weight at the rear, where it serves to counterbalance the weight and pull of the shovel or backhoe. This effect is usually increased by placing the engine and pumps across the rear.

Counterweight.   In addition to counterbalancing by arrangement of parts, there is usually a massive counterweight of shaped iron attached to the rear of the deck outside the cab. This may be very heavy, up to almost one-fifth of total excavator weight.

The swing axis is centered in the travel unit, so that the front of the deck is set back from the front of the tracks, and the rear edge overhangs. This overhang is increased by the counterweight.

FIGURE 13.23 Hydraulic excavator (backhoe). (*Courtesy of Case Corporation.*)

There is even greater overhang at the side, when the top is swung 90°. This overhang or tail swing must be allowed for in placing the excavator for work.

It is sometimes desirable to remove counterweight, to suit a lighter attachment, to reduce tail swing, or to lighten the machine to move it across a bridge of doubtful strength or to avoid overloading a trailer.

Most of the larger machines provide standard or optional means to handle counterweight by engine power. There may be a pair of hydraulic cylinders mounted on the back, with connections to a valve and pump.

Engine.   Standard power is diesel, from less than 100 horsepower to over 1,000. While bigger machines have bigger engines, there does not seem to be a standard ratio between power and either machine weight or bucket capacity.

Drives.   One or more hydraulic pumps may be driven directly by a shaft behind the clutch, in line with the engine crankshaft. But it is perhaps more usual for the pumps to be offset from this shaft, in a group of two or more, with drive-through gears. See Fig. 13.25.

Most of these machines have hydraulic drive to all functions. Cylinders operate parts directly, or through simple levers. Motors drive through reduction gears and sometimes through more or less extensive mechanical arrangements also.

Figure 13.4 shows hydraulic-mechanical arrangements for swing and propel in a hydraulic power system.

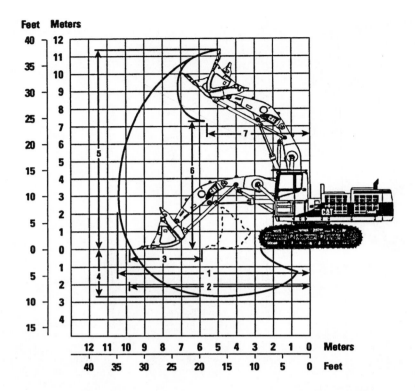

		3.48 m	11 ft 5 in
Stick Length		3.48 m	11 ft 5 in
Boom Length		4.6 m	15 ft 1 in
Bucket Capacity		5.2 m³	6.8 yd³
1	Maximum Reach	10 350 mm	33 ft 11 in
2	Maximum Reach at Ground Level	9890 mm	32 ft 5 in
3	Maximum Level Crowd Distance	3920 mm	12 ft 10 in
4	Maximum Digging Depth	2840 mm	9 ft 4 in
5	Maximum Digging Height	11 270 mm	37 ft 0 in
6	Maximum Dump Height	7440 mm	24 ft 5 in
7	Reach at Maximum Dump Height	5690 mm	18 ft 8 in
Bucket Breakout Force*			
	(SAE)	468 kN	105,210 lb
	(ISO)	538 kN	120,950 lb
Stick Breakout Force*			
	(SAE)	443 kN	99,590 lb
	(ISO)	429 kN	96,440 lb

* Forces shown are for 5.2 m³ (6.8 yd³) rock bucket

FIGURE 13.24   Hydraulic front dipper shovel. (*Reprinted courtesy of the Caterpillar company.*)

Cab. The cab may be almost identical with that for a cable shovel. It is always of the circle vision type, with the roof higher than the machinery covering. Windows have shatterproof glass for weather protection and vandal resistance. Some windows may be fixed, others may be moved by cranks or on slides, or removed after loosening bolts.

There may be a skylight or roof window, which is helpful in watching out for wires and tree branches.

Standard cab location is on the right front corner of the deck, beside the boom. However, it may be on the left side. Controls for all phases of machine operation, plus gauges and warning lights, are grouped here.

FIGURE 13.25   Triple pump set.

HYDRAULIC SYSTEM

Pumps and Circuits. Most of the engine power is usually delivered to a set of hydraulic pumps, which provide pressure flow of fluid to power the functions of the machine.

Three types of pump are used—gear, vane, and axial piston (hydrostatic)—in various combinations. Some machines have only two pumps—one for power and one for cooling—and others have as many as five.

Multiple pumps are desirable to provide a fixed and dependable flow of fluid for the use of each of several functions, so that one cannot be robbed (deprived of flow and pressure) by the opening of another circuit.

On the other hand, a pump is expensive, it consumes some power (and therefore fuel) when it is idling, and may be inefficient when run at part capacity. It is therefore economically desirable to provide it with as much useful work as it can handle.

These opposed needs are met in a variety of ways, including the use of variable-output pumps and proportioning valves, and the design of sequences to feed circuits that are important, and starve others that are less important at the moment.

The most desirable arrangement for both production and simplicity of operation is a separate circuit for each function that takes care of its normal needs, plus means to temporarily borrow fluid from another circuit. (See Fig. 13.26.)

Differences are too numerous and changeable to justify detailed discussion here. But the operator who is changing from one machine to another should be prepared to find differences in response to controls, particularly when two or more are used at the same time.

Layout. A revolving excavator has a rather complex hydraulic system, but fortunately it has enough space in which to arrange it. The engine and pumps are usually across the back; the swing motor and its vertical shaft must be above the big swing gear in the carbody. If the control valves are directly operated, they must be in the cab. (See Fig. 13.27.)

Otherwise, components can be arranged for efficiency, access, and convenience.

Lines. The basic piping on the deck and along the boom and stick is apt to be high-strength tubing, factory-formed into the lengths and curves needed. Where parts move in relation to each other during work, lines connecting them must be made of flexible hose. Hose may be used in other parts of the piping, for ease in installation, or to avoid stresses if parts shift slightly in position.

Stick and bucket lines must allow for very sharp bends. The length needed at sharp angles forms loops as the bends are straightened or reversed.

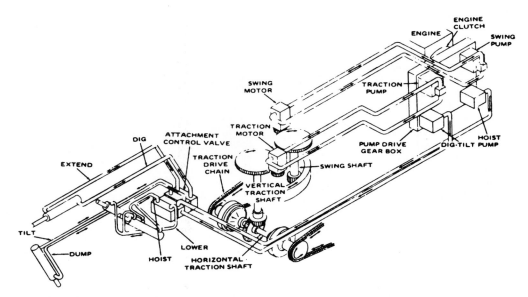

FIGURE 13.26   Hydraulic circuits.

If hydraulic motors or other units are in the travel unit, a swing joint is installed at the center of revolution. This device, described in Chap. 12, provides for flow of hydraulic fluid regardless of the rotation of the upper unit. It requires a flexible or cushion mounting for one or both sleeves, to avoid distortion when digging strains cause the deck to rock slightly on the turntable.

Controls.   A bewildering variety of operator's controls are offered. They vary in arrangement of levers and pedals, and in the means by which they operate the valves. (See Fig. 13.28.)

Levers are usually self-centering, that is, they return to neutral or Hold position automatically when released. A propel stick may have a detent to hold it in full-on position. Brakes may have finger-released latches to hold them in On positions. (See Fig. 13.29.)

Simple pedals are spring-returned to Off or neutral, like an automobile's accelerator or foot brake. Swing and traction brake pedals may have hold-down latches.

Rocker pedals are hinged at the center. Pressing down on the front opens a valve to cause a member to move. Pressing the back moves the same part in the opposite direction. A spring returns it to neutral from either position.

Cylinders are either naturally self-locking when flow of fluid is blocked, or can be made so by an automatic check valve. Hydraulic motors may or may not have a braking effect when shut off, but they cannot be depended upon to hold parts against creep. Positive braking may be supplied by a friction brake. This may be operated by a pedal. If automatic, it is engaged by spring pressure, and released by hydraulic pressure when the motor is used.

The control lever, pedal, or button may be connected mechanically to the valve that it operates, by cable or rod-and-clevis linkage. More often, a fluid system is used. In small machines, this may be static like a hydraulic brake system, with or without power boosters. Larger machines may use power from a constantly running hydraulic pump, or air, or combinations of static hydraulic with air or pumped hydraulic. Different methods may be used in one machine.

Direct connection is simple and prompt. If valves are excellently balanced and leverage is adequate, this is a highly satisfactory arrangement.

Power controls should make the fineness of balance of the operating valves less important, but their own smoothness may be questionable. However, a power system usually means less operator effort, and less lever or pedal motion for a given effect.

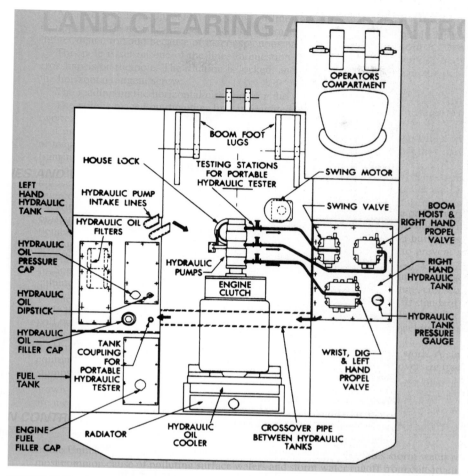

FIGURE 13.27 A typical deck layout.

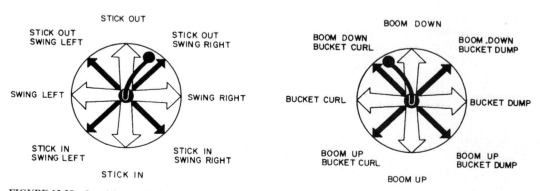

FIGURE 13.28 Joystick controls, hydraulic excavator.

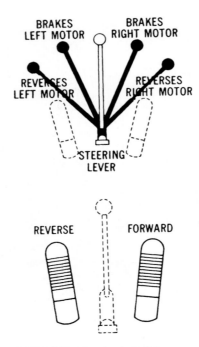

FIGURE 13.29 Travel and steering.

Power Management Systems. Most of the medium-to-large excavators now have a power management system so that the operator can control the power his or her machine applies for the work. The operator selects the mode that matches the kind of work to be done.

There may be three to five modes in the system to produce the most efficient hydraulic power distribution and fuel economy for the operation. One may be the heavy-duty mode for speed and optimum productivity, another mode may be standard for general excavation and fuel economy, still another mode may be for light work, where control and fuel economy may be most important, and finally, a mode for extra fine work: lifting, sloping, and grading.

For example, the light or extra fine mode might be set for deep digging using laser instruments to automatically dig to a preset depth and slope. Sensors set between the bucket and stick, between stick and boom, and between the boom and machine measure the angles. By monitoring these angles the laser controls send commands many times per second to electrohydraulic valves installed on the machine. These valves meter the precise hydraulic oil requirements to the cylinders. The operator follows the guidance of the laser controls unless he or she disengages the system.

The use of the different modes are controlled by electronics onboard set by the operator. The electronics can also set the environment in the cab for the operator's comfort while doing the work at hand.

TRAVEL UNIT

Most full-revolving hydraulic excavators have crawler mountings, as seen in Fig. 13.23. However, they can be adapted to truck carriers and self-propelled wheel mountings.

Carbody.   The carbody is a massive frame that includes the turntable and the dead axles or cross members that transmit its weight to the track frames. It carries the large ring gear that engages with the swing pinion extending downward from the deck frame.

Tracks.   Track frames are single or double beams welded to the outer ends of the dead axles in the carbody. Either of two types of track may be used. One is the traditional linked-shoe construction. The other follows the crawler tractor design of a roller chain with bolted-on shoes. Shoes are usually of the semigrouser type with three low cleats.

Drive wheels, idlers, and rollers conform to the type of track.

Propel.   The propel, traction, or travel drive to the tracks may come from a pair of live axles set across the center of the carbody, or from a pair of reversible hydraulic motors fastened to the track frames.

Axles are usually driven by a hydraulic motor on the deck, through reduction gears, a vertical shaft, and bevel gears. However, a hydraulic motor may be located in the axle housing, and turn them through a reduction-type transmission.

In the axle mechanism, a pair of brakes and jaw clutches provide for steering while traveling, and for holding the machine while working. Control is usually hydraulic. Friction brakes may be used instead of jaws.

Hydraulic motors, usually of the hydrostatic (axial piston) type, can be mounted in any convenient location on the inner surface of the track frames, and connected to the bull wheel axle by sets of reduction gears.

Direct-mounted hydraulic motors permit independent track movement on two sides. This makes possible counterrotation on the tracks for the spin turns discussed in Chap. 12. This feature makes it easy to turn in the length of the machine with little ground disturbance, and to maneuver accurately in very restricted spaces.

BACKHOE ATTACHMENT

Most hydraulic excavators are designed for hoe use only. The hoe might therefore be considered as an integral part of the machine.

However, it can be taken off and put back on, there are often options in stick length, and it may carry working tools other than hoe buckets.

Present capacity of buckets ranges from less than ½ yard to over 20 yards.

Construction.   All hydraulic hoe attachments are made of three strong structural members: the boom, the stick, and the bucket. These are hinged to each other, and the boom is hinged to the excavator deck. Movement at each of these hinges is controlled by two-way hydraulic cylinders.

Boom.   The boom is almost always of the bent or gooseneck type, concave toward the ground. It usually has one bend or angle, but may have two.

This shape serves three purposes. It allows space to pull the bucket closer to the machine, permits deeper digging without interference from the tracks, and enables the operator to see past it more easily when it is raised.

The boom foot is hinged to massive trunions 2 or more feet back from the deck edge. They are usually in front of the swing center, but may be behind it.

If there are two points of attachment to the boom, the upper one is used for maximum digging depth, the lower for maximum dump height.

The outer end of the boom is usually prolonged into a two-piece bracket, in which the stick is held by a heavy hinge pin or pins. The stick cylinder is mounted on the boom top.

Stick.   The stick, dipper stick, or arm is hinged to the end of the boom, and is connected to the stick cylinder rod at its upper (back) end, and to the bucket and bucket dump arms at the bottom or front. It is usually one-piece, but may extend and retract by telescoping.

The stick's connection to the boom is much nearer the top than the bottom of the stick. The proportion between the two sections varies widely in different makes and models.

The bucket teeth will be moved by the stick cylinder 4 to 8 times faster and farther than its piston moves, with one-fourth to one-eighth the force.

Some machines provide two places for the boom-to-stick hinge. The one that is closer to the bucket will supply more power for hard digging; the other will provide more speed in easier work.

The motions of the stick are variously described. Extending the cylinder forces the bucket in toward the machine, crowding it into the digging.

Bucket Mounting. The bucket, which will be described below, is connected to the lower end of the stick by a hinge pin, and to a triangular set of paired dump arms because the bucket has such an extended arc of rotary movement around the stick hinge.

When the cylinder is extended, the bucket teeth move inward in a curling or digging motion. When it is retracted, the bucket opens or extends.

Several sets of holes may be provided, so that bucket action can be changed by moving hinge pins. The choice is between a combination of greater speed and range of movement in bucket control, or slower motion and greater digging and breakout force. Selection depends on the work being done and the operator's preference, and is likely to be changed only under unusual conditions.

Buckets. The bucket is sometimes called a dipper or a tool. The primary use of a backhoe is digging ditches, although it is also well adapted to digging basements as discussed in Chap. 4, and general excavation. See Fig. 13.30. For efficiency in ditches, the bucket should cut the full required width on every pass. Therefore, buckets are usually supplied in a number of widths, ranging generally from 30 to 48 inches but available for some models down to 24 inches and up to 5 feet.

Narrow buckets tend to be deep in proportion to width, and may fill poorly in chunky or rocky digging. If width is the same, reducing depth from the front edge reduces capacity, but may increase

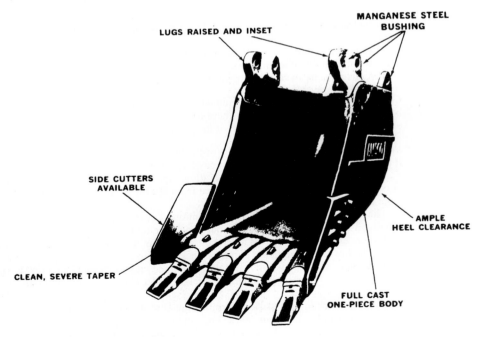

FIGURE 13.30 Backhoe bucket details.

efficiency in loading enough to compensate. A standard-width bucket intended for very hard digging might be made smaller (shallower) so that it could be reinforced without too great weight.

The bucket is usually slightly wider at the open or front end, to reduce friction at the sides and to allow for easier dumping. Additional clearance from trench walls may be obtained, and bucket cutting width increased by 2 to 8 or more inches, by installing sidecutters. They may be fixed-width or adjustable, smooth-edged or toothed.

Sidecutters are useful in accommodating a bucket to a wider trench, cramming more dirt into a narrow bucket, reducing drag in sticky soil, and reducing wear on the front edges.

Wide buckets may have poor penetration. General-purpose buckets for basements and pits are usually intermediate in width and capacity.

A bucket may be replaced by a single ripper tooth. This is intended for loosening and breaking, rather than excavating, but in certain soil types it may be able to cut a very narrow ditch for cable installation.

The digging edge is almost always equipped with teeth, which are removable for reversing, sharpening, or replacement.

Other Attachments.   Excavators are made to be very versatile machines with the variety of attachments that can be used on them. Many of the attachments are for excavation type operations. These include: the hoeclamp for handling stumps and boulders (Fig. 13.31A), and the hydraulic breaker (Fig. 13.31B) for breaking up rocks or other hard objects. Some other commonly used attachments to show the versatility of the hydraulic excavator for excavation operations are shown in Fig. 13.31C.

GRADALL

Figures 13.32A and 13.32B show the Gradall, a special type of full hydraulic pull shovel. It resembles a conventional shovel only in the construction of the turntable and swing rollers, and in having the engine mounted on the rear of the deck.

The engine drives a three-unit tandem pump that supplies pressure through control valves actuated by hand levers and foot pedals, to hydraulic cylinders and motors which power all working functions.

The boom is a hollow box girder, one section of which telescopes inside the other in response to the push and pull of a long two-way ram. It is mounted in a heavy steel cradle that serves for support, control, and counterweight. Cradle and boom are raised and lowered by two double-acting cylinders based on the platform. The boom can be lowered to a vertical position, for 25-foot digging depth, and raised to dump at over 18 feet. It can also be tilted (rotated on its long axis) in either direction from the horizontal by a hydraulic motor.

Swing is controlled by hydraulic motors driving reduction gears (See Fig. 13.3). These control such movements as extending or retracting the boom, and rotating the boom and bucket to do digging, as shown in Fig. 13.32B, and reaching the bucket out to dig and pulling it in to get a full load. With a crawler-mounted excavator, there are controls for moving the tracks individually, together, or in opposite directions for pivoting at a specific spot.

All hydraulic valves are spring-loaded so that they return to a neutral Hold position as soon as they are released. Affected parts stay locked in position until the controls are moved again.

Operation.   The normal digging procedure is to swing the machine and extend the boom until the bucket is over the spot where digging is to start, lower the bucket in digging position, pull it inward with the boom retract until it fills, rotate it to closed position, then lift the boom and swing it while extending or retracting as necessary to position the bucket over the dump spot. The bucket is then opened to dump, and the boom swung back to the digging.

The first few passes may be made by starting with the teeth at an angle of about 45° with the ground, and flattening it as it is pulled in. If the slice is too thin, a steeper angle may be used; if too thick, a flatter one. It is good operation to fill the bucket in the shortest pull that does not cause the engine to lug down or the carrier to shift.

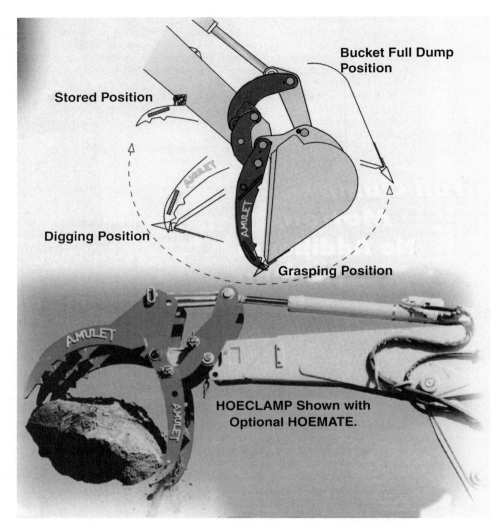

Bucket Full Dump Position

Stored Position

Digging Position

Grasping Position

HOECLAMP Shown with Optional HOEMATE.

FIGURE 13.31A Hoeclamp or power grip for use on excavators. (*Courtesy of Amulet Manufacturing Co., Inc.*)

The bucket should be closed when full and lifted out of the ground. Swing must not be started until it has lifted clear. If spoil is being dumped right alongside the ditch, it may not be necessary to close the bucket fully, and swing will be very short. If dumping is into a truck, it should be as near the ditch as possible, and swing may not be started until the bucket is high enough to clear the truck.

The ditch is deepened and cut toward the carrier by a series of similar passes. The far end can be trimmed to a vertical slope for a short distance down by opening the bucket wider; but if the ditch is deep, there will be a curve at the bottom that cannot be reached from this position. However, the carrier can be moved backward far enough to cut a square end (if it is needed), as there is ample clearance between the nearest part of the ditch and the carrier.

The bottom of the ditch is finished by making the final pass with the bucket level, using the hydraulic down pressure of the boom. This cleans up soil spill, evens out irregularities and tooth marks, and leaves an excellent surface for pipe or granular material.

FIGURE 13.31B   Hydraulic breakers used with excavators. Smaller hydraulic hammers are available for carriers such as backhoe-loaders, while larger units fit excavators in excess of 100,000 pounds.

If a boulder or heavy root is encountered, it may not be practical to remove it by direct pull of the boom retract. Much greater force can be applied by putting the bucket in a level or down-tilted position, getting the teeth under the obstacle, and closing the bucket. The floor will act as a fulcrum to aid the teeth in prying the object up, providing a powerful breakout force without dragging or tipping the machine.

When a boulder is too large to be pried out in this manner, it is necessary to widen the ditch until it is freed. A ramp is dug from it to the ground surface, the carrier is moved ahead a short distance, and the retract used to drag or roll the stone to the surface, where it can be rolled to the side.

Heavy lifts and any side pushing or pulling should be done with the boom retracted to obtain additional force and reduce strain.

Grading.   These machines are particularly well adapted to finish grading. The retractable boom makes it possible to draw in a bucket or grading blade in a smooth line that is not disturbed by hinge action of a boom or the effect of bumps on wheels or tracks. Its reach ranges from 30 feet from center pin in the small machine with a standard boom to almost 45 feet in the large model with a boom extension for mine or tunnel scaling work.

When doing light grading up or down a slope, the bucket or blade is kept in light contact with the ground as it is drawn in by the telescoping boom. When the ground is below the level of the boom pivot, retracting tends to pull the bucket up, so that for a level grade the boom-lowering valve must be kept partly open to compensate. If the ground slopes steeply upward, the boom may have to be lowered as the bucket moves in, to keep a straight-line contact.

The bucket or blade may be tilted to follow an uneven contour, or to work an even contour from a side angle.

Hydraulic control permits keeping the bucket at its most efficient digging angle at any point in its reach, and dumping just where desired. With the boom tilt it gives precise control, and allows exact cutting of floors and sloped or vertical walls in cellars and ditches. It is adept at

Rippers

Function: Attaches to or replaces bucket for lighter ripping applications. *Used with:* Backhoe-loader, hydraulic excavator, skid-steer loader.

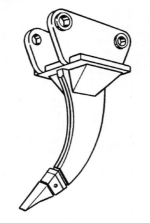

Compactor, Wheel

Function: Compacts material in trench. Clamps to bucket or replaces bucket, depending on model. May be equipped with hydraulic power to lock and unlock. *Used with:* Backhoe, excavator.

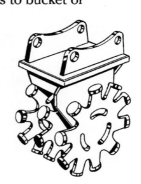

Brush cutter

Function: Cuts down heavy brush and small trees using rotary cutters with cutting teeth. Can be angled to cut in culverts or hillsides. *Used with:* Backhoe-loader, excavator.

Compactors, Vibratory Plate

Function: Compacts dirt, sand or gravel. Reaches into confined areas or below grade. Sometimes used to drive sheeting, piles or posts. *Used with:* Backhoe-loader, excavator.

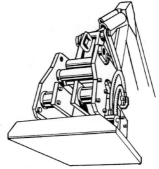

FIGURE 13.31C   Other attachments for hydraulic excavators. (*Courtesy of Construction Equipment.*)

shaping and fine-grading roadsides under even difficult conditions, and can stay safely on the road as it works.

Pull-type buckets are available in widths from 15 to 72 inches, and the digging action of push buckets resembles that of a front-end loader bucket with remote control. There is an 8-foot scraper blade for grading where dirt does not have to be picked up, and a ripper for breaking pavement or loosening hard dirt. (See Fig. 13.33.) Boom extensions and offset booms can be obtained.

FIGURE 13.32A   Multipurpose hydraulic excavator. (*Courtesy of The Gradall Company.*)

FIGURE 13.32B   Side sloping. (*Courtesy of The Gradall Company.*)

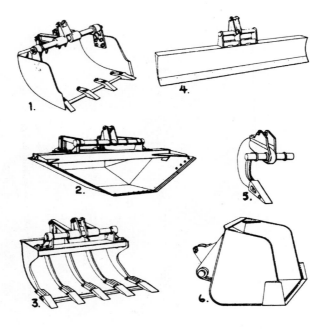

1. 36" Bucket
2. 72" Formed ditch digging bucket
3. 47" Pavement removal bucket
4. 8'0 Scraper blade
5. Single tooth ripper
6. 36" Push-type rehandling bucket

FIGURE 13.33 Buckets and attachments. (*Courtesy of The Gradall Company.*)

This machine can dig deeply in a small area, so that it can connect ditch sections from the side (see Fig. 5.3), and dig for septic tanks and manholes with little or no overcutting. It can push as well as pull and has the advantage of being able to reach under overhead structures.

COMPACT OR MINI-EXCAVATORS

The compact excavators with zero tail swing configuration first appeared in Europe and Japan but now are widely used in the United States. Basically, the upper body of the machine rotates within the footprint of the tracks, except for the boom. Refer to Fig. 13.34A.

The overall compact design of the mini-excavator allows it to move into places where larger machines can not go. The width of tracks of the smallest machine can be retracted to a 29 inch width so it can go through a 36-inch wide gate for work in a backyard. It can operate in narrow easements or even inside building basements.

The boom of a mini-excavator rotates independently of the upper body so that it can work close to a wall or foundation. See Fig. 13.34B. And an existing machine can have the frame tilted by up to 15 degrees, as seem in Fig. 13.34C, in case the supporting ground is not level.

FIGURE 13.34A   Mini-excavator with dozer blade for stability.

The tracks of a mini-excavator can be expanded to give it more stability. A dozer blade on front can be lowered to the ground improving the stability of the machine for digging. See Fig. 13.34A. Rubber tracks help minimize damage to landscaped areas and walkways.

The compact excavator is able to travel on muddy, wet ground better than a heavier backhoe-loader because it delivers a lower pounds-per-square-inch force on the ground.

The mini-excavator is small enough that it does not need a heavy trailer or large truck to move it long distances. It can be loaded on a trailer with other pieces of equipment.

It is said that a compact excavator is easier to operate than a backhoe-loader, which is discussed in Chap. 16, because it has simple and logical controls for quick learning. Manufacturers have added powerful engines and load sensing hydraulics. There are standard auxiliary hydraulic lines that allow the use of a variety of hydraulically driven attachments like those discussed earlier.

Mini-excavators are made in sizes weighing from 1,500 to 12,000 pounds and can dig from 5 feet to over 12 feet deep, just short of the common 14-foot depth for a standard sized hydraulic excavator.

DRAGLINE

A dragline attachment is shown in Fig. 13.35. It has a long light crane boom, with a fairlead set at its foot, and a bucket attached to the machine only by cables.

Boom.   The boom is of lattice construction. It may be of welded steel angles, angle corners and tubular braces, or all tubular. Very long booms sometimes use aluminum sections near the tip.

FIGURE 13.34B   Miniexcavator boom swing separate from cab.

Each boom is made up of at least two sections, tapering from their bolted center connection toward the end. The bottom is reinforced to hinge on the boom foot pins, and the top to hold the point sheaves.

Additional sections, usually in lengths of 5 or 10 feet, can be placed between the upper and lower sections to obtain extra reach or dumping height.

If the boom is intended for dragline work only, it carries one large sheave on the point; but if it is to be used for a clamshell also, it has two. A smaller sheave is carried on each side for the boom support line.

Fairlead.   The fairlead is a device mounted on the boom or boom foot, which lines up the drag cable to spool smoothly onto the drum, even when the bucket is far off to the side.

A common type, Fig. 13.36, has the front pulleys mounted in a frame on a vertical hinge, and a pair of vertical sheaves in a frame behind them.

Bucket.   Figure 13.37 is a picture of an Erie general-purpose bucket, which is somewhat similar to a pull shovel bucket without stick fastenings and side cutters. A pair of drag chains is attached to the front of the bucket, through brackets by which the pull point may be moved up or down. The upper position is used for deep or hard digging as it pulls the teeth into a steeper angle.

The drag chains converge in a drag yoke to which the drag cable is fastened. The hoist (bail) chains are attached to pivot (trunion) pins toward the rear of the bucket sides, rise vertically to a spreader bar, then converge to fasten to the dump sheave housing, which in turn is fastened to the end of the hoist cable.

Dragline buckets are made in various weights: light ones for digging soft earth, and rehandling stockpiles of material; medium-weight for general work, and heavy and extra heavy for deep and rocky digging. A light bucket means less weight to be lifted each cycle; a heavy one has better penetration and wear resistance. Light buckets may sometimes be obtained with a toothless cutting edge which is excellent for stripping soft topsoil, grading, and cleaning up.

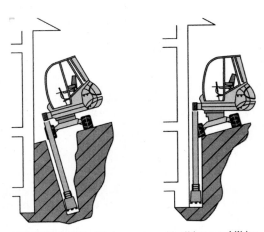

FIGURE 13.34C   Miniexcavator with tilting capabilitity.
Frame-tilt allows vertical sidewall digging.

FIGURE 13.35 A large dragline.

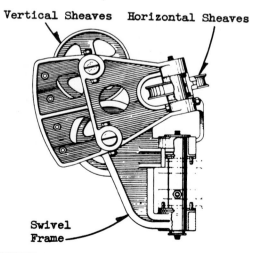

FIGURE 13.36 Four-sheave fairlead.

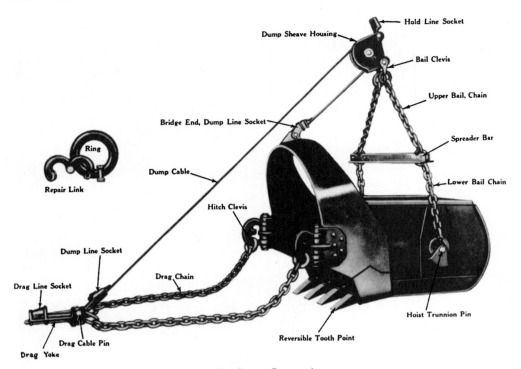

FIGURE 13.37 Dragline bucket. (*Courtesy of Erie Strayer Company.*)

Perforated or sieve buckets are standard buckets with a number of holes cut in the back and sides. These are useful in wet digging as water is pushed through the holes by incoming dirt and any remainder drains out while the bucket is being lifted. Water can be almost entirely manipulated out of a standard bucket if it is possible to take a deep bite so that a massive chunk of earth will push it out, particularly if the bucket can be pulled up a steep bank to dump any remaining water. However, if it is not possible to get a good bite, the perforations are necessary to avoid profitless carrying of water and sloppy spoil piles or loads. Very thin mud or fine dry soil may be lost through the holes, but most digging, wet or dry, can be handled.

Some operators weld $\frac{3}{8}$- or $\frac{1}{2}$-inch chain in the rear corners of solid or perforated buckets, as the slapping of the loose ends helps to dump sticky soil and to clean out thin layers remaining on the bucket sides and bottom and in corners after dumping.

The effectiveness of penetration of a dragline bucket decreases with depth below the machine, as the drag cable then pulls in a more upward direction, raising the teeth out of the soil. This can be compensated for in part by reversing or sharpening the teeth; by using a longer boom which, by permitting digging farther from the shovel, decreases the upward angle of the drag cable; or by fastening the drag cable higher on the bucket. Larger and heavier buckets dig much better at the same depth and distance.

Choice of bucket size is determined by the materials to be handled and the length of the boom. For example, a $\frac{3}{4}$-yard machine usually has a 40-foot boom and uses a $\frac{3}{4}$-yard general-purpose bucket. However, if the material to be dug is very heavy or tends to come up in amounts greater than the bucket capacity, if a longer boom is used without extra counterweight, or if digging is so hard or abrasive that a heavy-duty bucket is needed, then $\frac{5}{8}$-yard capacity should be more satisfactory. The same machine might use a $\frac{7}{8}$- or 1-yard light bucket on a standard boom in handling coal or dry humus.

Reeving. The dump cable runs from the top of the bucket arch over the dump sheave and forward to the drag yoke.

The boom line is a standard four-part rigging, similar to the dipper boom support, except that a longer cable is needed. The hoist line runs from the hoist drum over a large boom point sheave and down to the dump sheave case. The drag cable runs from the drag (digging) drum through the fairlead to the drag yoke. See Fig. 13.38.

Some machines have a light multiple line from the boom hoist drum to a hanging padlock sheave set, and a heavier two-part line from there to the boom point.

This shortens the inner cable, which is most subject to wear. One inner line can be used with different boom lengths.

Bucket Action. If the bucket is lifted with the hoist while the drag cable is slack, it will hang in fully dumped position. If tension is then put on the drag cable, it will pull on the dump cable before the slack is out of the drag chains. The dump cable will pull the front of the bucket up, toward the dump sheave, as seen in Fig. 13.38. Releasing the drag cable will allow the dump cable to run back over the sheave, and the bucket will return to dump position, pivoting on the hoist chain pins.

If the bucket is then lowered to the ground, it will turn to a horizontal position, or will rest on its teeth and arch, depending on its balance. A pull on the drag cable will now tip the bucket forward or backward onto its teeth, and the teeth and lip will dig in as it is dragged toward the shovel. If the

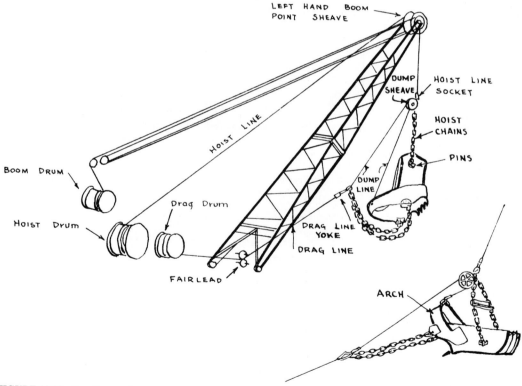

FIGURE 13.38 Dragline reeving.

pull is continued with the hoist line slack, the bucket will cut to a depth determined by its weight, the angle and sharpness of its teeth, and the resistance of the soil. If so deep a cut is not wanted, some tension is put on the hoist line, raising the bucket slightly. If the dump cable is long, the bucket will be raised in the rear by the hoist chains. A short dump line will cause an upward pull on the arch, raising the front of the bucket as much as or more than the rear. In either case the depth of cut is reduced. A further pull on the hoist will raise the bucket clear of the ground, as seen in Fig. 13.39.

Whether the bucket will remain in the carrying position, or partially dump while being raised, depends on opposing forces acting through the dump cable. The weight of the bucket front pulls down on the arch end of the cable, and the tension between the hoist and drag cables tends to stretch its drag-yoke-to-dump-sheave section and pull the bucket up. In effect, the dump cable must pinch the other two cables together in order to obtain slack to drop the front of the bucket.

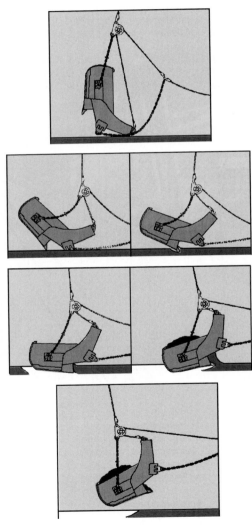

FIGURE 13.39   Bucket action.

This pinching requires comparatively little force when the angle between the cables is small and becomes increasingly difficult as the angle flattens out. Also, if the dump cable is long, it will not have to pull as strongly on the two cables to obtain slack as if it were short. The action of the bucket will therefore depend on the angle between the hoist and drag cables, the length of the dump cable, and the weight and distribution of the bucket load.

A wide angle between hoist and drag cables can be had when picking the bucket out of the soil, either by pulling the bucket close to the shovel, or by keeping the boom at a low angle. Bringing the bucket all the way in usually wastes time and causes wear on bucket and chains which might be avoided if it were picked up as soon as full. A low boom has a tendency to tip the shovel when heavily loaded, and often cannot be used because of obstructions or height of dump. A short dump cable makes it difficult to dump except directly under the boom point. If a live boom is used, it is possible to dig low and dump high, but this takes extra time and work.

The technique used will depend on the job and on the operator's preference. There is generally at least one good method of handling any situation.

Digging.   In an ordinary dragline digging cycle, the bucket is not thrown or cast. It is lowered into the pit with both lines taut, the hoist brake being almost wholly released, then reapplied smoothly as the bucket is about to strike the ground, and the drag brake is released enough to allow the bucket to drop straight instead of following an arc centering on the fairlead.

When the bucket rests on the ground, the hoist cable is slackened slightly and the drag clutch engaged. The drag cable pulls the bucket, with the teeth digging in and cutting a slice of dirt which piles inside the bucket. If the hoist brake is locked, the bucket will move up in an arc centering on the boom point, and on level ground may pivot so the teeth dig more sharply but no longer have the full weight of the bucket to force them in.

Ordinarily, the hoist brake is released enough to let the bucket cut level or follow the pit contour. If the pit slopes up toward the shovel, which is the most favorable digging condition, it may be necessary to partially engage the hoist clutch to avoid digging in, or to prevent the hoist line from becoming too slack and allowing the chains and dump sheave to slump into the bucket.

Hoisting.   When the bucket is filled, the hoist clutch is fully engaged and the bucket is lifted clear of the ground. If the bucket has a tendency to dump, the drag may be left engaged until the angle between drag and hoist cables is sufficiently wide to hold it. The drag clutch is then released and the hoist is continued, with the drag brake applied just enough to allow the hoist to pull the bucket forward and upward under the boom, without slackening the lines enough to dump it. If the drag brake is applied too tightly, the bucket may hit the boom.

The swing is started as soon as the bucket is clear of the ground. This is the period of heaviest load in the dragline cycle, as the machine is simultaneously lifting a load, pulling against the drag brake with the hoist clutch, and swinging. If overloads are picked up in the bucket, they may so slow the line and swing speeds that less dirt will be moved per hour than if smaller bites were taken.

Dumping.   Hoisting is discontinued as soon as the bucket is high enough to be dumped. When the swing is completed, the drag brake is released, partially or wholly, and the bucket swings out and dumps. A long dump cable will permit normal dumping inside the boom point, a short one under it only. Raising the boom will bring the dump point closer, lowering it puts it at a greater distance.

It is poor operation to raise the bucket higher than necessary before dumping. Clearance should be allowed so as not to strike trucks or other receptacles, but piles can be barely cleared. This saves time on a short swing and fuel and wear and tear under any circumstances.

Casting.   If the bucket is to be thrown or cast, it is pulled close in by the drag cable during the swing to the pit, with the hoist line held at a length that will keep the bucket above any obstructions during the cast, as in Fig. 13.40. When the swing is completed, the drag brake and clutch are released and the bucket swings outward like a pendulum. When it is just short of the farthest point of this swing, the hoist brake is released and the bucket falls downward and outward (B). It is

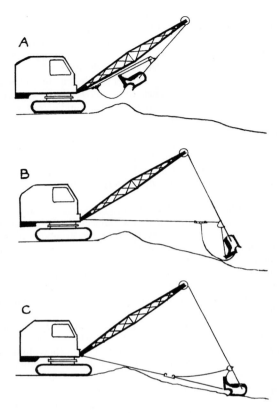

FIGURE 13.40   Casting the bucket.

checked by gradual application of both brakes just before hitting the ground, as otherwise it might be damaged, overturned, or tangled in its chains. It should then rest on the ground in the same position as if lowered and respond to drag pull similarly, except that if the pit floor is level, the hoist cable slackens as the bucket approaches the boom point and tightens again after it has passed under it.

A swing throw may also be used. During the swing from dump to pit, the drag cable is left slack and the hoist held so that the bucket is high enough to clear obstructions. The centrifugal force of the swing will pull the bucket outward, and the hoist brake should be released at the proper point so that the bucket will land where it is wanted. The swing should be checked as soon as the hoist brake is released, and hoist and drag brakes applied gently as or just before the bucket strikes. This swing throw requires much more expert operation than the other type and may damage the shovel seriously if improperly done.

Another technique is to make a pendulum throw while the shovel is swinging so that the centrifugal force adds to the outward sweep of the bucket.

Throwing to dump is a similar process, but usually the height of the piles prevents the use of a long enough hoist line to obtain much distance. The weight of the load, however, causes the bucket to cast farther than it could on the same length line if empty. A combination pendulum and swing throw is most effective, particularly if the shovel is dumping at 180° from the pit and can revolve in a full circle so that it can dump without stopping.

If the load sticks in a bucket which is thrown to dump, it may overturn the shovel.

The distance the bucket can be thrown is affected by the skill of the operator, the length and angle of the boom, the weight of the bucket, the depth of the pit, and even the wind. Casting onto a surface level with the tracks, the teeth ought to reach farther than the boom point would if lowered to a horizontal position. Much greater distance is attained when casting into a pit.

Manufacturers are wary about recommending or discussing throwing of buckets, because a careless operator may thus bang a good bucket into scrap iron in a short time, and the swing throw has possibilities for wrecking the boom as well. Beginners will do well to thoroughly master ordinary digging before practicing throws, particularly if they have trouble with tangled cables.

Throwing the bucket usually slows the digging cycle by several seconds, and reduces accuracy of work, so that digging should be done inside the boom point when practicable.

Tangles. If a drag cable becomes tangled, it is a good procedure to throw the empty bucket, as often the whole tangle can be unwound in this way, and at worst very few wraps will be left on the drum to be straightened. If the drag cable becomes bent or kinked so that it does not spool on smoothly, the bucket should be dragged in all the way a few times with a load to straighten it. If the bucket is swinging in the air, the drag should be wound in while it is swinging out and holding the cable under tension, and held while it is swinging in and the cable is slack. If the cable is running smoothly, this precaution need not be taken.

If the hoist cable becomes tangled, the bucket should be rested on the ground, the hoist brake released, and the machine swung. When the hoist cable is unwound past the tangles or to the anchor, the hoist clutch should be engaged and the swing clutch disengaged. The cable should then reel back properly, unless crushed or kinked, in which case tension should be put on it as it is wound in, by partially engaging the swing clutch to keep the boom from swinging back easily over the bucket.

The cables should not be allowed to run all the way out in a throw, for if the momentum of the bucket is stopped abruptly by the cable anchor, the anchor may tear out of the drum. Manufacturers often supply and specify a drag cable which is so short that it is unsafe for casting.

Novice Operators. To the novice, the dragline is apt to seem a very loose, rough, and contrary machine. Special provisions have to be made in every movement to keep the bucket from jerking and swinging, but these soon become automatic and very precise control can be obtained eventually. The beginner will obtain the best results by keeping the bucket fairly close in, except when actually digging or dumping, by making slow starts and slow stops when swinging, and dragging the brakes slightly to avoid spinning out the cables.

Bucket Wander. As the bucket is pulled in, it is liable to be deflected by irregularities and find a path considerably to one side of a direct line. Also, the shovel may swing by gravity during the haul-in if it is not standing level. In either case the drag cable will not be directly under the boom. The fairlead sheaves will put it in correct alignment for the drum, but if the angle is sharper than the fairlead pivot can meet, the cable will be dragged across the fairlead guard plate and will wear the cable, and wear or possibly tear off the guard. The shovel boom should be kept in line with the cable by the use of the swing lever.

Boom Twist. As a loaded bucket is lifted, the boom will tend to swing over it. However, if the boom is held to one side by the swing clutches during hoisting, or if a heavy, oversize load such as a stump is partially lifted, then dragged along the ground by the swing, a powerful twisting force is applied to the boom.

In order to give a long reach without prohibitive weight, a dragline boom is of light skeleton construction that is not intended to withstand heavy side and twisting strains. Sometimes a boom so strained will collapse, but more often will twist slightly, bending some of the cross braces particularly on the lower side. Once twisted, normal loads may increase the damage and failure will follow if the boom is not straightened.

A good ironworker is needed to properly repair such a boom; but if the bent members are angle irons, it may be kept in service quite a while by straightening them with a jack. A stout plank is placed inside the boom across the corner angles as a support for an automobile jack that can push the pieces straight.

This type of repair should not be attempted on tubular members as they lose their strength if flattened.

Applications.   The dragline does not have the positive digging force of the front shovel and the backhoe; as the bucket is not weighted or held in alignment by rigid structures and can therefore bounce, tip over forward, or drift sideward when it encounters hard material. This weakness increases with depth, and is particularly noticeable in small machines, as the weight and bulk of a large bucket are insufficient to give considerable stability and penetration.

The dragline experiences its greatest difficulty in cutting down and is able to continue a deep cut opened by other machines or by blasting in much harder material than it could dig from the surface.

The outstanding advantage of a dragline over most other rigs is long reach for both digging and dumping, plus the ability to dig below the tracks. It has the further good point of a high cycle speed, being second only to the dipper stick in this regard. It will be preferred to the front shovel for truck loading where the earth is not too tough, and where the original grade is better than the new, because of water, mud rock outcrops, steep ramps, or other problems within the excavation. Because of its greater reach, it will be preferred to the backhoe in any situation where it is capable of digging the soil effectively, where precisely cut vertical sides are not required, and where there is room for it to swing.

The dragline is the only practical attachment for extensive digging in mud, as its reach enables it to handle a wide area from a single stand and the sliding motion of the bucket avoids trouble with suction. It is also the best machine for many stripping operations in which the spoil is high-piled away from the pit, but is rivaled by long-boomed dipper stripping shovels if the spoil is to be moved across a narrow pit as in some strip mining.

Draglines with skillful operators do an excellent job of topsoil stripping, grading, and spreading piles of earth, but except under wet conditions they do not usually do as well as bulldozers of comparable ability.

Production.   In easy digging, dragline production should be between 60 and 90 percent of that of a front shovel of the same size if trucks are loaded on the same level. If trucks are in the pit, output may range from 70 to 100 percent.

In hard digging, dragline yardage falls off much more rapidly than shovel yardage because of weaker penetration. High banks also slow them when loading on their own level or sidecasting, because of the extra time required for hoisting.

In sidecasting, production on a volume-distance basis may be much higher than that of a front shovel.

Operation Suggestions.   The following are suggestions for dragline operation:

1. Keep bucket teeth sharp and built up to proper size.
2. Keep dump cable short so load can be picked up well out from the dragline.
3. Dig in layers, not ditches.
4. Keep digging surface sloped up toward shovel.
5. Do not drag in so far that dirt or rocks pile up in a ridge.
6. Keep drag cable from working in dirt.
7. Pick up bucket as soon as it is full.
8. Do not pull drag yoke into fairlead.
9. Inspect bucket chains frequently, especially at ends, and have them built up or replaced if worn thin.
10. Inspect fairlead frequently—worn bushings or spacers may let sheaves wear and cut cable.

11. Salvage short pieces of hoist or drag cable for dump cable. If dump anchors are too small, hammer the cable flat or change the anchors.

12. Do not guide bucket by swinging the boom while digging—you may twist the boom.

13. Do not swing until bucket or load is clear of the ground—you may twist the boom.

14. For heavy loads use high boom and swing slowly.

15. If the shovel starts to tip dangerously, drop the bucket. Do not try to dump it as its outward swing increases the tip.

16. Do not pick up overloads in bucket if they slow the hoist and swing, or if they tip the shovel.

17. Do not slap bucket against boom while hoisting.

18. Work with machine level. If you cannot, load trucks at either the top or bottom of the swing. Loading is sloppy at the side as the counterweight swings the boom uphill as the load is dropped.

19. Do not work with a cable that is cross-wound on the drum.

20. Do not travel uphill with a high boom.

WALKING DRAGLINES

In both parts of this book, mention is frequently made of crawler machines "walking," but this is with apologies to the true walking draglines (walkers), as shown in Figs. 13.41 and 13.42.

These machines are made in sizes of 5 yards and up, and are so slow and bulky that they must be taken apart for shipment by freight in most moves from one job to another. On the job, however, they are easily maneuvered, and have extremely small ground pressure.

They have a working base which consists of a wide, flattened cylinder called a tub, which rests on the ground and carries the turntable on its upper side. The walking mechanism is operated by cams or cranks, one on each end of a large-diameter, multiple-part shaft mounted across and extending beyond the sides of the upper frame and the tub. They are connected to the walking frames and the walking shoes. When the shaft is rotated, the shoes contact the ground, raise the machine, and support its weight as a movement or step is made. The rear lifts clear, and the front drags.

When the dragline is in normal operation, it sets on its base and the shoes are elevated as shown in (D). To move, the walking shaft is rotated clockwise, thus turning the crank from its upright position and bringing the walking shoes in contact with the ground, as in (A). The continued rotation of the shaft in the same direction transfers the weight from the base to the shoes and results in the leading (rear) edge of the base being moved upward. Continued rotation of the shaft causes the base to move backward and downward to again contact the ground, thus completing the step. The center of gravity of the machine is slightly forward of the shaft, so that only the back is raised off clear. Most of the weight is taken off the front, but it retains a steadying ground contact. The setting down is more gradual than the rise, to avoid jarring.

The process is very similar to that by which a person on their hands and knees progresses by resting their weight on their hands and swinging their body forward with feet dragging, then rests on their knees and toes while they moves their hands forward again.

Steering is done by lifting the shoes and turning the revolving unit until its rear faces the desired direction, then stepping off as described. This is probably the only steering device in heavy equipment which takes negligible power, does not tear up the ground, and permits turns of any desired sharpness under any footing conditions.

This easy steering and the light ground pressure made possible by the large area of the base are the special advantages of this mounting. They enable it to work in places dangerous or impossible for crawler machines because of difficult terrain, soft footing, or nearness to loose downslopes.

The disadvantages are that they are slow, with top speed around $\frac{1}{5}$ mile per hour, and are very wide, 30 feet or more. They must be disassembled to ship from one job to another, and are at a disadvantage in the bottom of a pit, because of both their width and the fact that they must walk away from their boom, which might be blocked by a hill.

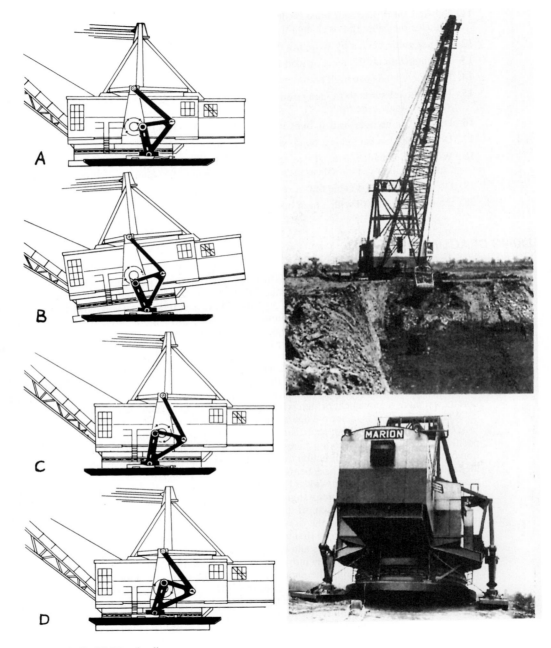

FIGURE 13.41 Walking dragline.

FIGURE 13.42   Walking draglines stripping coal.

CLAMSHELL

The clamshell front is a jack-of-all-trades. It can do most of the jobs the other rigs will do, although less efficiently, and in addition has its own specialties of digging deep, narrow, straight-sided excavations, and neat rehandling of materials.

Figure 13.43 shows this type of rig. It consists of a boom similar to that used for the dragline, with two boom line sheaves and two operating sheaves at the point, and a bucket hung from the operating sheaves by two cables.

Buckets.   A number of different types of clamshell buckets are available. Figure 13.43 shows a center pull construction, and 13.44 an Erie lever arm type. Each consists of two jaws hinged on a movable bar, the main shaft, and hinged on their outer ends to brackets extending to an upper bar, the head shaft parallel to the main shaft. The head shaft is supported by the head, or head beam, and the hoist cable. In Fig. 13.43, sheaves are fastened to each shaft and the digging cable is reeved through them and finally anchored to the head beam. In Fig. 13.44, the lower sheaves are on the lever arm, which is an upward extension of one jaw. If the digging clutch is engaged, the cable will pull the two sets of sheaves together, powerfully raising the center hinges of the jaws. Since the corner arms are fixed and will not allow the outer ends of the jaws to rise, the jaws pivot on their outer hinges and rotate inward and upward until they meet. The bucket is now closed, and will pick up its load if raised. If the digging line is released while the bucket is held in the air by the hoist or holding line, gravity will cause the main shaft to move down, pushing the jaws downward and outward and dumping the load.

These buckets, and certain other designs, can be obtained with various types of teeth with flat or curved lips, with toothless lips, with and without side cutters, with ballast plates for weight, and in various weights, widths, and shapes for special conditions.

FIGURE 13.43   Clamshell rig.

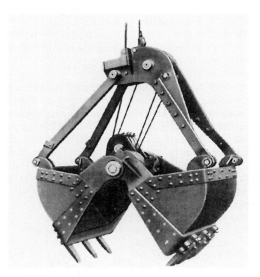

FIGURE 13.44   Lever arm bucket.

The shovel manufacturer or dealer can supply a table of lifting capacities, and the weight of the bucket and load should not exceed their recommendation. See Fig. 13.45. Figure 13.46 shows the ways in which bucket capacity is measured. Line of plate is the usual rating.

The more resistant the earth to be dug, the heavier the bucket must be in relation to its capacity. A good, extraheavy-duty clamshell will dig almost anything except solid rock, but its closing action when fully reeved is very slow, and its massive weight reduces the payload. A light bucket, suitable for handling soft or loose dirt, will have faster action and a minimum of deadweight, but it will not penetrate hard materials, and will suffer damage if repeatedly banged and scraped on them. For miscellaneous work including both hard and soft, a medium-weight, general-purpose bucket is usually selected. Some of these have provision for adding or subtracting plates to change weight and strength.

A shovel can usually just handle a medium bucket of its own rated capacity; that is, a ¾-yard machine uses a ¾-yard bucket. In heavy-duty work, the next smaller size is used, and for very light rehandling, a size larger. A small bucket must be used in mud as suction holds it, greatly increasing the force required to start it upward.

OPERATING RADIUS	TYPE OF SERVICE	40 FT. BOOM			60 FT. BOOM			80 FT. BOOM			100 FT. BOOM			Max. Jib Capacities With Outriggers
		Without Outriggers Over Side	Without Outriggers Over Rear	With Outriggers	Without Outriggers Over Side	Without Outriggers Over Rear	With Outriggers	Without Outriggers Over Side	Without Outriggers Over Rear	With Outriggers	Without Outriggers Over Side	Without Outriggers Over Rear	With Outriggers	
12'-0"	Crane	40500	52760	70000	.....	.....	.....	.....	.....	.....	.....	.....	.....	.....
15'-0"	Crane	30900	38480	70000	.....	.....	.....	.....	.....	.....	.....	.....	.....	.....
20'-0"	Crane	23600	26750	43900	30100	37680	69600	.....	.....	.....	.....	.....	.....	.....
25'-0"	Crane	17600	20200	31400	22800	25950	43100	22000	25150	42300	21200	24350	41500	.....
30'-0"	Crane	13100	15700	23200	16800	19400	30600	16000	18600	29800	15200	17800	29000	.....
35'-0"	Crane	10150	12750	18200	12300	14900	22400	11500	14100	21600	10700	13300	20800	.....
40'-0"	Crane	8400	10850	14500	9350	11950	17400	8550	11150	16600	7750	10350	15800	10000
45'-0"	Crane	.....	.....	.....	7600	10050	13700	6800	9250	12900	6000	8450	12100	9000
50'-0"	Crane	.....	.....	.....	6200	8400	11500	5400	7600	10700	4600	6800	9900	8000
55'-0"	Crane	.....	.....	.....	5350	7250	10200	4550	6450	9400	3750	5650	8600	7000
60'-0"	Crane	.....	.....	.....	4670	6310	8830	3870	5510	8030	3070	4710	7230	5700
65'-0"	Crane	.....	.....	.....	4050	5560	7800	3250	4760	7000	2450	3960	6200	4700
70'-0"	Crane	.....	.....	.....	.....	.....	.....	2740	4140	6140	1940	3340	5340	3700
75'-0"	Crane	.....	.....	.....	.....	.....	.....	2300	3610	5430	1500	2810	4630	2800
								2000	3200	4800	1200	2400	4000	2000

Radius is horizontal distance from center of rotation to hook.

All crane capacities based on 85% tipping load and were derived from tests made with the machine mounted on a P&H crane carrier with dual rear axle and pneumatic tires, fulcrum point of outriggers 7 ft. from center of truck and with the machine standing on a firm level uniformly supporting surface, using a Baldwin Southwark Load Cell and Brown Electronik Indicator for load measurement. Capacities without outriggers depend upon proper inflation, capacity and condition of tires.

Capacities shown include weight of hook, block, chains, etc.

Backstops are recommended for crane booms exceeding 50 ft. in length.

Maximum boom length for use without outriggers, 70 ft.

For clamshell and magnet ratings deduct 20% from crane ratings. Limit on clamshell rating is 10,000 lbs. Combined weights (bucket or magnet, etc., plus contents) should not exceed clamshell capacities.

Maximum length of boom for clamshell operation is 60 ft.

Single part hoist line for loads up to 12,000 lbs.

Two part hoist line for loads up to 24,000 lbs.

Three part hoist line for loads up to 36,000 lbs.

Four part hoist line for loads up to 48,000 lbs.

Six part hoist line for loads over 48,000 lbs.

Above jib capacities for 30 ft. jib on all booms up to 100 ft. maximum, using outriggers.

Deduct 1,500 lbs. from main hook capacities when boom is equipped with jib.

FIGURE 13.45   Capacities, 35-ton crane..

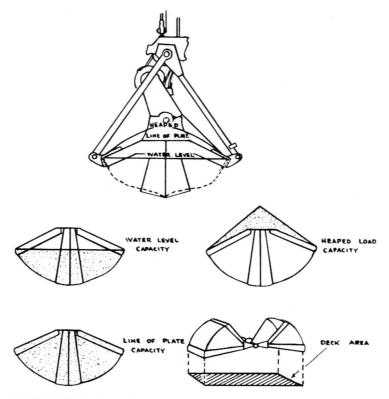

FIGURE 13.46 Bucket ratings.

In general, teeth should be easily detachable, upper and lower sheaves should be shielded against dirt and should turn on high-grade bushings or antifriction bearings; the closing line should be guarded against sharp edges where it enters the bucket head; and it should be possible to reeve the bucket without using all the sheaves, and without throwing the digging line off center.

For other uses than excavation there may be additional factors to consider. One of the commonest uses for clamshells is for rehandling loose material such as coal, gravel, cinders, sand, etc., either in piles or in transfer to or from barges, freight cars, trucks, hoppers, and other receptacles. In this work the bucket should be able to remove practically all the loose material from a hard surface, and should therefore be wide and equipped with a straight, toothless lip. If the dumping point is high in relation to the boom length, a bucket with minimum top-to-bottom measurement must be used for maximum efficiency.

Clamshells may be used to handle large rocks, cordwood, and other bulky objects, as the pinching effect of the jaws gives them an excellent grip. There are special buckets made for these jobs which may properly be considered multiple-jaw clamshells.

Grapples. When a clamshell has more than two jaws it is called a grapple, star, or orange peel bucket. The jaws may be made to fit accurately to each other when closed, Fig. 13.47, upper, so either loose dirt or large objects can be lifted; or as independent tongs that will hold big pieces only, as in the lower cut. Both show a four-jaw Owen that is constructed so that each of these jaws can work independently, making it possible for each of the four points to get a firm grip on an irregular or off-center object.

FIGURE 13.47   Orange peel bucket and tong grapple.

The multitine grapple is made with loose joints, so that the tines are free to feel their way as they close. Loads can be gripped by their sides as well as by the points.

Reeving.   Figure 13.48 indicates the reeving system. The hoist (holding) and digging (closing) lines are carried over the center pair of boom point sheaves and descend directly to the bucket. The hoist line is anchored to the head beam; the closing line goes through a guide in the head beam to the sheave sets. The bucket need not be fully reeved, as the full power of all the lines is not always needed. The cable may be anchored after rounding fewer sheaves for quicker but less powerful closing.

Tagline.   The tagline is a light cable running from the boom to the bucket that serves to prevent the bucket from twisting or spinning in the air. It is kept on a light tension by a weight sliding on a track inside the boom, as in Fig. 13.48, or by a spring-loaded drum, such as the electric dipper trip previously described, or the boom-mounted, spring-wound, Rud-O-Matic tagline drum shown in Fig. 13.49.

The cable is ordinarily fastened to two corners of the bucket by chains. If both are on one jaw, the hinge line of the jaw will be at right angles to the cable. If one corner of each jaw is held, the opening will be in line with the cable. The bucket can also be held in a diagonal position by attaching to only one corner.

The tagline pull is usually light enough that a person can guide the bucket by hand or with a stick as it is lowered to place it exactly as desired. This is done when it is necessary to cut in a position other than that in which the tagline holds it.

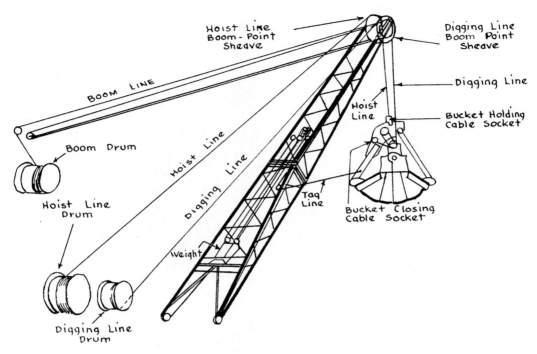

FIGURE 13.48 Clamshell reeving.

The same manufacturer builds a tagline with an extra drum for electric cable, for use with a magnet. Contact with the electric source is through brushes. The electric line is set slightly slack, the tagline taut.

There is also a magnet reel that has electric line only, which is taut at all times.

Digging. In digging, the bucket is placed over the work by swinging the boom, and either moving the shovel or raising or lowering the boom to obtain correct distance. The digging (drag, closing, or crowd) brake is released, causing the bucket to open, and the hoist brake is then released, allowing the bucket to contact the ground on its teeth. If the ground is soft, the hoist brake will be only partially released, or reapplied just before the bucket lands. If it is hard, the bucket will be allowed to fall freely, so that its weight will drive it into the ground for a good bite. In either case both brakes must be applied as soon as it has hit to avoid unspooling of the cable.

The digging clutch is then engaged to pull the jaws together. They first push the dirt inward, then curve and close under it. If the material is not too resistant and the bucket has proper weight, a full or heaping load will be gathered. The digging line will lift the bucket as soon as the jaws are tightly closed. The hoist clutch should then be engaged so that the hoist (holding) cable will not become slack while the bucket is being raised.

Dumping. The swing is started as soon as the bucket is clear of obstructions. The distance of the bucket from the machine may have to be adjusted by raising or lowering the boom if precise dumping is being done. This, of course, is practical only if the shovel is equipped with a live boom.

When the bucket is properly positioned, the digging brake is released, the jaws open by their own weight and that of the load, and the earth is dumped. The bucket is swung back to the pit in the open position and lowered or dropped for another bite.

FIGURE 13.49 Drum-type tagline.

Centrifugal force of the swing puts the bucket out beyond the boom point, offering some increase of reach for both digging and dumping. Stopping the swing will allow the bucket to swing inside the boom point for close digging.

Chopping. If the bucket does not fill at the first closing, it may be opened, hoisted, and dropped again, in which case the earth scraped together the second time will be added to that loosened the first time. This process may be repeated until a full bucket is obtained.

Signals. Clamshell work is often done at such depths that the operator cannot see the digging point. In this case it is best to have someone positioned to see the work and signal the operator. This is essential if there are workers in the pit where they might be hit by the bucket.

Hand Labor. If the bucket will not dig satisfactorily, either because it is too light in construction or slides off slopes, hand labor or explosives may be employed to loosen the dirt and the clamshell used to lift it out afterward.

Deep Digging. In deep ditches or shafts, it is difficult to keep the walls perpendicular, particularly if the earth is stony, as with each bite the bucket is edged a little away from the wall, thus causing the pit to grow narrower with depth. This tendency may be combatted by swinging the bucket against the wall as it drops, by hand trimming, or by making the top of the shaft enough oversize so minimum width will still be had at the bottom. It is helpful to equip the bucket with side cutters or corner teeth for this work.

Applications.   A clamshell has the unique advantages of being able to stand on either the new grade or the old and to excavate at its own level, to a depth limited only by the length of cable its drums will carry, and at a height limited only by its boom length. It is the best of the rigs for handling bulky objects such as boulders, stumps, and logs, as chaining is usually unnecessary. It is the best tool for piling and burning loose brush and trees. It does not push or pull loose material to other positions as it digs. There are very few excavating jobs which cannot be done with a clamshell, and it is an excellent utility and odd-job rig.

But it is slow. It ordinarily moves fewer yards per hour than any other rig which can do the particular job. This arises from the time consumed in closing the bucket and from the fact that the operator has minimum control over the position of the bucket, which is always directly under the boom point unless swung elsewhere. The bucket can be moved toward or away from the shovel by raising or lowering the boom, around it by swinging the shovel, outwardly by centrifugal force, and inward as a pendulum so that there is no point in its digging at a range it cannot reach. But careful operation is required to get it where it belongs.

A number of different buckets are required for best handling of a variety of digging, and there is often loss of efficiency in the use of an unsuitable or compromise type of bucket.

A clamshell has the same reach as a dragline, except that the bucket cannot be cast as effectively. Since most digging is done directly under the boom point, maximum depth in "diggable" soil is determined by the amount of cable the drums will carry. Dumping height is controlled by the height of the bucket, which is variable, under the boom point.

Operating Suggestions.   Suggestions for clamshell operation are:

1. Keep bucket teeth sharp and built up to size.

2. Do not use more parts of line in the bucket than you need.

3. Be sure footing is solid.

4. Do not travel with a high boom—a bump may tip it back on the cab.

5. Do not swing uphill with a high boom.

6. Keep back from the edge of deep, wide cuts.

7. If machine tips dangerously, release both holding and closing lines.

8. Do not hit boom with bucket.

DIPPER SHOVEL ATTACHMENT

The original shovel attachment, or front end, is the dipper. A very large one is shown in Fig. 13.50. It is commonly called a front shovel.

General Construction.   Reference is made to Figs. 13.1 and 13.8. The boom is a massive beam, hinged to the deck at the boom foot, and extending diagonally upward. It is supported by a four-part boom line. It may contain a center slot, in which a saddle block and shipper drum are pivoted on the shipper shaft, or it may carry a divided stick in side blocks.

The dipper stick, or handle, slides back and forth through the saddle block and pivots on the shipper shaft.

Bucket.   A typical bucket is a welded steel box open at the top, and closed on the bottom by a hinged door. Digging is done by the front top edge, which is reinforced by a lip. The lip contains

FIGURE 13.50   Bucket capacity, 140 yards.

tapered sockets to hold a set of teeth. The reinforcing ridges running down from the sockets are called tooth bases.

Construction is heavy. The whole bucket, or at least the lip and other parts subject to severe wear, is of alloy steel.

Teeth are usually of manganese steel. They have tapered shanks to fit into the sockets. A simple cotter pin through the shank prevents the tooth from falling out. Digging stresses force it into the socket, so the pin is not under strain.

The teeth take the brunt of the digging, and each one should be strong enough to take the entire power of the shovel. When they wear dull, they may be reversed to partly restore cutting ability. Eventually they must be built back to size and edge by welding on caps, or building up with rod; or be replaced with new ones. It is good practice to keep a spare set of teeth on hand.

The digging ability of the shovel and its fuel economy are both diminished, and maintenance costs are increased, by using dull teeth.

The door is hinged high on the back of the bucket to allow it to swing wide when it opens. A sliding latch fastens it to the bucket front. It is opened by a sharp pull on a light cable. The trip mechanism will be described below.

The top of the bucket is pinned to the end of the stick, and it is supported near the door hinge by two pitch braces bolted to the stick. The braces can be moved up and down the stick to change the digging angle of the teeth.

The padlock sheave is hinged to the bucket just above the stick, and carries the hoist line. Larger buckets may use a hinged yoke to fasten the sheave to the bucket.

Stick and Boom Drums.   The one-piece stick rests on the shipper drum and is held in alignment by wear plates in the saddle block. It moves freely back and forth in response to cable pull at its ends.

The boom foot drum, Fig. 13.51, revolves on the centerline of the boom foot pins, so that raising and lowering of the boom do not change its position relative to either the shipper drum or the deck machinery. It includes two grooved cable drums, and a center sprocket. It is turned in either direction by the crowd-retract chain from the front drum on the deck, and controls the crowd and retract cables.

FIGURE 13.51   Front view, boom foot drum.

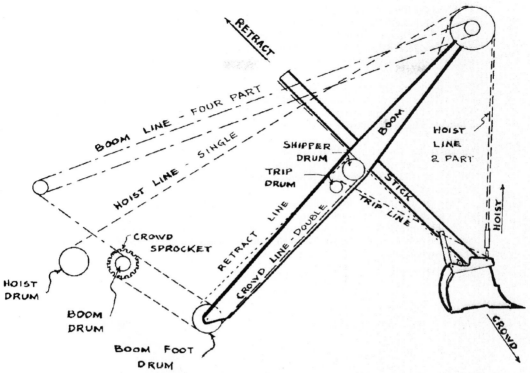

FIGURE 13.52   Dipper shovel reeving, cable crowd.

Reeving.   Figure 13.52 shows the reeving for a front shovel attachment. The boom line runs from the front of the boom drum, up and back over a sheave on the left side of the A-frame top, forward to a small sheave on the left side of the boom point, back to a horizontal sheave hinged to the A-frame cross member, forward to the top of the right-hand small sheave on the boom point, and back to an anchor on the A-frame.

The hoist cable takes off from the top of the hoist drum, runs upward and forward over the large left-hand sheave on the boom point, around the padlock or bucket sheave on the top front of the bucket, and over the large right-hand boom point sheave to an anchor on the boom.

The crowd cable is anchored and wrapped on the outer side of the left-hand boom foot drum, runs from the bottom of this through the hollow center of the boom to the bottom of the left groove in the shipper sheave, back along the stick to the end, around a curved holding plate and forward along the stick to the shipper sheave, around its right groove and to the bottom of the outer side of the right boom foot drum, where it is wrapped and anchored.

The retract cable is anchored to the inner side of the right-hand boom foot drum, takes off from the top, runs over the boom to the shipper sheave, around the center groove and forward along the bottom of the stick to an eyebolt anchor that has a threaded adjustment by which the cable can be tightened.

The boom foot drum is turned either way by the roller chain from the opposite-acting crowd and retract clutches. If it winds in the crowd cable, it will pull on the back of the stick, and force it to slide forward or downward through the saddle block. This will pull the retract cable, which unwinds from the drum at the same speed as the crowd winds onto it, so that the two cables remain tight and balanced against each other. If looseness develops, it may be removed by tightening the eyebolt anchor of the retract.

When the foot drum rotation is reversed, the retract will wind onto the drum, pulling the stick back through the saddle block and unwinding the crowd cable.

The crowd normally pushes the bucket into the digging, and the retract pulls it out. The crowd is therefore geared down for greater power, and a double cable is used for greater strength. Retract may be 50 or 60 percent faster than the crowd.

Chain Crowd. On some front shovels an endless chain is used. Chain crowd saves the expense and time loss of broken crowd cable, and suffers less damage from bending around small sheaves than cable of the same strength. However, it is noisier, an extra item to lubricate, and costs more to replace when it eventually wears out.

Big Stripping Shovel. The big Marion strippers use a special crowding mechanism, shown in Fig. 13.53. The stick is in two pieces, hinged to each other and to a stiff leg pivoted on the deck behind the boom. The boom has a wide center opening, allowing the front stick to work through it without touching.

The rear stick or crowding handle has rack teeth, and is crowded and retracted by shipper shaft pinion gears driven by electric motors mounted in the gantry.

The boom carries hoisting strains only, and therefore can be lighter in construction than if it had to line up and crowd the stick as well. The shovel shown in Fig. 13.54 has a standard framing for a large machine.

FIGURE 13.53 Knee-action crowd.

FIGURE 13.54   Stripping for coal.

Bucket Trip.   The dump line for the bucket door may be operated by mechanical or electric power.

The trip in Fig. 13.55 is mounted on the end of a live shaft in the deck machinery. A drum rides freely on the shaft and can be connected to it by a brake-type clutch or a ratchet which is held out of engagement by a spring.

The drum contains a spring which keeps a tension on the dump line sufficient to pull in slack when the bucket is retracted. Engaging the clutch or pawl reels in the line with a snap sufficient to pull the latch and allow the door to open.

An electric trip consists of a small starter-type electric motor equipped with an eccentric sheave. Actuating the sheave through a pushbutton switch controlling the motor produces a pull on the cable to release the latch.

The unit may be mounted on the lower boom near the shipper shaft, or on the shovel deck.

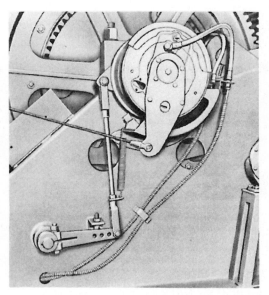

FIGURE 13.55   Hydraulic control dipper trip.

ELECTRIC SHOVELS

Electrics and Diesel-Electrics.   All-electric shovels and draglines normally operate on three-phase alternating-current (ac) power furnished at between 2,000 and 8,000 volts by utilities (high lines) or by generator sets assembled on the job.

Electric units are used to serve in the place of the internal-combustion engine, and of the clutches, brakes, mechanical controls, and to eliminate some of the gears, chains, and cables, so that construction is quite different from that of the diesel-powered rig.

A diesel-electric shovel carries a diesel engine and a generator to supply current to electric motors and controls that perform all operating functions. A part electric machine has diesel power, with mechanical drive to some power trains, and electric to others. Electric swing clutches may be used in a machine that is otherwise all-mechanical.

In excavators used by contractors, electric and diesel-electric drives begin to appear in the 3- and 4-yard class, and are standard in very large machines.

In the development of electric power for large shovel excavators, Harnischfeger found that what was needed is a system for static, solid-state electronic conversion of pit ac power to dc main power supply, with controlled delivery of this main power for the motion drive motors. They developed the patented Electrotorque Control system, which eliminated the need for the familiar ac-to-dc rotary type of motor-generator set. Also it essentially eliminated inductance, which is the main impediment to quick response from electrical systems. To do this required motion drive motors with much better transient dynamic response characteristics than were then available from any source.

Figure 13.56 shows a forward hoist motor and helical cut teeth of the first of two reductions to hoist rope drum gear. The hoist brake mounts on an involute splined extension of the pinion shaft.

In many ways electricity is an ideal power for excavators. Since it allows packing a lot of power in a small place, motors can be placed so as to eliminate complex gear trains and chain drives; maintenance and fueling of an internal-combustion engine are eliminated, exhaust gas is not a problem indoors or underground, and operation is smooth and quiet. Controls are easy to operate.

But first cost is higher, particularly in small units, and in any size the necessity of being near a power source and of having a cable connection can hamper activities severely. There are also technical problems involved in conveying the heavy current into the revolving frame, and in controlling and distributing it.

FIGURE 13.56   Electric hoist motor. (*Courtesy of P&H Mining Equipment.*)

The diesel-electric is independent of outside power, and has the smoothness and some of the simplicity of electric drive. Engine fueling, maintenance, and noise are about the same as with mechanical drive.

Controllers.   The main operations of the electric excavator depend on three master controllers at the operator's station. Two of them, the hoist and crowd-retract in the dipper shovel, or the hoist and drag in the dragline, have hand levers. The swing is operated by a pair of foot pedals.

By means of these, circuits carrying light current can be opened, closed, or "throttled" to regulate flow of current to a desired amount. There are several types of controllers now used in excavators, of which three will be described.

One, the rheostat or resistance type, includes a number of circuits of different resistance, connecting a power source and the generator fields. A rotating switch provides means to close any one of these to permit current to flow through it. The contacts are arranged so that in moving from the Off position, the controller closes the circuit of maximum resistance first, and those of diminishing resistance in succession, until the final full-flow connection is reached.

The P&H electric mining shovel motors have interpole field poles of a full-length design. The full length assists in the development of desirable optimum interpole field circuit inductance. The winding shown in Fig. 13.57 is used for large-horsepower motors. Bare conductor turns are edge-wound into a single coil and then furnace-annealed. The wide copper turns, being separated by the air gap, act as heat-exchanger fins for transfer of heat into the ventilating air.

Ward-Leonard System.   Most electric excavators use a drive and control system that makes it possible to control the heavy working current by varying a light current to generator fields. Figure 13.58 indicates schematically the principle of operation, without in any way representing the actual mechanical and electrical layout.

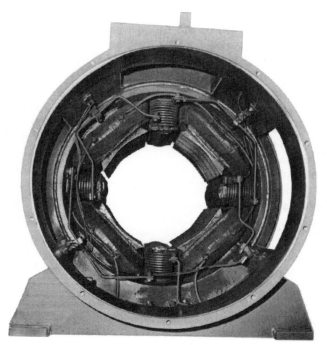

FIGURE 13.57   Motor interpole winding. (*Courtesy of P&H Mining Equipment.*)

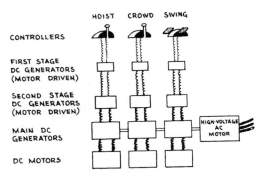

FIGURE 13.58   Block diagram of Double Ward-Leonard system.

One or more ac motors (one shown) turn a set of three generators, that supply direct current to the hoist, crowd, and swing motors. The output of each generator, and therefore the speed and power of the motor it powers, is regulated by the flow of exciting current through its field coils.

This current, in turn, is regulated by the position of the controller lever at the operator's station. Thus an easily handled low-voltage small-amperage current is stepped up to regulate the high voltage and very heavy amperage needed to provide the power required at the hoist drum and at the crowd and swing pinions.

The main ac motor turns at full speed whenever it is switched on, but current consumption is light when the generators are not charging and the dc motors are idle.

The swing and crowd-retract generators have means to reverse the direction of current flow. The crowd motor is reversible, and the swing may use either one or two reversible motors geared to individual swing pinions.

The hoist motor is also reversible, so that it lowers by power at a controlled speed, instead of allowing the bucket to drop by gravity. A mechanical clutch (friction or pneumatic) is usually put between it and the drum, adjusted so that there will be slippage to absorb shock if the dipper stops suddenly.

A gearshift, or moving of jaw clutches, permits the hoist motor to be used to propel the shovel.

DC is usually selected for motor drive because of its ability to develop high torque when lugged down by heavy loads and to use variable voltage without damage.

MINING SHOVEL

The P&H electric shovels of 22 to 56 cubic yard capacity, such as the machine in Fig. 13.8, are representative of a class of excavator that is widely used in pits, quarries, and mines; usually in loading blasted rock and ore into hauling units.

Power Line.   The power line is carried overhead, often on portable or temporary poles, to the immediate vicinity of the shovel; then along the ground in a flexible, heavily insulated tail cable to a terminal box in the lower frame.

A ring-type high-voltage collector is located at the base of the vertical propel shaft. This consists of three pairs of rings in an oil-filled moistureproof case. One ring of each pair is stationary and connected to one of the three input wires; the other revolves with the upper frame and is connected to one of three lines going up through a protecting sleeve inside the hollow propel shaft and to the control panel. The rings keep constant contact in any swing position of the shovel.

Power requirement of the 22-yard shovel is a minimum 2,000 kilovoltamperes from a high line, or more from a diesel generating set.

Control Panel. The control panel contains the master switches and fuses both for the incoming alternating current and for the dc circuits from main generators to motors; selenium rectifiers that convert ac to dc to operate brakes, dipper trip motor, and main motor fields; and low-voltage ac from the controllers into dc for generator field excitation.

There are also transformers to step down ac voltage for use in the controllers, lights, auxiliary motors, and the crowd-propel circuits. For instance, for the 22-yard shovel mentioned above, the transformer for the main armature is 1,500 kilovoltamperes, for the auxiliaries is 250 kilovoltamperes, and for the relays and lighting supply is 45 kilovoltamperes.

The approximate path of current through various parts is indicated in Fig. 13.59, and the operator's controls are shown in Fig. 13.11 or 13.60 and 13.61.

Motors. The identical crowd and swing dc generators are mounted on the extended shaft of the main motor. They are driven at constant speed, and output is varied from zero to maximum by the current supplied to the field coils.

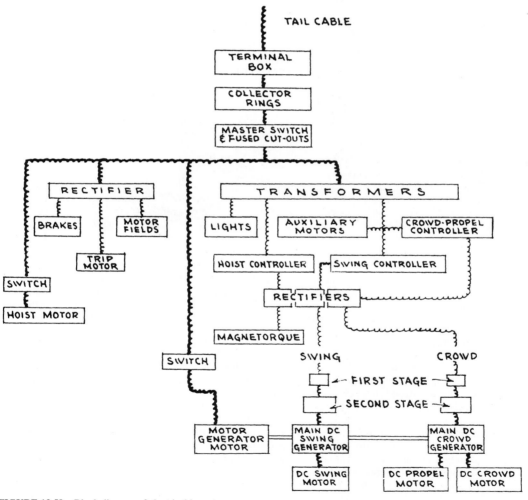

FIGURE 13.59   Block diagram of electrical layout.

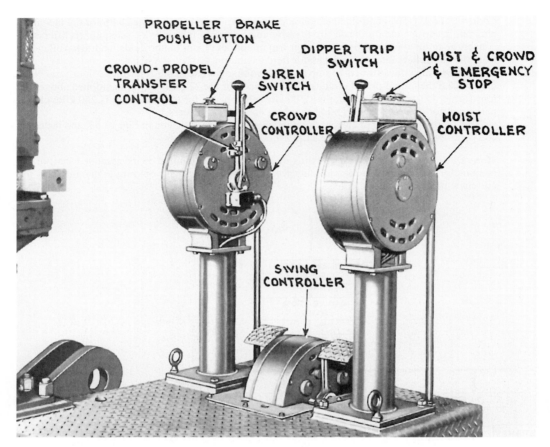

FIGURE 13.60   Operator's station.

The reversible crowd-retract motor, Fig. 13.62, is mounted on the boom just below the shipper shaft. Crowding or retracting of the straddle-type sticks is accomplished through a worm, worm wheel, and a set of spur gears, the combination of which links the crowd motor to the gear racks on the underside of the dipper sticks. The motor is cooled by filtered air from a blower.

Brakes.   Each motor or gear train is equipped with a powerful brake that is held out of engagement by an electromagnet, so that if power is cut off, either deliberately or accidentally, the unit "freezes." Current is shut off and brakes are applied by pushbuttons at the operator's station.

The pushbutton or automatic control is used to hold the bucket and sometimes the swing while waiting for arrival of trucks, or for other developments. The automatic application protects the machine and its surroundings against damage in the event of power failure.

Ventilation.   A blower in the top of the cab draws in filtered air, as shown by the flow arrows in Fig. 13.63, and maintains the cab above atmospheric pressure. Some air leaks out, but most of it is drawn out by blowers, one to each motor, that prevent overheating by keeping a current of air moving through them.

The main filter keeps the air clean so that abrasive dust is not blown into the motors.

There is a door between the operator's station and the main cab, to use or avoid the airflow, as desired. If this door is open, the windows must be closed.

FIGURE 13.61   Operator's switch panel.

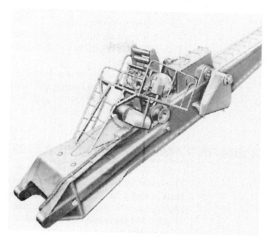

FIGURE 13.62   Crowd motor.

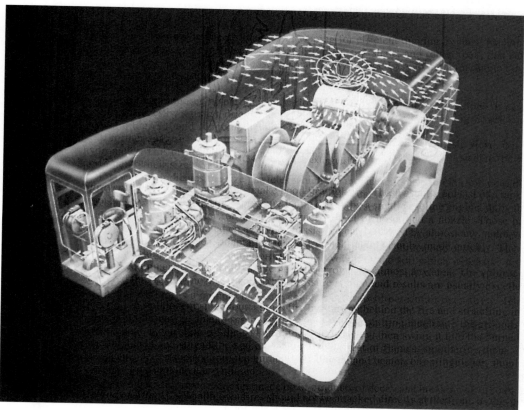

FIGURE 13.63   Ventilation system.

Propel.   Current from the crowd generator can be diverted from the crowd motor to the propel motor by flipping a switch on the crowd lever (controller handle). No gears or clutches are involved in this instantaneous one-finger shift. The automatic brake locks the crowd while current is used for propelling.

The propel motor is mounted vertically on the deck, and is permanently geared to the vertical propel shaft. This drives a horizontal shaft, which turns the live axles extending across the rear of the travel unit.

The propel shaft brake holds the machine while digging, and at any time that it is not disengaged by current activating the propel motor. The fact that the power train is permanently geared to one axle prevents any possibility of a runaway—a good consideration in an excavator of such weight, up to 1400 tons.

OPERATING AND MAINTAINING A SHOVEL

The techniques used for operating a front shovel, especially those for crowding the bucket into a bank, cannot be described well enough to be helpful to the reader. See Fig. 13.8. The movements and feel for the work of the machine have to be experienced to understand them. Therefore, it is recommended the inexperienced person seek the opportunity to work with an experienced operator to learn techniques.

There are some maintenance problems in dealing with the cables that will be described for the benefit of an inexperienced operator or manager of equipment maintenance. These will be discussed in the following paragraphs.

Tangled Cable.   If the hoist clutch and brake are released when the bucket is in the air, it will fall, unwinding the hoist cable from the drum rapidly. When the bucket strikes the ground, it will stop pulling on the cable but the drum will continue to revolve with its own momentum, pushing the cable. This will result in the cable's leaving the drum and looping itself all over the machinery, or the drum's revolving inside the wraps of cable, thus loosening all of them back to the anchor, or both. If the hoist clutch is engaged and the cable wound back onto the drum, the wraps will cross each other and tangle. If a heavy load is picked up by the cable in this condition, it will be strained, deformed, or crushed. In addition to having its life shortened it will then be difficult to spool onto the drum smoothly unless under enough load to pull its kinks straight as they come in.

The easiest way to straighten out this mess is to manipulate the bucket so that it will pull most of the cable away from the drum. It is first necessary to engage the hoist clutch and reel in the disordered cable, then hoist the bucket, dumping it first if loaded. The bucket may then be crowded all the way out and rested on the ground, dropped in a hole, or pulled back close to the tracks. There will now be only a few wraps of cable on the drum, which can be straightened out with the gloved hand, or gently with a screwdriver and a ball peen hammer. The engine clutch should be disconnected before handling the cable, as an inequality in the hoist clutch might cause it to tighten the cable with a jerk even while disengaged. Leather-palm gloves should be worn because even a preformed cable may have projecting broken wires that cut like needles.

If the cable has been loaded at all while tangled, it should be put under heavy tension while fully extended to straighten out kinks.

Tangling the cable in this manner is one of the principal bugaboos for beginners. It can be avoided almost entirely by riding the brakes, that is, by keeping a light pressure on the foot brakes at all times—not enough to stop the drum from being turned, but sufficient to stop free spinning. This is bad operation as it gives the shovel a constant drag to overcome, increases lining wear, and heats the drums. However, none of these drawbacks are serious enough to counterbalance the wasted time and labor, the loss of confidence, and the damage to the cable resulting from tangles. The new operator who develops good coordination with the controls can then train himself or herself to release the brakes fully instead of partially.

The crowd-retract cables should not tangle unless loose. They are straightened by prying with a screwdriver, and working the stick back and forth.

Cable Breaks.   Shovel cables are subject to shock, heavy loads, sharp bending, rapid motion, and exposure to weather; and drag cables may be exposed to friction with earth and rock. They may last a few hours or years, but sooner or later they will break unless replaced. If the break occurs during work, time is lost until another cable can be obtained and installed. If the break occurs at the wrong time, it may cause injury or death to personnel and damage to property.

A cable that is abused, is too small for its work, or is of poor quality or defective may break suddenly before showing any signs of wear. Usually, however, a wire rope will not break until weakened, and inspection will show thinned and broken wires on its surface. Such a cable may last a long time, but it is good operation to change it during maintenance time in order to eliminate work stoppage and possible damage when it fails.

Cables often break one strand at a time, giving the operator warning before parting completely. However, the fact that 50 cables may break in this manner is no proof that the fifty-first will not snap suddenly, either under load or reaction from load.

Torque converter drive prolongs rope life and reduces the likelihood of sudden failure.

Replacing Cable.   For safety's sake, the engine should be stopped before a cable is replaced. Some clutches may jerk a line even when they are disengaged, or a helper or a bystander might accidentally lean on a lever at just the wrong time. Drums may be turned and cable reeled in during work by using the starter, or by starting the engine, running it while necessary, and then shutting it off again.

A line that is being replaced is usually partially or wholly in place. If the machine is an unfamiliar one, it is great importance to memorize its exact layout. If a service manual is on hand, the diagram should be checked to make sure it is both right and understandable. If there is no diagram, a rough sketch may be made as a memory help.

It is very embarrassing to pull a cable out and then not remember how it was reeved. Fortunately, shovels are less complicated in this way than scrapers.

A typical hoist line reeving is shown in Fig. 13.64. The line runs from the hoist drum on the shovel deck over a boom point sheave, around the padlock or bucket sheave, over the other boom point sheave, and to an anchor on the boom.

Any part of the old line that is still wound on the drum is stripped off by pulling with gloved hands, with the engine shut off and the hoist brake and clutch disengaged. The old cable should be coiled and tied securely to avoid danger.

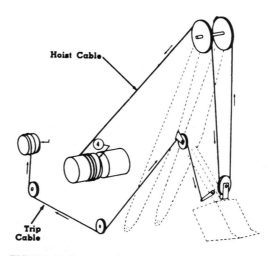

FIGURE 13.64 Dipper hoist and trip lines.

Most cable-to-drum anchors are of the wedge type, Fig. 13.65, that can be loosened by placing a driving punch against the small flat end and hammering until the wedge is loose. This will release the cable, which can then be pulled out of the socket with a pry bar or a screwdriver. The other anchor is reached by climbing up the boom or a ladder, or by standing in the raised bucket of a tractor loader, and it is loosened in the same manner.

The new cable, cut to proper length ahead of time or by measuring the old line, is drawn over the proper boom point sheave and down to the hoist drum. The end is pushed through the small end of the anchor socket and bent back into a loop, in which the wedge is placed. The slack is pulled back through the small end by hand, drawing the wedge into the socket. It is forced in by hammer blows, just enough to hold it in place until tension on the cable seats it firmly.

The drum is turned to reel in the line until the other end is near the bucket sheave. That end is then reeved through this sheave, over the empty boom point sheave, and is anchored to the boom. Putting the cable on the drum and then drawing it off in this manner is a precaution against kinking it.

With the line fastened at both ends, the engine is started and idled, and the hoist clutch is engaged to wind the slack cable onto the drum. It should wrap smoothly, with each loop lying against the previous one. A person working alone will guide it from near the drum with a gloved hand; a helper can pull the slack down to the bucket sheave and keep light tension on the line from there. Some resistance or back pull helps to make it wind snugly.

The anchor clamps are tightened by running the empty bucket down and out until the cable is stripped from the drum, and stopping it abruptly with the brake. This should be done two or three times. It is important to keep the cable from running slack for the first hour or two of operation, as it is most likely to cross loops on the drum before it is broken in.

Badly worn or damaged cables should be replaced in nonworking time before they break. Downtime on a shovel loading a string of trucks costs much more than the cable and the labor of changing it.

Sometimes a cable is cut to the wrong length by mistake, or it is convenient to use a precut line of slightly different length. One that is too long may be hard to fasten securely to the drum, as the extra wraps make it hard to put enough tension on the anchor to seat the wedge. If the drum is exactly the right size to hold the correct length of line, the extra will start a second layer, causing severe wear at the point where it snaps over.

If the drum is small enough that there is overlap with standard length line, or wide enough to take the extra on the first layer, this causes no problem.

A hoist line that is too short may limit bucket movement, and is likely to cause a damaging jolt by the anchor stopping it if it is run out too far. An instruction book may specify a cable that is too short for comfortable operation.

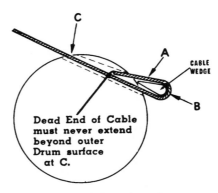

FIGURE 13.65   Wedge anchor to drum.

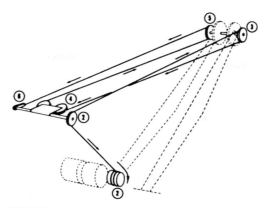

FIGURE 13.66   Dipper boom reeving.

The hoist is the most frequently broken line on a dipper shovel. Under severe conditions it may break several times a week, but this much trouble usually indicates overloading or other abuse. A too-large cable will be damaged by pinching in sheaves; one that is too small breaks from lack of strength.

Procedure is similar for a boom cable, except that the boom must be rested on the bucket before taking off the old cable, and reeving between the A-frame or gantry and the boom point is usually a four-part line that requires more care in getting it on right. See Fig. 13.66.

In order to change a crowd or retract cable, the turnbuckle fastening the retract to the bucket must be loosened. The new cable or cables are cut carefully to proper length, and installed to be snug with the turnbuckle loose. Tightening will put it under proper tension.

Safety Precautions.   An operator should lower the bucket to the ground before leaving the machine, as the brakes may lose their grip as they cool and let the bucket down, gradually or with a rush. There is also the possibility of a would-be operator accidentally releasing the hoist brake while an admiring friend stands under the bucket.

Workers and spectators should always keep beyond the farthest reach of the shovel, if possible, but they seldom do.

The engine clutch should always be disengaged during maintenance work, except when it is necessary to move parts for access.

Booming Up and Down.   The boom is normally held in a fixed position during digging. However, lowering the boom enables the bucket to dig deeper and farther away, and raising it permits a higher dump point. If desired, the boom latch may be locked out of engagement, the boom lowered to dig, raised during the swing to dump, and lowered again on the swing back. This uses extra power to lift the boom and may slow the digging cycle, but often simplifies the work. It is only possible with a live boom.

Booms should always be lowered slowly and stopped with the brake, not the ratchet.

Applications.   The front shovel is a good machine for truck loading. See Fig. 13.8. It is accurate because of the three-way control of the bucket and the ability to dump in any position. It is fast because only a small part of its weight—the stick and bucket—is involved in the digging motion, and because of its ability to dig in any spot it can reach with minimum waste motion when approaching and leaving the cut.

The dipper is also the attachment for hard, heavy digging—hardpan, boulders, ledge, and blasted rock. The weight of the boom, backed to a varying extent by the weight of the shovel itself, holds the bucket to its work. The effective cutting angle of the teeth in a bank, aided by the variable direction of pressure provided by the hoist and crowd, enables it to cut highly resistant material and to break up cracked and fissured rock formations. The shape of the bucket enables it to pick up objects much larger than itself without chaining. Larger shovels are increasingly effective at this type of work.

However, it is necessary that the shovel be able to move into its work constantly in order to retain its effectiveness. Solid rock projecting from the pit floor will prevent getting at the bank unless ramps are built over the rock, or bulldozers used to push the bank to the shovel. In digging below the tracks, the downslope should not be much over 20 percent if the machine is to follow the work. If it is to stand at the edge and dig, it can cut down steeply 5 to 7 feet, but is limited as to width of cut, and unless digging against a bank may push away about as much material as it picks up. Refer to Figs. 13.24 and 13.67.

A shovel does not work efficiently at its maximum reach as it loses in lift, in ability to fill the bucket in a short sweep, in tooth angle, and in accuracy of control. Good digging practice is to keep the shovel close to the work and trucks down at its level for easy, accurate loading. Special situations often arise, however, where every inch of reach must be utilized.

Swing speeds are quite variable, ranging from 4 to $5\frac{1}{2}$ revolutions per minute. This difference is not very important, as the gain in speed usually means a loss of power. A slow-swinging shovel may swing from digging to close dumping position more quickly than a fast-swinging one of the same power because of more rapid acceleration, but will be slower on a long swing.

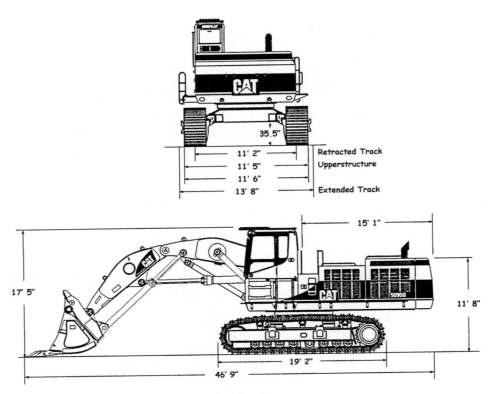

FIGURE 13.67   Caterpillar's 5090FS hydraulic front shovel.

Operating Suggestions.   The following is a list of suggestions for efficient dipper stick operation:

1. Keep bucket teeth sharp and built up to proper size.
2. Do not crowd bucket so deep into the bank that it slows the hoist.
3. Pull bucket out of the bank as soon as it is full.
4. Keep close to the bank unless it is likely to slide.
5. If bank is very hard, or material needs mixing, make passes in it with bucket door open.
6. Place shovel so that the drive chains are away from the digging.
7. Spot trucks as close to digging as safety permits to save swing time. See Fig.13.68.
8. Spot trucks on both sides if possible to save waiting time while trucks move.
9. Spot trucks so that they line up with arc of swing, or back them directly toward the shovel.
10. Do not swing over truck cab if you can avoid it.
11. Move up while waiting for trucks.
12. In mud, load driest material in bottom of truck, sticky stuff on top, for easier dumping.
13. Do not let tracks sink deep in ground—bury mud with dry fill or use poles or platforms.
14. Work smoothly—slamming and jerking are hard on machine and operator.

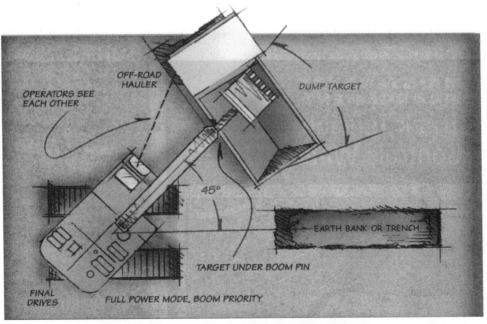

FIGURE 13.68   Best truck position for loading from shovel. The ideal setup for loading spoil in trucks puts haul units close to the cut where bucket swing is short and through the end gate. Dump as the boom swings, smoothly filling the truck body and unloading the excavator before it must reverse the swing direction. Place dump targets just off the end of the boom, so that extending the stick to keep the bucket over the truck bed will also position the stick to begin the next cut. You'll seldom see all of these conditions on a trenching site, but using any of these elements will boost productivity.

PARTIAL-SWING SHOVELS

Dredges.   Partial-swing cable shovels in large sizes are still used as dredges. They are mounted on barges or ships, for underwater work in which bottom hardness, spoil disposal, or other conditions are unsuitable for the hydraulic dredges described in the next chapter.

Figure 13.69 shows a large dipper dredge and Fig. 13.70 a grapple dredge. Loading is done on barges moored at either side, or on shore if it is close enough. Occasionally the spoil is dumped back in the water at the side.

Barges are towed to the job and maneuvered on it by winch lines attached to anchors. Pointed steel pipes, called spuds, are driven into the bottom and used as braces against digging thrust, wind, and waves.

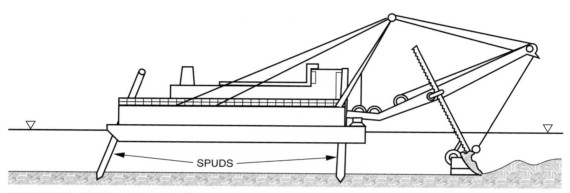

FIGURE 13.69   Dipper dredge.

FIGURE 13.70   Grapple dredge.

Dipper dredges are suited to deepening harbors and channels where the space is too restricted or the bottom too hard or coarse for hydraulic dredges. They are also used for digging canals through swamps where there is too much standing water to permit the use of land machinery.

Grapple dredges are best suited to work at depths beyond the economic limit of other types, and for cleaning up loose rock. Hydraulic dredges will be discussed in the next chapter.

OTHER ATTACHMENTS

Depth Gauge.   A backhoe of any type may have a built-in device by which the operator can check the depth of excavation from the control station.

The Accu-Depth, Figs. 13.71 and 13.72, is made up of a sensor or transmitter fastened to the stick near the bucket, a fluid-filled transmitter tube, and an indicating head or gauge.

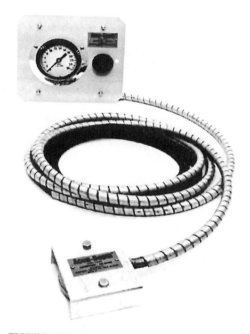

FIGURE 13.71   Accu-Depth unit.

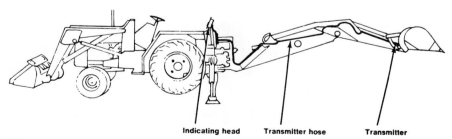

Indicating head   Transmitter hose   Transmitter

FIGURE 13.72   Accu-Depth installation. (*Courtesy of Amtek/U.S. Gauge.*)

To set it for a reading, the operator places the bucket on a reference point, depresses a valve button on the head, then rotates its dial until zero lines up with a pointer. The operator then places the bucket, in the same position, on the bottom of the trench, and opens the indicator valve again. The pointer swings to indicate the vertical distance (depth) of the second position (ditch bottom) below the reference level.

The fluid in the system is supported by atmospheric pressure. Its weight at the sensor increases in direct proportion to the vertical distance between sensor and head, creating a vacuum whose variation moves the pointer. Maximum working depth is 28 to 30 feet, but within this range, error is said to be less than $\frac{1}{2}$ percent.

The reference point may be the ground surface beside the ditch, a section of completed bottom, a grade stake, or a stringline.

Other methods of checking depth are discussed under Operation.

Clamshell.   Many hydraulic excavators may be fitted with a clamshell bucket, which replaces the hoe bucket or immobilizes it. The clam jaws are opened and closed by a two-way cylinder. This may be mounted horizontally across them, or be part of an attaching stick vertically above them. Closing power is supplied from the hoe bucket circuit.

The clam is usually attached to the end of the hoe stick by a universal mounting, which allows it to hang straight regardless of stick position or machine slope. There may be part or full rotation, controlled by the operator. See Fig. 13.73.

Attachment may be direct to the mounting, or through one or more vertical extension rods. Extensions provide greater digging depth but less dumping height. Maximum length is limited by dumping requirements, or by clearance needed to lift the bucket out of the ground.

This type of clamshell is a powerful digging tool, in which down pressure may be added to jaw closing power. However, its applications are limited by restricted lift as compared with a cable-operated clamshell, and it is slower than a hoe. Some additional lift may be obtained by using a power-controlled hinge in the boom gooseneck angle.

Demolition Head.   A demolition head, operated either by the excavator's hydraulics or by a separate air compressor, may be substituted for the bucket. This may be used for breaking boulders, weak bedrock, pavement, and structures, and driving trench bank supports. These devices are described in Chap. 19.

FIGURE 13.73   Clamshell on double-hinged boom. (*Courtesy of Liebherr-America.*)

Even with a standard bucket, a big hydraulic hoe is a powerful demolition tool because of its fairly long, high reach, with ability to both push and pull forcefully.

OPERATION OF FULL-SWING HOES

Some aspects of backhoe operation were discussed in Chaps. 4 and 5, in connection with basement excavations and ditches.

Operation of a full-revolving hydraulic excavator or backhoe is under complete control because of two-way power and "wrist-action" flexibility in the digging and dumping angle of the bucket.

Effect of Controls.   In one control system, the boom hoist lever actuates the cylinder(s) that supports the boom. Moving the lever forward forces the boom down, pulling it back raises the boom.

It may be unwise to move this lever all the way forward when the boom is raised, as in some machines it will allow the boom (and the whole attachment) to fall with dangerous speed and force.

There may be a finger button or some other device to divert fluid from the swing circuit to the boom circuit, to raise it more rapidly from a deep excavation.

Tilting the stick (crowd) pedal forward retracts the stick cylinder, pulling the top of the stick back and swinging the bucket forward. Back-tilting the pedal reverses these motions, crowding the bucket inward toward the cab.

Since the stick is fastened to the end of the boom, its position in space at any time depends on both the boom and the stick cylinders.

The bucket pedal, when tilted forward, causes the bucket cutting edge to move out, opening and dumping the bucket. When tilted back, it brings the edge inward, closing or curling the bucket.

The swing lever rotates the upper works to the left (counterclockwise) when pushed forward, and to the right when pulled back. In most machines, an automatic brake stops the swing smoothly when the lever is centered. There should be a lock-down pedal brake also, to lock the swing during travel.

In this example, the travel or propel lever moves the machine in the same direction it is moved. The right-hand steering lever controls the clutch and brake for a right turn, the left lever is for the left controls.

All these controls are (or should be) the metering type that can be partially opened to obtain fractional speed, for performing delicate operations slowly without reducing engine speed. But accuracy of response of part-open valves is quite variable.

Also, all levers except possibly the travel control are spring-loaded to return to neutral (center), and automatically lock their unit when centered by holding fluid in cylinders. In motors, there is usually a brake that is applied by springs in neutral, and released automatically by fluid pressure.

Other Systems.   Probably no two makes of excavator have exactly the same arrangement of levers and pedals. The type that is most conspicuously different from the example described is the so-called joysticks for attachment control, and separate hydraulic drive to the two tracks. See Figs. 13.28 and 13.29.

One joystick controls both the stick and the swing. The stick valve is operated by the front-to-back motion of the stick, the swing by its side-to-side movement. The other stick moves forward and back to move the boom in the same directions, and moves left to curl the bucket and right to dump it. The two motions do not interfere with each other, and the valves operate in the same manner as when responding to individual levers or pedals.

Hydrostatic propel can be controlled in several different ways. In this example there are two foot pedals. Pushing the left one, accelerator fashion, moves the machine back; the other one moves it forward. Speed is regulated by pressure on the pedal, from zero in full-up position to full speed when depressed to the floor.

A single steering lever is moved to the right for right turns, to the left for left. There are three stages. In the first, nearest center, one track is driven faster than the other. The next locks one track while drive (regulated by the foot pedal) is continued to the other. The third stage, reached by pushing the lever past a bumper spring, causes the tracks to rotate oppositely, for a spin or spot turn.

Attachment Motions.   Assume that the excavator is parked with bucket drawn in under the boom, as in Fig. 13.74, and that it is properly located on the centerline of a proposed ditch.

The bucket is freed from the ground by raising the boom, and moved toward digging position by extending the stick. The two movements are made at the same time. Until the stick reaches a vertical position, its arc of swing is down, and must be compensated by continued raising of the boom.

When the stick has passed the vertical position, its continued motion will take it in an upward arc. If it is to be kept near the ground, the boom is now lowered as the stick is moved outward.

When the stick is extended three-fourths or seven-eighths of maximum reach, the bucket is opened until the floor is about vertical (teeth straight down). It is then lowered (not dropped) to the ground by pushing the boom lever forward.

The stick may now be crowded inward, while the bucket is rotated gradually toward a half-closed or digging position. At the same time the boom is raised slowly, so that the downward arc of stick motion will not bury the bucket too deeply.

The bucket curl is adjusted for best penetration and filling, which is likely to be at an angle of 15° to 45° to the surface being penetrated. Except when finishing off, the bucket should usually be filled in as short a digging pass as is feasible, without slowing the engine or dragging the shovel. Long passes take longer, and usually wear the bucket more.

Digging angle may be varied during digging. There should always be at least a slight angle, so that the bucket floor will not be dragged on undug dirt. Such dragging consumes power and causes excessive wear. Small changes in angle help the teeth in "hunting" for easiest penetration.

If the digging pass continues under the boom point, the bucket path will curve upward, and the boom will require lowering instead of raising if it is to be kept level.

To guide the operator, a device like Spectra Precision's BucketPro (see Fig. 13.75) would be very helpful. A control screen next to the operator's seat shows the current position of the bucket relative to the bottom of the excavation to be dug. The final depth of the excavation can be set on the screen.

Down pressure can be applied to the bucket to improve penetration, by not raising the boom or by pushing it down, depending on bucket position.

When the bucket is full, it should be closed or curled, the crowd motion should then be stopped, and the bucket lifted out of the ditch by pulling back the boom hoist lever. See Fig. 13.76.

As soon as it is clear of the ground, it is swung to the side, and dumped by pushing the bucket lever. Generally, height is figured so that the teeth will just clear the pile. A higher position wastes time in excess hoisting, a lower one exposes bucket and stick to impact and twist.

However, if the pile will be big enough to threaten to slip into the excavation, the dump may be made in the pile, so that opening the bucket and extending the stick will push soil away, to make room for more. The swing should be fully stopped before entering the pile, to avoid twisting strains.

The bucket is swung back to the excavation as soon as it is dumped and clear of the pile. During the swing it is brought back to digging position, in both degree of curl and in-and-out position on the stick control. It is lowered to the ground, and pulled in to refill.

To avoid building a pile of earth at the near edge of the excavation, always rotate the bucket toward a closed position, before bringing it out of the ground.

FIGURE 13.74   Bucket tucked in.

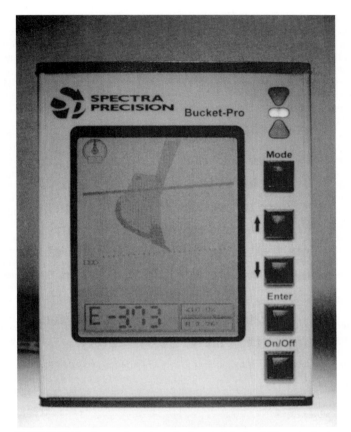

FIGURE 13.75   Spectra Precision's depth control device. With grade checking in the cab, the operator never loses the "feel" for grade or obstacles.

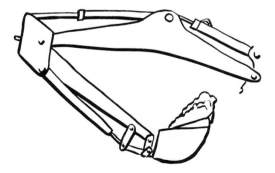

FIGURE 13.76   Bringing up a full bucket.

Finishing. The ditch is dug in a series of slices or chunks, down to the bottom. Depth may be measured by the operator or a helper. But if the trench is being dug to a fixed depth below ground surface, it is more convenient to put a mark on the stick that, when it is vertical and the bucket is flat on the ground, will indicate the correct depth. Or a device like the Accu-Depth described previously can be helpful.

The far wall can be trimmed by forcing down the bucket with the floor vertical, keeping it vertical as it goes down by curling out just enough to compensate for the inward curve of its descent. When the bottom is reached, the bucket may be pulled in with a slight lift until clear of the back wall, then closed into its load.

Pipe may be set in the trench under laser control, close to the digging. A mark may then be made on the back of the bucket, to line up with the beam at correct depth. This provides absolutely accurate control.

The ditch floor is trimmed by keeping the bucket curled so that its bottom is either at right angles to the floor, or flat on it, and crowding it in with the stick while moving the boom up and then down to keep the bucket path level. Curl will need to be adjusted as the stick swings on its hinge.

These two bucket positions share the advantage of having no suck (tendency to dig in). Teeth-down leaves soil looser and may be harder to keep level.

To trim the near wall (this is usually not necessary), pull the bucket to it along the ditch bottom, with its floor flat. Curl it up into the bank, then raise it with the boom hoist.

The boom will tend to pull it away, so crowd the stick in just enough to compensate. This will increase the curl of the bucket, so keep extending (flattening) it to keep the teeth in cutting position.

After practice, you can balance these three movements so that the bucket will come straight up on the desired line, at a good cutting angle.

Boulders. To dislodge a boulder, dig down on both its far side and its near side. Then place the bucket just beyond it, with the floor sloping at a steep angle. Apply boom down pressure and move the bucket lever back and forth rapidly.

When the teeth have penetrated to the underside, crowd and curl the bucket. Crowding keeps it close to the rock, and curling exerts great breakout force. See Fig. 13.77.

A boulder too large to lift in the bucket may be rolled out by ramping the near side, then backing away, pulling and rolling it up the slope.

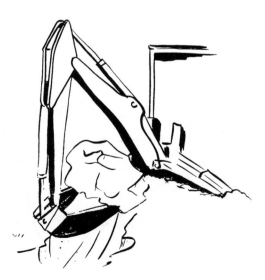

FIGURE 13.77 Dragging up a boulder.

Stalling.   The bucket may be stalled by digging resistance, either alone or in combination with a full bucket. Immediate action must be taken, either to reduce the resistance or to cut off the power. Otherwise, oil will overheat and parts will be strained.

Another effect of too much resistance is a tendency to drag the machine toward the bucket. This problem will be discussed in connection with tractor-mounted hoes, in which it is more serious.

Chopping.   Small quantities or layers of frozen ground, soft rock, or other resistant material may be loosened by lifting the attachment and dropping it, teeth down, a technique known as chopping. It is hard on the machine, and should be done only when absolutely necessary.

To avoid serious damage, be sure that no cylinders are at the ends of their strokes. Lack of cushioning by the hydraulic system in such positions causes more severe shock to the parts.

Combining crowd and down pressure with wiggling of the bucket control often gives more effective penetration than chopping does.

Undercutting Pipe.   To undercut a pipe, dig on both the far and near sides, keeping at least a few inches away from it. Go 1 foot or preferably 2 feet below the pipe, then undercut with a flat bucket from the far side, and a widely opened one from the near side. See Fig. 13.78.

Temporary Support.   When making a wide excavation in sandy soil, or above a soft, muddy subsoil, it might be advisable to rest the front of the tracks on a platform, plank or pole, to avoid digging close to the tracks.

Applications.   The hydraulic full-revolving backhoe is efficient at truck loading. Its typical cycle, digging at moderate depths and loading trucks (or piling) at its own level, is slower than that of a front shovel working in a bank, but is usually faster than a front-end loader. It varies from 20 to 40 seconds or more.

The cycle may be shortened by at least 25 percent if trucks can be spotted in the pit for loading. See Fig. 13.68. Time in raising the bucket to the surface is saved, and swing can usually be made shorter.

In addition to typical backhoe operations such as ditching, and digging basements and sunken pits, these machines are valuable for mud excavation. They may use platform supports, described in Chap. 3, or depend on their own tracks, either standard or extrawide, for support. See Fig. 13.79.

In such work, they have a great advantage in their ability to apply force to the bucket to pull, push, or raise themselves out of trouble.

TRACTOR-MOUNTED BACKHOE

A hydraulic hoe attachment similar to that described for revolving excavators can be mounted on the back of a wheel tractor, a crawler tractor, a drag trencher, or a truck. A two-wheel-drive tractor equipped with a front loader is the standard type of carrier.

It is usual to install permanent mounting brackets on the tractor, then use pins, hooks, and/or tie bars for more or less quick installation and removal of the hoe unit. The quick-detachable model in Fig. 13.80 is designed for four-wheel-drive units, and includes a seat. On two-wheel-drive units, the tractor seat may swivel to give access to the hoe.

There are also integral units, in which the tractor has mountings for both hoe and loader built into its own frame. These units are almost always smaller and lighter than full-revolving rigs, but they are nevertheless powerful and capable machines.

Boom, stick, and bucket are similar to those on the larger machines, but there are many differences in detail.

Swing.   The boom and associated digging parts are mounted on a small turntable that swings through only about half a circle, 175° to 210° on different models. Full swing is not practical because the tractor is in the way.

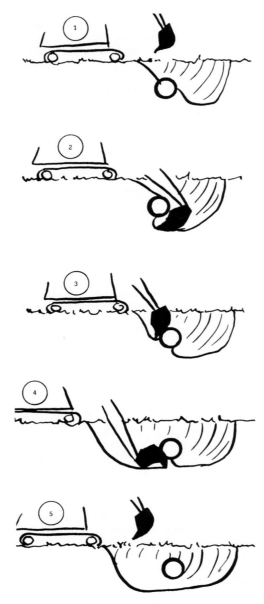

FIGURE 13.78 Undercutting a pipe.

FIGURE 13.79   Wide-track hoe working in mud.

FIGURE 13.80   Quick-detachable hoe. (*Courtesy of Ware Machine Works.*)

The pivot is usually a pair of pins, upper and lower. Power is from a pair of hydraulic cylinders based on the nonrevolving frame. The rods may be connected to brackets, or to opposite ends of a roller chain meshing with a sprocket on the rotating section. A partial-swing vane motor may replace the pistons.

Swing is usually very fast, as there is no heavy deck full of machinery to move. Acceleration and deceleration are rapid, allowing 90° swings to be completed in as little as 3 seconds. Action is likely to be jerky.

Partial swing causes little difficulty in straight, open work, like digging a ditch across a field. But it creates serious problems with complicated jobs.

Stabilizers. A pair of heavy stabilizer arms or outriggers is hinged to the sides of the turntable base. They are raised and lowered by two-way cylinders.

Cleated shoes on their outer ends are forced down against or into the ground while digging, to increase stability against tipping, and against being dragged by the bucket.

Hydraulics. The pump and reservoir are usually in the tractor. The most convenient arrangement is to have one pump for the backhoe and another for the loader. It is more usual to have one pump, and a diversion valve to route the flow to whichever unit is in use.

A backhoe and loader are not used for digging at the same time, but it is frequently necessary to make position adjustments in one while working with the other.

A pair of hoses with quick-detachable couplings brings the fluid to the valve bank in the hoe.

There are six operating valves, and usually six levers. But four functions may be performed by two joystick levers, that move to the sides as well as back and forth.

Figure 13.81 shows a standard type of control with a usual but not necessarily standard arrangement of functions. There are also universal controls, whose levers actuate cables that can be switched from one valve to another, so that the operator can make up her or his own pattern. See Fig. 13.82.

Pressure in different machines is variable, and is not always stated in specifications. Many models range from 2,000 to 3,000 pounds per square inch, with flow 15 to 35 gallons per minute.

Buckets. Buckets are similar in construction and linkage to those of the big machines, but differ in size. The smallest standard models seem to be 12 inches wide, with a capacity of 3 cubic feet; the largest, 36 inches wide with more than ½-yard capacity. In addition, there are specialty buckets on miniature hoes as narrow as 6 inches.

Even a standard-size tractor hoe may replace the bucket with a ripper tooth as narrow as 3 inches. This is primarily designed for penetrating very hard, frozen, or bouldery soil, but it may dig a slot for cable or conduit installation.

Applications. The tractor-mounted hoe is a small, powerful, and fairly economical package. It can be driven between jobs, or carried easily on a light trailer.

As a ditch digger, it can work in places that are difficult or impossible for larger machines, cross lawns without damage (except under unusually soft conditions), make narrow ditches, and show a high production rate.

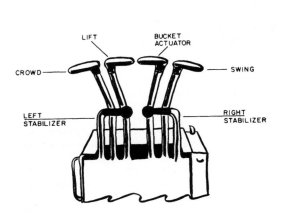

FIGURE 13.81 Tractor hoe controls.

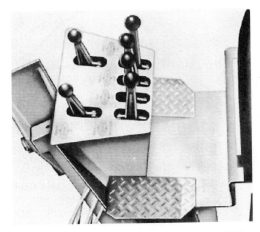

FIGURE 13.82 Universal controls. (Courtesy of Ware Machine Works.)

It does not trench as fast as the continuous-type ditchers discussed in the next chapter, but it can handle special situations that are difficult for them, can work in rocky soil, and can handle oversize pieces.

It is also a handy utility tool. It can dig out stumps and boulders so big that it cannot possibly lift them, can load trucks with either the hoe or the loader end in emergencies, and can serve as a light-duty (but very jerky) crane.

OPERATION

The boom, stick, and bucket are similar in action to full-swing attachments. Differences in operation include managing the tractor as a travel unit and counterweight, use of stabilizers, arranging work so that it can be done with limited swing, and doing heavy digging with a light machine.

Positioning for Digging.   The tractor is driven to the work spot, and maneuvered so that it is centered on the centerline of a ditch, the rear wheels toward the starting point, and about ¾ of the hoe's maximum reach from it.

The operator centers the steering wheel, puts the tractor in neutral, and locks its brakes. If one hydraulic pump supplies both loader and backhoe, the operator sets a diversion valve to deliver flow to the hoe. If there are separate circuits, he or she lowers the loader bucket to the ground.

The bucket will hold the tractor most effectively if it is put in fully dumped position, and forced down against the ground. See Fig. 13.83.

The operator then flips the seat over or swings it around, and sits rearward on the tractor, facing the hoe and its controls. In the terminology accepted by the trade, the operator is still facing forward, and hoe motions are described accordingly.

Stabilizers.   The two outside controls are used to push the stabilizers outward and downward, pressing them firmly against the ground so as to take just a little weight off the tractor's tires.

FIGURE 13.83   A dumped loader bucket helps stability.

If the ground is uneven, one is extended farther than the other, to make similar pressure on each side. If the ground slopes to the side, the low one is pushed down harder, to reduce or eliminate tractor tip.

Occasionally, you may wish to put a block under the stabilizer on the low side, or level the ground by some superficial digging.

The standard stabilizer shoe has a ridge or cleat on its undersurface, to penetrate and grip the ground. This is likely to tear up a lawn, not so much from original penetration as from dragging and twisting during digging. Very rarely, you may prefer to try digging without setting stabilizers, for this reason.

On soft ground, a plank may be placed to support the stabilizers.

Digging.   Digging motions are the same as those described earlier for the full-revolving excavator. However, this machine is probably very much lighter in proportion to digging power, and is more likely to be dragged into the work.

This dragging occurs as you pull or crowd the bucket inward against digging resistance. It is strongest at ground level, and reduces somewhat with depth.

Cutting a thin slice generates less pull than a thick one, but it takes longer and wears the bucket more. Rapid, small changes in bucket angle as it is brought in reduces resistance.

Curling or closing the bucket provides powerful digging and breakout force at the bucket teeth, with little or no pull on the tractor. See Fig. 13.84.

Starting a deep cut, then curling the bucket as soon as resistance builds up, will usually provide good digging with little dragging.

In general, you try to dig for production, but lighten the cut every time you feel a drag-the-tractor force developing. The reaction is a split-second one, which is difficult at first but soon becomes practically automatic.

Moving.   To move when a section of digging is finished, the stabilizers (and the loader bucket if it is down) are raised. The bucket is placed on the ditch bottom with its floor at a slight angle, with the stick almost vertical, but leaning slightly toward the tractor.

Pushing the hoist and crowd levers slowly forward will now lift the back of the tractor off the ground, and roll it forward on its front wheels. If it does not go far enough, the maneuver is repeated. Then the stabilizers are put down, and digging is resumed.

The tractor may be moved to the side in somewhat the same manner. The bucket is curled halfway, placed in the ditch a little forward (away from the tractor) of the boom point, and forced down by the boom cylinder until the rear (big) wheels are clear of the ground. The hoe is then swung until the tractor is in the desired position. See Fig. 13.85. Then it is lowered, and the stabilizers are set for digging.

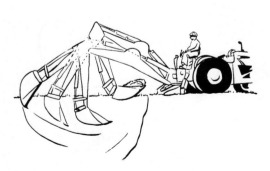

FIGURE 13.84   Digging motion.

FIGURE 13.85   The hoe moves the tractor.

Stability.   The total weight of a tractor-hoe, including its loader, is much smaller in proportion to digging power than that of a full-revolving machine. Stability is therefore more of a problem.

Down pressure on the bucket tends to lift the rear wheels off the ground. Upward pull lifts the front wheels. Rearward (toward the tractor) pull may either dig the dirt or drag the tractor.

Tractor weight must be increased to its maximum by putting water and calcium chloride, or perhaps mineral dust, in all four tires. Refer to Fig. 12.139. The loader bucket may be filled with dirt, or piled with rocks or other heavy objects. It may be held in the air for counterbalance (and for convenience in moving), or may rest on the ground for resistance to dragging.

Stabilizers should be put down far enough to take a little of the weight off the rear tires, as will be indicated by a slight change in shape at the bottom. Tires must never be lifted clear of the ground, or even so that most of the weight is off them, except one side in leveling on a slope. On ordinary ground, a ridge on each stabilizer pad sinks all the way in.

Rear wheels (larger ones in figures shown) are locked by the tractor brakes.

Even with these precautions, weight and grip on the ground are small in proportion to the digging forces—crowd and curl—which may be more than 6 tons. The operator must limit the force he exerts to the ability of the machine to keep its position.

Digging without pulling the tractor may be the hardest thing for the beginner to learn, as was discussed earlier, but it ordinarily offers few problems to the experienced operator. It is largely a matter of learning manipulation of the bucket to obtain penetration without excessive pull.

When the tractor is dragged toward the digging, it is usually necessary to reposition it with the bucket. If the movement has not been straight, it may be necessary to drive the tractor to get it in proper alignment.

Stability against tipping is ordinarily not a major problem in digging. But if the tractor is not level, or is being used to lift heavy objects with a chain, this must be considered. It usually has least stability when the load is at a 60° angle from center, on either side if level, on the low side if not level. Tipping tendency is reduced by bringing the load in toward the tractor, or stopped by setting it down.

Slopes.   A side slope affects stability of a hoe, making digging motions difficult to control, and produces a ditch that is out of plumb (side not vertical).

A slight slope may be ignored, or corrected with the stabilizers by putting pressure on the downhill one until it levels the tractor, or adding a block.

For a steeper slope it is advisable to make a cut or shallow trench for the uphill stabilizer and wheel, using the spoil as fill for the downhill side if necessary. This cut may be made by the loader bucket, preferably pushing downhill, or by the hoe. See Fig. 13.86.

A shallow leveling cut may be made by the hoe on fairly steep side grades, as little material is dug and accuracy is not important. This may extend the full length of the hillside, or just be scooped out for digging position.

When the tractor is tipped, it is much more stable dumping uphill from the ditch than downhill, and this side is generally (but not always) favorable for backfilling.

If there is a choice between trenching up a hill or down it, work with the tractor heading downhill. This reduces the principal problem of dragging the tractor toward the digging, and is much more stable against tipping when dumping the bucket.

An exception to this advice is that in wet ground it may be desirable to dig uphill to avoid ponding of water in the digging area.

Close Work.   Tractor mounting is better than full revolving for working close to buildings or trees, as it is not necessary to allow space for a tail swing. One wheel can be rubbed against the obstruction, if necessary.

It is sometimes possible to get closer to a wall by backing the machine up to it at an angle for short cuts. The front wheels may be turned parallel to the wall, so that the tractor can be boosted by the bucket from one digging position to another.

If the tractor has a frame-tilt capability, as seen in Fig. 13.34C, that mechanism can be used to advantage in close work.

FIGURE 13.86   Sidehill trenching.

CHAPTER 14
CONVEYOR MACHINERY

CONVEYORS

CONVEYOR BELTS

The belt conveyor is a transporting, elevating, or distributing machine made up of an endless wide belt which carries a load on its upper surface. It operates between a head and a tail pulley and is supported by idlers, which in turn are supported by a frame or by steel cables. Conveyors are made in small portable elevating units that are loaded with hand shovels, and as giants that carry millions of tons of earth for many miles.

As independent units, they are well suited for rapid transportation of loose material. They have less mobility and flexibility than trucks and scrapers, and are therefore used chiefly where large volumes of material are to be moved along one route. They are particularly applicable where the load must be lifted steeply, or carried across rough country where road construction would be difficult. They are desirable as feeders for processing plants because they provide an even and continuous flow. They simplify traffic problems where hauling space is restricted, as in tunnels and in busy pits. However, they are not adapted to hauling big chunks which clog hoppers, damage the belt, and are likely to fall off in transit.

Their mechanical efficiency is high, as very little deadweight must be moved with the load, friction is at a minimum, and power-consuming starts and stops are rare.

A conveyor can be kept in touch with a receding pit face by fitting in portable frame extensions and splicing in extra belt, by installing a feeder conveyor between the hopper and the face, or by trucking from the excavator to the belt.

In addition to their use as independent and semi-independent haulers, conveyor belts are used as parts of loading, ditching, and processing machines.

The large permanent type of belt conveyor is almost unique among machines used in connection with excavation in that it is usually custom-built and rules of design and construction are flexible, so that it can be "tailored" to the job.

In an average installation, the belt is about half of the original construction cost, and its repair and replacement are the biggest maintenance charge. Proper care of it is therefore very important.

The belt extends between a head pulley (which may be the drive pulley) and a tail or return pulley, and carries its load on its upper surface, usually toward the head pulley. Its upper strand is supported by idler sets whose three rollers are arranged to shape it into a trough, and the lower strand is supported at wider intervals by flat rollers called return idlers.

Figure 14.1 shows the parts of a simple conveyor and Fig. 14.2 shows one in action. There is a frame which keeps the working parts in position, an engine that turns a drive pulley that moves the belt by friction, a tail pulley to reverse the direction of belt travel, and idlers and return idlers.

Frame. Many frames are of the sectional type. A head and a tail piece must always be used, and as many intermediate sections as are required to obtain the desired length. Sections are bolted to each other.

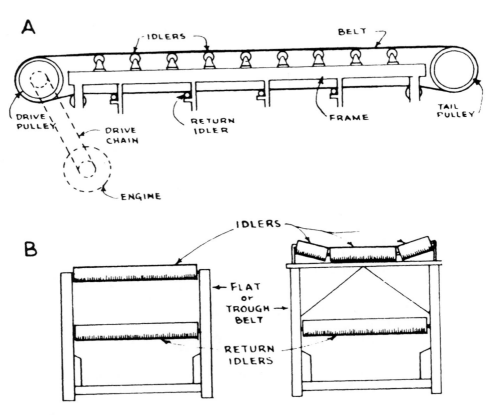

FIGURE 14.1   Belt conveyor parts.

Figure 14.3 shows several ways of mounting rigid frames. Wheel and caster carriers are suitable only for light machines. Pillar and floor are usually permanent, but may be made so that they can be disassembled for moving. Stationary installations usually include a catwalk from which the belt and rollers can be serviced.

Conveyors may also be carried on two strands of wire rope with movable head and tail sections, on skids, and in other ways.

Belt.   The belt is an endless flat strip of rubber-covered cotton or rayon fabric laid up in plies. Very long or heavily loaded belts may be reinforced with steel cable. The type of fabric, the number of plies, and any reinforcement which is present determine the strength of the belt. The rubber cover serves only to protect the fabric from abrasion and weather. Its thickness and quality are varied to suit different types of service. Several constructions are shown in Fig. 14.4.

There is no definite limit to the length of a single belt. More friction surface for the drive and stronger belt construction are needed as the unit is made longer, the climb steeper, or the load heavier. Common practice limits the carrying distance to ¼ mile, but belts with a carry of 1 mile or more have been constructed.

Any distance can be crossed by a series of belts, each of which dumps on the next through a hopper or over a baffle plate. The length of each conveyor is figured between the centers of the head and tail pulleys.

FIGURE 14.2   Belt conveyor in action.

Belts which move dirt or any loose material are usually run with a trough in the upper surface, which centers the load and reduces spill off the sides. Very short conveyors may carry a bigger load by using a flat surface with fixed side skirts.

Belts may be from 8 inches to 8 feet in width, but standard belts range from 12 inches to 60 inches with 30 inches very common.

Drive.   Power carries from the drive pulley to the belt by friction. If the resistance of the belt to moving is greater than the friction, the pulley will spin or slip inside the belt, with resultant loss of power, and will wear on both surfaces. The amount of friction or traction is determined by the nature of the surfaces, the slack side tension on the belt, and the area of contact.

The load or tension on the carrying strand of the belt, which tends to cause slippage, is made up of the pull of gravity on this strand and its load, friction in idlers, pulleys, the belt, and its load; and the inertia of the whole system when starting or accelerating.

The drive pulley surface may be bare metal, or covered by smooth face or grooved rubber lagging. Such lagging may be bolted or vulcanized in place. It increases traction, particularly when the belt is wet or frosty, and prevents pulley wear.

The belt is held in full contact with the pulley by its tension on the slack or low-tension side. This tension is normally regulated by some form of gravity takeup, in which a hanging weight exerts a pull on the tail pulley, or a special takeup pulley, which moves outward if the belt slackens and inward if it tightens. If the incline is steep, the weight of the slack side may maintain sufficient tension. Very short conveyors may have threaded adjustments to move the tail pulley in or out.

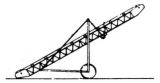

**Mast Truck Mounting
with Power Hoist**

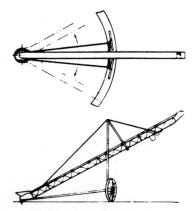

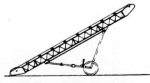

**V-Truck Mounting
with Hydraulic Hoist**

**Swivel Wheel
Radial Mounting
with fixed pivot**

Horizontal Four-Caster

**Horizontal 4-Wheel or Caster
adjustable discharge height**

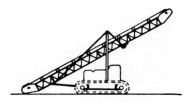

**Self-Propelled
Crawler Mounted
Power Raise**

**Rigid Axle 4-Wheel
for Shuttle Installation**

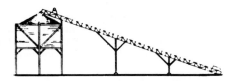

Two-Wheel Mounting

**Stationary Mounting
to suit requirements**

FIGURE 14.3   Conveyor mountings.

The amount of drive traction which may be obtained by increasing tension is limited by sharply increased power requirement and shortened life of a too-tight belt.

The area of contact is determined by pulley diameter, the arc of contact, and the number of pulleys. A thicker pulley not only increases the contact area, but also reduces flexing strain on the belt. Its disadvantage is the cost of the pulley itself and of changes in frame and layout to accommodate it.

If an existing drive pulley is replaced by a larger one, the belt will then be driven at higher speed with less power, unless the gearing or the motor is changed. Putting on lagging has the same effect.

STANDARD

Multiple layers of suitable duck of the same thickness and ply across the entire Belt. Standard is suitable for nearly all types of Conveyor service.

SHOCK PAD

Substantially the same as Standard but with a reinforced top *Cover* consisting of an abrasion-resistant tire-tread stock on the top surface, backed by a thick pad of resilient rubber. The pad yields to sudden, extreme impacts and pressures—protecting the cover from puncture or breakage and preventing rupture of the carcass.

STEPPED PAD . . . Stacker-type Cover

Substantially the same as Standard construction except that it is moulded with the *Cover* having an additional thickness standing out in relief in the center Belt area. It is recommended when the loading of abrasive material is concentrated in the center of the Belt with only slight abrasion at the cover edges. Stepped Pad Belting is not recommended for two-pulley or internal drives.

STEPPED PLY

A smooth-top construction having a heavier cover in the center of the Belt than at the edges. This is accomplished by moving the middle portion of one or two of the top fabric plies to the sides and filling in the extra space with cover stock. A Stepped Ply Belt has more crosswise flexibility and troughs more easily. The extra thickness of cover stock gives longer life to the Belt under loading conditions where abrasion is concentrated in the center Belt area.

FIGURE 14.4   Belt cross sections. (*Courtesy of Hewitt-Robins Incorporated.*)

A belt whose strands are parallel will have 180° of contact with the drive pulley. This contact can be increased by a snubbing pulley on the slack side. Increasing the degree of wrap in this manner is the cheapest way to increase contact area.

Head pulley drives are adequate for short conveyors, and for those so steeply inclined that the weight of the return strand maintains a high tension on the slack side. For conditions requiring greater traction, tandem pulleys are used; see lower right example in Fig. 14.5. Up to 440° of wrap can be obtained in this manner.

A belt which is pulling a load changes shape as it goes over a drive pulley. It is stretched thin where it first contacts it, and then fattens up as its tension is reduced. See Fig. 12.29. It moves fastest where it is thinnest, in the same manner that water is accelerated in going through a restricted place in its channel. The change in belt thickness and speed is quite small, but requires that the second of the tandem pulleys turn a bit more slowly than the first, if extra stress is to be

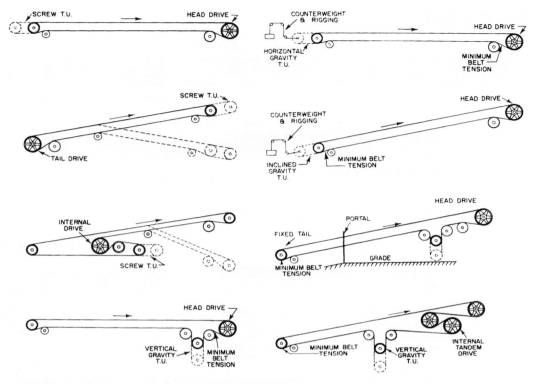

FIGURE 14.5   Drives and takeups. (*Courtesy of Hewitt-Robins Incorporated.*)

avoided. The amount of difference varies with the load. Where electric drive is used, separate motors on the two pulleys can be made to automatically adjust their speeds to each other.

A possible disadvantage of tandem drive is that one pulley works on the load-carrying surface of the belt, which may be wet and slippery so as to afford a poor grip; or gritty so that excessive wear of both belt and pulley will occur. This may be avoided by using the head and lower pulley for drive, and the intermediate one as an idler.

A number of drive and takeup arrangements are shown in Fig. 14.5.

Short reversible belts may have drive pulleys at both ends, connected by roller chain.

Idlers.   Idler rollers support the upper or working surface of the belt. They are usually of the troughing type, in which a horizontal center roll supports the loaded part of the belt, and a pair of outer rolls turn up the edges to create a trough cross section which keeps the load from spilling off the sides.

At loading points, shock and wear to both belt and idler can be reduced by using rubber idlers as in Fig. 14.6. The idlers turn on ball or roller bearings. Lubricant is usually renewed only at long intervals, and may be sealed in.

In excavation work, the flat idler and belt are used only on skirted conveyors.

Adjustment.   Short belts commonly have a screw adjustment or takeup on the tail pulley, which slides in and out on a track. Care should be taken to adjust both sides equally to keep the axle at right angles to the direction of belt travel. Any cocking will make the belt tighter on one side than the other, so that it will tend to climb the pulley and run off the tight side.

In long belts a fixed tension is not satisfactory, as the length of the frame is affected by changes in temperature, and the belt is affected by both temperature and moisture.

FIGURE 14.6   Cushion idler. (*Courtesy of Hewitt-Robins Incorporated.*)

There are two common types of automatic takeup which keep the belt under constant tension. In the horizontal gravity or counterweight-and-rigging method, tension is controlled by a tail pulley in a track, which pulls it outward by means of a weight hanging from a pulley. With a vertical gravity takeup, a fixed tail pulley is used, and the weighted pulley hangs between two return idlers, preferably at or near the point of minimum belt tension. These constructions are indicated in Fig. 14.5.

If the belt stretches too far to be adjusted, a piece is cut out, the ends are stapled or vulcanized together, and the adjustments are reset.

Alignment.   The belt is sensitive to tiny changes in frame and pulley alignment, which will cause it to wander out of a straight line. Internal changes in belt tension, or a splice which is beginning to pull apart, may have the same effect. Trouble with wandering can be greatly reduced by making both pulleys and frames wide enough that the belt does not have to run absolutely straight to keep out of trouble. The wider construction is more expensive, of course, but through the years it should more than pay for itself in longer belt life, and in reduced checking and adjusting.

If the framework is out of line, the carrying strand may still track well enough, being steered by the load seeking the idler troughs. The return strand, however, will find the shortest path, or allow itself to be influenced by sloping idlers, so that it will rub against stationary parts. Unfortunately, it is almost standard practice to carry the return strand inside the framework, where it is very difficult to see just what it is doing. It is much better to hang it below the frame members, where misbehavior can be readily observed and necessary corrections made before serious damage is done.

Troughed idlers will steer the belt if they are tipped in the direction it is moving. However, only a very slight tilt can be used, because if it is overdone, it will set up a drag against the bottom of the belt which will wear it rapidly and consume extra power.

Self-aligning or training idlers are mounted on a center swivel, and have vertical spools set at each edge. See Fig. 14.7. If the belt rubs against a spool, it tilts and presses a lined brake shoe against the adjoining roller, slowing it, swinging the idler, and shifting the belt back toward center. This device creates very little drag, and if placed every 50 feet in place of regular idlers, will keep a belt in line under any ordinary conditions. Two placed at 30-foot intervals ahead of the tail pulley will line the belt up properly to go under the loading point.

Holdbacks.   An inclined conveyor tends to run backward when power is cut off, because the weight of the load pulls the upper part of the belt downhill. This tendency can be overcome by means of a brake of sufficient size, but it is often more convenient to use a device which will automatically lock it against turning backward, without interfering with normal movement of the belt.

The holdback shown in Fig. 14.8 consists of a ratchet wheel keyed to the drive pulley shaft and held between a pair of side plates which are hinged to a toggle and tooth pawl anchored on the conveyor frame. When the belt and drive wheel are turning in the normal direction, friction between the ratchet and the side plates lifts the unit on the hinge, holding the pawl clear of the ratchet. If the drive wheel starts to turn backward, the hinge is pulled down and the pawl meshes with the ratchet.

This device cannot be used on a conveyor transporting downhill, as its load tends to move the belt forward.

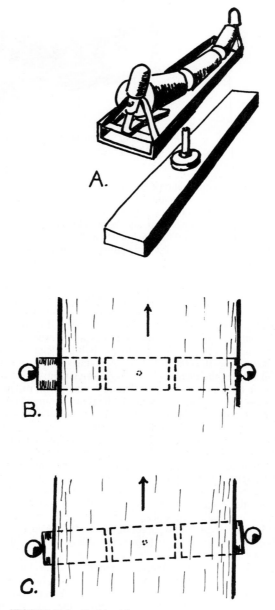

FIGURE 14.7 Training idler.

Hoppers and Loading. Hoppers are of two principal types: the primary or loading hopper, and the transfer. The primary usually has enough storage capacity that it can supply material steadily to the belt, although it is loaded intermittently by bucket or truck loads. See Fig. 14.9. It may have a screen or grizzly to keep out oversize pieces.

Loading hoppers must have a restricted chute capacity, or a gate or feeding device to regulate the rate of delivery to the belt. In some installations, increased flow is obtained by merely raising

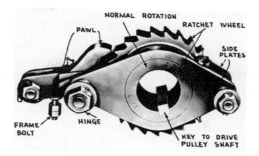

FIGURE 14.8 Holdback. (*Courtesy of Hewitt-Robins Incorporated.*)

FIGURE 14.9 Loading hopper and conveyor. (*Courtesy of Felco Industries, Ltd.*)

the chute, so that more material can pass under its forward edge. For any rate of feed, the load per foot of belt will be increased by slowing the belt, and decreased by accelerating it. Regulating devices must be so constructed and used as to allow through the largest lumps which get into the body of the hopper, and to avoid jamming by several smaller lumps.

Various feeding devices are available to ensure even flow from the hopper to the belt. A short feeder conveyor (transition belt) may be used, either for convenience of location or to save the long belt the extra wear of taking a part in measuring out material. Another common device is a pan which moves back and forth under the hopper chute, carrying and pushing a fixed amount of material toward the belt with each move. Other models use shaking or vibrating plates, or rotating rolls or vanes.

A conveyor belt is loaded by a chute from a hopper or another conveyor, or by a feeding device. The material always has some vertical drop, and may be moving in a different direction and at a different speed than the belt. As a result, there is impact to the belt in stopping its fall and giving it its proper speed and direction. Chutes and baffles should be so constructed that they convert as much as possible of vertical and cross movement into movement in the direction of belt travel, and so that they deposit their load squarely in the center of the belt. If the material is abrasive, the chute may be so constructed that a layer of the material piles up on its slope and protects the belt. When the main conveyor is a big one, it may pay to put in a short feeder conveyor (transition belt) to ensure feed at the proper speed and in the right direction. This transfers wear from an expensive belt to an inexpensive one.

Coarse, lumpy materials, particularly when they include sharp edges, require greater precautions at the loading point than fine or soft ones do. Impact damage to the belt can be reduced by using thicker rubber on both the top and bottom surfaces, and by using closely spaced rubber-cushioned idlers at the loading point.

A certain amount of belt wear from impact and abrasion occurs each time the belt goes over an idler, as it is a high point with a sag on each side. The slight but steady wear at these points can be reduced by minimizing the sag, which is affected by belt tension and idler spacing. Maximum smoothing effect can be obtained from any given number of idlers by using wide spacing near the head, where tension is greatest, and progressively narrower spacing toward the tail pulley.

If the load slides back on itself while going up an incline, a lighter load may be more efficient. If it slides on the belt, a belt with a cleated surface may be required. Often a change in the moisture content of the earth will cause or cure this type of trouble.

Spillage. There is usually very little spillage off a belt which is wide enough for its load. However, many belts are overloaded in bulk if not in weight, and the jouncing over idlers, movements in the load, and changes in alignment result in pieces rolling and bouncing off the sides. If decking is placed under the full length of the belt, or sometimes under the ends and the loading point, this material will be kept from spilling onto the return belt. A clean belt and clean pulleys and idlers mean longer belt life.

It is important that a diagonal cleaner bar or plow be placed on the return strand just above the tail pulley, as a large sharp object might otherwise fall on the belt and catch between it and the pulley, so as to puncture the belt or even cut it into strips.

Wet dirt and many other materials stick to belts and will build up to considerable amounts if permitted to do so. A scraper may be placed just after the discharge point, to clean the belt and drop the scrapings into the same receptacle as the mainstream.

This cleaner may consist of a rubber or steel blade, or a series of them, or be a serrated rubber roll or bristle brush which revolves oppositely to the belt. See Figs. 14.10 and 14.11. Proper pressure against the belt and adjustment for wear are provided by counterweight or springs.

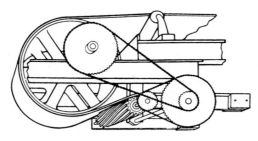

FIGURE 14.10 Belt cleaner. (*Courtesy of Hewitt-Robins Incorporated.*)

FIGURE 14.11 Brush and rubber roll cleaners. (*Courtesy of Hewitt-Robins Incorporated.*)

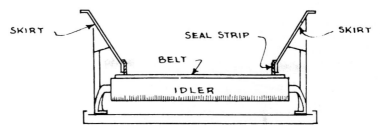

FIGURE 14.12   Skirted belt.

Troughs and Skirts.   Most belts that carry loose material—and this includes excavated soil and blasted rock—are troughed so that the center is lower than the sides. The side idlers are usually at a slope of 20°. However, there is a trend toward steeper slopes, and 35° and 45° may now be used to increase load and/or reduce spillage.

Skirts are used to prevent spillage off the sides of a belt at a loading point, and to increase the capacity of short belts. They usually consist of vertical or sloped faces of metal or wood, with a lower piece of flexible rubber which makes light contact with the belt. This rubber should be lowered as it wears, and replaced whenever necessary, to avoid damage from material wedging under it. See Fig. 14.12.

Both troughs and skirts may be used at a loading point. The trough tends to keep material from reaching or pressing against the skirt and working under it. Trough slopes as steep as 50° may be used.

Trippers.   A conveyor belt may discharge into any of several bins or piles. For a discharge point not at the end of the belt frame, a tripper may be used to dump the load into a sidecasting chute at any point. As shown in Fig. 14.13, it consists primarily of a pair of straight idler pulleys, one of which turns the belt under and dumps it, and the second returns it to its original direction.

The model illustrated is moved along tracks by a hand crank, and is anchored in position by a clamp. Other types can be propelled by the belt or by a separate motor, and can be under manual, automatic, or remote control.

Safety Devices.   A belt conveyor will run for long periods without attention. Idlers require lubrication twice a year or less (if of the modern type with bearings and seals), a well-founded and constructed frame will keep the parts in line, and properly designed and protected chutes may never clog. However, sudden accidents happen, which can be very costly if their results are not controlled, and it is impossible to keep a worker always on the alert for events which may not happen for years, or ever. Controls should therefore be installed which will automatically react to emergencies.

If the power fails on an inclined belt, it will tend to run backward, jamming its load into and around the loading chute. This movement can be prevented by a holdback ratchet on the head or drive pulley, which will allow free working motion, but immediately lock against backsliding.

If objects jam between the belt and a pulley, if discharged material backs up into the belt, or if the belt runs off its rollers, the power requirement will be sharply increased. If drive is electric, an automatic overload switch in the line can be set to cut off the power, and prevent or limit the resulting damage.

It is also a good plan to use a motor which is not too big for its job. A 200-horsepower motor can drag a belt into a lot more trouble than a 100-horsepower can, without increasing current requirement as sharply. The oversize motor will also put greater stress on the belt when starting it with a load.

A moderate rise in power consumption without increase in load indicates increasing friction. It may be in dry or broken idlers, result from the belt rubbing the frame, or be caused by too much tension. A record of current consumption will often reveal such conditions before they would be observed otherwise.

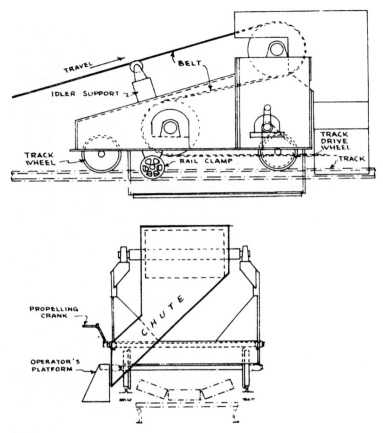

FIGURE 14.13   Tripper.

Inclines.   Conveyors are used for uphill, horizontal, and downhill (decline) transporting. See Fig. 14.14 for a bucket conveyor on a steep incline..

A belt will lift dry sand on an incline of about 15°, and wet sand up to 20°. Some cohesive materials can be carried on grades as steep as 28° on standard belts. The limiting factors are slippage between the belt and the load, and sliding of the load on itself.

Rough-surface belts may be used on somewhat steeper inclines if the load is thin. Belts with metal cleats will carry loads on inclines of up to 35° or 45°. Belts or chains with attached buckets can lift at any slope or vertically.

Capacities.   Persons used to seeing dirt moved by the bucketful or truckload have difficulty appreciating the volume production represented by the thin ribbon of dirt on a belt.

Tables showing conveyor sizes, inclines, and recommended speed will be found in the Appendix. As one example, a 30-inch belt carrying gravel weighing 3,400 pounds per yard, including occasional stones up to 14-inch diameter at 450 feet per minute, will move about 520 yards per hour.

Output is usually figured in tons per hour, and conversion to yards made by means of average-weight data. Production is determined by the width of the belt, the speed of the belt, and the height to which it is piled. Power needs are proportioned to tonnage and lift.

Those variables make design of a conveyor system a complex matter, with plenty of room for difference of opinion. In general, increasing belt width means more expensive construction

FIGURE 14.14   Incline bucket conveyor.

throughout. Increase of speed beyond specifications shortens belt life, and may involve lost power through slippage between the belt and the load. Heavier loading may call for stronger frame and idler construction, and heavier belting. In addition, a problem of spilling off the sides may be encountered.

If capacity of an existing installation is to be increased, the most economical, although often not the soundest, way is to increase power and speed and to maintain original load.

Recommended speeds for various materials range from 300 to 1,200 feet per minute, with dirt, sand, and gravel in the higher rates.

Conveyors many miles in length are often used in dam construction to bring material in from borrow pits. See Fig. 14.15. They can operate on grades much too steep for trucks, so that in rough country original construction may not be substantially more expensive than that of a haul road of equal capacity. Once built, a conveyor can show a much lower operating cost per yard than truck fleets.

Distribution of the fill may be handled by conveyors, trucks, or both.

Shiftable Frames.   A conveyor used in an open pit may require frequent sideward moves to keep near a bank that is being cut back. Dismantling a conventional unit and setting it up again is likely to be costly in labor and in lost production.

A German construction technique, shown in Fig. 14.16, is to mount the entire conveyor system on ties or skids that rest on a strip of land graded as if for a road. Idlers supports are grouped on

FIGURE 14.15 Cross-country conveyor. (*Courtesy of Hewitt-Robins Incorporated.*)

FIGURE 14.16 Side-shifting a conveyor.

frames about 15 feet long that are not directly connected to each other. The frames are supported at each end by a pivot mounting on a cross tie or skid.

The ties are connected by one or two 90-pound railroad rails. A heavy four-wheel-drive tractor, equipped with a side boom, carries a roller clamp that holds the top of one of the rails. Shifting is done by lifting the clamp and rail slightly, and running the tractor parallel with the conveyor and

FIGURE 14.17   Extensible conveyor.

at a slight distance from it. This will pull the rail and the whole conveyor 3 to 6 feet to the side. A return trip in reverse moves it the same distance again. The process is repeated until the conveyor reaches its new location. The rail is flexible enough for the bending involved.

The head and tail sections are usually mounted on skids or pontoons, and held by tightening anchor lines by hand winches. Sideward movement involves towing with the tractor, or pushing with a dozer. Very large drive stations may have pontoons that are equipped with hydraulic-powered legs that can be used for tension adjustment and for side movement.

Extensible Conveyor.   An extensible conveyor is a frame unit usually with mobile, crawler-mounted head and tail units. The head or drive unit holds up to 200 feet of extra belt, moving back and forth over a series of takeup pulleys. The tail unit carries two reels of rope held by brakes.

In coal mining work, the tail unit may follow a continuous miner, as in Fig. 14.17, receiving its discharge directly or across a short loading bridge conveyor. It can also be loaded by shuttle cars or in any conventional manner.

During operation the tail unit travels as much as is necessary to keep in contact with the excavator, while the head unit is kept stationary at a point where it discharges into a main conveyor, haulers, or a processing plant.

Both the ropes and the belt pay out automatically as the distance between the head and tail is lengthened. It is only necessary to place extra idler frames to support the belt.

When the reserve belt has all been paid out into the working stretch, the conveyor is stopped and an extra piece of belt is spliced in. This should be long enough to cover the working distance and refill the storage rolls. The rope reels have sufficient capacity to take care of several belt extensions. A reserve belt 200 feet long will permit a conveyor advance of 100 feet.

STACKER

A conveyor may end in a pivoted boom to allow placing material in separate piles or to facilitate building piles or embankments to a desired shape or size. The machine shown in Fig. 14.18 is building a levee along the Mississippi River. The long conveyor in the background feeds it through a hopper. Such a conveyor might be loaded in a pit miles away, or by trucks at a nearby transfer point.

Stackers are usually custom-made. They can be stationary, towed, or self-propelled. Booms vary from 60 to 200 feet in length, belt width from 24 to 54 inches, and capacity from 300 to over 1,000 tons per hour.

They are frequently used in stacking of processed material in mine and factory yards. They also can load materials from a pit floor directly into trucks or trains on the high wall.

FIGURE 14.18 Levee stacker.

BELT LOADERS

A belt loader may be any of a number of machines that load haulers or sidecast by means of a conveyor belt. The material may be pushed or dumped into a receiving hopper by other machines, or dug by a knife, blade, plowshare, disc, or bucket wheel that is part of the loader and is moved by it. In all these machines, the belt keeps the digging parts from being choked by accumulated cuttings, and is in itself the most economical type of elevator. As a result, these machines are a highly efficient type of loader under favorable circumstances, and are capable of very high volume of production in relation to power used. See Fig. 14.19.

In general, they do best in materials of a fairly even texture that can be cut readily by their feeding devices. They are inconvenienced by stones and lumps that are large in proportion to belt width.

PORTABLE BELT LOADERS

This type of belt loader does not propel or load itself, but can usually be moved around fairly readily. It consists of a heavy loading box or trap which is filled to the point of burial by bulldozers pushing material to it, a wide conveyor belt that carries the material to an elevated discharge point, an engine, and various accessories. See Fig. 14.20.

The trap must be very heavily built. Its opening may be either fixed or adjustable. If fixed, there is a reciprocating-plate feeder under it, to deliver more or less steady flow of material to the belt. If adjustable, it permits material to fall directly onto the belt, a desirable feature in sticky soil.

Box height, from base to the high end of the trap opening, may be 10 to 18 feet. The conveyor is comparatively short—40 to 60 feet—and 4 to 6 feet wide. Speed is moderate, about 350 feet per minute.

The tail end is in the loading box. The discharge or head end is carried by a cantilever frame over the loading point. The head may be fitted with a vibrating grizzly or screen to carry oversize away from the hauler.

The belt and any accessory units are driven through chains and intermediate shafts by a diesel engine of 100 to over 200 horsepower.

The whole machine is a single unit, supplied with a detachable pneumatic-tire undercarriage, wheels, and a kingpin attachment for use with a fifth-wheel trailer hitch. In addition, the bottom

FIGURE 14.19   Belt loader in action.

of the trap is fitted with heavy skid plates, so that it can be dragged on the ground without damage.

The discharge end ordinarily extends over a truck roadway. Discharge clearance may be 10 to 12 feet. The road may be excavated if more is required.

Since a standard machine has a fixed discharge position, a truck that has been loaded must move away before an empty one can be placed. Since the belt must be idle during the exchange, the reduced spotting time in drive-through is much more efficient than backing in.

Oversize.   Loose rock in the bank may be in pieces that are too large for the trap, the belt, or the fill. In general, pieces that will go through the trap opening will be carried by the belt, so the first two situations are a single problem.

Fortunately, any rock big enough to cause difficulty is likely to be noticeable in front of the blade, before reaching the trap. Operators can be instructed to separate oversize pieces from the feed before they reach the trap.

The trap may be protected by welding on heavy grizzly bars, but rocks retained on them may be difficult to reach to get them out of the way.

For protection of fill quality, a belt loader may be equipped (preferably at the factory) with a shaking or vibrating screen at the conveyor discharge. This can separate oversize from acceptable material, with minimum operating cost.

With a screen, there should be two discharge points. The soil will fall through the screen close to the conveyor tip; the rejects will be carried farther, to the end of the screen. The two discharge streams must be sufficiently separated so that two trucks or other haulers can stand side by side, one under each discharge stream.

FIGURE 14.20   Portable belt loader in action. (*Courtesy of Athey Products Corporation.*)

Haulers.   A belt loader can discharge into any type of hauler that has an open top. This includes trucks and trailers of rear dump, bottom dump, and side dump construction; either highway or off-highway models (discussed in Chap.18); and nonelevator scrapers (discussed in Chap.17).

The principal consideration is that the haulers should be big—the bigger the better. Time for spotting a truck, figured from stopping the conveyor to restarting it, may average 15 seconds. A 60-inch belt is rated at 1 cubic yard per second. Haulers carrying 45 yards (loose and heap) would keep the loader busy $\frac{3}{4}$ of the time, ones with 15-yard capacity would work it only half the time.

Belt loading is kind to haulers, as the material is poured into them, instead of being dumped in big chunks from a bucket. With portable loaders, one loading spot is likely to be used for concentrated traffic for several days, so that it is likely to be well prepared and maintained. Scrapers are spared the heavy stresses of being pushed through a digging run.

Production.   A 60-inch belt loader is rated at 3,600 yards per hour. A loss of one-fourth or more due to hauler spotting may be assumed, reducing this to 2,700 yards per hour. Average production on a 45-minute per hour basis would be almost 2,000 yards, which is a lot of dirt.

Loaders with 48-inch belts are rated at 2,000 to 2,800 yards per hour, and 72-inch units at 4,800. These figures are subject to the same markdowns for spotting time and 45-minute hours.

WHEEL EXCAVATOR WITH BOOM

Wheel excavators have been widely used in other countries, particularly Germany, for many years, but have been uncommon in the United States until recently. Most of them are big, electric-powered machines designed for large-scale removal of soil overburden in strip and open-pit mines.

Booms.   The wheel excavator has two conveyor-equipped booms. The digging boom or ladder carries the cutting wheel, and the discharge boom or stacker disposes of the cuttings. The two booms are mounted on a revolving superstructure. They may be fixed in line with each other, or may have partial swing independent of each other. In-line booms balance each other; independent ladders and long stackers must be equipped with counterweight structures. The two booms usually have separate hoists.

The ladder may be raised and lowered on a fixed pivot, or it may have a slide on which it can be crowded into the digging or pulled back from it.

Independent Loading Unit. Figure 14.21 shows a very large wheel excavator, digging a 100-foot bank and discharging spoil across a bridge conveyor into a separate mobile loading unit with a two-train capacity. This machine can dig 100 feet above its tracks and 16 feet below. Its daily output is about 80,000 bank yards. The excavator is mounted on three pairs of crawler tracks without leveling devices.

FIGURE 14.21   Excavator and car loader.

Wheel.   The digging wheel rotates on an axle at the tip of the ladder. It carries a number of toothed buckets, that will vary in size, number, and design for the material to be dug. The curved part may be made of link chain to reduce sticking.

The buckets are open against the wheel, so that their loads spill as they pass over the top; see Figs. 14.22 and 14.23. The material falls on a chute that slides it to a side conveyor belt.

The smallest wheels are less than 5 feet in diameter. They are usually double, and discharge to a center conveyor. The largest wheel under erection is 71 feet in diameter at the bucket lips. It carries 18 buckets with a capacity of 8.2 yards each, and is turned by four 845-horsepower electric motors.

Production is figured by multiplying the revolutions per minute of the wheel by the number of buckets by the load in each bucket. The largest machines may move 21,000 yards per hour.

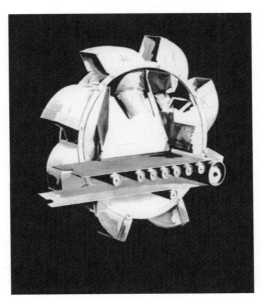

FIGURE 14.22   Big digging wheel.

FIGURE 14.23 Big digging wheel in action.

Operation without Crowd.   A wheel excavator with a noncrowding boom uses its track propel mechanism for crowd-retract.

Soil banks are cut to a stable slope, usually between 45° and 70°. The angle selected depends on the character of the soil, the length of time the bank will have to stand, and sometimes the season.

BLADE-FED BELTS

Elevating Graders.   These machines dig with a plowshare or blade that cuts a slice of earth and slides it onto a conveyor belt, which elevates and dumps it. Models that are to be used for loading loose windrowed material may also have a set of chain-driven paddles to move the dirt onto the conveyor, as seen in Fig. 14.24.

The belt keeps the share from being choked with accumulated cuttings, and is in itself the most economical type of elevator. As a result, these machines are the most efficient type of loader under favorable circumstances, and are capable of a very high volume of production. They cannot be used in rock, and are severely inconvenienced by boulders.

The disc is adjusted to secure the desired depth, and width is regulated either by side shift or by steering the machine. Digging is done alongside the machine rather than under it. Height of bank and width of slice taken vary with the material. Generally, a slice 20 to 30 inches high and 15 to 20 inches wide is good procedure.

Low gear is standard for digging to obtain the power and stability for heavy cuts, and for easy synchronization of speed with trucks being loaded. However, second gear is often used when cuts are light and the spoil is sidecast.

FIGURE 14.24   Elevating belt on grader.

FIGURE 14.25   Force-feed loader. (*Courtesy of Athey Products Corporation.*)

FIGURE 14.26   Hydraulic trencher loader. (*Courtesy of Rivinius.*)

A more complex adaptation enables a machine to make a quick, neat job of excavating a slot for road widening.

Force-Feed Loader.   Athey's force-feed loader, as pictured in Fig. 14.25, is a self-propelled, rubber-tire machine that is designed primarily for loading windrows of loose material including dirt, gravel, snow, leaves, and possibly broken pavement.

The plowshare wings rest on the ground and funnel loose material toward the center by action of the augers and the curved paddles as the loader moves into the material. The moldboard and feeder blades are raised together by a hydraulic ram and are lowered under control by gravity and hydraulic pressure. The discharge conveyor, which can have various angles from the centerline of the machine, is hydraulically driven, and its belt is controlled for variable speed.

Rear loading of trucks can permit the entire operation of digging and hauling to be in one lane. With the swiveling conveyor there is the option of loading or casting the material to either side of the loader. This machine can load up to 10 cubic yards per minute of loose material or as much as 20 cubic yards of snow per minute.

The relative speeds of the loader and the feeder and conveyor depend on the transmission gear selected. The speed should be high enough to keep the feeder well filled without choking it. If a truck is being loaded, it must move at the right speed to keep the discharged material filling its bed. When the truck is filled, both the driving and digging clutches of the loader are released. The loaded truck drives away, and another truck is moved into position to be loaded.

Hydraulic Trencher Loader.   The hydraulic trencher loader manufactured by Rivinius and seen in Fig. 14.26 can be quickly attached to, or detached from, a prime mover for shoulder shaping, berm cleaning, windrow loading, borrow pit loading, or trenching. The rotary-spade, paddle blade feeders take rapid bites of material, which is moved up the feeder belt to a discharge cross conveyor, directed to the right or left for loading into trucks or into a windrow.

MACHINERY FOR USE UNDERGROUND

Continuous Miner.   Coal, salt, potash, and other minerals are mined underground by machines that can cut them directly out of the solid vein, drop the material on a built-in chain conveyor, and load it into electric-powered, rubber-tire shuttle cars or continuous belt conveyors at the rear.

The Joy continuous mining machine in Figs. 14.27 and 14.28 has cutters mounted on sets of rotating wheels. It travels on reversible crawler treads. It is electric-powered and hydraulically controlled.

The cutting heads are swung by steering the whole machine, and in addition can be tilted and retracted by special controls. The discharge end of the conveyor can be swung to either side.

The crawler tread-gathering loader is designed to load coal that has been blasted from the seam. As the entire machine moves forward, the gathering arms on the loading head scoop up the coal and move it back to a built-in chain conveyor. See Fig. 14.29.

The conveyor can be swung at an angle of 40° to either side to permit loading wide areas with minimum maneuvering. The machine is built in models for work in low, intermediate, or high seams.

Mucking Machines.   When there is sufficient space, almost any excavator can be used for underground loading of blasted rock, minerals, or ore, known under the general term *muck*. However,

FIGURE 14.27   Boring-type continuous miner.

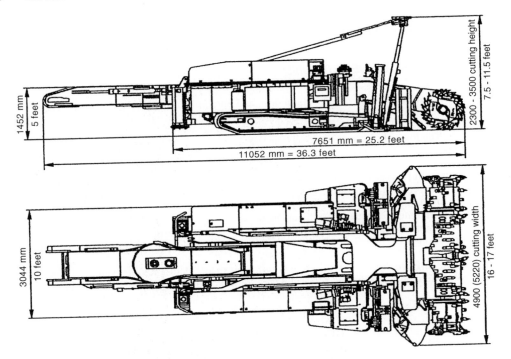

FIGURE 14.28   Two views of continuous miner.

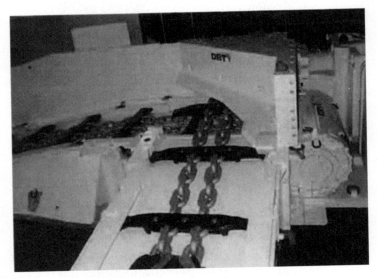

FIGURE 14.29   A chain conveyor.

FIGURE 14.30   Conveyor-loader.

FIGURE 14.31   Downhill loading.

FIGURE 14.32   Use of portable conveyor in car loading.

there are a number of machines designed specially to work in small spaces, and to operate with air or electric power to minimize air pollution.

The Joy Loader, Fig. 14.30, is essentially a chain flight conveyor mounted on crawlers, equipped with a gathering head which is pushed into the base of the pile.

The Eimco Rocker Shovels have a front-mounted bucket that is loaded by pushing into the pile, and lifted over the machine to dump into a car coupled to the rear.

Small models are rail-mounted, and can swing the whole upper works to reach the sides. They can be used on steep down-grades, as in Fig. 14.31. A conveyor belt can be added for longer reach.

Any of these loaders can be used in conjunction with portable conveyor belts, which discharge far enough to the rear that a whole train of cars can be run underneath the frame. When the front car is loaded, the locomotive moves ahead to put the next car under the discharge, until the last one. The loader and belt are then stopped, and the train pulls away. This setup is shown in Fig. 14.32.

In addition to their mucking work, loaders may be used as locomotives in switching cars at the heading, and in moving the jumbo and other equipment.

DITCHERS

WHEEL DITCHERS

Construction.   The Trencor model 1080, shown in Fig. 14.33, is representative of this class of machine. It consists of a crawler mounting, an engine on a front overhang frame, a cutting wheel, a conveyor belt, a shoe post, and the driving and control mechanisms.

The wheel, shown in greater detail in Fig. 14.34, is of channel construction with the buckets riveted or bolted to outside brackets. Both sides of the wheel are toothed and are driven by identical gears on the driveshaft (number one shaft). The wheel has no axle and is supported by idler wheels on an internal frame.

FIGURE 14.33   Large-wheel ditcher in action. (*Courtesy of Trencor, Inc.*)

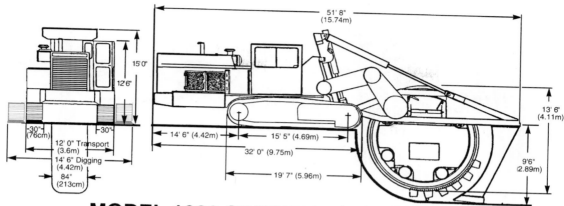

MODEL 1080 SPECIFICATIONS (Partial)

APPROXIMATE WEIGHT: 170,000 to 225,000 lbs. (77 112-102 060 kg)

DIGGING CAPACITY: 9½ ft. (2.89m) Maximum Depth. 84" (213cm) Maximum Width. Available with 72, 64, 58, 52 and 46-inch buckets.

ENGINE: Two engine options are available. The Caterpillar® Model 3408B DITA diesel, rated 503 hp @ 2,100 rpm or the Caterpillar® Model 3412 DITA diesel, rated 750 hp @ 2,100 rpm. Both engines have a two-stage air cleaner system, 24-volt electric starting and accessory system, cooling system for 120-degree ambient temperature operation, muffler, tachometer and engine-driven air compressor. Fuel consumption is approximately 25 to 38 gph

digging attachment is removed for transportation.

DIGGING WHEEL: Digging wheel rims are made of 2" steel plate. Multiple holes are drilled for various numbers of buckets to accommodate various soil conditions. Heavy-duty, heat-treated alloy steel drive segments are welded to both sides of the rim to provide four driving points on the wheel.

MAIN & BOOM FRAME: Experience-proven heavy-duty weldment designed for maximum resistance to applied stresses.

FIGURE 14.34   Specifications for wheel ditcher. (*Courtesy of Trencor, Inc.*)

Various types of buckets are available. Solid-bottom buckets can be used in any soil that is not sticky, and must be used when the dirt is loose and powdery. The slat type is suitable for any soil cohesive enough not to fall between the bars.

Spring-loaded bars set on the frame at the dump point push the dirt out of slat buckets so that even the stickiest soils can be dug. Large stones in the bucket will force the bars back against the springs.

The conveyor belt carries dirt dumped by the buckets to either side and leaves it in a windrow. The belt is reversible and can be shifted sideward.

The shoe post and crumber is hinged at the top to the wheel frame and held in position by two turnbuckles. It serves to support the back of the wheel, pushes spillage ahead of it until it is picked up again, and rounds and smooths the bottom.

Power Units.   The Trencor Jetco 1080 transmission is a three-element torque converter and constant-mesh planetary gearing with hydraulically applied clutches providing five forward speed ranges and one high-reduction reverse range. The crawler drive has a separate hydrostatic planetary speed reducer for each crawler driven by a two-speed hydraulic motor. This provides an infinite range of speeds to 1.4 miles per hour. The wheel hoist is all hydraulic. The digging wheel drive is through a mast-mounted digging differential providing power to a unique radial arm drive.

Conveyor.   The conveyor on a Trencor Jetco 1080 wheel ditcher is a high-volume, arc-type with hydrostatically powered end pulleys providing an infinite range of belt speeds to 1,000 feet per

FIGURE 14.35A Side-to-side shift.

FIGURE 14.35B Wheel tilt.

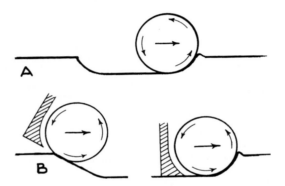

FIGURE 14.36 Starting wheel cut.

minute in either direction. The 48-inch-wide belt is hydraulically, transversely shiftable, with automatic antidrift brakes for precise placement of spoil.

Other Wheel Ditchers. A variety of wheel ditchers provide a range of digging depths from 4 to 10 feet with bucket speeds of 120 to 300 feet per minute. Their ground travel speeds may vary from 2 to more than 100 feet per minute.

Some wheels can be side-shifted, as shown in Fig. 14.35A, to work near obstructions, and/or tilted, as seen in Fig. 14.35B, to make vertical ditches on side slopes.

Operation. The buckets cut the earth and carry it forward to the top where they dump most of it on a conveyor which carries it to the side and piles it.

In starting a cut, as the wheel is lowered to the ground, the buckets will start to dig. The machine is stationary. Enough weight should be allowed to rest on the buckets that they will fill heaping without gouging deeply enough to slow the wheel.

If no shoe is used, the wheel can be lowered to cut bottom grade before walking the machine. The ditch will have a rounded start, as in Fig. 14.36(A). Position should be such that the center of the wheel is over the starting point of the full-depth ditch.

If a shoe is used, the rear hoist is pulled high and the wheel lowered in the position shown in (B). When the buckets cut so far that the shoe rests on the ground, the ditcher is moved forward.

The rear supports are held so that the rear weight rests on the shoe and only enough force is kept on the front support to prevent the buckets from biting too hard.

The wheel will cut to an increasing depth as the supporting shoe is pulled into the ditch, and the front hydraulic lift allows the wheel to sink.

When bottom grade or the limit of depth of cut is almost reached, the wheel is held from further down-cutting by holding the front supports, while continuing motion. The shoe will move down the ramp until it levels out. The rest of the cutting at that depth is done with rear support held and front hydraulic cylinder tight, unless it is the full depth to which the machine will dig, in which case both supports can be slack.

Control of the machine is complicated by the fact that the digging center is behind the tilting and turning center of the tractor. If the tractor starts up a grade or over a bump, a shoeless wheel will dig down and will have to be raised to keep level. If supported by a shoe, it will not cut down and may be raised slightly.

In starting down a grade, the tractor will pitch forward. The wheel will tend to rise out of the ground, and either the front or rear support should be relaxed to drop it to grade. If a shoe is used, a slight pitch might not require any attention, as the back will lower itself on its support. Greater pitch would involve dropping the wheel slightly with the front hoist with extra relaxing of the rear supports.

Digging.   When the wheel is at the correct depth, the machine is moved forward just fast enough to keep the buckets reasonably full. Crowding too hard will overwork the engine and put a heavy strain on the digging parts without adding particularly to the output. If there is a safety clutch on the digging mechanism, it may slip excessively and heat under overload.

Selection of the right speed range for various types of digging must be based on experience. If either the machine or the soil conditions are unfamiliar, it is good policy to err on the side of underfilling rather than overcrowding.

Soft rock usually responds best to a high wheel speed with a very slow walking speed. If dirt is very soft, it can sometimes be crowded so that material in excess of the bucket capacity piles on each side of the ditch without damage.

Obstructions.   Where boulders, heavy roots, or pipelines are liable to be met, both walking and wheel speeds should be low and the engine should be throttled down at critical points.

Soft boulders will be cut through by the teeth. Hard ones may be pulled up to the surface. This is less likely to happen if the boulder is deeply buried, because the deeper boulder not only is held down by a greater weight of dirt, but the direction of the tooth contact tends to force it forward rather than up.

The wheel will usually ride over a deep stone it cannot move, so that it will be lifted above grade. If a big stone is near the top, it may stop the forward motion of the machine, in which case power should be cut off promptly.

If a boulder is pulled to the surface, it is liable to land in a very inconvenient spot, forward of the wheel and between the tracks. It may be necessary to lift the wheel into transporting position, walk forward until clear of the rock, push it out of the way, and back up until the wheel can be lowered to the ditch bottom. If the boulder is too large for the wheel to clear, the wheel drive clutch may be released so that the wheel can turn as it crosses it.

It is sometimes easier to cut and repair tile lines than to work over them. Where they are expected, tile of proper sizes and joint fillers should be kept on the job.

Turning.   Turning must be done with great caution while digging, as the whole machine revolves around the center of one track, which causes the wheel to move sideward in the earth. If the buckets are equipped with long side teeth, or side cutting bars are used, and the earth is reasonably soft, a gradual turn can be made without damage. However, too sharp a turn may bend the wheel frame or the wheel itself, or pull the wheel frame off the vertical track.

FIGURE 14.37   Chain trencher with laser guidance. (*Courtesy of Trencor, Inc.*)

If a shoe is used, it further limits the ability to follow a curve, and if very long, such as a tile-laying shoe, may prevent any significant turning unless it is loose enough, or hinged, so that it can trail instead of being swung wide.

Line and Grade.   In most ditching work, it is important to keep the machine both accurately in line and working at the proper depth, although the operator cannot see the bottom.

The ditch is first surveyed, and the course and the depth of cut are ascertained. A line is then established at a fixed and constant distance above the bottom grade, and offset from the center beyond the track line of the ditcher. This can be done with a laser set up as described in Chap. 2 and seen in Fig. 14.37.

A rigid bar is fastened to the front of the power unit of the ditcher, with one end over the string when the ditching wheel is centered on the ditch line. A plumb bob or other weight is fastened to the bar so that it will hang directly over the string. If a laser system is used, a target is mounted on the machine, as seen in Fig. 14.38.

The operator can then keep the machine on correct line by keeping the plumb bob just over the string or the laser beam on the target. If the ground is irregular, the cord holding the plumb bob can be run through eyes or pulleys so that the operator can reach an end of it to raise and lower it when necessary.

The same apparatus fastened on the side beam of the wheel can be used with a fixed length of string to determine both line and depth.

Trapezoidal Ditchers.   Permanent irrigation ditches usually have a flat bottom and sloped sides. A wheel ditcher may be fitted with side cutters, called cones, that enable the wheel to produce this shape in one pass.

FIGURE 14.38   Trapezoidal ditcher. (*Courtesy of Trencor, Inc.*)

Two types of cones are used. One is driven directly by the wheel through its axle. These cannot be adjusted for different slopes, but they can be set out with spacers for a wider ditch. See Fig. 14.38. The others are turned by hydraulic motors. Their vertical angle can be increased or decreased by adjusting support rod length.

Both types move the cuttings down the slope to the wheel buckets, which carry them to the conveyor for discharge to the side.

In soft, caving soil, cones may be replaced by sloper bars, which are rigid blades held in side frames. They slice the dirt, which falls down to the buckets. The bars are simpler and cheaper, but are too hard to pull in many soils.

CHAIN OR DRAG TRENCHERS

The class of ditchers which pulls by a chain or cutter blades and drags the cuttings to the surface, rather than lifting them in buckets, may conveniently be called chain or drag trenchers. (See Figs. 14.39 and 14.40.) Many are small, 3 tons, but the chain trencher can be very large, as seen in Fig. 14.37. Some carry other related equipment such as a cable-laying plow, a cable reel, or a backfill blade and/or a backhoe on the other end of the tractor.

Depending on size and arrangements, the operator may stand beside the trencher and steer it with handlebars, sit on it sidesaddle, or have a regular tractor-style seat and control station.

FIGURE 14.39 Chain trencher in action. (*Courtesy of Vermeer Manufacturing Company.*)

In the digging, the machine should be propelled forward fast enough to keep the cutters loaded with soil, but not so fast as to slow or stall them. Since the resistance of soil may change frequently, it is necessary to keep alert and be ready to make adjustments often.

Usually the cutters are run at a uniform rate, and the ground speed is adjusted as necessary. This is particularly the case with machines having hydrostatic drive to the wheels.

When the cutters are moving rapidly, they throw cuttings out of the trench with considerable force. Most of the pieces strike baffles and fall harmlessly to the sides, but they can be a nuisance and a danger to a person close to the discharge point.

They are usually moved back from the edge by rotation of a pair of auger flights, to prevent sliding back into the slot. Cutters are usually carried on a roller chain. For cutting slots in rock or pavement, however, a solid wheel with carbide teeth may be used. Overcrowding can be very expensive with these, as it can cause breakage of a number of teeth in one turn of the wheel.

Cable or small, conduit-laying plows are discussed in Chap. 21.

FIGURE 14.40 Drag trenching. (*Courtesy of Vermeer Manufacturing Company.*)

PUMP EXCAVATORS

HYDRAULIC DREDGES

Hydraulic dredges consist of floating pumps, which suck in mixtures of soil and water and deliver them to a disposal point through a pipeline. They can excavate underwater soil or waterside banks, transport spoil through pipes, and rough-grade it on the fills. A single machine, with proper accessories, may perform all three functions, or it may dig without transporting, or a dredge-type pump may transport from a hog box without digging the soil itself.

General Construction.   The heart of the hydraulic dredge is the main centrifugal pump with its suction and discharge lines. Except for hog box work, the pump, power plant, and accessories are mounted in a floating hull. The suction line is carried in a live boom called the ladder. Its position and the working movements of the hull are controlled by winches and spuds on the hull, and by anchors on the bottom or on shore.

Plain suction dredges, such as the Ellicott machine shown in Fig. 14.41, erode the bank by the force of water flowing into the suction head.

High-pressure jets of water may be set around the rim of the head to loosen the soil. These are suitable for dredging loose, free-flowing material such as sand and some gravel deposits. They should not be used where there are many pieces too large to go through the pump, or where the bottom must be cut to a definite grade.

As a rule, suction dredges are moved around the job by winch lines to bottom or shore anchors, and they dig a succession of overlapping pits.

Cutter dredges, shown in Fig. 14.42, loosen material by means of revolving blades or chains. Position control is through two spuds on which the machine pivots, and a pair of swing anchors. While digging, the cutter is kept in almost continuous horizontal motion back and forth through an arc.

Dredges are usually rated as to size by the diameter of the pump discharge line, which is usually smaller than the suction pipe. Range is 6 to 30 inches.

FIGURE 14.41   Hydraulic dredge in action. (*Courtesy of Ellicott Machine Corporation.*)

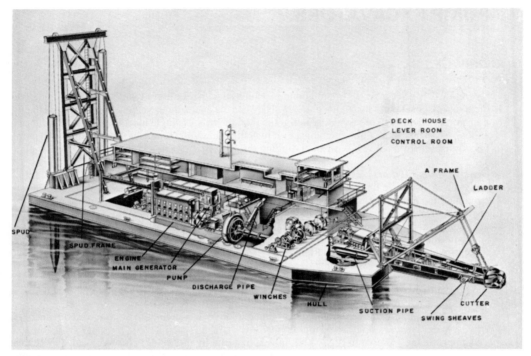

FIGURE 14.42   Cutter dredges. (*Courtesy of Ellicott Machine Corporation.*)

Hull.   The standard hull shape is rectangular, but the bow (front) may be tapered for work in narrow channels. The ladder is hinged at the deck edge, or set back into a slot or well for better stability.

Modern practice is to make hulls of steel divided into compartments. The steel is desirable to withstand the severe vibration of the dredging. The compartments prevent sinking from a single leak, and make it possible to trim (level) the dredge by selective pumping or filling.

The barge may be made of a number of pontoons bolted together. This construction is used when it is likely to be necessary to dismantle it for moves between jobs, or when it must be shipped overland to its first job.

The main machinery should be mounted in a single pontoon or section when possible, to avoid delay and damage during taking apart and reassembling.

Hull size is determined first by the weight of machinery to be floated, and second by the need to support and counterbalance the overhang of the ladder.

Cross section is varied according to special requirements. Dredging shallow channels requires little draft, so that length and width must be increased to provide support at the surface. Narrow channels call for reduced width with greater depth or length.

Large dredges may contain hoppers for storage of spoil, and may be equipped with screening, crushing, and washing plants for processing it.

Seagoing dredges are often of the hopper type, and may have a tapered hull propelled by their own engines. Other dredges are towed from job to job, or taken apart and shipped.

Power Plants.   Power may be diesel or electric. The dredge in Fig. 14.42 has two diesel engines. One drives the pump, the other turns the main generator that supplies current to run the cutter, winches, and accessories. Two engines are used so that the pump can work steadily at full capacity, without being affected by the variable requirements of the other equipment.

All-electric drive is frequently used in dredges that work within reach of power lines. They may also be used with a shore generating set. Current is brought from shore by special submarine cable, or supported on the discharge pipe floats.

Electric power makes a cleaner and quieter dredge and eliminates the fueling problem. However, caring for the power lines may be difficult because of danger from high voltage, movements of the dredge, and change in length of the discharge line.

Pumps.   The main or suction pump is of the centrifugal type, with wide clearances which permit pumping a high proportion of solids and passing of large pieces. Construction is varied according to the depth of digging, the length and rise of the discharge line, and the coarseness or abrasive qualities of the spoil. The pump must be able to maintain a suction velocity of at least 12 feet per second.

These units are designed so that most of the wear is taken by replaceable liners or tips. White iron, and manganese and nickel alloys are extensively used for these parts. Shafting is extraheavy to minimize shock damage from jamming on stones or logs.

Because of the clearances required, efficiency is usually only about 50 to 65 percent, and that is in a rather narrow speed range. If the engine has a different speed, pulleys or gears must be used to adapt it. Belt drives are often used because of their shock-absorbing nature. Power requirement is greatest when pumping heavy volume against a low head.

The pump may be mounted partly or wholly below outside water level, so that it will prime by gravity. If it will not, a valve may be closed in the discharge line and air pulled out by means of a vacuum pump or exhaust blower. Other units prime by raising the suction line and filling it by means of an auxiliary pump.

The auxiliary pump, usually a centrifugal, is used for wetting stuffing boxes and water-lubricated bearings, priming, flushing decks, and miscellaneous jobs. If a high-pressure type, it can supply water for jetting deposits around a plain suction line.

Suction Boom.   A plain suction dredge boom may consist of a pipe, supported by cables attached to a jack boom, which is raised and lowered by a winch line. Rubber hose of special construction is used at the upper end where it meets the hull, and heavy metal pipe for the balance. A suction hood is flared out and equipped with guards to keep out oversize rock. If agitation is used, the opening is smaller and the high-pressure jets are arranged around its rim. The pressure water line is fastened to the outside of the suction line.

A lattice boom or ladder may also be used to support the pipes.

Ladder.   A cutter must be supported by a ladder boom strong enough to carry its weight, withstand side stresses, and keep the driveshaft in alignment. The jack boom is fixed, and a multiple-winch line runs from it to the ladder point. The swinging lines are run from winches around swing sheaves near the cutter end, and outward to side anchors.

The open end of the suction line is carried in the suction head on the cutter end. This head contains the end bushing for the driveshaft and the supporting flange for the cutter.

Digging Depth.   The downward angle of the ladder during digging varies from a few degrees to a standard maximum of 45°. The greatest depth to which a dredge will dig is therefore about seven-tenths of the length of the ladder. Large machines ordinarily carry longer ladders than small ones.

Increasing the digging depth of a dredge adds greatly to its cost. The longer ladder demands additional buoyancy at the front to support it and weight at the back to counterbalance it, which means a substantial increase in the size and weight of the hull. The pump may also need redesign and additional power to develop the extra suction needed.

A dredge with an oversize ladder, hull, and pump engine with a special pump will have only about the output of a standard dredge of the same size, but the cost might be greater than that of a larger dredge with standard digging depth.

Under very special conditions, ladder angles up to 60° are sometimes used for increased depth.

Plain suction lines or ladders are much lighter and can be extended downward at less cost. A cutter machine, under necessity of digging a loose deposit below its reach, may have the cutter head removed and an extension suction pipe installed.

Depth of cut can sometimes be increased by partially draining the water body in which the dredge works or by diverting the waste water away from it, so that the digging will serve to pump out water and lower its surface level.

Forty feet is the deepest cut ordinarily recommended, but with special constructions and under favorable conditions depths down to 100 feet below water surface have been excavated. Fifty feet usually makes an adequate channel for oceangoing vessels.

Cutter Heads. The cutter head, Fig. 14.43, is driven through a shaft from a power plant mounted on the deck or on the upper part of the ladder. The lower end of the shaft is supported in a long bushing which is lubricated and flushed by clean water supplied by the auxiliary pump. The shaft may be fitted with a weak section that is easy to replace, or with a shear pin, for protection of the machinery if the cutter jams.

Cutter drives are occasionally reversible to aid in cleaning off vegetation caught in the blades.

The cutter head is fastened by a key and nut and is sometimes steadied by a backing flange. It encloses the suction head and serves to slice, chop, and stir up material so that it is easily picked up by the water flowing between the blades into the pipe.

There are several types of cutter heads. Each is made in a variety of sizes, and many in several weight classes as well. One-piece construction is the cheapest, but when worn, will have to be built up by welding, or scrapped.

Others use detachable blades or renewable blade edges. Teeth can be welded on the blades or sockets welded to carry renewable teeth. The proper number, shape, and protection of blades for various types of digging will have an important effect on dredge output and cutter life.

—Artist's view of the cutter of a hydraulic pipe line dredge.

—A spiral cutter with seven blades.

—For special work, cutters may be provided with teeth of the so-called "shovel" type welded to the blades.

FIGURE 14.43 Cutter heads. (*Courtesy of Ellicott Machine Corporation.*)

Large dredges with toothed cutters can dig cemented gravel, hard clay, and soft rock.

When in doubt about the proper cutter head, it is usually best to take a heavy type, as reduction in maintenance and downtime usually more than offsets the higher price.

In hard, abrasive digging it is good practice to keep a complete spare cutter head, as replacing wearing parts may take considerable time, particularly if the head has been sprung or damaged by rough usage.

Winches.   Plain suction dredges require a power winch to raise and lower the suction line and two to four power or hand winches for anchor lines and miscellaneous lifting.

Cutter dredges need at least five drums under individual clutch and brake control. Two control the swinging of the dredge, one the raising and lowering of the ladder, and two the spuds. Additional power or hand winches may be provided for handling barges, shore lines, and materials.

A five-drum winch set is shown in Fig. 14.44. This is of the layshaft design, the drums being carried on bearings on fixed shafts. Each drum flange carries a spur pinion that meshes with a spur gear that turns freely on the drive or jackshaft, running parallel to the drum shafts. The drive gear can be rotated by engaging it to its shaft by a friction clutch, or stopped by a band brake. This construction provides full control of the drum without subjecting its bearings to strain when it is stationary under load.

Control levers may be at the winch, as shown, or grouped in a pilot house. Mechanical, hydraulic, or air controls are used.

Spuds.   The spud wells, Fig. 14.45, are two pairs of guide collars mounted on the rear, stern, or spud end of the dredge hull. They are of two-piece construction so that they can be opened for convenient installation of the spuds. They are heavily made and securely fastened to reinforcements on the hull, as they have to withstand the reaction from the swinging of the dredge.

The spuds, Fig. 14.46, are heavy steel tubes, pointed at one end and provided with a lifting band and sheave block at the other. Hoist lines from the winch pass over sheaves in the spud gantry, around the spud sheaves, and are anchored on the tower.

Spuds must be long enough to reach the bottom without passing through the upper spud well. They are used to hold the dredge against digging resistance, wind, and tide and to advance it while working.

One, called the working spud, usually nearest to the discharge line, is down during digging, and provides stability against thrust and a pivot on which the dredge swings. The other, the walking spud, is used as a pivot in moving the dredge.

Canal Dredge.   Figure 14.47 shows a dredge made specially for construction and maintenance of narrow, shallow canals where spoil can be jetted over a bank or pumped through a short discharge line.

During digging this dredge is held in position by four spuds. The ladder and cutter move independently of the hull. Swing, hoist, and down pressure are hydraulic.

Advancing into the cut requires raising the front spuds, and rocking the rear ones alternately with a walking motion.

Sizes range from 6 to 12 inches, cutting widths in single passes are 12 to 20 feet, and maximum depth varies from 8 to 12 feet.

FIGURE 14.44 Five-drum dredge winch. (*Courtesy of Ellicott Machine Corporation.*)

FIGURE 14.45 Spud wells—and dredge factory. (*Courtesy of Ellicott Machine Corporation.*)

FIGURE 14.46 Spuds and shop carrier. (*Courtesy of Ellicott Machine Corporation.*)

Bumboats and Anchors. When working in narrow channels, dredges may be manipulated by lines to objects on shore or to anchors dropped in dug holes. Access to the shore may be by the discharge pipe, catwalks, or a rowboat. Floating pipe may be handled by a shore crane.

Under most working conditions, however, a small barge or boat with a hoist adequate to handle anchors and pipes, and an engine to move it around, is required. Sometimes the hoist will be on a barge which is moved by a small motorboat, but it is more efficient to combine the two.

FIGURE 14.47   Canal dredge. (*Courtesy of Ellicott Machine Corporation.*)

A heavily made boat on the style of a large flat-bottom rowboat with an outboard motor for propulsion and a hand winch for the hoist will be adequate for small and many medium dredges. Larger units may require a working deck, inboard power plant, and a power winch.

Anchors are usually of the fluke type, weighing 300 to 400 pounds for small dredges. When used for swinging, they are placed beside the ladder, and well off to each side, making certain the cutters will not foul their lines at the ends of the swing. When the dredge has moved forward so that the lines pitch back too much for efficiency, they are picked up and moved forward. If the dredge path is curved, the outside anchor is advanced farthest.

Floating Discharge Line.   If discharge is into a barge or hopper, a very short pipe with an open flap valve is used.

If the spoil is to be used as fill, heavy pipe takes it to the rear (spud end or stern) of the dredge. A lighter pipe on floats carries it to land where a still lighter shore pipe takes it to the discharge point.

Floating pipe is made of high carbon or alloy steel, generally in 20- to 40-foot lengths that have raised rims at the ends. Pieces are connected by short pieces of rubber hose with screw clamps. This arrangement provides flexibility to allow for motion from wind, waves, and advancing of the dredge, combined with great resistance to pulling apart.

Floats may be composed of oil drums with a metal or steel frame, of steel tanks as in Fig. 14.48, or small wooden barges. Each float normally carries one pipe section. An A-frame or cherry picker hoist with a hand winch may be built into the float nearest to shore to partly support that pipe so as to reduce the vertical angle with the shore pipe. Another hoist may be kept at the point behind the dredge where the line is opened to install extra sections.

The floats may have a light hand railing for use in walking along the pipe. Telephone and electric cables may be carried beside the pipe, or wires on short poles above the rail.

FIGURE 14.48   Float and discharge line section.

The pipe wears fastest on the bottom, and its life can be prolonged by rotating the straight pieces one-third of a turn occasionally.

"Floating" pipe is sometimes laid on the bottom, underwater, to leave a channel open for shipping.

Shore Lines.   Fixed shore lines leading to a permanent discharge point such as a gravel plant are customarily made of heavy pipe with bolted or threaded flanges.

Shore pipe of the standard movable type is of light welded construction, with one end expanded and the other tapered so that adjoining pieces make a sliding fit. Metal is high carbon or alloy steel, 10 to 16 gauge.

Each end carries a pair of shallow hooks to permit fastening the connection with wire or weaving wire rope along a number of sections.

The tapered joints are somewhat flexible so that the pipe can be laid and used slightly out of line. This causes additional back pressure and wear, but is often unavoidable near the discharge end.

The stream may be allowed to spurt directly out of the last pipe, or may be broken up and spread out by various types of baffles or Y or T ends which reduce pitting and make a wider fill.

Window valves are adjustable openings in the bottom of sections of pipe. They are used to distribute fill along the line of a pipe instead of concentrating it at the end, in order to produce a smoother grade.

Shore pipe ordinarily lies on the ground or fill surface, but is carried on a trestle of some sort if window valves are used, if a deep fill is to be made from a single discharge point, or if the discharge point is changed by removing pipe sections rather than adding them.

Y-shaped diversion valves operated by a hand crank are used to change the flow from one branch pipe to another.

Output.   The output of a dredge, Fig. 14.49, table A, depends on the capacity of the pump, the depth of digging, the height of discharge, the line friction, and the percentage of solids. Pump capacity is expressed in terms of yards per hour of solids. Ability to maintain production against a high head requires increase in power over that needed for a low head.

The vacuum required to raise the load depends on depth and percentage of solids. Aside from friction, it must be able to lift the excess of weight in the suction pipe over the weight of the same volume of plain water, which is the underwater weight of the solids.

The height of discharge and the friction head combine to produce back pressure on the pump. Pumps may be designed to work against either high or low heads.

Recommended maximum lengths of discharge line for nominal heights are given in table B. Much longer lines may be used, but with diminishing output or increased power requirement.

A. Dredge capacity			B. Maximum length of line		
Pipe size of dredge, inches	Cubic yards at 10%	Cubic yards at 20%	Size of dredge, inches	Heavy material, feet	Light material, feet
6	30	60	6	800	1,500
8	60	120	8	1,000	1,800
10	90	180	10	1,400	2,500
12	125	250	12	1,800	4,000
16	225	450	16	3,500	6,000
18	285	570			
20	350	700			
24	500	1,000			
28	700	1,400			
30	800	1,600			

FIGURE 14.49   Dredge output tables. (*Courtesy of Ellicott Machine Corporation.*)

If the required line length of lift is beyond the capacity of the dredge pump, a booster pump of the same characteristics should be installed in the discharge line.

The most critical factor in output of any particular dredge is the proportion of solids carried. Except under conditions of overload, there is little difference in volume or cost between pumping low and high percentages of solids. Dredge output is usually calculated on a basis of 10 percent, but if 20 percent solids can be carried, production is doubled.

Proportion of solids can be increased first by the proper use of the correct type of cutter, and second by increasing velocity of flow by the elimination of sharp bends, reducing the size of the discharge pipe, or speeding up the pump.

The limit on percentage of solids is the plugging point of the pipe. This will be reached more quickly at low velocities, and with smaller loads of coarse material such as gravel, than with fine soil such as silt, or light stuff such as humus. Plugs are expensive to remove and must be avoided.

OPERATION

Advancing.   The dredge is advanced into the digging by the pull of the swing winches and the pivot action of the spuds. One of these, called the working spud, is down while the dredge digs; the other is used for advancing. The swing lines go around sheaves near the end of the suction ladder to anchors placed at each side.

In Fig. 14.50, the dredge is headed directly toward the bank with the working spud down. Pull on the left (port) swing line, as in (B), moves the cutter in an arc to the left, pivoting the dredge on the working spud. Pull to the right (starboard), (C), pivots the dredge to the right.

These movements are made with the pump operating and the cutter head revolving. Firm material is taken in slices from the top down by lowering the ladder at the end of the swing. Flowing or sliding soil may be dug at the toe only.

When the material within reach of the cutter has been dug to bottom grade, the ladder is lifted until the cutter is above the top of the bank and the dredge swung to starboard, in the (C) position. The walking spud is dropped, the working spud raised, and the dredge pulled to the left, as in (D). This causes the dredge to pivot on the walking spud, moving the working spud forward a distance determined by the extent of the side swing in (C) and (D).

The working spud is then dropped, the walking spud raised, and digging resumed.

Spuds must penetrate the bottom sufficiently for a firm grip. If the bottom is hard, extraheavy spuds may be required, and it may be necessary to raise and drop one of them several times in one place in order to sink it.

Anchors must be raised and moved forward occasionally to avoid backpull from the swing lines.

Cutting.   The standard dredge does its hard digging while swinging toward starboard, with the cutter blades slicing up into the bank. The swing back to port is usually made with the cutter at the same level, cleaning up loosened material.

The revolving of the cutter causes it to act as a driving wheel, pushing the ladder to the left. This push will be weak in loose sand and strong in hard formations. On the right swing, the winch must overcome this resistance in addition to crowding the blades into the digging.

When swinging to the left, the cutter may pull the ladder too rapidly for effective cleaning and cause the winch line to run slack, and foul on the cutter. In this case the right-hand winch brake is applied enough to hold the cutter back to line speed.

Dry banks are undermined by cutting below the water level so that they slide or cave into the water where they can be picked up by the suction. The height that can be safely reduced this way depends on how readily it slides and the size of the unit. If too much material comes down at one time, it may bury the ladder or damage the dredge.

It is usual for a dredge to cut to bottom grade at each stand, regardless of the depth of material to be moved. Moving the dredge, with the attendant work of shifting anchors and extending discharge lines, is laborious.

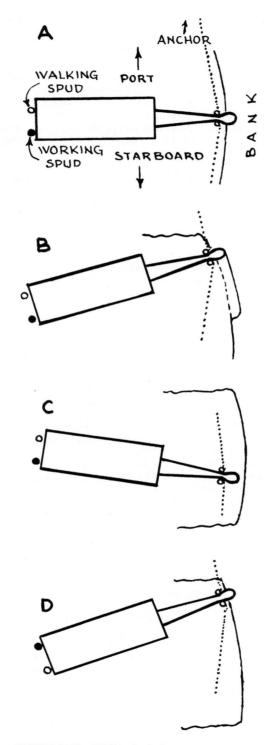

FIGURE 14.50 "Walking" a dredge.

The percentage of solids moved is regulated by the size of the slice taken with each pass and the speed with which it is cut. The slice can be enlarged by advancing the dredge farther toward the bank or by lowering the ladder. Cutting speed depends primarily on the hardness of the material and is controlled by the speed of the swing line, which can be reduced from maximum by throttling the engine, shifting to a lower gear, or slipping or intermittently disengaging the clutch. If the winch cannot pull at full speed, production may be increased by turning the cutter faster.

The dirt cut or stirred up by the blades falls inside the cutter head or is carried around the outside. Water pulled in from the top and the left side by suction of the pump further breaks up the pieces and carries them up the suction pump.

A skillful operator will keep the intake of solids as high as their nature and the velocity of the discharge line permit without plugging. But the operator should not allow the line to plug, as cleaning it out is liable to be a long and tedious job involving downtime that will far outweigh the gain of extra few percent of solids that caused it. A blocked line is particularly serious in freezing weather.

A vacuum indicator connected to the suction line indicates roughly the percentage of solids, since these are heavier than water and a higher vacuum is needed to lift them. A plugged intake will cause vacuum to rise sharply. This reading may also be affected by the angle of the ladder, being somewhat lower when taking off the top of a high bank than when working on the toe.

A pressure gauge attached to the discharge pipe indicates the head against which the pump is working. This will include the lift from the pump to the discharge opening, and the friction against the pipe walls, which is ordinarily a function of the length of the pipe and the velocity of the stream. If the line threatens to plug, discharge pressure will rise abruptly as the stream backs up behind the obstruction.

A threatened plug may be cleared out by alternately speeding and slowing the pump, or raising or backing off the cutter so that clean water or a leaner mixture will be pumped. When working with heavy loads, it may be good practice to run the pump at somewhat less than full speed so that extra force will be available to break through obstructions.

If the pipe does clog, the plug may be located by observing the small leaks usually found in shore pipe connections, or by tapping the tops of pipes with a light hammer. They will have different tones when full of water under pressure, or mud, and when partly empty.

Whenever a part of the pipe is to be cut off from the dredge to be taken apart, or when the dredge is to be shut down for more than a few minutes for any reason, plain water should be pumped until the discharge line is clean. Mud in a pipe will make it difficult or impossible to handle, and may dry out and cake in an unused line so that water will not move it again.

If a single discharge opening is used, the dredge pump must be stopped whenever a new section is to be added.

The dredge operator is usually unable to see much of the discharge area, but must fit his or her actions to its needs. Some quick signaling system must be used for efficient operation. Flags, telephone, and two-way radios are suitable for different conditions.

A dredge is generally run on a two- or three-shift basis, because of its high original cost, and a desire to spread its overhead over as many work hours as possible. It will often be run through lunch periods by a split crew.

Clearing.   Vegetation should be cleared away from the path of a dredge as thoroughly as possible. Cattails, reeds, and other swamp growth may wind around the cutter head, causing constant delays in removing them. Stumps and logs may jam or break the cutter, or block the pump or line.

When the cut is across dry land, brush, trees, and stumps should be completely removed and heavy sods plowed and disked. If it is in tree- or stump-filled water, it should be preceded by a snag boat, generally a grapple dredge. This will raise the stumps and logs, blasting them when necessary, and swing them outside of the cutting line or load them in barges for removal.

Discharge Pipe Patterns.   Dredge discharge lines can be rather readily extended at either end but are difficult to move. Floating pipe is unwieldy to manage in long towed sections, particularly in a wind, and it is a slow job to disassemble and relocate it piece by piece. Shore pipe must always be taken apart for moving. It should be flushed out with clean water first to lighten it.

As the dredge advances, the discharge pipe must be lengthened. This is done by disconnecting at a joint immediately behind the hull, placing another section by means of a light floating crane, and refastening.

In view of the difficulty of moving and the desirability of keeping the number of curves to a minimum, patterns involving a few right-angle turns and considerable extra footage of pipe are in general use. These are shown in Figs. 14.51 and 14.52.

Wherever possible, two or more discharge openings, with control by diversion valves, are used. This system permits adding pipe without losing work time by stopping the dredge.

For simplicity, a balanced amount of cut and fill is assumed in these diagrams, so that the fill advances about the same distance as the dredge moves. This favorable condition will be found when the bottom is used merely as a borrow pit so that depth or width of cut can be varied to keep pace with the fill, and when the disposal area is large or deep enough to easily accommodate the spoil from specified channel work.

If a natural balance does not exist, the depth of the fill may be varied to provide for disposal of greater or less yardage. This is a matter that should be carefully calculated in advance, so that seawalls or berms (earth dikes) may be built to correct height. Maximum depth of a single layer of hydraulic fill is limited by the pressure exerted on the dikes.

If the spoil is coarse or mixed, layers may be added by using the technique to be described for building dams. The extra height of the upper layers may put additional load on the pump, which would have to be speeded up to maintain full output. However, the lift may be more than balanced by the use of lines shorter than those needed to spread the same material over a larger area.

Handling Pipe.   The method of handling discharge pipe depends largely on the firmness of the fill surface. On quick-draining sand or gravel, a pile of pipes may be pushed up to the discharge end by a bulldozer, which can also place each pipe in a position to be engaged by hand. Front-end loaders, or cranes or cherry pickers mounted on crawlers or half-tracks, can also be used for handling.

On softer fills, pipe may be dragged in place by winch lines or handled by a dragline on platforms. More often, workers in hip boots roll pipes along the discharge line to its end and place them by hand. If footing is very bad, it may be necessary to lay boards on each side of the line to support the workers. Supporting boards may also be placed under the pipes.

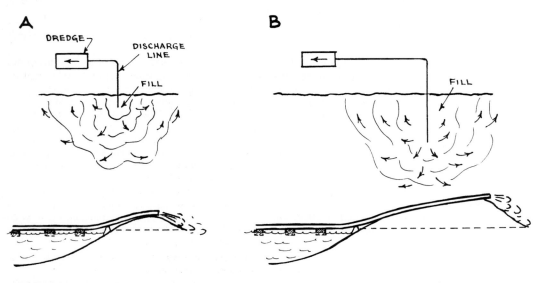

FIGURE 14.51   Open hydraulic fill.

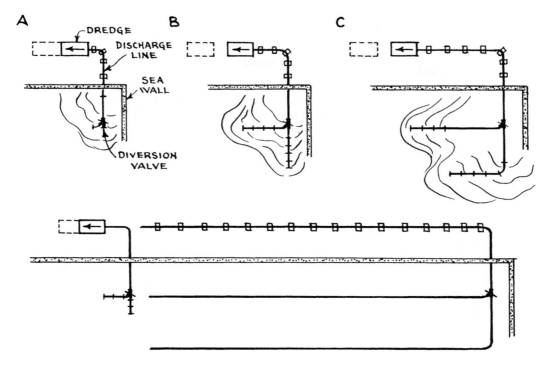

FIGURE 14.52 Discharge pipe pattern.

Fills are likely to be somewhat firmer in the immediate vicinity of the discharge than farther away, because coarser particles tend to settle out first.

Pipe joints are often rather out of line when connection is first made. They are straightened out two or more lengths back from the discharge opening, usually by prying into better position and blocking up. A straight line reduces friction on the load and wear on the pipe.

Leaks at joints are quite common. They can sometimes be stopped by tightening the connection or prying the pipes into better position. They can be checked by forcing in the thin ends of shingles. Those on the bottom block themselves with spoil without requiring attention unless the pipe is on a trestle.

In general, pipes lying on the ground do not need to be hooked together as the friction in the joint and with the ground holds them together. However, on curves and at changes in grade, fastening is necessary.

Individual joints are fastened by winding wire around the pipe hooks, which are placed in line with each other, and twisting it tight with a short stick which is left caught against the pipe. On curves only the outside hooks need fastening.

If the line is on a trestle, each joint must be fastened on both sides.

A number of joints may be fastened at a time with a cable. It is looped and clamped, or fastened in any manner to both ends of the section to be strengthened, put under the hooks on one side of the first joint, over the top of the pipe to be caught, under the opposite side on the next, and on alternate sides to the other anchor. The cable becomes tighter with each joint, and if the correct slack has been allowed, can be slipped over the last few joints only by means of a bar. Another method is to leave it loose enough to go over all hooks easily and then use a load binder to tighten it.

Light cable, $\frac{3}{8}$- or $\frac{1}{2}$-inch, is most satisfactory, but heavy, worn-out swing lines are often used.

HYDRAULIC FILLS

Hydraulic fills may be built by a dredge as wasting operations to dispose of unwanted spoil from channel dredging, they may be a secondary but valuable part of the work, or the dredging may be done merely to get the fill.

Open Fill.   The simplest type of open hydraulic fill is shown in Fig. 14.52. Spoil is being wasted on an area large enough that the water deposits practically all its burden when allowed to flow freely away from the pipe. A small percentage of the soil is fines which remain in suspension and drain back into the dredging area.

The end of the discharge pipe is supported on a low, light horse, or a log, or any object that will keep it above the fill. It may or may not have baffles to spread the stream. The stream will first erode slightly where it strikes the ground, but deposit the heaviest particles immediately around it and lighter ones farther away. In this manner a cone will be built up. Average slopes may be as steep as 15 percent for chunky material or as flat as 3 percent for silt. A high content of solids will give a steeper slope than a light burden of the same material. The action of the flowing water will sort and pack in the pieces.

When the fill has built up to grade or to a point where it threatens to interfere with the stream, another horse is set up, a pipe length is brought forward, the stream is diverted to another line, or a signal is sent to the dredge operator to stop the pump, and the pipes are connected.

The surface may be a slope up from the start of the fill, or level. In either case it should be made as smooth as possible to simplify keeping pipe gradient and to save future grading work. If the material is so coarse that it tends to build up a series of cones, the pipe may be set on a trestle and window pipes used. If the openings can be closed, a series of them can be used with open ports only in the last one or where needed. If the ports are simply slots, the window pipe must be taken off and replaced at the end of the line when it is lengthened.

Settling Basins.   If the proportion of suspended particles in the runoff water is large enough to constitute a serious waste or to cause undesirable silting where it is discharged, or if the edge of a fill is to be steeper than the natural slope, then the water must be retained or ponded in the fill area long enough to permit settlement. This is done by construction of dikes across the natural direction of flow. These may be permanent seawalls or berms thrown up by a dragline or dozer. See Fig. 14.53.

Berms must be of ample height and heavy enough not to be pushed out by pressure, even when softened by soaking. Dirt must be thoroughly tamped or puddled around the spillway.

The amount of settlement obtained is determined by the turbulence in the settling pool, the length of time the water remains in it, and the fineness of the particles. The first two factors are largely fixed by the proportion between the area and depth of the pool, and the rate of inflow. As the fill progresses, the area will be reduced. This may be counteracted for a while by raising the water level by putting more boards in the outlet gate, but ultimately additional land will have to be inundated if settlement is to be obtained.

Where the bottom being dredged includes known areas of fine and coarse soils, the fines may be worked first while a large settling area is available, and coarse deposits used to complete the fill.

Dams.   The velocity of the soil-water mixture discharged from a dredge pipe drops rapidly after it hits the ground because it spreads over a widening area and the water soaks into the fill. The stream may immediately lose its ability to carry cobbles, drop the gravel a few feet farther on, then the sand. Varying quantities of silt and clay may be carried after the water comes to rest in a settling basin.

This results in a gradation of material, from coarse at the pipe through finer sizes with increasing distance. The separation is not clean, as some fines are trapped near the pipe and some coarse particles are rolled or carried by occasional currents well into the edges of the cone.

An earth dam must have the impenetrability of clay and the slump resistance of gravel. One way to meet these requirements is the construction shown in Fig. 14.54. A water-resistant core of clay, silt, and fine sand is supported and protected by gravel sides.

FIGURE 14.53 Filling behind a wall. (*Courtesy of Ellicott Machine Corporation.*)

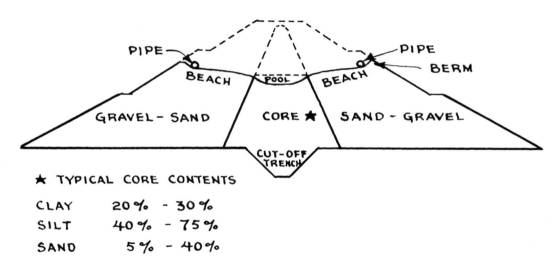

FIGURE 14.54 Cross section of earth dam.

This cross section can be obtained by the proper selection and placing of hydraulic fill. The bank material must have a higher proportion of fines than needed in the dam structure to allow for the amounts left in the coarser material and that carried away in the wastewater.

The dam is built up in a series of low fills. Berms are built by dragline or dozer along the outer edges of the fill, and discharge lines run inside them. The spoil builds up in a slope or beach from the outer line down to a central settling basin called the core pool. Coarse particles remaining in the stream are ordinarily deposited immediately upon entering this calm water. Occasionally, concentrated runs of water may build sand or gravel tongues out into the pool. This possibility is guarded against by anchoring beams or other floats just offshore, where they will break up such formations.

The core pool water is drained into a pipeline or flume, which is usually on one of the outer slopes and is built up as the dam level rises. The width of the core is regulated by that of the pool.

CLEANUP DREDGE

The Mud Cat, Figs. 14.55 and 14.56, is made in two models designed especially for cleaning mud deposits and weed growth out of shallow ponds and canals.

Its carrier is a specially designed barge, 30 or 39 feet long and 8 feet wide, which draws only 21 to 27 inches of water. A pair of side pontoons, connected by a floor and end bulkheads, are filled with polyurethane foam. The space between them is occupied by a 175-horsepower diesel engine, a dredge pump with an 8-inch inlet and a 6-inch discharge, and accessories, controls, and a cab.

The suction boom has a hydraulic hoist. It carries an 8-foot-wide auger driven by a hydraulic motor. The auger is partly enclosed while working by a mud shield. Single layers up to 18 inches thick may be removed at one pass. Maximum working depth is $10\frac{1}{2}$ feet for the smaller machine, 15 feet for the larger.

The dredge is propelled by a capstan winch and a $\frac{5}{16}$-inch-diameter cable. The cable is placed ahead of time, and anchored. After each pass it is moved 6 to 8 feet and reanchored. This arrangement minimizes the equipment needed on the dredge, and can ensure complete coverage of the area.

The suction line and the mud shield together ensure that most of the material loosened by the auger or the shield will be drawn up into the dredge, so that little turbidity (which may be injurious to fish and other pond life) is caused, in comparison with draglines and other excavators.

The auger is able to chop up water weeds and reasonable amounts of trash. If jammed, it can be lifted to the surface and cleaned. The shield tends to support it against digging too deeply into the bottom.

The pump has a capacity of 2,000 gallons per minute against a 155-foot head, and can transport $\frac{1}{2}$ mile or more without a booster. Removal of solid material may be at the rate of 50 to 120 yards per hour.

FIGURE 14.55   Cleanup dredge.

FIGURE 14.56   Suction head auger and mud shield.

SPOIL SEPARATION

In dredging cleanout of ponds in built-up areas, two critical problems may be to get enough water to supply the pump and float the carrier, and to obtain adequate sites for settling basins. Both problems may be reduced by use of mechanical and/or chemical means to accelerate separation of water and solids.

Water Supply.   An enclosed body of water may be sucked dry by a dredge, unless it is large or is fed by generous stream flow, or its water is returned to it after separation of the solids.

Reuse or clarification of water may be desirable or required in order to avoid damage to drainageways or other property from overflow of a settling pond.

Processing that accelerates separation of water and solids favors both efficient recycling of the water, and disposal of any surplus.

Cyclone.   The cyclone, Fig. 14.57, is a machine that is widely used in mineral processing plants to dewater finely ground material during or after processing. It has no moving parts, and is operated by the force of the water-solids mixtures that are pumped into it.

In the design shown, mixed water and solids enter a cylindrical chamber at a tangent. Pressure of about 22 pounds per square inch is sufficient to give the fluid high rotary speed. Centrifugal force presses solids against the outer wall, along which they move in a downward spiral, through a conical section, to an underflow or discharge gate. Water, with any solids too fine or too light to separate from it, overflows through a pipe at the top.

Adjustments for sharpness of separation and/or thickness of underflow mud may be made either manually or automatically during operation. Efficiency in separating fine or light particles diminishes as machine size increases.

Figure 14.58 shows data on four standard models. It will be noticed that capacity is small in relation to dredge output. For example, it would take four 18-inch cyclones to process spoil from an 8-inch dredge with a 2,000-gallon-per-minute flow.

Their separation range of 42 to 55 micrometers would enable them to remove all gravel and sand, a substantial fraction of silt, and some clay, with a variable proportion of organic material.

FIGURE 14.57   Cyclone, cutaway.

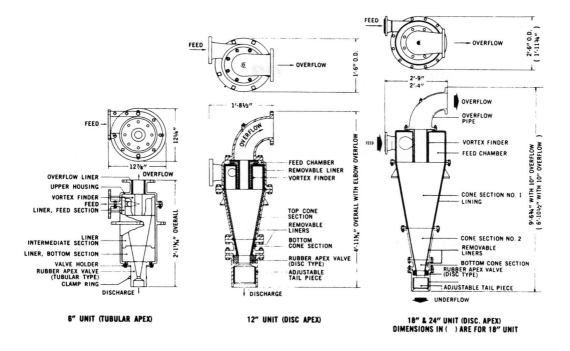

6" UNIT (TUBULAR APEX)   12" UNIT (DISC APEX)   18" & 24" UNIT (DISC. APEX)
DIMENSIONS IN ( ) ARE FOR 18" UNIT

OPERATING DATA

DIAMETER OF UNIT* INCHES	CAPACITY GPM	NORMAL OPERATING PRESSURE PSI	SEPARATION RANGE** MICRONS
6"	47-110	22	28-40
12"	185-390	22	36-46
18"	520-1100	22	42-55
24"	710-1520	22	50-68

*36" and 48" units are available on special request.
**Separation range is based on 2.7 specific gravity solids at a feed concentration below 25% by weight.

FIGURE 14.58 Cyclone details.

Unfortunately, they tend to leave in the water precisely those materials that take longest to settle, and which demand the largest quiet water area for separation.

However, under many conditions they remove most of the bulk, outside disposal of which can make it possible to use a settlement basin for a much longer time.

Solids from the underflow usually contain enough water to make a flowing mud, which may have to be channeled, dragged, or pumped to a drying area, or hauled away in tight-bodied dump trucks.

Clotting.   Organic mud, with or without included fine mineral particles, takes days and sometimes weeks to settle out of water naturally. But the addition of certain chemicals, called polymers, causes such material to form into relatively large clots which sink promptly, leaving clear water in 20 to 30 minutes.

Excellent results appear to follow injecting the polymer into the dredge outlet pipe about 20 feet short of its discharge point.

HYDRAULICKING

Hydraulicking may be said to be a dry-land dredging operation. Banks are washed down by high-pressure water jets and the mixed spoil and water moved by gravity, pumps, or both, to a fill. It is used in building hydraulic fill dams when sufficient material cannot be obtained by dredges, for breaking down banks, for cleaning rock for blasting, and for various special conditions.

Plant.   Hydraulicking equipment consists of water pumps, suction and discharge lines, monitors, and a flume or sluiceway to carry the mixed water and soil.

The centrifugal water pumps may be high-pressure jet types, or medium-pressure types, set in series with one pump's discharge supplying the intake of another. Usual pressure is between 80 and 150 pounds, but up to 275 may be used.

Steel or iron pipe is used except near the monitors, where flexible hose is required.

Monitors, also called giants, are large, high-pressure nozzles mounted in swivels on heavy skid frames. Small sizes may be aimed by hand, large ones have a movable baffle which utilizes the power of the stream of water to turn the nozzle.

The sluiceway may be a gully of controlled width in the pit floor, a flume of wood or metal, or dredge discharge pipe. A gully should have a slope of at least 4 percent, but flumes can be somewhat flatter. A narrow, deep waterway will give maximum velocity and transport the greatest bulk of solids.

The high-pressure water may do the transporting, or additional water may be supplied by low-pressure pumps or gravity flow.

The sluiceway runs into a distributing pipe which takes it to the fill. This pipe should start high enough on the slope to give sufficient head to push the spoil through it.

Hog Boxes.   If the grade is not sufficient to carry the spoil from the pit to the fill, it is poured into a hog box, which is a concrete trough. From there a dredge-type pump moves it through a pipeline to a fill.

When gravity feed is not possible and the hydraulic fill method is required on the dam, dirt may be dumped through grizzlies into the hog box by trucks, washed into a sump by monitors, then pumped to the fill.

Operating a Monitor.   The primary requirement of dam hydraulicking is the correct selection of material. Definite proportions of fines, sand, and gravel are usually required, and these are often not found in a single bank.

Several monitors may have to work in different sections, or in different pits, and increase or decrease their part of the output on the direction of the job inspectors.

If the bank will cave if undermined, the monitor should be as close to it as safety permits, and the stream should be directed at the base. If most of the cutting can be done at an angle, efficiency is increased and danger from slides is reduced.

If the bank will not cave, it is best cut by standing at such a distance that the stream will just start to bend downward before hitting it.

As the bank recedes, the monitor must be moved forward. The flexible hose should be long enough to be laid in loops so that considerable movement is required before an extension is needed.

Sometimes draglines, bulldozers, or other machines are used to break up the soil and push it into the sluiceway.

Water jets will erode very hard soils, but not efficiently. Light blasting of clay will greatly increase production.

Cleaning Rock.   Fire hose and nozzles supplied by small centrifugal pumps can often be used to advantage in cleaning thin layers and pockets of dirt off rock which is to be drilled. Pressures of 20 to 100 pounds may be used. Several workers will be required to handle one nozzle and hose.

CHAPTER 15
TRACTORS AND BULLDOZERS

TRACTORS

CRAWLER TRACTORS

Crawler (track-laying) tractors have a weight range from 4 or 5 tons to almost 150 tons. Maximum horsepower is over 1,000.

Figure 15.1A shows a typical direct-drive crawler tractor. The detail view (Fig. 15.1B) shows a center section or chassis that contains the engine, transmission, and steering units; and two track frames that supply traction and support. The Caterpillar tractors started the bull wheel sprocket above and slightly ahead of the rear track idler wheel, as seen in Fig. 15.2, to keep out of the dirt and environment, for better servicing and durability, and to improve balance and stability.

FIGURE 15.1A Typical tract-type tractor. (*Reprinted courtesy of Caterpillar Inc.*)

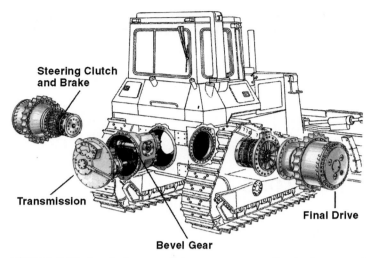

FIGURE 15.1B   Detail of power train in tract-type tractor. (*Reprinted courtesy of Caterpillar Inc.*)

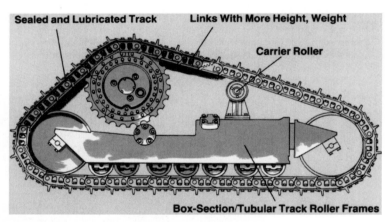

FIGURE 15.2   Elevated bull wheel sprocket. (*Courtesy of Caterpillar Inc.*)

Engine.   The engine is usually diesel, although in the smaller sizes, it may be gasoline-fueled. Power is rated first as net engine horsepower, meaning the net horsepower at the flywheel with the engine driving all accessories normal to tractor operation.

The second standard of measurement is drawbar pull or horsepower. This is a figure smaller than the flywheel horsepower and represents usable power at the drawbar under certain set conditions, after deducting losses in friction and slippage. This is the usual rating for tractors with direct mechanical drive.

Power may also be measured in terms of pounds of drawbar pull, a factor that is limited by traction and that may be increased by mounting a bulldozer or any extra weight. Crawlers with power shift and/or torque converters are generally rated in pounds of drawbar pull at a given speed.

Clutch.   Crawlers with mechanical drive have an engine or flywheel clutch with one or two discs, that are used to cut off the drivetrain when stopping the machine or shifting gears. When the machine is to be used to carry a loader or for any other very heavy service, the discs may be set with long-wearing, heat-resistant ceramic discs; or be kept in a circulating bath of oil that reduces wear and takes away heat.

The clutch is usually controlled by a hand lever at the operator's left. Large machines have a hydraulic booster to reduce operator effort and ensure sufficient engagement pressure.

Torque Converter.   Most large crawlers have a torque converter instead of, or in addition to, an engine clutch. This device, described in Chap. 12, provides shock-absorbing slippage between engine and tracks, multiplies torque so that fewer transmission gear ratios are needed, and allows on-the-move power shifting among those ratios that are used.

Center Frame.   Rigidity of the center section is obtained by making the steering clutch and transmission housings in one heavy casting or weldment with internal braces, together with heavy construction of all cases forward to the radiator. In addition, a pair of heavy side beams may run forward from the transmission case, supporting the engine, the crankcase guard, and the radiator base. See Fig. 15.3.

A front pull hook may be bolted to the crankcase guard, which is a plate protecting the bottom of the engine. Use of this hook for heavy pulls is a greater strain on the tractor than use of the drawbar.

Transmission—Manual Shift.   This type of transmission, used in machines with friction clutches, is compact, with very heavy construction. Shifting may be done by sliding spur gears, but newer machines may have constant-mesh helical gears with synchronized shifting.

The number of gear speeds varies in different models and may be from two to eight forward speeds, and one to six reverse speeds.

The engine clutch must be disengaged for any conventional gearshift. There may be a lock that will prevent any gear from disengaging unless the clutch is fully released.

There is a universal joint in the shaft from the clutch to the transmission to prevent damage if these two units should get slightly out of alignment.

The driveshaft from the rear of the transmission ends immediately in a bevel gear that drives the live axle ring gear.

The transmission and bevel gears operate in a sealed box that contains sufficient oil or fluid grease to lubricate all the gears and bearings by dip, splash, or pressure spray.

Transmission—Power Shift.   Tractors equipped with torque converters usually have power shift (shift-on-the-go) transmissions with two or more speeds in each direction. See Fig. 15.4.

FIGURE 15.3 Underside of tractor.

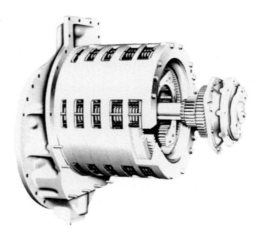

FIGURE 15.4   Power shift transmission.

The shifting is done by pairs of friction clutches connecting and disconnecting gears in the drive line, or by brakes controlling planetary gear sets. The shifting lever may control these by mechanical linkage, or hydraulic means.

Power Takeoff.   The standard power takeoff is a connection that will turn a shaft inserted through the rear wall of the gear case. It is used to power accessories such as cable control units, a winch, or a hydraulic pump. If the transmission is compound, the takeoff may have two gear ratios; otherwise its speed is controlled only by the engine. It is engaged by releasing the engine clutch, meshing a sliding jaw, and reengaging the clutch.

The takeoff usually turns more slowly than the engine. It operates in neutral or any gear, but not when the engine clutch is disengaged. It is not affected by the steering clutches.

Lack of power in the takeoff when the engine clutch is released, or when the output shaft of a torque converter is slowed by heavy load, results in inefficiency in many operations. There are therefore an increasing number of constant-running "live" power takeoffs, that can be operated whenever the engine is running, regardless of clutch position or converter action.

Rear-Drive Assembly.   The rear-drive or live axle assembly for clutch and brake steering is shown in the cutaway drawings in Figs. 15.1B and 15.5.

At each side of the ring gear, the axle extends through a section of the case which is kept free of grease by means of seals on the axle and by a drain hole in the bottom. The holes must be plugged while working in mud or water. In each of these compartments is a multiple-disc clutch and a band brake. This pair of clutches and brakes is used for steering in the same manner as those on the shovel live axles; but, being always of the friction type, they are smooth in operation.

Next to the steering clutch compartment on each side is the final drive. This is a dip-lubricated gear set of either single or double reduction construction. The large outer gear is attached to a short axle which turns the bull wheel (sprocket) that drives the track. Grease seals are used where the axle enters the final drive case, and where the bull wheel drive leaves it.

Dead Axle.   The dead axle, or pivot shaft, is a hinge pin which runs across the back of the tractor, through or forward of the final drive cases. It ties the track frames and center section together but allows them to oscillate vertically. In addition, it usually serves as an axle for the bull wheel.

Bull Wheel.   The bull wheel is a big sprocket of very heavy construction. The wheel itself is usually a flat disc, widening to shallow teeth at the rim. Hollows between them mesh with the track pin bushings, providing positive drive to the track.

FIGURE 15.5 Rear axle and final drive.

Tracks.   The track frame and mechanism is similar in general appearance to that described in Chap. 13. See Fig. 15.6*A*.

The live axles turn large toothed wheels, called drive sprockets or bull wheels, that are at the rear of the track frames or elevated as seen in Fig. 15.2. The frames rest on the small truck or track rollers. The idlers, which are smoothed flanged wheels similar in size to the drive sprockets, are mounted on spring-cushioned yokes at the front of the frames, or both front and back with an elevated bull wheel sprocket. One or two small support rollers are mounted above each frame, except in very small machines, to prevent excessive sagging in the upper track section.

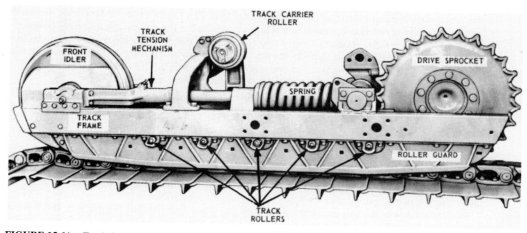

FIGURE 15.6A Track frame assembly.

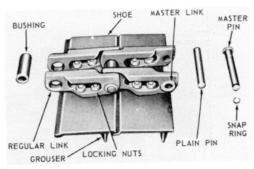

FIGURE 15.6B   Track links and shoes.

The track itself consists of a true roller chain and bolted-on shoes. The parts of a track chain are shown in Figs. 15.6A and 15.6B.

Certain types of stiff mud, and wet snow, may build up in the track and in the sprocket hollows so that the sprocket will spin, usually with abrupt and damaging stops and starts, and will probably make the track overtight at the same time. This condition may make work difficult or impossible, as repeated hand cleaning may be required. Ice and mud shoes usually have openings in the center that permit the sprocket teeth to force the snow out through them, leaving the inner parts comparatively free. There are also cutaway sprockets that allow mud to squeeze through openings in the tooth bottoms.

This is one of the reasons that Caterpillar Inc. introduced the elevated sprockets, as shown in Fig. 15.2, for their crawler tractors.

There are a great many types of track shoes, a few of which are shown in Fig. 15.7. The standard construction is a flat plate with a single high cleat or grouser across it. This affords good traction and protection against sideslipping under most conditions, but will not grip on ice or frozen ground, and tears up surfaces on which the machine works. If the tracks spin, each grouser acts as a bucket on a ditching machine, taking dirt from beneath and piling it at the rear. As a result, the machine may dig itself down into trouble very rapidly on soft ground when heavily loaded. In addition, such piles make backing up in the same path a very rough trip.

Flat shoes are used on machines that usually carry rather than push, and those that work in unpaved yards that would be cut up by the cleats. They have been used on bulldozers and shovel dozers, but usually do not give enough traction and permit a dangerous amount of sideslipping.

Rubber-face pads are used by some crawler tractors, sometimes called utility tractors, that work inside buildings and on paved roads. Traction is better than with flat shoes on hard surfaces, and scuffing and scarring are reduced to a minimum. They are usually not rugged enough for heavy pushing.

Semigrousers are flat shoes that carry two to three low cleats. They are the most satisfactory equipment for front loaders, as they do not dig up the ground in spinning and turning as much as full grousers, without reducing traction and stability as severely as flat shoes.

Snow and ice plates usually feature cutout or skeleton shoes, and high cleats.

A wide-track machine has much better stability on side slopes; in carrying high loads, it fouls less with mud; and it is possible to mount wide shoes on it for swamp work. It is heavier, is generally more clumsy to maneuver in restricted spaces, but can turn with a load with less difficulty than a standard model, and does not get stuck as readily.

Track Wheels.   The rollers and idlers are flanged in order to keep the track in line. The idler customarily has a wide center flange that fits between the track links. The track and support rollers have outer flanges that are on each side of the track rail. They may also have an inner flange. On the bottom, it is usual to alternate single- and double-flanged rollers.

Rollers and idlers revolve on fixed axles. They may have tapered roller bearings, or solid-sleeve types made of bronze or special metals. Good seals, to keep lubricant in and dirt out, are very important.

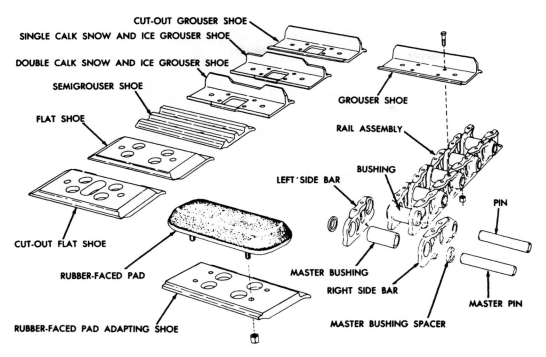

FIGURE 15.7 Various track links and shoes.

The most successful type, the "positive" seal, consists of two finely machined rings, one attached to the axle and one to the hub, which are pressed against each other by springs. These rings fit so perfectly that neither dirt nor grease can get past them, and they are hard enough to outlast other wearing parts.

Figure 15.8 shows a track roller of single-flange type, with roller and solid bearings. Figure 15.9 is a set that includes single- and double-flange construction mounted in the track frame.

FIGURE 15.8 Single-flange track roller.

Adjustment.   The big idler is held in position by a sliding yoke backed by a spring, or by a cylinder of compressed nitrogen, which pushes it forward in order to keep tension in the track. If the track collides with something, or an object is caught between the track and the idler, or sand or snow builds up on the sprocket or rails, the idler can move back by compressing the spring, thus absorbing the shock, or relieving the tension on the track. See Fig. 15.10.

Hydraulic track adjusters usually consist of a piston in a tight cylinder that has a grease fitting. The piston rod is an extension of the idler

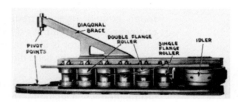

FIGURE 15.9 Track frame and rollers.

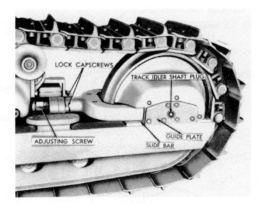

FIGURE 15.10   Mechanical track adjuster.

yoke. Pumping grease into the cylinder pushes the idler forward and tightens the track; bleeding grease through the fitting or a relief valve allows the idler to move back and release the track.

Hydraulic adjusters are very easy to use, but any leakage in them is likely to put the tractor out of action.

The tractor should be moved backward and forward a few times during adjustment to equalize upper and lower tension. A properly adjusted track should sag a little at the top when the machine moves forward.

Drawbar.   The drawbar is a heavy steel tow bar, fastened under the center of the tractor, extending backward across a support bracket, and projecting to the rear. It can swing horizontally and is held in the desired position by a pin or bolt through the bracket or by a pair of bolts in a clamp. It is desirable that the anchor of the drawbar be as low and as far forward as the construction of the tractor will permit for best distribution of stress.

The one shown in Fig. 15.11 has a safety catch to hold both the swing and the load pins.

The biggest tractors use a nonswinging drawbar rigidly attached to the frame.

Controls.   The standard operating controls for a small gearshift tractor consist of the main or flywheel clutch lever, which is held in the forward disengaged position by a spring, and locks over center in engaged position when pulled all the way back; the right and left steering clutch levers which are held in forward engaged position by springs, are released by pulling back, and do not lock over center; the gearshift lever; and the right and left brake pedals, which are depressed against springs to engage the brakes, and which may be locked down by moving the lock levers forward while the brakes are engaged.

Larger machines have hydraulic boosters on some or all of the controls to offset the greater effort required by large clutches and brakes.

A decelerator pedal permits the operator to leave the hand throttle at full-speed position, and still slow the engine as much as he or she wishes for accurate work, safety or comfort crossing rough ground, or noise reduction when receiving instructions. Pressing the pedal slows the engine; releasing it allows the engine to return to the speed required by the throttle setting.

The decelerator is used in crawlers because their low ground speed makes it safe and usual to operate them with the hand throttle set at full working engine speed, which has to be cut back to idling only occasionally.

Wheel tractors doing the same work have accelerators instead. An operator must develop discrimination when changing from one to the other, in regard to when the pedal should be depressed.

FIGURE 15.11   Drawbar, with pin lock.

OPERATION—MECHANICAL DRIVE

Starting the Engine.   Instructions for starting diesel engines will be found in Chap. 12. The special problem in a tractor (as well as in many other machines) is no-start interlocks.

Most tractors have a live starter circuit only when the gearshift is in neutral. If there are two shift levers, either one may be the key.

There may also be a switch in the battery case that must be in the connected position to allow the starter to turn.

Allow the engine to warm up at partial speed for 2 to 3 minutes before operating. You can spend this time checking the tractor for condition and signs of abnormal wear or behavior.

Starting the Tractor.   A conventional crawler tractor may be operated as follows: Release the engine or master clutch by moving it forward, shift into the gear to be used, set the throttle for the desired engine speed, pull back on the clutch lever until the machine starts to move or the engine starts to work, then pull slowly until the machine is moving and the clutch is carrying the full engine power without slipping, then pull it back rapidly until it locks over center. Starting with heavy loads, or in high gears, requires a longer, slow pull than loads easily moved. The pull should be slow enough to avoid jerking or jumping the tractor, but any hesitation beyond this minimum will increase wear.

Some engines will operate efficiently only if the throttle is wide open or nearly so. Others will perform reasonably well at any speed. On the whole, fast engine and tractor speeds are suited to rough, heavy work and fast traveling on smooth ground, and slow to moderate speeds for precise work, moving in close quarters and walking on rough ground.

The top speed of most crawlers is from 4 to 7 miles per hour on a level, but this is in a high gear which has little reserve power for climbing hills or pulling loads. It is expensive to operate at top speed, as wear on the track chains is greater per mile of travel at high speeds, but the time saved or work performed will often more than justify the cost.

Gear Selection.   For precise work, the lowest gear is desirable, not only because the slow pace gives more time to steer and make adjustments, but because the bulldozer blade or other unit will move faster in proportion to the speed of the tractor. This is because the blade speed is proportional to the speed of the engine, while the tractor speed is varied by the gear ratio. The blade can therefore cut a grade more accurately in low gear than in second, even if the tractor speed is the same.

For less exacting work the gears should be the highest that can be used without lugging the engine below its governed speed, and which will give the amount of control needed.

Gear shifting is slow and cumbersome in some tractors because of the spur gearing used, the hand throttle setting which makes double clutching impractical, and the fact that friction in the tracks slows it so rapidly that considerable operator skill is required to complete a shift before the machine stops. The gear used throughout is therefore usually low enough to start the load, although a higher gear might move it once it is underway.

An increasing number of machines use constant-mesh gears which shift more easily when standing, and which can be shifted on the move.

Speed may be increased or decreased by opening or closing the throttle, or by choice of gears. A decelerator pedal will slow the engine below throttle setting, when pushed down against a spring.

Reversing.   To change direction of travel of a standard crawler, first bring it to a full stop. With the clutch released it will stop by itself in a few feet, except when going down a grade. It may be stopped in inches by its load, or by applying a brake.

In an ordinary transmission, you move the shift lever through neutral into the desired gear, then reengage the clutch. See Fig. 15.12.

The reversing transmission may use sliding spur gears, which are usually somewhat clumsy or even difficult to shift, but should have helical gears with a sliding hub, which are faster and easier; or a pair of friction clutches.

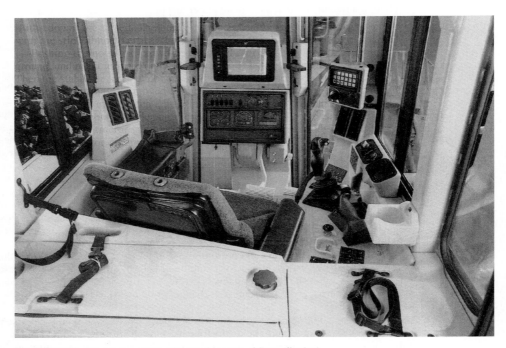

FIGURE 15.12   Operator's controls. (*Reprinted courtesy of Caterpillar Inc.*)

Steering.   Five factors are involved in steering a conventional crawler tractor—the effects of the steering clutches, of the steering brakes, of the footing, of the grade, and of the load being pulled or pushed.

If one steering clutch lever is pulled back while the tractor is moving forward, the track on that side will run idle and the other will continue to drive. The amount of turn in an unloaded tractor will depend on the resistance of the idle track to moving forward. Its resistance is made up of internal friction, which by itself will produce only very gradual change in direction and ground drag. As a result, on a smooth, hard surface the machine will turn very gradually. In soft ground it will turn more sharply. If the nondriving track collides with a stump, it will swing abruptly. See Fig. 15.13.

Rolling resistance in the free track can be increased to any desired degree by applying its steering brake. If the track is locked and traction is good, the machine will spin around it. Brake application

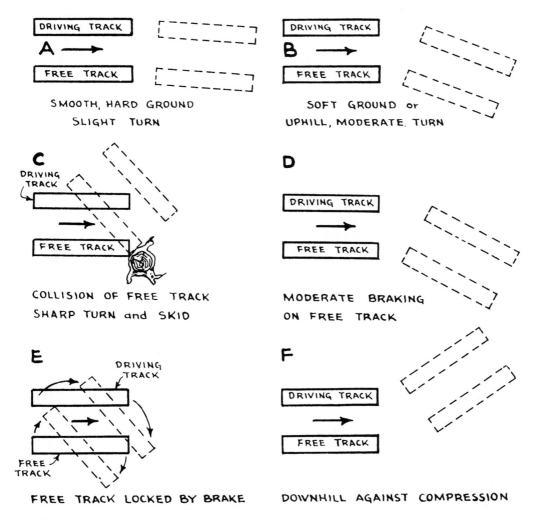

FIGURE 15.13   Steering with clutch and brake.

can produce any degree of turn between this and the nonbraked gradual change in direction. Braked turns are seldom smooth, as direction is usually changed in a series of steps.

When moving uphill, the free track is held back by gravity, and the weight of the machine also increases the tendency to turn, as shown in (*B*). Other conditions being equal, less braking will therefore be required for steering when going uphill than on the level.

If the ground is soft, both tracks will dig in. The driving track will tend to spin and dig; the braked track to push earth at the front in the direction of turn, and oppositely at the rear. At best, turning consumes some power, and under soft conditions the resistance built up by the displaced earth may be great enough to prevent the turn and will put severe strain on the tracks. Under such conditions turns are best made in a series of short jerks, to give the tracks a chance to work away from the dirt ridges before turning farther.

Since the machine drives from one track while being steered, only half of its normal traction is available. This complicates maneuvering in soft ground and with a load.

A heavy load, either on the drawbar or in front of a blade, has somewhat the effect of an upgrade in supplying a turning force when drive is from one side only; the load itself resists being turned. In general, a loaded tractor will steer more readily than a free one in response to a steering clutch only, and with greater difficulty in response to clutch and brake. Very sharp turns are usually difficult or impossible, but the location and character of the load influence this.

Even with all clutches engaged, a tractor with a heavy off-center load will tend to turn toward the loaded side.

A tractor so loaded that it is using nearly its full traction in straight going will have difficulty in turning, as the driving track will be inclined to spin. It may be necessary to relieve the load by backing in order to change direction. This is usually fairly easy with a pushed load but difficult with a pulled one.

Steering by means of clutches alone works backward in going down a hill steep enough to make the tractor try to roll faster than the engine drives it. The effect of releasing a steering clutch in this case is to allow its track to roll free, and move faster than the other track, which is braked by the engine, so that the machine will turn away from the released clutch. Firm application of that brake will cause the machine to turn toward it in the normal manner. If precise steering is required, the brake should be applied before the clutch is released.

Steering in reverse is the same process as steering forward.

Use of either steering brake while its steering clutch is engaged will slow, stop, or hold the machine, but has no steering effect, as while the clutches are engaged the axle shaft acts as a solid unit from track to track.

There are crawler tractors which differ from the type described in important respects.

OPERATION—WITH TORQUE CONVERTER

Figure 15.14 shows a typical but older set of controls for a crawler tractor with torque converter and power shift.

In this example, steering clutches and brakes are similar in location and action to those on a mechanical drive machine.

Shift.   This gearshift (speed selector) lever moves in a U-shaped slot, opening to the rear. The cross slot is neutral, the right side is forward, the left side reverse. See Fig. 15.15.

An additional lever, on the side of the U, is moved forward to lock the shift lever in neutral. This lock should be set whenever the operator gets off the machine. An interlock prevents use of the starter when this lever is in operating position.

To start work, first start the engine. Set the hand throttle to desired engine speed (indicated on the tachometer on the dash), which in most jobs is between three-quarters and full speed.

Unlock the speed shift, and move the shift lever to the desired direction and speed. This can be done at full throttle in most tractors. If this is undesirable, either because of instructions to the contrary, or

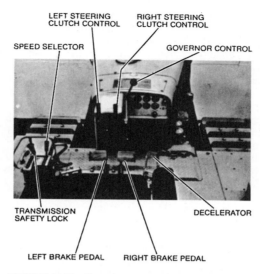

FIGURE 15.14 Controls, converter drive.

FIGURE 15.15 U-shift.

because the machine is going through a jerky shift phase, slow the engine with the decelerator pedal during the shift.

Steering is done in the same manner as in a gear drive model.

Varying Speed. A torque converter causes the relative speeds of tractor and engine to vary widely. Under full throttle a heavy load may slow the machine to a creep or even stop, and with a light load or no load it will reach its maximum speed in the gear range being used.

The loss of speed involves an increase in pulling or pushing power, because of the torque multiplication explained in Chap. 12. It saves the operator the responsibility of shifting to a lower gear, lightening the load, or even slipping the clutch, which the operator would do in a gear drive model. Under most conditions, this slowing is an efficient and desirable response to an increase in load.

Speeding up under light load is also efficient, but it is not always as welcome, as it may make it difficult or impossible to do precise work with a blade, bucket or implement. Fortunately, it can be controlled with the decelerator pedal, which will slow both engine and tractor to the desired rate of motion.

A machine might have an output shaft governor on the torque converter. This will keep the tractor, or a mounted unit such as a winch, at a steady speed regardless of load, as long as its requirement does not exceed engine power.

For this, there are two throttles, one controlling a standard engine governor, the other the output shaft governor. It is usual to keep the engine throttle wide open, and use the output governor to regulate speed of operation.

Proportioning Power.   The proportioning of engine power between the tractor and its attachment, which is usually hydraulically powered, is largely governed by the slippage factor in the torque converter.

Slowing the engine permits increased slippage in the converter, so that at idling speed little or no power is transmitted. But the parts driven by the hydraulic pump will be slowed only in almost direct proportion to the engine, and will keep nearly full power.

As a result, decelerating will usually stop tractor motion while allowing attachment action to continue at reduced speed. The uses of this will be discussed in Chap. 16, under Loader Operation.

DIFFERENT CONTROL SYSTEMS

Control levers and pedals may be arranged in a number of different ways in various makes and models, as in the combination levers below.

There are also several basically different control systems, which call for changes in operation.

Combination Controls.   Small tractors, and those with power boosters, may have a steering clutch lever or pedal also operate the steering brake. The first half or two-thirds of its travel disengages the clutch, then further movement contacts and applies the brake linkage.

This simplifies operation, but in many machines makes it impossible to use the brakes to slow the tractor without sacrificing the braking effect of the engine.

Hydrostatic.   A line of John Deere crawlers introduced a hydrostatic system for both fully automatic transmission control and independent track control for steering. The system is a dual-path hydrostatic drive. See Fig. 15.16.

Engine power is transmitted through a cold-weather disconnect clutch to a splitter drive, which provides power for the dual-path hydrostatic track drive, equipment hydraulic system, and a winch drive.

Track drive is through two driveshafts and two reversible, variable displacement hydrostatic pumps which drive two hydrostatic motors. The motors are also variable displacement, and together with pump variation, provide a stepless speed range of 0 to 6.5 miles per hour, forward and reverse. The two pump-motor pairs make up the transmission.

Infinitely variable travel speed, identical for both sides, is regulated by a transmission speed selector lever to the left of the operator. Its center position is neutral. Moving the lever forward causes the tractor to move forward, with travel speed increasing with distance from neutral. Reverse is selected by moving the lever rearward from neutral, with the travel speed increasing with the distance from neutral. As with conventional machines, slow speeds yield maximum push and faster speeds yield less push and less exact control.

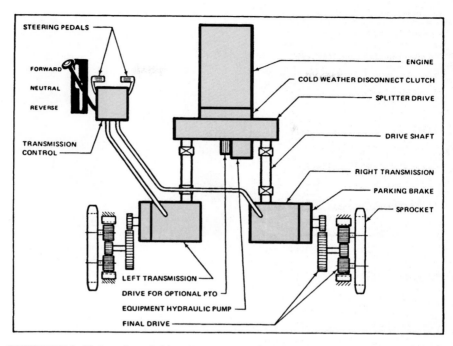

FIGURE 15.16   Hydrostatic track drive. (*Courtesy of Deere & Company.*)

The transmission is fully automatic. The speed selector lever is set at the maximum travel speed desired, and the automatic control regulates the speed, reducing travel speed as the load increases and increasing travel speed as the load decreases, maintaining engine rpm and horse-power near the selected maximum, keeping the tractor moving at the quickest rate possible for the load. The operator who wishes to go slower reduces maximum speed with the transmission speed selector lever. Adjustable detents are provided.

The hydrostatic transmissions also provide dynamic braking when the selector lever is in neutral, and will stop and hold the tractor on all but the steepest slopes. In addition, there are multidisc, wet-type brakes in the gear train to each track. These are spring-applied for parking when the engine is stopped, and are released by pressure from the transmission charging pump when the engine is running. They may also be applied by a foot pedal.

The tracks are driven by the dual transmission through separate gear reductions. There are no steering clutches or brakes. Steering is accomplished by independently varying the speed and direction of the right and left transmission. The optional steering controls are either two hand levers or two foot pedals, one for the right transmission and one for the left. Since the transmissions are infinitely variable in forward and reverse, steering is stepless, enabling live power turns of any degree of sharpness as well as counterrotation. Both give added productivity in some operating conditions. Counterrotation is obtained by moving the steering control past the neutral position. Increased resistance is felt on the control when in the counterrotating position.

Remote Control.   Small remote controlled excavators have been developed to operate in con-fined spaces where workers can not go. The Microtraxx loader can be outfitted with a dozer blade, a front-end shovel, or other attachments. The machine may weigh several tons and has a 20-hp diesel engine. See Fig. 15.17. The remote control links the operator to the machine only by radio frequency.

With a track width of just 42 inches and a frame height of less than 39 inches, this miniature loader can reach into most box culverts and storm drains to clear sediment.

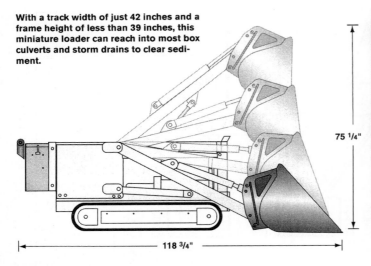

75 1/4"

118 3/4"

Radio remote control allows the operator to stand in the sunshine and send a Microtraxx MT 436 into box culverts with the snakes.

FIGURE 15.17   Remote control excavating machine.

WHEEL TRACTORS

Crawler tractors are the most compact, powerful, all-purpose pulling and pushing machines that have been developed. However, they are handicapped by low operating speed and by the large number of track parts which are subject to wear. In addition, unless equipped with special shoes, they wear and damage the pavements on which they work. There also seems to be a gradually rising limit on their maximum size, imposed by the limitations of the alloy steels commercially available and by problems of transportation between jobs.

Crawler construction is disproportionately expensive in small models, and rate of wear may be very high in any size.

For these reasons, the crawler tractor and other track-mounted machines are often replaced by equipment mounted on rubber tires. This may be described as wheel-mounted, rubber-mounted, or pneumatic-tired.

There are two principal types of wheel tractor. The basic or original style drives from large rear wheels only, and has much smaller front tires. The other drives through all four wheels, which are of equal size.

The second type is of greater importance in earthmoving, and will be discussed first.

FOUR-WHEEL DRIVE

The four-wheel-drive tractor occupies a position between the crawler and the two-wheel-drive. Traction is not usually as good as with tracks and grousers, but it is sufficient for most needs.

Such a machine needs more power and weight than a crawler to do the same class of work; it has great advantages over the crawler in speed and ability to use and work on highways; and it is more stable and comfortable on rough, hard terrain. It is not well adapted to mud work because of higher ground pressure and spinning of wheels.

Four-wheel drive is rather ineffective at dealing with peak loads such as are found in dislodging stumps and boulders, but very good when a heavy load can be picked up gradually without loss of speed, as in dozer or scraper work in smooth material.

Big tires and heavy tracks are both expensive, and under certain conditions will have similar maintenance costs. Tracks wear less than rubber on sharp-edged rock, but are rapidly cut to pieces by silica sand which does not damage tires.

Further discussion of these tractors will be found in Chap. 16.

Importance.   The four-wheel-drive tractor has become a substantial factor in earthmoving. Its most successful application is in front-end loaders, with bucket capacities up to 33 yards (a top of 8 to 14 yards is usual) with a moderate, SAE heap. They are the standard machines for truck loading in most yards, and in many pits, quarries, and highway cuts.

Most of these units are in the medium-weight or heavyweight class, ranging from 80 to over 700 horsepower. Weight with loaders varies from 8 to 100 tons, and even larger models.

Four-wheel-drive dozers have also found wide acceptance in scraper pushing, light to medium grading, pit cleanup work, and stockpiling. In addition, the front-end loaders do dozer work in their spare time.

There are also lighter skid-steer units. At least one of these weighs less than a ton, complete with loader. They are so different from the big ones that they will be discussed separately, in Chap. 16.

Construction.   In their early years, most of these machines had rigid frames and rear-wheel steering, as in Fig. 15.18. Current models have the center-articulated construction shown in Figs. 15.19 and 15.20, and described in Chap. 12.

The engine is usually at the rear, with considerable overhang behind the axle to increase its value as counterweight for the bucket or blade. Loader columns and arm pivots are behind the front axle; dozer columns somewhat farther forward.

The front driveshaft has a pair of sharp-angle universal joints for pivoting.

FIGURE 15.18   Rear-steer tractor and power train.

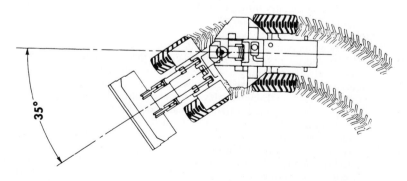

FIGURE 15.19   Articulation, full turn. (*Courtesy of Caterpillar Inc.*)

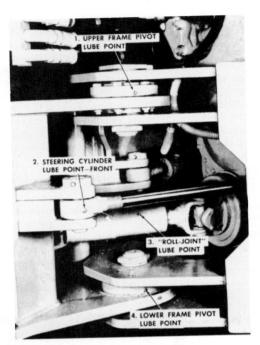

1. UPPER FRAME PIVOT LUBE POINT

2. STEERING CYLINDER LUBE POINT—FRONT

3. "ROLL-JOINT" LUBE POINT

4. LOWER FRAME PIVOT LUBE POINT

FIGURE 15.20   Center articulation parts.

The operator's station is usually out front on rear-steer machines, and over the pivot (on an extension of either frame) in articulated models. In either case, view of the bucket is good, but spill over its back may be dangerous.

Tires are large, and are weighted with calcium chloride solution or mineral dust to increase traction and stability. Wheelbase usually is short, so front and rear tires may be quite close to each other.

Transmission.   Practically all big four-wheel-drive tractors have torque converters and power shift. Many of them have the twin turbine converter that teams a two-speed manual shift with a two-speed automatic shift, giving four speeds forward with two reverse.

Smaller models are more likely to have hydrostatic or reversing clutch drives that are separate to the two sides.

Depending on instructions with the tractor, shifts may be made under any conditions or at any throttle position, or may be limited to closed throttle. Some models require that the machine be at a standstill before making a shift between forward and reverse. Instructions *must* be followed, if unnecessary and expensive breakdowns are to be avoided.

There may be a disconnect lever, to cut out drive to one axle. It is comfortable and economical to use two-wheel drive for light loads and traveling.

The tractors should not be either pushed or pulled to start, or towed from one place to another, unless the instruction book states specifically how it may be done. Removal of both driveshafts is usually required for towing on the road.

Lubrication is usually supplied by a pump driven by the input shaft, which does not turn unless the engine is running. Towing will work the transmission without lubricating it, unless the transmission is disconnected from the drive wheels by a disconnect lever for one end, and driveshaft removal from the other.

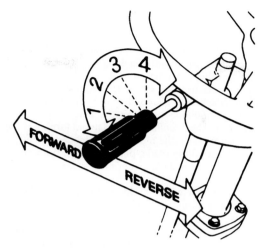

FIGURE 15.21   Twist shift.

Shift among the four or more speed ranges may be in a straight line, with a jog or obstacle at neutral; or a U pattern with forward on one side, reverse on the other, and the cross-slot neutral (see Fig. 15.15).

Another shifting arrangement, Fig. 15.21, utilizes a short lever under the left side of the steering wheel. It is pushed forward for forward and pulled back for reverse. Speed ranges are selected by twisting a handgrip on the lever. Detents mark each position.

Brakes and Clutch.   Brakes are on four wheels, and may be of any design. Vacuum-hydraulic is popular in smaller models, and air-over-hydraulic in big ones.

The torque converter may include a friction input clutch. This is usually released automatically by light pressure on a brake pedal. It disconnects the drive line to the transmission and wheels, without affecting the power takeoffs and pumps.

This effect is usually optional with the operator. There may be two brake pedals, only one of which operates the disconnect. Or there may be one pedal, with a separate lever whose position determines whether or not the brake will actuate the clutch.

The brake-operated clutch is particularly useful in loaders, and will be discussed in Chap. 16.

An input clutch may also be programmed to release automatically during any shift between reverse and forward.

Accelerator.   Wheel tractors usually have a stay-in-position hand throttle and a foot accelerator. The throttle may be set at low or moderate speed (low idle or fast idle), and full engine speed for heavy digging or fast travel is then obtained by pressing the accelerator.

This follows the automobile and truck pattern, and is opposite from crawlers doing the same work. They ordinarily operate on a high-speed throttle setting, with a decelerator pedal to slow them down.

The difference recognizes the speeds of the machines. Wheel jobs may travel up to 20 or even 40 miles per hour, and thus are safer if they slow down automatically on releasing an accelerator.

Steering.   There are several different steering systems, all power-operated.

Most new four-wheel-drive tractors in medium and large sizes have articulated or pivot steering, with a kingpin or hinge halfway between the axles.

Power is by hydraulic rams, but there is usually a follower valve that makes them respond to the wheel in normal automotive fashion. One set of wheels always follows the other directly, as seen in Fig. 15.18, so there is no problem of watching out for side swing of the wheels while turning.

There is loss of stability at sharp steering angles, requiring caution on side slopes or in loader operation.

The next most popular, but on-the-decline system, rear steer, causes the rear of the tractor to swing outside of the track of the front wheels, when on a curved path. This is the same problem seen with a car or truck while backing up. It must be kept in mind whenever working close to obstructions or other machines.

Skid steer is discussed in Chap. 16. Front steer for big tractors is unusual.

Whatever system is used, control over direction with heavy pushed or pulled loads will often be inferior to that obtained with crawler machines. It is hard to turn with a heavy centered load, either pulled or pushed, and difficult to avoid turning if the load is off-center. This is largely due to a failure of traction at the most heavily loaded wheels, and will vary widely with the footing.

If a dozer has a tilting blade, put it down on the side toward which you want to turn. A slight lift on the bucket will make a loader more steerable.

Traction.   Most four-wheel drives have more than enough power to spin the wheels readily. Such spinning is a normal part of operation, although it should be kept to a minimum in abrasive or rocky soils.

Both torque converter and wheel spin prevent you from applying shock loads to the work (and to the tractor). To some extent, you make up for this by utilizing momentum. Speed, even in the lowest range, gives the weight of the machine a substantial digging power, as it takes a lot of resistance to slow and stop it.

Calculate your depth of cut with a bucket to get maximum effect in the few seconds when momentum is added to tractive power. But keep in a low enough gear to carry through with straight digging.

In bulldozing, you will usually do best by cutting thin slices over a long enough strip to build up the load gradually, rather than tearing out a quick chunk as you would with a crawler. Under ordinary circumstances, if the wheels are spinning steadily, you are wasting too much power.

TWO-WHEEL DRIVE

Most wheel tractors used in farming and for light construction work have two-wheel drive. The rear or drive wheels and tires are much larger than the front ones, which serve for support and steering. Heavy-duty rear tires are needed in almost any construction work; while heavy front ones may be needed only with front loader equipment.

Industrial.   Models used in construction are called industrial, to differentiate them from farm tractors, which are usually not as strongly built. Figure 15.22 shows a standard type, although doing a farm job.

The gasoline or diesel engine at the front drives a disc clutch, a heavy-duty reduction-type transmission, a differential, and a pair of live axles which turn a pair of disc wheels equipped with large, low-pressure tires. The radiator base, engine, flywheel housing, transmission case, and rear axle housing are heavily built and make up the frame of the tractor. At the front, two small wheels and tires support an I beam front axle, which is hinged to the radiator baseplate. The axle will normally oscillate on rough ground without imparting twisting strains to the tractor. If a loader is to be carried, the axle may be rigidly fastened to the plate for greater stability.

The rear hubs may be designed to each carry one or two wheels. The use of dual wheels increases stability, traction, and weight, but makes the machine clumsy by adding greatly to the width. This disadvantage, and the extra stress it causes in the drive line by eliminating wheel spin, limit its use to special conditions.

A number of different transmission designs are used. There may be 5 to 10 forward speeds, controlled by one lever or by two. There are usually two reverse speeds, sometimes four. If there is a reversing gear or clutch, all forward speeds may be usable in reverse also.

A power takeoff is standard equipment. It turns a shaft to the rear of the axle housing, which can be used to power attached implements. It can be engaged and disengaged by moving a lever with the clutch released.

FIGURE 15.22 Two-wheel-drive tractor. (*Courtesy of New Holland North America, Inc.*)

It may be driven from the transmission gears, in which case it is controlled by the main clutch to the drive wheels. In modern machines, however, it is more often driven by a separate shaft from the engine through this clutch, and controlled by its own friction clutch. The two clutches are operated by the same pedal. Pushing it halfway down releases the propelling clutch only. Further movement cuts off the power to the takeoff also.

If there is a three-point hitch, its pump is controlled by the takeoff clutch, but it is not disconnected by disengaging the power takeoff lever.

Torque converters and power shift are common in the models that carry loaders or hoes, or handle three-point equipment.

Steering is primarily by automotive mechanical linkage to the front wheels. They angle very sharply, so the turning radius is short. A hydraulic booster is standard on large machines, and optional on small ones. It is a necessity with front loader equipment.

With manual steering, the wheel must be held firmly on rough ground, as otherwise a bump may cause it to spin with sufficient force to break a wrist. This danger is probably negligible with power steering, but the precaution is still a good one.

Brakes are on the rear wheels, and might be separately controlled by a pair of foot pedals, side by side. Then a single brake may be applied on the side toward which a turn is being made. The tractor tends to pivot on the braked wheel, thus making it easier for the angled front wheels to steer it around. This one-sided braking often makes it possible to turn sharply with loads that would cause unassisted front wheels to skid uselessly.

If one wheel spins on poor footing, and there is no differential lock, application of the brake on the spinning wheel will force the other one to turn.

The two brakes are applied together to stop. It is usually possible to lock the two pedals together, to simplify braking during travel without load.

Controls.   Controls vary in details and placement. Figure 15.23 shows an example.

The steering wheel is in top center, the clutch is on the floor at the front left, brake pedals front right, and the accelerator behind the brakes.

This machine has three gearshift levers. The tall center stick manipulates the main transmission through three speeds, the small right-hand one controls the high-low shift (and the starter interlock), and one high on the left is for forward and reverse. Other makes or models may have only one gear lever, or two.

Another lever engages the power takeoff, which is under clutch control when connected. In this illustration a lever engages the differential lock when pulled back.

Top speed of utility industrial tractors may be 12 to 20 miles per hour, with 15 or less being usual.

Traction.   The principal weakness of two-wheel drive is that it has poor traction on soft and slippery footing, particularly in reverse. This problem is reduced by putting the maximum possible share of the weight of the tractor and its attachments on the drive wheels, and by use of proper tires (usually high-cleated ones, in the largest recommended size).

Any rear-drive unit has better traction moving forward than backward. This is largely due to reaction from turning the drive axle, as the whole machine tends to rotate in an opposite direction from the axle.

FIGURE 15.23   Tractor controls.

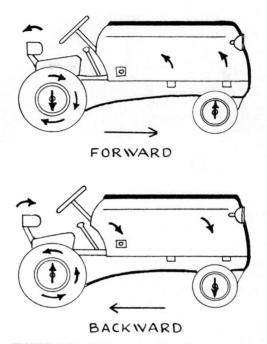

FORWARD

BACKWARD

FIGURE 15.24   Driving torque and traction.

Referring to Fig. 15.24, it will be seen that if the tractor is driven forward, the front of the drive wheels turn downward, and the equal force of reaction from the force applied tends to lift the front wheels up and back. The weight is thus concentrated on the drive wheels.

When the tractor is moved backward, the reaction from turning the axle forces the front wheels down, increasing their load, decreasing that on the drive wheels, and reducing traction.

If the tractor is moved up a hill, this shift in weight will be affected by a shift in the center of gravity. If the hill is very steep, there is danger that a tractor climbing it forward will overturn when the power needed to lift the front wheels becomes less than that needed to turn the rear wheels.

These effects will be more noticeable when the center of gravity is near the axle, or when the tire is large so that its contact with the ground is distant from the axle. Since wheel tractors normally have short wheel bases and big tires, it follows that reduction in traction while backing will be more severe than in cars or trucks.

Weight and Counterweight.   With two-wheel drive, traction problems are affected by (even determined by) weight distribution.

Liquid or powder in the drive tires increases their weight and traction. Wheel weights do the same. Balance is not greatly affected, and the proportionate loss of traction in reverse remains.

An attachment projecting to the rear of the tractor is 40 to 60 percent more effective than wheel weight in improving traction. This is because it tends to raise the front of the tractor, with the drive axle as a pivot or fulcrum, and thus adds some of the front weight to the rear wheels.

A tractor carrying a couple of tons of rear-mounted hoe has traction so improved that the forward-reverse problem can usually be ignored, and very heavy front loads can be handled. Loader performance becomes excellent, with improvements in traction, balance, and extra weight with momentum driving the bucket deep in the pile.

FIGURE 15.25   A mower is good counterweight.

A big rotary mower is an effective although lesser counterweight, and may be less in the way (Fig. 15.25). Mounted rippers or plows, or a drawbar-carried drum of sand or set of metal weights, are useful but lack most of the leverage advantage.

But a heavy weight attachment projecting to the rear may cause the tractor to become too light in the front for effective steering when going up a steep hill, and may even give it a tendency to turn over backward.

Steering is then done by cautious application of a wheel brake. Overturning may be prevented by moving slowly and smoothly, holding the attachment low for possible support, or backing up if there is sufficient traction.

Uses in Construction.   The two-wheel-drive tractor is the cheapest type of mobile power unit and attachment carrier that a contractor can buy. It is also reasonably fast, highly maneuverable, comparatively nondamaging to its footing, and relatively easy to hook up to attachments. It is often driven on roads between jobs.

As a result, its uses are too numerous to be detailed here. The two most important are the carrying and powering of backhoes and front loaders. Rear and side attachments include mowers and light grader blades.

The backhoes, which were discussed in Chap. 13, are used or are at least usable on almost every construction job. They also improve the traction of the tractors that carry them.

A front loader adds greatly to the usefulness of a tractor. Except in a few models, its efficient work with two-wheel drive is limited to soft digging and rehandling of loose material, but it does this work well. It can protect the tractor and clear the way for it.

The loader also serves as a motorized wheelbarrow, a work platform of adjustable height, a jack or hoist for heavy objects and disabled equipment, and a counterweight for a hoe.

THREE-POINT HITCH

Most farm tractors, and many industrial models, have a built-in mechanism to provide semiautomatic depth, draft, and position control to implements mounted at the rear. The device is called a three-point hitch, or sometimes a hydraulic drawbar. See Fig. 15.26.

The mechanisms differ from most hydraulic controls in that they regulate position rather than movement of the lift arms. An ordinary hoist valve such as is found in a bulldozer or a dump body, has UP, DOWN, and NEUTRAL or HOLD positions. In UP it will cause the ram or rams to raise the load until it meets the end of its travel, in DOWN it will allow it to lower or force it down as far as it will go, and in HOLD it will lock it in any position.

In three-point hitch controls there are no specific UP or DOWN lever positions. The top or forward end of the quadrant usually means DOWN, and the bottom or rear, UP, but any intermediate

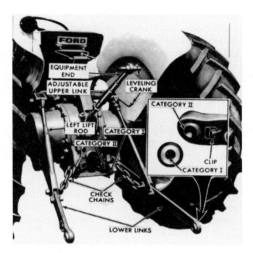

FIGURE 15.26   Three-point hitch.

position may result in raising, lowering, or holding the attached implement, depending on its position at the time, its weight, and the working resistance it is encountering.

In position control, and in depth control while conditions remain unchanged, a lever position corresponds to a specific implement position. The quadrant may have a movable stop, which can be locked to mark a desired setting, to which the lever can be returned accurately without looking at it.

Hoist and Linkage.   The hoist is operated by a hydraulic pump, valve, and cylinder unit inside the casing of the rear axle and transmission. The pump may be driven by an ordinary power takeoff, but more often (and preferably) is separately driven so that it turns whenever the engine is running. However, its power is usually cut off by disengaging the power takeoff clutch.

The cylinder turns a cross shaft which raises a pair of lift arms, one on each side of the differential case. Rods extend from these down to the lower links hinged to the axle case.

The linkage is made in two strength classes. Category I is standard equipment. Category II, usually offered at extra cost, is designed for heavy attachments and more rugged service.

The length of the top link determines the front-to-rear pitch of the implement. This may be very important in adjusting the suck of a grader blade or a plow, and may affect mower performance also. Shortening the link rotates the implement forward, increasing the suck of a plow but decreasing that of a blade. If it is too long, it will reverse these effects.

Too much suck causes diving and pitching, too little means no penetration.

Some links have hole-and-pin adjustment, but threads are easier to use and permit more accurate adjustment. If adjustments are required frequently or on the move, this top link may be replaced by a hydraulic cylinder.

One of the lift rods that connect the hoist lift arms to the lower links is equipped with a threaded length adjustment operated by a crank. This permits raising or lowering one side of the hoist for convenience in hitching, and to tilt an implement relative to the tractor. If the tool is a grader blade, both lift rods may be adjustable to make possible a steeper tilt.

The hitch is limited in side-to-side swing by a pair of chains anchored on the power takeoff cover plate.

Depth Control.   Semiautomatic depth control is ordinarily used for hitch-mounted implements that have edges or points that work below the ground surface. These include light grader blades, light rippers (subsoil plows), and moldboard and disc plows.

If the tractor is standing still, the hydraulic pump is operating, and the implement is resting on the ground; moving the depth-control lever to the top of its quadrant will cause the implement to rise until the ram reaches its limit of travel. If the lever is not quite to the top, the flow of oil may or may not be cut off by the movement of the ram before it reaches its limit, depending on the design of the particular unit.

If the front wheels of the tractor should start up a ridge, the tractor will assume a climbing position which tends to drop the implement deeper into the ground. This change in angle causes it to exert additional pressure on the top link so that fluid is supplied to the ram at higher pressure, thus raising the plow.

As the rear wheels cross the ridge, the tractor will nose down and tend to pull the teeth out of the ground. This will relieve pressure and perhaps pull against the top link, opening the drain port and allowing the plow to sink until balance is reestablished. In this way the plow is kept at an approximately constant depth, regardless of the pitching of the tractor.

The regulating pressure on the top link also acts to push the rear axle and wheels down, adding to traction. However, if the plow strikes something solid enough to stall the tractor, the shock load against the top link will drive it past the valve port opening position to another port that will drain the ram, taking the weight of the implement off the tractor wheels so that they can spin instead of stalling the engine.

This control arrangement maintains a fairly constant load on the tractor, prevents damage from shock, and keeps the implement near proper depth on rough ground. But it tends to vary working depth with soil resistance, skimping hard areas and overpenetrating in soft ones. The operator can counteract this behavior by adjusting the control lever for greater or less depth (draft resistance) as necessary. There may also be means to easily adjust the valve for greater or less responsiveness.

The one-way cylinder does not provide down pressure, so there is no way to force the implement down to cut soil that is too hard for it to penetrate.

Position Control.   Position control may be provided by the same lever after setting a selector, or by a second quadrant and lever alongside the depth control. Automatic response to changes in resistance is eliminated. The relationship between lever position and implement height is fixed, unless the ground or other surfaces raise the implement above its setting.

This control is used for attachments that work on or above the ground, not in it.

If the lever is set fully forward, the hoist does not operate, and the mower rests on the ground by its own weight, usually with additional rear support from a small caster wheel, skid shoes, or a height-regulating bar. It will stay in contact regardless of any pitching of the tractor while crossing ridges or obstructions.

In this position, a mower may tend to dig into the ground, cut too close, and scalp high spots. This can be prevented by raising its front slightly with the control. However, it will then be supported, and will lift clear of the ground as the tractor crosses a ridge, unless the operator lowers it.

The unused lever may have to be left in full UP position, to allow the other to function properly.

Hooking Up.   An implement is attached by backing the tractor to line up its hitch with the pins, raising or lowering the lower links until they are at the right height. The lower links have considerable free motion from side to side. One of them is slipped onto the corresponding pin, and the lock pin dropped and locked into place. The other link can now be made to line up by the adjustment crank, by raising or lowering, by moving the tractor, or moving the implement until the link ferrule can be slipped over the pin and locked.

The top link may originally be fastened to either the implement or the tractor, in either case it pivots freely up and down, but it may be either too short or too long to connect. If it is too long the lower links are raised so that the implement tips back and increases the space. If it is too short, the control lever is put at the bottom of the quadrant. If this does not bring it into position, the operator may stand on the links to push them down, raise the back of the implement, or tow it forward until it tilts into correct position. The ferrule is then slipped over the pin and locked. This may be done from the ground or the tractor seat.

The length of an adjustable link may be changed to attach it, then readjusted for operation.

Attaching is easy if the implement is light enough to be moved easily, or if the operator has been skillful or lucky in getting into just the right position relative to a heavy one. In general if a heavy tool is to be attached, the operator should have at least a strong back, and preferably a helper or a crowbar.

Detaching is easy. With the control lever down there should be little tension on the links so that they can be readily slid or hammered off the pins. A narrow unit, such as a one-bottom plow or a subsoiler, may fall over unless supported by blocks, and must be handled carefully.

BULLDOZERS

GENERAL DESCRIPTION

The bulldozer is the short-range member of a class of excavators which ordinarily dig dirt and transport it to the dumping point, and often grade both the cut and the fill in the same operation.

Bulldozers (dozers for short) are tractors equipped with a front pusher blade, which can be raised or lowered by hydraulic control and is used for digging and pushing. (See Figs. 15.27 and 15.28.) Angling dozers are bulldozers which have blades that may be set at angles to cast dirt to either side while the tractor moves forward. When their blades are set straight, they do the same work as straight dozers.

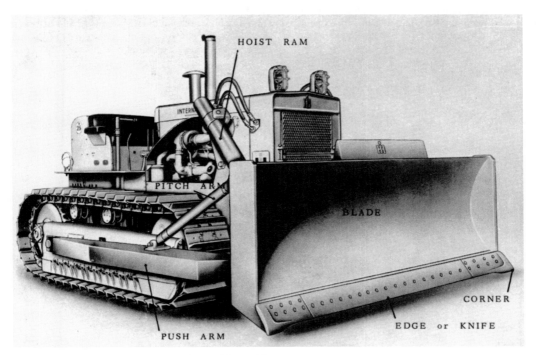

FIGURE 15.27   Direct lift hydraulic bulldozer, older model.

FIGURE 15.28   Single-ram hoist. (*Courtesy of Terex Division, General Motors Corporation.*)

Crawler tractors, known more familiarly as cats (when capitalized, "Cat" is a trademark of Caterpillar Inc.), carry most of the bulldozers intended for excavation or heavy pushing, but an increasing number are mounted on special four-wheel-drive, rubber-tire tractors. Dozers mounted on two-wheel-drive tractors, trucks, and graders are used chiefly for spreading and backfilling loose material.

Blade.   The blade is a massive structure that has a rectangular base and back. The leading edge of the base is a flat blade or knife of tough, hard steel which projects ahead of and below the rest of the blade. The front of the blade is called the moldboard and is concave and sloped back. See Fig. 15.29.

As this blade is pushed into the ground, the knife cuts and breaks up the dirt, which is then pushed up the curve of the moldboard until it falls forward. Material being pushed ahead of the blade is thus kept more or less in rotary movement, which tends to even up the load and offer less

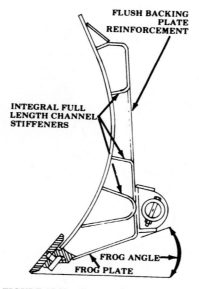

FIGURE 15.29   Cross section of dozer blade.

friction and a larger load than would be obtained with a flat vertical moldboard. The weight of the dirt first helps penetration, and then, as a full load is obtained, it pushes against the upper curve of the moldboard and "floats" the blade so that it cuts less deeply or not at all.

The edge is usually in three pieces, a wide center and two corners, which are bolted to the blade. They are reversible, so that they can be removed and turned upside down when worn. The corners wear fastest and are sometimes built up by welding until the whole edge is so worn as to need reversing or replacing.

Push arms should be attached to the blade near its outside edges, as the greatest strains normally occur at the corners. An exception to this is scraper pushing, where the center is subject to violent collision contacts again and again.

A blade that is to be used for scraper pushing is likely to cave in the center unless it is heavily reinforced. An extra plate or cup may be welded on the outside, or the inside may be braced with channels and plate steel.

Push Arms. Push arms are heavy, hollow beams extending from a hinged connection with the tractor to the bottom of the blade. Originally, dozers had arms that were mounted on the outside of the track frames, but more recently they have inside arms.

Outside arms are easier to design and to fasten to the tractor, as they need not get involved in the narrow clearances between tracks and center frame. Since blades are usually only slightly wider than the push arms, a wider blade and valuable extra clearance for the tracks in the slot cut by the blade are provided.

Inside arms are used in the majority of dozers when a track-width blade is wanted, when the dozer is mounted on a loader frame, or when the dozer will be used chiefly for scraper pushing and the narrow blade offers less danger of ripping tires.

Diagonal horizontal braces may extend from the inside of the push arms to the center of the blade.

Pitch (or tip) is the angle that the cutting edge or the blade itself makes with the ground. It is changed by rotating the blade around hinges on the front of the push arms.

Tilt is the side-to-side angle of the blade with the horizontal.

Pitch Arms (Pitch Braces). These are diagonal members between the push arms and the blade top. They brace the blade against loads above the line of the push arms, and may provide means for regulating its pitch and tilt. See Fig. 15.30.

Lengthening both pitch arms will tip the blade forward, shortening them will bring it back. The effect of such adjustments varies with digging conditions and the shape of the blade and the cutting edge. Usually, tipping forward will increase the suck of the edge and improve penetration in

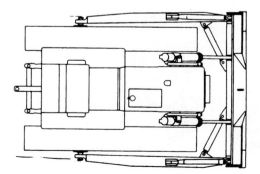

FIGURE 15.30 Top view of bulldozer with outside arms.

hard soils. It will also increase the upward pressure against the blade of loosened material coming up from the edge. This may be sufficient to float or carry the blade so that it will not be able to cut.

Since the blade is usually carrying a full load by the time there is enough upward pressure to float it, this does little harm. It is a convenience in pushing loose loads, as it automatically keeps the blade full.

Tipping the blade back reduces penetration, and may prevent the edge from cutting down except under heavy pressure. Such a setting is good in cleaning loose material off a firm, level surface, or in cutting off humps or knobs without regard to the whole surface.

Blade pitch adjustments are seldom made or needed with hydraulic lift bulldozers.

Mountings.   The direct-mounted, two-ram hoist shown in Fig. 15.27 is now the most popular type for large machines. The rams are carried in swivel brackets on the sides of a heavily reinforced hood and/or radiator guard structure. The piston rods are pinned to the back of the blade.

Direct mounting is best suited to tractors that have stabilizer bars or rigid frames instead of front springs. Otherwise the tractor springs will get involved in the stresses of lifting and lowering the blade. The engine will be pushed up by use of down pressure, and pulled down when the blade is lifted; the weight of tracks cannot usually be put on the blade by down pressure, grading hard ground is less accurate, and the springs make it difficult to raise the tracks out of mud by pushing the blade down.

The single-ram hoist substitutes a single thicker central ram for the two side rams. This arrangement allows all twisting strains to be absorbed by the blade and arms. It is better suited to pusher work than to general bulldozing.

Direct lifts may also be based on columns between the tracks and the hood.

Figure 15.31 shows an assembly that includes the radiator guard, a front-mounted hydraulic pump, a valve controlled by mechanical linkage from the seat, hood-mounting braces which contain the hydraulic oil reservoir, the hoist rams and their swivels, and the tubing and lines.

Track mountings, as shown in Fig. 15.32, place the weight as well as the push on the track frames, and permit spring mounting of the central frame without interference with dozer action. The leverage through the bell crank permits use of a shorter and thicker ram than with direct lift, with a corresponding reduction in hydraulic service problems.

The track frame mounting has lost out largely because the hood mounting is simpler, with fewer wearing parts.

Hydraulic Lift.   Dozers use a hydraulic system to raise and lower the blade. It may also provide from-the-seat adjustment of pitch, tilt, and even angle.

FIGURE 15.31   Front hydraulic unit.

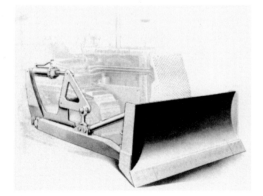

FIGURE 15.32   Outside lift arm bulldozer.

Such a system includes a pump, a control valve, cylinders, a reservoir, lines, and filters. General design is discussed in Chap. 12.

The hydraulic pump turns whenever the engine does. It is usually either gear or vane construction. Pressure may be 1,000 to over 3,000 pounds per square inch.

Blade Tilt.   When ground that is being dug is harder on one side than on the other, a bulldozer blade will tend to cut down on the soft side, particularly if it is a plastic soil that pulls it down. On a side hill the blade will tend to hang a little low on the downhill side because of shift in its center of gravity, and will cut low on that side.

There are a number of operational techniques which will produce flat grades under adverse conditions, or convert flats into side slopes where desired, but these usually involve extra skill, time, and work.

Mechanical Tilt.   Some blades can be tilted mechanically to cut more deeply on one side than on the other. Pitch braces, if adjustable, can be used for this purpose by making one longer than the other. The blade will be low on the side of the long brace. The twist of this adjustment is taken up by connections at the front and rear of the push arms.

FIGURE 15.33 Hydraulic tilt.

Mechanical tilting involves stopping the dozer and working on the blade for several minutes. It is usually not practical to make these adjustments frequently.

Power Tilt. The nuisance of mechanical adjustments of blade tilt is avoided by putting a two-way hydraulic ram in one of the pitch braces, as in Fig. 15.33, or in one or both of the push arm hinges, and controlling tilt by a valve at the operator's station. This permits rapid and effortless adjustment during operation.

Power tilting adds greatly to the ability of the machine to cut hard ground, to cut grades accurately up to walls or other obstructions, to crown roads, and to do accurate grading of slopes and curves.

If both pitch arms are adjustable, they can be used together to adjust blade pitch, in order to regulate penetration and pushing behavior. See Fig. 15.34.

Control Valve. The control valve may be a reciprocating or (rarely) a rotary type, operated by a hand lever. The operating position at the front is FLOAT, in which all the valve chambers are open (see the diagram, Fig. 12.85), so the pressure line empties into the exhaust line, and the ports to the fronts and backs of the rams are open so that no pressure is exerted on the bell cranks, and the blade rests by its own weight on the ground.

The next position of the valve control lever is DOWN, in which the pressure line feeds into the tops (backs) of the cylinders, and the bottoms (fronts) drain into the exhaust, causing the pistons to force the blade down to the limit of its travel, which may be 5 to 20 inches below the line of the tracks. If the blade meets resistance, the front of the tractor will be raised off the ground. Sometimes called neutral, the HOLD position closes the ports to the rams and opens the pressure line into the exhaust. This locks the blade in position without loading the pump. The rear position is UP, in which the pressure is directed into the bottom cylinder sections, while the upper sections empty, so that the pistons pull their rods to raise the blade.

The piping may be standard plumbing or special tubing, except where the line is expected to bend with the movements of the machine. Flexible high-pressure hose, woven of metal, fabric, and neoprene rubber, is used in such places.

Use of FLOAT. Most bulldozer control valves include a FLOAT position. Ordinarily FLOAT is used to smooth areas by backdragging, and in pushing dirt on pavements or other surfaces which are hard enough to support the blade, but which will be damaged if it is forced down. If the moldboard is so shaped and tilted that a full load tends to lift the edge to the surface, transporting can be done in FLOAT.

If the blade will not float, much of this work requires a high degree of operator skill and concentration, with results generally inferior to those obtained automatically by floating.

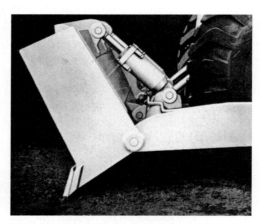

FIGURE 15.34 Hydraulic blade pitch control.

Replacement of a nonfloating with a floating valve is likely to be expensive. However, it usually is possible to install a pipe connecting the lines which carry pressure to the backs and fronts of the rams. A high-pressure, quarter-turn valve in this pipe will permit the blade to float when it is open, and to operate normally under control of the main valve when closed.

PUSHERS

Pushing of scrapers is an important use for heavy tractors equipped with dozer blades or special devices.

An ordinary dozer working as a pusher is likely to cave in the center of its blade unless it is reinforced. Reinforcement may be a plate or a cup welded outside, that will also serve to hold the scraper bumper from sideslipping, or it may be internal bracing.

Fixed plates, as shown in Fig. 15.35, are cheaper than dozers and are usually very rugged. However, they may not line up with bumpers on different models of scrapers, and they may not keep proper contact when pitching on rough ground. Similar plates may be mounted on angle dozer C-frames, where they are under as effective control as a blade, and do not cause as much hazard of ripping tires with their corners.

If two pushers are used in tandem behind one scraper, at least one of them should have a rear push bar or plate, with a frame that will carry the thrust directly to the front blade or plate.

If the tractor has conventional dozer push arms hinged to the outside of the track frames, the rear pusher should also be fastened to the track frames, to carry force directly. The rear pusher thrust will then not be carried by the central frame and gear cases.

If the front pusher is mounted on the central frame, the rear pusher is mounted on the steering clutch case, so that push is taken through the tractor in a direct line.

For efficient operation a pusher should make rapid contact with the scraper, that may be either stationary or moving. Any miscalculation of speed and/or distance may result in a collision that will damage both machines if repeated too often.

This situation is even more serious when tandem pushers are used. Doubling up the machines triples the likelihood of miscalculation and rough engagement. The front push blade or block must carry the power of two tractors, and therefore has less reserve strength.

Cushion. Use of spring-cushioned blades and/or rear plates will permit much faster engagement and make any damage unlikely. A blade may be hinged to a special push frame at the bottom, with rubber discs or compressed nitrogen to absorb shock in the middle, where scraper contact is made.

A rear cushion plate is valuable in a center mounting in protecting the gear cases from shock blows.

FIGURE 15.35   Pusher plate.

BULLDOZER OPERATION

References.   This section assumes familiarity with tractors. It is recommended that the sections on starting engines and on general operation procedures in Chap. 12, and the material on tractor operation earlier in this chapter, be reviewed.

Also, the operator's instruction manual for the particular machine should be studied, if it is available. It should provide control diagrams, operating instructions, and necessary information on lubrication and checking, which are part of the operator's responsibility on most jobs.

Digging.   A tractor bulldozer of any type is worked by moving the tractor forward, or less commonly, backward, and raising and lowering the blade to contact material to cut, spread, or transport it.

As a dozer moves forward and digs, some of the soil cut by the blade will pile up in front of the blade and move with it, and some of it will drift off the sides, forming ridges or windrows. Resistance to the machine's movement is made up of the power absorbed in cutting and breaking up soil, and in friction in the loosened dirt. If the blade is lowered, more work will be done and resistance will increase, as a thick slice requires more digging power than a thin one, and the total amount of dirt resisting the blade is increased. If the blade is raised, the slice will thin or disappear, and the amount of earth being pushed will decrease, so both work and resistance are reduced.

In heavy digging, efficient operation involves pushing the most dirt without losing too much speed by engine slowing, slippage in the torque converter, or spinning the tracks. The operator starts cutting a slice which should give this result, and if the machine slows, the operator raises the blade slightly; if it is not working to capacity, he or she lowers it. The upward blade movement should be made as gradually as possible to avoid leaving a bump in the path of the tracks.

If the blade is set to cut an even depth, digging resistance will remain about the same through the pass, but that of the loosened and transported dirt will increase steadily. This increasing resistance does not slow the machine at first, as it causes the governor to open the throttle to maintain tractor speed. Once the engine is wide open, further increase will slow it, so that the blade should be raised gradually to the surface of the ground where it can push the loose dirt without digging more. Sometimes the blade is "pumped" during this lift by being dropped and raised quickly, cutting out a bit of extra dirt each time it is lowered.

The dozer digs and transports much more effectively downhill than on a level or uphill, and work should be arranged to work down a grade when it can be done.

Breaking Piles.   A pile of dirt may be knocked down by walking into it with the blade at the desired grade, after which it may be spread or piled elsewhere. If the heap is too large or hard for the machine to take at one pass, or if it is to be spread in more than one direction, the first pass may be made to cut away part of the pile to grade, Fig. 15.36(*A*), or to cut the top of it partway down, as in (*B*). If the second method is to be used but the dozer cannot move the part it cannot reach, a ramp may be made by loosening the soil by pushing then backblading with down pressure, Fig. 15.37, so that the blade can contact the heap at a higher level.

If the pile is very large or hard in proportion to the power of the dozer, the side cut should be repeated from different angles, Fig. 15.38(*A*), in order to shorten the cut required for each pass. If the digging leaves a high face that might fall on the machine, the dozer should be turned toward it occasionally and driven into it with the blade held high. This should cause the bank to fall or slide without burying the side of the dozer in it, as it might in Fig. 15.36(*A*).

If the sides are not accessible, a center cut may be made by first ramping up and cutting a slot down to grade, widening it to both sides. This slot, and any cut more than a few inches in depth, should be made somewhat wider than the blade to avoid jamming the dozer between the walls. This may happen in very narrow cuts through rocks or roots in the sides being turned by grazing contact with the blade so that the rocks or roots project, or by creeping or falling in of the sides. Dozers whose blades are track width or very little wider are particularly subject to getting jammed in this manner.

A narrow cut also does not leave room to maneuver to get at a rock or other obstacle encountered in the floor.

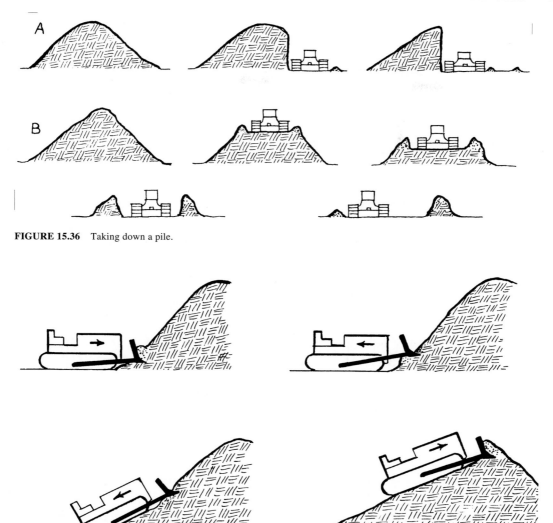

FIGURE 15.36   Taking down a pile.

FIGURE 15.37   Starting a slot.

Suck.   Plastic (rubbery silt or clay) soils will pull the blade down as it is pushed through them, dragging down the front of the tractor at the same time. An adjustable blade may be set for less penetration, and overdigging may sometimes be avoided by making a number of very thin slices. More often, digging is done in the regular way, and gouges made below grade are refilled with loose material.

Particularly cranky in such soil are dozers which have the cutting edge set at a pronounced forward angle and set well below the blade for a maximum suck, those with the hoist on the hood of a tractor with front springs, and any with spring cushions on the lift rods.

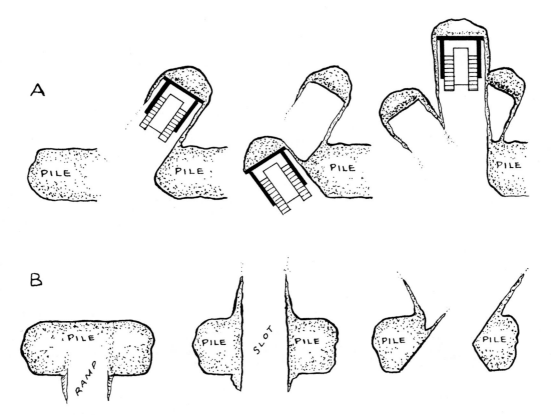

FIGURE 15.38 Spreading the pile from the center.

Road Cuts. A cut which is to have steep-sloped sides, as for a highway, should be started full width, and necessary measurements made to ensure cutting the slope correctly, as it may be difficult or impossible to get the machinery up on it afterward.

Side Slopes. A bulldozer not having a tilting control for the blade will cut deeper on the side which is downhill or in softer soil. On a slope, this tendency may be overcome in several ways. A shelf may be built by pushing downhill, as in Fig. 15.39(A) to (C), so the dozer can start its side cut level or tipped oppositely to the slope. Or the dozer may cut and turn downhill, raising the blade, as in (D) to (F), thus cutting a more or less level shelf which can then be enlarged or graded off as in (G) and (H).

Shallow stripping may be done by starting at the top of the slope so that in each pass except the first, the upper track can be walked in the hollow made on the previous pass. This works best in pushing two ways from a central point. Stripping downhill is more efficient, but it is not always possible.

Figure 15.40 shows a series of steps by which a short, wide shelf may be cut in a hillside, working from the side only. The material is piled in the background.

Uneven cutting may often be corrected by taking advantage of soil windrows, rocks, or other high spots to tilt the tractor up on the side where it cuts too deeply. In narrow cuts, the tractor may be backed up on one side of the cut opposite to the high or hard spots, or may turn so as to cut from the opposite direction. If no natural helps are available, the tractor may be walked onto boards or other lifts placed by hand at the low side, to start the cut at the desired slope.

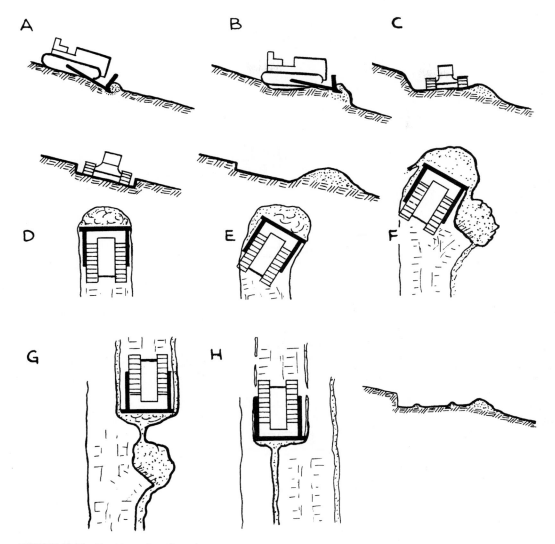

FIGURE 15.39   Notching a slope from above.

Transporting.   During transport of material the dirt which flows off the sides of the blade must be checked or replaced to keep a full load. It may best be checked by moving repeatedly in the same path, so that the ridges built up in earlier passes prevent dirt from leaving the blade. It may be replaced by digging down through the length of the push, just enough to replace the wastage, or by centering on a windrow left by a previous pass, which will replenish losses from the sides.

If the ground is smooth, the blade is held to just touch it or dig slightly. If the ground is uneven, an effort is made to cut enough in the humps to replace the dirt that is lost in the hollows or drifts off at the sides.

Wastage through side spill may be reduced by having two or more dozers work side by side, with the blades touching, so that little or no material can be lost between them.

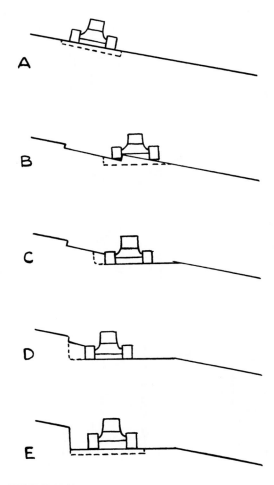

FIGURE 15.40 Notching a slope from the side.

Piling. Figure 15.41 shows two common ways of piling dirt with a dozer. Figure 15.41(A) is somewhat more efficient, as it does not involve pushing so much dirt up, only to have it slide down again at the back, but the difference would be important only if very large amounts of material were involved. Often a pile started according to method (A) is continued in the style of (B), because of the pile building back too near the digging.

If the earth is being pushed off a bank, the blade should be slowly lifted before it reaches the edge. The loose fill sinks under the weight of the machine. The extra height built at the edge supplies a safety factor, and should be enough to keep it level after compaction.

Each push to the edge should be on a slightly different path, if space allows, in order to distribute its packing action.

Spreading. When spreading material, the blade is held somewhat above the original surface so that dirt can slip under it in a smooth layer on which the tractor can walk. A thin layer may be spread to the desired grade, but a thick layer should be built higher to allow for compaction. If there is not enough dirt ahead of the blade to reach the end of the area to be covered, it saves time

A   B

FIGURE 15.41   Piling with a dozer.

to stop pushing as soon as the load is light and go back for more. The next bladeful will be pushed through the spot and can easily take the remnants of the first load with it.

It is best to vary the path used in spreading, as it is easier to keep track of the grade if no heavy windrows are built up.

Turning.   A bulldozer has difficulty turning while pushing a heavy load. More power is needed to swing the load than to push it straight ahead. As shown in Fig. 15.42, action is that of a bell crank lever pivoting on the braked track. Turns are easier with a wide-gauge machine, as the power arm is lengthened, and harder with an angle dozer or other dozer with blade carried a distance ahead of the tractor, as that lengthens the work arm, reducing the leverage.

It is often easier to break the curve up into two or more straight lines separated by angles; to pile the soil at the angle points; and to make a separate process of pushing it along the next line. The heaps at the angle points are best moved frequently instead of being allowed to pile up until difficult to dig.

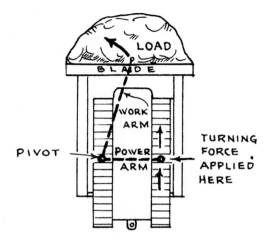

TURN TO LEFT

FIGURE 15.42   Leverage in swinging a load.

Scalloping.   Accurate grading is difficult on rough ground, or in rocky soil, as any pitching of the tractor is exaggerated in the movement of the blade. The blade control is not fast enough in most machines to be kept level while the tractor oscillates. If it is allowed to dig in on a drop, the material is apt to be left in a pile just beyond, and the tractor, in walking into the hole and up the heap, will pitch even more sharply so that a series of scallops are made. Once this process starts, only expert operators can level out without going back to the beginning and doing it over.

This scalloping is the bane of beginners and is liable to be troublesome until a feeling for the balance of the machine is acquired. An experienced operator can tell when the machine is about to either rear up or pitch down, and the operator can start moving the blade to compensate. If the change promises to be very abrupt, the operator can slow the tractor by cutting the throttle or slipping the clutch and using brakes, if necessary, to get more time to change the height of the blade.

A difficulty with hydraulic dozers is that the drop below the line of the tracks is limited and usually is not enough to enable it to work over a peak and start a grade down the other side; or to make a sharp down cut from a level grade, except by preparing a heap or ramp to back up on, to tilt the dozer down.

Backblading.   After an area has been graded, it may look a bit rough because of small windrows of loose dirt, grouser prints, and piles left where the dozer turned. These may be smoothed down by backing over the area with the blade floating. It acts as a drag, smoothing off humps and filling hollows, but does not move enough dirt to change the grade. This is called "Okie dozing."

Soil that has been pushed into places where the dozer cannot get behind it can be dragged in the same manner. Hard soil or a large quantity of loose dirt is better pulled with down pressure.

Rocks—in Cuts.   Rocks of flat or irregular shape may catch under the blade knife and be pushed along with it, increasing the resistance or floating it out of the ground. The blade may sometimes be freed by shaking it up and down, but it is often necessary to stop and back up in order to get behind and under it.

Firmly embedded rocks even of rather small sizes may roll or slide the blade up so that a bump is left. It may be necessary to back up, then move forward, forcing the blade into the ground deeply enough to get a grip on the stone that will roll or push it out. It may then be pushed ahead a few feet, with the blade high enough to let dirt slide under it, and that dirt backbladed into the gouge.

In digging out large rocks and stumps, the blade action usually is a combination of push and lift. The push is faster and more powerful, but objects may require chiefly upward motion.

In torque converter-equipped machines, the converter usually takes care of the drive slippage without resort to clutches.

Drive may be weakened in proportion to hoist by slowing the engine. No extra wear is involved, but full hoist speed is not maintained.

Rocks—in Fills.   Coarse material, such as rocks, lumps of sod, and other debris, make grading difficult or impossible. If a high fill is being made or a hole is dug to bury the trash, it can be worked over the front of the fill and buried. As the blade is raised in spreading dirt, the coarse material has a tendency to stay ahead of it to the last, although some will slip under and sometimes force the blade up so that the grade is lost. Such pieces can be moved along a bit farther with the next pass, and eventually gotten over the edge, although it is often quicker to get off the tractor and move such small, difficult pieces by hand.

If the stones are not to be buried but are to be left at the side for other disposal, they can often be worked over without making special passes for the purpose. Referring to Fig. 15.43(A), it will be seen that if a dozer pushes a bladeful of dirt into a loose rock, the rock will tend to drift off to the nearest side. If, during a series of pushes, each one is aimed to slide the rock in the same direction, it may be moved out of the area without any direct contact with the blade.

While the dozer is backing up, it can move a rock sideward by the maneuver shown in (B). When the side of the blade touches the rock, the steering clutch on the opposite side from the stone is released and the brake is applied hard. The dozer will spin on the braked track, the side of the blade will push the rock, and it may move several feet. The rear of the track is sometimes used in the same manner while going forward to move a rock a short distance, when restricted space makes it difficult to get at it otherwise.

Very heavy rocks, or those embedded in soil, should not be moved in this manner because of excessive strain on the tracks and blade mechanisms.

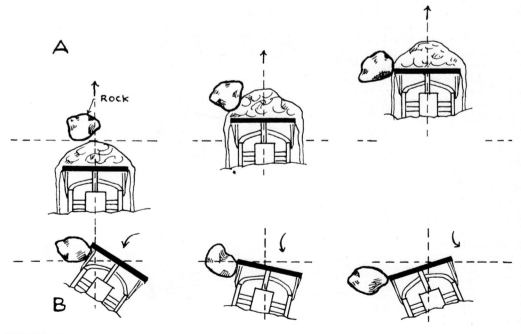

FIGURE 15.43   Side-shifting a boulder.

Pitching.   Tracked vehicles pitch badly in walking over ridges, stones, or poles. Figure 15.44 shows a series of positions assumed by a dozer walking over a small log. After overbalancing, the machine generally falls with a crash, which is very damaging to both tractor and operator. Such a log or bump should be pushed out of the way or avoided if possible. Crossing, if necessary, should be done slowly and at an angle, so that one side of the machine crosses the top and starts down while the other is still climbing. This slows the fall, and avoids any danger of turning over backward. If the bump is a soft ridge, turning the tractor sharply while crossing it will cut it down.

When a bulldozer is digging at capacity, the tracks often spin a little, then grip and move, and then spin again. Each time they spin, they build up piles of dirt at the back. If the machine backs up in the same track, these piles will have the effect of the log in the illustration, causing the back to rear up and then fall. This jolting can be avoided by keeping a wide enough work strip that it is not necessary to use exactly the same path backward as forward. A tractor is not much affected if only one track crosses a bump.

Tracks will carry the machine over a narrow ditch without any pitching, if the movement of the tracks does not break down the banks. A pair of quite frail boards will often be sufficient to prevent caving, as they will protect the bank against the backward push of the grousers. Crossing

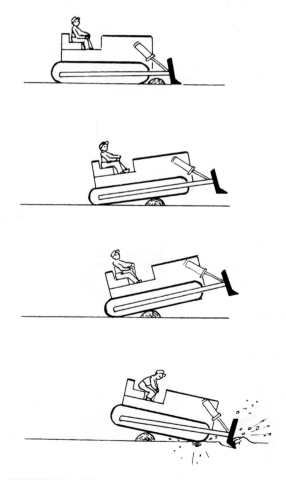

FIGURE 15.44   Wrong way to cross a log.

of a ditch wider than one-quarter to one-third of the length of the track on the ground is liable to be unsafe unless it is quite shallow.

Gears.   Bulldozing may be done in low or second gear. Low is less taxing for the machine in heavy pushing, and easier for the operator in precise work. Second gear is considerably faster, and in clean material good loads can be moved and smooth grades maintained. However, the higher gear makes it harder to cope with stones, hard spots, and other difficulties. Loose gravel may float a loaded blade so that it keeps a good grade automatically.

The amount of traction often determines whether a particular job can be done in second gear. A machine with narrow tracks, flat shoes, or worn grousers, or a standard machine on loose soil, spins the tracks easily, so that when too deep a bite is taken or an obstacle is hit, the load or shock to the engine is cushioned by slippage in the tracks, giving the operator time to raise the blade or disengage the clutch before stalling, even in a high gear.

Dozing Cycle.   Most bulldozer digging is done in shuttle fashion with the machine facing in one direction through the dig, push, spread, and return parts of the cycle. This is because the distances covered are usually quite short and turns, particularly in soft dirt, take time and spoil the grade, so that it is quicker and easier to back to the cut than to make two turns in order to use a higher gear. On pushes of 100 feet or longer the turns may be better unless the machine has a fast reverse.

Many tractors have as many reverse gears as they have forward gears and speeds. A bulldozer needs reverse speeds higher than forward ones. Backing is the unloaded part of the cycle, where lack of work should allow high speed, but some reverse gears must be powerful enough to climb steep grades and pull out of mud holes, and so cannot be fast.

The backup part of the cycle may be put to work by using back-ripper teeth on the blade to loosen soil for the next push.

Speed in reverse may also be limited by the quality of grading done during the push. In making heavy cuts it may not be efficient to take time and skimp loads to make a level floor with each pass, particularly if the soil is coarse. Gouges and bumps may be left by the blade and humps made by spinning tracks. The result is that a slow return may be made to avoid pitching even when a higher gear is available.

Hill Work.   Dozers may be used on moderate side slopes, and wide-track models on steep ones up to 30° or more. However, they are quite likely to overturn unless care is used. A machine which appears to have an ample margin of safety may be suddenly flipped on its side by running over a stone with the higher track, at the same moment that the lower track enters a hollow or soft ground. This is less apt to occur if the machine is pushing than walking, as it will then be moving slowly, will have the blade close to the ground, and will be steadied by the load. It also obtains some support from the windrow spilling from the downhill side of the blade.

Working on frozen slopes is hazardous, as the grousers may act as skates and allow the machine to slide uncontrollably downhill, regardless of the direction in which it is facing or trying to move. Sharp ice cleats will hold in such conditions, but dirt grinds their points off very rapidly.

A similar danger is encountered on rock slopes, particularly shale with beds parallel to the surface.

Slopes on soft fills are very treacherous, as the tip will be increased by the lower track sinking more deeply than the upper one.

If a machine starts to roll over slowly, it can sometimes be saved by turning downhill and lowering the blade.

A slope which is too steep to be safely worked sideward may sometimes be graded by running the dozer along it diagonally. If it is too steep for this, soil may be pushed straight down from the high spots, moved along the bottom, then pushed up to the low places.

Dozers can safely negotiate very steep up- and downgrades. Digging and pushing efficiency are much greater downhill and taper off to zero on steep upgrades. Steering is apt to be tricky on steep slopes, whether up or down, because of track slippage and shift in the center of gravity. Very steep grades of 25° or more should be climbed forward rather than in reverse, because of better balance and traction.

Cutting should be done downhill whenever possible, and in very hard ground it may be advisable to dig it downhill, even if the spoil must then be pushed up the same hill for disposal.

The engine oil-pressure gauge should be watched closely on steep work, as some engines do not get proper lubrication when tilted steeply, especially at compound angles, and a low oil level which still gives adequate lubrication on a level may leave the pump dry on either up- or downgrades.

Where a run includes a downslope, the operator may push several loads to the top, then push most of the resulting pile down in a single pass.

Unless the ground is loose, a dozer can push much more of it than it can cut and move while cutting. Here again it may be good technique to drop one or more loads at the end of the cut, pushing the final load all the way through, along with the bulk of dirt piled previously.

Cutting Hard Ground.   If the blade refuses to cut down, it should usually be tipped forward by means of the pitch rods. This helps it to cut when it is empty, but a load can float it out of the ground more readily. If it will not cut into humps or a bank, which is less usual, it should be tipped back. For general pushing the blade should be centered or back, as it rolls the material most effectively in that position, reducing friction. See Fig. 15.45.

If the blade will not cut after adjusting the pitch, a limited amount of ground can sometimes be cut up by spinning the tractor, first on one track then on the other, on the area to be dug. The grousers may chew up the ground sufficiently that it can be bladed off readily.

Digging in hard or stony ground will be easier if the work is arranged so that cuts are made by only part of the blade. One corner of the blade should be tilted down if possible. If not, one cut can be ground down to a depth of a few inches and overlapping cuts made at the sides. If half the blade is in the air over the cut, weight per inch of edge will be doubled for the part on the ground and it will cut more effectively. The dirt it gets under will add weight to the blade, so that it will probably do some digging in the original cut in addition to removing a substantial bit at the side. Effective work may also be done by working outward from a cut, herringbone fashion.

If any considerable quantity of hard dirt must be bulldozed, it should be loosened by back-ripper teeth or by a separate ripper, or dug with hydraulically controlled front-end crawler tractors.

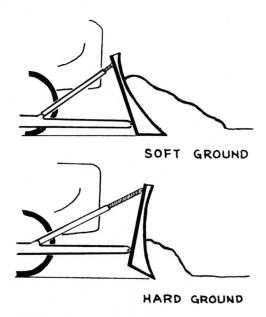

SOFT GROUND

HARD GROUND

FIGURE 15.45   Dozer blade pitch.

Output.   Dozer output falls off in almost direct ratio with increase in distance. Use of scrapers, or shovel and trucks, should be considered for pushes of much over 100 feet.

If space is ample, the large dozer will move dirt at lower cost per yard than a small one, and its advantage is increased in hard or rocky digging. In land clearing, the large machine is even more economical.

In restricted quarters, as in landscaping, backfilling trenches in narrow spaces, and working inside buildings, small dozers may show larger production and much lower cost than large ones.

Dozer output varies more than that of any other excavator. Production tables can be found in the *Caterpillar Performance Handbook,* but the figures should not be used in estimating without careful checking.

SPECIAL CONSTRUCTIONS

U-blade.   The U-blade, Fig. 15.46, has the sides advanced farther than the center. This makes it possible to transport a larger load by reducing side spill. The pointed corners assist penetration in hard soil and under stumps and boulders. It functions well in rough pioneering on sidehills, and is ideal for handling coal, garbage, and other loose materials.

The edges will cut more deeply than the center if the blade is below track level.

Bowldozer.   The Bowldozer, Fig. 15.47, is the highest-capacity dozer blade. It utilizes a standard dozer mounting with tilt control.

The rear wall is similar to a regular blade, except that it has a scraper-type edge with an advanced center section. The sideboards extend forward from $4\frac{1}{2}$ to 7 feet, and are held rigid by a cross beam connecting the lower front corners. This includes another cutting edge, set at a sharp angle.

This unit is bottomless. It is raised and backed at the end of its push. It can handle most types of material, but may be damaged in rough digging. It is specially adapted to light material, such as coal, or high-volume work such as supplying a belt loader.

Angle Dozers.   Angle dozers, known under various trade names such as Angledozer and Gradebuilder, are bulldozers with blades that can be angled to the left or right, in addition to a center straight-across position. They may have either hydraulic or mechanical hoists. See Fig. 15.48.

The angling blade is desirable in pioneering roads across rough country, and is superior to the straight blade in light trench backfilling and some other jobs. Its drawbacks are somewhat greater weight, cost, and upkeep; clumsiness in restricted spaces; difficulty in turning with a load; and looseness in the joints. Many of its functions can be served equally well by a power tilting dozer.

FIGURE 15.46   U-blade handling garbage. (*Courtesy of Balderson Inc.*)

FIGURE 15.47   Bowldozer. (*Courtesy of Balderson Inc.*)

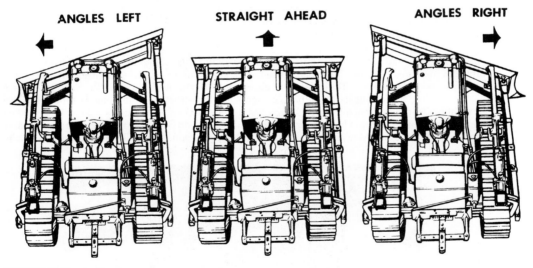

FIGURE 15.48   Angling dozer.

Cushion Blade.   The Balderson Cushion blade is used on Caterpillar's larger (D8 through D11) tractors for on-the-go push loading. It is called the C-blade and has rubber cushions, which allow the dozer to absorb the impact of contacting a scraper push block. The narrow width of the C-blade increases the machine's maneuverability in congested cuts and reduces the possibility of cutting tires, which is a problem for the wider U-blades. When the C-blade is not push loading, it can be used for cut maintenance and other general dozing jobs.

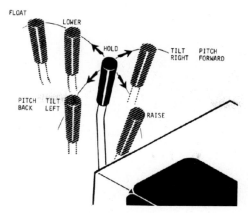

FIGURE 15.49   Hoist-tilt-pitch control.

Hydraulic Angle-Tilt-Pitch Dozer.   The most versatile dozers perform all blade adjustments by hydraulic power, with on-the-move adjustments by valve controls within reach of the operator.

The control lever for one such dozer is shown in Fig. 15.49. The main dozer frame is a C-frame, conventionally hinged near the tractor dead axles, and fastened to the center of the back of the blade by a universal bearing.

The C-frame is raised and lowered by a pair of cylinders mounted to the trunnions on the front wrapper. The blade is conventional, being about as high as a straight dozer.

Angling is effected by a pair of horizontal hydraulic cylinders mounted above the C-frame, and a control lever at the right side of the operator's compartment.

The angle cylinders are connected to a pair of struts extending forward to the top and near-bottom of the blade. The upper struts are hydraulic cylinders that lengthen or shorten to regulate tilt and pitch of the blade.

Blade lift and tilt are controlled by a single hand lever to the left of the angling lever. Pitch is controlled by the same lever in conjunction with a foot pedal. The lever is a standard four-position control, with FLOAT at the front and RAISE at the rear. The lever is pushed sideward to tilt the blade.

If the valve foot pedal is depressed while simultaneously pushing the hand lever sideward, blade pitch is obtained rather than blade tilt. Maximum pitch can only be attained when the tilt adjustment is centered. A separate hand lever controls the angling function of the blade.

Clearing Blades.   There are a number of special blades and attachments used chiefly or entirely in land clearing. These will be described in Chap. 21. They include stumpers, rake and rock blades, and tree dozers.

CHAPTER 16
FRONT-END LOADERS

FRONT LOADER

A machine developed from the bulldozer might be called the front-end tractor shovel. This machine may also be called a shovel dozer, dozer shovel, tractor loader, front-end loader, or just loader. It is used for digging, loading, rough grading, and limited hauling.

A typical loader is shown in Fig. 16.1 including a support frame on a tractor, a hydraulic system, a pair of push or lift arms (the boom) hinged to the top of the support frame, a tractor-width bucket hinged to the front of the arms, and a pair of dump arms hinged both to the push arms and to the bucket. Nontractor movements are under the control of two pairs of hydraulic cylinders.

The loader may be carried by any type of tractor. Crawler and four-wheel-drive tractors are used for heavy service, and two-wheel-drive tractors for lighter work.

CRAWLER MOUNTING

Crawler tractors that carry loaders are usually specially designed for them, and differ from standard models. Tracks may be wide gauge and made extra long, with an additional track roller on each side. The idler and sometimes the front roller are of extra heavy-duty design. Width is necessary for stability against side tipping when carrying high loads. The long tracks move the center of balance forward so that heavier loads may be broken out and carried. The heavier idler and roller construction is required by the heavy front loads.

Tractors redesigned to carry loaders do not have springs. Most of them have a rigid connection between the track frames and main frame at the front, thus improving stability at the expense of some operator comfort and grading control.

If drive is mechanical, the engine clutch must be rugged, as a loader is very hard on it. It is likely to have ceramic discs instead of lining, or operate in a circulating and cooling oil bath.

However, most loader tractors have torque converters, teamed with power shift transmissions. This construction avoids the problem of slipping a clutch, and improves lugging qualities in the bank. Power shift improves flexibility and shortens cycles.

Hydrostatic drive may replace the converter and transmission.

Loader Frame. The frame is composed of a massive weldment fastened to the track frames and/or the central frame. It carries the pivot or hinge pins for the push and dump arms and their hydraulic rams, and transmits the weight, thrust, and twisting strains of the loader to the tractor.

The frame connections or stress points should be inspected periodically, and tightened or welded as required.

Arms. The push or lift arms are hinged to the top of the columns or tower on the frame. They extend forward to hinges near the bottom of the bucket. A cross beam or other linkage braces them near the front. The arms-and-brace assembly may be called a boom.

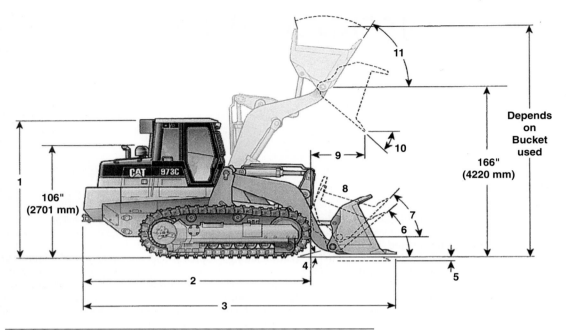

Overall machine width without bucket		
with standard track — 500 mm (19.7" shoes)	2580 mm (102")	
with wide track — 675mm 26.6"shoes)	2755 mm (108")	
Ground clearance from face to shoe	456 mm (17.9")	
Grading angle	69°	
1 Machine height to top of cab	3450 mm (136")	
2 Length to front of track	5163 mm (203")	
3 Overall machine length	depends on bucket used	
4 Carry position approach angle	15°	
5 Digging depth	depends on bucket used	
6 Maximum rollback at ground	GP 43°/MP 46°	GP = general purpose or
7 Maximum rollback at carry position	GP 50°/MP 52°	MP = multipurpose bucket
8 Bucket in carry position	—	
9 Reach at full lift position	depends on bucket used	
10 S.A.E. specified dump angle	45° (46° max.)	
11 Maximum rollback. fully raised	59°	

FIGURE 16.1   Large crawler-mounted front loader. (*Reprinted courtesy of Caterpillar Inc.*)

Lift arms are raised and lowered by means of two-way cylinders in the bases of the columns.

In principle, the dump arms are a connection between the back of the bucket, above the lift arm hinges, and the columns. This connection includes hydraulic cylinders, whose lengthening dumps the bucket, and whose shortening rolls it back.

The actual mechanism is more complicated, to supply changes of leverage, stability, and/or a mechanical (parallelogram) linkage to automatically level the bucket floor as it is lifted. This last function may be taken over by hydraulic valve arrangements and controls.

Bucket.   The bucket is of simple box construction. A heavy cutting edge of tempered steel runs along the front and partway up the sides. The upper back curves forward. It is placed as close to the tractor as possible, for stability.

Buckets are about the same width as the outside of the tracks. Present size range for standard weights ranges from 1 to around 4 yards for crawlers. Wheel tractors have both smaller and larger sizes, from 5 cubic feet to over 30 cubic yards.

Buckets are made in different sizes and weights for various types of material and work conditions. Light material buckets for handling humus, sawdust, or snow may be from 40 to 100 percent larger than standard buckets. Rock buckets are heavily reinforced. Slat buckets are used for handling loose rock or wood. They allow unwanted dirt to fall away through the slots. See Fig. 16.2.

Teeth are standard equipment on rock buckets, and optional on standard weight. They help greatly in hard digging, and in handling rock, stumps, and brush, but they interfere more or less with grading. Their cost is partly offset by the protection they give to the cutting edge.

Design must be a compromise. The bucket should be strong enough to take any punishment the tractor can give it, but light enough to raise a big load without overbalancing the tractor and without absorbing too much of its lifting power.

Bucket Action.   The standard bucket has three working motions. It is raised and lowered by two-way rams controlling it through the push arms, it is tilted or rolled between carrying and dumping position by the dump rams and linkage, and it is crowded and retracted by the forward and reverse travel of the tractor.

Dumping height is the elevation above ground level of the lip of the bucket in dumped position. It may be several feet below the height of the lip in carrying position, and $1\frac{1}{2}$ to $2\frac{1}{2}$ feet below the bucket hinges. Maximum dump height varies from 7 to 10 feet, being greatest in the larger and newer machines.

The bucket will usually tilt nearly 100° between dump and full-back positions. See Fig. 16.3. At maximum height the dumping slope of the bottom is 45 to 50°. At ground level a bucket may be kept fully dumped for float-grading while moving forward.

A bucket is said to be rolled, tilted, or curled when the floor is tilted so as to retain a load.

The standard Society of Automotive Engineers' (SAE) breakout force of a tractor to dig material is less than its operating weight. This is necessary; otherwise, the tractor would be lifted out of its position. There are several aspects of the tractor that lead to a strong breakout force. Obviously, the engine and hydraulic power contribute. Another is the linkage between the push arm and the dump arm of the bucket. It is claimed that a Z-shaped linkage, shown in Fig. 16.4, can produce better breakout force. This linkage is compared to the parallel linkage seen in Fig. 16.3, which does a good job of keeping the bucket in a level position while it is being raised and improving its reach.

When a bucket is rolled back during penetration into soil, it pivots on its heel (rear of the floor) as a fulcrum, usually developing much more breakout force than can be provided by the hoist.

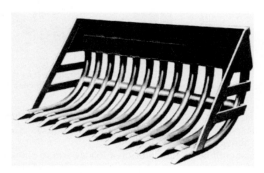

FIGURE 16.2   Slat loader bucket.

DIMENSIONAL DATA MICHIGAN L70
Tires: 17.5 - 25 (12PR) L-2

Where applicable, specifications and dimensions are in accordance with SAE Standard J 732 c, J 742 b and J 818 b.
Liquid ballast in rear tires only recommended for stabilizing purposes in timber and pallet handling on hard and level ground.

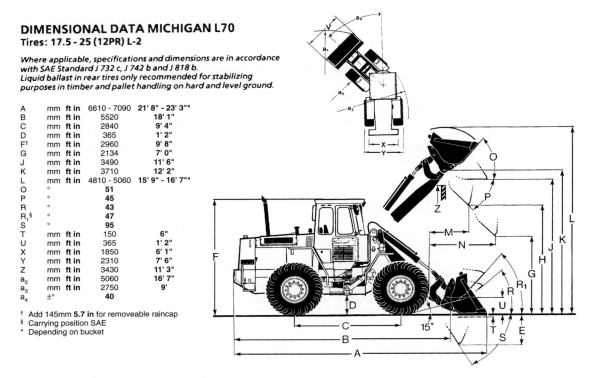

A	mm **ft in**	6610 - 7090	21' 8" - 23' 3"*
B	mm **ft in**	5520	18' 1"
C	mm **ft in**	2840	9' 4"
D	mm **ft in**	365	1' 2"
F†	mm **ft in**	2960	9' 8"
G	mm **ft in**	2134	7' 0"
J	mm **ft in**	3490	11' 6"
K	mm **ft in**	3710	12' 2"
L	mm **ft in**	4810 - 5060	15' 9" - 16' 7"*
O	°	51	
P	°	45	
R	°	43	
R₁§	°	47	
S	°	95	
T	mm **ft in**	150	6"
U	mm **ft in**	365	1' 2"
X	mm **ft in**	1850	6' 1"
Y	mm **ft in**	2310	7' 6"
Z	mm **ft in**	3430	11' 3"
a₂	mm **ft in**	5060	16' 7"
a₃	mm **ft in**	2750	9'
a₄	±°	40	

† Add 145mm **5.7 in** for removeable raincap
§ Carrying position SAE
* Depending on bucket

FIGURE 16.3 Representation of medium-size articulated wheel loaders.

FIGURE 16.4 Z linkage between lift boom and bucket. (*Courtesy of Samsung Construction Equipment Co.*)

Rollback is also useful in slicing upward in hard or heavy banks. It makes it easier to pick up heavy, heaped, or sloppy loads, and oversize objects. They can then be carried at a safe level, 2 or 3 feet above the ground.

There should be an indicator on a loader arm to show the tilt of the bucket, as this is usually difficult for the operator to observe directly. Its usefulness may be improved by a paint mark, a strip of bright tape, or a weld tack at level-on-the-ground position.

Automatic control of rollback will be discussed under OPERATION.

Rolling back is useful in slicing upward in hard or heavy banks, and in picking up and carrying heavy, heaped, or sloppy loads, and oversize objects. It improves balance by moving the load toward the tractor.

Hydraulic Systems.   Loader hydraulic pumps are designed for flow capacities varying from 12 gallons per minute (gpm) for small loaders up to more than 100 gallons per minute for the largest loaders. The relief pressure is from 2,000 to 4,500 pounds per square inch. Most loaders, regardless of size, have a load-sensing system for both the hydraulic systems and transmission. For instance, a fully automatic transmission will have preset shift points, so that shifting occurs at optimum torque.

The reservoir, holding more than the maximum gpm required of oil, is equipped with filters to remove outside dirt and products of wear. It may be a closed system, or an open one with a filtered air vent. It is important to keep oil at the proper level. Too little will allow the system to suck air and perform jerkily, too much may cause squirting out of the vent or building up of damaging pressure when lowering a loaded bucket.

Hoist and dump are controlled by two-way-type levers at the right of the cockpit. The hoist lever may have four positions, FLOAT at the front, then DOWN pressure, HOLD, and UP at the back. To control the bucket, a lever has three positions: a center position to lock it against change in tilt, one to dump it, and one to rotate it back.

The control levers, which produce electronic signals, may be just 3 inches tall. These signals are received by the machine's microprocessor, which activates the required hydraulics.

The control bank may carry an additional valve and lever, for use in operating a multipurpose bucket or a rear-mounted ripper.

Counterweight.   On a small machine, 500 to 1,500 pounds will permit lifting and carrying heavier (perhaps too heavy?) loads, and will improve traction when carrying a load. Larger machines can carry proportionately greater rear weights.

Counterweight may be a deadweight of metal or concrete attached to the back of the tractor, but is then most in the way.

The weight may also be a working part, such as a rear-mounted ripper, a power control unit, or a towing winch. The ripper enables the machine to break up soil in advance of grading or loading. However, the 7- to 30-inch layer of soil that can be reached by a ripper is not deep enough for efficient loading.

A counterweighted machine may be more difficult to steer with a load, as keeping the back down increases the track contact with the ground.

Track Shoes.   Most crawler loaders have the semigrouser (three-cleat) shoes depicted in Fig. 15.7. Some have two-cleat types.

When shoe surfaces are flat, the tracks will spin rather readily on many footings, giving the effect of a slipping clutch. This tendency to spin cushions all parts of the tractor against shock loads, but it often interferes with steering and traction and prevents the full power of the machine from being applied to its work. Semigrouser and other special, semiflat shoes give better traction but will still spin rather freely under slippery conditions. Full grousers grip well and aid in digging but make the machine very touchy and apt to stall, build up ridges behind the tracks, and subject machine and bucket to shocks and overloads that may shorten bucket life materially.

The chief objection to the use of grousers on a loader is that they tear up the ground when it turns, causing it to work itself down into holes that make work slow and sometimes dangerous. Flat shoes do this damage more slowly, or not at all if the ground is firm. Dirt loosened in this manner is easily smoothed off but will dig up again on the next turn. Under such conditions truck positions may have to be changed frequently to keep the loader on good footing.

WHEEL MOUNTINGS

Wheel-mounted front loaders work on the same basic principles as crawlers, but there are a number of differences in size and structure. Uses are strongly affected by type of carrier.

The most important units for heavy work are mounted on four-wheel-drive tractors. Smaller models are on skid-steer or conventional two-wheel-drive machines.

Dump rams and arms may be hinged on the columns above or below the lift arms, as in the crawler machines. But they may also be on the same hinge pins, or based on the lift arms themselves. In the two last cases, the bucket remains at a fixed angle with the lift arms as they are raised or lowered, unless changed hydraulically by automatic controls.

Four-Wheel Drive.   The medium-size and large loaders on rubber are on articulated, four-wheel-drive tractors. See Figs. 16.5 and 16.7. Loader frame columns are forward of the swivel. Lift arms are therefore shorter than on the crawlers, and rise to a steeper angle for the same bucket height.

Most of these machines are equipped with torque converters and easy-shift or power-shift transmissions. It is important that the forward-reverse shift be of the clutch or power shift type, to avoid delays in meshing gears or synchronizing clutch and gears. See Fig. 16.6.

Operators sit high and forward, where they have a good view of the bucket, but are exposed to danger of objects falling off its back. The engine is in the rear.

Even though the articulation turns to about 40°, turning radius is much longer than that of crawlers, so more space is required to maneuver from bank to truck and back. Faster travel and usually faster shifting keep cycle time as short as that of a conventional crawler.

In common with crawler-mounted loaders, wheel loaders can handle large boulders by carrying or pushing and can keep their pit area cleaned and leveled. See Fig. 16.7.

Wheel loaders have the outstanding advantage of quick and easy moving from one part of a job to another. They can also be driven from one job to another, but the bigger ones are so wide as to require special permits, and slow enough (30 miles per hour or less) that trailers may be preferred on long hauls.

Very large tires are generally used. They serve to provide excellent flotation, permitting work on most footings. Ground pressure is still much higher than with crawlers, but the packing effect of the tires and the more gradual turns make it possible to work easily on sandy ground that would tear up under crawlers, and cause excessive track wear. But slippery surfaces may cause loss of both traction and steering accuracy.

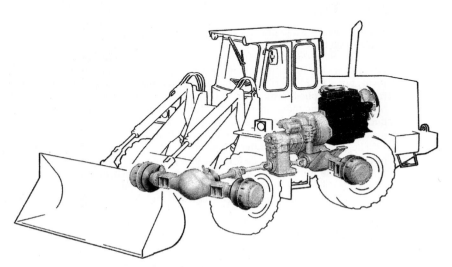

FIGURE 16.5   Power train, articulated loader.

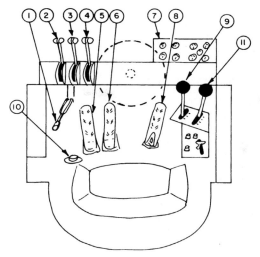

1. Parking brake
2. Hand throttle
3. Range lever
4. Directional lever
5. Brake pedal with
   drive cutout
6. Brake pedal
7. Instrument panel
8. Accelerator
9. Bucket control
10. Horn
11. Boom control

FIGURE 16.6 Controls, four-wheel drive.

FIGURE 16.7 Medium-size wheel loader in action.

SKID STEER

Skid-steer loaders are small and very compact. See Fig. 16.8. Weights range from less than a ton to more than 4 tons, with bucket capacities from 5 cubic feet up to $1\frac{1}{2}$ yards for light material. Dumping clearance is 6 to 10 feet. Drive is usually hydrostatic, but may be by belts or chains.

The lift arms are pivoted to triangular columns behind the rear wheels, and extend alongside the tractor for its full length. The cab includes steel-mesh sides, to protect against operator entanglement with the arms. The entrance is at the front, across steps provided on the bucket.

Travel is controlled by a T-bar controller or a pair of hand levers, each of which causes its side of the machine to move forward or back, as the lever is moved. The loading part is operated by rocker foot pedals, Fig. 16.9. The center pedal is installed for auxiliary hydraulically controlled equipment.

Skid steer enables these units to turn in approximately their own lengths. Combined with short wheelbase, this enables them to work comfortably in areas so restricted that many standard loaders could not even enter.

Pushing a lever forward causes the wheels on that side to rotate to move the machine forward; pulling it back reverses the movement. When both levers are moved the same amount in the same direction, the machine will move in a straight line. If one lever is moved more than the other, the machine will usually turn toward the side of lesser movement. See Fig. 16.10.

Ergonomic features are improving to help the operator of a skid steer more comfortably. These include low-effort controls, a comfortable arm rest, a retractable seatbelt, and a suspension seat.

If the levers are moved oppositely, one side will propel forward and the other backward, so that the machine will spin horizontally. In general, a machine with a loaded bucket will tend to slide the back wheels, with an empty one it will slide the front wheels. The actual movement is rather complicated, but the result is a U-turn in little more than overall machine length. See Fig. 16.11.

FIGURE 16.8 Skid-steer loader in action. (*Courtesy of New Holland North America, Inc.*)

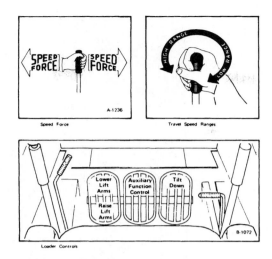

FIGURE 16.9   Travel and loader controls.

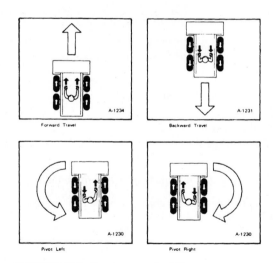

FIGURE 16.10   Steering controls, skid steer.

Small size and sharp turning enable these machines to work in very restricted areas. The drive to all wheels gives them good pushing power in proportion to weight.

Attachments.   One of the important attractions for a skid steer unit of equipment is the variety of attachments it can handle. For earthwork the attachments include

- Front-end loader bucket
- 4-in-1 bucket (see figures later in this chapter)

FIGURE 16.11   Skid-steer turn.

- A variety of other buckets
- Dozer blade
- Hydraulic breaker
- Hole drilling augur
- Trencher
- Backhoe excavator mounted on back of machine

Quick coupling devices for changing buckets or other front attachments are available or may be standard with the skid-steer machines.

TWO-WHEEL DRIVE

Two-wheel-drive loaders depend on good traction to dig competently. A large machine, equipped with a backhoe or a heavy rear implement, can give an excellent account of itself in medium digging, if the ground is neither slippery nor loose sand. But poor ground or really hard digging make its work difficult and uneconomical.

These machines, as seen in Fig. 16.12, are excellent units for general helping and cleanup work. In yards, they can load crushed rock and other stockpiles just as well as four-wheel-drives.

Design is usually simple. Side frames are fastened to the rear axle housing and the sides of the radiator base, and columns, cross-braced to each other, are erected on the frames. Lift arms hinged on the column tops extend forward of the radiator guard to the bottom hinge of the bucket. These are between the hood and the front wheels when lowered. Double-acting hoist rams are hinged on the column bases and push upward and forward against the arms.

FIGURE 16.12   Two-wheel-drive loader and backhoe. (*Courtesy of Deere & Company.*)

The dump rams are hinged to the lift arms. Their piston rods are extended into dump arms that hinge to the upper back of the bucket. The pump is driven from the front of the engine. A single two-section control valve regulates movements of the hoist and dump rams. The standard bucket is box type, straight-edged with an option of teeth, and as wide as the rear wheel tread. Other widths are usually available.

DIGGING

Bucket Angle.   Most loader digging is started with the floor flat or tilted to a slight downward angle. This position gives maximum penetration into banks and high spots, and cuts a smooth path on which the tracks can follow.

A rollback bucket is fitted with an indicator by which the operator can tell its angle of tilt. The operator may adjust it to flat position by reading the indicator, or by observing the bucket directly.

Cutting Down.   For cutting down into a level surface, tip the bucket downward 10 to 30°. When it has penetrated to a depth of 2 to 6 inches, it should be turned up to a flat or almost flat position, while the forward motion of the tractor is continued until the bucket is filled.

This tilt sequence combines good original penetration, sturdiest position of the bucket for most of the pass, and a powerful prying effect during the change in angle. Under some soil conditions continual minor adjustment of the angle while digging helps penetration.

The flat position is best for pushing a quantity of loose dirt, but the bucket should be turned down steeply for spreading and grading it, so that dirt will flow freely off the floor into holes, and so it will not be pulled down by sticky soil. Care should be taken not to hook into solid obstructions at a steep angle, as the bucket is then in its weakest position, and leverage against the dump mechanism is at its maximum.

The depth of cut may be regulated either by the hoist or the dump rams. See Fig. 16.13. For any given position of the push arms the edge will be highest when flat and 2 or more feet lower when fully dumped.

Penetration.   The loader bucket has much weaker penetration in proportion to size than a front shovel, because it is larger in relation to the power and weight of the machine, has a wider edge in digging, and may lack teeth. Another digging difficulty is that the hoist is slow in proportion to the speed of the tractor, so that the bucket tends to get under more dirt than it can break loose and lift.

Digging Banks.   A standard method is to use low gear range to force the flat bucket into the bank toe at ground level. When resistance slows the tractor, the bucket is rolled back gradually and hoisted, while crowding with the tractor is continued. See Fig. 16.14.

Rolling back the bucket as it rises in the bank increases cutting efficiency by aligning the edge with its upward movement, and by retracting it for a thinner slice. Its own suction, and crowding by the tractor, tends to make the cut thicker.

The proper balance among these forces varies with the machine, the bank, and the position and momentum of the bucket. It is the operator's responsibility to balance them so as to get a good bucket load in minimum time.

FIGURE 16.13   Starting a shallow cut.

FIGURE 16.14   Bucket movement in bank.

If the bank is hard, a thin slice may be most productive. If it is vertical, no crowding may be needed after initial penetration. If it slopes away from the loader, more crowding and less rollback are needed.

In general, a nearly vertical cut face in a bank not more than a foot or two higher than the lift arm hinges is most efficient for fast loading. This is called the optimum depth.

If the bank is higher than this, and overhangs, the upper part should be nudged with the full bucket occasionally to bring it down. See Fig. 16.15. But if the overhang is substantial, dig somewhere else.

If the loader has a self-leveling device, it may be necessary to cut it off, or to override it, to keep proper tilt in the bank. If it does not have it, you must remember not to give the bucket a high lift in full curl-back position, as this is likely to spill part of the load over the back—and probably on you.

If the bank is only 2 or 3 feet high, it may be dug by keeping the bucket at final grade and running beside the bank, cutting into it as much as possible without slowing the tractor excessively. The cutting will be done by the side and floor of the bucket which is in the bank. The soil will roll along the bucket and fill it, although the load will be heavier on the bank side.

The tractor with a torque converter meets resistance in the bank that will affect tractor speed much more than hoist speed. Even with the tracks or wheels stalled, the engine loader's hydraulic pump should maintain good speed.

This has the favorable effect of slowing crowd speed relative to hoist. However, the crowding force is increased by multiplication in the converter, so the bucket is still likely to be crowded into more dirt than it can lift.

FIGURE 16.15 Loader working in high bank.

Converters cannot be operated at full throttle under stall or near-stall conditions for more than a few seconds without excessive heat buildup and strain.

If the throttle is cut back to idling speed, the torque converter will exert almost no drive force. The loader pump will be slowed, but it will continue to exert full pressure to lift the bucket at a slower speed. This combination favors breaking out an overloaded bucket. Part throttle weakens drive without stopping it.

The wheel-mounted loader may be operated with the hand throttle closed, so that the engine governor setting is controlled by a foot accelerator. The operator cuts engine speed by lifting his or her foot.

Some four-wheel-drive loaders have an engine (input) clutch in addition to the torque converter. This clutch is power-released when you press one of the brake pedals lightly, or touch a separate pedal. This permits disconnecting the drive (crowd) without slowing the engine or the hoist.

The crawler usually works with the hand throttle wide open. A decelerator (opposite of accelerator) pedal is depressed to slow it down. This is the pedal to push when you start to get stuck in the bank.

The separate responses of push and lift to load and throttle are the chief features that distinguish torque converter operation.

Payload. The amount picked up in the bucket varies with the type of bucket, the power behind it, the traction, the nature of the material, and the skill of the operator.

For each yard of bucket capacity, you might pick up anything from $\frac{1}{2}$ to $1\frac{1}{2}$ yards, and should average $\frac{7}{8}$ yard in medium digging. A bucket must be rolled back to carry its maximum load.

When dumping close to the digging point, as in sidecasting, or loading a properly placed truck, keeping up a fast cycle is usually more important than getting maximum loads with each pass. As distance to the dump point increases, capacity loads become more important than the time taken to get them.

If the load must be carried over rough ground or backward up a slope, it should be limited to the weight the machine can carry easily without tipping.

If the bucket does not fill sufficiently, the tractor should be backed up, the bucket lowered to floor level, and another pass made. If the load is one-sided, the second cut should be made at an angle to the bank so that the empty side will penetrate first.

It is often difficult for the operator to judge the amount in the bucket, unless the digging is easy enough to permit dirt to be forced over the top. It is a good idea to cut a row of holes, about $1\frac{1}{2}$ inch in diameter, in the upper back of the bucket, through which the operator can see whether it is filled. Spillage through such holes is negligible in ordinary digging.

Ramping Down.   If the digging is downward, as in cutting a basement or a ramp, hard material may require pitching the bucket floor at a 20 to 30° angle to the line of the tracks, and cutting in thin slices. The downward pitch of the ramp should be gradual, as the machine is nose-heavy with a loaded bucket.

Gouging.   A common difficulty in digging heavy soils is that the penetration of the bucket is too good, so that it will be pulled down by the slice it has dug, either raising the back of the tractor or pulling the front of the tracks down into the ground. This may be combatted by keeping the bucket as flat as possible, or by tipping the floor into a nearly vertical position, which cures the difficulty but puts extra strain on the bucket.

It is often best to let it gouge, and make an extra pass to grade off the area as often as necessary.

Transporting.   If the ground between digging and dumping is hard and smooth, backing, turning, and dumping can all be done at speed with safety. If the ground is rough, the machine must move slowly, as going over a bump or ridge with a heavy load may cause it to fall forward, the bucket dropping to the ground and the operator's seat rising into the air. If part of the load dumps, the tractor will settle back. In crossing rough ground the operator should be alert to lower or dump the bucket if an upset starts. Lowering has the effect of taking the bucket weight off the tractor long enough to enable it to recover its balance.

Dropping a loaded bucket and stopping it abruptly in the air may burst a hoist ram hose if it does not overbalance the machine.

With crawlers, you cross ridges at such an angle that one track will be partway across before the other reaches it. If the ridge is soft, it may be possible to cut a more level path through it by turning sharply while on it.

If the bucket is carried 2 to 4 feet above the ground, the consequences of overbalancing are unlikely to be serious. If it is high, the weight is not quite so far forward, so tipping is less likely; but if it occurs, it will give the operator a worse toss. A high-held bucket also involves the danger of less likely but far more serious side tipping.

The bucket should not be given a high lift if there are rocks or lumps projecting over the back, as they might fall and injure the operator or the tractor, as seen in Fig. 16.16. Normally such things fall to the ground, or on the front of the tractor without serious damage, but they might roll down the arms or bounce back off the hood. This danger is particularly to be guarded against when the operator's station is at the front of the tractor.

TRUCK LOADING

This discussion is based chiefly on popular-size loaders with buckets from 1- to $2\frac{1}{2}$-yard capacity. See Fig. 16.17.

Procedures for the actual digging in the bank are the same as those already described.

FIGURE 16.16 Too much tipback.

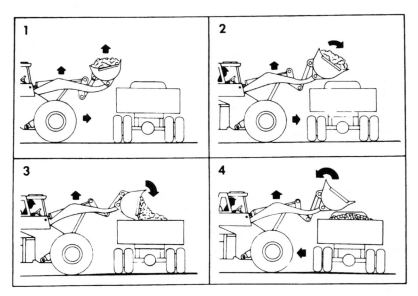

FIGURE 16.17 Loading a truck.

Maneuvering. The normal cycle for truck loading is made up of digging in the bank, backing out and making a partial turn toward the truck, then moving forward to the side of the truck while completing the turn and raising the bucket to clear the truck body.

The bucket should be high enough that the downward movement of the bucket lip during dumping will not cause it to strike the truck, and it is good practice to have it high enough to clear the side to avoid an accident while backing (contrary to the bucket height in part 3 of Fig. 16.17). The hoist is usually completed before the truck is reached. The control is then moved to HOLD, and the tractor moved so that the bucket is over the truck body.

A good procedure is to time the lift so that the bucket will just safely clear the truck as it is moved over it and the control can be left in UP using the dump. These levels can be automatically controlled by preset electronic controls.

The tractor is walked forward until the bucket is as far over the body as desired or until the radiator guard touches the truck body or tire. The main clutch is released (or, with nonclutch converter drive, the throttle closed), the tractor held with a brake, and the control moved forward to dump the load. The first bucket or two are best dumped slowly to reduce shock to the truck. If the soil is sticky, the bucket may be shaken by banging against the dump stops by moving the dump valve lever rapidly back and forth.

The loader is shifted into reverse and backed away, the bucket lip being raised to clear the body if necessary. When clear of the truck, the machine is stopped, put in the forward gear for digging, and headed toward the bank, the bucket being put in digging position and lowered during the return trip.

Front-end-loader cycle time varies all the way from 20 to 50 seconds. Average with small machines is about 25 seconds, with big ones 35 or more. But in rough digging a big machine may have the shorter cycle.

Easy digging, good truck spotting, and power shift favor fast cycles.

Figure 16.18 shows the result of a study of cycle time in big wheel loaders.

Loader Size, Cubic Yards	Loading Cycle Element	Well-Blasted Rock Average Cycle Element Time, Seconds	Poorly-Blasted Rock Average Cycle Element Time, Seconds	Common Earth Average Cycle Element Time, Seconds
8 or less	Bucket Load	10		6
	Maneuver Loaded	10		9
	Dump	5	No Data	3
	Maneuver Empty	9		9
	Total Cycle Time	34		27
8 to 10	Bucket Load	8		5
	Maneuver Loaded	13		11
	Dump	5	No Data	4
	Maneuver Empty	11		10
	Total Cycle Time	37		30
10 or more	Bucket Load	10	15	8
	Maneuver Loaded	11	15	12
	Dump	5	6	6
	Maneuver Empty	9	12	11
	Total Cycle Time	35	48	37

FIGURE 16.18 Cycle time, big front loaders. (*Courtesy of U.S. Department of Transportation.*)

Spotting Trucks.   There are many possible patterns of digging and dumping, some of which are shown in Fig. 16.19. Figure 16.19(A) is the most-used method, in which the side of the truck is at right angles to the face of the bank. This involves a quarter turn twice in each digging cycle. The turning stress is high.

In (*B*) the truck is parked at an angle of about 45° to the bank, so the loader's turns are only half as sharp. In this position some loads can be swung from the bank onto the truck with a very short backward movement, thus increasing loading speed. This is the best system for tracks, but because of driver resistance and indifferent supervision, it is little used.

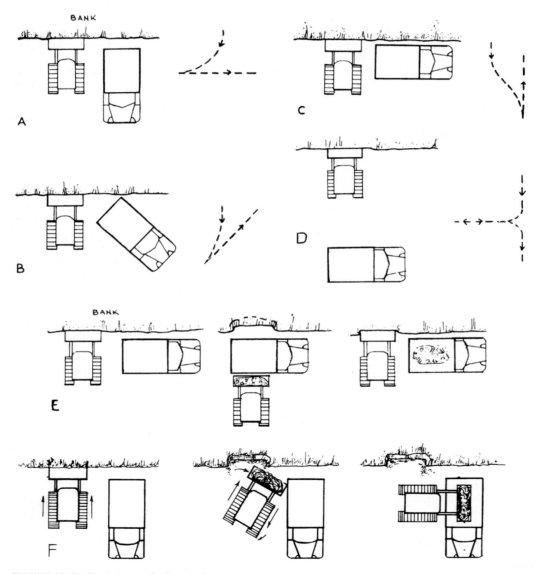

FIGURE 16.19   Truck patterns for front loader.

In (C) the truck is parked parallel with the bank, and the digging is done just behind it. This involves about the same amount of turning as (B), with a greater amount of walking.

In (D) the truck is parallel to the bank but at a distance from it, so that the tractor must make a 180° turn each time to put the load in the truck, and 180° to head back to the digging. More walking is required than in other methods. This is the slowest and most unsatisfactory of the arrangements suggested, but is often used in muddy or sandy pits where trucks are restricted to certain drives.

In (E) the truck is again parallel to the bank, but it comes to the loader for each bucketful. The machine fills its bucket from the bank and backs straight away. The truck backs in front of it, the tractor advances and dumps into the truck, the truck moves forward, the loader takes another bite, backs up, and waits for the truck to come back.

This method involves no turning, is fast and efficient if properly coordinated, and keeps wear on the tractor to a minimum. It is particularly valuable when the soil is loose and very abrasive, so that it would wear the tracks rapidly if it were worked into them by steering. However, the method is so unpopular with truck drivers that it is seldom used.

Figure 16.19(F) shows the effect of counterrotating tracks (or wheels) in shortening travel.

For wheel loaders, extra travel distance from bank to truck, up to 50 or more feet, may not consume time because of more maneuver space and use of higher gears.

Loader output can usually be increased substantially by having a spotter place the trucks, or by training the drivers to be alert to the needs of the machine so that they will not only take a convenient position, but be ready to move if the machine works away from them. This last is particularly important where the digging is shallow and truck bodies are large.

Dumping in Body.   The width of standard buckets varies with machine width, usually from about 6 to 11 feet, but a few up to 30 feet. Truck bodies (including trailers and off-the-road models) vary from 7 to over 25 feet.

Dumping height of the bucket, measured from the ground to the lip of the bucket, held at maximum height with the floor inclined downward at 45°, varies from 8 to 12 feet, in a few cases going as high as 20 feet. Highway dumpers have side heights of from 5 to 10 feet; off-the-road loaders may be over 15 feet.

Ordinarily, loaders are matched in size to the trucks that they fill. However, it is often necessary to cope with a mismatched set.

If the loader is too big for the truck, care must be taken to pour the load gently into the body, and allow excess to spill off the back.

A well-matched body can be fully loaded by dumping in the center only. A long body is loaded by dumping alternately in the front and rear from the side, although there is a tendency to pile up too much in the center and skimp the corners.

Such a body may also be loaded from the side at the front, and finished by filling from the rear, over the tailgate.

For convenience and efficiency in side loading, the sideboard of the truck should be a foot or two lower than the lip of the bucket in full-lift dumped position. This makes it easy to place the load in the center or even in the far side of the body.

High Trucks.   If a body too high for convenience is loaded, a heaping pile of soil is built up on one side. This heap may be moved toward the other side by pushing it with a loaded bucket held with its floor parallel with the ground and just clearing the sideboards. The bucket is then dumped and its load pushed over by the next bucketful, this process being repeated until the body is full.

Dirt may also be worked over by putting the bucket in fully dumped position, dropping it just inside the body, and rotating it partway toward flat position.

If many big trucks are to be filled, it may save time to dig slots 2 to 3 feet deep into which they can back while the loader operates on the higher level. If the pit floor must not be torn up, a few bucketsful will build a ramp up which the loader can walk to get an easy working height. The ramp should be made so that the machine is not tipped up steeply while dumping, as this reduces reach and increases the effort of holding position.

If there is uneven distribution of earth front and rear while loading, it may be corrected by cutting into the bank at an angle so that the side of the bucket which will dump over the low spot will get the heaviest part of the load. Dirt already dumped can sometimes be rearranged with the bucket in dumped position by pushing, pulling, or turning.

Distances.   The loading is fastest if the truck is so close to the digging that the machine has just comfortable room to turn, although the hoist is not fast enough to lift the load to the required height in a very short travel distance. When a pit is too wet or sandy for truck operation, it can dig in the pit, carry material onto firm ground, and there put it in the truck.

Output.   Figure 16.18 shows theoretical no-delay possibilities for large and very large wheel machines, based on Federal Highway Administration Report No. FHWA-RDDP-PC-520, "Production Efficiency Study on Large-Capacity, Rubber-Tired Front-End Loaders."

Conditions get unfavorable much more easily for loaders than for shovel excavators, as their production diminishes more rapidly as the digging becomes harder or coarser, or the footing gets softer. On the other hand, an alert foreman and a good operator can often step up loader production way above average, simply by using sound procedures that are generally ignored.

When compared to a shovel of similar rate of production, the loader has the advantages of moving around more readily, cleaning up the pit floor and moving boulders without assistance, picking up bigger rocks without chaining, and use as a dozer while not loading. The shovel can handle harder digging, work on softer floors, and has lower repair and maintenance costs because its tracks and rollers have much less use.

OPERATION—TWO-WHEEL-DRIVE

A front loader that is mounted on a rear-drive tractor reduces its traction when it is carrying a load. The front axle acts as a balance point, and any weight ahead of it will counterbalance the weight on the rear wheels, sometimes to the extent of raising them off the ground. Any reduction in weight on the driving wheels reduces their traction, which in any case is not as good backing as going forward because of reaction from driving torque.

As a result of these factors, the machines cannot carry good loads on loose, slippery, or soft ground, especially in reverse. This difficulty may be reduced by heavy counterweights on the rear wheels or on the back of the tractor, by using dual wheels, or by attachment of a rear-mounted tool such as a ripper, a winch, or a scraper blade.

The rear tires should be filled with a water solution of calcium chloride, as discussed in Chap. 12. This is necessary for efficiency in any tractor work, and is the first step in counterweighting a loader.

If the load is carried very high, more of the weight will be on the rear wheels, although danger of side tipping is increased. Extra traction may be obtained for a moment by lifting the load high and letting it drop. While it is falling, the tractor is almost free of its weight, and the wheels may grip enough to get the machine moving. Sometimes the machine will be able to drag the loaded bucket backward on the ground or on skids.

Another consideration is that the front wheels normally carry less than half the tractor weight; but when the load is heavy, they carry most of the tractor and all of the load. This results in hard turning, particularly with large machines, in which power steering is a necessity. Front tires must be the heaviest offered with the tractor.

The two-wheel-drive depends on its momentum to drive the bucket deeply enough into the pile to pick up a good load. The speed required may be greater than would be safe for turning and dumping, and it is therefore desirable to have a foot accelerator by which the operator can speed it up as it approaches the pile without changing the hand throttle adjustment.

The pit must be arranged so that the machine will not have to carry a load while it backs uphill or across rough or sandy ground, as it is likely to lose time or be unable to work due to poor traction.

Direct digging of hard or firm soil can be made much easier by breaking up the ground with a ripper or subsoil plow, and cutting only to the depth the tool penetrates. Such a machine may be mounted on the rear of the tractor and serve for counterweight, or be a separate unit attached and pulled by the drawbar and detached during digging.

LOADER HAULING

The large rubber-tire loaders may carry loads in complete-operation digging, hauling, and spreading, over distances up to ¼ mile.

Bucket capacity is smaller than in scrapers or trucks of comparable price, travel is slower, and the tipping influence of a loaded bucket limits them to well-graded routes.

They have an advantage over scrapers in being able to dig into banks from the floor of a pit or a cut, and in the ease with which they can build steep-sided stockpiles or dump over banks or into hoppers, without the help of other equipment.

A loader can supply a low or medium bin, saving either the expense of a truck ramp and hopper, or rehandling by clamshell. If the edge is a little too high, the loader can build a steep ramp of dirt or aggregate, as in Fig. 16.20.

If hauling is done by trucks, the loader (or an equivalent machine) is needed to fill them. A comparison graph such as that in Fig. 16.21 can be worked out, to determine the relative cost per yard of hauling by truck or by loader. If truck hauling requires a spreading or piling machine at the dump, its cost must be included.

FIGURE 16.20   Ramp up to a hopper.

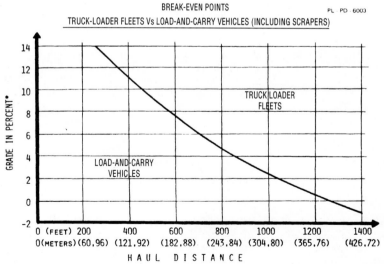

FIGURE 16.21   Load and carry, efficiency curve.

The cost is not the whole story, however. The work may require more material than the loader alone can deliver, and there could be many reasons why the acquisition of an additional loader would not be justified.

Job simplification may also be important. Where one unit can do the work of several, problems of scheduling are cut. This factor may be particularly important in a small operation. The loader might be able to do this hauling, and also take care of other work, so that trucks could be eliminated without adding other costs.

Carry or Push.   If material is to be moved a short distance, 200 feet or less, and the quantity is large, it may be economical to push rather than carry it. A flat bucket floating on the ground, moving between windrows of spill, will move two to three times as much dirt or rock as it will carry. Loads may be even better if the push is downhill.

Disadvantages include probable use of a lower gear, poor loads until windrows build up, rocks rolling behind the bucket into the tires, possible damage to the push route, and time spent cleaning it afterward.

Pushing is standard procedure in supplying belt loaders, even over quite long distances. See Fig. 14.20.

Underground.   The load-and-carry method is firmly established in subsurface mining, where it may be called load-haul-dump (LHD) transport.

The Eimco 920 LHD, Fig. 16.22, is one of the largest machines of this type. It weighs about 43 tons, is 10 feet wide and 38 long, is rated to carry $16\frac{1}{2}$ tons in its bucket (8- to 17-yard buckets for various weights of material), has almost 400 diesel horsepower, and is only $6\frac{1}{2}$ feet high with the bucket in low carrying position. A water bath conditioner reduces exhaust fumes to a permissible level.

Operation is usually shuttle fashion, without turns. The machine self-loads at the digging face, rolls back the bucket, backs past a chute or hoist, and usually makes a small forward turning movement to dump into it. It then straightens out, and goes back to the heading.

Maximum speed is 15 miles per hour, but underground haulage roads usually demand slower movement. It has four-wheel drive with articulated construction, and can climb a 40 percent grade with a load.

Dumping height is the same as machine height, $6\frac{1}{2}$ feet. An ejector bucket can dump a couple of feet higher.

The 911 LHD is the smallest in the line, with 4-foot width and a height of only 44 inches. It may have either diesel or electric power, and can work in a tunnel as small as 6 × 6 feet. See Fig. 16.23.

FIGURE 16.22   Load-haul-dump unit, large.

FIGURE 16.23 Load-haul-dump unit, small.

HANDLING OVERSIZE

The floor of a truck body should be protected by a layer of dirt or other cushioning material, before placing big rocks on it. This is particularly important if they are to be dumped rather than lowered. See Fig. 16.24.

Loader buckets are not well adapted to picking up large, loose objects on the ground. They tend to push ahead of the edge unless it is dug into the ground under them, and the overhang of the back makes it hard to balance anything bulky on the floor.

Rollback buckets can pick up many objects by getting the edge under them, then tilting back the bucket until they fall in. Larger pieces, such as stumps or boulders, may be crowded against a steep bank which will prevent them from falling out while the bucket is rolled back enough that the object will slide in. Two loaders can lift the object between them until it settles into one of the buckets.

If the object is too large to be picked up in this way or there is nothing to crowd against, it can be maneuvered onto the floor and held from falling out by a chain to the top back of the bucket, as in Fig. 16.25(A). This method is especially effective with loose stumps resting upright.

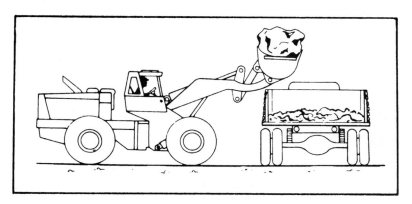

FIGURE 16.24 A dirt cushion protects the truck.

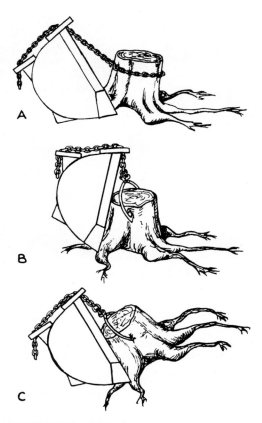

FIGURE 16.25   Picking up stumps.

A load usually settles back as it is raised or tilted back, so that it is easy to unhitch before dumping. If it settles forward and puts so much tension on the chain that it cannot be released, its overhang may be lowered so as to be supported by the truck sideboard or the pile. The chain may then be removed and the load boosted into position.

The bucket may also be lowered over a loose stump in dumped position, and attached by a chain or chain and tongs, as in (*B*). The bucket is curled, prying the stump up to a forward carrying position, as in (*C*).

For efficient odd-job use, or for crane work, it is almost essential to have a chain grab bracket, or a hook or hooks, fastened onto the top or back of the bucket. For use in dumped position, a second bracket on the rear beam is desirable.

A chain should not be anchored on a push arm and passed over the top of the bucket, as these change their relative positions during a lift, and the chain is likely to be stretched and broken.

BULLDOZER WORK

A regular wide bulldozer blade may be substituted for the bucket. Its use is advisable if the machine is to be used as a pusher for scrapers, or if it is desired to equip it with grousers and do heavy digging in rocky soil. The blade is more rugged than the bucket. A wide blade enables the machine to turn in its path without climbing onto the edges of the cut. However, the standard width bucket will do most bulldozing jobs better than a bulldozer, and is immediately available for loading or other special work.

This blade is somewhat more effective than on a regular bulldozer, because the digging angle can be changed with the dump control to meet any change in the digging, and because the blade can be dumped at the top of steep piles, where part of the load has a tendency to ride back down the pile on a standard blade. In addition, it can be floated forward in full-dump position as well as backward to smooth over loose soil.

Bucket Manipulation.   In dozer work, the bucket is used most efficiently by cutting with it in a digging position and spreading with it more or less dumped. In the latter position the bucket floor will have a tilt similar to that of a blade. Its advantages over a blade are the greatly increased cutting efficiency due to the knifelike edge; its ability to push bigger loads of soil because a good pushing load is contained in the bucket where it causes so little load or friction that nearly a full blade load can be pushed ahead of the bucket; faster grading work because of greater transporting capacity and ability to carry dirt to hollows without disturbing surfaces in between; the facility with which material can be backed out of banks; and the exactness with which it can be deposited where needed.

The transition from forward digging to spreading is most easily made by putting the lift lever at UP, and dumping the bucket a little at a time by rapid back-and-forth movements of the dump lever. Each dumping movement should lower the edge just enough to compensate for the amount it was hoisted while the dump lever was in neutral. As the hoist will not operate while the dump ram is moving, it is possible after practice to keep the edge running an even grade this way. The dumping operation may be stopped as soon as the floor has enough tilt to spill the entire load. Hoisting may be interrupted occasionally if a thin layer is wanted.

If the dirt is to be placed in a pile, the same method may be used, starting the dump with the edge kept down soon enough that the bucket will be in dumping position at the pile. The bucket can be allowed to rise while the shift into reverse is made; or can be left stationary during the shift and lifted slightly while backing out.

Another system is to leave the bucket in pushing position until the place to dump is reached, putting the hoist in UP, shifting into reverse, then dumping and backing away.

If a high pile is being built, either method may be used, simply lifting the bucket to conform to the slope of the pile while climbing it. It is a good plan to knock the top off the pile occasionally, by forcing the bucket down in dumped position in the loose dirt a few feet short of the top. Several yards may be pushed over at a time in this way, thereby shortening the climb to dump the next few buckets.

In backfilling against a wall, you may fill the bucket, put it in a flat position, and push a large amount of material ahead of it. Let this fall into the hole, then go back with the full bucket for more. This avoids the danger of hitting the wall with the upper part of the bucket. See Fig. 16.26.

Reach-down.   A loader has greater effective reach below itself than any bulldozer. In the vertical position the bucket edge will reach more than 2 feet below the tracks, and it will cut a sheer

FIGURE 16.26   Backfilling against a wall.

wall to that depth. Because of their wide bases no dozer blades can be forced straight down or can reach such a depth in a short distance, even when capable of such a drop.

The ability to reach down enables the loader to cross sharp ridges without the bucket loader losing contact with the ground, a matter of great importance in clearing work. The bucket may also be used to ease the machine down lesser drops by holding it a foot or so above the bottom and walking very slowly off the edge. When the tractor overbalances, it will be supported by the bucket, which can be gradually hoisted, thus lowering the front of the tracks to the bottom. This process may crush the bank enough that the machine can then be walked off it, or it may be necessary to put small blocks to reduce the drop of the back.

Backdragging.   The bucket is efficient at backdragging because the floor, when tilted straight down, can penetrate fairly hard material vertically, and because the clear space between the bucket and the tracks is sufficient to accommodate a lot of dirt. When backdragging in order to smooth loose dirt, the operator has a choice of tilting it to full dump, where it will cut (see Fig. 16.27); through intermediate positions where it will not cut and will pull decreasing quantities of earth; to flat, where it will move hardly any. In the flat or near flat positions, down pressure can be applied so that loose dirt may be compacted.

General Considerations.   But it can be seen that the operator usually needs two controls more or less continuously, versus one in a bulldozer, in order to bulldoze with a bucket. Some operators may dislike the machine for this reason.

Another drawback is that the location of the push arms between the tracks and the narrow path cut by the bucket enables stones and dirt to roll between the bucket and the track, making it necessary to lift the bucket and back far enough to get behind them in order to get through or to keep a grade. Also, the top of the bucket may be in the way of pushing close to a building.

A more serious consideration is the possibility of breaking up the bucket. It is not practical to make it as heavy or as strong as a bulldozer blade, and maintenance costs in heavy digging, particularly among rocks, may be higher. This is especially true if the operator does not keep the bucket at a proper angle when digging or if full grouser track shoes are used.

The use of the heavily built rock bucket for hard digging will materially reduce maintenance costs, but it necessitates the lifting of several hundred pounds of extra weight when loading.

The dumping rams may be used to push or pull the tractor when lack of traction or breakdown prevent it from walking in the regular way. The bucket is pushed down in firm ground or on a mat of poles or planks. Dumping the bucket will pull the tractor forward; closing it will push it back.

This control over the bucket edge can also be used in pulling stones back out of water or marsh, and in reducing the length of the machine to fit it in a short space.

In general bulldozing work where traction is good, a front-end-loader bucket should produce from 25 to 50 percent more work than a blade. It shows its greatest advantage in hard ground and among obstructions.

FIGURE 16.27   Backdragging with bucket.

MULTIPURPOSE (4-IN-1) BUCKET

This special bucket has a back that is similar to a dozer blade, and a separate floor attached to upper hinges by the bucket sides. A pair of double-acting rams on the rear of the bucket can lift this clamshell fashion, or clamp it firmly against the bottom of the blade. It is equipped with a cutting edge at the base of the back, and another on the bucket lip.

Figure 16.28 shows some ways to use its special features. When the floor is raised all the way, the back of the bucket can be used as an ordinary dozer blade. If it is lifted slightly, it acts as a float or depth gauge to regulate depth of cut, and as a bowl for holding the cuttings.

If fully back, operation is the same as with a one-piece bucket. In addition it can be used clamshell fashion for picking up loose material without pushing it around.

It can also grip and raise substantial pieces, such as tree trunks and boulders. Such objects should be held at the center whenever possible, to avoid twisting the loader frame. Such clamped loads can be released only by opening the clam.

Loads lying in the bucket may be dumped through the bottom in this way, or by tilting the whole bucket forward, in the conventional way.

Bottom dump permits the loading of very high trucks, as the bucket lip does not swing down when discharging the load. However, it tends to fill only the near side.

OTHER SPECIAL BUCKETS

Ejector Bucket. A rock ejector bucket has a back wall that slides forward to completely clear the bucket of its load. Its movement is controlled by the tilt mechanism through a mechanical linkage. In any position less than full height, it advances automatically as the bucket floor is tilted downward.

At full height, however, the bucket floor stays level when the tilt lever is moved to DUMP, so that the back moves forward horizontally. Dumping is complete, and can be done at a greater height than is possible with a standard bucket, whose edge must pivot downward.

Side Loader. Side dump buckets as special equipment for regular front loaders were mentioned earlier. Such buckets are standard equipment on a number of small to medium-size crawler-mounted loaders designed for underground use. See Fig. 16.29.

In narrow tunnels, it is usually necessary for a mechanical loader to hoist the load overhead and dump it behind onto a belt or into a hauler.

A side loader can dig straight-on at the face, back alongside the dump point, and spill the load into it without turns. It then moves (trams) forward to refill.

Dumping is faster than with an overhead, and less roof clearance is needed. All cars of a train (instead of just the nearest one) can be loaded without shuffling.

Side loaders are usually $4\frac{1}{2}$ and $8\frac{1}{2}$ tons in weight, with buckets from 15 to 35 cubic feet capacity. Width varies from 3 to 5 feet. Power is usually air or electric, but may be diesel.

OVERSHOT MUCKER

The overshot mucker digs at the front in the same manner as a front-end loader. The filled bucket is then lifted entirely over the tractor, and dumped behind it. In spite of increasing competition from load-and-carry models, the overshot mucker is extensively used in mines and tunnels. See Fig. 16.30.

The bucket is mounted on a rocker frame that raises it as it is pulled back by a pair of leaf chains that wind on rear winch drums. After dumping, it returns to digging position by air or gravity.

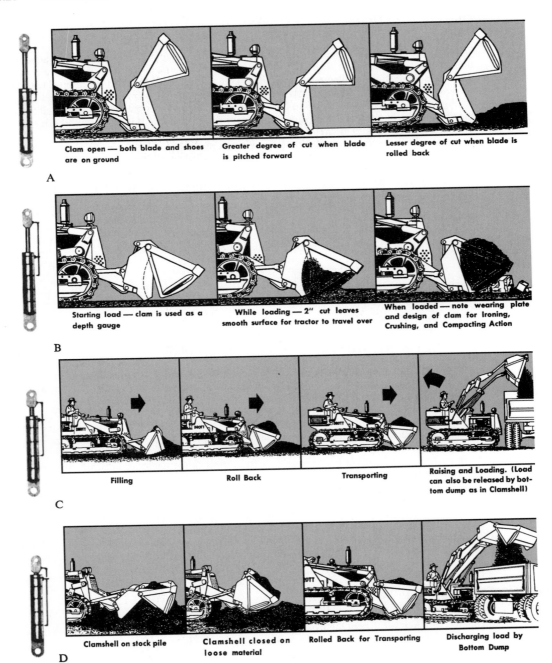

A

Clam open — both blade and shoes are on ground

Greater degree of cut when blade is pitched forward

Lesser degree of cut when blade is rolled back

B

Starting load — clam is used as a depth gauge

While loading — 2" cut leaves smooth surface for tractor to travel over

When loaded — note wearing plate and design of clam for Ironing, Crushing, and Compacting Action

C

Filling

Roll Back

Transporting

Raising and Loading. (Load can also be released by bottom dump as in Clamshell)

D

Clamshell on stock pile

Clamshell closed on loose material

Rolled Back for Transporting

Discharging load by Bottom Dump

FIGURE 16.28 Ways to use multipurpose bucket.

FIGURE 16.29   Side loader.

FIGURE 16.30   Overshot mucker.

FIGURE 16.31 RockerShovel with conveyor.

The loader can be obtained in three different loading heights, and can be converted from one to another in the field. The high discharge can load trucks as high as 11 feet, and standard high railroad gondolas. It requires a headroom of 7 feet—too high for many underground jobs. The low discharge will load a 7-foot truck, and has a maximum height of only 12 feet.

The manufacturer of this kind of equipment also makes air and electric rocker shovels for use in tunnels and mines where space is more restricted. Model 21 runs on the same rails that carry the cars it loads. It can dig straight ahead, or swivel about 30° to either side. The loaded bucket is passed over the top of the machine and dumped into a car directly behind. The operator stands on a side platform. A great many models are made to suit particular requirements.

The controls of Model 21 are shown in Fig. 16.31. The operator stands on the step plate and holds the two rubber-covered handles. Moving the left (front) one forward moves the machine forward; moving this same lever backward backs the machine. The rear curved lever can be moved forward to raise the bucket, backward to lower it, and to center (neutral) to lock it in any position.

The loading cycle is only 6 seconds, so this small machine can move a lot of muck in a shift.

Model 40H, Fig. 16.31, dumps its bucket onto a conveyor belt that carries the spoil well back from the machine. The front lever controls locomotion (crowd-retract) when moved forward and backward, and swings the bucket and deck when moved sideward. A small knee-operated lever controls the conveyor motor.

CHAPTER 17
SCRAPERS

PLACE IN EARTHMOVING

The bottom-dump scrapers discussed in this chapter are also known as carrying scrapers or pans, and by various trade names. They are highly mobile excavators with a centrally located bowl that digs, carries, and spreads loads. There are a wide range of types and sizes. Struck capacity is usually between 6 and 56 yards, but there are larger and smaller units also.

The typical modern scraper is a self-powered, rubber-tire unit. Controls are usually hydraulic or a combination cable-hydraulic.

Standard or conventional self-powered models have power and traction sufficient for most hauling needs, but require the help of pusher tractors or other machines in order to dig efficiently.

Self-loading scrapers, which ordinarily do not need pusher help, may have two engines with separate drive axles for extra power, or an elevator that reduces loading resistance.

There are also full trailer scrapers, which are towed by a drawbar attached to a separate tractor. These were once the dominant type, and are still useful for farming and in special situations. They are self-loading if the tractor is large in proportion to scraper capacity.

Scrapers are of primary importance in earthmoving. They are the standard tool for alternating cuts and fills, under the wide range of conditions where the cut is firm enough to support them, and the soil (including rock that is soft, ripped, or blasted) is digable.

The scraper digs, hauls, and spreads in a single cycle. It works in thin layers in the cut and on the fill, without limit as to the number of layers, so that its efficiency is not particularly affected by depth of cut or height of fill. Its use causes considerable compaction of fills, and favors proper use of rollers.

In various models, its economical haul distance ranges from less than 1,000 feet to nearly a mile. Where conditions are favorable, it can move earth for those distances at lower cost per cubic yard than any other type of earthmover.

It is not only an excellent machine for bulk earthmoving, but a precision finishing tool as well. The cutting edge is carried between front and rear wheels, so that it is unaffected by pitching, and the operator can control its position very accurately. If job conditions give enough time, the operator can cut or fill accurately to grade, and when space is wide enough for maneuvering, can build crowns and slopes as well.

TRACTOR-SCRAPER

The standard or conventional scraper is a single-engine self-powered machine, as shown in Fig. 17.1. It is made in two distinct sections, tractor and scraper, which are connected by a swivel hitch and hydraulic lines. In steering, the two parts pivot on this swivel.

FIGURE 17.1   Standard single-engine scraper. (*Courtesy of Terex.*)

SCRAPER

The scraper has three basic operating parts: the bowl, the apron, and the ejector or tailgate. In addition, it includes the gooseneck and the scraper (trailer) wheels. (See Fig. 17.2.)

Gooseneck.   In front, the gooseneck or yoke has a vertical swivel connection with the tractor, which is usually in two parts with two pivots, upper and lower. It permits turns of 85 to 90° to each side of center. See Fig. 17.3. In addition, there are horizontal links that permit the two sections to tip independently through a limited angle.

Behind the swivel, it arches up to allow space for the tractor wheels to roll under it on turns, then widens into a very massive crossbeam, and is finally a pair of side arms extending backward and somewhat downward to trunnion fastenings on the sides of the scraper bowl.

The gooseneck carries the steering cylinders, the lift cylinder and lever arm for the apron, and a pair of hoist cylinders for the bowl. All of these may have two-way action, or be one-way with return by gravity, springs, or counteracting cylinder.

Bowl.   The bowl, Fig. 17.4, is the principal member and carries the cutting edge. It is substantially a box with rigid sides, with the apron forming a movable front and the ejector a movable back. Extensions of the sides converge behind the rear axle, forming a case for the ejector cylinder, and support for a bumper by which the machine may be pushed.

The bowl is supported at the rear by the rear or scraper axle, at the center by trunnions on the ends of the draft arms, and at the front by a pair of hydraulic cylinders suspended from the gooseneck.

The pull of the tractor is applied through the gooseneck. Most of it is transmitted by the trunnions at the bowl center, but a variable amount comes through the lift cylinders, depending on their position.

The floor is cut forward of the centerline and fitted with a cutting edge, often called a knife. This edge is usually very hard steel plate in three pieces—a wide center one and narrower ends,

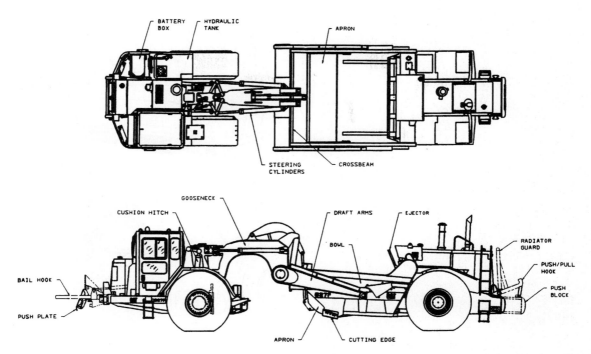

FIGURE 17.2 Scraper parts identified.

fastened with plow bolts with smooth sides up. The sections can be removed, inverted, and reinstalled when worn on one side.

For most work, the center piece is set farther forward than the sides. It is mounted back flush with the ends only when the job is grading, working light cuts, or in sand.

Teeth may be bolted to the center section, to improve penetration in difficult ground. They interfere with dumping and spreading and are laborious to install and remove, particularly on a worn edge. They are used more often on elevating scrapers than on standard models. See Fig. 17.27.

The fronts of the bowl sides, at the bottom, usually have bolted-on wear plates, called side cutters. These receive less wear than the bottom edge, but eventually need replacement.

Apron. The apron forms the forward side and a variable amount of the bottom of the scraper assembly. When in a down or closed position, it rests against the scraper bowl at the cutting edge. When lifted, it moves upward far enough to leave the whole front of the bowl open.

It is lifted, lowered, and in some models clamped down forcibly, by a hydraulic cylinder which usually is linked near the base of a lever hinged to the gooseneck.

Since apron movement calls for travel through a considerable distance without (usually) the need for much brute force, this third-class lever arrangement is efficient.

When the scraper is digging (loading), the apron is held at the moderate distance forward of and/or above the cutting edge. Dug material moves both backward into the bowl and forward onto the apron. In sand, the apron may be kept in a digging position for compaction; in chunky or boulder-filled soil it must be kept up and forward to be out of the way.

Occasionally, down pressure is used on the apron to clamp bulky objects against the bowl, in order to carry them in a half-loaded position.

There may be a mechanical adjustment in the apron lift, to obtain a larger opening at the expense of tight closing. The higher position may be useful in loading big pieces.

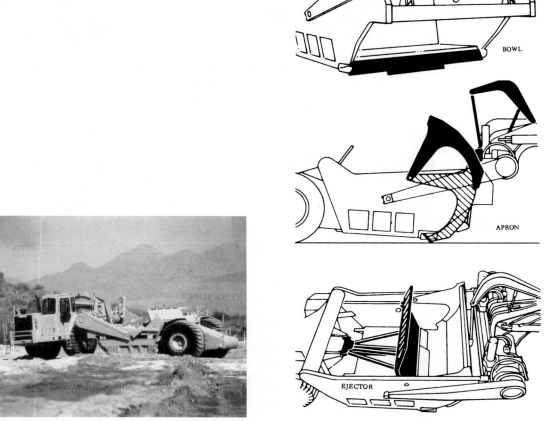

BOWL

APRON

EJECTOR

FIGURE 17.3   Scraper making tight turn. (*Reprinted courtesy of Caterpillar Inc.*)

FIGURE 17.4   Bowl edge, apron, and ejector.

Ejector (Tailgate).   The ejector is the rear wall of the bowl. It is usually a sliding or bulldozer type, that moves forward horizontally, forcing the load out of the bowl, over the cutting edge. It is supported by rollers riding on the floor, and on tracks welded to the sides of the bowl.

Power is a two-way hydraulic cylinder inside the rear pusher block (bumper) frame. Machines of the larger sizes may have two cylinders, and they may be telescoping in design, to increase the length of push in proportion to casing length.

Controls.   A typical arrangement of controls for scraper movements is shown in Fig. 17.5. The levers operate independently unless some optional combination control is used.

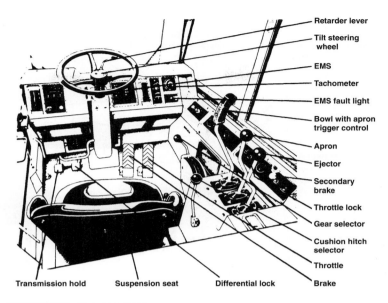

FIGURE 17.5 Scraper controls.

The bowl lever has three standard positions, RAISE, HOLD, and LOWER (DOWN). LOWER is with pressure, so that on hard soil continued movement will raise the scraper, with weight resting on the cutting edge. The DOWN position is useful for emergency slowing or stopping on a downgrade.

The apron lever also has the three positions, to raise, hold, or lower with pressure. It may also have a FLOAT position, to cause the apron to rest by its own weight.

The ejector control has three standard positions: FORWARD (EJECT), HOLD, and RETURN. There may be a FAST RETURN also. There is often an automatic control, so that it can be held in RETURN by a detent until the return is complete. It then kicks itself into HOLD. This cycle is canceled by the operator's moving any control.

TRACTOR

The tractor includes the engine, drivetrain and drive wheels, hydraulic pumps, and operator's station. It is more or less permanently attached to the scraper by a swivel or articulated joint.

Two-Axle (Overhung).   A typical two-axle scraper, Fig. 17.1, is powered by a tractor having only a drive axle. Stability is supplied by the attached scraper. The engine projects forward, a position called overhung, with its entire weight on the drive wheels.

The articulated connection between the tractor and the scraper gooseneck varies in construction, but usually includes separate upper and lower hinges, and a pair of steering cylinders based on the gooseneck.

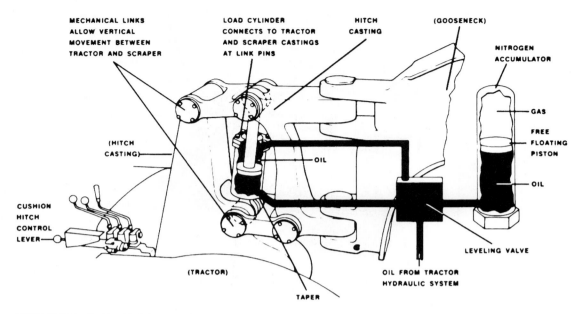

FIGURE 17.6   Cushion hitch for two-axle scraper.

There may be a cushioning arrangement, such as that in Fig. 17.6, to reduce a riding defect called loping. Oil displacement from this cylinder is regulated by a leveling valve, and resisted by a nitrogen accumulator, so that a motion-dampening balance of pressure is obtained.

The articulated connection and the high arch of the gooseneck permit very sharp turns, up to 180° from one side to the other. As a result the unit is highly maneuverable, and can often turn within its own length.

Drivetrain.   Practically all self-powered scrapers have torque converters and power shift. The shift may be partly automatic.

The transmission is usually behind the drive axle, as in Fig. 17.13. Input is by a long shaft from the converter to the top of the transmission; output is a shorter shaft from the bottom forward to the differential. There may be anywhere from 4 to 10 forward speeds, and one or two reverse. An eight-speed box has converter drive in the two lowest ranges, and automatic shift with direct drive in the top six. The automatic goes only up to the range set by the shift lever position.

Shift pattern may be straight line or U-shaped. If straight, the lever is in a slot with toothed edges, which indicate effective position for each range, and may make it necessary to move the lever only one position at a time. See Fig. 17.7.

With automatic shift, a hold pedal may be provided. When depressed, it keeps the transmission in the range being used at the time. The operator may use it to stay in a low ratio to increase hydraulic power by maintaining a high engine speed, or to keep in a low direct drive for safety on downgrades.

A retarder in the transmission may be either optional or standard. It saves wear on brakes and prevents overspeeding down long grades. It is not for quick use, as it takes 2 to 6 seconds to become effective after it is engaged.

The differential may be a nonspin design or a standard type that allows one wheel to spin. The standard unit may be fitted with a lock, applied by a foot pedal, that forces the two wheels to turn

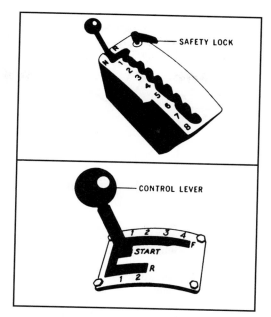

FIGURE 17.7   Scraper shifts.

together. The throttle should be cut to stop wheel spin before engaging it, and the machine should not be turned while it is in use. It releases automatically when traction becomes equalized.

Final drives are reduction gearing of the planetary type.

Brakes.   Brakes may be of any kind suited for heavy service. Smaller units may have booster hydraulic, larger ones pressure-hydraulic, air-over-hydraulic, or full air.

A sequence valve may apply the rear (scraper) brakes a little before the front ones to keep the rig running straight. This is a precaution against jackknifing.

In air systems, there is likely to be an emergency tank that provides power to automatically apply and lock the scraper brakes if air pressure falls below 40 pounds.

Steering.   Two-axle scrapers steer by swinging around the kingpin during steering, with a feeling similar to that of turning a tractor-trailer truck. The rear or scraper wheels do not follow the front wheels accurately on curves, but track inside them.

In forward driving, there is the need to take precautions against jackknifing. There is danger that if an abrupt stop is made, particularly with the front wheels turned, or when the front brakes are more effective than the rear ones, the scraper will swing forward beside the tractor, with total loss of control over direction of travel.

The steering can be used to walk the tractor out of soft spots, with or without blocking. The diagrams in Fig. 17.8 show a succession of moves which depend on the fact that when the tractor is turned while stationary, one wheel moves forward, the other back, thus putting them both on a new footing. If necessary, the forward wheel can then be blocked and the machine steered in the opposite direction, so that the other wheel will move forward, pulling the scraper after it. This technique may also be used without blocks while moving steadily forward. While standing still the strain is sufficient to make the steering system overheat. The technique should be used only in emergencies, and then slowly or intermittently.

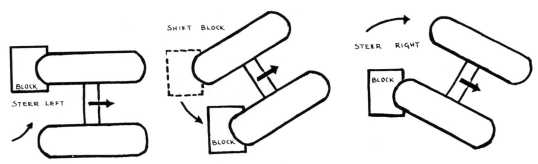

FIGURE 17.8   Steering out of mudhole.

PUSHER

A conventional scraper with only one pair of driving wheels does not have sufficient power or traction to load itself. It can usually pick up a partial load of cubic yards, then the weight of this partial load prevents entrance of more dirt. Two-engine and trailer scrapers do not have adequate self-loading ability for many conditions.

Additional power is usually supplied by one or more pusher tractors as seen in Fig. 17.9 (or, more rarely, pull or snatch tractors), crawler or four-wheel drive, which are discussed in Chaps. 8 and 15. They supply the force needed to pick up a full load in an efficient time.

Push-Pull.   Scrapers may be used to assist each other in loading. See Fig. 17.10. The two are connected during loading by bumpers (push) or by hooks or brackets (pull). The power of the two machines is concentrated on loading one of them. When it is full its bowl is lifted, and all power is diverted to the other.

The push-pull device in Fig. 17.11 is fitted to a minimum of two tractors, preferably of the same model. Each carries a pusher block and a hook at the rear, and a cushion block and a loose-fitting, hydraulically controlled bail in front.

FIGURE 17.9   Scraper with pusher. (*Reprinted courtesy of Caterpillar Inc.*)

FIGURE 17.10   Scrapers in push-pull mode. (*Reprinted courtesy of Caterpillar Inc.*)

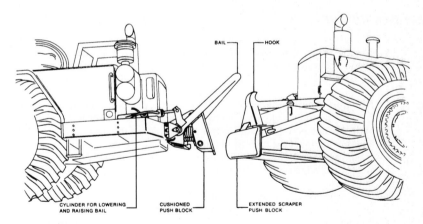

FIGURE 17.11   Push-pull connectors.

The front scraper starts loading; the second comes up behind it and establishes pushing contact between the bumpers, then lowers the bail onto the hook. It is a loose fit, so that exact lining up is not necessary. Loading of the front scraper is continued until the load appears adequate to the operator of the rear scraper, who then signals the front operator that he or she has dug enough.

The front scraper lifts its bowl, the rear one lowers its. The hitch opens so that the bumpers lose contact, and pull is exerted through the hook and bail. When the second load is complete, the bowl is lifted, and tension on the hook is decreased so that the bail can be lifted automatically. The two scrapers then operate independently to the fill, and hook up again on their return to the cut.

This arrangement has great advantages in job efficiency, but the total traction of a pair of conventional scrapers may not be sufficient for good loads under unfavorable conditions. However, the hitch can be used on the two-engine tractors described below, and should then provide excellent loading wherever tires can find a grip.

TWO-ENGINE SCRAPER

A scraper's power may be doubled, and its traction often more than doubled, by mounting a second engine at the rear, above the ejector cylinder, so that it can drive the scraper wheels. The unit may be called two-engine, tandem-powered, or four-wheel-drive.

The engines must be compatible with each other in controls and performance, and equipped with torque converters and power shift transmissions. They are controlled from the operator's station in front, and under ordinary conditions are coordinated in speed and transmission ratio.

The second engine and drive give better loading than push-pull, as all wheels are drive wheels, and power-weight ratio is much greater. The machine has substantial advantages in maneuverability and acceleration, in ability to climb steep or slippery grades, and in independent (instead of paired) activity.

These two-engine scrapers can usually work alone in easy to medium digging. In hard digging they need, or at least benefit by, pusher help. When paired as push-pulls, they can usually handle any type of scraper loading without outside help, except on slippery ground.

Even when pushers are used for loading, the second engine often pays it way in making it possible to pull loads up steep, loose, or muddy grades, to improve acceleration, and increase both loaded and empty speeds on soft ground and moderate grades. See Fig. 17.12.

ELEVATING SCRAPER

The elevating scraper is a truly self-loading machine. Pusher help is not required and is not even useful under ordinary conditions. If a pusher is used, it is usually a crawler on ground too slippery for tires, and it must be operated with care to avoid damage to the elevator. See Fig. 17.13.

Elevator.   In these machines, the apron is replaced by an elevator made up of two roller chains carrying a number of crossbars called flights. Its foot is near the bowl cutting edge, it slopes back 40 to 45° and is somewhat higher than the sides of the bowl. It is driven through reduction gearets, which are connected by a rigid cross shaft.

In digging, the elevator is rotated as shown in Fig. 17.14. The flights may cut soil ahead of the cutting edge or above it, as in the figure. The arrangement varies among different makes and models.

The backward motion of the flights (which is considered to be "forward" rotation) carries the cut soil into the bowl, along with that loosened by the edge.

The material falls or is thrown into the bowl as the loaded flights move upward. It cannot fall out because of the narrowness of the space between the edge and the elevator, and the continued motion of the flights. Angle and height are such that a full load, usually sloping from the elevator down to the tailgate, can be obtained.

The elevator is held in line at the bottom by a pair of brackets which may be spring-loaded to permit the elevator to ride up and over boulders entering the bowl. There may be a mechanical adjustment of clearance, to suit different types of digging.

The elevator may have a single or several forward speeds, usually between 100 and 300 feet per minute. Means to reverse it may be provided as standard or optional equipment.

FIGURE 17.12   Two-engine scraper moving up grade.

FIGURE 17.13   Elevating scraper, cutaway. (*Courtesy of Deere & Company.*)

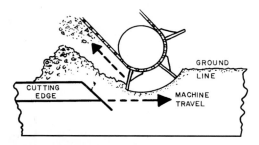

FIGURE 17.14   Elevator cutting action.

During loading, which is done in forward speed, the elevator may use half the power of the engine. An overload valve or switch will stop it if it jams.

Reverse is used for dislodging obstacles or sticky soil, and occasionally to prevent soil from entering the bowl during light grading.

The allocation of available power to the elevator or to the drive wheels is usually determined with efficiency by the torque converter delivering slower motion with multiplied torque to the wheels, while a power takeoff or generator delivers full-speed power to the elevator.

This difference enables the elevator to retain its speed while the forward motion slows in hard digging, to give it an opportunity to cut and/or scoop up the soil in thinner bites.

Bowl and Ejector.   The general shape of the bowl is similar to that in a conventional scraper, but its structure must be different to allow for ejection.

It is not practical to lift the elevator up and forward in the manner of an apron, because of lack of space under the gooseneck. An opening for discharge of load is made by pulling the front half of the bowl floor to the rear. This dumps part of the material.

Then the rear wall of the bowl is moved forward, usually in conventional ejector fashion, pushing the rest of the load out the gap left by the backsliding floor.

During unloading, the machine is moving forward continually, and the discharged material is struck off by the cutting edge. The spread is usually quite uniform, as the flights break lumps and aerate soil as they load it.

The double action of ejection moving first the floor and then the rear wall, is accomplished automatically in one position of the control lever. In some machines, one cylinder or a pair do both jobs through mechanical linkage, in others two or four cylinders operate separately in obedience to a sequence valve.

Ejection is sometimes started and stopped with an empty bowl, to move the edge back far enough to make it possible to pick up a boulder. The scraper is moved over it at full height until it is behind the elevator. The bowl is then lowered, and the object scooped up by the retracted edge, which is then slid forward to retain it. See Fig. 17.15.

Working Characteristics.   The self-loading principle of these machines is that the elevator flights continuously remove the material that they dig, and that is dug by the bowl edge, so that it does not impose a back pressure or load on further digging.

The last yard in a load is therefore as easy to dig as the first, although it does take some extra power to hoist it to the top of the heap.

The problem of loading is reduced to one of cutting or chopping a slice of the ground. There is usually sufficient traction to permit this in fairly hard soil, although depth of cut may have to be reduced. Teeth bolted to the center section of the edge help digging, but are a nuisance in spreading.

Self-loading makes each scraper an independent unit that can work alone or as a member of a large fleet, with equal efficiency. The cost and the planning and supervision problems of pushers

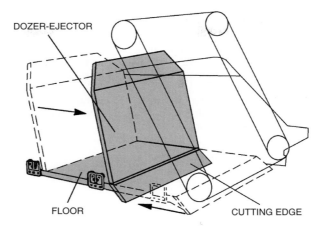

FIGURE 17.15 Ejector and floor movements.

are eliminated. Coarse or hard soil is broken into small lumps or pulverized, reducing voids in the load and problems in the spreading of it.

There are drawbacks, of course. The elevator is expensive, and requires maintenance. It is heavy, and must be carried on the haul, with resulting loss of acceleration and of speed either up grades or on soft footing. It cannot be used where there are large rocks, and repairs are apt to be increased greatly by small ones and even by hard soil.

The balance of advantages against disadvantages is definitely favorable, as is indicated by the increasing share of the scraper market being taken over by the elevating models.

Auger Scrapers. Another self-loading machine, with a broad material appetite, the auger scraper has a hydraulically powered auger assembly in the center of the bowl, as seen in Fig. 17.16. Incoming material is augured up and does not have to force itself through material already loaded. This reduces the cutting-edge resistance and required rim pull, and increases tire life. There are fewer voids, and the broken material is easily retained by the tight-closing apron. The ejector, with angled wing extensions, moves forward to completely clean out the area on each side of the auger for complete emptying of the bowl.

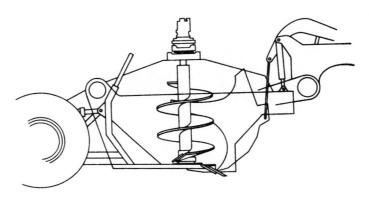

FIGURE 17.16 Auger scraper mechanism.

Two-Engine.   The two-engine auger scraper combines the two separate approaches to self-loading. Doubled power helps loading chiefly by permitting a thicker slice of earth and thereby shortening the time to get a full load. The front engine may put the majority of power into the auger, while the rear one supplies push for the scraper.

The principal advantage of two engines is the increased ability to work, or to work more rapidly under unfavorable conditions, such as slippery footing in the cut, or on the haul road, and/or steep grades and soft ground.

Economics.   The auger adds weight to a scraper on the haul and is more costly to own and operate than standard scrapers. However, its work-alone capability makes the auger scraper cost-effective on many tough material applications.

TOWED SCRAPER

The towed or full trailer scraper, pulled by a crawler tractor, was the dominant type for years. It has been replaced by the faster and more flexible self-propelled machines, and is now of secondary importance.

However, it is still efficient under many conditions of short haul, difficult circumstances, and only-occasional need, and will continue to be used for years. See Fig. 17.17.

The frame structure is that of a low-slung, pneumatic-tired wagon. The rear axle is rigid. The narrower front axle, if there is one, is fastened to the frame yoke or gooseneck by a hinged kingpin which permits it to turn and to oscillate.

FIGURE 17.17   Towed scraper in use.

The bulk of the scraper consists of two parts: the front gooseneck, which includes a crossbeam and a pair of arms extending backward from it; and the gooseneck body, which is hinged to the gooseneck arms, pivots on the rear axle, and includes the bowl.

Bowl.   The bowl is substantially a box open at the top. The bottom and sides are part of the body. The front, or apron, is hinged near the center of the bowl so that it can be lifted away from the bottom. The front of the floor is equipped with a cutting edge or blade. The back, known as the tailgate or ejector, moves forward to the edge and back again from spring pressure, and keeps close sliding contact with the bottom and sides.

The cutting blade is composed of three pieces of wear-resistant steel bolted onto the bowl bottom. The center piece may be held forward of the wide pieces for improved penetration, or, more rarely, the pieces may be in one straight line for smooth-finish grading.

The bowl is attached to the rear axle inside of the wheels, which are positioned so that they will roll in the path made by the bowl edge.

Control Units.   The traveling, steering, and digging power of this machine is provided by the movements of the towing tractor. Individual motions of the bowl, apron, and ejector are powered by hydraulic cylinders connected to the hydraulic system of the tractor.

Tilting Ejector.   Some makes of scraper dump their loads by tilting the bowl floor and back upward and forward. See Fig. 17.18.

The tilting ejector is usually a curved piece that makes up the back and most of the floor of the bowl. It is hinged to the bottom just behind the cutting edge. During the dump it is rotated forward

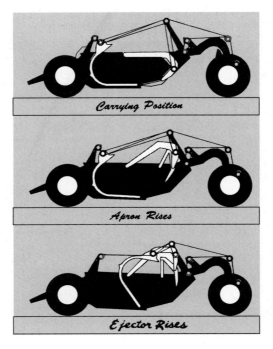

Carrying Position

Apron Rises

Ejector Rises

FIGURE 17.18   Tilting ejector with cable (old) controls.

on this hinge, scraping dirt from the sides of the bowl and spilling the load over the front of the edge. The highest angle is steep for the floor and overhanging for the back. Return is started by spring action and finished by gravity.

LOADING

Conventional.   The conventional scraper needs the help of a pusher to obtain a worthwhile load. However, to avoid confusion, the scraper operation will be described separately first. The cooperation of the pusher in the work will be described afterward.

For normal loading, the tailgate is always fully retracted. In rough or bouldery soil, the apron is held almost all the way up; in most digging it is just high enough not to drag, and in sand it may be allowed to drag for a compacting effect. Up to one-third of the load in the bowl may rest on the apron when it is kept fairly low.

The bowl is lowered so that its edge will cut a slice of ground. In hard ground, it might be forced down to lift the scraper wheels in the air.

Digging is done in low or second gear. The throttle is adjusted at the highest engine speed that will not cause the wheels to spin.

Depth of cut is regulated by raising or lowering the bowl. Except when working near final grade, the cut should be as deep as the machines can handle without spinning drive wheels. If speed is maintained, a deep cut fills the bowl fastest.

A spinning wheel causes expensive wear to its tire, and delivers less push than a nonspinning one.

There may be a differential lock pedal, which you push down if only one of a pair of wheels spins. They will then turn together. But release the accelerator a moment to stop spinning before using it. Do not steer while the differential is locked. The pedal will come back up when traction becomes equal on both sides.

The dirt piles up in the bowl, part of it falling forward on the apron. See Fig. 17.19. Incoming dirt must be forced to rise through an increasing depth in the bowl. Loading rate slows, and a point of refusal is likely to be reached where power is not sufficient to force any more dirt into the bowl, or when the soil will not take the thrust, so that it breaks up and is pushed ahead, or drifts off the sides.

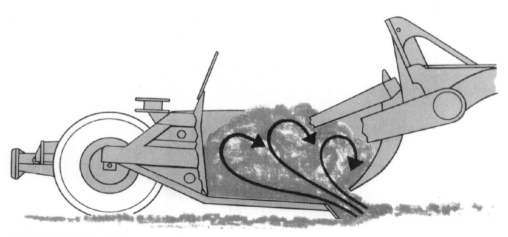

FIGURE 17.19   Dirt boiling into scraper bowl.

If a full load has not been obtained, some may be added by pumping. The edge is raised sufficiently to decrease the draft and allow the tractor to regain its full speed, then dropped several inches below the normal cutting depth. Some soil will be forced into the bowl. The knife is lifted as the engine starts to labor, and the process is repeated if desired.

The advantage obtained is partly the momentum of the accelerated scraper, and partly the ram-rod effect of the thicker layer of dirt punching its way up through the load.

This punching effect is important. It will be found that clay and other heavy soils may be loaded most effectively in thin layers which reduce the cutting power necessary, without sacrificing too much thrust. Sand, however, must be taken in deeper cuts and requires pumping with a much smaller load. It is sometimes helpful to set the apron to drag on the surface of loose soil to compact it while loading.

When a satisfactory load has been obtained, the apron is lowered and the bowl edge raised an inch or two above the ground. This position is held for several feet before lifting to carrying position, to spread any loose material in front of the blade and leave the cut smooth.

Scraper cuts, loading, and pushers were discussed in Chap. 8 in connection with road building.

Pushers. For the minute or two of loading, the scraper needs several times as much power as it does at any other time. It is efficient to have one pusher, as in Fig. 17.20, provide the power for several scrapers, rather than have it part of the scraper and unused most of the time.

Pushers are usually tractors, either crawler or four-wheel-drive, equipped with dozer blades or pusher plates.

Because of their greater speed, rubber-tire pushers can service a maximum number of scrapers under average conditions. However, they do not exert as heavy a push in proportion to weight as a crawler can, and their usefulness declines rapidly if the cut becomes slippery.

On some jobs and under some conditions, the scraper moves into position to dig, then waits for a pusher. At other times the scraper will start its run, loading a few yards before being pushed.

The pusher is driven up behind the scraper, stopped or moving, and contact made with its bumper as smoothly as possible. The two machines each exert as much power as possible without excessive spinning of wheels or tracks. Most of the digging power is supplied by the pusher, and the scraper engine may be cut back to one-half or two-thirds speed.

The push must be in a straight line, as a blade at an angle may cut scraper tires, or cause jack-knifing or steering problems.

When a full load is obtained, the pusher drops back to service another machine, and the scraper goes into a higher gear and leaves. The pusher operator has a better view of the load, and may signal when it is ready.

In general, it does not pay to struggle to get a heaping load, as diminishing returns may make the last yard a very expensive one. This situation is discussed later in this chapter.

FIGURE 17.20 Pushing.

HAUL

The loaded scraper is driven to the fill area at the highest speed allowed by safety and reasonable comfort. Some of the driving problems are discussed later, under Return.

Scrapers can move loads successfully over soft and rough ground, but always at the expense of speed and maintenance expense. Yardage, both in the load and in the number of loads, can be greatly increased by providing a smooth, hard surface for rubber-tire tractors. If a very large quantity is to be hauled, it pays to oil or blacktop the road.

In general practice, the haul route is dirt unless bad weather is expected. Then bank gravel, shell, or other material not seriously softened by water should be used. The road should be maintained by a grader, but if none is available, it can be kept in shape at somewhat higher cost by the scrapers. In dry weather, watering equipment will be required. See Fig. 17.21.

Upgrades slow the movement of loaded scrapers, the amount of loss depending largely on the ratio of the power of the tractor to the weight of the load. Two-engine rigs will get up the grades faster than single-engine ones of the same type. On severe grades, level loads may make sufficiently better time than heaped loads to give higher production. See Fig. 17.22.

FIGURE 17.21   Watering equipment works with scrapers.

FIGURE 17.22   Two-engine scraper working at a coal field. (*Reprinted courtesy of Caterpillar Inc.*)

Soft footing absorbs a lot of power and may make a critical difference in uphill hauling.

On graded haul roads, the highest gear that will move the load without lugging down is used, and on rough-going, the highest which will give enough stability and control.

Bowl Position.   A scraper bowl is carried high to avoid obstacles, and low to keep the machine stable. The actual height will therefore vary with route conditions.

A good haul road maintained by graders should permit keeping the bowl within 2 or 3 inches of the ground. But such a road is usually good enough that there is little danger of upsetting if it is held higher.

On a rough or soft road the bowl must be fairly high to avoid colliding and dragging, but the stability of a low bowl may be needed. Under these conditions, a number of up and down adjustments may be made during the haul.

The bowl may be deliberately dragged to slow the machine descending a steep hill, or in an emergency stop. But the wear on the edge makes this form of braking as expensive as it is inefficient.

The bowl should never be held fully up. In this position the hoist cylinders become rigid frame members, and must absorb damaging stresses on rough ground.

Shifting.   The modern scraper shifts very easily. You just move a lever, and clutches inside the transmission make the shift.

There may be as few as four or as many as nine forward ranges. In the upper group they may be in pairs such that each has one set of gears, with drive through the torque converter in one range and direct mechanical in the other. When you start up a grade or get into a soft spot, the converter will allow the road speed to drop but maintain engine speed. In direct drive, road speed will hold up better at first, but the engine will slow in proportion to any drop.

If the resistance continues or becomes heavier, direct drive will cause the engine to lug down so you have to shift, while the converter might keep going comfortably at reduced speed. In this last case, the next-lower direct drive range might give better performance.

In some semiautomatics, the two lowest (digging) speeds are manual. From second you can move the lever directly to the highest gear you want to use, and the transmission will do its own figuring and shifting. You can override the automatic with the lever anytime you want. There may be a pedal you can push down to hold it in whatever gear it is in, until the pedal is released.

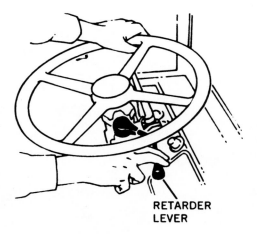

RETARDER
LEVER

FIGURE 17.23 Retarder control.

Shifting is guided mostly by the feel of the machine and your estimate of conditions immediately ahead. But it may be controlled by dashboard gauges. You must manage to keep the engine within its efficient operating speed range. If it is on the fast side, shift into a higher gear. If it is near the bottom, shift into a lower gear.

If the torque converter fluid is too hot, use a lower gear.

Retarder. Many torque converters include a hydraulic retarder. This is a drag or resistance mechanism which runs freely when empty, but when filled with oil it slows the driveshaft. Partial filling for partial effect may be possible. See Fig. 17.23.

The retarder is used to slow the machine on downgrades. You have to think in advance to take advantage of it, as it needs from 2 to 6 seconds to take in enough fluid to start working.

The energy of motion which it absorbs is changed into heat, which is dissipated by the converter cooling system. On a long descent it is useful in keeping the converter warm.

Brakes. Big scrapers usually have air brakes; smaller ones have hydraulic of either the automotive or the pressure-pumped type. All of these are absolutely dependent on tight, nonleaking systems. Most or all air brakes need more than 60 pounds of pressure. If the gauge does not show that much, stop immediately and investigate.

There is usually an arrangement that puts on the rear (scraper) brakes sooner and/or harder than the front (tractor) set, to ensure straight stops.

Air brakes should not be pumped (put on and off rapidly), as this exhausts the air supply.

DUMPING AND SPREADING

Dumping and spreading is one operation, as the scraper dumps only while moving forward, and the material falls under its edge and is automatically spread by it.

In dumping a conventional scraper, the bowl is lowered until the edge will just allow a layer of dirt the desired thickness to slide under it. The apron is then raised enough that the falling dirt will supply the knife sufficiently to make a continuous, smooth layer, without dropping excess which would heap in front, or drift to the side to make windrows. See Fig. 17.24.

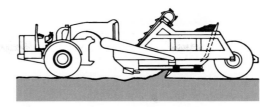

Move the ejector gate lever to "FORWARD"
(toward operator) and hold there as necessary.
(An "ON" and "OFF") operation of this lever will
give the best results until the gate is fully forward.

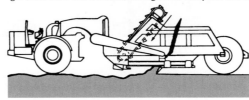

After the sliding bowl floor is all the way back,
the ejector gate will move forward pushing
material out of the bowl. (To remove sticky
material from the bowl, place the elevator
control lever in "UNLOAD".) (lever away
from operator)

FIGURE 17.24 Dumping an elevating scraper.

When the apron is fully raised, the tailgate is moved forward gradually, pushing or dumping the dirt out of the bowl. Too fast a crowd will supply too much dirt to the knife, and will place unnecessary strain on the cylinders and linkage of the ejector mechanism. If the dirt is sticky, good results are obtained by advancing the gate 1 foot, allowing it to slide back 6 inches and advancing it again when required. A liftgate may be banged against its stops to knock it clean. Dirt resting on the edge after the gate is fully forward may often be dumped by moving the gate back and forth.

When the dump is complete, the grate is fully retracted, the apron dropped, and the bowl raised to carrying position.

Spreading is usually done in thin layers, as it provides better compaction and eliminates or reduces the need for other grading equipment. It also keeps a more even grade on the dump, reducing the hazards of high-speed work.

Dumping and spreading is ordinarily done heading away from the cut and at medium or high speed. In finishing a grade and under some other conditions, a low gear may be used to give more exact control.

The fill should be started at its outer edges and kept built out to finish the slope all the way up. This is important, as the type of scraper ordinarily used has no way of dumping over edges, and any patch fill made could not be readily compacted so as to support the machinery building additional layers.

It is also desirable to have the fill slope up toward the edge, as this arrangement will place the larger part of the weight of both tractor and scraper on the side away from the slope, and will reduce danger of caving. See Fig. 17.25. However, if rain is expected, the fill may be crowned enough to shed water at the end of the day.

The closeness with which a scraper can approach a high edge is determined by the type of soil, the degree of compaction, the slopes, and the weight of the machines, and to some extent the skill

TO MAINTAIN FILL SLOPE

1 MAKE FILL HIGH ON THE OUTSIDE

2 THIS PREVENTS SCRAPER FROM SLIDING OVER SLOPE

3 ACCURATE SLOPES CAN THUS BE MAINTAINED TO DESIRED HEIGHTS ELIMINATING NECESSITY FOR HANDWORK

4 IF WET CONDITION PREVAILS ARRANGE FOR DRAINAGE TO PREVENT WATER POOLING IN CENTER OF FILL

RESULT OF INCORRECT METHOD

1 SCRAPER WILL SLIDE OVER SIDE OF HILL

2 DAMAGE TO SLOPE WILL BE CAUSED

3 IMPOSSIBLE TO MAINTAIN ACCURATE DEGREE OF SLOPE, TENDENCY TO WORK AWAY FROM EDGE OF FILL

FIGURE 17.25   Cross section of fill. (*Courtesy of U.S. Army Engineers.*)

of the operator. In general, fine-grained soils which are not in a muddy condition are safer than sandy soils. Working right to the edge or overhanging it slightly with a tamping roller will give good compaction.

If conditions are such that the scrapers cannot safely go near enough to the edge to build a proper slope, they can drop the material a few feet in and a grader can distribute it.

The slope should be checked frequently for proper grade.

Elevating Scraper.   The elevator may stay in place during the dump. An opening for unloading is made by pulling back the knife and front half of the floor, sliding on guides and/or rollers under the rear half. The tailgate is then moved forward.

A substantial part of the load falls by gravity through the floor opening. The gate and floor controls are interlocked, so that gate movement does not start until the floor is completely open. The control may be returned to HOLD for a few seconds after floor motion is complete, so as not to overload the opening.

RETURN

After completing the spread, the scraper is turned, either on the fill or after going on to a turning point, and driven back to the cut at the highest safe speed.

Safe speed might be the maximum of the scraper at full governed engine speed, but usually there are limiting factors. The machine will be automatically slowed by upgrades and soft footing, and should be slowed for rough spots, curves, close approaches to other machines, downgrades, and other hazardous conditions.

Accuracy of steering and efficiency of brakes must be taken into account. Two-axle machines may have a tendency to rhythmic bouncing, called loping, which is most uncomfortable and may reduce control dangerously. Special cushion hitches may reduce this, but you should immobilize them during digging and spreading, and may forget to reconnect. Otherwise, a small increase or decrease in speed may steady it.

A loaded scraper (or truck) has the right of way over an empty unit.

TRAILER SCRAPER OPERATION

Connecting. Towed scrapers are usually stored with the bowl blocked up and the hydraulic hoses in place. To reconnect, the tractor is backed to within a foot or two of the tongue, and each hose is connected to the tractor's hydraulic system.

Hydraulic two-axle scrapers depend on counterbalancing or jacking the tongue, or leaving it blocked at correct height when disconnecting. One-axle hydraulic scrapers allow manipulation of hitch height by applying down pressure to the bowl.

Hydraulically operated scrapers may use either two or three controls. The third valve and lever permit independent operation of the apron and the ejector.

Overturning. On sideslopes always move slowly, keep the blade low, avoid going over rocks or bumps with the upper wheels, or hollows or soft spots with the lower ones, and be extra careful when turning.

A full trailer scraper should not be used on steep sideslopes unless the operator is very skillful, or other equipment capable of righting it is on the job. The narrow front tread, oscillating front axle, and tendency to jackknife when backed up all make it liable to side-tipping. It is also unsafe to use it close to the edge of poorly compacted fills.

If sidehill work cannot be avoided, a scraper with minimum overhead structure should be used and the bowl should be carried low. Turns should be made uphill where possible, and downhill turns should be made gradually, preferably with the apron fully down and the bowl scraping on the ground.

Danger of tipping is particularly severe if the soil under the upper side of the machine is harder than that under the downhill wheels or if the upper side is forced up by running over boulders.

Servicing. Servicing a towed scraper is dangerous, unless the operator or mechanic has experience and takes proper precautions.

When changing blades or doing other work under the scraper, both the bowl and the apron should be blocked, so that they cannot come down.

If work is being done behind a forward-positioned or raised tailgate (ejector) it must be blocked from moving back. The blocking must be heavy and positive, and should be checked by releasing the hydraulics, then relocking them.

The tractor should be immobilized by locking, or providing some well-understood indication at the controls that the scraper is being serviced.

The dumping mechanism is returned to loading position, or at least started back, by heavy springs or torque rods. These can be dangerous if their tension is accidentally released by work done on their cases.

Do not leave scraper parts in positions from which they would move if the control unit brakes were released.

Hydraulic. Hooking up a hydraulically controlled model is simplified by counterbalancing or jacking the tongue, or leaving it blocked at correct height when disconnecting. It may or may not be necessary to connect the hydraulic lines before the draw tongue, to move it into a good position.

Disconnect couplings should be wrapped in plastic when they are opened, and checked carefully for cleanliness when reconnecting.

GRADING

Scrapers are sometimes used to smooth and patch haul roads, and in grading and leveling areas where the quantity of dirt to be moved is not enough to fill the bowl, and where distances are too short for the loading-dumping cycle. See Fig. 17.26.

For grading, the center piece of the knife should be flush with the side pieces. Otherwise, finish work will be limited to the width of the center. See Fig. 17.27(*A*).

The location of the knife between the front and rear axles provides good stability for grading, and controls are usually sufficiently sensitive for medium-fine work. But there is no means to shift dirt from side to side, grader fashion, so that crowning a road or shaping a slope is a laborious process.

On the other hand, the scraper can move substantial amounts of material along a road, and can bring in borrow, or dispose of surplus off the road.

When cuts and fills are shallow and closely spaced, the usual working position is with the apron fully lifted, and the tailgate most of the way forward. It is bad to have it all the way forward, as it is then subject to twisting strains from the frame and the bowl.

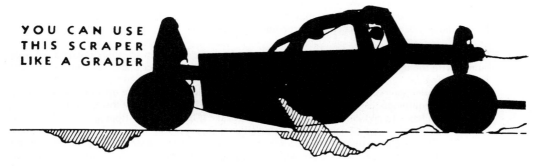

FIGURE 17.26   Grading with scraper.

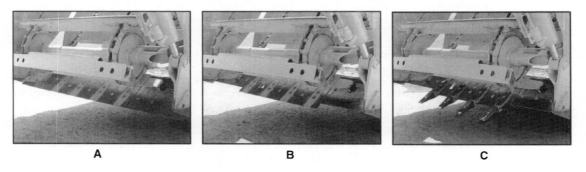

	Cutting Edge	Penetration Capability	Resistance To Breakage	Typical Application Range
A	Straight	Low	High	Finish / General
B	Stinger	Medium	Medium	General
C	Teeth	High	Low	General

FIGURE 17.27   Scraper cutting edges. (*Reprinted courtesy of Caterpillar Inc.*)

The bowl is lowered or raised to give the desired depth of cut or thickness of fill, and acts much as a very steady dozer blade.

If a cut is deep or long enough to provide more spoil than can be carried in front, the tailgate can be retracted enough to admit some inside the bowl, and then used to push it out again when a fill is reached.

An elevating scraper must have the elevator operating in reverse, or the floor partly retracted, to avoid picking up loads. In either case, there is more than normal strain on the elevator flights, and the scraper should not be used in this way to cut hard surfaces.

The self-loading feature makes the elevating scraper ideal for grading work that requires a moderate amount of off-site borrow. It makes a good combination with a grader for this.

There is no definite line of separation between this light grading and heavy cuts and fills.

STUMPS AND BOULDERS

Scrapers are not designed to dig or transport bulky objects, but they may be used for the purpose when more suitable equipment is not available.

Large objects can usually be dug and loaded rather readily after practice, unless the size is excessive or the shape ridiculous. But getting them out again is occasionally a real problem, calling for gate-jiggling, crowbars, and ingenuity, and perhaps finally a crane. Either the digging or ejecting of a stump or boulder might strain or bend parts of a scraper, so the work should be done carefully.

If the object is low enough for the tractor to walk over it, raise the apron to full height, drop the bowl 2 or 3 feet behind it, apply down pressure, and move forward. If the edge hooks into it, lift the bowl while inching forward. If it slips off, back for another try. You may have to try from other directions in order to get a grip.

If the piece is to be picked up and carried, the tailgate is kept in normal full-back position. But if you want to just loosen and shove it, keep the gate forward.

Often, or perhaps usually, the stump or boulder cannot be picked up on the same pass that loosened it, so you have to back off and take another bite. Try to capture it before it is entirely loosened, as you might then have trouble catching up with it.

The apron can sometimes be used to push the object back into the bowl, or clamp it to the edge so that it can be carried.

If a stump is too high to walk across with the tractor, pass it very closely, and then steer sharply in front of it. This should bring the bowl into position to get its edge into or under it.

Scrapers, particularly of those models which have very high apron lifts, can pick up large stones or stumps. However, these sometimes turn or catch when in the bowl, and become jammed when an effort is made to discharge them. They can be worked out by moving the gate back and forth and the bowl up and down; by digging enough dirt to shift position; by putting the machine on a favorable slope or over a hole, prying with bars or saplings, lifting with a loader or a crane, or reducing the objects with chainsaw, axes, chisels, sledges, or pneumatic drills, then dumping the pieces.

Oversize rock is liable to cause damage by denting, bending, or straining parts. Damage may also be done by accidental collision with boulders during ordinary digging or transporting.

JOB FACTORS

Operating Efficiency.   The self-powered scraper is more sensitive to weather delays than most excavators are. Rain quickly makes the surface of the cut too wet to use in the fill, stopping work. On resumption of work it may be necessary to waste the top layer, or stockpile it to dry.

Also, scraper drive tires quickly lose traction on rain-wet surfaces, particularly if they have been dusty.

A U.S. Bureau of Public Roads study covering 2,660 hours of scraper operation in the southern states during a rainy year showed that scrapers averaged a loss of 54 percent of available working time because of rain, a substantially worse record than other equipment. Under the same conditions, crawler-drawn scrapers lost only 25 percent of their time because of weather.

Job Delays.   In the same rainy-year study, self-powered scrapers showed a loss of about two-fifths of the time the weather allowed them to work, because of other delays. These delays were divided about equally among the following five causes:

Waiting for opening of cuts

Waiting for pusher

Maintenance and repair

Lack of operators

Miscellaneous

The net result was that these scrapers worked about 35 minutes per hour during working periods, and an average of 17 minutes per hour for the whole job.

Other studies, not quite so complete or carefully recorded, indicate somewhat better performance. In this discussion, an efficiency of 75 percent or a 45-minute work hour will be assumed. However, each estimator must of course use his or her own judgment and figures.

Scrapers are not particularly subject to stoppage from breakdown of other equipment. A pusher can usually be replaced quickly by odd-job dozers, scrapers can sometimes load each other, or they can operate for a while at partial load. They can maintain their own haul road, at a price, and can continue to build a fill for quite a while in the absence of grading and compacting equipment, unless supervision is very exacting.

Ownership Expense.   The self-powered scraper, as a busy contractor's first-line equipment, may be used from 600 to 1,500 hours per year, with the probability that it will be less than 1,000.

For depreciation purposes it is considered to have a 5-year life. With reasonable care it may last much longer, many still-active units being over 10 years old. On the other hand, rough treatment and lack of care can finish the useful life of a scraper in a year.

Insurance and property tax costs are average. Some of the smaller units may have occasional use on highways and require licensing, but most are too wide.

Pusher costs are those for dozers, medium to hard service.

Figure 17.28 gives a sample scraper cost setup including both fixed and operating expenses. This illustration is only a framework, in which the contractor inserts figures for his or her own conditions and costs of equipment.

Operating Expense.   Self-powered scrapers are expensive machines. Their work is usually hard, and they get punishing treatment trying to keep up their output, particularly in being forced through resistant material by too-powerful pushers and in high-speed travel on rough ground.

Fuel and Oil.   Engines operate at full and nearly full load a greater part of the time than most excavators, as the highest possible gear is used most of the time, whether loaded or empty. Medium loading occurs in long, level runs on firm roadways, light loads in downhill travel.

With equipment in excellent condition, fuel consumption is likely to average .05 to .06 gallon per rated brake horsepower per hour. In ordinary operating conditions, figure .06 to .07. Chassis lubrication is usually once per shift of 8 or 10 hours. Oil changes vary with manufacturers' recommendations and dust conditions, but are average for construction machinery.

Lacking specific information, figure lubrication at one-third of fuel cost.

Maintenance.   Bowl cutting edges need periodic reversal or replacement, always before they are so far gone that they let the bowl wear.

Ownership and Operating Cost
SELF-PROPELLED SCRAPER
Struck Capacity ___ Yards
___ horsepower Empty Weight ___ tons

PURCHASE

Price at factory $

Extras (cable reel, hardfacing, spare parts

Sales tax, %

Freight
    Purchase price, or total cost $

Deduct, 4 tires @ $ each, fleet price
    Net cost to be depreciated

Repair factor 1/10,000 of net cost $

Average annual investment, 5 year basis, 6/10 net cost $

OWNERSHIP

Hourly depreciation, 6,000 hours total use, no salvage $

Carrying charges, per cent of average annual investment

Interest	0 to 11.3%	assume at	%	
Insurance	.5 to 2.5%	assume at	%	
Taxes	0 to 4.5%	assume at	%	
Storage	0 to 5.0%	assume at	%	

    Total   for example ⟶ 8.0%

Hourly carrying charges, .08 x $\dfrac{\text{av. ann. inv.}}{1{,}200 \text{ work hours per year}}$ = $

Total ownership costs per hour $

OPERATING

	Hourly Cost		
Type of service, normal range	Light	Medium	Heavy
Original tire purchase, $ $			
Tire replacements, 4 recaps @ $			
3 recaps, 1 new			
3 recaps, 4 new			
Tire repairs, 15% of combined cost			
Mechanical repairs and maintenance,			
Diesel fuel @ ¢ a gallon,			
Lubrication, 1/3 of fuel			
Total non-labor operating cost per hour			
Ownership cost, above			
Total non-labor cost			
Operator's wages			
Total hourly cost			
Cost per minute, 45 minute hour			
Cost per 1,200 hour year $			

FIGURE 17.28 Scraper cost data format.

17.27

The contractor who has detailed cost records can compute the cost of these items separately. Otherwise they are lumped together with repairs.

Scrapers are often parked outdoors for long periods between jobs. They are likely to suffer serious rust damage unless they are carefully protected by paint and grease.

Repairs.   Repair costs increase with weight of load, length of struggle to get loaded, roughness and coarseness of ground, grades, and age of the machine. Severe damage may be done by boulders, ledge, and ripped rock, particularly with powerful pushers. Some makes and models are more subject to rock damage than others are.

For rule-of-thumb calculation when lacking specific records, a scraper's nontire maintenance and repair will equal 80 percent of its purchase price under ordinary operation for 5,000 hours. In light service, repair costs may run only 50 percent of purchase, in heavy service 150 percent or more.

Tires.   Tire repair, recapping, and replacement may make up one-third or more of the nonlabor operating expense.

A particular problem with drive tires is that scraper low gears are faster than the low gears of crawler pusher tractors. If the scraper is kept at full throttle and the pusher is in low, considerable spinning of the drive wheels must occur. This is wasted effort and wasted rubber, as the scraper usually supplies a modest fraction of the loading power, and a spinning wheel has less push than one whose tread is engaging the ground.

Spinning is most apt to occur when loading is kept up too long. The scraper operator feels the struggle of the pusher and lack of load response, and has a tendency to tramp on the throttle in an effort to help.

Drivewheel spin in the cut may be reduced by driver education, running pushers in second gear, using rubber-tire pushers, increasing pusher power, or having scrapers equipped with torque converters.

Scraper drive tires start to slip and spin on loose or muddy haul routes on grades as low as 3 to 5 percent. Under severe conditions of abrasive mud and steep grades, a tire can be worn smooth in 500 hours of operation.

Labor.   A scraper has one operator, and no helpers or hand laborers are ordinarily needed. Pusher expenses, including labor, are divided by the number of scrapers they service, and added to the cost of each scraper.

Supervision.   Policies on supervision vary. A grading foreman may be assigned to one cut or to one fill, or may divide her or his time among a number of cuts and fills. In the absence of a foreman, a cut may be supervised by the pusher operator and a fill by a grader operator.

Experienced scraper operators need little supervision. The critical items are proper loading, efficient haul speed, cutting sideslopes correctly and finishing them before they get out of reach, keeping the pit floor reasonably smooth, and accurate finishing of the floor.

Operators may need a foreman to secure help or services for them when they are needed.

HOW BIG A LOAD?

Heaped capacity ratings of scrapers are for advertising purposes only. A rule of thumb is that even under good digging conditions with plenty of pusher power, a scraper should not be expected to pick up more than its rated struck capacity in bank yards. But it might be more profitable to leave the cut sooner with a yard or two less.

Rate of Loading.   During a loading run, material first enters the empty scraper rapidly. As the load increases, the rate of loading drops. Refer to Fig. 17.19. When the weight of the dirt above the opening balances the digging force, the scraper does not pick up any more material.

The block graph in Fig. 17.29 shows the amount that might be expected to be loaded during each 10-second period. It will be noted that this machine picked up 8 yards in the first 10 seconds, but only about $\frac{1}{20}$ yard in the last 10. Loading cost of the last cubic foot is over 100 times as much as for the first one in the bowl.

Line 1 in Fig. 17.30 shows the increase in average cost per yard of a load as the scraper keeps digging.

It is obvious that 2 minutes is too long to keep this scraper in the cut. The question is, What is the most economical time to pull out? The answer depends partly on the hauling cost.

Hauling Cost. Increase of load size decreases hauling cost per yard, unless the speed and acceleration of the scraper are seriously reduced. Sometimes a moderate increase in load may reduce

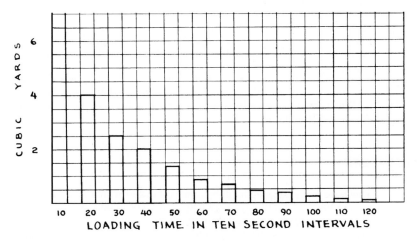

FIGURE 17.29   Scraper load increase in 10-second intervals.

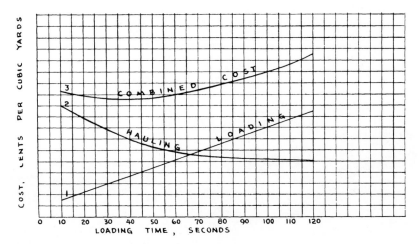

FIGURE 17.30   Cost per yard and loading time.

speed by forcing the machine into the next-lower gear, but under slightly different conditions the same increase might have no effect.

Any increase in load will increase time required for acceleration and will have a small effect on dumping time.

In general, a light load can be hauled rapidly but the cost is divided among few yards, as the cost per yard is high. Larger loads move more slowly, but costs are shared by more yards, so are lower on a per yard basis.

Line 2 shows the cost per yard of hauling each of the 10-second loads 2,500 feet, one way. Note that the last $1\frac{1}{2}$ yards that added so much cost to loading do not lower the hauling cost much.

Most Efficient Time.   Line 3 shows the loading and hauling costs combined. It is a sagging curve, whose low section is the most economical time of loading. For this particular problem, it is 40 seconds, with 50 seconds so close that there is no practical difference. A range from 30 to 60 seconds might be tested in the field.

Maximum Production.   If scrapers loaded themselves without help, the most efficient loading time would also give the greatest production. However, the pusher cost makes the most efficient time a little short of best production. Usually 10 to 20 seconds longer push will boost production, at a slightly higher cost.

Aiming for maximum production is the right idea where working time is limited. It may prove economical as well, if the pusher has to wait for the hauling units.

When scrapers are waiting for the pusher, it does not pay to keep them in the cut longer than the most efficient time.

LAND LEVELERS

These are towed scrapers with bottomless buckets which are chiefly used in leveling farmland as seen in Fig. 17.31, although they may be used profitably in other large-scale grading operations. They include drag levels and the long-chassis land levelers.

Drag levels do the heavier part of the work, and the land levelers perform the final smoothing. The latter consist of a frame from 20 to 80 feet in length, which rests on wheels or skids on each end and carries a bottomless bucket in the middle. The bucket can be moved up or down by either hand wheels or hydraulic controls.

Some land levelers are made up of a drag scraper bolted into the long frame. Others are made so that they can be taken apart immediately behind the bucket and used as a drag scraper.

There are a number of makes and types of these machines, but a few examples of each will suffice to show the type of construction used.

FIGURE 17.31   Land leveler, single or tandem.

Agricultural Drag Scraper.   The scraper like the one shown in Fig. 17.32 may be 5 to 20 feet wide and holds up to 10 yards. It is towed by tractors of 70 to 160 horsepower.

The full-width bucket or bowl is shallow with no floor, and has a cutting knife or blade at the bottom. It is rigidly fastened to a drawbar or tongue extending forward to the tractor and is hinged to a frame supported by a pair of pneumatic wheels behind the bucket.

A two-way hydraulic ram is mounted on the top of the bucket. Its piston rod is hinged to an upward extension of the wheel frame. When the ram is extended, the wheel frame rotates on the axles, raising the bowl through the hinge connection. When the ram is retracted, the bowl is lowered or forced down.

There are several working differences from a dozer that may be noted. Since the bucket is hung between the tractor and the rear wheels, it has little tendency to scallop, and can run a good grade readily. No work is done ahead of the tractor, so that an area cannot be worked unless the tractor can get through it or a dozer smooths its path. The blade is much wider in proportion to the power of the tractor, and capacity is larger. Sharp-sided cuts, ditches, and steep dump piles may be left, as the tractor does not follow. It ordinarily does not dig hard soil as well.

As compared with a carrying scraper, this machine can grade a greater width with each pass, will penetrate more rapidly, and may be dumped completely while standing still to build dikes or to avoid bogging down. Its capacity is smaller and speed is lower, because the load is dragged on the ground rather than carried.

The machine acts as a semitrailer, and after some practice can be backed readily. It is possible, therefore, to use it either in a shuttle or in any of the scraper patterns, the shuttle being favored for short hauls.

Land Plane.   The land plane, shown at work in Fig. 17.33, has its bucket or blade centrally mounted under a long girder frame. Its front and rear sections (extensions) can be telescoped inward to reduce its wheelbase, for decreased pull and better handling in rough ground, and for convenience in moving between jobs.

Maximum frame length varies in different models from 35 to 90 feet, with bucket widths from 10 to 16 feet.

The drawbar, or tongue, is fastened to a hitch which allows for lateral movement in making 90 or 180° turns. The hitch does its pulling from diagonal braces fastened to the main frames close to the center of draft near the bucket, and extending downward to the front end of the machine.

Wheels may be all steel with flat rims, or equipped with single rubber pneumatic tires.

These machines are intended to be used on land that is fairly level or that has been made so by scrapers of the carrier or drag type. Their principal feature is length. The longer the framework, the more accurate the leveling operation will be. The effective leveling length is the distance between the front and rear wheels. The bucket suspended midway between the wheels acts as a plane to shave off any high spots that appear in the field, and pushes and carries this dirt until a corresponding low place is found elsewhere in the field.

FIGURE 17.32 Agricultural drag scraper.

FIGURE 17.33   Automatic land leveler.

The bucket gives its load a rolling motion, pulverizing clods so that the fine dirt automatically filters out of the bottom of the blade into low places. The tractor operator sets the blade at the start of the operation to the point where there is enough dirt in the bucket to keep it from running empty during the operation. It is not necessary or desirable to raise and lower the bucket dozer or grader fashion, but if it is found to be running empty, the bucket may be lowered to gather more dirt from the high places in the field.

On extremely rough ground or during the first planing operation, it may be necessary to work with the bucket blade higher than normal, to work down the rough spots without making cuts so deep that they stall the tractor. When the high places have been cut down, the tractor operator can lower the bucket with the manual screws to the proper cutting depth. Normally, the bucket blade is set to a depth that would be level with the bottom of the wheels, assuming that the wheels were resting on firm, unworked ground; however, it is necessary to compensate for the settling of wheels in soft ground by raising the bucket.

In very loose or sandy ground such as desert lands, the air pressure may be reduced in dual rubber tires to around 12 pounds, which will provide greater surface area and flotation to carry the weight of the machine. Under extreme sandy conditions, oversized dual tires are used to provide the necessary flotation.

These machines will turn sharply in their own wheelbase, particularly if the bucket is partly filled with dirt so that the machine will pivot on it. The field may be worked in strips or in any pattern that seems most desirable to efficiently level the area. The manufacturer recommends that the planing on farmed fields be done diagonally to the plowing or cultivating for the best results and easiest pulling. It is also recommended, where irrigation is practiced, that the last operation be done with the flow of water.

They are sometimes useful to contract levelers of desert land to level fields in both directions before the survey stakes are set, in order to move as much dirt as possible at the least expense and bring the field to a more uniform level before actually moving dirt to the proper grade. They are valuable for final smoothing after the stakes are pulled.

Correction leveling on an annual basis is advisable to restore smooth level surfaces on fields which are roughened by wind and water erosion, ditches, levees, furrows, and ground settling.

Other uses include leveling snow on airport runways in Greenland.

Land Leveler.   The Automatic Land Leveler, Fig. 17.33, has the center wheels mounted on a crank axle that is supported by the mainframe and that supports the front of the bowl subframe. Linkage is arranged so that when the wheels move down in relation to the mainframe, the bowl is moved up, and when the wheels come up, it is lowered.

When the center of the machine is passing over a ridge, these wheels ride up on it while the frame, supported by lower ground at the ends, will not. The crank axle will turn as the wheels move up, thereby lowering the blade to make a deeper cut. In crossing a hollow, the wheels drop relative to the mainframe and the blade is lifted, thereby spreading a thicker layer of fill than a standard leveler would.

The bowl is raised to dump or to clear obstructions by means of a hydraulic ram. An indicator gauge tells the operator its position, and aids him or her in getting it back into automatic operation setting after use. The gauge pointer is not affected by bowl movements caused by the axle.

The ram is also used to relieve weight while making manual adjustments.

Working depth is set by a crank handle and screw at the hitch block connection between the draw tongue and the front frame member. The bowl should average a half load, filling in ridges and emptying in hollows. A high setting may be needed for the first working of uneven ground, or when the tractor has barely enough power to pull the leveler.

The rear of the bowl is supported by springs. An adjustment permits reducing tension when more weight is needed on heavy, cloddy ground, or increasing it to hold the bowl up in light, loose, or damp soil.

The forked rear section swivels on a vertical pin for ease in steering. It may be removed when the machine is being used chiefly as a dirt-moving scraper. It, and the tongue and dolly, may be loaded on the mainframe for transporting.

This machine is 32 feet long and 12 wide, is used chiefly in land smoothing and seedbed preparation, and requires a three-plow tractor to pull it.

Eversman levelers are available in sizes up to 70-foot length with a 14-foot blade.

Towed Ditcher.   Irrigation and other shallow ditches may be made and cleaned by the pull ditcher in Fig. 17.34. It has two plowlike moldboards or wings flared out from a single share. It is raised and lowered by a hydraulic ram, or by an optional screw crank.

FIGURE 17.34   Towed ditcher, working position.

FIGURE 17.35 Mounted ditcher-plow.

A pin and quadrant adjustment on the towbar regulates the angle at which the wings enter the ground. When the bar is at the bottom of the quadrant, their full length will enter the ground and a wide, flat ditch is made. If the bar is higher, the ditcher will nose down and dig with only the front part, making a narrower, steeper-sided cut.

Mounted ditchers are also made to fit most three-point hitch tractors. See Fig. 17.35.

LIGHT SCRAPERS

A number of types of scrapers are designed for use behind light tractors. These may be carried on the tractor, on wheels, or dragged on the ground. Many of them are controlled through ropes and trip levers.

They are usually economical to buy and maintain, and they work well in small places. However, their capacity is too small for general work, and they will not cut hard soil unless it is plowed or rooted. They are primarily farmers' tools and will be used by the contractor chiefly for spaces or quantities too small to justify the use of heavier equipment.

Miskin.   A closed-bottom scraper, Figs. 17.36 and 17.37, is attached to a wheel tractor by a three-point hitch, which adjusts its working or carrying level. Manipulation of tilt for digging or dumping is controlled by an auxiliary, two-way cylinder.

The scraper bowl is hinged to the front bar and side plates. A two-piece hinged rod runs from the top hitch connection to the rear of the bowl. A vertical cylinder based on the bar lifts or pulls down the rod to control digging and dumping angles of the bowl.

When empty, the scraper is lifted clear of the ground for transport, by the lift arms of the hitch. When loaded, it is usually dragged along the ground, resting on hardened steel skid plates.

An 8-foot-wide bucket has a capacity of over 2 yards.

FIGURE 17.36   Closed-bottom scraper.

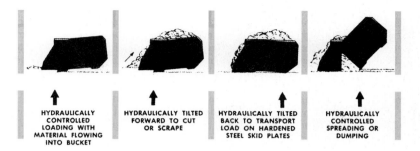

| HYDRAULICALLY CONTROLLED LOADING WITH MATERIAL FLOWING INTO BUCKET | HYDRAULICALLY TILTED FORWARD TO CUT OR SCRAPE | HYDRAULICALLY TILTED BACK TO TRANSPORT LOAD ON HARDENED STEEL SKID PLATES | HYDRAULICALLY CONTROLLED SPREADING OR DUMPING |

FIGURE 17.37 Loading and dumping, closed-bottom.

FIGURE 17.38 Rear-lift scoop.

Dearborn Scoop. The tractor scoop is carried by the three-point hydraulic lift drawbar used on tractors. Digging depth is regulated by lowering the drawbar, and transporting height by raising it. See Fig. 17.38.

Pulling a cord attached to a trip lever releases a catch and allows the bucket to dump by gravity. Another catch will hold in a vertical position for spreading and leveling loose soil, dozer fashion.

The bowl can be rotated into any desired position by allowing it to drag on the ground, while moving the tractor forward or backward. Capacity is about $\frac{1}{3}$ yard.

CHAPTER 18
TRUCKS

DUMP TRUCKS

The dump truck is probably the most familiar of the machines used for excavation. However, its structure is rather complex, and it is so important that a detailed description is in order.

It is composed of four major assemblies. The chassis includes the frame, bumper, springs, dead axles, wheels, and tires. The power train, which is supported by the chassis, consists of engine, clutch, transmission, driveshaft, differential, and live axles. The cab is the driver's compartment. The body assembly, which includes the carrying box, tailgate, cab shield, and the hydraulic system and controls, is an entirely separate unit, usually made by a different manufacturer and adaptable to different makes of truck.

The dump truck in Fig. 18.1 may be considered typical of the medium-size trucks that can operate on highways.

FIGURE 18.1 On-highway dump truck. (*Courtesy of Marmon-Herrington Co.*)

ON-HIGHWAY TRUCKS

Frame.   The frame, shown in Fig. 18.2, consists of two parallel pressed-steel channels with cross braces, some of which serve as supports for the engine and transmission. The front cross member is extended to the sides and serves as a bumper.

The frame side members behind the cab have about a 3-foot spacing. The length of the chassis from the back of the cab to the rear axle may be 5, 6, 7, or $8\frac{1}{2}$ feet. The width and lengths of this section are standardized for most makes of trucks—for convenience in mounting bodies.

For dump use, the side members are cut immediately behind the rear cross member.

Pull hooks should be fastened to the top of the frame members just behind the front bumper, and on the rear cross member. A rear pintle or clevis hook is useful in towing other machines.

Springs.   Springs are of the leaf type, and for a single-drive axle are shown in Figs. 18.3 and 18.4. They are fastened to the frame by two shackles, one of which is a single pin hinge, the other a U hinge that takes care of the increase in length of the spring as it is compressed.

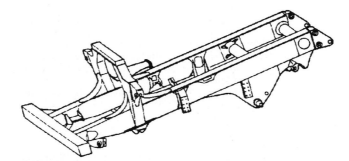

FIGURE 18.2   Sturdy truck main frame.

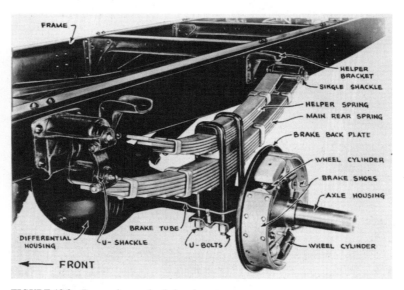

FIGURE 18.3   Rear springs and axle housing.

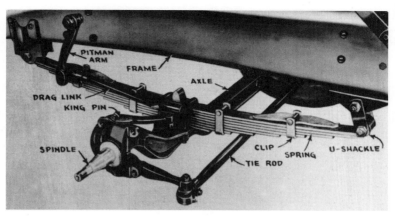

FIGURE 18.4   Front spring and axle.

The rear springs carry the largest part of the load and are proportionately heavy. A helper spring is placed just above the main spring under frame brackets which rest on its ends when the main spring is partly compressed. The helper adds sufficient strength to carry heavy loads without increasing stiffness under light loads. The helper could be rubber suspensions.

Each spring is fastened to the axle by a pair of U bolts and by the spring center bolt, the head of which fits into a socket in the top of the axle. The braking power on all four wheels and the driving power in the rear are transmitted to the frame through the springs, so it is important to keep their fastenings very tight. If the U bolts are loose, the center bolt may shear and the axle move out of line.

The front axle is a drop center I beam, the rear a hollow casing which carries the differential and the axle shafts.

The front wheel hubs pivot on nearly vertical kingpins held by the ends of the axles. The steering mechanism is similar to that used in automobiles.

Brakes.   The foot brakes on most small and medium trucks are hydraulic, with the type of vacuum booster described below. Each brake shoe has a separate wheel cylinder. Some details of construction are shown in Fig. 18.5. For air brakes on heavy trucks, see page 18.15.

The parking brake is a single mechanical unit on the driveshaft behind the transmission. Its effective grip is multiplied by the rear-end gearing. It is not designed to be used for stopping the truck.

Vacuum Brakes.   The vacuum in an engine intake manifold can be used to apply brakes and do odd jobs around a machine. Sometimes it is given so much work that an auxiliary vacuum pump is installed to ensure an ample and steady suction. Diesels do not create an intake vacuum.

It should be understood that vacuum is a minus quantity—an air pressure that is lower than that of the atmosphere—and that work is really done by pressure of atmospheric air trying to force its way into the vacuum. It is convenient to speak of a vacuum as having suction, but this suction is really pressure going the other way.

Vacuum brakes involve a cylinder or chamber with a movable piston or diaphragm. Atmospheric air on one side of the piston pushes the piston into the vacuum on the other side of it to apply the brakes. The vacuum cylinder is usually used as a booster to make application of air-assisted hydraulic brakes easier, but it may also supply all the braking effort through direct mechanical connections.

Figure 18.6 shows a cutaway view of a vacuum-hydraulic booster, and Fig. 18.7, a brake system. There are three principal parts in the booster: the control valve, the slave cylinder, and the vacuum cylinder. When the brake is off, the vacuum piston is pushed full to the left by a spring, and has vacuum on both sides of it.

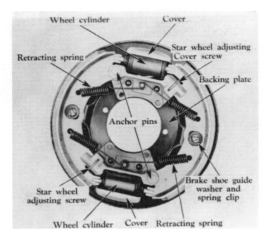

FIGURE 18.5 Hydraulic brake.

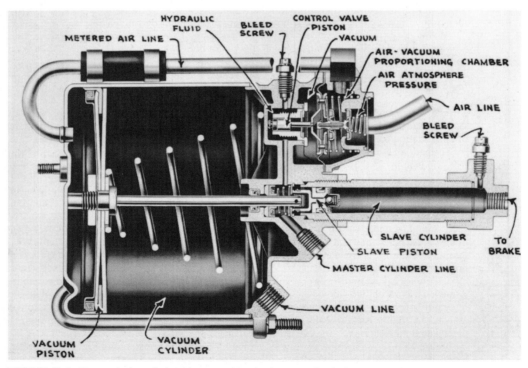

FIGURE 18.6 Vacuum brake cylinder. (*Courtesy of Bendix Commercial Vehicle System LLC.*)

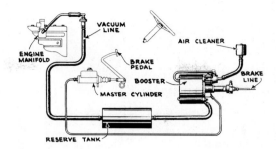

FIGURE 18.7 Vacuum brake lines. (*Courtesy of Bendix Commercial Vehicle System LLC.*)

The control valve consists of four chambers which, from left to right in the illustration, contain hydraulic fluid, vacuum, a variable proportion of vacuum and air, and atmospheric air. The proportioning chamber is connected to the left side of the vacuum piston by tubing.

When the brake pedal is depressed, hydraulic pressure from the master cylinder moves the control valve piston to the right, closing the vacuum poppet valve, and then opening the atmospheric valve which allows air to flow from the air chamber to the proportioning chamber. The resulting increase in pressure (drop in vacuum) closes the atmosphere valve by its action on the diaphragm. Increased hydraulic pressure will reopen it. Decreased hydraulic pressure will allow it to remain closed and will open the vacuum valve that will drain air from the proportioning chamber away into the vacuum system until a new balance is reached. The amount of air pressure in the chamber and in the left end of the vacuum cylinder is therefore exactly regulated by the hydraulic pressure, which in turn depends on the amount of push exerted on the brake pedal.

An air drier, to eliminate moisture in the intake air, is desirable to ensure that the system operates efficiently.

Pressure from the brake cylinder also enters the left portion of the slave cylinder. It pushes the piston to the right, forcing the fluid beyond it into the brake lines. At the same time, air metered by the control valve pushes the vacuum piston to the right, adding its force to that of the slave piston. The brakes thus receive direct pressure from fluid and booster pressure from the vacuum mechanism. The initial movement of the vacuum piston closes a check valve in the slave cylinder piston so that higher pressures may be maintained in the lines on the brake system side of the piston.

The booster reduces the required pedal pressure from 30 to 70 percent, depending on its model and the pedal ratio used.

If the vacuum system does not function, the vacuum piston will be held to the left by its spring, and if the brake is applied, fluid will flow directly from the master cylinder lines to the brake lines through the slave cylinder piston check valve, which is open when the piston is in its leftward position.

Trailer brakes are designed so that in case the trailer becomes disconnected from the tractor, the broken vacuum lines will automatically cause the trailer brakes to be applied. This prevents the trailer from running free. Trailers should be blocked when left disconnected, as brakes may release because of leakage.

Wheels and Hubs.   Brake drums are anchored to the hubs by the same studs that hold the wheels. Rear hub and drum construction is shown in Fig. 18.8. Tires are tubeless in new trucks.

Six identical wheels are used. These are steel, of either cut-out disc or spoke construction, with a lock ring to hold the tire. The front wheels and the inner rears are mounted with the convex side out, and the outer rears have that side in, in order to meet the hub.

The wheel stud is pressed through the hub from the back so that its head holds the drum. The inner wheel is fastened by five hollow lugs with inner and outer threads. The outer wheel is then mounted and fastened with large nuts which screw onto the outer threads of the lugs.

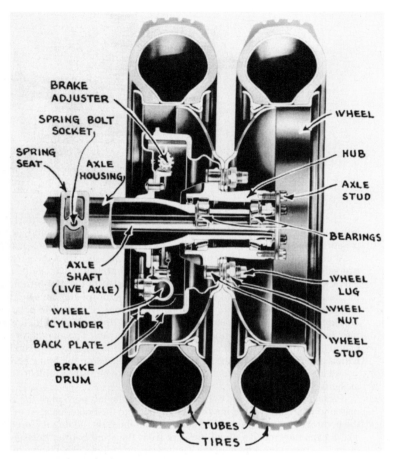

FIGURE 18.8 Rear hub.

It is very important that both inner and outer fastenings be tight, as the driving power and stopping power of the truck are transmitted through them. If any looseness develops, they will wear and ultimately break.

Studs, lugs, and nuts on the right-hand hubs have the usual right-hand thread, but those on the left have a left thread. The part of the lug or nut nearest the rim moves faster than that near the center, and during an abrupt stop will tend to revolve on the stud in the direction the wheel turns. The thread arrangement causes this force to tighten the connection.

TIRES

Tire construction is described in Chap. 12. The table of wear factors given there does not apply to highway use.

Tires built for highway use wear faster with increasing speed, but the rate of wear does not increase as rapidly as with the heavier off-the-road types. Forty miles per hour is considered moderate highway speed.

Speed is damaging chiefly when combined with curves, overuse of brakes, or rough or littered roads. Underinflation and overloading may cause severe damage. For the dump use, the largest tires usually are the best investment because of extra traction and resistance to abuse. Tires of the desired size should be supplied at the factory, as it is more economical than to change over. In addition, a change of rear tire size necessitates changing the speedometer gear.

The front wheels usually carry much less weight than the rear wheels, and can often safely use smaller and lighter tires. However, when all tires are of the same size, only one spare is needed, and the life of rear tires can be prolonged by a rotation program that puts them on the front for part of the time. Large front tires may rub on the frame when steered sharply.

From a safety standpoint, the front wheels should have good tires, as a front blowout on a loaded truck may put it out of control, while a blown-out rear will be carried temporarily by the tire next to it.

It is usually not practical to carry a spare tire on a dump truck unless the body is specially made to accommodate it. It cannot be mounted on the side of the body because of the 8-foot-width limit. It can be placed on a reinforced cab shield, but its weight is too great for one person to handle it there. However, it is sometimes possible to make a rack for it under the body.

Generally, if an empty truck has a rear flat, it can return to its base. If the trip is long, it is advisable to remove the flat tire. A front flat can be removed, and one of the outer rears substituted for it. If the truck is loaded, however, it should be parked and a tire brought to it.

Outer rear tires may be changed without a jack by running the inner wheel up on a block so that the outer one will be held clear of the pavement.

Tires are expensive and are quickly destroyed by neglect or abuse. Proper inflation, which can be determined only by a pressure gauge, is extremely important when heavy loads are carried. If the tire bulges prominently at the bottom when resting on a smooth surface, it is either soft or overloaded, and running it will develop destructive heat, weakening the fabric.

Dual Tires.   Dual tires are two tires mounted on two wheels that are bolted to the same hub. All highway dump trucks and most off-the-road rear dumps use dual-drive tires. Dual front tires are very rare.

The two tires of the set work as a unit. They must be the same size and ply rating, have nearly the same amount of tread, and carry the same pressure. Otherwise the larger or harder tire will carry more than its share of the load, and is likely to be damaged.

New tires of the same size but of different makes may differ in outside diameter. Tires of the same size and make but of different ply ratings are likely to differ in either the outside diameter of the loaded radius, or both.

Scuffing.   The two tires of a dual pair do not travel the same distance on a curve. If a truck equipped with dual 8.25-20 tires spaced 11 inches on centers makes a U-turn between curbs 60 feet apart, the outer outside tire will travel 94.2 feet, and the inside one of the pair 91.4, a difference of 2.8 feet. The difference in travel distance in the inner pair is similar.

Travel distance differences between tires that are locked together cause slippage and scuffing. Because of leverage and road crown, the outside tires slip more than the inside, so they wear faster. As they get smaller, they carry less of the vehicle weight, are less firmly pressed against the road surface, and do a larger share of the slipping.

The more heavy and powerful the truck, the bigger the tire, the wider the spacing, and the more pronounced the wear.

It is important to rotate the tires before the inside tires become overloaded and the difference becomes too great for proper matching. The maximum permissible differences, measured in inches, are as follows:

Tire size	Diameter	Circumference
8.25 and smaller	$\frac{1}{4}$	$\frac{3}{4}$
9.00 and larger	$\frac{1}{2}$	$1\frac{1}{2}$

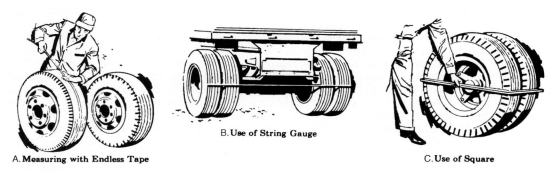

A. **Measuring with Endless Tape**   B. **Use of String Gauge**   C. **Use of Square**

FIGURE 18.9   Matching dual tires. (*Courtesy of Rubber Manufacturers Association.*)

Where differences exist that are within these limits, the truck owner has a choice between two recommended practices. He or she may put the two larger tires on the outside wheels to conform to the crown of the road, or put both of them on the right side and let the differential take care of the differences.

Measurement.   Sets of duals should be checked for size differences at least every 1,000 miles, after making sure that the air pressure of the four tires is exactly the same. Replacement tires should be measured and compared before they are put on the truck.

Figure 18.9 shows three ways of checking sizes. In (*A*), the circumference is measured with a tape, either before mounting or after jacking up the wheel. In (*B*) a straightedge or a taut string checks both pairs at once. In (*C*) a "square" made of two 1-inch by 2-inch wood strips rigidly fastened to make an exact right angle can be laid along the side of the outside tire and across the treads.

Rotation.   Systematic rotation of tires will prevent damaging size differences from developing. The simplest system is to move the right tires to the left side and the left tires to the right side, putting inside tires on the outside and outside tires on the inside, as shown in Fig. 18.10.

Tandems.   Tandem drives, to be described later, have two sets of axles, one in front of the other, each equipped with dual tires.

Scuffing is much greater with this arrangement. The outer tires wear more than the inner, and all tires are dragged sideward on turns. The side drag is hardest on the rear set.

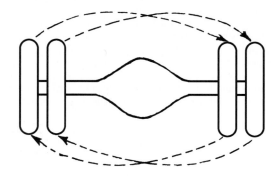

FIGURE 18.10   Rotating position of dual tires.

Many tandems do not have any differential or power divider between axles, so that all eight tires rotate at the same speed. In this case it is very important that the tires be matched so that the average diameter of those on one axle is within $\frac{1}{4}$ inch of the average diameter on the other axle.

Front Tires.   Truck front tires should last much longer than the rears, as they do not transmit driving force and carry much lighter loads. However, they are subject to excessive wear from being run out of line.

Front wheels do not roll exactly parallel to each other, as steering is more stable if they toe in slightly (fronts of the tires closer together than the rears), have a slight camber (bottoms of the tires closer together than the tops), and a slight caster, or backward tilt of the kingpin.

Too much toe-in shows a feathered edge at the inside of the tread design, while toe-out feathers the outside edge. Too much camber concentrates smooth wear on the outside of the tire, negative or reversed camber wears the insides and makes it look as if the axle is sagging. Wrong caster may result in cupping wear.

Misalignment of front wheels may result from poor adjustment, but more often it is the result of bending caused by glancing blows against curbs or banks. Severe misalignment may be felt by the driver, but if it is slight, it will show up only in tire wear. The two precautions are careful driving and frequent tire inspection.

Wobble.   Rapid tire wear, with or without cupping, may be caused by wheel wobble. The trouble may be a bent wheel, in which case only the tire is damaged.

A loose or broken bearing will allow a wheel to wobble. Unless repair is made, the wheel may come off. Defective bearings are a common cause of rear axle breakage.

Maintenance.   One of the best ways to provide tire maintenance is to use, or call for the use of, tracking software provided by the tire manufacturer.

POWER TRAIN

Engine.   Trucks may have diesel or gasoline engines, with compression ratios that permit using regular gasoline, and that can be converted to burn LPG (liquefied petroleum gas).

Clutch.   The standard clutch is a dry single-plate type located in the engine flywheel housing. It is held in engagement by springs, and released by pushing a foot pedal.

Transmission.   The transmission may have either four or five forward speeds, with one reverse. The five-speed may have direct drive in either fourth or fifth. All gears except first are synchromesh.

Four-wheel drive may be provided by factory installation of a two-speed transfer case behind the transmission, from which power may go to the rear driveshaft only, or to both front and rear shafts, and a live front axle. See Fig. 18.11.

The power takeoff is in the left side of the transmission, and, for dump use, drives a shaft to the body hoist pump. The takeoff is operated by engaging its drive gear in the transmission. The clutch must be depressed while engaging it, but, except when under load, it can be disconnected without clutching.

The driveshaft from the rear of the transmission supports a drum for the mechanical parking brake, and continues through a bearing in the center frame cross member. Behind this the shaft is fitted with two universal joints, one of which includes a sliding connection. This makes it possible for the axle to move up and down relative to the frame without damage to the shaft.

Automatic Transmission.   Optional automatic transmissions are similar to those described in Chaps. 12 and 17.

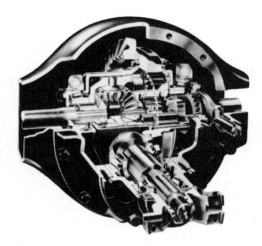

FIGURE 18.11 Two-speed differential.

The friction clutch is replaced by a torque converter. The Allison transmission has four or five forward ranges and one reverse. Gears are constantly engaged, and power is directed through them by hydraulically shifted disk clutches.

Five speeds are used when a very low or creeper gear is required, as in dump or transit mix work.

In lower ranges, the converter acts to give variable speed-power ratios within the range; upper gears have an automatic lockup for nonslip drive.

In operation, drivers use the shift lever to select the highest range they wish to use. Shifting will then be automatic up to and down from that range. The lever can be moved at any time.

Automatic shifting is regulated by speed, torque, and throttle position.

Final Drive.   The live axles are splined into the axle gears in the differential and are bolted to the hubs, so that they revolve on the bearings supporting these parts and carry no weight. See Fig. 18.12.

The bolts which fasten the axle shaft flange to the hub must take the full drive torque of the truck, and must be kept very tight to avoid shearing.

The power is transmitted from the hub through studs and friction contact to the wheels. The wheel and tire act as a unit as long as sufficient air pressure is maintained. Friction between the tires and the road converts the rotary movement of the axles into straight forward or backward motion of the axle housings and the truck.

This horizontal push of the tires is transmitted through the hub bearings to the axle housings, and through the rear springs and their front shackles to the frame.

The rear springs have four functions: to carry the weight of the truck, to absorb road shocks, to steady the axle housing against twisting in reaction to the turning of the wheels, and to keep the housing from shifting forward or backward in response to driving and braking forces.

The term "final drive" is sometimes limited to reducing gear sets that carry power from axles to hubs.

Cab.   The standard cab resembles a passenger car coupe body, except that it is cut off immediately behind the driver's seat in most trucks.

The clutch, brake, and accelerator pedals are identical in location and function with those in standard passenger cars. The gearshift lever may be on the floor or on the steering column. A typical dashboard of controls is shown in Fig. 18.13.

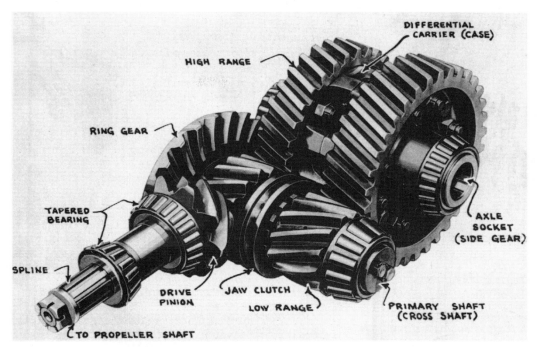

FIGURE 18.12   Double-reduction two-speed differential.

Rear-vision mirrors, projecting beyond the body line, are required on both sides. There must also be blinker-light turn signals, backup light, and possibly an automatic backup horn signal.

Such items as heater, windshield wiper and washer, adjustable seat, seat belts, and trim are similar to or identical with automotive equipment.

Cab over Engine.   In this truck construction, which is usually known by the abbreviation C.O.E., the frame is shortened and the cab mounted over the rear of the engine.

Advantages are shorter turning radius, better load distribution, less overall length, and improved vision from the cab. Disadvantages are awkwardness in servicing the engine, extra climb up into the cab, and difficulty in keeping the cab cool.

Electronics.   The use of computers on board trucks has many purposes. They are designed to control certain functions of the mechanism just as used on other pieces of construction equipment described in previous chapters. In trucks the electronics from the computers give readings for fuel use and economy. They help maintenance people scan the on-board computer's memory for active fault codes to see what is currently out of specification on the machine. When they scan for inactive fault codes, they get a list of any events in which the machine temporarily operated outside normal parameters. The mechanic can maximize the electronic resource by checking for fault codes at regular intervals to find out how the truck is being used in the field.

For example, the computer records each time the antilock braking system was engaged or when the engine reached an overspeed situation. This kind of input can help identify abusive drivers or abusive conditions. Electronics can indicate all the times when a truck overheated or when the oil pressure ran low, even when the event was only momentary. The information is valuable in helping to establish a maintenance schedule that can prevent breakdowns.

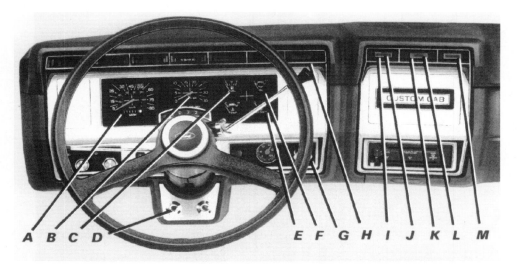

(A) Electric speedometer and
odometer (U.S. or metric)
(B) Tachometer*
(C) Fuel gauge
(D) Choke (gas engine)/ether cold start
control* (diesel engine)
(E) Charge indicator light or gauge*
(F) Engine oil pressure light or gauge*
(G) Parking brake control
(H) Brake indicator light

(I) Low oil pressure indicator light
(J) High coolant temperature indica-
tor light
(K) Low coolant level indicator light
(diesel)
(L) Water-in-fuel indicator light (diesel)
(M) Automatic transmission oil tem-
perature indicator light (with
optional automatic transmission)
*Optional

FIGURE 18.13 Dump truck control panel.

BODY

A dump body unit consists of the box or body proper, the tailgate, body hardware such as chains and pins, and optional equipment such as cab guards. The hoist, which is often sold as a separate unit, includes a subframe, pump, valve, cylinder, and the controls.

A very wide range of body and hoist constructions are available for every truck. These units are usually not made by the truck manufacturer.

Figure 18.14 shows a heavy-duty body. A hoist subframe extending backward from the cab is bolted onto the truck frame. This is attached at the rear by heavy hinges to the body frame, which consists of two beams that rest on the subframe, and cross pieces to support the floor and sides.

The sides are sheet-metal-reinforced by flanges at top and bottom, and V-type or pyramidal side braces, and are welded to the flanged front wall. At the rear, heavy corner posts and the rear frame member combine to make a structure rigid enough to resist bending outward.

The front and rear corners have slots or gusset pockets in which sideboards can be placed. These usually consist of planks 1½ inches thick, which may be as high as desired. They may be used to increase the capacity of the body or to prevent spillage off the sides.

Body capacity is figured level with the sides and is stamped on a plate on a front corner. Sideboard capacity and heap must be calculated as explained in Chap. 2.

The double-acting tailgate is somewhat higher than the sides, and usually has offset hinges at the top to increase clearance for dumping bulky objects and to make closing more positive. It is made of steel plate with box reinforcing.

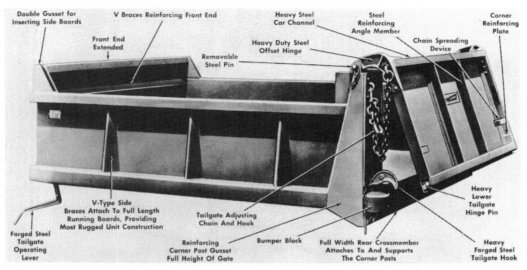

FIGURE 18.14 Heavy-duty dump body.

The upper hinges are equipped with removable pins. The lower hinge pin is a fixed part of the gate, but the hinge itself can be opened by means of a lever at the left front corner of the body, within reach of the driver.

If the body is flat on the subframe and the latch is open, the gate will hang in a closed position, with the lower hinge pins lying in the hollow of the latches. When the tailgate lever is moved up, it moves the latches upward and forward against the post, locking the bottom of the gate tight to the body.

The body and the gate are held in this position for loading and transporting. To dump, the gate lever is pushed down and the body raised in the front, pivoting on the rear hinges. Its own weight and the pressure of the load sliding against it cause the gate to swing outward on the upper hinges. When the load is fully dumped, the body is lowered, the gate swings into closed position, and then is clamped there by pulling the lever up.

Each rear corner post contains two keyhole slots of such size that a chain fastened to the top of the gate can pass through the upper part of the hole freely, but any link dropped in the lower slot will be caught.

The tailgate can be held at any desired angle by unfastening it at the top hinges, keeping it clamped at the lower ones, and passing the chains through the upper slots. Adjustment is made by raising the tailgate by hand to take its weight off the chain, then lifting the chain out of the slot and moving it in or out to the desired length. The other chain is then adjusted to as near the same tension as possible.

The tailgate may also be fastened at the top hinges, released at the bottom, and restricted in opening by the chains. Each is passed down the back of the gate through a hole in its slide flange, and forward to the lower slot. The gate opening is restricted in order to spread thin, even sheets of free-flowing material.

The cab guard, which is an optional extra, is a sheet of reinforced steel, curving or angling upward and forward over the cab. This is almost a necessity for trucks that are to be loaded overhead. Holes should be cut in it to allow the driver to see into the body through the rear window of the cab.

Hoist.   A direct-type hoist is shown in Fig. 18.15. It consists of a hydraulic pump, a valve, and a cylinder.

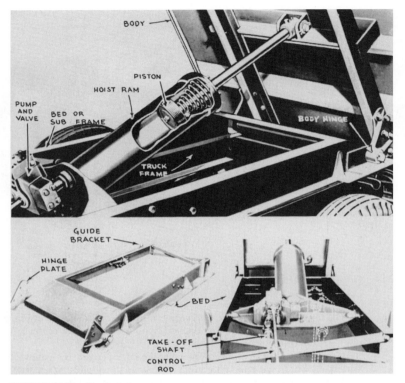

FIGURE 18.15   Simple hoist.

The pump is driven from a transmission power takeoff through shafts and universal joints, and works only when both the engine clutch and the takeoff gear are engaged.

The valve in many hoists is built into the pump body and has three positions:, UP, HOLD, and DOWN. It is controlled from the cab, through either a floor lever or a knob and wire on the dash. The lever is sturdier but takes up floor space, interferes with insulating the cab, and may not permit the operator to watch the load as carefully.

The single-acting ram is bolted to a cross member which is hinged to the body subframe, and the piston rod is hinged to a crossbeam of the body. A spring may be placed between the piston and the ram head to cushion the piston when forced to the limit of its travel and to help to start the body down when pressure is released.

When the body is down, the ram slopes up a little from the horizontal. When it is expanded, it pushes both back and up. The body hinge pins are made strong enough to resist the backward pressure so that the body is forced up. Leverage is lowest and load greatest at the start of the dump. As the body rises, a large part of the weight is transferred from the ram to the rear hinges.

A number of hoist linkages are offered in which the leverage is greatest at the beginning of the dump, so that the body moves slowly at first and more rapidly as it approaches the top.

HEAVY TRUCKS

The U.S. Department of Transportation requires annual inspections of all trucks with gross vehicle weight (GVW) of 26,000 pounds or more. Large trucks for use on highways are very similar in design to lighter ones, except that all parts must be stronger and heavier. Either gasoline or diesel engines can be obtained. Air brakes are standard. The transmission contains more speeds, and an auxiliary high-low box may be provided. Standard, rock, or quarry type bodies may be used.

Power Steering.   A typical power steering unit, the Vickers hydraulic steering booster, is shown in Fig. 18.16. It consists of a two-way ram with built-in valving and is powered by an engine-driven hydraulic pump. The ram is bolted to the back of the drag link, and the piston rod is fastened to the frame by a ball joint. The pitman arm of the steering gear is clamped in the control valve between the ram and the drag link, and can move slightly either way against light springs before being stopped by the valve case.

If the steering wheel is turned so as to move the pitman arm to the right, it moves the valve spool with it, opening a passage for oil under pressure from the pump to the rod end (right side in the illustration) of the piston. The piston rod is anchored to the frame, so this causes the ram, valve casing, and drag link to move to the right until the valve casing catches up with the spool and closes the passage.

If the wheel is turned the other way, the pitman moves the valve spool forward to open a passage to the front of the piston. The ram will then move forward until the passage is closed.

If there should be no pressure in the system, a check valve opens which permits fluid to move from one side of the piston to the other. Pressure of the pitman arm will then move the valve case and ram in either direction mechanically, and will operate the drag link as if the ram were not present.

Road shocks are transmitted from the drag link through the ram to the truck frame without affecting the steering wheel.

This type of control may also be used in graders, tractors, and other machines; but usually not in rollers or scrapers.

Air Brakes.   Brakes on heavy trucks and wheel tractors are usually applied by compressed air. Figure 18.17 contains a diagram of a system that is used in heavy trucks.

The air is supplied by a compressor constantly driven by the engine, which unloads, or stops pumping, when full pressure—usually 100 pounds—is reached in the reservoir or receiver and resumes when it falls to 90 pounds.

A valve operated by a foot treadle or conventional brake pedal allows exact control of air pressure in lines leading to the brake chambers, where it acts against diaphragms that move rods and

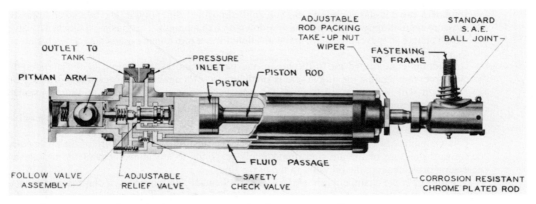

FIGURE 18.16   Hydraulic steering booster. (*Courtesy of Vickers Incorporated.*)

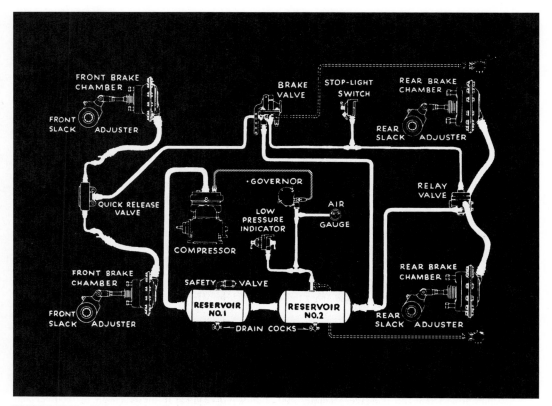

FIGURE 18.17   Air brake piping diagram.

levers called slack adjusters which apply the brakes. The front brake lines are provided with a quick-release valve which drains them rapidly when released to prevent any lag that might interfere with steering or vehicle balance. The rear brakes, and trailer brakes if used, may have a relay valve that feeds air direct from the reservoir into the lines.

The brake valve is shown in Fig. 18.18 in applied position. In applying, the treadle compresses the pressure-regulating spring, which pushes down on a diaphragm. This pushes the exhaust valve seat down on its ball valve, closing it, and with slight further movement pushes the inlet valve down off its seat, opening it.

This admits compressed air to the brake line passage and to the bottom of the diaphragm. When its pressure against the diaphragm is sufficient to lift it against the mechanical pressure on the spring, it will move upward to close the inlet passage so that the brake pressure is sealed against the movement of air in or out.

If the treadle is released, the air will push the diaphragm higher, unseating the exhaust valve and allowing the air to flow out of the brake lines.

This system is always in a state of balance between the downward pressure of the spring and the upward pressure of the air, and immediately responds to any movement of the treadle by increasing or decreasing the pressure in the brake lines.

One type of brake chamber is shown in Figs. 18.19 and 18.20. It contains a diaphragm attached to a push rod, which is moved in a direction to apply the brake by air pressure, and is released by coil springs. The force applied to the push rod depends on the pressure supplied from the brake

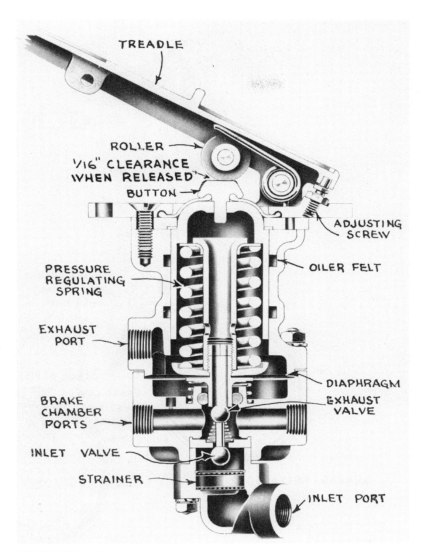

TREADLE

ROLLER

1/16" CLEARANCE
WHEN RELEASED

BUTTON

ADJUSTING
SCREW

PRESSURE
REGULATING
SPRING

OILER FELT

EXHAUST
PORT

DIAPHRAGM

EXHAUST
VALVE

BRAKE
CHAMBER
PORTS

INLET VALVE

STRAINER

INLET PORT

FIGURE 18.18   Air brake valve.

valve and the size of the diaphragm area against which it pushes. The force applied to the brake is further affected by the length of the slack adjuster lever.

When the brakes are applied, inlet air pressure forces the diaphragm down so that it closes the exhaust port, and allows air to flow into the brake lines. When the brakes are being held, inlet and brake line pressure is balanced, the exhaust port remains closed, and the inlet is closed by the rim of the diaphragm. When the brakes are released, pressure drops in the inlet, and pressure in the brake lines forces the diaphragm body upward so that the air can flow from the brake lines through the exhaust port.

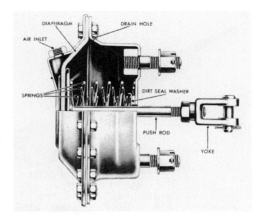

FIGURE 18.19   Brake chamber.

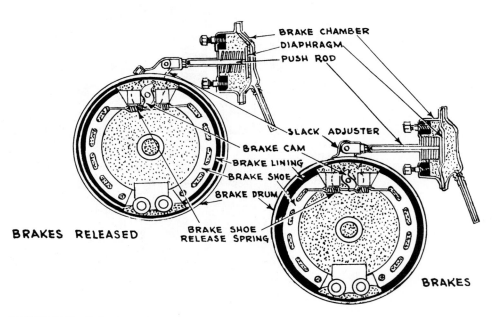

FIGURE 18.20   Brake shoe action.

This allows the brakes to release more rapidly than if the escaping air had to follow the pressure line back into the brake valve.

When a truck is equipped with air brakes, it is absolutely essential that it not be operated when there is not enough pressure to apply them. Warning of low pressure may be given by a buzzer or other noisemaker, but this may not be heard over engine and road noise.

The visible warning shown in Fig. 18.21 is simple and effective. It hangs directly in the driver's line of vision except when pushed out of the way by air pressure.

FIGURE 18.21 Low-pressure warning signal.

TANDEM DRIVE

The load-carrying capacity of a two-axle truck of any size can be increased by installing an extra axle in the rear. For dump use this should be a driving axle. A truck so equipped may be called a six-wheeler, a ten-wheeler, or a tandem. The double-axle unit itself may be called a tandem or a bogie.

Tandem drive permits carrying much heavier loads in proportion to tire size and axle strength by distributing both weight and driving strains over twice as many units. It improves traction but not nearly as much as a driving front axle does.

There are a number of different makes and types of tandem drives. Figure 18.22 is a side and a detail view of a tandem drive unit.

Two identical axle housings are held parallel to each other by eight torque rods. The upper four rods are hinged to the axle housings and to the side frame brackets, the lower set to the bottom of the housings and to the bottom of a heavy cross member rigidly fastened to the frame. These rods hold the axles against twisting and absorb power and braking thrusts.

The weight of the truck is carried on two leaf springs, fastened on an oscillating collar on the lower frame cross member. The ends of the springs are held on the tops of the axle housings by spring pads and rebound brackets, which allow the axles to move forward and back and to tip, without putting extra strain on the spring.

If the load on the truck is increased, the spring will flex and the frame move down relative to the axles. As the frame settles, the axles will first be forced apart by the torque rods, and then pulled together. In and out movement of the axles also occurs, as the tires hit bumps in the road.

If worn tires are used on one axle and new ones on the other, the rotation of the new one will be slower, and fight between the two axles, with resultant tire scuffing and gear wear, will occur.

This difficulty may be avoided either by using a full set of tires with about the same amount of tread, or by putting the good treads on one side and the worn ones on the other so that the average diameters of the tires on each axle are the same. The axle differentials can then adjust the differences. The right side is preferable for the good tires to offset the crown of the road.

This company's current models have an interaxle differential to equalize such differences.

The Mack bogie or tandem, shown in Fig. 18.23, uses a parallelogram of springs, instead of torque rods. The springs are anchored to the axles in rubber shock insulators (pillow blocks) enclosed in steel boxes under preload. The upper and lower boxes are carried on trunnions, so that any twisting of the double pairs of springs is avoided.

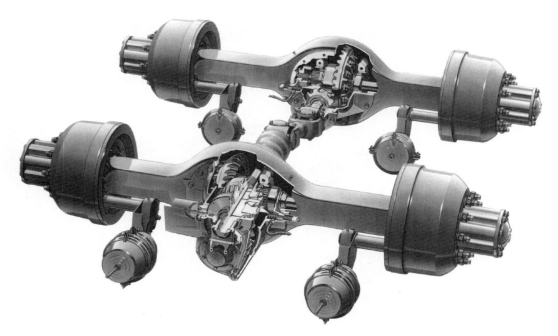

FIGURE 18.22 Tandem drive cutaway view.

FIGURE 18.23 Double-spring tandem drive.

In any of these tandem drives, some tire scuffing will occur because neither axle has any provision for steering. As the truck turns, the rear bogie wheels have a tendency to slide toward the outside of the curve, and the front set toward the inside.

If a bogie is improperly designed, it may tend to nose down in front when stopping and rear up when accelerating, with resultant bouncing and excessive wear on parts.

Six-wheelers are more difficult to steer on slippery surfaces than standard trucks, because of the resistance of the extra axle to sideward movement.

Highway Limitations.   Vehicles to be used on highways must not be more than 8 feet wide, and they are limited in length, gross weight, and weight on any one axle. These restrictions vary in different states, but a limit of 9 tons for the load on any one axle is usual.

Dual-drive tires are always used, with either single or tandem axles. The size and weight-bearing capacity of tires are limited by width restriction and the space required by the frame.

As a result of both weight and width regulations, the highway dump truck is usually limited to a maximum carrying capacity.

Uses.   The highway dumper usually has a combined weight of chassis and body equal to maybe three-quarters of its rated payload, as against a usually higher ratio for off-highway haulers in comparable sizes. Highway trucks may go over 70 miles per hour, nearly double the speed of other haulers.

The result is that the highway truck is a highly efficient hauler when operating conditions are suitable. It should not be used constantly in rough or soft pits or on poorly maintained haul roads, or on excessively steep grades, or under big shovels loading coarse rock.

However, the highway dumper is the preferred hauler for jobs where distances are long and roads well maintained, and loading and ground conditions in the pit and on the fill are not too rough.

The highway dumper is necessary to the contractor whose hauling must be done on streets and highways. This includes most excavating for foundations, city and suburban roads, supplying sand, gravel, fill, and topsoil, and most other small to medium-size work. On major highway construction it is usually possible to obtain permits to operate off-highway vehicles over short sections of public road, but the contractor is likely to have to rebuild them after using them.

OFF THE ROAD

Trucks that are built to operate in mines or pits, or in other types of excavation in which the use of public roads is not required, are not subject to any legal restrictions in size or weight.

Off-highway rear-dump trucks may be 10 to 30 feet wide, up to 50 feet long, over 20 feet high, with loading height (body sides) between 8 and 17 feet. Capacity range is from 20 to more than 400 tons, with body ratings of up to 300 cubic yards struck.

Gross vehicle (empty) weight may equal 1.8 times the payload capacity, down to as low as four-fifths of it.

Construction is heavier than in highway trucks, in order to stand up under conditions of rough footing, heavy loads, and short hauls. Substantial amounts of high-strength steel may be used.

Top speed is usually 35 to 50 miles per hour. Road conditions and tire wear limit practical speed.

Construction.   Figures 18.24 to 18.27 are representative of this type of truck. Components tend to be massive and comparatively simple.

Road shocks may be absorbed by conventional leaf springs, coil springs, nitrogen-and-oil (air-over-hydraulic) cylinders, rubber discs, oscillating bars, and various combinations of these.

Horsepower range is from about 125 up to 2,500. Most of these haulers have torque converters and power shift transmissions, with hydraulic retarders being either standard or optional. A majority of these have mechanical drive, but some of the largest use a generator and electric wheel motors.

Differentials may be either standard or limited action. They are usually single reduction, but may be double. Further gear reduction is obtained through planetary final drives in the wheel hubs.

There usually are two axles, with drive through dual wheel sets on the rear. There are also tandems and all-wheel drives.

Tires are among the largest made, up to 59/80 R63, with the same size front and rear.

FIGURE 18.24 Off-the-road dump truck.

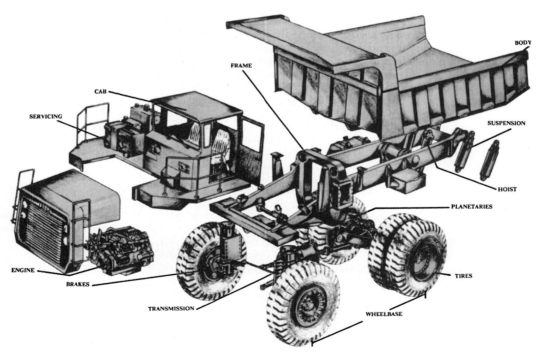

FIGURE 18.25 Off-the-road dump truck, exploded view. (*Courtesy of Terex Corporation.*)

FIGURE 18.26   Articulated truck frame.

FIGURE 18.27   Telescoping hoist cylinders.

Steering usually is of the automotive type, with front wheels swiveling on a rigid axle, but there are also articulated models.

Body.   Bodies are of the heavy-duty rock or quarry type. The sides may flare out to make a larger loading target. The floor is usually a single, heavily reinforced plate. Air ducts may be provided for internal heating of both floor and sides with exhaust air, to prevent loads from freezing down during subzero hauls.

There is no tailgate. The body floor may have a continuous upward slope from front to back to retain the load, or it may be flat or nearly so, with an upturned chute in the rear.

Standard hoist construction is a pair of direct-acting, three-stage telescoping cylinders, with power up and partway down. Power-down permits raising the body to a very steep dumping angle, 55° or more, then pulling it back until gravity can lower it the rest of the way. See Fig. 18.27.

Most materials can be dumped readily, even when backing up a grade.

ELECTRIC DRIVE

Some of the biggest off-the-road trucks have electric drive. A diesel or a turbine engine drives a generator, which supplies current to electric drive motors. These may be located in each wheel, or in differentials between pairs of wheels.

The motors use direct current on account of its superior lugging performance. The generator may produce dc directly, or it may be an alternator whose alternating-current (ac) output is converted to dc in a rectifier. The motors operate through reducing gears, as in Fig. 12.38, and are reversible.

Electric drive eliminates all gearshifts and speed ranges, automatic or otherwise. There is smooth, stepless transition from zero to maximum speed, which is determined by the amount of reduction in the motor gearing, and by grade and rolling resistance. It may be 24 to 36 miles per hour.

Drive motors also supply dynamic braking, to slow the hauler on downgrades and before stops.

TRACTION

Soft Footing.   Truck efficiency drops rapidly on soft ground. The wheels sink in, so that they must constantly climb a slope to move the truck horizontally. As the front wheels sink, they have an increasing tendency to push in the manner of a sled runner instead of rolling up and out, so that the force required to move them is greatly increased.

If the power required to roll a driving wheel up the slope in front of it is greater than that needed to spin the wheel in the hole, it will spin and usually dig the hole more deeply and polish its slope. The ordinary differential will deliver its full drive power to the spinning wheel, so that the standard truck will not be able to move even if only one side lacks traction.

Another difficulty is that the front wheels, if turned to climb out of a rut, are liable to stop revolving and skid sideward so that the truck cannot be steered.

When the truck is reversed, the reaction to the twist of the driving axles tends to lift the back of the truck and to push down the front, thus reducing the weight on the driving wheels and their traction.

Similar difficulties are encountered in sand, which allows all wheels to sink somewhat and, although not slippery, will first compress under the push of a driving wheel, then shear, allowing the wheel to spin a part turn, carrying the sand with it, then compress to hold it momentarily before shearing again. This produces a succession of shocks to the power train, while digging the wheel in more deeply each time it moves.

These factors severely limit the use of trucks in many circumstances and may be combatted in a number of ways.

In muddy pits the use of big tires is advisable, because of their larger area which increases contact and reduces sinking. It may be said that a tire will sink in mud until a large enough area is in contact to carry its weight. If the tire is small, it will sink much farther to develop enough contact; therefore, it will have a higher slope to climb to get out of the hole and have more side friction to hold it in.

Tires with coarse block treads afford better grip on most muds but are at a disadvantage on hard slippery surfaces and in sand. Tire chains are helpful in mud and on slippery surfaces, but not in sand. Block treads and chains on front wheels help in steering out of ruts.

Tandem drives are helpful, particularly if tires are kept as large as those which would be used with a single axle. Limited-action differentials keep all wheels driving instead of wasting power in spinning one wheel or pair of wheels.

Live Front Axle.   The most effective way to combat bad footing is to supply driving power to the front axle as well as to the rear. The increase in traction will vary between 50 and 200 percent, depending on the number of rear driving wheels, ground conditions, load distribution, and steepness of climb.

The advantages obtained are conversion of the front wheels from a dead obstacle to a driving force, ability to pull around a turn instead of having only a straight, forward push, tendency of front wheels to climb out of a rut instead of sliding in it, and good traction in reverse.

The two principal problems connected with the all-wheel drive are driving the front wheels through sharp turning angles, and making allowance for the fact that the front wheels go farther than the rear wheels on curves, both backward and forward.

Figure 18.28 shows why the front wheels have longer travel. The rear wheels take a smaller circle on a curve, so that they need to turn fewer times in the same truck travel distance. On loose surfaces this difference adjusts itself, but means should be provided to allow the front and rear

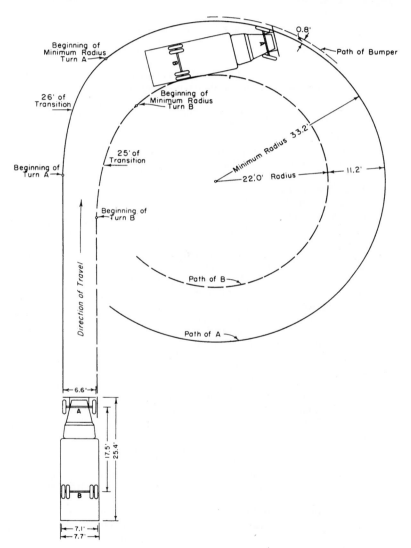

FIGURE 18.28   Right turn, two-axle truck. (*Courtesy of Department of Public Works, State of California.*)

FIGURE 18.29   Steerable drive axle. (*Courtesy of Marmon-Harrington Co.*)

driveshafts to revolve at slightly different speeds on hard pavements in order to prevent excessive wearing of tires and strain on shafts, pressure on gear teeth, and waste of power.

The truck or other vehicle is customarily kept in two-wheel drive. The front shaft is engaged whenever extra traction is needed for either moving or stopping.

The front wheels are driven through constant-velocity universal joints protected by ball sleeves. Figure 18.29 is a detail of a steerable drive axle.

Drive to four wheels greatly increases efficiency of brakes on slippery surfaces. Front wheels do not lock as readily, so that steering control is better maintained.

Freewheeling Hubs.   The front hubs may be fitted with freewheeling attachments, so that the axles will turn the wheels forward but not backward, and the wheels will not turn the axles when the vehicle is moving forward.

This allows the front hubs, axles, differential, and driveshaft to stop whenever drive is disconnected in the transfer case, reducing wear and noise.

But in four-wheel drive, hubs in freewheeling will not transmit power in reverse, or permit the steadying effect of front drive on braking.

When hubs are locked, a vehicle can be shifted in and out of four-wheel drive instantly at any speed, which is a safety factor in intermittently slippery conditions. With hubs freewheeling, it is necessary to stop to shift into four-wheel drive.

The hubs are easily shifted into a solid connection by turning their caps with the fingers. This adjustment can be made readily if it is necessary to engage them to back out of a mudhole. Control from the driver's seat may be available.

It is common practice to keep the hubs in freewheeling position except when bad footing is expected. Sometimes they are locked for all winter driving.

Positive Drive.   All the systems described thus far have the weakness that if one front and one rear wheel have no traction, they will spin and the truck will be unable to move.

The Walter power train overcomes this difficulty by the use of a set of three self-locking differentials employing worm gears instead of bevel spiders. These will permit sufficient differential action for steering but will lock against spinning.

Figure 18.30 shows the power train. The engine is located ahead of the front wheels. The engine transmission inputs into a transfer case. The transfer case includes a center differential. Output shafts from the transfer case direct drive to the front and rear axles. This drive is connected at all times.

The front and rear differentials are rubber-mounted and suspended from the frame, so that they are not subject to moving with the wheels and do not form part of the unsprung weight. All wheels are carried on beam-type dead axles with attachment to the chassis through conventional leaf springs.

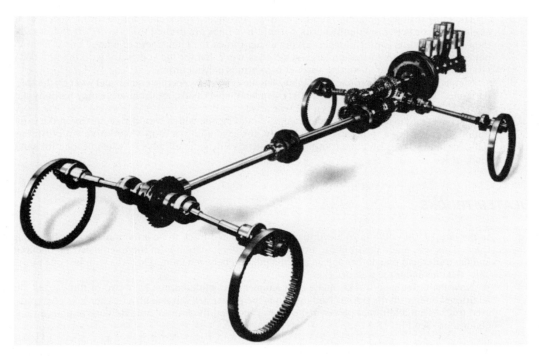

FIGURE 18.30 Walter power train. (*Courtesy of Walter Motor Truck Co.*)

The live axles each include a pair of constant-velocity universal joints, and carry brake drums close to their differentials. An additional pair of brakes is located in the rear hubs.

The live axles end in small spur gears, which turn large internal tooth ring gears in the hubs.

This construction permits the use of lighter axles than could be used if the speed reduction were in the differentials only, or if the axles carried the weight of the truck.

6 × 6.   Many of the all-wheel-drive systems discussed have been of the two-axle, or 4 × 4, type. Any of these may be expanded to a three-axle, or 6 × 6, drive, by substituting a bogie for the rear axle and extending and reinforcing the frame. The extra axle improves flotation, traction, and carrying capacity, but increases cost and makes the truck less maneuverable.

General Considerations.   All-wheel drive is of great value wherever mud, sand, or snow is encountered. Trucks so equipped can keep a job going under conditions that would be impossible for two-wheel or tandem drive trucks.

On highways, the extra traction is desirable for snowplowing and for operating under-frame or towed graders, and sometimes in towing trailers.

However, the traction and flotation obtained cannot be compared with those obtained by crawler equipment. Mud must have a fairly solid bottom near the surface for any truck to operate, and very slippery surfaces require chains no matter how many wheels are driving.

All-wheel drive has several disadvantages. It is expensive, particularly when it is installed as an accessory rather than in the original assemblyline. The extra cost may be anywhere from 25 to 100 percent of the price of the standard truck.

In return for this, the truck can carry heavier loads under much worse conditions than before conversion, so that the cost is justified in many lines of work.

The most serious operating fault is steering. A standard truck can turn the front wheels at an angle of 35° or better, while the various front-drive hubs are limited to 28 to 30°. It follows that their turning radius is much longer, giving a small truck the clumsiness of a large one. This may not be particularly important on large-scale open work, but for the contractor working in woods, around residences, and in storage yards it is a serious consideration.

Maintenance costs are somewhat higher, but the extra parts decrease strain and wear on the standard ones. If a truck is to be consistently given heavy overloads, the front drive may partially pay for itself in reduced maintenance on the rear drive.

It seems that the ideal mechanical hookup would be one which would give the driver the triple option of direct drive to the rear wheels only, for easy going; a four-wheel drive with differentials, for heavy loads on good footing; and a positive drive to all wheels, when needed for sand, mud, or ice.

ARTICULATED TRUCKS

In the late 1950s Volvo introduced their first articulated truck to be a more maneuverable dump truck. This was made possible for a tractor-trailer combination with the tractor part on one axle and the trailer part on another and an articulated joint between them. That was with just one power axle, the one under the tractor.

Now the articulated trucks, made by a number of manufacturers, have two or three axles, the additional ones with the trailer. And there may be power delivered at all axles such as a 6x6 powered truck. That additional power makes the articulated truck more able to work in poor ground conditions. See Fig. 18.31.

Specifications.   Articulated dump trucks (ADT) used in the United States are generally of sizes with 11- to 40-ton capacity. The ADTs have engines that deliver 100 to more than 400 horsepower.

FIGURE 18.31   An articulated dump truck.

The smaller ones, which might be used in tunnel work, may be permitted to transport over a highway, but the larger ones are more restricted. The ADTs might range from less than 30 to over 40 feet in length. The trailer bed generally is raised like a dump truck to be unloaded. That would require a 20-foot, or so, height clearance. Where there is a height limitation, or if the material is very sticky, there are some ADTs equipped with an ejector style body part similar to a scraper's part. Refer to Fig. 17.15. The ADT outside turning radius is up to 30 feet maximum.

Excavation Uses.   The articulated dump truck is useful on sites that may call for scraper-type earthmovers. Of course, the ADT cannot load itself and needs some form of excavator or front-end loader to do that part of the work. But if there is a haul distance of half a mile up to two miles, where a scraper would lose its economy of use because of its lower speed, an equipment combination with ADTs would be a good choice. Their speeds are 25 to 40 miles per hour.

As with any loader-hauler combination, the rule of thumb is for the loader to take 3 to 5 bucketfuls to load an ADT. More than 5 and the hauler is spending too much time being loaded, and less than 3 requires too many haulers to keep the loader busy.

TRAILERS AND WAGONS

Dump bodies may be mounted on semitrailers and trailers, with either standard or special constructions.

A semitrailer is a frame which has supporting axles and wheels at the rear and rests on the prime mover or tractor at the front. See Fig. 18.32. Tractors for highway-type semitrailers are equipped with a connecting device called a fifth wheel, one variety of which is shown in Fig. 18.33. To connect, the trailer is held at a proper height by jacks or wheeled standards, the wheels are blocked, and the tractor is backed into it. The upper fifth wheel, or trailer hitch, is hit by the tractor fifth wheel and wedged slightly up until its knob reaches the socket, into which it slips and locks automatically. The pin and socket take the pull and thrust of towing and stopping, while the fifth wheel surface carries the weight. The knob can be released from the hitch when not under load by moving a hand lever.

Off-the-road semis may use a kingpin arrangement, such as that described in the preceding chapter.

Full trailers, often called wagons, are equipped with supporting axles and wheels at both ends, so that none of their weight rests on the prime mover. See Fig. 18.34. The front axle may swivel or

FIGURE 18.32   Dump bodies on semitrailers.

FIGURE 18.33   Fifth wheel for truck trailer. (*Courtesy of Holland Group Inc.*)

FIGURE 18.34   Full trailer chassis.

may have steering linkage. It may be fastened to a drawbar that is attached to the prime mover or it carries a fifth wheel or other support for the trailer frame.

Tractors may be short coupled trucks or have more of the characteristics of large wheel tractors, there being no sharp distinction between the types. The two-wheel or four-wheel tractors used with self-powered pans may be used with semitrailers.

Trailers of either semi or full construction must be equipped with brakes. These are operated by air, vacuum, fluid, or electricity from the tractor, and are usually fixed to go on automatically if the connection is broken. This is to prevent the trailer from rolling free if it breaks away from the tractor. Brakes may be synchronized with the tractor system, controlled separately, or both.

Trailer lights are controlled from the tractor cab. Current is conducted along a multiple-wire cable with plug-ins for both tractor and trailer slotted so that connection cannot be made incorrectly.

Rear-dump bodies are usually installed on semitrailers rather than full trailers, and may be dumped hydraulically as in Fig. 18.32. The hydraulic dump conforms to the systems already described but may be much larger and heavier. Bodies may be very long and require level ground and good overhead clearance. The pump and valve may be in the tractor and connected to the hoist ram or rams by pipes and flexible tubing.

A semitrailer can be operated in more restricted areas than a truck of the same capacity. Since the drive wheels are not at the rear, it can be backed somewhat closer to the edges of fills and farther onto soft ground than a truck. However, traction is not as good, and more driving skill is required.

BOTTOM AND SIDE DUMPS

Bottom dumpers are usually semitrailers, occasionally full trailers. Tapered bodies are shaped to dump through a full-length bottom opening controlled by one or two pairs of clamshell doors. See Fig. 18.35. The doors are operated by either hydraulic or air cylinders, to slide or pivot sideward.

FIGURE 18.35   Bottom-dump trailer.

Dumping is done with the unit moving forward, without any particular limit to travel speed. The load is deposited in a windrow. Its cross section is determined by the width to which the doors are opened, and the speed of movement. Maximum height is limited by the rear clearance of the body, but this may increase as the rear wheels ride up on the dumped material.

Body walls are much lower than those of rear dumps of comparable size, which makes loading easier. Dumping on the move saves time, and windrows are easier to spread than full piles.

Disadvantages are inability to dump off edges or close against obstructions. Receiving hoppers must be made specially long, with drive-over provisions to accommodate them.

Capacities range from about 20 tons or 20 cubic yards to more.

Side Dump. Side dumps are used in building out edges of fills which are long enough to allow a broadside approach. They are lower and more stable with a raised body than a rear dump, and can be unloaded at higher speeds. See Fig. 18.36.

TRUCK SIZES

A desired hauling capacity can be obtained by a few large trucks, or a larger number of small ones. Truck fleets may contain only one size unit or a variety.

A big truck should move dirt more cheaply than a small one of the same speed, particularly on long hauls. Its purchase and maintenance costs usually are lower on a per yard basis, and there is a definite saving in driver's wages. Insurance premiums and registration costs are variable.

A big body presents a larger target for loading by a big shovel or dragline, and may increase its production as much as 20 percent in better cycle time and reduced spillage. However, a high-sided body will make loading by a small shovel or front-end loader difficult or impossible.

The large truck requires substantially more room to maneuver and, if it does not have it, may waste so much time getting placed to load or dump that its production will be smaller and more costly than that of a small truck.

One 50-yard truck in a fleet may break up a haul road that would last indefinitely under 20-yard units. Damage is particularly likely when the road surface rests on soil which supports moderate loads but either flows or compresses under great weights. Costly additional fill or surfacing may be required for protection. This problem is affected both by ground pressure of the individual tires and by the total weight of the whole machine.

Fleets made up of big trucks do not have traffic problems as frequently as more numerous small ones, and in general offer fewer problems to the dispatcher; but breakdown of one unit will cripple the fleet more seriously.

FIGURE 18.36 Side-dump bodies. (*Courtesy of Smith Co. Manufacturing.*)

Overloads.   Trucks are generally capable of carrying much heavier loads than are recommended by their manufacturers. However, overloading uses up the reserve of strength that is built into a truck's parts, and causes accelerated wear and more frequent breakdowns. Brakes and the holding power of engine compression will not be as effective or safe under additional tonnage. Overloads will slow acceleration, reduce pulling power, and lower speeds on rough ground, so that under severe conditions they will decrease production.

Conditions favoring overloads include smooth roads, hard ground in loading and dumping areas, level or easy grades, long hauls, and careful operators. Rough or soft surfaces and steep grades will slow and strain a heavily loaded vehicle, and rough or careless operation will damage it far more than if it carried fewer tons.

If the decision is to overload a light truck, and it is bought new, it should be equipped with the largest and heaviest tires the manufacturer offers and have a two-speed, heavy-duty axle. If springs are not adequate, extra leaves should be added.

The frame should be reinforced. The point of greatest stress is between the body and the cab. The usual flat reinforcing fishplate may not be satisfactory because it reinforces against vertical stresses only, and twist is an important factor in failure of frame members. Use of angles or channels produces better results. A channel inserted and welded in the frame channel is the preferred method. Fishplating should extend along at least 2 feet of frame, preferably 3 or more, and should be securely welded top and bottom. Bolting or riveting weakens both the frame and the plate.

The rear cross member may require strengthening also. Any type of fishplating, or welding a heavy pipe between the outer frame members immediately ahead of it, should give sufficient support.

The hoist should be of the largest capacity that will fit the body. Overloaded hoists need frequent repairs, which are much more costly than the extra price of the bigger hoist.

An expensive preparation for overload is installation of tandem drive, together with lengthening and reinforcement of the frame. Some makes of trucks give excellent service with tandems while others have frequent transmission and clutch trouble.

If soft or slippery ground is an important factor, conversion to all-wheel drive will increase the carrying capacity, but only at considerable expense. Wider rims and oversize tires are often installed and frames reinforced in connection with this change.

DRIVING A DUMP TRUCK

Operating a conventional highway-type dump truck is somewhat like driving a car. However, the truck is much heavier and more bulky, has very poor visibility to the rear, and usually requires special methods of shifting gears and climbing and descending hills, because of greater weight in proportion to engine power and braking capacity.

In addition, checking the truck, spotting it to load and dump, and operation of the body are special techniques not related to any experience gained in driving a car.

Checking.   It is particularly important that tire inflation and condition be checked at least daily, as the tires are the most vulnerable and most expensive wearing parts of a truck. Running them soft with a load may ruin them in a few minutes. The driver should stand clear while the truck is being loaded.

If the truck has air brakes, it should not be moved until the dash gauge shows safe operating pressure.

Direction signals and stop lights must be in working condition for driving on highways or on heavily traveled haul roads.

Stones stuck between dual tires should be removed immediately, as they may be thrown behind the truck with deadly effect. A stone can usually be pried out with a bar or board, but it is occasionally necessary to let the air out of one tire to free it.

Each load should be checked for rocks or other objects which might fall off. These may be moved to safer positions or be pushed off in the pit.

Loading.   Truck drivers may add substantially to shovel production by getting promptly in proper position. If a spotter is present or the shovel operator is giving signals, directions should be followed exactly regardless of the driver's ideas of where the truck should be. If a spot log is used, the rear wheels should be backed squarely against it.

If the shovel has not moved, a truck will generally be spotted in the same place as the last one loaded. Following tire tracks or observing the location of spilled material makes it easy to get in the same position.

In general, a truck hauling from a revolving shovel is placed to require the shortest practical swing from the digging. As in Fig. 18.37, it faces directly away from the shovel, so that the shovel can reach from back to front of the body by use of the crowd mechanism, or at a right angle so that the swing controls can be used to distribute the spoil along the length of the body.

When large front shovels are loading rock, the truck is sometimes placed facing the shovel. The open door of the bucket tends to prevent rock from rolling toward the shovel and therefore serves to protect the cab and the operator.

The distance from a front shovel should be such that the center of the body is in the middle of the crowd arc of the bucket. With draglines and clamshells it should be directly under the boom point.

A backhoe or a dragline loads most efficiently when it is on the bank and the trucks are on the floor of the pit. For draglines and hydraulic hoes, the difference in cycle time is the number of seconds needed to hoist the load to the higher level.

When being filled by a front-end loader, the truck should usually be backed against the bank at an angle of 45°, so that the loader need make only one-eighth of a turn to dump. See Fig. 18.38. The truck should be close, and the loader walking time should not be greater than the time required to lift the bucket. Wheel tractors require more space to maneuver than crawlers.

It is often necessary to move a truck while it is being filled. This is particularly true with belt or bucket loaders, which may themselves be moving, or if stationary, may not distribute the load throughout the body.

Gearshift.   On-highway trucks generally are equipped with manual transmissions, but they can be provided with automatic transmissions. In either case, the clutch is used to get into an operating gear to begin moving. The operator knows pretty well what gear or gear range is best for the travel to be done

FIGURE 18.37   Back toward a hoe for loading.

FIGURE 18.38   Good truck placement.

and so shifts into that gear. When more power is needed, the engine will be revved up to overcome the load, then the accelerator is relaxed to let the chosen gear function normally. To save braking excessively going downhill, the clutch should be used to shift into a low gear so the engine helps to reduce the speed for braking the motion. If the vehicle has a retarder gear, that helps going downhill.

Practically all off-the-road trucks have automatic transmissions with torque converters and power shifts, like the other off-highway construction equipment. The operator selects the gear range most appropriate for the travel to be done and leaves the control up to the automatic features of the transmission.

Axle Shift.   Trucks can be equipped with two-speed axles providing two gear ratios in each gear. This allows for perhaps one-third more power in each gear for differing conditions of operation. The shift to the higher power setting is done by a lever or control switch in the operator's cab. The power for this shift may be vacuum, air, or electricity.

Operator's Controls.   A typical operator's control setup is shown in Fig. 18.39. The electronic monitoring system (EMS) message center has alert indicators that light up to show various conditions to help the operator maintain the vehicle in the conditions set for successful operation. One indicator light may show overheating in the engine or transmission, another may show the need for certain maintenance. These are like the indicators found on most modern passenger cars. Other dashboard or EMS indicators may show the engine's total number of operating hours, travel distances, engine speed, number of loads since last reset by the operator, and a variety of faults the message center has detected.

Hills.   A truck with a gross weight of 16,000 pounds may use the same engine as a car weighing 3,500. Brakes, although larger, are not increased in size in proportion to the extra weight which must be controlled.

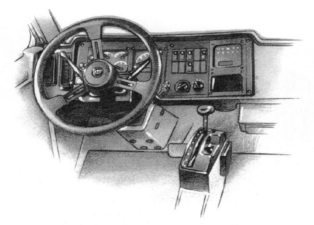

EMS Message Center

FIGURE 18.39 Operator's control center.

Hills, therefore, present two problems to trucks: pulling up and getting down safely. It is sometimes possible to rush small hills and get over them in high gear, and to descend moderate grades in high gear with the use of brakes.

Generally, however, the gearshift must be used. When the truck slows below the efficient torque point of the engine (usually between one-third and two-thirds of top speed in the gear being used), a shift should be made to the next-lower gear. If the shift is delayed too long, the next-lower gear may not be able to pull so that another shift down must be made immediately.

If the truck is overloaded or the upgrade very steep, the truck may not be able to start moving from a dead stop, so that unless gears are shifted properly it may become necessary to back the truck to the bottom again or to get help. Under such circumstances the amateur truck driver may do well to put it in low before he or she starts up.

Failure to shift properly going downhill may have far more serious consequences. If too high a gear is used, the brakes may become overworked so that the drums heat and expand away from them. If the brakes are loose, the pedal may strike the floor or require "pumping," before it can make an effective contact. If they are tight, there is still the danger that the curve of the shoes will no longer fit the expanded drums, so that not enough lining will be in contact to stop the truck. Heat glaze may destroy the effectiveness of the lining.

If a shift down is attempted when the truck is moving too fast, it will be difficult to complete because of both the speed and its mental effect, so the gears may remain in neutral. A truck going downhill out of gear is even more out of control than when in high gear.

Proper procedure is to slow to a near stop before starting down a grade and to shift into a gear low enough to hold the truck to a safe speed with only occasional braking. If the speed still increases too much, a near or full stop should be made before a shift into a lower gear.

Downhill speed should be kept within the normal range of the gear being used. If the engine is allowed to "wind up" past its maximum operating speed, severe damage may result. If the truck is allowed to coast rapidly in a low gear with the clutch out, the clutch may be wrecked by centrifugal force. See Fig. 18.40.

Hill problems are directly related to load. An empty truck needs only a little more care than a passenger car, while a severely overloaded one may be unsafe even with the most careful handling.

Curves. Curves must be taken more slowly with a load than without. The truck should be slowed before entering the curve and accelerated or held at an even speed while in it.

When rounding a curve, the centrifugal force tends to pull the truck toward the outside, and this force is resisted by the friction of the tires on the road. If traction is good, the truck will tilt,

FIGURE 18.40   Hauling out of a mine.

compressing the outside springs and tires. Increase in load will both raise the center of gravity and increase the force tending to tip or slide the truck.

The force required to turn the truck is exerted by the front wheels, which slow the truck and cause it to nose down very slightly. The outside wheel does more than half of the turning work because the tip of the truck adds to the weight it is carrying. This wheel is therefore under the greatest strain of any part of the truck.

When brakes are applied, the forward momentum of the truck is resisted by the wheels. This transfers weight forward, compressing the front springs and relieving the rear ones. This increases the burden of the front wheels, particularly the outside one which may already be overloaded.

Compression of the outside front spring by both turning and braking forces throws the truck off balance and makes it difficult to control. The overloaded wheel may roll its tire, hop, or skid.

Accelerating the engine causes the rear wheels to push forward against the inertia of the truck, which tends to make the truck nose up, transferring some weight from the front wheels to the rear ones and improving stability.

If the road is banked, the centrifugal force acts to pull the truck to the pavement so that stability is improved. Reverse banking, as in a left curve on a crowned road, will cause it to pull the truck away from the road.

Skids.   In normal operation the direction in which a wheel and tire are rolling controls the course of the vehicle, and considerable resistance exists against forces tending to slide the wheel sideward. This resistance decreases if the wheel is turning either faster or more slowly than required by the speed of the truck. If the wheel is locked, it moves sideward almost as readily as along its plane of rotation.

If the front wheels are locked while rounding a curve, they will no longer turn the truck, which then moves straight ahead. This type of skid most often occurs when brakes are applied too hard, but on very slippery surfaces the wheels may lock when turned sharply without brakes. When the brake is released, the wheel may revolve again, or may remain locked until turned in the direction of the skid so that its friction with the road tends to revolve rather than to twist it sideward or bend it.

When the front wheels are again revolving, they will substantially control the direction of the movement of the front of the truck until again locked by brakes or too sharp a turn.

If the rear wheels lock on a curve and the front ones do not, centrifugal force causes the rear of the truck to slide outward, pivoting on the front axle. The brakes should be immediately released and the front wheels straightened, so that the side pull of the turn will be stopped.

If the rear wheels have skidded too far sideward to be started rolling in this manner, the tendency of the truck to spin may be checked by steering sharply in the direction of skid. The front wheels will travel in the same direction as the back is sliding, and the truck will move diagonally sideward. The front wheels will tend to get in the line of skid, and the rear wheels to start turning. If the engine stalls, it should be started immediately.

Should the rear wheels slide due to spinning, traction can usually be restored by releasing the accelerator and pressing it down gradually.

Sometimes it is more important to avoid collision with a person or a particular object than to straighten out the skid. Also, once a truck has started to "pinwheel," straightening it out is a matter of skill and luck rather than rules. The safest system is to avoid skidding by driving slowly and cautiously when roads are slippery.

Backing. Most fills using rear-dump trucks require turning and backing to the edge of the fill. If possible, a fill should be arranged so that the turn is made near the dump spot; the turn spot should be wide enough that reverse gear need be used only once; turns in reverse should be toward the driver's left; and the truck should be level or facing uphill while dumping.

If the truck has a two-speed axle or an auxiliary transmission, two reverse speeds are available. Low should be used when the load is heavy, the ground soft or rough, or complicated steering is required.

Vision to the rear is blocked off by the body. The cab guard may be solid, so as to render the rear window of the cab useless. If there is no guard, or it has eyeholes, view of the ground will still be blocked by the load or the tailgate.

A view to the left can be obtained by watching the mirror, or better by leaning out of the cab past the side of the body. Unless the body is very narrow or the cab very wide, it may be necessary to open the door and stand on the left-hand running board. The gas and brake are controlled by the right foot, and the steering is done by both hands or the right hand alone. To stop, the driver swings back onto the seat in order to use both feet for clutch and brake.

The driver can see to the rear of the right side of the truck in a right-hand mirror, or directly by partly raising the body to look between it and the frame.

Under many circumstances, such a view to the right is not sufficient for accuracy. The route must then be plotted ahead of time, either before or while backing, so that steering can be done in reference to the left wheels with enough space left to keep the right ones out of trouble.

Dumping. When dumping off the edge of a fill, the driver should back so that both rear wheels will be the same distance from it, Fig. 18.41(A), rather than at an angle to it, as in (B). If one wheel sinks in more deeply than the other, it may not be possible to either dump the truck safely or pull out with the load.

Safe distance from the edge is determined by circumstances and the judgment of the driver. If the fill is shallow or of firm material or the truck has tandem or all-wheel drive, a very close approach is possible. Certain rear-dump semitrailers can be backed over the edge. On the other hand, it may be necessary to keep 6 or more feet back if the fill is soft, slippery, sandy, or otherwise treacherous. Very high fills should always be treated with respect.

Another factor is the means used to spread. If one or more large bulldozers are spreading and easily keeping ahead of the loads, the truck need not risk a close approach and the danger is reduced as it can be pulled out readily if stuck. If spreading is done by hand or by overworked machines, dumping over the edge will speed the job, but it will be less convenient to extract the truck if stuck.

When the truck is in position to dump, the tailgate latch is released; the transmission power takeoff engaged, with use of the clutch and a hand control; and the hoist valve placed in UP position. The engine is accelerated to a moderate speed but not raced.

Most hoists are constructed so that they can be left in the UP position to hold the body at its high position. However, if this appears to strain the mechanism, the valve may be moved to an intermediate HOLD position as soon as the body is at its steepest angle.

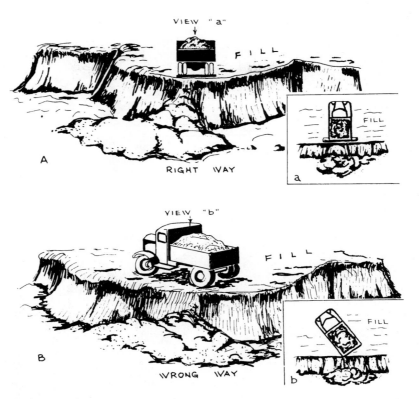

FIGURE 18.41 At the dump.

As the body rises, the load slides backward along the floor and under the tailgate, Fig. 18.42(*A*). Unless right at the edge of a bank, the load will tend to pile up until it blocks the gate, as in (*B*). The truck is then placed in low and moved forward without disturbing the hoist controls until there is space for the remainder of the load to slide out.

If part of the load sticks in the body, the truck may be backed into the pile to shake it loose. This may have to be done several times if the load is sticky or the dumping is uphill. The clutch should be released just before the wheels hit the pile to avoid shock to the power train. If the truck rolls away from the pile on the rebound, the shaking out can be done without shifting out of reverse; otherwise low gear is used to drive forward a few feet and reverse to hit the pile.

If any considerable part of the load cannot be shaken out, it must be dislodged with a hand shovel or other tool. Most of the body can be reached from the pile and tailgate, but the upper end may require climbing up on the body running boards. Sticking can be reduced if the shovel operator will put the driest material in the bottom of the truck so that it will slide out, carrying the wetter dirt.

If the load includes rocks, stumps, or mats of trash large enough to have difficulty going under the tailgate, the gate should be left latched at the bottom and unfastened at the top so that it can drop down and the load slide over it. However, it is not safe to back against the pile to shake out the load with the gate down, as it will be bent or knocked off. If much bulky material is to be carried, it is good practice to take the gate off and use higher sideboards, so that a full load can be piled toward the front.

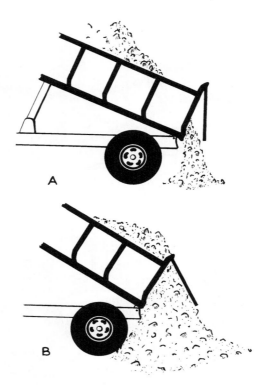

FIGURE 18.42 Dumping a load.

If the tailgate becomes twisted, it may jam the upper hinge pins so that they cannot be removed. In this case, the bottom hinges should be released. This will relieve tension on the upper pins which can be removed while the gate is held in position by hand. The bottom is then relatched, and the gate is lowered.

Loads put more strain on the hoist when heaped in front than when evenly distributed.

If dumping is done with the gate hinged at the top in the usual manner, and the load jams instead of sliding under it, the upper part of the body usually empties and the load piles up at the back.

The weight of dirt makes it difficult to remove the obstruction or the tailgate. Sometimes the body can be lowered by driving forward and applying the brakes sharply, with the control in LOWER.

A stuck load may overbalance the truck, so that the front wheels will rise several feet off the ground. This usually does no particular harm, but gives the driver quite a surprise.

A truck in this position will usually settle down if towed in the conventional manner by front bumper pull hooks. If it does not, the line should be attached to a front corner of the body. If the load is dislodged, the truck may right itself suddenly.

A low lift makes it difficult or impossible to dump uphill, so the truck should be on a level or faced uphill (so the load will slide downhill) whenever it can be arranged.

Trouble in dumping also comes from defects in the hoist. If it acts normally except that it will not lift to full height, the trouble is apt to be lack of oil. In this case, the body should be blocked up and oil poured into the system. A quart or two generally is sufficient.

If the lift is slow and tends to creep back down if the engine is idled, the hoist control lever may have slipped. Sometimes it can be made to work properly in a different position, and at other times the linkage must be adjusted.

A loaded body should not be raised into dumping position unless the rear wheels are nearly level. Any slant will cause the center of gravity of the body to shift downhill as it is raised. The twisting force exerted may break the body hinges, tear the subframe off the chassis, or overturn the truck, as shown in Fig. 18.43.

When the dump is completed, the body is lowered and the latch closed. Often this is forgotten, with resulting embarrassment in the pit when part of the next load spills through. If the gate is sprung out of shape, or material sticks in its way, it may not latch on the first attempt.

Spreading.   The load may be spread over a considerable space instead of being dumped in a pile. The procedure is to start the dump in the manner described, and when the body is high enough to begin to spill, put the truck in gear and drive forward while continuing the hoist. If low gear is used and the hoist is fast, the spread will be short and heavy. If the hoist is slow, or is so constructed that it can be slowed by opening the valve only partially, the material will be spread more thinly.

Use of higher gears for the truck will also thin the spread. If the hoist cannot be set partially open, and if the highest gear that will move the truck is not fast enough for the spread desired, the hoist valve can be opened and closed as the truck is driven forward.

This type of dumping is liable to result in a series of piles rather than a smooth sheet. If the load will flow freely and contains no lumps, a much smoother spread can be obtained by chaining the bottom of the tailgate to the body so that it will open only a few inches.

When the body is raised so that the load starts to pour out under the gate in a sheet, the truck is driven forward. The body is hoisted just enough to keep the opening fully supplied, as raising it to full height might cause part of the load to slide over the gate or might overbalance the truck.

With any one material, the thickness of spread will be determined by the gate opening and the speed of the truck. A larger opening will increase the thickness, a higher speed decrease it.

Surface Dumping.   Dumping on the surface of a fill is usually arranged so that it can be done without backing. Rear-dump trucks may be used, although their height in full-dump position makes it advisable to keep on smooth ground and move slowly. They may spread the load or dump in piles.

Such surface building is usually done by scrapers, or by bottom- or side-dump trucks or wagons.

The bottom-dump wagon moves rapidly while dumping. The thickness of the spread is determined by the relation between the speed and the width of the door opening. The maximum height of the windrow is determined by the clearance of the wagon.

The wagon may be followed immediately by a dozer, which spreads the material in a single pass, or a number of windrows may be built both behind and beside the front one. The latter way

FIGURE 18.43   Dumping on a sidehill.

is economical of dozers but is liable to slow down the wagons and might cause upsets because of the necessity of traveling on the rough fill.

A side dump may be used on the surface of the fill in the same manner as a bottom-dump wagon, or dump off an edge which is long enough to permit it to get broadside to it.

TRAILER OPERATION

Semitrailer operation differs from truck driving in backing, and in precautions needed when steering sharply and when stopping.

Very sharp turns may be made in proportion to the length of the whole unit, particularly when the tractor is a short coupled, off-the-road type, and the trailer yoke is high enough to allow space for the tractor rear tires under it. However, turns must be made carefully to avoid overrunning and jackknifing.

In a solid truck only the front wheels are turned, and the momentum of the truck pushes against them at their center of revolution. The light weight and fairly positive control of the wheels, impossibility of turning the wheels beyond their stops, and the low line of push make this a rather safe and stable arrangement.

A sharp fast turn with a semi is dangerous. When the tractor is turned, the weight of the trailer pushing straight ahead tends to skid the driving wheels to the outside of the turn, increasing its sharpness, so that the unit may jackknife with the two parts, still connected, trying to move in different directions. This may do no more than cause a tangle that will be difficult to straighten without a tow, but it can result in serious damage, overturning both units and tearing the hitch apart.

Jackknifing will also occur as a result of a fast stop in which the tractor brakes are stronger or find better traction than the trailer brakes.

In rounding a curve, the tractor rear wheels will track toward the inside of the curve from the front, and the trailer wheels will track still farther in, as shown in Fig. 18.44. Swinging turns wide enough to keep the trailer clear and avoiding stops sudden enough to cause jackknifing are special steering requirements for forward driving.

Jackknifing is a common occurrence when backing a semi, but is then frustrating rather than destructive.

When a semi is backed around a curve, the trailer wheels will move toward the same side as the front wheels, and any difference in direction between the two sections increases as the tractor is backed straight. Jackknifing is more apt to occur if the trailer turns sharply and the tractor slowly, rather than when the reverse is true.

If the swivel point between tractor and trailer is any noticeable distance behind the tractor rear axle, the swivel moves in an arc as the tractor turns and accentuates the steering effect on the trailer.

In backing down a hill the trailer brakes should be released and those of the tractor applied sufficiently to create a drag. The trailer has greater tendency to roll straight when pulling than when being pushed.

Full trailers are much more difficult to control than semis. Most drivers cannot back them more than a few feet.

In general, full trailers should not be used where backing is required. Emergency measures for getting one out of a blind street include unhooking so as to tow it backward.

Rear-Dump versus Bottom-Dump Trucks.   Use rear-dump trucks when

1. The material hauled is large rock, ore, shale, etc., or a combination of free-flowing and bulky material.

2. Dumping is into restricted hoppers or over edges of a waste bank or fill.

3. The hauling unit is subject to severe loading impact when under a large shovel, dragline, or loading hopper.

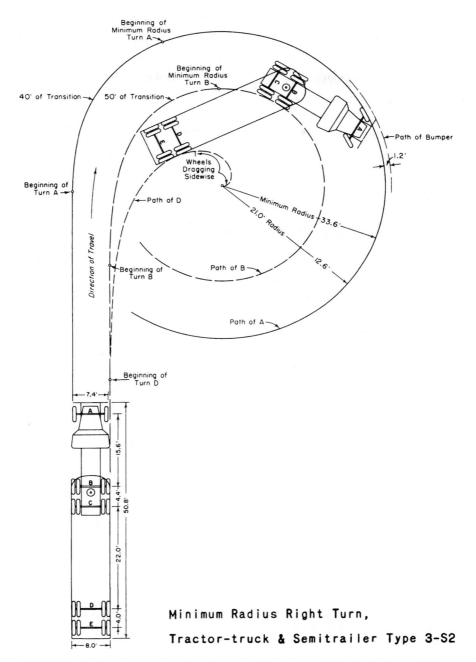

FIGURE 18.44 Tire tracks of semitrailer. (*Courtesy of Department of Public Works, State of California.*)

4. Maximum gradeability and rapid spotting in a restricted area are required.

5. Maximum flexibility is required for hauling a variety of materials under variable job conditions.

Use bottom-dump semitrailers when

1. The material is free-flowing.

2. Maximum flotation of the large single tires is required.

3. The haul is level, allowing high-speed travel.

4. Dumping is unrestricted into a drive-over hopper, or the load is spread in windrows.

5. Long adverse grades do not exceed 3 to 5 percent. This is recommended as a general rule for optimum performance, but is by no means a measure of the actual gradeability of bottom-dump semitrailer trucks.

MACHINERY TRAILERS

CONSTRUCTION

Most excavating machines are not designed to travel under their own power on highways, particularly for long moves, and are carried from job to job on machinery trailers. These are manufactured in a wide variety of sizes and types, of which only a few can be described here.

A semitrailer has a rigid drawbar supported by the towing truck. This may carry the direct weight of more than half the trailer and load, if the wheels are in the rear, or only serve to stabilize against rocking, if the axle is under the middle of the load. Full trailers are usually supported in front by a swiveling axle and connected to the truck by a comparatively light hinged draw tongue.

Any trailer should be equipped with brakes and lights operated from the truck or tractor cab. Even a small one can produce an unpleasant mess by jackknifing during an abrupt stop, and the side shift of its weight may put the truck out of control.

A trailer must have a deck or carrying space large enough to support or hold the machine to be carried; it must be strong enough for the job; and it must provide for loading and unloading with a minimum of difficulty and danger. It should be pulled by a truck or tractor amply powerful to climb any grade it may meet, and have brakes adequate for stopping it going down.

In general, a trailer should be purchased which is of sufficient size for the biggest and heaviest machine to be regularly carried. For occasional use most trailers will take considerable overloads, particularly if oversize or extra-ply tires are installed, but frequent overloading is almost sure to result in sagging beams which are very expensive to repair. Braking effectiveness may be dangerously reduced by too much weight.

A limiting factor in trailer size is often the power of the tractor or truck that is available. It is a far too common practice to use small units to haul disproportionate loads, with resultant rapid wear and breakage in the power train, embarrassing failures to get up hills when the engine is a little below par, and the danger of disastrous runaways on downgrades. Big, rugged trailers are heavy, particularly when in the full-trailer deck construction, and their own extra weight may make a critical difference in the ability of tractors to handle the load.

Another factor is the size and weight of ramps. In deck jobs, height increases as the span from rear axle to front support gets longer, and as load capacity increases. Larger tires also raise the deck. Higher decks require longer and heavier ramps or more dependence on loading from a bank. A crane or loader, or two to four workers, may be required to place and remove ramps to a high trailer.

Maneuverability may be critical. From this standpoint semitrailers are much preferred. When in small sizes, an experienced operator can spot them almost as readily as the truck. Large ones may find themselves in difficulties in country roads and residential districts.

A small trailer is much cheaper than a big one to buy, maintain, license, and insure; it gets around more easily, does not require as much towing power and weight, and can be more readily equipped with a low deck, or the means for reducing or eliminating the need for ramps. Under some circumstances it may be good practice to own one of them and to hire a big one for infrequent use. At other times, the small unit may pay its own way by saving a big one from being worn out running around carrying small machines.

Deck Trailers.   Figure 18.45 shows a medium-weight trailer in both level deck and drop deck models. Capacity range is 15 to 40 tons.

The trailer proper consists of a girder frame supporting a flat wood deck, usually hardwood planks 2 inches thick. Standard width is 8 feet. It is supported at the rear by four dual wheel sets (eight tires), and at the front by a welded gooseneck.

When the unit is used as a semitrailer, the gooseneck is attached to the "fifth wheel" of a tractor truck. For full trailer operation, it rests on a dolly.

The wheels may be mounted on a pair of walking beams, left and right, to allow front to rear oscillation to maintain road contact and reduce jolting. The tops of the tires are even with the top of the deck on level ground. Cutaway deck sections permit them to go higher as they oscillate. See Fig. 18.46.

In another design, there are two axles in tandem, with leaf springs and torque arms. See Fig. 18.47.

Wheels are 15 inches. Tire sizes are 8:25 up to 11:00, 10 to 16 ply. Level deck height may be 28 to 36 inches, differences depending mostly on tire size. Drop decks are 6 to 9 inches lower.

Air or vacuum brakes are used on all four hubs. These are controlled from the truck cab through piping and flexible hoses with snap-on couplings. They are designed so that if the hoses become disconnected, the trailer brakes will lock on automatically to prevent a runaway. They can be released by bleeding the air or vacuum tank, or may release very gradually by leakage.

The dolly consists of a towing platform, spring-mounted on a single axle. Two perforated discs are used in the swivel connection to the gooseneck to hold grease, to increase bearing surface, and to eliminate sway or whip. See Fig. 18.48.

FIGURE 18.45   Deck trailer. (*Courtesy of Rogers Brothers Corporation.*)

1. Bearing plates   4. Walking beam
2. Bronze bushings   5. Wheel bearings
3. Lube reservoir   6. Brake shoe
        7. Detachable drums

FIGURE 18.46 Walking beam suspension. (*Courtesy of Rogers Brothers Corporation.*)

FIGURE 18.47 Tandem wheel oscillation. (*Courtesy of Rogers Brothers Corporation.*)

FIGURE 18.48 Trailer dollies. (*Courtesy of Rogers Brothers Corporation.*)

The draw tongue is hinged to the base of the platform. A pair of breakaway chains are attached to the frame, and can be hooked to the back of the truck to hold the trailer if the tongue should become disconnected.

If the trailer is to be used both with and without a dolly, the dolly can be equipped with a tractor-type fifth wheel instead of the standard turntable. The tongue is then fitted with a bracket so that it can be made to hold the dolly upright during the change.

Dollies are also available with two rocking axles, each carrying two or four tires.

Level deck construction is shorter, provides more unencumbered deck space, and is somewhat easier to load and unload from the back. The drop deck lowers the overall height of the loaded machine 6 inches or more and is easier to load from the side, but reduces road clearance.

The sloping deck section behind the wheels, called a beavertail, may be standard or optional equipment. It reduces ramp requirements for loading from the rear. It has a bracket on its rear face, usually full width, to hold metal angles on the ends of ramps. If there is no tail, a similar bracket is built into a rear bumper.

Heavier trailers may have three or more axles. The model shown in Figs. 18.49 and 18.50 is convertible between two and three. The rear and the frame and deck above it are hinged to the main frame. In working position the rear axle carries its full share of the load. When not needed, it can be unlocked, then swung up, pivoting on top pins, so that it lies upside down on the deck. It can be disconnected and removed by pulling the pins and uncoupling the brake and light lines.

FIGURE 18.49   Flip-up third axle. (*Courtesy of Rogers Brothers Corporation.*)

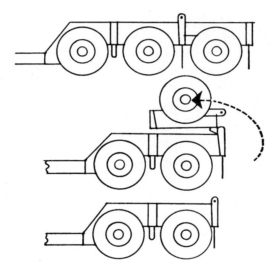

FIGURE 18.50   Flip-up axle changes. (*Courtesy of Rogers Brothers Corporation.*)

Reduction of overall length in this way reduces the need for or cost of overlength permits, makes the unit more maneuverable, and reduces tire wear.

Detachable Gooseneck.   Some medium to heavy deck trailers are made so that the gooseneck can be removed, leaving the front of the frame resting on the pavement or ground. Machinery can then be loaded readily using very short ramps (usually hinged to the deck) or small blocks. See Fig. 18.51.

The gooseneck and frame are held in alignment by removable pins or other safety locks. With these released, and brake and electric lines opened, the two units are lowered to the ground by a hydraulic jack in the gooseneck, or by a line from a winch mounted on the tractor. The gooseneck is then detached from the frame, and carried or dragged a short distance by the truck.

FIGURE 18.51 Detachable gooseneck. (*Courtesy of Talbert Manufacturing, Inc.*)

Ramps are flipped over to rest on the ground, and the machine is driven up onto the trailer. The gooseneck is backed into place, attached, lifted, and locked, lines are reconnected, and the ramps are folded back onto the deck. See Fig. 18.52.

Girder or I-Beam Trailers. One of this type of trailer is shown in Fig. 18.53. It is often not practical to transport shovels on deck trailers, because the combined height of trailer and shovel is too great to go under bridges on the route. This difficulty may be reduced by the use of a girder-type trailer which supports the frame of the shovel and allows the tracks to hang within a few inches of the ground.

Girder construction also eliminates the necessity of using long ramps, which may be so heavy as to require a crew to handle them, and which are dangerous to use when slippery.

The tracks of a shovel on a girder trailer hang very low and may rest on the road on sharp humps or hill crests. If the tracks are not locked, they will reduce resulting drag by turning. If the machine hangs up badly, it can be walked forward while the trailer is towed at the same speed until the trailer frame will raise the tracks clear of the ground again.

Loose tracks can be prevented from hanging below the rollers by raising the upper section with a jack on the truck frame, or by working pieces of wood in between the track and the bull wheel by turning it.

Occasionally the tracks must be removed to obtain sufficient clearance from the ground when the machine is carried in the conventional manner.

Low-Bed Trailers. Figure 18.54 shows two sizes of low-bed semitrailers which are used to carry machines weighing 10 tons or less. In order to keep the overall width down to 8 feet, the deck must be limited to a width of 6 feet 4 inches.

In the lighter model, the wheel stub axles and brake backplates are welded directly to the deck rails. The tandem wheels in the larger size are mounted on rocker arms, allowing front and rear oscillation.

The draw tongue is adjustable for height, and is fastened to the truck by a vertical pin through a bracket on the rear cross member. Connection is made by jacking the trailer and tongue to the

FIGURE 18.52   Trailer bed with ramps folded back. (*Courtesy of Talbert Manufacturing, Inc.*)

required height, backing the truck so that upper and lower holes in its tow bracket line up with the tongue hole, and then placing and locking the pin. The pin is a loose fit in the hole, so that the trailer can tip somewhat without binding on it.

A built-in hydraulic jack can be obtained as optional equipment. Ordinarily the trailer is blocked up when left, so that it need not be jacked to reconnect.

Deck height is 14 inches. Length is 14 to 16 feet. Load is about 60 percent on the pintle hook of the towing truck.

Ramps are not required for loading crawler machines, and very short ones are needed for wheeled vehicles. A pair of blocks placed behind the trailer will cause tracks to rear up so as to pass over the rear edge of the deck in loading, and will ease them down in unloading.

Tilt Trailer.   In a tilt trailer the deck is hinged to a subframe that connects the gooseneck or drawbar to the axles. The deck can be tilted so that the back edge rests on the ground. See Fig. 18.55. A machine being loaded can walk directly up the sloped deck as if it were a ramp. As its center of gravity passes the hinge, the deck automatically goes back to level position.

To load one of the simplest and smallest of these units, the trailer is attached to the truck, the deck lock is released, and the front of the deck is pulled up by hand until the rear rests on the ground. The machine is then driven from the ground directly up the sloping deck. When the machine has moved past the hinges, the deck will swing into carrying position and will lock automatically.

The machine must be moved very slowly as it starts to overbalance the deck, which is liable to come down so rapidly as to strike a heavy blow on the lower frame, and to subject the towing truck to shock. Damage to the truck can be prevented by supporting the tongue on blocks or a standard during loading.

If blocks are jammed under the tongue, moving the machine to the back of the trailer, with the deck locked, will release them. For transportation, the machine may be moved well forward so that the deck will remain level even if the lock fails.

FIGURE 18.53   Loaded girder trailer, with dimensions. (*Courtesy of Talbert Manufacturing, Inc.*)

Most modern tilt trailers are equipped with a hydraulic ram or other snubbing device, so that the deck will move smoothly from loading to travel position. Heavy units may start the tilting motion by hydraulic power. Otherwise, they may be tilted by standing on the rear.

Large tilt trailers (maximum capacity about 25 tons) may have a regular gooseneck attachment to a dolly or fifth wheel. Smaller units have a rigid two-bar attachment to a special bracket on the rear of a dump or other truck. The bar usually includes a jack for raising it to the truck, and for supporting it during parking and loading.

There may be a full-width solid deck over the wheels, requiring a deck height of 34 to 41 inches. Slope in loading position may be 15 to 18°, too steep for a weak machine or a slippery deck. This angle can be reduced by putting blocks under the edge before lowering it, and putting blocks or very short ramps behind it.

LOADING

Rear Loading.   It is standard practice to drive self-propelled machinery onto deck trailers over the back with the aid of ramps, blocks, or banks, or combinations of them.

FIGURE 18.54 Low-bed trailer.

FIGURE 18.55 Narrow-tilt deck.

Ramps. Ramps may be made of planks of oak or other strong wood, 2 to 5 inches thick (three is usual), and 10 to 16 inches wide. A metal angle is fastened under one end of the plank, and rests in an angle on a shelf on the rear of the trailer.

Ramps may also be made of steel box or channel construction, or of wood reinforced with steel plates or angles.

They are ordinarily so heavy that it takes at least two people to handle them, so that it is often advantageous to place them with the machine that is to be loaded. However, they are usually not strong enough to take the weight of a heavy machine unless supported between the ends. Short pieces of heavy timbers, railroad ties, or heavy sawhorses can be used to block them up.

In loading crawler machines, the point of greatest strain is near the ground where the tracks are forced upward into a climb. In unloading, bracing is particularly needed where the machine strikes the ramp as it pitches off the trailer deck. In addition to these special points, it is good policy to have blocking under one or two more points to reduce the length of unsupported spans.

Thorough blocking will make possible the use of lighter ramps and add materially to their life.

Ramps supplied with a trailer are designed to load most machines on level ground. Length should be at least 3 feet for each foot of trailer height. The incline should be considerably less than the machine is capable of climbing.

Rubber-tire machines impose a concentrated but fairly even strain on ramps. Rollers must have very slight gradients and need help from dirt piles or banks to get on a deck trailer.

Crawlers may be loaded either forward or backward, but are under better control when backing uphill, particularly if the tracks are loose.

Ramps are set to line up with the tracks or wheels of the machine, which is placed so that it can move up them without turning. If the ramps are wider than the tracks, ramps are set so that the operator can see an edge and steer accordingly. If they are narrow so that the operator cannot watch them, a competent worker should stand on the trailer or the ground and guide the operator by signals.

The machine should be in its lowest gear and should move slowly, but with sufficient throttle to avoid any chance of stalling. It should not be steered sharply or stopped except in emergency. A loading sequence is shown in Fig. 18.56.

When the top is reached, the upper end of the tracks will move upward above the tires and deck, and fall when the machine overbalances. A jarring fall can be avoided by moving very slowly. A sloping extension on the rear of the trailer will provide an intermediate grade between the ramp and deck, which reduces the abruptness of the fall.

If the ramp is short, it can be extended by blocking, as in (E).

Wet ramps are dangerous and should be sprinkled with sharp sand. Steel surfaces may be too slick to climb even when sanded. Machines should never be loaded up ramps when there is snow or ice on the ramps, the deck, the tracks, or the ground where the machine is standing.

Since the machine must walk over the tires after leaving the ramp, tracks should be inspected and any sharp pieces of metal removed. Grousers will not damage tires, but some varieties of ice cleat might when new.

If the trailer can be backed against a bank, or into a deep gutter, it may be possible to load it without ramps, and perhaps even without blocks. Lower banks or shallower gutters may be useful in making it possible to ramp up at an easier gradient than from a level.

Pieces which are wider than the trailer may be jacked and blocked up, the trailer backed under them, and the blocks knocked out.

Protection of Pavements.   When crawler machines are loaded or walked on paved streets, the road surface must be protected. Damage can be caused by side-scuffing during turns, and by digging-in of grousers.

Crushing is particularly likely on oiled gravel when the subgrade is wet, or on any blacktop in hot weather. It is guarded against by laying planks or boards as in Fig. 18.57, to spread weight over a wider area. Note the extra boards where the front of the tracks tends to dig in just below the ramps.

In many localities, it is unlawful to walk crawler machines on streets, but nothing is said if no visible signs of the move are left. Bruising of pavements, and scuffing on turns, can be prevented by laying thin boards under the tracks. A single thin board, narrower than the track, may protect hard pavements, while softer ones may need wide planks, or narrow planks in pairs.

If the walk is long, the same boards can be used repeatedly.

Digging-in of grousers may be prevented by walking on boards in the same manner.

Side Loading.   It is often more convenient to load from the side. Normal curb and sidewalk heights above the gutter permit use of shorter and lighter ramps. Drop-deck models are lower at the side than at the back. In addition, the machine may have to enter or leave the job through a driveway or other narrow access. By side loading, it can be walked directly into the drive, whereas rear ramps will put it on the street pavement and necessitate a right-angle turn, often under tight conditions. Also, it frequently is possible to pull up alongside a roadside bank, where backing into it would not be practical.

Trailers are not ordinarily equipped with ramp-supporting rails on the sides, but they can be welded on easily. There is usually enough recess in the girders to do this without increasing width.

Unless it is blocked up, the trailer will tip sideward under the weight of the machine as it climbs on. This simplifies loading from blocks or a low bank, and may make it possible for a large

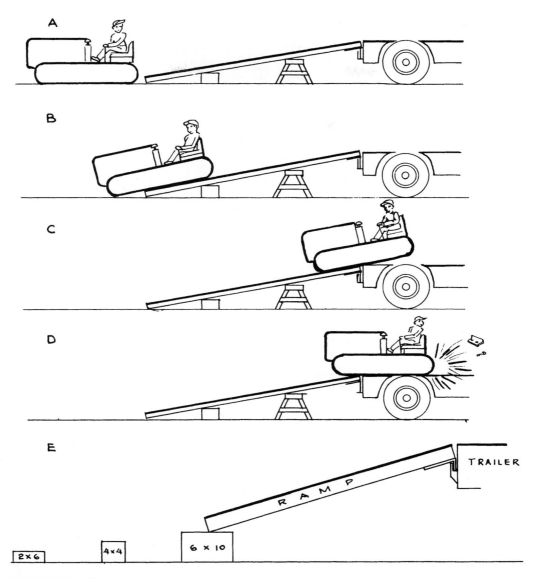

FIGURE 18.56 Climbing a ramp.

tractor to walk on from ground level, pulling the trailer down with its grousers. But tipping may make unblocked loading dangerous or impossible if the deck is slippery with rain.

The machine can be carried crosswise or turned on the deck. If the machine is large in proportion to the deck space, turning may be difficult or impossible. Machines narrower than the deck can turn rather readily, unless the planking is wet and slippery.

The front corners of the deck should be blocked up before turning, but the blocks must not be so high that the deck will rest on them when loaded.

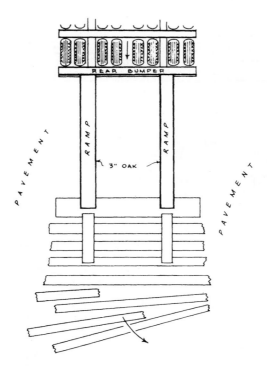

FIGURE 18.57   Protecting pavement.

Transporting.   The amount of load on rear trailer tires, and on those of the dolly or tractor, can be adjusted by placing the machine toward the front or rear of the deck. Placement will also be affected by the need to rest the bucket or blade, and the direction in which the boom is to be carried.

No part of the load should rest over the tires, unless high enough to allow them to oscillate without hitting.

A front shovel bucket may be tucked under the boom on the deck, rested on the gooseneck or between the rear tires, or, if the tractor is a dump truck, in its body. Long booms are usually faced to the rear and carried low, with the bucket or hook resting on the deck. It may be necessary to have the operator stay with a large shovel in transit to raise, lower, or swing the boom to avoid obstacles.

When the machine is correctly positioned, it is left in gear with the clutch engaged and the brakes locked. The ramps and blocks are loaded onto the deck, and the machine is secured against sliding by blocks or chains, or both.

Chains may be fastened to the four corners of the machine, and to the "D" loops on the trailer sides. At least two of these chains should be tightened and locked with load binders. Sometimes a machine can be lashed by passing a chain from a loop through or over it to a loop on the other side, and tightening it.

Blocks should be placed at each end of both tracks. The ramp blocks or the ramps themselves can be used, but triangular blocks or ones curved to fit the track or tire are much more effective. These may be wedged in place, or nailed down lightly.

Many machines are moved without any chaining or blocking, particularly on short trips. Crawler tractors tend to stay put very well. Shovels, because of their height, are less stable. Rubber-tire vehicles tend to bounce and creep.

Blocking is recommended for all machine transportation, since a minor accident may become a serious one if it slides a heavy unit off a trailer. In some places, truckers are required to block and chain trailer loads.

In most states, truck and trailer widths are limited to 8 feet. Wider vehicles must obtain special permits, which may be limited as to route and time of day, and which may also be void while streets are wet or icy. In practice, many localities do not enforce width restrictions for local moves. However, if an accident occurs to an overwidth vehicle lacking a permit, the operator may be in a difficult position.

Drivers of full trailers should be sure of their route so as not to get into blind streets. The average driver cannot back such a rig any distance, or turn it around in restricted quarters. It may be necessary to disconnect the towing truck, turn that around, and tow the trailer out backward. One or two people might be able to steer the trailer by manipulating the drawbar. If the load is a revolving shovel, it can pick up the tongue with a chain and steer by swinging. A jeep or wheel tractor following the trailer can sometimes manipulate the drawbar successfully.

Another method of getting out of a blind street is to unload the machine and drag the empty trailer sideward until it faces the exit.

Trailer routings must also be planned carefully to avoid underpasses too low to get through, and bridges too weak to cross over. Long, low trailers can also get hung up crossing banked highways or railroad tracks and on sharp hilltops on country roads.

Trucks.   Machinery may also be transported on trucks with either flatbed or dump bodies. This gives better maneuverability but involves the use of longer and heavier ramps or dependence on banks. It has the disadvantages of the labor for handling the ramps, the danger in using them, and the greater height of the machine when loaded.

When a platform body is used, it is helpful to extend and curve it down at the rear. This decreases the ramp length or bank height required, and increases stability by reducing the angle between deck and ramp.

A dump body can be extended by making braces to support the tailgate. These braces should be made to slide or swing out of the way when not in use, to allow hanging the tailgate down if desired. Tailgate clamps cannot support machines safely.

Hoist Hitches.   Rubber-tire machines are often transported by raising either the front or the rear and attaching it by a swivel to the rear of a truck or trailer.

If the front wheels are on the road, they must be securely fastened to roll straight.

Serious damage may be done by towing with the rear wheels on the road and the engine stopped. Tractor transmissions may depend on turning of the clutch (input) shaft for lubrication, and this does not turn with the rear wheels in neutral.

Towing speed may be so much faster than that normal to the tractor, that power train parts may burn out or even fly apart.

Machines with automatic transmissions must not be towed with drive wheels on the ground unless the driveshaft is removed.

Cleated tires wear irregularly on hard pavements, and may overheat and blow at ordinary highway speeds.

The usual way to make connection is to place a pair of inclined blocks behind the truck, and then drive the machine up on them until the hitch can be locked. The truck is then backed a few inches, and the blocks are picked up. Unhooking is done by pulling the machine up on blocks.

The hitch may not allow a sharp angle between truck and tow, so caution must be used in making turns and in backing.

CHAPTER 19
GRADING AND COMPACTING MACHINERY

GRADERS

MOTOR GRADER

The grader, motor grader, or motor patrol, is a machine used principally in shaping and finishing, rather than in digging or transporting. It is available in sizes ranging from 35 to over 600 horsepower, in a variety of constructions.

Except for a few light models that are rear-mounted tractor accessories, a grader consists of a wide, controllable blade (moldboard) mounted at the center of a long-wheelbase, rubber-tire prime mover. The blade is usually held at an angle to travel direction. It will cut, fill, and sidecast. These functions can be performed separately or at the same time, in any combination.

A grader may be used as a prime mover or carrier for a number of accessories. These include snowplows and side wings, belt loaders, arm-mounted clearing tools, bank slopers, trailer drags, and disc harrows, in addition to scarifiers or rippers.

The typical grader works for more than 15 years. The average machine is rebuilt twice. The first time is for the transmission after about 8,500 hours of use, whereas the engine should not need an overhaul until 12,000 to 14,000 hours of service.

HEAVY GRADER

Most heavy graders are tandem drive units of the type shown in Fig. 19.1. The diesel engine, with 125 to 275 horsepower, is mounted at the rear, and drives four single wheels in tandem pairs through gears and chains. The frame connection to the front axle is long and high, to allow space for carrying and manipulating the moldboards or blade under it.

The central location of the blade, and the long wheelbase of the machine, provide natural stability for the cutting edge. Smooth-acting, multidirectional control provides precision finishing ability.

Weight range is from 13 to 20 tons. Lighter grading machines will be discussed in other sections. Grader terms are given in Fig. 19.2.

Clutch. In many types of grader operation, the engine clutch may be used frequently under heavy load. It is required to be always smooth in operation. A clutch may be retained even when the machine is equipped with torque converter.

Transmission. Graders need slow, powerful gearing for heavy or precise work, moderate speeds for lighter or less fussy jobs, and also travel speeds up to 20 to 30 miles per hour. There

FIGURE 19.1   Heavy motor grader. (*Reprinted courtesy of Caterpillar Inc.*)

must also be a choice of reverse speeds, as some heavy work may be done backing up, but most reversing is for return travel only, and should be brisk. There may be six to nine forward speeds, and two to nine reverse, in either direct drive or power shift.

An eight-speed, direct-drive, power-shift transmission pattern is shown in Fig. 19.3.

If there is a converter, it is usually single-stage. It may be equipped with an output shaft governor, to keep ground speed uniform in spite of changes in load. There may be an input clutch in the converter, or a modulating pedal to momentarily disengage the transmission (and power to the wheels) while moving the blade. The clutch is also used for small, exact machine movements.

Hydrostatic Drive.   The CMI tandem graders use an all-hydraulic drive system that replaces clutch, torque converter, transmission, and a parallel shafting. There is a fixed-displacement piston pump driven by the engine, which supplies power to a pair of motors geared to a two-speed rear axle.

Power Train.   Figure 19.4 is a top rear view of the rest of a power train in a typical transmission-equipped grader. A bevel gear on the transmission output shaft is meshed with a bevel on an intermediate shaft, which turns the inner, live, or drive axles through spur gears.

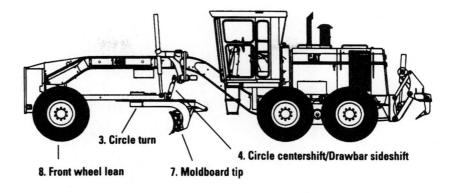

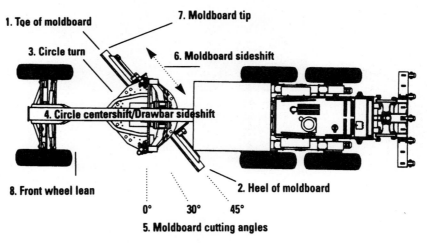

FIGURE 19.2 Motor grader terms.

An option available for some graders is variable horsepower. With that variable, when the grader is shifted to higher gears, an electrical system on the transmission moves an automatic stop to a higher setting on the fuel governor. More fuel is injected, giving more power for high-speed, light-load work. For example, Caterpillar's 140H grader moves from a net 150 to 185 horsepower in gears 4 to 8 forward and 3 to 6 reverse.

The tandem drive case carries the outer axles, supports the weight of the machine, and provides protection and lubrication. It is pivoted on the inner axle housing, to permit both wheels to follow ground contour.

Tires. Tires are of a special heavy-duty design for grader work. They are preferably the same size all around, but front ones may be smaller and/or lighter, unless it is an all-wheel-drive grader. Traction tread is usual, but the fronts may be ribbed.

Size range is from 13.00-24, 12 PR up to 23.5 × 25, 12 PR.

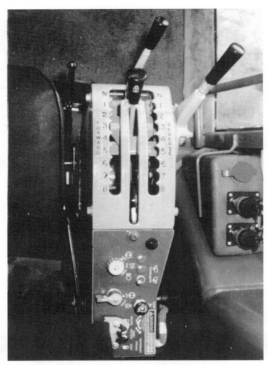

FIGURE 19.3 Power shift controls. (*Reprinted courtesy of Caterpillar Inc.*)

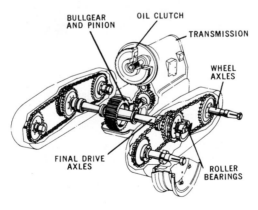

FIGURE 19.4 Grader power train.

Brakes. There are usually brakes on each of the four drive wheels, but none on the fronts. They may be either shoe or multiple-disc design, and are applied by booster-assisted hydraulic pressure. Or there may be a single disc brake in the transmission, acting through the four wheels.

The parking brake is usually on the lower transmission shaft, on the end opposite the drive pinion, and is operated by a hand lever.

FIGURE 19.5   Grader frame.

Throttle.   Basic governed engine speed is set by a hand throttle, which may be a lever on the console.

There is usually an accelerator for speeding the engine beyond throttle setting, and a decelerator for slowing it. These may be combined in one rocker pedal, providing acceleration for toe pressure and deceleration for heel pressure.

Frame and Front Axle.   The rear of the frame is a pair of beams that support the engine and the power train. Forward of the operator's station these slope upward and converge into a single beam of reinforced box cross section. This may slope down to the front axle hinge, or end in a column above it. See Fig. 19.5.

The front of the frame may be a mounting plate for front-end accessories, and/or project forward far enough to serve as a narrow bumper.

The front axle is compound. The lower section usually carries weight of the machine, may be arched high in the center for clearance over windrows, and ends below the wheel spindles. It oscillates on a central pin, and is hinged to wheel spindle brackets just below the kingpins.

The upper section, of lighter construction, is substantially a straight bar (lean bar) hinged to the tops of the hub brackets. It can be moved from side to side by a hydraulic cylinder, or by mechanical means. When moved off-center, it causes the front wheels to lean sideward. The angle may be as much as 18°. See Fig. 19.6.

Wheels are leaned to increase their resistance to sliding sideward on the ground because of load on the blade, or steering stresses.

Steering may be mechanical with a booster, or by a hydraulic cylinder located behind or above the axle, and operating through a more or less conventional linkage. The cylinder control is equipped with a follower valve that keeps a direct relationship with the position of the steering wheel. Special arrangements may be made to avoid interference between leaning and steering.

BLADE AND MOUNTING

Blade.   The blade is the principal working part of a grader. Most of it is a curved piece of steel called a moldboard. The distinction is chiefly that "moldboard" emphasizes its function in causing dirt to roll and mix as it is moved.

Standard moldboard lengths in tandem graders are 12, 14, and 16 feet, with some 10-footers in small machines. Two-foot extensions are sometimes added for light work or long side-reach. Height is usually 24 inches, but may be as much as 31.

The cross section is such that dirt being pushed has a rotary movement, rising at the bottom and falling forward at the top. This characteristic, combined with the usual sideward drift caused by working it at an angle, makes spillage over the top very unusual.

The bottom is fitted with a removable and reversible wearing edge, with separate pieces (end bits) at the corners.

The blade is supported and held in position by a pair of heavy curved brackets, called circle knees. They are attached to the underside of a rotatable ring, called a circle. See Fig. 19.7.

FIGURE 19.6 Leaning front wheels. (*Reprinted courtesy of Caterpillar Inc.*)

Tilt (Pitch). The blade may be fastened to the knees by a hinge bar at the bottom, and a curved slide at the top. This construction permits changing the tilt (forward lean) of the moldboard.

The adjustment may be manual, with clamp nuts to be loosened and then retightened, as in Fig. 19.7, or by hydraulic cylinders controlled from the operator's station, as in Fig. 19.8.

The blade is usually kept near the center of its tilt adjustment, but may be tipped up to 44° forward and 6° back.

Blade Sideshift. Many grader operations can be made more efficient by extending the blade a greater-than-normal distance to the side. This move can usually be made with the circle controls, but they disturb the side-to-side slope. It is often better or easier to shift the blade itself, by sliding it along the knees. For extreme reach, both the circle and the blade adjustments may be used.

With power sideshift, the move is made by holding the blade clear of the ground and obstructions, and sliding it sideward by means of the control valve for its hydraulic cylinder.

Drawbar and Circle. The drawbar is a V-shaped (or sometimes T-shaped) connection between the front of the grader frame and the circle. It is rigid with the circle, and fastened to the frame by a ball and socket, or a universal fitting, that allows limited angular movement from side to side and up and down.

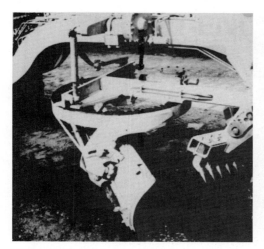

FIGURE 19.7   Circle and moldboard.

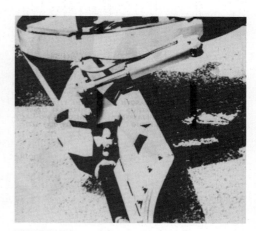

FIGURE 19.8   Moldboard with power tilt.

The drawbar carries the full horizontal load on the blade, as other connections provide only vertical and side support.

The circle is a toothed ring that is rotated in or on a supporting frame by a circle turn mechanism. Pads and shims provide for low-friction movement, and for adjustment of clearances.

The circle is turned by a spur pinion gear meshing with teeth cut all around it, usually in the inside. The pinion may be turned by a shaft driven by the engine power takeoff or by a hydraulic motor, or by a direct-mounted hydraulic motor and reducing gears. In each case, movement is controlled by a lever in the control console, at the operator's station.

It is important that the blade angle, which is controlled by circle rotation, be capable of exact adjustment, and that it remain in place during work, without drifting or creeping.

Means to keep it locked against hydraulic creep include drive through an irreversible worm gear, automatic clamping brake on the shaft, or clamping the circle itself. The clamps are applied by springs, and released by hydraulic pressure whenever the control lever is used.

Rotation in full-size graders is usually continuous in both directions. However, a full circle can be turned only when the blade is in such a position that it will not strike the ground or any part of the machine as it is turned. Some machines have very limited clearances, in others they are generous. Scarifier teeth must usually be removed.

In general, for rotation into reverse blading position, the blade should be level, a few inches above the ground, and approximately centered.

Lift.   The blade is lifted and lowered through the circle by a pair of arms or cylinders, supported by the frame and fastened to the sides of the drawbar at the rear.

Direct lift by the pair of two-way cylinders, Fig. 19.9, is superficially the simplest method. But cylinder rods are subject to change of position caused by seepage of fluid past the piston, or in the valve. Such creep can be prevented, or compensated, but corrective devices involve loss of simplicity.

Hydraulic cylinders can usually be made long enough to take care of full range of position without manual adjustments.

Circle Sideshift.   The circle sideshift, or lateral shift, swings the circle and blade to the side, usually to the right.

When the move is mechanically powered, a curved track is mounted on the underside of the frame. A long, curved gear section slides on this, being moved by a pinion gear based on the frame. An arm is fastened by ball and socket joints to the right side of the gear section and the left side of the drawbar, for reach out to the right.

FIGURE 19.9   Circle lift, direct hydraulic. (*Courtesy of John Deere Company.*)

Turning the pinion for right shift moves the gear outward and upward to the right, pulling the drawbar and circle with it. For left shift, the arm is fastened to the left side of the gear and the right side of the drawbar, and the pinion rotated oppositely.

Sideshift to the right raises the blade increasingly on the right side and lowers it somewhat on the left. The lift levers must be manipulated if it is to be kept level.

A hydraulic cylinder mounted on one side of the frame, with its piston connected to the opposite side of the drawbar, can be used to shift the circle.

Blade lift cylinders may be based on a hydraulic saddle or cross piece on the grader frame. This may be designed to rotate on the frame when locks are released. When free, it can be turned by action of one lift cylinder, then locked in a side position.

The principal use of circle or lateral sideshift is to position, or start to position, the blade at a steep or vertical angle offside, for slope work. On many machines, manual adjustment of telescoping arms is required also.

A combination of circle and blade sideshifts will put the blade a long way out, as seen in Fig. 19.9, to work in spots too soft to support the grader.

Blade Vertical Position.   The moldboard blade can be positioned for shaping a bank or embankment slope, as seen in Fig. 19.10. This is done by moving the circle out and at an angle with a horizontal plane as shown in that figure. The extreme angle is for the blade to be vertical or at a 90° angle from the horizon. This is possible either to the right or left side of the motor grader. The movements are done entirely by hydraulics.

FIGURE 19.10   High reach with blade.

CONTROLS

The control levers for the tools or working parts of the grader are usually arranged in a more or less straight line on a console on the dash.

There are five basic controls that are always present: the two blade lifts (right and left), circle turn, circle shift, and wheel lean. Common additional controls are for scarifier, and blade shift. A dozer or a snowplow might have another lever, or use the scarifier control. A rear ripper will have one or two levers.

Location of controls is not completely standardized. Figure 19.11 shows the control station of a Caterpillar grader, and Fig. 19.12 indicates location of the major control levers in four other machines. Individual graders may have lever locations switched to suit operator preferences. They are arranged for the operator's convenience as shown in Fig. 19.13.

All these levers are two-way, operating in forward and back positions, and holding in the center. Manufacturers have not introduced the multifunction joystick, like those in excavators and loaders, in graders, for fear of interfering with operator control.

Hydraulic.   Many graders use all-hydraulic control for some or all of their functions. The three-position valves are similar to those used in other hydraulic equipment, but must be of the metering type to allow gradual and/or partial engagement whenever necessary for slow and precise movement.

Precautions against creep in these systems include irreversible worm gearing, and clamps that hold shafts from turning or piston rods from sliding, when valves are in neutral. They are pressure-released whenever valves are engaged to move the parts.

FIGURE 19.11   Grader operator's controls. (*Reprinted courtesy of Caterpillar Inc.*)

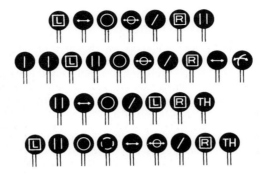

BLADE LIFT, LEFT   SCARIFIER
BLADE LIFT, RIGHT   RIPPER
MOLDBOARD SIDE SHIFT   MOLDBOARD TILT
CIRCLE REVERSE   THROTTLE
CIRCLE SIDE SHIFT   DOZER
LEAN WHEELS

FIGURE 19.12   Four grader control sets.

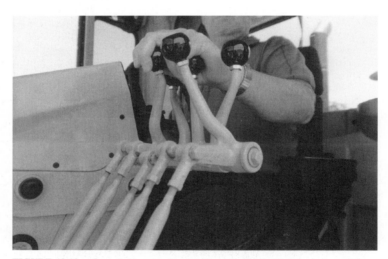

FIGURE 19.13   Operator's convenience with controls. (*Reprinted courtesy of Caterpillar Inc.*)

OPTIONAL EQUIPMENT

Scarifier.   The scarifier is a set of teeth used for breaking up surfaces too hard to be readily penetrated by the blade. The teeth, consisting of rather slender shanks with replaceable tips, are set in a bar with a flattened V shape, as seen in Fig. 19.14, that is narrower than the grader; or in a wider straight bar. It is pulled by a pair of beams hinged to the bottom front of the frame, and is usually raised and lowered by a hydraulic cylinder. Angle of penetration may be adjustable from the operator's station.

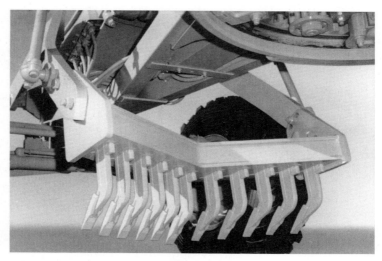

FIGURE 19.14 V-scarifier mounted under circle.

The shanks are wedged or clipped in place, and may be readily adjusted for height, or removed. A full set, often eleven in number, is used for shallow penetration and light work, and in material that crumbles into fine pieces. For deeper penetration, or slabby material, every other tooth or two out of three may be removed.

Individual teeth are not designed to take the full push of the grader, and caution must be used in hard material and among rocks to avoid bending or breaking them.

Scarifiers used to be almost standard equipment but are rather less used now. One can speed up grader work, and greatly reduce wear on the blade. It may be used in the same pass with the blading, or separate passes may be made for loosening and for shaping.

It is usually necessary to remove the shanks, or set them high in the bar, in order to reverse the blade. They may interfere with cutting flat-bottom ditches and handling high windrows.

Ripper-Scarifier. This unit, Fig. 19.15, which mounts on the back of a grader, may be fitted with 11 teeth for scarifying depths up to 9 inches, or with 5 heavier ripper teeth for penetration to 14 inches.

It is raised and forced down by a hydraulic cylinder. The lift frame is of the parallelogram type that keeps the teeth at the same vertical angle, whether up or down.

A rear scarifier or ripper will handle much heavier work than a standard front-mounted one, it can process a strip the full width of the machine, and it is not in the way of the blade. However, it adds to machine length so that it is awkward in close quarters, usually produces coarser and more difficult-to-grade pieces, and cannot break ground on the same trip that it is bladed.

Front Attachments. The front of many grader frames can be built out to form a narrow vertical bumper, forward of the tires. This may have limited use for protection in straight-on collisions, and for occasional pushing of other vehicles.

It may carry a dozer blade, Fig. 19.16, raised and lowered by the scarifier cylinder or by separate controls, which can be convenient in rough work, particularly in leveling piles.

A snowplow, preferably of the V type, is often fitted to the heavier machines, to provide possible work in the winter months. Chains should easily be put on the drive wheels for snowplowing.

Blade Stabilizer. The stabilizer is a shoe or skid plate that can be rested on or pressed against the ground behind the blade. See Fig. 19.17.

FIGURE 19.15   Rear-mounted ripper. (*Reprinted courtesy of Caterpillar Inc.*)

FIGURE 19.16   Dozer attachment.

FIGURE 19.17   Blade stabilizer and bowl.

In high-speed, shallow work a grader blade may develop various kinds of vibration, including a long-period rhythmic bounce or lope, or a short-period chatter. These cause a wavy, rough, or washboard surface on the work, discomfort to the operator, and extra machine wear.

The stabilizer, which is under separate hydraulic control, can be put in a dragging position where it will steady the blade and dampen or eliminate vibration. It can also be used as a support shoe to limit the cutting depth of the blade.

Blade Bowl.   A blade bowl consists of two sideboards, a spreader bar, and a cutting edge. It is attached at the ends of a standard moldboard, and is supported from the circle by an adjusting bolt. The edge is ordinarily a little lower than the moldboard edge.

This attachment converts the blade into a bottomless scraper bowl. The leading edge cuts and shatters material, which is held by the sides and caught and moved forward by the blade. When a depression is reached, dirt falls into it and is smoothed off by the moldboard edge.

The bowl is used only with the blade set straight across, and where somewhat more dirt must be moved than is normal to grader operation.

ARTICULATED (FRAME) STEERING

A majority of graders have frames that are hinged just forward of the engine, with pivoting controlled by a pair of hydraulic cylinders. This permits the front frame section, which carries the circle and blade, to be turned at an angle of 20° or more to the tandem power section, as in Fig. 19.18. In addition, normal sharp-angle steering of the front wheels is retained.

With two points of turn, these machines can make a U-turn in less than two times overall length, a remarkable feat for a tandem grader. A differential may be required between the drive axles to minimize scuffing.

Independent steering of wheels and frame may provide operational advantages. As indicated in Fig. 19.19, the front end can be offset to the side by steering front and rear in the same direction (crab steering), so that its wheels will not have to tangle with the windrow or other obstruction, while the tandem rear is kept squarely behind the blade. Front wheels can still be leaned or turned as necessary to resist any remaining side-thrust.

ALL-WHEEL DRIVE AND STEER

Some motor graders have been equipped with a mechanism to power-drive and steer on all four or six wheels. An example of the power delivery is shown in Fig. 19.20.

FIGURE 19.18   Tight turn with articulated grader. (*Courtesy of John Deere Company.*)

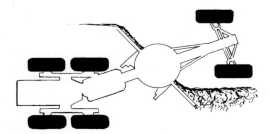

FIGURE 19.19 Crab steering with articulation.

FIGURE 19.20 Four-wheel drive and steer.

The engine is rear-mounted in the standard manner. A torque converter with an input clutch delivers power to a four-speed, power-shift transmission with a separate forward-reverse lever, and a high-low range auxiliary, providing eight speeds in both directions. A transfer case drives front and rear propeller shafts.

The front shaft is in an under-the-frame arch, with multiple universal joints to enable it to conform to its curve.

The all-wheel-drive option increases the machine cost by 20 percent or more.

Applications. As in other wheel-mounted equipment, all-wheel drive makes the full weight of the machine available for traction, whereas rear drive can only utilize the weight on the drive wheels. In conventional graders, this is about 70 percent.

As a finishing tool, the grader should not often be called upon to work in mud. But it does have to work in loose soil or sand, and it may be urgently needed to skim slippery films off haul roads. A grader may also be exposed to poor footing on turnarounds, between work strips, and in traps such as loosely filled ditches. In any of these cases, the reserve traction of all-wheel drive is valuable.

The pulling ability of powered front wheels improves ability to resist the side-thrust of heavy blade loads, whether on straightaways or turns.

Extra traction provides for more effective handling of accessories such as scarifier, dozer blade, and snowplow.

AUTOMATIC CONTROLS

Automatic Blade Control.   Several automatic systems are available for control of hydraulically hoisted grader blades. Slope regulation is the most widely used.

With this there is a pendulum device mounted on the grader blade, that remains perfectly vertical in a right-to-left or cross plane, regardless of movements of the grader. It is coordinated with transistor computer circuits. See Fig. 19.21.

The operator sets a dial at the cross slope wanted, which usually is from job specifications. He or she then operates the grader, keeping the leading edge of the blade at the desired grade.

The instrument, operating on battery power and through solenoid valves, raises and lowers the heel of the blade to keep the correct slope. Corrections are made at the rate of 10 or more per second, so a high degree of accuracy can be obtained.

It is usually necessary to keep the blade at a predetermined angle with the grader's direction to have the slope the same as that set on the dial. A sharper angle reduces the effective width, and therefore changes the amount of slope.

Some controls include a sensor on the blade circle, which corrects for changes in angle and allows more flexibility.

The next step up is to add a sensor to keep the leading edge of the blade on grade. This is carried by an arm projecting beyond the side of the machine. It may terminate in a finger (sensing fork) guided by a stringline, as in Fig. 19.22, or a small wheel rolling on pavement or on a previous pass by the grader.

FIGURE 19.21 Sonar sensor grade control.

FIGURE 19.22   Stringline for grade control.

The grade and the slope corrections are usually separate, with grade controlling the leading edge and slope the trailing one.

Such controls are by no means a sleeping license for operators. They have to steer, work the throttle (so far), and be sure not to run into anything. If a too-heavy cut lugs down the grader, the operator must override the control and raise the blade.

Laser.   An almost-final step in automation is grade control by laser. See Fig. 19.23. It is adapted only to graders that already have slope control.

The background of this system is a laser, a beam of intense (usually red), highly concentrated light, that can maintain an almost pencil-thin beam for many hundreds of feet.

Lasers, and many of their uses, are described in Chap. 2.

For mobile machines, the laser shines down on a prism, which casts a basically horizontal beam. This can be adjusted to job grades, slope up or slope down. The prism revolves rapidly, casting the beam over a circular pattern.

A mast is fastened to the leading edge of the blade, and kept vertical by a pendulum and hydraulic minicylinders. It carries a set of laser-sensitive cells that activate solenoids in the lift valve to lift or lower the blade whenever necessary.

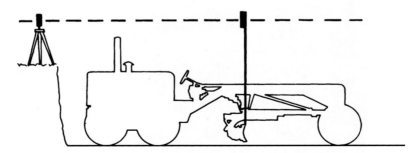

FIGURE 19.23   Laser grade control.

Load.   Automatic grade regulation devices usually know nothing about load. Any one of them will stall a machine if a cut to grade is too heavy for it, or will try to fill a hollow without enough material in front of the blade. They are therefore limited to finishing sections that are already almost on grade, and must be closely supervised by the operator and the foreman.

HEAVY GRADER OPERATION

General.   The following discussion refers to the tandem rear-drive grader with hydraulic control, unless otherwise indicated.

The grader is normally moved forward while working. It can work in reverse by using the back of the blade for spreading or smoothing; or by reversing the blade so that its cutting edge faces to the rear.

In forward operation the grader is kept in line by steering, and by leaning the front wheels away from the direction of thrust or toward the direction of turn. The wheels may also be leaned to avoid rubbing vertical banks.

Adjustments.   The blade is controlled in a number of ways. The ends can be raised or lowered independently of each other, or together. It may be positioned across the line of travel, parallel to it, or at any angle. It can be shifted to the side and into a vertical position by hydraulic power. There are also mechanical adjustments for extending its range.

The blade is ordinarily kept near the center of the tipping adjustment so that the top of the blade is directly over the edge. Increasing the lean forward decreases cutting ability and causes the blade to ride over its load rather than to push it. It diminishes the likelihood of catching on solid obstructions and may be used for rapid, light planing of rather regular surfaces and for mixing operations. When leaned back, the blade cuts readily but tends to let the load ride over its top, and to dig into obstructions. In machines not carrying a scarifier this tilt may be used to cut hard surfaces.

Bulldozing.   A grader blade may be used as a bulldozer to a limited extent, often in spreading piles of loose material. If there is space to work beside the pile, the blade should be extended well to the side, and the pile reduced in a series of cuts, as shown in Fig. 19.24.

If there is not room enough to do this and the piles are not too high, the front wheels may be driven over them. The front axle will push the top off, and the blade will cut as much more as power permits.

The blade should be kept well below its highest position so that if the machine gets hung up on it and loses traction, it will be possible to raise the blade to restore weight to the wheels.

Piles to be distributed by a grader should be spread as much as possible while being dumped.

The grader can also be used for light cut-and-fill work in building and regrading roads.

The load to be pushed is limited by the power and traction of the machine and will usually be much less than by a crawler of the same weight, although it will be moved faster. The blade itself is quite low, but being more concave than the dozer blade impacts a more pronounced rolling action to the load so that a large quantity can be pushed without spilling over the top.

If the blade is lowered on the right only, the other lift arm will remain stationary and the circle and blade will pivot around it, causing the left end of the blade to rise about one-quarter of the distance the right side lowers. If the left corner is to be held in position, it must be lowered intermittently while the right side is lowered steadily. This effect occurs on either side.

Keeping track of both ends of the blade is necessary and may be the hardest technique for the beginner to learn.

If the blade is raised to its full height and the controls are left engaged, the lift arms will continue to rotate inward until stopped by the frame. The blade will be lowered during the movement from top center to the frame, although the controls are in the RAISE position. If the controls are moved to DOWN, the blade will rise until the arms are past center, after which it will respond normally.

If much dozer work is to be done, it may be advisable to install a front dozer blade.

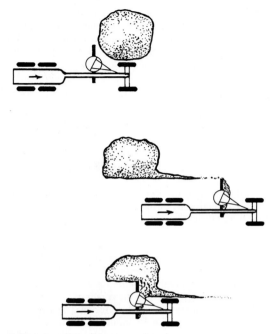

FIGURE 19.24   Spreading a pile.

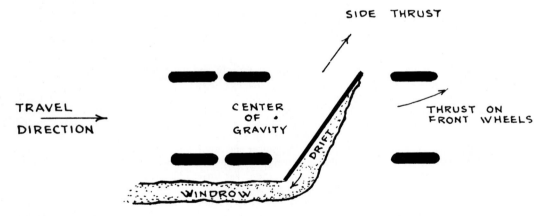

FIGURE 19.25   Sidecasting.

Sidecasting.   When the blade is set at an angle, the load pushed ahead of it tends to drift off to one side, as in Fig. 19.25. The rolling action caused by the curve of the moldboard assists this side movement. As the blade is angled more sharply, the speed of the side-drift increases, so that the dirt is not carried forward as far and a deeper cut can be made.

The sideward movement of the load exerts a thrust against the blade in the opposite direction, which tends to swing the front of the grader toward the leading edge. This thrust is handled by leaning the front wheels to pull against the side-drift, and steering enough to compensate for any side-slipping which occurs in spite of setting the front wheels to lean.

The most usual way in which to describe blade setting is to say that a blade set straight across, as in Fig. 19.24, is at zero, and all other settings are described by their angular distance from that position. Most road shaping and maintenance are done at a 25 to 30° angle, with straighter settings for distributing windrows and sharper ones for hard cuts and ditching.

The angle of the blade is regulated by the circle reverse control. The mechanism is self-locking and can be turned any desired amount. In some makes it can be adjusted only while the blade is empty or doing light work, in others while pushing a heavy load.

Sideshifting the circle from center changes the slope of the blade, requiring compensating lift adjustments. Sideshifting the blade itself does not affect slope.

Planing. If the blade is set at an angle, it can be used to plane off irregular surfaces by lowering them sufficiently that enough material will be cut off the humps to fill the hollows. Enough extra material should be cut to keep a partial load in front of the blade. The forward and sideward movement of the loosened dirt serves to distribute it effectively. If a windrow is left at the trailing edge of the blade, it is picked up on the next pass. On the final pass a lighter cut is made, and the trailing edge of the blade is lifted enough to allow the surplus material to go under rather than around it, to avoid leaving a ridge.

This type of light planing will produce a smooth surface under favorable conditions, but the fill in the hollows is liable to settle or be compressed below the cut sections. Also the blade may chatter in a very shallow cut, particularly if the mechanism is loose or worn.

A more thorough method is shown in diagrams in Fig. 19.26. A series of cuts are made across the area to a depth sufficient to reach the bottoms of the holes, or at least to 2 inches. The large windrow of loosened dirt is then spread back evenly over the area.

It is easier to get a smooth surface with this method because of the advantage of working loosened material and the more uniform distribution with a full blade. The surface will tend to remain smooth after settling or rolling.

It is desirable to vary the blade angle during this work, the first cuts being taken with a straighter blade than the later ones, and the first spreading pass having a sharp angle which will be reduced on each following pass as the size of the windrow is reduced.

Windrows should not be piled in front of the rear wheels, as they will interfere considerably with grading accuracy and traction.

Crowning. When the piece to be smoothed is a dirt or gravel road, it is generally crowned so that water will flow off to the sides. Figure 19.27 shows a sequence of passes in a crowning operation. Road material is bladed inward from the shoulders or ditches, and the top of the crown is cut with the blade at zero angle, or at a slight angle which will sidecast some material to either side that may require it. The windrows are then spread by putting the blade at an angle of 10 to 25° toward the center, and using a fast working speed. The blade is held above the level of the undisturbed surface so as to avoid collision with solid objects. The speed causes the loose material to be thrown from the blade, so that it will feather out and blend at the top. Any ridge built in the center is then spread out at speed with a straight blade. This should finish the job, but it may be desirable to backblade or rework some sections where proper crowning was not obtained.

If the road is gravel or other imported material and the ditch or shoulder is loam or clay, the road may be made muddy by blading in too much from the edges. Since the road must be finished so as to drain off the sides, it may be necessary to blade off high spots on edges outward.

Sods and other debris brought up on the road from the ditches interfere with grading, as lumps catch under the blade, leave ruts, and block the sideward drift of the dirt. If labor or raking equipment is available, it is best to remove the debris from the side windrows before making the center cut. If the grader is working alone, the operator may occasionally climb down to remove a particularly annoying piece, but most of it will be left on the road to dry out and be broken up by traffic.

Stones are a more serious nuisance, both in making smooth grades difficult or impossible and in causing damage to machine and operator during cutting. If the cut is shallow or the road dirt compacts readily, it is possible to blade loose stones out of the road while grading it.

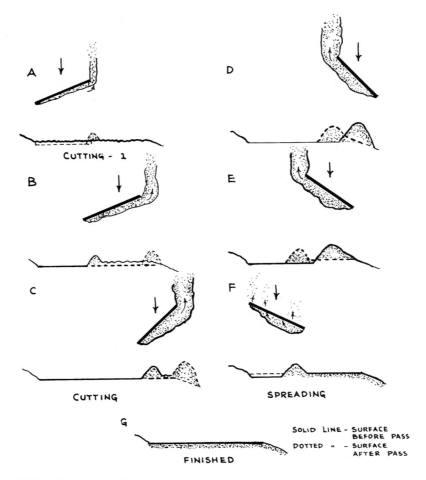

A

CUTTING - 1

B

C

CUTTING

D

E

F

SPREADING

G

FINISHED

SOLID LINE - SURFACE
                  BEFORE PASS
DOTTED "   - SURFACE
                  AFTER PASS

FIGURE 19.26   Refinishing a flat section.

Large rocks or those firmly embedded in roads or shoulders are dangerous. If the blade hooks into one too solid to move, the grader will jump sideward or stop abruptly. This imposes severe shock on the blade, circle, and power train, and is liable to throw and injure the operator. Danger of serious consequences increases greatly with higher speeds, so that all cutting should be done in low gear and often at part-throttle where buried rocks or roots are expected.

Work Patterns.   The following discussion will concern work done by single graders. However, it is often efficient to have two or more of the machines working together, each performing one step in the work sequence. This speeds the job, produces better results than working small sections with individual machines, and reduces or eliminates blocking of roads with windrows.

Road building and grading may be done on three general patterns. One is to work the two sides alternately, turning at the end of the strip. Another is to do one side at a time, working in both directions by means of a reversible blade. The third is to do one side at a time, with the reverse trips nonworking or utilized for light work with the back of the blade.

The pattern used is determined by the length of the strip being worked, the turning space and footing, and the reversibility of the blade. The machine does its best work going forward, and this increased efficiency must be balanced against the time, labor, and risks involved in turning. In a long run even difficult turning conditions may take an insignificant part of the working time, where easy turns may not be justified on a short run. A general rule is that a grader should not be turned if the strip is less than 1,000 feet long.

Turns.   In turning, the front wheels are leaned all the way over in the direction the front of the frame is to turn, and are left in this position for both forward and reverse movements until the turn is completed. If ditches or rough ground must be used, the machine should be backed into or

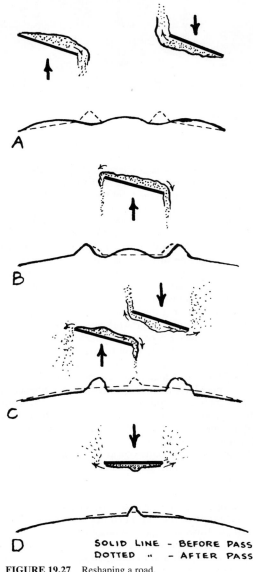

SOLID LINE - BEFORE PASS
DOTTED " - AFTER PASS

FIGURE 19.27   Reshaping a road.

FIGURE 19.28   Turning on rough ground.

across them. The oscillating tandem drive will readily climb ditches or obstructions which would be difficult to cross with front wheels.

Turning may be a serious problem. See Fig. 19.28. When the grader lacks a differential, the rear wheels drive straight forward and the tandem arrangement gives them considerable resistance to sideward movement. Weight on the front wheels is light, and even when leaning properly, the tires may not have enough traction to turn the machine on loose material, but will skid and slide sideways.

If the front tires do grip enough for a sharp turn (minimum radius is about 25 feet), the rear wheels on the inner side of the curve must spin or those on the outside drag enough to compensate for the different distances they travel. Turning a grader sharply therefore results in scuffing and gouging, which may be sufficient to damage soft surfaces.

It is therefore often necessary to make a gradual turn either by swinging in a wide circle or by jockeying backward and forward in order to avoid damaging or loosening the turn spot. It is sometimes possible to do most of the turning on ungraded areas where no damage will be done.

Other factors affecting the choice between turning and working in reverse may be idle travel distance between the end of the work and the first possible turning area, interference with traffic while turning, and chances of getting stuck.

Tandem drive affords powerful traction, but a careless operator may hang the blade or circle up on ridges or rocks, and the front wheels particularly may sink into mud with the same result. Recently filled ditches which have become water-soaked make grader traps. Repeated sharp turns on sandy soil may loosen it enough that the rear wheels will spin in it.

If the lifts are in good condition, many graders can be unstuck by putting blocks under the blade and applying down pressure. This will normally lift the front wheels off the ground so that they can be shored up, and the blade can then be lifted clear. If the pressure is applied to the rear corner of a sharply angled blade, it may lift the rear wheels on one side. The scarifier will lift the front only.

Large front tires—the same size as on the drive wheels—give better front-end flotation and are more effective in climbing ditches and banks and in holding the machine to its course on slopes, turns, and loose ground.

Road Building.   A grader, without assistance from other machines or hand work, can shape up a road across a field by digging a pair of parallel ditches and using the soil to build the road crown. See Fig. 19.29. However, sod can make the finishing operation tedious and unsatisfactory, as it tends to ball up under the blade and catch and pull out of loose surfaces. For this reason, the strip should be thoroughly disked before grading is commenced.

FIGURE 19.29   Excavating side ditch for roadway. *(Reprinted courtesy of Caterpillar Inc.)*

The outer ditch lines are marked by stakes or by the edge of the disked strip. The first cut on each side is made about 2 feet inside the edge, Fig. 19.30(*A*). The blade is held at a very sharp angle, perhaps 50 or 60°, with the leading edge just outside the wheel track and the windrow rolled off under the grader. The cut is a light one, made primarily to mark the edge of the work and to hold the wheels against sideslipping.

The next cut is made at about a 25° angle, as in (*B*), casting the spoil beyond the inner wheels. If the windrow is large enough, it is spread toward the center, as in (*C*). Otherwise, additional ditch cuts are made until sufficient material is piled to justify a spreading pass.

Ditch cuts, alternated with casting or spreading, are continued until the ditch is the proper depth. The outer slope is then cut, as in (*D*), and the spoil moved up the inner slope, as in (*E*), from whence it can be spread over the road.

The other side is done in the same manner, and the fills are blended at the top. Ditch cuts, except the first one or two, can be made in either forward or reverse. Light casting and spreading can be done in either direction, but forward is more efficient if the windrow is heavy.

Manufacturers recommend doing forward ditch cuts and other heavy grader work in second gear, at a speed of 3 or 4 miles per hour. Blading windrows and similar handling of loose soil can often be done in third gear at speeds up to 6 miles per hour. However, when there is loose rock, lower speeds will pay off in improved quality of work. In the presence of buried obstructions heavy enough to stall the grader, very slow movement may be required for protection of both the operator and the machine.

If a wide-bottom ditch is required, the further operations shown in (*F*) and (*G*) are undertaken. Slices are cut from the inside slope of the ditch, cutting down to the level of the original bottom and leaving a ridge. This is cut out by running the grader with its outer wheels in the original

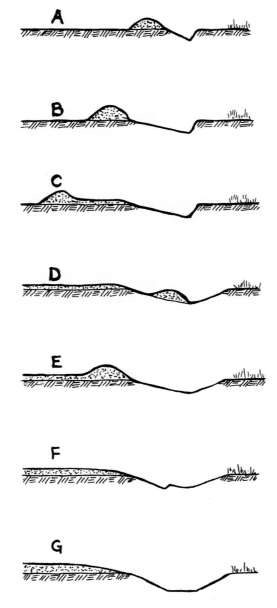

FIGURE 19.30   Ditching for fill.

bottom, and setting the blade with its leading edge even with the outer tires and at a sharp angle so that it will cut only the width of the desired flat bottom. The spoil is piled on the bottom of the inner slope, from whence it is cast onto the road and spread by following passes.

The number and sequence of passes are affected by the depth of the ditch, the width of the road, and the resistance of the soil.

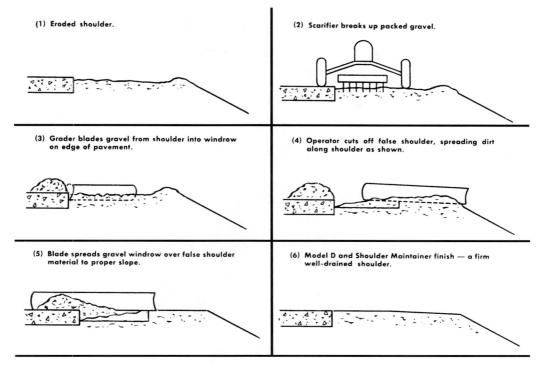

(1) Eroded shoulder.

(2) Scarifier breaks up packed gravel.

(3) Grader blades gravel from shoulder into windrow on edge of pavement.

(4) Operator cuts off false shoulder, spreading dirt along shoulder as shown.

(5) Blade spreads gravel windrow over false shoulder material to proper slope.

(6) Model D and Shoulder Maintainer finish — a firm well-drained shoulder.

FIGURE 19.31   Reshaping a road shoulder.

Figure 19.31 indicates a series of steps that can be taken to restore a shoulder on a paved road.

All-Wheel Drive.   Traction on all wheels makes it possible to push larger loads with the same weight grader, and working in mud or sand too soft for conventional drive.

In working, the front and rear axles can be offset from each other to either side, as in Figs. 19.29 and 19.32. This steadies the grader against sidethrust, and enables it to increase its effective sidereach by running a front wheel up on a bank, or in a soft gutter, while the rear wheels stay on more level or firm ground. It also allows sideshifting of the trailing wheels to avoid running on a windrow.

LIGHT GRADERS

The graders discussed so far have been of the high-powered, heavy-duty type, or lighter models patterned directly after the big units.

There are also lighter machines, 10 tons or less, that are simplified in structure. The most conspicuous difference is that the blades cannot rotate in a full circle. Circle sideshift may be limited or lacking. See Fig. 19.33.

They have a net 35- to 90-horsepower engine with standard single or tandem drive with the engine over the drive wheels. Length is less than 24 feet, height is 10 feet or less, and turning radius is generally less than 20 feet. Their maximum speed is 20 miles per hour, or less.

The blade is generally 10 feet long, and is supported by a $^3/_4$-turntable resting on the circle. It is rotated through less than a half-circle by a hydraulic motor and self-locking reduction gears.

FIGURE 19.32   Balancing load with four-wheel drive.

Blade lift is by two direct-acting cylinders cradled on frame brackets. Circle sideshift is manual, with pin-and-hole adjustments of a telescoping arm. Blade sideshift may be either manual or hydraulic.

Steering and leaning front wheels are hydraulically controlled. A scarifier can be mounted behind the blade, or at the rear.

Comparison with Heavy Graders.   Light graders do not provide the full range of blade positions, blade angle usually cannot be changed while working, and total weight and power not only are much less, but are even smaller in relation to blade width.

This disproportion is necessary because chassis width cannot be reduced far without loss of stability, and a moldboard should be long enough to reach past the wheel line even when sharply angled.

The result is that full-width cuts must be shallow, except in loose material, and the operator should not expect to load the blade as full or move it as fast as with the heavier machines.

These graders cannot slope banks too steep for them to walk on, dig flat-bottom ditches narrower than the tread, cut in reverse, or move heavy windrows in one pass.

On the other hand, they can work in narrow, crooked roads and driveways, around buildings, and among obstructions where the large machines would be at a disadvantage. They can often do maintenance and light construction jobs more economically.

Drawbar Grader.   Light grader blades can be readily mounted on the back of wheel tractors equipped with a three-point hitch, as described in Chap. 15.

A simple model is shown in Fig. 19.34. The blade can be locked straight across, at seven different angles to either side, or reversed, by pulling a lockpin and rotating it by hand. It can be

FIGURE 19.33 Medium-light grader. (*Courtesy of New Holland Construction.*)

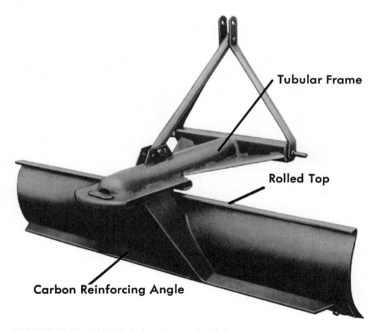

Tubular Frame

Rolled Top

Carbon Reinforcing Angle

FIGURE 19.34 Light blade for a three-point hitch.

given a moderate cross slope by raising or lowering one of the lower hitch arms. Suck is adjusted by shortening or lengthening the upper arm.

More elaborate models have a greater choice of working angles, built-in adjustments for slope, and means to offset the blade to obtain greater reach on one side.

As the blade is entirely behind the tractor, it lacks the stability of the under-frame mountings, tending to dig when the tractor starts to climb, and to rise when the tractor starts to tip forward. This is partly compensated for by the use of an automatic leveling device in the drawbar control valve. When this is used, the position of the hoist is controlled partly by a hand lever and partly by the pull or pressure exerted by the grader on the upper link. Under favorable circumstances, an even grade may be obtained by setting the hand lever and letting it take care of itself. For more exacting work, the automatic mechanism can be cut out and direct control used.

There are a number of useful accessories, such as rollers or shoes behind the blade to provide more stability and/or to regulate cutting depth, box sides to increase dirt-dragging capacity, and ditching points to help penetration of the leading edge.

Rake Blade.   A light grading blade may be made up of a set of curved teeth fastened to a cross-bar frame, as in Fig. 19.35. This model has alloy steel teeth 30 inches long, 1 inch wide, and spaced a little more than 1 inch apart.

The caster wheels are removable for doing rough work. The cross frame just behind the tractor wheels can be fitted with ripper teeth. There is a low grader blade that can be folded down across the bottoms of the teeth when rake action is not desired. A three-point hitch provides for lift and adjustments.

The principal use of the rake is to sift stones out from gravel on roads or from soil being shaped and planted. The stones are sidecast into a windrow, along with other oversize objects, while loose soil shifts between the teeth.

Stones may be sidecast off the area by a succession of passes, pulled into piles with the rake set straight across, or picked up by a loader or by hand.

FIGURE 19.35   Rake blade. (*Courtesy of York Modern Corp.*)

Rakes are also used to spread crushed stone and blacktop; to remove trash, break up lumps, distribute soil in seedbed preparation; and to cover seed.

TOWED GRADERS

Towed graders depend on a tractor or truck for movement. The grader controls may be hydraulic, mechanical, or manual. Those with power controls may carry a small gasoline engine, or be operated by a valve and tubing from the tractor. The tractor operator may run the grader also, but it is more usual to have two operators.

Steering, or tracking, is controlled by angling the draw tongue and shifting the frame along the rear axle.

Towed graders are made in quite small sizes, and are sometimes equipped with stone rakes, smoothing drags, and other accessories. Sometimes the front axle is removed, and the gooseneck mounted directly on a hydraulic lift tractor drawbar. See Fig. 19.36. This change makes it possible to back the machine as a semitrailer—an important convenience in redoing short pieces and in repeated scraping of hard spots. The weight transfer improves wheel tractor traction.

A towed grader with two axles can have a blade and a scarifier as well as the rake. The actions resulting are shown in Fig. 19.37.

A towed grader can also be used to advantage behind a motor grader to avoid blocking roads and to eliminate the need for return trips. When a ditch is being cleaned out toward the road, the resulting windrow may be a serious obstacle to traffic. If the patrol tows a grader behind it, offset toward the center of the road, it can spread the windrow in the same trip.

If a tractor operator works the towed grader controls also, he or she must alternate between looking ahead for steering and behind for grading. Either the quality or the speed of work usually suffers as a result, but it is still a satisfactory practice under many conditions.

FIGURE 19.36   Towed rake scraper. (*Courtesy of York Modern Corp.*)

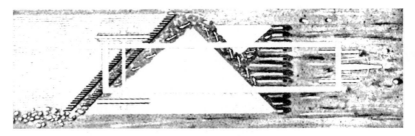

FIGURE 19.37   Blade and rake action. (*Courtesy of York Modern Corporation.*)

FIGURE 19.38   Preparing a subgrade. (*Courtesy of CMI Corporation.*)

AUTOMATIC LANE PROCESSORS

There are quite different automatic leveling, spreading, and trimming machines that are used in final preparation of highway lanes for paving. These represent the most elaborate and efficient development of the grading machine. They are also readily changed over to paving machines, for either concrete or asphalt.

Autograde.   The CMI Dual Lane Autograde Trimmer-Spreader, Figs. 19.38 to 19.45, is the original and apparently the most successful type and will be used for description.

The basic machine consists of two augers and two strikeoff blades under a wide, short frame that is supported and propelled by four crawler track units, at the ends of horizontal legs.

The augers and blades are under both hydraulic and mechanical control from the operator's station. Leveling of the machine by means of hydraulic jacks in the leg supports, and steering, are under both manual and automatic control.

References used for automatic control are stringlines and/or completed grades or pavements.

FRAME

Mainframe.   The mainframe of this model is a boxlike unit, about 28 feet wide and 10 feet long.

Its flat deck supports the power unit, which is made up of a diesel engine and a set of hydraulic pumps; the operator's station with its control console or panel; and a two-section tank containing fuel and hydraulic fluid.

The augers and blades, called collectively the subframe components, are held within and beneath the deck by slide plates or bars, and are supported by hydraulic cylinders and mechanical stops.

The flooring is partly an open grating that permits the operator to watch the behavior of the units beneath. A waist-high railing surrounds the deck. An opening and climbing rungs are provided at each side. A cab is optional.

Legs.   This size machine is supported by four legs, extending forward and backward to crawler drive units. Single-lane models may have three legs.

Each leg is normally attached to the mainframe by a pair of horizontal hinge pins at the bottom, and the piston rod of a hydraulic cylinder near the top. When in working position, the cylinder is retracted to clamp the leg firmly in place.

Extending the cylinder forces the leg to pivot downward. Control is by two manual valves, one for front and one for rear. This action raises the frame for loading on a trailer, or for convenience in inspection and repair under it.

Each leg is attached at its outer end, through a steering strut fork, to a vertical hydraulic cylinder whose piston rod is supported by a bracket on the crawler frame. This cylinder raises or lowers the leg to establish and maintain the level of the mainframe.

A light outrigger frame may be mounted on the outer side of each leg, to support a sensing unit for automatic controls.

Crawler Units.   The crawler units have tractor-type frames and tracks, with the three-cleat pads used on loaders. Any of several track lengths, up to 10 feet, and widths up to 2 feet, are chosen according to machine size and expected firmness of ground.

When steered, the crawler frames swivel in a limited arc around the centerlines of the leg elevation cylinders.

Drive is by means of hydrostatic (axial cylinder) motors, usually mounted on the inboard sides of the frames, with speed-reducing roller chains to sprockets on short axles that turn the bull wheels and tracks.

The rear crawler motors have fixed displacement, the front ones variable displacement. Drive is to all four during work, but may be shifted to the rear wheels only when traveling.

Steering.   The crawlers are swiveled for steering, with front and rear sets under separate control.

The same mechanism is used front and rear. A two-way cylinder is mounted horizontally across the outside of the mainframe. The piston rod is anchored to the frame. The cylinder is fastened to a bracket-supported slide bar.

In Fig. 19.39, expanding the cylinder moves the bar to the left, retracting it moves it right. The bar is connected at both ends to link rods that move wide, rigid steering arms under the legs. The arms swivel the crawlers through yokes.

Steering is usually controlled automatically by front and rear sensors and a stringline while working. During travel, and work without a line, it is operated by a pair of switches on the console. See Fig. 19.40.

Turning front and rear tracks to the same angle in opposite directions enables the machine to follow curves accurately. Turning them equally in the same direction sideshifts it without any change in centerline direction. Intermediate positions make possible very delicate maneuvers.

SUBFRAME COMPONENTS

The work of the Autograde is done chiefly by two full-width, horizontal, 30-inch-diameter augers, backed by two high, full-width blades or moldboards. These units are known collectively as subframe components.

Each of them is divided into right and left sections, with a hydraulic lift cylinder at each outer end and one in the center. Mechanical depth indicators extend from near the lift points up through the deck.

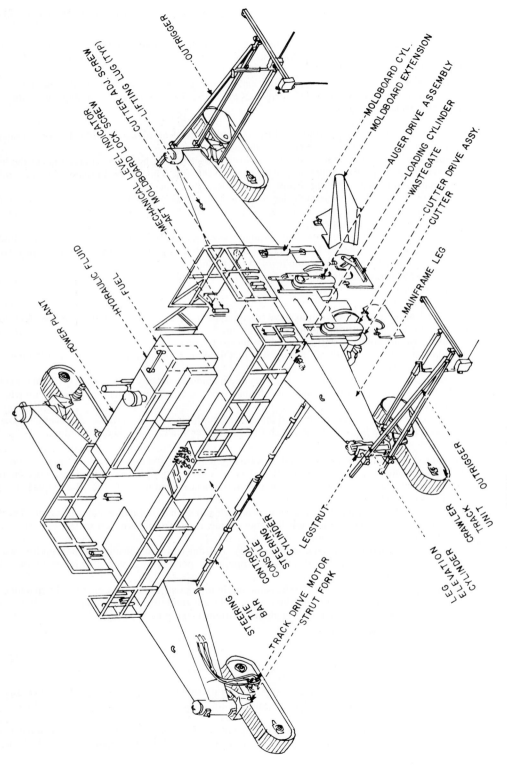

OUTRIGGER

MOLDBOARD CYL.
MOLDBOARD EXTENSION
AUGER DRIVE ASSEMBLY
LOADING CYLINDER
WASTEGATE
CUTTER DRIVE ASSY.
CUTTER

LIFTING LUG (TYP)
CUTTER ADJ. SCREW
AFT MOLDBOARD LEVEL LOCK INDICATOR SCREW
MECHANICAL LEVEL LOCK ADJ. SCREW

MAINFRAME LEG

POWER PLANT
HYDRAULIC FLUID
FUEL

OUTRIGGER

STEERING TIE BAR
CONTROL CONSOLE
STEERING CYLINDER
LEGSTRUT
TRACK DRIVE MOTOR
STRUT FORK

ELEVATION CYLINDER
ELEVATOR
CRAWLER TRACK UNIT

FIGURE 19.39 Automatic two-lane processor. (*Courtesy of CMI Corporation.*)

FIGURE 19.40   Crawler unit. (*Courtesy of CMI Corporation.*)

Cutter.   The front auger is called a cutter. It takes care of initial digging, pulverizing, and spreading operations.

There are two 14-foot, 12-inch-diameter tubes. They carry one welded-on spiral of helicoid flighting, and a spiral of digging teeth. The tooth shanks are welded into the tube, and have replaceable carbide points.

Spirals are usually oppositely pitched on the two halves, so that if they are turned in the same direction, all material will be moved outward to both sides, or inward. If rotated in opposite directions, material will be shifted to one side.

The front moldboard is close behind the cutter, and holds material in contact with it until it passes under the edge or off to the sides, ensuring a constant supply for filling hollows.

Discharge to the sides is controlled by the cutter doors. These are hinged plates which are raised manually to allow free flow of material, or lowered for various degrees of restriction, and secured in position by pins.

Auger.   The rear auger has only smooth helicoid flights, with no teeth. The flighting is installed in short hinged sections, wrapped and bolted on an inner cylinder.

Otherwise, general construction is similar to that of the cutter. It is split into right and left sections with separate drive, with lift cylinders at the center and sides. Drive motors and mechanisms are identical.

The flow of material out the sides is controlled by waste gates. These are similar to the cutter doors, but are hydraulically adjusted from the operator's station.

Moldboards.   The front and rear moldboards generally are similar in construction. There are separate right and left halves, with a shared lift mechanism in the center and separate lifts at the sides. See Fig. 19.41. They are about 43 inches high, only slightly concave to the front. Wearing edges are standard grader blades, bolted in place.

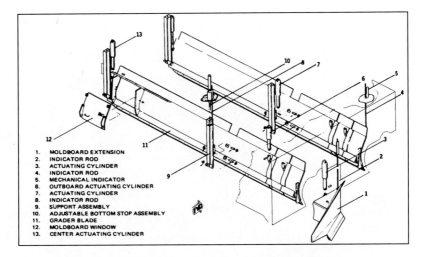

1. MOLDBOARD EXTENSION
2. INDICATOR ROD
3. ACTUATING CYLINDER
4. INDICATOR ROD
5. MECHANICAL INDICATOR
6. OUTBOARD ACTUATING CYLINDER
7. ACTUATING CYLINDER
8. INDICATOR ROD
9. SUPPORT ASSEMBLY
10. ADJUSTABLE BOTTOM STOP ASSEMBLY
11. GRADER BLADE
12. MOLDBOARD WINDOW
13. CENTER ACTUATING CYLINDER

FIGURE 19.41   Moldboards, right side. (*Courtesy of CMI Corporation.*)

Each lift includes a hydraulic cylinder, and a slide bar that moves up and down inside a square channel secured to the mainframe. This mechanism braces the moldboard against working stresses.

Each of the rear moldboard sections has a large, square opening or window near the inner end. This is normally closed by a removable section, which is kept on hangers on the back of the moldboard when the window is open.

These windows are opened when surplus material is to be left in a windrow in the center of the work strip for later removal, or is to be fed into a reclaimer, which will be described below. When windows are open, the auger sections are rotated to move material toward the center. Some adjustment of flighting sections may be advisable for maximum efficiency.

There are three adjustable bottom stops for the rear moldboard—one center and two sides. The adjusting screws are contained in pipes that extend above the deck.

The stops are set when the machine is originally lined up with the stringline at the beginning of a work run, and the moldboard normally rests on them. They permit the operator to raise the moldboard hydraulically if necessary, and then return it to its exact setting without recalibration.

Base Reclaimer.   The subgrade or base reclaimer is an accessory mounted across the rear Autograde legs. It picks up surplus material discharged through windows in the rear moldboard on a conveyor belt.

This belt discharges onto another belt that can direct the material anywhere within a 180° arc. It is high enough to load following trucks over the cab. See Fig. 19.42.

The belts are driven and controlled hydrostatically by two extra pumps mounted on the forward part of the Autograde engine. Swing and height of the second conveyor are adjusted by hydraulic cylinders connected to the constant-pressure circuit.

Controls are at the operator's stations on the Autograde.

HYDRAULIC SYSTEM

Pumps.   The hydraulics are powered by six pumps. There are five axial-piston hydrostatic pumps driven from the rear of the engine, and one pressure-compensated pump at the front. These, with the electrical system, use the full power of the engine, as there are no mechanical drives.

FIGURE 19.42   Base reclaimer. (*Courtesy of CMI Corporation.*)

All of the hydrostatics are driven by a gearbox mounted at the rear of the engine. Each of them is a cam-regulated, reversible, variable-displacement unit operating in a closed loop with its motor or motors. However, a single 35-gallon reservoir, a cooler, and a three-filter system are used by all of them.

Four pumps supply fluid separately to the four motors at the outer ends of the cutter and auger sections. The fifth usually takes care of all four of the travel motors.

The front pump is for all the hydraulic cylinders that steer and level the machine, and raise and lower the subframe members and extensions.

The output of each hydrostatic pump is controlled by the pitch of a cam (see Chap. 12) that is positioned mechanically by a lever on the console. Center position is neutral or nonpumping. Moving away from neutral increases speed, crossing neutral reverses direction.

Additional hydrostatic pumps may be mounted in order to power attached equipment such as a belt reclaimer.

Track Drive.   Each track has an axial piston drive motor, with fluid circulation through high-pressure lines in the leg. All motors have fixed displacement, so that torque and speed are regulated by the pressure and volume received from the pump.

Cylinders and Valves.   All cylinders are double-acting. Wherever necessary, they are linked with slide bars or other devices that ensure straight-line motion of the parts they control.

Valves in the steering and leveling circuits are electrically controlled, and respond either to automatic sensors actuated by stringlines, or to manual control switches on the panel.

The rest of the cylinders, for subframe and outboard units, have electrically actuated valves controlled by panel switches only.

Gauges.   Each hydraulic system has a pressure gauge. These are located with the engine gauges on the left side of the console panel. Control levers are below them, and each is aligned with its own gauge.

A single gauge reports fluid temperature at the outlet to the reservoir. During normal operation the temperature should be 180°F, and it should never be permitted to go over 200°F, where thermal decomposition of the oil becomes likely.

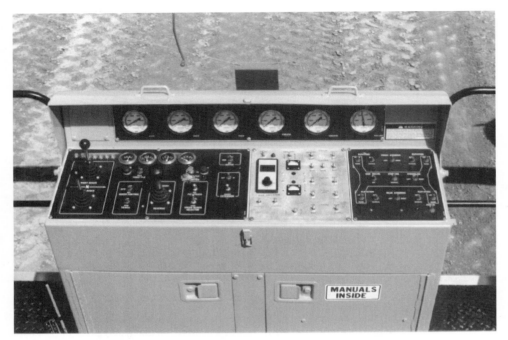

FIGURE 19.43 Control console. (*Courtesy of CMI Corporation.*)

MANUAL CONTROLS

The operator's control console, Fig. 19.43, has three panels or groups. Included are what might be called rotation control levers, two for the left and right cutter sections, two for the auger, one for the tracks, and one for the engine throttle.

Except for the throttle, these levers are neutral when centered, increase speed as they move away from neutral, and change direction of drive when crossing neutral.

One square panel controls the height of the working units relative to the frame. Rubber-covered toggle switches actuate valves in solenoids that raise and lower cutter, auger, and moldboard sections, and the waste gates and any extensions.

The other panel has the same type of switches, to raise, level, and lower the whole machine, either for work or for loading on a trailer. The working-leveling section is labeled TRACERS. It also controls the track drive selection, steering, and the master electric switch.

Leveling and steering switches have signal lights that indicate when a circuit is active. Their principal value is to keep the operator informed about the actions of the automatic controls.

AUTOMATIC CONTROLS

CMI's standard Hydra-Mation, a patented automatic control system, is a fully proportional all-hydraulic system. Information from a preset stringline, or lines, is transmitted by aluminum wands attached to direct-acting, servo-valve-proportional sensors. Individual flow control at the sensors provides for fine-tuning. The standard elevation control may be operated either manually or automatically. The standard control system includes equipment for four-point control (dual stringline), lock-to-grade, or automatic reference control of elevation.

Outriggers are adjustable as required.

Stringline.   The general principles of setting up a stringline were discussed in Chap. 2. It is a continuous line of taut string, or occasionally wire, that is exactly parallel horizontally and vertically with an edge of a proposed grade or surface.

For satisfactory results with the Dual Lane Autograde, the string should be between 18 and 30 inches above desired grade, and 5 inches above obstructions.

Distance between a pair of stringlines may be 32 to 64 feet, depending on the outrigger equipment. If the machine has a cross-slope system, one line, on either side, is sufficient.

Since automatic controls cause a machine to follow a stringline exactly, it is absolutely essential that it be accurate.

Tracer.   A tracer or wand is usually an aluminum rod, extending about 18 inches from the sensor. An extension across the sensor carries a counterweight that presses the tracer lightly upward against the string. This keeps it in contact, and tends to counteract any slight sag of the line.

Steering wands are balanced so as to press lightly outward.

Other Grade Guides.   Leveling may be regulated by the surface of an existing grade or pavement, instead of by a stringline.

One method is to attach a probe frame to the outrigger, to slide along the grade to be matched. A string is stretched in the frame, and the wand is placed in contact with it, as in Fig. 19.44. As the probe moves along the surface, the frame and string are raised or lowered by any change, and the information is transmitted to the sensor.

FIGURE 19.44   Probe tracer control. (*Courtesy of CMI Corporation.*)

Another method is to attach a small wheel directly to the tracer, and allow it to roll on the finished surface.

Sensor Box.   The tracer and its counterweight extension are both secured by setscrews to a rotary switch shaft. A shock-absorbing spring protects the shaft from damage from tracer collisions.

If the stringline raises a tracer, this indicates that the machine is too low at that point. The tracer rotates the switch, making a contact that sends current to the RAISE side of a solenoid valve, and also lights a signal on the operator's console.

The valve directs pressured fluid to the lower side of the leg cylinder, the leg and sensor are raised, lift on the wand ceases, the switch rotates back to null (neutral), the valve closes, and the signal goes out.

If the wand dipped while following a stringline, a too-high grade would be indicated. The switch would be turned in the reverse direction, opening the LOWER side of the valve, and lighting a different signal.

Signal lights tell the operator what the controls are doing, which is important if any malfunction is suspected. The lights are also important while adjusting controls at the beginning of a job.

The steering sensors respond to movement toward or away from the inside of the string, opening and closing the proper side of the steering valve and showing signals.

The automatic changes are usually each too small to be felt, and are positively indicated only by the signals.

A horn blows to warn the operator if any tracer jumps or otherwise loses contact with the string.

Overriding.   The operator can use the toggle switches on the control panel to override the automatic action, and lift, lower, or turn the machine if desired. As soon as he or she releases a manual control, the automatic returns the machine to its preset condition.

Hydraulic Sensor System.   Movements of the wand open and close ports in a very small and precise hydraulic valve. Fluid in a pressure line is blocked, or directed through either of two lines to small cylinders that operate the raise or lower (or left or right) ports in the main control valve.

Subframe Components.   The four subframe tools—cutter, auger, and two blades—are lifted, lowered, and leveled as a group by the automatic leg-elevation mechanism. But they remain largely under individual manual control.

The rear moldboard is set with threaded bottom stops, so that it cannot be dropped below grade. But it can be raised to clear an obstacle or for some other reason, although then it should be immediately returned to bottom.

The other three units are not usually limited, as they have less (or no) effect on final grade, unless manipulated so as to starve the rear moldboard of material.

The cutter should be adjusted upward if tooth marks show behind the machine, or downward if the rear auger or moldboard has to grind through hard, unbroken material.

OPERATING CHARACTERISTICS

Starting a Grade.   The relationship between stringline setting and the grade and centerline varies from job to job and section to section. The grade changes with each successive cutting and spreading pass in the same strip. It is therefore necessary to set the automatic controls for each work run.

First, a section large enough for the Autograde is graded to a reasonably smooth finish and approximate grade. This may be done by other equipment, or by the machine itself under manual control. Stringline(s) is (are) set or rechecked carefully.

The Autograde is then operated manually to place it in precise level and side-to-side position. The rear moldboard is set so that its bottom edge is exactly at grade, and preferably flush with the bottom of the side frame.

Threaded bottom stops are set manually from the deck to hold it from going lower.

The sensors are adjusted, one at a time, to be in a null or neutral position when in proper contact with the string. The machine is then operated for a short stretch, the grade it produced checked by instrument, and any necessary adjustments made.

The finished strip should be checked by engineers at frequent intervals during work.

Rough Ground.   These machines are usually operated on finished or semifinished surfaces, where corrections to final grade are relatively small. However, the machine may also be operated on irregular or rough ground that is free of large stones and vegetation. But heavy work may interfere with accuracy, so that it must be brought back for a second or finishing pass.

If the rough work includes filling hollows, a second pass must be made after compaction.

Whenever a second trip will be needed, the first grade should be left high, so that the machine will have something to work with when it comes back.

Compaction.   An optional compactor screed converts CMI Corporation's TR-6004 trimmer reclaimer to a base material compactor. Vibratory units are hydraulically operated with a variable frequency of 0 to 4,500 vibrations per minute. A towed window box is available for controlled spreading of base and subbase materials.

All-cut grading calls for disposal of large amounts of surplus material. Extension moldboards may push it well back, but unless used for shoulder or other-lane fill, it must be removed eventually.

A base reclaimer may be used to pick up material wasted through rear moldboard windows, to load into following trucks, or windrow it well off to the side.

Superelevations and Crowns.   Superelevated (banked) curves have a straight but tipped cross section. This may be indicated by two stringlines. If these are properly set, the machine will automatically take care of transitions from flat cross section to banked slope and back again. The CMI cutter assemblies are fixed at the outboard ends, and will crown 7 inches up and 3 inches down. The basic cutter is equipped with 226 carbide-tipped teeth.

If the straight sections are crowned, careful operator attention is required to set the crown originally, and to make proper transitions from crown to superelevation and back to crown.

In the original setup, the outer edges of the rear moldboard are set at zero on the indicators, and the center is raised hydraulically by the amount indicated on the plans.

In making a transition to a curve, the operator must have the engineers' instructions about the stations at which it begins and ends, and divide that part of the run into a number of small bits of lowering of the center, to make a smooth and precise adjustment.

As the curve ends, the crown is gradually restored by lifting the center.

If the operation is on one stringline with a cross-slope system, the operator must also keep track of variations in transverse slope as the roadway enters and emerges from the most steeply banked stretch.

LOADING ON TRAILER

The Autograde is placed on a trailer by raising it on its legs, backing the trailer under it, then raising the legs. See Fig. 19.45. First, select a reasonably firm and level spot, where the low-bed trailer can be backed under the machine from the side. Put all rotary drives in OFF position. Raise the subframe components all the way. Adjust the constant-pressure pump to 1,800 to 2,000 pounds.

Using the four LOADING switches in the upper-left corner of the right-hand column, lower all the legs, taking care to keep the frame fairly level as the legs push it up. When raising is finished, put the loading switches in neutral.

The trailer needs a clear deck space, 10 feet wide and more than 28 feet long, and must be accurately aligned under the machine. When it is in place, 2 × 6 timbers are set under the ends of the mainframe, and it is lowered onto them by raising the legs. Reduce the hydraulic pressure to 1,000 to 1,100. Lower the cutter and auger until they rest lightly on the trailer bed.

FIGURE 19.45   Straddling a trailer. (*Courtesy of CMI Corporation.*)

If there is sufficient clearance, the trailer may be able to carry the machine to another worksite on the job with its legs extended. If this is not possible, or if the rig must use a highway, the legs must be removed.

This job is done with the help of a crane or other hoist that can handle more than 7,000 pounds. With the weight of a leg and crawler unit supported by the hoist, two bottom pins and locks are taken from the bottom of the leg, and the steering and loading ram connections are opened. The leg is then removed, and placed on a second trailer. This is done with each of them.

At the new work area the machine is reassembled, and lifts itself off the trailer by reversing the loading process. It is usual to make a thorough inspection, and lubricate all points, before putting it back to work.

COMPACTORS

COMPACTION

Compaction is an absolute requirement in at least some part of most earthmoving projects, and is usually worth its cost in filling and backfilling even if it is not required. Its underlying principles were discussed in Chap. 3.

Soil may be compacted naturally (settled) by time and weather. If it is porous, settlement may be speeded by soaking it and allowing it to dry, a method called puddling. But nature is very slow, and neither process can be depended upon to produce the high densities required in modern construction.

This discussion will be limited to mechanical means of obtaining high-density compaction promptly. There are three basic methods: rolling, vibrating, and hammering. The first two are the most widely used, and are often combined. Adequate rolling can sometimes be obtained by systematic routing of hauling equipment.

Compaction Effects.   The primary effect of a roller is usually to compress material under it by dead or static weight. There is also a kneading effect, which usually is small under smooth steel rolls and more important with tamping and rubber-tire rollers.

A vibrator shakes soil particles individually, causing a tendency to move into closer contact with each other, and to displace excess water. Loose soil such as sand or clean gravel may be sufficiently compacted by vibration alone, but usually it is desirable, and often essential, to have weight (sometimes great weight) on the vibrating surface.

Most vibrators used in compaction of earth or asphalt are rollers, but a few are flat plates. Tubular internal vibrators, such as are used in placing concrete, may be useful in stabilizing mud pockets in granular or mixed granular soil.

Hammers are made in a number of different designs, including slow and heavyweight droppers and jump rammers that crush particles together, as well as air and hydraulic hammers with such rapid strokes that part of their work is done by vibration. They are used chiefly in trenches, and in places where obstructions prevent use of rollers.

STEEL WHEEL ROLLERS

Smooth steel rolls are used in consolidation of most blacktop surfaces, and in rolling gravel roads and road bases, and some subgrades. They produce a smooth, solid surface under favorable conditions, but may fail to compact hollows narrower than the roll, and do not compact deeply in proportion to their weight.

Many smooth steel rollers are two- or three-wheel machines. Three-wheel rollers have a pair of large drive wheels in the rear and a smaller, wider, steerable wheel in front. There are also models that have one smooth roll, with drive from tires. Any steel drum that is not a drive wheel can be equipped with a vibrator. The mechanism and its effects will be discussed in a later section.

Two-Wheel Smooth Rollers. The two-wheel roller is also known as a tandem roller. The standard two-wheel roller has same-width rollers in front and in the rear. It may have equal-diameter wheels; or the front, steering one, may be a little smaller in diameter. Weights are usually between 2 and 15 tons, but both smaller and larger rollers are available. The rear drive wheel carries about 60 percent of the roller machine's total weight. An example of this compactor, which may incorporate vibratory action (explained later), is shown in Fig. 19.46.

FIGURE 19.46   Two-wheel static roller. (*Courtesy of Compaction America, Inc.*)

Weight can usually be increased substantially by filling the roller wheels with water or damp sand ballast. If the descriptive rating for a machine has two figures, for example, 10-14 ton, the first represents empty weight and the second the maximum weight with ballast added. The drive roller, which carries more weight, will produce considerably greater compression on the surface the machine is compacting. Compression produced is measured in pounds per linear inch of roll width.

Engines are center-mounted between the rollers and may be either parallel to, or at right angles to, the direction of travel. The frame of the machine is outside the rolls, so that the roller cannot work against vertical walls or other obstructions high enough to hit the frame. The vertical clearance is 17 to 20 inches.

Clutch and Transmission.   A roller drivetrain must be designed so that starting, stopping, and reversal of direction can be done easily and with great smoothness. Rough or sudden response to any of these actions tends to cause shifting of material and overcompaction under the drive rolls.

In gear drive, a reversing clutch is teamed with four speeds. Models with torque converters have two speeds, whereas hydrostatic drive has either two speeds or one speed. All speeds are reversible. The upper gear(s) is (are) generally only for travel. The brake is usually inside the transmission. With hydrostatics, most stopping is done automatically by the drive transmission.

Hydrostatic Drive.   In hydrostatic drive, a variable-displacement, reversible piston pump is mounted on the engine output shaft. It drives a similar, but usually fixed-displacement, motor geared to a simplified transmission, which may be one-speed or two-speed.

This mechanism, which was discussed in Chap. 12, provides smooth reversing of direction and a stepless change of speed ratios, from zero to maximum speed at the throttle setting, by movement of a single lever. It also provides dynamic braking when in neutral.

Guide Roll and Steering.   The front or guide roll is made up of two sections which turn independently. It is connected by a yoke and a horizontal hinge to an overhead kingpin. See Fig. 19.47. The weight of the frame is carried on roller bearings on the kingpin shoulder, and the kingpin itself is kept in line by tapered roller bearings. Steering is by means of a two-way hydraulic ram, acting against a lever clamped to the kingpin. Extending the ram turns the steering roll to the left, retracting it, to the right.

FIGURE 19.47   Guide roll.

The rolls are equipped with scraper blades, two to a wheel, which are held in light contact with the surface by springs unless locked out of contact. These are essential to prevent material from sticking to the roll and building up so as to spoil the smooth surface.

The roller may be equipped with a sprinkling system. This consists of a tank, valves, and piping to the wheels. At each outlet the water trickles over a cocoa mat, a fabric of wood fibers that distributes the water as a film of moisture over the entire roll surface as it turns against it. The mats are swung out of contact when not in use.

Scarifier. A scarifier is a common accessory. Its base frame is bolted to the rear of the roller. The teeth are mounted in individual arms, fastened to a shaft which is raised or lowered mechanically or by a two-way ram hinged to the carrying frame. See Fig. 19.48. A pair of wheels which ride on the surface limit the depth to which the unit can be depressed. Depth of penetration is also regulated by setting the teeth up or down in their clamps, and by holding the wheels above the ground.

A roller has rather weak traction, so in order to do effective scarifying, it usually is necessary to put teeth or lugs in the driving wheels, in tapered sockets provided for the purpose. These sockets are normally filled with plugs flush with the roll. Use of spike teeth not only gives good traction, but damages the surface so that the scarifier has an easier job breaking it.

Portables. There are a number of light tandem rollers, 6 tons or smaller, that are "portable," that is, they can be towed as semitrailers. They have a pair of rubber-tire wheels that are held up out of the way during work, but can be lowered to support them when towed behind a truck.

In the simpler models, the wheels are moved from upper to lower positions manually, after driving one roll up a ramp or bank. In others, they are shifted hydraulically.

In either case, the wheels carry part of the roller weight when they are lowered, causing the guide roll (which is the lighter end) to rise off the ground.

Before lowering the wheels hydraulically, a towbar at the drive roll end is attached to the pintle (tow) hook of a truck. It is forced down hydraulically to lift the drive roll, so that the roller is supported on the rubber tires and the towbar. It can then be towed at moderate highway speeds.

Trench Rollers. Trench rollers are used for compaction of backfills in ditches and for road widening when the strip is narrow. The rollers may be 15 inches to 3 feet wide and have a static weight of 1,000 to nearly 3,000 pounds. They may not be smooth rolls, as seen in Fig. 19.49. To get more compactive effort, this type of roller generally has a vibratory mechanism to produce the equivalent of 1 to 6 tons of force. The vibratory mechanism will be discussed in a following section.

FIGURE 19.48   Hydraulic-lift scarifier.

FIGURE 19.49 Trench roller compactor. (*Courtesy of Compaction America, Inc.*)

OPERATION

Controls. So far as the controls are concerned, roller operation is simple. With either torque converter or hydrostatic drive, there is only one gear for working speeds, and probably another for travel. The forward-reverse lever, the steering bar or wheel, the throttle, and sometimes the brake are all you have to use.

Modern machines have hydraulic steering. If there is a steering wheel, control resembles that of an automobile.

Units steered by a tiller or steering bar are turned by moving the bar in the direction of turn until the front roll is at the angle desired. The bar is returned to center until the turn is completed, moved to the other side until the front roll is straightened out, and is then returned to center. Any adjustments in sharpness of turn may be made by moving the bar one way or the other.

The error to be avoided is that of leaving the bar in the side position, as one would a car's steering wheel. This causes the ram to continue to move, making the angle sharper and sharper until the stops are reached.

A differential lock is not used during compacting work. It is intended to give additional traction when scarifying, or walking over rough or steep ground.

The sprinkling system cannot be used in compaction of subgrades, as dirt or gravel will stick to the wet rolls and come up in chunks, spoiling the surface. If water is needed for the compaction, it should be supplied by water wagons sufficiently in advance that it can dry slightly before the roller reaches it.

Stopping. Stopping and starting should be done gradually in order to avoid scuffing of the surface. It is best to disengage the clutch before reaching the end of the run, so that the roller can drift to a stop without using the brake or reversing the clutch. When it is stopped, the clutch for movement in the opposite direction is gradually engaged for a smooth start.

Obstructions.   Manholes or other obstructions in the road interfere with regular rolling patterns. The way they are handled will vary greatly with their height, construction, and location. In rolling up to an obstruction head-on, the clutch should be released or partially released before coming to it, so that the roll will just touch it without having momentum enough to break or climb it.

The pattern is rolled the first time as nearly normally as possible, and curving passes are made later. Spaces which the roller cannot reach without excessive maneuvering should be hand-tamped, or a smaller roller should be used in them.

Sequence.   Rolling speeds are slow. One and a half to three miles per hour is usual.

The rear wheels of a roller do most of the compacting, particularly on the three-wheel type. The smaller and lighter front roll serves to "work" and stabilize the soil.

In rolling deep, loose material such as fill or gravel, all passes in a series except the first should be overlapped at least half the width of the drive roll. Gradual extension of the roller into the unrolled area makes possible greater concentration of weight on local ridges and high spots, and keeps the rolls running at a truer grade.

In rolling a graded area with a sideslope, as a crowned or banked road, work should always be done from the bottom up. The lower edges of the rolls have a tendency to push downhill, which can be best resisted by compacted material. In working uphill, the creep of soil away from the upper edge helps to preserve the slope.

A crowned road is rolled according to the pattern in Fig. 19.50, starting at one edge and working up until the center is reached by the upper roll, then moving diagonally to the opposite side and working up from there. Each rerolling is started at the bottom in the same manner.

Banked or superelevated curves are rolled from the inner edge to the outer edge, Fig. 19.51. The transition from crown to bank is made by a diagonal from center to low side. From bank to crown the move is from either edge straight into the adjoining low side. The meeting of these two types of grade is a convenient place to end a rolling section, if the continuous-advance system is not being used.

Rolling should be continued until no advantage is noted from successive passes. Presence of too much water in the subgrade may make its compaction impossible, but long rolling will at least bring much of the water to the top where it can evaporate more readily. The waterlogged condition results in a rubbery action of the ground, in which it goes down under the rolls and springs back into nearly its original condition when they have passed. This condition may not be apparent at the

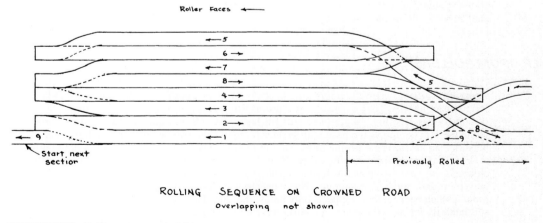

ROLLING   SEQUENCE   ON   CROWNED   ROAD
Overlapping   not   shown

FIGURE 19.50   Rolling sequence, straight.

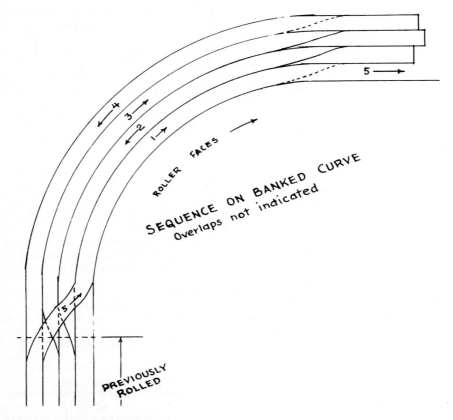

ROLLER FACES

SEQUENCE ON BANKED CURVE
Overlaps not indicated

PREVIOUSLY ROLLED

FIGURE 19.51 Rolling banked curve.

start of the work, as the larger airspaces in the unconsolidated soil may be adequate to hold the water. As these spaces are reduced, however, the water is forced out of them and becomes a lubricant between all the particles.

OTHER DRUM ROLLERS

Tamping Rollers. Towed tamping or sheepsfoot rollers were for a long time the standard tool for compacting fills. They consist of steel drums fitted with projecting "feet" and towed by means of box frames. See Fig. 19.52. On a soft fill layer, the roll will compact the surface somewhat while the feet compress the base with greater force. As the soil becomes packed, the feet do not penetrate as far, and first lift the roll clear and finally walk themselves almost out of the ground.

Feet are usually from 7 to $9\frac{1}{2}$ inches long. Two types are used. The original sheepsfoot has an enlarged, off-center sole and a straight shank. The tapered foot diminishes from base to sole and may be round, square, or angular. The sheepsfoot is easier to pull, but is liable to tear up the ground when backed, and compacts only below the sole. The tapered foot kneads and compresses soil laterally for its full length, and works as effectively backward as forward.

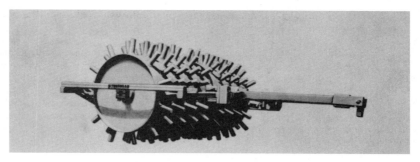

FIGURE 19.52 Tamping rollers.

Two or more units can be combined in multiple frames. These are hinged to oscillating bars in the rear, and to tow bars in front. The towing tongue should be spring-cushioned to minimize shock to tractor and roller.

Cleaner bars are usually put on the rear of the frame to remove dirt caught between the teeth. If the roller is to be used extensively in reverse on sticky soils, it should have front cleaners also. Clogging destroys the effectiveness of the machine.

Individual rolls may be from 4 to 6 feet wide and 3 to 5 feet in diameter, not including the feet. They can be filled with ballast of water or sand, and may carry sandboxes in addition. Foot pressures range from about 150 to 750 pounds per square inch, depending on the size of the unit and the amount and kind of ballast used.

A bulldozer may tow a tamping roller while spreading fill, so that both grading and compaction can be done by one operator.

Self-propel. Tamping rollers are now self-propelled. They may be specially designed machines, or four- or two-wheel-drive tractors with special wheels. These machines are usually center-articulated, with arrangements for rear wheel feet to step accurately between prints of the leading wheels, in forward or reverse movement.

Traction with four-wheel drive is commonly by rear axle oscillation, and it often is profitable for the machine to carry a dozer blade, for use in rough grading while compacting.

Self-propelled tamping rollers may have a working speed of 4 miles per hour forward and reverse. The long feet tend to kick out and loosen soil at over 6 miles per hour, and thus limit working speed.

Shorter feet, or the pads described below, allow higher speed during compaction.

There are many tamping roller combinations. A single machine, usually articulated, may have a pair of rubber drive wheels and a tamping drum, a tamper and a smooth wheel, or a tamper and a vibrating roll. Usually, each has its own special uses, with an additional range of general applications.

Pads and Crushers. Steel drums may be obtained with a variety of surfaces, some of which are shown in the accompanying illustrations.

The pad roller carries wider, closer-spaced feet than a tamping roller, for more concentrated compaction effect to a lesser depth. It has increased effectiveness in loose soil.

Feet may be combined into parallel ridges around the drum, with height varying according to a pattern. Used on self-powered, center-articulated tractors, these are designed so that the rear ridges track between the front ones, even while turning. They are effective crushers of miscellaneous material, including soft rock, old pavement, garbage, and vegetation.

RUBBER-TIRE ROLLERS

The rubber or pneumatic tire roller substitutes a number of wheels and tires for the steel drums discussed so far. They add to their downward pressure a kneading effect, as material is pressed toward spaces between tires.

Wheels are mounted in pairs that can oscillate, or singly with spring action, so that tires can move down into soft spots that would be bridged by a drum. There may be means to lock out individual movement for special work.

The typical rubber-tire roller is a weight box mounted on two rows of tires in tandem. See Fig. 19.53. The rear row may have an even number, the front an odd number of wheels. They are aligned so that the rear tires more than cover the spaces left between front tire tracks.

FIGURE 19.53 Rubber-tire roller in action.

FIGURE 19.54   Fifty-ton Rollopactor. (*Courtesy of Wm. Bros Boiler & Manufacturing Co.*)

Ballast in the boxes is usually dry, and almost anything will do. Sand, gravel, stone, and/or chunks of metal are used.

These rollers may be trailers. More often, they are self-propelled with (usually) torque converter, reversing clutch, and two to four speeds each way. Final drive may be gear drive or hydrostatic. Top speeds may be 15, or over 20, miles per hour. Weights range from about 5 to 35 tons.

Rubber (pneumatic) tire rollers are successful in most types of compaction work, from raw fills to asphalt surfaces. The tires, which are usually smooth tread, may be fitted with scrapers for sticky soil.

Heavy Compactors.   There are larger rollers, made as trailers only, in which there is a single row of two or four tires across the center of the weight box. See Fig. 19.54. Stability is provided by the hitch to the towing tractor. There may be a vibratory attachment.

With ballast, these weigh from 50 to 200 tons. They are used for compaction of earth fills in layers up to 2 feet thick, and even thicker layers of broken rock.

These big jobs are not nearly as popular as the two-axle types. They are criticized as being slow and cumbersome, and prone to getting stuck, or tipping. Weak fills may be damaged by shearing, due to extreme weight. But under a number of conditions, they are highly efficient tools.

VIBRATORY ROLLERS

Vibration.   Vibration of a tool in contact with the ground produces a rapid series of impacts that develop pressure waves which penetrate the soil, setting its particles in motion. If the tool exerts static pressure, the combined effect will be to rearrange the particles, and force them into a compact structure with a minimum of voids.

Vibration is usually produced by rapid rotation of an axle carrying off-center weights, called eccentrics. It is transmitted to the vibrator parts through ordinary bearings, and insulated from nonvibrating parts of the machine by pads or hangers of flexible material, called isolators.

Measures of vibratory motion are frequency and amplitude. Frequency is in cycles per minute, and depends on the rotation speed of the eccentric axle. There may or may not be means to change frequency during operation. See Fig. 19.55.

Frequency available in rollers ranges from a few hundred vibrations per minute (vpm) to 3,200 or more, but in any one machine, the range usually is much smaller.

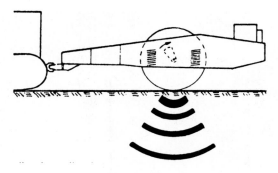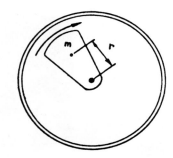

FIGURE 19.55   Vibration, and vibrator.

Amplitude is the distance of movement of the vibrating surface from a central position. It is half the total movement. It varies with the elastic and damping properties of the soil, but may be given a nominal rating, neglecting soil effect.

In some machines, amplitude may be adjusted by transfer of fluid between rotating chambers, one of which adds to off-center weight, while the other counterbalances it.

In general, increase in a compactor's ground speed calls for faster vibration. Otherwise, there is a great deal of contradictory evidence about the relative advantages of various speeds.

Most soils have a resonant frequency at which they vibrate most readily. A vibrator in contact with the soil will develop its greatest amplitude at that frequency. Some authorities consider that best compaction is obtained at resonant frequency, while others consider it less important.

Dynamic force is the centrifugal force produced by the eccentrics. In a general way, it represents the power behind the vibratory stroke.

Vibration in Rollers.   A roll vibrates if it is mounted on a vibrating axle. The axle turns in end bearings supported in vibration-damped brackets on the frame, and in another set of bearings on which the roll revolves.

The roll vibrates with the eccentric, and may be called the vibrating mass. The frame and the engine have little vibration, but their weight is essential to the effectiveness of the roll. They may be called the surcharge.

An eccentric may also be turned by a hydraulic motor and V-belt, substituting flexible hose for belts and couplings, and providing means for control of frequency independent of engine or travel speed, and for reversing rotation of the eccentrics. This reversal may be made automatic when direction of travel is changed, to reduce sinking and scuffing.

Many vibratory rollers are of articulated construction, with an engine and control section driving by means of a pair of large wheels, which usually are rubber-tired but may be steel drums, smooth or with tamping or pad feet. The other section is a drum roller, smooth or otherwise, with a vibratory axle turned by a hydraulic motor.

Vibrating rollers usually give best results when moving slowly, from $1\frac{1}{2}$ to 7 miles per hour. They will provide some compaction at higher speeds, but must usually do the same area a greater number of times. See Fig. 19.56.

Vibration greatly increases roller effectiveness on loose, granular soils, but makes little difference on silt and clay. The rule of thumb is: The larger the grain, the greater the effect. Comparisons between vibratory and static rolling are therefore of little value unless the soil type is specified.

Under favorable soil conditions, a vibrating roller may produce compaction equivalent to that of a static roller 2 to 4 times as heavy. Some of them are therefore very small, even so small that the operator walks alongside instead of riding.

FIGURE 19.56   Vibratory tamping compactor. (*Courtesy of Compaction America, Inc.*)

TAMPERS AND VIBRATORS

It is often not practical to use rollers against walls, between manholes, in ditch bottoms, and many other places where space is restricted.

Soil in such spots is compacted by tamping. Considerable benefit can be obtained by walking or stamping down 2- to 3-three inch layers, or pounding with sledges or other tools.

Hand-Operated Rammer.   Mechanical tampers, which may be operated by gasoline, air, or electricity, may operate by jump pounding, by leaving the shoe on the ground and hammering it, by vibration, by combined hammering and vibration, or by weight dropping.

The rammer, Fig. 19.57, includes a single-cylinder, two-cycle engine. The working piston is attached to the foot. This type of machine weighs more than 100 pounds and is operated at the rate of 800 to 3,000 blows per minute, with jump strokes of $1\frac{1}{2}$ to 3 inches off the ground. Tilting it slightly in the direction of the desired motion while working will cause it to walk in that direction.

Compaction Meter.   There is a patented Compaction Meter available from MBW that takes the guesswork out of backfilling a trench to the required compaction. Using a very inexpensive, disposable sensor placed at the bottom of the trench with an attached wire long enough to reach the surface, the backfilling can begin. As the earth backfill is placed in lifts and compacted, the readout meter attached to the upper end of the wire tells by a red light when the lift has been compacted to a set design level, say 95 percent of optimum density. The required design level of compaction can be field-calibrated to any specified level for the backfill. When the backfilling is completed to the top,

FIGURE 19.57   Rammer compactor. (*Courtesy of Vibromax America Inc.*)

the wire is cut off at the surface and only the sensor at the bottom of the trench plus the depth of wire is not recovered.

Vibrating Plate Tamper.   The hand-operated plate tamper, like the one in Fig. 19.58, has a gasoline engine that operates a reciprocating-type vibrator, with frequencies between 4,000 and 6,000 vibrations per minute. It has a very short stroke.

The machines weigh from 400 to 1,500 pounds. Working depths depend largely on soil characteristics. Compaction may be obtained for some distance around the plate, by vibratory settling. The machine is designed to move forward or in reverse as controlled by the operator.

Traveling Vibrator.   This flat vibrating unit, Fig. 19.59, is primarily intended for compaction of blacktop, loose granular soils, clean gravel, and crusher rock.

These machines move themselves along the working surface at speeds of up to 30 feet per minute, when hand-operated. They can also be mounted in gangs held in a lift frame on a tractor, which is also equipped with a generator to provide current for them.

Compactor Selection.   Since soils are usually mixtures of granular and fine-grained soils, compactor machine selection becomes more of a problem. Generally, the machine used should be the type

FIGURE 19.58   Vibrating plate tamper. (*Courtesy of Compaction America Inc.*)

best suited for whichever soil material predominates. The graph in Fig. 19.60 shows application of this approach.

Vibratory Probe.   Sandy soil may be compacted to depths of up to 60 feet by means of an open pipe clamped in a pile-driving vibrator. This is sunk and withdrawn vertically at calculated intervals, usually 4 to 10 feet, in a square grid pattern over the area to be stabilized. See Fig. 19.61.

Vibration packs the soil particles closely together, and forces excess water out of the top of the formation. If the soil type is suitable, compaction can be accomplished underwater.

Power-Down Hammer.   Newer portable hammers may use hydraulic pressure to supply power for the downstroke, as well as for the lift. The hammer in Fig. 19.62 may operate on the rear of either a truck or a tractor.

The truck is much faster and more convenient when moving between jobs, while the tractor is more maneuverable.

FIGURE 19.59   Vibrating compactor.

SOIL

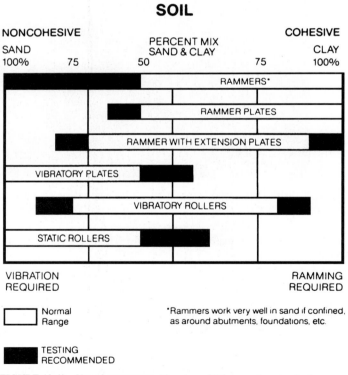

FIGURE 19.60   Use of compactors. (*Courtesy of Vibromax America Inc.*)

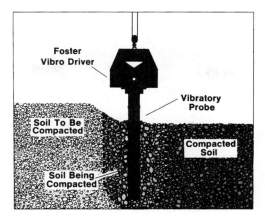

FIGURE 19.61   Vibratory probe.

FIGURE 19.62   Tractor hammer.

This model has a downstroke force that is adjustable by the operator to an impact force of from 80 to 16,000 pounds. Cycle can be varied from 0 up to 110 strokes per minute. This tool is used to compact and stabilize deep, narrow fills, as in trenches. It can be fitted with tools for breaking pavement (illustrated) or boulders.

Dynamic Deep Compaction.   The process of heavy weight dropping in order to modify the load-bearing condition of the ground is called dynamic deep compaction. On one site, two cranes lifted and dropped 10- to 20-ton weights in a grid pattern covering a 83,000-square-foot building area.

The dynamic deep compaction procedure densified the soil so much that the ground surface after debris recompaction was about 3 feet lower than the original surface.

The dynamic impact of a heavy weight dropped from heights of up to 150 feet transmits shock waves downward through the soil. Dropping weights like this on the surface is practical for soil depths of up to 100 feet. Most soil types, even silts and some clays, can be helped by this process. For treatment of cohesive soil, careful planning and field control are mandatory. Between weight drops, high pore pressures must be allowed to dissipate. This can be accelerated by using sand wick drains.

CHAPTER 20
COMPRESSORS AND DRILLS

COMPRESSORS

Light and medium drilling and breaking tools are usually powered by air supplied by portable compressors that are driven by standard types of industrial engines. These machines draw in atmospheric air, compress it, and deliver it through hose to the tools. The compressed air provides a power medium somewhat similar to steam in performance, with the important difference that it can be used cold.

Types of Construction. Compressors used in construction work are one of three types. They may be of the original reciprocating type, the newer rotary vane type, or the newest rotary screw type. Most portable compressors used in construction are of the rotary vane or screw type of air pumps with one or two stages.

In all manufactured forms, it is important that incoming air be filtered to prevent dust from causing excessive wear, fouling moving parts, and choking passages. Intake may be through one or several passages or manifolds.

The diagram in Fig. 20.1 shows the airflow in a single-stage compressor. The inlet passage receives atmospheric air through a cleaner. The discharge is at full working pressure, usually 100 pounds per square inch (psi), and opens into a line to a storage tank (air receiver) from which it is piped to the tools.

The two-stage machine, Fig. 20.2, has one or more primary or low-pressure cylinders which draw in atmospheric air, compress it to about 30 psi, pass it through a cooling radiator (intercooler) to

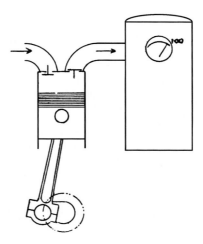

FIGURE 20.1   Single-stage compression.

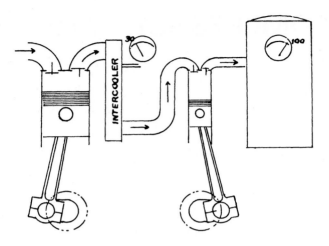

FIGURE 20.2 Two-stage compression.

a secondary or high-pressure cylinder, where pressure is increased to 100 psi and discharge is to the receiver.

Rating.   Compressors are rated according to the number of cubic feet of atmospheric air they take in each minute, when working at maximum governed speed, with a specified pressure, usually 100 psi, at the discharge. This rating is abbreviated to cfm or CFM. The "actual cfm" is this intake measurement, less any losses by leakage in the machine.

Actual cfm, or actual air delivery, is the volume of compressed air coming from the receiver outlet, transformed back to standard atmospheric pressure at sea level. Some garage and industrial compressors are rated according to piston displacement per minute at full speed. This figure will always be larger than the cfm in the same compressor, as air left in the space above the piston on the compression stroke expands and partly fills the cylinder on the suction stroke so that the full displacement cannot be filled with new air. See Fig. 20.3.

The relationship between cfm and piston displacement is called volumetric efficiency. This figure may be between 65 and 80 percent, and depends largely on the airspace above the piston and the increase in pressure attained in the cylinder. In two-stage machines, it is calculated on the basis of the displacement in the primary cylinders only.

Pressure Control.   Air-operated automatic controls are used to keep the pressure in the receiver within certain limits, usually 90 to 100 psi. In addition, the engine must have a mechanical governor to prevent it from racing when air pressure is low.

During operation the engine and compressor continue turning at all times. When the receiver pressure reaches maximum, the compressor intake valves are held open so that air that enters on the suction stroke is forced out again on the compression stroke. In this condition the compressor is said to be unloaded. The engine is throttled down at the same time the unloading takes place. In rotaries, the throttling of the engine automatically unloads the compressor.

Heat.   Compression of air produces heat. The working parts of the compressor tend to become very hot and must be cooled by fins, water jackets, or an oil bath.

The ability of air to hold moisture is decreased by pressure and increased by heat. As the hot compressed air enters the receiver, these effects are often balanced, but as the air cools in the receiver or in the lines, extensive condensation may occur, particularly on damp days.

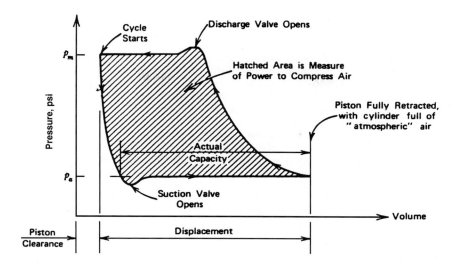

FIGURE 20.3 Cycle of air compression.

Compressed air loses some pressure as it cools, so that the compressor must pump air to compensate for shrinkage, as well as to supply the tools.

Reciprocating Compressors.   The reciprocating type of compressor, which was the form originally used, will be discussed first because it is easier to understand. The reciprocating-type machines have cylinders, each of which takes in air during a suction stroke, and discharges it through a check value at higher pressure during a compression stroke. These may be in the same block as the engine cylinders, but more often are in a separate block with a disk clutch connection between crankshafts.

The compressor proper consists of four or six cylinders in line in a single block. Standard industrial engine types of pistons, rods, crankshaft, bearings, and cooling system are used, and lubrication is by force feed through passages in the shaft.

The intake valve, Fig. 20.4, is a poppet type which opens during each downstroke of the piston. The push rod consists of two opposed pistons, sliding in an open cylinder. These pistons are normally in contact with each other and transmit the thrust of the cam to the valve stem.

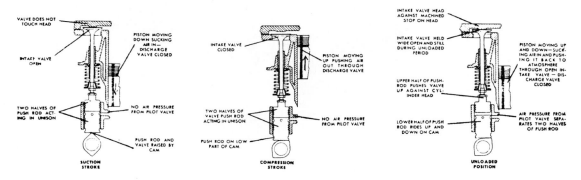

FIGURE 20.4 Unloading-type poppet valve.

The compressed air is cooled by water jackets around the exhaust passages, which are called the aftercooler.

Cylinders and Pistons.   Compressor pistons may have either flat or convex tops.

Single-stage compressors ordinarily have a number of identical cylinders and pistons in line in a water-cooled block. In two-stage units, the low-pressure stage requires more cylinder capacity than the high-pressure, so that if a single cylinder of each type is used, the primary is much larger. Two or three first-stage may be used for each second-stage cylinder so that all cylinders can be of similar or identical size.

Such cylinders are arranged radially, as shown in Fig. 20.5, so as to operate off one throw of the crankshaft. Large models may have two or more sets, each on a separate throw.

Two-stage cylinders may be either air- or water-cooled. Air cooling employs the thin cylinder walls and outside fins illustrated. The intercooler fan keeps air moving over the fins.

Heads and Valves.   On two-stage machines, each head contains a discharge valve and passage, and usually an inlet valve and passage also. The primary stage inlet valves are fitted with an unloading device.

Valves are of the automatic check type. Suction valves are similar or identical to discharge valves, but are inverted to act oppositely. See Fig. 20.6.

When the compressor unloads (stops compressing), air is admitted to the chamber above a diaphragm, forcing the diaphragm and plunger down until the fingers press the valve away from the seat. Air drawn in on the suction stroke is now discharged back through the inlet valve on the compression stroke, and no work is done. When air pressure above the diaphragm is released, springs push the fingers out of engagement.

This type of unloader may also be worked by air pressure against a piston attached to the plunger, rather than against a diaphragm.

Valves are usually accessible for service by removal of a cover plate, without need for disturbing other parts of the machine.

Intercooler.   The intercooler, used only in two-stage machines, is a tubular cooling radiator connecting the first-stage discharge passages with the second-stage inlet. A fan mounted on the rear

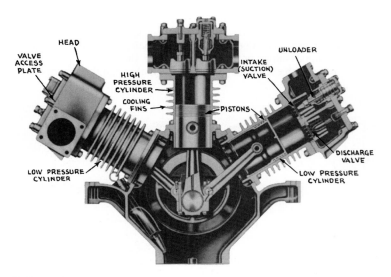

FIGURE 20.5   Two-stage compressor cylinders.

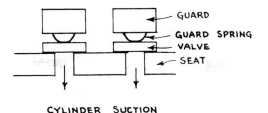

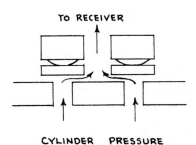

FIGURE 20.6   Exhaust valve action.

of the compressor provides air circulation around both these tubes and the compressor cylinders. Intercooler efficiency varies, but it is expected to reduce the temperature of the contained air to within a few degrees of that of the atmosphere.

Intercooler pressure has a natural relationship to the receiver pressure, generally in the neighborhood of 30 to 100 psi. If intercooler pressure rises disproportionately, leakage in high-pressure suction valves is often indicated.

The intercooler is fitted with a spring-loaded safety valve set for about 35 psi. This can be tripped by hand, and should be blown daily to avoid the accumulation of material which might clog it.

A drain valve in the bottom should also be opened daily, or more often, under pressure, to blow out accumulations of water or oil.

Intercoolers are sometimes fitted with an unloader or automatic drain, which consists of a blowoff valve that is opened by air pressure in the control tubing.

Air Receiver.   The air receiver is a cylindrical tank with convex ends, usually mounted horizontally on the rear of the frame. It must conform with various federal and state safety regulations, and National Board certification is normally stamped on it.

The receiver acts as a small reservoir between the compressor and the tools, which reduces the frequency of unloading in light use. It serves to separate moisture and oil from the air and provides a place for draining them.

It is equipped with an inlet from the compressor, and with one or more outgoing air lines with shutoff valves. A smaller air line goes to the pressure gauge and the automatic controls.

The following fittings are required for safe and convenient operation:

Safety valve of the spring type that will open when air pressure rises above the highest operating pressure. This must have the capacity to let air out faster than the compressor can put it in. It should have a hand trip mechanism by which it can be opened and any sediment blown out daily.

A gauge indicating pressure in the receiver. It is the standard practice to use a 200-psi face, so that the needle will be vertical at normal high pressure of 100 psi.

Compressor Size	50	105	210	315	420
Safety Valve	1"	1 1/4"	1 1/2"	2"	3"
Pressure Gauge	3 1/2"	3 1/2"	3 1/2"	6"	6"
Drain Valve	1/2"	1/2"	1/2"	1/2"	1/2"
Fusible Plug	3/8"	3/8"	3/8"	3/8"	3/8"
Air Line Valve	3/4"	1"	1 1/2"	2"	3"

FIGURE 20.7   Minimum size of fittings.

A drain cock or valve at the low point in the receiver. This is used for draining or blowing out water, oil, and sediment.

A fusible plug will melt if the air gets hot enough to be near the flash point of lubricating oil vapor. The table in Fig. 20.7 gives recommended safe sizes for these attachments.

Air Pressure Controls.   Automatic controls are necessary to keep sufficient pressure in the receiver to operate the tools and to prevent building up of excessive pressure.

Air from the receiver is admitted to the control system by the pilot or trigger valve, two types of which are shown in Fig. 20.8. Receiver air contacts a disc or a ball held on a finely machined seat by a spring. When pressure becomes high enough to force it off its seat, a much larger area can be reached by the compressed air so that the push against the spring is greatly increased, and the disc or ball is snapped back far enough to allow air to escape quickly into a side passage to the controls.

When pressure drops sufficiently, the disc will be pushed down by the spring. When it passes the side passage, air from the passage presses against its upper surface, equalizing the receiver pressure below, and allows the spring to push the disc back on its seat. The air in the passage then leaks out through the top of the valve.

The engine throttle is either fully open or in idling position, and is liable to move frequently from one to the other. This wastes fuel because of the nonproductive time while unloaded and the extra consumption during acceleration. The change of speed and load also increases wear.

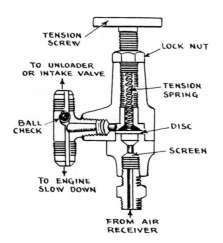

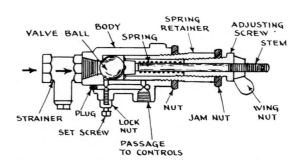

FIGURE 20.8   Pilot or trigger valve.

Most manufacturers offer devices that will proportion engine speed to receiver air pressure. Various constructions are employed, but in principle they consist of a diaphragm or piston acted upon directly by air from the receiver, which pushes against the spring-held throttle lever, so that increased pressure will partly close the throttle, and falling pressure open it. Adjustment to the desired pressure-speed relationship may be made by adjusting the tension of the spring.

These "fuel savers" may show substantial economies when the average consumption of the tools used is below the capacity of the compressor, as they will permit continuous operation on partial throttle. However, when the air demand is intermittent, or as great as or greater than the compressor output, they are of little use.

Gauges and Controls.   The instrument panel should carry the indicators for the compressor oil pressure and receiver air pressure, in addition to the usual engine gauges. A two-stage compressor should have a gauge for intercooler pressure also.

The clutch control is usually an over-center lever, but in one make it is a hydraulic jack which is pumped to disengage the clutch and released to engage it. Auxiliary levers are provided for hand operation of the throttle and the unloading mechanism.

Capacities.   Air requirement of tools varies with weight, model, and condition. In general, a 60- or 85-cfm compressor can supply two light hand drills or paving breakers, or one of medium weight. A 105- or 125-cfm can take care of a heavy hand drill, or two or more light ones. A wagon drill requires a 210- or 315-cfm, a light crawler drill a 600-cfm one.

When tools are used intermittently, a compressor may be able to take care of a larger number than is indicated by its capacity, but it will do so at the risk of reduced pressure and delays.

Mountings.   All makes of portable compressors may be had in a variety of mountings.

Rubber-mounted compressors with a capacity of 60 cfm and smaller can be moved short distances by hand; 105-cfm compressors may be towed by a car or shifted by hand on level pavements. Larger models should be handled with a truck or tractor.

Operation.   Before starting a compressor, a routine check should be made of oil level in engine and compressor, fuel, and cooling system. The receiver drain cock should be opened and the clutch disengaged. See Fig. 20.9.

After starting, the engine is run for several minutes to warm, then the clutch is engaged. Soon after it turns the compressor without choking or laboring, the drain cock may be closed.

If the engine is not thoroughly warm or is in poor condition, it may not be able to build up full receiver pressure immediately. If it threatens to stall, the clutch should be disengaged or the compressor unloaded by a hand control, and extra warm-up time allowed.

During work the drain in the bottom of the receiver should be opened occasionally to get rid of water from condensation. In hot, damp weather this can amount to several gallons per day, and it is important to keep it down.

The safety valves on the receiver and the intercooler should be tripped by hand at least once per day to blow out any carbon or sludge deposits which might prevent them from working in an emergency.

The filters in the air intake for both engine and compressor, and any in the control system, should be cleaned as often as necessary.

Once started, the operation of the compressor is completely automatic, except for the occasional attentions listed above. The oil and air pressure gauge should be checked occasionally.

A spare valve assembly should be kept on hand.

Carbon and Explosions.   Air temperatures are high enough to cause some vaporization of lubricating oil. The nonvolatile residue, in combination with any dirt in the air, is liable to build up hard or gummy deposits that will interfere with valve action, and sometimes will choke air passages. Since thin oils leave less residue than heavy ones, it is advisable to use the lightest oil that will give proper lubrication, and which will not be pumped into the cylinders in excessive quantities.

Thorough cleaning of incoming air is essential.

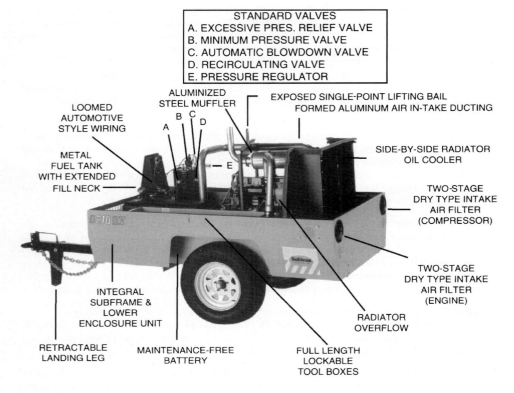

STANDARD VALVES
A. EXCESSIVE PRES. RELIEF VALVE
B. MINIMUM PRESSURE VALVE
C. AUTOMATIC BLOWDOWN VALVE
D. RECIRCULATING VALVE
E. PRESSURE REGULATOR

ALUMINIZED STEEL MUFFLER

EXPOSED SINGLE-POINT LIFTING BAIL
FORMED ALUMINUM AIR IN-TAKE DUCTING

LOOMED AUTOMOTIVE STYLE WIRING

METAL FUEL TANK WITH EXTENDED FILL NECK

SIDE-BY-SIDE RADIATOR OIL COOLER

TWO-STAGE DRY TYPE INTAKE AIR FILTER (COMPRESSOR)

TWO-STAGE DRY TYPE INTAKE AIR FILTER (ENGINE)

RADIATOR OVERFLOW

INTEGRAL SUBFRAME & LOWER ENCLOSURE UNIT

RETRACTABLE LANDING LEG

MAINTENANCE-FREE BATTERY

FULL LENGTH LOCKABLE TOOL BOXES

FIGURE 20.9   Portable compressor without hood.

If exhaust valves leak, some of the hot compressed air which has just been forced out of the cylinder will come back in and be recompressed, resulting in very high pressure locally, and excessive temperatures which will boil off the lubricant.

Under these conditions enough oil may be present in the air to cause explosions, most often in the cylinder (dieseling) or the passages, but occasionally in the receiver or even in the tools.

This danger is avoided by replacing piston rings in the compressor as soon as they allow pumping of oil, keeping valves free of carbon and dirt, and replacing broken or worn parts.

Multiple Units.   Air demand too large for a single compressor may be supplied by two or more of any size, discharging into a single supply system.

The hookup can be arranged so that one machine does most of the work, or so that two or more work equally.

The first method is most satisfactory when one machine can supply the normal load but cannot keep up with peak demand. The compressor which is to carry the full burden has its pilot valve set for normal pressures, probably loading at 90 psi and unloading at 100. The auxiliary has both settings about 2 psi lower. Discharge lines from both receivers should be flexible hose leading into a common manifold, from which air is taken by the tool lines, as illustrated in Fig. 20.10.

When the machines are started, both will run until a pressure of 98 psi is built up in the receivers and manifold. The auxiliary will then cut out and will be unloaded as long as the main machine can keep pressure above 88 psi.

The compressor manufacturers need to be consulted for the piping and value arrangements to team multiple units.

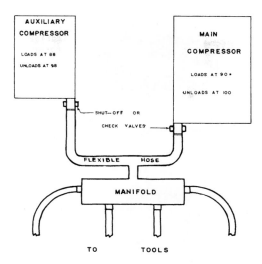

FIGURE 20.10   Piping for main and auxiliary compressors.

ROTARY COMPRESSION

The early rotary compressors operated on the principle of a vane-type air motor, as shown in Fig. 20.11. A cylindrical rotor is mounted off-center inside a larger cylindrical casing. Sliding vanes are fitted into lengthwise slots in the rotor. Centrifugal force causes them to keep in contact with the casing wall whenever the rotor is turning rapidly.

The vanes divide the space between the rotor and the casing into a series of compartments. Because of the off-center mounting, these are very small as they pass one side of the casing, and large on the opposite side.

Slots in the casing allow air to enter the compartments as they expand. As they contract, the air is compressed and is forced out an exhaust passage just before the point of closest contact between rotor and casing.

Rotary Screw-Type Compressors.   The rotary screw form of compressor uses a pair of cylinders with matching screws, as shown in Fig. 20.12, to compress the air in a single operation. There are no sliding vanes or metal-to-metal contact. Oil provides sealing and cooling. The efficient two-stage oil separation and Air-Glide regulation provides smooth, stepless control of compressor capacity, from full load to no load in response to demand of the Ingersol-Rand compressors. See Fig. 20.13.

Servicing.   A rotary compressor of either type requires little attention, except to make sure it has clean air and clean oil of the type and viscosity recommended by the manufacturer. The air filter must be serviced frequently, condensation water must be drained out of the oil filter daily, the oil filter must be in excellent condition, and oil must be changed when it starts to break down.

The immediate result of allowing the oil to become dirty is excessive wear on the vanes or screws. They are pressed against the back or trailing edge of the rotor slots by air pressure ahead of them, and each vane may slide in and out a million times in 12 hours of operation. It is easy to understand that rapid wear will be caused by lubrication failure.

In some machines it is fairly easy to remove vanes for inspection; in others it is difficult. Contractors who are very careful about keeping air and oil clean may get more than 6,000 hours from a set, while others who are less careful may have to replace them every 1,500 hours or else run serious risk of breakage.

Rotor with nonmetallic sliding vanes.

Air is gradually compressed as pockets get smaller.

As Rotor turns, air is trapped in pockets formed by vanes.

Compressed air is pushed out through discharge port.

FIGURE 20.11   Rotary vane-type compressor cylinders.

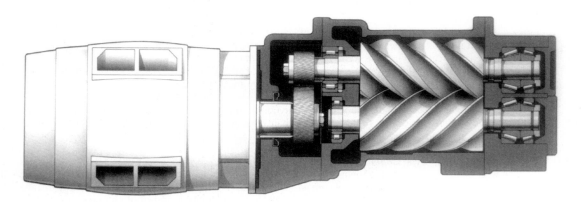

FIGURE 20.12   Rotary screw-type compressor cylinders. (*Courtesy of Ingersoll-Rand Company.*)

Because of maintenance problems with the vanes, the majority of rotary compressors now are of the screw type.

Tractor-Compressors.   Many small contractors have only occasional use for a compressor, with the result that the machine stands idle so much of the time that it cannot make an adequate return on investment, and tends to deteriorate rapidly because of excessive idle time, usually exposed to the weather. However, this occasional use may be sufficiently important to justify the ownership of the machine. Otherwise, renting the equipment is the way to go.

Air compression may be stopped during tractor operation by unloading the intake valves.

FIGURE 20.13   Large rotary screw-type compressor. (*Courtesy of Ingersoll-Rand Company.*)

The Skyworker attachment shown in Fig. 20.14 is a pair of 13-foot hinged booms which carry a basket, air hose, and tools. It can be used to place a worker high in the air and in a wide radius around the tractor, to facilitate drilling, scaling, and other work in hard-to-reach places. It is steadied by hydraulic outriggers.

The tractor can be equipped with all standard accessories and with other special drill mountings.

AIR LINES AND ACCESSORIES

Metal Pipe.   If the compressor cannot be brought close to the work, ordinary plumbers' pipe may be used to carry the air. The pipe should be at least as large as the receiver discharge line, and preferably should be a size larger.

Angles in the line should be kept to a minimum. If the line is reduced in size, reducing couplings instead of couplings with bushings should be used for smoother airflow. If air may be required later between the compressor and the end of the line, tees with plugs may be used instead of couplings between lengths of pipe, so that additional outlets can be added without opening and reconnecting the line.

The drop in pressure in an air line depends on

1. The size and length of the pipe
2. Pipe bends, and type of fittings used
3. The volume of air flowing
4. The pressure of air as it enters the line

FIGURE 20.14   Pneumatractor and Skyworker. (*Courtesy of Schramm, Inc.*)

The table in Fig. 20.15 gives the drop of pressure per 100 feet of straight pipeline, and the equivalent in feet of straight pipe for 90° turns (elbows or tees), and of shutoff valves. If the tools use less than full compressor output, the drop will be proportional to their consumption rather than to the capacity of the compressor.

Hose and Fittings.   Connection between the receiver or metal lines and the tools is made by flexible hose or tubing of rubber and fabric. The rubber should be neoprene or of some other oil-resistant type. See Fig. 20.16.

Most hoses have between three and seven plies, or one to three braid layers, and may be of either wrapped or molded construction. The molded type with rayon braid fabric is lighter and more flexible for its size and strength. Heavy-duty hose with tough covers is required for mining, for quarrying rock which breaks with sharp edges, and where the hose is subject to abuse from machines, tools, or rock falls. For less severe conditions the lighter hose is both more economical and easier to handle.

Hose should not be used after the inner tubing starts to deteriorate, as broken pieces will clog filters or valves in the tools. See Fig. 20.17. Such pieces are often so small as to be mistaken for carbon.

Hoses are fastened to each other and to other units by threaded, quarter-turn, or snap-on couplings.

TABLE I *

AIR PRESSURE DROP
per Hundred Feet of Straight Pipe

Air Flow C.F.M.	Diameter of Air Line						
	3/4"	1"	1 1/4"	1 1/2"	2"	2 1/2"	3"
50	2.51	.7	.16	.07	—	—	—
105	—	3.1	.71	.31	.09	—	—
210	—	—	2.82	1.26	.34	.13	—
315	—	—	—	2.73	.76	.29	—
420	—	—	—	—	1.35	.53	.17

TABLE II **

EQUIVALENTS OF FITTINGS *in feet of straight pipe*

Type Fitting	Diameter of Pipe Fitting						
	3/4"	1"	1 1/4"	1 1/2"	2"	2 1/2"	3"
Elbow or Tee	1.2	1.6	2.2	2.6	3.6	4.4	5.7
Globe Valve	3.5	4.7	6.5	7.8	10.6	13.1	17.1

* Adapted from Schramm Data Sheet C-50 B
**Copied from Schramm Data Sheet C-50 B

FIGURE 20.15   Air-pressure-drop data.

FIGURE 20.16   Air line connectors.

The threaded connections are best suited to connections which are seldom changed during the work. They take more time to assemble and disassemble than the other types, but usually require less servicing. No gaskets are needed.

The quarter-turn or quick-detachable are usually obtainable only for medium and small hose. The connectors in any one make are all the same size, and different-size hose can be connected without the use of bushings. Both surfaces and the gasket should be kept out of the dirt when apart and should be cleaned before coupling. Spare washers should be kept, as these may be lost or damaged.

Air pressure should be turned off and the line bled before opening either type of connection, for ease of operation and because of danger from the hose whipping and possible damage to eyes from blowing dirt.

The snap-on couplings are assembled by pulling back a sliding collar on the socket, inserting the plug on the other hose piece, and releasing the collar. They are opened by pulling back the collar and pulling the plug out. The socket automatically shuts off the air when the plug is out, and should be installed on the side toward the receiver.

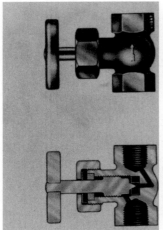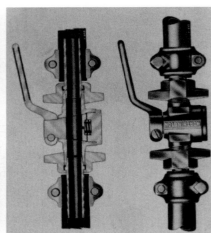

FIGURE 20.17   Air line valves.

At present, these are available only in small sizes. They are quick and convenient to use. The sockets should be wrapped in a cloth or otherwise protected from dirt if they are to be left open on the ground.

Reels.   Hose can be wrapped around the compressor or coiled in tool boxes when not in use. However, keeping it on a reel is good policy, as it keeps it away from contact with sharp or heavy objects and avoids kinking.

A dead reel, similar to the type used for garden hose, may be used for storage. If the inside end of the hose is left projecting near the axle, it may be connected to the receiver so that the necessity of removing it entirely from the reel is avoided.

A more convenient device is a live reel permanently mounted on the compressor. Air from the receiver is admitted to the axle through a rotating, pressure-sealed connection, and from the axle to the hose by an ordinary coupling.

The outer end of the hose should be tied to the reel when the compressor is moved to avoid unspooling.

Manifolds.   When several working lines are to be taken off one receiver or supply line, a manifold or "pig," such as those shown in Figs. 20.18 and 20.19, is used.

Oilers.   Most air hammers are equipped with oiling systems supplied by small reservoirs in the tool itself. However, these are often neglected, need frequent attention, and may not function satisfactorily in a worn tool.

A line oiler, illustrated in Fig. 20.20, is a reservoir which feeds into the airstream through a needle valve. The oil is blown into a fine spray which is carried into the tool to keep it lubricated. Oil will feed only under pressure.

Oilers are built into track drills and other heavy air-operated equipment. Otherwise they should be as close to the tool as possible, as the air and oil may separate almost completely in 10 or 15 feet. However, sufficient distance must be allowed for easy manipulation of the tools, and some lubrication will be afforded at long distances by the airstream pushing condensed oil along the inside of the hose.

Some line oilers will operate in any position, while others must be upright. They should always be placed in the line so that the air moves in the direction of the arrow or the case. Hose between the oiler and the tool must be oil-resistant.

FIGURE 20.18   Line manifold.

FIGURE 20.19   Tank manifold. (*Courtesy of Schramm, Inc.*)

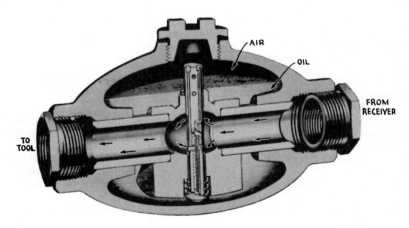

FIGURE 20.20   Line oiler.

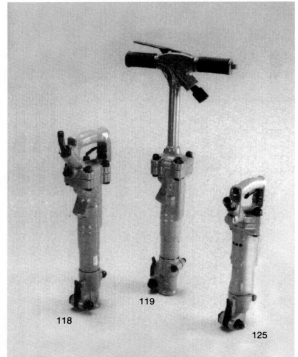

Model	Chuck Size	Bore
118	7/8 x 3-1/4 • 1 x 4-1/4	1-3/4
119	7/8 x 3-1/4 • 1 x 4-1/4	1-3/4
125	3/4 sq. x 2-3/4 • 7/8 hex x 2-3/4 or 3-1/4	1-11/16

FIGURE 20.21   Hand-operated diggers. (*Courtesy of American Pneumatic Tools, Inc.*)

BREAKERS AND DIGGERS

Most of the light, handheld breakers, sheeting drivers, and tampers are air hammers that do not rotate the steel. An anvil block may be used to transmit the blow of the piston to the steel. See Fig. 20.21.

Hydraulic.   Hammers may also be hydraulically operated. Figure 20.22 shows one that is powered by a loader, by means of hoses to tees inserted in its bucket lift and lower lines. In absence of other equipment, it may be run by an engine-driven pump on a small trailer.

Requirement in two sizes ranges from 4 to 7 gallons per minute, at 800 psi pressure. Since machines with such hydraulic systems are on most jobs, hooking up is usually much easier than bringing in a compressor. No muffler is needed, as it has no external exhaust.

Hydraulic Splitters.   Hydraulic splitters apply the plug and feathers mechanism (see Fig. 9.37) to very heavy work.

The splitter in Fig. 20.23 has a pair of hard shims (feathers) at the base of a two-way hydraulic cylinder. They are rounded on the outside and slightly enlarged at the bottom to fit a straight

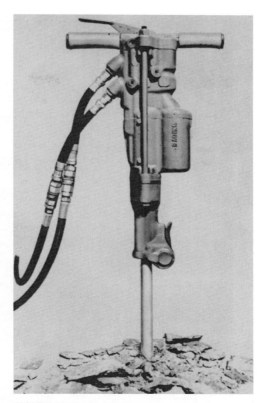

FIGURE 20.22   Hydraulic hammer.

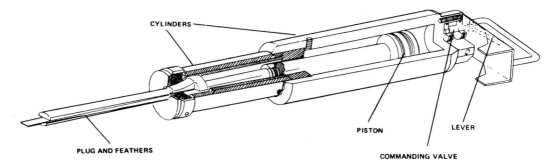

FIGURE 20.23   Hydraulic rock splitter.

drilled hole $1\frac{3}{16}$ to $1\frac{5}{8}$ inches in diameter. The inner, facing sides are flat with a gradual taper, and are smoothly finished.

The plug is a slender wedge, also very hard, that is forced between the feathers by the hydraulic piston. An enormous side pressure, of up to 400 tons, is exerted against the sides of the hole.

This model is supplied with 7,000-psi hydraulic pressure from a separate engine-driven pump, with hose connections. Another make uses a hand pump on the cylinder.

FIGURE 20.24   Machine-held rock splitter. (*Courtesy of Stanley Hydraulic Tools.*)

Almost any sound rock can be cracked by proper use of these devices. Boulders usually break up, but for solid rock excavation, a backhoe-mounted demolition hammer may be used to loosen and reduce the pieces before loading them. See Fig. 20.24.

The outward force should be exerted at right angles to the proposed crack. As in blasting, there must be a free face toward which at least one side can move, for the pressure to be fully effective.

The primary use for splitters is rock breakage in areas where explosives would be very dangerous to people or property, or where the noise of blasting could not be tolerated.

PERCUSSION ROCK DRILLS

Most pneumatic rock drills (air drills) used in excavation are percussion machines which both hammer and turn a cutting bit, which is usually threaded onto the end of a hollow steel rod.

Weights of the drills range from about 10 to 500 pounds. The light ones, up to 30 pounds, are always hand-operated. Medium drills, up to 80 pounds, are generally handheld but may be mounted on frames equipped with hand or power feeding mechanisms. Heavier models are almost always frame-mounted and power-fed.

Hand rock drills are called sinkers, hammers, and jackhammers (jack hammers). They are air-operated, and except for a small type called a plug drill, they rotate and hammer a hollow drill steel to which a bit is attached. Weight is 15 to 70 pounds.

To minimize vibration felt by the operator, Atlas Copco has patented a spring-loaded design for the tool that eliminates 80 percent of the vibration typically created by the air breakers. The patented spring-loaded design allows the handle to move with the tool's movement instead of against it. According to the company, the resulting decrease in vibration increases productivity by enabling the operators to work as much as 4 times longer without fatigue.

Hammering chips, flakes, or crushes the rock, rotation gives the bit fresh striking surfaces, and exhaust and direct blower air through the steel remove the cuttings.

Handheld. A 55-pound medium-weight drill is shown in Fig. 20.25. The three principal parts of the drill body are the upper end or back head, the cylinder, and the front head. These are machined to fit each other accurately without gaskets, and are held together by a pair of alloy steel bolts, called assembly rods.

Air reaches the drill through flexible hose, then a curved metal tube with a swivel connection to the back head, and the throttle valve. See Fig. 20.26. The throttle can be set in closed, wide-open, and several intermediate positions, so that the speed of drill action can be regulated.

The piston is moved rapidly up and down in the cylinder by compressed air. On the downstroke its stem strikes the upper end of the drill steel. It is kept in alignment by the cylinder bore, rifle bar, stem bearing, and chuck.

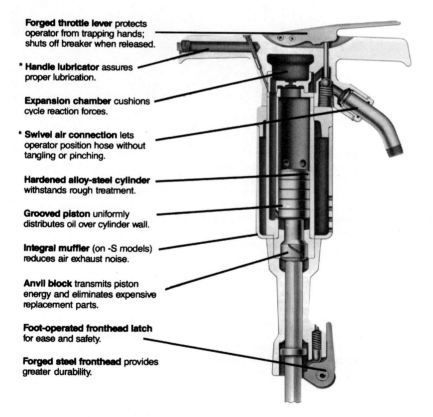

Forged throttle lever protects operator from trapping hands; shuts off breaker when released.

* **Handle lubricator** assures proper lubrication.

Expansion chamber cushions cycle reaction forces.

* **Swivel air connection** lets operator position hose without tangling or pinching.

Hardened alloy-steel cylinder withstands rough treatment.

Grooved piston uniformly distributes oil over cylinder wall.

Integral muffler (on -S models) reduces air exhaust noise.

Anvil block transmits piston energy and eliminates expensive replacement parts.

Foot-operated fronthead latch for ease and safety.

Forged steel fronthead provides greater durability.

FIGURE 20.25 Hand-operated air hammer. (*Courtesy of Ingersoll-Rand Company.*)

FIGURE 20.26   Air hammer drilling rock. (*Courtesy of Ingersoll-Rand Company.*)

The rifle bar is splined and slightly twisted. Four pawls attached to its upper end allow it to rotate in the direction of the twist. The rifle nut, in the top of the piston, has matching splines.

As the piston moves up, the rifle nut splines exert a twisting force on both the rifle bar and the piston. The rifle bar is held by the ratchet, so the piston turns. On the return stroke, the rifle bar turns, having less rotation resistance than the piston, which drives straight down. Each up-and-down cycle of the piston therefore rotates it an amount determined by the twist in the rifle bar and the spacing of the ratchet teeth.

Rifle bars with matching chuck nuts can be obtained in several pitches, and sometimes with a reverse twist which will give rotation on the downstroke.

The lower end of the piston, called the stem, is splined into the chuck driver that is threaded into the chuck, which is made in upper and lower jaws. The lower jaw has a socket into which the shank of the steel fits. Rotation is carried from the piston stem through the chuck parts into the steel.

Chucks may be obtained to fit the hexagonal shank, octagonal designs, or round steel with lugs. Steel diameters range from $\frac{7}{8}$ to $1\frac{1}{2}$ inches. One inch is generally standard for medium drills.

The steel puller is a clamp that holds the steel in position during work. It is released by foot pressure on the projecting lever.

The steel is shown in its highest position, with the chuck resting on the steel collar and the piston near the bottom of its stroke. As the piston completes its downstroke, it drives the steel down (or the drill up), until the collar reaches the steel puller. The piston is then forced up by air entering the bottom of the cylinder through the automatic valve located around the upper part of the rifle bar, and the weight of the drill causes it to slide down along the shank until the chuck rests on the collar again.

It is important that the shank be the correct length, as the drill will not operate properly on a longer or shorter stroke than that for which it is designed. Standard length is $3\frac{1}{4}$ inches from the collar to the end.

From the throttle, air goes through a drilled passage in the back head, past the pawls on the rifle bar, into the valve, which directs it alternately to the top and bottom of the cylinder. Air exhausted from the top of the cylinder passes into the open air while the hammer is working. Exhaust from below the piston goes through a drilled passage into the piston stem bearing, and downward along the chuck driver splines into the space below the piston stem. Here it enters the hole in the center of the steel, which takes it to the bit where it serves to blow rock chips out of the hole during drilling.

In deep holes or soft rock, the amount of air provided by this puff blowing may not keep the hole clean. All the air that comes through the throttle can be turned down the steel by moving the blower or throttle valve to the blow position.

A bypass in the back head can be connected to divert part of the air into the steel while leaving the bulk of it to run the hammer. This provides more cleaning in proportion to drilling.

Lubrication is provided by a reservoir around the piston stem bearing, which is filled through a plug in the side. This oil is forced by air pressure to points where it directly lubricates some surfaces, and it is carried by the airstream to others. However, it is safer to use a line oiler.

Drifters. Percussion drills that weigh 65 pounds or more are usually supplied with supports, and with automatic devices to feed them into the work. They are called drifters or drifter drills, that is, drills intended for mounting on "drifter" or power feeds, such as are carried by crawler drills or jumbos.

Drifters are rated both by cylinder bore and weight. Figure 20.27 gives specifications for a manufacturer's line. The smallest has a 3-inch bore and weighs 65 pounds, the largest a $5\frac{1}{2}$-inch bore and a weight of 460 pounds.

Recommended hole diameter is usually smaller than cylinder bore.

Rotation may be powered by rifle bar, as described for the smaller drill, rifle bar with double sets of pawls to permit reversing, or by a reversible air motor turning the chuck through reduction gears. Rifle bar rotation is simple, and economical of air, but it lacks flexibility.

Drifter	Type of Rotation	Bore	Weight	Chuck (Std.)	Air Consumption	Hole Sizes
			ENGLISH			
D300A	Rifle Bar	3''	65 lb.	$\frac{7}{8}$'' x $4\frac{1}{4}$'' Hex	220 cfm	$1\frac{1}{4}$''-$1\frac{3}{4}$''
DC35	Rifle Bar	$3\frac{1}{2}$''	111 lb.	$1\frac{1}{4}$'' Round	270 cfm	$1\frac{1}{2}$''-2''
URD350	Air Motor	$3\frac{1}{2}$''	156 lb.	$1\frac{1}{4}$'' Round	355 cfm	$1\frac{1}{2}$''-2''
X71WD	Rifle Bar	4''	164 lb.	$1\frac{1}{4}$'' Round	271 cfm	$1\frac{3}{4}$''-$3\frac{1}{2}$''
D40	Rifle Bar	4''	156 lb.	$1\frac{1}{4}$'' Round	328 cfm	$1\frac{3}{4}$''-$3\frac{1}{2}$''
D475A	Rifle Bar	$4\frac{3}{4}$''	239 lb.	$1\frac{1}{2}$'' Round	425 cfm	$1\frac{3}{4}$''-$4\frac{1}{2}$''
URD475U	Air Motor	$4\frac{3}{4}$''	335 lb.	$1\frac{1}{2}$'' Round	520 cfm	$1\frac{3}{4}$''-$4\frac{1}{2}$''
URD550	Air Motor	$5\frac{1}{2}$''	460 lb.	2'' Round	800 cfm	$2\frac{1}{2}$''-6''
			METRIC			
D300A	Rifle Bar	76.2 mm	29.6 Kg	22 mm x 108 mm Hex	6.2 M³/min.	32.0 - 45 mm
DC35	Rifle Bar	85.8 mm	50.4 Kg	32 mm Round	7.6 M³/min.	38.0 - 51 mm
URD350	Air Motor	85.8 mm	71.0 Kg	32 mm Round	10.0 M³/min.	38.0 - 51 mm
X71WD	Rifle Bar	101.6 mm	74.4 Kg	32 mm Round	7.7 M³/min.	44.5 - 89 mm
D40	Rifle Bar	101.6 mm	70.8 Kg	32 mm Round	9.3 M³/min.	44.5 - 76 mm
D475A	Rifle Bar	121.0 mm	109.0 Kg	38 mm Round	12.0 M³/min.	44.5 - 104 mm
URD475U	Air Motor	121.0 mm	152.0 Kg	38 mm Round	14.7 M³/min.	44.5 - 104 mm
URD550	Air Motor	139.7 mm	209.0 Kg	51 mm Round	22.6 M³/min.	63.5 - 152.4 mm

FIGURE 20.27   Drifter specifications. (*Courtesy of Ingersoll-Rand Company.*)

Reverse rotation is used in unscrewing (breaking) steels and bits, and in pulling back out of holes.

The drill assembly includes a striking bar, which is a short piece of hollow steel. The upper end receives the blows of the piston; the lower end is threaded for coupling to the drill steel. It is held in line by a chuck bushing and other parts, and in place by a threaded front head or chuck housing cap, which usually has lugs to allow loosening it by hitting it with a hammer or other tool.

Fit must be good. Play to the side, caused by an undersize striking bar or by worn drive ring or chuck bushing, will cause extensive damage.

Blow air is carried by an air tube from the back head through the center of the piston (in the rifle bar, if there is one) and into the striking bar. The tube is a sliding fit in the bar. If it is loose, it will leak high-pressure air and cause "short stroking." If too tight, it will probably break.

Some blow air, both from the high-pressure line and from the exhaust, feeds through the tube during drilling. Full airflow is turned into it in the BLOW position of the control valve.

The throttle controls for drilling, blowing, and rotation may be mounted on the back head, but are usually at an operator's station where other movements of the machine may be controlled also.

Down Hole. A down-the-hole drill is a simple, heavy pneumatic unit in which the piston blow is delivered against the bit shank, without any steel or rod between. The drill is very slender in proportion to its weight and strength. It has no rotation mechanism in itself. Air is exhausted through the bit, and carries chips to the surface, around the outside of the rods. See Fig. 20.28.

Diameter of piston hammers varies from $2\frac{1}{2}$ to 6 inches, and they weigh from 13 to 175 pounds. Strokes may be as short as $\frac{7}{8}$ inch or as long as 5 inches. Blows per minute range from 500 to 2,700 in standard units, and to 3,800 in high-pressure designs.

The theoretical limit on frequency of strokes is about 6,000 per minute. Lubrication is supplied by heat-resistant drill oil added to the air from the compressor.

The drill rod above the drill supplies air, keeps it in contact with the rock at proper pressure, and rotates it. The rotation, usually at the rate of 15 or 25 revolutions per minute, is supplied by an air or hydraulic motor that has the position on the chain or other feed that is usually occupied by a drill.

A carbide insert bit is held in the drill chuck by a split collar and a nut. In taking out the bit, the nut is loosened or "broken" by a hydraulic wrench, and then spun off by the rotary mechanism.

Down-hole drills are made to operate at the usual construction machinery air pressure of 100 pounds per square inch. High-pressure models may run at 250 up to 500 psi. There is a tendency to increase pressures for greater efficiency.

A pressure of 100 psi at the receiver usually means about 90 psi at the drill. Back-pressure from the restricted openings in the bit and from the rising column of air and chips around the rods is likely to be about 20 psi, leaving an effective pressure of only 70 psi to operate the drill. If receiver pressure were 300 psi, losses would be the same 30 psi or less, so the efficiency would be much greater. Higher pressure is usually accompanied by decrease in stroke and increase in frequency and drilling rate.

As with all "souping-up" operations, the possible damages must be weighed against the increase in production.

The down-hole drill is used for making holes of 5- to $7\frac{1}{2}$-inch diameter in hard and medium rock. Since none of its striking force is absorbed by the rod (steel for big and deep working drills is usually called rod or pipe), its working depth is limited only by the ability of the airstream to keep cuttings blown out of the hole, and the capacity of the rotary bearings to carry the weight of a long string of rods.

Down-pressure or pulldown on these drills is much less than with rotaries. It seldom exceeds 2,500 psi even in very hard rock.

The exhaust air rising from the bit should have a velocity of almost 3,000 feet per minute to clean the hole. In rapid drilling with coarse chips 5,000 feet per minute may be needed.

When the hole is large in proportion to rod diameter, air velocity will be low because it has a wide passageway. It can be speeded up by using an auxiliary compressor, decreasing the size of the bit, or using bigger rods.

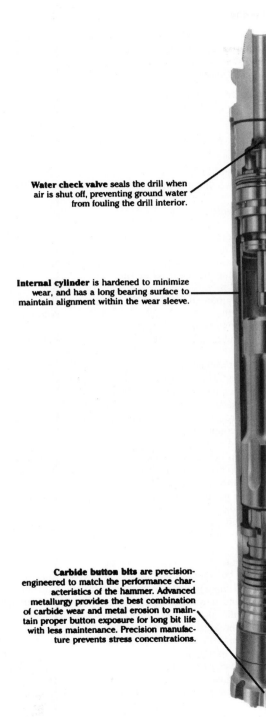

Backheads are available in a variety of popular thread sizes.

Water check valve seals the drill when air is shut off, preventing ground water from fouling the drill interior.

Wearsleeve is thicker and hardened for long life. It can be replaced economically. On most sizes it can be reversed end-for-end, doubling its life.

Internal cylinder is hardened to minimize wear, and has a long bearing surface to maintain alignment within the wear sleeve.

Piston functions as a sliding valve to control the operating cycle. No separate valves or valve chest assembly are required. Large piston bearing area minimizes wear on piston and cylinder.

Carbide button bits are precision-engineered to match the performance characteristics of the hammer. Advanced metallurgy provides the best combination of carbide wear and metal erosion to maintain proper button exposure for long bit life with less maintenance. Precision manufacture prevents stress concentrations.

Continuous hole cleaning is provided by the drill exhaust air which passes out through the bit. The bit exhaust tube can be replaced in the field, and eliminates the need for a piston stem and stem bearing.

FIGURE 20.28   Down-hole drill assembly. (*Courtesy of Ingersoll-Rand Company.*)

A principal disadvantage is the danger of losing a whole drill as a result of a rock fall, or formation of mud collars. For this reason, use may not be prudent in badly fractured formations, or in wet shale or other muddy strata.

Volume needed for blowing only can be worked out approximately by using the formula:

$$(\text{Hole diam.})^2 - (\text{rod diam.})^2 \times 16.4 = \text{cfm}$$

STEELS, RODS, AND PIPES

Steels and rods are the connectors between non-down-hole percussion drills and their bits. If only one piece is needed, it is usually called a steel; if two or more pieces are fastened together, they may be called either sectional steel or rods.

Rotary drill rods, when made with tapered threads, are usually called stems or pipes, sometimes rods, but almost never steels.

In moderate to deep drilling, the set of connectors and bit is called the string. In percussion work it carries hammering and rotation with light to moderate down-pressure; in rotaries it is subjected to rotation and moderate to very heavy pressure. For both, it carries air under pressure to blow cuttings out of the hole. In larger rotaries air may be replaced by fluid.

Steels.   Hollow drill steels for hand- and machine-mounted percussion drills are made in outside diameters of ⅞ to 2 inches. There are standard lengths of 1 to 10, 12, 14, and 20 feet. Many odd lengths are produced by cutting down during repair work.

Cross section may be hexagonal, octagonal, or round, or the whole steel may have a continuous, rounded thread.

One end may be forged into a shank, which fits in the chuck of a hand drill. For drifters, the shank is usually a short, separate piece, called a striking bar. The opposite end of shank pieces, and both ends of most other steels, carries a male thread which may or not be backed by a raised shoulder. Such threads are coarse, left-handed, and made in several nonmatching types.

A steel may have one tapered end to carry a tapered bit, or be forged into a cutting bit. Sections of steel (sectional steel) are fastened together by couplings, which are short, strong tubes with internal threads. Bits have female threads also, for direct fastening to steels.

Most of the hammering impact shock is carried between rod ends, which butt against each other inside the coupling, or from rod end to bit socket (bottom). If there are shoulders on the steel threads, they carry shock through the coupling, or the shoulder of the bit.

If a connection is loose, or ends or shoulders are not properly shaped, some or all of the shock will be carried by the threads, with resulting rapid deterioration.

Threads and tapers are designed to be tightened by the rotation of the drill, to jam tightly enough so that reverse twisting will not loosen them, but to loosen when struck sharp blows when free of pressure or twist.

Threads need to be well lubricated by special (heat-resistant) grease that prevents direct metal-to-metal contact. Without this, full tightening may be difficult, and impact may create tiny weld points that both prevent natural tightening while drilling and make loosening excessively difficult.

Threads and couplings deteriorate in service, and require every-use inspection and frequent repair. Forging machines are available that will make (upset) new threaded ends, with shoulders when required, or reshape old ones. Continuously threaded rod is cut to a new flat face, but must have the rim chamfered for the final ⅛ inch, to about 30°. See Figs. 20.29 and 20.30.

Every steel must have a hollow center to carry air to the bit for hole cleaning and cooling. Care must be taken not to plug or narrow it with mud, trash, or repair mistakes.

Steels are long and thin, and not nearly as strong as they look. They are easily bent by accident, by rough handling, and particularly by being blasted out of stuck holes. Their life in service may be limited by metal fatigue, and by cracks starting at dents and scratches on the surface. In general, long steels and thin steels are the ones most often and most severely damaged.

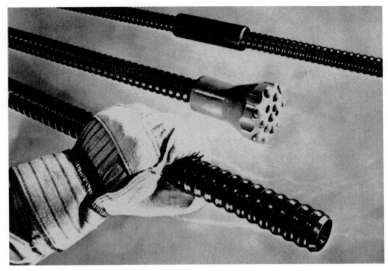

FIGURE 20.29 Spiral steel. (*Courtesy of Ingersoll-Rand Company.*)

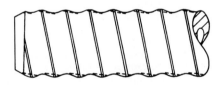

Cut it off before it's worn out

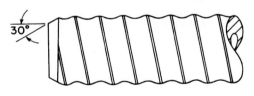

Chamfer carefully and evenly

FIGURE 20.30 Trimming spiral steel. (*Courtesy of Ingersoll-Rand Company.*)

A steel should *never* be used as a crowbar or a lever, even if the load seems light. Tubes have almost no strength for such work.

A blacksmith shop, if available, can usually straighten and rethread steel for less than the cost of replacement, unless it is the carburized type.

A set of hand steels is a series in graduated lengths, from starter to the longest piece required. For hand work, a 2-foot increase with each change is usual. The length difference between successive pieces is called the steel change.

Shanks. Three types of shank are hexagonal, quarter octagonal (square, with chamfered corners), and round lugged. The round shanks with two lugs are largely used with the heavier drills and thicker steels. Shank diameter is normally the same or only slightly larger than steel diameter. See Fig. 20.31.

The shank collar serves to limit the upward penetration of the steel into the chuck, and enables the puller to hold the steel from sliding out of the drill.

Shank lengths, from the upper side of the collar to the end, are $3\frac{1}{4}$ inches for small steels and $4\frac{1}{4}$ inches for the larger sizes. It is important that this length not vary more than $\frac{1}{16}$ inch, and that the shank end be kept straight across.

In drifter drills, the shank is replaced by the striking bar, which is basically a short shank with a threaded lower end, which is kept clamped in the drill.

Drill Pipe. Down-hole and rotary drills use a heavier type of steel, called rod or pipe, that is designed to carry torque rather than impact. It uses A.P.I. (American Petroleum Institute) steeply tapered threads, male on one end, female on the other.

Standard lengths for deep drilling are 21, 29, and 43 feet. For mobile rigs they are made up to suit tower height and steel change, in lengths of 20, 25, or more feet.

Metal. There are four principal types of steel used in making drill steels (or rods): plain carbon, high carbon, alloy, and carburized.

Plain carbon steel (.80 to .85 percent carbon) is the oldest type, and is still standard for hand drills and other light hammers. It is the least expensive, usually has shoulders behind the threads, and can be rethreaded and repaired in a small shop. Outside diameters range from $\frac{7}{8}$ to $1\frac{1}{4}$ inches.

High carbon (.95 to 1.05) steel is used in the same sizes, but without shoulders, for somewhat heavier service with light drifters. Skill and good heat-treating equipment are needed for proper repair, unless this consists only of cutting the damaged tip off a long thread.

Various steel alloys such as manganese-chrome-molybdenum are made into $1\frac{1}{4}$- and $1\frac{1}{2}$-inch diameters, which are harder, up to 45 Rockwell C scale, and highly fatigue-resistant. They can be rethreaded in the field with carbide tools, without heat treatment. Other repairs require special shops.

Carburized steel is low carbon, but surface-treated for great hardness, both inside and outside, after forming and threading. It is expensive but high-performance and long-lasting. It cannot be rethreaded or repaired in the field.

Outside diameters range from 1 to 2 inches, for use with the heaviest drills.

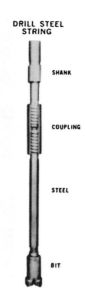

DRILL STEEL STRING

SHANK

COUPLING

STEEL

BIT

FIGURE 20.31   Drill steel string.

PERCUSSION BITS

Design. Percussion rock bits chip or crush rock with hammer blows. Rotation changes the area of impact with each blow. Compressed air, or air and water, supplied through the center passage in the steel, and through one or more holes in the bit, blows the chipped and powdered rock away from the bit and out the drill hole.

Bits are usually made of deep hardened alloy steel. They may have inserts of silicon carbide at impact areas. Most of them can be restored by grinding when worn.

Left-hand female threads, or smooth off-center taper, are provided to match one or more styles of steel.

The work of smashing the rock is usually done by four, but occasionally two or six, ridges that are radial to a center hole, with their faces at right angles to the centerline of the steel.

Two ridges (four wings or points) at right angles to each other are called a four-point cross, the most common construction. See Fig. 20.32. If not at right angles, it is an X bit. A six-wing (six-point) design is called a rose bit.

FIGURE 20.32 Multiple-use steel bit.

FIGURE 20.33 Duro Rock drill bit with exhaust hole. (*Courtesy of Ingersoll-Rand Company.*)

The wings may be replaced by round buttons of carbide set in an almost continuous, slightly convex surface. See Fig. 20.33.

The hollows between the wings are flutes. The width across the face of a pair of wings (the width of the bit) is the gauge or size.

The bit is widest at the face, and tapers to the back. The taper is called the gauge angle. It is usually 3° in carbide bits.

Bits are selected according to hole size (the same as bit gauge), size and type of thread, type of rock, and hole footage to be expected.

Airholes. The number and placement of airholes in percussion bits are quite variable.

Small bits for light drills may have a single hole in the center of the face. This is the most effective position for instant removal of cuttings, but it is vulnerable to plugging. Downward movement presses it into a core that is not cut directly. This may be forced into the hole in soft or tough material.

Most bits have this center hole. In addition, there may be one to five holes in the volutes, or recesses between the wings. In general, bits designed for soft and/or wet rock have more holes than those meant for hard, dry formations.

A venturi bit has one or more holes to direct air back up the hole at an angle. Reverse airholes tend to break up mud collars either while they are forming or when the bit is being pulled out.

Smooth-face button bits may have a front hole somewhat off center, and other holes in volutes are cut into the bit to provide an escape route for the cuttings.

Carbide Inserts. The drilling speed and service life of bits can be greatly improved by carbide inserts.

The material used is sintered tungsten carbide, a mixture of 9 to 12 percent cobalt (CO) and 91 to 88 percent tungsten (WC), with a hardness of 91 to 87 on the Rockwell A scale. Fine powders of these materials are mixed in whatever proportion is desired, and cemented into a solid by heat and pressure.

Maximum hardness is obtained with maximum tungsten, minimum cobalt, and finest grain size. Increase in cobalt and grain size reduces hardness but increases toughness. These variations are within a narrow range, but may be important in performance.

The harder inserts will give maximum footage between regrinds, but may chip or break under certain conditions. The softer ones will wear more rapidly (but far, far more slowly than steel), but are less susceptible to breakage.

Excessive rotation speed and heavy down-pressure are likely to break or chip inserts of any grade.

A carbide bit will cut faster than an all-steel bit of similar design, and may drill 100 times as much footage before requiring reconditioning or replacement.

Carbide bits are expensive, so that their use is not practical unless loss in holes, or by careless-ness or theft, can be controlled. Under average drilling conditions, however, they will usually prove more economical than steel bits.

Their use is particularly indicated in deep holes, where their negligible taper saves the waste of drilling oversize at the top; in hard and abrasive rock, where taper, lost time, and expense of bit replacement are important; and where transportation and reconditioning of bits are difficult.

In fissured rock, the large investment in carbide bits may be too risky, because of the danger of sticking.

Any pieces of carbide loose in a hole, as a result of chipping or falling-out of a bit, are extremely destructive to the rest of the bit, and to other bits. If they cannot be blown or flushed out of the hole, it should be abandoned, and a new hole drilled alongside.

Bit Action.   With the downstroke of the drill piston, each edge of the bit is driven into the rock like a cold chisel hit with a hammer, a tiny fraction of an inch from where it hit with the previous stroke.

If the bit is sharp and rotation rapid enough, it will tend to cut off a wedge-shaped slice. If it is dull, it is more likely to crush the edge to powder to a shallower depth—a much less efficient operation.

For one quality of rock frequency of stroke, and size and sharpness of bit, the thickness of the slice will depend on the speed of rotation. The depth of cut is governed by the force of impact, plus down-pressure, through the drill string.

Most drills have a fixed blow frequency at proper air pressure and full throttle. Rifle bar drills have a fixed rotation rate, so with them the operator has only control over down-pressure. This is light in handheld hammers, but may be regulated up to a ton or more in drifters and down-holes. Too little pressure may reduce depth of chipping and therefore footage (rate of drilling). Too much down-pressure prevents the bit from getting back up on top of the rock for a clean stroke, and may reduce footage even more. In addition, it subjects the whole mechanism to damaging strain.

The depth of penetration by each blow is usually very slight. The wings follow each other closely, overlapping their work, and cut a round hole with smooth sides.

However, in soft rock the bit may penetrate so rapidly so that each wing cuts a spiral thread in the wall, which is not entirely removed by the following wing. This process, which is called rifling, results in a threaded, undersize hole, from which it may be difficult or impossible to remove the bit.

Rifling may be reduced or eliminated by supporting the drill to reduce down-pressure, or by using dull bits, rose bits, or X bits.

A dull bit spreads blows over too large a surface, and does no chipping and very little crushing. In addition, it sets up a powerful dragging effect, which results in overtightening thread connections, and greatly increases strain on the drill string and the rotation mechanism.

Efficient drilling is absolutely dependent on removal of rock chips and dust as fast as they are formed. This important subject will be discussed under Hole Cleaning.

Reconditioning.   There are a few throwaway bits that are designed to be used only until worn, and then discarded. These include steel bits with surface hardening only, and bits with shallow carbide inserts or buttons.

One-use bits may have a pilot consisting of advanced cutting edges near the center. These facil-itate starting a hole, cut more rapidly in some formations, and avoid the nuisance of rehandling.

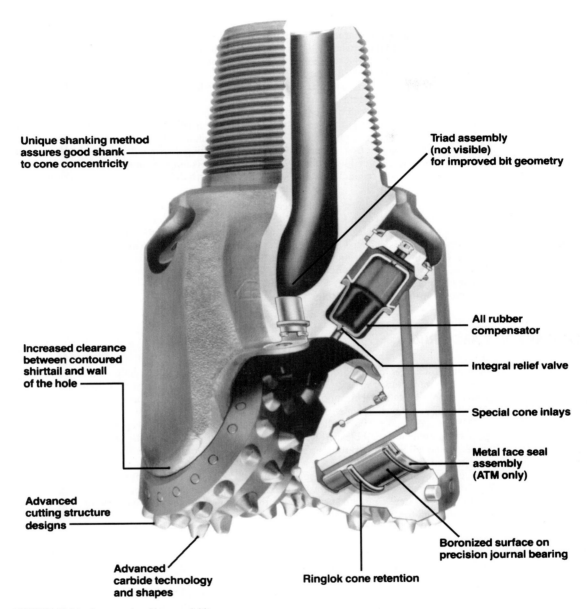

Unique shanking method
assures good shank
to cone concentricity

Triad assembly
(not visible)
for improved bit geometry

All rubber
compensator

Integral relief valve

Increased clearance
between contoured
shirttail and wall
of the hole

Special cone inlays

Metal face seal
assembly
(ATM only)

Advanced
cutting structure
designs

Boronized surface on
precision journal bearing

Advanced
carbide technology
and shapes

Ringlok cone retention

FIGURE 20.34   Inner works of large rock bit.

Most bits, however, are designed to be reconditioned several times. See Fig. 20.34, showing inner works of larger bit.

As a bit is used, it wears along the cutting edges and on the sides of the wings. The side-wear is the more serious, as it soon reverses the taper so that the edge becomes narrower than the part behind it. This causes the bit to drive into a hole that is too small for it, so that it sticks. The degree of wear which causes sticking varies with the type of rock.

Worn bits of the multiuse type can be reconditioned in special machines. Generally it is not economical to attempt to sharpen them by hand.

A deep hardened bit can usually be ground several times. It is cut back from the sides until the reverse taper is removed, and on the face to restore the edge.

A shallow hardened bit will be softened by deep grinding but may be hot-milled. It is heated, reshaped, and rehardened in a special forging machine. Many deep hardened bits will also give better service if hot-milled than if ground.

The gauge size is reduced by reconditioning.

Carbide bits may be reconditioned by hand or machine grinding, using a silicon carbide wheel.

In soft, abrasive rock it is important to use bits with pronounced taper to avoid excessive sticking and possible frequent loss. In fissured rock where sticking does not depend on bit design, the large investment in carbide bits is a risky proposition, and is not recommended.

Carbide bits may be reconditioned by hand or machine grinding, using a silicon carbide wheel.

Hole Taper. As a bit wears, its hole becomes smaller. In soft rock, wear may be so slight that the same bit can be used on several steels in succession. In hard or abrasive rock, it may wear badly in a few inches. Holes taper smoothly with bit wear and in steps at bit changes.

Whenever a bit is changed in a hole, the new bit must be smaller to avoid sticking. If all new bits are used, the reduction must be $\frac{1}{8}$ inch. Used bits can be compared by setting their edges together, and very small reductions can often be made.

The starting bit must be large enough to allow for the taper and still have a big enough hole at the bottom for the quantity of explosive needed.

ROTARY BITS AND DRIVE

The rotary bit cuts, chips, and/or grinds, without hammering. In construction work, it is usually carried by mobile tower drills, described in a following section. A rotating table or head above ground supplies rotation. This motion, down-pressure, and cleaning air are carried by hollow drill pipe with threaded connections.

The bits are usually of the roller or tricone type discussed under Deep Rotary Drilling, with special adaptions to being cleaned by air instead of mud. Rotation of the bit against the bottom causes the cones to rotate, setting up a chipping and crushing action, with a variable amount of scraping and abrading.

In general, large teeth with wide spacing in each row are used for soft formation, as shown in Fig. 20.35(*B*), and tooth size is progressively reduced for harder drilling. Teeth are either steel or full carbide inserts. Steel teeth are surfaced with tungsten carbide or other hard facing. Full carbide inserts are used for the hardest and most abrasive formations.

In standard bits, the airstream comes from the bits behind the cones, tending to clean the hole bottom and blow the cuttings out of the drilled hole. This construction is best for all formations.

The jet bit directs the airstreams directly on the bottom. This avoids sandblasting the teeth and cones with fine particles sucked into the airstream. Part of the stream of clean air may go through the bearings to cool and lubricate them.

The blast hole rotary drive is usually a rotary head, fastened through a chuck to the top of the drill rod. It is turned by an air, hydraulic, or electric motor with variable speeds from zero up to 150 or more revolutions per minute. The head is raised and lowered with the drill string by the feed mechanism, which is described in a later section.

Rigs built primarily for down-hole drilling can usually be changed over to use rotary bits. However, rotation may not be as fast, or down-pressure as great, as in machines that are built primarily for rotary drilling.

Bit footage may be as little as 5 feet, but the usual expectation is from several hundred to several thousand. Life is affected by the drillability of the formation, the design of the bit, and the manner of use and abuse.

A   B

FIGURE 20.35   Examples of rock bits (*A*) for hard, abrasive rock; (*B*) for soft formations.

HOLE CLEANING

As the bit reduces rock to chips, sand, and dust, these particles must be removed promptly, or they will form a layer that will prevent the bit from striking the rock.

The basic method of removal is by a current of compressed air, entering the hollow steel at the drill, and emerging from the bit in holes at the front or bottom (one) and sides (none to four). Some of the air is exhausted from below the piston (puff blowing) at moderate pressure, and part is blow air admitted directly from the high-pressure chamber supplying the drill, by putting the throttle in BLOW position.

In a properly balanced rig, the air that cleans cuttings from the bottom of the hole has sufficient volume and velocity to carry them up the hole and out the top. There the chips and coarse sand will pile in a ring around the hole, fine sand will go a little farther, and dust will drift with the wind.

Chips and sand must occasionally be pushed back or removed from the edge of a deep hole, so as not to slide back in. This is a minor nuisance. The material itself may be useful as stemming for explosives, grit on slippery surfaces, and perhaps in other ways.

Dust. The dust, however, is a major nuisance. It cannot be tolerated underground. Even in the open, it is unhealthy to inhale for both people and engines. It contaminates exposed lubricant and lubricated surfaces, often turning them into grinding tools. Its suppression has become required by law.

There are three principal methods to control dust in air exhausted from drill holes: vacuum separation, wet or damp (detergent) drilling, and foam.

Vacuum (Filter) Collectors. A hood may be placed over the drill hole, with a loosely gasketed fit around the rod, and connected by flexible tubing to a container. See Fig. 20.36. A vacuum

FIGURE 20.36   Effect of dust collector.

apparatus pulls the air out of this through dry filters and/or cyclonic separators, leaving dust and chips behind. The clean air is discharged to the atmosphere, and the ground rock is removed as often as necessary.

There may be considerable leakage of outside air into the hood, particularly on rough ground. This is not serious, as even a weak vacuum will remove the dust completely, leaving harmless coarse particles behind. The hood may depend principally on the drill rod for support on steep slopes.

The collector is usually built into a mobile tower drill and often into crawler drills. Skid- or wheel-mounted units are used for smaller crawlers, and for hand drills. See Fig. 20.37. One collector may have several intake hoses to service a number of holes at the same time. Power may be gasoline or electric.

FIGURE 20.37   Vacuum dust collector.

There are also venturi systems for hand drills whose vacuum is created by the flow of compressed air.

A separate vacuum apparatus is suited to quarries with fairly regular work surfaces and drilling patterns, but may be an almost hopeless nuisance in rough and pioneer work. Its great advantage is that it permits bit and hole cleaning to be done efficiently by dry air.

Detergent System.   Dust may be suppressed by dampening it at the bit.

Water in a tank is mixed with a small amount of detergent, and fed slowly into the hole-cleaning air. The detergent breaks the surface tension of the water, enabling it to coat dust particles thoroughly, so that they stick together in little balls that are much too heavy to float in the air.

Unfortunately, the dampening slows the escape of the material from the face, so part of it is still there for the next blow. This regrinding absorbs power, weakens impact on the rock, and wears the bit. The loss in footage in damp drilling as compared with dry drilling may be 10 percent in hard rock.

The detergent is necessary for efficiency. Without it, much more water is needed (wet drilling), and action is less uniform. The extra water turns part of the dust to mud, which may build up mud collars that make pulling the bit difficult.

In winter the detergent-water needs antifreeze. Alcohol (methanol) is standard for this. Fuel oil, which may be kept in suspension by the detergent, is sometimes used.

Detergent dampening is a widely used dust-suppression system. Any crawler drill can easily carry a tank of 30- to 50-gallon capacity, enough for a full shift of drilling. Skid-, trailer-, or even wheelbarrow-mounted tanks can be used for smaller units. Aside from refilling, little maintenance is required. The principal disadvantage is reduced production, with increased bit wear.

Foam.   A foam system uses the same equipment as the detergent method, with change only in chemicals.

The detergent in the water is replaced by a special foaming agent, plus some diesel fuel. When drawn into the airstream, this generates a stiff foam of fine air bubbles which carries dust and chips out of the hole in a dampened and harmless condition.

The bubbles hold and buoy up rock particles of all sizes, so that the particles do not tend to sink or fall down against the current. Much less air is therefore required to clean the hole, so that full-blow air may be needed seldom or not at all. The saving in air may be 150 to 350 cfm, reducing compressor requirement and cost.

There is a loss in penetration and increase in bit wear compared to dry drilling, but the loss is probably less than with detergent. On the other hand, the chemical is more expensive and may not be obtained as readily.

NOISE SUPPRESSION

Percussion drilling is an inherently noisy operation. Aside from the compressor, which was considered separately, the drill admits and ejects high-pressure air at a rapid frequency, while it also hammers metal against metal and metal against rock. Exhaust from air motors may add to the din.

There seems to be little hope of converting this to a quiet operation, but there are methods of lowering the noise level noticeably.

One means is to install mufflers, usually called silencers, on direct exhausts of pressure air from the drill piston, as in the tool seen in Fig. 20.38, and from any air motors. These are usually simple devices, involving abrupt changes in direction of airflow. They are designed to cause minimum back-pressure. On small drills it is a major problem to keep them small and light enough, and sufficiently out of the way, so as not to interfere seriously with the use of the tool.

Exhaust air coming out of the hole has greatly lowered velocity, and is not an important noise problem in itself.

The blow of drill piston on striker bar or shank is the most difficult noise to reduce. Some improvement can be obtained by reducing the resonance of the parts involved, as by a vibration-absorbing

FIGURE 20.38   Muffler in hand-operated breaker. (*Courtesy of Ingersoll-Rand Company.*)

clamp. But the only way to substantial reduction would be to encase the entire drill in a sound-absorbent structure. This would create extreme problems for the operator and service technician.

Bit-on-rock noise is diminished by depth, and is further reduced by foam dust suppression, as the bubbles are an effective absorber of noise.

A down-hole drill becomes much quieter as soon as it enters the hole, and might be considered on noise-sensitive jobs for just that reason.

HAND DRILLING

Starting.   The rock drill can be carried by hand or truck to the work area and connected to a flexible air line from the compressor. A short starter steel, 1 foot to 30 inches in length, is fitted with a bit and clamped in the chuck. This bit should be the largest size which is to be used in the hole.

The area surrounding the hole should be free of dirt and litter that would interfere with starting the hole or might slide into it during drilling.

If the rock surface is horizontal, the drill is held vertically and the throttle partly opened. If the bit wanders on the surface instead of cutting it, the bit can be held in position with a foot pressed against the lower end of the steel. If the rock slopes, the steel should be held at a right angle to the slope instead of straight up and down, until it has dug a pit that will hold it.

When the bit has cut in enough that it is not likely to bounce out, the hole is said to be collared. The throttle can then be fully opened. The drill can be simply supported against leaning over, or if it tends to bounce, it can be pushed down with one or both hands.

In soft rock it is not advisable to push down on the drill, as this increases any tendency toward rifling or binding. It is sometimes necessary to pull up slightly to avoid these difficulties. In hard rock, extra weight is helpful, particularly if the drill is light.

While the hole is shallow the automatic puff blowing should keep it clean. However, it is good practice to turn the valve to BLOW occasionally. The amount of extra chips and dust raised will indicate whether the puff blowing has been doing the job.

Dust around an air drill is a damaging nuisance, and its suppression is required on many jobs.

Deepening.   When the hole is cut so the drill is close to the ground, the throttle should be closed and the drill lifted until the bit and steel are out of the hole. The steel is taken out, and another steel about 2 feet longer is locked in the chuck. If it is not worn much, the starting bit may be put on the second steel.

The bit is detached by holding the steel firmly and striking the bit upon any projection so as to unscrew it. The steel may be held under the foot and the bit struck with a hammer. Some operators prefer to use wrenches—a pair of Stillsons, or special wrenches—for changing bits. In installing a bit, it is hand-tightened then tapped to seat it firmly.

It should be remembered that the thread is left-handed.

If another bit is used, it should be slightly smaller than the starter. The easiest way to compare sizes is to put the two bits together.

Drilling is resumed with the second steel. When it is mostly in the hole, it is replaced with a longer steel, using the same or a smaller bit. This process is repeated until the hole is finished. Its depth is readily measured from the length of the steels.

As the hole deepens, it becomes increasingly necessary to blow it out. Some hammers can be adjusted to send down an extra stream of air while drilling, which reduces the need to stop drilling to blow.

If air and dust stop coming out of the hole, it should be blown immediately. If it will not blow, the steel should be pulled out. The trouble is usually a plugged bit. The cuttings can be removed with a thin punch, a nail, or an ice pick.

An attempt to drill without a flow of air will overheat and spoil the bit and probably cause a collar of compacted cuttings to form just over it, which will make it very difficult to pull.

Whenever possible, holes should be drilled vertically. This puts the weight of the hammer on the bit instead of on the operator and simplifies keeping the hole straight.

Jamming.   There are several conditions under which getting the steel and bit out of the hole is a greater problem than the drilling. No drill runner is so good that he or she never sticks a steel.

Generally, the principal cause of sticking is the use of worn bits. In most ground the bit gives a warning that it is reaching a danger point before it jams. This may be slower penetration, production of fine rock dust instead of chips, twisting of the drill in the hands, or failure of the steel to rotate. Condition of the hole may be roughly checked by raising the drill a few inches occasionally to see if it is free.

In hard or brittle rock a bit may function satisfactorily which is so worn that it would stick immediately in a soft formation. However, worn bits are also dull bits, and drilling is slowed by their use. It is also poor economy to use a worn bit because excessive wear may make it impractical to recondition it, particularly by grinding.

In soft rock a bit may rifle the hole. This trouble may be expected whenever the drilling rate is very rapid. New, sharp bits are worse offenders than worn ones. Precautions which can be taken include use of special X or rose bits, using a lighter drill or partly supporting the weight of a heavy one, grinding back the cutting edges of the bit, or reconditioning bits by grinding the sides to restore taper without sharpening the edges.

Seams of dirt or finely disintegrated rock in a formation will cause plugging of the airhole. At the same time there is rapid penetration and dirt falls on the top of the bit, where it may compact into a hard, tight collar.

When the speed of penetration increases suddenly, drilling should be stopped immediately and the hole blown. Drilling can then be resumed on partial throttle with frequent, thorough blowing. If air stops coming out of the hole, the steel should be pulled and the hole in the bit opened.

If the ground is composed of sloping layers of rock of varying hardness, or seams or cracks slope across the line of drilling, the hole will tend to drift down the slope. The curved hole that results will bind the steel.

Drilling on a dip at right angles to the seams should prevent this trouble. Often an intermediate direction will be satisfactory. In vertical drilling, curving can often be prevented by using light pressure and sharp bits and arranging for the drill to get maximum air pressure. This may involve cleaning the drill air filter, shortening and straightening the line from the compressor, or temporarily shutting down other units using air from the same source.

Pulling Steels. In taking a steel over 3 feet long out of a hole, the usual procedure is to unclamp the steel puller, set the hammer aside, and lift the steel by hand. However, if the steel is stuck, the hammer should be left attached and lifted with the throttle alternately in drilling and blowing position. The heavy vibration and rotation of the steel and the air pressure help to break it free.

Two people—one on each handle—can lift the drill much more effectively than one person.

If the sticking is caused by spiraling (rifling), it may be possible to extract the steel by turning it against the direction of rotation while lifting. If two people are working, one person can lift the drill with throttle closed while the other person turns the steel and drill with a wrench. If the operator is alone, she or he should remove the drill, place the wrench just under the shank collar, and pull the wrench upward while turning it.

This may cause the steel to unscrew from the bit. If the hole is full depth, a steel bit can be sacrificed economically. If it is not full depth, leaving the bit in will make it necessary to drill another hole. If the bit is carbide, every effort should be made to salvage it by direct pull.

Usually steel and bit can be raised together by jacking. No really suitable jack is generally available, but a chain jack or a regular bumper jack with a chain grab fastened to the lifthook can be used. The jack is placed on the edge of the hole, the end of a light chain given a double turn around the steel and hooked, and the chain pulled tight and caught in the jack head. The best grip is just under the collar, but a small wood wedge will generally enable it to hold anywhere on the steel.

This jack does not pull straight up, and jack and steel will tip as the steel rises. This may cause bending of the steel or the jack if lifting is continued too long on one hold. It is a good plan to release the chain and fasten it again farther down before bending is severe.

A chain fastened to wagon drill feeds, loader buckets, bulldozer blades, cable control units, or to any type of shovel or crane may be used for steel pulling.

If all methods of extraction fail, a parallel hole is drilled nearby and the steel picked out of the loosened rock after the blast. It is generally bent by the explosion, and unless it lands on top, may be further twisted by a shovel.

Care of the Drill. A drill is built of special steels and is very finely machined throughout. It therefore costs many times as much per pound as a shovel or dozer, and it deserves much better care than it usually gets.

The oil reservoir is built into the bottom of the cylinder and may require filling from one to four times a day, depending on the make and condition of the drill. This oil is forced or sucked into lubricating passages, and into the airstream during work. The oil mist in the exhaust air gives an indication of the amount being used.

A line oiler should be used instead of or in addition to the reservoir for best results, as it has a larger capacity and is more reliable.

The frequency with which either a reservoir or an oiler must be checked is best determined by experience. If exhaust air ceases to carry oil, the cause should be found immediately.

Air drills have very slight clearances and have a severe temperature drop. Air often enters them at a temperature of 100°F or more, and is exhausted close to freezing.

In addition, water condensing from the air or leaking from a watertube may wash oil off the parts which need it. Special oils designed for these conditions will give better service than standard lubricants. However, any oil is better than no oil.

Every possible precaution should be taken to keep the oil clean. This is a difficult matter when the air is full of rock dust, but it is important. Dirt may clog oil or air passages or score sliding parts.

Use of an inlet air filter may prevent trouble with scoring or clogging, but it should be cleaned frequently. Most filters are so small that a little debris can reduce the air passage enough to starve the drill.

A drill needs periodic cleaning, the interval depending on the quality of air it receives. An oil-pumping compressor, very hot air, or deteriorated hose may foul it in a day or less. With clean air it may function well for weeks.

A quick cleanout can be done by disconnecting the air hose, pouring a cup of kerosene or fuel oil into the drill, reconnecting the air, and running the drill idle for about a minute. The hose is again disconnected, a cup of rock drill oil poured in, and the drill operated again.

If it is very dirty, several doses of a mixture of three parts kerosene and one part oil can be used.

At much longer intervals, it is good practice to take the drill to the shop (drill doctor), to be disassembled, thoroughly cleaned, and checked for worn parts.

POWER FEEDS

Drills that weigh 75 pounds and more are usually supplied with supports and with automatic feeds.

Air Leg.   The air leg in stopers provides both support and automatic feed for light overhead drilling. It is used almost entirely in underground work. The drill is supported by an air cylinder, whose piston ends in a stinger point or some other device to hold its position on rock. Air is admitted to the leg cylinder by a control valve, causing it to expand sufficiently to keep the bit in contact with the rock as it cuts into it.

Chain Feed.   Chain is the most usual type of feed in open air work. A reversible piston type of air motor drives a roller chain through reduction gears. It both feeds the drill bit into the rock and, when reversed, pulls it out of the hole.

The chain goes around idler sprockets at the top and bottom of the mast. The drill is fastened to the chain, and moves along the shell on lubricated guides.

Hydraulic.   Hydraulic feed is used with a hydraulic motor operating in the same manner as an air motor to power a chain feed, or a two-way hydraulic ram that may work through a chain, a cable, or a direct fastening to the drill.

MOUNTINGS

Stoper.   In the standard stoper, Fig. 20.39, the leg and the drill make up a single straight-line unit. The operator places bit and steel in the drill, places the pointed bottom of the leg, called the stinger, on the floor or in a wall crevice, and opens the valve to extend the feed leg until the bit is held against the overhead rock in drilling position. During drilling, air is fed to the leg cylinder so as to keep an even pressure on the bit.

When the steel is all in the hole or the leg is fully extended, the leg is retracted, pulling the steel out of the hole, and a longer steel is substituted.

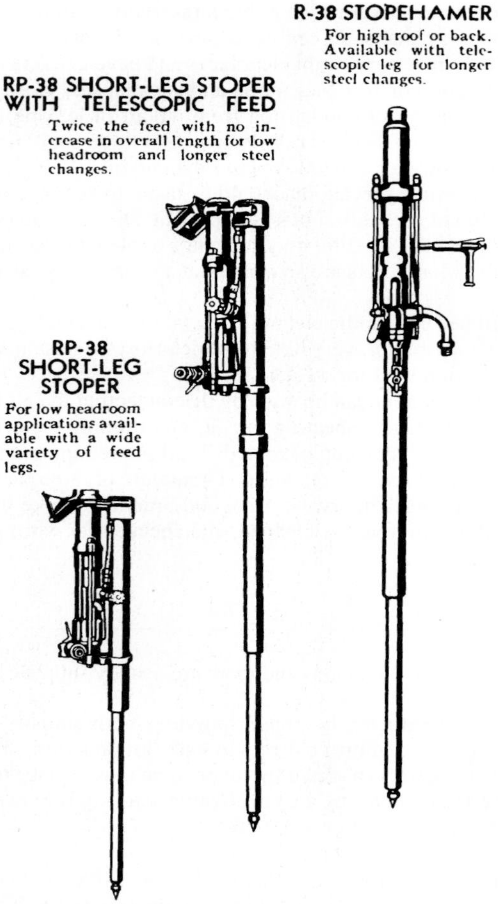

R-38 STOPEHAMER

For high roof or back. Available with telescopic leg for longer steel changes.

RP-38 SHORT-LEG STOPER WITH TELESCOPIC FEED

Twice the feed with no increase in overall length for low headroom and longer steel changes.

RP-38 SHORT-LEG STOPER

For low headroom applications available with a wide variety of feed legs.

FIGURE 20.39 Stopers.

Stoper operation is tricky. The point may slip out of position and even jump onto a foot (safety shoes are a *must* for stoper operators), the moderately heavy machine must be supported until the hole is collared, and collaring may prove difficult because of the strain of keeping the stinger down and handling the drill.

Some deluxe or safety stopers have a double stinger so that they can lock at both ends to support the entire weight during collaring.

The stoper is intended primarily for overhead drilling, as in other positions its weight is more difficult to handle. Holes should be inclined away from the vertical, and the operator should stay on the uphill side, out from under the hole.

Jackdrill. The air feed leg shown in Fig. 20.40 is hinged to the drill. This arrangement permits convenient use in horizontal and angle drilling, as well as vertical.

This unit may be set up with the leg and drill almost in line, so that the feed supplies both support or push, or at a sharp angle where the leg supports and the operator pushes. The feed tends to push the steel against the top of the hole, but a slight down-pressure on the drill will keep it in line.

FIGURE 20.40   Jackleg.

FIGURE 20.41   Column drill support.

The operator should keep a firm grip on the handle of an air leg machine, because if the steel sticks, it may whip out of the operator's hand and cause injury.

Column.   A hydraulic feed drifter drill in underground use may be supported by a pipe column. The column, of 3- or 4-inch diameter, usually extends from floor to ceiling of a tunnel, as in Fig. 20.41, but it may be placed directly against the rock or timbering, or have foot and head blocks. Columns are usually tightened by means of threaded sections. Pneumatic columns are extended by air pressure in the manner of an air leg, and may carry a stinger point on each end.

The drifter is supported by a saddle on the column itself, or on a side arm. It can swivel vertically and horizontally. The saddle can be moved along the arm, and the arm can be moved along the column.

Column mountings are somewhat clumsy to use. They are being replaced by air legs in light work, and by jumbos in big jobs.

BOOMS

Drill Booms and Jibs.   Air drills and feeds of the type carried by crawler rigs may be mounted on hydraulically controlled booms. See Fig. 20.42. Such booms can be mounted in any desired

FIGURE 20.42 Drifter drill with hydraulic feed. (*Courtesy of Atlas Copco.*)

number, from one up, on any type of base large and strong enough to carry them. See Fig. 20.43 for typical construction.

The boom is attached to its support by a universal or two-hinge bracket. A hoist ram raises and lowers the boom, and a side ram swings it to right and left. The two controls can be used together to place it in a wide range of positions.

The drill tower or mast is fastened to the outer end of the boom by a double-hinge clamp connection called a cone. There may be a hydraulic ram that can move the mast back and forth along this clamp to adjust the reach, and there may be hydraulic means to lower (dump) and tilt (roll) it.

Booms are often mounted in pairs on old 20-ton tractors which are no longer suitable for heavy work. A compressor may be mounted on the same machine.

Jumbos.   Modern rock tunneling involves drilling a number of holes in the face at the same time. The drills and their operators are supported on a moveable platform, called a jumbo. These devices are made in great variety, including single or double platforms that merely support drill runners and hand drills, arrangements of columns and swivels to support either jackhammers or automatic feed drills, hand-jacked carriers, or full hydraulic control as the Hydra-Boom machine, which is shown in Fig. 20.44.

Where muck is hauled away from the face by railcars, the jumbo is usually rail-mounted. It may use the same tracks, or much wider ones that permit it to straddle the trains without blocking them. It may also carry a crane for transferring cars.

Rubber-tire or crawler-mounted jumbos are now widely used, particularly where haulage is by truck.

TRACK-MOUNTED DRILL

This section deals with the lighter models of track-mounted drills, which have both a mast and a boom. This is the dominant type of drill rig for hole diameters of $2\frac{1}{2}$ to $4\frac{1}{2}$ inches, up to 80 or perhaps 100 feet deep. Some models can drill larger holes, and most of them can go deeper, but at lower efficiency than the big mobile tower or hinged-mast machines.

In smaller holes, and depths of 20 feet or less, they compete with handheld drills.

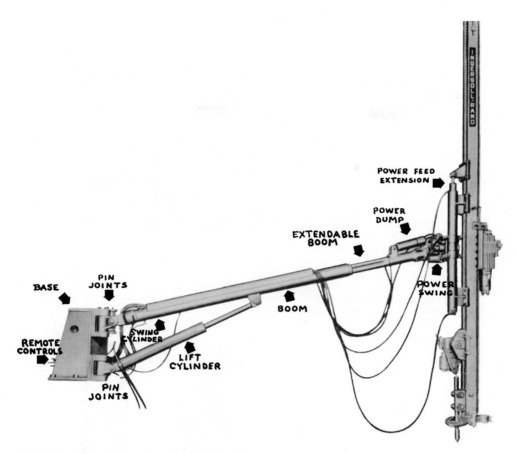

FIGURE 20.43 Hydra-Boom. (*Courtesy of Ingersoll-Rand Company.*)

FIGURE 20.44 Hydra-Boom jumbo. (*Courtesy of Ingersoll-Rand Company.*)

General Description.   A crawler drill has a pair of tracks; a body and turret mounted between them; a boom that is based on the turret and can be raised, lowered, and swung in a partial circle; a mast on a universal mounting at the boom point; and a percussion drifter drill that can be moved along the mast by a power feed. See Fig. 20.45.

Power is supplied by a separate compressor, towed by the drill rig. Some functions are air-operated, others are hydraulic, with a pump driven by an air motor.

Travel.   The tracks are the type used on crawler tractors, except that they are narrower and lighter, usually 10 inches wide. Single grouser shoes are standard. See Fig. 20.46.

Travel or propel is called tramming, a term derived from mining. Each track is driven independently, forward or backward, by an air motor. Maneuverability is excellent.

Brakes usually go on automatically when tramming controls are put in neutral. These brakes can (and must) be locked out of operation by a manual control if the machine is to be towed.

Travel speeds are slow—2 or $2\frac{1}{2}$ miles per hour. Speed while being towed must not exceed 5 miles per hour.

Before moving, the mast foot must be lifted sufficiently to clear the highest spots on the ground to be crossed. If the move is longer than between adjacent holes, it is good practice to lower the boom, and tilt the mast back into a carrying position. The drifter should be directly over the turret.

The track frames are pivoted to the body at the rear, and support it on a walking beam (hinged cross member) near their centers. The mechanism may provide for extremes of oscillation, up to 24° difference in vertical angle between the tracks.

FIGURE 20.45   Track-mounted drills, range of sizes. (*Courtesy of Ingersoll-Rand Company.*)

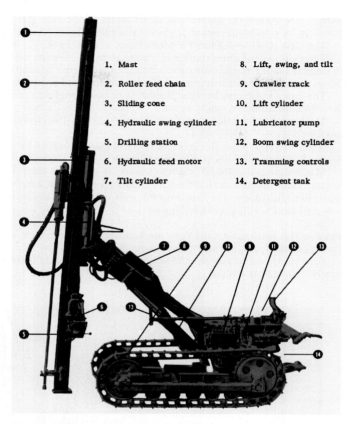

1. Mast
2. Roller feed chain
3. Sliding cone
4. Hydraulic swing cylinder
5. Drilling station
6. Hydraulic feed motor
7. Tilt cylinder
8. Lift, swing, and tilt
9. Crawler track
10. Lift cylinder
11. Lubricator pump
12. Boom swing cylinder
13. Tramming controls
14. Detergent tank

FIGURE 20.46   Track-mounted drill, labeled.

These machines are designed to tow a compressor of appropriate size (600 to 900 cfm) except on extremely rough or steep ground. For frequent use under such conditions, an air-powered winch may be installed in the rear. The crawler can then go ahead, by either track drive or winch power, until the connecting hose is fully extended. It can then stop, be braced, and winch the compressor to itself.

Body.   The body, supported by the rear axle and the walking beam, contains or carries the air manifold and lubricator, tramming motors, hydraulic pump, the detergent tank, and air and hydraulic controls for all machine functions except drifter movements.

The body serves as the connection between the tracks, and as the support for the pedestal or turret that carries the boom.

Boom.   The boom is a beam, hinged to the top of the turret, and supported by a two-way hoist cylinder that can move it in a vertical arc of about 80°. It can be swung horizontally about 100°, either by another cylinder or by rotation of the turret.

The boom may be one piece, with fixed length, or made extendable by having two hydraulically telescoping sections. The extendable feature adds greatly to the area that can be drilled from one position of the tracks.

A straight line of closely spaced holes may be drilled by extending or retracting the boom as it swings from one to another, without need to adjust the vertical angle between boom and mast each time.

Mast.   The mast is the support and slide (guide) for the drill. It is attached to the right side of the boom tip by a multidirectional, hydraulically controlled connector called a power cone.

The cone provides for three separate types of movement.

The mast may be dumped, that is, swung so as to move the lower end or foot forward or back in the plane of the boom. This movement makes it possible to keep the mast vertical with different boom angles, and to drill at other angles within an arc of about 180° relative to the boom.

The dump angle may be used together with boom angles to provide for drilling in almost any direction, including vertically upward.

The mast may be rolled, that is, tilted to right or left. The arc varies from 80 to 180° in different models.

It can also slide several feet on the cone, so that its foot can be extended or retracted (raised or lowered, if vertical) without affecting its angle with the boom or the ground.

Mast height is usually sufficient to handle 10- or 12-foot steel changes, and starter steels 2 feet longer. There are also models that can handle 20-foot steels 2 inches thick, but these high, heavy masts limit flexibility. They are desirable only on fairly level ground. They need hydraulic or air-powered steel hoisting and handling devices.

Feed.   The mast carries the feed mechanism, which supports, crowds, and retracts the drill. Power is usually air, but may be hydraulic. See Fig. 20.47.

Feed pressure and pull may be varied from zero up to 3,000 or 4,000 pounds. Speed may be nearly 200 feet per minute.

FIGURE 20.47   Tractor-mounted drill for 3- to $4\frac{1}{2}$-inch holes. (*Courtesy of Atlas Copco Construction and Mining N.A.*)

Drifter.   The drifter (drill), also described earlier, may have either rifle bar or independent rotation. There are a range of sizes, usually from 4½-inch bore and 290 pounds up to 6-inch bore and 460 pounds. Smaller and larger sizes may be used. See Fig. 20.48.

The drill is slidably fastened to the mast, and clamped to the feed cable, by a bolted-on backing plate called a slab back.

If a down-hole drill is used, the drifter is replaced by, or converted to, a rotary unit that turns the rod, and also transmits the lowering and raising movements of the feed, but does not hammer.

Recommended drill bit diameters:
76–115 mm, 3"–4½"

Compressor
Atlas Copco, screw type compressor

Working pressure	10,5	bar	152	psi
FAD	143	l/s	303	cfm

Engine
Mercedes Benz
Watercooled diesel OM441A

Rating at 2000 rpm	170	kW	230	HP

Fuel tank

Capacity	380	l	approx. 100	gal.

Feed

Feed length tot.	7 300	mm	24'	
Travel length	4 600	mm	15'1"	
Feed rate, max.	0,92	m/s	180	ft/min
Feed force, max.	16	kN	3 520	lbf

Tramming

Travel speed, max.	3,6	km/h	2,2	mph
Traction force	147	kN	33 000	lbf
Hill climbing ability, without winch	20°		20°	
Hill climbing ability, with winch	35°		35°	
Track oscillation	±10°		±10°	
Ground clearance	405	mm	16"	

Hydraulic rock drill
COP 1838HE, COP 1840HE, COP 1850

Impact power	20	kW	26,8	HP
Working pressure max.	230	bar	3 335	psi
Torque	980	Nm	722	lbf.ft
Weight approx.	190	kg	420	lb
Includ. hydr.drill steel				
Extractor approx.	250	kg	550	lb

Transport dimensions
Length

folding boom version	10 850	mm	35'6"
single boom version	11 000	mm	36'
Width	2 420	mm	8'

Height

folding boom version	3 500	mm	11'6"
single boom version	3 350	mm	11'

Weight

folding boom version	14 000	kg	31 000	lb
single boom version	13 700	kg	30 200	lb

FIGURE 20.48 Specifications for tractor-mounted drill shown in Fig. 20.47. (*Courtesy of Atlas Copco.*)

Air System.   Compressor size is determined by the needs of the drifter, including the rotation motor if it has one. The tramming motors and the hydraulic system (except feed, if it is hydraulic) are shut off during drilling. Their requirements, when in use, are much smaller than those of the drilling.

The compressor usually has a 600- to 900-cfm rating at 100 pounds, but one machine requires a 1,200-cfm size.

Recommended main or bull hose is 2-inch-diameter and 100 feet long, with surface reinforcement against scuffing. Larger hose may be too stiff for good maneuverability. Length must be sufficient to allow a good working area for the drill rig from one stand of the compressor, and to allow the compressor to be parked away from the dust if drilling is dry and open.

Air enters through a main shutoff valve, a manifold, and a lubricator. From there it feeds the valves and motors for travel and the hydraulic pump, and a main supply hose runs forward to the mast-and-drill control station.

It is essential that the compressed air be clean. It is filtered at the compressor, and with tight hoses, there is no problem. However, any hose that is opened should be blown out before replacement. This particularly applies to the bull hose between the compressor and the drill rig. This often brings complications that may be solved by disconnecting, and it is right down in the dirt.

Always disconnect at the crawler drill end of the hose, and always blow it out thoroughly before reconnecting.

Most of these machines are meant to operate at 100 pounds of air, and do so badly at less than 90 that they should be stopped if pressure falls that far. The problem might be using too small a compressor, or, in one of proper size, either engine or air trouble.

Hydraulic System.   Hydraulics are used instead of air in all position adjustment functions because air, being compressible, does not hold position against changes in load.

A small hydraulic pump, protected by a relief valve, is driven by an air motor. The hydraulics usually control setup functions after the machine is in position, and before drilling starts. The air is therefore shut off and the pump stopped during travel, parking, and drilling.

A typical control bank is shown in Fig. 20.49. It has all the hydraulic valve levers, plus air controls for tracks and the pump. This bank may be on a swivel, so the operator can have access to it while walking behind the machine during travel, or when standing beside it while positioning the mast. Standing two-way cylinders may be used for boom rotation and lift.

Drilling Controls.   The control panel in Fig. 20.50 has four rotary valves, each of a metering type to allow fine adjustment of opening.

FIGURE 20.49   Tramming and hydraulic controls.

FIGURE 20.50   Swivel panel for drilling controls.

The two left-hand valves are one-way, on or off. DRILL controls the percussion or hammering action, BLOW the volume of cleaning air.

The two right handles govern reversible action—FEED down or up, and rotation for drilling (FWD), or reversing (REV) for bit or steel removal, or getting out of bad holes.

If the drifter had a rifle bar, the ROTATE handle would control direction only, not speed.

TRACK-MOUNTED DRILL OPERATION

Travel. Crawler drills are moved between jobs on trailers, or if distances are short, may be towed very slowly. On the job they move themselves, towing the compressor that provides their power.

For self-propelling (tramming), the operator walks behind or beside the machine, within reach of the track controls. There usually are two levers, one for each side. Moving a lever forward turns that track forward; backward moves it back. An automatic brake goes on when the lever is centered.

Steering is done by varying the relative speed and/or forward-reverse direction of the two tracks. When they go in opposite directions, the machine will spin-turn in its own length. Very precise maneuvering and spotting are possible.

The rig usually pulls the compressor with it by means of a pintle hook or drawbar connection on any but the shortest moves. Then, and when moving independently, the operator has a major responsibility to avoid damaging the hose by stretching, kinking, or walking over it.

Tramming is done with the mast above any possible obstructions, and preferably folded back to a near-horizontal position, for maximum stability.

Setting Up. To set up for a single vertical drill hole, choose the most convenient and stable position for the machine. If ground conditions permit, try to drill in the area between the line of tracks and just forward of them.

Raise the boom to an angle of about 25 or 30°, and set the mast vertically. Position the foot of the mast so the hole marker lines up with the steel centralizer, by swinging the boom and extending or retracting it (if it is extendible), or by raising or lowering it. Recheck the mast for vertical, then slide it down until the foot rests solidly on the ground.

The crawlers give the machine a good grip on most types of ground, their brakes lock automatically, and it is heavy enough to stay in place. But if there is any question of its sliding even a fraction of an inch, block it with wood or rocks.

Once the boom and mast are set for drilling, shut off the air motor that drives the hydraulic pump, unless feed is hydraulic.

Starting the Hole. Move the drifter (drill) to the top of the mast by using the feed motor. Make sure that all threads are the same—steels, couplers, and striking bar.

Grease both threads on the first length of drill steel. Screw a good bit onto one end and a coupling on the other, finger tight. Remember, these are left-hand threads. Swing the centralizer arms open, then stand the steel (or rod) on its bit in the centralizer.

Move the drill down with the feed so that the striking bar threads engage the rod coupling. With the rotation motor, turn the bar counterclockwise until it is firmly seated in the coupling. Be sure *not* to start hammering.

Lower the feed until the bit rests firmly on the ground. Close and latch the centralizer. Using the drifter's rotation motor, turn the striking bar counterclockwise until the rod and bit turn with it, and threads appear firmly seated.

Start part throttle hammer action of the drill, continue rotation at moderate speed, and feed it down slowly. When the bit has penetrated any overburden and has made a good socket (collared the hole) in rock, open the centralizer arms wide.

Drilling. The interaction of bit and rock, and some of the effects of varying pressure and feed, was discussed under Bit Action on page 20.28.

Rotation should be fairly rapid when penetrating overburden or other loose material, but should be slow to moderate in rock. Fast rotation keeps giving the edges thicker slices of rock than they can break efficiently, therefore slowing penetration while using more air. It is also likely to cause the drill string to whip and vibrate, and break prematurely.

Pressure should hold the bit firmly against the bottom, but pressure in excess of that does not allow proper rebound from the strokes. This hampers the bit in the same manner that a low ceiling interferes with using a sledgehammer.

Insufficient pressure, which usually means partly supporting the weight of the string on the feed, prevents the bit from following through enough on its strokes, reducing penetration. It also loosens threads and heats up the bottom of the drifter.

A good stream of blow air is essential. It is the most important factor in penetrating overburden or soft rock, as these can readily plug up the holes in the bit. After this, there is nothing to do but pull it out for cleaning—promptly.

There should be a continuous stream of chips, sand, and dust coming out of the hole. If you are drilling with detergent, the dust should be in little balls, but it should still come out. The volume and coarseness of this material give you a reading on bit action.

The positive indication of progress, of course, is how fast the rod goes down into the hole.

Adding Steel. If the hole is to be deeper than the length of the first rod, you have to add another. This is done when the striking bar coupling has almost reached the centralizer.

You start by checking the hole. Turn off the hammer, continue rotation slowly, and raise and lower the bit a foot or so to make sure it is not binding.

Then turn off the rotation, lower the bit to the bottom, and alternately start and stop the hammer action of the drifter a few times, to loosen the top-of-the-steel coupling threads. Lifting the string so it bounces may help. The ringing note of the steel changes when loosening occurs.

It is important to keep rotation shut off while hammering to loosen.

Operate the drill rotation motor in reverse, to turn the striking bar clockwise and unscrew it from the coupling. Grease the striking bar threads, and raise the drill to the top of the mast.

Grease the threads on another section of steel, turn a coupling onto one end by hand, and raise that end to line up with the striking bar. Fit the bottom of the steel into the coupling on the piece that is in the hole, and turn to catch the threads. Lower the drifter so that the bar threads enter the upper coupling.

Tighten the threads by turning the rotation counterclockwise slowly. Blow air and rotate some more, then start the hammer action after both threads are thoroughly seated.

This operation is repeated for each steel added.

Withdrawing Steel. When the hole has been drilled to required depth, turn blow air on full, rotating the string slowly without hammering, until air coming out of the hole is clean.

Stop the rotation and blowing, and loosen the threads as described under Adding Steel. Stop the hammer action, and raise the string by running the drifter up to the top of the mast.

Next, close and latch the centralizer arms, then lower the string until the coupling below the top steel section rests on the centralizer, so that it supports the weight of the string.

It is absolutely essential that the centralizer be latched, and that the coupling be above it. Otherwise, the whole string might be lost in the hole.

Rotate the drill slowly in reverse (clockwise) and raise it until the striking bar thread is loosened from its coupling. Stop the rotation, and unscrew the steel by hand from its bottom coupling, which is resting on the centralizer. Remove the steel section.

Lower the drifter until the striking bar thread engages the coupling on the centralizer. Rotate the bar to thread it loosely into the coupling, sufficiently to support it. Stop rotation and raise the string to take its weight off the centralizer.

Open the centralizer arms. Run the drifter up the mast until the next coupling is above the centralizer. Close and latch the centralizer, and lower the string until the coupling rests on the centralizer. Then repeat the steps described above, for each section of steel in the hole.

The bit is usually removed after the steel has been laid aside. It should be inspected for wear, and for chipping or loss of inserts. If unscrewed over the hole, the hole should be covered by a board.

The steels should be laid out in a definite pattern. Their positions should be changed in the string on the next hole, so that throughout the job wear and fatigue will be about equal throughout the set being used.

Problems. The drill string does not always come out readily. Occasionally, it does not come out until a loader picks it out of the blasted rock.

If the ground is wet, or if water is being added to blow air to suppress dust, the drill chips and dust may combine with the water to build a mud collar above the bit. See Fig. 20.51. Sometimes it is thick enough to just allow space for the rod, but a thickness of $\frac{1}{16}$ inch may be enough to cause trouble.

The bit must have a taper so that it can follow the cutting edges without dragging and binding on the sides of the hole. If the hole becomes partly clogged, this taper causes a wedging action that compresses and often hardens mud to increase its resistance, as the bit is pulled up.

The operator can usually get through this resistance by some combination of force (the feed can usually lift a ton or more), rotation, and hammering, accompanied by maximum blowing of air. Only experience can fill in the details of just how to do it.

The most important advice is not to get stuck in the first place. Most times, the collar results from improper drilling. The most common error is not supplying enough blow air in proportion to rate of penetration, so that an overburdened airstream allows accumulation just above the bit.

Another cause of sticking, most common in big holes, is pieces of rock falling out of the wall onto the top of the bit, jamming it. The operator should not blame himself or herself for this, but must struggle to overcome the resistance in much the same way as described above.

The stream of air and cuttings out of the hole should be checked frequently, as changes in its character usually call for changes in technique. Finer grain and/or reduced output means harder rock or a dull or broken bit, calling (perhaps) for another bit, or greater pressure. Coarser grain and/or increased penetration rate means softer rock, with possible need for faster rotation and/or decreased pressure to reduce the danger of a mud collar.

Chips of carbide detached from a bit will grind up the rest of the bit, or any other bit. Repeated blowing might get them out, or the hole may have to be abandoned.

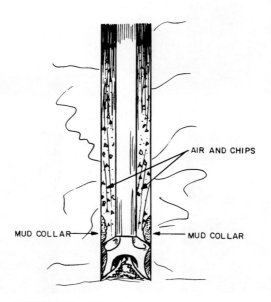

FIGURE 20.51 Mud collar.

Angle Drilling.   Setup procedures for angle drilling are the same in principle as for vertical holes, but are harder to visualize. Arranging for the first hole may take extra time and figuring. See Fig. 20.52.

If the direction of the hole is above horizontal, the rods will have a tendency to slide out of the hole. The coupling is therefore kept on the opposite side of the centralizer from the drill, while adding or withdrawing steels.

If the angle is nearly horizontal, the blow air may travel along the top, leaving excessive drilling debris on the bottom. It is a good plan to pull the steel back toward the mouth of the hole occasionally, with full blow air, to drag this material toward the opening.

Alignment.   It is necessary that the hole and the line of motion of the drill be in close alignment. Careful attention to instructions on starting a hole should ensure a correct beginning. But if tracks should move, the boom sink, or the mast tilt during drilling, alignment will be lost and problems will arise.

Drilling out of alignment gets worse the more deeply the hole is drilled. It puts severe strain on the drill string, and on the mast and its positioning mechanism. Drilling force will be lost in flexing of rods, and in friction against the sides of the hole. The life of both rods and couplings will be shortened by bending stresses. Failure may occur immediately, or at some later time. It is generally the operator's fault.

When misalignment is detected promptly, the operator may have a choice of methods to correct it. The operator may decide to reposition the mast by using the boom and cone controls, or may move the whole machine.

FIGURE 20.52   Drilling positions. (*Courtesy of Gardner-Denver Company.*)

However, if the hole already is crooked, it will probably be necessary to abandon it, and drill another hole close alongside, with greater care.

Lubrication.   Chassis lubrication is conventional. There are a number of high-pressure grease fittings and a few gear case reservoirs that need daily or weekly attention, according to type of use and a schedule supplied with the machine.

The drifter and other air-operated units are lubricated by oil in the airstream. A special rock drill oil is kept in a 5-gallon tank, from where it feeds through a regulating needle valve into the manifold, where it is picked up by the airstream.

One recommendation for lubricator adjustment is to turn the needle valve handle to closed, then open it two turns. Then open the main air hose at the drifter, let air blow until it shows an oil mist, then reconnect.

Visible mist is the standard and dependable indicator of the presence of oil in the air. But it does not tell how much, except to the expert. Correct proportion is indicated in many of these machines by a slow seepage of oil out of the bottom of the chuck housing and down the striker bar while the drill is operating.

If this area is dry, there is not enough oil and the lubricator valve should be opened further. Oil running down calls for cutting down the feed.

The oil tank must be full at the start of the shift, and it should be checked at least once during the shift. Many expensive components can be ruined by working briefly without oil.

MOBILE TOWER DRILLS

Larger drill rigs usually have a higher mast or carriage, often called a tower, hinged directly to the travel base, with hydraulic cylinders to raise it to working position or to lower it for travel. See Fig. 20.53.

Most units drill only with the tower vertically fixed to the carrier, but means are sometimes provided to compensate for uneven ground, or to drill at angles.

Tower height may be 30 to 60 feet, or more, to provide feed length or stroke of 20 to 55 feet.

The tower is usually at the rear of the carrier, which then becomes the "front" of the whole unit. This is the simplest construction, permits accurate spotting, and enables the whole machine to keep farther away from dangerous edges. Central mounting permits using a greater proportion of machine weight for down-pressure, and is used only in very heavy units.

The mounting may be on a special carrier with individually controlled shovel-type tracks, on a big crawler tractor, or on a heavy-duty truck similar to a crane carrier. When crawler tracks are driven by separately reversible air or hydraulic motors, it can maneuver accurately in limited space. However, it is not suited to rough or steep ground. Truck mounting is preferred where job locations are changed frequently, as in water well drilling.

The machine is leveled up before drilling by means of three hydraulic jacks separately controlled from the operator's station. A bubble or pendulum level on the tower shows when it is vertical. Check valves hold pressure in the jacks if the hydraulic system fails. The hydraulic pump may be driven mechanically or by an air motor.

Drill rod is ordinarily rotated from the top by an air, hydraulic, or electric motor, geared down. There may be two or three speeds, or infinitely variable rates from zero to 80, 100, or higher rpm.

The rotary head must include a swivel head through which compressed air enters the rod. Oil field units may use the mechanically driven fixed-height rotary table described later under deep rotary drilling.

The feed mechanism may be an air motor and chain, a two-way hydraulic ram connected directly or operating a chain through a pulley, or a cable and drum crowd-retract with electric drive for hoisting and lowering and hydraulic motor drive for feed control.

The feed keeps the bit in contact with the rock and provides proper drilling pressure (pulldown). This may be as much as 25 tons in these mobile machines, and much more with stationary deep-well

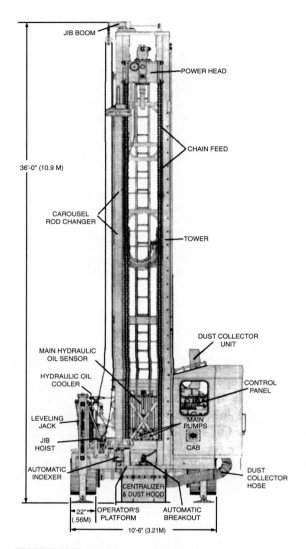

FIGURE 20.53 Mobile tower drill. (*Courtesy of Ingersoll-Rand Company.*)

rigs. The feed is reversed to pull rods and bit out of the hole when it is finished or when trouble develops.

The tower has one side open for handling drill stems. It has stem storage and hydraulically controlled devices for adding them to the drill string or for pulling it out.

Air requirement is 250 to 1300 cfm, mostly for blow air. The compressor is usually mounted on the tractor or truck that carries the drill, making a completely self-contained unit. Small or special models may tow a compressor, or use piped air from a stationary or distant source.

Power may be supplied by the carrier engine through a power takeoff, by a separate diesel or gasoline engine, or by electric motors.

The compressor or compressors may be the only power source, with all drill functions operated by air motors. Some machines are powered by the carrier engine through a power takeoff, or by a separate diesel engine, with various combinations of air, hydraulic, and mechanical drive.

One large unit with electric drive uses 50 horsepower for travel, 125 for air compression, 15 for a hydraulic pump, and 10 for air compression. Another uses two 125-horsepower motors for air, and two 25-horsepower motors for hoist and rotary drive.

These rigs may be equipped with down-hole drills for holes up to 9-inch diameter, and with rotary bits for diameters up to 15 inches. Augers are sometimes used. Models designed primarily for down-hole equipment may have slower rotation and less down pressure than those meant for rotary work. The down-hole drill may require more air capacity and higher pressure than the rotary. However, most machines are convertible to some extent between types.

Conventional drifter drills of $4\frac{1}{2}$ or $5\frac{1}{4}$ inches may be used for drilling holes up to 4 inches in the conventional manner. Rotation may be provided by either the head or the drill. Depth is limited by the absorption of hammer impact by long drill steel. Efficiency starts to decline at 30 feet in hard rock, and at somewhat greater depths in soft material. They are sometimes used down to 100 feet or more.

Down-hole and rotary drills are limited in depth by the ability of the rotary table bearings to carry the weight of long rods, and by the ability of the airstream to keep the cuttings cleaned out of the hole. Down holes are intended to work down to about 200 feet, but have been drilled to 800 feet or more. An extra compressor may be hooked in to help clean the hole at great depths.

If a rotary drill is used, and an optional mud-circulating system substituted for air blast cleaning, there is no exact limit to depth.

AUGER DRILLS

Augers are rotary drills that have a helix or screw thread on drill rods that are called augers or flights. Drilling resistance and/or control of the rate of feed prevents the thread from penetrating in proportion to its turning speed, so that material cut by the bit is gripped by the threads and forced out of the hole by a screw conveyor action.

The flights are made in sections proportional to the feed length of the drill unit. They are connected to each other by bolting or pinning.

Boring Head.   The boring heads are bits of the drag or fishtail type that cut by rotary scraping. There are a great variety in design and material to meet different conditions. See Fig. 20.54. There is usually an advanced center or pilot cutter. The teeth, called fingers, are generally detachable. They may be set in a separate head or in the leading edge of the auger flight. Cutting edges are of steel hardened by various processes, or of tungsten carbide. Worn steel teeth may be built up with borium or carbide hardfacing.

The head should be slightly larger than the auger flights, so that they will not bind in the hole.

Capacity.   Augers are made in standard diameters from 2 inches to 6 feet. Some of them are easily operated by one person, others require rigs weighing 35 tons or more. Vertical holes are usually limited to a depth of 100 feet or less, while horizontal holes may extend over 250 feet.

Most auger drills will handle a variety of sizes. With any one machine, increasing the size of the hole decreases the speed of penetration and maximum depth obtainable, and limits the hardness of material that can be drilled.

The rule of thumb is that maximum depth is in inverse proportion to auger diameter. That is, if a machine will drill down 50 feet with a 12-inch bore, it should not be expected to go much deeper than 25 feet with a 24-inch auger.

Capacity diminishes in regard to both diameter and depth of hole when the going gets tough.

Drilling rates vary with the power of the unit, the hole size, the material, and the skill of the operator. A big machine may put a small bit down in firm soil so rapidly that most of the working time is spent latching on additional augers.

FIGURE 20.54   Auger boring heads.

Uses.   Augers are primarily earth drills, but are also adapted to penetrating soft to medium rock such as shale, soft limestone, and sandstone.

Augers are widely used for soil testing, prospecting for minerals, blast holes, drainage holes, putting pipes and conduits under fills, pavements or obstructions, placing deep footings, setting fenceposts and utility poles, mining, and many other purposes.

Construction.   Units for vertical auger drilling are somewhat similar to blast hole rotaries, in that they have a rotating head that grips the top of the rod, and a power feed mechanism. However, the feed or carriage stroke is usually quite short, and the rotation mechanical instead of air or hydraulic. Wagon or crawler drills may be used by substituting an air-driven rotary head for the drill.

The truck-mounted drill in Fig. 20.55 is designed to handle augers of 8-inch and larger diameters to depths of 30 to 100 feet. It has a feed stroke of 8 feet and uses 6-foot sections. Maximum carriage speed in feet per minute is 40 downward and 34 up. Down-pressure is about 7 tons, and lift capacity 8 tons.

Rotary drive is mechanical with eight ratios, ranging from 13 to 270 revolutions per minute at full engine speed. Leveling jacks may be mechanical or hydraulic. Maximum torque at the bit is about 3,000 pounds.

In horizontal units the drill frame makes a bed or cradle for supporting the auger as it is fed into the hole.

Vertical Drilling.   The machine is carefully leveled by means of its jacks. If the ground is soft, planks or blocks should be used to spread the weight, as any movement of the carrier during drilling is likely to damage or break the augers. See Fig. 20.56.

The drill chuck is raised to the top of its stroke. A cutting head is fastened to an auger, and the other end of the auger is fastened in the chuck.

The cutting head is lowered to the ground. A guide or centralizer may be latched around the auger just above it; the drill clutch or hydraulic valve is engaged to rotate the auger, which is then pushed into the ground with the feed control.

FIGURE 20.55   Truck-mounted vertical auger drill.

FIGURE 20.56   Vertical auger with tower.

As the bit cuts the ground, the spoil is brought to the surface on the auger thread. Most of it is thrown off by centrifugal force and makes a circular pile around the hole. A shield may be used to prevent it from building so close to the pile that it will tend to fall in when the auger is removed. Cuttings may be removed with a hand shovel from time to time.

When the auger is mostly underground, the rotation is stopped, the guide removed, the auger unlatched, the chuck raised all the way. Another auger is attached to the one in the ground and to the chuck, then drilling is resumed. The guide is used only for the first length.

When full depth is reached, the feed is stopped and the auger revolved until the hole is clean of cuttings. The chuck is raised all the way, bringing the top section clear out of the ground and showing the next one. An auger fork or some other device is caught around the lower auger to prevent it from sliding down.

The upper section is then detached at both ends and set aside. The chuck is lowered, fastened to the top of the next auger, and raised. Sections are removed one by one in this manner.

Some drills have an auxiliary tower with a cable hoist that permits raising several sections at a time.

The head should be inspected for wear or breakage of the bits after every hole. If the bits are not sharp and full size, they will cut an undersize hole, causing excessive wear on the head and auger flights, and causing the augers to bind and possibly break off.

The positions of the auger sections should be changed with each hole. The section immediately behind the head wears more rapidly, and if used in this place repeatedly, it will wear to a smaller gauge than the others and will pack cuttings against the wall instead of bringing them up.

Bits and wearing parts of augers can be built up by hardfacing with borium or carbide rod.

Posthole Digger.   The powered posthole digger, Fig. 20.57, is an attachment that mounts on a tractor three-point hydraulic lift, and is driven by a propeller shaft from the power takeoff.

The auger can pivot on the support arms so that it will drill straight down even if the tractor is headed up- or downhill. It is adjusted for side tilt by the leveling arm in the drawbar.

It is lowered into the ground by the drawbar control. Spoil is raised and spun off by the rotation of the auger. It may be necessary to raise and lower a few times, or to stop lowering and continue rotation to avoid carrying too heavy a load.

Standard auger diameter is 9 inches, but widths of 4 to 24 inches are obtainable. Except for the smallest sizes, they have removable cutting edges. The auger is attached to the drill shaft by a soft bolt that will shear if the auger jams.

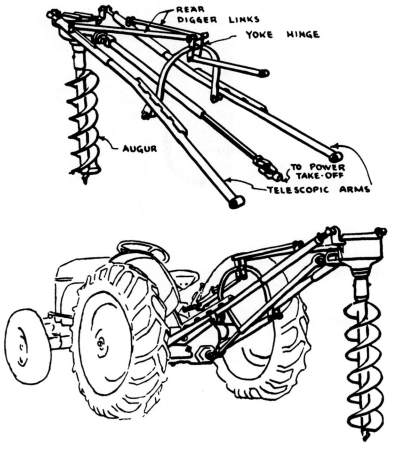

FIGURE 20.57   Posthole digger.

This machine is used chiefly for making holes in which to set posts for fences and for planting seedlings. It can also be used for test holes, to find depth and quality of topsoil or other shallow surface layers.

Horizontal Auger Drills.   The horizontal earth-boring drill shown in Fig. 20.58 is used largely in drilling horizontal blast holes in strip mine overburden.

When the machine is in position, jacks are used to set it at the proper height and angle for the hole.

A machine like this might operate as follows. The feed carriage is moved all the way back, and an auger and head are placed so that the secondary cutters in the auger flight are in front of the guide. The feed is then moved forward and the auger attached to the chuck. The feed is moved forward until the drill head makes contact.

The drill transmission is placed in first speed, and its clutch engaged. When the head has penetrated 2 or 3 feet, the guide is removed.

When the drill cradle reaches the forward end of its track, the chuck is disconnected from the auger and the cradle is moved to the rear.

Another auger section is then latched to the rear of the front auger and inserted in the chuck. Except for the first few feet, which are drilled in low gear, any one of four transmission speeds may be used.

When the hole is finished, the drill is put in third or fourth speed and the auger turned for 2 or 3 minutes. It is sometimes kept turning while being pulled out also.

Augers are withdrawn one section at a time, uncoupled, and the cradle moved forward to engage the next one. The positions of the auger sections should be changed with each hole.

Coal Recovery Auger.   The coal recovery drill, Fig. 20.59, is a horizontal auger that uses sections from 16 to 54 inches in diameter, and up to 24 feet long. Penetration may be 60 to 250 feet.

The compound drilling head, with inner and outer cutters, is designed to remove soft coal in lumps, with minimum fine breakage. The cuttings are carried back by the auger flights to a conveyor belt that loads into trucks.

Sections are handled by a crane that may be hydraulic or hand-powered. The machine moves itself hydraulically from hole to hole on mobile skids.

This drill is used chiefly for mining coal from the edges of strip pits. Standard practice is to cut parallel holes, as shown, leaving thin pillars or ribs between them to support the overburden. Sometimes a single hole is cut for a centerline tunnel, and coal from each side is drilled and blasted into it.

Such auger drilling may be a final operation, where the coal seam has become too thin or the overburden too heavy to justify further stripping. On other jobs parallel slots or keyways are cut at intervals, and the coal seam between them is removed by augers.

HORIZONTAL DIRECTIONAL DRILLING

The use of open cuts for installing utilities or removing contaminated soil has been the common practice in the past. This was done clumsily with excavators, backhoes, or trenchers of various forms. Excavation of that sort has always raised complaints from nearby citizens and angered environmentalists. So manufacturers have developed methods to avoid surface disruption and the associated traffic delays. Simply stated, trenchless methods are now available for underground drilling, punching, or auguring more horizontal boreholes into which pipe or cable can be pulled. This use of the methods was discussed briefly in Chaps. 1 and 5.

The first directional drilling machines were introduced in the late 1980s. Their development has made rapid advancements in the ensuing years. Now these machines pack high thrust-pullback and torque in relatively small packages. Many self-contained models have convenience features such as self-leveling, pipe handling, and pipe makeup and breakout systems, all of which make them easier to use and more productive. One such model is Vermeer's Navigator, shown in Fig. 20.60.

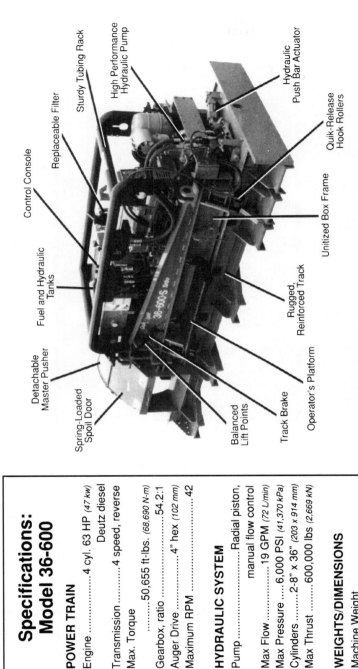

Specifications:
Model 36-600

POWER TRAIN

Engine ....4 cyl. 63 HP *(47 kw)*
Deutz diesel

Transmission.........4 speed, reverse

Max. Torque
.............50,655 ft-lbs. *(68,690 N-m)*

Gearbox, ratio 54.2:1

Auger Drive ....4" hex *(102 mm)*

Maximum RPM 42

HYDRAULIC SYSTEM

Pump...........Radial piston,
manual flow control

Max Flow ....19 GPM *(72 L/min)*

Max Pressure ....6,000 PSI *(41,370 kPa)*

Cylinders ....2-8" x 36" *(203 x 914 mm)*

Max Thrust ....600,000 lbs *(2,669 kN)*

WEIGHTS/DIMENSIONS

Machine Weight

Base Unit ....5,800 lbs *(2,635 kg)*

Power Pack ....2,800 lbs *(1,275 kg)*

Master Pusher ....1,000 lbs *(455 kg)*

Master Track...........2,400 lbs *(1,090 kg)*

Track Section ....2,000 lbs *(910 kg)*

Work Range ....12" - 36" *(305-914 mm)*
casing diameter

Centerline ....22½" *(572 mm)*

Height x Width ..52" x 52" *(1.32 x 1.32 m)*

Track: Width x Length
............60" x 120" *(1.52 x 3.05 m)*

FIGURE 20.58 Horizontal earth-boring machine. *(Courtesy of American Augers Inc.)*

FIGURE 20.59   Coal recovery drills. (*Courtesy of Salem Tool Inc.*)

Drilling Equipment.   There are nearly a dozen manufacturers making horizontal directional drilling machines. They range in power from less than 100 horsepower (hp) to as much as 1500 hp. This much power produces a drilling thrust of 16,500 to more than 1,000,000 pounds (lb) and pullback of 27,000 lb to more than 1,000,000 lb with the largest machines.

These drilling machines are generally mounted on tracks as seen in Fig. 20.60, but a few are mounted on trailers.

The units that are used in residential areas are the smaller ones that have somewhat over 100 hp and deliver a drilling thrust of around 20,000 lb and a pullback force of around 30,000 lb. For drilling they deliver a rotational torque of 3,000 to 5,000 pound-feet and have an operating weight somewhat less than 20,000 lb.

All the directional machines use special drill rods to allow flexibility through the length of the bore so that the operator can steer a path to avoid obstacles. Special equipment, such as pipe pullers, pulling harnesses, and back reamers may be used to assist the pullback operation.

Once the pilot bore is completed, the backreamer is attached to the drill stem string and pulled back—enlarging the bore wall to take the conduit or pipe being installed.

Electronic Guidance.   The directional drilling machine has guidance by having a transmitter in the drill head which sends out a radio signal to a receiver. Originally, the receiver was handled by a worker moving directly over the drill. That worked all right if the drill was under solid ground. But it did not work if the drilling was under a waterway.

FIGURE 20.60   Directional drilling machine. (*Courtesy of Vermeer Manufacturing Company.*)

Now, the drill operator can guide the bore by adjusting the roll angle of the drill head, its pitch, i.e., horizontal inclination, its location, and depth. All of this information is received and adjusted on a laptop computer connected to the machine's guidance system. This sort of guidance was essential for the crossing under the Mississippi River for a more than mile long gas line that had to be more than 200 feet below the top of levees on both sides of the river.

Drilling Mud.   Rotational torque is required to cut the hole opening and to mix the excavated soil with the proper drilling fluids. The mud mixture can range from polymer to bentonite mixtures with viscosities as high as 60 SSU. Polymers, bentonite, and other new combinations have allowed contractors to simplify mixing techniques and generate gel strength for optimum downhole conditions. Self-contained machines have hydrostatic systems. Pressures up to 6,000 psi improve efficiency, allowing the manufacturer to reduce the size of components. Horsepower is no longer lost by pumping oil through hundreds of feet of hose. True self-contained directional boring machines carry the high-pressure fluid pump. The trailer unit is left with fluid storage and a small power source to mix the mud and transfer it to the drill unit.

The necessary development to do directional drilling through solid rock formations is progressing. Borrowing technology from the oil fields, drillers are working on rock using mud motors. A mud motor is a down-hole tool powered by drilling fluid. Because drilling fluid drives the mud motor, its operation requires a higher volume of mud than most small directional drill systems can provide. In addition to the mud motor itself, separate mud pumping and recirculating systems are necessary. The pump must provide the flow necessary to operate the motor. Cuttings are screened out of the fluid during recirculation so the fluid can be reused. However, not all directional drilling machines can support a large enough fluid rate to run the mud motor. The amount of flow determines the size of mud motor that a machine can operate.

PNEUMATIC UNDERGROUND PIERCERS

An air-powered mole, Fig. 20.61, may be used instead of an auger for small underground horizontal bores.

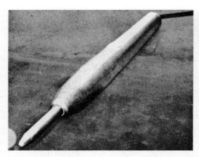

FIGURE 20.61   Hole Hog and launcher. (*Courtesy of Allied Construction Products, Inc.*)

This is usually a cylinder, 4 to 5 feet long and $3\frac{1}{2}$ to 5 inches in diameter, tapered at the front and attached to a 1-inch air hose at the back. Weight may be 60 to 190 pounds.

It contains a valve and a two-way cylinder chamber. Whenever it has air pressure, a piston moves back and forth in the cylinder, striking heavy blows against an anvil that is rigidly fastened into the tapered nose. Striking force may be 150 or more foot-pounds, with a blow frequency of up to 400 a minute.

When the unit is embedded in soil, each blow tends to drive it forward. The tapered nose makes an opening, displacing material to make a hard-packed tube. Slipping back between strokes is prevented by friction, and by roughening the cone with spiral threads.

Working air pressure is about 90 psi. Requirement is 60 to 150 cfm in various models. Lubrication is provided by a line oiler near the compressor, or by occasionally opening the air hose and pouring in a small quantity of a transmission fluid.

It is recommended that the hole be at least 3 feet below the surface in ordinary soils, and 4 feet in sand or gravel. Otherwise, displaced dirt may tend to move chiefly toward the surface, causing the hole to incline upward.

Some units have a reversible valve. It is shifted by turning the hose counterclockwise one-quarter turn. Air hammering will then bring the tool out. The hose must be pulled by hand to avoid binding.

If there is no reverse, there should be a bracket on the unit for attachment of a light retrieval cable, which follows it into the hole and can be used to pull it back if necessary.

An oversize hole may be made by adding a collar. A coupling may be fastened to the front to enable the tool to drive pipe or conduit ahead of it. Such pipe must be fitted with a tapered nose cone.

Operation.   A starting pit is dug, similar to that for an auger, but smaller. It must be slightly deeper than the proposed hole level. If a starter or launching frame is used, the pit must accommodate its extra width also.

An air hose is attached to the back, provided with lubrication, and coupled to a compressor. Air is turned on and off at the compressor. The hose should be marked so that length of penetration (and therefore, the location or at least the distance of the tool) is always known.

A pit is also dug where the hole is to finish. It is usual to dig one or more slots across its path for check purposes, at least until one has become accustomed to the tool.

Place the mole in its pit and against the face to be drilled. See Fig. 20.62. Level and align it carefully, by instrument or otherwise. Open the air valve partway, and push on the cylinder or the hose to help it get in. When about one-third is underground, stop and recheck alignment and grade, and check again before it disappears. When the full length has penetrated, open the air valve fully.

Additional lengths of air hose may be added at the compressor end when necessary. Speed may be 1 to 4 feet per minute, depending on the model and ground conditions. These are mostly used for rather short holes, but might go several hundred feet under favorable circumstances.

When the exit pit is reached, shut off the air, disconnect the hose, lift out the mole, and pull the hose back to the compressor.

If the hole dead-ends because of an obstacle, wandering off course, or because a blind hole is wanted, the tool must be brought back by reversing it, or by pulling a retrieval cable or the hose. If, for any reason, it cannot be moved either backward or forward, it must be dug up.

FIGURE 20.62   Starting and finishing a hole. (*Courtesy of Allied Construction Products, Inc.*)

DEEP ROTARY DRILLS

The rotary drills to be discussed in this section are primarily used in boring deep holes for oil. They employ a toothed bit which is rotated against the bottom of the hole by a drilling string driven by a power plant. Mud or, occasionally, water or air is pumped through the hollow drilling string and bit and rises outside of them.

A diagram of a large stationary unit for deep work is shown in Fig. 20.63.

Derricks and Masts.   The derrick shown is built on the job out of precut and drilled angle iron on poured concrete footings. Other constructions include prefabricated sections of angle or tubing, which are assembled and bolted; complete derricks assembled on the ground and raised by winches; and hydraulic or cable-controlled telescoping or folding masts, transported to the job on large trailers which are used as both support and substructure. The derricks are generally centered on the hole and are self-supporting, while the masts are on one side of the hole, lean over it, and are supported by guys.

The derrick or mast serves as a support for a pair of hoist blocks which handle the drill pipe and casing. Its strength must be sufficient to easily carry the weight of the longest string of pipe or casing that might be used on the job, and it is therefore designed with reference to the estimated drilling depth.

Its height is determined by the length of pipe to be pulled as a unit. Economic drilling of very deep wells requires removing sections 120 feet or more in length, whereas shallow holes may justify handling 30 feet or less.

Derrick structures and many other items of oil drilling equipment should be approved by the A.P.I. (American Petroleum Institute), which has set up standards for promotion of safety and efficiency.

Substructure.   Substructures are massive supports and platforms of steel and wood which carry the power plants, drilling machinery, and pipe racks. They are usually separate from the derrick and can be removed without disturbing it.

The drilling floor is inside the derrick legs and high enough above the ground to provide easy access to the space under it, called the basement.

Shallow holes cased with pipe protruding a few feet above the derrick floor are drilled opposite the draw works. One of these, called the "rat hole," supports the kelly and swivel when not in use. The other, called the "mouse hole," supports a single piece of drill pipe ready to be added to the drilling string.

When the surface soil is full of boulders, when subzero weather might freeze the kelly into the conventional rat hole, and under some other conditions, a mechanical rat hole is used. It is a 40-foot track, pivoted like a seesaw, which bears a dolly that supports and facilitates moving the kelly. It is placed at the corner of the derrick.

Power Plant and Controls.   Power requirements are heavy and may exceed 1,500 horsepower. Multiple engines are commonly used and may be high-pressure steam, electric, or internal combustion with electric, mechanical, or fluid drive.

The type of internal combustion engine is usually decided by the fuel available locally. Standard diesels or, less often, gasoline engines are used, or adaptions of either type to use crude oil, natural gas, or butane.

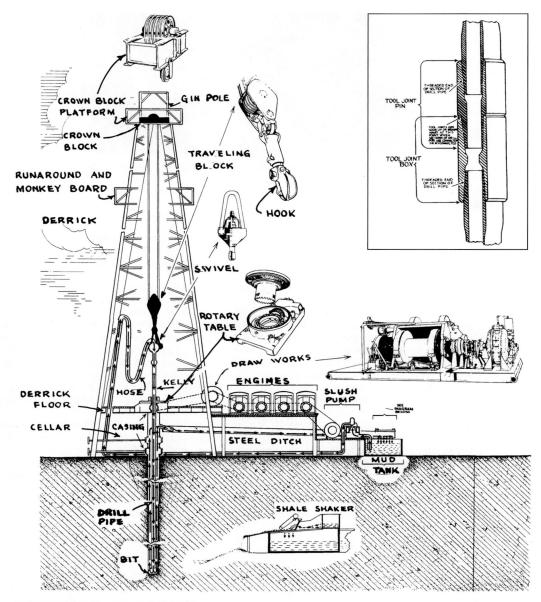

FIGURE 20.63 Structures for rotary drill.

During drilling the rotary machinery absorbs about one-third of the power expended and the pump the balance. The hoisting mechanism requires as much power as the two combined. In small units one engine may drive the pump and a smaller one the rotary table, with provision made for hooking them both to the hoist winch.

Large plants may use as many as six engines, driving into a transmission box from which the power of any one engine, or any group of them, can be applied to any function. Engines can be disconnected and stopped when not needed.

Engines driving the rotary are subject to severe shock, and it is good practice to protect them with torque converters.

Draw Works. The draw works is the power distribution and control mechanism. It includes one or more variable-speed transmissions; clutches and brakes operated mechanically, or by air or fluid; a heavy-duty winch for the main hoist; one or more cat heads or light winches for handling pipe, miscellaneous lifting, and coupling and uncoupling pipe; and chain belt or shaft drives to the rotary table and the pump.

The instrument panel has engine gauges and indicators for the weight of the drilling string, rotary speed, and mud pressure.

Rotary Table. The rotary table consists of a stationary base with a machined cup, and a rotary, also called turntable or table, which rests on a wide ball bearing in the cup, and is turned by bevel gears.

The rotary is pierced by a large, square-shouldered hole in which the two-piece outer or table bushing rests. This has a square hole through it which holds the inner or kelly drive bushing. This is also of split construction, and has a square hole with smoothly machined sides.

This arrangement transmits rotary drive to the kelly, a square shaft which is free to move vertically in the inner bushing. The hole can be enlarged for passage of the drill string by lifting the inner bushing; and further widened to pass the bit by removing the outer one.

A soft rubber wiper collar is fastened to the lower side of the base, to clean mud off the drill string as it is raised.

Kelly. A simple drilling string consists of a kelly, lengths of drill pipe, joints, some heavier drill collars, and a bit. A reamer may be required under special conditions. They are all made of high-grade alloy steels.

The kelly or grief stem is a hollow square shaft (hexagonal or splined cross sections are rare) which slides up and down in the inner table bushing. The ends are round and are upset, or hot-hammered, into a heavier cross section. Tapered A.P.I. threads are used, left-hand on the top, right-hand on the bottom. A recess is machined into the upset upper portion to provide a grip for the hoist swivel.

The lower end is subject to constant connecting and disconnecting of pipe, so that threads wear rapidly. The kelly is therefore protected by a joint protector, called a sub, and this in turn is protected by another, the sub saver. These are more economical to repair or replace than the kelly.

A kelly should have a square section somewhat longer than the drill pipes which are to be used.

Drill Pipe. The drill pipe is made in standard lengths of 27 to 33 feet, known as ranges. Standard outside diameters are 4½, 5¾₆, and 6⅝ inches. In the field, lengths will be variable because of shortening due to repair work. Construction is seamless steel tubing. The ends are upset to give thicker walls. Taper right-hand male threads are used on each end.

Protecting joints, used for attachment to other pipes or collars and for protection of the drill pipe threads, are heated before screwing on so as to form a very tight connection, and are sometimes welded in addition.

Pins and connectors have a larger outside diameter than the pipe.

This brief introduction to deep rotary drilling is not part of moving the earth in the strict sense, but is included because it is similar to deep drilling for blast holes or pilot bore for a shaft.

CAISSON DRILLS

Caisson drills are specialized auger or rotary rigs that drill holes from 12 inches up to 10 feet in diameter, to depths up to 100 or 200 feet, either vertically or at a slant. They may be equipped with devices to bell out the bottoms to 3 to 5 times the hole diameter.

Their principal use is in making holes for poured concrete piers. They have largely replaced the driven pile in large areas where deep clay soils are found, and they are used in many other soil types as well.

They are not particularly suited for drilling in loose soil that tends to cave, or for boulders or solid rock that is too hard to be dug with a pick. However, they can sink holes in loose ground when casing is used, and boulders may be broken by lowering a person into holes 2 or more feet in diameter, to hand-break or blast them.

For continuous hard digging, many of the rigs can use the rotary shaft cutters described in a later section, or rotating "bundles" of down-hole drills.

In addition to pier work, caisson drills may be used for preboring for driven piles, sinking ventilation shafts, soil and mineral sampling, and digging for manholes, cesspools, and shallow wells.

When used for sampling, they can take big chunks of formation for thorough checking, and permit lowering a person to inspect the undisturbed formations in the wall.

Figure 20.64 shows a heavy-duty foundation drilling machine on a special truck carrier, with the mast folded down into transport position. It has a rotary table at the foot of the mast, that is driven by a transmission with four forward and two reverse speeds. This receives engine drive through a pair of input clutches, one low speed for drilling, the other high speed for spinning off cuttings.

The kelly or drive stem telescopes. The inner stem is $4\frac{1}{2}$ inches square, the outer one 7 inches. The two bars are linked by an automatic latch release that permits application of both down-pressure and hoist pull.

Figure 20.65 shows a caisson drill unit mounted on a standard $1\frac{1}{2}$-yard crane. Its frame is supported by the boom foot and by tierods to the roof structure. It carries its own engine, transmission,

FIGURE 20.64   Caisson drill with tower folded down.

FIGURE 20.65   Crane attachment and drilling bucket.

FIGURE 20.66   Caisson drill buckets.

and rotary table. Torque converter drive is used, and manual or power shift transmission is optional. The kelly may be solid, telescoping, or "triplescoping."

Drilling may be done by augers with short flights that are lifted straight to the surface to unload by spinning, or by buckets that are lifted and swung for dumping to the side. See Fig. 20.66.

Buckets may be fabricated out of steel tubing, with convex hinged bottoms. Cutters and slots provide for filling the bucket by rotating it on the bottom.

Hole bottoms may be belled out by side cutters that are kept folded into the bucket walls while drilling hole, or by various types of undercutting reamers.

SHAFT SINKING

Rotary drills are being used to an increasing extent to sink ventilation and escape shafts into underground mines. The diameter of the shaft that can be drilled is now limited more by the sizes needed, than by limitations of the machinery or the technique.

Such holes are drilled by conventional deep rotary drill rigs, with rotary tables altered to accommodate the huge bits and drill collars needed.

Most big-hole work is done in two stages. First a pilot hole is drilled, centered on the final shaft center, several diameters deeper than the intended shaft bottom. The big bit has an extension or stinger that is guided by the pilot hole to keep exact alignment. See Fig. 20.67. Rock cutting is done by rows of rotary shaft cutters bolted to the body of the bit.

Big bits require very heavy weights to keep them cutting. A 90-inch ventilation shaft was drilled in New Mexico with three lead-filled collars 40 inches in diameter weighing 21 tons each above the bit. The bit carried a total weight of about 50 tons, after deducting buoyancy of the mud and tension support from above.

Rotation was 12 revolutions per minute, average penetration rate 2.15 feet per hour, with a maximum of 5 feet. Maximum mud-circulation rate was 8,000 gallons per minute.

A drilled hole may be lined by lowering reinforced-steel casing, welding each joint before it goes in, or by pouring concrete from the top into forms that are lowered in stages to follow the bit.

Costs of drilling and digging shafts are somewhat similar. The drilled shaft has the advantages of speed, safety because no personnel need to work in it until it is finished and lined, minimum difficulty with water flow or caving because of mud pressure, simplicity of setting lining, and better chance of using the shaft without lining.

In unlined ventilating shafts, the smooth-walled circular hole cut by machine may have the same air capacity as a blasted and dug rectangular hole with twice its cross section.

FIGURE 20.67   Shaft drill.

Shaft Sinking Machines.   There are also drilling units that operate entirely inside the shaft. The design shown in Fig. 20.68 cuts a 76-inch-diameter hole by enlarging a 12¼-inch pilot hole that has been drilled into an opening of a mine beneath.

The machine is suspended by the rotary drilling rig used to put down the pilot hole. It is in three sections, with a rotary drive motor case at the top, a rotating cutter head at the bottom, and a works cylinder in the middle.

The machine is lifted out of the hole by the surface rig when inspection and servicing are needed.

Three 60-horsepower electric motors turn the cutter head at 6 revolutions per minute, through a central shaft and reducing gears. The head is set with rolling cutters of design suitable for the formation being drilled, and has a stinger projecting down into the pilot hole.

The cylinder is held in position against drilling thrust by hydraulic jacks powered by an electric pump. They push shoes outward against the shaft walls. The upper directional jacks keep the cylinder lined up with the pilot hole.

The drilling unit is lowered by the surface rig until the cutters rest on the hole bottom. The holding jacks are set and the directional jacks are adjusted. The cutter head is then rotated, and pushed down by sets of hydraulic jacks that can exert thrust of up to 450,000 pounds.

A flush system blows cuttings down the pilot hole into the mine, where they are removed by mine haulage equipment, or blown up a nearby ventilating hole.

When the thrust jacks have reached the end of their stroke, rotation of the head is stopped, holding and directional jacks are released, the hoist line to the surface is slackened, and the cylinder is lowered by releasing the thrust jacks. The cylinder is then locked in position, and drilling is resumed.

Cutters are available to permit drilling any formation from soft shale to granite and quartzite.

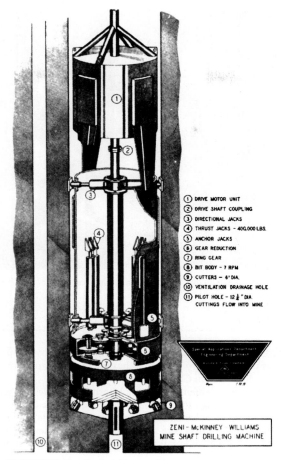

① DRIVE MOTOR UNIT
② DRIVE SHAFT COUPLING
③ DIRECTIONAL JACKS
④ THRUST JACKS - 400,000 LBS.
⑤ ANCHOR JACKS
⑥ GEAR REDUCTION
⑦ RING GEAR
⑧ BIT BODY - 7 RPM
⑨ CUTTERS - 6' DIA.
⑩ VENTILATION DRAINAGE HOLE
⑪ PILOT HOLE - 12½" DIA.
    CUTTINGS FLOW INTO MINE

ZENI - McKINNEY WILLIAMS
MINE SHAFT DRILLING MACHINE

FIGURE 20.68   Shaft sinking machine.

A limiting factor on use of this machine, and the raise borers described below, is that there must be access to the bottom of the proposed shaft, to provide for removal of cuttings. Raising them to the surface with air or mud is more difficult and costly when hole diameter is large.

Raise Borers.   A machine working at the top can enlarge a pilot hole from the bottom. The unit first cuts an 11-inch hole down to a lower-level passage, using a 10-inch drill pipe in 5-foot sections that weigh 710 pounds each. Cuttings are washed up out of the hole with air, water, or drilling mud. Bits may be steel tooth or carbide insert, depending on the formation.

When the drill breaks through into the passage, the bit is removed and replaced by a much larger back reamer, 4 to 12 feet in diameter. Steel or carbide insert teeth, or steel disc cutters, are mounted upon its upper surface. See Fig. 20.69

The reamer is rotated and pulled, with a force that may be as high as 200 tons, and penetrates at rates from 2 to 20 feet per hour. Cuttings fall down the enlarged opening, and are removed by other equipment at the bottom.

The bored raise may be vertical, or pitched at any slope steep enough to drop the cuttings down away from the bit. Maximum depth may be 1,000 to 2,000 feet, depending on the machine and the job conditions.

FIGURE 20.69   Drilling head of raise borer.

TUNNELING MACHINES

An increasing share of tunneling work is being done by boring or tunneling machines, often called borers or moles. The problems that they encounter, or perhaps the circumstances under which they work, are discussed in Chap. 9.

The principal elements in a tunnel-boring machine (TBM) are usually a full-size cutting head that rotates or oscillates to cut the ground, a complete or partial tubular casing or shield, a crowding or propelling mechanism with provision for steering, a muck disposal system, and electric and hydraulic power units. There are usually means to place tunnel lining and/or rock bolts.

Diameter may be as small as 6 feet, or over 36 feet, but both extremes are rare. Sizes from 10 to 25 feet are common.

Machines are preferably assembled on the outside, at portals, to dig their own way underground. However, it may be possible to lower one in sections down a large shaft, and assemble it in a drill-and-blast stub tunnel.

Cutterhead.   Cutterheads are made in a variety of designs, to suit different conditions, and to work out the ideas of their engineers, and of the contractors who use them.

TBMs, like that pictured in Fig. 20.70, rotate on a central structural member. They are fabricated to hold a number of cutters. There are slots or windows to take cuttings through the head, away from the face, and buckets or scoops to gather and load them into a conveyor system.

The TBM head revolves as a single unit. This results in difference in speed between cutters near the center and those at the rim. This may be compensated by variations in spacing, and in type of cutter. In some machines, the center is driven separately at higher rpm. It often has a forward extension or stinger that helps to prevent wandering off course.

Oscillating heads of other tunneling machines have a motion like a windshield wiper for each of a number of radial cutter arms. Their arcs overlap, and distance of motion can be adjusted. They are used chiefly in soft or variable ground.

Cutters.   Cutters may be simple steel teeth set in rigid sockets, if the ground is soft. More often, they are cones, spools, or discs, as in Fig. 20.71, whose contact with the face causes them to rotate with a dragging, crushing, or fracturing effect on the rock formation.

Their principle of operation is generally similar to the rotary drill bits described earlier, but individual cones may be widely separated. Teeth may be smaller for harder formations. Carbide inserts or disc-type cutters are used for both medium and hard rock.

Steel cutters primarily cut and chip, carbide inserts grind. Each cutter should make a set of grooves and ridges. The following cutter will break out or slice the ridges in making its own grooves. As the rock becomes harder, the cuts get shallower, and the efficiency of the operation declines.

Shield. Except when rock is unusually stable, a borer is likely to be protected by a tubular shield or casing, or at least by a curved roof section. Shield length and characteristics are determined largely by the stability of the ground.

The shield may be in two longitudinal sections. The front one moves with the cutterhead. The rear one is anchored to the tunnel or its lining, and serves as a thrust block for hydraulic cylinders that move the front section and cutterhead slowly forward into the digging.

In soft, caving, or flowing ground the cutterhead may operate inside the shield, whose top, and sometimes sides and bottom, penetrate the soil before the cutters reach it. In extreme conditions, the head may function more as a barrier against admitting too much material, than as a cutting tool. Flow of material into the machine may then be regulated by adjustable openings, called windows. Shields may also be operated without cutterheads.

In hard ground, the cutterhead is out in front of the shield, and cuts a circle large enough for it to follow without binding.

A rearward extension of the shield may be used for setting lining for the tunnel. A system suitable for very unstable ground is to fabricate the lining inside the shield from curved sections, building each ring onto a previously completed one. Such a lining is necessarily smaller than the tunnel bore. The annular space is filled by immediately pumping in fillers such as pea gravel and/or cement grout.

If the ground is judged to be stable enough to maintain itself without support for a few hours, the lining may be fabricated or expanded behind the shield, so that it may be full tunnel size and require less filler behind it.

On other jobs, roof bolts and wire mesh may be placed instead of lining. The rear section of the shield can have ports or slots through which holes can be drilled and bolts inserted.

Tunnels that are left bolted or unsupported during excavation are often lined later as a separate project.

Propel (Crowd). The tunneling machine is usually moved forward into the digging by bracing or locking the rear portion to the tunnel walls or lining, as shown in Fig. 20.70, then moving the front portion forward by means of hydraulic cylinders. When these reach the end of their stroke, the front portion may be locked to the walls, and the rear drawn up to it with the same cylinders. The cycle is then repeated. The resetting part takes from ½ up to 3 minutes.

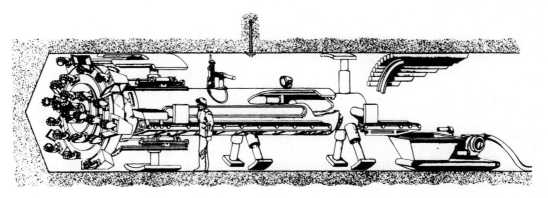

FIGURE 20.70 Tunneling machine.

FIGURE 20.71   Revolving head of a TBM. (*Courtesy of Atlas Copco Robbins, Inc.*)

The stroke of the pistons may be anywhere from 2 to 6 feet in various models. Crowding force varies up to 8,000 tons.

Power.   Power is usually electric and hydraulic. The cutterhead is driven directly by one or more electric motors, with a total of 100 to over 1,000 horsepower, depending on machine size and expected digging resistance.

Other functions are powered by hydraulics, with electric-driven pump or pumps. Big machines carry these pumps with them, small ones usually tow them behind.

The electricity is preferably supplied by power lines from outside. If this is not practical, it is generated by a diesel-electric set towed or self-propelled behind the borer.

Muck Disposal.   Cuttings are scooped up by buckets in the rim of the cutterhead, and carried up to dump on a conveyor belt. See Fig. 20.72. This carries them back through the machine, and a substantial distance to the rear. It usually dumps into railcars, a string of which is pushed into the tunnel by a locomotive, as seen in Fig. 9.52. As each car is loaded, the train of empty cars moves the next one into position.

FIGURE 20.72   Conveyor system.

FIGURE 20.73   Inside a drilled tunnel.

In dry ground or under any dusty conditions, there are water sprays at the cutterhead, and at each transfer point on or off the belt, to keep down dust. In addition, the ventilation system is usually of the suction type, so that any escaped dust, fumes, or rock gas is continuously drawn out of the work area. See Fig. 20.73.

Muck cars may be hauled out of a portal, hoisted up a shaft by a crane, or dumped within reach of hoisting machinery. The spoil is usually then reloaded into surface-hauling units for removal from the area.

CHAPTER 21
AUXILIARY EQUIPMENT

TRACTOR-MOUNTED RIPPER

A ripper is a tooth, or set of teeth, which is usually mounted in a frame hinged to the back of a crawler tractor, with hydraulic hoist. It is pulled through rock or hard soil to break and loosen it.

The mounting may be a radial type, in which the teeth move in an arc, so that they tend to point downward in carrying position, and swing forward as they are lowered. A parallelogram or double mounting has two pairs of hinge arms, upper and lower, that keep the shank(s) at a constant angle to the ground as it is raised or lowered. See Figs. 21.1 and 21.2.

In an adjustable parallelogram, the upper arms are replaced by a pair of hydraulic cylinders which can be used to regulate shank angle, either before or during ripping.

Adjustment is useful in selecting an efficient point angle for a particular formation. It allows using a steep angle for entry, and a flatter, point-saving one for ripping. Dull points may sometimes be sharpened during ripping by changing from a steeper to a flatter angle.

The hydraulic pump and controls are built into the tractor, which should carry a bulldozer. This should have hydraulic tilt, for efficiency in handling oversize pieces of rock.

Shanks and Teeth. A tooth is made up of a long shank and the tooth proper, which may be called a point, a tip, or a cap.

A shank is usually fastened to a ripper bracket by two removable pins. If the bracket is open at the rear, teeth not being used may be stored as shown in Fig. 19.15.

Shanks may be straight or curved. Straight ones are generally used for massive or blocky formations; curved ones for bedded or laminated rock or road pavement that is further shattered by the lifting action of the bottom of the curve. Brackets may be loosely pinned to allow the tooth to swivel slightly to "hunt" weaknesses in rock, and to reduce side strain when the tractor is turned. They may allow for adjustment of depth of cut and/or angle of penetration.

Teeth are detachable, and may be reversible. See Fig. 21.3. They may be built up with hard-face rod when worn, but this may spoil their shape and efficiency. It is better practice to weld on a forged cap.

Tooth points may have a service life of 30 minutes to 1,000 hours, most of the difference depending on the abrasive qualities of the rock. The operator should glance at them every time they come out of the ground. Dull teeth are inefficient, and worn-out or broken ones allow destruction of the shank.

Most heavy-duty rippers may be used with three teeth, with the center tooth only, or with the side teeth only. Use of all the teeth requires maximum power, works a strip somewhat wider than the ripper, and produces most thorough breakage where full penetration can be obtained. But it may limit depth and thoroughness, and even make any penetration impossible. Trouble may occur with slabs and boulders pushing ahead of the shanks.

Ripping with only the outer teeth increases efficiency of penetration, reduces power required, and usually produces coarser pieces than with three teeth. Breakage may be poor or lacking in the center, so that overlapping passes may be required.

FIGURE 21.1   Ripper with parallelogram mounting. (*Reprinted courtesy of Caterpillar Inc.*)

Most very heavy ripping work is done with a single center tooth. All the weight and pull available can be applied to the single tooth, and it can hunt and follow weaknesses in the rock more readily.

One mistake to be avoided is running the tractor for extended periods "on tiptoe." Down-pressure on the ripper may raise the rear of the tractor, so that only the front of the tracks provides traction. See. Fig. 21.4. This reduces production, wears the undercarriage, and provides the operator with the roughest kind of a ride.

Rippability.   The factors involved in the resistance of rock to ripping are discussed on page 3.5.

Penetration.   Maximum penetration of general-purpose heavy rippers varies from about 10 to 60 inches. Overall width between outer teeth may be 5 to 11 feet, to correspond with tractor width.

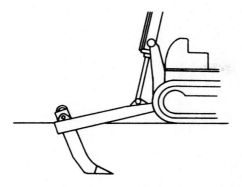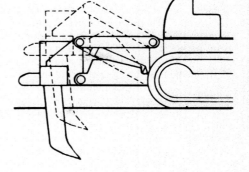

FIGURE 21.2   Radial and parallel mountings.

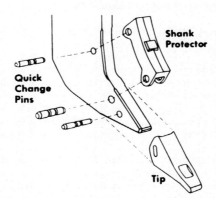

FIGURE 21.3   Ripper tooth details.

FIGURE 21.4   Tiptoe operation is poor operation.

A pair of D9 tractors can put as much as 60,000 pounds of down-pressure on a ripper tooth, making it possible to fracture very resistant rock.

There are also special-purpose rippers used in pipeline work, loosening coal, and subdrainage that may have single shanks permitting penetration as deep as 8 feet.

Heavy-duty rippers may be designed for operation by tandem tractors. The front machine carries and pulls the ripper; the second pushes it with a dozer blade held by a shelf or socket at the rear of the bracket or the shank.

The front tractor may do the whole job wherever it can produce good results alone, and be helped by the pusher only when necessary. The pusher blade may be allowed to float most of the time, with down-pressure only if the ripper tries to ride up.

Depth. Except when breaking hard or tough ground (pavement or frost) over soft material, teeth usually should be near full depth, with the pull beam approximately horizontal.

Shallow ripping tends to be irregular in depth, and causes tooth wear out of proportion to production.

However, there are a number of conditions that may call for part-depth ripping. Full depth may create too heavy a load, or may require so much down-pressure that the tractor is on the front of the tracks only (on tiptoe), with loss of traction and production, with increase in wear. Or the rock may have a natural horizontal weakness partway down, that should be utilized.

In any of these cases, it may pay to make a shallow cut. But if better depth can be reached easily by removing a tooth, or teeth, they should probably come off.

Direction. Any direction is suitable in soil, frost, rock with bedding planes parallel to the surface, and any material without a definite structure.

On slopes it is safest to rip up- or downhill, as large chunks might tend to lift and overturn the tractor. Difficult ground might be ripped downhill only.

When bedding planes or joint structure is at an angle with the surface, primary ripping is usually done against the grain, so that the slope of the beds tends to pull the tooth down. However, if this results in excessive pulling up of big slabs, another direction may be used.

If bedding planes are perpendicular to the surface, ripping should be across them. If parallel or nearly parallel, the tooth or teeth may tend to cut steep-sided grooves with unbroken ribs between them.

Whenever it is practical, tractors doing double duty as scraper pushers and as rippers, rip in the direction of scraper travel, to avoid turns.

Irregular depth and poor breakage can often be corrected by cross-ripping; that is, ripping the same area again at right angles to the original direction. The tractor must walk on rocks turned up the first time, and may find the going very rough. It may be necessary to push some of the pieces off to the side or to smash them before cross-ripping.

Spacing. Wide spacing means fewer passes, and therefore increased production on an area basis. However, close spacing may be needed, either to obtain reasonably even bottom breakage, or to produce pieces fine enough for removal by the equipment used.

With full penetration, 3 feet between one-tooth passes is often satisfactory. Slab material may be loosened on spacings of 8 or even 10 feet, if the pieces can be handled. Crumbly material may call for wide spacing to reduce fines.

Slabs. Many formations, frozen ground, strong or rubbery rock with weak bedding planes, and concrete pavements tend to break out in big slabs that are pushed ahead by the ripper shanks, so that they lift the ripper out of the ground and/or jam against the tractor. A number of small slabs may combine to block multiple teeth.

A big slab may make it necessary to lift the teeth high, drive forward to clear it, lower the teeth to the ground surface, push the slab back until the teeth reach broken ground, then force them down to resume ripping. If the teeth are simply raised to clear the slab and forced in again at the other side, an unbroken pinnacle of rock will be left.

Slabs may be broken by forcing the tooth or teeth down on them, by climbing them with the tractor and turning, or by a separate machine with a drop ball or pile driver.

Operation.   Most ripping is done in low gear, at 1 to 1½ miles per hour. Higher speed tends to increase wear out of proportion to added production.

A torque converter reduces repair cost, as it cuts down both track spin and shock to the gear train. However, it may also reduce production substantially, particularly in rough work, as shock breaks even more rock than it does tractor parts.

Ripper operation in hard rough rock is the toughest service ever required of a crawler tractor. Hoist pull, down-pressure, and a portion of the drawbar pull are taken by the transmission case. An exceptionally solid rock mass may bring the tractor from walking speed to a dead stop in a fraction of a second. Tracks spin on smooth slabs, then are stopped abruptly by a grouser catching an edge. The underframe and power train cases may be hammered and jammed by pieces of rock too high for clearance, turned up by the tracks, or forced forward by the ripper. Reripping an area may involve constant pitching over loose rock, even after bulldozing the biggest pieces to the side.

Damage to the tractor and ripper is greatly affected by operator skill and attitude. A rough operator on such a rough job can wreck a new crawler in a season or less. A very careful person might be able to obtain almost as much production, and keep the machine running for its normal span of years, although repair costs will be high in this work. The careful operator will last longer, as bouncing a big tractor around on rock is rough on an operator's insides.

Ripper Uses.   The ripper has two principal uses in earthmoving. One is to make a diggable soil easier to dig, the other is to compete with explosives in loosening otherwise undigable material. Secondary uses include laying underground cable, cutting tree roots, and rolling out boulders.

The first use is a very old one, being chiefly employed in aiding scrapers and dozers to dig hard soil. The machine often has the double job of ripping, and pushing scrapers or bulldozing.

The use of rippers in soil being dug is discussed in Chaps. 8 and 19. The particular problem is that a very fine or loose soil is harder to load than one that is a little too hard for easy cutting. Ripping of a borderline soil may have the effect of decreasing production, or of not increasing it enough to justify the expense of ripping. Fine breakage with three teeth may cause loads to be smaller than if one or two teeth are used. Many contractors consider that if three teeth can be pulled through a soil at full depth, it does not need ripping.

A bulldozer does not have this problem, as it will almost always do better in loose soil than in formations that are even moderately hard. The worst that ripping can do here is to be a waste of effort. It cannot harm the digging and it usually helps it.

Production.   A ripper designed for mounting on a 20-ton crawler tractor usually has an overall width of 9 feet when carrying three teeth, and should loosen a 12-foot strip. Maximum penetration is 2 feet or more. In loosening soil or decomposed rock that is free from hard ledge or boulders, the ripper might be worked at full depth in low gear at a bit better than 1 mile per hour, say 30 linear yards per minute, less 5 for turn time, leaving a net of 25.

Multiplying this by the 4-yard strip width, it covers 100 square yards per minute. At 2-foot depth, this would yield 66 bank yards per minute, or 3,000 per hour, enough to keep a whole spread of scrapers in full-time operation.

Even after deducting one-third for contingencies, there is a very respectable 2,000 yards per hour. Potential production such as this is the reason that contractors try to assign ripping as just a part-time job for pushers.

Difficult soil containing boulders or frost or interrupted by ledges is loosened much more slowly. And all-rock ripping may go at a rate as low as 100 yards per hour before it is given up. In very hard work, the work hour may be figured at 20 minutes.

In ripper work, most decreases in production caused by difficult ground conditions also increase machinery costs. There is probably no other earthmoving tool that shows such a wide range of output and unit cost.

Ripping is most apt to be profitable where the power is ample or even excessive for the job. A 10-ton tractor with a ripper might tear itself to pieces with little production in a formation that a 20-ton machine would walk through, loosening substantial yardage with only moderate strain on itself. This larger machine might in its turn be nearly helpless in a harder formation that would yield easily to tandem tractors or a single 30-ton unit. The heavier and higher-powered units usually provide better and more uniform breakage than the overworked smaller ones. (See Fig. 21.5.)

Ripping Cost—Rock.   A thin, weak-bedded shale might be ripped at little more cost than hard soil. However, a really tough rock might prove so resistant that only one tooth could be used, penetration would still be poor, and breakage so difficult and irregular that production would be cut down to 100 yards per hour.

A tractor ripping difficult rock is likely to be worn out within 3,000 hours if it keeps at it. Repair costs can easily be double those in ordinary heavy-duty service. If the right ripper is

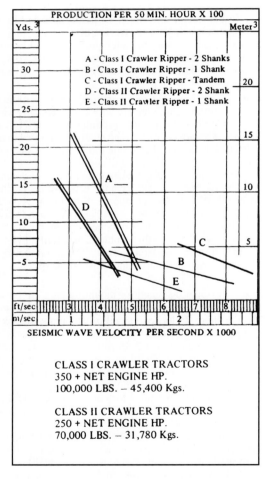

FIGURE 21.5 Ripper production graph. (*Courtesy of ATECO.*)

bought for the job, it should last at least as long as the tractor. Some heavy-duty rippers outlast two or three of the tractors that carry them.

Production may be substantially increased in many formations, and machine life lengthened, by light shake-up blasting before ripping.

If the ripper were strong enough, a pusher tractor with a down-pressure blade could be used, doubling both pull power and penetration rate. Assuming that the rock had good breaking qualities and just needed more force, output might be increased to 400 or even 600 yards per hour.

The tractor carrying the ripper would be under less strain, but the ripper would be under much more. The net effect would usually be a substantial reduction in repairs, and increase in tractor life. But even if no savings were obtained in this way, the pusher would add only 50 percent to costs, and would be paid for over and over by increased production.

Other Costs.   Any comparison of ripping and blasting costs must take into account the method by which the broken rock is to be moved.

In a sidehill cut, it may be possible to sidecast the ripped rock with a dozer at a fraction of the cost of using a shovel in blasted rock. This would be true even if ripper breakage were coarse and irregular.

A dozer, often the same one that carries the ripper, may be used economically for pushing ripped rock up to 200 or 300 feet along the road route to a fill. Production would be much better with fine breakage than with coarse.

For longer hauls the scraper is the preferred machine. If the material is suitable for scraper loading, it can probably move it for less than a shovel and trucks could. In the numerous formations that can be scraper-loaded after ripping, but not after blasting, the combined saving of ripping instead of blasting, and scraper instead of shovel, may be very substantial.

Where the rock cut is shallow, ripper-scraper combinations are even more desirable, because of the relative inefficiency of both blasting work and shovels or loaders in low banks.

But a scraper is not designed to handle coarse rock. Where ripper breakage is poor, scraper loading is likely to be slow, repair costs high, and useful life short. Scrapers may also be damaged at the dump by riding over rocks, and in struggles to eject oversize pieces.

On slopes, both fine and coarse ripped rock may be bulldozed downhill to belt loaders that put it into trucks, as shown in Fig. 14.20.

If the rock cannot be ripped into sizes for scrapers or belts, and there are too many big pieces to make it economical to break them with a drop ball, pile driver, or explosives, it will probably be necessary to bulldoze the rippings into piles for a front loader or power shovel.

The loader can do its own dozing. Work conditions are apt to be very rough, production low, and costs high.

If ripped rock can be pushed off a nearby face, the yardage from several layers can be heaped for efficient shovel loading. The cost of ripping can then be added to that for dozing, to get a figure to compare to that for drilling and blasting.

If there is no face to push to, or when repeated cutting of the top destroys it, the loose rock may be pushed up into windrows for the shovel. The shovel will not work efficiently, as side material will tend to slide away from it, and dozer help will be needed to keep the pile trimmed. For this condition the ripping method should be charged both with the piling expense and the lowered shovel efficiency.

Light Mounted Rippers.   Figure 21.6 shows a ripper of lighter construction mounted on a heavy two-wheel-drive tractor equipped with a front loader. It is similar in design to the heavy units, except that it has four teeth and one ram.

This unit is primarily designed to loosen soils to be loaded by the same machine, but can do medium-duty rooting on many projects. It also serves as counterweight.

The John Deere subsoil plow may be carried on a hydraulic lift drawbar. It is essentially an agricultural tool, but will break up quite hard soils, and can prove a valuable aid to a light loader or dozer. It is inexpensive, readily mounted and removed, and does not interfere with many other uses for the tractor.

FIGURE 21.6   Multiple-shank ripper on a wheel tractor.

CABLE LAYERS

Either mounted or towed rippers can be used to lay cable and flexible conduit underground, without ditching. The soil is preferably even-textured. It is essential that the ripper be able to penetrate it and hold a reasonably even depth.

The cable may be mounted on a reel carried on the tractor; or may be laid on the ground and carried over or under the tractor by sheaves or tubes. See Fig. 21.7. One end is fed through a conduit or around sheaves to take it down the back of the single ripper shank and curve backward near the heel of the tooth.

This end is anchored at the beginning of the work strip. The tractor moves forward, the ripper makes a slot in the ground, the cable (or conduit) slides through its guides to be left near the bottom of the slot. The cut in the ground may heal itself, or be pressed together by one passage of the tires of a car or truck.

The ability of a cable-laying ripper or plow to penetrate ground, keep in line, and make minimum surface disturbance is improved by high-frequency vibration of the shank. However, the vibration may have an unfavorable effect on some types of cable.

The cable plow in Fig. 21.8 has a vibrating shank (saber) attached to a cable static chute by vibration-absorbing hinges. It is mounted on a category II, three-point hitch on a heavy wheel tractor with hydrostatic drive or creeper gears.

The special cable plow machine in the next illustration has the saber mounted under the center of a short-wheelbase four-wheel-drive tractor. It can swing 90° either forward or backward from its vertical plowing position, for starting and finishing a line. It can work up close to an obstruction on either end.

The cable/pipe plow in Fig. 21.9, for mounting on crawlers over 70 horsepower, can slide and swivel to both sides to permit cable laying either inside the track path, or at an offset to either side.

WINCHES

A winch is a steel spool or drum that is capable of powerful rotary movement, and that has means to attach a line (usually cable, officially called wire rope) that winds in as it turns.

FIGURE 21.7   Cable-laying machinery.

FIGURE 21.8   Cable layer with vibrating shank.

FIGURE 21.9   Side-pivoting cable layer. (*Courtesy of ATECO.*)

The operating drums in cable-powered machines such as shovels and cranes answer to this description, but the term "winch" is usually limited to units that normally exert their power outside the machine that carries them.

An exception to the definition is the capstan or cathead winch, described later, which is not attached to the hemp rope to which it provides power.

A winch must have power to wind in its line, and means to release the line to be pulled to its next load. Pull out may be arranged by releasing the drum from the drivetrain, so that it will turn and unwind as the line is pulled; or by providing a reverse gear. If there is a reverse, it may be twice as fast as the inpull.

A winch is said to underwind when the cable is reeled on and off the bottom of the drum, and to overwind when it is at the top.

A few winches are hand-operated, but most of them are powered by the engine of the carrier. Drive usually is mechanical, through a power takeoff, but may be hydraulic or electric.

Power.   A winch is usually power-rated on the basis of the maximum pull it can exert on its line. Specifications should also give the speed of the line, in feet per minute (fpm). These figures vary with drum diameter, in inverse ratio, if drive is mechanical.

A small drum winds in less cable per revolution than a large one turned with the same power and speed. It therefore exerts more pull, but at a lower line speed.

The effective diameter of a drum increases with each full layer of line it wraps around itself, as additional line must wind on top of it. This means that pull will decrease as the drum fills, and line speed will increase.

A drum 8 inches in diameter carrying $\frac{7}{8}$ cable might have a line pull of 28,000 pounds bare, and about 16,000 on a fifth layer. But line speed would increase from 80 to 140 feet per minute. Additional effects of multiple layers are discussed later.

Line speed specifications vary with the tractor or other carrier. Pull is assumed to be fixed, and should not be exceeded even if power permits.

Torque Converter.   If the winch is on a power shift tractor, its drive should be through the torque converter. This device confuses the speed-power relationship, as indicated by the graph in Fig. 21.10.

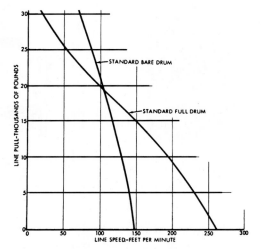

FIGURE 21.10   Line speed and pull with torque converter.

When pull on the line is light, the full drum shows the relatively fast line speed that would be expected. But increasing load slows down a full drum much more than it does a bare drum, because of greater leverage against the converter. At peak load (30,000 pounds in this example) the line moves faster on the bare drum than on the full one, as reduced slippage in the converter provides that much faster drum speed.

Brakes.   A full-service winch should have a brake that can lock the drum against full load. It may be necessary to hold a suspended or springy load when an engine stalls or direction of pull must be adjusted.

The brake may also be needed to hold the drum when the tractor is pulling the load.

The brake may be an external band, contracting on a drum on an intermediate or sideshaft in the winch case. It may be applied manually, hydraulically, and/or automatically.

In automatic position, a pawl locks the brake whenever a line that is being wound in starts to move out and unwind.

In free spooling, when line is being pulled off the drum to extend it to a load, there is a danger that the drum will spin (that is, turn faster than the removal of the line), with the resulting loosening and tangles. This is prevented by setting the brake to drag slightly, so that the drum will turn only the amount that the line pulls it.

Reversing clutches may be used as a brake by engaging both of them while a winch input clutch disconnects power.

A light winch may also not have a brake. In this the gearing is usually irreversible, and can be turned only by the takeoff shaft. The drum is connected to it by a jaw clutch. If the clutch is engaged and the power takeoff disengaged, the drum will be locked. In spooling out, spinning can be avoided by holding a foot against the drum, leaning a log against it, or snubbing it with a wood wedge.

Towing Winch.   The towing or logging winch is usually mounted on the rear of a crawler tractor, and driven by its power takeoff. Standard rotation direction is overwinding.

If the takeoff is clutch-controlled, the winch may have a jaw or sliding gear to connect the drum to the power train, and depend on the tractor clutch for starting and stopping of rotation, and for shifting gears or jaw clutches. For both safety and convenience, such a unit should have duplicate clutch and throttle controls that can be reached from behind the winch. See Fig. 21.11.

A live or constant running takeoff in a power shift tractor may drive the winch through a pair of multiple-disc clutches on a reversing shaft, so that the drum may be power-rotated in either direction. Clutch control may be manual, or hydraulically actuated.

Most winches now sold are power-shift models.

In either type, several stages of gear reduction are provided in the winch between takeoff drive and the drum shaft, to provide for slow and powerful rotation.

A towing winch drum has rounded flanges, which turn inside closely fitted and rounded casing parts. This construction prevents, or at least reduces, damage to line pulled at such an angle that it drags over these edges.

If the work is such that pulling at angles is frequent, additional protection can be obtained from a fairlead. See Fig. 21.12. Three rollers line up the cable with the drum from the sides and top. But it still is harmful to reel it with great force at sharp angles, because of bending and crushing stresses in the cable.

A towing winch line can be reeved over a boom to provide a tractor crane. However, only a fraction of its power can be used in such an application.

Wheel Tractor Mounting.   Wheel tractors may be fitted with winches mounted either on the front or rear. Rear winches should underwind and their tractors should be anchored by a line from the front for heavy loads, to avoid danger of overturning backward.

The winch shown in Fig. 21.13 could get a slow 50-ton pull from the engine of a light wheel tractor. The winch case had a connection for an anchor line extending forward under the tractor.

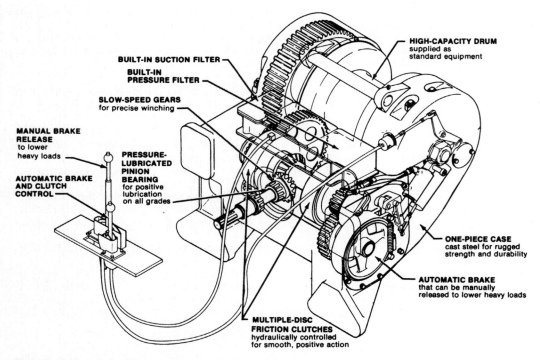

BUILT-IN SUCTION FILTER

BUILT-IN PRESSURE FILTER

SLOW-SPEED GEARS
for precise winching

HIGH-CAPACITY DRUM
supplied as standard equipment

MANUAL BRAKE RELEASE
to lower heavy loads

AUTOMATIC BRAKE AND CLUTCH CONTROL

PRESSURE-LUBRICATED PINION BEARING
for positive lubrication on all grades

ONE-PIECE CASE
cast steel for rugged strength and durability

AUTOMATIC BRAKE
that can be manually released to lower heavy loads

MULTIPLE-DISC FRICTION CLUTCHES
hydraulically controlled for smooth, positive action

FIGURE 21.11   Winch with clutch shift.

FIGURE 21.12   Towing winch with fairlead.

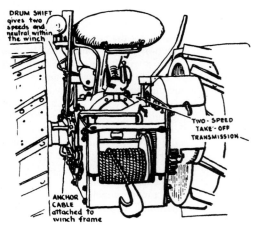

DRUM SHIFT gives two speeds and neutral within the winch

TWO-SPEED TAKE-OFF TRANSMISSION

ANCHOR CABLE attached to winch frame

FIGURE 21.13   A 50-ton pull on a wheel tractor.

Applications.   Winch operation was discussed in Chap. 1.

A towing winch provides accessible storage for a long piece of cable, which enables the pull of the tractor to be readily exerted at a distance, and it also enables the tow line to be adjusted to the exact length required.

A long cable permits use of snatch blocks and anchors to build up the line pull to any power required. Snatch block applications and other winch uses will be discussed later.

The winch provides a pull equal to or greater than that of the tractor drawbar, and is largely independent of traction. In muddy or sandy conditions, or in working on roads or landscaped ground, much less disturbance is made by winch than track pull. If the tractor is dragged backward even with the brakes set, it may be anchored by a line to a stump or tree, or braced by placing a log behind it. If there is no objection to tearing up the ground, the tractor can try to pull the load with tracks spinning until a mound of earth is built up behind them. This will act as a block for the tractor when the winch is used.

The cable is pulled off the drum by hand to attach it to the load. If the drum is disconnected from the gearing so that the pull on the cable will turn it, the brake should be applied sufficiently to prevent it from spinning free and tangling the cable when the pull stops. If two people are available, one person can use the tractor engine and winch controls to turn the drum to unspool while the other pulls away the cable.

Large winches require heavy cable, and dragging hundreds of feet of it is laborious, even for several people. If many loads are to be brought out of one place it may be advisable to pull the line out by means of another winch with lighter cable reeved around a pulley behind the loads.

The second winch may be a towing winch, a cable control unit on another machine, or a second drum on the same winch.

TRUCK MOUNTING

Winches in a range of sizes may be installed at the front or rear of the frames of trucks, pickups, and jeeps, or on platform bodies. Rear frame mountings may interfere with body type, or with access for loading, and are seldom used.

Mechanical drive is generally through a power takeoff on the transmission, and a propeller shaft with universal joints, but may be hydraulic from an engine pump.

A takeoff may be one speed in one direction, one speed each in forward or reverse, or two or more speeds in one or both directions. Vehicles with four-wheel drive operate the winch through the main transmission with its choice of speeds, with the auxiliary shift in neutral.

Ordinarily, the only control on the winch itself is a jaw clutch. The driver at the cab controls cannot watch the line, which means that two workers are required for efficient and safe operation. However, it is usually fairly easy to add clutch and throttle controls beside the winch, so that one person can operate it safely, although not always conveniently. It must be possible to lock the extra clutch lever in disengaged position.

Figure 21.14 contains a diagram of a front-mounted winch and controls. When the jaw clutch in the drum is engaged, the winch can be connected or disconnected at the power takeoff, and controlled through the engine clutch and throttle.

Truck winches may have either manual or automatic holding brakes. The drum is released or reverse gear is used to strip the cable. See Fig. 21.15. Recent-model winches are equipped with drag brakes that prevent overspooling and snarling of cable when the clutch is released for free spooling.

Hydraulic.   A winch of almost any kind, in any position, may be powered by a hydraulic pump and motor set, instead of a power takeoff shaft. Such a motor has a speed-reducing gearbox, and may have a stepdown chain drive as well. If design is adequate, performance should be satisfactory.

Hydraulic drive puts some additional weight and bulk at the winch or close to it, but avoids the often-difficult problem of getting a shaft drive to it. Hydraulic hoses can reach almost anywhere,

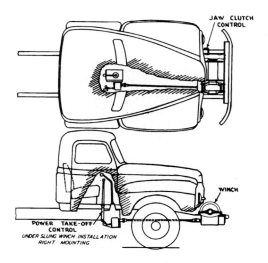

JAW CLUTCH
CONTROL

WINCH

POWER TAKE-OFF
CONTROL
*UNDER SLUNG WINCH INSTALLATION
RIGHT MOUNTING*

FIGURE 21.14   Front winch on light truck.

FIGURE 21.15   Winch on front of pickup truck.

although care must be taken to wrap or cushion all points where they rub or even might rub against moving or vibrating parts.

Electric.   Electric winches, powered by the battery of the carrying truck, tractor, or car, are usually made in small sizes, with pull of 4 tons or less.

These units are compact, offer minimal problems in installation, and may be economical in price. The principal problem is that the average good battery may provide peak power for only about a minute. Many winch jobs, even getting out of a ditch, take longer than that. See Fig. 21.16.

There is usually a hand crank that will turn the drum after the battery refuses to. It is often able to provide enough pull to complete a job, although the chief purpose is to relieve cable tension if power should fail during a pull.

The engine should be running during a pull, fast enough to provide maximal help from the alternator. Also, the winch may run down the battery so that the starter will not work.

Booster cables may be used to provide extra winch action from another battery.

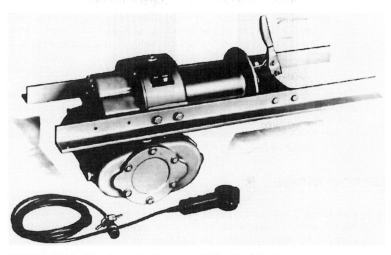

FIGURE 21.16   Electric winch. (*Courtesy of Tulsa Winch Inc.*)

HAND WINCHES

Winches with hand-crank power may be used for occasional jobs that do not justify the cost of a power unit, and in places that are not accessible to machinery. They are made in 2-, 5-, and 15-ton sizes. The two smaller ones are the most popular.

The winch, Fig. 21.17, consists of a spool drum with a double reduction set of gears operated by a hand crank. The handle may be inserted in the small outer gear, as shown, for full power, or in the larger gear for faster cranking and lighter pull. Gear ratios are 24 to 1 and 4 to 1.

FIGURE 21.17   Hand winch.

A spring-held ratchet may be engaged to prevent slipping back while changing the handle or resting, and the load can also be held or controlled with a friction brake.

The frame can be bolted to a deck or skids or attached to a tail yoke for anchoring by a chain or cable.

The 2-ton winches of the same make are similar in construction, and also have two gear ratios, 22 to 1 and 4 to 1.

Either size may be carried by two workers and readily set up in remote places. If the unit is clean and well lubricated, it may be able to pull stumps or rescue machinery better than a power winch of the same capacity, because of sustained pull and ability to take up a little at a time.

Substituting a ratchet handle for the standard type will greatly reduce the labor of winding in. It is often possible to rig a hand winch for power drive.

LAND-CLEARING TACKLE

Tow Lines.   Standard types of cable (wire rope) are used on winches. A 6 × 19 construction is most common. Independent wire core is recommended by manufacturers because of its greater strength, but many operators prefer fiber center because of its flexibility.

Wire core resists mashing, but if deformed it will not straighten out well under load so that it may become difficult to spool on the drum or feed through tackle.

Use of heavy cable decreases the length that can be wound on the drum and increases the labor of handling. However, land clearing which includes stump pulling usually requires the heaviest cable recommended for the winch. If lighter lines are used, the tractor should not be anchored and maximum pulls should be avoided.

Strength of cable is given in a table in Chap. 12.

Light and medium cables are ordinarily attached to a tail chain and round hook. The chain is usually made of stock of the same diameter as the cable, and may be from 18 inches to 8 feet long. It should be made of high-strength or alloy steel, to give it substantially greater strength than the wire rope.

If the chain is long enough to make a complete choker around most of the loads or anchors, it may end in a standard slip or round hook. If it is so short that it functions merely to relieve the cable end of twists and kinks, it should have a hook with a wide, rounded inner surface which will do minimal damage to the rope wires. It is also desirable to have the hook end turned in, to minimize catching on obstructions when pulled in empty. See Fig. 21.18.

Winch lines of $\frac{7}{8}$-inch diameter and larger may be fastened directly to a hook. Attachment may be by cable clips, wedge and socket, or poured fitting. Cable clips are light, easily obtained, and inexpensive, but are a nuisance to install, hard to remove, difficult to salvage, make a clumsy joint, and are apt to damage the cable.

FIGURE 21.18   Cable-choker hooks.

Wedge clamps are comparatively easy to install and to take apart, can be used many times without replacement or repair, and do not damage cable. However, they are heavy and bulky and may be put out of action by loss of an improperly installed or tightened wedge.

Poured fittings are compact, light, and strong but require special materials and techniques for installation or removal.

A winch line may also be ended in a shackle, a cluster, a loop or a thimble, or a cable takeup with or without a swivel.

It is good practice to use separate attachment or choker lines. They are more readily repaired or replaced than the winch cable, and will serve to save it from the severe wear resulting from contact with the choker hook and load.

Chain.   Towing chain is made in a variety of sizes, strengths, and designs, four of which are shown in Fig. 21.19. Size is designated by the approximate diameter of the round bar stock used in making the links, and ranges from ¼ to 2 inches or larger.

Strength is determined by the bar size, by the quality and treatment of the steel, and to a smaller extent by the shape of the links. The table in Fig. 21.20 gives data for various types. The safe load shown is about one-quarter of the average breaking strength.

Alloy steel chain is very expensive, but is desirable for land-clearing work and load handling because of its light weight in proportion to strength. For example, $\frac{3}{8}$-inch alloy chain is 30 percent stronger than $\frac{5}{8}$-inch proof chain and weighs only two-fifths as much. In addition, the lighter

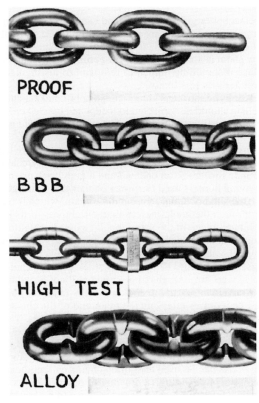

FIGURE 21.19   Chain.

Size	Approximate weight (lbs) per foot	Safe working load—pounds*				
		Dredge iron	Proof coil	BBB coil	High test	Alloy
$\frac{1}{4}$	.8	—	900	1100	2500	2750
$\frac{5}{16}$	1.1	—	1400	1700	4000	—
$\frac{3}{8}$	1.6	—	1900	2300	5100	6600
$\frac{7}{16}$	2.1	—	2500	3100	6600	—
$\frac{1}{2}$	2.8	—	3300	4100	8200	11250
$\frac{9}{16}$	3.4	—	4100	5100	—	—
$\frac{5}{8}$	4.3	6930	5000	6300	11500	16500
$\frac{3}{4}$	6.2	10140	7100	8900	16200	23000
$\frac{7}{8}$	8.5	14000	9600	12000	—	28750
1	10.5	18600	12400	15500	—	38750
$1\frac{1}{8}$	12.6	23400	15600	19500	—	44500
$1\frac{1}{4}$	15.4	28800	19200	24000	—	57500

*Safe working load is about one-fourth of average breaking strength.

FIGURE 21.20   Chain weight and strength.

chain gets a better grip on objects and can be passed through narrow spaces. Substantial savings in labor and machine time are realized from its use, but only in the hands of workers who will not abuse it or lose it.

Alloy chain is almost standard for permanent attachment to winch cables. When used as a separate piece it should be dipped in red paint to distinguish it from less strong and valuable chain and to render loss less likely.

Short links are stronger than long ones but make a chain more balky and inclined to kink.

No chain should be subjected to load when twisted or kinked, as the links will be deformed and weakened. Twisting can be reduced if swivels are used in the chain or at the end hooks.

Hooks and Rings.   The most-used chain fastenings are hooks and rings, some types of which are shown in Fig. 21.21. The standard logging or utility chain, Fig. 21.22, is of variable length, and has a round (slip) hook on one end and a grab hook on the other.

The round hook is used to form a choker that will tighten under pull. The chain is put around the load, the slip hook placed around the chain, which is pulled from the grab end. The chain tends to slide through the hook, increasing the tightness of its hold in proportion to the force exerted.

The hinged piece in the safety hook swings back automatically to admit a line, but it must be manipulated in order to release it. This prevents the hook from falling off the line when it is slack, avoiding delay and damage.

Rings, which should be of heavier stock than the chain, are used in the same manner as round hooks. They are stronger than hooks and will not lose their hold when the chain is slack, but they are less convenient. A stump choker may be made by pulling chain through the ring, but anchoring to a tree requires releasing the other end of the chain and threading it through.

"Rings" are made in a number of different shapes. The three illustrated are the most popular. If two are used on a chain, one should be small or narrow enough to slide through the other.

The grab hook is used to form a chain loop that will not tighten. It slides over any link in the proper-size chain, as in Fig. 21.23, but will not slide along the chain.

Grab hooks can be used to shorten chains by lengthening the loop or by blocking the chain so that it will not slide through a drawbar or other opening.

Shackles, Fig. 21.24, can be used in place of rings or hooks for many purposes, and are handy means of attaching lines to objects. They can be used for emergency chain repair.

HOOKS   SWIVELS

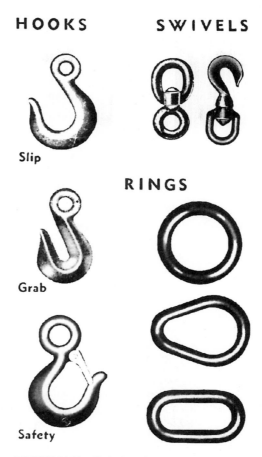

Slip

RINGS

Grab

Safety

FIGURE 21.21   Chain fastenings.

Load binders, Fig. 21.25, are used chiefly in tightening chains. The binder is expanded, the slack pulled out of the chain, and the hooks are attached to links on each side of the slack. The lever is pulled to shorten the binder, and locks over center. This process may have to be repeated on successive links. The handle should be tied down when left as part of a load lashing, as otherwise any slackening of the chain will allow it to fall open.

Load binders which do not have springs can be used with a chain to move heavy objects.

Repairs.   It is common practice to overload winch and land-clearing chains to the point of destruction, with resulting heavy costs in repair and replacement. This is in part because the extremely heavy and variable pulls required to uproot tough stumps or to free jammed logs call for a chain which in ordinary quality is too heavy to handle and to use. High investment in alloy chain may be justified.

Broken chains can be repaired by hot forging of new links or by using special repair or connecting links, two types of which are shown in Fig. 21.26.

Such links are purchased assembled and separated by driving a chisel or a very sharp screwdriver between the pieces. This is most conveniently done in the shop.

FIGURE 21.22   Log chain.

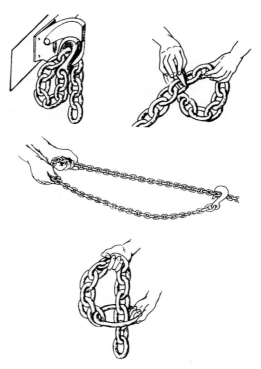

FIGURE 21.23   Grab hook uses, and stump choker.

FIGURE 21.24   Shackles.

FIGURE 21.25   Load binders.

If the links to be connected have pulled out of shape, it may not be possible to get the repair pieces through them. Such a link can be opened up by placing it on a block with a hole in it, and driving a big punch through it.

A good repair link is somewhat stronger than a standard chain link.

Snatch Blocks.   A snatch block is a pulley or set of pulleys on an axle, which is held in a portable case that can be attached to one or two pull lines.

These blocks are generally employed to increase the force exerted by a line. They are used for hoisting, dragging heavy loads, changing direction of pull, and land clearing. The principles involved are described in Chap. 12, and their use in land clearing is discussed in Chap. 1.

Figure 21.27 shows several types of blocks. Single-sheave blocks should have latch fastenings that permit opening to insert or remove cable, but double and triple sheaves have solid cases.

The pulleys (sheaves) turn on a single axle. Self-lubricating bronze bushings are commonly used, but bushings greased through the axle and sealed antifriction bearings are also obtainable.

These general-purpose blocks can be obtained having a loose hook, a swivel hook, or a rigid or swivel shackle. The shackle is the strongest unit in proportion to weight, but is also the least convenient to use. Swivels have the advantage of allowing the block to adjust itself automatically to line of pull, so that rubbing of the lines against the case is kept to a minimum.

The latched blocks are much more efficient than the solid ones. The average hook or other attachment is too large to pass between the pulley and a fixed case so that it must be removed from the line for reeving. If it will pass, it is still necessary to thread the end of the cable through the block and to pull it out the same way. This takes much more time than inserting and removing the cable from the side, particularly when a number of blocks and lines are used.

Fixed blocks are often left rigged during moving, and crossing and tangling of lines may result, particularly if they are not pulled snugly together.

Figures 21.28 and 21.29 illustrate some heavy-duty blocks designed specially for land clearing. The sling block permits use of two parts of light choker line instead of a single heavy one.

MISSING LINK

Swing Link

FIGURE 21.26   Repair links.

When a choker cable or chain is permanently fastened to a block, the combination is called a power choker. One type shown is attached to the choker line by a shackle, and the other by a poured fitting. The shackle or loop on the opposite end makes it possible to attach an extra line, and to use the block as part of a direct line when the extra power of the pulley is not required.

Block and Fall.   A block and fall is a set of light pulley blocks using fiber rope. See Fig. 21.30. In small sizes, pull is by hand. In larger units, several workers, a light tractor, or a horse may provide power.

They range in size from very small, 4 pounds or less complete with $\frac{3}{8}$-inch rope, to heavy-duty models with 1-inch rope.

In the smallest sizes they are used chiefly for tightening fence wire. In general, they are useful in countless situations where required pull is greater than can be exerted by a worker, and machine power either cannot be readily applied, or might be too strong, fast, or jerky for the job.

Nylon rope should be avoided, because it stretches, wasting space and energy.

Root Hook.   This is a very heavily constructed hook used for pulling small stumps, and individual roots on large ones. It is designed more for strength than for gripping power, and it is often necessary to notch a stump or dig behind a root, in order to give it a grip.

Tongs.   Tongs are often used in place of chokers to grip loads that are to be hoisted or pulled. The type shown in Fig. 21.31 has points which are pulled into the load and which tighten automatically. It is sometimes necessary to tap the points in order to obtain the first grip. They can often be shaken loose after the load is placed, without hand work.

These are used for picking up or dragging loose stumps, logs, and similar objects. They are seldom used for pulling "solid" stumps, as considerable weight would be required to obtain sufficient strength, and the points have a tendency to tear through wood under extreme stress.

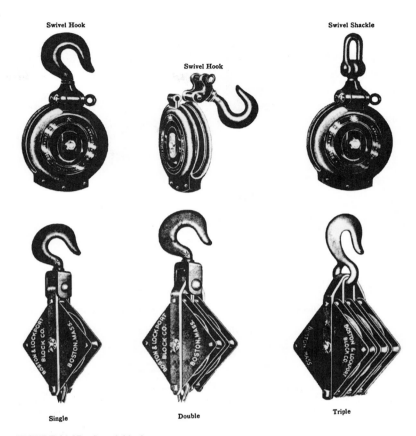

FIGURE 21.27   Snatch blocks.

They are of particular advantage in handling objects which are so placed or piled that it is difficult to get chains under them, or to remove chains after they are placed.

The Johnson grubbing tongs shown in Fig. 21.32 grip with the inner surfaces of the arms. They are heavily built for rugged service in pulling stumps and trees up to 10- or 12-inch diameters, and can also be used for hoisting. Under many conditions they are more readily attached and detached than chains, and are of particular advantage when the pulling is done by a shovel dozer or some other machine which can carry them to the point of attachment.

In heavy work, it is important to keep the line of pull straight through the tongs, as a side pull imposes excessive strain.

DOZER EQUIPMENT

The bulldozer is an important land-clearing tool in itself. But its performance in this field can be extended and improved by replacing the standard connectors with special blades and devices, the more important of which are described below.

Perhaps the most necessary special construction for clearing is protection devices, for both operator and machine. A strong cab, OSHA-approved, should include an extra strong roof and

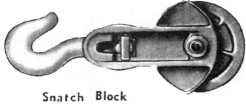

Snatch Block

Sling Block

FIGURE 21.28   Heavy-duty blocks.

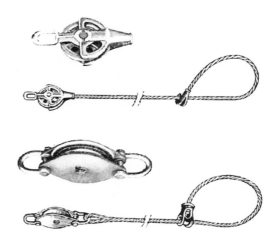

FIGURE 21.29   Power chokers.

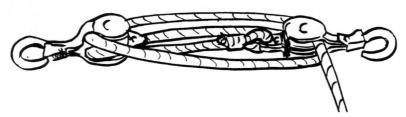

FIGURE 21.30   Block and fall.

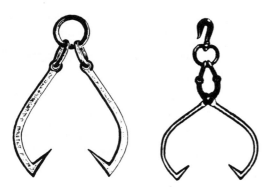

FIGURE 21.31   Grapple hook and skidding tongs.

FIGURE 21.32   Grubbing tongs.

verticals, and a rear protection of heavy screening. See Fig. 21.33. The radiator must have a strong, small-mesh guard. The engine must have side guards. Fine-mesh cab screening for operator protection against insects may be needed.

The operator must be constantly alert for falling trees and branches, for high stumps, rocks, and pits concealed by vegetation, for poles thrusting toward her or him, and for buildup of dangerous tensions in pushed material.

Stumper.   The stumper in Fig. 21.34 is designed to fit over and be pin-fastened to the center swivel of an angling dozer C frame, or to a special frame operated by a bulldozer hoist.

These tools are usually $2\frac{1}{2}$ feet wide, and have a drop below the C frame of 2 or more feet. Construction is very massive. A serrated edge assists penetration in earth or wood, and keeps it from skidding to the side.

The stumper is used for pushing over trees and stumps, driving under stumps to boost them out, and digging around them when necessary. It is also effective in digging out boulders, knocking dirt off loose stumps, digging up railroad ties, ripping up shale and old paving, and making shallow ditches.

It concentrates the full power of the tractor on a narrow front. It makes possible cutting and lifting underground roots with minimum soil disturbance, and without wasting power in unnecessary digging of a wide strip of soil. It is not subject to the twisting strains which shorten the life of full blades used for stumping.

It is not good at piling or transporting loose stumps, or at backfilling holes or clearing brush. It should be teamed with a dozer or used alternately with a rake or blade on the same tractor.

Tree Pusher (Knockdown Beam).   A tree pusher is usually a heavy tractor equipped with a dozer, stumping, or angle or V-blade, and a higher push frame with longer reach. See. Fig. 21.35. The push frame is preferably under separate control, but may be linked to the blade so that it is raised and lowered by its controls. Sometimes a pusher is carried without a blade.

Trees are pushed by the upper frame. This may uproot them, or just put them under heavy tension, so that the blade can drive under them readily to tear them out of the ground.

A V or angled blade can cast the tree to the side. Disposal is usually handled by other equipment.

V-Tree Cutter.   A V-cutter for trees, Fig. 21.36, consists of a V-shaped, dozer-mounted blade fitted along the bottom with horizontal, scalloped cutting edges, and a center-mounted splitting point or stinger. Floats keep the edges from digging in.

The V-cutter is designed to shear off all vegetation at ground level, whether it is large or small, and to cast the debris to each side, where it is left for other equipment.

FIGURE 21.33   Cab guard.

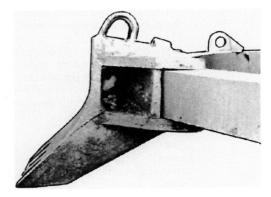

FIGURE 21.34   Stumper.

FIGURE 21.35 Tree dozer in action.

Trees that are too large to be sliced from the side are rammed and split by the stinger. The machine might then be able to slice the halves without stopping, or might have to back up to make additional passes.

Operation is most efficient on even-surfaced ground where the edges maintain good contact, where it is firm enough to hold the roots in position while the trunk is cut, and where there are no rocks hard or large enough to damage the cutting edges.

Under favorable conditions, cleared areas will be left entirely firm, and free of projecting stumps. The cut trees and brush may be pushed off by a dozer, preferably equipped with a rake blade.

Angle-Blade Cutter. This somewhat similar device has the stinger mounted on the forward end of a long blade set at a 30° angle. The full-length cutting edge is straight. See Fig. 21.37.

This arrangement permits dropping and pushing all trunks to one side and doing after-cutting windrow piling. In very heavy work, there may be problems with off-center loads.

The blade can be tilted downward to use the stinger as a stumper, or the whole blade in cutting shallow ditches.

Rake Blade. A rake blade may replace either a dozer blade or a front loader bucket, or be fastened by brackets and pins to a standard blade. It is made up of a set of tines that may be more or less vertical at top and center, and curve forward at the bottom. See Fig. 21.38. An upward extension, as seen in Fig. 1.2, is called a brush guard. There may be a solid center section to protect the radiator.

FIGURE 21.36   V tree cutter.

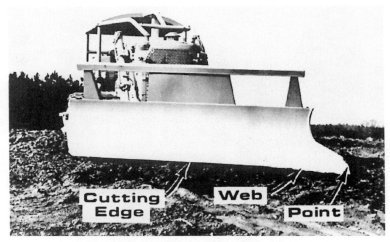

FIGURE 21.37   Angle-blade tree cutter. (*Courtesy of Rome Plow Co.*)

FIGURE 21.38   Rake blade.

The tines may be operated below ground level to bring up roots and boulders, without moving solid masses of soil. However, separation does not usually occur automatically, except perhaps in dry sand.

Wet, sticky, or lumpy dirt may build across the teeth, blocking the slots between them, either by itself or after becoming matted with brush or roots. Under such conditions, it takes patience, skill, and experience to take growth (and rocks) and leave the soil behind.

Rakes vary greatly in weight, strength, and tooth spacing, to suit various types of work. The heaviest ones, designed for grubbing out stumps, heavy roots, and boulders, must have tines so strong that any one of them can take the full push of the tractor without bending. Tooth spacing is usually wide in this type.

The lightest ones are intended for raking up loosened material on the surface, and for underground removal of light or weak root system. Tines are lighter and more closely spaced. They may be badly damaged by heavy work.

Heavy soils, such as clay and silt, are much more troublesome than sand, gravel, or loam. Almost any soil is easier to separate when dry than when wet.

Tree Shear.   Tractor-mounted, hydraulic tree shears are presently able to cut hardwood trees up to 20-inch diameter, and softwood over 20 inches. They snip or shear the tree a few inches above the ground. Capacity is reduced if the wood is frozen.

The unit in Fig. 21.39 is shown in open and closed positions. It is supported by a dozer (or loader) frame and lift. The left side (looking at it in this picture) is a massive fixed jaw or anvil, the right side is a hinged knife (cutter blade) that is moved by a powerful hydraulic cylinder.

The tractor is maneuvered until the tree trunk is between the knife and the anvil, which usually rests on the ground. The cutter is then forced through the wood. Unless badly out of balance, the tree falls across the anvil, away from the knife.

FIGURE 21.39   Tree shear.

The shear may crush the bottom few inches of the trunk. A slice would have to be trimmed off saw timber, but not from a log to be pulped. No loss of timber is involved, as hand cutting is usually done much higher, and would leave more than that section on the stump.

Such a shear may be teamed with a handling device mounted on the same tractor. The accumulator in Fig. 21.40 is a double clamp that can hold one large tree, or a number of small ones, both while being cut, and while moving on to the next tree. When the arms are full, the tree or the stack is laid (piled down) on the ground, for dragging (skidding) away, or to be picked up by a chipper.

Root Plow.   A root plow, or root cutter, is usually a horizontal knife, straight or V-shaped, supported by vertical standards or shanks at each side, and carried in a rear-mounted frame hinged to the track frames of a big crawler.

The model illustrated in Fig. 21.41 has a cable lift, but hydraulic operation is optional. Working depth is 8 to 20 inches, depending on growth size, soil condition, and tractor power.

Angled fins slide roots up to the surface, but cleave through soil with little disturbance.

A fringe benefit from root plowing is almost complete loosening of soil, which increases water absorption. This good result may be accomplished with very little, if any, increase in erosion, which is retarded by innumerable fragments of vegetation left to rot in the disturbed soil.

Root Rake.   Roots and stumps cut loose by the plow may be brought to the surface and piled by either a rake blade or a root rake. They are most effective in sandy soil and with coarse pieces. Heavy, wet soil, and fine matted roots, may cause almost constant clogging.

FIGURE 21.40   Accumulator to hold cut trees. (*Courtesy of Rome Industries.*)

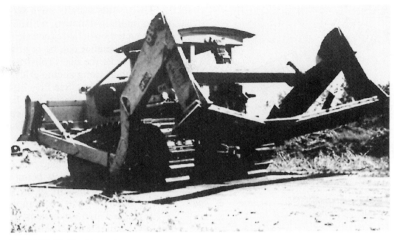

FIGURE 21.41   Root plow.

Towed Chain.   Trees and stiff brush may be uprooted by a chain towed between two dozers of 180 horsepower or more. Chains must be very heavy, with bar diameter in the links being 2 to 3 inches, and weight up to almost 90 pounds per foot. They should be about three times as long as the spacing between tractors, and 2½ times the height of the tallest tree.

One or more steel balls, from 3 to 10 feet in diameter, which may be hollow, filled, or solid, may be fastened in the chain by universal connectors. They should be centered or equally distributed in the center half of the chain. Balls may serve to hold the chain up to slide over stumps and ridges, and to add momentum to overcome sudden resistance.

The chain should include several links, preferably one at each end and two in between, to prevent development of damaging twists. A supply of quick repair links should be available.

A third tractor with a pusher may be needed to follow the chain, and assist with stubborn trees, or lifting over obstacles.

It is often necessary to cover the area twice, in opposite directions, to complete the uprooting. But there is danger of loosened trees moving with the chain, and creating a massive, tangled pile.

CHOPPERS, SHREDDERS, AND CHIPPERS

The actual bulk of vegetation in any area is far less than it appears to be when it is standing, or even after if has been cut, uprooted, or piled. The difference is usually greater with brush and saplings than with trees with substantial trunks.

As a result, sufficient clearing for many purposes can often be accomplished by chopping vegetation into small pieces, without removing it. Such pieces may be partly or wholly buried by the chopping process, or left scattered on the surface.

When chopping involves partial burial, as by a disc harrow, the debris cannot be removed. Shredding and surface scattering by a rotary mower permit raking up for burning or removal, with increased difficulty in getting all the pieces. Chips are delivered through a spout, may be piled or scattered, but can conveniently be put directly in trucks for removal and possible use elsewhere.

Disc Harrow.   Weeds, brush and small saplings may be knocked down, chopped or mangled, and partially buried by a heavy disc harrow. See Fig. 1.5.

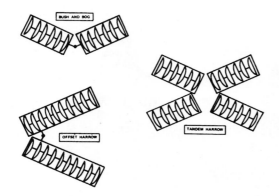

FIGURE 21.42   Disc arrangements. (*Courtesy of Rome Plow Co.*)

These units usually have six or more concave discs, 2 feet or more in diameter, with sharp, scalloped edges. They are mounted on an axle (or axles) whose angle to the direction of movement can be regulated. If it is parallel to the tractor's axle, the discs roll freely; if at an angle, they cut down with a strong slicing effect, and mix and overturn the soil and vegetation. See Fig. 21.42.

At the best, the result will be well-chopped vegetation so well mixed with loosened earth that only a small part will be visible on the surface. The surface will usually have low parallel ridges, but might be almost smooth, and ready for planting with a cover crop, preferably some legume.

If the ground is too hard or rocky, or the vegetation is too coarse for the size harrow being used, there will be spots or areas where one or more of the functions—chopping, loosening, or burial—was not complete.

A single harrow sidecasts, somewhat in the manner of a plow, so that it leaves a furrow on one side of each pass and a ridge on the other. If the field is worked in straight lines back and forth, the line where the two directions meet will be left as a trough or a ridge. Skipped spots will create rough spots for the same reason.

In a rough field these irregularities probably will not matter. In cropland, they can be corrected at the next plowing, But they should be kept in mind when planning the work.

A bog harrow has two axles, which angle to both cast outward, creating a center trough and side ridges. These are smoothed by overlapping passes by a half-width.

In tandem construction, with two axles, one behind the other, the discs in the second set throw the opposite way from the first, thus eliminating troughing of the dirt except on turns.

Rolling Chopper.   A rolling chopper is usually a big drum fitted with cutting blades, set in a tow frame with a draw tongue. See. Fig. 21.43. It is towed behind a tractor, usually a crawler, of 60 horsepower or more. Weight can be adjusted by putting water in the drum, or draining it.

A fully ballasted drum may put $1\frac{1}{2}$ to 2 tons of weight on each foot of blade, providing a tremendous crushing and cutting force. Widths are available up to 16 feet. Smaller units may be towed in tandem, with different blade spacing on front and rear drums, or in triangular groups of three.

FIGURE 21.43   Rolling chopper.

Choppers are usually pulled through standing trees and brush. The tractor should have a dozer blade or, for heavy growth, a tree pusher.

These units are generally used on jobs where the chopped vegetation can be left on and in the ground to decay. If the purpose is to produce rangeland for grazing, grass seed may be broadcast directly behind the towing tractor, so that the chopper will mix in and cover some of it.

A variable amount of material is mixed into the soil and buried, but this is usually not considered an important part of the result.

If the downed vegetation is to be piled, this may be best done by rake blades working at right angles to the direction of chopper travel.

The rolling chopper is most effective on ground that is firm enough to support sticks under the cutter, but soft enough to allow the cutter itself to sink in. Quality of work increases with speed, so it is advantageous to tow with a tractor powerful enough to use a high gear. Small stones do not bother it, but large ones interfere with the work and damage the blades.

The size growth that can be handled is limited by the ability of the tractor to knock it down, and by the resistance of the wood. Results vary with roller weight, tree type, and ground conditions. A few of the larger trunks on a job may be left almost intact.

Rotary Mixer. The rotary mixer is a machine that is used chiefly in mixing and stabilizing road bases and surfaces, but is also an excellent tool for clearing and mulching brushland. See Fig. 21.44.

A rotor assembly (Fig. 21.45), which consists of a shaft, tine-holding plates, and tines, is mounted across the direction of travel under a mixing chamber which controls the movement of materials so that two things are accomplished: first, a high percentage of the materials is deflected ahead of the rotors, so that constant remixing cycles are established; second, the material is at the final stage controlled in such a way that coarse and fines are mixed, blended, and placed so that aggregate segregation is entirely corrected. All sizes of the material it is working, from dust to the largest gravel, are uniformly distributed, the aggregate is keyed and interlocked, and the voids are filled with fines which create the most solid mass.

A rolled plate at the trailing edge of the mixing chamber acts as a strike-off to provide a smooth surface and to partially compact the mix. See Fig. 21.46.

Rotor speeds are controlled by a multi-speed transmission and by the engine throttle. Depth of cut is regulated by raising or lowering the rotor relative to mixing chamber. Adjustable springs carry part of the rotor weight to permit it to ride over obstructions.

FIGURE 21.44 Rotary mixer.

FIGURE 21.45 Rotor assembly.

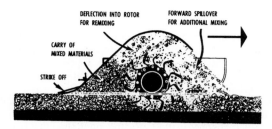

FIGURE 21.46   Rotary mixing action.

The tine-holding plates are driven by the shaft through individual friction clutches, designed to slip momentarily under shock loads. A variety of interchangeable tines can be obtained for different working conditions.

Smaller units of 3-foot working width may be purchased for mounting directly on most wheel tractors of two-plow capacity or over, and still smaller sizes for garden tractors, but these are not adapted for industrial work. Principles of operation are similar, but there are wide differences in ruggedness and in details of construction.

All sizes are useful in landscape work. They pulverize topsoil, mix it thoroughly with subsoil, sand, fertilizer, humus, or any other material desired, smooth it off, and leave it in ideal condition for planting. They will chop up sod and weeds and mix them with soil, increasing its bulk and making its appearance more attractive.

The large motorized and self-propelled mixers do a good job of clearing brush and palmetto, and will handle trees up to $2\frac{1}{2}$ to 3 inches in diameter. Light brush is cut up and completely buried; larger branches and trunk sections will be partly or wholly buried if the rear section of the hood is left down, or scattered on the surface for removal if it is raised. It is unsafe to walk behind the machine when the plate is up, as objects are thrown out with great force.

Rotary Mower.   Rotary mowers, which in small sizes are the homeowners' favorite for cutting the lawn, are also made in big, rugged models capable of mowing tall weeds, thick brush, and small saplings.

A heavy-duty rotary, such as the Bush Hog, Fig. 21.47, has a flat disc blade holder or flywheel fastened to the mower axle. Two heavy blades or flails are fastened to it by hinge pins.

When the blade holder is turning at high speed, up to 776 revolutions per minute, centrifugal force causes the blades to extend straight out, in cutting position, even against considerable cutting resistance.

But if a blade strikes something it cannot cut, it simply folds back on its hinge. It extends again immediately, and yields again at the next collision.

This arrangement entirely prevents shock damage to the shaft or drive, and limits blade damage to dulling or chipping. It also allows continued effort to cut or grind through the obstruction.

Most of these mowers are designed for mounting on the rear of medium-size wheel tractors with a three-point rear hitch. This allows positive (but not rigid) in-line fastening, and hydraulic lift. Drive is from the tractor's power takeoff, through a shaft and universal joints. The shaft has a protective shield sleeve to reduce winding up of brush and vines.

In ordinary operation, the mower rests on the ground, supported at the rear by a swivel wheel that may be adjustable for height. In front it may rest on narrow skid shoes, and/or be supported by the hydraulic lift. Lower gears are used, the choice depending on the power available and the heaviness of cut. The engine is run at three-quarters to full throttle.

Quality of cutting is somewhat better going forward than backward, but the difference is usually not important. If the tractor has a reversing transmission, or a suitable reverse gear, mowing may be done neatly in straight adjoining lines, backward and forward. Otherwise the area may be cut by working inward from all sides, with some removing of skips in the corners.

FIGURE 21.47   Heavy mowing for a rotary.

If cutting becomes very heavy, the engine will lug down and efficiency will drop sharply. It may be restored by raising the mower, if close cutting is not required. Or the tractor may be put into a lower gear.

In general, a heavy-duty rotary will chop up anything that its towing tractor can easily go over. For a 40-horsepower tractor and a 6-foot mower, that would mean almost any thickness of brush, densely growing saplings up to $1\frac{1}{2}$ inches in diameter, and occasional trunks up to 3 inches.

Heavy stands often cannot be done in reverse, as they block the mower box. But for small pieces of cleanup at edges, or to chop piles of brush, the mower may be raised, backed into or over the material, and then lowered with the tractor standing still and the blades turning.

Most vegetation cut by this mower is left in rather short pieces, with considerable shredding, and it can usually be left on the ground to rot. However, a certain number of long pieces escape by lying flat on the ground. Sapling stumps may not be cut off flush, and might require repeat cuttings. Work is usually not as thorough and neat as the shredder's.

Vines may wind around the driveshaft, and have to be cut off with hand clippers or a heavy knife. This trouble is greatly reduced by making first passes in reverse.

Rocks damage the blades slightly, and repeated contacts will wear them away or make them impossibly dull. The rocks themselves are often smashed.

The machine is dangerous, as it may throw rocks and other hard pieces with great force for more than 50 feet, chiefly to the rear. It should always be stopped if someone approaches it.

Rotary brush choppers are versatile and economical clearing tools, for work within their capacity. And they make excellent junior partners for bigger machinery, when controlling areas that have resprouted after clearing.

Brush Chipper.   The brush chipper or grinder is the standard tool for processing hand-cut brush, saplings, and branches, wherever burning is impractical. See Fig. 21.48.

This type of vegetation is reduced to small chips by an engine-driven toothed cylinder or disc turning at high speed. The unit is usually mounted on a light trailer, and towed by a truck into which it can discharge the chips. It may be moved by a tractor on rough ground, or where complicated maneuvering is required.

FIGURE 21.48   Brush chopper loading truck. (*Courtesy of Vermeer Manufacturing Company.*)

The chips have only a tiny fraction of the bulk of the brush that produces them. Where brush must be hauled away, 10 to 20 truckloads may be reduced to one load of chips.

On rough ground, in or bordering on fields and woods, the chips can be scattered to save hauling away. Small quantities are inconspicuous, and their decay adds compost to the soil. But piles or thick layers produce barren areas that may persist for years.

Cut material is fed into the machine by hand, in single stems or in bundles. The roller or its feed mechanism pulls them in. Great care must be taken not to allow hands to follow the brush. More time is usually needed than would be required to throw the material on a fire.

The chipper can usually be kept close to the cutting or gathering, so less carrying time is needed than with fires. Problems of building and maintaining fires, preventing spreading, and putting them out when leaving the area are all avoided.

Disadvantages include a high noise level, cost of purchase and maintenance, and consumption of fuel.

It is most efficient when kept close to the work, so that pieces can be ground as they are cut. If this is not practical, brush should be piled with all the butt ends facing the side where the chipper is to stand.

Tree Chipper.   There are also chippers that can process full-size trees, up to 30-inch diameter or more. The model in Fig. 21.49 is mounted on a semitrailer with a fifth wheel for connection to a highway-type tractor.

Its operator can pick up trees (or bunches of small trees) with a grapple, and place them on a chain conveyor and into the grip of vertical and horizontal compression rollers. These force it against a rotating 75-inch, three-knife chipping disc. See Fig. 1.29.

FIGURE 21.49   Self-feeding tree chipper. (*Courtesy of Morbark Industries.*)

The chips are screened to separate fines (mostly bark dust) and oversize pieces. A conveyor and chute take them from there, and can be adjusted to load them into trucks, pile them, or scatter them.

The compression rolls are powerful enough to pull most trees into the cutter complete, without need to trim any branches.

The chipper accepts only cut trees, or uprooted ones with the stump cut off and discarded. Stumps are almost sure to carry dirt to dull the knives, and are likely to include rocks to chip or break them.

In many areas, it is possible to sell chips to a paper mill, or to other wood product factories. Value depends on quality in relation to the processors' requirements; possible profit may depend largely on haul distance. But it often pays to sell chips for less than cost, when other disposal methods would be more expensive. And utilization is greatly preferable to wasting.

If there is no market for chips, they may be sprayed on slopes for erosion control, usually in conjunction with tree planting, or spread thinly over the ground for decomposition into humus. But thick layers and piles should be avoided except in wasteland, as they may make the covered areas barren for years.

Stump Chipper.   It is often difficult or impossible to get to a stump with equipment big enough to pull or dig it out. Even more often, the tearing up of the area and creation of a big hole are unacceptable. Hand digging is less destructive, but it is laborious and prohibitively expensive, and leaves a hole to be filled.

A stump chipper of appropriate size can chew almost any stump to chips, with little ground disturbance outside of its cutting arc. The work is done by a cutting wheel equipped with teeth, preferably carbide. Working depth is 6 to 24 inches in various models.

Figure 21.50 shows one of the larger and more sophisticated machines. It has a wheel carrying 48 teeth, some of which are on the rim, but most on both sides. This wheel can be lifted and lowered, and moved back and forth sideward continuously at a controllable rate during digging.

FIGURE 21.50   Eliminating a stump. (*Courtesy of Vermeer Manufacturing Company.*)

The unit is a trailer, attached to a light towing vehicle by a drawbar that can be swung from side to side by a hydraulic cylinder for accurate machine placement. Another cylinder extends and retracts the bar to move the unit and its wheel backward and forward during work.

The hole made by chipping a stump is much smaller than the one made by digging or pulling, as its lower parts are left in place. Chips may be used for temporary backfill, but they are loose, barren, and shrink with decay. Soil, or a mixture of soil and chips, should be put in for a longer-lasting repair.

It is economical to cut the stump off as close to the ground as is practical, before chipping it.

Stump chippers are used in land clearing, where stumps are to be removed after tree cutting, and in parks and built-up areas. Teeth are likely to be extensively damaged by contact with buried rock, so the wheel should be operated at slow speed and with great caution if its presence is suspected. Visible rock should be removed if possible.

Log Movers.   There is a great variety of specialized equipment designed to move tree trunks and shorter logs from the place they are cut to a nearby processing or transporting facility.

A rear-mounted towing grapple has a pair of pincer arms that are closed hydraulically on the butt ends of a log or pile of logs, then lifted to tow them. See Fig. 21.51. A universal mounting allows the grapple and its load to swing and tilt, to accommodate turns and rough ground. The hoist may be either a direct or a parallelogram type, similar to those in rippers.

A grapple does not require the fastening of individual choker lines to the logs, but it may require that they be prepiled in position to be picked up.

The vertical and horizontal accumulators used for carrying sheared trees were mentioned earlier.

There are several clamp designs that replace loader buckets. They carry one or more logs, up to the limits of clamp size and loader stability, at right angles to travel direction. These need plenty of clear space, and there may be problems of catching ends against the ground, and of side tipping on slopes and rough ground.

FIGURE 21.51   Towing grapple.

CHAIN SAWS

A chain saw is a handheld power tool designed to cut and trim trees. Generally, it has a high-speed, two-cycle, rope-start gasoline engine.

There is an automatic centrifugal clutch. This turns a sprocket, usually by direct connection, but sometimes through a speed-reducing gearbox.

The sprocket meshes with a toothed cutting chain that slides in a groove around a flat bar. The bar is sometimes fitted with a very thin sprocket or wheel in its outer end.

The only operating control is the throttle. Cutting is done at top engine speed. The clutch disconnects the chain when the engine drops toward idling speed.

Power Head.   A chain saw, minus its bar and chain, may be called a power head, saw body, or just body.

There is a case of sheet metal that encloses the engine and the reservoirs. In the Homelite models in Fig. 21.52 and 21. 53, the engine is horizontal, with the cylinder at the rear, and the crankshaft across the center. Reservoirs for fuel and chain oil are at the front left.

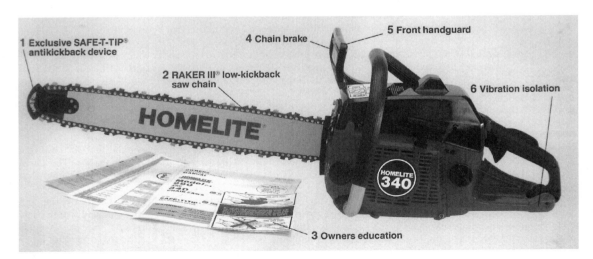

FIGURE 21.52   Chain saw with safety devices.

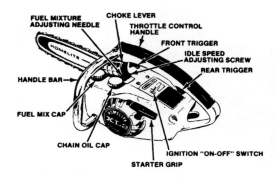

FUEL MIXTURE
ADJUSTING NEEDLE

CHOKE LEVER

THROTTLE CONTROL
HANDLE

FRONT TRIGGER

IDLE SPEED
ADJUSTING SCREW

REAR TRIGGER

HANDLE BAR→

FUEL MIX CAP

CHAIN OIL CAP

IGNITION "ON-OFF" SWITCH

STARTER GRIP

FIGURE 21.53   Chain saw parts and controls.

The case also encloses the carburetor and ignition system, and carries the rewind rope starter unit on the left side, opposite the sprocket.

There is a curved handle, running from front to back of the body. This carries the throttle trigger or triggers, and a throttle latch or hold-down, and is often called the throttle handle.

The handlebar is at right angles to the throttle handle, at the front of the saw.

The throttle handle should be held in the right hand, and is the principal holding member, while the handlebar is used to guide, tilt, and steady the saw. Both must be held at all times during operation.

The engine is a two-cycle design turning at high speed, 8,000 revolutions per minute, in this model. It requires a mixture of regular gasoline with oil for fuel. The type and proportion of oil will be discussed under Lubrication.

To the left side of the bar, looking toward its tip or nose, the body protrudes in a slightly rounded structure called the bumper. In ordinary cutting this is rested against the near side of the log, and the saw is pulled tightly against it by the pull of the cutting action.

The bumper is used as a fulcrum in pivoting the saw in a number of the cuts described in the next section. Its effectiveness is greatly increased by fastening a set of spike teeth to it. This can usually be obtained as standard or optional equipment.

Chain.   A section and an exploded view of a typical chain are shown in Fig. 21.54. A chain is made in three interlocked narrow sections or strips. The center is a succession of identical drive links, which have bottom tangs (projections) for meshing with the drive sprocket, keeping alignment in the bar groove, and cleaning sawdust out of it.

Sides of the chain are called right and left, looking in the direction of motion.

Each cutter link also carries a depth gauge, which prevents the cutter from gouging in too deeply.

The chain moves away from the saw body along the top of the bar, and returns along the bottom. Most cutting is done at the bottom. Chain speed is about 3,500 feet per minute in many direct-drive saws, and is much lower in gear drives.

Cutters may be sharpened in the field, with a round file and a guide, but it is a job that must be done just right. An inexpertly sharpened saw is likely to cut in a curve. It is safer to take it to a shop, for sharpening by machine, until experienced instruction can be obtained.

Saw Bar.   The saw bar is a flat piece of steel with rounded ends, with a narrow groove to accommodate the chain drivers. Length may be 10 inches to over 3½ feet, in various models. It is clamped into the saw body, in line with the sprocket, and is drilled or cut to allow for clamping and adjustment.

It is possible to obtain a bar having either a wheel or a sprocket in the outer end (nose), to reduce friction as the chain is pulled around it. These items must be extremely thin.

Chain Adjustment.   The chain must be loose enough to move freely around the bar, and tight enough so that the drive tangs cannot jump out (or be flipped out) of the groove. The chain changes in length, as it is longer when it is hot than when it is cold, and it becomes longer as it wears.

Tension is adjusted by moving the bar outward, away from the drive sprocket, to tighten; and inward to loosen. This bar is moved by means of a pin or dog on an adjustment screw in the saw body or drive case, which is parallel to the bar and is turned by a screwdriver, clockwise for tightening. The pin fits into a square or round hole in the bar.

This bar is held in alignment by the adjustment pin, and by a fixed pin that engages a long, straight slot. It is clamped in place by one or two bolts.

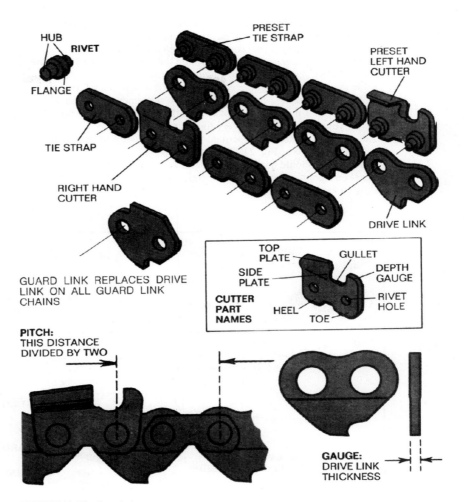

FIGURE 21.54 Saw chain parts.

Instructions vary with different saws and chains. Lacking information to the contrary, adjust a cold chain so that it is nearly snug in the groove all around, and a warm chain so that the tangs at the bottom center hang about halfway out of the groove. Let a hot chain cool.

Adjustment is made by loosening the clamp bolts, supporting the tip of the bar so that it will not slump downward, and turning the screw. When proper tension has apparently been obtained, pull the chain out of the groove and let it snap back at several points, to relieve any possible kinks. Recheck the tension and correct if necessary. Tighten the bolts, still supporting the end of the bar.

A too-tight chain absorbs power, runs hot, and wears itself and the bar rapidly. A loose chain chews its tangs and the sprocket, and also causes extra wear by slapping against the bar. It is vulnerable to coming off the bar, most often because of twigs or splinters getting under it.

A loose chain is more apt to get in trouble during horizontal cuts, or random cutting of small branches, than in vertical down-cut. It is a major cause of kickbacks.

Replacement.   To replace the chain, remove the clamp nuts and take off the sprocket housing, and the sawdust guard if present. Pull the bar off the adjusting pin, and slide it toward the sprocket. The chain will then be very loose, and can be taken off easily.

Wrap the sharpened or new chain around the bar and the sprocket, making sure that the cutters on top of the bar have their sharp edges away from the saw body.

Slide the bar away from the sprocket, and fit it onto the adjusting pin. It is often necessary to move the pin by turning the screw, before it will engage. Put back the guard and the housing, being careful that the bar does not come off the adjuster while you do it. Turn down the clamp nuts with your fingers only, or very lightly with a wrench.

Adjust chain tension as described above, being sure to snap it to relieve kinks. Tighten the clamp nuts firmly. See Fig. 21.55.

A new chain should have extra lubrication, usually from an oil can, before it is run on the saw. The first half hour of work should be light cuts, while the chain and bar are wearing in. Tension is adjusted as required.

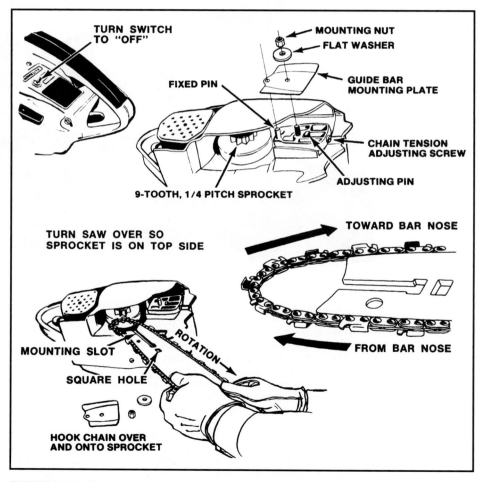

FIGURE 21.55   Chain installation. (*Courtesy of Homelite.*)

Engine Lubrication.   The standard chain saw requires regular lubrication only for the engine and the chain.

Engine oil is mixed with the gasoline. Manufacturers recommend special two-cycle (SAE 30) motor oil in the ratio of 1 part oil to 16 parts of regular gasoline. This is $\frac{1}{2}$ pint of oil per U.S. gallon, or a proportion of 6 percent oil.

If no two-cycle oil is available, regular good-quality motor oil, SAE 30, may be used in the 16-to-1 ratio. However, ordinary oils are not recommended by saw manufacturers, and detergents and other additives in diesel and multigrade lube oils tend to foul up a two-cycle engine.

Chain Oiling.   The chain is lubricated from a reservoir, which should be filled each time the machine is refueled. There are special bar and chain oils which are formulated to flow even in cold weather, and to cling to the chain.

If these are not available, any clean motor oil of SAE 30 weight may be used. In cold weather, weight 20 or 10 may be better, or 30 cut with 1 part of kerosene or fuel oil to 4 parts of oil. Multigrade oil may be used.

In most saws the oil is transferred to the chain by a tiny pump. This may be engine-driven if lubrication is automatic, or be operated by a plunger that is pressed with the right thumb at intervals, if manual.

There should be sufficient oil fed to the chain to keep it damp around the connecting links, and to allow a very fine spray to be thrown off the front at full speed. Such throwoff may be much less when special chain oil is used.

More oil is needed in heavy cutting than in light work. The hand pump allows the operator to increase the amount, but it also makes it possible to underlubricate, or even not lubricate.

A few machines have both automatic chain oiling and a hand pump. With this arrangement, the operator has the option of pumping more oil when he or she thinks it desirable, without having full responsibility.

With manual oiling, a stroke may be given at intervals that vary between 10 seconds and 1 minute of actual cutting time. If a little oil is left each time the fuel tank goes dry, the chain is probably being properly lubricated.

Running without oil, or with too little, will wear out both the chain and the bar groove, and may cause the chain to break.

Indications of a dry chain are a dry appearance, too-quick stopping when the throttle is closed, and sometimes thin smoke, tightening, and engine stalling.

The bearing in the clutch hub usually has enough lubricant to last between overhauls. A nose sprocket or wheel should be lubricated with a special miniature grease gun every time the fuel tank is filled, or at least once a day. This should be done when it is warm, so that dirt will come out with the old grease. Lubrication at the end of the day fills the chamber and prevents condensation of moisture in it.

Controls.   The saw is usually equipped with a throttle trigger, a latch (for starting) that holds the throttle partly open, perhaps a chain oil pump handle, an ignition switch, and a choke. Some models have two throttle triggers for convenience in changing hand holds on the throttle bar.

There may be a lock to prevent accidental opening of the throttle. If located on the top of the throttle bar, it is kept disengaged by pressure of the palm of the hand while gripping the bar.

CHAIN SAW OPERATION

The need for chain saw work is also discussed in Chap. 1.

Starting.   Since tank capacity is very small, in the interest of light weight, it is customary to fill with oil-gas for the engine and oil for the chain before each use.

Even if reservoirs were left filled after the last saw work, they should be checked, in case of leakage.

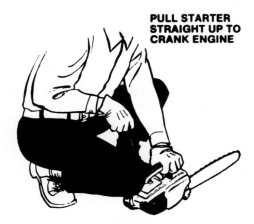

**PULL STARTER
STRAIGHT UP TO
CRANK ENGINE**

CHAIN MUST BE IN THE CLEAR

FIGURE 21.56   Starting a chain saw.

Usual procedure is to flip on the ignition, latch the throttle in starting position, and pull out the choke. Details vary among different makes and models. See Fig. 21.56.

The starter is a rewind pull rope. You hold the saw down firmly with one hand, and brace it with a foot or knee; grasp the starter handle in the other hand, pull a few inches until you feel the starter engage, then pull briskly to give the engine a fast spin. Do not pull all the way against the end stop, as this may damage the starter or the rope. Keep hold of the handle, relaxing enough to let the starter springs pull the handle back down. Then pull again.

Repeat the pull several times, until the engine fires. It may simply kick without taking hold. This is usually a signal that it has had enough choke. Push the choke button in (or half in). The engine should start on the next pull. If not, rechoke and pull some more.

It may or may not need half choke for a few seconds to keep running. Let it run on the latched throttle briefly, then release the latch by squeezing the throttle, so the engine can idle. It may or may not need more warm-up time before working.

Cutting.   Hold the saw with the left hand by the cross bar (handlebar), and with the right hand by the throttle or sidebar. The throttle, and the chain oil pump if there is one, should be within reach of your right fingers.

To cut a log lying on the ground, the bar is lowered onto the cutting line, at right angles (or any desired angle) to the log, with the bumper just touching its near side. The throttle is opened just before the cutters reach the log.

Enough down-pressure should be applied to cause a moderate to rapid cut, but not enough to make the engine slow down or the clutch slip. A substantial stream of coarse sawdust or chips should spray out of the cut.

Moderate down-pressure can be exerted and satisfactory cuts made by pressing straight down. However, it it much more efficient and more usual to pivot the saw in the cut, holding it with the cross bar, and raising the rear of the throttle bar so that it pivots on the bumper, forcing the bar down by leverage.

The bottom of the cut may be kept horizontal, or shaped as desired, by making a moderate rocker cut, then drawing the saw back to slice off the ridge left on the near side, then pivoting it in again, as in Fig. 21.57. Cutting can then be resumed on the far side.

Varying the angle of approach shortens the line along which cutting is taking place, and makes it easier for the cutters to bite in. It is helpful under almost any conditions, and is necessary with big logs or dull chains. But a dull chain should be sharpened or changed.

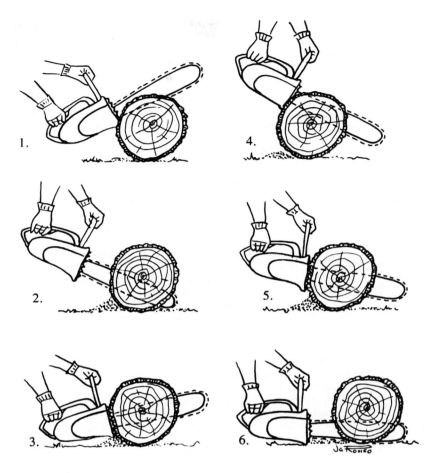

FIGURE 21.57   Working through a thick log.

Pressure required, and cutting rate obtained, vary widely with conditions. The most important factor is chain sharpness. A very dull chain might not cut through a log, but still slice a 2-inch stick easily. A moderately dull chain may cut a stick rapidly but a log slowly or not at all. Fine sawdust shows the chain is too dull for its job, or, more rarely, that its depth gauges need filing.

Procedures in cutting a standing tree were discussed in Chap. 1.

Hardwood such as oak requires a sharper saw and cuts more slowly than a softwood. Cutting with the grain (rip sawing) is slower and more difficult than across the grain, and requires more than normal lubrication. Low cuts in stumps are slowed by very dense wood and cross-graining.

Direct-drive saws, which are the most common type, have very high-speed chains and require (and will accept) very little pressure. The old gear-drive saws usually needed heavy pressure for heavy cuts.

Ground Damage.   If a log that is being cut is on the ground, or close to it, it is difficult but important to prevent the chain from touching the ground. Even fine, soft dirt causes rapid wear, multiplied by the fact that it is drawn into the cut in the wood. A stone can dull and damage, and perhaps ruin, a chain in a fraction of a second.

If the cut must be completed in place, it is done by using only the nose (boring), pointing it downward, using light pressure, and being alert to pull it back as it goes through.

If a number of cuts are being made in one trunk, it may be necessary to complete only one of them. The cutoff log can then be rolled over, with levers if necessary, to put the uncut sections on top of the rest of the cuts. They can then be cut easily and safely.

Stresses.   Cut or fallen trees and their branches are usually under bending stress of some kind. An important exception is straight trunks lying on smooth ground. When a stressed piece of wood, a long log for example, is being cut, the two pieces will tend to move at the point of weakness being developed. This movement may tend to close the cut, to widen it, or to twist the two pieces around it. See Fig. 21.58.

If the log is supported at the ends, it will tend to sag in the middle and close a cut made in the top, and open one in the bottom. If supports are on one side of the cut, it will open if made from the top or close if it is made from the bottom. Twist, which is usually found in trunks with branches attached, or in branches themselves, may close a cut made from any direction.

A saw will be caught and jammed if left in a cut as it closes. It may be freed by jacking or prying up the log, driving in a wedge, or cutting it out with another tool. If the pressure against it is very heavy, the chain may be damaged.

A cut that is opening will not jam the saw, and the tension makes the wood easier to cut. But the weight of the free section may split the other part, and its direction of fall could be hazardous.

When both top and bottom are accessible, it is good practice to cut the log one-fourth to one-third through on the side that is expected to close, then complete the cut from the other side. This avoids working in wood that is under heavy stress, and prevents splitting.

When twisting is present, cuts should be made first at spots where the direction of stress is simple, and preferably at the thinnest accessible part. Two parallel cuts, 1 inch or 3 inches apart, may be worked in alternately, to reduce pressure on one spot.

Undercutting.   Undercutting means using the upper side of the bar for cutting upward. It is standard procedure where a top cut would bind, and there is space to get the bar underneath the log.

Its principal drawback is that the resistance of the wood to the saw tends to force it against the operator, who must brace himself or herself to withstand a moderate force. Any relaxation lets the chain push itself and the saw out of the cut. This usually results in a kickback, which will be discussed below.

A secondary problem is that it is difficult to use a pivoting movement for pressure. With gravity against the saw too, it takes more effort (usually) to cut up than to cut down.

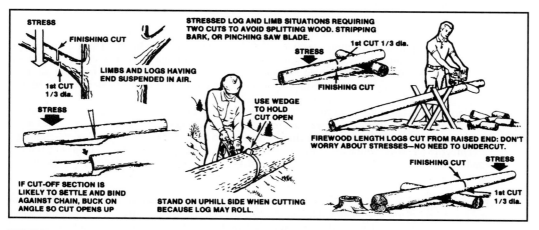

FIGURE 21.58   Stress in wood must be considered when cutting it. (*Courtesy of Homelite.*)

Kickback.   In normal cutting position the resistance of wood to the cutters simply pulls the saw bumper more tightly against the log. In undercutting it is a straight horizontal force that can be resisted by straight pressure.

But when the force is at the tip or nose, or the upper side, leverage is completely against the operator. If the nose is not held by an obstruction, it will flip upward at nearly the speed of the chain, about 60 feet per second, with enough force to swing it through a considerable and possibly dangerous arc. This may be a truly shocking experience when it is unexpected.

Figure 21.59 shows three conditions under which kickback often occurs. The angle of approach made the difference between a successful cut and a shock. In the other two examples, the nose extended farther than expected, and touched an unnoticed obstacle.

There is a special chain construction (guard link) which is supposed to reduce kicking. A safety guard, as shown in Fig. 21.52, can be used to protect against kickback. But avoiding kickback is chiefly a matter of the operator's being vigilant and using good judgment.

BRUSH BURNER

Under many circumstances, burning is still the preferred and often the only practical way to dispose of cut or uprooted vegetation.

Trained crews with adequate equipment can usually burn clearing debris under almost any conditions, wet or dry. But the time (including lost time) and labor involved in getting new fires started under difficulties may be excessive.

Substantial help may be obtained from brush burners, such as that in Fig. 1.32. These trailer-mounted devices have a small (3- to 12-horsepower) engine, an airplane-type propeller, and a fuel oil sprayer.

The air blast, delivered off the back of the trailer, may be enough to invigorate a fire, and to push it into the material to be burned, which may be brush, logs, or stumps. With help of a mist spray of fuel oil, injected into the airstream as needed, it also provides whatever additional heat is required.

However, the basic principles of properly starting and maintaining fires, as given in Chap. 1, should be understood if this type of machine is to do its best work.

WATER PUMPS

The pumps most used by contractors are of the centrifugal and diaphragm types.

Centrifugal.   The centrifugal pump, Fig. 21.60, operates by throwing water outward from a center by means of rapidly revolving vanes. The vacuum thus created at the center is constantly filled

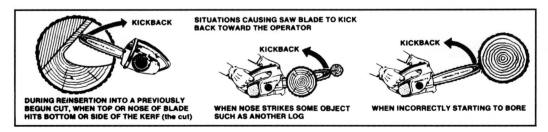

FIGURE 21.59   Kickback situations. (*Courtesy of Homelite.*)

FIGURE 21.60   Centrifugal pump supported by safety raft.

by water "sucked" (forced by atmospheric pressure) into it through the inlet passage. The velocity of the water thrown off the tips of the vanes creates pressure which drives the water out of the pump through the outlet passage.

In the construction of a centrifugal pump, as shown in Fig. 21.61, the water enters the system through a strainer at the bottom of a hose fastened to the inlet. A foot valve may be installed above the strainer to prevent the hose from draining when the pump is shut down.

The inlet hose is made of heavy plies of stiffened fabric and rubber which will not flatten under atmospheric pressure. It is usually made in 10- to 20-foot sections. Each section ends in a metal coupling. Four-inch-diameter and smaller lines usually have pipe thread couplings, while larger sizes have bolted flanges.

The upper end of the hose may be led directly into the pump body or through an elbow, usually with an angle of 30 to 45°, which can be set in different positions by unbolting and rotating.

The hose-to-pump connections each include a small threaded and plugged hole. A vacuum gauge should be put in the inlet. This can be loosened or removed to admit air in order to drain a hose not equipped with a good foot valve, or to allow the pump to run without pulling water. The outlet fitting will take a pressure gauge.

The connector is fastened to the pump with a gasket, and hose flanges are fastened to each other in the same manner. Any threaded connections are smeared with sealing compound before connecting. Even very small leaks on the inlet side will substantially reduce the amount of water handled by the pump, and may stop it from working at all.

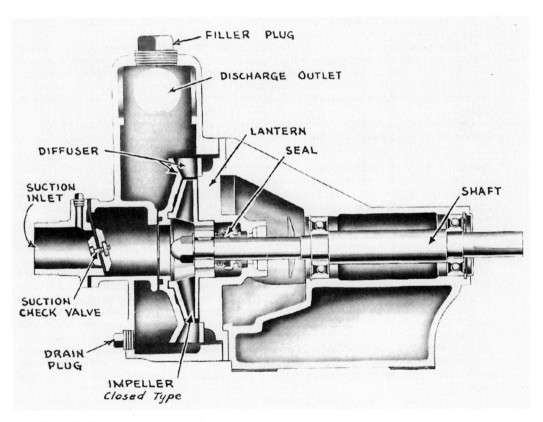

FIGURE 21.61   Cross section of centrifugal pump.

The pump body is a rigid casing which serves as a support for the pumping mechanism and as a tank to supply priming water.

The inlet check valve serves to prevent water in the tank from flowing back down in the inlet hose when the pump is stopped, will stop the outlet hose from draining if its end is underwater, and prevents siphoning of water out of a higher discharge point.

The diffuser or liner forms the inner shell of the pump. It provides an inlet passage by which water reaches the center of the impeller vanes, a plate which limits leakage of water or air past the back of the vanes, and peeler passages which take the water from the tips of the vanes and convert its velocity into pressure. It is the most widely used device to make a centrifugal pump self-priming.

The impeller consists of vanes curved back from the direction of rotation so as to minimize shock and turbulence where they hit the water at the center, and so as to give maximum outward velocity at the tips. They add a substantial wedging action to the centrifugal force.

Three types of impeller are in general use—the open, semiopen, and closed. Efficiency, maximum pressure, and ability to handle air are ordinarily greatest with the closed types, while the open ones are best able to pass solids.

The impeller is backed by the lantern which closes the front of the casing, carries the driveshaft bearing a seal, and, with open impellers, prevents leakage past the front of the vanes.

In some makes, the diffuser and lantern are equipped with renewable wear plates facing the impeller.

The discharge or outlet tee is bolted to the top of the body and can be turned. The filler plug is removed in order to pour in water for priming a standard pump.

The outlet hose is of lighter construction than the inlet and will flatten out when empty. Couplings are identical.

The driveshaft extends forward from the impeller which is splined or keyed to it, and through the lantern to a direct connection with the engine driveshaft. Portable pumps usually have combustion engines, while fixed ones have electric power; but there are many exceptions to this generality. Electricity is more satisfactory where power is available because of ease of arranging for automatic operation and avoidance of the nuisance of fueling, lubricating, and adjusting the engine.

The driveshaft is usually on antifriction bearings, although bushings may be used. The rear bearing may be lubricated by the water in the pump, by an outside grease fitting, or may be sealed.

Priming.   A self-priming centrifugal pump must be filled with water (primed) in order to function. The tank should contain enough water to cover the impeller. Water acts as a sealer of the clearance spaces next to the impeller, mixes with air to give it enough weight to be affected by centrifugal force, and serves as a check valve to prevent air on the discharge side of the pump from working back to the inlet. In most machines it provides lubrication for the shaft bushings and seals.

Contractors' centrifugal pumps are of the so-called self-priming type. The pump will not prime itself—that is, draw the water into itself—when dry, but once it is filled with water, it will remain full and prime and reprime itself through any number of starts and stops. And as long as water is in the tank, and the pump is turning, it tends to move air and water from the inlet to the discharge.

If the inlet hose fills with air instead of water, the pump is said to have lost its prime.

There are two general methods of circulating the priming water so as to pump air. The diffuser method depends on gravity forcing the tank water between the outer or discharge tips of the vanes, without its being able to reach the center because of their rotation. The water at the outside and the air at the inside mix where they make contact, and air bubbles are thrown from the impeller tips into the diffuser passages, from which they rise to the surface of the tank and escape through the discharge pipe, while the water continues to press against the vanes. This process is illustrated in Fig. 21.62.

As air is removed in this manner, a vacuum is created in the inlet pipe that causes water to rise in it until it enters the pump. The priming process is then complete. Water is forced through the pump hundreds of times faster than air, and vacuum increases sharply.

The recirculation method is to open a passage by which water can flow from the tank, which is on the outlet side, into the inlet passage. This mixes with air as it is caught by the impeller. On the discharge side the air bubbles to the surface, and the water returns to the inlet. Since the movement of air is quite slow, a very small leak may delay or prevent building up sufficient vacuum.

The height of the pump above water is an important consideration, as higher lifts require higher vacuums. A new pump may be able to pull water up 28 feet (although volume handled drops rapidly with increase in lift), but a badly worn unit may have difficulty raising it 5 feet. Another factor is that under high vacuum, warm or hot water may evaporate so rapidly that the pump will be kept busy pumping water vapor.

Even a badly worn pump can draw water to a considerable height, once the air is exhausted. In emergencies, defective pumps may be primed by forcing water into the inlet hose from another pump, or siphoning water from a higher point through the discharge line, with the pump stopped. When the inlet is filled, the pump is started, and the direction of flow reversed.

FIGURE 21.62   Diffuser priming action.

In general, it is cheaper to overhaul a pump when it first shows signs of weakness, than to delay or lose work through its inefficiency.

Portables.   In addition to the heavy wheel-mounted pumps, there are light hand-carry models (45 to 105 pounds) in 1½-, 2-, and 3-inch sizes, that are made of aluminum and magnesium alloys, and driven by high-speed two-cycle engines. One is shown in Fig. 21.63. Electric drive is also available.

Their output is standard for their hose size, and their extreme portability makes them invaluable where access is difficult, and when jobs are small and scattered. They are handy where frequent changes of position must be made to follow a receding water level, and for keeping down inflow during work.

Several may be used in one excavation if there is too much water for one. They are ordinarily used on jobs needing intermittent rather than steady round-the-clock use, but contractors often keep them running for days at a stretch. An auxiliary gas supply can be rigged up to feed automatically into the pump tank, by a float valve in the tank, and a line to a drum or other container.

As with all two-cycle gasoline engines, it is absolutely essential that lubricating oil be mixed with the gasoline.

Jetting Pump.   Jetting pumps usually are centrifugal pumps of heavy construction, so designed that they can deliver water at very high pressure but which have comparatively weak suction.

They may be of the self-priming construction described or may use an auxiliary vacuum pump for priming.

Their special uses include sinking foundation piles by water pressure, hydraulic excavating and flushing, making test holes, fire fighting, and dewatering where the discharge point is too high for the ordinary pump.

Well Point Pumps.   These are centrifugal pumps which are able to handle water with a large proportion of air because of special internal design, or by means of auxiliary vacuum pumps.

FIGURE 21.63   Hand-carry pump. (*Courtesy of Homelite.*)

They are used for removing underground water from the soil in excavation areas, and were discussed in Chap. 5.

One pump will be supplied from numerous points or inlets. These consist of pipes with gauze or slotted screens of fine mesh which are driven or jetted into the ground. The well points are connected with the surface by pipes which are connected to a main or header pipe to the pump through hose or pipes.

Diaphragm Pump.   A diaphragm pump consists chiefly of a movable diaphragm in a closed chamber between two check valves, as seen in Fig. 21.64. The diaphragm moves vertically. On the upstroke it causes a suction which makes the inlet ball rise off its seat and pulls the contents of the inlet pipe. On the downstroke, the inlet valve is seated and the outlet valve forced open, discharging part of the contents of the chamber into the outlet pipe.

The action of this pump is positive, so that it will handle any material which will flow through the pipe. It is not subject to air lock, but because of its elasticity air moves through it slowly, particularly on high lifts.

It can be slowed or put out of action by trash lodging in the valves. For this reason, the valve chambers are made quickly accessible for easy removal of foreign matter.

The maximum lift is somewhat less than that of the centrifugal pump, because some pressure in the intake line is required to open the inlet valve, and power is not provided for high outlet pres-

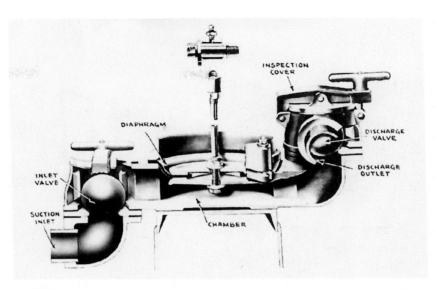

FIGURE 21.64   Diaphragm pump cross section.

sure. On high lifts it is good practice to set the pump about halfway between the inlet and discharge levels so as to equalize the resistance to the up- and downstrokes. See Fig. 21.65.

SUBMERSIBLES

Problems of priming, of diminished output with height above water, and of impossibility of pulling water up more than 25 or 28 feet may all be eliminated by placing the pump under the water, and providing power from above through a shaft, hydraulic lines, or electric wires.

Submersible pumps can be left unattended for hours, quietly working "overtime." They can operate totally or partially submerged, so there is no need to move the pump as the water level changes. The pumps are designed to run dry for reasonable lengths of time without seal damage. Submersible pumps have no suction or priming problems. They have no suction hose. Just submerge the pump, and it is ready to go to work, providing higher heads for greater lifts than above-ground pumps of comparable size.

The pumps range in size from $1\frac{1}{2}$ to 12 inches, with capacities up to 7,400 gallons per minute (gpm) and head to 600 feet. The discharge of one pump can be connected through hose and couplings to the suction of another for tandem operation, which effectively doubles the head at a given flow.

Electrics.   Small, submersible electric-powered pumps are commonly used for household water supply, where there is electricity readily available. For construction or mining sites, the electricity may have to be generated on the site. In either case, electric submersible pumps are very effective where large solids are not a problem.

A wide, solid base, as shown in Fig. 21.66, helps prevent the pump from turning into the ground or pumping its way into a hole. They are made in sizes from 2- to 12-inch discharge lines, with the largest having a capacity to 7,400 gallons per minute and a total head of 135 feet. An 8-inch model can handle a 600-foot total head. Other models and sizes have maximum total heads of 100 to 300 feet and maximum capacities of several thousands of gallons per minute.

There are also slimline submersible pumps for use in drilled wells, narrow cofferdams, and hard-to-reach places. They are more lightweight and easier to handle than the wide-base variety.

FIGURE 21.65 Diaphragm pump ready for dewatering. (*Courtesy of Gorman-Rupp Company.*)

The slimline design for a 2-inch pump, for instance, may be only $7\frac{1}{2}$ inches at the widest. These pumps are provided in 2- to 6-inch sizes, with capacities of several hundred to several thousand gallons per minute and total heads up to 300 feet.

Hydraulics. Hydraulic submersible pumps are designed for solids-handling construction dewatering applications and for pumping sludges and slurries, such as from mining operations. These pumps are powered by a gas or diesel engine-driven hydraulic power unit which provides up to 2,900 pounds per square inch. Pump speed can be varied to suit job requirements, thereby allowing maximum efficiency and minimum fuel consumption. This makes them more economical to operate than generator-driven electric submersible pumps, which must operate at constant speed.

The hydraulic submersible pumps are available in 3-, 4-, and 6-inch discharge sizes with capacities up to 1,600 gallons per minute and heads of up to 175 feet. They will handle up to 5-inch-diameter solids depending on the model. The 4- and 6-inch sizes are driven by a standard 63-horsepower diesel engine.

PUMP OPERATION

If the water is small in volume or contains a heavy load of mud or other solids, a diaphragm pump is preferred. For larger quantities of water a centrifugal pump is needed. Both light hand-carry and heavier wheel models of centrifugal pumps are available.

The most satisfactory results are obtained when the capacity of the pump or pumps is substantially greater than the inflow, particularly when there is a large volume of standing water.

Setting Up. The pump should be level and placed as near the water level as possible, as centrifugal pumps can push more strongly than they can pull. The table of output, Fig. 21.67, shows

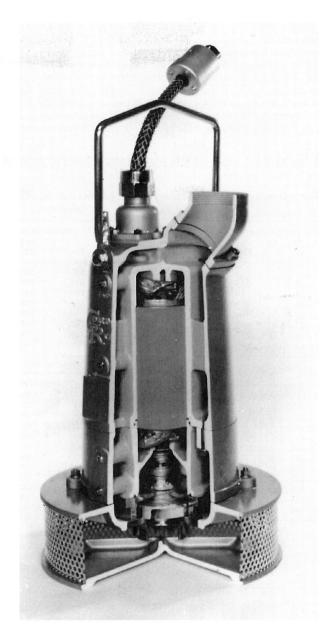

FIGURE 21.66   Submersible pump, cutaway. (*Courtesy of Gorman-Rupp Company.*)

the loss in volume which occurs as a pump is raised above the water. The capacity of a pump is greatest when total lift is low.

The pump should be supported on a platform, or on boards as in Fig. 21.60. Otherwise vibration, and softening or washing of the ground, may cause it to settle off level, or even to fall into the pit.

A strainer should always be used when there is a possibility of sucking up stones or other objects which might damage or clog the line; and it is a good practice to use one whenever possible.

Use of a foot valve is optional. If the hose diameter is small, the lift low, the pumping steady, and the inlet opening unlikely to be exposed, it is not needed. On high lifts, particularly where the pump is worn and the inlet valve defective, a foot valve is very desirable.

STANDARD TABLES FOR SELF-PRIMING CENTRIFUGAL PUMPS

The following tables give capacities in gallons per minute

Model 5-M, 1-1/2 inch

Total Head Including Friction	Height of Pump Above Water In Feet			
	10	15	20	25
15 feet	85	-	-	-
20	84	68	-	-
25	82	67	-	-
30	79	66	49	35
40	71	60	46	33
50	59	52	41	28
60	42	40	32	22
70	22	22	20	12

Model 10-M, 2 inch

Total Head Including Friction	Height of Pump Above Water In Feet			
	10	15	20	25
25 feet	166	-	-	-
30	165	140	110	-
40	158	140	110	75
50	145	130	106	70
60	126	117	97	68
70	102	100	85	60
80	74	74	68	48
90	40	40	40	32

Model 17-M, 3 inch

Total Head Including Friction	Height of Pump Above Water In Feet			
	10	15	20	25
25 feet	284	-	-	-
30	278	245	203	158
40	260	239	198	155
50	236	224	187	150
60	204	200	172	138
70	164	164	149	120
80	122	122	118	98
90	77	77	77	70
95	54	54	54	33

Model 20-M, 3 inch

Total Head Including Friction	Height of Pump Above Water In Feet			
	10	15	20	25
30 feet	333	280	235	165
40	315	270	230	162
50	290	255	220	154
60	255	235	205	143
70	212	209	184	130
80	165	165	157	114
90	116	116	116	94
100	60	60	60	60

Model 30-M, 4 inch

Total Head Including Friction	Height of Pump Above Water In Feet			
	10	15	20	25
30 feet	500	435	350	250
40	495	430	345	250
50	475	415	340	245
60	450	400	325	240
70	415	370	300	230
80	355	325	270	210
90	250	240	215	175
100	100	100	100	100
105	20	20	20	20

Model 40-M, 4 inch

Total Head Including Friction	Height of Pump Above Water In Feet			
	10	15	20	25
25 feet	665	-	-	-
30	660	575	475	355
40	645	565	465	350
50	620	545	455	345
60	585	510	435	335
70	535	475	410	315
80	465	410	365	280
90	375	325	300	220
100	250	215	195	145
110	65	60	50	40

FIGURE 21.67 Pump capacities. (*Courtesy of Contractors' Pump Bureau.*)

Model 90-M,   6 inch

Total Head Including Friction	Height of Pump Above Water In Feet			
	10	15	20	25
25 feet	1500	-	-	-
30	1480	1280	1050	790
40	1430	1230	1020	780
50	1350	1160	970	735
60	1225	1050	900	690
70	1050	900	775	610
80	800	680	600	490
90	450	400	365	300
100	100	100	100	100

Model 125-M,   8 inch

Total Head Including Friction	Height of Pump Above Water In Feet			
	10	15	20	25
25 feet	2100	1850	1570	-
30	2060	1820	1560	1200
40	1960	1740	1520	1170
50	1800	1620	1450	1140
60	1640	1500	1360	1090
70	1460	1340	1250	1015
80	1250	1170	1110	950
90	1020	980	940	840
100	800	760	710	680
110	570	540	500	470
120	275	245	240	240

Model 200-M,   10 inch

Total Head Including Friction	Height of Pump Above Water In Feet			
	10	15	20	25
20 feet	3350	3000	-	-
30	3000	2800	2500	1550
40	2500	2500	2250	1500
50	2000	2000	2000	1350
60	1300	1300	1300	1150
70	500	500	500	500

FIGURE 21.67   (*Continued.*)

Foot valves are not particularly dependable, being more subject to jamming by trash or mud than parts of the pump proper. In addition, the end of the inlet hose is frequently horizontal or nearly so, and some valves do not function well in this position.

Inlet Protection.   The inlet should be well below the surface of the water—about six times the inside diameter of the hose when possible. This is frequently not practical, particularly when the place must be pumped dry.

At lesser depths a whirlpool may form over the inlet, and air enter through its center in sufficient quantity to form an air lock and cause the pump to lose its prime. Such a vortex will not form if the end of the hose is vertical, and is most likely to occur if it is horizontal.

The whirlpool may be prevented by digging a sump pit to lower the inlet; by floating a square or round piece of wood two or more feet in diameter over the inlet, or by bolting a roof over the strainer.

The inlet should not be allowed to rest in soft mud. It is liable to pick up abrasive material which will cause rapid wear of the pump; and may sink in sufficiently to reduce the intake or choke it off entirely. It is particularly apt to be choked when the pump is not running, and mud can slump into the hollow it has dug by sucking in soil with the water.

This can be prevented by resting the strainer in a wood box, or on a wide board. A light metal plate may be fastened permanently to the underside of the strainer.

When leaves, grass, or other vegetable debris is in the water, a strainer may become plugged every few seconds. The best cure for this is to make an additional strainer of wire of sufficiently fine mesh to catch the bulk of the trash, with a large enough area that the water will move through it at too low a velocity to hold the trash against it.

Even if too small for self-cleaning, such an additional screen will clog less often, and be more easily cleaned, than the standard strainer. Two constructions were illustrated in Chap. 5.

A riskier but less laborious method is to remove the hose strainer, and allow the trash to go through the pump.

Air in Inlet.   A contractor's centrifugal pump is built to handle solid water, and pumps air with difficulty. When the pump and inlet hose are empty, the pump must be filled with water (primed) before starting, and will then work for a while, slowly drawing the air out of the inlet pipe. When the water is sucked into the pump, its efficiency rises abruptly. If air is permitted to enter the intake, the pump will lose its grip, and have to again slowly work up enough vacuum to lift the water into it. This process is fairly quick in new pumps, but in worn ones is often a slow process if the lift is at all high.

Very often, failure of a pump to develop sufficient vacuum to prime itself is due to air leaks in the inlet line. These are most often past the gaskets at hose couplings, but may be at any spot on the inlet side. This includes the hoses themselves, the gasket between the inlet elbow and the body, cracks in the inlet or body, threaded fittings, loose or worn shaft packings, or lack of grease in a shaft bushing.

Such leaks are hard to detect, as, if the pump is worn or the lift is high, a very small leak will do the damage. Wet clay should be smeared over all suspicious spots on the hose, couplings, gaskets, or casing parts. This makes a good temporary seal for small leaks and will show up larger ones by disappearing into them. Leaks along the shaft will generally drip when the pump is stopped.

Gauges.   The pump casing should be checked for cracks. Gaskets and connector threads should be smeared with pipe dope or Permatex #2. Installation of a vacuum gauge on the inlet side is an excellent idea, as it will immediately indicate the nature of the trouble. Low vacuum indicates leakage, a worn pump, or lack of prime water, and high vacuum a plugged inlet—usually the strainer, but sometimes a stuck valve. A fairly normal vacuum, with little or no production, indicates back-pressure from too high an outlet or a plugged line.

A vacuum gauge may be calibrated in pounds or inches. If pounds, it measures the difference between air pressure inside and outside of the inlet chamber in pounds per square inch. If inches, it indicates the height of a column of mercury which could be supported by this difference in pressure.

Figure 21.68 contains data for conversion of inches of vacuum into the number of feet that water should rise in the inlet hose, and for some other calculations.

If the hose is full, vacuum will rise very rapidly to full pumping level.

Outlet or Discharge Line.   If the outlet hose is not long enough to reach a disposal point, a level or downhill extension may be made by means of a wood flume, a ditch, or parallel dikes.

If the discharge hose end is underwater, the pump is stopped, and the inlet and foot valves leak or are lacking, water will siphon from the higher level through the pump and out the inlet. On a high lift, the force may be sufficient to turn the pump and engine backward. Otherwise the flow is leakage past the vanes.

TABLE CONVERTING INCHES VACUUM INTO FEET SUCTION

To convert inches vacuum into feet, multiply by 1.13.

Inch Vac.	Feet	Inch Vac.	Feet	Inch Vac.	Feet
1	1.13	11	12.47	21	23.81
2	2.27	12	13.61	22	24.95
3	3.41	13	14.74	23	26.08
4	4.54	14	15.88	24	27.22
5	5.67	15	17.01	25	28.35
6	6.80	16	18.14	26	29.48
7	7.94	17	19.28	27	30.62
8	9.07	18	20.41	28	31.75
9	10.21	19	21.55	29	32.89
10	11.34	20	22.68		

CONVERSION FACTORS FOR PRESSURE AND VACUUM

Feet head (water) $\times$ .433 = pounds pressure

Pounds pressure (water) $\times$ 2.31 = feet head

Feet head (brine, sp. gr. = 1.2) $\times$ .52 = pounds pressure

Pounds pressure (brine) $\times$ 1.92 = feet head

Feet head (gasoline, sp. gr. = .75) $\times$ .325 = pounds pressure

Pounds pressure (gasoline) $\times$ 3.08 = feet head

Inches of mercury $\times$ 1.132 = feet head of water

Feet suction lift of water $\times$ .882 = inches of mercury

FIGURE 21.68 Conversion tables. (*Courtesy of Contractors' Pump Bureau.*)

If the discharge point is high, 30 or more feet above the pump, it may be advisable to install a check valve just above the pump, to relieve the pump of destructive water hammer when shut down.

If the outlet hose is long or crooked, pump capacity may be increased by using oversize hose. This reduces the velocity and friction of the water. Sharp turns or abrupt changes in size should be avoided through use of taper fittings and 30 or 45° elbows.

It is frequently desirable to reduce the output of the pump, as when the water level has been lowered and pumping need only remove an inflow which is less than the pump capacity. Gasoline power permits throttling the engine down. If this does not afford sufficient reduction, and it is not convenient to stop and start the pump frequently, a gate valve may be put on the outlet connection, and output regulated by opening or closing it. This does not add materially to the strain on the pump, whereas throttling the intake line does.

Freezing.   Special precautions are needed to operate a pump in freezing weather. Whenever the pump is to be shut down, the body and hoses should be drained.

The drain plug at the bottom of the case, and the filler plug, should be removed. If the end of the discharge hose is above the pump and underwater, it should be lifted out to avoid siphoning water down to the pump.

The discharge hose can be emptied of water by holding it so that there is a continuous slope to the ends. If this is not practical, water can be worked out of it by raising it at the high end, and walking toward the other end, holding it up in a loop high enough to force the water over other high points. If the line is metal, it should be opened at its low points, and blown out if air is available.

If the inlet hose is light, it can be disconnected, pulled out of the water, and drained. If it is heavy, and has no foot valve, it may drain itself if the pump check valve is not airtight. If it does not drain, an opening should be made in the metal at its upper end. A petcock may be provided, or a vacuum gauge removed, or a new hole cut and threaded. The air admitted in this way will allow the water inside the hose to drop to the surrounding water level.

This opening must be sealed when the pump is used again.

If freezing is apt to be severe enough to form a plug of ice at water level in the hose, the opening at the top should be made large enough to permit pouring sufficient alcohol or fuel oil into the hose to protect it. If the hose is deep in the water, one dose should give protection for a long time if the pump is not used. Should the opening in the hose be close to the surface, alcohol may have to be renewed every few days.

If a foot valve is required for the pumping, it will be necessary to lift the lower end of the hose out of the water to open the valve to drain it unless a "tickler" is installed. Any homemade device which will unseat the flap in the foot valve by pulling on a line or a rod will be adequate.

If the pump body does not drain completely, it may be necessary to tip it to get all the water out, or to pour in enough alcohol to prevent the residue from freezing.

The use of salt or calcium chloride as an antifreeze is not recommended, because of probable damage to the pump.

The engine should be protected by antifreeze in the regular manner.

In freezing weather, it is advisable to start the engine before priming the pump, although it should not be allowed to run dry longer than a minute or two. If the temperature is below 0°F, it may be advisable to use hot water or a weak antifreeze mixture for priming.

If the pump is frozen, the engine will not turn over. Filling the tank with hot water or with a strong antifreeze solution should free it.

The outlet hose should be checked for ice before starting. Even if it has been drained, a small amount of water may collect in a low spot and glue the walls together. A hose blocked in any manner is liable to blow out under working pressure. Such ice can be readily broken by tapping the hose with a hammer or a block of wood.

Thawing an ice-blocked inlet hose is a major operation, and methods will depend on the facilities on hand. A gasoline-burning heater equipped with a blower and a warm air duct is most effective.

Under extreme conditions, a pump may freeze up while running. Provided the inlet is deep in the water, this may be prevented by wrapping hoses and pump body with sacking or blankets, or by applying heat (engine exhaust, bonfire, or large blowtorch) to the case. Flame should not be used if there are gasoline leaks anywhere, and caution should be used if the engine is oily or dirty.

GRIZZLIES AND SCREENS

Screens of various types are used to eliminate oversize stone from fill or crusher feeds, and to separate mixtures into uniform sizes.

In order for separation to take place, the raw material must be moved or shaken on the screen surface. Sticky or wet materials and fine openings require the maximum amount of movement.

Movement may be accomplished by gravity flow along an inclined screen, by rotating, shaking, or vibrating the screen, or less commonly, by raking devices. Some combination of gravity and other methods is usual.

Screens may be made of welded bars; sheet steel or rubber with round, square, slot, or octagonal holes; or wire cloth. Cloth is the most popular material in a wide range of applications.

A cloth may be described in terms of "space" or "mesh." The difference is illustrated in Fig. 21.69. Space is the actual dimension of the clear opening between adjacent parallel wires. Mesh is the number of openings per inch, measured from center to center of parallel wires.

A square opening, whether in cloth or plate, will allow larger material through it than will pass a round hole of the same diameter. Figure 21.70 contains a table of equivalent sizes of square and

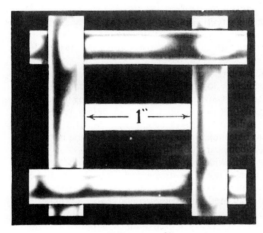

"Space"

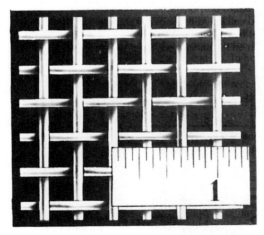

"Mesh"

DIAMETER OF ROUND HOLE	SIZE OF SQUARE OPENING	DIAMETER OF ROUND HOLE	SIZE OF SQUARE OPENING
1/8	3/32	2 1/4	1 7/8
3/16	5/32	2 3/8	2
1/4	3/16	2 1/2	2 1/8
5/16	1/4	2 3/4	2 1/4
3/8	5/16	3	2 1/2
1/2	3/8	3 1/4	2 3/4
5/8	1/2	3 1/2	3
3/4	5/8	3 3/4	3 1/8
7/8	3/4	4	3 5/16
1	7/8	4 1/4	3 1/2
1 1/4	1	4 1/2	3 3/4
1 3/8	1 1/8	4 3/4	4
1 1/2	1 1/4	5	4 1/4
1 5/8	1 3/8	5 1/4	4 1/2
1 3/4	1 1/2	5 1/2	4 3/4
1 7/8	1 5/8	5 3/4	4 7/8
2	1 3/4	6	5
—	—	7	5 3/4
—	—	8	6 3/4
—	—		

FIGURE 21.69   Wire cloth measurements. (*Courtesy of Hewitt-Robins.*)

FIGURE 21.70   Equivalent sizes, square and round.

round openings. As an example of its use, if a screen with round holes 2 inches in diameter is separating the aggregate properly, a square-hole cloth to replace it should have spacing of $1\frac{3}{4}$ inches.

Rectangular openings pass larger material than a square of their short dimension, but smaller than a square of the long dimension. Each shape rectangle will have different characteristics.

The wires in the cloth are crimped together—that is, bent under extreme pressure to a permanent set. Crimping may affect only one set of wires, or both. Various special arrangements are made to obtain extra tightness of connection, or to have the cloth smooth on one side.

Manufactured screen cloth does not need to be welded, as the crimps hold the wires in place. Homemade cloth—usually made of bars for rough grizzly work—must be welded in order to keep it in alignment.

Rectangular openings are used when a steep screen slope reduces the effective length of the squares. Nonblind cloth may have the longer wires of a smaller size than the cross ones, so that they vibrate and tend to dislodge and pass pieces sticking between them. "Blinding" is the partial or complete clogging of the screen with inert material.

Figures 21.71 and 21.72 show four of the many screen cloth constructions that are available.

Gravity.   A grizzly is generally a primary screen with fairly coarse openings that rejects boulders or large fragments from run-of-pit material.

Portable gravity grizzlies are shown in Fig. 21.73(*A*), a light unit which is placed on a truck body before loading. Gravel or soil is deposited on the top of the slope and pours and rolls down along the surface, the undersize falling through the spaces and the oversize rolling clear. After loading, the shovel picks up the grizzly and places it on the next truck. A helper is needed to steady it during placement.

When used for stockpiling, it can be bolted to the truck body, as in (*B*). It must then not be high or heavy enough to overbalance the body in dumped position.

Figure 21.73(*C*) is a similar unit, built up on a skid frame. The truck backs under it to be loaded. It often becomes partly buried, and the frame should be strong enough to withstand heavy pulls. It may be braced between the skids also.

Figure 21.74 shows a fixed installation that may feed into trucks, a conveyor, a crusher, or a small stockpile.

These units may have screen surfaces of parallel steel bars or coarse mesh, supported by heavy cross members. The bars may be made of triangular, square, diamond, or round cross section, or of inverted rails. Mesh construction works more accurately than parallel bars.

A major difficulty with this type of screen may be slightly oversize rocks sticking between the bars. This trouble is least with triangular pieces with a flat side up, and greatest with round sections. A gradual increase in slot width toward the bottom is helpful.

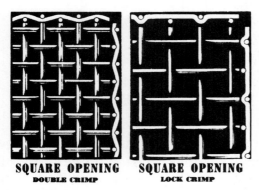

SQUARE OPENING
DOUBLE CRIMP

SQUARE OPENING
LOCK CRIMP

FIGURE 21.71   Wire cloth, square openings. (*Courtesy of Hewitt-Robins.*)

RECTANGULAR

NONBLIND

FIGURE 21.72   Wire cloth, rectangular openings. (*Courtesy of Hewitt-Robins.*)

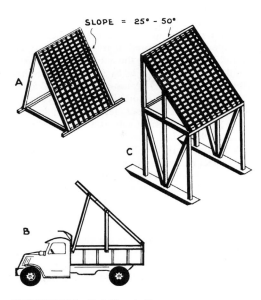

FIGURE 21.73 Portable grizzlies.

FIGURE 21.74 Fixed grizzly.

Slot-type surfaces pass oversize flat rocks readily. If the underbracing is placed at such close intervals that the slots are reduced to squares, this difficulty is decreased. However, a wider bar spacing, or a longer slide, is then required to give the same capacity.

For a given material and bar spacing the amount of material passing through the grizzly will be increased by lengthening the screen or by reducing the slope. A long screen occupies more space both vertically and horizontally. A too-flat slope will allow material to rest on the surface and clog the slots. A too-steep or too-short slope will reject undersize material.

Spacing may be from 2 inches to 2 feet, depending on the type of service.

Loose dry dirt will separate satisfactorily at a grade of 30 to 40°. Wet gravel or damp loam requires slopes up to 50°. Wet sticky materials cannot be satisfactorily processed.

Increase in width of slots will usually make it possible to reduce the slope or the length. Grizzlies used only to reject occasional large boulders, using bar spacings of 2 feet, may be flat if some means of removing the boulders is provided.

Use of these units is restricted by inaccuracy of sizing, their tendency to clog, or the space required. They are desirable in that they can be built cheaply on the job, often out of scrap material, and are easy to operate and repair.

The slope of a grizzly can be reduced and the accuracy of sizing and capacity increased by shaking or vibrating the screen. Best results are obtained by purchasing a made-up unit, such as the one shown in Fig. 21.75, in which the screen has a limited movement relative to the frame and is connected to a shaking or vibrating device. Performance of a standard grizzly may often be improved by fastening a vibrator tightly to a side or top frame.

Revolving.   Revolving screens, such as the Telsmith model shown in Fig. 21.76, are usually cylinders of wire cloth or perforated plate set at a slope of 5 to 7°. They are rotated at a speed of 15 to 20 revolutions per minute.

Material placed inside the upper end of the cylinder is rotated up the side of the screen until its weight overcomes adhesion and centrifugal force and it falls to the bottom. The upward movement is slightly toward the lower end and the fall is straight, so material tends to move from the higher to the lower end in a series of short steps.

The amount of rise is affected by the stickiness and speed of the particles. Sticky pieces on a fast-turning screen will cling to it for full revolutions, clogging or "blinding" the holes, while hard smooth particles on a slow screen will appear to move straight down the bottom. In general, speed should be adjusted to carry the load less than one-third of the way up the screen.

FIGURE 21.75   Vibrating bar grizzly. (*Courtesy of Nordberg.*)

FIGURE 21.76   Rotary screen. (*Courtesy of Telsmith, Inc.*)

Two concentric cylinders may be used. The first screens on the inner or main cylinder are coarser than the finest products required. Material passing through them is further classified by the finer-mesh outer screens which do not have to carry any rough material. This type is well adapted to sand and gravel.

Screen frames may be made up with hexagonal or other polygonal cross sections. These allow the use of flat sections of mesh or plate and agitate the load more.

Blinding.   Blinding, or clogging of screen holes, may be a serious problem. It is most severe when the material is sticky; when most of the load is a size close to that of the screen opening; when the screen has insufficient slope, or wires that are rough with rust; or when it is overloaded.

Corrective steps include shaking or vibrating the screen, drying the material, changing particle size by adjusting the crusher, heating the screen wires or the aggregate, or tapping or scraping from below.

Shaking.   Shaking screens are flat and rectangular, are suspended on loose or flexible attachments, and are shaken longitudinally by connecting rods or other eccentrics. See Fig. 21.77. Slope is usually around 16 to 18°. The throw may be as much as 1 foot, but 6 inches is more usual. The number of strokes per minute is usually the quotient of 600 divided by the length of throw in inches. This would be 100 strokes for a 6-inch throw.

Material dumped on the upper end works its way down, most of the motion being on the back- or upstroke. Small particles fall through the screen, and the rejects move off the end.

Shaking screens set up heavy vibrations. These are reduced by arranging them in oppositely moving pairs, which preferably should carry equal loads.

Vibrating.   Screens may be vibrated mechanically or electrically to agitate the material lying on them, to increase the rate of passage through the holes, to separate particles which are stuck together, and to move all the rejects off the discharge end promptly.

Mechanical vibrators include eccentric, cam, bumper, and unbalanced flywheel types. Electric vibrators are electromagnets. Motions imparted to the screen may be reciprocating or rotary. Frequency ranges from 600 to 3,600 per minute, and throw up to $\frac{1}{2}$ inch. Slopes vary from horizontal to 40°. Standard widths are 1 to 6 feet, and lengths $2\frac{1}{2}$ to 16 feet.

Vibrating screens are usually mounted on springs or cushions, or suspended from cables. Solid contact would dampen the screen and vibrate the building.

Scalping.   Scalping screens are coarse vibrating screens used instead of the smaller sizes of grizzly. Screen construction is very heavy, as they are fed pit-run material that may include big pieces. Occasionally they are protected by widely spaced grizzly bars.

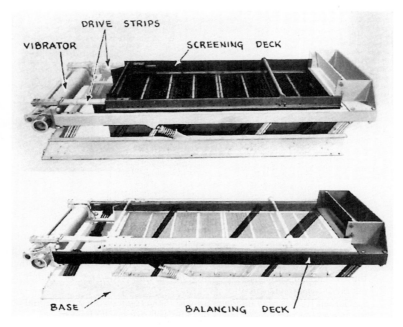

FIGURE 21.77   Counterbalanced level deck screen. (*Courtesy of Nordberg.*)

Deck.   Deck screens consist of two or more vibrating screens placed one above the other. The lowest screen may be short, and if so, is called a half deck. They are usually vibrated through the frame so that motion is the same on each deck.

The top deck has the coarsest grid. Material passed through it is further separated into sizes by each succeeding layer. The rejects move down the screen surfaces to chutes or conveyors.

The number of decks can often be changed in the field by adding or removing one or more screens.

Figure 21.77 shows a screen of the counterbalanced level type.

WASHERS

Aggregate (usable material) frequently comes to the plant mixed with clay or other impurities which have to be washed out. The amount and kind of impurity and the water supply largely determine the method used.

Two principal methods are immersion and spray. Either may be assisted by vibration, which tends to separate stuck particles.

Figures 21.78 and 21.79 show cutaway views of a rotary scrubber. A water spray screen is illustrated in Fig. 21.80.

ROCK CRUSHERS

Rock (or stone) crushers reduce rocks to smaller and more uniform sizes. They are chiefly used in connection with blasted rock, and for cobbles and boulders found in gravel deposits.

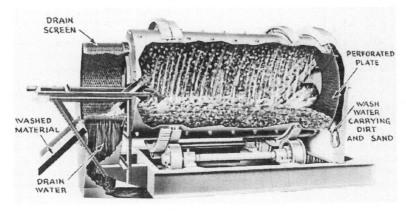

FIGURE 21.78   Rotary washer. (*Courtesy of Telsmith, Inc.*)

FIGURE 21.79   Scrubber drum.

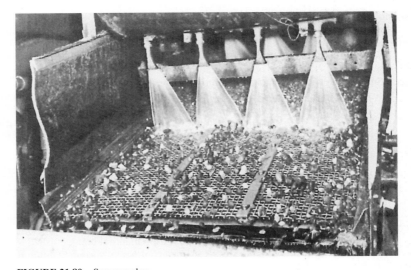

FIGURE 21.80   Spray washer.

The reduction may be accomplished by pressure, impact, shearing, or combinations. The crushers must be of heavy construction, and surfaces in contact with the stone should be renewable plates of manganese or other special alloys. Flat or corrugated surfaces are used for different conditions.

Most crushers are provided with overload reliefs, so that "tramp" iron (iron scrap) or other uncrushable material will not seriously damage the machine. These may be springs, shear pins, or easily accessible under-strength parts.

Crushers are roughly classified as primary and secondary, but these types overlap. The primary work on run-of-pit material has relatively large hopper openings and produces a coarse product. Secondary crushers will accept only small stone from the primaries, and turn out a finer and more uniform product. Both types are commonly protected against both oversize and undersize pieces by grizzlies and/or screens. Primary and secondary functions may be combined in one crusher, or in two or three crushers in series.

The ratio of reduction of a crusher is the difference in size between the largest pieces it will process readily and the product it obtains from them.

Figure 21.81 contains tables in general use in the industry, showing the capacities of shovels and certain crushers in size of pieces and in hourly production. Production figures should not be taken as a basis for estimates, as improvement in machines and differences in field conditions cause wide variation.

Jaw.   Jaw crushers are simple and economical in construction and require minimal power.

The usual construction is called swing-jaw. In this, one jaw is movable and the other is fixed in position. In twin-jaw models both of them move.

Figure 21.82 shows a cross section of a typical jaw crusher. The crushing surfaces consist of two jaws which do not quite touch at the bottom, and which are widely separated at the top. Jaw faces may be flat or convex.

One jaw is fixed, the other is attached to the pitman or eccentric arm, which is hinged on a toggle and spring fulcrum. At the top the pitman is mounted on an eccentric shaft running between two heavy balance wheels. Rotation of the shaft causes the pitman and jaw to gyrate, leaning first toward the fixed jaw, then away from it, with some vertical rubbing motion. The pitman acts as a second-class lever, with longest movement of $\frac{3}{8}$ to $\frac{1}{2}$ inch at the top, and greatest power toward the bottom.

The toggle bar runs the full length of the jaw. It is held in replaceable sockets by heavy springs. It is notched to produce a line of weakness at which it will break before other crusher parts if tramp iron (iron chunks) gets between the jaws.

Adjustment for fineness of product is made by moving the toggle forward, usually by means of spacers placed behind its frame socket.

Rocks resting in the V between the jaws are crushed by pressure, then allowed to drop as the jaw moves back. This process is repeated until the rocks are reduced to pieces small enough to pass through the narrow space at the bottom of the jaws.

Jaw crushers are made in considerable variety of construction and in sizes up to a jaw opening of $5\frac{1}{2}$ by 7 feet. They are classified according to size of opening. For example, a 10-inch by 30-inch crusher has a 10-inch opening at the top, and the jaws are 30 inches long (or wide).

The maximum diameter round rock that will be accepted or "nipped" is about 80 percent of the width of the opening.

Gyratory.   These machines have a conical or domed crushing member called the cone, head, or sphere, which moves in a small circle around a vertical axis, inside a fixed bowl or mantle. A Telsmith construction is shown in Fig. 21.83.

The cone may be relatively stationary at the top and move at the bottom only, may gyrate equally at top and bottom, or may be mounted so that the head can wobble as well as gyrate.

The crushing head is free to turn under thrust from material being crushed. Modern secondary units provide relief from breaking strains by means of heavy springs holding the mantle in place against normal loads. Fineness of product is adjusted by raising or lowering the mantle.

TABLE III.—APPROXIMATE CAPACITIES OF JAW CRUSHERS

Material weighing 100 lbs. per cu. ft.
Discharge opening, closed, in inches.

Feed Opening Size (in.)	Smallest Discharge Opening (in.)	Capacity (tons per hr.)	Largest Discharge Opening (in.)	Capacity (tons per hr.)	Horsepower	Height (ft. and in.)
16 by 10	1½	15	4	45	15	2—4
24 by 15	2	30	5	80	35	3—6
36 by 24	3	75	6	160	75	4—11
42 by 40	4	130	8	250	125	7—9
48 by 36	5	175	8	275	150	7—6
48 by 42	5	175	8	275	150	8—6
60 by 48	5	240	9	450	200	9—2
84 by 56	8	350	12	600	200	10—7
84 by 60	8	350	12	600	250	10—7
84 by 66	8	400	12	600	250	11—8

TABLE I.—SIZES OF SHOVEL DIPPERS AND CORRESPONDING CRUSHERS

The dimensions of the largest stone that a given dipper will handle are those of a stone which will pass through the dipper. The crusher recommended is the smallest that will handle the stone without regard to capacity.

Rated Capacity of Dipper (cu. yd.)	Size Stone to Pass Dipper (in.)	Size of Jaw Crusher (in.)	Size of Gyratory (in.)
¾	32 by 35	36 by 48	13 or 16
1	33 by 38	36 by 48	16 or 20
1¼	33 by 40	42 by 48	20 or 26
1½	30 by 36	36 by 42	20 or 26
1¾	33 by 45	42 by 48	20 or 26
2	33 by 45	42 by 48	20 or 26
2½	36 by 48	48 by 60	36 or 42
3	40 by 48	48 by 60	42 or 48
3½	44 by 50	48 by 60	42 or 48
4	48 by 57	56 by 72	48 or 60
5	48 by 60	66 by 86	60 or 72

TABLE IV.—APPROXIMATE CAPACITIES OF GYRATORY CRUSHERS

Material weighing 100 lbs. per cu. ft.
Discharge opening—open side

Feed Opening Size (in.)	Smallest Discharge Opening (in.)	Capacity (tons per hr.)	Largest Discharge Opening (in.)	Capacity (tons per hr.)	Horsepower	Height (ft. and in.)
2½	⅜	.5	½	.75	3	1—1
8	1	15	2	40	20	5—6
12	2	40	2¾	70	45	7—5
16	2½	100	4½	160	80	8—3
20	3	150	5	250	125	10—1
26	3½	225	6	400	150	11—0
30	4	250	6½	450	175	12—2
36	4½	370	7	600	225	14—0
42	5	420	7½	700	250	15—5
48	5½	750	9	1,200	300	18—6
54	6¼	900	9½	1,600	350	19—0
60	7	1,200	10	2,000	400	20—0
72	9	2,000	12	3,000	500	20—0

TABLE II.—SHOVEL CAPACITIES

Based on stone with a specific gravity of 2.6.

Size of Dipper Cu. yd.	Tons	Swings per Minute	Tons Loaded per Hour, 70% Efficiency
½	.45	2.4	46
¾	.675	2.36	67
1	.90	2.23	86
1¼	1.125	2.15	103
1½	1.35	2.10	120
1¾	1.575	2.00	136
2	1.80	1.97	150
2¼	2.025	1.91	165
2½	2.25	1.85	177
2¾	2.475	1.80	190
3	2.70	1.75	202
3¼	2.925	1.70	213
3½	3.150	1.65	222
3¾	3.375	1.61	233
4	3.600	1.57	240

TABLE V.—APPROXIMATE CAPACITIES OF SINGLE-ROLL CRUSHERS

Material weighing 100 lbs. per cu. ft.

Feed Opening Size (in.)	Capacity in Net Tons per Hour 4 in.	6 in.	10 in.	Horsepower	Height (ft. and in.)
21 x 48	125–150	200–250	300–375	75–100	3–10
21 x 60	150–175	225–400	325–400	100–125	3–10
24 x 48	150–175	225–275	325–400	100–125	4– 5
24 x 60	175–200	250–300	350–425	125–150	4– 5
30 x 48	175–200	250–300	350–425	125–150	5– 0
30 x 60	200–225	275–325	375–450	150–175	5– 0
36 x 48	200–225	275–325	375–450	150–175	6– 0
36 x 60	225–250	300–350	400–475	175–200	6– 0
42 x 48	225–250	300–350	400–475	175–200	7– 0
42 x 60	250–275	325–375	425–500	200–225	7– 0

FIGURE 21.81 Shovel and crusher capacities. (*Courtesy of Pit and Quarry Publications.*)

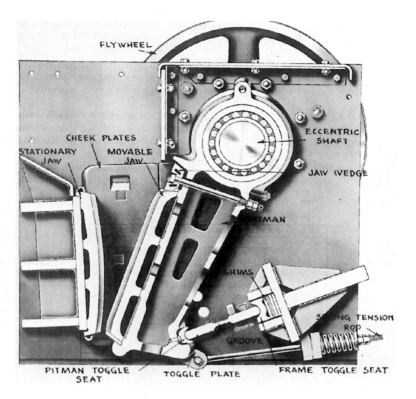

FIGURE 21.82   Jaw crusher.

The crushing chamber is annular (ring-shaped), and wedge-shaped in cross section. Rock fed into the top falls between cone and mantle and is crushed as the opening narrows with the movement of the cone. When it widens again, the pieces fall farther, to be further crushed on its return. The action is somewhat similar to that of the jaw crusher, but the squeeze comes from the side rather than the bottom, and the curve of the chamber breaks up flat pieces.

Figure 21.84 shows a succession of steps in the reduction of a large stone.

Cone speed and distance of travel must be carefully synchronized. A wide space allows pieces to fall more freely than a narrow one, and if coupled with slow movement would allow pieces to fall too far before the next impact. Fast gyration and short travel would not allow them to fall far enough, and would waste power.

Gyratory crushers are more expensive than jaw types, require more power, and need more vertical space for installation, but have greater production and will produce a finer and more uniform product.

Most jaw and cone crushers are not suited to soft moist rock, which tends to cake under pressure, or to rock mixed with clay.

Swing Hammermills.   These machines, one of which is illustrated in Fig. 21.85, break by impact. Rapidly revolving flails, with tip speeds often over 2 miles per minute, strike stones as they slide in from the hopper and throw them repeatedly against a breaker plate. The flails then sweep the broken pieces across a bar grate, through which they fall if they are small enough. Any that are oversize are carried around to the breaker plate again for recrushing.

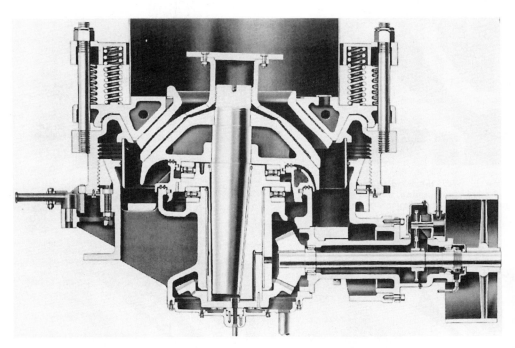

FIGURE 21.83   Gyratory crusher, secondary type. (*Courtesy of Telsmith, Inc.*)

The grate openings may be all the same width, or may be in several sets, with narrow openings near the breaker plate and progressively wider ones away from it. The latter arrangement permits the use of several hoppers under the grate, and the separation of crushed stone into different sizes.

These machines are subject to extreme peak loads when encountering large or hard rocks, and require heavy flywheels to maintain momentum.

Hammermills have the largest ratio of reduction of any type crusher, and in weak rock of favorable structure may reduce 48-inch cubes to 1-inch cubes in one pass. They do primary crushing in soft and medium rock, secondary in any type, and are also garbage shredders.

Their product tends toward cubic shape more than that of the pressure-type crushers. Fineness of crushing can be obtained by setting the breaker plate close to the hammers; but fineness of product is determined by the setting of the grate bars.

Some crushing is done against the bars, but this is kept to a minimum as they are not as sturdy as the breaker plates.

Roll Crusher.   The single-roll crusher consists of a toothed or fluted roll which revolves close to a crushing plate. The teeth or projections are called sluggers and act as sledges in breaking large stones. Smaller pieces are dragged between the roll and the plate and crushed by dragging pressure.

This does its best work on stratified or laminated rock that is not particularly abrasive. It passes clay or other sticky material with little difficulty. Product usually is coarse.

Double-roll crushers consist of two power-driven rolls which rotate in opposite directions, with their top surfaces moving toward each other. Stone is pulled down between them by gravity and by friction with the roll surfaces. The rolls may be smooth, corrugated, or toothed. A toothed model is shown in Fig. 21.86.

In general, the ratio of reduction for feed materials over 1 inch in diameter is restricted to 4 to 1, but smaller pieces may be reduced by as much as 10 to 1.

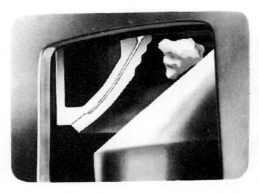

Position 1. With the head in the position of maximum opening, a large stone is just entering the crushing cavity.

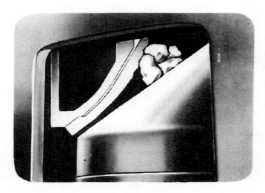

Position 2. As the head moves to the closed side, the stone receives its initial impact and is broken into several smaller particles.

Position 3. The broken particles fall vertically toward the head as it recedes from under the stone when moving to the open side.

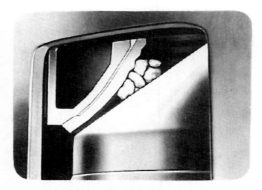

Position 4. The head has once more moved to the closed side and, receiving another impact, the particles are again reduced in size.

Position 5. The particles again take a vertical path, spread across the head and advance farther into the crushing cavity.

FIGURE 21.84 Gyratory crushing action.

Position 6.   With another crushing impact, a further reduction in size takes place corresponding to the opening of the cavity at this point.

Position 8.   Again a reduction takes place with those particles now in the parallel zone reduced to the size to which the crusher is set.

Position 7.   The material has traveled farther on its downward path and is just entering the wide parallel zone at the bottom of the head.

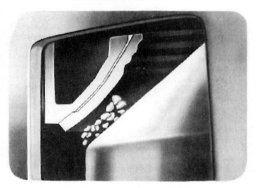

Position 9.   Here the head is again at the open side with all the material now in the parallel zone.   Note the large opening created for the discharge of the crushed fines.

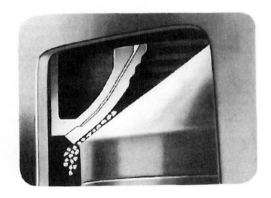

Position 10.   All particles are now reduced in size, having received five impacts from the head in passing through the crushing cavity.

FIGURE 21.84   *(Continued)*

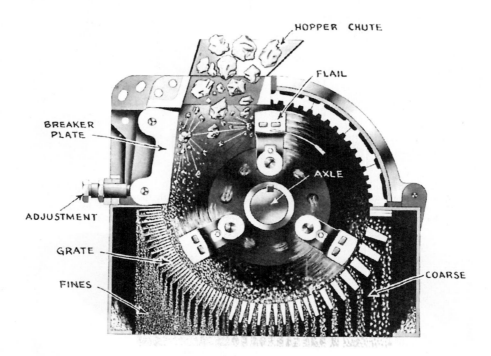

FIGURE 21.85 Swing hammermill.

FIGURE 21.86 Toothed roll crusher.

The size of rock that can be processed depends on the angle of nip and friction between the stone and roll surfaces.

The angle of nip is found by drawing lines from the roll centers to the points of first contact with the stone, and drawing tangents to those lines. The angle at which the tangents intersect is the nip. It should not be over 31° for general use of smooth rolls.

This angle is reduced by using smaller rock or larger rolls, or by pulling the rolls apart so that product will be coarser. The friction will be affected by the hardness or slipperiness of the rock and the surface of the roll. A toothed, pitted, or corrugated surface will increase gripping power.

Roll surfaces tend to wear in grooves, so that a side shifting adjustment is often provided to equalize wear. This may be manual or automatic.

Mountings. Heavy crushers are installed on concrete foundations. Smaller sizes may be set in masonry, in steel frames for semipermanent installations, or mounted on mobile chassis of either self-powered or trailer types.

The engine may be permanently hooked up to the crusher or be in a separate unit connected by a shaft or belting. Diesel, gasoline, or electric power is used.

PROCESSING PLANTS

Complete plants for the processing of blasted rock, or of pit-run gravel containing oversize stone, perform the function of taking in a coarse and variable material and turning out one or more classes of graded and uniform product of greater value.

Equipment includes a hopper into which spoil is dumped by shovel, conveyor or haul units, screens to separate pieces according to size both before and after crushing, from one to three crushers to reduce oversize pieces, and hoppers or chutes and conveyors for individual handling of different sizes separated by the screens. Provision must be made for removal or storage of the products.

The plant units should be protected by a grizzly (coarse primary screen) on the hopper, which would prevent rocks too large for the passages or primary crusher from entering them. Feed from the hopper to the primary crusher or screens may be by gravity, apron conveyor, or belt conveyor. Such conveyors may be a few feet or several hundred feet long. They serve both to transport the material and to regulate the amount entering the crusher.

Bypassing and Reprocessing. A crusher is most efficient when all pieces in the feed are too large to pass through it without crushing. It is therefore desirable to separate and direct the raw material so that the fines will go around the crusher rather than through it.

The product of the primary crusher will have a number of pieces small enough to fall through the secondary crusher, and it is also usual to cause the fines which bypass the primary to mix with its product. A further separation is therefore required to avoid overloading the secondary with small sizes.

Any crusher will pass some oversize stone, particularly of material that tends to splinter into long thin pieces. Crushed material is therefore returned to the feed stream from which the finished pieces will be screened out and delivered to output hoppers, intermediate sizes routed to the secondary crusher, and oversize sent back to the primary crusher.

This rehandling of material gives the plant an appearance of complexity and duplication of work. However, conveyors and screens have lower power and maintenance requirements than crushers and pay for themselves in saving crusher size and wear, in addition to making possible more uniform and valuable products.

The use of two or three crushers in series may be economical even where a single crusher could be obtained which would complete the reduction itself. The ratio of reduction—that is, comparative size of the feed and the product—determines both the power requirement and the output of a crusher. The table of single-roll crusher capacity in Fig. 21.81 indicates the increase in output obtained by increasing spacing to obtain a coarser aggregate. Also, when the opening is large, a greater percentage of the feed can be bypassed, again increasing the volume of unsorted material handled.

If a secondary crusher is used, the primary may have quite a low reduction so that a comparatively small, light, and inexpensive crusher can be substituted for a heavy one. Also, if a third stage is used, the secondary unit will be relieved of part of its burden and will have to reduce that which it does get much less.

The size of the primary crusher in a stationary plant is determined largely by the production required, the size stone to be handled, and the cost.

Required production may be obtained by installing one crusher large enough to do the whole job, a series of crushers that will do it in stages, or two or more separate units each of which processes a part of the material.

Size. Usually, crushers are purchased to provide output required by estimated demand, and excavators are obtained in sizes to keep them supplied. Occasionally, however, a crusher is selected to match the output of the excavators. The capacity tables can be used in calculating sizes required in any case, but because of the very wide variations in proportion of fines, and in production of both types of machine, results should be checked against field experience.

The primary crusher will not have to handle all material supplied to the plant, as much of it will be undersize, and will be screened out for direct delivery to the secondary. The proportions will vary in different pits, in different sections of the same pit, and in the same section for successive blasts.

Large crusher openings reduce both blasting and loading costs. If big pieces can be processed, fragmentation need not be so fine so that holes can be more widely spaced or lightly loaded. Or drilling and loading can be standard and savings obtained in reduced secondary blasting. Shovels can be less selective about the size rocks they pick up.

Big shovels are more efficient than small ones in rock piles, and where sufficient output is required will give substantially lower digging costs. However, they pick up big rocks and should be teamed with crushers large enough to take anything that will go through the bucket. The tables indicate the size crushers required for various bucket capacities. This does not allow for oversize balanced on top of the dipper.

In large operations, several shovels and crushers will be used. Cost of plant may be reduced by including one very large crusher that will take rock rejected by grizzlies on smaller units. This arrangement will require either occasional or constant use of some machine capable of collecting the boulders and feeding them to the big crusher.

In general, rock which is too large to crush should not be hauled unless the plant is near an abandoned pit into which oversize can be rolled or pushed. Secondary blasting near the plant must be done under mats and is more expensive and risky than when done in the pit.

Mobile Crushing Plants. The portable plant, shown in cross section in Fig. 21.87, is a mobile, single-unit, double-reduction rock crushing and sorting plant designed for use in limestone quarries.

The hopper may be protected by a bar grizzly to prevent the entrance of boulders too large for the primary crusher. Blasted rock may be dumped in the hopper by a shovel, conveyor, or trucks on a platform or bank.

An apron feeder carries the rock forward to fall on a grizzly, which passes pieces small enough not to require primary crushing, and slides the larger ones into the 18-inch by 24-inch jaw crusher.

This crusher dumps on a conveyor belt, which also carries the pieces that dropped through the sorting grizzly. Transfer is made to a second belt, which runs up an incline and discharges into a rotary elevator made up of a series of buckets carried inside a revolving drum which is set across the frame. This Rotovator elevates the load and drops it on an upper conveyor which discharges onto a three-deck screen.

Stones which fail to pass the coarse mesh screen on the top deck move forward into the secondary crusher, a hammermill. Product of the hammermill falls on the forward bottom belt, which takes it back to the screen.

Pieces passing the upper deck, but failing to go through either the second or third deck, slide forward into a chute opening onto the rock delivery conveyor. Movable gates are arranged so that when smaller sizes are required, the rejects from the second, or from both the second and third

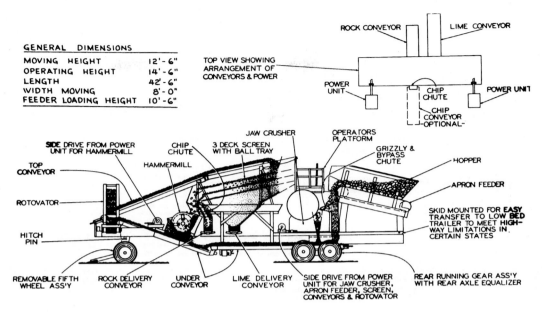

FIGURE 21.87 Portable crushing and screening plant.

decks, may be directed into the secondary crusher. An extra "chip" conveyor can be installed with an additional chute to allow side delivery of two sizes of crushings.

Fine particles which go through all three screens fall into a hopper that feeds the lime delivery conveyor.

The side delivery conveyors for rock, lime, and chips fold up against the sides of the machine for transit. The hopper and primary crusher are mounted on a subframe and skids so that they can be removed for separate transportation, if necessary.

Two engines are used. These are mounted separately on rubber tires. They are positioned beside the main unit and drive through propeller shafts or belts. One operates the hammermill, the other the primary crusher and the rest of the machinery.

The use of two engines permits separate control of the speed of the secondary crusher. High speed produces a greater percentage of fines for agricultural limestone; low speed, pieces for road metal.

The power unit for each section is mounted permanently on its chassis to facilitate moving.

The Universal 880 Gravelmaster, Fig. 21.88, serves to reduce and size run-of-pit gravel. In this the feed is taken from the hopper by conveyor belt to a $2\frac{1}{2}$-deck gyrating screen, which sends rejects from the top deck into the primary jaw crusher, from the secondary deck into the secondary roll crusher, and from the half deck onto the delivery or output conveyor.

Material passing all the screens enters the sand hopper. An adjustable gate permits adding any desired quantity of this sand to the gravel on the output conveyor. A sand conveyor can be installed to stockpile it or to sideload it onto trucks.

The product of both crushers falls on a conveyor, which takes it through the bucket drum and back to the screens.

Trailer-mounted portable plants may be completely self-contained, including engines, or use separate wheel- or skid-mounted engines, hoppers, and conveyors. Conveyors can be self-powered or driven by the plant engine.

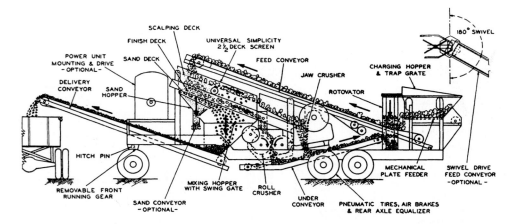

FLOW OF MATERIAL

From the charging hopper the pit material is uniformly fed by the mechanical feeder onto the feed conveyor which conveys it to the 2½-deck screen. Material retained on the top scalping deck is chuted to the primary jaw crusher. Material passing top deck and retained on second deck flows to the roll reduction crusher. Material passing sec-ond deck and retained on bottom half-deck sand screen flows to the mixing hopper. A swinging gate permits desired amount of fines to be mixed with the finished product. Material from both crushers falls onto the under conveyor where it is conveyed to the Rotovator which elevates it back onto the feed conveyor which again takes it to the screen. Finished material is delivered from the mixing hopper to truck by conveyor.

FIGURE 21.88 Material flow diagram.

Fixed or Permanent Plants.   Fixed plants generally are heavier and more spacious than mobile units, and usually are custom-built to include a selection of standard crushers. They may be more economical to operate because of the weight of the units and the ease of getting at parts needing service. Moving is difficult and expensive.

Figure 21.89 shows a layout of a fixed plant that includes several conveyor-connected buildings.

LUBRICATION EQUIPMENT

High-Pressure Fittings and Couplings.   There are at least six different fittings which are attached to machinery to permit pumping grease. Of these, the most popular is the hydraulic fitting, Fig. 21.90. The gun nozzle or coupling, Fig. 21.91, contains jaws which are wedged apart by forcing them over the top of the fitting, and which clamp below the bulge. High pressure developed in the gun pushes against the head of the fitting and tends to push the gun away, but it also presses the tops of the jaws outward, wedging the bottoms inward for a tight grip. If grease passes freely and pressure is light, the gun can be removed by direct pull. If pressure is high, it is taken off by tilting sideward until the nozzle touches the shoulder and pries the jaws off.

The ball and spring should prevent grease from leaking back from the bearing. They are lacking in some small fittings.

These two types have the advantage of speed, with the hydraulic providing the more positive seal. If the machine is exposed to dirt or dust, the fitting should not be cleaned off after greasing, as the dab of grease sticking to it will prevent dirt from getting in the neck of the opening.

Standard threads are ⅛-inch P. T. (pipe thread), ¼-inch P. T. for oversize fittings, and ¼ to 28 inches for small ones. A variety of threads are available for special uses. Many small fittings have no threads and are driven in. These may not stand high pressures.

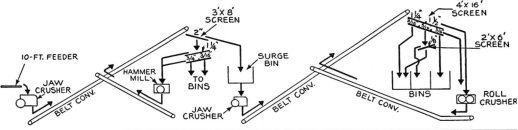

Flowsheet of material through plant from primary crusher to finished sizes

FIGURE 21.89   Stationary crushing and processing plant.

Large ($\frac{1}{4}$-inch P. T.) hydraulic fittings are sometimes made of a small fitting screwed or driven into a larger base. The small part may come out when the coupling is pulled away. If much trouble is experienced, one-piece units, or $\frac{1}{8}$-inch P. T. fittings in $\frac{1}{8}$- to $\frac{1}{4}$-inch bushings should be substituted.

Hydraulic fittings are made in a variety of angles and lengths. Special extensions can be made up with $\frac{1}{8}$-inch pipe material.

If grease leaks at the fitting-coupling connection, it usually indicates that one of these units, most often the fitting, has been scarred or damaged so that the coupling cannot seat properly. The defective part should be replaced.

The coupling may connect to it by pushing, pulling, or sideward motion. Leakage of grease around it usually indicates that the neoprene face seal is damaged or worn. The coupling cannot be pulled off while the grease in it is under high pressure.

The pin type is the original Alemite fitting. The coupling locks on it with a quarter-turn. This fitting projects farther than the others described, and is more subject to accident. Making connection is slower than with the hydraulic. It is no longer in wide use.

The Dot is made in standard and Mogul sizes. The gun is screwed on a full turn, usually forcing some grease into the fitting. Any additional amount required is then pumped in.

The flush type is used where the fitting cannot be allowed to project. It has limited use in earth-moving equipment because of the difficulty of cleaning.

Hand Grease Guns.   The majority of hand grease guns now in production are the lever type. Figure 21.92 shows a typical example in cross section. The cylindrical barrel has a smooth inside

FIGURE 21.90 Hydraulic fittings.

finish. The metal piston with leather seals is pushed toward the head by a light spring. The follower rod is used to pull the piston back against the spring when refilling by suction. The collar groove in the rod permits locking it in the back position.

The head contains a fitting through which grease can be pumped into the reservoir, and a passageway from the reservoir into the nozzle tube. A piston actuated by the hand lever moves up and down in this passage in which a ball check is located.

When the piston is pulled up, grease or air in the tube is prevented from following it by the check. This leaves a vacuum so that when the passage to the reservoir is opened, grease is sucked into the passage. The grease is urged in by the pressure of the follower spring, and by atmospheric air entering the barrel around the follower rod in the back cap.

When the piston is moved down, it blocks the reservoir passage, then forces the grease down and compresses the check spring so that the grease can flow past the ball into the tube. When the piston is raised, the ball reseats itself and the passage refills from the reservoir.

The small piston and the comparatively long lever enable this gun to develop pressure up to 10,000 pounds per square inch. Most nozzles and fittings are designed to take 20,000 pounds per square inch pressure. However, the seals and casings of the parts being lubricated will often bend or break at less pressure, so caution must be used when forcing grease into them.

The gun may be filled in three ways. The easiest is to pump grease into it through the fitting in the head casing from a loader pump. Precautions should be taken against pumping in air with the grease.

If there is no loader fitting, or no loader pump is available, the head casing is unscrewed, the head of the barrel cleaned and pushed well down in the grease supply, and the follower arm drawn back slowly. If the grease is thin enough to flow, it will be sucked into the barrel. It may be necessary to move the gun around in the container to prevent air from entering. When the follower is fully back, it is locked with a sideward motion, the head screwed on, and the follower released. It is good practice to keep the grease in a warm place to keep it soft enough to flow.

If the grease cannot be pumped or sucked into the gun, it may be put in with a small paddle and air kneaded out of it. It is difficult to avoid air pockets with this method.

Air in the grease may form a pocket in the passageways that will prevent the gun from working. The block may be temporary until the air is worked out, or permanent if the gun is worn enough to allow it to return beside the piston into the reservoir.

Air takes much longer to get through the head than the same bulk of grease. However, many times a gun is said to be air-locked when the trouble is partly or wholly foreign matter which prevents the ball check from seating properly. This allows grease or air to be sucked into the cylinder from the outlet tube on the upstroke, and pushed back into it on the downstroke.

In either case the cure is to disassemble the unit, clean it, and pack it with fresh grease.

The outlet tube consists of a piece of $\frac{1}{8}$-inch pipe, or of equivalent size flexible hose. Pipe thread is used. Any of the standard couplings can be attached to this.

Thick grease may not feed properly in this type of gun, as combined atmospheric and spring pressure may not be enough to make it flow. It may be persuaded to work by tapping or by heating. Best results may be obtained if the follower piston is so built that the rod can be locked to it (as in this example), so that pressure can be applied while pumping.

A special gun may be obtained in which the grease is forced along the gun by twisting a threaded follower.

The push-type gun, Fig. 21.93, is convenient for narrow spaces and high spots. The newer models may be filled through the loader fitting shown, but for older types it is necessary to remove the backplate, pull out the follower, and pack it by hand. The amount of lubricant in the gun is indicated by the distance the follower handle projects.

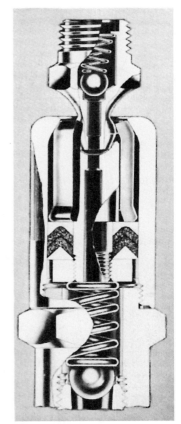

FIGURE 21.91 Hydraulic nozzle and fitting.

The gun is held by the handle and pushed against the fitting. After the coupler grips, the end of the gun slides on a plunger which pushes the grease out of the stem into the fitting. Pressure is then released, the spring forces the gun back, the stem fills with grease, and it is ready for the next push. A ball check in the coupling prevents air from entering the stem.

Screw-type guns are of simple construction. The follower rod is threaded, and screwing it into the rear cap forces the follower piston against the grease, which passes directly into the coupling.

Bucket Pumps.   As used here, this term includes all large-capacity hand-operated grease pumps. Pressure varies from a few pounds for transfer pumps up to 15,000 pounds per square inch (psi) in heavy-duty models.

Such pumps generally consist of a pail, a pail cover, a hand pump fastened in the cover, a flexible hose, and a coupling or nozzle. An Alemite bucket pump is shown in Fig. 21.94, good for pressures from 2,500 to 5,000 pounds per square inch and 35 pounds or 5 gallons of lubricants.

The pressure release lever opens a bypass so that grease pumped is returned to the tank, and the hose can drain back into the tank also. This allows the release of any pressure binding a coupling on a fitting, allows the handle to be pushed down to carrying position without forcing grease out

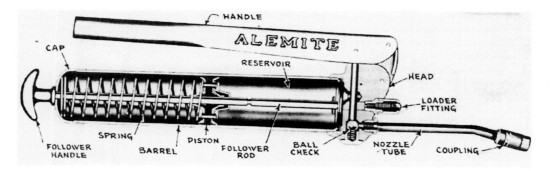

FIGURE 21.92 Lever-type hand grease gun.

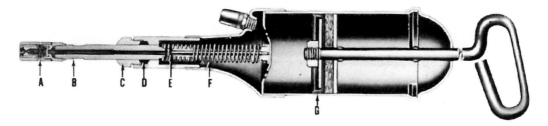

A. Nozzle or coupler

B. Stem

C. Slide cup

D. Packing

E. Piston

F. Main spring

G. Follower piston

FIGURE 21.93 Push-type hand grease gun.

FIGURE 21.94 Multipressure bucket pump. (*Courtesy of Alemite Corporation.*)

of the coupling, and allows priming the pump after refilling, without forcing air into the hose.

Air-Operated Pumps. Wherever possible, grease is pumped by compressed air. Many pumps are intended for use in a standard 55-gallon (400-pound) drum.

A single vertical rod connects the piston of a one-cylinder, double-acting air motor to the grease pump piston. This is hollow and carries a ball on a machined seat at the bottom. When the piston moves down, grease unseats the valve and fills the piston. When it moves up, the ball seats and the grease is forced up the space around the piston rod and out toward the nozzle.

The mechanical primer is a large plunger which moves grease toward the ball check with each upstroke.

When this unit is connected to the air line, it pumps until full pressure is built up in the grease line. This pressure then shuts off the air by closing the passage through the trigger valve. The unit is then said to be balanced. When grease pressure drops, air is admitted again and pumping is resumed.

These pumps are available in high- and low-pressure constructions, and similar units dispense oil in large volume and at very low pressure.

Grease Trucks.   A contractor who has ten or more pieces of equipment in the field may consider the advisability of buying or building a mobile greasing unit to take care of the lubrication requirements. Such units are shown in Figs. 21.95 and 21.96. These are only samples of possible arrangements. A wide variety of equipment and many layout plans are available.

The usual carrier is a flatbed truck with a capacity of 2 tons or more. A beefed-up pickup may carry a small unit. A van-type body may be used when the climate is cold or rainy. Four-wheel-drive trucks are used for rugged conditions. Semitrailers are suited to part-time use, as they can be parked and the tractor used for other jobs.

A grease truck may carry from 2 to 10 drums of grease and oil, each weighing about 400 pounds; a gasoline engine that drives an air compressor, an air pump in each active drum (spares may be carried), and a reel of hose for each lubricant.

The reels receive grease through their axles and feed it into a connection to the hose. They may be hand-wound or equipped with pullback springs that allow the hose to be pulled out, then wind it back when it is released.

The nozzles may be plain or equipped with meters. Pressure in the hoses is automatically maintained by the air pumps.

The truck may carry drums of fuel, antifreeze, and/or water in addition to the lubricants. If it is not needed full time for greasing, it may be equipped with a small electric generator, drills, lights, vise, and tools, so that it can be used for field repairs.

Lubrication may be done by one or two workers. The time per machine is highly variable. Some have only 1 to 3 points for daily greasing, others may have 30 or more. In addition to daily lube

FIGURE 21.95   Grease truck.

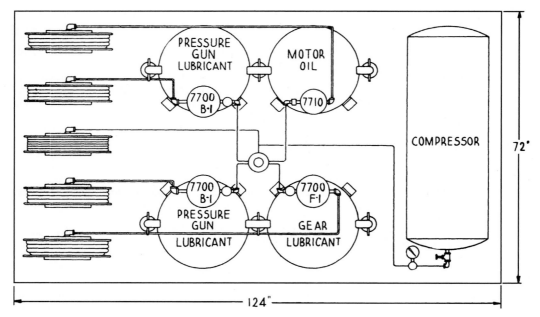

FIGURE 21.96   Grease truck layout.

jobs there are more extensive lubrications required weekly or monthly, and changes of engine, transmission, final drive, and hydraulic oil are made according to hourly or seasonal schedules.

Machines that are lined up together at the end of a shift may be lubricated by one person at the rate of 2 to 12 per hour, with a long-term average of 3 to 5. Those that have to be found where they are scattered around the job require additional travel and search time. If the system is to catch and stop machines at work, time will be longer and interference with work may be costly.

Time spent by the grease crew, or the number of lube trucks and crews required, may be greatly reduced by having operators take care of light lubrication themselves. A few contractors go further and leave all lubrication to operators and oilers, paying overtime for the work.

When lubrication must be done during workhours, trucks can be most effectively done by a stationary unit, or a mobile unit parked along the haul route.

A survey made by the magazine *Construction Methods and Equipment* indicates that contractors use one grease truck for 7 up to 110 pieces of equipment, with the usual range being between 20 and 80. Some, with as many as 80 nonhighway units, relied entirely on hand greasing and/or fixed grease racks.

TIRE HANDLING

With mobile construction equipment, such as scrapers and trucks, getting bigger and bigger, it is essential that their tires be handled by heavy machinery. That is the role of machinery seen in Fig. 21.97. The IMT TireHandler offers a choice of loader-mounted, forklift-mounted, or hydraulic crane-mounted models. These can handle tires from 25 to 164 inches in diameter and up to 25,000 pounds.

The TireHandler provides a natural, more efficient way for an operator to hydraulically grip, pull, lift, turn, and position a tire. The operator in the loader, forklift, or crane can do it all from the safety of the control console. The tire body can be rotated up to 390° in a vertical plane, rotated to a horizontal plane for carrying with an unobstructed view, and moved laterally so the operator does not have to jockey the equipment for a more perfect approach for installing the tire. See Fig. 21.98.

FIGURE 21.97 Handling tires of equipment. (*Courtesy of IMT.*)

FIGURE 21.98 Movements of tire handler. (*Courtesy of IMT.*)

APPENDIX

This section is made up of technical, statistical, and advisory material largely contributed by others, that has been reproduced without change or comment.

Working conditions vary so widely that averages such as may be found in these tables and examples cannot be safely applied to any specific job. Changes and refinements in equipment may have changed production characteristics by the time this is read.

Because of these factors, the figures serve for general information only, and should not be used directly in estimates of production or cost, or in comparison of different makes and models of equipment.

The following is a guide to the major subjects in this section:

Power shovel yardages									
Shovel bucket capacity in cu. yds.									
Class of material	$\frac{3}{8}$	$\frac{1}{2}$	$\frac{3}{4}$	1	$1\frac{1}{4}$	$1\frac{1}{2}$	$1\frac{3}{4}$	2	$2\frac{1}{2}$
Moist loam or light sandy clay	3.8'	4.6'	5.3'	6.0'	7.0'	7.4'	7.8'	8.4'	
	85	115	165	205	250	285	320	355	405
Sand and gravel	3.8'	4.6'	5.3'	6.0'	6.5'	7.0'	7.4'	7.8'	8.4'
	80	110	155	200	230	270	300	330	390
Good common earth	4.5'	5.7'	6.8'	7.8'	8.5'	9.2'	9.7'	10.2'	11.2'
	70	95	135	175	210	240	270	300	350
Clay, hard, tough	6.0'	7.0'	8.0'	9.0'	9.8'	10.7'	11.5'	12.2'	13.3'
	50	75	110	145	180	210	235	265	310
Rock, well blasted	40	60	95	125	155	180	205	230	275
Common, with rocks and roots	30	50	80	105	130	155	180	200	245
Clay, wet and sticky	6.0'	7.0'	8.0'	9.0'	9.8'	10.7'	11.5'	12.2'	13.3'
	25	40	70	95	120	145	165	185	230
Rock, poorly blasted	15	25	50	75	95	115	140	160	195

Conditions: (1) Cu. yds. bank measurement per hour. (2) Suitable depth of cut for maximum effect. (3) No delays. (4) 90° swing. (5) All materials loaded into hauling units.

Haul units needed to spot under shovel per hour in medium digging					
Size excavator dipper, yd	Minimum haul unit capacity at 4 times dipper size, yd	Approx. shovel cycle time in seconds 90° swing...no delays loading on grade	Loading time for 4-dipper truck in sec.	Time to synchronize loading spot one truck, min	No. spots by hauling units at. shovel needed per hr.
$\frac{3}{8}$	$1\frac{1}{2}$	19	76	1.26	48
$\frac{1}{2}$	2	19	76	1.26	48
$\frac{3}{4}$	3	20	80	1.33	45
1	4	21	84	1.4	43
$1\frac{1}{4}$	5	21	84	1.4	43
$1\frac{1}{2}$	6	23	92	1.53	39
2	8	25	100	1.66	36
$2\frac{1}{2}$	10	26	104	1.73	35

Short boom dragline performance, cu. yds.									
Class of material	$\frac{3}{8}$	$\frac{1}{2}$	$\frac{3}{4}$	1	$1\frac{1}{4}$	$1\frac{1}{2}$	$1\frac{3}{4}$	2	$2\frac{1}{2}$
Light moist clay or loam	5.0'	5.5'	6.0'	6.6'	7.0'	7.4'	7.7'	8.0'	8.5'
	70	95	130	160	195	220	245	265	305
Sand or gravel	5.0'	5.5'	6.0'	6.6'	7.0'	7.4'	7.7'	8.0'	8.5'
	65	90	125	155	185	210	235	255	295
Good common earth	6.0'	6.7'	7.4'	8.0'	8.5'	9.0'	9.5'	9.9'	10.5'
	55	75	105	135	165	190	210	230	265
Clay, hard, tough	7.3'	8.0'	8.7'	9.3'	10.0'	10.7'	11.3'	11.8'	12.3'
	35	55	90	110	135	160	180	195	230
Clay, wet, sticky	7.3'	8.0'	8.7'	9.3'	10.0'	10.7'	11.3'	11.8'	12.3'
	20	30	55	75	95	110	130	145	175

Note: Top figure denotes optimum depth of cut. Bottom figure denotes yards per hour.

Source: Courtesy of Power Crane and Shovel Association.

Excavator Production Before Delays

Theoretical Excavator Production in Loose Yards, for 60 minute hour without delays, for various bucket loads and cycle times

NOT FOR DIRECT USE IN ESTIMATING

Cycle time, Seconds	12	13	14	15	16	17	18	19	20
Cycles per Hour	300	276 9	257.1	240	225	211.8	200	189.5	180
Bucket Load, Cubic Yards									
3/8	111.5	103.8	96.3	90	84.3	79.4	75	71.1	67.5
1/2	150	138.4	128.5	120	112.5	105.9	100	94.7	90
3/4	225	207.6	192.6	180	168.6	158.8	150	142.2	135
1	300	276.9	256.8	240	224.8	211.2	200	189.5	180
1-1/2	450	415.2	385.2	360	337.2	317.7	300	284.4	270
2	600	554	514	480	450	424	400	379	360
2-1/2	750	692	642	600	562	530	500	474	450
3	900	831	770	720	674	634	600	568	540
3-1/2	1050	970	899	840	786	740	700	663	630
4	1200	1108	1028	960	900	847	800	758	720
5	1500	1384	1284	1200	1125	1059	1000	947	900
6	1800	1662	1540	1440	1350	1271	1200	1137	1080

Cycle Time, Seconds	21	22	23	24	25	30	35	40	50
Cycles per Hour	171.4	163.6	156.5	150	145	120	102.9	90	72
Bucket Load, Cubic Yards									
3/8	65.5	61.3	58.7	56.2	54.3	45	38.6	33.5	27
1/2	85.7	81.8	78.2	75	72.5	60	51.4	45	36
3/4	128.5	122.7	117.3	112.5	108.7	90	77.2	67.5	54
1	171.4	163.6	156.5	150	145	120	102.9	90	72
1-1/2	256.9	245.4	234.7	225	217.5	180	154.3	135	108
2	343	327	313	300	290	240	205.4	180	144
2-1/2	429	409	391	375	362	300	257	225	180
3	514	491	470	450	435	360	309	270	216
3-1/2	600	573	548	525	507	420	360	315	252
4	686	655	616	600	580	480	412	360	288
5	858	818	782	750	724	600	515	450	360
6	1028	982	940	900	870	720	617	540	432

Weight and Measure Tables

Measure of Length
1 Mile = 1760 Yds. = 5280 Ft. = 63,360 Inches
1 Mile = 8 Furlongs = 80 Chains
1 Furlong = 10 Chains = 220 Yds.
1 Chain = 4 Rods = 22 Yds. = 66 Ft. = 100 Links
1 Rod = 5.5 Yds. = 16.5 Ft.

Measure of Length—English to Metric
1 Mile = 1.609 Kilometer
1 Yard = 0.9144 Meter
1 Foot = 0.3048 Meter = 304.8 Millimeters
1 Inch = 2.54 Centimeters = 25.4 Millimeters

Measure of Length—Metric to English
1 Kilometer = 0.6214 Mile
1 Meter = 39.37 Inch = 3.2808 Ft. = 1.0936 Yd.
1 Centimeter = 0.3937 Inch
1 Millimeter = 0.03937 Inch

Square Measure
1 Sq. Mile = 640 Acres = 6400 Sq. Chains
1 Acre = 10 Sq. Chains = 4840 Sq. Yds. = 43,560 Sq. Ft.
1 Sq. Chain = 16 Sq. Rods = 484 Sq. Yds. = 4356 Sq. Ft.
1 Sq. Rod = 30.25 Sq. Yds. = 272.25 Sq. Ft. = 625 Sq. Links
1 Sq. Yd. = 9 Sq. Ft.
1 Sq. Ft. = 144 Sq. Inches
An Acre is equal to a Square 208.7 Feet per Side

Square Measure—English to Metric
1 Sq. Mile = 2.5899 Sq. Kilometers
1 Acre = 0.4047 Hectare = 40.47 Ares
1 Sq. Yard = 0.836 Sq. Meters
1 Sq. Foot = 0.0929 Sq. Meters = 929 Sq. Centimeters
1 Sq. Inch = 6.452 Sq. Centimeters = 645.2 Sq. Millimeters

Square Measure—Metric to English
1 Sq. Kilometer = 0.3861 Sq. Mile = 247.1 Acres
1 Hectare = 2.471 Acres = 107,640 Sq. Ft.
1 Are = 0.0247 Acre = 1076.4 Sq. Ft.
1 Sq. Meter = 10.764 Sq. Ft. = 1.196 Sq. Yd.
1 Sq. Centimeter = 0.155 Sq. Inch
1 Sq. Millimeter = 0.00155 Sq. Inch

Cubic Measure
1 Cubic Yd. = 27 Cu. Ft.
1 Cubic Ft. = 1728 Cu. Inches
1 Cord = 128 Cu. Ft.
1 Gallon = 0.1137 Cu. Ft. = 231 Cu. Inches
1 Cubic Ft. = 7.48 U. S. Gallons
1 U. S. Gallon = 0.83268 Imperial Gallon
1 Imperial Gallon = 1.2009 U. S. Gallons

Cubic Measure—English to Metric
1 Cubic Yd. = 0.7646 Cubic Meters
1 Cubic Ft. = 28.316 Liters
1 Cubic Inch = 16.38 Cubic Centimeters
1 U. S. Gallon = 3.785 Liters
1 U. S. Quart = 0.946 Liters
1 U. S. Pint = 0.473 Liters
1 Imperial Gallon = 4.542 Liters

Cubic Measure—Metric to English
1 Cubic Meter = 35.314 Cu. Ft. = 1.308 Cu. Yd. = 264.2 U. S. Gallons
1 Cubic Centimeter = 0.061 Cu. Inch
1 Liter = 0.0353 Cu. Ft. = 61.023 Cu. Inches
1 Liter = 0.2642 U. S. Gallon = 1.0567 U. S. Quart

Measures of Weight—English and Metric
1 Long Ton = 2240 Lbs. = 1016.05 Kilograms
1 Short Ton = 2000 Lbs. = 907.18 Kilograms
1 Metric Ton = 2204.6 Lbs.
1 Kilogram = 2.2046 Lbs.
1 Lb. = 0.45359 Kilograms

Specific Gravity—is a number indicating how many times a certain volume of material is heavier than an equal volume of water.

ENGLISH SYSTEM—If a material has a specific gravity of 2.7 for instance, multiply this by 62.4 lbs. (weight of 1 cu. ft.) of water to obtain the weight in lbs. per cu. ft. of the material in question.

METRIC SYSTEM—If a material has a specific gravity of 2.7 for instance, multiply this by 1000 kilograms (weight of 1 cu. meter of water) to obtain the weight in kilograms per cu. meter of the material in question.

Equivalents of Density—English and Metric
1 Lb. per Cu. Yd. = 0.5933 Kg. per Cu. Meter
1 Kg. per Cu. Meter = 1.6856 Lbs. per Cu. Yd.

Equivalents of Pressure—English and Metric
1 Lb. per Sq. Inch = 0.0703 Kg. per Sq. Centimeter
1 Kg. per Sq. Centimeter = 14.244 Lbs. per Sq. Inch

Weights of Diesel Fuel
1 U. S. Gallon = 7 lbs. average.
1 U. S. Gallon = 3.17 kilograms.

Material Weights

Material	Weight in Bank per Cubic Yard	Percent of Swell	Swell Factor	Loose Weight per Cubic Yard
Ashes, Hard Coal.................	700-1000 lbs.	8%	.93	650-930 lbs.
Ashes, Soft Coal with Clinkers.....	1000-1515 lbs.	8%	.93	930-1410 lbs.
Ashes, Soft Coal, Ordinary.........	1080-1215 lbs.	8%	.93	1000-1130 lbs.
Bauxite........................	2700-4325 lbs.	33%	.75	2020-3240 lbs.
Brick...........................				2700 lbs.
Cement, Portland.................	94 lbs. per bag			
Cement, Portland.................	2970 lbs. (packed)	20%	.83	2450 lbs.
Coke, Lump, Loose...............				620-865 lbs.
Coke, Solvay, Egg, Chestnut or Pea...				840 lbs.
Coke, Gas, Egg, Chestnut or Pea.....				785 lbs.
Coke, Gas Furnace...............				730 lbs.
Concrete.......................	3240-4185 lbs.	40%	.72	2330-3000 lbs.
Concrete, Mix Wet...............				3500-3750 lbs.
Copper Ore.....................	3800 lbs.	35%	.74	2800 lbs.
Gasoline, 56° Gaume..............	6.3 lbs. per gallon			
Granite.........................	4500 lbs.	50 to 80%	.67 to .56	3000-2520 lbs.
Iron Ore, Hematite...............	6500-8700 lbs.		.45	3900 lbs.
Iron Ore, Limonite...............	6400 lbs.			
Iron Ore, Magnetite..............	8500 lbs.			
Kaolin.........................	2800 lbs.	30%	.77	2160 lbs.
Lead Ore, Galina................	12550 lbs.			
Lime...........................				1400 lbs.
Limestone, Blasted...............	4200 lbs.	67 to 75%	.60 to .57	2400-2520 lbs.
Limestone, Loose, Crushed.........				2600-2700 lbs.
Limestone, Marble...............	4600 lbs.	67 to 75%	.60 to .57	2620-2760 lbs.
Mud, Dry (Close)................	2160-2970 lbs.	20%	.83	1790-2460 lbs.
Mud, Wet (Moderately packed).....	2970-3510 lbs.	20%	.83	2470-2910 lbs.
Oil, Crude......................	6.42 lbs. per gallon			
Phosphate Rock.................	5400 lbs.			
Sand, Dry......................	3250 lbs.	12%	.89	2900 lbs.
Sand, Wet......................	3600 lbs.	14%	.88	3200 lbs.
Sandstone......................	4140 lbs.	40 to 60%	.72 to .63	2980-2610 lbs.
Shale, Riprap...................	2800 lbs.	33%	.75	2100 lbs.
Slag, Sand.....................	1670 lbs.	12%	.89	1485 lbs.
Slag, Solid.....................	4320-4860 lbs.	33%	.75	3240-2640 lbs.
Slag, Crushed..................				1900 lbs
Slag, Furnace, Granulated.........	1600 lbs.	12%	.89	1430 lbs.
Slate..........................	4590-4860 lbs.	30%	.77	3530-3740 lbs.
Trap Rock......................	5075 lbs.	50%	.67	3400 lbs.

Wood & Lumber

Beechwood..............	3250 lbs. per cord	Hemlock...............	2200 lbs. per cord
Chestnut...............	2350 lbs. per cord	Hickory...............	4500 lbs. per cord
Elm...................	2350 lbs. per cord	Pine, Norway or White...	2000 lbs. per cord
		Poplar................	2350 lbs. per cord

Miscellaneous Information

The COMPACTION TEST FOR OPTIMUM MOISTURE CONTENT is similar in purpose to the STANDARD PROCTOR TEST. The two tests differ in details of procedure as to the number of dirt layers and thickness of dirt, weight of the tamper used for compacting and the distance through which the tamper is moved.

MEASURE OF ANGLES

Degrees	Rise in Inches per ft.	Rise in Inches per ft.	Degrees and Minutes	Per cent Rise in 100 ft.	Degrees and Minutes	Per cent rise in 100 ft.	Degrees and Minutes
1	.210	¼	1° 11′	1	34.4′	36	19° 48′
2	.419	½	2° 23′	2	1° 8.7′	37	20° 18′
3	.629	¾	3° 35′	3	1° 43.1′	38	20° 48′
4	.839	1	4° 46′	4	2° 17.5′	39	21° 18′
5	1.050	1¼	5° 56′	5	2° 51.8′	40	21° 48′
6	1.261	1½	7° 7′	6	3° 26.0′	41	22° 18′
7	1.473	1¾	8° 18′	7	4° 0.3′	42	22° 47′
8	1.686	2	9° 28′	8	4° 34.4′	43	23° 16′
9	1.901	2¼	10° 37′	9	5° 8.6′	44	23° 45′
10	2.116	2½	11° 46′	10	5° 42.6′	45	24° 14′
11	2.333	2¾	12° 54′	11	6° 16.6′	46	24° 42′
12	2.551	3	14° 2′	12	6° 50.6′	47	25° 10′
13	2.770	3¼	15° 9′	13	7° 24.4′	48	25° 38′
14	2.992	3½	16° 15′	14	7° 58.2′	49	26° .6′
15	3.215	3¾	17° 21′	15	8° 31.9′	50	26° 34′
16	3.441	4	18° 26′	16	9° 5.4′	51	27° 1′
17	3.669	4¼	19° 30′	17	9° 38.9′	52	27° 28′
18	3.900	4½	20° 33′	18	10° 12.2′	53	27° 55′
19	4.132	4¾	21° 36′	19	10° 45.5′	54	28° 22′
20	4.368	5	22° 37′	20	11° 18.6′	55	28° 49′
21	4.606	5¼	23° 38′	21	11° 51.6′	56	29° 15′
22	4.848	5½	24° 37′	22	12° 24.5′	57	29° 41′
23	5.094	5¾	25° 36′	23	12° 57.2′	58	30° 7′
24	5.313	6	26° 34′	24	13° 29.8′	59	30° 32′
25	5.596	6¼	27° 31′	25	14° 2.2′	60	30° 58′
26	5.853	6½	28° 27′	26	14° 34.5′	61	31° 23′
27	6.114	6¾	29° 22′	27	15° 6.6′	62	31° 48′
28	6.381	7	30° 16′	28	15° 38.5′	63	32° 13′
29	6.652	7¼	31° 8′	29	16° 10.3′	64	32° 37′
30	6.928	7½	32°	30	16° 42.0′	65	33° 1′
31	7.210	7¾	32° 51′	31	17° 13.4′	66	33° 25′
32	7.498	8	33° 41′	32	17° 44.7′	67	33° 49′
33	7.793	8¼	34° 30′	33	18° 15.8′	68	34° 13′
34	8.094	8½	35° 19′	34	18° 46.7′	69	34° 36′
35	8.403	8¾	36° 5′	35	19° 17.0′	70	35° 0′

USEFUL CONVERSION FACTORS FOR RAPID APPROXIMATION

Feet	X	.00019	= Miles
Links	X	.66	= feet
Feet	X	1 5	= links
Square inches	X	.007	= square feet
Square feet	X	.111	= square yards
Acres	X	4,840.	= square yards
Square Yards	X	.002066	= acres
Width in chains	X	8.	= acres per mile
Cubic feet	X	.04	= cu. yds. (Ap.)
Cubic inches	X	.00058	= cu. ft.
U. S. bu.	X	.046	= cu. yds.
U. S. bu.	X	1.244	= cu. ft.
U. S. bu.	X	2,150.42	= cu. in.
Cubic feet	X	.8036	= U. S. bu.
Cubic inches	X	.000466	= U. S. bu.
U. S. gals.	X	.13368	= Cu. ft.
U. S. gals.	X	231.	= cu. in.
Cubic feet	X	7.48	= U. S. gals.
Cubic inches	X	.004329	= U. S. gals.
Cylindrical feet	X	5.878	= U. S. gals.
Cylindrical in.	X	.0034	= U. S. gals.
Pounds	X	.009	= cwt. (112 lbs.)
Pounds	X	.00045	= tons (2,240 lbs.)

Example: Given seven acres of land. To find number of square yards multiply seven by 4,840. Answer: 33,880 square yards.

TABLES OF USEFUL ENGINEERING DATA
MECHANICAL-ELECTRICAL EQUIVALENTS

Power

1 horsepower (hp	= 550 foot-pounds (ft.-lb.) per second (sec.)
	= 33,000 ft.-lb. per minute (min.)
	= 1,980,000 ft.-lbs. per hour (hr.)
	= .275 ft.-tons per sec.
	= 16.5 ft.-tons per min.
	= 990 ft.-tons per hr.
1 horsepower-second (hp-sec.)	= 550 ft.-lb.
	= .275 ft.-tons.
1 horsepower-minute (hp-min.)	= 33,000 ft.-lb.
	= 16.5 ft.-tons.
1 horsepower-hour (hp-hr.)	= 1,980,000 ft.-lb.
	= 990 ft.-tons
1 horsepower (hp)	= 746 watts (w)
	= .746 kilowatts (kw)

Energy

1 horsepower-hour	= 2544 BTU
	= .746 KW-hr.
1 Kilowatt-hour	= 3413 BTU

Pressure

1 lb. per sq. in.	= 2.0360″ of mercury at 32° F.
	= 27.71″ of water at 32° F.
	= 2.3091 ft. of water at 60° F.
	= 144 lb. per sq. ft.
1 in. of mercury	= .491 lb. per sq. in.
1 in. of water	= 5.2 lb. per sq. ft. = .0361 PSI.

PERCENT OF SEA LEVEL HORSEPOWER AVAILABLE FOR A FOUR CYCLE, GASOLINE OR DIESEL ENGINE FOR VARIOUS ALTITUDES

Altitude in feet	Temperature in degrees Fahrenheit							
	110	90	70	60	50	40	20	0	—20
0	95.4	97.1	99.1	100	100.8	101.8	103.9	106.2	108.5
1000	92.0	93.7	95.5	96.4	97.4	98.4	100.3	102.5	104.8
2000	88.7	90.4	92.1	93.0	93.8	94.8	96.8	98.8	101.0
3000	85.5	87.2	88.8	89.6	90.5	91.4	93.3	95.2	97.4
4000	82.5	84.0	85.6	86.5	87.3	88.2	89.9	91.8	93.8
5000	79.5	80.9	82.5	83.3	84.2	84.9	86.7	88.5	90.4
6000	76.7	78.1	79.5	80.3	81.1	82.0	83.6	85.3	87.2
7000	73.8	75.2	76.7	77.5	78.2	79.0	80.6	82.3	84.0
8000	71.2	72.5	73.9	74.6	75.4	76.2	77.6	79.3	81.1
9000	68.6	69.9	71.3	72.0	72.7	73.4	74.8	76.4	78.2
10000	66.2	67.5	68.7	69.3	70.7	70.7	72.2	73.7	75.3

PERCENT OF SEA LEVEL HORSEPOWER AVAILABLE IN TRACTORS AT VARIOUS ALTITUDES POWERED BY G.M.C. TWO-CYCLE DIESEL ENGINE (APPROXIMATE ONLY)

Altitude in feet	Percent of Horsepower Available	Altitude in feet	Percent of Horsepower Available
0	100.0	6000	96.0
1000	100.0	7000	95.3
2000	99.1	8000	94.7
3000	98.2	9000	94.2
4000	97.5	10000	93.6
5000	96.8		

Miscellaneous Information (*Continued*)

RULES OF THUMB

The following "Rules of Thumb" are approximately only.

ROUND TRIP HAUL TIME IN MINUTES for one way haul in ft.

$$\text{tractor-scraper} = \frac{}{100} + 1\tfrac{1}{2}.$$

GRADE RESISTANCE is equal to twenty pounds per ton of tractor weight for each 1% of grade.

THE MAXIMUM POUNDS DRAWBAR PULL of a crawler tractor is equal to 90% of its weight.

REPAIRS AND REPAIR LABOR COSTS for a crawler tractor will amount to about 100% of the delivered price of the machine based on a 5-year life of 10,000 hours.

SHEEPSFOOT COMPACTION OF SUBGRADE—continue passes until tamper "Walks itself out."

TO CORRECT ENGINE HORSEPOWER RATING FOR ALTITUDE:

1. For a gasoline or four stroke cycle engine deduct 3% for each 1000 ft. of altitude above sea level.

2. For a two stroke cycle engine deduct 1% for each 1000 feet above 1000 feet.

TO CORRECT ENGINE HORSEPOWER FOR TEMPERATURES:

1. Deduct 1% of rated power at 60°F. for each 10° temperature rise.

2. Add 1% of rated power at 60°F. for each 10° temperature drop.

MILES PER HOUR IN FEET PER MINUTE AND FEET PER SECOND

Miles Per Hour	Feet Per Minute	Feet Per Second
1	88	1.46
2	176	2.94
3	264	4.4
4	352	5.87
5	440	7.33
6	528	8.8
7	616	10.26
8	704	11.73
9	792	13.2
10	880	14.67
11	968	16.13
12	1,056	17.6
13	1,144	19.07
14	1,232	20.52
15	1,320	22.00
16	1,408	23.47
17	1,496	24.93
18	1,584	26.4
19	1,672	27.86
20	1,760	29.33
21	1,848	30.8
22	1,936	32.26
23	2,024	33.72
24	2,112	35.2
25	2,200	36.67
26	2,288	38.14
27	2,376	39.6
28	2,464	41.04
29	2,552	42.50
30	2,640	44.00

ANGLES OF SLOPES

Slopes ½ to 1	= 63° 30'
Slopes ¾ to 1	= 53° 00'
Slopes 1 to 1	= 45° 00'
Slopes 1¼ to 1	= 38° 40'
Slopes 1½ to 1	= 33° 42'
Slopes 1¾ to 1	= 29° 44'
Slopes 2 to 1	= 26° 35'
Slopes 3 to 1	= 18° 25'
Slopes 4 to 1	= 14° 2'

TABLE FOR CONVERTING PRESSURE PER SQUARE INCH INTO FEET HEAD OF WATER

Pounds per Sq. In.	Feet Head	Pounds per Sq. In.	Feet Head	Pounds per Sq. In.	Feet Head
1	2.31	45	103.90	140	323.26
2	4.62	50	115.45	150	346.34
3	6.93	55	126.99	160	369.44
4	9.24	60	138.54	170	392.53
5	11.54	65	150.08	180	415.62
6	13.85	70	161.63	190	438.71
7	16.16	75	173.17	200	461.80
8	18.47	80	184.72	225	519.52
9	20.78	85	196.26	250	577.25
10	23.09	90	207.81	275	634.97
15	34.63	95	219.35	300	692.70
20	46.18	100	230.90	325	750.42
25	57.72	110	253.99	350	808.15
30	69.27	120	277.08	375	865.87
35	80.81	125	288.62	400	923.60
40	92.36	130	300.17	500	1154.45

TABLE FOR CONVERTING FEET HEAD OF WATER INTO PRESSURE PER SQUARE INCH

Feet Head	Pounds per Sq. In.	Feet Head	Pounds per Sq. In.	Feet Head	Pounds per Sq. In.
1	.43	55	23.81	190	82.37
2	.87	60	25.98	200	86.60
3	1.30	65	28.14	225	97.42
4	1.73	70	30.31	250	108.25
5	2.17	75	32.47	275	119.07
6	2.60	80	34.64	300	129.90
7	3.03	85	36.80	325	140.72
8	3.46	90	38.97	350	151.55
9	3.90	95	41.13	375	162.37
10	4.33	100	43.30	400	173.20
15	6.50	110	47.63	500	216.50
20	8.66	120	51.96	600	259.80
25	10.83	130	57.29	700	303.10
30	12.99	140	60.62	800	346.40
35	15.16	150	64.95	900	389.70
40	17.32	160	69.28	1000	433.00
45	19.49	170	73.61	....	......
50	21.65	180	77.94	....	......

TRAVEL TIME IN MINUTES
TRAVEL DISTANCE IN FEET

Speed in Miles Per Hour	100	200	300	400	500	600	700	800	900	1000
1.0	1.136	2.275	3.405	4.540	5.67	6.82	7.95	9.08	10.22	11.36
2.0	.568	1.136	1.705	2.275	2.84	3.41	3.98	4.55	5.12	5.68
3.0	.379	.758	1.136	1.515	1.89	2.27	2.66	3.03	3.41	3.79
4.0	.284	.568	.853	1.136	1.42	1.70	1.99	2.27	2.56	2.84
5.0	.227	.454	.682	.910	1.14	1.36	1.59	1.82	2.04	2.27
6.0	.189	.378	.568	.758	.95	1.14	1.32	1.51	1.70	1.89
7.0	.162	.324	.486	.648	.81	.97	1.14	1.30	1.46	1.62
8.0	.142	.284	.427	.570	.71	.85	.99	1.14	1.28	1.42
9.0	.126	.252	.378	.505	.63	.76	.88	1.01	1.14	1.26
10.0	.114	.227	.341	.455	.57	.68	.80	.91	1.02	1.14
11.0	.103	.206	.310	.414	.52	.62	.72	.83	.93	1.03
12.0	.095	.189	.284	.379	.47	.57	.66	.76	.85	.95
13.0	.087	.174	.262	.349	.44	.52	.61	.70	.79	.87
14.0	.081	.162	.244	.325	.41	.49	.57	.65	.73	.81
15.0	.076	.152	.227	.303	.38	.46	.53	.61	.68	.76
17.5	.065	.129	.194	.259	.32	.39	.45	.52	.58	.65
20.0	.057	.113	.170	.227	.28	.34	.40	.45	.51	.57
22.5	.050	.101	.151	.202	.25	.30	.35	.40	.45	.50
25.0	.045	.090	.136	.181	.23	.27	.32	.36	.41	.45

EXAMPLE: To estimate time required to travel 550 feet at 6.0 MPH.

First establish time for 500 ft. at 6.0 MPH .... .95

50 ft.—½ of time shown for 100 ft. at 6.0 MPH .... .09
_____

1.04 **Minutes**

Enter 1.04 min. for travel time for 550 ft. at 6 MPH.

Tire Load and Inflation Table

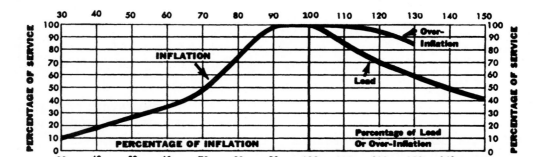

Source: Courtesy of Rubber Manufacturers Association.

General industry practices

The following is for information purposes only to explain the general industry practices of equipment rental, and is not to suggest or influence the rates or terms of rental of any item as these are matters which must be determined between the lessee and lessor. To avoid disagreements, the rental terms should be agreed upon prior to rental and, especially where larger equipment is involved, the terms should be spelled out in a written agreement signed by both the lessor and lessee.

Time basis of rates

It is the general practice in the industry to base rates upon one shift of 8 hours per day, 40 hours per week, or 176 hours per month of a 30 consecutive day period.

If the equipment is rented by the day, the rate for overtime is one-eighth of the daily rate for each hour in excess of eight. If it is rented by the week, the rate for overtime is 1/40 of the weekly rate for each hour in excess of 40. If it is rented by the month, the overtime rate is 1/176th of the monthly rate for each hour in excess of 176 hours in any one 30 consecutive day period.

Many distributors do not rent by the day or by the week, especially in the case of large equipment, even though daily and weekly rates appear in this compilation.

Costs of repairs

Tractor equipment: In the case of tractor equipment and/or rubber-tired hauling equipment, the difference between "normal" and "abnormal" wear and tear is not easily discernible and can be the cause of unpleasant relationships between the lessor and lessee. Most lessors, therefore, require the lessee, in their lease agreements, to bear all costs of repair to such equipment, regardless of the cause. Many distributors measure tread wear to determine charges to lessee for tire wear. Tractor equipment would include crawler as well as rubber-tired.

Non-tractor equipment: The lessor, under the lease agreement, usually bears the cost of repairs due to normal wear and tear on non-tractor equipment, and the lessee bears all other costs. **Cranes and shovels:** Rubber-tired or crawler mounted cranes or shovels would not be included in the category of either tractor equipment or rubber-tired hauling equipment and generally the lessor agrees to bear the cost of repairs due to normal wear and tear. In individual cases (especially where the equipment may be subjected to unusual abuse or excessive wear and tear), lessors insist on the lessee bearing all costs of repair on equipment regardless of its type or classification or charge a higher rental rate.

"Normal wear and tear"

"Normal wear and tear" is that which would be expected to result from the use of the equipment under normal circumstances, provided the equipment is properly maintained and serviced. This is a question of fact which must be settled by the interested parties.

Fuel and Lubricants

In all cases, the lessee bears costs of fuel and lubricants.

Operator

In no case does the average rental rate in this publication include the cost of operator.

Condition of equipment

The equipment rented is to be delivered to the lessee in good operating condition, and is to be returned to the lessor in the same condition as delivered, less normal wear, unless there is an agreement which may vary this general practice. Where excess cleaning or repairs are necessary, many lessors have made additional charges for this cleaning and repair.

Freight charges

The rates in the schedule are all f.o.b. the lessor's warehouse or shipping point. The lessee pays the freight or drayage charges from shipping point to destination and return. When loading, unloading, dismantling or assembling is necessary, the lessee pays this additional charge.

Rental period

On local rentals the rate starts when equipment leaves the lessor's warehouse and stops when it is returned to such warehouse. On out-of-town shipments, the rental starts on the date of the bill of lading of shipments and stops on the date of the return bill of lading.

Payment, taxes and insurance

Rentals are payable in advance and subject to the terms and conditions of the lessor's rental contract. No insurance, license, sale or use taxes are included in these rates.

Courtesy of Associated Equipment Distributors (AED)

EXPLANATION OF TABULATED ITEMS

ALL PERCENTAGE RATES ARE OFFERED AS A GUIDE AND SHOULD BE REVIEWED BY THE INDIVIDUAL EQUIPMENT OWNER.

DEPRECIATION AND OBSOLESCENCE

The straight-line method of depreciation based on the useful life of the machine is best suited to a compilation of this type. A uniform percentage is charged off yearly during the economic life of each piece of equipment.

Obsolescence is an element whose effects are hidden but nonetheless real. When recently purchased equipment is outmoded by advanced models or new methods it is apparent that the purchaser's business potential is affected.

OVERHAUL, MAJOR REPAIRS, PAINTING

This section covers the cost of periodic major repair and reconditioning work, usually performed in the owner's shops. Broadly speaking it is the cost of offsetting the effects of the heavy wear-and-tear to which construction equipment is subjected, and of repairing damage which may be discovered only during a period of overhaul. The percentage factors listed for major repairs are for average working conditions and should be re-negotiated if abnormal conditions exist.

GENERAL OVERHEAD

This broad heading incorporates a number of items which are constant for all items of equipment and consequently have been grouped together. The total of these is 10% and they are divided into four classifications:—

INTEREST. An interest rate of 4% has been used in this schedule.

STORAGE. This expense consists of the cost of providing land and buildings for the storage of equipment when not in use, starting or running machines in storage, the maintenance and servicing of the property and any direct overhead involved. An average figure for this expense is 1%.

INSURANCE. An allowance of 1% is made to cover the cost of general policies covering equipment.

OVERHEAD. An allocation of 4% is made to cover the general expense of doing business. Overhead includes all business operating costs such as office space, supplies and office equipment, salaries and wages and other essentials for the conduct of a business.

TOTAL PERCENTAGE

This is the total of the percentages of the first three columns. Under average conditions this figure covers the cost of doing business and ensures funds to purchase new equipment.

AVERAGE USE MONTHS PER YEAR

To convert the total to a monthly rental percentage rate it is necessary to determine, for each piece of equipment, the average number of months of operation annually. Dividing the total percentage by the average use in months per year produces the monthly percentage rate. There is little need to stress the extent to which the number of working months is affected by the vagaries of demand, job nature, climate and other factors, and the fact that the monthly rental rate varies accordingly.

MONTHLY RENTAL RATE

The monthly rental percentage rate is based on a minimum period of not less than one month, and operational time of not more than 200 hours monthly. A month shall be considered from the date of commencement of the rental period up to, but not including, the same date in the next calendar month. After the minimum rental period has expired, the rent payable for a fraction of any succeeding month shall be the proportionate part of the monthly rental rate in dollars according to the number of calendar days in such fraction.

CURRENT REPLACEMENT VALUE

The current replacement value of new equipment including Provincial and Municipal sales taxes must be used as the basis for calcuating equipment rental rates, and not the original purchase price. The necessity for this is apparent when it is appreciated that changes in the purchase price of new equipment are normally accompanied by corresponding changes in the cost of all the factors which comprise the rental rate. When the monthly rental percentage rate is applied to current replacement value there is automatic compensation for such changes. This would not

Courtesy of Canadian Construction Association

be the case if the original purchase price were used. Consequently, all percentages in this schedule are expressed in terms of current replacement value.

Current Replacement Value ($) x Monthly Percentage (%) = Monthly Rental ($).

WEEKLY AND DAILY RATES

Regardless of the length of the rental period it is necessary for an owner to prepare equipment for delivery in good working order and, upon its return, to inspect and restore it as nearly as possible to its original condition. When the rental period is short, it is apparent that the cost of inspection and handling would be out of proportion to the earned rental if the monthly rate were used.

This condition is adjusted by the use of a weekly rate which is ⅓ of the monthly rate, and by a daily rate which is 1/12 of the monthly rate.

The weekly rental rate is based on operational time of not more than 60 hours in seven consecutive days, while the daily rate has a basis of not more than 10 working hours in a twenty-four hour period.

OVERTIME

Since rental rates are based on a maximum number of working hours in a given period it is necessary to charge for the additional wear-and-tear on equipment due to overtime operation. Wear-and-tear increases in proportion to the overtime while other factors may undergo little or no change.

Overtime should be charged by the hour at 66⅔% of the straight-time rate. For example, when equipment is rented on a monthly basis the hourly

cost to the lessee is $\dfrac{Monthly\ Rental\ in\ Dollars,}{200}$

and overtime hours should be charged at 66⅔% of this figure. Similarly, the overtime hourly rate

on weekly rentals is 66⅔% of $\dfrac{Weekly\ Rental\ in\ Dollars}{60}$

and, for daily rentals, 66⅔% of $\dfrac{Daily\ Rental\ in\ Dollars.}{10}$

Courtesy of Canadian Construction Association

"ALL-FOUND" RENTALS

It is the practice of some owners to provide equipment on an "operated" basis. Under this arrangement it is the responsibility of the owner to pay all expenses in connection with normal operation of the equipment, with the usual exception that the lessee pays transportation costs.

The rates listed in this schedule are bare "machine" rates and will not be adequate under such conditions. A separate and additional rate must be calculated for those items which may be classified as job or operating costs. This rate, in dollars, is added to the bare "machine" rate in dollars to produce the required "all-found" rental.

Typical of the job costs which must be incorporated into an "all-found" rental are wages for operators and other labour; unemployment insurance; holiday pay allowance; workmen's compensation; public liability insurance; equipment insurance; fuel, lubricants and supplies; labour and material for field repairs; unloading; erecting; dismantling; loading; and an allowance for overhead.

RENTAL AGREEMENTS

While the terms of agreement may be subject to negotiation between lessor and lessee it may be of value to enumerate briefly those conditions which are commonly accepted.

Equipment to be rented is assumed to be in good mechanical condition, clean and with all standard attachments and accessories. Rental starts when the equipment leaves the lessor's shipping point and ends on and includes the date of actual delivery of the equipment to the Lessor or at any other equidistant point if instructions to do so are given by the Lessor. Rental is normally f.o.b. the lessor's premises and the lessee pays the cost of transportation both ways, plus unloading and loading costs at the job. Costs of erecting, dismantling and operating (including field repairs) are borne by the lessee. Rental is payable in advance by the month.

In the interests of uniformity it is strongly recommended that standard rental agreement forms be used, a copy of which is included on pages 63 to 65.

SOIL CLASSIFICATION

In order to describe soils, the Public Roads Administration has investigated various soil types which exhibit characteristic field behavior. On the basis of this study soils have been divided into eight distinct classes. These classifications are sufficiently detailed so that characteristics such as compressibility, elasticity, capillary action, cohesion, shrinkage and moisture content—all extremely vital considerations to a good subgrade—can, when considered with local climatic and usage conditions, give a good index to the adequacy of the soil for a desired purpose. These eight soil classifications are as follows:

A-1—Well graded material, coarse and fine, excellent binder. Highly stable under wheel loads irrespective of moisture conditions. Functions satisfactorily when surface treated or when used as a base for relatively thin wearing courses.

A-2—Coarse and fine materials, improper grading or inferior binder. Highly stable when fairly dry. Likely to soften at high water content caused either by rains or high capillary rise from saturated lower strata, when an impervious cover prevents evaporation from top layer, or to become loose and dusty in long continued dry weather.

A-3—Coarse material only, no binder. Lacks stability under wheel loads, but is unaffected by moisture conditions. Not likely to heave because of frost, nor to shrink or expand in appreciable amounts. Furnishes excellent support for flexible pavement of moderate thickness and for relatively thin rigid pavements.

A-4—Silt soils, without coarse material, and with no appreciable amount of sticky colloidal clay. Has a tendency to absorb water very readily in quantities sufficient to cause rapid loss of stability even when not manipulated. When dry or damp presents a firm riding surface which rebounds but very little upon the removal of load. Likely to cause cracking in rigid pavements as a result of frost heaving, and failure in flexible pavements because of low supporting value.

A-5—Similar to Group A-4 but have highly elastic supporting surfaces with appreciable rebound upon removal of load even when dry. Elastic properties interfere with proper compaction of macadams during construction and with retention of good bond afterwards.

A-6—Clay soils without coarse material. In stiff or soft plastic state absorb additional water only if manipulated. May then change to a liquid state and work up into the interstices of macadams or cause failure due to sliding in high fills. Furnish firm support essential in properly compacting macadams only at stiff consistency. Deformations occur slowly and removal of load causes very little rebound. Shrinkage properties combined with alternate wetting and drying under field conditions are likely to cause cracking in rigid pavements.

A-7—Similar to Group A-6 but at certain moisture contents deforms quickly under load and rebounds appreciably upon removing of load, as do subgrades of Group A-5. Alternate wetting and drying under field conditions leads to even more detrimental volume changes than in Group A-6 subgrades. May cause concrete pavements to crack before setting and to crack and fault afterwards. May contain lime or associated chemicals productive of flocculation in soils.

A-8—Very soft peat and muck incapable of supporting a road surface without being previously compacted.

To classify a given soil, a sample is run through a series of tests to determine into which of the above groups it most closely falls. The tests to determine its classification are as follows:

1. SIEVE ANALYSIS TEST—This test determines the per cent of total quantities that will pass through seven different size sieves. Certain further checks are made to determine the distribution of material passing through a No. 40 sieve.

2. MOISTURE EQUIVALENT TEST—This test determines the per cent of weight difference between a dry sample and a moist sample.

3. LIQUID LIMIT TEST—This test is defined as the per cent of moisture at which soil changes from a plastic to a liquid condition. The test is conducted by thoroughly mixing a sample with water, smoothing it out, marking a groove in the sample and then determining the number of controlled shocks necessary to close the groove. By repeated tests it is determined what moisture content will permit the groove to close with twenty-five shocks. This moisture content is the liquid limit.

4. PLASTIC LIMIT TEST—This is defined as the per cent of moisture at which the soil changes from a solid to a plastic condition. Test is conducted by moisting a sample and rolling it into a $\frac{1}{8}$" diameter thread with the palm of the hand. The moisture content at the time the thread begins to crumble determines the Plastic Limit.

5. PLASTICITY INDEX—The numerical difference between the liquid limit and plastic limit.

6. SHRINKAGE TEST—This test determines the "Shrinkage Limit" and the "Shrinkage Ratio". Test is conducted by putting a sample in a test bowl, drying out, and noting volume change.

$$\text{The Shrinkage Limit} = (\% \text{ moisture content}) - \frac{(\text{Volume of dish} - \text{volume dry soil})}{\text{Weight dry soil}} (100)$$

$$\text{The Shrinkage Ratio} = \frac{\text{Weight of dry soil}}{\text{Volume of dry soil}}$$

7. FIELD MOISTURE CONTENT—Minimum moisture content, expressed as a percentage of the weight of the oven dried soil, at which a drop of water placed on a smoothed surface of the soil will not immediately be absorbed, but will instead spread out over the surface and give it a shiny appearance.

8. SOIL ACIDITY OR ALKALINITY—Determine pH value with colorometric test equipment. One purpose is that a lime content has certain beneficial characteristics.

When the above tests have been made the results are compared by use of charts and the soil classed accordingly. Many soils will be border line cases as to classification.

Although the soil tests and resulting classification will usually give a good index to the behavior of a soil, it does not fill the need for practical soil classification terminology required by the engineer out on the job. Under field circumstances he may be able to test the soil only by visual examination. One of the common classification methods used by many engineers in the field is grouping soils by texture and structure. The terms are general and the range in any one group may be great.

These groups are as follows:

SANDY SOIL—Loose and granular soil, the individual grains of which can readily be seen or felt and may range from very fine sand to coarse sand.

CLAY SOILS—Clay soil is a fine-textured soil which forms hard lumps or clods when dry.

LOAM—A loam is a soil having a relatively even mixture of sand, silt and clay.

SANDY LOAM—A soil containing much sand but having sufficient silt and clay to render it coherent.

SILT LOAM—When this class of soil is dry and powdered, it is often called "rock flour." It is a soil having a moderate amount of fine sand and clay, over half the particles being of the size called "silt". The dry lumps are easily broken and then feel soft and floury.

CLAY LOAM—A fine-textured soil having a large percentage of clay. When dry, the clods are hard and difficult to break.

GRAVELLY OR STONY SOILS—All the above soils, if mixed with a considerable amount of pebbles, are classed as gravelly sand loams; sand clay loams; sandy clay soils, etc.

SOIL COMPACTION

The primary objective of compacting soil by sheepsfoot rollers, flat wheel rollers, pneumatic tired units, or other means is to obtain a soil of a specific density in order that it will carry specified loads without undue settlement. Much has been written on this subject, but soil types, equipment, operating conditions and moisture content are so variable that it is not practical to attempt to state definitely what work is required and what equipment is needed to get certain definite results from compaction.

The work necessary to get the desired compaction on a specific job should be determined by actual test on the job.

Soil settlement occurs under load for two reasons: (1) Air and water are expelled from the earth due to compression; and (2) The earth is forced out laterally into the surrounding soil.

Compaction operations attempt to do these things artificially by means of various types of rollers or tampers so that settlement after construction work is completed will be held to a minimum. To do this, two principles of action are involved.

1. It is necessary to place the earth in layers sufficiently thin to permit air and water to be expelled efficiently and easily. Some soils, depending upon their permeability, may be put down in thicker layers than others. For example, clay must be placed in thin layers whereas a sandy soil could be rolled in thick layers.

2. The second principle to consider is that the compression of soil particles requires movement of the individual particles in order to fit them together and fill in the voids. Before movement can take place friction must be reduced. Lubrication of the soil particles by means of moisture will help to overcome friction. Too little moisture will not materially reduce friction; too much moisture only means that the excess water must be expelled. There is, then, an optimum or ideal moisture content.

Tests have been developed for determining the adequacy of soil compaction. There is some difference in the exact procedure of tests as used by the Army Engineers and the various States, but the fundamental principles remain the same. The tests generally used are based on procedures established by the American Association of State Highway Officials (AASHO). Three main tests are used to test soil for proper compaction.

1. Moisture-content test.
2. Unit-weight determination or density test.
3. Compaction test for optimum-moisture content.

The MOISTURE CONTENT TEST (Similar to Public Roads Administration "Moisture Equivalent Test") is used to determine the ratio of the weight of the water contained in a given sample to the dry weight of the sample. The answer is expressed in per cent. The test is conducted by weighing a moist sample of earth, drying it in an oven, then noting the loss in weight due to the water evaporation. The weight of water lost divided by the weight of the dry sample and multiplied by 100 equals the per cent of moisture content.

The UNIT WEIGHT determination is a test for determining the weight of a unit volume. The answer is expressed in pounds per cubic foot.

The COMPACTION TEST FOR OPTIMUM MOISTURE CONTENT (Modified ASSHO method) is an important test used to determine what quantity of moisture in earth will permit the greatest compaction. If too much water is present more work must be done to expel the excess water. If insufficient water is present the dirt will not compact easily. This test is made by compacting in a standard test machine a quantity of the sample dirt which has been thoroughly mixed with water. After compaction the weight per unit volume of the compacted material is determined. Next, samples of the compacted earth are taken and the moisture content is determined as in the MOISTURE CONTENT TEST discussed above. From this information the moisture content for a unit weight of dirt is now known. This same procedure is repeated on several samples with varying amounts of water added until the addition of more water does not give any weight increase for a given volume. The moisture content which results in the greatest weight per volume is the OPTIMUM MOISTURE CONTENT.

Compacted Cubic Yards of Run-of-Bank Gravel = Square Yards x Depth in inches x .033

Square Yards	1"	2"	3"	4"	5"	6"	8"	10"
1	.033	.066	.099	.132	.165	.198	.264	.330
2	.066	.132	.198	.264	.330	.396	.528	.660
3	.099	.198	.297	.396	.495	.594	.792	.990
4	.132	.264	.396	.528	.660	.792	1.056	1.320
5	.165	.330	.495	.660	.825	.990	1.320	1.650
6	.198	.396	.594	.792	.990	1.188	1.584	1.980
7	.231	.412	.693	.924	1.155	1.386	1.848	2.310
8	.264	.528	.792	1.056	1.320	1.584	2.112	2.640
9	.297	.594	.891	1.188	1.485	1.782	2.376	2.970
10	.330	.660	.990	1.320	1.650	1.980	2.640	3.300
20	.660	1.32	1.98	2.64	3.30	3.96	5.28	6.60
30	.990	1.98	2.97	3.96	4.95	5.94	7.92	9.90
40	1.32	2.64	3.96	5.28	6.60	7.92	10.56	13.20
50	1.65	3.30	4.95	6.60	8.25	9.90	13.20	16.50
60	1.98	3.96	5.94	7.92	9.90	11.88	15.84	19.80
70	2.31	4.62	6.93	9.24	11.55	13.86	18.48	23.10
80	2.64	5.28	7.92	10.56	13.20	15.84	21.12	26.40
90	2.97	5.94	8.91	11.88	14.85	17.82	23.76	29.70
100	3.30	6.60	9.90	13.20	16.50	19.80	26.40	33.00
200	6.60	13.20	19.80	26.40	33.00	39.60	52.80	66.00
300	9.90	19.80	29.70	39.60	49.50	59.40	79.20	99.00
400	13.20	26.40	39.60	52.80	66.00	79.20	105.60	132.00
500	16.50	33.00	49.50	66.00	82.50	99.00	132.00	165.00
600	19.80	39.60	59.40	79.20	99.00	118.00	158.40	198.00
700	23.10	46.20	69.30	92.40	115.50	138.60	184.80	231.00
800	26.40	52.80	79.20	105.60	132.00	158.40	211.20	264.00
900	29.70	59.40	89.10	118.80	148.50	178.20	237.60	297.00
1000	33.00	66.00	99.00	132.00	165.00	198.00	264.00	330.00
2000	66.00	132.00	198.00	264.00	330.00	396.00	528.00	660.00
3000	99.00	198.00	297.00	396.00	495.00	594.00	792.00	990.00
4000	132.00	264.00	396.00	528.00	660.00	792.00	1056.00	1320.00
5000	165.00	330.00	495.00	660.00	825.00	990.00	1320.00	1650.00
6000	198.00	396.00	594.00	792.00	990.00	1180.00	1584.00	1980.00
7000	231.00	462.00	693.00	924.00	1155.00	1386.00	1848.00	2310.00
8000	264.00	528.0(	792.00	1056.00	1320.00	1584.00	2112.00	2640.00
9000	297.00	594.00	891.00	1188.00	1485.00	1782.00	2376.00	2970.00
10000	330.00	660.00	990.00	1320.00	1650.00	1980.00	2640.00	3300.00

Example: 546 square yards x 6 inches deep:

500 s.y.	99.000 cu. yd.
40 s.y.	7.920 cu. yd.
6 s.y.	1.188 cu. yd.
	108.108, say 108 cu. yd.

For crushed stone, add 10% to figures in table
For loose yards of gravel (truck measure) add 30% to figures in table.

"DOS AND DON'TS"
Adopted by The Institute of Makers of Explosives

Definitions

1. The term "explosives" as used herein includes any or all of the following: dynamite, black blasting powder, pellet powder, blasting caps, electric blasting caps and detonating cord.

2. The term "electric blasting cap" as used herein includes both instantaneous electric blasting caps and all types of delay electric blasting caps.

3. The term "primer" as used herein means a cartridge of explosives in combination with a blasting cap or an electric blasting cap.

When Transporting Explosives

1. DO obey all federal, state and local laws and regulations.

2. DO see that any vehicle used to transport explosives is in proper working condition and equipped with a tight wooden or non-sparking metal floor with sides and ends high enough to prevent the explosives from falling off. The load in an open-bodied truck should be covered with a waterproof and fire-resistant tarpaulin, and the explosives should not be allowed to contact any source of heat such as an exhaust pipe. Wiring should be fully insulated so as to prevent short circuiting, and at least two fire extinguishers should be carried. The trucks should be plainly marked so as to give adequate warning to the public of the nature of the cargo.

3. DON'T permit metal, except approved metal truck bodies, to contact cases of explosives. Metal, flammable, or corrosive substances should not be transported with explosives.

4. DON'T allow smoking or unauthorized or unnecessary persons in the vehicle.

5. DO load and unload explosives carefully. Never throw explosives from the truck.

6. DO see that other explosives, including detonating cord, are separated from blasting caps and/or electric blasting caps where it is permitted to transport them in the same vehicle.

7. DON'T drive trucks containing explosives through cities, towns or villages, or park them near such places as restaurants, garages and filling stations, unless it cannot be avoided.

8. DO request that explosive deliveries be made at the magazine or in some other location well removed from populated areas.

9. DON'T fight fires after they have come in contact with explosives. Remove all personnel to a safe location and guard the area against intruders.

When Storing Explosives

10. DO store explosives in accordance with federal, state or local laws and regulations.

11. DO store explosives only in a magazine which is clean, dry, well ventilated, reasonably cool, properly located, substantially constructed, bullet and fire resistant, and securely locked.

12. DON'T store blasting caps or electric blasting caps in the same box, container or magazine with other explosives.

13. DON'T store explosives, fuse, or fuse lighters in a wet or damp place, or near oil, gasoline, cleaning solution or solvents, or near radiators, steam pipes, exhaust pipes, stoves, or other sources of heat.

14. DON'T store any sparking metal, or sparking metal tools in an explosives magazine.

15. DON'T smoke or have matches, or have any source of fire or flame in or near an explosives magazine.

16. DON'T allow leaves, grass, brush, or debris to accumulate within 25 feet of an explosives magazine.

17. DON'T shoot into explosives or allow the discharge of firearms in the vicinity of an explosives magazine.

18. DO consult the manufacturer if nitroglycerin from deteriorated explosives has leaked onto the floor of a magazine. The floor should be desensitized by washing thoroughly with an agent approved for that purpose.

19. DO locate explosives magazines in the most isolated places available. They should be separated from each other, and from inhabited buildings, highways, and railroads, by distances not less than those recommended in the "American Table of Distances."

When Using Explosives

20. DON'T use sparking metal tools to open kegs or wooden cases of explosives. Metallic slitters may be used for opening fiberboard

Courtesy of E. I. du Pont de Nemours & Co. (Inc.)

cases, provided that the metallic slitter does not come in contact with the metallic fasteners of the case.

21. DON'T smoke or have matches or any source of fire or flame, within 100 feet of an area in which explosives are being handled or used.

22. DON'T place explosives where they may be exposed to flame, excessive heat, sparks, or impact.

23. DO replace or close the cover of explosives cases or packages after using.

24. DON'T carry explosives in the pockets of your clothing or elsewhere on your person.

25. DON'T insert anything but fuse in the open end of a blasting cap.

26. DON'T strike, tamper with, or attempt to remove or investigate the contents of a blasting cap or an electric blasting cap, or try to pull the wires out of an electric blasting cap.

27. DON'T allow children or unauthorized or unnecessary persons to be present where explosives are being handled or used.

28. DON'T handle, use, or be near explosives during the approach or progress of any electrical storm. All persons should retire to a place of safety.

29. DON'T use explosives or accessory equipment that are obviously deteriorated or damaged.

30. DON'T attempt to reclaim or to use fuse, blasting caps, electric blasting caps, or any explosives that have been water soaked, even if they have dried out. Consult the manufacturer.

When Preparing The Primer

31. DON'T make up primers in a magazine, or near excessive quantities of explosives, or in excess of immediate needs.

32. DON'T force a blasting cap or an electric blasting cap into dynamite. Insert the cap into a hole made in the dynamite with a punch suitable for the purpose.

33. DO make up primers in accordance with proven and established methods. Make sure that the cap shell is completely encased in the dynamite or booster and so secured that in loading no tension will be placed on the wires or fuse at the point of entry into the cap. When side priming a heavy wall or heavy weight cartridge, wrap adhesive tape around the hole punched in the cartridge so that the cap cannot come out.

When Drilling And Loading

34. DO comply with applicable federal, state and local regulations relative to drilling and loading.

35. DO carefully examine the surface or face before drilling to determine the possible presence of unfired explosives. Never drill into explosives.

36. DO check the borehole carefully with a wooden tamping pole or a measuring tape to determine its condition before loading.

37. DO recognize the possibility of static electrical hazards from pneumatic loading and take adequate precautionary measures. If any doubt exists, consult your explosives supplier.

38. DON'T stack surplus explosives near working areas during loading.

39. DO cut from the spool the line of detonating cord extending into a borehole before loading the remainder of the charge.

40. DON'T load a borehole with explosives after springing (enlarging the hole with explosives) or upon completion of drilling without making certain that it is cool and that it does not contain any hot metal, or burning or smoldering material. Temperatures in excess of 150° F. are dangerous.

41. DON'T spring a borehole near another hole loaded with explosives.

42. DON'T force explosives into a borehole or through an obstruction in a borehole. Any such practice is particularly hazardous in dry holes and when the charge is primed.

43. DON'T slit, drop, deform or abuse the primer. DON'T drop a large size, heavy cartridge directly on the primer.

44. DO avoid placing any unnecessary part of the body over the borehole during loading.

45. DON'T load any boreholes near electric power lines unless the firing line, including the electric blasting cap wires, is so short that it cannot reach the power wires.

46. DON'T connect blasting caps, or electric blasting caps to detonating cord except by methods recommended by the manufacturer.

When Tamping

47. DON'T tamp dynamite that has been removed from the cartridge.

48. DON'T tamp with metallic devices of any kind, including the metal end of loading poles. Use wooden tamping tools with no exposed metal parts, except non-sparking metal connectors for jointed poles. Avoid violent tamping. Never tamp the primer.

49. DO confine the explosives in the borehole with sand, earth, clay, or other suitable incombustible stemming material.

50. DON'T kink or injure fuse, or electric blasting cap wires, when tamping.

When Shooting Electrically

51. DON'T uncoil the wires or use electric blasting caps during dust storms or near any other source of large charges of static electricity.

52. DON'T uncoil the wires or use electric blasting caps in the vicinity of radio-frequency transmitters, except at safe distances. Consult the manufacturer or the Institute of Makers of Explosives pamphlet on "Radio Frequency Hazards."

53. DO keep the firing circuit completely insulated from the ground or other conductors such as bare wires, rails, pipes, or other paths of stray currents.

54. DON'T have electric wires or cables of any kind near electric blasting caps or other explosives except at the time and for the purpose of firing the blast.

55. DO test all electric blasting caps, either singly or when connected in a series circuit, using only a blasting galvanometer specifically designed for the purpose.

56. DON'T use in the same circuit either electric blasting caps made by more than one manufacturer, or electric blasting caps of different style or function even if made by the same manufacturer, unless such use is approved by the manufacturer.

57. DON'T attempt to fire a single electric blasting cap or a circuit of electric blasting caps with less than the minimum current specified by the manufacturer.

58. DO be sure that all wire ends to be connected are bright and clean.

59. DO keep the electric cap wires or leading wires disconnected from the power source and short circuited until ready to fire.

When Shooting With Fuse

60. DO handle fuse carefully to avoid damaging the covering. In cold weather warm slightly before using to avoid cracking the waterproofing.

61. DON'T use short fuse. Know the burning speed of the fuse and make sure you have time to reach a place of safety after lighting. Never use less than two feet.

62. DON'T cut fuse until you are ready to insert it into a blasting cap. Cut off an inch or two to insure a dry end. Cut fuse squarely across with a clean sharp blade. Seat the fuse lightly against the cap charge and avoid twisting after it is in place.

63. DON'T crimp blasting caps by any means except a cap crimper designed for the purpose. Make certain that the cap is securely crimped to the fuse.

64. DO light fuse with a fuse lighter designed for the purpose. If a match is used the fuse should be slit at the end and the match head held in the slit against the powder core. Then scratch the match head with an abrasive surface to light the fuse.

65. DON'T light fuse until sufficient stemming has been placed over the explosive to prevent sparks or flying match heads from coming into contact with the explosive.

66. DON'T hold explosives in the hands when lighting fuse.

In Underground Work

67. DO use permissible explosives only in the manner specified by the United States Bureau of Mines.

68. DON'T take excessive quantities of explosives into a mine at any one time.

69. DON'T use black blasting powder or pellet powder with permissible explosives or other dynamite in the same borehole in a coal mine.

Before And After Firing

70. DON'T fire a blast without a positive signal from the one in charge, who has made certain that all surplus explosives are in a safe place, all persons and vehicles are at a safe distance or under sufficient cover, and that adequate warning has been given.

71. DON'T return to the area of any blast until the smoke and fumes from the blast have been dissipated.

72. DON'T attempt to investigate a misfire too soon. Follow recognized rules and regulations, or if no rules or regulations are in effect, wait at least one hour.

73. DON'T drill, bore, or pick out a charge of explosives that has misfired. Misfires should be handled only by or under the direction of a competent and experienced person.

Explosives Disposal

74. DON'T abandon any explosives.

75. DO dispose of or destroy explosives in strict accordance with approved methods. Consult the manufacturer or follow the Institute of Makers of Explosives pamphlet on destroying explosives.

76. DON'T leave explosives, empty cartridges, boxes, liners, or other materials used in the packing of explosives lying around where children or unauthorized persons or livestock can get at them.

77. DON'T allow any wood, paper, or any other materials employed in packing explosives to be burned in a stove, a fireplace, or other confined space, or to be used for any purpose. Such materials should be destroyed by burning at an isolated location out-of-doors and no person should be nearer than 100 feet after the burning has started.

ADDITIONAL "DOS AND DON'TS" PARTICULARLY APPLICABLE TO SEISMIC PROSPECTING

1. DO post "Explosives" signs conspicuously near explosives magazines. These signs should be so placed that a bullet passing through them at right angles cannot strike a magazine.

2. DO provide separate storage compartments for the dynamite and electric blasting caps on the shooting truck, where it is permitted to transport them on the same vehicle. These compartments should be lined with some soft material such as wood or rubber. If detonating cord is to be used, it should be carried in the dynamite compartment.

3. DON'T make up more charges than needed to be loaded and fired in one shot.

4. DO place the cap near the top of a dynamite charge, either midway in the side of the top cartridge or in the top of the second cartridge. When side-priming, wrap adhesive tape around the hole punched in the cartridge so that the cap cannot come out.

5. DO use sufficient half-hitches with the cap wires to secure the electric blasting cap in the cartridge. One half-hitch may be inadequate to prevent tension on the wires from pulling the cap out of the cartridge.

6. DO make sure, particularly in dry holes, that the hole is cool and there are no hot pieces of drill steel in the hole. If in doubt, pour in enough water or dirt to cover the bottom of the hole or wait at least one hour before loading it. Temperatures in excess of 150°F. are dangerous.

7. DON'T drop into a borehole an explosive charge that contains an electric blasting cap. DON'T drop the next explosive unit if any must be placed on top of the one containing the cap.

8. DO make certain that the charge is securely placed at a safe depth in the hole. Use shot anchors if there is any chance that the charge may "float," such as in heavy drilling mud or in water containing marsh gas.

9. DO securely anchor any casing, if there is a possibility that it may blow out of the borehole.

10. DON'T approach any explosives that have been thrown out of the borehole until it is evident that they are not burning.

11. DON'T return to any borehole until the smoke and fumes from the blast have dissipated and it is certain that the borehole has ceased to "blow."

Courtesy of E. I. du Pont de Nemours & Co. (Inc.)

Belt Speed and Capacity Tables

SUGGESTED SPEEDS WHICH ARE TODAY CONSIDERED GOOD PRACTICE FOR VARIOUS WIDTHS OF BELT HANDLING VARIOUS MATERIALS

KIND & CONDITION OF MATERIAL HANDLED	WIDTH OF BELT										
	14"	16"	18"	20"	24"	30"	36"	42"	48"	54"	60"
UNSIZED COAL, GRAVEL STONE, ASHES, ORE, OR SIMILAR MATERIAL	300'	300'	350'	350'	400'	450'	500'	550'	600'	600'	600'
SIZED COAL, COKE OR OTHER BREAKABLE MATERIAL	250'	250'	250'	300'	300'	350'	350'	400'	400'	400'	400'
GRAIN, WET OR DRY SAND..............	400'	400'	500'	600'	600'	700'	800'	800'	800'	800'	800'
CRUSHED COKE, CRUSHED SLAG OR OTHER FINE ABRASIVE MATERIAL .	250'	250'	300'	400'	400'	500'	500'	500'	500'	500'	500'
LARGE LUMP ORE, ROCK SLAG OR OTHER LARGE ABRASIVE MATERIAL..	—	—	—	—	350'	350'	400'	400'	400'	400'	400'

TABLE 2 — CAPACITIES OF TROUGHED BELT CONVEYORS BASED ON SPEED OF 100 F.P.M. FOR VARIOUS WEIGHTS OF MATERIAL — **TABLE 2**

BELT WIDTH	MAXIMUM LUMPS		CAPACITY PER HOUR AT SPEED OF 100 F.P.M.								
	SIZED	UNSIZED	30# cu.ft.	50# cu.ft.	90# cu.ft.	100# cu.ft.	125# cu.ft.	150# cu.ft.	160# cu.ft.	180# cu.ft.	200# cu.ft.
14"	2"	2½"	9T	15T	28T	31T	39T	46T	49T	56T	62T
16"	2½"	3"	13T	21T	38T	42T	52T	63T	67T	75T	83T
18"	3"	4"	16T	27T	48T	54T	67T	81T	86T	97T	107T
20"	3½"	5"	20T	33T	60T	67T	83T	100T	107T	120T	133T
24"	4½"	8"	30T	50T	90T	100T	125T	150T	160T	180T	200T
30"	7"	14"	47T	79T	142T	158T	197T	236T	252T	284T	315T
36"	9"	18"	70T	117T	210T	234T	292T	351T	374T	420T	467T
42"	11"	20"	100T	167T	300T	333T	417T	500T	534T	600T	667T
48"	14"	24"	138T	230T	414T	460T	575T	690T	736T	828T	920T
54"	15"	28"	178T	297T	534T	593T	741T	890T	948T	1070T	1190T
60"	16"	30"	222T	369T	664T	738T	922T	1110T	1180T	1330T	1480T

Courtesy of Hewitt-Robins Incorporated

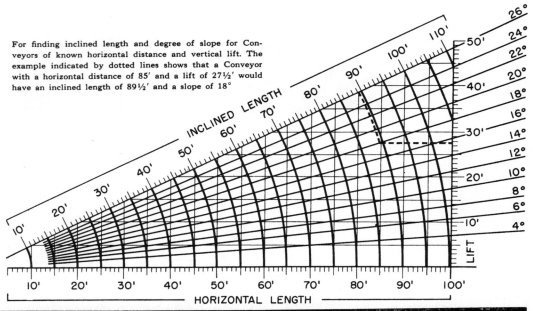

For finding inclined length and degree of slope for Conveyors of known horizontal distance and vertical lift. The example indicated by dotted lines shows that a Conveyor with a horizontal distance of 85' and a lift of 27½' would have an inclined length of 89½' and a slope of 18°

MAXIMUM SAFE INCLINATIONS OF TROUGHED BELT CONVEYORS FOR HANDLING VARIOUS BULK MATERIAL								
MATERIAL	ANGLE	RISE PER 100 FT.	MATERIAL	ANGLE	RISE PER 100 FT.	MATERIAL	ANGLE	RISE PER. 100 FT.
CEMENT – LOOSE	22°	40.4'	EARTH – LOOSE	20°	36.4'	PACKAGES – PAPER WRAP	16°	28.6'
CLAY – FINE DRY	23°	42.4'	GLASS – BATCH	21°	38.4'	ROCK – FINE CRUSHED	22°	40.4'
CLAY – WET LUMP	18°	32.5'	GRAIN	16°	28.6'	ROCK – MIXED	20°	36.4'
COAL – MINE RUN	18°	32.5'	GRAVEL – BANK RUN	18°	32.5'	ROCK – SIZED	18°	32.5'
COAL – SIZED	18°	32.5'	GRAVEL – SCREENED	15°	26.8'	SALT	20°	36.4'
COAL – BIT SLACK	23°	42.4'	GYPSUM – POWDERED	23°	42.4'	SAND – DRY	15°	26.8'
COAL – ANTHRACITE	16°	28.6'	LIME – POWDERED	23°	42.4'	SAND – DAMP	20°	36.4'
COKE – OVEN RUN	18°	32.5'	LIMESTONE	18°	32.5'	SAND – TEMP'R'D FOUNDRY	24°	44.5'
COKE – SIZED	18°	32.5'	ORE – FINE	22°	40.4'	SULPHUR POWDERED	23°	42.4'
COKE – BREEZE	20°	36.4'	ORE – CRUSHED	20°	36.4'	WOOD – CHIPS	27°	50.9'
CONCRETE – WET	15°	26.8'	ORE – SIZED	18°	32.5'			

The horsepower required at the prime mover to drive a Belt Conveyor is the sum of these integral parts:

1. Power to move the empty Belt over the Idlers.
2. Power to move the load horizontally.
3. Power required to lift or lower the load.
4. Power to turn the pulleys.
5. Power required by Trippers.
6. Drive losses.

This may be expressed in a basic formula:

L = Length of Conveyor (feet)
S = Belt Speed (fpm)
T = Capacity (short tph)
H = Height of Lift or Drop (feet)
F_P = Pulley Friction
C = Idler Friction Factor*
Q = Weight of Moving Parts Per Ft. of Conveyor*

$$\text{Hp at Motor Shaft} = \frac{.03\,LSCQ}{1000}_{\text{Empty Belt}} + \frac{CLT}{1000}_{\text{Load Horiz.}} + \frac{TH}{1000}_{\text{Load Lift}} + F_p + \text{Trippers} + \text{Drive Losses}$$

Courtesy of Hewitt-Robins Incorporated

LENGTH

Unit		Metric Equivalent		U.S. Equivalent	
millimeter	(mm)	0.001	meter	0.03937	inch
centimeter	(cm)	0.01	meter	0.3937	inch
decimeter	(dm)	0.1	meter	3.937	inches
METER	(m)	1.0	meter	{ 39.37	inches
				{ 3.28	feet
dekameter	(dkm)	10.0	meters	10.93	yards
hectometer	(hm)	100.0	meters	328.08	feet
kilometer	(km)	1000.0	meters	{ 3280.8	feet
				{ 0.6214	mile

WEIGHT OR MASS

Unit		Metric Equivalent		U.S. Equivalent	
milligram	(mg)	0.001	gram	0.0154	grain
centigram	(cg)	0.01	gram	0.1543	grain
decigram	(dg)	0.1	gram	1.543	grains
GRAM	(g)	1.0	gram	15.43	grains
dekagram	(dkg)	10.0	grams	0.3527	ounce avoirdupois
hectogram	(hg)	100.0	grams	3.527	ounces avoirdupois
kilogram	(kg)	1000.0	grams	2.2	pounds avoirdupois

CAPACITY

Unit		Metric Equivalent		U.S. Equivalent	
milliliter	(ml)	0.001	liter	0.034	fluid ounce
centiliter	(cl)	0.01	liter	0.338	fluid ounce
deciliter	(dl)	0.1	liter	3.38	fluid ounces
LITER	(l)	1.0	liter	1.05	liquid quarts
dekaliter	(dkl)	10.0	liters	0.284	bushel
hectoliter	(hl)	100.0	liters	2.837	bushels
kiloliter	(kl)	1000.0	liters	264.18	gallons

AREA

Unit		Metric equivalent		U.S. Equivalent	
square millimeter	(mm²)	0.000001	square meter	0.00155	square inch
square centimeter	(cm²)	0.0001	square meter	0.155	square inch
square decimeter	(dm²)	0.01	square meter	15.5	square inches
square meter (also CENTARE)	(m²) (ca)	1.0	square meter	10.76	square feet
square dekameter (also ARE or AR)	(dkm²) (a)	100.0	square meters	0.0247	acre
square hectometer (also HECTARE)	(hm²) (ha)	10,000.00	square meters	2.47	acres
square kilometer	(km²)	1,000,000.00	square meters	0.386	square mile

VOLUME

Unit		Metric Equivalent		U.S. Equivalent	
cubic millimeter	(mm³)	0.001	cubic centimeter	0.016	minim
cubic centimeter	(cc, cm³)	0.001	cubic decimeter	0.061	cubic inch
cubic decimeter	(dm³)	0.001	cubic meter	61.023	cubic inches
cubic meter (also STERE)	(m³) (s)	1.0	cubic meter (or stere)	1.308	cubic yards
cubic dekameter	(dkm³)	1000.0	cubic meters	1307.943	cubic yards
cubic hectometer	(hm³)	1,000,000.0	cubic meters	1,307,942.8	cubic yards
cubic kilometer	(km³)	1,000,000,000.0	cubic meters	0.25	cubic mile

SYMBOLS

G	10^9 times (1,000,000,000): giga-	m	10^{-3} times (1/1000 or .001): milli-
M	10^6 times (1,000,000): mega-	μ	10^{-6} times (1/1,000,000 or .000001): micro-
k	10^3 times (1,000): kilo-	n	10^{-9} times (1/1,000,000,000): nano-
h	10^2 times (100): hecto-	$\mu\mu$	10^{-12} times (1/1,000,000,000,000):
dk	10 times: deka-		micromicro-
d	10^{-1} times (1/10 or .1): deci-	Å, λ	Angstrom unit
c	10^{-2} times (1/100 or .01): centi-	$\mu\mu$	micromicron
		μ	micron

TEMPERATURE CONVERSION

Fahrenheit and Celsius (Centigrade) temperatures may be converted into each other by use of either of the following formulas:

$$\text{Temperature, Fahrenheit} = \frac{9 \times \text{Temperature, Celsius}}{5} + 32$$

$$\text{Temperature, Celsius} = \frac{5 \times (\text{Temperature, Fahrenheit} - 32)}{9}$$

This conversion works out to whole numbers every 5 degrees on the Celsius scale and every 9 degrees on the Fahrenheit scale, as in the table below, where equivalent temperatures on the two scales are placed beside each other.

Fahrenheit	Celsius	Fahrenheit	Celsius	Fahrenheit	Celsius
−40	−40	122	50	302	150
−22	−30	140	60	320	160
0	−17.7	150	65.5	350	176.6
14	−10	176	80	356	180
32	0	200	93.3	392	200
50	10	212	100	400	204.4
68	20	230	110	428	220
86	30	250	121.1	450	231.1
100	37.7	275	135	464	240
104	40	300	148.9	482	250
				500	260

WIND-CHILL INDEX

How you find Wind-Chill Temperature: Locate actual temperature in top row and wind speed in left hand column. Equivalent (wind-chill) temperature is found where these two intersect. Example: with temperature of 10° and wind speed of 10 mph, the Wind-Chill Temperature is −9°.

(From the U.S. Weather Bureau)

Wind Speed (mph)	THERMOMETER READING						
	30	20	10	0	−10	−20	−30

EQUIVALENT (WIND-CHILL) TEMPERATURE (F.)

Wind Speed	30	20	10	0	−10	−20	−30
calm	30	20	10	0	−10	−20	−30
5	27	16	7	−6	−15	−26	−35
10	16	2	−9	−22	−31	−45	−58
15	11	−6	−18	−33	−45	−60	−70
20	3	−9	−24	−40	−52	−68	−81
25	0	−15	−29	−45	−58	−75	−89
30	−2	−18	−33	−49	−63	−78	−94
35	−4	−20	−35	−52	−67	−83	−98
40	−4	−22	−36	−54	−69	−87	−101

APPENDIX

Name __O & K__ Excavating   Date __October 5,__ ____

Address _____, __Ohio__

Operation __Route 303 Relocation__   Production req'd __600 BCY/hr.__ Location __Hinckley__

Model __S-24__   Struck Capacity, cu. yds. __24__   Capacity lbs. __80,000__

Material __Sandy earth__   Bank Yd. __3200__ lbs. Swell Factor __0.85__ Loose Yd. __2720__ lbs.

Pay Load per Cycle; Loose Cu. Yds. _____ Bank Cu. Yds. __25__ lbs. __80,000__

Type of Loading Unit __Pushed by__ _____ cut   Bucket Size __--__   No. of Passes to Load __--__

Loading Conditions __well maintained__ Loading Production _____ Tons or Bank Cu. Yds./Hr.

A. LOADING TIME __Average Conditions__   __0.75__ Min.

Loaded Haul—Total Length __2400__ _____ Ft. Elevation __500__ ____ Ft.

Road Section	Length in Ft.	Rolling Resist.	Per Cent Grade	Trans. Gear	Max. Speed	Speed Factor	Average Speed	Hauling Time in Min.
	200	5	0	4	14.3	.33	4.7	.48
	600	3	1	5	19.5	.65	12.7	.54
	1000	3	0	6	22.5	.75	16.9	.67
	400	3	8	2	7.5	1.20	9.0	.51
	200	5	0	4	14.3	.40	5.7	.40

B.   TOTAL HAULING TIME __2.60__ Min.

Return Empty—Total Length __2400__ ____ Ft.

Road Section	Length in Ft.	Rolling Resist.	Per Cent Grade	Trans. Gear	Max. Speed	Speed Factor	Average Speed	Return Time
	200	5	0	4	29.5	.40	11.8	.19
	400	3	-8	6	31.5	.72	22.7	.20
	1000	3	0	6	29.5	.81	23.9	.47
	600	3	-1	6	29.5	.77	22.7	.30
	200	5	0	4	29.5	.40	11.8	.19

C.   TOTAL RETURN TIME __1.35__ Min.

D. Turning and Dumping—Conditions __Average__   Turning and Dumping Time __0.60__ Min.

E. Spot at Loading Machine—Conditions __Average__   Spotting Time __0.60__ Min.

F. TOTAL TIME PER COMPLETE HAULING CYCLE (A+B+C+D+E) __5.90__ Min.

G. Average Trips per Hour $= \dfrac{\text{50 Min. Prod. Hr.}}{\text{(F) Total Cycle Time}} =$ __8.48__ Trips per Hour

H. Hourly Production = (G) Trips per Hour × Pay Load = __8.48 x 25 = 212__ Tons or Bank Yds. per Hour

J. Number of TEREX Req'd $= \dfrac{\text{Hourly Production Req'd}}{\text{(H) Bank Yds. or Tons per TEREX per Hour}} =$ __600/212 = 2.83 or 3__ TEREX

Fleet Production per Hour = J × H = __3 x 212 = 636__ Bank Yds. ~~or Tons~~

Hauling Cost per Bank Yd. or Ton

  Hourly Cost of Owning and Operating _____ TEREX @ _____ each = _____

  Hourly Cost of Owning Spare TEREX _____ @ _____ each = _____

K. Hourly Cost for Fleet of _____ TEREX _____ Total _____

ESTIMATED HAULING COST PER YD. OR TON $= \dfrac{K}{\text{Fleet Production}} =$ _____

or $\dfrac{K}{\text{Req'd Production}} =$ _____

*This is an important checking figure. Its relation to the rated capacity of the unit determines the need for top extensions or smaller body, should unusually light or heavy materials be handled.

GLOSSARY

abrasion   Wear by rubbing of coarse, hard, or sharp materials.

acre   Unit for measuring land, equal to 43,560 square feet; or 4,840 square yards; or 160 square rods.

adhesion   The soil quality of sticking to buckets, blades, and other parts of excavators.

A-frame   An open structure tapering from a wide base to a load-bearing top.

aftercooler   Any device which will cool compressed air after it is fully compressed.

aggregate   Crushed rock or gravel screened to sizes for use in road surfaces, concrete, or bituminous mixes.

air receiver   The air storage tank on a compressor.

air waves   Airborne vibrations caused by explosions.

alloy steel   Steel compounded with other metals to improve its quality.

ampere   The intensity of electric current produced by one volt acting through a resistance of one ohm.

angle   The difference in direction of two lines which meet or tend to meet. Usually measured in degrees.

angling dozer (Angle dozer)   A bulldozer with a blade which can be pivoted on a vertical centerpin, so as to cast its load to either side.

annular   Ring-shaped.

anvil block   In a rock breaker, a movable piece of steel between the air piston stem and the steel.

A.P.I.   American Petroleum Institute.

apron   The front gate of a scraper body. A short ramp with a slight pitch.

assembly rod   An external bolt holding a machine together.

atmospheric pressure   Pressure of air enveloping the earth, averaged as 14.7 pounds per square inch at sea level, or 29.92 inches of mercury as measured by a standard barometer.

auger   A rotating drill having a screw thread that carries cuttings away from the face.

auxiliary   A helper or standby engine or unit.

avalanche protectors   Guard plates that prevent loose material from sliding into contact with the wheels or tracks of a digging machine.

axis   A straight line around which a shaft or body revolves. The centerline of a tunnel.

axle, dead   A fixed shaft functioning as a hinge pin. A fixed shaft or beam on which a wheel revolves.

axle, live   A revolving horizontal shaft.

babbitt   A soft antifriction metal composed of tin, antimony, and copper in varying proportions.

backfurrow   (Land). The first cut of a plow, from which the slice is laid on undisturbed soil.

backhoe   A revolving shovel-mounted excavator, or a tractor-mounted attachment, which digs with a boom and bucket pulling the loading bucket toward the machine.

backhaul   A line which pulls a drag scraper bucket backward from the dump point to the digging.

backfill   The material used in refilling a ditch or other excavation, or the process of such refilling.

backfire   A fire started to burn against and cut off a spreading fire. An explosion in the intake or exhaust passages of an engine.

backhoe   A hoe or pull shovel.

bail   A hinged loop used for lifting. A hoist yoke or bracket.

bailer   A hollow cylinder used for removing rock chips and water from churn drill holes.

ballast   Heavy material, such as water, sand, or iron, which has no function in a machine except increase of weight.

ball joint   A connection, consisting of a ball and socket, which will allow a limited hinge movement in any direction.

bank   Specifically, a mass of soil rising above a digging or trucking level. Generally, any soil which is to be dug from its natural position.

bank gravel   A natural mixture of cobbles, gravel, sand, and fines.

bank measure   Volume of soil or rock in its original place in the ground.

bank yards   Yards of soil or rock measured in its original position, before digging.

barrel   The water passage in a culvert.

baseline   (Traversing) The main traverse or surveyed line running through the site of proposed construction, from which property lines, street lines, buildings, etc., are located and plotted on the plan.

batter   Inward slope from bottom to top of the face of a wall.

batter boards   Horizontal boards placed to mark line and grade of a proposed building.

battery   A storage battery or dry cell. In blasting, often a blasting machine.

bearing   A part in which a shaft or pivot revolves.

bearing, antifriction   A bearing consisting of an inner and outer ring, separated by balls or rollers held in position by a cage.

bearing, needle   An antifriction bearing using very small-diameter rollers between wide faces.

bearing, pilot   A small bearing that keeps the end of a shaft in line.

bearing, solid   A one-piece bushing.

bearing, throwout   A bearing that permits a clutch throwout collar to slide along the clutch shaft without rotating with it.

bed   A base for machinery.

bedding   Ground or supports in which pipe is laid.

bedding plane   A separation or weakness between two layers of rock, caused by changes during the building up of the rock-forming material.

bedrock   Solid rock, as distinguished from boulders.

bell   An expanded part at one end of a pipe section, into which the next pipe fits.

bell crank   A lever whose two arms form an angle at the fulcrum, or a triangular plate hinged at one corner.

belt conveyor   An endless pulley-driven belt supported on rollers, which transports material placed on its upper surface.

bench   A working level or step in a cut which is made in several layers.

benchmark   A point of known or assumed elevation used as a reference in determining and recording other elevations.

bench terrace   A more or less level step between steep risers, graded into a hillside.

bends (Caisson disease) A cramping disease induced by too rapid decrease of air pressure after a stay in compressed atmosphere, as in a caisson.

bent (set) In tunnel timbering, two posts and a roof timber.

berm An artificial ridge of earth.

bid To make a price on anything; a proposition either verbal or written, for doing work and for supplying materials and/or equipment.

binder Fines which hold gravel together when it is dry. A deposit check that makes a contract valid.

bit The part of a drill which cuts the rock or solid.

bit, carbide A bit having inserts of tungsten carbide.

bit, chopping A bit that is worked by raising and dropping.

bit, coring A bit that grinds the outside ring of the hole, leaving an inner core intact for sampling.

bit, diamond A rotary bit having diamonds set in its cutting surfaces.

bit, drag A diamond or fishtail bit. A bit that cuts by rotation of fixed cutting edges or points.

bit, fishtail A rotary bit having cutting edges or knives.

bit, multiuse A bit that is sharpened for new service when worn.

bit, plug A diamond bit that grinds out the full width of the hole.

bit, roller A bit that contains cutting elements that are rotated inside it as it turns.

bit, throwaway A bit that is discarded when worn.

black powder Gunpowder. A mixture of carbon, sodium or potassium nitrate, and sulfur.

blade Usually a part of an excavator which digs and pushes dirt but does not carry it.

blanket Soil or broken rock left or placed over a blast to confine or direct throw of fragments.

blast To loosen or move rock or dirt by means of explosives or an explosion.

blast hole A vertical drill hole 4 or more inches in diameter, used for a charge of explosives.

blasting gelatin A jellylike high explosive made by dissolving nitrocotton in nitroglycerin.

blasting machine (battery) A hand-operated generator used to supply firing current to blasting circuits.

blasting mat A steel blanket composed of woven cable or interlocked rings.

bleed To remove unwanted air or fluid from passages.

blinding Compacting soil immediately over a tile drain to reduce its tendency to move into the tile. Clogging of a screen.

block A pulley and its case.

block, crown A sheave set suspended at the top of a derrick.

block, snatch A sheave in a case having a pull hook or ring.

block, sling A frame containing two sheaves mounted on parallel axles, so that they will line up when pulled from opposite directions.

block, traveling A frame for a sheave or a set of sheaves that slides in a track.

blockholing Blasting boulders by means of drilled holes.

blue tops Grade stakes whose tops indicate finish grade level.

BM Benchmark.

body The load-carrying part of a truck or scraper.

body, quarry A dump body with sloped sides.

body, rock A dump body with oak planking set inside a double steel floor.

bogie (Tandem) (Tandem drive unit) A two-axle driving unit in a truck. Also called tandem drive unit or a tandem.

boom   In a revolving shovel, a beam hinged to the deck front. Any heavy beam which is hinged at one end and carries a weight-lifting device at the other.

boom, lattice   A long, light shovel boom fabricated of crisscrossed steel or aluminum angles or tubing.

boom, live   A shovel boom which can be lifted and lowered without interrupting the digging cycle.

booster   An auxiliary device that increases force or pressure.

booster pump   A pump that operates in the discharge line of another pump, either to increase pressure, or to restore pressure lost by friction in the line or by lift.

boring   Rotary drilling.

borrow pit   An excavation from which material is taken to a nearby job.

boulder   A rock which is too heavy to be lifted readily by hand.

bowl   The bucket or body of a carrying scraper. Sometimes the moldboard or blade of a dozer.

box   A transmission. A dump body.

box girder   A hollow steel beam with a square or rectangular cross section.

box thread   The female side of A.P.I. tapered thread.

brake   A device for slowing, stopping, and holding an object.

brake, disc   A brake which utilizes friction between fixed and rotating discs, or between discs and shoes.

brake drum   A rotating cylinder with a machined inner or outer surface upon which a brake band or shoe presses.

brake, friction   A brake operating by friction between two surfaces rotating or sliding on each other.

brake horsepower   The horsepower output of an engine or mechanical device. Measured at the flywheel or belt, usually by some form of mechanical brake.

brake, self-energizing   A brake that is applied partly by friction between its lining and the drum.

brake, tooth (jaw brake)   A brake used to hold a shaft by means of a tooth or teeth engaging with fixed sockets. Not used for slowing or stopping.

braze   To solder with brass or other hard alloys.

break   To twist open or disconnect. A short rest period.

breast board   A temporary barrier to prevent the digging face from caving or flowing into a tunnel.

breast timber   A leaning brace from the floor of an excavation to a wall support.

bridge   In an electric blasting cap, the wire that is heated by electric current so as to ignite the charge. Sometimes the shunt connection between the cap wires.

bridle cable   An anchor cable that is at right angles to the line of pull.

bridle hitch   A connection between a bridle cable and a cable or sheave block.

Brinell test   A method of determining the hardness of metal by the indentation of a standard steel ball of known hardness under a definite load.

British imperial gallon   A fluid gallon equal to 1.2 U.S. gallons (4.5 liters) approximately; contains 277.42 cubic inches. There are 6.23 such gallons per cubic foot.

bucket   A part of an excavator which digs, lifts, and carries dirt.

bucket loader   Usually a chain bucket loader, sometimes a tractor loader or shovel dozer.

bucket sheave (padlock sheave)   A pulley attached to a shovel bucket, through which the hoist or drag cable is reeved.

bucket, slat   An open-work bucket made of bars instead of plates, used in digging sticky soil.

bucking   Sawing a long log into shorter pieces.

buffer   A pile of blasted rock left against or near a face to improve fragmentation and reduce scattering from the next blast. A movable metal plate used in tunnels to limit scattering of blasted rock.

bulkhead   A wall or partition erected to resist ground or water pressure.

bull clam   A bulldozer fitted with a curved bowl hinged to the top of the front of the blade.

bulldozer   A tractor equipped with a front pusher blade. A cleaning blade that follows the wheel or ladder of a ditching machine.

bull gear   A toothed driving wheel which is the largest or strongest in the mechanism.

bumboat   A small boat equipped with a hoist and used for handling dredge lines and anchors.

bumper (guard)   A slotted or perforated plate that holds a check-type air valve near its seat.

burden   The distance from a drill hole to the face, or the volume of rock to be moved by the explosive in a drill hole.

burn   To cut with a torch. To pulverize with very heavy explosive charges.

burn cut   A narrow section of rock pulverized by exploding heavy charges in parallel holes.

bushing   A metal cylinder between a shaft and a support or a wheel, that serves to reduce rotating friction and to protect the parts.

bushing, split   A bushing made in two pieces, for ease of insertion and removal.

butt joint (open joint)   In pipe, flat ends that meet but do not overlap.

cab guard   On a dump truck, a heavy metal shield extending up from the front wall of the body and forward over the cab.

cable   Rope made of steel wire.

cable, backhaul   In a cable excavator, the line that pulls the bucket from the dumping point back to the digging.

cable, drag   In a dragline, the line that pulls the bucket toward the shovel.

cable excavator   A long-range, cable-operated machine which works between a head mast and an anchor.

cable, inhaul (digging line)   In a cable excavator, the line that pulls the bucket to dig and bring in soil.

cage   A circular frame that limits the motion of balls or rollers in a bearing.

cairn   A pile of stones used as a marker.

caisson   A box or chamber used in construction work underwater. A foundation element.

cam   A rotating or sliding piece, or a projection on a wheel, used to impart exactly timed motion to light parts.

camber   Vertical convex curve in a culvert barrel or a structural member. Outward lean of the front wheels of a motor vehicle.

cantilever   A lever-type beam that is held down at one end, supported near the middle, and supports a load on the free end.

cap   A detonator, set off by electric current or a burning fuse. A pipe plug with female threads. The roof or top piece in a three-piece timber set used for tunnel support.

cap, delay   An electric blasting cap that explodes at a set interval after current goes through it.

cap, millisecond delay (short delay)   A detonating cap that fires from 20 to 500 thousandths of a second after the firing current passes through it.

capillary attraction   The tendency of water to move into fine spaces, as between soil particles, regardless of gravity.

capillary movement   Movement of underground water in response to capillary attraction.

capillary water   Underground water held above the water table by capillary attraction.

capstan (cat head)   A nonwinding winch used with soft rope.

carbide   Tungsten carbide, a very hard and abrasion-resistant compound used in drill bits and other tools.

carbide bit   A steel bit which contains inserts of tungsten carbide.

carbon steel   Usually a hardened steel not alloyed with other metals.

carriage   A sliding or rolling base or supporting frame.

carrier   A rotating or sliding mounting or case.

cartridge   A wrapped stick of dynamite or other explosive.

casing   A pipe lining for a drilled hole.

casing spider   A frame and wedge set that supports the top of a casing string while new sections are added.

caster   A wheel mounted in a swivel frame so that it is steered automatically by movements of its load. In an automotive vehicle, the toe-in of the front wheels.

Cat   A trademarked designation for any machine made by the Caterpillar tractor company. Widely used to indicate a crawler tractor or mounting of any make.

cat head   A capstan winch.

catskinner   Operator of a crawler tractor.

catwalk   A pathway, usually of wood or metal, that gives access to parts of large machines.

cave (caving)   Collapse of an unstable bank.

Celsius   A temperature scale on which the freezing point of water at sea-level atmospheric pressure is indicated as 0° and its boiling point as 100°. Degrees Celsius (°C) equals Fahrenheit (°F) minus 32 multiplied by $\frac{5}{9}$.

center of gravity   That point in a body about which all the weights of all the various parts balance. It is found experimentally by balancing on a knife edge or a point. The center of mass of a cut or a fill.

center of mass   In a cut or a fill, a cross-section line that divides its bulk into halves.

centerpin (center pintle)   In a revolving shovel, a fixed vertical shaft around which the shovel deck turns.

centralizer   A device that lines up a drill steel or string between the mast and the hole.

centrifugal force   Outward force exerted by a body moving in a curved line. It is the force which tends to tip a car over in going around a curve.

centripetal force   The force or restriction exerted inward to keep a body moving in a curved line. The force which keeps a car from being thrown out of a curve about which it is moving is centripetal force.

cetane number   An indication of diesel fuel ignition quality. The cetane number of a fuel is the percentage by volume of cetane in a mixture of cetane and alpha-methylnapthlalene which matches the unknown fuel in ignition quality. American diesel oil usually varies from 30 to 60 cetane.

cfm   Cubic feet per minute. A standard capacity or performance measurement for compressors.

C-frame   An angling dozer lift and push frame.

chain   A tow line or drive belt made of interlocked links. A surveyor's steel tape measure.

chain, breakaway (safety chain)   A chain that holds a tractor and a towed unit together if the regular fastening opens or breaks.

chain bucket loader (bucket loader)   A mobile loader that uses a series of small buckets on a roller chain to elevate spoil to the dumping point.

chain, leaf   A silent chain designed for low-speed, heavy-duty work.

chain, logging   A chain composed of links of round bar pieces curved and welded to interlock, with a grab hook at one end and a round hook at the other.

chain, roller   Generally, any sprocket-driven chain made up of links connected by hinge pins and sleeves. Specifically, a chain whose hinge sleeves are protected by an outer sleeve or roller that is free to turn.

chain, silent   A roller-type chain in which the sprockets are engaged by projections on the link sidebars.

chain, stud type   A roller chain in which the inner (block) links are connected solidly by nonrotating bushings.

chamfer   To bevel or slope an edge or corner.

channel terrace A contour ridge built of soil moved from its uphill side, which serves to divert surface water from a field.

check dam A dam that divides a drainageway into two sections with reduced slopes.

check valve Any device which will allow fluid or air to pass through it in only one direction.

cherry picker A small derrick made up of a sheave on an A-frame, a winch and winch line, and a hook. Usually mounted on a truck.

chip blasting Shallow blasting of ledge rock.

chipping Loosening of shallow rock by light blasting or air hammers.

chock A block used under and against an object to prevent it from rolling or sliding.

choker A chain or cable so fastened that it tightens on its load as it is pulled.

choker hook (round hook) A hook that can slide along a chain.

chord A straight line connecting two points on a curve.

chuck The part of a drill that rotates the steel. A device that clamps a rod or shaft.

churn drill (spudding or well drill) A machine that drills holes by dropping and raising a bit and drill string hung by a cable.

circle In a grader, the rotary table which supports the blade and regulates its angle.

circle reverse The mechanism that changes the angle of a grader blade.

clam A clamshell bucket.

clamshell A shovel bucket with two jaws that clamp together by their own weight when it is lifted by the closing line. A shovel equipped with a clamshell bucket.

clay A "heavy" soil composed of particles less than $\frac{1}{256}$ mm in diameter.

clean Free of foreign material. In reference to sand or gravel, means lack of binder.

cleavage plane Any uniform joint, crack, or change in quality of formation along which rock will break easily when dug or blasted.

clevis A shackle. A split end of a rod, drilled for insertion of a pin through the two sections.

clinometer A hand instrument for measuring grades by sighting.

clod buster A drag that follows a grading machine to break up lumps.

closing line (digging line) The cable which closes the jaws of a clamshell bucket.

cloth, wire Screen composed of wire or rod woven and crimped into a square or rectangular pattern.

clutch A device that connects and disconnects two shafts that revolve in line with each other.

clutch, automatic A clutch whose engagement is controlled by centrifugal force, vacuum, or other power, without attention by the operator.

clutch brake A device to slow the jackshaft when a clutch is released, to permit more rapid gear shifting.

clutch, centrifugal A clutch that is kept in engagement only by centrifugal force, so that it automatically disconnects the power train when the engine idles.

clutch, denture A jaw clutch.

clutch, disc A coupling that can be engaged to transmit power through one or more discs squeezed between a backplate and a movable pressure plate, and that can be disengaged by moving the plates apart.

clutch, fluid A fluid coupling other than a torque converter.

clutch, jaw (positive or denture clutch) A toothed hub and a sliding toothed collar that can be engaged to transmit power between two shafts having the same axis of revolution.

clutch, lockup A clutch that can be engaged to provide a nonslip mechanical drive through a fluid coupling.

clutch, overrunning (freewheeling unit) A coupling that transmits rotation in only one direction, and disconnects when the torque is reversed.

clutch, slip (safety clutch)   A friction clutch that protects a mechanism by slipping under excessive load.

clutch, wet (oil clutch)   A clutch that operates in an oil bath.

cobble   Rounded stone with diameter of 4 to 12 inches.

cockpit   The part of a tractor or grader containing the operator's seat and controls.

cofferdam   A set of temporary walls designed to keep soil and/or water from entering an excavation.

cohesion   The soil quality of sticking together.

collar   A sliding ring mounted on a shaft so that it does not revolve with it. Used in clutches and transmissions. The open end of a drill hole.

collaring   Starting a drill hole. When the hole is deep enough to hold the bit from slipping out of it, it is said to be collared.

compacted yards   Measurement of soil or rock after it has been placed and compacted in a fill.

compaction   Reduction in bulk of fill by rolling, tamping, or soaking.

compensating drive   In a four-wheel-drive truck, a freewheeling unit in the front propeller shaft that allows the front wheels to go farther than the rear on curves.

compression   For steel wheel rollers, the compacting effect of the weight at the bottom of the roll, measured in pounds per linear inch of roll width.

compression ratio   The ratio of the volume of space above a piston at the bottom of its stroke to the volume above the piston at the top of its stroke.

compression roll (drive roll)   The drive wheel of a steel wheel roller.

compressor   A machine which compresses air.

concussion   Shock or sharp air waves caused by an explosion or heavy blow.

conduit   A pipe or tile carrying water, wire, or pipes.

cone of depression   The dried-up area of soil around a single underground suction point.

contour line   A level line crossing a slope.

conveyor   A device that transports material by belts, cables, or chains.

conveyor, apron   One or more endless chains carrying overlapping or interlocking plates that carry bulk materials on their upper surface.

conveyor belt   An endless belt of rubber-covered fabric that transports material on its upper surface.

conveyor, decline   A conveyor that transports downhill.

conveyor, feeder   A short conveyor belt that supplies material to a long belt.

conveyor, screw   A revolving shaft fitted with auger-type flights that moves bulk materials through a trough or tube.

corduroy   A road made of logs laid crosswise on the ground or on other logs.

cordwood   Wood cut in 4-foot or shorter lengths to be used as fuel.

core   A cylindrical piece of an underground formation, cut and raised by a rotary drill with a hollow bit.

core barrel   A hollow cylinder containing a socket and choker springs for holding a section of drilled rock.

core drill   A rotary drill, usually a diamond drill, equipped with a hollow bit and a core lifter.

corrosion   Wear or dissolving away through chemical action, as by rusting or acids.

countershaft   A shaft which receives power from a parallel mainshaft, and transmits it to another part of the mainshaft or to working parts.

counterweight   A "dead" or nonworking load attached to one end or side of a machine to balance weight carried on the opposite end or side. A working part attached or positioned partly for the purpose of improving machine balance.

coyote holes   Horizontal drilled holes in which explosives are packed for blasting a high rock face.

cradle   A support bracket with a hinged connection to its load. A carriage.

crane   A mobile machine used for lifting and moving loads without use of a bucket.

crankshaft   The engine shaft that converts the reciprocating motion and force of pistons and connecting rods to rotary motion and torque.

crawler   One of a pair of roller chain tracks used to support and propel a machine, or any machine mounted on such tracks.

creep   Very slow travel of a machine or a part. Unwanted turning of a shaft due to drag in a fluid coupling or other disconnect device.

crimp   A tight bend in metal made under pressure.

crosshair   A hair mounted horizontally in a telescope so as to divide the field of view into halves.

crosshead   A connection between a connecting rod and a piston rod which is guided so as to move in a straight line.

cross section   A profile taken at right angles to the centerline of a project.

cross-section paper   Paper ruled in squares for convenience in drawing and measuring.

crowd   The process of forcing a bucket into the digging, or the mechanism which does the forcing. Used chiefly in reference to machines which dig by pushing away from themselves.

crown   The elevation of a road center above its sides. The curved roof of a tunnel.

crown fire   A fire burning in treetops.

crumber   A "bulldozer" blade that follows the wheel or ladder of a ditching machine to clean and shape the bottom.

crusher   A machine which reduces rocks to smaller and more uniform sizes. See also **jaw crusher, gyratory crusher,** and **hammermill.**

culvert   A pipe or small bridge for drainage under a road or structure.

curtain drain (intercepting drain)   A drain that is placed between the water source and the area to be protected.

curve, vertical   A change in gradient of the centerline of a road or pipe.

cut   To lower an existing grade. An artificial depression. To stop an engine, or throttle it to idling speed.

cut and cover   A work method which involves excavation in the open, and placing of a temporary roof over it to carry traffic during further work.

cut, gross   The total amount of excavation in a road or a road section, without regard to fill requirements.

cut, net   The amount of excavated material to be removed from a road section, after completing fills in that section.

cutter (cutter head)   On a hydraulic dredge, a set of revolving blades at the end of the suction line.

cutting   Excavating. Lowering a grade.

cycle, digging   Complete set of operations a machine performs before repeating them.

cylinder   In hydraulic systems, a hollow cylinder of metal, containing a piston, piston rod and end seals, and fitted with a port or ports to allow entrance and exit of fluid.

cylinder, slave   A small cylinder whose piston is moved by a piston rod controlled by a larger cylinder.

dart valve   A drain for a well bailer that opens automatically when rested on the ground.

datum   Any level surface taken as a plane of reference from which to measure elevations.

dead axle   See **Axle, dead.**

dead furrow   The line in a field where two directions of plowing meet, and the slices are turned away from each other.

deadheading   Traveling without load, except from the dumping area to the loading point.

decking   Separating charges of explosives by inert material which prevents passing of concussion, and placing a primer in each charge. Several levels of screening in an aggregate plant.

deck screens   Two or more screens, usually of the vibrating type, placed one above the other.

decompression   The process of reducing high air pressure gradually enough not to injure personnel who have been working in it.

deflagration   Burn with sudden and startling combustion. Describes explosion of black powder, in contrast with more rapid detonation of dynamite.

degree of curve   The number of degrees at the center of a circle subtended by a chord of 100 feet at its rim. Occasionally in highway surveying it is defined as the central angle subtended by an arc of 100 feet.

delay   An electric blasting cap which explodes at a set interval after current is passed through it.

delay, short period (millisecond delay)   An electric blasting cap that explodes $\frac{1}{50}$ to $\frac{1}{2}$ second after passage of an electric current.

density   The ratio of the weight of a substance to its volume.

derrick   Usually a nonmobile tower equipped with a hoist, but may be used as a synonym for crane.

detail drawing   A large-scale drawing showing all small parts, details, dimensions, etc.

detergent   A chemical compound that acts to clean surfaces and to keep foreign matter in solution or suspension.

detonation   Practically instantaneous decomposition, explosion, or combustion of an unstable compound, with tremendous increase in volume.

detonator   A device to start an explosion, as a fuse or cap.

dewatering   Removing water by pumping, drainage, or evaporation.

diamond drill   A light rotary drill, most often used for exploratory work and blast holes.

diaphragm   A flexible partition between two chambers.

dieseling   In a compressor, explosions of mixtures of air and lubricating oil in the compression chambers or other parts of the air system.

differential   A device that drives two axles and allows them to turn at different speeds to adjust to varying resistance.

differential, nonspin (limited-action differential)   A differential that will turn both axles, even if one offers no resistance.

differential, two-speed   A differential having a high-low gearshift between the driveshaft and the ring gear.

diffuser   Inner shell and water passages of a centrifugal pump.

digging line   On a shovel, the cable which forces the bucket into the soil. Called crowd in a dipper shovel, drag in a pull shovel, and dragline and closing line in a clamshell.

dike   A long, low dam. A thin rock formation that cuts across the structure of surrounding rock.

dimension stone   Rock quarried in blocks to predetermined sizes, in such a manner as not to weaken or shatter it.

dip   The slope of layers of soil or rock.

dipper   A digging bucket rigidly attached to a stick or arm.

dipper trip   A device that unlatches the door of a shovel bucket to dump the load.

direction of irrigation   Direction of flow of irrigation water.

ditch   Generally, a long, narrow excavation. In rotary drilling, a trough carrying mud to a screen.

ditcher, ladder   A machine that digs trenches by means of buckets mounted on a pair of chains traveling on the exterior of a boom.

ditcher, wheel   A machine that digs trenches by rotation of a wheel fitted with toothed buckets.

diversion valve   A valve that permits flow to be directed into any one of two or more pipes.

dog   A heavy-duty latch.

dolly   A unit consisting of draw tongue, an axle with wheels, and a turntable platform to support a trailer gooseneck. A small wheeled carriage designed to support heavy machines.

donkey   A winch with two drums that are controlled separately by clutches and brakes.

dope   A viscous liquid put on pipe threads to make a tight joint.

double   In rotary drilling, two pieces of drill rod left fastened together during raising and lowering.

double-clutching   Disengaging and engaging the clutch twice during a single gearshift, in order to synchronize gear speeds.

downstream face   The dry side of a dam.

dozer   Short for bulldozer.

draft   Resistance to movement of a towed load.

drag   Pulling a bucket into the digging, or the mechanism by which the pulling is done or controlled.

drag brake   On a revolving shovel, the brake which stops and holds the drag (digging) drum.

dragline   A revolving shovel which carries a bucket attached only by cables, and digs by pulling the bucket toward itself.

drag scraper   A digging and transporting device consisting of a bottomless bucket working between a mast and an anchor. A towed bottomless scraper used for land leveling. Called "leveling drag scraper" to distinguish from cable type.

dragshovel   Backhoe.

drain, intercepting (curtain drain)   A drain that intercepts and diverts groundwater before it reaches the area to be protected.

drainage head   The farthest or highest spot in a drainage area.

draw   A small valley or a gully.

drawbar   In a tractor, a fixed or hinged bar extending to the rear, used as a fastening for lines and towed machines or loads. In a grader, the connection between the circle and the front of the frame.

drawbar horsepower   A tractor's flywheel horsepower minus friction and slippage losses in the drive mechanism and the tracks or tires.

drawbar pull   The pull a tractor can exert on a load attached to the drawbar. Depends on power, weight, and traction.

draw knife   A curved, two-handled knife used in digging clay.

draw pin   A removable pin that attaches a load to a drawbar.

drawpoint   A spot where gravity-fed ore from a higher level is loaded into hauling units.

draw tongue   A bar hinged to a towed machine, fitted with some device for attaching it to a tractor.

draw works   The power distribution and control machinery of a rotary drill.

dredge   To dig underwater. A machine that digs underwater.

drift   A small, nearly horizontal tunnel.

drifter   An air drill mounted on a column or crossbar, and used for horizontal drilling underground.

drill, auger   See **Auger.**

drill bit   See **Bit.**

drill, blast hole   A machine capable of drilling holes 4 inches or more in diameter to a depth of 100 or more feet.

drill, churn (spudding drill)   A drill that cuts its hole by raising and dropping a chisel bit.

drill collar   Thick-walled drill pipe used immediately above a rotary bit to provide extra weight.

drill core   A drill that cuts around a cylinder of rock or soil, and lifts it to the surface for inspection.

drill, diamond   A rotary drill that uses a diamond-studded bit.

drill doctor   A mechanic or shop that sharpens and services drill bits, tools, and steels.

drill, percussion   A drill that hammers and rotates a steel and bit. Sometimes limited to large blast-hole drills of the percussion type.

drill pipe   The sections of a rotary drilling string connecting the kelly with the bit or collars.

drill, quarry   A blast-hole drill.

drill steel   Hollow steel connecting a percussion drill with the bit.

drill string   In rotary drills, all revolving parts below the ground. In churn drills, the tools hanging from the drilling cable.

drill, well   A churn drill, mounted on a truck.

drilling, core   Exploratory drilling that includes cutting cylinders of rock or soil and bringing them to the surface for inspection.

drilling, directional (offset drilling)   Curving a rotary drill hole to avoid obstacles or to reach side areas.

drilling, solid   In diamond drilling, using a bit that grinds the whole face, without preserving a core for sampling.

drive   To dig or make a tunnel. To hammer down piling.

drive clamp   A collar fitted on a churn drill string to enable it to be used as a hammer to drive casing pipe.

drive, positive   A driving connection to two or more wheels or shafts that will turn them at approximately the same relative speeds under any conditions.

drum   A rotating cylinder with side flanges, used for winding in and releasing cable.

drum, spudding   In a churn drill, the winch that controls the drilling line.

dry well   A deep hole, covered, and usually lined or filled with rocks, that holds drainage water until it soaks into the ground.

dynamic   Forces tending to produce motion.

dynamic balance   A condition of rest created by equal strength of forces tending to move in opposite directions.

dynamite   A mixture of an explosive or explosives with relatively inert material.

dynamite, straight   A dynamite in which nitroglycerin is the principal or only explosive.

earth drill   An auger.

eccentric   A wheel or cam with an off-center axis of revolution.

ejector   A cleanout device, usually a sliding plate. The moving wall of a scraper bowl to eject earth from it.

elevating grader   A grader with a belt to move earth up and away from the blade.

elevation (surveying)   The height of a point above a plane of reference.

elevator   A cage hoist. A machine that raises material on a belt or a chain of small buckets.

embankment   A fill whose top is higher than the adjoining surface.

environment   The conditions of atmosphere, etc., surrounding an activity or operation.

erosion   Wear caused by moving water or wind.

excavation, unclassified   Excavation paid for at a fixed price per yard, regardless of whether it is earth or rock.

excavator, hydraulic   A revolving shovel with a boom and bucket, loaded by pulling toward the machine. A backhoe shovel.

exploit   Excavate in such a manner as to utilize material in a particular vein or layer, and waste or avoid surrounding material.

explosive   A chemical compound that can decompose quickly and violently.

explosive, high   A material that detonates, that is, explodes almost instantaneously.

face   The more or less vertical surface of rock exposed by blasting or excavating, or the cutting end of a drill hole. An edge of rock used as a starting point in figuring drilling and blasting. The width of a roll crusher.

factor of safety   The ratio of the ultimate strength of the material to the allowable or working stress.

fairlead   A device which lines up cable so that it will wind smoothly onto a drum.

false set (horsehead)   A temporary support for forepoles used in driving a tunnel in soft ground.

fast powder   Dynamites or other explosives having a high-speed detonation.

faulting   In geology, the movement which produces relative displacement along a fracture in rock.

feather   To blend the edge of new material smoothly into the old surface.

feed   A mechanism which pushes a drill into its work. The process of supplying material to a conveying or processing unit.

feeder   A pushing device or short belt that supplies material to a crusher or a conveyor.

feed travel   The distance a drilling machine moves the steel shank in traveling from top to bottom of its feeding range.

ferrule   A short unthreaded tube or bushing, shrunk or soldered onto a tube or line.

fifth wheel   The weight-bearing swivel connection between highway-type tractors and semitrailers.

fill   An earth or broken-rock structure or embankment. Soil or loose rock used to raise a grade. Soil that has no value except bulk.

fill, net   In sidehill work, the yardage of fill required at any station, less the yards of material obtained from the cut at that station.

fill, net corrected   Net fill after making allowance for shrinkage during compaction.

filter bed   A fill of pervious soil that provides a site for a septic field.

filter cake (mud cake)   A deposit of mud on the walls of a drill hole.

final drive   A set of reduction gearing close to or inside of a drive wheel.

fines   Clay or silt particles in soil.

finish grade   The final grade required by specifications.

fishing   The operation of recovering an object left or dropped in a drill hole.

fitting, poured   A wire rope attachment fastened by separating the wires, expanding them in a conical socket, and filling it with molten zinc.

fitting, wedge socket   A wire rope attachment in which the rope lies in a too-small groove between a wedge and a housing, so that pull on the rope tightens the wedge.

flail   A hammer hinged to an axle so that it can be used to break or crush material.

flame gun   A large blowtorch using kerosene for fuel.

flange   A ridge that prevents a sliding motion. A rib or rim for strength or for attachments.

fleet angle   The maximum angle between a rope and a line perpendicular to the drum on which it winds.

flight   The screw thread (helix) of an auger.

float   In reference to a dozer blade—to rest by its own weight, or to be held from digging by upward pressure of a load of dirt against its moldboard.

flotation   Separation of minerals by floating the lighter ones in a fluid. The weight-supporting ability of a tire, crawler track, or platform on soft ground.

flow gradient   A drainageway slope determined by the elevation and distance of the inlet and outlet, and by required volume and velocity.

fluid clutch   A hydraulic coupling which does not increase torque.

fluid drive   A connection between two shafts that transmits torque through a fluid.

flume   An artificial channel, often elevated above the ground, used to carry fast-flowing water.

follower   A piston that maintains a light pressure against a variable amount of fluid in a container.

foot   In tamping rollers, one of a number of projections from a cylindrical drum.

footing (foot wall)   A sill under a foundation. Ground, in relation to its load-bearing and friction qualities.

foot pin   The hinge that attaches the boom to a revolving shovel.

foot-pound   Unit of work equal to the force in pounds multiplied by the distance in feet through which it acts. When a 1-pound force is exerted through a 1-foot distance, 1 foot-pound of work is done.

foot valve   A check valve in the inlet end of a pump suction hose.

ford   A place where a road crosses a stream under water.

forepole   A plank driven ahead of a tunnel face to support the roof or wall during excavation.

fork   A two-pronged rod or yoke used to slide shifting collars along their shafts.

fork head   A wheel-guiding frame with a swivel connection to the machine or vehicle that rests on it. (A caster frame.)

foul air duct   A suction line in a tunnel ventilation system.

four by four (4 × 4)   A vehicle with four wheels or sets of wheels, all engine-driven.

four-part line   A single rope or cable, reeved around pulleys so that four lines connect the fixed and the movable units.

French drain (rubble or stone drain)   Covered ditch or pipes with open joints and covered by a layer of fitted or loose stone or other pervious material.

friction   Resistance to motion when one body is sliding or tending to slide over another.

front   The working attachment of a shovel, such as dragline, hoe, or front shovel.

front-end loader   A tractor loader with a bucket which operates entirely at the front end of the tractor.

front shovel   A name for the standard revolving shovel also known as a dipper shovel.

frost   Frozen soil.

frostline   The greatest depth to which ground may be expected to freeze.

fulcrum   A pivot for a lever.

full trailer   A towed vehicle whose weight rests entirely on its own wheels or crawlers.

fumes   Usually, smoke from an explosion.

fumes, excellent   Fumes that contain a minimum of toxic and irritating chemicals.

fumes, poor   Toxic or irritating chemicals produced by an explosion.

fuse   A thin core of black powder surrounded by wrappings, which, when lit at one end, will burn to the other at a fixed speed.

fuse, detonating   A stringlike core of PETN, a high explosive, contained within a waterproof, reinforced sheath. Primacord is the best known brand.

gantry   An overhead structure that supports machines or operating parts. An upward extension of a shovel-revolving frame that holds the boom line sheaves.

gauge (gage)   Thickness of wire or sheet metal. Spacing of tracks or wheels.

gauge size   The width of a drill bit along the cutting edge.

gear A toothed wheel, cone, or bar.

gear, bevel A gear made of teeth cut in the surface of a truncated cone.

gear, bull A gear or sprocket that is much larger than the others in the same power train.

gear, cluster Two or more gears of different sizes made in one solid piece.

gear, helical A gear with straight or curved teeth cut at an angle of less than 90° to the direction of rotation.

gear, herringbone A gear with V teeth.

gear, idler A gear meshed with two other gears that does not transmit power to its shaft. Used to reverse direction of rotation in a transmission.

gear, pinion A drive gear that is smaller than the gear it turns.

gear, planetary set A gear set consisting of an inner (sun) gear, an outer ring with internal teeth, and two or more small (planet) gears meshed with both the sun and the ring.

gear, rack A toothed bar.

gear, sprocket A gear that meshes with roller or silent chain.

gelatin, blasting A high explosive made by dissolving nitrocotton in nitroglycerin. It is the strongest and highest-velocity commercial explosive.

general drawing A drawing showing elevation plan and cross section of the structure, also the borings for substructure and the main dimensions, etc.

giant (monitor) In hydraulicking, a large high-pressure nozzle mounted in a swivel on a skid frame.

glory hole A vertical pit, material from which is fed by gravity to hauling units in a shaft under the pit bottom.

gooseneck An arched connection, usually between a tractor and a trailer.

grade Usually the elevation of a real or planned surface or structure. Also means surface slope.

grader A machine with a centrally located blade that can be angled to cast to either side, with independent hoist control on each side.

grade stake A stake indicating the amount of cut or fill required to bring the ground to a specified level.

gradient Slope along a specific route, as of a road surface, channel, or pipe.

grapple A clamshell-type bucket having three or more jaws.

gravel Rock fragments from 2 to 64 millimeters (0.8 to 2.5 inches) in diameter. Or a mixture of such gravel with sand, cobbles, boulders, and not over 15 percent of fines.

grease Thick oil. A solid or semisolid mixture of oil with soap or other fillers.

grid A set of surveyor's closely spaced reference lines laid out at right angles, with elevations taken at line intersections.

grief stem A kelly, i.e., a hollow square shaft used in vertical drilling.

grizzly A coarse screen used to remove oversize pieces from earth or blasted rock. A gate or closure on a chute. (May be spelled "grizzlie.")

ground pressure The weight of a machine divided by the area in square inches of the ground directly supporting it.

ground waves Vibrations in soil or rock.

grouser A ridge or cleat across a track shoe, which improves its grip on the ground.

grout A cementing or sealing mixture of cement and water, to which sand, sawdust, or other fillers may be added.

grubbing Digging out roots.

guard (bumper) In a compressor check valve, a backing or retaining plate for the movable part.

gudgeon A reinforced bushing or a thrust-absorbing block.

guy   A line that steadies a high piece or structure by pull against an off-center load or other guys.

gypsy spool (cat head)   A capstan winch.

gyratory crusher   A crusher having a central conical member with an eccentric motion in a circular chamber tapering from a wide-top opening.

half track   A heavy truck with high-speed crawler track drive in the rear and driving wheels in front.

hammermill   A rock crusher or a shredder employing hammers or flails on a rapidly rotating axle.

handle (stick)   In a dipper shovel or hoe, the arm that connects the bucket with the boom.

hand level   A sighting level that does not have a tripod, base, or telescope.

hardpan   Hard, tight soil. A hard layer that may form just below plow depth on cultivated land.

harrow   An agricultural tool that loosens and works the ground surface.

haul, average   The average distance a grading material is moved from cut to fill.

haul distance   The distance measured along the centerline or most direct practical route between the center of the mass of excavation and the center of mass of the fill as finally placed. It is the distance that material is moved.

haul, station yards of   The number of cubic yards multiplied by the number of 100-foot stations through which it is moved.

haul, free   The distance every cubic yard is entitled to be moved without an additional charge for haul.

haul, over   The distance in excess of that given as the stated haul distance to haul excavated material.

haulageway   A main tunnel connecting underground excavation areas with an exit.

haulaway   An excavation method that involves hauling the spoil away from the hole.

hauler   A carrier piece of equipment filled to move earth materials.

haunch   In pipe, the sides of the lower third of the circumference.

head   Height of water above a specified point. The back pressure against a pump from a high outlet.

heading   In a tunnel, a digging face and its work area.

head mast   In a cable excavator, the tower that carries the working lines.

headwall (sidewall)   A culvert sidewall. Sometimes only the upstream wall.

heap   The soil carried above the sides of a body or bucket.

heel   A floor brace or socket for wall-bracing timbers. The trailing edge of an angled blade.

heeling in   Temporary planting of trees and shrubs.

helical   Spiral.

H.I.   Height of instrument.

high line   A high-tension electric line. Electric power supplied by a utility.

high wall   A face which is being excavated, as distinguished from spoil piles. Undisturbed soil or rock bordering a cut.

hinge   A connection that allows swinging motion in one plane.

hitch   A horizontal shelf along the side of a rock tunnel, which supports roof timbers. A connection between two machines.

hoe (backhoe, pull shovel)   A shovel that digs by pulling a boom-and-stick-mounted bucket toward itself.

hog box   A concrete box in which water and dirt are mixed to be pumped to a fill.

hoist   The mechanism by which a bucket or blade is lifted, or the process of lifting it.

hood   A casing on the end of a suction line that causes it to pick up material from the bottom only. A curved baffle that prevents scattering and separation of material discharged by a conveyor belt.

hook, cable   A round hook with a wide, beveled face.

hook, grab   A chain hook that will slide over any one link, but will not slide along the chain.

hook, pintle   A towing bracket having a fixed lower part and a hinged upper one, which when locked together make a round opening that can hold a tow ring.

hook, round (slip hook)   A hook that has a smooth inner surface, and will slide along a chain.

hook, safety (lock-on hook)   A round hook with a hinged piece across the opening, that allows a line to enter it readily, but requires special manipulation to remove it.

hook, swivel   A hook with a swivel connection to its base or eye.

hopper   A storage bin or a funnel that is loaded from the top, and discharges through a door or chute in the bottom.

horizon   A horizontal layer.

horsehead (false set)   A temporary support for forepoles used in tunneling soft ground.

hp   Horsepower.

horsepower   A measurement of power that includes the factors of force and speed. The force required to lift 33,000 pounds 1 foot in 1 minute.

horsepower, drawbar   Horsepower available to move a tractor and its load, after deducting losses in the power train.

horsepower, indicated   The horsepower developed in the cylinders. Determined by use of an indicator gauge. Does not include engine friction losses.

horsepower, rated   Theoretical horsepower of an engine based on dimensions and speed. Power of an engine according to a particular standard.

horsepower, shaft (flywheel or belt horsepower)   Actual horsepower produced by the engine, after deducting the drag of accessories.

holdback   An automatic safety device that prevents a conveyor belt from running backward.

holding line   The hoist cable for a clamshell bucket.

hot mill   To heat metal, then shape it.

housing   A heavy case or enclosure for rotating parts.

hub   The strengthened inner part or mounting of a wheel or gear.

hull   The substructure and deck of a ship or dredge.

humus   Decayed organic matter. A dark, fluffy swamp soil, composed chiefly of decayed vegetation, that is also called peat.

hunting tooth   A sprocket and roller chain combination in which one has an odd number of contacts and the other an even number, so that no tooth will contact the same pin twice in succession.

hydraulic dredge   A floating pump that sucks up a mixture of water and soil, and usually discharges it on land through pipes.

hydraulic fill   Fill moved and placed by running water.

hydraulic gradient   The slope of the surface of open water flow or the theoretical free water surface of water under pressure in a full pipe.

hydraulicking   Excavating on dry land by means of water jets.

hydrometer   A device (usually a float in a glass tube) for measuring the specific gravity of fluids.

hydrostatic   Relating to the pressure or equilibrium of fluids.

hygroscopic   Water absorbed from the atmosphere.

hypoid   A pinion-and-ring gear set transmitting rotation through a right angle by means of teeth having a structure intermediate between a bevel and a worm set.

ICC   Interstate Commerce Commission in the United States.

idler   A wheel or gear which changes the direction of rotation of shafts, or the direction of movement of a chain or belt.

impeller   A rotary pump member using centrifugal force to discharge a fluid into outlet passages.

impervious   Resistant to movement of water.

inclined plane   A slope used to change the direction and speed-power ratio of a force.

inertia   The property of matter by which it will remain at rest, or in uniform motion in a straight line, unless acted upon by an external force.

inhaul   The line or mechanism by which a cable excavator bucket is pulled toward the dump point.

injector   In a diesel engine, the unit that sprays fuel into the combustion chamber.

instrument   A telescopic level, such as a transit or a builder's level.

intercepting drain   Curtain drain.

intercooler   A radiator in which air is cooled while moving from the low-pressure to high-pressure cylinders of a two-stage compressor.

intermediate shaft   A shaft which is driven by one shaft, and drives another.

interruptions   Secondary cutters in auger drills.

invert   The inside bottom of a pipe or tunnel.

jack   A mechanical or hydraulic lifting device. A hydraulic ram or cylinder.

jack boom   A boom which supports sheaves between the hoist drum and the main boom in a pull shovel or a dredge.

jackhammer   An air drill that hammers and rotates a hollow steel and a bit, and that can be operated by one person.

jackknife   A tractor and trailer assuming such an angle to each other that the tractor cannot move forward.

jackleg   An outrigger post.

jackshaft   A short driveshaft, usually connecting a clutch and transmission.

jars   A tool in the churn drill string which contains slack to allow hammering upward to free a stuck bit.

jaw   In a clutch, one of a pair of toothed rings, the teeth of which face each other. In a crusher, one of a pair of nearly flat faces separated by a wedge-shaped opening.

jaw clutch   A clutch consisting of two toothed jaws, one of which slides along its shaft to engage or disengage from the other.

jaw crusher   A fixed and a movable jaw widely spaced at the top and close at the bottom, with means to move one jaw toward and away from the other.

jetting   Drilling with high-pressure water or air jets.

jetty   A long fill or structure extending into water from the shore, that serves to change the direction or velocity of water flow.

jib boom   An extension piece hinged to the upper end of a crane boom.

jig   A guide used in shaping pieces of wood or metal.

journal   That part of a rotating shaft or axle which turns in a load-supporting bearing.

jumbo   A number of drills mounted on a mobile carriage, and used in tunnels.

kelly   A square or fluted pipe which is turned by a drill rotary table, while it is free to move up and down in the table. Also called grief stem.

key   A hard steel strip inserted in matching grooves (keyways) in a shaft and a hub to make them turn as a unit.

keyhole slot   A slot enlarged at one end to allow entrance of a chain or bolt that can then be held by the narrow end.

keyway   A square-edged, lengthwise slot in a shaft or hub.

kill   Cut off electric current from a circuit. Stop an engine.

kilowatt   An electrical unit of work or power. Equal to 1000 watts, 1.34 horsepower, and 1.18 kVA.

kingpin (king pin)   A vertical swivel or hinge pin, usually supported at both top and bottom.

knife   The dirt-cutting edge of a digging machine.

kVA (kilovoltampere)   Approximately $^{89}/_{100}$ kilowatt.

lacing   Small boards or patches that prevent dirt from entering an excavation through spaces between sheeting or lagging planks.

ladder   The digging boom assembly in a hydraulic dredge or chain-and-buckets ditcher.

ladder ditcher   A machine that digs ditches by means of buckets in a chain that travels around a boom.

lag   Delay in one action following another. To install lagging, or increase the diameter of a drum.

lagging   The surface or contact area of a drum or flat pulley, especially a detachable surface or one of special composition. In a tunnel, planking placed against the dirt or rock walls and ceiling, outside the ribs. Boards fastened to the back of a shovel for blast protection.

lagging, split   Drum lagging made in two pieces to allow changing it without dismantling the drum.

land leveler   A towed scraper with a bottomless bucket centrally mounted in a long frame. Used chiefly in agricultural grading.

land tile   Porous clay pipe with open (butt) joints for a French drain.

laminated   In thin parallel layers.

lantern   In a centrifugal pump, a hollow casing on the engine side of the pump body.

lapped   Overlapped and fitted together.

lay   The direction of twist in wires and strands in wire rope.

lay, lang   A wire-rope construction in which the wires are twisted in the strands in the same direction as the strands are twisted in the rope.

lay, regular   A wire-rope construction in which the direction of twist of the wires in the strands is opposite to that of the strands in the rope.

layshaft   A fixed shaft supporting revolving drums.

lead wires   In blasting, the heavy wires that connect the firing current source or switch with the connecting or cap wires.

leg   A side post in tunnel timbering. A wire or connector in one side of an electric circuit.

level   To make level or to cause to conform to a specified grade. Any instrument that can be used to indicate a horizontal line or plane.

leveling rod (surveying)   A telescoping rod marked in feet and fractions of feet, and fitted with a movable target or sighting disc.

lever   A bar that pivots so that force applied at one part can do work at another, usually with a change in the force-distance ratio.

lever, first-class   A bar having a fulcrum (pivot point) between the points where force is applied and where it is exerted.

lever, second-class   A lever whose force is exerted between the fulcrum and the point where it is applied.

lever, third-class   A lever to which force is applied between the fulcrum and the work point.

lift   A step or bench in a multiple-layer excavation.

line   A cable, rope, chain, or other flexible device for transmitting pull. To line pieces up in order to couple them together.

line, drilling   In a churn drill, the cable that supports and manipulates the tools.

line oiler   An oil reservoir and metering device placed in a compressed-air line to lubricate air tools.

line, spinning   A line wrapped around a threaded pipe, so that a pull will rotate the pipe to fasten or unfasten it from another.

lip   The cutting edge of a bucket. Applied chiefly to edges, including tooth sockets.

liquid limit   Minimum moisture content which will cause soil to flow if jarred slightly.

load   To place explosives in a hole. To transfer material to a hauling unit or hopper.

load binder   A lever that pulls two grab hooks together, and holds them by locking over center.

load, deck   Charges of dynamite spaced well apart in a borehole, and fired by separate primers or by detonating cord.

load factor   Average load carried by an engine, machine, or plant, expressed as a percentage of its maximum capacity.

loader, belt (elevating grader)   A machine whose forward motion cuts soil with a plowshare or disc and pushes it to a conveyor belt that elevates it to a dumping point.

loader, bucket   A machine having a digging and gathering rotor, and a set of chain-mounted buckets to elevate the material to a dumping point.

loader, front-end   A tractor loader that both digs and dumps in front.

loader, paddle   A belt loader equipped with chain-driven paddles that move loose material to the belt.

loader, reversed   A front-end loader mounted on a wheel tractor having the driving wheels in front and steering at the rear.

loader, swing   A tractor loader that digs in front, and can swing the bucket to dump to the side of the tractor.

loader, tractor   A tractor equipped with a digging bucket that can dump into hauling equipment. See **Front-end loader.**

loam   A soft, easily worked soil containing sand, silt, and clay.

lock   In a compressed-air system, a chamber that can be opened to pressured air at one end, and to atmospheric air at the other.

logging tongs   Tongs with end hooks that dig in when the tongs are pulled.

loose yard   Measurement of soil or rock after it has been loosened by digging or blasting.

low bed   A machinery trailer with a low deck.

lug down   To slow down an engine by increasing its load beyond its capacity.

MA (mechanical advantage)   Increase in force obtained at the expense of speed or distance.

machined   A smooth surface finish on metal. Shaped by cutting or grinding.

magazine   A structure or container in which explosives are stored.

manifold   A chamber or tube having a number of inlets and one outlet, or one inlet and several outlets.

masonry   Cast-in-place concrete or molded cementitious units.

mass diagram   A plotting of cumulative cuts and fills used for the engineering computation of highway earthwork jobs.

mass profile   A road profile showing cut and fill in cubic yards.

mass shooting   Simultaneous exploding of charges in all of a large number of holes, as contrasted with firing in sequence with delay caps.

mast   A tower or vertical beam carrying one or more load lines at its top.

mastic   A soft sealing material.

mat   A heavy, flexible fabric of woven wire rope or chain used to confine blasts. A wood platform used in sets to support machinery on soft ground.

mechanical efficiency   As applied to engines, the ratio of the useful horsepower available at the flywheel or power takeoff to the horsepower developed in the engine cylinders, expressed in percent.

mesh   In wire screen, the number of openings per lineal inch.

metering pin   A valve plunger that controls the rate of flow of a liquid or a gas.

metric ton   Approximately 1.1 tons.

millisecond delay (short-period delay)   A type of delay cap with a definite but extremely short interval between passing of current and explosion.

millwright   A mechanic specializing in installation of heavy machinery in permanent plants.

mining   Usually, removal of soil or rock having value because of its chemical composition.

misfire   Failure of all or part of an explosive charge to go off.

mixed face   In tunneling, digging in dirt and rock in the same heading at the same time.

mm   Millimeter.

moldboard   A curved surface of a plow, dozer, or grader blade, or other dirt mover, which gives dirt moving over it a rotary, spiral, or twisting movement.

mole (mole ball)   An egg-shaped device pulled behind the tooth of a subsoil plow to open drainage passages.

monitor (giant)   In hydraulicking, a high-pressure nozzle mounted in a swivel on a skid frame.

motor grader   See **Grader.**

mouse hole   In a rotary drill substructure, a socket that holds a single piece of drill pipe ready to be added to the string.

muck   Mud rich in humus. Finely blasted rock, particularly from underground.

mud   Generally, any soil containing enough water to make it soft. In rotary drilling, a mixture of water with fine drill cuttings and added material, which is pumped through the drill string to clean the hole and cool the bit.

mudcapping   Blasting boulders or other rock by means of explosive laid on the surface and covered with mud.

multiuse bit   A detachable drill bit that can be sharpened and reshaped when worn.

multiple lines   A single line reeved around two or more sheaves so as to increase pull at the expense of speed.

net cut   In sidehill work, the cut required, less the fill required, at a particular station or part of a road.

net fill   The fill required, less the cut required, at a particular station or part of a road.

New York rod   A leveling rod marked with narrow lines, ruler fashion.

nip   The seizing of stone between the jaws or rolls of a crusher.

nip, angle of   In a roll crusher, the angle between tangents to the roll surfaces at the widest point at which they will grip a stone.

nipple   A short piece of pipe with male threads on each end.

nipple, close   A nipple so short that its two sets of threads meet in the middle.

nitroglycerin   A powerful liquid explosive that is dangerously unstable unless combined with other materials.

normal haul   A haul whose cost is included in the cost of excavation, so that no separate charge is made for it.

octane number   Percent of iso-octane by volume in a mixture of iso-octane and normal heptane that has the same antiknock character in a standard, variable-compression Cooperative Fuel Research test engine as the fuel under test. Octane has antiknock characteristics. A mixture having 75% octane and 25% heptane is said to have an octane rating of 75.

offset digging   In a ladder ditcher, digging with the boom not centered in the machine.

ohm   Unit of electrical resistance to current flow. It is equal to a fall in potential of 1 volt when a current of 1 ampere flows.

oil   Any fluid lubricant. Any liquid petroleum derivative that is less volatile than gasoline.

one on two (1 to 2)   A slope in which the elevation rises 1 foot in 2 horizontal feet.

one-part line   A single strand of rope or cable.

open cut   A method of excavation in which the working area is kept open to the sky. Used to distinguish from cut-and-cover and underground work.

optimum   Best.

ore   Rock or earth containing workable quantities of a mineral or minerals of commercial value.

oscillation   Independent movement through a limited range, usually on a hinge.

outrigger   An outward extension of a frame that is supported by a jack or block. Used to increase stability.

overbreak   Moving or loosening of rock as a result of a blast, beyond the intended line of cut.

overburden   Soil or rock lying on top of a pay formation.

overhang   Projecting parts of a face or bank.

overhaul   In many highway contracts, a movement of dirt far enough that payment, in addition to excavation pay, is made for its haulage.

overhead loader   A tractor loader which digs at one end, swings the bucket overhead, and dumps at the other end.

overtopping   Flow of water over the top of a dam or embankment.

overwinding   A rope or cable wound and attached so that it stretches from the top of a drum to the load.

pad (shoe or plate)   Ground contact part of a crawler-type track.

pan   A carrying scraper.

parallel   An arrangement of electric blasting caps in which the firing current passes through all of them at the same time.

parallel series   Two or more series of electric blasting caps arranged in parallel.

parts of line   Separate strands of the same rope or cable used to connect two sets of sheaves.

pass   A working trip or passage of an excavating or grading machine.

pawl   A tooth or set of teeth designed to lock against a ratchet.

pay formation   A layer or deposit of soil or rock whose value is sufficient to justify excavation.

peat (humus)   A soft, light swamp soil consisting mostly of decayed vegetation.

peeler   One of a set of blades that pick up and channel water moved outward by the impeller of a centrifugal pump.

peg point (steady point)   A pointed bar in a slide clamp. Used to brace a machine during work.

pellet powder   Black powder made up into hollow cartridges.

perched water table   Underground water lying over dry soil, and sealed from it by an impervious layer.

permissible   Low-flame explosive used in gassy and dusty coal mines.

petcock   A small drain valve.

pH   Percentage of free hydrogen ions. A measurement of soil acidity. pH 7 is neutral, smaller readings increasingly acid.

Philadelphia rod   A leveling rod in which the hundredths of feet, or eighths of inches, are marked by alternate bars of color the width of the measurement.

pi (π)   A number, approximately 3.1416 or $3\frac{1}{7}$, which when multiplied by the diameter of a circle, will give the circumference.

pig   An air manifold having a number of pipes which distribute compressed air coming through a single large line.

pilot valve   In a compressor, an automatic valve which regulates air pressure.

pillow block   A metal-cased rubber block that allows limited motion to a support or thrust member.

pin, master   The only pin in an integrated crawler track that will open the track when driven out.

pin, taper   A straight-sided pin that is smaller at one end than at the other.

pin thread   The male side of A.P.I. tapered thread.

pin, track   A hinge pin connecting two sections or shoes of a crawler track.

pintle   A vertical pin fastened at the bottom that serves as a center of rotation.

pintle hook   A towing device consisting of a fixed lower jaw, a hinged and lockable upper jaw, and a socket between them to hold a tow ring.

pioneering   The first working over of rough or overgrown areas.

pioneer road   A primitive, temporary road built along the route of a job, to provide means for moving equipment and personnel.

piston displacement   The amount of air displaced by moving all pistons of an engine or compressor from the bottom to the top of their stroke.

piston, free-running   A piston not connected with a rod, that does its work by hammerlike blows.

piston, slave   A small piston having a fixed connection with a larger one.

piston speed   Total feet of travel of a piston in one minute.

pit   The slope of a surface or tooth relative to its direction of movement. In a roller or silent chain, the space between pins, measured center to center.

pit, dig-down (sunken pit)   A pit that is below the surrounding area on all sides.

pitch   Any mine, quarry, or excavation area worked by the open-cut method to obtain material of value.

pitch arms (pitch braces, pitch rods)   Rods, usually adjustable, which determine the digging angle of a blade or bucket.

pitman arm   An arm having a limited movement around a pivot.

pivot   A nonrotating axle or hinge pin.

pivot tube   A hollow hinge pin.

pivot shaft   A tractor dead axle, or any fixed shaft that acts as a hinge pin.

planimeter   A device that measures an area on a map when run around its edges.

plastic limit   The minimum amount of water, in terms of percent of oven-dry weight of soil, that will make the soil plastic.

plastic soil   A soil that can be rolled into $\frac{1}{8}$-inch-diameter strings without crumbling. A soft, rubbery soil.

plate, pressure   A flywheel-driven plate that can be slid along a clutch shaft to squeeze a lined plate against the flywheel.

platform   A wood mat used in sets to support machinery on soft ground. Also called a pontoon. An operator's station on a large machine.

plenum   Use of compressed air to hold soil from slumping into an excavation.

plug   A stoppage in the discharge line of a dredge, or in an underground drain.

plug and feathers   A set of two half-round pieces of hard steel and a gradual-taper wedge, used for splitting drilled boulders.

plug, magnetic   A drain or inspection plug magnetized for the purpose of attracting and holding iron or steel particles in lubricant.

plumb bob   A pointed weight hung from a string. Used for vertical alignment.

plumbers' dope   A soft sealing compound for pipe threads.

ply   One of several layers of fabric or of other strength-contributing material.

pneumatic   Powered or inflated by compressed air.

point, well   A pipe having a fine-mesh screen and a drive point at the bottom. Used for pumping out groundwater.

pond   A small lake. In dredge work, an area where discharge water is held long enough to allow fine soil particles to settle.

pontoon   A float supporting part of a structure, such as a bridge. A wood platform used to support machinery on soft ground.

poppet valve   A valve shaped like a mushroom, resting on a circular seat, and opened by raising the stem. Standard automotive equipment.

portal   A nearly level opening into a tunnel.

pothole   A small, steep-sided hole, usually with underground drainage.

poured fitting   A connecting device that is fastened to the end of a cable (wire rope) by inserting the cable end into a funnel-shaped socket, separating the wires, and filling the socket with molten zinc.

powder   Black powder or gunpowder. General term for explosives including dynamite, but excluding caps.

powder, black   A mixture consisting mostly of carbon, sodium or potassium nitrate, and sulfur, used as an explosive.

power arm   The part of a lever between the fulcrum and the point where force is applied to the lever.

power control unit   One or more winches mounted on a tractor and used to manipulate parts of bulldozers, scrapers, or other machines.

power-divider   A nonspin differential.

power takeoff   A place in a transmission or engine to which a shaft can be so attached as to drive an outside mechanism.

power train   All moving parts connecting an engine with the point or point where work is accomplished.

preform   In wire rope, to shape the wires so that they will lie in place.

preselective   An arrangement by which a gear lever can be moved, but the resulting speed shift will not take place until the clutch or the throttle is manipulated.

pressure plate   In a clutch, a plate driven by the flywheel or rotating housing, which can be slid toward the flywheel to engage the lined disc or discs between them.

Primacord   Trademark name for a detonating fuse.

primary excavation   Digging in undisturbed soil, as distinguished from rehandling stockpiles.

prime   To provide means to start a process, as to supply sufficient water to a pump to enable it to start pumping. In blasting, to place a detonator in a cartridge or a charge of explosive.

primer   Usually the combination of a dynamite cartridge and a detonating cap.

prime mover   A tractor or other vehicle used to pull other machines.

profile   A charted line indicating grades and distances, and usually depth of cut and height of fill for excavation and grading work. It is commonly taken along the centerline.

projected pipe   A pipe laid on the surface before building a fill that buries it.

propagation   Spread of an explosion through separated charges by concussion waves in water or mud.

propeller shaft   Usually a main driveshaft fitted with universal joints.

propel shaft   In a revolving shovel, a shaft which transmits engine power to the walking mechanism.

prospecting, seismic   Underground exploration conducted by measuring vibrations caused by explosions set off in drill holes.

protractor   A device for measuring angles on drawings.

psi   Pressure in pounds per square inch.

P.T.   Pipe thread. Point of tangent on a roadway curve.

puddle   To compact loose soil by soaking it and allowing it to dry.

puff blowing   Blowing chips out of a hole by means of exhaust air from the drill.

pull   To loosen the rock around the bottom of a hole by blasting. Usually used with a negative to describe a blast which did not shatter rock to the desired depth.

pulley   A wheel that carries a cable or belt on part of its surface.

pull shovel   See **backhoe.**

pulpwood   Wood to be used in making paper.

pump, centrifugal   A pump that moves water by centrifugal force developed by rapid rotation of an impeller.

pump, diaphragm   A pump that moves water by reciprocating motion of a diaphragm in a chamber having inlet and outlet check valves.

pump, jetting   A water pump that develops very high discharge pressure.

pump, mud (slush pump)   The circulating pump that supplies fluid to a rotary drill.

pump, well point   A centrifugal pump that can handle considerable quantities of air, and is used for removing underground water to dry up an excavation.

pumping   Mechanical transfer of fluids. Alternately raising and lowering a digging edge to increase the volume of dirt being transported.

pusher   A tractor that pushes a scraper to help it pick up a load.

quadrant   One-quarter of the circumference of a circle. A curved guide for a lever. A curved scale for measuring angles.

quarry   A rock pit. An open-cut mine in rock.

quarter octagonal   A square shaft with corners cut back.

quicksand   Fine sand that is prevented from settling firmly together by upward movement of groundwater. Any wet, inorganic soil so unsubstantial that it will not support any load.

quill shaft   A light driveshaft inside a heavier one, and turning independently of it.

races   The inner and outer rings of a ball or roller bearing.

radial   Lines converging at a single center.

radius   Horizontal distance from the center of rotation of a crane to its hoisting hook.

rail   A piece of railroad-type track. The chain or inner surface of a crawler track.

raise   A shaft being dug upward from a tunnel.

rake blade   A dozer blade or attachment made of spaced tines.

rake, brush   A rake blade having a high top and light construction.

ram   A hydraulic cylinder.

rake, rock   A heavy-duty rake blade.

ram, one-way or single-acting   A hydraulic cylinder in which fluid is supplied to one end so that the piston can be moved only one way by power.

ram, two-way or double-acting   A hydraulic cylinder in which fluid can be supplied to either end, so the piston can be moved by power in two directions.

ramp   An incline connecting two levels.

range pole   A pole marked in alternate red and white bands 1-foot high.

ratchet   A set of teeth, vertical on one side and sloped on the other, which will hold a pawl moving in one direction, but allow it to move in another.

rat hole   In a rotary drill substructure, a socket that supports the kelly and swivel when they are not in use.

ratio of reduction   The relationship between the maximum size of the stone that will enter a crusher and the size of its product.

reamer   A cutting device that enlarges or straightens a hole.

reamer shell   A cutter just above a diamond bit, used to ensure a full-size hole.

rearing   Rising of the front of a tractor when pulling a heavy load.

receiver   The air tank or reservoir on a compressor.

reciprocating   Having a straight back-and-forth or up-and-down motion.

reclaiming   Digging from stockpiles. Reprocessing previously rejected material.

reduction, double   Two sets of gears in series that both reduce speed and increase power.

reduction, single   A gear set that causes one shaft to turn another at reduced speed.

reel   A revolving rack used for storage of hose and cable. In a churn drill, the winches are usually called reels.

reel, bull (spudding reel)   The churn drill winch that lifts and lowers the drill string.

reel, calf (casing reel)   The churn drill winch used for handling casing and for odd jobs.

reel, dead   A storage reel.

reel, live   A reel that supplies air, water, or electricity to the inner end of the hose or wire wound on it.

reel, sand   In a churn drill, the high-speed winch that lifts the bailing cylinder.

reeving   Threading or placement of a working line.

relay   A valve or switch that amplifies or restores original strength to an air, hydraulic, or electrical impulse.

relief holes   Holes drilled closely along a line, which are not loaded, and which serve to weaken the rock so that it breaks on that line.

relief valve   A valve that will allow air or fluid to escape if its pressure becomes higher than the valve setting.

retaining wall   A wall separating two levels.

retract   The mechanism by which a dipper shovel bucket is pulled back out of the digging.

reverse bend   To bend a line over a drum or a sheave, and then in the opposite direction over another sheave.

reversing clutch   A forward-and-reverse transmission that is shifted by a pair of friction clutches.

revetment   A wall sloped back sharply from its base. A masonry or steel facing for a bank.

revolving shovel   A digging machine in which the upper works can revolve independently of the supporting unit.

rheostat   A device that regulates flow of electricity by varying the amount of resistance in the circuit.

rib   A ridge projecting above grade in the floor of a blasted area.

ridge terrace   A ridge built along a contour line of a slope to pond rainwater above it.

rifle bar   A cylinder with curved splines.

rifle nut   A splined nut that slides back and forth on a rifle bar.

rifling   Forming a spiral thread on the wall of a drill hole, which makes it difficult to pull out the bit.

rig   A general term, denoting any machine.

riparian rights   Rights of a landowner to water that naturally flows on or bordering his or her property, including right to prevent diversion or misuse of upstream water.

ripper   A towed machine equipped with teeth, used primarily for loosening hard soil and soft rock.

riprap   Heavy stones placed on a slope or at water's edge to protect soil from waves or current.

rob   To remove part of an installation for use elsewhere. To take out supporting pillars or walls of pay rock in a mine.

rock   The hard, firm, and stable parts of the earth's crust. Any material that requires blasting before it can be dug by available equipment.

rocker arm   A lever resting on a curved base so that the position of its fulcrum moves as its angle changes. A bell crank with the fulcrum at the bottom.

rocking   Pushing a resistant object repeatedly, and backing or rolling back between pushes to allow it to reach or cross its original position.

rod stock   Round steel rod.

roll   The wheel of a roller.

roll, compression   The drive wheel of a roller.

roll, guide   The front or steering wheel of a roller.

roller, hook   In a revolving shovel, a swing roller attached by a bracket to the revolving section, and contacting the lower face of a circular track on the travel unit.

roller, support   In a crawler machine, a roller that supports the slack upper part of the track.

roller, swing   In a revolving shovel, one of several tapered wheels that roll on a circular turntable and support the upper works.

roller, track   In a crawler machine, the small wheels that rest on the track and carry most of the weight of the machine.

roller, truck   A track roller.

root buttress   A root that is above ground where it joins the trunk.

rooter   A heavy-duty ripper.

root hook   A very heavy hook designed to catch and tear out big roots when it is dragged along the ground.

rotary (rotary table)   In a rotary drill, the unit that turns the kelly and drill string.

rotary tiller   A machine that loosens and mixes soil and vegetation by means of a high-speed rotor equipped with tines.

rotation firing   Crushing a small piece of rock with a first explosion, and timing other holes to throw their burdens toward the space made by that and other preceding explosions. Or row shooting.

rotor   Any unit that does its work in a machine by spinning, and does not drive other parts mechanically.

round   A blast including a succession of delay shots.

row shooting   In a large blast, setting off the row of holes nearest the face first, and other rows behind it in succession.

rpm   Revolutions per minute.

rubble drains   French drains.

rule of thumb   A statement or formula that is not exactly correct, but is accurate enough for use in rough figuring.

run levels   To survey an area or strip so as to determine elevations.

running   Operating, particularly a drill.

saddle block   In a dipper shovel, the boom swivel block through which the stick slides when crowded or retracted.

safety clutch   A clutch that slips instead of transmitting loads beyond the capacity of the machine.

safety factor   The ratio between breakage resistance and load.

sand   A loose soil composed of particles between $\frac{1}{16}$ and 2 millimeters in diameter. Rock chips and other waste produced by drilling action.

sandhog   An underground worker who works in compressed air.

scaling   Prying loose pieces of rock off a face or roof to avoid danger of their falling unexpectedly.

scalping screen   Usually, a vibrating grizzly.

scarifier   An accessory on a grader, roller, or other machine, used chiefly for shallow loosening of road surfaces.

scavenge   To clean out thoroughly. To pick up surplus fluid and return it to a circulating system.

schematic   Display of principles of construction or operation, without accurate mechanical representation.

scoria   Cinderlike lava filled with bubbles. Bricklike material formed by volcanic heating of clay beds.

scour   Erosion in a streambed, particularly if caused or increased by channel changes.

scraper (carrying scraper, pan)   A digging, hauling, and grading machine having a cutting edge, a carrying bowl, a movable front wall (apron), and a dumping or ejecting mechanism.

scraper, bottom-dump   A carrying scraper that dumps or ejects its load over the cutting edge.

scraper, drag   A digging bucket operated on a cable between a mast and an anchor, that is not lifted off the ground during a normal cycle. A two-wheel, tractor-towed scraper, equipped with a bottomless bucket.

scraper, rear-dump   A two-wheel scraper that dumps at the rear.

scraper, self-powered   A scraper built into a single unit with a tractor.

scraper, two-axle   A full, trailer-type carrying scraper.

screen   A mesh or bar surface used for separating pieces or particles of different sizes.

screen, deck   Two or more screens placed one above the other for successive processing of the same run of material.

screen, scalping   A coarse primary screen or grizzly.

screen, shaking   A screen that is moved with a back-and-forth or rotary motion to move material along it and through it.

screen, vibrating   A screen that is vibrated to move material along it and through it.

seam   A layer of rock, coal, or ore.

section   An area equal to 640 acres or 1 square mile. A part of a work area or strip.

seepage   Movement of water through soil without formation of definite channels.

seize   To bind wire rope with soft wire, to prevent it from raveling when cut.

selective digging   Separating two or more types of soil while digging them.

semigrouser   A crawler track shoe with one or more low cleats.

semitrailer   A towed vehicle whose front rests on the towing unit.

series   An arrangement of electric blasting caps in which all the firing current passes through each of them in a single circuit.

serrated   (An edge) cut into a line of teeth.

set, timbering   A tunnel support consisting of a roof beam or arch, and two posts.

sewer tile   Glazed waterproof clay pipe with bell joints.

shackle   A connecting device for lines and drawbars, which consists of a U-shaped section pierced for a cross bolt or a pin.

shaft   A round bar that rotates or provides an axis of revolution. A vertical or steeply inclined tunnel.

shaft, cam (camshaft)   A shaft carrying cams which open and close valves.

shaft, counter (countershaft)   A shaft that allows one end of a (main) shaft to drive the other through reduction gears.

shaft, crank (crankshaft)   The main shaft of a piston-type engine, that converts reciprocating motion to rotation.

shaft, idler   A shaft that carries a gear that reverses direction of rotation in a transmission.

shaft, input   The shaft that delivers engine power to a transmission or clutch.

shaft, jack (jackshaft)   A short driveshaft, usually connecting a clutch and a transmission.

shaft, lay   A fixed shaft supporting rotating drums or gears.

shaft, main (mainshaft)   The transmission shaft forming a continuation of the input shaft.

shaft, output   A shaft that transmits power from a transmission or clutch.

shaft, reversing   A shaft whose direction of rotation can be reversed by use of clutches or brakes.

shaking screen   A suspended screen which is moved with a back-and-forth or rotary motion with a throw of several inches or more.

shale   A rock formed of consolidated mud.

shale pit   A dumping place for coarse material screened out of rotary drill mud.

shale shaker   A screen in the mud-circulating system of a rotary drill.

shank (standard)   The connecting bar between a ripper or scarifier tooth and the frame. The part of a drill steel that fits into the drill.

sheave (pronounced "shiv")   A grooved wheel used to support cable or change its direction of travel.

sheave, padlock   The bucket sheave on a dipper or backhoe. A sheave set connecting inner and outer boom lines.

sheave, traveling   A sheave block that slides in a track.

sheave block   A pulley, and a case provided with means to anchor it.

sheepsfoot   A tamping roller with feet expanded at their outer tips.

sheet erosion   Lowering of land by nearly uniform removal of particles from its entire surface by flowing water.

sheeting   Tongue-and-groove board. Planks used in shoring and bracing.

sheeting driver   An air hammer attachment that fits on plank ends so that they can be driven without splintering.

sheeting jacks   Push-type turnbuckles, used to set ditch bracing.

sheet piling   Steel strips shaped to interlock with each other when driven into the ground.

shift   A work period.

shift, graveyard   Work "day" from midnight to 8:00 A.M.

shift, swing   Work "day" from 4:00 P.M. to midnight. Occasionally refers to the midnight to 8:00 A.M. shift.

shipper shaft   In a dipper shovel, the hinge on which the stick pivots when the bucket is hoisted.

shoe   A ground plate forming a link of a track, or bolted to a track link. A support for a bulldozer blade or other digging edge to prevent cutting down. A cleanup device following the buckets of a ditching machine.

shoe, tile   A box towed behind a ditching machine, in which tile can be laid on the ditch bottom.

shot rock   Blasted rock.

shoot   Blast.

shoring   Temporary bracing to hold the sides of an excavation from caving.

shoulder   The graded part of a road on each side of the pavement. The side of a horizontal pipe, at the level of the centerline.

shovel   A digging and loading machine or tool.

shovel, dipper (front shovel)   A revolving shovel that has a push-type bucket rigidly fastened to a stick that slides on a pivot in the boom.

shovel, hoe (dragshovel, pull shovel, ditching shovel, backhoe)   A revolving shovel having a pull-type bucket rigidly attached to a stick hinged on the end of a live boom.

shovel, hydraulic   A revolving shovel in which drums and cables are replaced by hydraulic rams and/or motors.

shovel, part-swing   A revolving shovel that cannot swing through a full circle.

shovel, revolving   A digging machine that has the machinery deck and attachment on a vertical pivot, so that it can swing independently of its base.

Shrinkage   Loss of bulk of soil when compacted in a fill. Usually is computed on the basis of bank measure.

shunt   A connection between the two wires of a blasting cap that prevents building up of opposed electric potential in them.

shuttle   A back-and-forth motion of a machine that continues to face in one direction.

sidecasting   Piling spoil alongside the excavation from which it is taken.

sidehill   A slope that crosses the line of work.

sidehill cut   A long excavation in a slope that has a bank on one side, and is near original grade on the other.

sidewalls   Walls, usually masonry, at each end of a culvert.

silicosis   A lung disease caused chiefly by inhaling rock dust from air drills.

silt   A soil composed of particles between $\frac{1}{256}$ and $\frac{1}{16}$ millimeter in diameter. A fine-grained soil intermediate between clay and sand.

silting   Filling with soil or mud deposited by water.

silt trap   A settling hole or basin that prevents waterborne soil from entering a pond or drainage system.

siphon   A tube or pipe through which water flows over a high point by differential pressure.

six by six (6 × 6)   A truck having drive to the front wheels and to tandem rear wheels.

six-wheeler (ten-wheeler including dual rear tires)   A truck with two sets of rear axles.

skewed   On a horizontal angle, or in an oblique course or direction.

skip   A nondigging bucket or tray that hoists material.

skirt (skirt board)   A vertical strip placed at the side of a conveyor belt to prevent spillage or to increase capacity.

skiving   Digging in thin layers.

slack adjuster   In air brakes, the connection between the brake chamber and the brake cam.

slackline (slackline cableway)   A cable excavator having a track cable that is loosened to lower the bucket, and tightened to raise it.

slag   Refuse from steelmaking.

slave unit   A machine that is controlled by or through another unit of the same type.

sleeper   In corduroy roads, a cross log or timber supporting the stringers (longitudinal supports).

slick hole   A hole column loaded with explosive, without springing.

slick sheets   Thin steel plate spread on a tunnel floor before a blast, to make hand mucking easier.

slide   A small landslide.

slide coupling   A slip joint.

sling   A lifting hold, consisting of two or more strands of chain or cable.

sling block   A frame in which two sheaves are mounted so as to receive lines from opposite directions.

slat bucket   A digging bucket of basket construction, used in handling sticky, chunky mud.

slip joint   A splined connection loose enough to allow its two parts to slide on each other to change shaft length.

slow powder   Black powder, often called gunpowder. Also, some of the slow-acting dynamites.

sludge samples   Samples of mud from a rotary drill, or sand from a churn drill, used to obtain information about the formation being drilled.

slugger   A tooth on a roll-type rock crusher.

sluice   A steep, narrow waterway.

slurry   Cement grout.

slusher   A mobile drag scraper, with a metal slide to elevate the bucket to dump point.

slush pump   The mud pump for a rotary drill.

smart aleck   A limit switch that cuts off power if a machine part is moved beyond its safe range.

smoother bar   A drag that breaks up lumps behind a leveling machine.

snag boat   A boat equipped with a hoist and grapple for clearing obstacles from the path of a dredge.

snake hole   A hole driven into a toe for blasting, with or without vertical holes.

snakeholing   Drilling under a rock or face in order to blast it.

snaking   Towing a load with a long cable. Inserting a tow or hoist line under an object without moving the object.

snatch block   A pulley in a case, which can be easily fastened to lines or objects by means of a hook, ring, or shackle.

soil   The loose surface material of the earth's crust.

soil, heavy   A fine-grained soil, made up largely of clay or silt.

solid loading   Filling a drill hole with all the explosive that can be crammed into it, except for stemming space at top.

space   In a screen, the actual dimension of the clear opening between adjacent parallel wires or bars.

spaced loading   Loading so that cartridges or groups of cartridges are separated by open spacers that do not prevent the concussion from one charge from reaching the next.

spacing   The distance between drill holes along a line parallel to the face.

spall   To break off from a surface in sheets or pieces.

specific gravity (solids or liquids)   The ratio of the mass of a body to an equal volume of water.

spider gear (carrier pinion)   A differential gear which rotates on its shaft in a rotating case.

spile (forepole)   A plank driven ahead of a tunnel face for roof support.

spillway   An overflow channel for a pond or a terrace channel.

spinning line   A chain or rope used as a wrench in attaching and detaching drill pipe sections.

spiral cleaner   A device for removing dirt from a conveyor belt.

spiraling rifling   A drill hole twisting into a spiral around its intended centerline.

spirit level   A glass tube containing fluid and an air bubble.

spline   A set of parallel grooves running lengthwise down a shaft.

split sprocket   A two-piece sprocket that can be assembled on a shaft without removing the shaft bearings.

spoil   Dirt or rock that has been removed from its original location.

spool   The movable part of a slide-type hydraulic valve. To wind in a winch cable.

spot (trucks and other haulers)   To direct to the exact loading or dumping place.

spot log   A log or marker, placed to show a truck driver the spot where to stop to be loaded.

spotter   In truck or scraper use, the person who directs the driver into loading or dumping position.

spring, helper   On a truck rear axle, an upper spring that carries no weight until the regular spring changes shape under load.

springing   Enlarging the bottom of a drill hole by exploding a small charge in it.

spring line   The meeting of the roof arch and the sides of a tunnel.

spring-loaded   Held in contact or engagement by springs.

sprocket   A gear that meshes with a chain or a crawler track.

spudding drill (churn drill)   A drill that makes a hole by lifting and dropping a chisel bit.

spuds   On a dredge, steel tubes pointed at the bottom and provided with lifting tackle at the top which are used to hold and to move the dredge.

spud well   On a dredge, a pair of guide collars for a spud.

spur   A rock ridge projecting from a sidewall after inadequate blasting.

spur valley   A short branch valley.

squib   A detonator consisting of a firing device, and a chemical that will burn with a flash which will ignite black powder.

stab   In adding to a drill string, the action of lining up and catching the threads of the loose piece.

stabilize   To make soil firm and to prevent it from moving.

stacker   A large, mobile elevating belt.

stadia   Measurement of distance by proportion to the space on a vertical rod seen between upper and lower instrument crosshairs. The usual proportion is 1 vertical to 100 horizontal.

stake, side   On a road job, a stake on the line of the outer edge of the proposed pavement. Any stake not on the centerline.

stake, slope   A stake marking the line where a cut or fill meets the original grade.

starboard   Right side of a boat.

starter   In drilling, a short steel used to start a drill hole.

static balance   A condition of rest created by inertia (deadweight) sufficient to oppose outside forces.

static load   A load that is at rest and exerts downward pressure only.

station   Any one of a series of stakes or points indicating distance from a point of beginning or reference. A distance of 100 feet along a roadway.

station, minus   Stakes or points on the far side of the zero point from which a job was originally laid out.

stator (reactor)   In a torque converter, a set of fixed vanes that change the direction of flow of fluid entering the pump or the next stage turbine.

steady point (peg point)   A pointed steel bar that can be locked in a clamp, and is used to brace a drill frame against the ground.

steel   In air hammers, the hollow or solid steel bar that connects the hammer with the cutting tool.

steel, alloy   Steel compounded with other metals to increase strength, wearing, or rust resistance, or to obtain other desired qualities.

steel centralizer   On a wagon drill, a guide to hold the starting steel in proper alignment.

steel changes   The difference in length between successive steels used in drilling one hole.

steel puller   A hinged clamp on the bottom of a hand drill.

steering brake   A brake that slows or stops one side of a tractor.

steering clutch   A clutch that can disconnect power from one side of a tractor.

stemming   Dirt or other inert material placed in parts of a drill hole instead of explosives.

stick or handle   In a dipper shovel or backhoe, a rigid bar hinged to the boom and fastened to the bucket.

stockpile   Material dug and piled for future use.

stone boat   A flat steel sled with an up-curved front.

stope   An underground excavation that is made in a series of steps or benches.

stoper   A hand-operated air drill mounted on a column or other support.

street ell   A pipe elbow with female threads on one end, male on the other.

strength   In an explosive, the energy content in relation to its weight.

stress   The force per unit area. When the force is one of compression, it is known as "pressure." It is an internal force that resists an external force.

string (of tools)   In a churn drill, the tools suspended on the drilling cable.

stringer   A beam running lengthwise down a bridge or wood road.

string level   A spirit level equipped with prongs so that it can be hung from a string.

string loading   Filling a drill hole with cartridges smaller in diameter than the hole, without slitting or tamping them.

strip   Remove overburden or thin layers of pay material.

stripping   Removal of a surface layer or deposit, usually for the purpose of excavating other material under it.

stripping shovel   A shovel with a specially long boom and stick that enable it to reach farther and pile higher.

strut   An inside brace.

stud   A bolt having one end firmly anchored.

stuffing box   A space around a shaft filled with soft packing so as to prevent fluids or gases from leaking along it.

stumper   A narrow heavy-dozer attachment used in pushing out stumps.

sub (joint protector)   A threaded thread protector used with drill pipe.

subgrade   The surface produced by grading native earth, or cheap imported materials, which serves as a base for a more expensive paving.

subsoil plow (pan breaker)   A one-tooth ripper designed for agricultural work.

subsaver   A protector for the thread protector on the kelly of a rotary drill.

suck   The shape of the bottom of a cutting edge or tooth, which tends to pull it into the ground as it is moved.

suction   Atmospheric pressure pushing against a partial vacuum. The "pull" of a pump. Adhesion of a mass of mud to the underside of an object being lifted out of it.

sump   A low spot to which water is drained, and from which it is removed by a pump.

sun gear   The central gear in a planetary set.

sun gears   A planetary gear set consisting of a central gear, an internal-tooth ring gear, and two or more planet gears meshed with both of them.

supercharger   A blower that increases the intake pressure of an engine.

surge bin   A compartment for temporary storage, which will allow converting a variable rate of supply into a steady flow of the same average amount.

surveying   To find and record elevations, locations, and directions, by means of instruments.

sweat   To unite two closely fitting pieces by enlarging the outer one by heat.

swell (growth)   Increase of bulk in soil or rock when it is dug or blasted.

swing   In revolving shovels, to rotate the shovel on its base. In churn drills, to operate a string of tools.

swing angle   The distance in degrees that a shovel must swing between digging and dumping points.

switchback   A hairpin curve.

swivel head   In a diamond drill, the mechanism that rotates the kelly and drill string.

synchromesh   A silent-shift transmission construction, in which hub speeds are synchronized before engagement by the contact of leather cones.

tagline   A line from a crane boom to a clamshell bucket that holds the bucket from spinning out of position.

tail   The rear of a shovel deck.

tail anchor   The anchor for a track cable, or the turn point for a backhaul line in a cable excavator.

tailblock   The boom foot and idler sprocket assembly on a ladder ditcher.

tailboard   Tailgate.

tailgate   The hinged rear wall of a dump truck body. The hinged or sliding rear wall of a scraper bowl. (Ejector)

tailings   Second-grade or waste material separated from pay material during screening or processing.

tail swing   The clearance required by the rear of a revolving shovel.

talus   Loose rock or gravel formed by disintegration of a steep rock slope.

tamp   Pound or press soil to compact it.

tamper   A tool for compacting soil in spots not accessible to rollers.

tamping roller   One or more steel drums, fitted with projecting feet, and towed with a box frame.

tandem   A double-axle drive unit for a truck or grader. (A bogie.) A pair in which one part follows the other.

tandem drive   A three-axle vehicle having two driving axles.

tangent   A line that touches a circle and is perpendicular to its radius at the point of contact.

taproot   A big root that grows downward from the base of a tree.

target rod   A leveling rod.

tee   A pipe fitting that has two threaded openings in line, and a third at right angles to them.

telescope   To slide one piece inside another.

terrace   A ridge, a ridge and hollow, or a flat bench built along a ground contour.

terrain   Ground surface.

ten-wheeler (six-wheeler)   A truck with tandem rear axles.

three-part line   A single strand of rope or cable doubled back around two sheaves so that three parts of it pull a load together.

thrible   Three sections of drill pipe handled as a unit.

through cut   An excavation between parallel banks that begins and ends at original grade.

throw   The longest straight distance moved in the stroke or circle of a reciprocating or rotary part. Scattering of blast fragments.

throwout bearing   A bearing, sliding on a clutch jackshaft, that carries the engage-and-disengage mechanism.

thrust arm   A cable-controlled bar that can slide by power in two directions.

thrust washer   A washer that holds a rotating part from sideward movement in its bearings.

tight   Soil or rock formations lacking veins of weakness. Blasts or blast holes around which rock cannot break away freely.

tile   Pipe made of baked clay.

tile, land   Short pieces of porous pipe with butt (open) joints, used for underground drainage.

tile, sewer   Glazed clay pipe with bell joints.

tile shoe (tile box)   A device that permits laying tile directly behind a ditcher.

tilth   Soil condition in relation to lump or particle size. Tilled, cultivated land.

timber   Wood beams larger than 4×6.

timbering   Wood bracing in a tunnel or excavation.

toe   The projection of the bottom of a face beyond the top.

tongs   A pair of curved arms pivoted to each other, scissor fashion, so that a pull on a ring or chain connecting the short ends will cause the long ends to close to grip an object between them.

tongue   Drawbar of a towed vehicle.

tooth base   The inner part of a two-piece tooth on a digging bucket.

topographic map   A map indicating surface elevation and slope.

topping   Fine material forming a surface layer or dressing for a road or grade.

topsoil   The topmost layer of soil. Usually refers to soil containing humus that is capable of supporting a good plant growth.

torque   The twisting force exerted by or on a shaft, without reference to the speed of the shaft.

torque converter   A hydraulic coupling that utilizes slippage to multiply torque.

torque rod   A bar having the function of resisting or absorbing twisting strains.

track   A crawler track. A railroad-type track.

track, crawler   One of a pair of roller chains used to support and propel a machine. It has an inner surface that provides a track to carry the wheels of the machine, and an outer surface providing continuous ground contact.

tilting dozer   A bulldozer whose blade can be pivoted on a horizontal center pin to cut low on either side.

track frame   In a crawler mounting, a side frame to which the track roller and idler are attached.

track roller   In a crawler machine, the small wheels that are under the track frame and that rest on the track.

traction   The total amount of driving push of a vehicle on a given surface.

tractive efficiency   A measure of the proportion of the weight resting on tracks or drive wheels that can be converted into vehicle movement.

tractor   A motor vehicle on tracks or wheels used for towing or operating vehicles or equipment.

tractor loader (front-end loader)   A tractor equipped with a bucket that can be used to dig, and to elevate to dump at truck height.

trailer (full trailer)   A towed carrier that rests on its own wheels both front and rear.

trailer, semi (semitrailer)   A towed carrier that rests on the tractor in front, and on its own wheels in the rear.

tramp iron   Scrap metal entering a crusher.

transfer case   In an all-wheel-drive vehicle, a transmission or gear set that provides drive to the front shaft.

transfer point   Turning point.

transit   A surveying instrument that can measure both vertical and horizontal angles.

transite   Cement-asbestos pipe.

transition belt (feeder conveyor)   A short belt carrying material from a loading point to a main conveyor belt.

transmission   A gear set that permits change in the speed-power ratio and/or direction of rotation.

transmission, clutch-shifted   A constant-mesh transmission in which power is directed through geartrains by engagement of friction clutches.

transmission, compound   A gear set in which power can be transmitted through two sets of reduction gears in succession.

transmission, reduction-type   A transmission whose output shaft (usually the countershaft) always turns more slowly than the input shaft.

transmission, reversing   A transmission that has only a forward and reverse shift.

tread   The ground contact surface on a tire or track shoe.

treadle   A foot pedal hinged to the floor at one end.

trench   A ditch.

trestle   A bridge, usually of timber or steel, that has a number of closely spaced supports between the abutments.

trickle drain   A pond overflow pipe set vertically, with its open top level with the water surface. A nearly horizontal trough to carry low flow.

trim holes (relief holes)   Unloaded drill holes, closely spaced along a line to limit the breakage of a blast.

trip   A release catch.

tripod   A three-legged support for a surveying instrument.

tripper   A double pulley that turns a short section of a conveyor belt upside down in order to dump its load into a side chute.

troughing   Making repeated dozer pushes in one track, so that ridges of spilled material hold dirt in front of the blade.

truck, bottom-dump (dump wagon)   A trailer or semitrailer that dumps bulk material by opening doors in the floor of the body.

truck, dump   A truck or semitrailer that carries a box body with a mechanism for discharging its load.

truck, platform (rack body truck)   A truck having a flat, open body.

truck, rear-dump (end dump)   A truck or semitrailer that has a box body that can be raised at the front so the load will slide out the rear.

truck frame   Track frame.

trunnion (walking beam or bar)   An oscillating bar that allows changes in angle between a unit fastened to its center, and another attached to both ends. A heavy horizontal hinge.

trussed   Braced by an assembly of members into a rigid unit.

T.U. (takeup)   A mechanism for adjusting belt or chain tension.

tub   The base of a walking dragline.

tunnel   A more or less horizontal passageway underground for automotive traffic, railroads, or water.

turbine   A rotary engine driven by pressure of liquid or gas against its vanes.

turn angles   Angles between directions, taken with a surveying instrument.

turning point (transfer point)   A point whose elevation is taken from two or more instrument positions to determine their height in relation to each other.

turntable   A base that supports a part and allows it to rotate or swing. In a shovel, the upper part of the travel unit.

two-part line   A single strand of rope or cable, doubled back around a sheave so that two parts of it pull a load together.

universal joint   A connection between two revolving shafts that allows them to turn or swivel at an angle.

upset   To enlarge an end of a bar by shortening it.

vein   A layer, seam, or narrow irregular body of material that is different from surrounding formations.

venturi   A pressure jet that draws in and mixes air.

vernier   A device permitting finer measurement or control than standard markings or adjustments. In a spudding drill, a brake adjustment that permits the line to pay out automatically as the hole deepens.

vertical curve   The meeting of different gradients in a road or pipe.

vertical drains   Usually, columns of sand used to vent water squeezed out of humus by weight of fill.

vibrating screens   A screen that is vibrated to separate and move particles resting on it.

viscosity   The resistance of a fluid to flow. A liquid with a high viscosity rating will resist flow more readily than will a liquid with a low viscosity. The Society of Automotive Engineers (S.A.E.) has developed a series of viscosity numbers for indicating viscosities of lubricating oils.

vitrify   Glaze during heat treatment.

volt   The electromotive force that will cause a current of 1 ampere to flow through a resistance of 1 ohm.

voltage (potential)   Electromotive force (emf).

volumetric efficiency   In compressors, the relationship between cubic feet per minute and piston displacement.

wadding   Paper or cloth placed over explosive in a hole.

wagon   A trailer with a dump body.

wagon drill   A wheeled frame holding a pneumatic drill and a mechanism for feeding it into the rock and retracting it.

walker   A walking dragline.

walking bar   A trunnion or walking beam.

walking beam (trunnion)   A rigid member whose ends rest on supports that may move up and down independently, and whose center is hinged to the load it carries.

walking dragline   A dragline shovel that drags itself along the ground by means of side-mounted shoes.

wash boring   A test hole from which samples are brought up mixed with water.

waste   Digging, hauling, and dumping of valueless material to get it out of the way; or the valueless material itself.

watershed   Area that drains into or past a point.

water table   The surface of underground, gravity-controlled water.

watt   The power of a current of 1 ampere, flowing across a potential difference of 1 volt.

wedge   A piece that tapers from a thick end to a chisel point.

weld   To build up or fasten together metals by bonding with molten metal.

weldment   A base or frame made of pieces welded together, as contrasted with a one-piece casting or a bolted or riveted assembly.

well   A slot in the front of a hydraulic dredge hull in which the digging ladder pivots. A wall around a tree trunk that protects it from fill.

well drill   A churn drill used for water wells. It usually has a limited depth capacity and a truck or trailer mounting.

well point   A pipe fitted with a driving point and a fine-mesh screen, used to remove underground water. A complete set of equipment for drying up ground, including well points, connecting pipes, and a pump.

wetland   A condition of the earth's surface between dry land supporting plant life and water supporting aquatic life.

wetting agent   A chemical that reduces the surface tension of water so that it soaks into porous material more readily. Example: synthetic soap powder.

whaler   A horizontal beam in a bracing structure.

wheel, bull   A driving sprocket for a crawler track.

wheel, track   One of a set of small flanged steel wheels resting on a crawler track and supporting a track frame.

wheel ditcher   A wheel equipped with digging buckets, carried and controlled by a tractor unit.

winch   A drum that can be rotated so as to exert a strong pull while winding in a line.

winch, capstan (cat head)   A revolving spool that exerts a pull by friction with one or more loops of fiber rope.

winch, donkey (yarder)   A two-drum towing winch.

winch, oil field   An extremely powerful low-speed winch on a crawler tractor.

winch, power control (power control unit)   A high-speed, tractor-mounted winch with one to three drums. Used chiefly on bulldozers, scrapers, and rooters.

winch, towing (logging winch)   A heavy-duty winch mounted on the rear of a crawler tractor.

window pipe   A dredge discharge pipe with one or more openings in the bottom.

windrow   A ridge of loose dirt.

wing   Projection on an air drill bit.

wing wall   A wall that guides a stream into a bridge opening or culvert barrel.

work arm   The part of a lever between the fulcrum and the working end.

working cycle   A complete set of operations. In an excavator, it usually includes loading, moving, dumping, and returning to the loading point.

working drawing   Any drawing showing sufficient detail that whatever is shown can be built without other drawings or instructions.

worm   A gear formed of a cylinder with spiral threads cut in its surface.

worm wheel   A modified spur gear with curved teeth that meshes with a worm.

wrist action   In a bucket, the ability to change its digging or dumping angle by power.

INDEX

* An italic page number indicates a detailed discussion, often continuing on the following pages.

ABOUT THE AUTHORS

HERBERT L. NICHOLS, JR., has devoted most of his business and professional life to earthmoving and heavy equipment. A longtime private consultant, he owned his own excavation and grading business in Connecticut for many years. In addition to his widely translated benchmark work *Moving the Earth,* he is the author of *Excavator Operation, Heavy Equipment Repair,* and numerous other books and manuals in the field. He is a member of the American Association of Cost Engineers, the American Institute of Mining Engineers, and the American Welding Society. He lives in Roswell, Georgia.

DAVID A. DAY, P.E., is a consultant with ABCO Engineering Corporation, with more than 50 years of experience in civil engineering and construction. A graduate of Cornell University and the University of Illinois with a master's degree in construction and structures, he is the author of *Construction Equipment Guide.* Mr. Day is also a Life Member of the American Society of Civil Engineers, where he helped start the annual Underground Technology Conference (UTC), and a Life Member of the National Society of Professional Engineers. He lives in Centennial, Colorado.